AF847436

Advanced Structured Materials

Volume 89

Series editors

Andreas Öchsner, Faculty of Mechanical Engineering, Esslingen University of Applied Sciences, Esslingen, Germany
Lucas F. M. da Silva, Department of Mechanical Engineering, University of Porto, Porto, Portugal
Holm Altenbach, Institute of Mechanics, Faculty of Mechanical Engineering, Otto-von-Guericke-University Magdeburg, Magdeburg, Saxony-Anhalt, Germany

Common engineering materials reach in many applications their limits and new developments are required to fulfil increasing demands on engineering materials. The performance of materials can be increased by combining different materials to achieve better properties than a single constituent or by shaping the material or constituents in a specific structure. The interaction between material and structure may arise on different length scales, such as micro-, meso- or macroscale, and offers possible applications in quite diverse fields.

This book series addresses the fundamental relationship between materials and their structure on the overall properties (e.g. mechanical, thermal, chemical or magnetic etc.) and applications.

The topics of *Advanced Structured Materials* include but are not limited to

- classical fibre-reinforced composites (e.g. class, carbon or Aramid reinforced plastics)
- metal matrix composites (MMCs)
- micro porous composites
- micro channel materials
- multilayered materials
- cellular materials (e.g. metallic or polymer foams, sponges, hollow sphere structures)
- porous materials
- truss structures
- nanocomposite materials
- biomaterials
- nano porous metals
- concrete
- coated materials
- smart materials

Advanced Structures Material is indexed in Google Scholar and Scopus.

More information about this series at http://www.springer.com/series/8611

Holm Altenbach · Joël Pouget
Martine Rousseau · Bernard Collet
Thomas Michelitsch
Editors

Generalized Models and Non-classical Approaches in Complex Materials 1

Springer

Editors
Holm Altenbach
Institut für Mechanik
Otto-von-Guericke-Universität Magdeburg
Magdeburg
Germany

Bernard Collet
Centre National de la Recherche Scientifique, UMR 7190, Institut Jean Le Rond d'Alembert
Sorbonne Université
Paris
France

Joël Pouget
Centre National de la Recherche Scientifique, UMR 7190, Institut Jean Le Rond d'Alembert
Sorbonne Université
Paris
France

Thomas Michelitsch
Centre National de la Recherche Scientifique, UMR 7190, Institut Jean Le Rond d'Alembert
Sorbonne Université
Paris
France

Martine Rousseau
Centre National de la Recherche Scientifique, UMR 7190, Institut Jean Le Rond d'Alembert
Sorbonne Université
Paris
France

ISSN 1869-8433 ISSN 1869-8441 (electronic)
Advanced Structured Materials
ISBN 978-3-319-72439-3 ISBN 978-3-319-72440-9 (eBook)
https://doi.org/10.1007/978-3-319-72440-9

Library of Congress Control Number: 2018934438

© Springer International Publishing AG, part of Springer Nature 2018
This work is subject to copyright. All rights are reserved by the Publisher, whether the whole or part of the material is concerned, specifically the rights of translation, reprinting, reuse of illustrations, recitation, broadcasting, reproduction on microfilms or in any other physical way, and transmission or information storage and retrieval, electronic adaptation, computer software, or by similar or dissimilar methodology now known or hereafter developed.
The use of general descriptive names, registered names, trademarks, service marks, etc. in this publication does not imply, even in the absence of a specific statement, that such names are exempt from the relevant protective laws and regulations and therefore free for general use.
The publisher, the authors and the editors are safe to assume that the advice and information in this book are believed to be true and accurate at the date of publication. Neither the publisher nor the authors or the editors give a warranty, express or implied, with respect to the material contained herein or for any errors or omissions that may have been made. The publisher remains neutral with regard to jurisdictional claims in published maps and institutional affiliations.

Printed on acid-free paper

This Springer imprint is published by the registered company Springer International Publishing AG part of Springer Nature
The registered company address is: Gewerbestrasse 11, 6330 Cham, Switzerland

Dedicated to the memory of a great creative spirit, G.A. Maugin

Foreword

Gérard A. Maugin

Gérard A. Maugin born at Angers (France) on December 2, 1944, married to Eleni Zachariadou in 1978, passed away in Villejuif (France) on September 22, 2016, 7 p. m.

He had retired from the University of Paris VI since 2010. His longstanding scientific activity in Continuum Mechanics and Continuum Physics is well known in the Community of Mechanics. In these fields he enjoys a well-established reputation. His interests covered almost all disciplines of Continuum Mechanics and his studies have been addressed fundamental problems of mechanics and electromagnetism and applications as well.

One of his first papers (1965) is concerned with *The Race Tidal Power Plant*, a topical subject in the current engineering applications. A few years later he published a series of papers in the Comptes Rendus de l'Académie des sciences, Paris (1970-71), on the macroscopic description of magnetic media in the relativistic framework. His striking versatility in scientific research, which emerged since the very beginning of his career, cannot be unnoticed. In April 1971 he defended his PhD dissertation thesis on micromagnetism, supervisor Prof. Cemal A. Eringen from Princeton University. The Princeton University Press has published the thesis with the title *Micromagnetism and Polar Media*, 1-294, (1971). Four years later, in May 1975, Prof. Maugin achieved his "habilitation" (Doctorat d'Etat en Sciences Mathématiques) in Paris, supervisor Prof. Paul Germain de l'Académie des sciences.

By 1975, Gérard Maugin already had a ripe scientific *curriculum studiorum* in Mechanics and Physics: 45 papers published in the most well known scientific journals of mechanics and mathematics (Ann. Inst. Henri Poincaré, J. of Physics, J.

of Mathematical Physics, General Relativity and Gravitation, J. de Mécanique and others).

His favourite topics in this period are the behaviour of electromagnetic materials in the relativistic framework and in the Galilean approximation as well. Specifically, the behaviour of deformable dielectrics, ferro-magnetic and ferri-magnetic bodies are examined and explored in such frameworks.

His training in relativity and in electromagnetism presumably developed in his mind a specific sensitivity toward the mathematical description of Continua with coupled-fields, Continua with structures and/or Microstructures.

In 1980 Gérard Maugin published the paper *The method of virtual power in Continuum Mechanics: Application to Coupled Fields*, Acta Mechanica, 35, 1-70, (1980). The energetic approach therein proposed represents one of the most powerful methods for describing complex materials from the viewpoint of continua. The method also provides the proper tools, with which to attack problems of structured continua, both, from the theoretical viewpoint and from the standpoint of applications. The method of virtual power, such as expounded in the aforementioned paper, is formulated in its most general form and is applied to electromagnetic materials in their various aspects (thermo-elastic dielectrics with polarisation gradients, dielectrics with quadrupoles, ferromagnets, liquid crystals in external electromagnetic fields, et cetera). This contribution of Prof. Maugin stands as a referential point to many searchers in continuum mechanics.

Wave propagation was also one of his favourite topics of interest. To this topic he devoted his attention and his studies since the very beginning of his studies. Due to the interesting results achieved in applied problems of wave propagation, he was awarded by a scientific prize, the Prize of Mechanics Doisteau-Blutet of the French Academy of Sciences in 1982. His interest in wave propagation never ceased nor decreased in the subsequent years, even when his main efforts were focused on other fields. As a result of the expertise that he had acquired in this field, Gérard Maugin was invited to deliver a course on *Physical and Mathematical Models of Nonlinear Waves in Solids*, in Udine, at the International Centre for Mechanical Sciences (CISM), in 1993. Springer-Verlag will publish the lecture notes of this course in the series CISM Courses and Lectures. Afterward, he also published the book *Nonlinear Waves in Elastic Crystals*, Oxford University Press (1999).

This specific attention to the dynamical problems in continua is often transferred to his graduate students. Some of them investigated the possibility of "soliton-propagation" in structured materials, under his advice. Interesting and unexpected results are shown in evidence by their studies, with the help of numerical techniques.

Gérard Maugin not only provided his students with an excellent professional training in Continuum Mechanics and Physics he also transferred to his students and co-workers enthusiasm in research along with motivations and scientific curiosity. These qualities represent the primary source of his prolific scientific activity.

An impressive number of papers and many books and monographs (published by Springer-Verlag, McGraw-Hill, Elsevier, Oxford University Press UK, Cambridge University Press UK) emerge from his Curriculum and numerous awards and honours.

A detailed list can be found on the website:
`http://www.dalembert.upmc.fr/home/maugin/`.

Gérard Maugin was a member of the editorial board of many scientific journals, among them

- International Journal of Engineering Science from 1976 to December 1995,
- Wave Motion since 1986,
- Journal of Thermal Stresses,
- Journal of Technical Physics of the Polish Academy of Science,
- International Journal of Applied Electromagnetics and Mechanics - one of its founders in 1990,
- Applied Mechanics Reviews - Associate Editor since 1985
- Journal of Non-equilibrium Thermodynamics,
- Egyptian Journal of Mathematics,
- Archives of Applied Mechanics (formerly Ingenieur-Archiev),
- Yugoslav Journal of Mechanics,
- Archives of Mechanics of the Polish Academy of Science,
- ARI - Associate Editor since its creation 1997,
- Proceedings of the Estonian Academy of Science since 1997,
- Mechanics Research Communications - one of the five Editors since July 1999, and
- the Marocain Revue de Mécanique Appliquée et Théorique.

He held a membership in scientific societies (in most of the cases as member of the executive committee or of the advisory board)

- Society of Natural Philosophy (SNP), USA,
- Society of Engineering Science, (SES), USA, Life member,
- Society for Industrial and Applied Mathematics (SIAM), USA,
- American Physical Society (APS), USA,
- American Mathematical Society (AMS), USA,
- Acoustical Society of America (ASA), USA,
- Gesellschaft für angewandte Mathematik und Mechanik (GAMM), Germany,
- International Society for the Interaction of Mathematics and Mechanics (ISIMM) - Member of the Executive Committee 1986-1990, 1997-2001,
- Société Française d'Acoustique (SFA), France,
- Association Française de Mécanique (AFM), France, and
- EUROMECH Society (European Society of Mechanics),

and was appointmented as consulting editor (for Springer, John Wiley & Sons, Kluwer, Oxford University Press) or as expert for research contracts and grants (in USA, Canada, UK, Belgium, France and other countries). In addition, he acted as a Series Editor of the CRC Series on Continuum Modelling and Discrete Systems (CRC, Boca Raton, Florida, USA) and Applied Electromagnetics and Mechanics (Elsevier, then I.O.S. Press, The Netherlands).

He was a visiting professor and visiting scientist at Princeton, Belgrade, Warsaw, Istanbul, at the Royal Institute of Technology in Stockholm, at the TU Berlin, Rome, Tel Aviv, the Lomonosov University, Kyoto, Darmstadt and Berkeley. In 2001 he received the Max Planck Research Award, was the 1991/92 Fellow of the Berlin Institute for Advanced Study, and in 2001 received an honorary doctorate from the Technical University of Darmstadt. In 1982 he received the mechanics Prize of French Academy of Sciences and in 1977 the Medal of the CNRS in physics and engineering. He was a member of the Polish Academy of Sciences (1994), of the Estonian Academy of Sciences and was awarded an honorary professorship by the Moscow State University. In 2003, he received the A. Cemal Eringen Medal.

I would like rather to emphasise his natural attitude as searcher and as teacher. This attitude combined with his skill in finding the proper (and often the simplest) mathematical tools, through which to expound and to clarify the physical nature of the phenomenon under consideration.

The so-called Configurational Mechanics or Material Mechanics is the "novel" field, to which Gérard Maugin devoted his main interest during the last decades. He initiated the search and the studies of configurational forces in elasticity, being concerned with the elastic energy-momentum tensor, a notion introduced by Eshelby in a few seminal papers in the fifties. It is not difficult to show that the Eshelby tensor naturally applies to defective materials and in fracture mechanics. For instance, based on this tensor, one is able to recover all known invariant integrals around a defect, including the celebrated J-integral around the tip of a crack. In addition, fracture criteria can be (and indeed, are) properly extended to elastic dielectrics and to elastic magnetised materials.

The early studies of Gérard Maugin and others in this field are also concerned with inhomogeneous materials. Specifically, Maugin and others re-proposed the Eshelby tensor in finite elasticity, basing on Noll's notion of homogeneity and uniformity. Such an extension of the Eshelby tensor shows in evidence important physical properties and relevant geometrical features, which are hidden in the linear framework. All these features eventually address the notion of configurational force. Gérard Maugin and others suddenly realised that the notion of configurational force confers to the Eshelby stress tensor a deeper physical meaning. They also realised that the notion of configurational (or material) force could not be confined to the in-homogeneities in the elasto-static framework. Hence, the important role of this force was enquired in dynamics. One of the relevant results is the natural relationship of the material force with the so-called material-momentum, or pseudo-momentum. Such a result also represents a turning point for the introduction of the so-called configurational mechanics, which now stands on firm bases. In addition, configurational mechanics is also shown to be the natural framework for thermodynamical transformations, such as solid-phase-transitions.

The notion of configurational force becomes even more powerful in complex materials and materials with structures. Based on this notion, Gérard Maugin (with a second author) contributed to disentangling the following quarrel in liquid crystals (Int. J. Engng. Sci., 33, 1663-1678, 1995): as to whether the Ericksen stress tensor should be regarded as related to a configurational force or to the classical traction. The point is that the Ericksen tensor for liquid crystals has the form of an energy-stress tensor, just like the Eshelby stress. Hence, one could be tempted to incorrectly identify the one with the other. It is worth noticing that the quarrel involved Ericksen and Eshelby themselves, along with Kröner and other prominent people.

Eventually, the interest arises in discriminating configurational forces from traction in the more general context of structured continua. This interest becomes a crucial need in the case of electromagnetic materials. In this regard, it is worth recalling that Eshelby was initially inspired by the Maxwell stress tensor of electromagnetism. The latter however, though possessing the form of an energy-stress, is undoubtedly related to the classical traction. In order to avoid misunderstandings, one envisages the existence of two meaningful energy-stress-tensors in continua and, more specifically, in electromagnetic materials. The introduction of the material energy-stress (namely, the Eshelby tensor) provides a novel standpoint, which allows one to enlighten unclear issues or rather obscure aspects of electromagnetic materials. One of these is the proper form of the electromagnetic momentum. Basing on a criterion established by Gérard Maugin and others, one is able to distinguish between momentum and *pseudo-momentum* or *crystal-momentum*, in the language of Solid State Physics. These themes are still nowadays open to further developments. New applications of these ideas are proposed from time to time in the community of continuum mechanics, in which a steadily increasing interest is recorded on this subject.

It should be noted that Gérard Maugin delivered his knowledge and new research results immediately to the PhD and post graduated students. One of his loveliest places for lectures was the International Center of Mechanical Sciences (CISM, Udine, Italy), where he presented not only the aforementioned course on wave propagation. He was involved, for example, in the following activities:

- Non-Equilibrium Thermodynamics with Application to Solids (coordinated by W. Muschik in 1992): lectures on "Non-Equilibrium Thermodynamics of Electromagnetic Solids",
- Nonlinear Waves in Solids (coordinated by A. Jeffrey and J. Engelbrecht in 1993): lectures on "Physical and Mathematical Models of Nonlinear Waves in Solids",
- Configurational Mechanics of Materials (coordinated by R. Kienzler and G.A. Maugin in 2001): lectures on "Elements of Field Theory in Inhomogeneous and Defective Materials" and "Material Mechanics of Electromagnetic Solids" (together with C. Trimarco),
- Surface Waves in Geomechanics: Direct and Inverse Modeling for Soils and Rocks (coordinated by K. Wilmanski and C.G. Lai in 2004): 6 lectures on waves on interfaces, in thin layers on linear media, on surfaces with curvature, grating and roughness, nonlinear surface waves, propagation of surface soliton packages, interactions with nonmechanical fields,
- Generalised Continua and Dislocation Theory. Theoretical Concepts, Computational Methods and Experimental Verification (coordinated by C. Sansour in 2007): 4 lectures on defects, dislocations and the general theory of material inhomogeneities,
- Mechanics and Electrodynamics of Magneto- and Electro-Elastic Materials (coordinated by R.W. Ogden and D.J. Steigman in 2009): 7 lectures on the basics of electromagnetics in matter, with emphasis placed on the notions of electromagnetic forces, momentum and stresses, on the general thermomechanical framework,

and on applications to magnetoelasticity at different scales, the notions of internal stresses, internal variables, homogenization, ferromagnetic polycrystals and configurational force,

- Generalized Continua from the Theory to Engineering Applications (coordinated by H. Altenbach and V. Eremeyev in 2011): 6 lectures on electromagnetism and generalized continua, ponderomotive couple, electromagnetic microstructure, resonance couplings with classical deformation, effects on configurational forces (fracture and phase transformation).

Gérard Maugin was also greatly attracted by researches in Epistemology and the History of Science. The naissance of fundamental concepts of Mechanics and Physics and their evolution through the centuries were fascinating topics for him. Toward these topics he had developed a unique sensitivity, since he was a young researcher. To them he devoted his efforts in the last years until the end of his life by writing a history of Continuum Mechanics in the following three volumes published by Springer, Solid Mechanics and Applications Series:

- Continuum Mechanics Through the Twentieth Centuries: A Concise Historical Perspectives *(2013)*,
- Continuum Mechanics Through the Eighteenth and Nineteenth Centuries: Historical Perspectives from John Bernoulli (1727) to Ernst Hellinger (1914) *(2014)*,
- Continuum Mechanics Through the Ages: From the Renaissance to the Twentieth Century *(2016)*.

Last but not least, he offered clear and reliable explanations of over 100 keywords in Continuum Mechanics for better understanding the fundamental concepts

- Non-Classical Continuum Mechanics - A Dictionary *(2017)*.

This book was published also by Springer in the Advanced Structured Materials Series (Series Editors: Andreas Öchsner, Lucas F.M. da Silva, and Holm Altenbach) as volume 51.

His memory will endure among his many friends and in the Scientific Community of Mechanics.

Università di Pisa, Italia, January 2018 *Carmine Trimarco*

Preface

At the beginning of February 2017 the invitation letters for a special remembrance book were sent to approximately 70 friends and colleagues of the great French scientist in the field of Continuum Mechanics (or more general Continuum Physics) Gérard A. Maugin who died on September 22nd, 2016. As usual in such case that the response is 50% sending a kind reply that they will submit a paper and finally one gets 15-20 papers. In the case of Gérard the resonance was overwhelming - the editors got finally approximately 60 papers and the decision was made to publish two volumes. This is the first one including 40 papers from authors living in more than 20 countries.

The scientific interests of Gérard are well reflected by variety of subjects covered by the contributions to this book including the following branches of Continuum Mechanics

- relativistic continuum mechanics,
- micromagnetism,
- electrodynamics of continua,
- electro-magneto-mechanical interaction,
- mechanics of deformable solids with ferroïc states (ferromagnetics, ferroelectrics, etc.),
- thermomechanics with internal state variables,
- linear and nonlinear surface waves on deformable structures,
- nonlinear waves in continua,
- Lighthill-Whitham wave mechanics,
- lattice dynamics,
- Eshelbian Mechanics of continua on the material manifold,
- geometry and thermomechanics of material defects,
- material equations and
- biomechanical applications (tissue and long bones growth).

In addition, he published several papers and books on the history of continuum mechanics. This was reason that the authors of this book have submitted so different papers with the focus on the research interests of Gérard.

We have to thank all contributors for their perfect job. Last but not least, we gratefully acknowledge Dr. Christoph Baumann (Springer Publisher) supporting the book project.

Magdeburg, Paris
January 2018

Holm Altenbach
Joël Pouget
Martine Rousseau
Bernard Collet
Thomas Michelitsch

Contents

List of Contributors

Moustafa S. Abou-Dina
Department of Mathematics, Faculty of Science, Cairo University, Giza 12613, Egypt
e-mail: moustafa_aboudina@hotmail.com

Ransés Alfonso-Rodríguez
Department of Mathematics, University of Central Florida, 4393 Andromeda Loop North, Orlando, FL 32816, USA
e-mail: ranses.alfonso@knights.ucf.edu

Vladimir I. Alshits
Shubnikov Institute of Crystallography, Russian Academy of Sciences, 119333 Moscow, Russia
e-mail: valshits@mail.ru

Holm Altenbach
Institut für Mechanik, Otto-von-Guericke-Universität Magdeburg, Universitätsplatz 2, D-39106 Magdeburg, Germany
e-mail: holm.altenbach@ovgu.de

Attila Askar
Koç University, Sarıyer, Istanbul 89010, Turkey
e-mail: AASKAR@ku.edu.tr

Jean-Louis Auriault
Université Grenoble Alpes-CNRS, 3SR Lab, F-38000 Grenoble, France
e-mail: jean-louis.auriault@3sr-grenoble.fr

Svetlana M. Bauer
St. Petersburg State University, 199034 St. Petersburg, Universitetskaya nab. 7-9, Russia
e-mail: s_bauer@mail.ru

Arkadi Berezovski
Department of Cybernetics, School of Science, Tallinn University of Technology, Akadeemia tee 21, 12618, Tallinn, Estonia
e-mail: berezovski.arkadi@gmail.com

Harald Berger
Institut für Mechanik, Otto-von-Guericke-Universität Magdeburg, Universitätsplatz 2, D-39106 Magdeburg, Germany
e-mail: harald.berger@ovgu.de

Guy Bonnet
Université Paris-Est, Laboratoire Modélisation et Simulation Multi Echelle, MSME UMR 8208 CNRS, Marne-la-Vallée, France
e-mail: guy.bonnet@u-pem.fr

Claude Boutin
Université de Lyon, LGCB/LTDS CNRS 5513, ENTPE, Vaulx-en-Velin, France
e-mail: claude.boutin@entpe.fr

Manfred Braun
Chair of Mechanics and Robotics, University Duisburg–Essen, Lotharstraße 1, 47057 Duisburg, Germany
e-mail: manfred.braun@uni-due.de

Julián Bravo-Castillero
Facultad de Matemática y Computación, Universidad de La Habana, San Lázaro y L, Vedado, CP 10400, La Habana, Cuba & Instituto de Investigaciones en Matemáticas Aplicadas y en Sistemas, Universidad Nacional Autónoma de México, Delegación Álvaro Obregón, Apartado Postal 20-126, 01000 CDMX, México
e-mail: jbravo@matcom.uh.cu, julian@mym.iimas.unam.mx

Renald Brenner
Sorbonne Université, Centre National de la Recherche Scientifique, UMR 7190, Institut Jean Le Rond d'Alembert, F-75005 Paris, France
e-mail: renald.brenner@upmc.fr

Noël Challamel
Université de Bretagne Sud, EA 4250, Institut de Recherche Dupuy de Lôme (IRDL), Centre de Recherche, Rue de Saint Maudé-BP 92116, F-56100 Lorient, France
e-mail: noel.challamel@univ-ubs.fr

Sanda Cleja-Ţigoiu
University of Bucharest, Faculty of Mathematics and Computer Science, str. Academiei 14, 010014-Bucharest, Romania
e-mail: tigoiu@fmi.unibuc.ro

David Christie
Physics department, Lancaster University, Lancaster LA1 4YB and The Cockcroft Institute of Accelerator Science, Keckwick Lane, Warington WA4 4AD, Unided Kingdom
e-mail: d.christie@lancaster.ac.uk

Bernard Collet
Sorbonne Université, Centre National de la Recherche Scientifique, UMR 7190, Institut Jean Le Rond d'Alembert, F-75005 Paris, France
e-mail: bernard.collet@upmc.fr

Oscar L. Cruz-González
Facultad de Ciencias Técnicas, Departamento de Matemática, Universidad de Matanzas, Varadero road, Km. 2 1/2, Matanzas, Cuba
e-mail: oscar.lcg93@gmail.com

Xiangyu Cui
Computational Engineering, Ruhr University Bochum, D-44780 Bochum, Germany
e-mail: 15120080178@163.com

Hui-Hui Dai
Department of Mathematics, City University of Hong Kong, 83 Tat Chee Avenue, Kowloon Tong, Hong Kong
e-mail: mahhdai@cityu.edu.hk

Francesco dell'Isola
Università di Roma "La Sapienza" & International Research Center on Mathematics and Mechanics of Complex System (M&MOCS), Universitá degli Studi dell'Aquila, Via Giovanni Gronchi 18 - Zona industriale di Pile, 67100, L'Aquila, Italy
e-mail: francescodellisola@uniroma1.it

Fabio di Cosmo
International Research Center on Mathematics and Mechanics of Complex System (M&MOCS), Universitá degli Studi dell'Aquila, Via Giovanni Gronchi 18 - Zona industriale di Pile, 67100, L'Aquila, Italy
e-mail: fabio.dicosmo.memocs@gmail.com

Serge Dumont
Université de Nîmes, Institut de Mathématiques Alexander Grothendieck, CNRS, UMR 5149, CC.051, Pl. E. Bataillon, 34 095 Montpellier Cedex 5, France
e-mail: serge.dumont@unimes.fr

Amr R. El Dhaba
Department of Mathematics, Faculty of Science, Damanhour University, Egypt
e-mail: amrramadaneg@gmail.com

Jüri Engelbrecht
Department of Cybernetics, School of Science, Tallinn University of Technology, Akadeemia tee 21, 12618, Tallinn, Estonia
e-mail: je@ioc.ee

Marcelo Epstein
University of Calgary, Canada
e-mail: mepstein@ucalgary.ca

Victor A. Eremeyev
Gdańsk University of Technology, ul. Gabriela Narutowicza 11/12, 80-233 Gdańsk, Poland
e-mail: eremeyev.victor@gmail.com

Samuel Forest
Mines ParisTech, Centre des Matériaux, CNRS, UMR 7633, France
e-mail: samuel.forest@ensmp.fr

Alexander B. Freidin
Institute for Problems in Mechanical Engineering of the Russian Academy of Sciences, Bolshoy pr., 61, V.O., St. Petersburg, 199178 & Peter the Great St.Petersburg Polytechnic University, Polytechnicheskaya st., 29, St.Petersburg, 195251, Russia
e-mail: alexander.freidin@gmail.com

Ahmed F. Ghaleb
Department of Mathematics, Faculty of Science, Cairo University, Giza 12613, Egypt
e-mail: afghaleb@sci.cu.edu.eg

Alain Gérard
Bordeaux University, I2M-MPI, CNRS, UMR 5295, 33600 Pessac, France
e-mail: ajr.gerard@gmail.com

Elena F. Grekova
Institute for Problems in Mechanical Engineering of the Russian Academy of Sciences, Bolshoy pr., 61, V.O., St. Petersburg, 199178, Russia
e-mail: elgreco@pdmi.ras.ru

Sonia Guarguagli
Università degli Studi di Torino, Via Giuseppe Verdi 8, Torino, Italy
e-mail: sonia.guarguagli@gmail.com

Raúl Guinovart-Díaz
Facultad de Matemática y Computación, Universidad de La Habana, San Lázaro y L, Vedado, CP 10400, La Habana, Cuba
e-mail: guino@matcom.uh.cu

Igor A. Gula
The Ioffe Institute of the Russian Academy of Sciences, 194021, St. Petersburg, Russia (current Department of Physics, Chemistry and Pharmacy, University of Southern Denmark (SDU), Campusvej 55, Odense M 5230, Denmark)
e-mail: wswggg@gmail.com

Dorin Ieşan
Department of Mathematics, Al. I. Cuza University and Octav Mayer Institute of Mathematics (Romanian Academy), Bd. Carol I, nr. 8, 700508 Iaşi, Romania
e-mail: iesan@uaic.ro

Shoji Imatani
Kyoto University, Sakyo-ku, Kyoto, Japan
e-mail: imatani@energy.kyoto-u.ac.jp

David Jou
Universitat Autònoma de Barcelona, Grup de Fisíca Estadística, 08193 Bellaterra, Catalonia, Spain
e-mail: David.Jou@uab.cat

Stanislava V. Kashtanova
St. Petersburg State University, 199034 St. Petersburg, Universitetskaya nab. 7-9, Russia
e-mail: kastasya@yandex.ru

Attila Kocsis
Department of Structural Mechanics, Budapest University of Technology and Economics and Engineering Center Budapest, Robert Bosch Kft., 1111, Budapest, Hungary
e-mail: kocsis@ep-mech.me.bme.hu

Marco Laudato
Dipartimento di Ingegneria e Scienze dell'Informazione e Matematica, Universitá degli Studi dell'Aquila, Via Vetoio (Coppito 1), 67100 Coppito & International Research Center on Mathematics and Mechanics of Complex System (M&MOCS), Universitá degli Studi dell'Aquila, Via Giovanni Gronchi 18 - Zona industriale di Pile, 67100, L'Aquila, Italy
e-mail: laudato.memocs@gmail.com

Khanh Chau Le
Lehrstuhl für Mechanik - Materialtheorie, Ruhr-Universität Bochum, D-44780 Bochum, Germany
e-mail: chau.le@rub.de

Frederic Lebon
Aix-Marseille Univ., CNRS, Centrale Marseille, LMA, 4 Impasse Nikola Tesla, CS 40006, 13453 Marseille Cedex 13, France
e-mail: lebon@lma.cnrs-mrs.fr

Yuanyou Li
Department of Mathematics, City University of Hong Kong, 83 Tat Chee Avenue, Kowloon Tong, Hong Kong
e-mail: yuanyouli2-c@cityu.edu.hk,

Wael Mahmoud
Department of Mathematics, Faculty of Science, Cairo University, Giza 12613, Egypt
e-mail: mwael@sci.cu.edu.eg

María-Belén Martínez-Pavetti
Kyoto University, Sakyo-ku, Kyoto, Japan (currently Universidad Nacional de Asunción, San Lorenzo, Paraguay)
e-mail: bmartinez.py@gmail.com

Raúl Martínez-Rosado
Tecnologico de Monterrey, Escuela de Ingenieria y Ciencias, Atizapan de Zaragoza, Estado de México, México
e-mail: rrosado@itesm.mx

Tiphaine Mathieu-Pennober
Ecole Polytechnique, Paris, France
e-mail: tiphaine.mathieu-pennober@polytechnique.org

Markus Mehnert
Chair of Applied Mechanics, University of Erlangen-Nuremberg, Paul-Gordan-Str. 3, 91054 Erlangen, Germany
e-mail: markus.mehnert@fau.de

Thomas Michelitsch
Sorbonne Université, Centre National de la Recherche Scientifique, UMR 7190, Institut Jean Le Rond d'Alembert, F-75005 Paris, France
e-mail: michel@lmm.jussieu.fr

Nikita F. Morozov
St. Petersburg State University, 199034 St. Petersburg, Universitetskaya nab. 7-9, Russia
e-mail: morozov@nm1016.spb.edu

Wolfgang H. Müller
Institute of Mechanics, Chair of Continuum Mechanics and Constitutive Theory, Technische Universität Berlin, Einsteinufer 5, 10587 Berlin, Germany
e-mail: whmueller1000@gmail.com

Wolfgang Muschik
Institut für Theoretische Physik, Technische Universität Berlin, Hardenbergstr. 36, D-10623 Berlin, Germany
e-mail: muschik@physik.tu-berlin.de

Konstantin Naumenko
Institut für Mechanik, Otto-von-Guericke-Universität Magdeburg, Universitätsplatz 2, D-39106 Magdeburg, Germany
e-mail: konstantin.naumenko@ovgu.de

Franck Nicolleau
Sheffield Fluid Mechanics Group, Department of Mechanical Engineering of the Russian Academy of Sciences, University of Sheffield, Mappin Street, Sheffield S1 3JD, United Kingdom
e-mail: f.nicolleau@sheffield.ac.uk

Jerzy P. Nowacki
Polish-Japanese Academy of Information Technology, 02-008 Warsaw, Poland
e-mail: nowacki@pja.edu.pl

Andrzej Nowakowski
Sheffield Fluid Mechanics Group, Department of Mechanical Engineering of the Russian Academy of Sciences, University of Sheffield, Mappin Street, Sheffield S1 3JD, United Kingdom
e-mail: a.f.nowakowski@sheffield.ac.uk

Alena E. Osokina
Institute for Problems in Mechanical Engineering of the Russian Academy of Sciences, Bolshoy pr., 61, V.O., St. Petersburg, 199178 & Peter the Great St.Petersburg Polytechnic University, Polytechnicheskaya st., 29, St.Petersburg, 195251, Russia
e-mail: aeosokina@gmail.com

Martin Ostoja-Starzewski
University of Illinois at Urbana-Champaign, Urbana, IL 61801, USA
e-mail: martinos@illinois.edu

José A. Otero
Instituto Tecnológico de Estudios Superiores de Monterrey CEM, Atizapán de Zaragoza, EM CP 52926, México
e-mail: j.a.otero@itesm.mx

Franco Pastrone
Accademia delle Scienze di Torino, Via Accademia delle Scienze 6, Torino, Italy
e-mail: franco.pastrone@unito.it

Tanel Peets
Department of Cybernetics, School of Science, Tallinn University of Technology, Akadeemia tee 21, 12618, Tallinn, Estonia
e-mail: tanelp@ioc.ee

Leslie D. Pérez-Fernández
Universidade Federal de Pelotas, Departamento de Matemática e Estatística, Instituto de Física e Matemática, Caixa Postal 354, CEP 96010-900, Pelotas, Rio Grande do Sul, Brazil
e-mail: leslie.fernandez@ufpel.edu.br

Alejandro Perez Riascos
Department of Civil Engineering, Universidad Mariana San Juan de Pasto, Colombia
e-mail: aaappprrr@gmail.com

Alexey V. Porubov
Institute for Problems in Mechanical Engineering of the Russian Academy of Sciences, Bolshoy pr., 61, V.O., St. Petersburg, 199178 & Peter the Great St.Petersburg Polytechnic University, Polytechnicheskaya st., 29, St.Petersburg, 195251, Russia
e-mail: alexey.porubov@gmail.com

Joël Pouget
Sorbonne Université, Centre National de la Recherche Scientifique, UMR 7190, Institut Jean Le Rond d'Alembert, F-75005 Paris, France
e-mail: joel.pouget@upmc.fr

Vít Průša
Faculty of Mathematics and Physics, Charles University, Sokolovská 83, Praha 8 – Karlín, CZ 186 75, Czech Republic
e-mail: prusv@karlin.mff.cuni.cz

Ramon Quintanilla
Department of Mathematics, ESEIAAT, Polytechnic University of Catalonia, Colón, 11, 08222 Terrassa, Barcelona, Spain
e-mail: ramon.quintanilla@upc.edu

Bharath V. Raghavan
University of Illinois at Urbana-Champaign, Urbana, IL 61801, USA
e-mail: braghav2@illinois.edu

Kumbakonam R. Rajagopal
Texas A&M University, Department of Mechanical Engineering, 3123 TAMU, College Station TX 77843-3123, USA
e-mail: krajagopal@tamu.edu

Lalaonirina R. Rakotomanana
Institut de Recherche Mathématique de Rennes, Campus Beaulieu, 35042 Rennes, France
e-mail: lalaonirina.rakotomanana-ravelonarivo@univ-rennes1.fr

Enaam K. Rawy
Department of Mathematics, Faculty of Science, Cairo University, Giza 12613, Egypt
e-mail: enaamkhalifa@yahoo.com

Liliana Restuccia
University of Messina, Department of Mathematical and Computer Sciences, Physical Sciences and Earth Sciences, Contrada Papardo, Viale Ferdinando Stagno d'Alcontres, 98166 Messina, Italy
e-mail: lrestuccia@unime.it

Reinaldo Rodríguez-Ramos
Facultad de Matemática y Computación, Universidad de La Habana, San Lázaro y L, Vedado, CP 10400, La Habana, Cuba
e-mail: reinaldo@matcom.uh.cu

Vitoriano Ruas
Sorbonne Université, Centre National de la Recherche Scientifique, UMR 7190, Institut Jean Le Rond d'Alembert, F-75005 Paris, France
e-mail: vitoriano.ruas@upmc.fr

Federico J. Sabina
Instituto de Investigaciones en Matemáticas Aplicadas y en Sistemas, Universidad Nacional Autónoma de México, Delegación Álvaro Obregón, Apartado Postal 20-126, 01000 CDMX, México
e-mail: fjs@mym.iimas.unam.mx

Andrus Salupere
Department of Cybernetics, School of Science, Tallinn University of Technology, Akadeemia tee 21, 12618, Tallinn, Estonia
e-mail: salupere@ioc.ee

Alexander M. Samsonov (†)
The Ioffe Institute of the Russian Academy of Sciences, 194021, St. Petersburg, Russia

Sushma Santapuri
Department of Applied Mechanics, Indian Institute of Technology Delhi, New Delhi 110016, India
e-mail: ssantapuri@am.iitd.ac.in

Boris N. Semenov
St. Petersburg State University, 199034 St. Petersburg, Universitetskaya nab. 7-9, Russia
e-mail: semenov@bs1892.spb.edu

Igor Sevostianov
Department of Mechanical and Aerospace Engineering, New Mexico State University, Las Cruces, NM 88003, USA
e-mail: igor@nmsu.edu

Leah L. Sharipova
Institute for Problems in Mechanical Engineering of the Russian Academy of Sciences, Bolshoy pr., 61, V.O., St. Petersburg, 199178, Russia
e-mail: sleah07@gmail.com

Mario Spagnuolo
CNRS, LSPM UPR3407, Université Paris 13, Sorbonne Paris Cité, 93430 Villetaneuse, France
e-mail: mario.spagnuolo@lspm.cnrs.fr

David J. Steigmann
Department of Mechanical Engineering, University of California Berkeley, CA 94720, USA
e-mail: dsteigmann@berkeley.edu

Paul Steinmann
Chair of Applied Mechanics, University of Erlangen-Nuremberg, Paul-Gordan-Str. 3, 91054 Erlangen, Germany
e-mail: paul.steinmann@fau.de

Kert Tamm
Department of Cybernetics, School of Science, Tallinn University of Technology, Akadeemia tee 21, 12618, Tallinn, Estonia
e-mail: kert@ioc.ee

Mireille Tixier
Département de Physique, Université de Versailles Saint Quentin, 45, Avenue des Etats-Unis, F-78035 Versailles, France
e-mail: mireille.tixier@uvsq.fr

Robin Tucker
Physics department, Lancaster University, Lancaster LA1 4YB and The Cockcroft Institute of Accelerator Science, Keckwick Lane, Warington WA4 4AD, Unided Kingdom
e-mail: r.tucker@lancaster.ac.uk

Péter Ván
Department of Theoretical Physics, Wigner Research Centre for Physics, H-1525 Budapest, Konkoly Thege Miklós u. 29-33., Department of Energy Engineering, Faculty of Mechanical Engineering, Budapest University of Technology and Economics, 1111 Budapest, Műegyetem rkp. 3., Montavid Thermodynamic Research Group, Budapest, Hungary
e-mail: van.peter@wigner.mta.hu

Elena N. Vilchevskaya
Institute for Problems in Mechanical Engineering of the Russian Academy of Sciences, Bolshoy pr., 61, V.O., St. Petersburg, 199178 & Peter the Great St.Petersburg Polytechnic University, Polytechnicheskaya st., 29, St.Petersburg, 195251, Russia
e-mail: vilchevska@gmail.com

Mathias Würkner
Institut für Mechanik, Otto-von-Guericke-Universität Magdeburg, Universitätsplatz 2, D-39106 Magdeburg, Germany
e-mail: mathias.wuerkner@ovgu.de

Chapter 1
Effective Coefficients and Local Fields of Periodic Fibrous Piezocomposites with 622 Hexagonal Constituents

Ransés Alfonso-Rodríguez, Julián Bravo-Castillero, Renald Brenner, Raúl Guinovart-Díaz, Leslie D. Pérez-Fernández, Reinaldo Rodríguez-Ramos, and Federico J. Sabina

Abstract The asymptotic homogenization method is applied to a family of boundary value problems for linear piezoelectric heterogeneous media with periodic and rapidly oscillating coefficients. We consider a two-phase fibrous composite consisting of identical circular cylinders perfectly bonded in a matrix. Both constituents are piezoelectric 622 hexagonal crystal and the periodic distribution of the fibers follows a

---

Ransés Alfonso-Rodríguez
Department of Mathematics, University of Central Florida, 4393 Andromeda Loop North, Orlando, FL 32816, USA
e-mail: ranses.alfonso@knights.ucf.edu

Julián Bravo-Castillero
Facultad de Matemática y Computación, Universidad de La Habana, San Lázaro y L, Vedado, CP 10400, La Habana, Cuba & Instituto de Investigaciones en Matemáticas Aplicadas y en Sistemas, Universidad Nacional Autónoma de México, Delegación Álvaro Obregón, Apartado Postal 20-126, 01000 CDMX, México
e-mail: jbravo@matcom.uh.cu, julian@mym.iimas.unam.mx

Raúl Guinovart-Díaz · Reinaldo Rodríguez-Ramos
Facultad de Matemática y Computación, Universidad de La Habana, San Lázaro y L, Vedado, CP 10400, La Habana, Cuba
e-mail: jbravo@matcom.uh.cu, guino@matcom.uh.cu, reinaldo@matcom.uh.cu

Renald Brenner
Sorbonne Université, Centre National de la Recherche Scientifique, UMR 7190, Institut Jean Le Rond d'Alembert, F-75005 Paris, France
e-mail: renald.brenner@upmc.fr

Leslie D. Pérez-Fernández
Universidade Federal de Pelotas, Departamento de Matemática e Estatística, Instituto de Física e Matemática, Caixa Postal 354, CEP 96010-900, Pelotas, Rio Grande do Sul, Brazil
e-mail: leslie.fernandez@ufpel.edu.br

Federico J. Sabina
Instituto de Investigaciones en Matemáticas Aplicadas y en Sistemas, Universidad Nacional Autónoma de México, Delegación Álvaro Obregón, Apartado Postal 20-126, 01000 CDMX, México
e-mail: fjs@mym.iimas.unam.mx

© Springer International Publishing AG, part of Springer Nature 2018
H. Altenbach et al. (eds.), *Generalized Models and Non-classical Approaches in Complex Materials 1*, Advanced Structured Materials 89,
https://doi.org/10.1007/978-3-319-72440-9_1

rectangular array. Closed-form expressions are obtained for the effective coefficients, based on the solution of local problems using potential methods of a complex variable. An analytical procedure to study the spatial heterogeneity of the strain and electric fields is described. Analytical expressions for the computation of these fields are given for specific local problems. Examples are presented for fiber-reinforced and porous matrix including comparisons with fast Fourier transform (FFT) numerical results.

1.1 Introduction

At the beginning of the nineteen nineties, different homogenization techniques were applied to investigate the macroscopic or effective properties of periodic piezoelectric composites (Galka et al, 1992; Maugin and Turbé, 1991; Telega, 1991; Turbé and Maugin, 1991). The initial studies of the effective dynamical properties of periodic piezoelectric composites by considering Bloch expansions were reported in Telega (1991); Turbé and Maugin (1991). The method of Γ-convergence was used to study the static effective properties without dispersive behavior. In Galka et al (1992), the two-scale asymptotic homogenization was applied for thermo-piezoelectric heterogeneous media.

In this framework, the computation of the effective properties depends on the solution of the so-called local problems. Many works have been devoted to the application of analytical and numerical techniques for solving the local problems, see, for instance, Berger et al (2003, 2006); Bravo-Castillero et al (1997, 1998, 2001); Galka et al (1996); Otero et al (2003); Rodríguez-Ramos et al (1996); Sabina et al (2001). In general, those efforts have been addressed to piezoelectric composites whose constituents exhibit a 6mm symmetry class which are of interest in smart materials applications.

The purpose of this work is essentially twofold. Firstly, to provide closed-form expressions for the effective coefficients of fibrous composites with piezoelectric components which belong to the 622 hexagonal symmetry (Nye, 1957) and with a rectangular distribution of the fibers. These results generalize those published in López-López et al (2005); Aguiar et al (2013) where the periodic cell is a square. Secondly, to describe a procedure to obtain analytical expressions for the components of both the local strain tensor and local electric field intensity vector.

These studies could be interesting for the modeling of biomaterials in bone mechanics applications (for instance, collagen is a natural substance which possesses the 622 symmetry, see Fukada, 1984). In Telega (1991), for the first time, the application of homogenization methods for finding the effective piezoelectric properties of compact bones was sketched. However, up to now, few papers on composites with 622 symmetry have been reported (Aguiar et al, 2013; Alfonso-Rodríguez et al, 2017; López-López et al, 2005; Sevostianov et al, 2014).

The paper is organized as follows. In Sect. 1.2, a family of boundary value problems for periodic piezoelectric media with rapidly oscillating coefficients is presented in matrix notation. The main steps of the asymptotic homogenization procedure to obtain the averaged problem, the local problems, the effective coefficients and the components of the local fields, are summarized. In Sect. 1.4, the homogenization model is applied to the case of unidirectional fibrous composites with 622 piezoelectric phases and a rectangular periodic cell. The relevant local problems are solved based on the theory of functions of a complex variable and closed-form expressions are derived for the corresponding effective coefficients. Analytical expressions are also explicitly given for the components of the strain and electric local fields associated with particular local problems. In Sect. 1.6, some numerical examples are presented and the accuracy of the results is assessed through comparisons with results derived from the FFT numerical scheme (Brenner, 2009, 2010).

1.2 A Boundary Value Problem of the Linear Piezoelectricity Theory

Let $\Omega \subset \mathbb{R}^3$ be a three-dimensional domain with infinitely smooth boundary $\partial\Omega$. The material properties of a piezoelectric body occupying Ω are described by elastic (c_{ijkl}), piezoelectric (e_{ijk}), and dielectric (κ_{ij}) coefficients. These coefficients are assumed to be differentiable, rapidly oscillating and εY-periodic functions in the local variable $y = x/\varepsilon$, where $x = (x_1, x_2, x_3) \in \Omega$ is the global variable, $\varepsilon > 0$ is the usual small geometric parameter, and Y is the periodic cell.

The material functions are defined by

$$c^{\varepsilon}_{ijkl}(x) = c_{ijkl}\left(\frac{x}{\varepsilon}\right), \quad e^{\varepsilon}_{ijk}(x) = e_{ijk}\left(\frac{x}{\varepsilon}\right), \quad \kappa^{\varepsilon}_{ij}(x) = \kappa_{ij}\left(\frac{x}{\varepsilon}\right), \qquad i,j,k = 1,2,3,$$

which are denoted in a unified fashion by $A_{jl} \equiv \left(a^{i'k'}_{jl}\right)_{i',k'=1,..,4}$, where

$$a^{ik}_{jl} = c_{ijkl}, \quad a^{i4}_{jl} = e_{lij}, \quad a^{4k}_{jl} = e_{jkl}, \quad a^{44}_{jl} = -\kappa_{jl}.$$

The material functions satisfy the usual symmetry conditions

$$a^{ik}_{jl} = a^{jk}_{il} = a^{il}_{jk} = a^{ki}_{lj}, \quad a^{i4}_{jl} = a^{j4}_{il}, \quad a^{4k}_{jl} = a^{4l}_{jk}, \quad a^{44}_{jl} = a^{44}_{lj}, \tag{1.1}$$

and we will assume that there exist a constant $\varkappa > 0$ such that, for any symmetric matrix $q = (q_{ij})$ and any vector $a = (a_i)$

$$a^{ik}_{jl}(x)\, q_{ij} q_{kl} \geq \varkappa q_{ij} q_{ij}, \quad a^{44}_{jl}(x)\, a_j a_l \geq \varkappa a_j a_j. \tag{1.2}$$

Note that the summation rule on the repeated indices will be used throughout the paper.

A boundary value problem for the system of equations of linear piezoelectricity can be written as

$$\frac{\partial}{\partial x_j}\left(A^{\varepsilon}_{jl}(x)\frac{\partial}{\partial x_l}U^{\varepsilon}(x)\right) = 0 \in \Omega, \tag{1.3}$$

$$U^{\varepsilon}(x) = \widehat{U}(x), \quad x \in \partial\Omega, \tag{1.4}$$

where $U^{\varepsilon}(x) = \left(u^{\varepsilon}_1(x), u^{\varepsilon}_2(x), u^{\varepsilon}_3(x), v^{\varepsilon}(x)\right)^T$ and $\widehat{U}(x) = (\hat{u}_1(x), \hat{u}_2(x), \hat{u}_3(x), \hat{v}(x))^T$ represent the unknown and the prescribed boundary conditions, respectively. The superscript T stands for transposition. Equation (1.3) represent a system of partial differential equations to find the mechanical displacement field $u^{\varepsilon} = \left(u^{\varepsilon}_k\right)$ and the electric potential v^{ε}. The problem (1.3)-(1.4) describes the piezoelectric state of a composite material that occupies the domain Ω and is free of external forces.

1.3 Homogenization, Local Problems and Effective Coefficients

In this section, the asymptotic homogenization method (AHM) (Bakhvalov and Panasenko, 1989) is applied to the family of problems (1.3)-(1.4). More specifically, the methodology used in Sixto-Camacho et al (2013) is followed.

The solution of (1.3)-(1.4) is sought in the form

$$U^{\varepsilon}(x) = U^{(0)}(x,y) + \varepsilon U^{(1)}(x,y) + \dots + \varepsilon^i U^{(i)}(x,y) + \dots, \tag{1.5}$$

where

$$U^{(i)} = \left(u^{(i)}_1, u^{(i)}_2, u^{(i)}_3, v^{(i)}\right)^T, \qquad i = 0,1,2,\dots,$$

being $u^{(0)}_k(x,y)$, $u^{(1)}_k(x,y)$,..., $v^{(0)}(x,y)$, $v^{(1)}(x,y)$, ... infinitely differentiable and $Y-$periodic functions with respect to y. Substituting (1.5) into (1.3)–(1.4), applying the differentiation chain rule and equating to zero the terms corresponding to equal powers of ε (from $\varepsilon^{-2}, \varepsilon^{-1}, \varepsilon^{0}, \dots$), a recurrent family of partial differential equations is obtained. From the term corresponding to ε^{-2}, it is possible to conclude that the non-perturbed terms of the asymptotic (1.5) are independent of y, that is $U^{(0)} = U^{(0)}(x)$. From the term that corresponds to ε^{-1} the local problems are obtained, which have a solution $U^{(1)}(x,y)$ in the class of $Y-$periodic functions with respect to y. Such a solution can be expressed using the method of separable variables as follows

$$U^{(1)}(x,y) = N^p(y)\frac{\partial U^{(0)}(x)}{\partial x_p}, \tag{1.6}$$

with

$$N^p(y) = \begin{pmatrix} \Xi^{pq}_k(y) & \Upsilon^p_k(y) \\ \Theta^{pq}(y) & \Pi^p(y) \end{pmatrix}_{k,q=1,2,3},$$

where the matrix $N^p(y)$ is a $Y-$periodic solution of

$$\frac{\partial}{\partial y_j}\left(A_{jp}(y) + A_{jl}(y)\frac{\partial N^p(y)}{\partial y_l}\right) = 0. \tag{1.7}$$

Based on the periodicity and ellipticity of the material coefficients it is possible to apply the theorem of the Appendix of Sixto-Camacho et al (2013) to prove that equations (1.7) have a unique $Y-$periodic solution up to an additive constant. The problems involving such equations are the so-called local problems. The solutions of such problems play an important role for the calculation of the effective coefficients. Usually, the condition of null average of the local functions (i.e., $\langle N^p(y)\rangle = 0$) on the periodic cell is imposed for uniqueness. The angular brackets denote the average per unit volume over the cell i.e.

$$\langle g(y)\rangle = \frac{1}{|Y|}\int_Y g(y)dy.$$

On the other hand, from the terms corresponding to ε^0, we obtain the homogenized problem

$$\underline{A}_{pq}\frac{\partial^2 U^{(0)}(x)}{\partial x_p \partial x_q} = 0, \qquad x \in \Omega, \tag{1.8}$$

$$U^{(0)}(x) = \widehat{U}(x), \qquad x \in \partial\Omega. \tag{1.9}$$

and the effective coefficients $\underline{A}_{pq}$, which are defined by

$$\underline{A}_{pq} = \left\langle A_{pq}(y) + A_{pl}(y)\frac{\partial N^p(y)}{\partial y_l}\right\rangle. \tag{1.10}$$

The terms $U^{(i)}(x,y)$ $(i > 1)$ of (1.5) can be also expressed in separable variables by

$$U^{(i)}(x,y) = N^{pp_1\ldots p_{i-1}}(y)\frac{\partial^i U^{(0)}(x)}{\partial x_p \partial x_{p_1}\ldots\partial x_{p_{i-1}}}. \tag{1.11}$$

where $N^{pp_1\ldots p_{i-1}}$ are $Y-$periodic solutions of certain partial differential equations which can be found in Eq. (4.13) of Sixto-Camacho et al (2013).

1.3.1 Explicit Form of the Homogenized Problem, Effective Coefficients and Local Problems

From (1.8)–(1.10), it is possible to obtain the explicit form of the homogenized problem

$$\underline{c}_{ijkl}\frac{\partial^2 u_k^{(0)}(x)}{\partial x_j \partial x_l} + \underline{e}_{mij}\frac{\partial^2 v^{(0)}(x)}{\partial x_j \partial x_m} = 0, \qquad x \in \Omega, \tag{1.12}$$

$$\underline{e}_{ikl}\frac{\partial^2 u_k^{(0)}(x)}{\partial x_i \partial x_l} - \underline{\kappa}_{im}\frac{\partial^2 v^{(0)}(x)}{\partial x_i \partial x_m} = 0, \qquad x \in \Omega. \tag{1.13}$$

$$u_k^{(0)}(x) = \hat{u}_k(x), \quad v^{(0)}(x) = \hat{v}(x), \quad x \in \partial\Omega, \tag{1.14}$$

and the effective coefficients

$$\underline{c}_{ijpq} = \left\langle c_{ijkl}\left[\delta_{kp}\delta_{lq} + \epsilon_{kl,y}(\Xi^{pq})\right] + e_{lij}\frac{\partial\Theta^{pq}}{\partial y_l}\right\rangle, \tag{1.15}$$

$$\underline{e}_{jpq} = \left\langle e_{jkl}\left[\delta_{kp}\delta_{lq} + \epsilon_{kl,y}(\Xi^{pq})\right] - \kappa_{jl}\frac{\partial\Theta^{pq}}{\partial y_l}\right\rangle, \tag{1.16}$$

$$\underline{e}_{pij} = \left\langle e_{lij}\left[\delta_{lp} + \frac{\partial\Pi^p}{\partial y_l}\right] + c_{ijkl}\epsilon_{kl,y}(\Upsilon^p)\right\rangle, \tag{1.17}$$

$$\underline{\kappa}_{ip} = \left\langle \kappa_{il}\left[\delta_{lp} + \frac{\partial\Pi^p}{\partial y_l}\right] - e_{ikl}\epsilon_{kl,y}(\Upsilon^p)\right\rangle, \tag{1.18}$$

where δ_{kl} is the Kronecker's delta and

$$\epsilon_{kl,\xi}(u) = \frac{1}{2}\left(\frac{\partial u_k}{\partial \xi_l} + \frac{\partial u_l}{\partial \xi_k}\right).$$

The local functions Ξ_k^{pq}, Θ^{pq}, Υ_k^p and Π^p are Y-periodic solutions of the following problems on the cell Y:

- Problem L^{pq}: Find the Y-periodic functions Ξ_k^{pq}, Θ^{pq} such that:

$$\begin{cases} \dfrac{\partial}{\partial y_j}\left\{c_{ijkl}\left[\delta_{kp}\delta_{lq} + \epsilon_{kl,y}(\Xi^{pq})\right] + e_{lij}\dfrac{\partial\Theta^{pq}}{\partial y_l}\right\} = 0, & \text{in } Y, \\ \dfrac{\partial}{\partial y_j}\left\{e_{jkl}\left[\delta_{kp}\delta_{lq} + \epsilon_{kl,y}(\Xi^{pq})\right] - \kappa_{jl}\dfrac{\partial\Theta^{pq}}{\partial y_l}\right\} = 0, & \text{in } Y. \end{cases} \tag{1.19}$$

- Problem L^p: Find the Y-periodic functions Υ_k^p, Π^p such that:

$$\begin{cases} \dfrac{\partial}{\partial y_j}\left\{e_{lij}\left[\delta_{lp} + \dfrac{\partial\Pi^p}{\partial y_l}\right] + c_{ijkl}\epsilon_{kl,y}(\Upsilon^p)\right\} = 0, & \text{in } Y, \\ \dfrac{\partial}{\partial y_j}\left\{\kappa_{jl}\left[\delta_{lp} + \dfrac{\partial\Pi^p}{\partial y_l}\right] - e_{jkl}\epsilon_{kl,y}(\Upsilon^p)\right\} = 0, & \text{in } Y. \end{cases} \tag{1.20}$$

1.3.2 Local Fields

Now boundary conditions (1.14) are given by linear functions of the type

$$\hat{u}_k(x) = \hat{\varepsilon}_{kl}x_l, \quad \hat{v}(x) = -\hat{E}_l x_l, \tag{1.21}$$

where $\hat{\varepsilon}_{kl}$ and $\hat{E}_l$ are the components of a constant strain tensor and a constant electric field intensity vector, respectively on the boundary of the composite. Under these conditions, the functions $u_k^{(0)}(x) = \hat{\varepsilon}_{kl}x_l$ and $v^{(0)}(x) = -\hat{E}_l x_l$ represent the solution of the homogenized problem (1.12)-(1.14). So the linearity of $U^{(0)}(x)$ implies that $U^{(i)}(x,y) = 0$ for $i > 1$. Consequently, the components of the asymptotic expansion (1.5) take the form

$$u_k^{\epsilon}(x) = \hat{\varepsilon}_{kl}x_l + \varepsilon\left[\hat{\varepsilon}_{pq}\Xi_k^{pq}(y) - \hat{E}_p\Upsilon_k^p\right], \tag{1.22}$$

$$v^{\epsilon}(x) = -\hat{E}_l x_l + \varepsilon\left[\hat{\varepsilon}_{pq}\Theta^{pq}(y) - \hat{E}_p\Pi^p\right]. \tag{1.23}$$

Therefore, it is possible to obtain the components of the local strain field

$$\varepsilon_{kl,x}(u^{\epsilon}) = \hat{\varepsilon}_{pq}\left[\delta_{kp}\delta_{lq} + \varepsilon_{kl,y}(\Xi^{pq})\right] - \hat{E}_p\varepsilon_{kl,y}(\Upsilon^p), \tag{1.24}$$

and the components of the local electric field intensity

$$E_l(v^{\epsilon}) = \hat{E}_p\left[\delta_{pl} + \frac{\partial \Pi^p}{\partial y_l}\right] - \hat{\varepsilon}_{pq}\frac{\partial \Theta^{pq}}{\partial y_l}, \tag{1.25}$$

where

$$E_l(v^{\epsilon}) = -\frac{\partial v^{\epsilon}}{\partial x_l}.$$

Note that the fields defined by (1.24) and (1.25) only depend on the local variable y.

1.4 Application to a Binary Fibrous Piezocomposite with Perfect Contact Conditions at the Interfaces

In this section, we apply the previously described method to a particular composite. We consider a two-phase fibrous composite consisting of identical circular cylinders embedded in a matrix. Both components are homogeneous piezoelectric materials. The axis of the fibers is parallel to the x_3-axis. The periodic distribution of the fibers follows a rectangular array as shown in Fig. 1.1. Perfect contact conditions on the interface Σ^{ε} between the fibers and the matrix are assumed. The application of the above described homogenization process leads to (1.12)–(1.20), with the addition of contact conditions on the interfaces.

The local problems (1.19)–(1.20) on the periodic cell Y can be written as (Bravo-Castillero et al, 2001; Sabina et al, 2001):

- Problem L^{pq}: To find the Y-periodic functions Ξ_k^{pq} and Θ^{pq} such that:

$$\sigma_{i\delta,\delta}^{pq(\gamma)} = 0, \quad \text{in} \quad Y_{\gamma}, \tag{1.26a}$$

$$D_{\delta,\delta}^{pq(\gamma)} = 0, \quad \text{in} \quad Y_{\gamma}, \tag{1.26b}$$

$$[\![\Xi_k^{pq}]\!] = 0, \quad \text{on} \quad \Sigma, \tag{1.26c}$$

$$[\![\Theta^{pq}]\!] = 0, \quad \text{on} \quad \Sigma, \tag{1.26d}$$

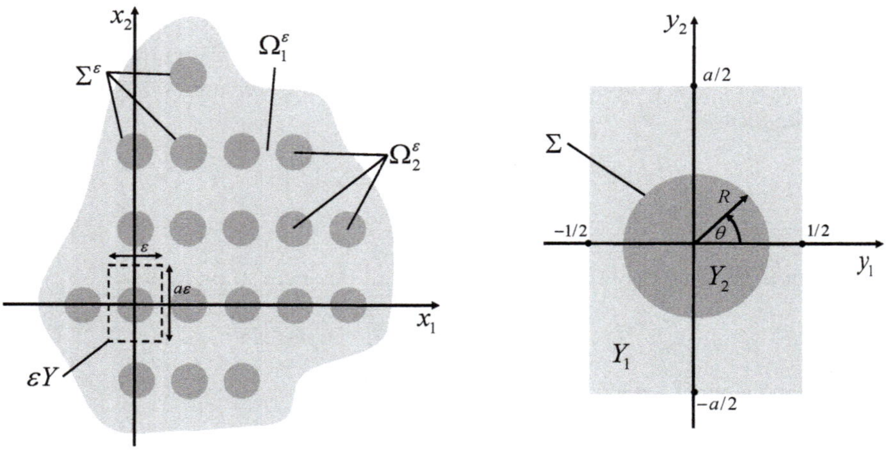

Fig. 1.1: Description of the cross-section of a two-phase fibrous periodic medium and the rectangular periodic cell

$$\left[\!\left[\sigma_{i\delta}^{pq} n_\delta\right]\!\right] = -\left[\!\left[c_{i\delta pq}\right]\!\right] n_\delta, \quad \text{on} \quad \Sigma, \tag{1.26e}$$

$$\left[\!\left[D_\delta^{pq} n_\delta\right]\!\right] = -\left[\!\left[e_{\delta pq}\right]\!\right] n_\delta, \quad \text{on} \quad \Sigma, \tag{1.26f}$$

$$\left\langle \Xi_k^{pq} \right\rangle = 0, \tag{1.26g}$$

$$\left\langle \Theta^{pq} \right\rangle = 0, \tag{1.26h}$$

with

$$\sigma_{i\delta,\delta}^{pq(\gamma)} = c_{i\delta k\lambda}^{(\gamma)} \epsilon_{k\lambda,y}\left(\Xi^{pq(\gamma)}\right) + e_{\lambda i\delta}^{(\gamma)} \Theta_{,\lambda}^{pq(\gamma)}, \tag{1.27a}$$

$$D_\delta^{pq(\gamma)} = e_{\delta k\lambda}^{(\gamma)} \epsilon_{k\lambda,y}\left(\Xi^{pq(\gamma)}\right) - \kappa_{\delta\lambda}^{(\gamma)} \Theta_{,\lambda}^{pq(\gamma)}. \tag{1.27b}$$

- Problem L^p: Find the Y-periodic functions Υ_k^p and Π^p such that:

$$\sigma_{i\delta,\delta}^{p(\gamma)} = 0, \quad \text{in} \quad Y_\gamma, \tag{1.28a}$$

$$D_{\delta,\delta}^{p(\gamma)} = 0, \quad \text{in} \quad Y_\gamma, \tag{1.28b}$$

$$\left[\!\left[\Upsilon_k^p\right]\!\right] = 0, \quad \text{on} \quad \Sigma, \tag{1.28c}$$

$$\left[\!\left[\Pi^p\right]\!\right] = 0, \quad \text{on} \quad \Sigma, \tag{1.28d}$$

$$\left[\!\left[\sigma_{i\delta,\delta}^{p} n_\delta\right]\!\right] = -\left[\!\left[e_{pi\delta}\right]\!\right] n_\delta, \quad \text{on} \quad \Sigma, \tag{1.28e}$$

$$\left[\!\left[D_\delta^{p} n_\delta\right]\!\right] = -\left[\!\left[\kappa_{\delta p}\right]\!\right] n_\delta, \quad \text{on} \quad \Sigma, \tag{1.28f}$$

$$\left\langle \Upsilon_k^p \right\rangle = 0, \tag{1.28g}$$

$$\left\langle \Pi^p \right\rangle = 0, \tag{1.28h}$$

with

$$\sigma_{i\delta,\delta}^{p(\gamma)} = c_{i\delta k\lambda}^{(\gamma)} \epsilon_{k\lambda,y}\left(\Upsilon^{p(\gamma)}\right) + e_{\lambda i\delta}^{(\gamma)} \Pi_{,\lambda}^{p(\gamma)}, \tag{1.29a}$$

$$D_{\delta}^{p(\gamma)} = e_{\delta k\lambda}^{(\gamma)} \epsilon_{k\lambda,y}\left(\Upsilon^{p(\gamma)}\right) - \kappa_{\delta\lambda}^{(\gamma)} \Pi_{,\lambda}^{p(\gamma)}, \tag{1.29b}$$

where n_δ are the components of the outer unit normal vector to the interface Σ, and $[\![\cdot]\!] = (\cdot)^{(1)} - (\cdot)^{(2)}$ denotes the contrast across Σ taken from the matrix to the fiber. $\sigma_{i\delta}$ and D_δ are the components of local stress tensor and electric displacement vector, respectively. The Eqs. (1.26a) and (1.28a) are the corresponding equilibrium relations of solids bodies; (1.26b) and (1.28b) are the quasi-static approximation of Maxwell's equations in the absence of free charges. Perfect contact conditions on the interface are represented by (1.26c)–(1.26f) and (1.28c)–(1.28f). Conditions for uniqueness are given by (1.26g)–(1.26h) and (1.28g)–(1.28h). Formulae (1.27a)–(1.27b) and (1.29a)–(1.29b) are the local constitutive relations. The Latin indices run from 1 to 3, and the Greek ones from 1 to 2. The comma denotes partial differentiation with respect to the local variable y.

1.5 Local Problems for Fibrous Composites with Constituents of 622 Hexagonal Class

In this work, we solve the local problems for the case corresponding to matrix and fibres made of piezoelectric materials with 622 hexagonal symmetry (Nye, 1957). These materials are characterized by eight independent constants (k, m, l, n, p, s', t and u), which are given by five elastic constants

$$2k = c_{1111} + c_{1122}, \quad 2m = c_{1111} - c_{1122} = c_{1212},$$
$$l = c_{1133} = c_{2233} \quad n = c_{3333}, \quad p = c_{1313} = c_{2323};$$

one piezoelectric constant

$$s' = -e_{123} = e_{213} = e_{231} = -e_{132}$$

and two dielectric permittivity constants

$$t = \kappa_{11} = \kappa_{22}, \quad u = \kappa_{33}.$$

Consequently, the local problems L^{11}, L^{22}, L^{33} and L^{12} are exactly the same purely elastic problems which were solved in Nava-Gómez et al (2012) to obtain the effective coefficients $\underline{c}_{1111}$, $\underline{c}_{2211}$, $\underline{c}_{3311}$, $\underline{c}_{2222}$, $\underline{c}_{3322}$, $\underline{c}_{3333}$ and $\underline{c}_{1212}$. On the other hand, from L^3 one obtains that $\underline{\kappa}_{33} = \langle u \rangle$. Therefore, only four local problems (L^{13}, L^{23}, L^1 and L^2) are relevant to obtain the remaining nonzero effective coefficients, which are $\underline{c}_{1313}$, $\underline{c}_{2323}$, $\underline{e}_{213}$, $\underline{e}_{123}$, $\underline{\kappa}_{11}$ and $\underline{\kappa}_{22}$.

In the local problems L^{13}, L^{23}, L^1 and L^2, the relevant constitutive relations (1.27) and (1.29) can be summarized as

$$\sigma_{31} = p\varepsilon_{13,y} - s'E_2, \tag{1.30}$$
$$\sigma_{32} = -s'\varepsilon_{23,y} + tE_1, \tag{1.31}$$
$$D_1 = p\varepsilon_{23,y} + s'E_1, \tag{1.32}$$
$$D_2 = s'\varepsilon_{13,y} + tE_2. \tag{1.33}$$

Thus, only three material properties are here involved: the longitudinal shear modulus p, the shear stress piezoelectric coefficient s' and the transverse permittivity constant t. The next subsections will be dedicated to the solution of these local problems and the further computation of the related effective coefficients.

1.5.1 Local Problems L^{23} and L^1

These two problems can be stated in a unified form as follows

$$\Delta\Xi^{(\gamma)} = 0 \text{ in } Y_\gamma, \tag{1.34a}$$
$$\Delta\Theta^{(\gamma)} = 0 \text{ in } Y_\gamma, \tag{1.34b}$$
$$[\![\Xi]\!] = 0 \text{ on } \Sigma, \tag{1.34c}$$
$$[\![\Theta]\!] = 0 \text{ on } \Sigma, \tag{1.34d}$$
$$[\![(p\Xi_{,1} - s'\Theta_{,2})\, n_1 + (p\Xi_{,2} + s'\Theta_{,1})\, n_2]\!] = An_2 \text{ on } \Sigma, \tag{1.34e}$$
$$[\![(s'\Xi_{,2} - t\Theta_{,1})\, n_1 - (s'\Xi_{,1} + t\Theta_{,2})\, n_2]\!] = Bn_1 \text{ on } \Sigma, \tag{1.34f}$$
$$\langle\Xi\rangle = 0, \tag{1.34g}$$
$$\langle\Theta\rangle = 0, \tag{1.34h}$$

where Δ is the two-dimensional Laplacian. Therefore, the solutions $\Xi^{(\gamma)} (\equiv \Xi_3^{23(\gamma)})$ and $\Theta^{(\gamma)} (\equiv \Theta^{23(\gamma)})$ are doubly periodic harmonic functions of the complex variable $z = y_1 + iy_2$ defined in the rectangular cell Y with periods $\omega_1 = 1$ and $\omega_2 = ai$. The values of A and B in the right hand side of equations (1.34e) and (1.34f) for L^{23} are $-[\![p]\!]$ and $-[\![s']\!]$ respectively. However, for the local problem L^1, these values are $A = -[\![s']\!]$ and $B = [\![t]\!]$, whereas $\Xi^{(\gamma)} (\equiv \Upsilon_3^{1(\gamma)})$ and $\Theta^{(\gamma)} (\equiv \Pi^{1(\gamma)})$.

The solution of (1.34) is sought as follows

$$\Xi^{(1)}(z) = \Im\left\{ -\frac{\delta_2}{\omega_2} a_1 z + \sum_{k=1}^{\infty}{}^{o} a_k \frac{\zeta^{(k-1)}(z)}{(k-1)!} \right\}, \tag{1.35a}$$

$$\Theta^{(1)}(z) = \Re\left\{ -\frac{\delta_1}{\omega_1} b_1 z + \sum_{k=1}^{\infty}{}^{o} b_k \frac{\zeta^{(k-1)}(z)}{(k-1)!} \right\}, \tag{1.35b}$$

$$\Xi^{(2)}(z)=\Im\left\{\sum_{k=1}^{\infty}{}^{o}\, c_k z^k\right\}, \tag{1.35c}$$

$$\Theta^{(2)}(z)=\Re\left\{\sum_{k=1}^{\infty}{}^{o}\, d_k z^k\right\}, \tag{1.35d}$$

where a_k, b_k, c_k and d_k are real and undetermined coefficients. $\Re\{z\}$ and $\Im\{z\}$ are, respectively, the real and imaginary part of the complex number z, and $\zeta(z)$ is the quasi-periodic Weierstrass Zeta function; whereas $\zeta^{(k)}(z)$ denotes the k-th derivative of periods ω_1 and ω_2. The superscript "o" indicates that the summation is carried out only over the odd indices. $\Xi^{(\gamma)}$ is an even function of θ, with $z=Re^{i\vartheta}$, and $\Theta^{(\gamma)}$ is an odd function of θ.

The expressions for the undetermined constants δ_α can be obtained from the quasi-periodicity of $\zeta(z)$

$$\zeta(z+\omega_\alpha)-\zeta(z)=\delta_\alpha, \tag{1.36}$$

where

$$\delta_\alpha=2\zeta\left(\frac{\omega_\alpha}{2}\right)$$

and Legendre's relation is fulfilled (see, for instance Lang, 1993).

The Laurent expansion about the origin for $\Xi^{(1)}$ and $\Theta^{(1)}$ can be written as

$$\Xi^{(1)}(z)=\Im\left\{\sum_{l=1}^{\infty}{}^{o}\, a_l z^{-l}-\sum_{k=1}^{\infty}{}^{o}\, a_k\sum_{l=1}^{\infty}{}^{o}\, k\eta_{kl} z^l\right\}, \tag{1.37a}$$

$$\Theta^{(1)}(z)=\Re\left\{\sum_{l=1}^{\infty}{}^{o}\, b_l z^{-l}-\sum_{k=1}^{\infty}{}^{o}\, b_k\sum_{l=1}^{\infty}{}^{o}\, k\eta'_{kl} z^l\right\}, \tag{1.37b}$$

with

$$\eta_{11}=\frac{\delta_2}{\omega_2},\qquad \eta'_{11}=\frac{\delta_1}{\omega_1},\qquad \eta_{kl}=\eta'_{kl}=\frac{(k+l-1)!}{k!\,l!}S_{k+l}\qquad \text{for}\quad k,l\neq 1, \tag{1.38}$$

and the lattices sum S_{k+l} is defined by

$$S_{k+l}=\sum_{m,n}{}'(m\omega_1+n\omega_2)^{-k-l},\quad k+l\geq 3, \tag{1.39}$$

where the prime on the summation means that it excludes the term $m=n=0$. Now we use contact conditions (1.34c)–(1.34f) to derive the following relations between the undetermined coefficients

$$R^l c_l=-\left(R^{-l}a_l+\sum_{k=1}^{\infty}{}^{o}\, k\eta_{kl}R^l a_k\right), \tag{1.40a}$$

$$R^l d_l = R^{-l} b_l - \sum_{k=1}^{\infty}{}^{o} k\eta'_{kl} R^l b_k, \tag{1.40b}$$

$$AR\delta_{1l} = \left(p^{(1)} + p^{(2)}\right) R^{-l} a_l - [\![p]\!] \left(\sum_{k=1}^{\infty}{}^{o} k\eta_{kl} R^l a_k\right) + [\![s']\!] \left(R^{-l} b_l - \sum_{k=1}^{\infty}{}^{o} k\eta'_{kl} R^l b_k\right), \tag{1.40c}$$

$$BR\delta_{1l} = [\![s']\!] \left(R^{-l} a_l + \sum_{k=1}^{\infty}{}^{o} k\eta_{kl} R^l a_k\right) - \left(t^{(1)} + t^{(2)}\right) R^{-l} b_l - [\![t]\!] \left(\sum_{k=1}^{\infty}{}^{o} k\eta'_{kl} R^l b_k\right), \tag{1.40d}$$

for $l = 1, 3, 5, \ldots$ Note that the coefficients a_k and b_k from (1.40c) and (1.40d) are solutions of an infinite system of linear algebraic equations.

In order to solve the system (1.40), it is convenient to introduce the following change of variables

$$\tilde{a}_l = \sqrt{l}R{-}la_l, \qquad \tilde{b}_l = \sqrt{l}R{-}lb_l, \qquad \tilde{c}_l = \sqrt{l}Rlc_l, \qquad \tilde{d}_l = \sqrt{l}Rld_l. \tag{1.41}$$

Thus, now we can write (1.40) as

$$(I + W)\mathfrak{V}_1 = \mathfrak{V}_3, \tag{1.42a}$$

$$(I - W')\mathfrak{V}_2 = \mathfrak{V}_4, \tag{1.42b}$$

$$\varphi_{11}^{(1)}\mathfrak{V}_1 + \varphi_{12}^{(1)}\mathfrak{V}_2 + \varphi_{11}^{(2)} W\mathfrak{V}_1 + \varphi_{12}^{(2)} W'\mathfrak{V}_2 = \widehat{V}_1, \tag{1.42c}$$

$$\varphi_{21}^{(1)}\mathfrak{V}_1 + \varphi_{22}^{(1)}\mathfrak{V}_2 + \varphi_{21}^{(2)} W\mathfrak{V}_1 + \varphi_{22}^{(2)} W'\mathfrak{V}_2 = \widehat{V}_2, \tag{1.42d}$$

where I is the identity matrix, and the components W and W' for $k = l = 1$ are

$$w_{11} = \frac{\delta_2}{\omega_2} R^2, \qquad w'_{11} = \frac{\delta_1}{\omega_1} R^2, \tag{1.43}$$

and, otherwise,

$$w_{kl} = w'_{kl} = \frac{(k+l-1)!}{(k-1)!(l-1)!} \frac{R^{k+l}}{\sqrt{k}\sqrt{l}} S_{k+l}. \tag{1.44}$$

So, both W and W' are real, symmetric and bounded; then we can use classical results from the theory of infinite systems (Kantorovich and Krylov, 1964). Furthermore,

$$\mathfrak{V}_1 = (\tilde{a}_1, \tilde{a}_3, \tilde{a}_5, \ldots)^T, \qquad \mathfrak{V}_2 = \left(\tilde{b}_1, \tilde{b}_3, \tilde{b}_5, \ldots\right)^T, \tag{1.45}$$

$$\mathfrak{V}_3 = (\tilde{c}_1, \tilde{c}_3, \tilde{c}_5, \ldots)^T, \qquad \mathfrak{V}_4 = \left(\tilde{d}_1, \tilde{d}_3, \tilde{d}_5, \ldots\right)^T. \tag{1.46}$$

and all components of $\widehat{V}_1$ and $\widehat{V}_2$ are zero except the first ones, which are equal to $R\chi_p$ and $R\chi'_t$, respectively, in L^{23}, and to $-R\chi_p$ and $-R\chi'_t$ in L^1

$$\chi_p = \frac{[\![p]\!]}{p^{(1)} + p^{(2)}}, \qquad \chi'_t = \frac{[\![s']\!]}{t^{(1)} + t^{(2)}}, \tag{1.47}$$

$$\chi_p' = \frac{[\![s']\!]}{p^{(1)}+p^{(2)}}, \qquad \chi_t = \frac{[\![t]\!]}{t^{(1)}+t^{(2)}}. \tag{1.48}$$

Moreover, matrices $\Phi^{(\delta)}$, of components $\varphi_{\alpha\beta}^{(\delta)}$, are non-symmetric matrices and can be defined as

$$\Phi^{(1)} = \begin{bmatrix} -1 & -\chi_p' \\ -\chi_t' & 1 \end{bmatrix}, \qquad \Phi^{(2)} = \begin{bmatrix} \chi_p & \chi_p' \\ -\chi_t' & \chi_t \end{bmatrix}. \tag{1.49}$$

Note that the knowledge of $\mathfrak{V}_1$ and $\mathfrak{V}_2$ is enough to solve the system (1.42). Equations (1.42c) and (1.42d) can be transformed into

$$\varphi_{11}\mathfrak{V}_1 + \varphi_{12}\mathfrak{V}_2 + W\mathfrak{V}_1 = V_1, \tag{1.50a}$$
$$\varphi_{21}\mathfrak{V}_1 + \varphi_{22}\mathfrak{V}_2 + W'\mathfrak{V}_2 = V_2, \tag{1.50b}$$

or, in matrix form

$$\Phi \begin{bmatrix} \mathfrak{V}_1 \\ \mathfrak{V}_2 \end{bmatrix} - \begin{bmatrix} W & O \\ O & W' \end{bmatrix} \begin{bmatrix} \mathfrak{V}_1 \\ \mathfrak{V}_2 \end{bmatrix} = \begin{bmatrix} V_1 \\ V_2 \end{bmatrix}, \tag{1.51}$$

where the O denotes the null matrix and, in L^{23}, only the first component of V_1 is nonzero, and equal to R; and in L^1, only the first component of V_2 is nonzero, and equal to $-R$. Besides, we have

$$\Phi = \left[\Phi^{(2)}\right]^{-1} \Phi^{(1)} = \frac{1}{\Lambda} \begin{bmatrix} -\chi_t + \chi_p'\chi_t' & -\chi_p'(1+\chi_t) \\ -\chi_t'(1+\chi_p) & \chi_p - \chi_p'\chi_t' \end{bmatrix}, \tag{1.52}$$
$$\Lambda = \chi_p\chi_t + \chi_p'\chi_t',$$

In order to solve the infinite system (1.50) it must be truncated as follows

$$H\mathfrak{V} = V, \tag{1.53}$$

where $\mathfrak{V} = (\mathfrak{V}_{1i}, \mathfrak{V}_{2i})^T$ and $V = (V_{1i}, V_{2i})^T$, for $i = 1, 3, \ldots, 2n_0 - 1$. The natural number n_0 denotes the truncation order.

The general form of the components of the principal matrix $H = (h_{ij})$ of (1.53) can be defined as follows:

$$h_{ij} = \begin{cases} \text{for } i \text{ odd} \begin{cases} h_{ii} = \varphi_{11} + w_{ii}, \\ h_{ij} = w_{ij}, \quad \text{if } j \text{ odd}, \\ h_{ii+1} = \varphi_{12}, \end{cases} \\ \text{for } i \text{ even} \begin{cases} h_{ii} = \varphi_{22} + w'_{i-1i-1}, \\ h_{ij} = w'_{i-1j-1}, \quad \text{if } j \text{ even}, \\ h_{ii-1} = \varphi_{21}, \end{cases} \end{cases} \tag{1.54}$$

From system (1.53) the values of $\tilde{a}_i$ and $\tilde{b}_i$ can be obtained using the inverse matrix method to solve systems; and then the values of a_i and b_i by reversing the change of variables described in (1.41), resulting in the following formulae

$$a_i = R\mathscr{H}_i U, \qquad b_i = R\mathscr{H}_{i+1} U, \tag{1.55}$$

where $\mathscr{H}_i$ is the i-th row of H^{-1}, which is the inverse matrix of H.

1.5.2 Effective Coefficients Related with the Local Problems L^{23} and L^1

The nonzero effective coefficients which can be computed from the local problems L^{23} and L^1 are

$$\underline{c}_{2323} = p_v + \langle p\Xi^{23}_{3,2} - s'\Theta^{23}_{,1} \rangle, \tag{1.56}$$

$$-\underline{e}_{123} = s'_v + \langle s'\Xi^{23}_{3,2} + t\Theta^{23}_{,1} \rangle = s'_v + \langle p\Upsilon^1_{3,2} + s'\Pi^1_{,1} \rangle, \tag{1.57}$$

$$\underline{\kappa}_{11} = t_v + \langle t\Pi^1_{,1} - s'\Upsilon^1_{3,2} \rangle, \tag{1.58}$$

where $p_v = c_1 p^{(1)} + c_2 p^{(2)}$, with $c_1 + c_2 = 1$ and $c_2 = \pi R^2/a$.

After that, the application of Green's theorem, the doubly periodicity of the local functions and the conditions on Σ leads to the following expressions (see, for instance, Sabina et al, 2001; Aguiar et al, 2013)

$$\underline{c}_{2323} = p^{(1)} \left(1 + \frac{2\pi}{a} a^{23}_1\right), \tag{1.59}$$

$$-\underline{e}_{123} = s'^{(1)} \left(1 + \frac{2\pi t^{(1)}}{a s'^{(1)}} b^{23}_1\right) = s'^{(1)} \left(1 + \frac{2\pi p^{(1)}}{a s'^{(1)}} a^1_1\right), \tag{1.60}$$

$$\underline{\kappa}_{11} = t^{(1)} \left(1 - \frac{2\pi}{a} b^1_1\right), \tag{1.61}$$

where only the residues a_1 and b_1, of $\Xi^{(1)}$ and $\Theta^{(1)}$, are relevant for computing such effective coefficients. The superindices on a_1 and b_1 indicate the local problem which is solved in order to use the formulas (1.55) for $i = 1$.

1.5.3 Local Problems L^{13} and L^2 and Related Effective Coefficients

The study of the local problems L^{13} and L^2 and the related effective coefficients is very similar as above. Then only the main results will be summarized in this subsection.

The simultaneous formulation of these problems consists in to find the doubly periodic functions $\Xi^{(\gamma)}$ and $\Theta^{(\gamma)}$ such that:

$$\Delta \Xi^{(\gamma)} = 0 \text{ in } Y_\gamma, \tag{1.62a}$$

$$\Delta \Theta^{(\gamma)} = 0 \text{ in } Y_\gamma, \tag{1.62b}$$

$$[\![\Xi]\!] = 0 \text{ on } \Sigma, \tag{1.62c}$$

$$[\![\Theta]\!] = 0 \text{ on } \Sigma, \tag{1.62d}$$

$$[\![(p\Xi_{,1} - s'\Theta_{,2})\, n_1 + (p\Xi_{,2} + s'\Theta_{,1})\, n_2]\!] = Cn_1 \text{ on } \Sigma, \tag{1.62e}$$

$$[\![(s'\Xi_{,2} - t\Theta_{,1})\, n_1 - (s'\Xi_{,1} + t\Theta_{,2})\, n_2]\!] = Dn_2 \text{ on } \Sigma, \tag{1.62f}$$

$$\langle \Xi \rangle = 0, \tag{1.62g}$$

$$\langle \Theta \rangle = 0. \tag{1.62h}$$

The solutions $\Xi^{(\gamma)}(\equiv \Xi_3^{13(\gamma)})$ and $\Theta^{(\gamma)}(\equiv \Theta^{13(\gamma)})$ are Y-periodic harmonic functions depending on $z = y_1 + iy_2$ with periods $\omega_1 = 1$ and $\omega_2 = ai$. The values of C and D in the right hand side of equations (1.62e) and (1.62f) for L^{13} are $-[\![p]\!]$ and $[\![s']\!]$, respectively. For the local problem L^1, these values are $C = [\![s']\!]$ and $D = [\![t]\!]$, whereas $\Xi^{(\gamma)}(\equiv \Upsilon_3^{2(\gamma)})$ and $\Theta^{(\gamma)}(\equiv \Pi^{2(\gamma)})$.

According to the interface conditions (1.62e) and (1.62f), the solution of (1.62) is sought as

$$\Xi^{(1)}(z) = \Re\left\{ -\frac{\delta_2}{\omega_2} a_1 z + \sum_{k=1}^{\infty}{}^{o}\, a_k \frac{\zeta^{(k-1)}(z)}{(k-1)!} \right\}, \tag{1.63a}$$

$$\Theta^{(1)}(z) = \Im\left\{ -\frac{\delta_1}{\omega_1} b_1 z + \sum_{k=1}^{\infty}{}^{o}\, b_k \frac{\zeta^{(k-1)}(z)}{(k-1)!} \right\}, \tag{1.63b}$$

$$\Xi^{(2)}(z) = \Re\left\{ \sum_{k=1}^{\infty}{}^{o}\, c_k z^k \right\}, \tag{1.63c}$$

$$\Theta^{(2)}(z) = \Im\left\{ \sum_{k=1}^{\infty}{}^{o}\, d_k z^k \right\}. \tag{1.63d}$$

Furthermore, using similar ideas to those discussed previously, the following formulae for the related effective coefficients can be obtained

$$\underline{c}_{1313} = p^{(1)}\left(1 - \frac{2\pi}{a} a_1^{13}\right), \tag{1.64}$$

$$\underline{e}_{213} = -s'^{(1)}\left(1 + \frac{2\pi t^{(1)}}{as'^{(1)}} b_1^{13}\right) = s'^{(1)}\left(1 + \frac{2\pi p^{(1)}}{as'^{(1)}} a_1^{2}\right), \tag{1.65}$$

$$\underline{\kappa}_{22} = t^{(1)}\left(1 + \frac{2\pi}{a} b_1^{2}\right), \tag{1.66}$$

where a_1 and b_1 are also the residues of $\Xi^{(1)}$ and $\Theta^{(1)}$.

Now, by using the change of variable (1.41) it is possible to arrive to the infinite system to compute $\tilde{a}_1$ and $\tilde{b}_1$:

$$(I-M)\mathfrak{V}_1 = \mathfrak{V}_3, \tag{1.67a}$$

$$(I+M')\mathfrak{V}_2 = -\mathfrak{V}_4, \tag{1.67b}$$

$$\psi_{11}^{(1)}\mathfrak{V}_1 + \psi_{12}^{(1)}\mathfrak{V}_2 + \psi_{11}^{(2)}M\mathfrak{V}_1 + \psi_{12}^{(2)}M'\mathfrak{V}_2 = \widehat{V}_1, \tag{1.67c}$$

$$\psi_{21}^{(1)}\mathfrak{V}_1 + \psi_{22}^{(1)}\mathfrak{V}_2 + \psi_{21}^{(2)}M\mathfrak{V}_1 + \psi_{22}^{(2)}M'\mathfrak{V}_2 = \widehat{V}_2, \tag{1.67d}$$

where the only difference to system (1.42) is that

$$m_{11} = \frac{\delta_1}{\omega_1}R^2, \qquad m'_{11} = \frac{\delta_2}{\omega_2}R^2, \tag{1.68}$$

and, otherwise,

$$m_{kl} = m'_{kl} = \frac{(k+l-1)!}{(k-1)!(l-1)!}\frac{R^{k+l}}{\sqrt{k}\sqrt{l}}S_{k+l}. \tag{1.69}$$

Besides,

$$\Psi^{(1)} = \begin{bmatrix} 1 & -\chi'_p \\ -\chi'_t & -1 \end{bmatrix}, \qquad \Psi^{(2)} = \begin{bmatrix} \chi_p & -\chi'_p \\ \chi'_t & \chi_t \end{bmatrix}, \tag{1.70}$$

and, as above, we find the matrix Ψ, which is given by

$$\Psi = \left[\Psi^{(2)}\right]^{-1}\Psi^{(1)} = \frac{1}{\Lambda}\begin{bmatrix} \chi_t - \chi'_p\chi'_t & -\chi'_p(1+\chi_t) \\ -\chi'_t(1+\chi_p) & -\chi_p + \chi'_p\chi'_t \end{bmatrix}. \tag{1.71}$$

Therefore, we can write the system in the following way

$$\Psi\begin{bmatrix} \mathfrak{V}_1 \\ \mathfrak{V}_2 \end{bmatrix} + \begin{bmatrix} M & O \\ O & M' \end{bmatrix}\begin{bmatrix} \mathfrak{V}_1 \\ \mathfrak{V}_2 \end{bmatrix} = \begin{bmatrix} V_1 \\ V_2 \end{bmatrix}, \tag{1.72}$$

where only the first components of V_1 and V_2 can be different from zero, and equal to R. In the case of problem L^{13}, the first component of V_1 is the one that is nonzero; in the case of problem L^2, only the first component of V_2 is equal to R.

System (1.72) can be written in the form (1.53) with the particularity that the components of the principal matrix $H = (h_{ij})$ are defined by

$$h_{ij} = \begin{cases} \text{for } i \text{ odd} \begin{cases} h_{ii} = \psi_{11} + m_{ii}, \\ h_{ij} = m_{ij}, \quad \text{if } j \text{ odd}, \\ h_{ii+1} = \psi_{12}, \end{cases} \\ \text{for } i \text{ even} \begin{cases} h_{ii} = \psi_{22} + m'_{i-1i-1}, \\ h_{ij} = m'_{i-1j-1}, \quad \text{if } j \text{ even}, \\ h_{ii-1} = \psi_{21}, \end{cases} \end{cases} \tag{1.73}$$

Finally, the corresponding values of a_1 and b_1 can be computed by using Eqs. (1.55), respectively.

1.5.4 On the Computation of the Local Fields from the Solutions of the Local Problem L^{13}

As we can note, the analytical solutions of the local problems are expressed in terms of the local coordinates ρ and θ. However, to study the behavior of the local fields (1.24) and (1.25) it is necessary to compute their derivatives with respect to the Cartesian coordinates y_δ, and then the chain's rule must be used

$$\frac{\partial}{\partial y_1} = \cos\theta \frac{\partial}{\partial \rho} - \frac{\sin\theta}{\rho}\frac{\partial}{\partial \theta}, \qquad \frac{\partial}{\partial y_2} = \sin\theta \frac{\partial}{\partial \rho} + \frac{\cos\theta}{\rho}\frac{\partial}{\partial \vartheta}. \tag{1.74}$$

For instance, the solutions $\Xi^{(\gamma)}(\equiv \Xi_3^{13(\gamma)})$ and $\Theta^{(\gamma)}(\equiv \Theta^{13(\gamma)})$, of the local problem L^{13}, can be written as

$$\Xi^{(1)}(\rho,\theta) = \sum_{l=1}^{2n_0-1}{}^{o}\, a_l \rho^{-l}\cos(l\theta) - \sum_{k=1}^{2n_0-1}{}^{o}\, a_k \sum_{l=1}^{2n_0-1}{}^{o}\, k\eta_{kl}\rho^{l}\cos(l\theta), \tag{1.75a}$$

$$\Theta^{(1)}(\rho,\theta) = -\sum_{l=1}^{2n_0-1}{}^{o}\, b_l \rho^{-l}\sin(l\theta) - \sum_{k=1}^{2n_0-1}{}^{o}\, b_k \sum_{l=1}^{2n_0-1}{}^{o}\, k\eta'_{kl}\rho^{l}\sin(l\theta), \tag{1.75b}$$

$$\Xi^{(2)}(\rho,\theta) = \sum_{k=1}^{2n_0-1}{}^{o}\, c_k \rho^{k}\cos(k\theta), \tag{1.75c}$$

$$\Theta^{(2)}(\rho,\theta) = \sum_{k=1}^{2n_0-1}{}^{o}\, d_k \rho^{k}\sin(k\theta), \tag{1.75d}$$

where the constants a_l, b_l, c_l and d_l $(l = 1,3,\ldots,2n_0-1)$ can be obtained from the solution of (1.67a)–(1.67d) for a truncation order n_0.

As an example, we consider the following homogeneous boundary conditions: $\hat{\varepsilon}_{13} = 1$ and $\hat{\varepsilon}_{pq} = 0$, for the strain tensor components, and $\hat{E}_p = 0$ for the components of electric field intensity vector. Then, Eqs. (1.24) and (1.25) take the form

$$\varepsilon_{kl}(y) = \delta_{k1}\delta_{l3} + \delta_{k3}\delta_{l1} + 2\varepsilon_{kl,y}\left(\Xi^{13}\right), \qquad E_l(y) = -2\frac{\partial \Theta^{13}}{\partial y_l}. \tag{1.76}$$

Finally, it is possible to compute the components of the local fields from the solutions of the local problem L^{13} as follows

$$\varepsilon_{13}(y) = 1 + \frac{\partial \Xi_3^{13}}{\partial y_1}, \qquad \varepsilon_{23}(y) = \frac{\partial \Xi_3^{13}}{\partial y_2}, \tag{1.77}$$

$$E_1(y) = -2\frac{\partial \Theta^{13}}{\partial y_1}, \qquad E_2(y) = -2\frac{\partial \Theta^{13}}{\partial y_2}. \tag{1.78}$$

which can be computed by using the expressions (1.75) and the chain rule. A similar procedure allows to obtain analytical expressions for the local fields related with the solutions of the local problems L^{23}, L^1 and L^2.

1.6 Numerical Examples

Some examples are presented, which include comparisons between the analytical results derived from the present model (PM) and those obtained with the fast Fourier transform (FFT) numerical method (Moulinec and Suquet, 1998; Michel et al, 2001; Brenner, 2009, 2010). For the case of a fiber-reinforced matrix, we have considered the data previously used in López-López et al (2005). They correspond to a collagen matrix with collagen-hydroxyapatite (HA) fibers. For the case of a porous piezoelectric matrix, we have considered the data used in Aguiar et al (2013) corresponding to bone material (Table 1.1).

In Figs. 1.2–1.5, the evolution of the effective coefficients with the fiber volume fraction c_2 is shown. The range of variation of c_2 goes from zero up to the percolation limit when two neighboring fibers or holes get in contact. The results are normalized by the properties of the matrix. In Fig. 1.2, a square periodic cell ($a = 1$) is considered and the results from López-López et al (2005) are reproduced. In Figs. 1.4 and 1.5, a rectangular periodic cell (for $a = 2$) is discussed. Finally, in Figs. 1.6 and 1.7, the spatial distribution of the local fields is illustrated.

Table 1.1: Piezoelectric properties used for the computations (López-López et al, 2005; Aguiar et al, 2013). ϵ_0 denotes the permittivity of free-space.

	Collagen	HA
Longitudinal shear modulus p (GPa)	1.400	2.697
Shear strain piezoelectric coefficient $d = s'/p$ (pC/N)	0.062	0.041
Transverse permittivity constant t/ϵ_0 (no units)	2.825	2.509
	Bone	
Longitudinal shear modulus p (GPa)	8.2	
Shear stress piezoelectric coefficient s' (N/Vm)	2.214×10^{-3}	
Transverse permittivity constant t/ϵ_0 (no units)	6.85	

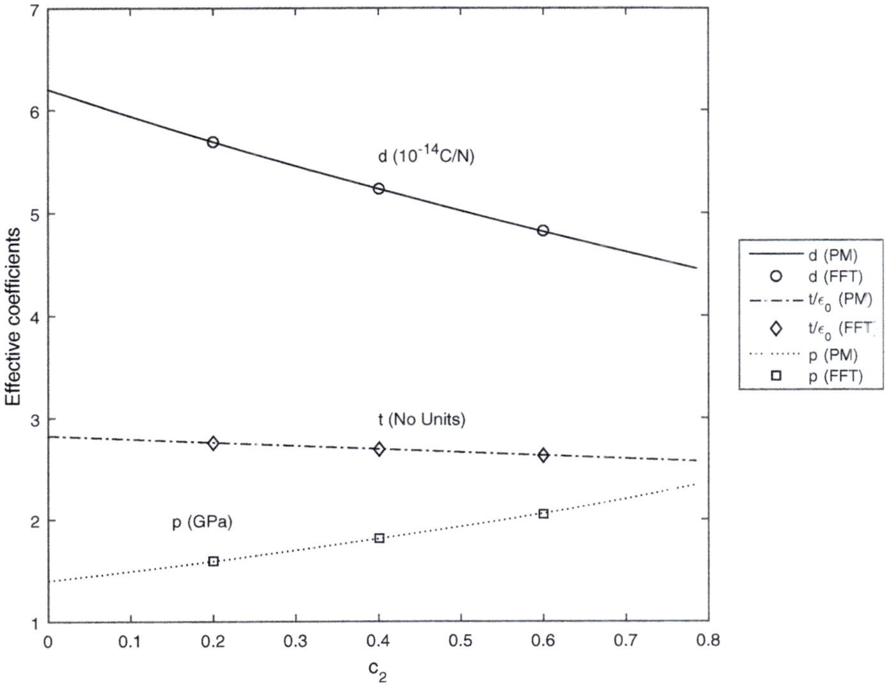

Fig. 1.2: The elastic, piezoelectric and dielectric effective coefficients for a two-phase fibrous piezocomposite with a square ($a = 1$) periodic distribution of fibres. Comparisons of the results obtained with the present model (PM) with those derived by the FFT

1.6.1 Square Array Distribution

In Fig. 1.2, numerical results are shown for a fiber reinforced material with square periodic cell ($a = 1$). In this case the following equalities were numerically verified: $\underline{c}_{1313} = \underline{c}_{2323}$, $\underline{\kappa}_{11} = \underline{\kappa}_{22}$ and $\underline{e}_{213} = -\underline{e}_{123}$. Then, for simplicity, p, t and s' are used to denote the effective elastic, dielectric permittivity and piezoelectric coefficients, respectively. Also, $d = s'/p$ denotes the effective shear strain piezoelectric coefficient. A truncation order $n_0 = 2$ is used for the computations with the present model.

A good agreement can be observed between the semi-analytical (PM) and the numerical results (FFT). The curves corresponding to the present model reproduce those published in López-López et al (2005).

The results shown in Fig. 1.3 correspond to a porous material and, as in Fig. 1.2, the periodic cell is square. Here, the equalities $\underline{c}_{1313} = \underline{c}_{2323} = \underline{\kappa}_{11} = \underline{\kappa}_{22}$ and $\underline{e}_{213} = -\underline{e}_{123}$ stand and were numerically verified. A good agreement between the PM results and those from the numerical FFT method can be observed whatever the

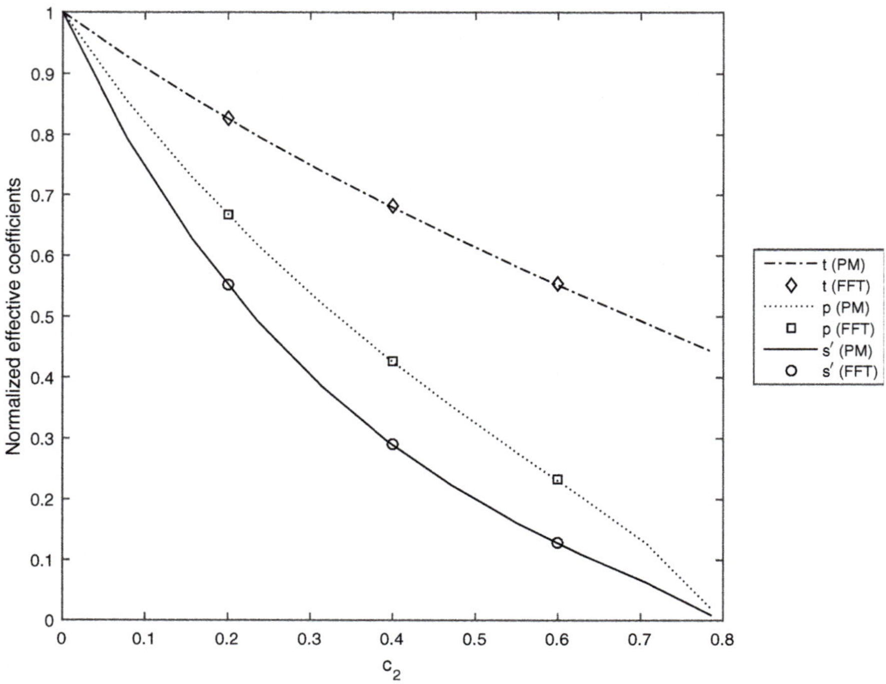

Fig. 1.3: The normalized elastic, piezoelectric and dielectric effective coefficients for a piezocomposite with a square ($a = 1$) periodic distribution of cylindrical holes. Comparisons between the results derived from the present model (PM) with those derived from the Fast Fourier Transform numerical method (FFT)

fiber volume fraction. The results shown for the present model were obtained using a truncation order $n_0 = 5$.

1.6.2 Rectangular Array Distribution

In Figs. 1.4 and 1.5, piezoelectric composites with a rectangular periodic distribution of the fibers are examined for $a = 2$. In Fig. 1.4, a two-phase composite with an orthotropic effective behavior is studied. In Fig. 1.5, a porous piezocomposite which preserves the symmetry properties of the matrix is considered. In both figures, an excellent agreement can be observed, for all the range of fiber area fractions, among the analytical (PM) and numerical (FFT) results and for all the normalized effective coefficients. Note that both models capture the expected global behavior for each type of composite. The truncation order used for the computations with the present model was $n_0 = 4$.

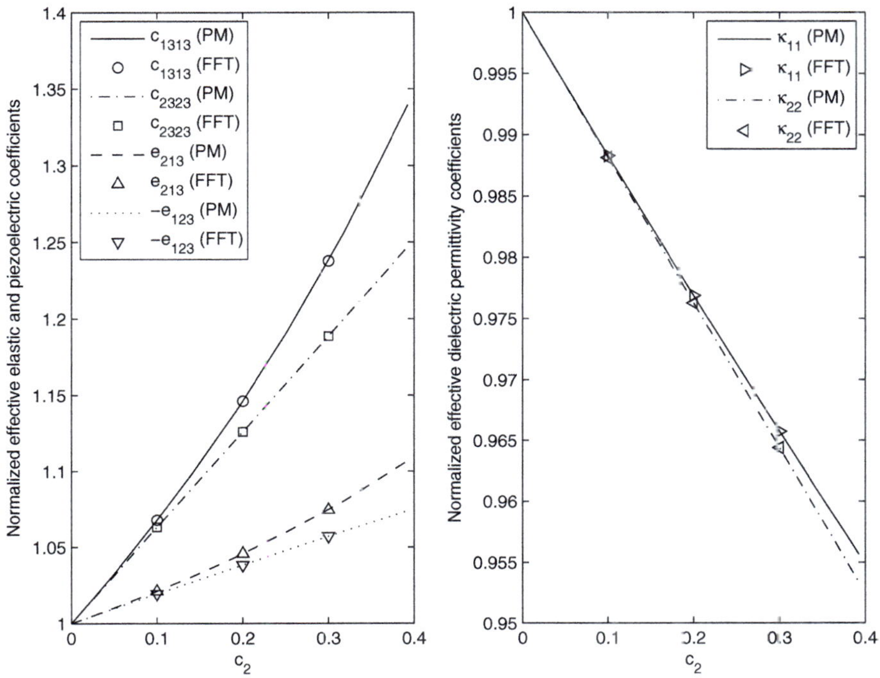

Fig. 1.4: The normalized elastic, piezoelectric and dielectric effective coefficients for a two-phase fibrous piezocomposite with a rectangular ($a = 2$) periodic distribution of fibres. Comparisons of the results derived from the present model (PM) with those derived by the Fast Fourier Transform numerical method (FFT)

1.6.3 Spatial Distribution of Local Fields

The local fields defined by (1.24) and (1.25) depend on the solution of all local problems. However, for specific homogeneous boundary conditions only one local problem needs to be solved in order to compute the local fields. For instance, as was shown above, the solutions of the local problem L^{13} are enough to compute the local fields given by (1.77)–(1.78). In Figs. 1.6 and 1.7, the distribution of the components ϵ_{13} and E_1, for a truncation order $n_0 = 7$ is plotted. In both cases the radius of the fiber is $R = 0.35$. A significant variation is predicted in the matrix at the fiber-matrix interface normal to the x_1 and x_2 axis respectively, and almost uniform fields are predicted within the fibers. The latter is an expected result. Let us consider, for example, the Eq. (1.77): using Eqs. (1.75c) and (1.74) it yields

$$\varepsilon_{13}(y) = 1 + \frac{\partial \Xi_3^{13}}{\partial y_1}, \qquad \varepsilon_{13}(y) = 1 + \sum_{k=1}^{2n_0-1}{}^{o}\, kc_k\rho^{k-1}\cos(k-1)\theta. \tag{1.79}$$

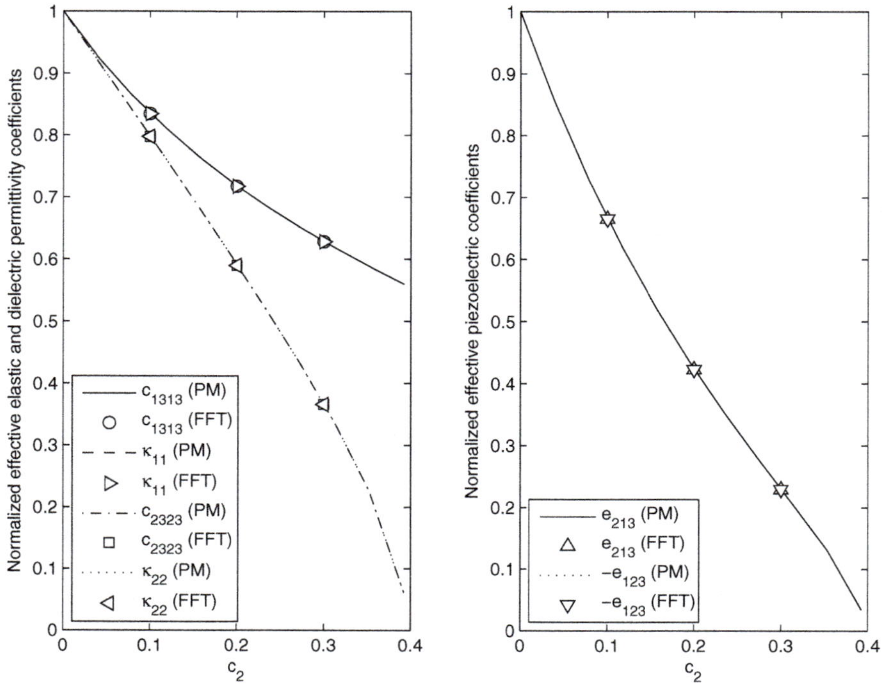

Fig. 1.5: The normalized elastic, piezoelectric and dielectric effective coefficients for a piezocomposite with a rectangular ($a = 2$) periodic distribution of cylindrical holes. The results derived from the present model (PM) and those derived from the Fast Fourier Transform numerical method (FFT) are compared

The sum in the right hand side of Eq. $(1.79)_2$ can be easily majored by the sum

$$|c_1| \sum_{k=1}^{2n_0-1} {}^{o}\, k \left(\frac{1}{2}\right)^{k-1} . \tag{1.80}$$

Hence, the aforementioned sum should have little to nil influence in the final value of the field inside the fiber, since the value of c_1 for the given data set is very small.

The results shown in Fig. 1.6 were obtained using the formula (1.77) corresponding to the solutions of the local problem L^{13}, for homogeneous boundary conditions of the type $\hat{\varepsilon}_{pq} = \delta_{1p}\delta_{3q}$ and $\hat{E}_p = 0$. This field is similar to that predicted via FFT, as can be seen in Fig. 1.6, where the relative error between the computation using Eq. (1.77) and the one using FFT is plotted, i.e. $|\mathrm{PM} - \mathrm{FFT}| / \mathrm{PM}$. It can be seen that in the matrix both methods coincide, while in the fiber they predict behaviors with about a 20% difference. This difference can be explained by the little influence of the sum in the right hand side of Eq. (1.79) in the actual value of the field inside the fiber.

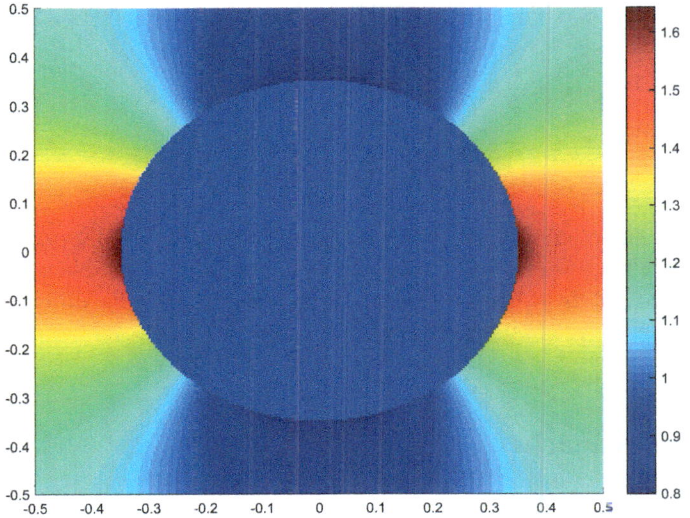

(a)

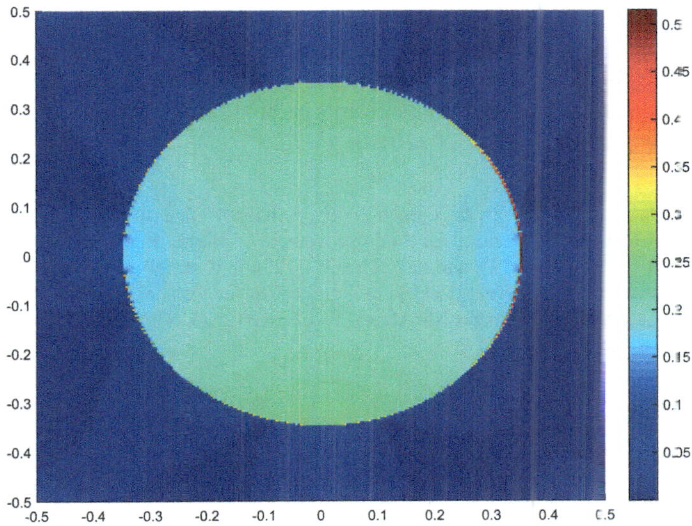

(b)

Fig. 1.6: (a) Component ϵ_{13} of the local strain field within the fiber composite with square arrangement for an axial shear strain load $\hat{\epsilon}_{13} = \hat{\epsilon}_{31} = 1$; (b) Absolute error between the computation in (a) using the present model (1.77) and the FFT numerical scheme

The variation shown in Fig. 1.7 was computed for homogeneous boundary conditions of the type $\hat{\varepsilon}_{pq} = 0$ and $\hat{E}_p = \delta_{1p}$. Such conditions transform (1.24) and (1.25) into

$$\varepsilon_{13}(y) = -\frac{\partial \gamma_3^1}{\partial y_1}, \quad \varepsilon_{23}(y) = -\frac{\partial \gamma_3^1}{\partial y_2}, \tag{1.81}$$

$$E_1(y) = 1 + \frac{\partial \Pi^1}{\partial y_1}, \quad E_2(y) = \frac{\partial \Pi^1}{\partial y_2}. \tag{1.82}$$

Consequently, only the solutions of the problem L^1 are involved in (1.81). Notice that again in Fig. 1.7 the relative error of the computations through the present method and FFT was plotted and they show coincidence in the matrix, while the differences inside the fiber are not greater than a 6%.

1.7 Concluding Remarks

Analytical formulae for the effective coefficients were obtained for binary fibrous composites with 622 hexagonal piezoelectric components and a rectangular distribution of the unidirectional circular fibers. These results contain as particular cases those reported in López-López et al (2005) and Aguiar et al (2013) where only the square periodic cell was considered. Analytical expressions to study the fluctuations of the components of the local strain and the local electric field intensity are explicitly given. For the binary and the porous piezoelectric materials, the present model has been successfully compared to the results obtained with the FFT numerical homogenization method (Brenner, 2009, 2010).

Acknowledgements Support provided by the French Priority Solidarity Fund (Campus France FSP Cuba 29896TD), the Brazilian Coordination for the Improvement of Higher Education Personnel (Project CAPES 88881.030424/2013-01), and the CoNaCyT 129658 is gratefully acknowledged. The authors are also grateful to Ana Pérez Arteaga and Ramiro Chávez for computational support. JB acknowledges the Cátedra Extraordinaria IIMAS and PREI-DGAPA, UNAM.

References

Aguiar AR, Bravo-Castillero J, Rodríguez-Ramos R, da Silva UP (2013) Effective electromechanical properties of 622 piezoelectric medium with unidirectional cylindrical holes. J App Mech 80(5):050,905

Alfonso-Rodríguez R, Bravo-Castillero J, D PL (2017) Local fields in two-phase fibrous piezocomposites with 622 constituents. Math Methods App Sci 40:3221–3229

Bakhvalov N, Panasenko GP (1989) Homogenisation: Averaging Processes in Periodic Media. Kluwer, Oxford

Berger H, Gabbert U, Köppe H, Rodríguez-Ramos R, Bravo-Castillero J, Guinovart-Díaz R, Otero JA, Maugin GA (2003) Finite element and asymptotic homogenization methods applied to smart composite materials. Comp Mech 33:61–67

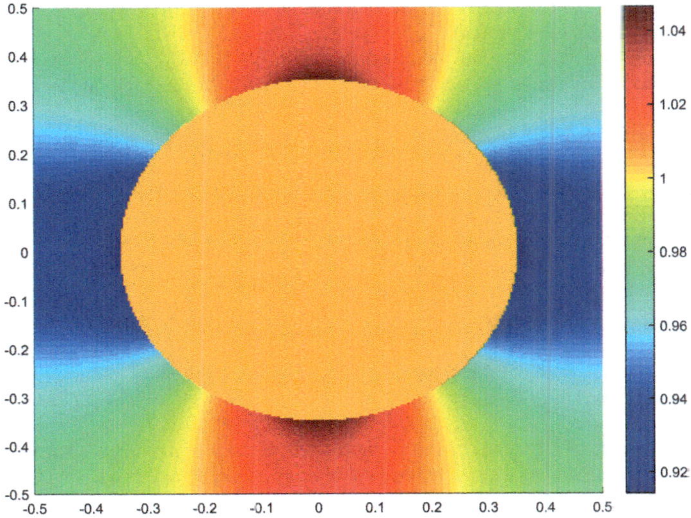

(a)

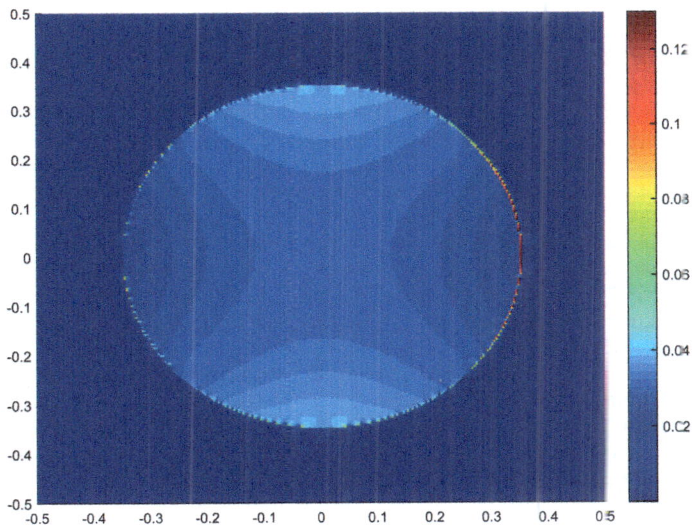

(b)

Fig. 1.7: (a) Component E_1 of the local electric field intensity within the fiber composite with square arrangement for an unitary uniaxial applied electric field $\hat{E}_1 = 1$; (b) Absolute error between the computation in (a) using the present model (1.82) and FFT

Berger H, Kari S, Gabbert U, Rodríguez-Ramos R, Bravo-Castillero J, Guinovart-Díaz R, Sabina FJ, Maugin GA (2006) Unit cell models of piezoelectric fiber composites for numerical and analytical calculation of effective properties. J Smart Mat Struct 15:451–458

Bravo-Castillero J, Guinovart-Díaz R, Otero JA, Rodríguez-Ramos R (1997) Electromechanical properties of continuous fibre-reinforced piezoelectric composites. Mech Comp Mater 33:670–680

Bravo-Castillero J, Otero JA, Rodríguez-Ramos R, Bourgeat A (1998) Asymptotic homogenization of laminated piezocomposite materials. Int J Solids Struct 35:527–541

Bravo-Castillero J, Guinovart-Díaz R, Sabina FJ, Rodríguez-Ramos R (2001) Closed-form expressions for the effective coefficients of fibre-reinforced composite with transversely isotropic constituents - II. Piezoelectric and square symmetry. Mech Mater 33:237–248

Brenner R (2009) Numerical computation of the response of piezoelectric composites using Fourier transform. Phys Rev B 79:184,106

Brenner R (2010) Computational approach for composite materials with coupled constitutive laws. Smart Mater Struct 61:919–927

Fukada E (1984) Piezoelectricity of natural biomaterials. Ferroelec 60:285–296

Galka A, Telega JJ, Wojnar R (1992) Homogenization and thermopiezoelectricity. Mech Res Comm 19:315–324

Galka A, Telega JJ, Wojnar R (1996) Some computational aspects of homogenization of thermopiezoelectric composites. Computer Assisted Mechanics and Engineering Sciences 3:133–154

Kantorovich LV, Krylov VI (1964) Approximate Methods of Higher Analysis. Wiley, New York

Lang S (1993) Complex Analysis, 3rd edn. Graduate Texts in Mathematics, Springer, New York

López-López E, Sabina F, Bravo-Castillero J, Guinovart-Díaz R, Rodríguez-Ramos R (2005) Overall electromechanical properties of a binary composite with 622 symmetry constituents. Antiplane shear piezoelectric state. Int J Solids Struct 42:5765–5777

Maugin GA, Turbé N (1991) Homogenization of piezoelectric composites via Bloch expansion. Int J Appl Electromagnetics in Mat 2:135–140

Michel JC, Moulinec H, Suquet P (2001) A computational scheme for linear and non-linear composites with arbitrary phase contrast. Int J Numer Meth Engng 52:139–160

Moulinec H, Suquet P (1998) A numerical method for computing the overall response of nonlinear composites with complex microstructure. Comput Methods Appl Mech Engrg 157:69–94

Nava-Gómez GG, Camacho-Montes HF, Sabina FJ, Rodríguez-Ramos R, Fuentes L, Guinovart-Díaz R (2012) Elastic properties of an orthotropic binary fiber-reinforced composite with auxetic and conventional constituents. Mech Mater 48:1–25

Nye JF (1957) Physical Properties of Crystals: Their Representation by Tensors and Matrices. Clarendon Press, Oxford

Otero JA, Bravo-Castillero J, Guinovart-Díaz R, Rodríguez-Ramos R, Maugin GA (2003) Analytical expressions of effective constants for a piezoelectric composite reinforced with square cross-section fibers. Arch Mech 55:357–371

Rodríguez-Ramos R, Otero JA, Bravo-Castillero J, Sabina FJ (1996) Electromechanical properties of laminated piezoelectric composites. Mech Comp Mater 32:410–417

Sabina FJ, Rodríguez-Ramos R, Bravo-Castillero J, Guinovart-Díaz R (2001) Closed-form expressions for the effective coefficients of fibre-reinforced composite with transversely isotropic constituents - II. Piezoelectric and hexagonal symmetry. J Mech Phys Solids 49:1463–1479

Sevostianov I, da Silva UP, Aguiar AR (2014) Green's function for piezoelectric 622 hexagonal crystals. Int J Eng Sci 84:18–28

Sixto-Camacho LM, Bravo-Castillero J, Brenner R, Guinovart-Díaz R, Mechkour H, Rodriguez-Ramos R, Sabina FJ (2013) Asymptotic homogenization of periodic thermo-magneto-electro-elastic heterogeneous media. Comp Math App 66:2056–2074

Telega JJ (1991) Piezoelectricity and homogenization. application to biomechanics. In: Maugin GA (ed) Continuum Models and Discrete Systems, Longman, Essex, vol 2, pp 220–229

Turbé N, Maugin GA (1991) On the linear piezoelectricity of composite materials. Math Meth Appl Sci 14:403–412

Chapter 2
High-Frequency Spectra of SH Guided Waves in Continuously Layered Plates

Vladimir I. Alshits and Jerzy P. Nowacki

Abstract The spectra of SH guided waves in an isotropic continuously layered plate with arbitrary profile of the limiting slowness $\hat{s}(y)$ across the plate are explicitly analyzed for high frequencies ω in the framework of "adiabatic" approximation. Dispersion equations and their solutions are analytically found for free, clamped or free-clamped faces of the plate. The positions of horizontal asymptotes for dispersion branches are determined by extreme points y_i of the dependence $\hat{s}(y)$ including also inflection points and "linear" (non-extreme) min/max points. In the vicinity of all asymptotic levels, apart from the upper one, spectra form specific ladder-like patterns. Explicit asymptotics of dispersion curves are derived for a series of particular local dependencies $\hat{s}(y)$ in the vicinity of points y_i.

2.1 Introduction

The subject of this paper, dedicated to G.A. Maugin, belongs to a circle of his interests and closely associates with his results. We mean an analytical explicit theory of SH waves in an elastic inhomogeneous plate. Thirty five years ago he published the well known paper (Maugin, 1983) about properties of SH waves in solids. In this paper and in the previous publication (Bakirtas and Maugin, 1982) the influence of continuous inhomogeneity on the dispersion of SH waves was numerically studied. Fifteen years ago he participated in the study (Alshits et al, 2003) of high-frequency spectra of guided waves in a homogeneous elastic plate of arbitrary anisotropy where the intermediate asymptotes in spectra with a non-trivial ladder-like behaviour of

---

Vladimir I. Alshits
Shubnikov Institute of Crystallography, Russian Academy of Sciences, 119333 Moscow, Russia,
e-mail: valshits@mail.ru

Jerzy P. Nowacki
Polish-Japanese Academy of Information Technology, 02-008 Warsaw, Poland,
e-mail: nowacki@pja.edu.pl

© Springer International Publishing AG, part of Springer Nature 2018
H. Altenbach et al. (eds.), *Generalized Models and Non-classical Approaches in Complex Materials 1*, Advanced Structured Materials 89,
https://doi.org/10.1007/978-3-319-72440-9_2

dispersion branches were first time analytically described. Then the general theory (Alshits and Maugin, 2005, 2008) of acoustic waves in arbitrary anisotropic layered plate was developed (see also Baron et al, 2003; Shuvalov et al, 2004, 2005). In the theory (Alshits and Maugin, 2005, 2008) ladder-like patterns again appeared in spectra of piece-wise homogeneous and continuously layered plates. In both cases the detailed predictions were given of number and positions of the asymptotes in spectra and of the character of approaching these asymptotes by the dispersion branches when the frequency ω increases. The analytical theory for stratified plates was rigorously formulated in terms of the propagator matrix. And the description of graded media was based on a semi-intuitive approach. In accordance with the latter consideration, asymptotic levels in such spectra must occur in all extreme points of the slowness profile $\hat{s}(y)$ across the continuously layered plate (of minimum, maximum and inflection type) and in non-extreme min/max points at the surfaces. The predictions of this theory were in agreement with previously obtained numerical data (Baron et al, 2003).

The further corroboration was given by Shuvalov et al (2008), who basing on the WKB method obtained the same asymptotic dependencies in the cases of extreme maximums for a monoclinic graded plate. They did not consider "linear" points, minimums or inflection points, but just noticed that they could hardly course "a prominent falloff of dispersion" in spectra not being "markedly flat". Such doubts contradicting to Alshits and Maugin (2005, 2008) sound reasonable and need checking.

In this paper we shall reconsider predictions of Alshits and Maugin (2005, 2008) including linear min/max points at the faces, as well as minimums and inflections of the slowness profile $\hat{s}(y)$. We shall use the adiabatic approximation for the propagator matrix. This approach introduced by Alshits and Kirchner (1995) is similar to the WKB method but seems to be more convenient for us being well adjusted to the propagator theory. We shall consider a purely isotropic plate with arbitrary transverse inhomogeneity. This will transform the 6D theory of Alshits and Maugin (2005) into a 2D formalism admitting explicit calculations. It will be even easier than the remarkably simple monoclinic 2D theory (Shuvalov et al, 2008). On the other hand, the presented in Shuvalov et al (2008) positive testing of results of WKB method by comparing them with exact computing will allow us to limit our analysis to adiabatic approximation.

2.2 Statement of the Problem and Main Equations

Consider the layered elastic plate of the thickness d with the faces parallel to the layers. The material constants involved into the analysis, the shear modulus μ and the density ρ, are supposed to be arbitrary continuous functions of the coordinate y along the normal $\mathbf{n}$ to the faces. The latter are intersected by this axis at $y = 0$ and $y = d$ (Fig. 2.1). The sagittal plane of wave propagation is xy, and the propagation direction is x. Certainly the theory is equally applicable to a layered hexagonal plate with the sagittal plane parallel to the plane of transverse isotropy. In this case $\mu = c_{44}(y)$.

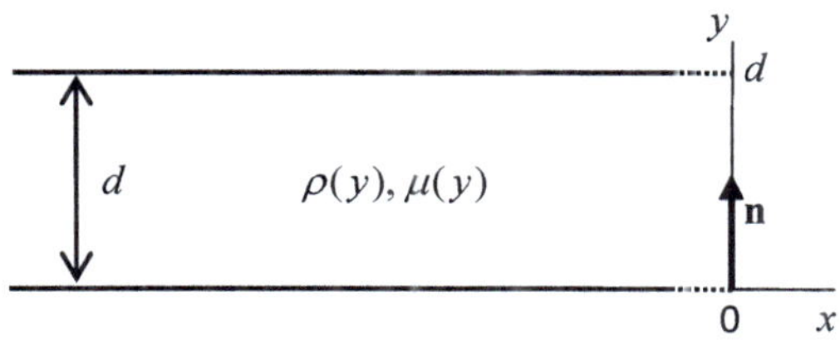

Fig. 2.1 The continuously layered plate with the surfaces parallel to the layers in the Cartesian coordinate system.

In such plate an SH wave field is characterized by the displacement ($\mathbf{u}$) and traction ($\sigma\mathbf{n}$) vectors which are both directed along z and can be described by the scalar steady-state fields

$$u(x,y,t) = A(y)\exp[ik(x-vt)], \tag{2.1}$$

$$\sigma(x,y,t) = -ikL(y)\exp[ik(x-vt)], \tag{2.2}$$

where k is the wave number, $v = \omega/k$ is the tracing speed and ω is the wave frequency. Let us form the two-component vector function

$$\boldsymbol{\eta} = \begin{pmatrix} A(y) \\ L(y) \end{pmatrix}. \tag{2.3}$$

In the considered case the standard 6D equation of Stroh (1962) acquires the 2D form

$$\frac{d\boldsymbol{\eta}}{dy} = ik\hat{N}_{SH}(y)\boldsymbol{\eta}, \tag{2.4}$$

where $\hat{N}_{SH}$ is the 2×2 analog of the Stroh matrix:

$$\hat{N}_{SH} = -\begin{pmatrix} 0 & 1/\mu \\ p^2\mu & 0 \end{pmatrix}. \tag{2.5}$$

Here

$$p^2 = (\hat{s}/s)^2 - 1, \tag{2.6}$$

with $s = v^{-1}$ for the slowness of the wave and $\hat{s}(y)$ for the limiting slowness profile of the plate:

$$\hat{s}^2 = \rho(y)/\mu(y), \tag{2.7}$$

which is supposed to be a known function.

The eigenvectors ($\xi_{1,2}$) and eigenvalues ($p_{1,2}$) of the $\hat{N}_{SH}$ matrix are equal (Alshits et al, 2005)

$$\xi_1(y) = \frac{i}{\sqrt{2}}\begin{pmatrix} (\mu p)^{-1/2} \\ -(\mu p)^{1/2} \end{pmatrix}, \quad \xi_2(y) = \frac{1}{\sqrt{2}}\begin{pmatrix} (\mu p)^{-1/2} \\ (\mu p)^{1/2} \end{pmatrix}, \quad p_{1,2}(y) = \pm p\,. \tag{2.8}$$

2.3 The Propagator Matrix and Its Adiabatic Approximation

Thus, we should solve Eq. (2.4) and then from boundary conditions find the dispersion branches $s_l(\omega)$, $l = 1, 2 \ldots$. The solution of (2.4) is conveniently expressed in the form (Alshits and Maugin, 2005, 2008)

$$\boldsymbol{\eta}(y) = \hat{W}^{(k)}(y|0)\boldsymbol{\eta}(0), \tag{2.9}$$

where $\hat{W}^{(k)}$ is the propagator matrix which in analogy with Alshits and Maugin (2005); Alshits and Kirchner (1995) may be presented in the form

$$\hat{W}^{(k)}(y|0) = \mathrm{Ord}\exp[ik\int_0^y \hat{N}_{SH}(y')dy']. \tag{2.10}$$

The Ord operator means the definite order of non-commuting matrices $\hat{N}_{SH}$ in the products $\hat{N}_{SH}(y')\hat{N}_{SH}(y'')\ldots\hat{N}_{SH}(y''')$ (with $y' > y'' > \ldots > y'''$ for $y > 0$ and $y' < y'' < \ldots < y'''$ for $y < 0$) in integrands of multiple integrals arising after expansion of the exponential operator in (2.10).

Such solutions of equations similar (2.4) are widely used in quantum mechanics with time-dependent Hamiltonian (Landau and Lifshitz, 1991), in mathematics (Gantmacher, 1989; Shilov, 1996) and in mechanics of inhomogeneous media (Nayfeh, 1995). In our paper we are interested by only high-frequency region of the acoustic spectrum, i.e. we need the asymptotic representation of the matrix $\hat{W}^{(k)}$ at large k. The proper analytical representation known as adiabatic approximation has the form (Alshits and Kirchner, 1995)

$$\hat{W}^{(k)}_{\mathrm{ad}}(y|0) \approx \sum_{\alpha=1}^{2} \xi_{k\alpha}(y) \otimes \hat{T}\xi_{\alpha}(0), \tag{2.11}$$

where

$$\xi_{k\alpha}(y) = \xi_{\alpha}(y)\exp[iky\overline{p}_{\alpha}(y)]\,, \quad \overline{p}_{\alpha}(y) = y^{-1}\int_0^y p_{\alpha}(y')dy'\,; \tag{2.12}$$

$$\hat{T} = \begin{pmatrix} 0 & 1 \\ 1 & 0 \end{pmatrix}. \tag{2.13}$$

It is easily checked that this approximate solution satisfies Eq. (2.4) at large k. Substituting (2.12) into (2.4) one obtains the criterion for this satisfaction:

$$|\xi'_{\alpha}(y)| \ll k|p(y)\xi_{\alpha}(y)|\,. \tag{2.14}$$

Thus the approximation is the better, the larger k and the less inhomogeneity of the material. However, inequality (2.14) is not the only limitation on the applicability of representation (2.11). Apart from differential equation (2.4), the solution should also satisfy some known criteria (Alshits and Kirchner, 1995) for the propagator matrix. For instance, as is seen from (2.10) the identity must be valid

$$[\hat{W}^{(k)}(y|0)]^* = \hat{W}^{(-k)}(y|0)\,. \tag{2.15}$$

This relation is satisfied by (2.11) only when the averaged factors $\overline{p}_{1,2}$ (2.12)$_2$ are either real, or imaginary (in the second case the two terms in sum (2.11) are complex conjugate). However, as follows from (2.6), (2.8) and (2.12)$_2$, the factors $\overline{p}_{1,2}$ might be also generally complex at some regions of slowness s. This difficulty may be overcome basing on the main propagator property

$$\hat{W}^{(k)}(y_3|y_1) = \hat{W}^{(k)}(y_3|y_2)\hat{W}^{(k)}(y_2|y_1)\,. \tag{2.16}$$

At any s one can always divide the interval $[0,y]$ on several subintervals $[0,y_1]$, $[y_1,y_2],\ldots,[y_N,y]$, in which the parameters $p_{1,2}$ (2.6) (and consequently $\overline{p}_{1,2}$) are either real, or purely imaginary. Then, with (2.16), one can present the asymptotic formula as the matrix product

$$\hat{W}^{(k)}(y|0) = \hat{W}^{(k)}_{\text{ad}}(y|y_N)\ldots\hat{W}^{(k)}_{\text{ad}}(y_2|y_1)\hat{W}^{(k)}_{\text{ad}}(y_1|0) \tag{2.17}$$

where all matrices on the right-hand side are given by (2.11)-(2.13). Of course, with changing the slowness s the positions y_l of the division points and their number N may vary (Fig. 2.2).

Substitution of (2.8) into (2.11), (2.12) determines the matrices $\hat{W}^{(k)}_{\text{ad}}$ in (2.17):

$$\hat{W}^{(\omega)}_{\text{ad}}(y_{l+1}|y_l) = \begin{pmatrix} \sqrt{\gamma_l/\gamma_{l+1}}\cos(\omega I_l) & -(i/\sqrt{\gamma_l\gamma_{l+1}})\sin(\omega I_l) \\ -i\sqrt{\gamma_l\gamma_{l+1}}\sin(\omega I_l) & \sqrt{\gamma_{l+1}/\gamma_l}\cos(\omega I_l) \end{pmatrix}. \tag{2.18}$$

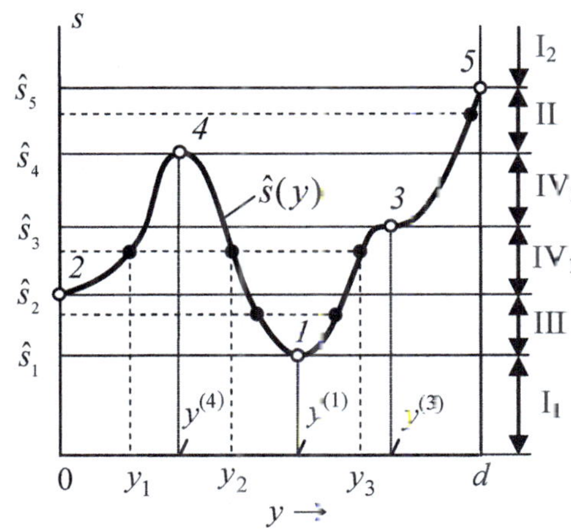

Fig. 2.2 Profile $\hat{s}(y)$ of the limiting slowness of the plate material with five min/max and inflection points (*1, 2, …, 5*) determining asymptotic levels $\hat{s}_1,\hat{s}_2,\ldots,\hat{s}_5$ in the SH wave spectrum. The regions I_i, II, III and IV_j of slowness s relate to different numbers of zeros of the function $p(y)$ (e.g. in the regions $\mathrm{IV}_{1,2}$ there are three such zeros: y_1, y_2 and y_3).

Here $k = \omega s$ is put with replacing the superscript $(k) \to (\omega)$ and the notations

$$\gamma_l \equiv \gamma_{y_l} = \mu(y_l)p(y_l), \quad I_l(s) \equiv I_{y_l y_{l+1}} = \int\limits_{y_l}^{y_{l+1}} dy\sqrt{[\hat{s}(y)]^2 - s^2} \tag{2.19}$$

are introduced. In these terms we shall imply below that $y_0 \equiv 0$ and $y_{N+1} \equiv d$.

It is easily checked that (2.18) satisfies identity (2.15) for both real and imaginary parameter $\overline{p}(y_l) = I_l/s$ (2.12)$_2$. And still, there remains one more trouble in solution (2.17)-(2.19): all the division points $y_1, y_2, \ldots, y_N$ exactly correspond to the limiting slowness $s = \hat{s}$ for which $p = 0$ (2.6) and the corresponding parameters $\gamma_1, \gamma_2, \ldots, \gamma_N$ vanish. In view of (2.18), this leads to singularities in all matrices of product (2.17). In order to exclude these singularities, let us introduce in the vicinity of each internal division point y_l the two additional points $y_l^{\pm} = y_l \pm \varepsilon$ where ε is chosen so, that $\varepsilon k \ll 1$. Basing on (2.18) one can prove that for small enough ε the matrix products $\hat{W}_{\rm ad}^{(\omega)}(y_l^+|y_l)\hat{W}_{\rm ad}^{(\omega)}(y_l|y_l^-)$ for any $l = 1, 2, \ldots, N$ can be made arbitrarily close to the identity matrix $\hat{I}$ with corrections linear in ε. Thus, (2.17) is reduced to the form

$$\hat{W}^{(\omega)}(y|0) = \lim_{\varepsilon \to 0}\left\{\hat{W}_{\rm ad}^{(\omega)}(y|y_N^+)\hat{W}_{\rm ad}^{(\omega)}(y_N^-|y_{N-1}^+)\ldots\hat{W}_{\rm ad}^{(\omega)}(y_2^-|y_1^+)\hat{W}_{\rm ad}^{(\omega)}(y_1^-|0)\right\} \tag{2.20}$$

which does not contain singularities. Indeed, the small parameters $\gamma_l^{\pm} \propto \varepsilon$ present in (2.20) only in the combination γ_l^-/γ_l^+ with the following limiting behaviour:

$$\lim_{\varepsilon \to 0}\left\{\gamma_l^-/\gamma_l^+\right\} = i\,\mathrm{sgn}\left\{\hat{s}'(y_l)\right\}. \tag{2.21}$$

2.4 Boundary Problems and Their General Solutions

As shown by Shuvalov et al (2008), the dispersion equations for SH waves in inhomogeneous plates are directly formulated in terms of scalar elements of the propagator

$$\hat{W}^{(\omega)}(d|0) = \begin{pmatrix} W_1 & W_2 \\ W_3 & W_4 \end{pmatrix}. \tag{2.22}$$

It allows a convenient description of the boundary problems for the plate surfaces which are both free (f/f), both clamped (c/c), the lower face free - the upper face clamped (f/c) and the lower face clamped - the upper face free (c/f):

$$W_3 = 0\ (\mathrm{f/f}),\ W_2 = 0\ (\mathrm{c/c}),\ W_1 = 0\ (\mathrm{f/c}),\ W_4 = 0\ (\mathrm{c/f}). \tag{2.23}$$

2.4.1 Spectral Regions Without Division Points

2.4.1.1 The Range $s < \min\{\hat{s}(y)\}$ (I_1 in Fig. 2.2)

In this case Eq. (2.6) gives real $p(y)$ in any layer of the plate which means that there are no division points in the region $0 \le y \le d$ and the propagator matrix is directly determined by (2.18):

$$\hat{W}^{(\omega)}(d|0) = \hat{W}^{(\omega)}_{\text{ad}}(d|0) = \begin{pmatrix} \sqrt{\gamma_0/\gamma_d}\cos(\omega I_{0d}) & -(i/\sqrt{\gamma_0\gamma_d})\sin(\omega I_{0d}) \\ -i\sqrt{\gamma_0\gamma_d}\sin(\omega I_{0d}) & \sqrt{\gamma_d/\gamma_0}\cos(\omega I_{0d}) \end{pmatrix}. \quad (2.24)$$

Then, by (2.22), (2.23), in this range one has the dispersion equations

$$\sin(\omega I_{0d}) = 0 \text{ (f/f and c/c)}, \quad \cos(\omega I_{0d}) = 0 \text{ (f/c and c/f)}. \quad (2.25)$$

With (2.25), asymptotics for inverse dispersion branches $\omega(s_l)$ $(l = 1, 2, \ldots)$ are given by

$$\omega(s_l) = \frac{\pi l}{I_{0d}(s_l)} \text{ (f/f and c/c)}, \quad \omega(s_l) = \frac{\pi(l - \frac{1}{2})}{I_{0d}(s_l)} \text{ (f/c and c/f)}. \quad (2.26)$$

2.4.1.2 The Range $s > \max\{\hat{s}(y)\}$ (I_2 in Fig. 2.2)

In this range the parameter $p(y)$ (2.6) is purely imaginary throughout the plate. Again there are no division points. And the propagator is given by the replacement $I_{0d} \to i|I_{0d}|$ in (2.24):

$$\hat{W}^{(\omega)}(d|0) = \begin{pmatrix} \sqrt{\gamma_0/\gamma_d}\cosh(\omega|I_{0d}|) & (1/\sqrt{\gamma_0\gamma_d})\sinh(\omega|I_{0d}|) \\ \sqrt{\gamma_0\gamma_d}\sinh(\omega|I_{0d}|) & \sqrt{\gamma_d/\gamma_0}\cosh(\omega|I_{0d}|) \end{pmatrix}. \quad (2.27)$$

No elements of this matrix can vanish which means that *at $s > \max\{\hat{s}(y)\}$ there are no dispersion branches for SH waves.*

2.4.2 Spectral Regions with one Division Point

In the intermediate range $\min\{\hat{s}(y)\} < s < \max\{\hat{s}(y)\}$ there must be regions for y where $p(y)$ (2.6) takes either real or imaginary values. We start from the situation when the equation $\hat{s}(y) = s$ has only one solution $y = a$ dividing the plate thickness into two parts: $0 < y < a$ and $a < y < d$. In this case the short wavelength asymptotics of the propagator matrix has the form

$$\hat{W}^{(\omega)}(d|0) = \lim_{\varepsilon\to 0}\left\{\hat{W}^{(\omega)}_{\text{ad}}(d|a^+)\hat{W}^{(\omega)}_{\text{ad}}(a^-|0)\right\} \tag{2.28}$$

With appropriate expressions from (2.18) for matrices $\hat{W}^{(\omega)}_{\text{ad}}$ in (2.28) the further analysis depends on the sign of the derivative $\mathring{s}'(a)$.

2.4.2.1 The Case $\mathring{s}'(a) > 0$ (the Range II in Fig. 2.2)

In this case the parameters γ_0 and I_{0a} are imaginary and the parameters γ_d and I_{ad} are real, and the propagator matrix acquires the form

$$\hat{W}^{(\omega)}(d|0) = \frac{\exp(\omega|I_{0a}|)}{\sqrt{2}}\begin{pmatrix} \sqrt{\dfrac{\gamma_0}{i\gamma_d}}\sin(\omega I_{ad}+\dfrac{\pi}{4}) & \dfrac{1}{\sqrt{i\gamma_0\gamma_d}}\sin(\omega I_{ad}+\dfrac{\pi}{4}) \\ \sqrt{i\gamma_0\gamma_d}\cos(\omega I_{ad}+\dfrac{\pi}{4}) & \sqrt{\dfrac{i\gamma_d}{\gamma_0}}\cos(\omega I_{ad}+\dfrac{\pi}{4}) \end{pmatrix}. \tag{2.29}$$

Here we used (2.21) and the high frequency approximation for hyperbolic functions

$$\sinh(\omega|I_{0a}|) \approx \cosh(\omega|I_{0a}|) \approx \frac{1}{2}\exp(\omega|I_{0a}|). \tag{2.30}$$

Found expression (2.29) gives dispersion equations for all boundary problems (2.23):

$$\cos(\omega I_{ad}+\frac{\pi}{4}) = 0 \text{ (f/f and c/f)}; \quad \sin(\omega I_{ad}+\frac{\pi}{4}) = 0 \text{ (c/c and f/c)}\,. \tag{2.31}$$

Solving these equations gives the inverse dispersion branches $\omega(s_l)(l = 1,2,\ldots)$:

$$\left.\begin{aligned} \omega &= \pi(l+\frac{1}{4})/I_{ad}(s_l) \text{ (f/f and c/f)} \\ \omega &= \pi(l-\frac{1}{4})/I_{ad}(s_l) \text{ (c/c and f/c)} \end{aligned}\right\}\ \mathring{s}'(a) > 0\,. \tag{2.32}$$

2.4.2.2 The Case $\mathring{s}'(a) < 0$

This time, vice versa, the parameters γ_d and I_{0d} are imaginary and the parameters γ_0 and I_{0a} are real. Accordingly, Eqs. (2.29) and (2.31) should be transformed by replacing there $\gamma_d \leftrightarrow \gamma_0$ and $I_{ad} \leftrightarrow I_{0a}$. And the inverse dispersion branches $\omega(s_l)(l = 1,2,\ldots)$ are

$$\left.\begin{aligned} \omega &= \pi(l+\frac{1}{4})/I_{0a}(s_l)\text{(f/f and c/f)} \\ \omega &= \pi(l-\frac{1}{4})/I_{0a}(s_l)\text{(c/c and f/c)} \end{aligned}\right\}\ \mathring{s}'(a) < 0\,. \tag{2.33}$$

2.4.3 Spectral Regions with two Division Points

Let us suppose that the equation $\hat{s}(y) = s$ has two roots: $y_1 = a$ and $y_2 = b > a$. In this case the propagator is the product of three submatrices:

$$\hat{W}^{(k)}(d|0) = \hat{W}^{(k)}_{\mathrm{ac}}(d|b^+)\hat{W}^{(k)}_{\mathrm{ad}}(b^-|a^+)\hat{W}^{(k)}_{\mathrm{ad}}(a^-|0). \tag{2.34}$$

Matrix product (2.34) leads to different dispersion equations for two types of sign combinations of derivatives $\hat{s}'(a)$ and $\hat{s}'(b)$ which will be considered separately.

2.4.3.1 The Case $\hat{s}'(a) > 0, \hat{s}'(b) < 0$

Now the parameter $p(y)$ (2.6) is imaginary in the intervals $(0,a)$ and (b,d), being real in the interval (a,b). Equations (2.34) and (2.18) give

$$\hat{W}^{(k)}(d|0) = \frac{1}{2}\exp[\omega(|I_{0a}| + |I_{bd}|)\cos(\omega I_{ab})\begin{pmatrix} \sqrt{\gamma_0/\gamma_d} & 1/\sqrt{\gamma_0\gamma_d} \\ \sqrt{\gamma_0\gamma_d} & \sqrt{\gamma_d/\gamma_0} \end{pmatrix}. \tag{2.35}$$

Thus, all boundary problems f/f, c/c, f/c and c/f are reduced to one equation

$$\cos(\omega I_{ab}) = 0 \tag{2.36}$$

and the inverse branches $\omega(s_l)$, $l = 1,2,\ldots$ are in this case also universal:

$$\omega = \pi(l - \frac{1}{2})/I_{ab}(s_l)\ . \tag{2.37}$$

2.4.3.2 The Case $\hat{s}'(a) < 0, \hat{s}'(b) > 0$ (the Range III in Fig. 2.2)

Now parameter $p(y)$ (2.6) is real in the intervals $(0,a)$ and (b,d), being imaginary in the interval (a,b). So, the propagator matrix acquires the form

$$\hat{W}^{(k)}(d|0) = \exp(\omega|I_{ab}|)\times$$

$$\begin{pmatrix} \sqrt{\frac{\gamma_0}{\gamma_d}}\cos(\omega I_{0a} + \frac{\pi}{4})\sin(\omega I_{bd} + \frac{\pi}{4}) & \frac{-i}{\sqrt{\gamma_0\gamma_d}}\sin(\omega I_{0a} + \frac{\pi}{4})\sin(\omega I_{bd} + \frac{\pi}{4}) \\ i\sqrt{\gamma_0\gamma_d}\cos(\omega I_{0a} + \frac{\pi}{4})\cos(\omega I_{bd} + \frac{\pi}{4}) & \sqrt{\frac{\gamma_d}{\gamma_0}}\sin(\omega I_{0d} + \frac{\pi}{4})\cos(\omega I_{bd} + \frac{\pi}{4}) \end{pmatrix}. \tag{2.38}$$

From here we obtain the following dispersion equations (2.23):

$$\begin{aligned}
&\text{f/f}: \quad \cos(\omega I_{0a}+\frac{\pi}{4})=0,\ \cos(\omega I_{bd}+\frac{\pi}{4})=0;\\
&\text{c/c}: \quad \sin(\omega I_{0a}+\frac{\pi}{4})=0,\ \sin(\omega I_{bd}+\frac{\pi}{4})=0;\\
&\text{f/c}: \quad \cos(\omega I_{0a}+\frac{\pi}{4})=0,\ \sin(\omega I_{bd}+\frac{\pi}{4})=0;\\
&\text{c/f}: \quad \sin(\omega I_{0a}+\frac{\pi}{4})=0,\ \cos(\omega I_{bd}+\frac{\pi}{4})=0.
\end{aligned} \tag{2.39}$$

Each of them describes a couple of branch series, $s_m(\omega)$ and $s_n(\omega)$. They are numbered separately $(m,n=1,2,\ldots)$. As before, we present the inverse functions:

$$\begin{aligned}
&\text{f/f}: \quad \omega=\pi(m+\frac{1}{4})/I_{0a}(s_m),\ \omega=\pi(n+\frac{1}{4})/I_{bd}(s_n);\\
&\text{c/c}: \quad \omega=\pi(m-\frac{1}{4})/I_{0a}(s_m),\ \omega=\pi(n-\frac{1}{4})/I_{bd}(s_n);\\
&\text{f/c}: \quad \omega=\pi(m+\frac{1}{4})/I_{0a}(s_m),\ \omega=\pi(n-\frac{1}{4})/I_{bd}(s_n);\\
&\text{c/f}: \quad \omega=\pi(m-\frac{1}{4})/I_{0a}(s_m),\ \omega=\pi(n+\frac{1}{4})/I_{bd}(s_n).
\end{aligned} \tag{2.40}$$

Until now for each boundary condition we obtained only one series of branches. In (2.40) we have a couple of branch series for each boundary problem which are independent and admit multiple intersections which are generally forbidden (Alshits et al, 2003; Alshits and Maugin, 2005, 2008; Shuvalov et al, 2008). In fact, these intersections arise due to our asymptotic approximation (2.30) which excludes very weak branch repulsion. Exact calculations would eliminate such degeneracies. However the higher is frequency the less is size of the repulsion region. Often in order to visualize this region one should greatly scale up the picture. This effect is one of reasons for formation of terracing patterns in spectra (Alshits et al, 2003; Shuvalov et al, 2008).

2.4.4 Extension for an Arbitrary Number of Division Points

2.4.4.1 An Odd Number of Division Points $N=2n-1$ $(n>1)$

As before, the results of an analysis depend on the sign of the derivative $\mathring{s}'(y_1)$. If $\mathring{s}'(y_1)>0$ the parameters I_m (2.19) are purely imaginary when their numbers are even $(m=2l)$, being real for odd numbers $(m=2l-1)$. The propagator matrix in this case takes the form

$$\hat{W}^{(\omega)}(d|0) = \frac{1}{\sqrt{2}}\exp\left[\omega\sum_{l=0}^{n-1}|I_{2l}|\right]\prod_{l=1}^{n-1}\cos(\omega I_{2l-1})\times$$
$$\begin{pmatrix} \sqrt{\frac{\gamma_0}{i\gamma_d}}\sin(\omega I_N+\frac{\pi}{4}) & \frac{1}{\sqrt{i\gamma_0\gamma_d}}\sin(\omega I_N+\frac{\pi}{4}) \\ \sqrt{i\gamma_0\gamma_d}\cos(\omega I_N+\frac{\pi}{4}) & \sqrt{\frac{i\gamma_d}{\gamma_0}}\cos(\omega I_N+\frac{\pi}{4}) \end{pmatrix}. \tag{2.41}$$

The corresponding dispersion equations (2.23) are

$$\begin{aligned} &\textstyle\prod_{l=1}^{n-1}\cos(\omega I_{2l-1})=0,\ \cos(\omega I_N+\frac{\pi}{4})=0\ (\mathrm{f/f,\ c/f}); \\ &\textstyle\prod_{l=1}^{n-1}\cos(\omega I_{2l-1})=0,\ \sin(\omega I_N+\frac{\pi}{4})=0\ \ (\mathrm{c/c,\ f/c}). \end{aligned} \tag{2.42}$$

So, for each boundary problem we obtain the n sets of inverse dispersion branches:

$$\mathrm{f/f,\ c/f}: \begin{cases} \omega=\pi(m-\frac{1}{2})/I_{2l-1}(s_m),\ m=1,2,\ldots,\ l=1,\ldots,n-1,\ n>1; \\ \omega=\pi(m+\frac{1}{4})/I_N(s_m),\ m=1,2,\ldots; \end{cases} \tag{2.43}$$

$$\mathrm{c/c,\ f/c}: \begin{cases} \omega=\pi(m-\frac{1}{2})/I_{2l-1}(s_m),\ m=1,2,\ldots,\ l=1,\ldots,n-1,\ n>1; \\ \omega=\pi(m-\frac{1}{4})/I_N(s_m),\ m=1,2,\ldots . \end{cases} \tag{2.44}$$

If $\hat{s}'(y_1)<0$, the set I_{2l} is real and I_{2l-1} is imaginary. Thus we obtain

$$\hat{W}^{(k)}(d|0) = \frac{1}{\sqrt{2}}\exp\left[\omega\sum_{l=1}^{n}|I_{2l-1}|\right]\prod_{l=1}^{n-1}\cos(\omega I_{2l})\times$$
$$\begin{pmatrix} \sqrt{\frac{i\gamma_0}{\gamma_d}}\cos(\omega I_0+\frac{\pi}{4}) & \frac{1}{\sqrt{i\gamma_0\gamma_d}}\sin(\omega I_0+\frac{\pi}{4}) \\ \sqrt{i\gamma_0\gamma_d}\cos(\omega I_0+\frac{\pi}{4}) & \sqrt{\frac{\gamma_d}{i\gamma_0}}\sin(\omega I_0+\frac{\pi}{4}) \end{pmatrix}. \tag{2.45}$$

Again for each boundary problem (2.23) there occur n sets of inverse branches

$$\mathrm{f/f,\ f/c}: \begin{cases} \omega=\dfrac{\pi(m-\frac{1}{2})}{I_{2l}(s_m)},\ m=1,2,\ldots,\ l=1,\ldots,n-1,\ n>1; \\ \omega=\pi(m+\frac{1}{4})/I_0(s_m),\ m=1,2,\ldots; \end{cases} \tag{2.46}$$

$$\text{c/c, c/f}: \begin{cases} \omega = \dfrac{\pi(m-\frac{1}{2})}{I_{2l}(s_m)}, \; m = 1,2,\ldots, \; l = 1,\ldots,n-1, \; n > 1; \\ \qquad \omega = \pi(m-\frac{1}{4})/I_0(s_m), \; m = 1,2,\ldots. \end{cases} \tag{2.47}$$

Found solutions (2.43), (2.44) and (2.46), (2.47) are valid for odd numbers N of division points starting from $N = 3$ and extend above solutions (2.32), (2.33) valid for $N = 1$.

2.4.4.2 An Even Number of Division Points $N = 2n$ ($n > 1$)

In the case $\hat{s}'(y_1) > 0$ one obtains

$$\hat{W}^{(\omega)}(d|0) = \frac{1}{2}\exp\left[\omega\sum_{l=1}^{n}|I_{2l}|\right]\prod_{l=1}^{n}\cos(\omega I_{2l-1})\begin{pmatrix} \sqrt{\dfrac{\gamma_0}{\gamma_d}} & 1/\sqrt{\gamma_0\gamma_d} \\ \sqrt{\gamma_0\gamma_d} & \sqrt{\gamma_d/\gamma_0} \end{pmatrix}. \tag{2.48}$$

The n sets of dispersion equations are common for all four boundary problems

$$\prod_{l=1}^{n}\cos(\omega I_{2l-1}) = 0 \quad (\text{f/f, c/c, f/c, c/f}) \tag{2.49}$$

which are solved by

$$\omega = \frac{\pi(m-\frac{1}{2})}{I_{2l-1}(s_m)}, \; m = 1,2,\ldots, \; l = 1,\ldots,n, \; n \geq 1. \tag{2.50}$$

The found general solution describes also the case $n = 1$ and coincides with (2.37).

For $\hat{s}'(y_1) < 0$ the propagator matrix is more bulky than (2.48):

$$\hat{W}^{(k)}(d|0) = \exp\left[\omega\sum_{l=1}^{n}|I_{2l-1}|\right]\prod_{l=1}^{n-1}\cos(\omega I_{2l})\times$$
$$\begin{pmatrix} \sqrt{\dfrac{\gamma_0}{\gamma_d}}\cos(\omega I_0+\dfrac{\pi}{4})\sin(\omega I_N+\dfrac{\pi}{4}) & \dfrac{-i}{\sqrt{\gamma_0\gamma_d}}\sin(\omega I_0+\dfrac{\pi}{4})\sin(\omega I_N+\dfrac{\pi}{4}) \\ i\sqrt{\gamma_0\gamma_d}\cos(\omega I_0+\dfrac{\pi}{4})\cos(\omega I_N+\dfrac{\pi}{4}) & \sqrt{\dfrac{\gamma_d}{\gamma_0}}\sin(\omega I_0+\dfrac{\pi}{4})\cos(\omega I_N+\dfrac{\pi}{4}) \end{pmatrix}. \tag{2.51}$$

This time we obtain $n+1$ sets of dispersion branches different for all boundary conditions:

$$\text{f/f}: \begin{cases} \omega = \dfrac{\pi(m-\frac{1}{2})}{I_{2l}(s_m)}, \; m=1,2,\ldots, \; l=1,\ldots,n-1, \; n>1; \\ \omega = \pi(m+\dfrac{1}{4})/I_0(s_m), \; m=1,2,\ldots; \\ \omega = \pi(m+\dfrac{1}{4})/I_N(s_m), \; m=1,2,\ldots; \end{cases} \tag{2.52}$$

$$\text{c/c}: \begin{cases} \omega = \dfrac{\pi(m-\frac{1}{2})}{I_{2l}(s_m)}, \; m=1,2,\ldots, \; l=1,\ldots,n-1, \; n>1; \\ \omega = \pi(m-\dfrac{1}{4})/I_0(s_m), \; m=1,2\ldots; \\ \omega = \pi(m-\dfrac{1}{4})/I_N(s_m), \; m=1,2,\ldots; \end{cases} \tag{2.53}$$

$$\text{f/c}: \begin{cases} \omega = \dfrac{\pi(m-\frac{1}{2})}{I_{2l}(s_m)}, \; m=1,2,\ldots, \; l=1,\ldots,n-1, \; n>1; \\ \omega = \pi(m+\dfrac{1}{4})/I_0(s_m), \; m=1,2,\ldots; \\ \omega = \pi(m-\dfrac{1}{4})/I_N(s_m), \; m=1,2,\ldots; \end{cases} \tag{2.54}$$

$$\text{c/f}: \begin{cases} \omega = \dfrac{\pi(m-\frac{1}{2})}{I_{2l}(s_m)}, \; m=1,2,\ldots, \; l=1,\ldots n-1, \; n>1; \\ \omega = \pi(m-\dfrac{1}{4})/I_0(s_m), \; m=1,2,\ldots; \\ \omega = \pi(m+\dfrac{1}{4})/I_N(s_m), \; m=1,2,\ldots. \end{cases} \tag{2.55}$$

These equations are valid for $N \geq 4$. They extend sets (2.42) of branches for the case $N=2$.

2.5 The Low-Slowness Approximation and the Cut-Off Frequencies

Below we shall study the found spectra in the vicinity of some levels of the slowness where the branches $s_l(\omega)$ manifest some specific features. We start from the range of low slowness $s \ll \min\{\hat{s}(y)\}$. In this limit the spectrum is easily found from (2.26) with

$$I_{0d}(s_l) = \int_0^d dy \sqrt{\hat{s}^2(y) - s_l^2} \approx d\left(\bar{\hat{s}} - \frac{1}{2} s_l^2 \overline{\hat{s}^{-1}}\right), \tag{2.56}$$

where the notations are introduced

$$\bar{\hat{s}} = \frac{1}{d}\int_0^d dy\hat{s}(y)\,, \qquad \overline{\hat{s}^{-1}} = \frac{1}{d}\int_0^d dy\hat{s}^{-1}(y)\,. \tag{2.57}$$

Combining (2.26) with (2.56) one readily obtains the spectrum. In particular, for the f/f boundary condition we have

$$s_l(\omega) \approx \sqrt{\frac{2\bar{\hat{s}}}{\overline{\hat{s}^{-1}}}}\left(1 - \frac{\omega_l^0}{\omega}\right), \tag{2.58}$$

where ω_l^0 are the so-called cut-off frequencies,

$$\omega_l^0 = \pi l / \bar{\hat{s}} d \qquad l = 1, 2, \ldots. \tag{2.59}$$

Each branch $s_l(\omega)$ exists only for frequencies ω exceeding its own cut-off frequency ω_l^0 for which $s_l(\omega_l^0) = 0$. As is seen from (2.58), the derivative $\partial s_l / \partial \omega$ at $\omega \to \omega_l^0$ tends to infinity which means that the branches $s_l(\omega)$ come out the cut-off frequencies ω_l^0 with vertical slopes.

A similar approximate description of this part of the spectrum was earlier presented by Shuvalov et al (2008) in somewhat more general statement. It was shown that Eqs. (2.58), (2.59) are in quite good fit with an exact computer spectrum in the vicinity of cut-off frequencies exceeding 10 rad/μs. Indeed, in this range of the spectrum we have the small slowness, $s_l^2 \ll \hat{s}^2(y)$, and accordingly the large parameter $p \approx \hat{s}(y)/s_l$ (2.6). Then the criterion (2.14) reduces to the inequality

$$\omega \gg \frac{(\mu\hat{s})'}{2\mu\hat{s}^2} = \frac{(\rho/\hat{s})'}{2\rho}, \tag{2.60}$$

which is satisfied for the example considered in Shuvalov et al (2008) (see Fig. 3 in those paper).

2.6 Example of Inhomogeneity Admitting an Explicit Analysis

Let us consider the example of slowness profile $\hat{s}(y)$ shown in Fig. 2.3:

$$\hat{s}(y) = \hat{s}_0\sqrt{1 + \delta\frac{y}{d}} \tag{2.61}$$

which can be explicitly analyzed basing on the above theoretical approach. As was shown in Subsects. 2.4.1 and 2.4.2, the spectral regions $s \leq \hat{s}_0 = \min\{\hat{s}(y)\}$ and $\hat{s}_0 \leq s < \hat{s}_m = \max\{\hat{s}(y)\}$ should be considered separately.

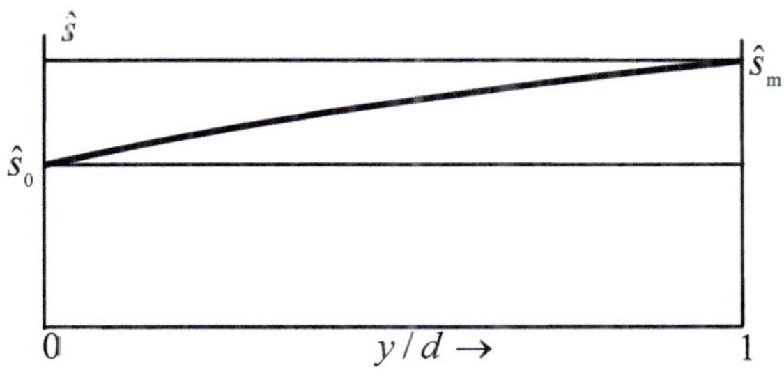

Fig. 2.3 Schematic plot of the function $\hat{s}(y)$ (2.61).

2.6.1 *The Region* $0 < s \le \hat{s}_0$

In this range the inverse branches $\omega(s_l)$ are given by (2.26) with

$$I_{0d} = \frac{2d}{3\hat{s}_0^2\delta}\left\{\left(\hat{s}_m^2 - s_l^2\right)^{3/2} - \left(\hat{s}_0^2 - s_l^2\right)^{3/2}\right\}. \tag{2.62}$$

For instance, for the f/f boundary problem one has

$$\omega = \frac{3\pi l\hat{s}_0^2\delta}{2d\left\{\left(\hat{s}_m^2 - s_l^2\right)^{3/2} - \left(\hat{s}_0^2 - s_l^2\right)^{3/2}\right\}} = \frac{\delta^{3/2}\omega_l}{\left(1+\delta - s_l^2/\hat{s}_0^2\right)^{3/2} - \left(1 - s_l^2/\hat{s}_0^2\right)^{3/2}}, \tag{2.63}$$

where

$$\omega_l = l\Delta\omega\,, \qquad \Delta\omega = \frac{3\pi}{2d\hat{s}_0\sqrt{\delta}}\,. \tag{2.64}$$

2.6.1.1 The Cut-Off Frequencies of the Spectrum

Each of branches $s_l(\omega)$ exists only for frequencies ω exceeding the cut-off frequency ω_l^0 for which $s_l(\omega_l^0) = 0$. Substituting into (2.63) $s_l = 0$ one finds the spectrum of cut-off frequencies

$$\omega_l^0 = l\Delta\omega^0\,, \qquad \Delta\omega^0 = \frac{3\pi\hat{s}_0\delta}{2d(\hat{s}_m^3 - \hat{s}_0^3)} = \frac{\Delta\omega\delta^{3/2}}{(1+\delta)^{3/2} - 1}\,. \tag{2.65}$$

As is seen from (2.63), at the points $s_l = 0$ the derivative $d\omega/ds_l$ vanishes which means that the branches $s_l(\omega)$ come out the cut-off frequencies ω_l^0 with vertical slopes (Fig. 2.4). This part of the spectrum related to the range $s_l \ll \hat{s}_0$ has the form

$$s_l(\omega) = \sqrt{\frac{\pi l\hat{s}_0\delta}{d(\sqrt{1+\delta}-1)}\left(\frac{1}{\omega_l^0} - \frac{1}{\omega}\right)}\,. \tag{2.66}$$

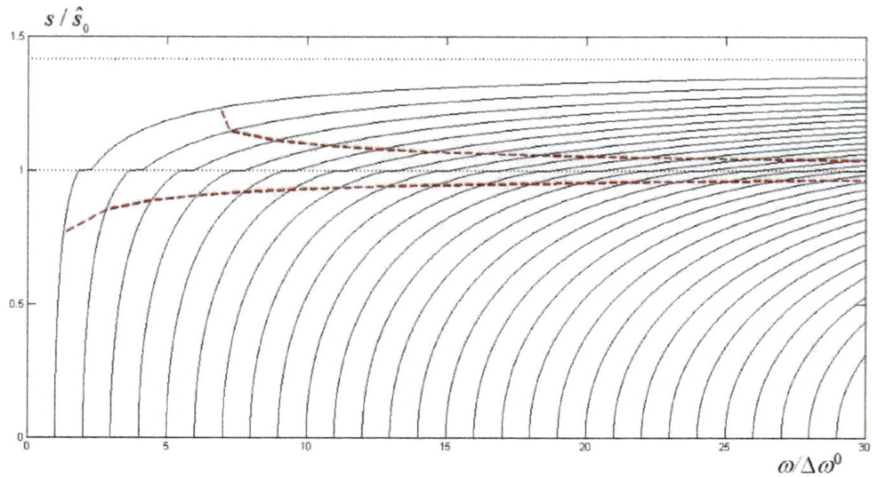

Fig. 2.4: The spectrum of the plate with slowness profile (2.61) for $\delta = 1$. Solid lines relate to branches $s_l(\omega)$. Dashed lines bound the forbidden band. Dotted lines indicate the levels $s = \hat{s}_0$ and $s = \hat{s}_0\sqrt{2}$.

At small inhomogeneity $\delta \ll 1$ this expression as well as the interval $\Delta\omega^0$ between the cut-off frequencies become independent of δ:

$$s_l(\omega) = \sqrt{\frac{2\pi l s_0}{d}\left(\frac{1}{\omega_l^0} - \frac{1}{\omega}\right)}, \qquad \Delta\omega^0 \approx \frac{\pi}{\hat{s}_0 d}. \tag{2.67}$$

2.6.1.2 Spectrum just Under the Level $s_l = \hat{s}_0$

Consider now the narrow band of the spectrum, $\Delta s_l = \hat{s}_0 - s_l \ll \hat{s}_0$, just below the level $s_l = \hat{s}_0$. Here one should be careful because the upper limit of this band $s_l = \hat{s}_0$ relates to a singular level of the spectrum where the eigenvectors $\xi_{1,2}$ (2.8) diverge due to vanishing of p (2.6). In the vicinity of this level the solution (2.63) satisfies the criterion (2.14) of applicability of the adiabatic approximation if

$$\omega \gg \frac{\hat{s}_0^2\delta}{2d(\hat{s}_0^2 - s_l^2)^{3/2}}. \tag{2.68}$$

Thus for large enough frequency ω Eq. (2.63) describes the spectrum as close to the singular level $s_l = \hat{s}_0$ as we need. Let us choose the arbitrarily narrow forbidden gap of the width $\Delta s \ll \hat{s}_0$ under the level $s_l = \hat{s}_0$ and estimate the lower bound l_0 of spectrum branch numbers, $l \gg l_0$, for which our approximation (2.68) works at $s_l = \hat{s}_0 - \Delta s$ and consequently throughout the region $\Delta s < \hat{s}_0 - s_l \ll \hat{s}_0$.

The frequencies ω_l^b where the dispersion lines intersect the bound,

$$s_l(\omega) = \hat{s}_0 - \Delta s,$$

are found from (2.63):

$$\omega_l^b = l\Delta\omega^b\,, \qquad \Delta\omega^b \approx \frac{3\pi\delta}{2d\hat{s}_0(\delta + 2\Delta s/\hat{s}_0)^{3/2}} \approx \Delta\omega\,. \tag{2.69}$$

Here we supposed for simplicity that $\Delta s \ll \hat{s}_0\delta/3$. In this approximation the found frequencies (2.69) are very close to the set (2.64) related to intersections with the singular level $s_l = \hat{s}_0$. Combining (2.69) with criterion (2.68) taken at $s_l = \hat{s}_0 - \Delta s$ one obtains

$$l \gg l_0 = \frac{(\hat{s}_0\delta/\Delta s)^{2/3}}{6\pi\sqrt{2}}\,, \tag{2.70}$$

i.e. $l_0 = 5$ when $\Delta s/\hat{s}_0 \approx 0.04\delta$. Thus, the width Δs_l^0 of the forbidden zone, where our adiabatic approach fails, depends on l and converges $\propto l^{-2/3}$ (Fig. 2.4):

$$\Delta s_l^0 \sim \frac{\hat{s}_0\delta}{(2\pi l)^{2/3}}\,. \tag{2.71}$$

For the considered interval $\Delta s_l^0 < \Delta s_l \ll \hat{s}_0$ Eq. (2.63) takes the form

$$\left(1 + \frac{2\Delta s_l}{\hat{s}_0\delta}\right)^{3/2} - \left(\frac{2\Delta s_l}{\hat{s}_0\delta}\right)^{3/2} = \frac{\omega_l}{\omega}\,. \tag{2.72}$$

In the region $\Delta s_l^0 < \Delta s_l \ll \hat{s}_0\delta$ this equation is solved by

$$s_l(\omega) \approx \hat{s}_0\left\{1 - \frac{\delta}{3}\left[\frac{\omega_l}{\omega} - 1\right]\right\}\,. \tag{2.73}$$

It follows from (2.73) that slopes of the branches $s_l(\omega)$ in points ω_l^b of their intersection with the level $s_l = \hat{s}_0 - \Delta s_l^0$ decrease with growing branch number as $1/l$:

$$s_l'(\omega)\mid_{\omega_l} \approx \frac{\hat{s}_0\delta}{3\omega_l}\,. \tag{2.74}$$

2.6.2 *The Region* $\hat{s}_0 < s \leq \hat{s}_m$

In this spectral interval we meet the situation described in the Sect. 2.4.2.1. The inverse spectrum now is described by equation (2.32) with

$$I_{ad}(s_l) = \frac{2\omega d}{3\hat{s}_0^2}(\hat{s}_m^2 - \hat{s}_l^2)^{3/2}\,. \tag{2.75}$$

For the f/f boundary condition the resulting spectrum is given by

$$s_l^2(\omega) = \hat{s}_m^2 - \hat{s}_0^2\delta(\tilde{\omega}_l/\omega)^{3/2} \tag{2.76}$$

where

$$\tilde{\omega}_l = \omega_{l+1/4} = \Delta\omega\left(l+\frac{1}{4}\right) . \tag{2.77}$$

As we shall see, the shift of the frequency $(l \to l+\frac{1}{4})$ in (2.77) as compared with (2.64) plays a key role in the spectrum pattern in the vicinity of the level $s_l = \hat{s}_0$.

2.6.2.1 Spectrum just Over the Level $s_l = \hat{s}_0$

We start from the narrow region $\Delta s_l^0 < s_l - \hat{s}_0 \ll \hat{s}_0$ just over the level $s = \hat{s}_0$. Here one obtains from (2.76)

$$s_l(\omega) \approx \hat{s}_0\left\{1+\frac{\delta}{2}\left[1-\left(\frac{\tilde{\omega}_l}{\omega}\right)^{2/3}\right]\right\} . \tag{2.78}$$

This is a continuation of branch (2.73) through the forbidden zone $|s_l - \hat{s}_0| < \Delta s_l^0$. Certainly, for large l its upper side is practically symmetrical to the lower one. If to forget for a moment about this zone, the points of intersections of lines (2.73) and (2.78) with the level $s = \hat{s}_0$ are shifted on the distance $\tilde{\omega}_l - \omega_l = \frac{1}{4}\Delta\omega$ (Fig. 2.4).

It is essential that the slope of curve (2.78) in the point $\tilde{\omega}$ practically coincides (for large l) with the slope (2.74) of the other curve:

$$s_l'(\omega)\,|_{\tilde{\omega}_l} \approx \frac{\hat{s}_0\delta}{3\tilde{\omega}_l} . \tag{2.79}$$

Thus, the interpolation line through the gap between curves (2.73) and (2.78) must contain an inflection point. With growing l and respective converging the gap, these imaginary fragments of the spectrum should tend to a horizontal line along the discussed level $s_l = \hat{s}_0$. Unfortunately, such tendency is concealed by the other tendency: with increasing l the slopes of all branches on both sides of the expected asymptote also decrease $\propto 1/l$ and remain identical (compare (2.74) and (2.79)). Therefore it would be difficult to notice manifestations of discussed anomaly in real spectra. This is evident from Fig. 2.4.

2.6.2.2 Spectrum Under the Asymptote $s = \hat{s}_m$

Dispersion branches $s_l(\omega)$ (2.76) never exceed the upper asymptotic level $s = \hat{s}_m$ (Fig. 2.4). In the vicinity of this level $(s_m - s_l \ll \hat{s}_m)$ one obtains

$$s_l(\omega) \approx \hat{s}_m \left\{ 1 - \frac{\delta}{2(1+\delta)} \left(\frac{\tilde{\omega}_l}{\omega} \right)^{2/3} \right\} . \tag{2.80}$$

We stress that asymptotic tending $s_l \to \hat{s}_m$ is accompanied by approaching the lower limit (a) of the integral I_{ad} to the upper one (d). This relates to convergence of the region admitting the propagation of bulk SH waves, i.e. to channelling of the wave field in the vicinity of the upper face of the plate.

Note in passing that approaching the curve $s_l(\omega)$ (2.80) to the asymptote $s = \hat{s}_m$ at growing frequency, $\hat{s}_m - s_l \propto \omega^{-2/3}$, coincides with the prediction (Alshits and Maugin, 2005, 2008) based on completely different consideration.

2.7 Levels Related to Extreme Points on the Slowness Profile

In this section we shall study behaviour of dispersion branches in the vicinity of levels associated with extreme points of the slowness profile. Here we shall return to the scheme of the function $\hat{s}(y)$ shown in Fig. 2.2 and consider the minimum (*1*), inflection (*3*) and maximum (*4*) types of levels. The non-extreme points *2* and *5* will be beyond our interest because these types of asymptotes are already studied in the previous Sect. 2.6.

2.7.1 An Absolute Minimum of the Function $\hat{s}(y)$

In the vicinity of the level $\hat{s}_1$ related to the minimum on the slowness profile in Fig. 2.2 it is convenient to present the function $\hat{s}(y)$ in the form

$$\hat{s}^2(y) = \hat{s}_1^2 \{1 + f_1(y)\} . \tag{2.81}$$

Close to its minimum point $y^{(1)}$ the function $f_1(y)$ will be approximated by the expression

$$f_1(y) \approx \lambda \left| \frac{y - y^{(1)}}{d} \right|^m , \tag{2.82}$$

where λ is a dimensionless parameter and the power m is supposed to be integer-valued: $m = 2, 3, \ldots$ Expression (2.82) for $f_1(y)$ will be implied applicable in the definite region $|y - y^{(1)}| < \Delta y \ll d$.

2.7.1.1 Spectral Features just Under the Level $s = \hat{s}_1$

The spectrum branches $\omega(s_l)$ under the level $s = \hat{s}_1$ is given by (2.26) with

$$I_{0d}(s_l) = \int_0^d dy\sqrt{\hat{s}_1^2 - s_l^2 + \hat{s}_1^2 f_1(y)} \, . \tag{2.83}$$

Let us divide this integral into two parts: $I_{0d} = I_1 + I_2$. The first one (I_1) is determined by the range (I): $|y - y^{(1)}| < \Delta y$, where approximation (2.82) works. And the integral I_2 is taken over the rest part (II) of the interval $[0,d]$. These integrals may be presented in the forms

$$\begin{aligned} I_1 &= 2\varepsilon_{1l}^{\frac{m+2}{2m}} \left(\hat{s}_1 d/\lambda^{1/m}\right) \int_0^{\frac{\Delta y}{d}\left(\frac{\lambda}{\varepsilon_{1l}}\right)^{1/m}} dx\sqrt{1+x^m} \, , \\ I_2 &= \int_{II} dy\sqrt{\hat{s}_1^2 - s_l^2 + \hat{s}_1^2 f_1(y)} = \hat{s}_1 \int_{II} dy\sqrt{\varepsilon_{1l} + f_1(y)} \, , \end{aligned} \tag{2.84}$$

with $\varepsilon_{1l} = (\hat{s}_1^2 - s_l^2)/\hat{s}_1^2$. For our asymptotic consideration we shall suppose ε_{1l} to be very small,

$$\varepsilon_{1l} \ll \lambda(\Delta y/d)^m \, . \tag{2.85}$$

In this limiting case one can obtain

$$I_{0d}(s_l) \approx \hat{s}_1 d\left(A_1 + B_1 \varepsilon_{1l}^{\frac{m+2}{2m}}\right) , \tag{2.86}$$

where $A_1 \sim 1$ and $B_1 \sim 1$ are dimensionless constants depending on λ.

Substituting (2.86) into (2.26), one can find asymptotic dispersion branches $s_l(\omega)$ just under the level $s = \hat{s}_1$ for any of boundary problems (2.23). In particular, for free faces one has

$$s_l \approx \hat{s}_1 \left\{1 - \kappa_1 \left(\frac{\omega_{1l}}{\omega} - 1\right)^{\frac{2m}{m+2}}\right\} , \tag{2.87}$$

where the notations are introduced

$$\kappa_1 = \frac{1}{2}(A_1/B_1)^{\frac{2m}{m+2}} \, , \qquad \omega_{1l} = l\Delta\omega_1 \, , \qquad \Delta\omega_1 = \pi/A_1\hat{s}_1 d \, . \tag{2.88}$$

The set of frequencies ω_{1l} approximately indicates points of intersections of the branches $s_l(\omega)$ with the level $\hat{s}_1$. However here one should be careful in the same way as in Sect. 2.6. In the vicinity of $\hat{s}_1$, solution (2.87) satisfies criterion (2.14) of applicability of the adiabatic approximation, only if

$$\omega \gg \frac{m\delta^{1/m}}{8\hat{s}_1 d\sqrt{2}} \varepsilon_{1l}^{-\frac{m+2}{2m}} \tag{2.89}$$

(compare with (2.68)). Thus, for large enough frequencies ω, formula (2.87) describes the spectrum as close to the singular level $s = \hat{s}_1$ as we need.

The further consideration is completely analogous to that in 2.6.1.2. Again close to the asymptote there is a forbidden zone of the width Δs_l^1 which depends on l and converges with its growth:

$$\Delta s_l^1 \sim \hat{s}_1 \left(\frac{m\lambda^{1/m} A_1}{2\pi l} \right)^{\frac{2m}{m+2}} . \tag{2.90}$$

For parabolic minimum $m = 2$ this gives $\Delta s_l^1 \propto 1/l$. By (2.87), the frequency ω_{1l}^b where the branch $s_l(\omega)$ intersect the bound $\hat{s}_1 - \Delta s_l^1$ is

$$\omega_{1l}^b \approx \frac{\omega_{1l}}{1 + \left(\Delta s_l^1 / \kappa_1 \hat{s}_1\right)^{\frac{m+2}{2m}}} . \tag{2.91}$$

For $\kappa_1 \sim 1$ and $\Delta s_l^1 \ll \hat{s}_1$ the frequencies ω_{1l}^b must be close to the set ω_{1l} (2.88).

It also follows from (2.87) that with increasing branch number l the slopes of the branches $s_l(\omega)$ in the points ω_{1l}^b (2.91) of the bound decrease as

$$\left.\frac{ds_l}{d\omega}\right|_{\omega=\omega_{1l}^b} \approx \frac{\hat{s}_1}{\omega_{1l}} \frac{A_1 m}{B_1 (m+2)} \left(\frac{2\Delta s_l^1}{\hat{s}_1} \right)^{\frac{m-2}{2m}} . \tag{2.92}$$

Note in passing that for $m = 2$ the last multiplier is equal 1. However one should be aware that even for $m > 2$ this multiplier containing a small parameter $2\Delta s_l^1/\hat{s}_1$ remains of order of unity at least for not very large m because of small index $\frac{m-2}{m+2}$ (see (2.90)).

The character of the relay race approaching the level $s = \hat{s}_1$ by the infinite set of branches is conveniently described by the decrease of distances $\hat{s}_1 - s_l > \Delta s_l^1$ in the discrete series of equidistant frequencies $\hat{\omega}_l \equiv \omega_{1(l-\Delta l)}$ with the growing l and the constant $\Delta l \ll l$. For large enough numbers $l - \Delta l$ when asymptotic formula (2.87) is applicable, it gives

$$\hat{s}_1 - s_l(\hat{\omega}_l) \approx \hat{s}_1 \kappa_1 \left(\frac{\omega_{1\Delta l}}{\hat{\omega}_l} \right)^{\frac{2m}{m+2}} \propto \hat{\omega}_l^{-\frac{2m}{m+2}} . \tag{2.93}$$

This dependence is similar to the prediction in Alshits and Maugin (2005, 2008).

2.7.1.2 Spectrum Features just Over the Level $s = \hat{s}_1$

For the other side from the level $s = \hat{s}_1$ the analysis is completely analogous. However, this time the line $s =$const $> \hat{s}_1$ intersects the curve $\hat{s}(y)$ in two points a and b on the left and right sides from the minimum $y^{(1)}$, respectively (region III in Fig. 2.2). The inverse spectrum for such case is found in 2.4.3.2. For the f/f boundary problem it is described by the two series of branches given by $(2.42)_1$. The integrals I_{0a} and I_{bd} determining these series are similar to (2.86):

$$I_{0a}(s_i) \approx \hat{s}_1 d\left(A_1' - B_1'|\varepsilon_{1i}|^{\frac{m+2}{2m}}\right), \quad I_{bd}(s_j) \approx \hat{s}_1 d\left(A_1'' - B_1''|\varepsilon_{1j}|^{\frac{m+2}{2m}}\right), \tag{2.94}$$

where $i, j = 1, 2, \ldots$ are numbers of branches, and the positive coefficients in (2.94) are estimated as $A_1', B_1', A_2', B_2' \sim 1$. With (2.94) and $(2.40)_1$, the new asymptotic branch series are

$$s_i \approx \hat{s}_1[1 + (\kappa_1'(1 - \omega_i'/\omega)^{\frac{2m}{m+2}}], \quad s_j \approx \hat{s}_1[1 + (\kappa_1''(1 - \omega_j''/\omega)^{\frac{2m}{m+2}}], \tag{2.95}$$

where $\omega > \omega_i', \omega_j''$ and

$$\begin{aligned} \kappa_1' &= \frac{1}{2}(A_1'/B_1')^{\frac{2m}{m+2}}, \ \omega_{1i}' = \Delta\omega_1'(i + \frac{1}{4}), \ \Delta\omega_1' = \pi/A_1'\hat{s}_1 d\,; \\ \kappa_1'' &= \frac{1}{2}(A_1''/B_1'')^{\frac{2m}{m+2}}, \ \omega_{1j}'' = \Delta\omega_1''(j + \frac{1}{4}), \ \Delta\omega_1'' = \pi/A_1''\hat{s}_1 d\,. \end{aligned} \tag{2.96}$$

One can prove that the parameters A_1' and A_1'' in (2.94) and (2.96) should approximately relate to A_1 in (2.86) as $A_1 \approx A_1' + A_1''$. Then the intervals $\Delta\omega_1'$, $\Delta\omega_1''$ (2.96) and $\Delta\omega_1$ (2.88) must relate to each other as

$$\frac{1}{\Delta\omega_1} \approx \frac{1}{\Delta\omega_1'} + \frac{1}{\Delta\omega_1''} \tag{2.97}$$

This relation has clear interpretation. Indeed, it is evident that at large enough frequency ω the total number of branches in the spectrum with frequencies less than ω coming to the level $s = \hat{s}_1$ from below, $l \approx \omega/\Delta\omega_1$, must be approximately equal to the sum of numbers of branches over the asymptote in the same frequency range, $i \approx \omega/\Delta\omega_1'$ and $j \approx \omega/\Delta\omega_1''$, i.e. $l \approx i + j$. But this leads directly to (2.97).

Thus, each branch l under the level $s = \hat{s}_1$ will be continued above it by the branch i, or j. And the intersection points of this pair with the level $s = \hat{s}_1$ must be shifted correspondingly by $\frac{1}{4}\Delta\omega_1'$ or by $\frac{1}{4}\Delta\omega_1''$ which reminds the situation of Sect. 2.6. But the slopes of all three lines from both sides of the level now differ from each other (ordinarily, the slope (2.92) of the lower l-line is less). Of course, the discussed discontinuities of functions (2.87), (2.95) and their derivatives occur nonmetering the forbidden gap around the singular level. However with increasing frequency the forbidden band converges faster than earlier (e.g., for $m = 2$ by (2.90) $\Delta s_l^1 \propto 1/l$) and the changes in the shape of the dispersion line should become more and more pronounced. Such relay-race asymptotic behavior of spectrum on the level related to an extreme minimum of the profile $\hat{s}(y)$ was expected in Alshits and Maugin (2005).

All mentioned features are seen in Fig. 3 of Shuvalov et al (2008) where the numerically obtained exact spectrum is presented for a continuously layered plate. The level $s \approx 0.29\mu$ s/mm just corresponds to such intermediate asymptote related to the minimum of the slowness profile at $y \approx -0.2$mm. One can see a tendency of decreasing slopes of branches under this level with growing frequency and then a rather sharp increase in their slopes on the upper side with two types of branches going up to different asymptotes. There are no comments in Shuvalov et al (2008) related to this part of spectrum.

2.7.2 The Level Related to an Inflection Point

The vicinity of the level $s = \hat{s}_3$ which relates to the inflection point *3* in Fig. 2.2 may be described similarly to the previous subsection. In this case the situation is simpler because the ranges just under the level $\hat{s}_3$ and just over it belong to regions $\mathrm{IV}_{1,2}$ (Fig. 2.2) of the same type (with 3 division points) where the spectra are given by the same relations (2.45) with $n = 2,\ l = 1$. In both cases they provide two series of dispersion branches, one asymptotic and the other not. So, in case of the f/f boundary problem we obtain for asymptotic branches,

$$\omega = \pi(p - \frac{1}{2}) \begin{cases} 1/I_+(s_p),\ s_p < \hat{s}_3\ , \\ 1/I_-(s_p),\ s_p > \hat{s}_3\ ; \end{cases} \tag{2.98}$$

$$I_\pm = \hat{s}_3 \int_{y^{(3)} \mp \delta y}^{d} dy \sqrt{f_3(y) \pm |\varepsilon_{3p}|}\ , \qquad \varepsilon_{3p} = \frac{\hat{s}_3^2 - s_p^2}{\hat{s}_3^2}\ . \tag{2.99}$$

Let us calculate the integrals in (2.99) for the function

$$f_3(y) \approx \lambda \left(\frac{y - y^{(3)}}{d} \right)^3\ . \tag{2.100}$$

In this case one can use in the lower limits of the integrals in (2.99) the small shift

$$\delta y = d\,(|\varepsilon_{3p}|/\lambda)^{1/3}\ , \tag{2.101}$$

and the above integrals acquire the form

$$I_\pm = \hat{s}_3 d \left(A_3 \pm B_3^\pm |\varepsilon_{3p}|^{5/6} \right)\ , \tag{2.102}$$

with

$$A_3 \approx 0.4\sqrt{\lambda} \left(1 - y^{(3)}/d \right)^{5/2}\ , \quad B_3^+ \approx 2.86/\sqrt[3]{\lambda}\ , \quad B_3^- \approx 1.04\sqrt[3]{\lambda}\ . \tag{2.103}$$

Combining (2.102) with (2.98) one obtains the following spectrum

$$s_p \approx \hat{s}_3 \begin{cases} 1 - \kappa_3^- (\omega_{3p}/\omega - 1)^{6/5}\ ,\ \omega < \omega_{3p}\ , \\ 1 + \kappa_3^+ (1 - \omega_{3p}/\omega)^{6/5}\ ,\ \omega > \omega_{3p}\ ; \end{cases} \tag{2.104}$$

where

$$\kappa_3^\mp = (A_3/2B_3^\pm)^{6/5}\ , \quad \omega_{3p} = \Delta\omega_3 (p - \frac{1}{2})\ , \quad \Delta\omega_3 = \pi/A_3 \hat{s}_3 d\ . \tag{2.105}$$

Thus, in the given case the dispersion branch $s_p(\omega)$ is continuous at the level $s = \hat{s}_3$. Let us prove that its derivatives at bounds of the forbidden gap are substantially different. These bounds $s = \hat{s}_3 \mp \Delta s_p^3$ and the frequencies $\omega_{3p}^{\mp}$ where the branch $s_p(\omega)$ intersect the bounds can be estimated similarly to (2.90), (2.91):

$$\Delta s_p^3 \sim \hat{s}_3 \left(\frac{3\lambda^{1/3}A_3}{2\pi p}\right)^{6/5}, \qquad \omega_{3p}^{\mp} \approx \frac{\omega_{3p}}{1+\left(\Delta s_p^3/\kappa_3^{\mp}\hat{s}_3\right)^{5/6}} . \tag{2.106}$$

And branch slopes at bounds of the forbidden gap are found similarly (2.92):

$$\left.\frac{ds_p}{d\omega}\right|_{\omega=\omega_{1p}^{\mp}} \approx \frac{\hat{s}_p}{\omega_{3p}}\frac{3A_3}{5B_3^{\pm}}\left(\frac{2\Delta s_P^3}{\hat{s}_3}\right)^{1/5} \approx \frac{\hat{s}_p}{\omega_{3p}}\frac{3A_3}{5B_3^{\pm}} . \tag{2.107}$$

Thus, with increasing branch number p the both slopes decrease $\propto 1/p$ and the slope under the level is $B_3^+/B_3^- \approx 3$ times less than that over the level. So, we again obtained asymptotic spectrum with branches tending in a relay-race manner to the level $s = \hat{s}_3$ from below. Then they, one by one, increase their slopes and after intersection of the level go up to the next asymptote. The character of approaching the level by branches from below with growing p is again well described by the dependence

$$\hat{s}_3 - s_p(\hat{\omega}_p) \approx \hat{s}_3\kappa_3^- \left(\omega_{3\Delta p}/\hat{\omega}_p\right)^{6/5} \propto \hat{\omega}_p^{-6/5} \tag{2.108}$$

in the complete analogy with the above equation (2.93) for $m = 3$ which fits the predictions in Alshits and Maugin (2005, 2008).

2.7.3 Asymptote Related to Maximum at the Profile $\hat{s}(y)$

We conclude this section by a short description of the spectrum under the asymptote $s = \hat{s}_4$ associated with maximum point 4 on the slowness profile in Fig. 2.2. We do it just for completeness since this problem is already solved in Shuvalov et al (2008).

Under this level there are three division points (region IV_2 in Fig. 2.2) and two series of dispersion branches described by equation (2.43) at $n = 2$ and $l = 1$. Only the first series is asymptotic and just relates to our case. For the f/f boundary condition the solution is given by

$$\omega = \pi(k - \frac{1}{2})/I_4(s_k) , \quad k = 1,2,3,\ldots; \tag{2.109}$$

$$I_4 = \hat{s}_4 \int_{y^{(4)}-\delta y}^{y^{(4)}+\delta y} dy\sqrt{\varepsilon_{4k} - f_4(y)} , \qquad \varepsilon_{4k} = \frac{\hat{s}_4^2 - s_k^2}{\hat{s}_4^2} . \tag{2.110}$$

In the considered asymptotic region $\hat{s}_4 - s_k \ll \hat{s}_4$ the function $f_4(y)$ will be chosen in the form similar to (2.82)

$$f_4(y) \approx \lambda \left| \frac{y - y^{(4)}}{d} \right|^m , \quad m = 2,3,4\ldots . \tag{2.111}$$

The value of δy in limits of the integral in (2.110) is found as a root of the integrand:

$$\delta y = d(\varepsilon_{4k}/\lambda)^{1/m} . \tag{2.112}$$

With (2.111) and (2.112) the integral I_4 is equal

$$I_4 = 2\chi_m \hat{s}_4 d \varepsilon_{4k}^{\frac{m+2}{2m}} \tag{2.113}$$

where χ_m is the dimensionless parameter

$$\chi_m = \lambda^{-\frac{1}{m}} \int_0^1 dt \sqrt{1 - t^m} \sim 1 . \tag{2.114}$$

Substituting (2.113) with $(2.110)_2$ into (2.109) one obtains for the studied asymptotic region $\hat{s}_4 - s_k \ll \hat{s}_4$:

$$s_k \approx \hat{s}_4 \left\{ 1 - \frac{1}{2} \left(\frac{\omega_{4k}}{\omega} \right)^{\frac{2m}{m+2}} \right\} \tag{2.115}$$

$$\omega_{4k} = \Delta\omega_4 (k - \frac{1}{2}) , \quad \Delta\omega_4 = \pi/2\chi_m \hat{s}_4 d . \tag{2.116}$$

The estimated asymptotic branches (2.115) coincide with the result of Shuvalov et al (2008) and fit the expectation of Alshits and Maugin (2005, 2008) that

$$\hat{s}_4 - s_k \propto \omega^{-2m/m+2} .$$

As was explained in the end of Sect. 2.6, this part of the spectrum relates to channeling of energy in the thin layer related to the position of the slowness profile maximum in the plate. This part of spectrum exists only under the asymptote. Above that there are branches of the mentioned second series which intersect this asymptote and go up to the last asymptote related to the point *5* in Fig. 2.2. Their "intersections" with the first series are spurious being the consequence of approximation (2.30). Small repulsion eliminates degeneracies and creates a ladder-like pattern in spectrum.

2.8 Conclusions

The presented approximate approach to an analytical analysis of acoustic spectra of transversely inhomogeneous plates appears to be quite convenient and efficient.

It allows one to study rather peculiar unexpected properties and nontrivial specific features of acoustic behavior created exclusively by inhomogeneity of the medium even when it is isotropic. We analyzed the spectral effects in the vicinity of slowness levels associated with extreme material characteristics of the inhomogeneous profile. As was predicted by Alshits and Maugin (2005, 2008), all such levels should be characterized by specific asymptotic behavior of dispersion branches. Indeed, we found that maximum points on the slowness profile of the plate provide channeling of energy in appropriate layers and create high frequency asymptotes for spectral branches. This coincides with data of Shuvalov et al (2008). But it was also expected by Alshits and Maugin (2005, 2008) that minimum and inflection points on the same profile would create specific intermediate asymptotes to which branches should approach in the relay-race manner. The Asymptotes of these two types were found. The non-extreme minimum on the surface was not included in Alshits and Maugin (2005, 2008) into the list of candidates for asymptotic behavior. Indeed, we found that the anomalies related to such sort of levels are hardly observable. They are not visible in the mentioned above numerical spectrum (Fig. 3 in Shuvalov et al, 2008).

Acknowledgements The authors are grateful to D.A. Bessonov for the help in computing.

References

Alshits VI, Kirchner HOK (1995) Elasticity of multilayers I. Basic equations and solutions. Philosophical Magazine A 72(6):1431–1444

Alshits VI, Maugin GA (2005) Dynamics of multilayers: elastic waves in an anisotropic graded or stratified plate. Wave Motion 41(4):357–394

Alshits VI, Maugin GA (2008) Dynamics of anisotropic multilayers. Wave Motion 45(5):629–640

Alshits VI, Deschamps M, Maugin GA (2003) Elastic waves in anisotropic plates: short-wavelength asymptotics of the dispersion branches $v_n(k)$. Wave Motion 37(3):273–292

Alshits VI, Deschamps M, Lyubimov VN (2005) Dispersion anomalies of shear horizontal guided waves in two- and three-layered plates. The J Acoust Soc America 118(5):2850–2859

Bakirtas AI, Maugin GA (1982) Ondes de surface SH pures en élasticité inhomogène. J Mècan Theor Appl 1:995–1013

Baron C, Poncelet O, Shuvalov A (2003) Calculation of the velocity spectrum of the vertically inhomogeneous plates. In: Proceedings of 5th World Congress on Ultrasonics, Paris, pp 605–608

Gantmacher FR (1989) The Theory of Matrices. Chelsea, New York

Landau LD, Lifshitz EM (1991) Quantum Mechanics. Pergamon Press, New York

Maugin GA (1983) Elastic surface waves with transverse horizontal polarization. In: Hutchinson JW (ed) Advances in Applied Mechanics, Academic Press, New York, vol 23, pp 373–434

Nayfeh AH (1995) Wave Propagation in Layered Anisotropic Media. North-Holland, Amsterdam

Shilov GE (1996) Elementary Functional Analysis. Dover, New York

Shuvalov AL, Poncelet O, Deschamps M (2004) General formalism for plane guided waves in transversely inhomogeneous anisotropic plates. Wave Motion 40(4):413–426

Shuvalov AL, Poncelet O, Deschamps M, Baron C (2005) Long-wavelength dispersion of acoustic waves in transversely inhomogeneous anisotropic plates. Wave Motion 42(4):367–382

Shuvalov AL, Poncelet O, Kiselev AP (2008) Shear horizontal waves in transversely inhomogeneous plates. Wave Motion 45(5):605–615

Stroh AN (1962) Steady state problems in anisotropic elasticity. J Math Phys 41(1-4):77–103

Chapter 3
Nonlinear Schrödinger and Gross - Pitaevskii Equations in the Bohmian or Quantum Fluid Dynamics (QFD) Representation

Attila Askar

Abstract The Quantum Fluid Dynamics (QFD) representation has its foundations in the works of Madelung (1929), De Broglie (1930 - 1950) and Bohm (1950 - 1970). It is an interpretation of quantum mechanics with the goal to find classically identifiable dynamical variables at the sub-particle level. The approach leads to two conservation laws, one for "mass" and one for "momentum", similar to those in hydrodynamics for a compressible fluid with a particular constitutive law. The QFD equations are a set of nonlinear partial differential equations. This paper extends the QFD formalism of quantum mechanics to the Nonlinear Schrödinger and the Gross-Pitaevskii equation.

3.1 Introduction

Schrödinger equation is the most conventional formalism for describing quantum mechanics (Wyatt, 2005; Styer et al, 2002). Another formalism is referred to equivalently as "the hydrodynamic analogy to quantum mechanics" or "the quantum fluid dynamics" (QFD) (Wyatt, 2005). This approach is partly motivated by Einstein's questioning of the completeness of the quantum theory that is dramatized by his famous statement: "God doesn't throw dice" while admitting to its internal consistency (Einstein et al, 1935; Bohr, 1935; Schilp, 1949).

The essence of the formalism is the polar representation of the complex wave function by its amplitude and phase as opposed to its Cartesian representation using the real and imaginary parts.

The approach leads to two conservation laws, one for "mass" and one for "energy" whose gradient becomes the "momentum" equation, similar to those in hydrodynamics. This formalism was derived originally by Madelung (1926), elaborated on by de Broglie in work spanning the period 1927 to 1967 (de Broglie, 1926, 1951, 1957,

Attila Askar
Koç University, Sarıyer, Istanbul 89010, Turkey
e-mail: AASKAR@ku.edu.tr

© Springer International Publishing AG, part of Springer Nature 2018

H. Altenbach et al. (eds.), *Generalized Models and Non-classical Approaches in Complex Materials 1*, Advanced Structured Materials 89,
https://doi.org/10.1007/978-3-319-72440-9_3

1967) and in the fifties to end of sixties by Bohm (1951); Bohm and Bub (1966). The initial attempts have been more in terms of the philosophy of interpreting quantum mechanics, trying in particular find a classically identifiable quantities or dynamical variables at the quantum level.

The QFD equations are a set of nonlinear partial differential equations. In this sense, QFD may be seen as a step in the negative direction as compared to the Schrödinger equation that is linear. However, in this scheme, the oscillatory real and imaginary components of the complex wave function are replaced by the usually significantly less oscillatory amplitude and phase. To appreciate this aspect of the procedure, consider a plane wave $\psi = \exp[i(kx-t)]$ as the simplest illustrative example. Its real and imaginary parts are the oscillatory sine and cosine functions, while its amplitude is a constant and its phase is linear in both the time and space coordinate. This aspect of QFD formalism is utilized advantageously to generate numerical algorithms for solving the time dependent Schrödinger equation for scattering problems (Wyatt, 2005; Weiner and Askar, 1971; Askar et al, 1980; Dey, 1998; Sales, 1999). The QFD formalism replaces the abstract wave function or operator representation by the concrete image of an ensemble of mass points in classically familiar terms for a more intuitive interpretation of the dynamics.

3.2 Polar Representation of the Wave Function

The (i) Schrödinger, (ii) Nonlinear Schrödinger and (iii) Gross-Pitaevskii equations are defined in the same structure respectively with (i) $k=0$, (ii) $V=0$ and (iii) with both $k \neq 0$ and $V \neq 0$ as:

$$i\hbar\frac{\partial\psi}{\partial t} = -\frac{\hbar^2}{2m}\left(\nabla^2\psi - k^2|\psi|^2\psi\right) + V\psi \tag{3.1}$$

These equations are treated in the paper with their physical forms to reinforce the structure of the QFD representation. Planck's constant $\hbar$, mass m and the wave number k are kept for dimensional convenience and reference to quantum mechanics.

The complex wave function is considered in the polar representation by its amplitude A and phase S as:

$$\psi(\mathbf{x},t) = A(\mathbf{x},t)\, e^{i[S(\mathbf{x},t)-Et]/\hbar} \tag{3.2}$$

Above by reference to quantum mechanics, E is singled out to be identified as at the total energy. The substitution of the representation of the wave function in the polar form in Eq. (3.2) into the Eq. (3.1) and the separation of its respectively imaginary and real parts yields:

$$A_t + \frac{1}{2}\,(\nabla S\cdot\nabla A + \frac{1}{2m}A\nabla^2 S) = 0 \tag{3.3}$$

$$A\left[\frac{\partial S}{\partial t}-E+\frac{1}{2m}\nabla S\cdot\nabla S+V\right]+\frac{\hbar^2}{2m}A^3-\frac{\hbar^2}{2m}\nabla^2 A=0 \tag{3.4}$$

The two equations above are rearranged in the following sections into both physically meaningful and computationally suitable forms.

3.3 Conservation Laws

3.3.1 Mass Conservation Equation

The rearrangement of the first equation above after multiplying it by A, introducing the definitions for ρ and v yields:

$$\rho=A^2 \qquad \mathbf{v}=\frac{\nabla S}{m} \qquad \rightarrow \qquad \frac{\partial\rho}{\partial t}+\nabla\cdot(\rho\mathbf{v})=0 \tag{3.5}$$

Equation (3.5) is identified as the conventional mass conservation equation law of fluid dynamics for a compressible gas with the density $\rho(\mathbf{x},t)$ and velocity field $\mathbf{v}(\mathbf{x},t)$. The phase function $S(\mathbf{x},t)$ plays the role of the velocity potential in view of the definition in Eq. (3.5) (Wyatt, 2005; Dey, 1998; Sales, 1999).

3.3.2 Energy Conservation Equation

The rearrangement of the expression in Eq. (3.4) after dividing it by A and introducing the definitions

$$V_{qu}=-\frac{\hbar^2}{2m}\left(\frac{\nabla^2 A}{A}\right) \qquad V_p=\frac{\hbar^2 k^2}{2m}\rho \tag{3.6}$$

yields:

$$\frac{\partial S}{\partial t}+\left(\frac{1}{2}m\mathbf{v}\cdot\mathbf{v}+V+V_p+V_{qu}\right)=E \tag{3.7}$$

Equation (3.7) is interpreted as an energy conservation law with S as an action whose time rate S_t is an energy;

$$\frac{1}{2}m\mathbf{v}\cdot\mathbf{v}+V$$

form the classical energy with its kinetic and potential energy components; V_p is a local "pressure potential" and V_{qu} as a non-local "quantum potential" as the Laplacian connects neighboring regions.

3.3.3 The Momentum Equation

A momentum conservation equation can be derived by taking the gradient of the expression in Eq. (3.7). With the use of the vector identity

$$\nabla(\mathbf{a}\cdot\mathbf{b}) = \mathbf{a}\cdot\nabla\mathbf{b} + \mathbf{b}\cdot\nabla\mathbf{a} + (\nabla\times\mathbf{a})\times\mathbf{b} + (\nabla\times\mathbf{b})\times\mathbf{a}$$

with $\mathbf{a} = \mathbf{b} = \nabla S$, the result is:

$$m\left(\frac{\partial\mathbf{v}}{\partial t} + \mathbf{v}\cdot\nabla\mathbf{v}\right) = -\nabla(V + V_p + V_{qu}) \tag{3.8}$$

The term $\mathbf{v}_t + \mathbf{v}\cdot\nabla\mathbf{v}$ is the acceleration in fluid dynamics and is composed of the sum of the local and the convected components (Landau and Lifschitz, 1959; Eringen, 1962).

3.3.4 Pressure Interpretation

The term $V_p + V_{qu}$ can also be interpreted as "pressure" in the fluid dynamics terminology. Consequently, again in the fluid dynamics terminology, a diagonal stress tensor can be introduced with the use of the unit tensor $\mathbf{I}$:

$$p = V_p + V_{qu} \qquad\qquad \boldsymbol{\sigma} = -p\,\mathbf{I} \tag{3.9}$$

With this interpretation and the definition of the external force density acting on the fluid as $\mathbf{g} = -\nabla V$, the momentum equation reads:

$$m\left(\frac{\partial\mathbf{v}}{\partial t} + \mathbf{v}\cdot\nabla\mathbf{v}\right) = \nabla\cdot\boldsymbol{\sigma} + \mathbf{g} \tag{3.10}$$

3.3.5 The Lagrangian Representation

The conservation laws in Eqs. (3.5), (3.7) and (3.8) can be rewritten in the convected coordinates (Wyatt, 2005; Landau and Lifschitz, 1959; Dey, 1998; Sales, 1999; Eringen, 1962), thereby leading to the Lagrangian representation. The total time derivative of a general function $f = f(\mathbf{x},t)$ is defined as:

$$\frac{df}{dt} = \frac{\partial f}{\partial t} + \mathbf{v}\cdot\nabla f \tag{3.11}$$

and the position vector of a particle is related to the velocity in the usual way as:

$$\frac{d\mathbf{x}}{dt} = \mathbf{v} \tag{3.12}$$

With the above basic definitions, the mass and momentum conservation equations become:

$$\frac{d\rho}{dt} + \rho\nabla\cdot\mathbf{v} = 0 \qquad (3.13)$$

$$m\frac{d^2\mathbf{x}}{dt^2} = -\nabla(V_p + V_{qu} + V) \qquad \rightarrow \qquad m\frac{d^2\mathbf{x}}{dt^2} = \mathbf{F} \qquad (3.14)$$

The representation above for the momentum conservation is in the familiar Newtonian form. The interpretation of the momentum equation in the Lagrangian formalism is most clear: the evolution of the wave equation is equivalent to calculating the trajectories of the ensemble of particles subject to the total potential made of the classical and quantum components. This form of the momentum equation in (3.14) along with the mass conservation in (13) has been exploited in computational algorithms with success (Wyatt, 2005; Dey, 1998; Sales, 1999).

The above momentum equation represented as a collection of trajectories, can also be written in the fluid dynamics terminology:

$$m\frac{d^2\mathbf{x}}{dt^2} = \nabla\cdot\boldsymbol{\sigma} + \mathbf{g} \qquad \boldsymbol{\sigma} \equiv -(V_p + V_{qu})\,\mathbf{I} \qquad \mathbf{g} \equiv -\nabla V \qquad (3.15)$$

Above, the stress σ and the external force $\mathbf{g}$ are as defined in Eqs. (3.9) and (3.10).

3.4 Adding a Dissipation Term as in Navier - Stokes Equation

The QFD equations conserve mass, momentum and energy. In this sense, they correspond to a perfect fluid, with a particular constitutive law for a pressure field: $p = V_{qu} + V_p$. Introducing the dissipative term as the rate of strain into the stress, in the manner of the Navier - Stokes equation, yields:

$$\boldsymbol{\sigma} \equiv -(V_p + V_{qu})\,\mathbf{I} + \mu\left(\nabla\mathbf{v} + (\nabla\mathbf{v})^T\right) \qquad (3.16)$$

Above, the superscript T represents the transpose in defining the rate of strain and thereby making the momentum equation invariant under rigid body rotations.

With the above stress, the momentum equations in (3.8) and (3.14), respectively for the Eulerian and Lagrangian representations, become:

$$m\left(\frac{\partial\mathbf{v}}{\partial t} + \mathbf{v}\cdot\nabla\mathbf{v}\right) = -\nabla p + \mu(\nabla^2\mathbf{v} + \nabla\nabla\cdot\mathbf{v}) + \mathbf{g} \qquad (3.17)$$

$$m\frac{d\mathbf{v}}{dt} = -\nabla p + \mu(\nabla^2\mathbf{v} + \nabla\nabla\cdot\mathbf{v}) + \mathbf{g} \qquad (3.18)$$

Above, p and $\mathbf{g}$ have the definitions in (3.9) and (3.10).

The mass conservation equation remains unchanged. However, the addition of the Laplacian of the velocity changes drastically the nature of the set of differential equations from hyperbolic to parabolic type.

3.5 Vorticity

Vorticity is defined as $\omega \equiv \nabla \times \mathbf{v}$. The QFD as well as its equivalent Schrödinger equations do not lead to vorticity. The reason follows simply from the identity $\omega \equiv \nabla \times \mathbf{v} = \nabla \times (\nabla S) = \mathbf{0}$. Figures 3.1 present numerical results of the application of the QFD formulation (Dey, 1998; Sales, 1999). These solutions are obtained by a discretization of $\rho(x, y, t)$ and $\mathbf{x}(t) = (x(t), y(t))$ within the Lagrangian representation in Eqs. (3.13) and (3.14). The purpose here is an illustration of the absence of vorticity.

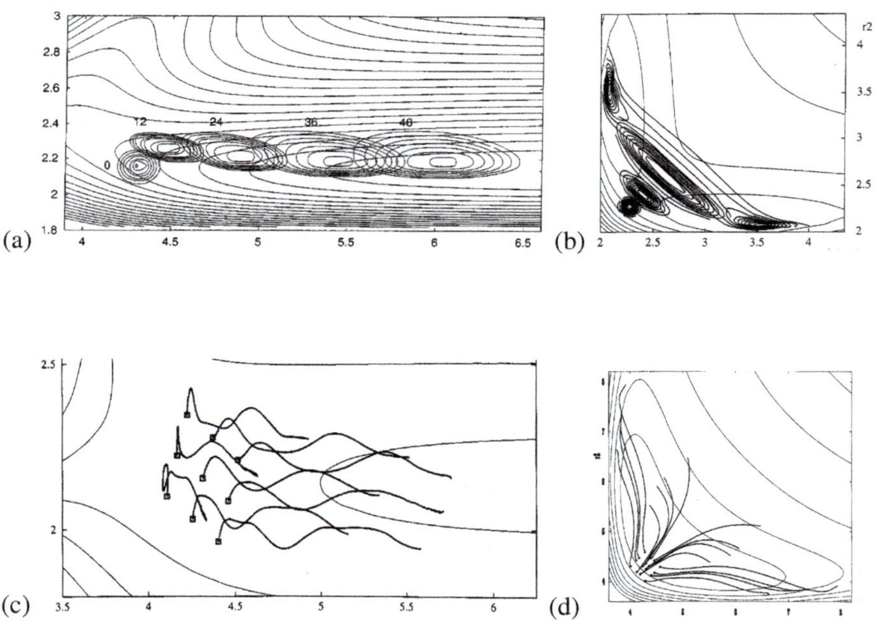

Fig. 3.1: Sample density and trajectory results. The figures on the top show the evolution of the initial wave packet through the display of $\rho(x_j, y_j, t_k)$ at the selected times t_k and the positions $\mathbf{x}_j$. (a) $NOCl \rightarrow NO + Cl$; (b) $NO_2 \rightarrow NO + O$ density contours. The figures in the second row are sample trajectories for selected discrete points $\mathbf{x}_j = \mathbf{x}_j(t)$ of the wave function: (c) $NOCl \rightarrow NO + Cl$; (d) $NO_2 \rightarrow NO + O$.

The figures on the top show the evolution of the initial wave packet through the display of $\rho(x_j, y_j, t_k)$ at the selected times t_k and the positions $\mathbf{x}_j$. The figures in the second row are sample trajectories for selected discrete points $\mathbf{x}_j(t)$ of the wave function, each corresponding to the evolution of the selected points of the wave packet above. In accordance with the prediction by the Schrödinger equation, no vortices are present in the plots. (The physics behind the figures is about a photo-dissociation problem with the linear Schrödinger equation. The figures on the left are for the photodissociation of $NOCl$ and those on the right are for the photodissociation of NO_2 (Dey, 1998; Sales, 1999).

Regardless of the nature of the physical problem, one would expect a vorticity field due to viscosity, as in the Navier - Stokes equation for dissipative fluids. In fact, the vorticity equation is obtained by taking the curl of the momentum equation in (3.8). With the curl of the gradients dropping out of the equation, $\nabla \times (\nabla f) = 0$ and the use of vector identities for $\nabla \times (\mathbf{v} \cdot \nabla \mathbf{v})$ we obtain:

$$m\nabla \times \left(\frac{\partial \mathbf{v}}{\partial t} + \mathbf{v} \cdot \nabla \mathbf{v}\right) = \nabla \times \left(-\nabla p + V\right) + \mu \nabla \times \left(\nabla^2 \mathbf{v} + \nabla \nabla \cdot \mathbf{v}\right) \tag{3.19}$$

$$\rightarrow \qquad m\left(\frac{\partial \omega}{\partial t} + (\mathbf{v} \cdot \nabla \mathbf{v})\omega - (\omega \cdot \nabla)\mathbf{v}\right) = \mu \nabla^2 \omega \tag{3.20}$$

Equivalently, in the Lagrangian representation this equation reads:

$$m\,\frac{d\omega}{dt} = -(\mathbf{v} \cdot \nabla \mathbf{v})\omega + \mu \nabla^2 \omega \tag{3.21}$$

Approaches to dissipation in quantum systems and chaos based on stochastic noise, as well as their relevance to physics can be viewed in Braun (2001).

The importance of vorticity lies in the heuristic - phenomenological explanation of the formation of turbulence in classical fluids. In the classical Navier - Stokes equation, it is stipulated that vorticities break into a hierarchy of smaller and smaller ones, eventually covering the whole space and thereby leading to "turbulence".

Further analytical and computational work is needed to determine weather the QFD equations with the viscosity term would lead to chaos and eventually "turbulence".

3.6 Closing Remarks

The representation here has the following set of purposes:

1. Benefit from the fluid dynamics analogy towards a concrete image of the complex valued equations and the function respectively as a fluid flow or set of trajectories of an ensemble of discrete set of points;
2. Present a platform for a natural extension to of the Schrödinger family of equations to multidimensions;

3. Present formulations with numerical advantages in discretized forms, including for the multidimensional extensions of the family of Schrödinger equations;
4. The nonlinear field equations with the dissipation in the Navier - Stokes manner transform the differential equations from hyperbolic to parabolic type, thereby opening a natural framework for studying fundamentally nonlinear phenomena;
5. With the dissipation term, the extended set of equations in the fluid dynamics analogy which lead to vorticity and thereby are likely to lead to chaos and eventually to turbulence.

The fluid dynamics equations derived here, including the dissipative case, are invariant under Galilean transformation of the coordinates, i. e. rigid body translations and rotations, as they should be.

References

Askar A, Cakmak AS, Rabitz H (1980) Nodal structure and global behavior of scattering wave functions. J Chem Phys 72:5287

Bohm D (1951) Quantum Theory. Prentice-Hall, Inc., New York

Bohm D, Bub J (1966) A proposed solution of the measurement problem in quantum mechanics by a hidden variable theory. Rev Mod Phys 38:453

Bohr N (1935) Can quantum-mechanical description of physical reality be considered complete? Phys Rev 48:696

Braun D (2001) Dissipative Quantum Chaos and Decoherence, Springer Tracts in Modern Physics, vol 172. Springer, Berlin, Heidelberg

de Broglie L (1926) Sur la possibilité de relier les phénomènes d'interférences et de diffraction à la théorie des quanta de lumière. Compt Rend Acad Sci Paris 183:447

de Broglie L (1951) Remarques sur la théorie de l'onde pilote. Compt Rend Acad Sci Paris 233:641

de Broglie L (1957) Idées nouvelles concernant les systèmes de corpuscules dans l'interprétation causale de la mécanique ondulatoire. Compt Rend Acad Sci Paris 244:529

de Broglie L (1967) Le mouvement brownien d'une particule dans son onde. Compt Rend Acad Sci Paris 264:1041

Dey B (1998) Multidimensional wave-packet dynamics within the fluid dynamical formulation of the schrödinger equation. J Chem Phys 109:8770

Einstein A, Podolsky B, Rosen N (1935) Can quantum-mechanical description of physical reality be considered complete? Phys Rev 47:777

Eringen AC (1962) Nonlinear Theory of Continuous Media. McGraw-Hill

Landau LD, Lifschitz EM (1959) Fluid Mechanics. Pergamon Press

Madelung E (1926) Quantentheorie in hydrodynamischer Form. Z Physik 40:322–326

Sales F (1999) Quantum fluid dynamics in the lagrangian representation and applications to photodissociation problems. J Chem Phys 111:2423

Schilp PA (ed) (1949) Albert Einstein, Philosopher-Scientist. Library of Living Philosophers, Evanston, Illinois

Styer DF, Balkin MS, Becker KM, Burns MR, Dudley CE, Forth ST, Gaumer JS, Kramer MA, Oertel DC, Park LH, Rinkoski MT, Smith CT, Wotherspoon TD (2002) Nine formulations of quantum mechanics. American Journal of Physics 70(3):288–297

Weiner JH, Askar A (1971) Time-dependent perturbation calculations based on the hydrodynamic analogy to quantum mechanics. J Chem Phys 54:1108

Wyatt RE (2005) Quantum Dynamics with Trajectories: Introduction to Quantum Hydrodynamics, Interdisciplinary Applied Mathematics, vol 28. Springer, New York

Chapter 4
The Stability of the Plates with Circular Inclusions under Tension

Svetlana M. Bauer, Stanislava V. Kashtanova, Nikita F. Morozov, and Boris N. Semenov

Abstract This paper deals with the problem of local buckling caused by uniaxial stretching of an infinite plate with a circular hole or with circular inclusion made of another material. As the Young's modulus of the inclusion approaches that of the plate, the critical load increases substantially. When these moduli coincide, stability loss is not possible. This paper also shows the difference between them when the inclusion is softer than the plate and when the inclusion is stiffer than the plate. Computational models show that instability modes are different both when the inclusion is softer than the plate and when the inclusion is stiffer than the plate. The case when plate and inclusion have the same modulus of elasticity, but different Poisson's ratio is investigated too. It is also discussed here the case when a plate with inclusion is under biaxial tension. For each ratio of the modulus of elasticity of plate versus inclusion it's obtained the range of the load parameters for which the loss of stability is impossible.

4.1 Introduction

The study of buckling elements, mechanisms and structures, together with the tasks of destruction are actual problems of modern mechanics. The stability of thin plates with holes and cracks under uniaxial tension was analyzed in Guz et al (1981); Bochkarev and Grekov (2015); Shimizu (2007).

It was noted in Guz et al (1981); Shimizu (2007) that compressive stresses can arise not only in plates subjected to compression, which is obvious, but also near the boundaries of holes in plates subjected to tension.

Svetlana M. Bauer · Stanislava V. Kashtanova · Nikita F. Morozov · Boris N. Semenov
St. Petersburg State University, 199034 St. Petersburg, Universitetskaya nab. 7-9, Russia
e-mail: s_bauer@mail.ru, kastasya@yandex.ru, morozov@nm1016.spb.edu, semenov@bs1892.spb.edu

© Springer International Publishing AG, part of Springer Nature 2018
H. Altenbach et al. (eds.), *Generalized Models and Non-classical Approaches in Complex Materials 1*, Advanced Structured Materials 89,
https://doi.org/10.1007/978-3-319-72440-9_4

In recent years, increasingly interest has arise in the various mechanisms using nano-objects. When calculating the deformation, stability, and failure of nanoscale objects, it is necessary also to take into account the surface effect. The local loss of stability of a plate with circular nanohole under uniaxial tension was considered by Bochkarev and Grekov (2014). In this case, the surface stresses on the circular hole boundary, which define more exactly the known solution of the Kirsch problem for small hole sizes, were taken into account in the subcritical state. It is noted that taking into account the surface stresses at the hole edge results in a loss of stability of the plate under a smaller load than in the classical formulation. The value of the load decreases by 5-7%.

However, it is clear that for nanoscale thicknesses of the plate, it is necessary to take into account also the surface stresses operating on the face surfaces. So in Bauer et al (2014) the problem was solved taking into account surface effects not only on the border of the hole, but also taking into account surface stresses on the facial planes, which change the bending stiffness of the plate. To determine the critical stress corresponding to the point of bifurcation, in which the plate loses the flat shape deformation, the energy method of S.P. Timoshenko is used.

The normal deflections of the plate was sought in the form

$$w(\rho,\varphi) = R\sum_{k=1}^{\infty}\sum_{l=0}^{\infty} A_{kl}\frac{\cos(l\varphi)}{\rho^k}, \tag{4.1}$$

Here R is radius of the inclusion hole, r, φ- polar coordinates centered in the center of a circular hole, $\rho = r/R$. These expressions satisfy the boundary condition of the symmetry of deformations and the deflection decrement at infinity.

It is shown that taking into account surface stresses for plates of nanoscale thickness leads to a significant increase in the flexural rigidity of the plate comparison with the classical value. The consequence is an increase in the critical load even when taking into account the surface stresses on the contour boundary. The problem considered in Bochkarev and Grekov (2014) was solved by the method of finite elements. It was built of the buckling modes and corresponding critical loads. The results demonstrated a good coincidence between the first critical loads constructed by the method of finite elements and by analytical method.

Compression stresses could appear also in the plates with the inclusion, and therefore these stresses may cause local buckling of the plate.

Loss of stability under uniaxial tension in an infinite plate with a circular inclusion made of another material was analyzed in Bauer et al (2017). The influence of the inclusion, or more exactly, the influence of the elastic modulus of the inclusion on the critical load was examined. The Poisson's ratio of the inclusions and plates were considered to be the same. The minimum eigenvalue corresponding to the first critical load was found as in Bauer et al (2014) by applying the variational principle. Computations were performed in Maple and were compared with results obtained with the finite element method in ANSYS .

4.2 Problem Statement

This part deals with the problem of the local buckling caused by uniaxial stretching of an infinite plate with a circular inclusion in the case when plate and inclusion have the same modulus of elasticity, but different Poisson's ratio. The problem on the deformation of a plate with a circular inclusion under uniaxial tension was solved by Muskhelishvili (1963); Kacharov et al (2003) in polar coordinates with the origin placed at the center of the inclusion. The solution of this problem in Cartesian Coordinate System is presented in Deryugin and Lasko (2012).

Suppose that E_1, ν_1 - are Young's modulus and Poison's ratio of the plate, and E_2, ν_2 - parameters of the inclusion, x, y - Cartesian coordinates. If we consider an infinite plate with a circular inclusion made of different material with applied at infinity load σ along the y-axis, the stresses in the plate will have the follow form:

$$
\begin{aligned}
\frac{\sigma_y}{\sigma} &= 1+\frac{(1-k_y+k_x)R^2}{2r^2}\left[1+\frac{3R^2+10y^2}{r^2}-F+G\right]-\frac{k_xR^2}{r^2}\left(1-\frac{2y^2}{r^2}\right),\\
\frac{\sigma_x}{\sigma} &= \frac{(1-k_y+k_x)R^2}{2r^2}\left[3-\frac{3R^2+18y^2}{r^2}+F-G\right]+\frac{k_xR^2}{r^2}\left(1-\frac{2y^2}{r^2}\right),\\
\frac{\sigma_{xy}}{\sigma} &= \frac{(1-k_y+k_x)R^2xy}{r^4}\left[3-\frac{2(3R^2+4y^2)}{r^2}+\frac{12R^2y^2}{r^4}\right]-\frac{2k_xR^2xy}{r^4}.
\end{aligned}
\tag{4.2}
$$

where

$$
r^2=x^2-y^2,\quad F=\frac{8y^2(3R^2+2y^2)}{r^4},\quad G=\frac{24R^2y^4}{r^6},
$$

R - radius of an inclusion, h - thickness. Coefficients k_x and k_y depend on elasticity modulus of the inclusion and the plate:

$$
\begin{aligned}
k_y &= \frac{E_2[(3-\nu_2)E_1+(5+\nu_1)E_2]}{(E_1+2E_2)^2-[\nu_2E_1+(1-\nu_1)E_2]^2},\\
k_x &= \frac{E_2[(3\nu_2-1)E_1+(1-3\nu_1)E_2]}{(E_1+2E_2)^2-[\nu_2E_1+(1-\nu_1)E_2]^2}
\end{aligned}
\tag{4.3}
$$

According to Eshelby's theorem (Deryugin and Lasko, 2012; Eshelby, 1957) the stress field inside the circular inclusion is homogeneous and symmetric with respect to the tension axis:

$$
\sigma_y=k_y\sigma,\quad \sigma_x=k_x\sigma,\quad \sigma_{xy}=0. \tag{4.4}
$$

It follows from (4.1) that the stresses σ_y/σ are mainly positive (negative zones arise as the limiting cases are approached, i.e., absolutely rigid inclusion or absent of an inclusion). As for the stresses σ_x/σ, then there are always appear negative values; moreover, the region of negative stresses and their absolute values are smaller for a stiffer inclusion. There are no negative stresses when the inclusion and the plate are made of the same material (and this is consistent with the classical solution). When the inclusion becomes stiffer than the plate, once again there appears a region of negative stresses and, hence, the loss of stability of the plate is again possible. Note that in the case of an inclusion stiffer than the plate, the region of negative stresses σ_x/σ is shifted by 90 degrees. Similar results are obtained if the problem is solved by applying the finite element method in ANSYS (see Figs. 4.1a and 4.1b).

4.3 Stability Loss

The numerical results and analysis in ANSYS showed that the loss of stability of plates with circular inclusion is possible if the elastic modulus of the inclusion is less than the modulus of elasticity of the plate $E_2 < E_1$ (inclusion softer than the plate) or when the elastic modulus of the inclusion larger than the modulus of elasticity of the plate $E_2 > E_1$ (i.e. inclusion is stiffer). Figures 4.2a and 4.2b show the stability loss modes for a plate with an inclusion subjected to tension along the plate's Y axis. Note that the stability loss modes are different for the case $E_2/E_1 < 1$ and $E_2/E_1 > 1$. In the case when the inclusion is stiffer than the plate, as was noted above, the zones of compressive stresses lie along the x axis (they are shifted by 90 degrees with respect to the case where the inclusion is softer than the plate).

If the Young's modulus of the inclusion is close to the modulus of elasticity of the plate than the loss of stability occurs either at very high loads or does not occur at all in the case of identical materials (Fig. 4.3).

4.3.1 Case with Different Poisson's Ratio

All results obtained in Bauer et al (2014) and shown in Fig. 4.3, were carried out for the same Poisson's ratio for inclusion and for plate $\nu_1 = \nu_2 = 0.3$. However,

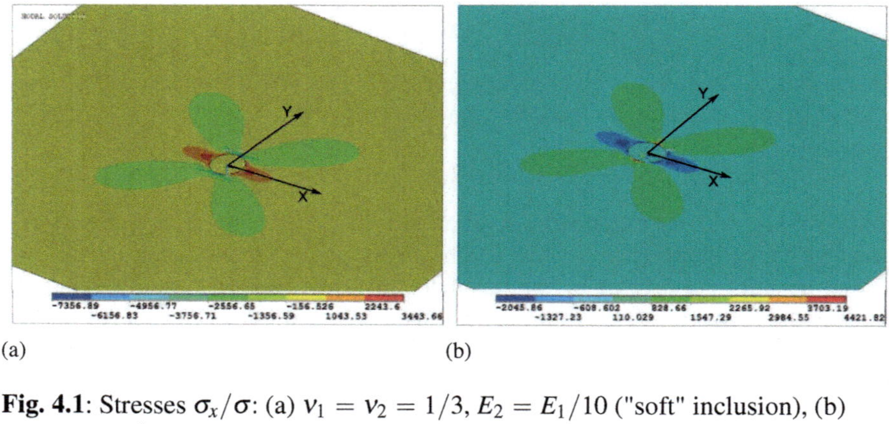

(a) (b)

Fig. 4.1: Stresses σ_x/σ: (a) $\nu_1 = \nu_2 = 1/3$, $E_2 = E_1/10$ ("soft" inclusion), (b) $\nu_1 = \nu_2 = 1/3$, $E_2 = 10E_1$ ("rigid" inclusion)

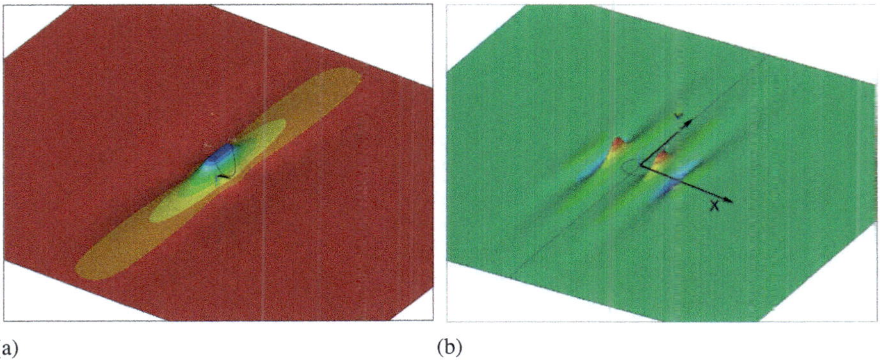

Fig. 4.2: Loss of stability: (a) $E_2 = E_1/10$ ("soft" inclusion), (b) $E_2 = 10\,E_1$ ("rigid" inclusion)

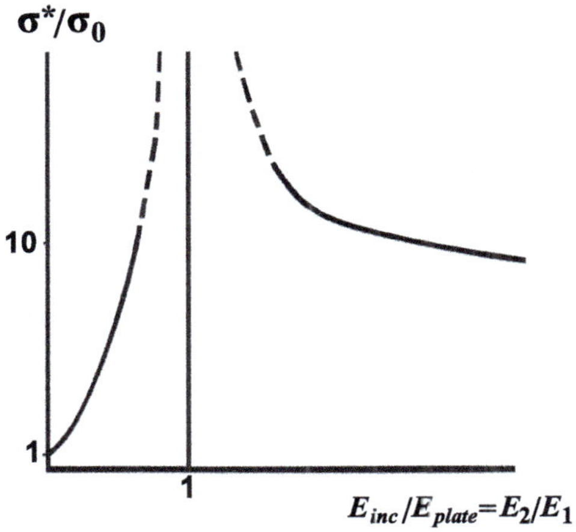

Fig. 4.3: Critical load as a function of the ratio of the inclusion modulus to the plate modulus (σ_0 is critical load for plate with a hole)

in the case of equal modules of elasticity of inclusion and plate, but with different Poisson's ratios can also be area of negative stress. Figures 4.4a and 4.4b represents stress distribution in the plate, obtained on the relations (4.1) for the case when $E_2 = E_1 = E$, but $\nu_1 \neq \nu_2$.

It means that in the case when the materials of the plate and inclusion have the same elastic moduli but different Poisson's ratios, the plate can also lose the stability under uniaxial tension. Figures 4.5a and 4.5b show the stress distribution before buckling and form of the plate stability loss in the case when $E_2 = E_1 = 2 \cdot 10^{11}$ Pa. Poisson's ratio of the plate $\nu_1 = 0.49$, Poisson's ratio of the inclusion $\nu_2 = 0.01$.

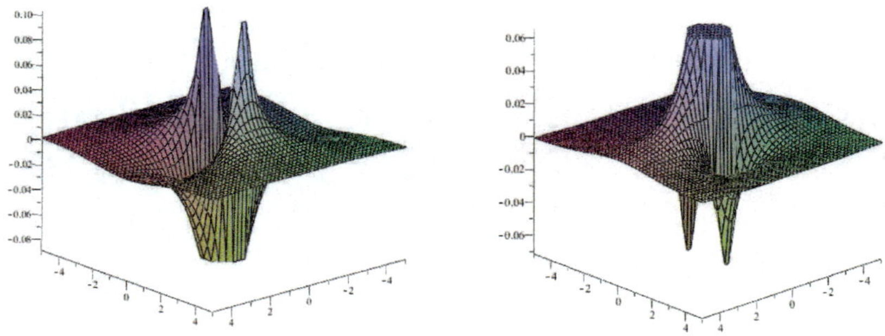

(a) (b)

Fig. 4.4: Stress distribution in the plate: (a) $\nu_1 = 2\nu_2, \nu_2 = 1/4$, (b) $\nu_1 = \nu_2/2$, $\nu_2 = 1/3$

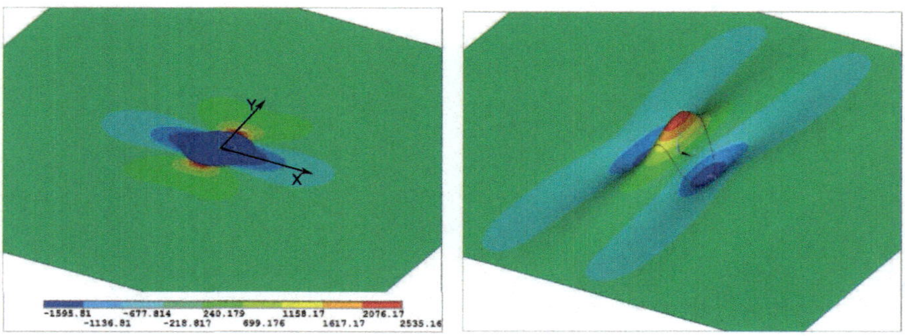

(a) (b)

Fig. 4.5: (a) Stress distribution before buckling $E_2 = E_1 = 2 * 10^{11}$ Pa, $\nu_1 = 0.49$, $\nu_2 = 0.01$, (b) Form of the plate stability loss $E_2 = E_1 = 2 * 10^{11}$ Pa, $\nu_1 = 0.49$, $\nu_2 = 0.01$

Figures 4.6a and 4.6b show the stress distribution before buckling and form of stability loss of a plate in the case when $E_2 = E_1 = 2 \cdot 10^{11}$ Pa. Poisson's ratio of the plate $\nu_1 = 0.01$, Poisson's ratio of the inclusion $\nu_2 = 0.49$.

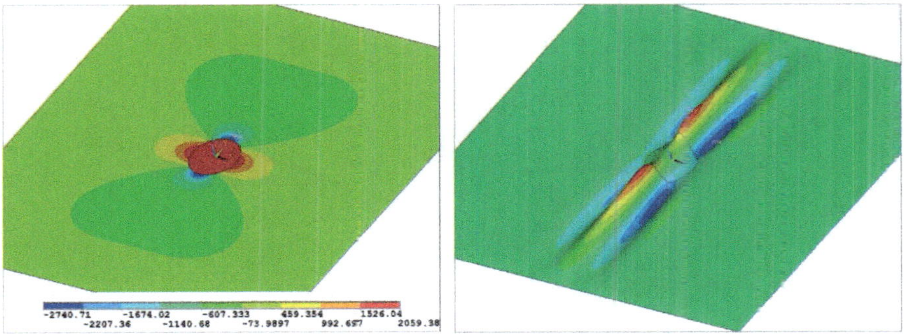

(a) (b)

Fig. 4.6: (a) Stress distribution before buckling $E_2 = E_1 = 2 * 10^{11}$ Pa, $\nu_1 = 0.01$, $\nu_2 = 0.49$, (b) Form of the plate stability loss $E_2 = E_1 = 2 * 10^{11}$ Pa, $\nu_1 = 0.01$, $\nu_2 = 0.49$

4.3.2 A Plate with a Circular Inclusion under Biaxial Tension

The case when the plate with inclusion is under tension not only along the Y-axis but also along the X-axis is also interesting to consider. If tension along the Y-axis is equal to the tension along the X-axis then all stresses in the plate will be positive and there is no loss of stability. But if the stresses applied along the X-axis is less than stresses applied along Y-axis then again occur negative stresses in the plate near inclusion and thus the stability loss of the plate is possible.

Under uniaxial tension negative stresses vanish only if the elastic moduli and Poisson's ratios of the plate and inclusion are equal (see Fig. 4.3). The area of negative stresses in the plate with inclusion under tension along the Y-axis decreases with the application to the plate stresses along the X-axis. By increasing the load along the X-axis we can discover the value of tension P_{comp} when the regions of compressive stresses in the plate disappear and loss of stability is impossible. If we continue to increase the load along the X-axis, starting with a certain value P_{comp} we can obtain new areas of compressive stresses. These areas are rotated on 90 degrees compared to the initial location of compressive stresses.

For each ratio of the modules of elasticity of inclusion and plate (E_2/E_1) there is an interval of stresses P_{comp} along X-axis, where there are no zones of compressive stresses, thus we determine the whole range of values for which there is no possibility for stability loss of the plate (between lines 1 and 2). In Figs. 4.7a and 4.7b curve 1 is the lower boundary of this zone, and the curve is 2 the upper bound. If E_2/E_1 goes to 1 then the upper boundary of the interval (line 2) goes to infinite.

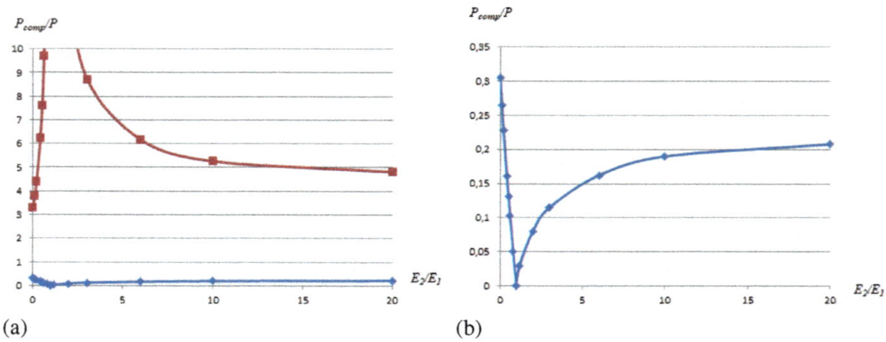

Fig. 4.7: (a) Additional stress curves bounding the stability region of the plate with inclusion. Curve 1 (blue) is the lower boundary of this zone, and the curve 2 (red) is the upper bound. P_{comp}/P is the ratio of the additional tension P_{comp} along the X-axis to the initial tension P along the Y-axis.

References

Bauer SM, Kashtanova SV, Morozov NF, Semenov BN (2014) Stability of a nanoscale-thickness plate weakened by a circular hole. Doklady Physics 59(9):416–418

Bauer SM, Kashtanova SV, Morozov NF, Semenov BN (2017) Stability loss in an infinite plate with a circular inclusion under uniaxial tension. Vestnik St Petersburg University, Mathematics 50(2):161–165

Bochkarev A, Grekov M (2015) On symmetrical and antisymmetrical buckling of a plate with circular nanohole under uniaxial tension. Applied Mathematical Sciences 9:125–128

Bochkarev AO, Grekov MA (2014) Local instability of a plate with a circular nanohole under uniaxial tension. Doklady Physics 59(7):330–334

Deryugin YY, Lasko GV (2012) Field of stresses in an isotropic plane with circular inclusion under tensile stresses. Engineering 4:583–589

Eshelby DE (1957) Definition of the stress field, which was creating by elliptical inclusion. Proceedings of the Royal Society A 241(1226):376

Guz AN, Dishel' MS, Kuliev GG, Milovanova OB (1981) Fracture and Stability of Thin Bodies with Cracks (in Russ.). Naukova Dumka, Kiev

Kachanov M, Shafiro B, Tsurkov I (2003) Handbook of Elasticity Solutions. Kluwer Academic, Dordrecht

Muskhelishvili NI (1963) Some Basic Problems of the Mathematical Theory of Elasticity. Noordhoff, Groningen

Shimizu S (2007) Tension buckling of plate having a hole. Thin-Walled Structures 45(10):827–833

Chapter 5
Unit Cell Models of Viscoelastic Fibrous Composites for Numerical Computation of Effective Properties

Harald Berger, Mathias Würkner, José A. Otero, Raúl Guinovart-Díaz, Julián Bravo-Castillero, and Reinaldo Rodríguez-Ramos

Abstract The paper presents an extension to viscoelastic composites of a former developed numerical homogenization procedure which was used for elastic and piezoelectric material systems. It is based on an unit cell model using the finite element method. In the paper a brief description of the basic equations and the homogenization algorithm with specific attention to the numerical model is given. The investigated composites consist of a viscoelastic matrix with unidirectional embedded cylindrical elastic fibers. Hence the homogenized behavior of the composite is also viscoelastic. Consequently the effective coefficients are time-dependent. The geometrical shape of the unit cell is rhombic which allows to analyze a wide range of nonstandard unidirectional fiber distributions. Otherwise it includes the special cases for square and hexagonal fiber arrangements which can be used for comparisons with other solutions. Here results are compared with an analytical homogenization method. Furthermore the influences of rhombic angle and fiber volume fraction on effective coefficients are investigated. In addition two limit cases are considered. One is with air as inclusions which is equivalent to a porous media and the other is the pure matrix without fibers.

Harald Berger · Mathias Würkner
Institut für Mechanik, Otto-von-Guericke-Universität Magdeburg, Universitätsplatz 2, D-39106 Magdeburg, Germany
e-mail: harald.berger@ovgu.de, mathias.wuerkner@ovgu.de

José A. Otero
Instituto Tecnológico de Estudios Superiores de Monterrey CEM, Atizapán de Zaragoza, EM CP 52926, México
e-mail: j.a.otero@itesm.mx

Raúl Guinovart-Díaz · Julián Bravo-Castillero · Reinaldo Rodríguez-Ramos
Universidad de La Habana, Facultad de Matemática y Computació, San Lázaro y L, Vedado, La Habana, CP 10400, Cuba
e-mail: guino@matcom.uh.cu, jb_castillero@yahoo.com.mx, reinaldo@matcom.uh.cu

© Springer International Publishing AG, part of Springer Nature 2018
H. Altenbach et al. (eds.), *Generalized Models and Non-classical Approaches in Complex Materials 1*, Advanced Structured Materials 89,
https://doi.org/10.1007/978-3-319-72440-9_5

5.1 Introduction

Nowadays composite materials have a very important meaning in many fields of industry. Due to their advantageous properties related to weight, strength, stiffness, functionality and resistance they belong to future orientated materials which replace more and more common materials. Furthermore the combination of components with multi-physical behavior (e.g. piezoelectricity) opens new ways in constructions.

But on the other hand composites are a great challenge to estimate their overall behavior in structural design. One possible way is to calculate effective material properties by appropriate homogenization techniques. Since many years scientists deal with this matter. Different types of composites and different material components have been considered. But there are still some issues which need a deeper investigation.

Also the homogenization of viscoelastic composites has been a matter of investigation since many years. There are analytical approaches (Yancey and Pindera, 1990; Nguyen et al, 1995; Tang and Felicelli, 2016; Cruz-González et al, 2018) and numerical approaches (Haasemann and Ulbricht, 2009; Nguyen et al, 1995; Pathan et al, 2017; Daridon et al, 2016; To et al, 2017). Unit cells (RUC) are a typical basis for homogenization models which assume an infinite regular distribution of inclusions with repeatable pattern. For random distribution of inclusions so-called representative volume elements (RVE) are used. A wide range of literature especially deals with the characteristics of such cells considering appropriate boundary conditions, influence of cell size related to inclusions, etc.

In this paper a former developed homogenization procedure based on a RUC model (Berger et al, 2006; Kari et al, 2007; Würkner et al, 2011) is extended to viscoelasticity in order to calculate effective material properties of unidirectional fiber-reinforced composites, where elastic cylindrical fibers are embedded in a viscoelastic matrix. The procedure allows the investigation of composites, where the fiber distribution can be characterized by a RUC with a rhombic cross section (see Fig. 5.1). By varying the rhombic angle and the fiber volume fraction the change in overall viscoelöastic material behavior can be studied for such types of composites.

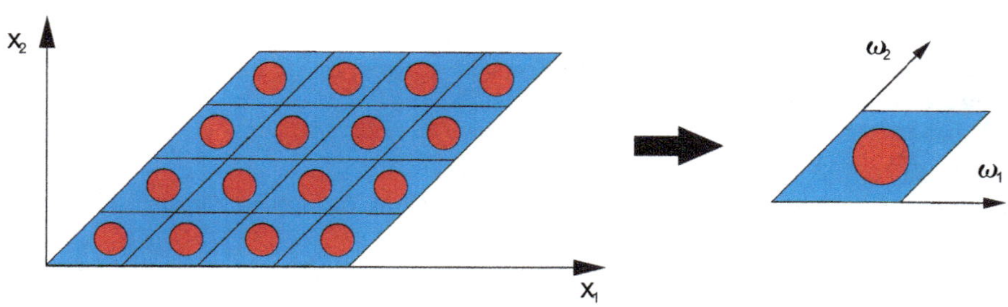

Fig. 5.1: Geometrical type of composite structure and extracted rhombic RUC

5.2 Linear viscoelastic relations

In general viscoelastic materials are characterized by a time-dependent change of stress or strain after loading. In case of a constant stress load the strain is increasing over time which is called creep. In case of a constant strain load the stress is decreasing over time which is called relaxation. Viscoelasticity means that the stress-strain relations can be divided in an elastic and non-elastic part. In case of a restriction to small deformations it leads to linear viscoelastic behavior.

The typical phenomenological behavior can be described by the well known models from Maxwell and Kelvin-Voigt consisting of springs and dashpots (Gutierrez-Lemini, 2014). The following basic equations are focused on solving a linear viscoelastic problem using a finite element software.

Derived from the generalized Maxwell relaxation model (Gutierrez-Lemini, 2014) the linear isotropic viscoelastic behavior is characterized by time-dependent material parameters and can be written in the hereditary integral formulation

$$\sigma(t) = \int_0^t 2G(t-\tau)\frac{de}{d\tau}d\tau + I\int_0^t K(t-\tau)\frac{d\Delta}{d\tau}d\tau. \tag{5.1}$$

It couples stresses σ with strain rates $de/d\tau$ and $d\Delta/d\tau$. In detail 'σ' are the Cauchy stresses, 'e' is the deviatoric part of the strains, 'Δ' is volumetric part of the strains, $G(t)$ is the shear relaxation function, $K(t)$ is the bulk relaxation function, t is the current time, ‚τ' is the relaxation time and I is the unit tensor. The time-dependent functions $G(t)$ and $K(t)$ which represent the viscoelastic material parameters can be expressed in terms of Prony series (Gutierrez-Lemini, 2014) in the following form

$$G(t) = G_\infty + \sum_{i=1}^{n_g} G_i e^{\left(-\frac{t}{\tau_i^G}\right)}, \tag{5.2}$$

$$K(t) = K_\infty + \sum_{i=1}^{n_k} K_i e^{\left(-\frac{t}{\tau_i^K}\right)}. \tag{5.3}$$

Here G_i and K_i are the shear elastic moduli and the bulk elastic moduli, respectively, and τ_i^G and τ_i^K are the corresponding relaxation times for each Prony component. G_∞ and K_∞ are the long-term moduli related to $t=\infty$. n_g and n_k are the numbers of Prony terms for the corresponding moduli.

Introducing the relative moduli

$$\alpha_i^G = \frac{G_i}{G_0}, \tag{5.4}$$

$$\alpha_i^K = \frac{K_i}{K_0} \tag{5.5}$$

with

$$G_0 = G_\infty + \sum_{i=1}^{n_g} G_i \,, \tag{5.6}$$

$$K_0 = K_\infty + \sum_{i=1}^{n_k} K_i \tag{5.7}$$

where G_0 and K_0 are the instantaneous shear elastic modulus and instantaneous bulk elastic modulus, respectively, Eqs. (5.2) and (5.3) can be rewritten in

$$G(t) = G_0 \left[\alpha_\infty^G + \sum_{i=1}^{n_g} \alpha_i^G e^{\left(-\frac{t}{\tau_i^G}\right)} \right], \tag{5.8}$$

$$K(t) = K_0 \left[\alpha_\infty^K + \sum_{i=1}^{n_g} \alpha_i^K e^{\left(-\frac{t}{\tau_i^K}\right)} \right]. \tag{5.9}$$

For isotropic materials G_0 and K_0 can be expressed by the Young's modulus E_0 and the Poisson's ratio ν_0 with

$$G_0 = \frac{E_0}{2(1+\nu_0)}, \tag{5.10}$$

$$K_0 = \frac{E_0}{3(1-2\nu_0)}. \tag{5.11}$$

Finally the necessary input values are the instantaneous Young's modulus E_0, Poisson's ratio ν_0 and the Prony terms τ_i^G, α_i^G and τ_i^K, α_i^K . The number of shear terms n_G and bulk terms n_K need not to be the same.

Inside the finite element algorithm the problem is solved by a time integration scheme. The solution is carried out by time steps until the target time is reached. The time step width must be carefully chosen to achieve a stable solution.

5.3 Numerical Homogenization Model

To find the effective material properties for the composite the elasticity tensor must be calculated with the developed numerical homogenization procedure. Only a brief description is given here. For more details see Berger et al (2006); Würkner et al (2011).

The constitutive relation at a discrete time t_i can be expressed in the following form

$$\sigma_{ij}(t_i) = C_{ijkl}(t_i)\, \varepsilon_{ij}(t_i) \tag{5.12}$$

where $C_{ijkl}(t_i)$ is the coefficient of the elasticity tensor at time t_i. and $\sigma_{ij}(t_i)$ and $\varepsilon_{ij}(t_i)$ are the stresses and strains, respectively. Although the constituents have isotropic behavior the overall properties of the composite can become anisotropic due to different types of fiber arrangements. In case of anisotropic material the elasticity tensor has 21 independent coefficients (here expressed in matrix notation which is used in Sect. 5.4 for presentation of calculated effective coefficients)

$$\begin{bmatrix} \sigma_{11} \\ \sigma_{22} \\ \sigma_{33} \\ \sigma_{12} \\ \sigma_{23} \\ \sigma_{31} \end{bmatrix} = \begin{bmatrix} C_{11} & C_{12} & C_{13} & C_{14} & C_{15} & C_{16} \\ & C_{22} & C_{23} & C_{24} & C_{25} & C_{26} \\ & & C_{33} & C_{34} & C_{35} & C_{36} \\ & & & C_{44} & C_{45} & C_{46} \\ & symm & & & C_{55} & C_{56} \\ & & & & & C_{66} \end{bmatrix} \begin{bmatrix} \epsilon_{11} \\ \epsilon_{22} \\ \epsilon_{33} \\ \epsilon_{12} \\ \epsilon_{23} \\ \epsilon_{31} \end{bmatrix} . \tag{5.13}$$

In the numerical homogenization procedure the RUC is subjected to 6 strain load cases (applied as prescribed displacement differences) in such a way that in every load case only one strain component is nonzero and all others are zero. These are three pure tension cases in all spatial directions and three pure shear cases.

Furthermore to guarantee that also after deformation the periodicity is ensured appropriate periodic boundary conditions must be applied. This can be achieved by fulfilling the following equation (Würkner et al, 2011)

$$u_i^{X_j^+} - u_i^{X_j^-} = \varepsilon_{ij}^0 (x_j^{X_j^+} - x_j^{X_j^-}). \tag{5.14}$$

The values $u_i^{X_j^+}$ and $u_i^{X_j^-}$ are the ith displacement components on the boundary surfaces of the cell, which are perpendicular to the x_j-axis ("+" for a positive normal direction, "-" for a negative normal direction, see Fig. 5.2). The locations, in which the values are calculated, are characterized by an offset in x_j-direction. In the finite element model this means that every opposite nodal pair is coupled by constraint equations corresponding to Eq. (5.14). This requires an identical mesh configuration on opposite surfaces which is ensured using a special meshing strategy. ε^0 represents the applied strain tensor and has the following form for the three tension load cases

$$\begin{pmatrix} \varepsilon_{11}^0 & 0 & 0 \\ 0 & 0 & 0 \\ 0 & 0 & 0 \end{pmatrix}, \begin{pmatrix} 0 & 0 & 0 \\ 0 & \varepsilon_{22}^0 & 0 \\ 0 & 0 & 0 \end{pmatrix}, \begin{pmatrix} 0 & 0 & 0 \\ 0 & 0 & 0 \\ 0 & 0 & \varepsilon_{33}^0 \end{pmatrix} \tag{5.15}$$

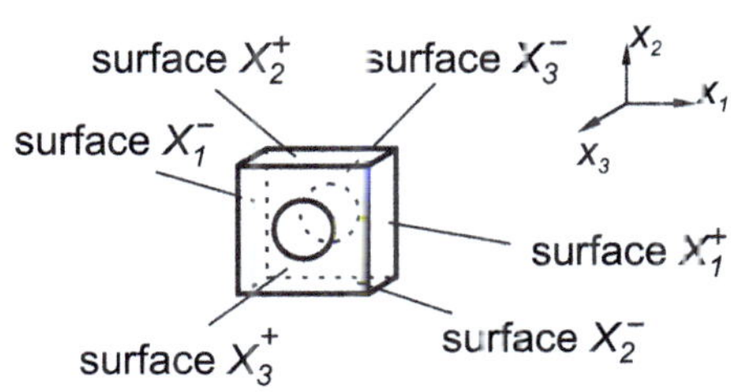

Fig. 5.2 Scheme for notation of the different surfaces on the RUC

and for the three shear load cases

$$\begin{pmatrix} 0 & \varepsilon_{12}^0 & 0 \\ \varepsilon_{12}^0 & 0 & 0 \\ 0 & 0 & 0 \end{pmatrix}, \quad \begin{pmatrix} 0 & 0 & 0 \\ 0 & 0 & \varepsilon_{23}^0 \\ 0 & \varepsilon_{23}^0 & 0 \end{pmatrix}, \quad \begin{pmatrix} 0 & 0 & \varepsilon_{31}^0 \\ 0 & 0 & 0 \\ \varepsilon_{31}^0 & 0 & 0 \end{pmatrix}. \tag{5.16}$$

ε_{ij}^0 can have an arbitrary value but is chosen with an unit value.

From the six calculated load cases all effective coefficients can be extracted by dividing the averaged stresses by the averaged strains considering every column of the matrix in Eq. (5.13) separately. For the calculation of averaged stresses $\hat{\sigma}_{ij}$ and strains $\hat{\varepsilon}_{ij}$ we take the averaged element values σ_{ij}^k and ε_{ij}^k, respectively, which are the arithmetic mean of Gauss point results, weight these values with the element volume V_k, calculate the sum over all elements and divide it by the RUC volume V_{RUC}

$$\hat{\sigma}_{ij} = \frac{1}{V_{RUC}} \sum_{k=1}^{n} \sigma_{ij}^k V_k, \tag{5.17}$$

$$\hat{\varepsilon}_{ij} = \frac{1}{V_{RUC}} \sum_{k=1}^{n} \varepsilon_{ij}^k V_k. \tag{5.18}$$

Then, for example from the first load case in Eq. (5.15) the following effective elasticity coefficients can be extracted

$$C_{11} = \frac{\hat{\sigma}_{11}}{\hat{\varepsilon}_{11}}, \quad C_{12} = \frac{\hat{\sigma}_{22}}{\hat{\varepsilon}_{11}}, \quad \text{etc.} \tag{5.19}$$

In case of a composite with a rhombic fiber arrangement the unit cell has the typical shape shown in Fig. 5.1. But for this shape with the oblique coordinate system ω_1, ω_2 it is difficult to apply the above mentioned periodic boundary conditions. The solution lies in extracting a rectangular cell which is rotated against the global coordinate system by the half of the rhombic angle α (see Fig. 5.3). Assuming unit length of rhombic edges the rectangular cell has the edge lengths

$$l_1 = 2\cos(\frac{\alpha}{2}), \tag{5.20}$$

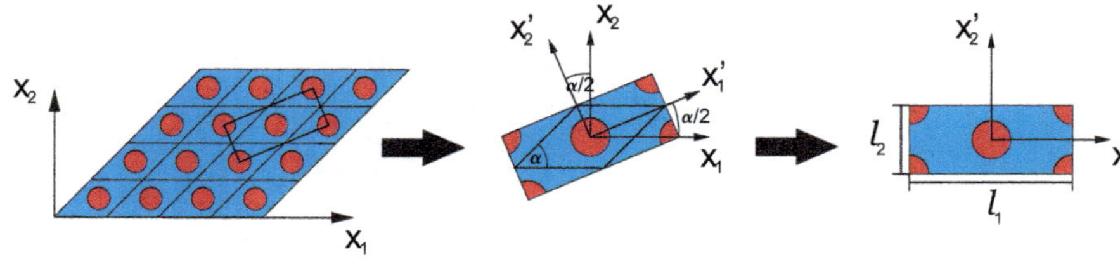

Fig. 5.3: Rectangular RUC with local coordinate system (Würkner et al, 2011)

$$l_2 = 2\sin(\frac{\alpha}{2}). \tag{5.21}$$

Then all calculations can be made in the local coordinate system x_i' and, if necessary, the results can be transformed to the global coordinate system x_i.

To calculate all effective coefficients of the elasticity tensor a three-dimensional RUC is used shown in Fig. 5.4. Due to unidirectional embedded fibers the thickness of the cell can be chosen arbitrarily and only one element in this direction is enough. 3D standard elements (hexahedrons and pentahedrons) with quadratic shape functions is used which provide sufficient accuracy with moderate mesh size like shown in Fig. 5.4. Since all calculations are made with the finite element package ANSYS many parts of model generation as well as extracting and processing results can be automated using the included ANSYS parametric design language APDL.

5.4 Results

Various calculations are made to consider different aspects of the developed procedure. First, for a validation results are compared with available values calculated by the asymptotic homogenization method for the case of hexagonal fiber arrangement. Furthermore calculations are made to study the change in the overall behavior of the composite by varying the rhombic angle and keeping a constant fiber volume fraction and then by varying the volume fiber fraction and keeping a constant rhombic angle, respectively. Finally investigations are made by considering the two limit cases: embedded air instead of fibers (porous media) and no inclusions (pure matrix).

For all calculations a glass fiber/epoxy composite with material properties for the constituents given in Table 5.1 is considered. The fiber is considered to be elastic isotropic. The matrix has viscoelastic isotropic properties. According to Sect. 5.2 the necessary input values are the instantaneous Young's modulus E_0, the Poisson's ratio ν_0 and the Prony series parameters. Here n^G is 3 and n^K is 0. That means 3 pairs of the shear part and no values for bulk part are used. For the relaxation time range a period of 60000 seconds is used. This is chosen in order to achieve the long-term material behavior.

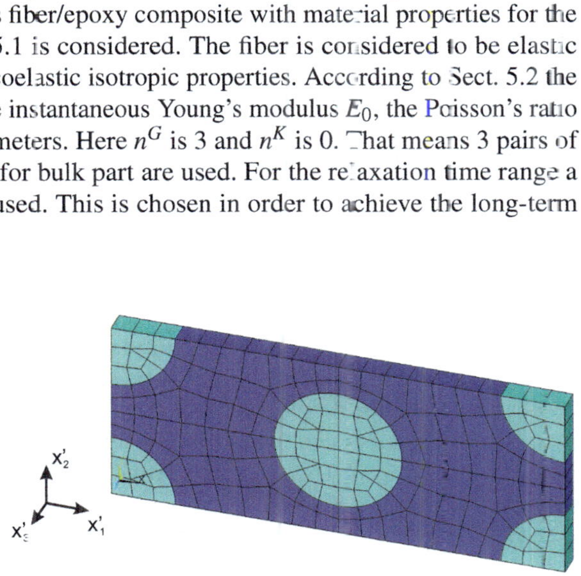

Fig. 5.4 Finite element mesh of the RUC with rhombic angle 45 ° and fiber volume fraction 0.3

Table 5.1: Material properties

Constituent	E (GPa)	ν				
Fiber	100	0.3				
Matrix	3.606 (E_0)	0.4(ν_0)	Prony series parameters	i	τ_i^G (s)	α_i^G
				1	9.6	0.03807
				2	372	0.0458
				3	9887	0.0668

For the validation of the presented homogenization procedure for rhombic cells a comparison is made with results obtained from the asymptotic homogenization method for the special case of hexagonal fiber arrangement (Cruz-González et al, 2018). This arrangement is identical with a rhombic angle of 60 degrees. Hexagonal fiber arrangement leads to transverse isotropic behavior of the composite. Hence only 6 coefficients in the matrix of Eq. (5.13) are nonzero. From these coefficients tangential ($x_1' - x_2'$ plane; index t) and axial (x_3' direction; index a) engineering constants can be derived. The fiber volume fraction is fixed with $V_{fiber}/V_{RUC} = 0.3$. Figures 5.5-5.10 show the six engineering constants over time as a comparison between the presented method (FEM) and the asymptotic homogenization method (AHM). A good agreement can be stated for all engineering constants. Although the curves show differences in the low time region the percentage difference is not more the 2%. All curves show the asymptotic behavior at the end of the chosen time range.

To study the influence of the rhombic angle different calculations are made by varying the angle and keeping the fiber volume fraction constant. For arbitrary angles the resulting behavior is orthotropic with respect to the local axes in Fig. 5.2. This leads to 9 independent coefficients in the matrix of Eq. (5.13). Only selected effective coefficients are considered for this study. Figures 5.11-5.16 show the results over time for all coefficients of the main diagonal of the matrix related to the local coordinate system. Four rhombic angles α are investigated: 45°, 60°, 75° and 90°. The fiber volume fraction is kept constant with $V_{fiber}/V_{RUC} = 0.3$. For $\alpha = 60°$, the hexagonal case, an expected transversal isotropy can be recognized considering the corresponding coefficients C_{11}, C_{22} (in-plane) and C_{55}, C_{66} (out-of-plane). The coefficient C_{33} for the axial direction depends only on the fiber volume fraction in correspondence with the simple rule of mixture and is nearly identical for all rhombic angles. But for the other coefficients the typical differences for rhombic angles of 45° and 75° can be seen. This is caused by the effective orthotropic behavior of the composite due to different fiber distances in the $x_1' - x_2'$ plane. It can be seen for instance in the case of 45° where the shear coefficient C_{55} is lower and C_{66} is higher then the transversal isotropic curve (angle 60°).

In the next investigations the influence of fiber volume fraction for a constant rhombic angle of 45° is studied. Like expected all effective coefficients increase with

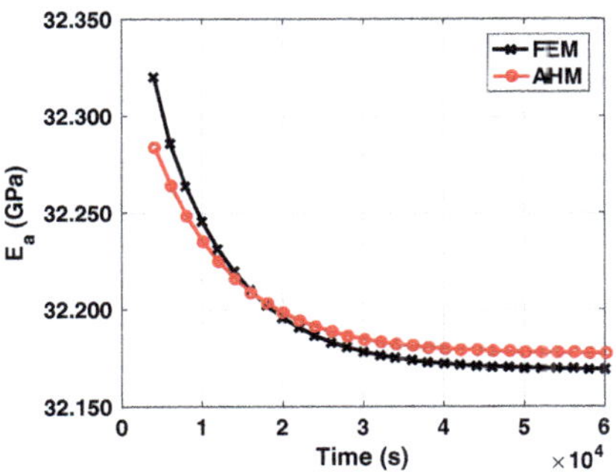

Fig. 5.5: E_a for rhombic angle 60 °

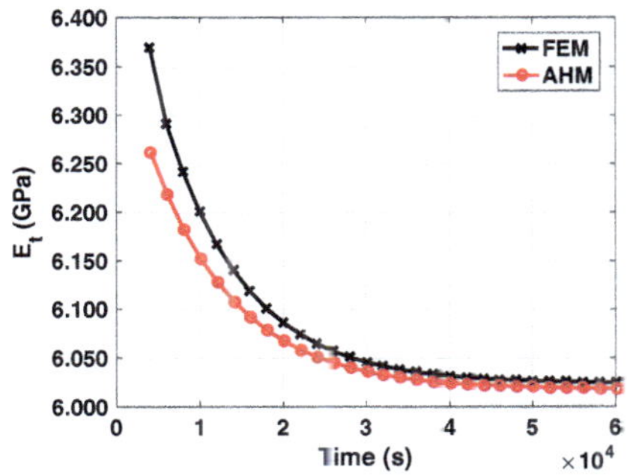

Fig. 5.6: E_t for rhombic angle 60 °

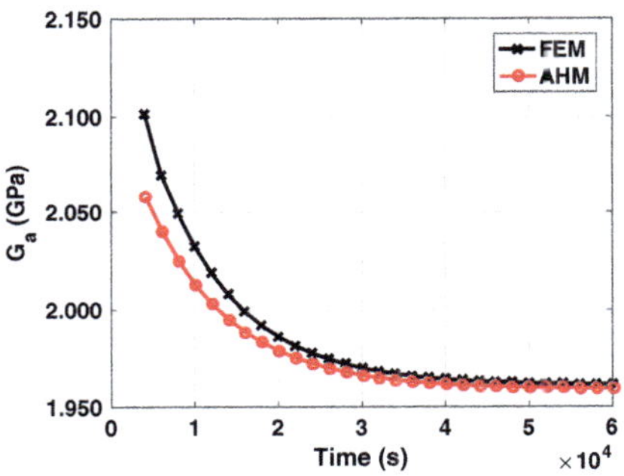

Fig. 5.7: G_a for rhombic angle 60 °

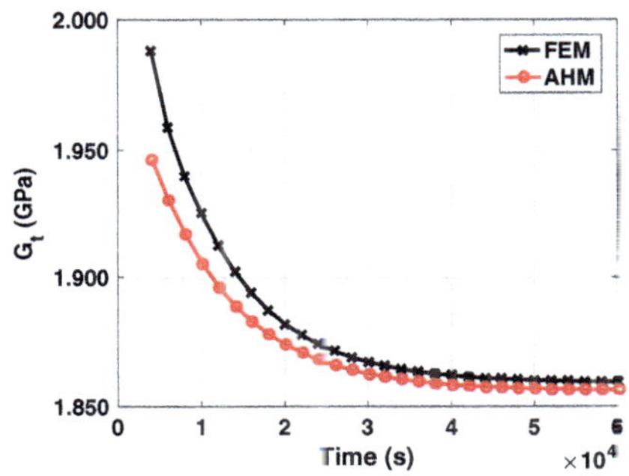

Fig. 5.8: G_t for rhombic angle 60 °

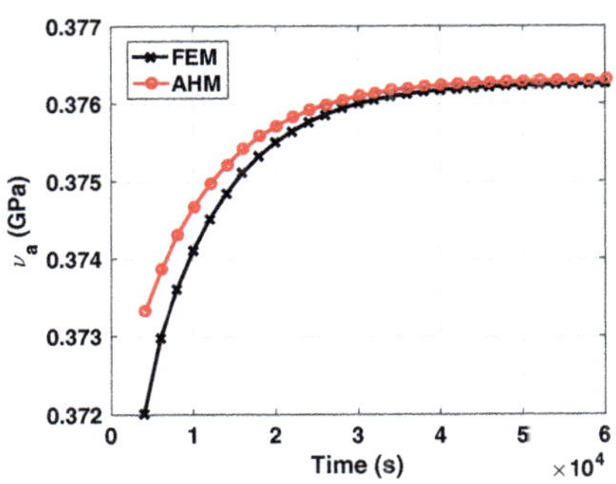

Fig. 5.9: ν_a for rhombic angle 60 °

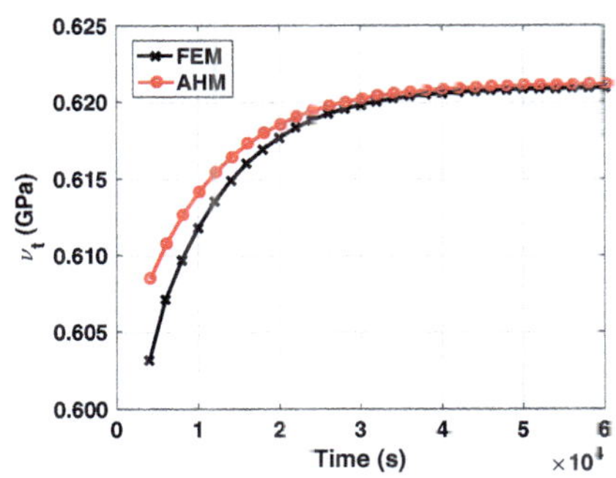

Fig. 5.10: ν_t for rhombic angle 60 °

an increase of fiber volume fraction due to a higher stiffness in comparison to the matrix phase. This can be clearly seen in Figs. 5.17-5.22.

The last consideration is dedicated to the limit cases. In the first case the fibers are replaced by air. This leads to the simulation of a porous medium. In the RUC the air is modeled as a phase with a very low Young's modulus related to the matrix material. The second limit case is a homogeneous material consisting only of the

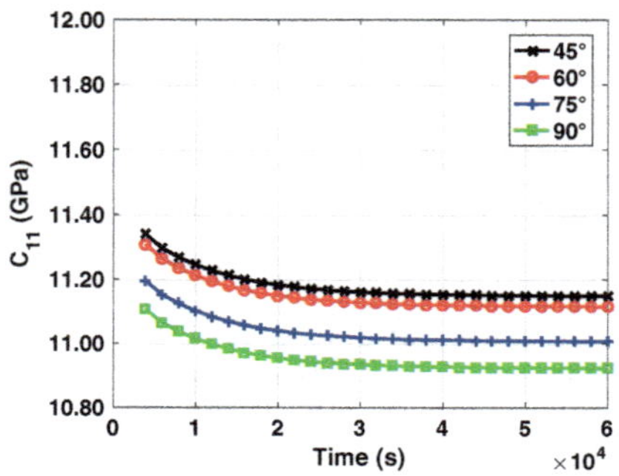

Fig. 5.11: C_{11} for different rhombic angles and constant fiber volume fraction of 0.3

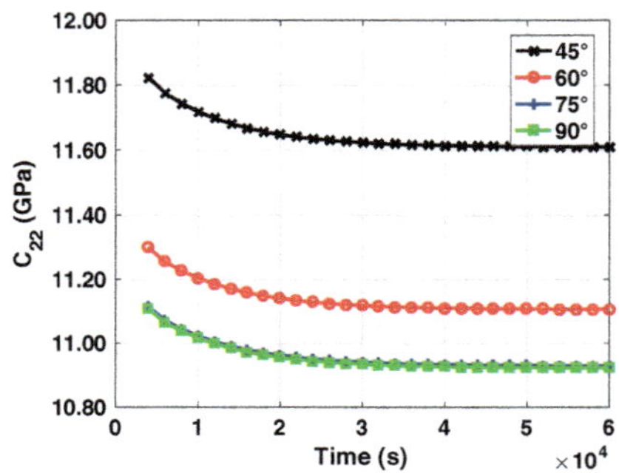

Fig. 5.12: C_{22} for different rhombic angles and constant fiber volume fraction of 0.3

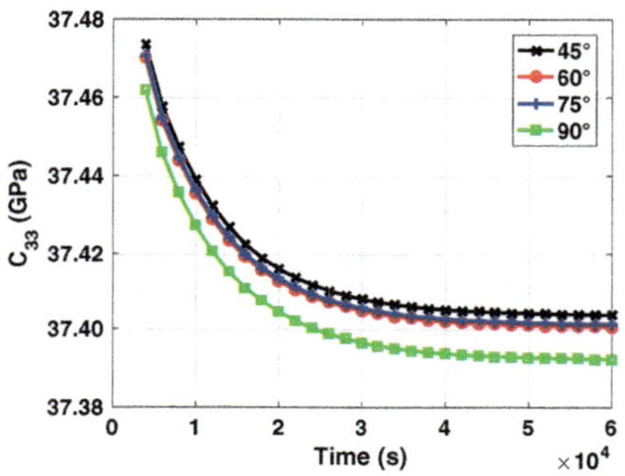

Fig. 5.13: C_{33} for different rhombic angles and constant fiber volume fraction of 0.3

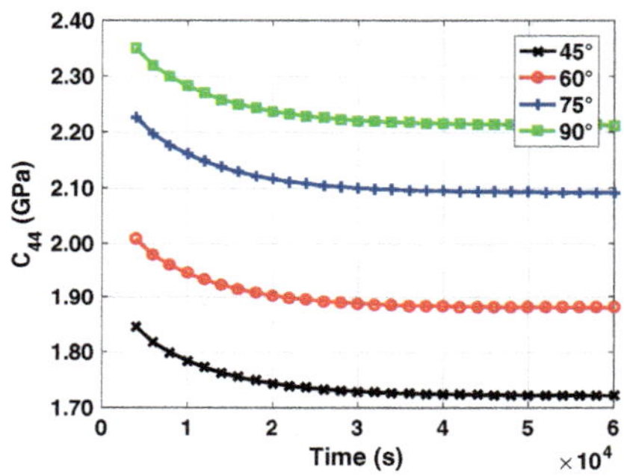

Fig. 5.14: C_{44} for different rhombic angles and constant fiber volume fraction of 0.3

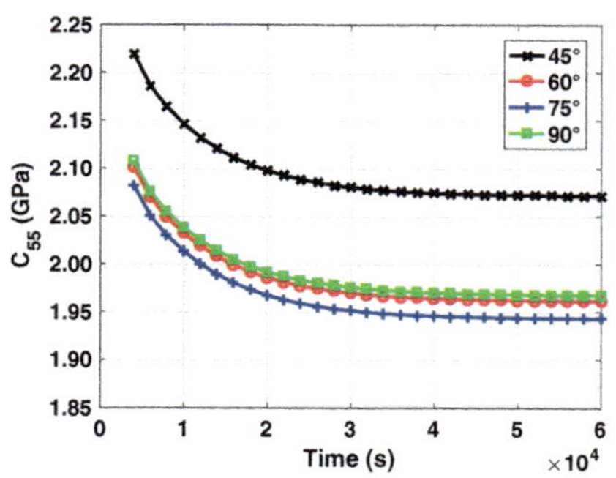

Fig. 5.15: C_{55} for different rhombic angles and constant fiber volume fraction of 0.3

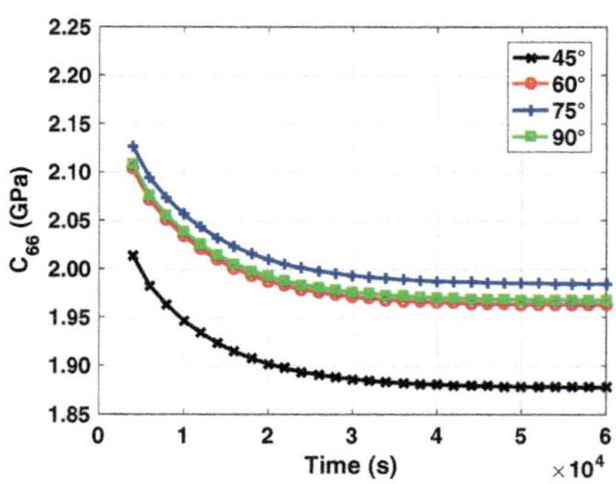

Fig. 5.16: C_{66} for different rhombic angles and constant fiber volume fraction of 0.3

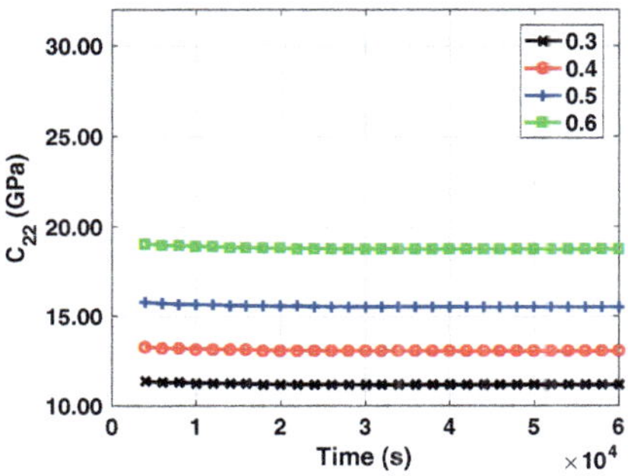

Fig. 5.17: C_{11} for different fiber volume fractions and constant rhombic angle of 45°

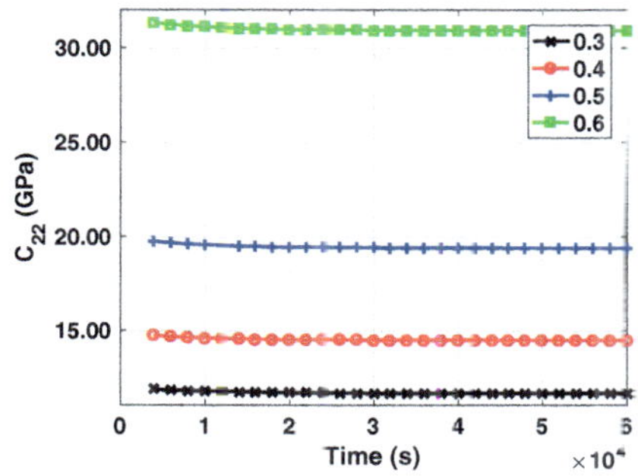

Fig. 5.18: C_{22} for different fiber volume fractions and constant rhombic angle of 45°

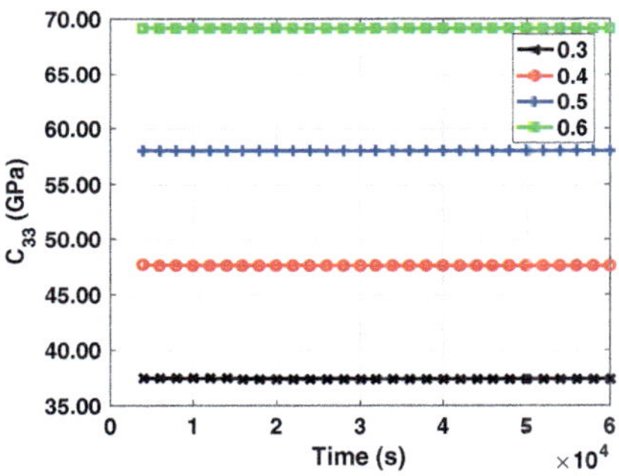

Fig. 5.19: C_{33} for different fiber volume fractions and constant rhombic angle of 45°

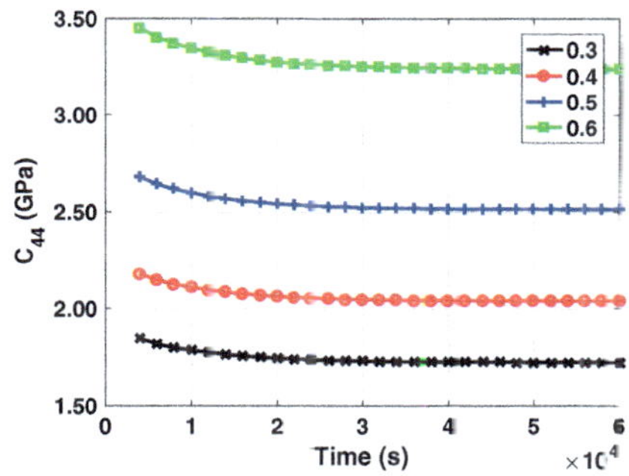

Fig. 5.20: C_{44} for different fibre volume fractions and constant rhombic angle of 45°

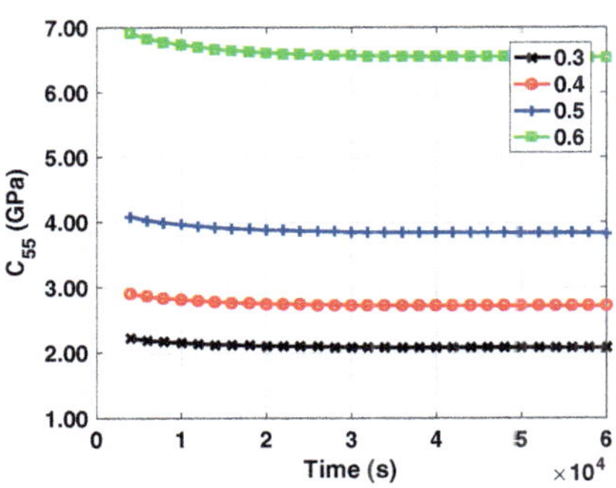

Fig. 5.21: C_{55} for different fibre volume fractions and constant rhombic angle of 45°

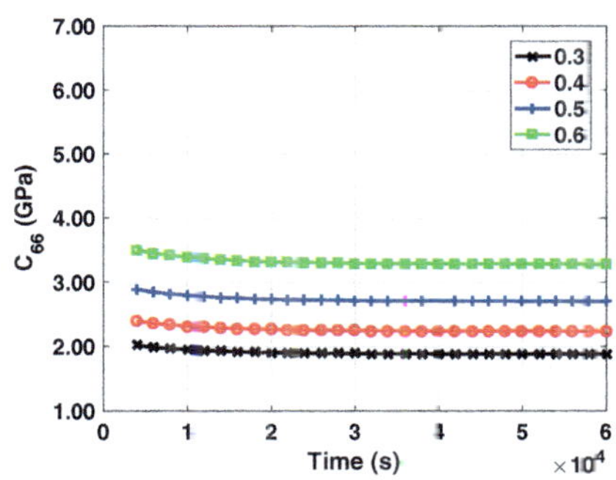

Fig. 5.22: C_{66} for different fiber volume fractions and constant rhombic angle of 45°

viscoelastic matrix phase which is denoted in the figures with "monolithic". Figures 5.23-5.28 show selected effective coefficients of the elasticity tensor over time. In addition to the limit cases the curves for a composite with 0.3 fiber volume fraction and a rhombic angle of 45° are included in the figures for a better exhibition of the tendency. The reinforcing influence of the fibers can be clearly observed. Like expected the porous case show the lowest stiffness. On the other hand the achievable effective stiffness of the composite depends mainly on the stiffness of the fiber in relation to the matrix and on the fiber volume fraction. Furthermore, with varying the rhombic fiber arrangement it is possible to construct composites which have a different elastic behavior in two perpendicular directions in the $x_1' - x_2'$ plane (see also Figs. 5.11-5.16). The orthotropic in-plane behavior can be valuable for optimal design of constructions.

5.5 Conclusions

This work is based on a former research of the authors in numerical homogenization. Here an extension to viscoelastic composites is presented. Furthermore it is applied to unidirectional fiber-reinforced composites with a rhombic fiber arrangement. The developed models are verified by values calculated with an analytical homogenization method. This has been realized for a special case of rhombic shape with an angle of 60° which is identical with a hexagonal arrangement. The correctness of the procedure can be stated by corresponding results.

Furthermore interesting points of the behavior by changing the rhombic angle, fiber volume fraction and material composition are revealed. Especially in this work it is shown that the former developed algorithms can be extended to time-dependent material behavior like viscoelasticity to get the effective time-dependent coefficients.

References

Berger H, Kari S, Gabbert U, Rodríguez-Ramos R, Bravo-Castillero J, Guinovart-Díaz R, Sabina FJ, Maugin GA (2006) Unit cell models of piezoelectric fiber composites for numerical and analytical calculation of effective properties. Smart Mater Struct 15:451–458

Cruz-González OL, Rodríguez-Ramos R, Otero JA, Bravo-Castillero J, Guinovart-Díaz R, Martínez-Rosado R, Sabina FJ, Dumont S, Lebon F, Sevostianov I (2018) Viscoelastic effective properties for composites with rectangular cross-section fibers using the asymptotic homogenization method. In: Altenbach H, Pouget J, Rousseau M, Collet B, Michelitsch T (eds) Generalized Models and Non-Classical Approaches in Complex Materials - Vol. 1, Springer, Singapore, Advanced Structured Materials, vol 92, pp 203–222

Daridon L, Licht C, Orankitjaroen S, Pagano S (2016) Periodic homogenization for Kelvin-Voigt viscoelastic media with a Kelvin-Voigt viscoelastic interphase. European Journal of Mechanics - A/Solids 58:163–171

Gutierrez-Lemini D (2014) Engineering Viscoelasticity. Springer, New York

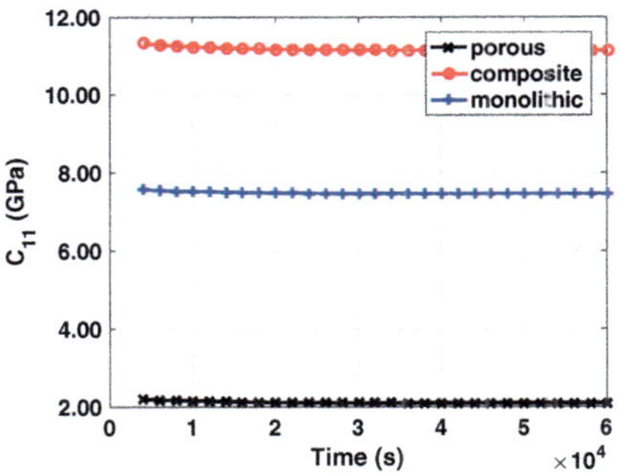

Fig. 5.23: C_{11} for different fiber material with rhombic angle 45° and volume fraction of 0.3

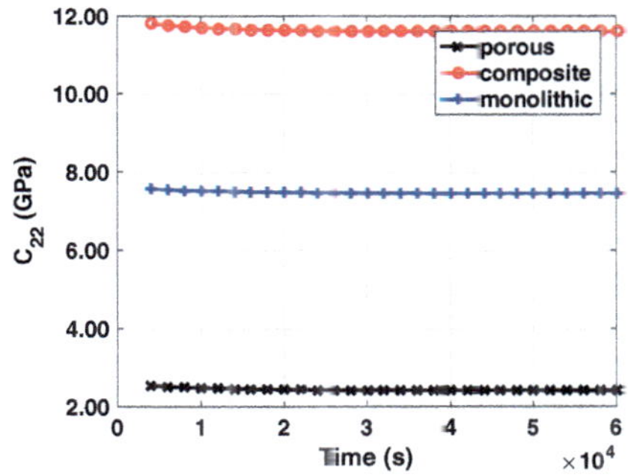

Fig. 5.24: C_{22} for different fiber material with rhombic angle 45° and volume fraction of 0.3

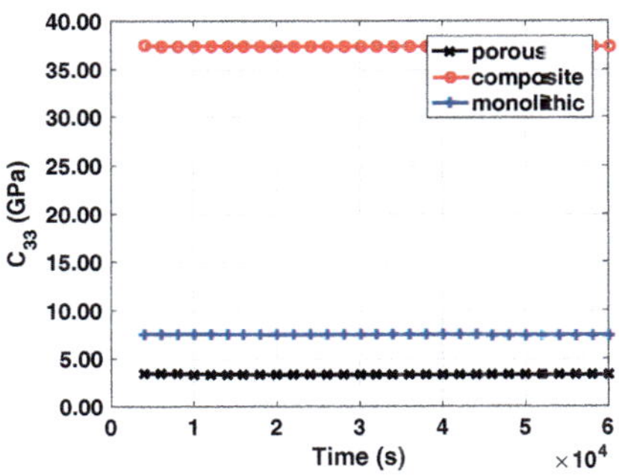

Fig. 5.25: C_{33} for different fiber material with rhombic angle 45° and volume fraction of 0.3

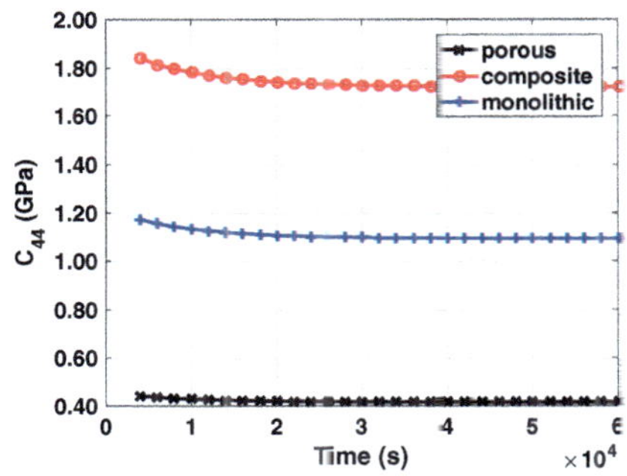

Fig. 5.26: C_{44} for different fiber material with rhombic angle 45° and volume fraction of 0.3

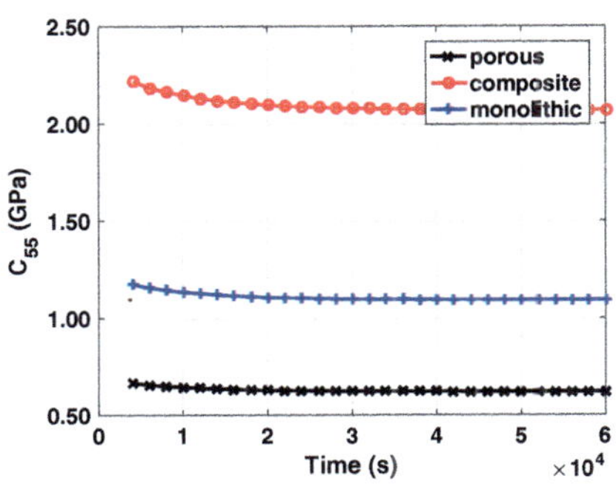

Fig. 5.27: C_{55} for different fiber material with rhombic angle 45° and volume fraction of 0.3

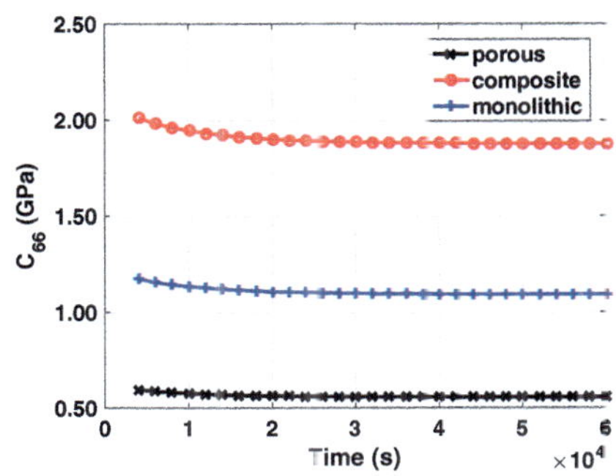

Fig. 5.28: C_{66} for different fiber material with rhombic angle 45° and volume fraction of 0.3

Haasemann G, Ulbricht V (2009) Numerical evaluation of the viscoelastic and viscoplastic behavior of composites. Technische Mechanik 30:122–135

Kari S, Berger H, Rodríguez-Ramos R, Gabbert U (2007) Numerical evaluation of effective material properties of transversely randomly distributed unidirectional piezoelectric fiber composites. Journal of Intelligent Material Systems and Structures 18(4):361–372

Nguyen H, Pastor J, Muller D (1995) A method for predicting linear viscoelastic mechanical behavior of composites, a comparison with other methods and experimental validation. European Journal of Mechanics - A/Solids 14:939–960

Pathan MV, Tagarielli VL, Patsias S (2017) Numerical predictions of the anisotropic viscoelastic response of uni-directional fibre composites. Composites Part A: Applied Science and Manufacturing 93:18–32

Tang T, Felicelli SD (2016) Effective creep response and uniaxial tension behavior of linear viscoelastic polymer composites. In: Sano T, Srivatsan TS (eds) Advanced Composites for Aerospace, Marine, and Land Applications II, Springer International Publishing, Cham, pp 335–345

To QD, Nguyen ST, Bonnet G, Vu MN (2017) Overall viscoelastic properties of 2d and two-phase periodic composites constituted of elliptical and rectangular heterogeneities. European Journal of Mechanics - A/Solids 64:186–201

Würkner M, Berger H, Gabbert U (2011) On numerical evaluation of effective material properties for composite structures with rhombic fiber arrangements. International Journal of Engineering Science 49(4):322–332

Yancey RN, Pindera MJ (1990) Micromechanical analysis of the creep response of unidirectional composites. Journal of Engineering Materials and Technology 112(2):157–163

Chapter 6
Inner Resonance in Media Governed by Hyperbolic and Parabolic Dynamic Equations. Principle and Examples

Claude Boutin, Jean-Louis Auriault, and Guy Bonnet

Abstract This chapter deals with the modeling and design of inner resonance media, i.e. media that present a local resonance which has an impact on the overall dynamic behaviour. The aim of this chapter is to provide a synthetic picture of the inner resonance phenomena by means of the asymptotic homogenization method (Sanchez-Palencia, 1980). The analysis is based on the comparative study of a few canonical realistic composite media. This approach discloses the common principle and the specific features of different inner resonance situations and points out their consequences on the effective behavior. Some general design rules enabling to reach such a specific dynamic regime with a desired effect are also highlighted The paper successively addresses different materials

- having different behaviours and inner structures as elastic composites, reticulated media, permeable rigid and elastic media,
- undergoing phenomena governed either by momentum transfer or/and mass transfer,
- in which the inner resonance mechanisms can be highly or weakly dissipative,
- in situation of inner resonance or inner anti-resonance.

The results related to different physical behaviours show that inner resonance requires a highly contrasted microstructure. It constrains the resonant constituent to respond in a forced regime imposed by the non resonant constituent. Then, the effective

---

Claude Boutin
Université de Lyon, LGCB/LTDS CNRS 5513, ENTPE, Vaulx-en-Velin, France
e-mail: `claude.boutin@entpe.fr`

Jean-Louis Auriault
Université Grenoble Alpes-CNRS, 3SR Lab, F-38000 Grenoble, France
e-mail: `jean-louis.auriault@3sr-grenoble.fr`

Guy Bonnet
Université Paris-Est, Laboratoire Modélisation et Simulation Multi Echelle, MSME UMR 8208 CNRS, Marne-la-Vallée, France
e-mail: `guy.bonnet@u-pem.fr`

© Springer International Publishing AG, part of Springer Nature 2018

H. Altenbach et al. (eds.), *Generalized Models and Non-classical Approaches in Complex Materials 1*, Advanced Structured Materials 89,
https://doi.org/10.1007/978-3-319-72440-9_6

constitutive law is determined by this latter while the resonating constituent acts as an atypical source term in the macroscopic balance equation. It is established that inner resonance governed by momentum (resp. mass) balance yields unconventional mass (resp. bulk modulus). Furthermore, inner-resonance in media characterized by hyperbolic or parabolic dynamic equations can be handled in a similar manner, leading however to strongly distinct effective features.

6.1 Introduction

This paper deals with the macroscopic description of inner resonance media, i.e. media that present a Representative Elementary Volume (REV) in which dynamic phenomena exist at micro-scale and lead to specific features of the overall behaviour at macro-scale. Such an occurrence of a dynamic regime coexisting at both micro and macro scales will be named "co-dynamics".

Such a "co-dynamics" regime is impossible in (nearly-)homogeneous media and can occur only in heterogeneous materials with sufficiently contrasted properties. This type of materials, or "metamaterials", are of prime interest for their atypical properties, that are seemingly impossible to reach with classical materials. Indeed, in presence of inner resonance, the description strongly departs from usual standard dynamics. The critical discrepancy is that effective parameters, as apparent mass or compressibility, are frequency dependent and can take negative or complex values in a frequency range related to the inner-resonance frequency.

In mechanics, the first study dedicated to inner-resonance media was developed in Auriault and Bonnet (1985) (in English Auriault, 1994) where the macroscopic behavior of highly-contrasted bi-composite elastic materials were rigorously derived through the homogenization method. It was shown that the local resonance within the soft medium leads to a frequency dependent effective tensorial density. Then waves are dispersive and frequency band-gaps occur around the series of local resonance frequencies, see Fig. 6.1. Note that the idea of local resonating elements was already considered by Maxwell and Rayleigh as an analogous mechanical system for explaining anomalous dispersion (Maugin, 1995), Fig. 6.2.

This subject receives now a great interest and a wide amount of works are now devoted to inner resonance in mechanics and acoustics. Among them, let us mention the theoretical studies based on asymptotic and/or physical approaches (Zhikov, 2000; Ávila et al, 2005; Babych et al, 2008; Chesnais et al, 2007; Milton, 2007; Smyshlyaev, 2009; Krynkin et al, 2011) and experimental investigations as Fang et al (2006); Liu et al (2000, 2005); Shanshan et al (2008); Sheng et al (2003), where band gaps related to local resonance are evidenced. A good state of art in acoustic metamaterials can be found in the recent review (Ma and Sheng, 2016). Among other subjects, this review paper presents some papers studying the coupling between local resonance and diffraction, which occurs when the elastic properties of inclusions are of the same order as the ones of the matrix. A main consequence of this coupling is that the effective behaviour is no more local in space, contrarily to the kinds of

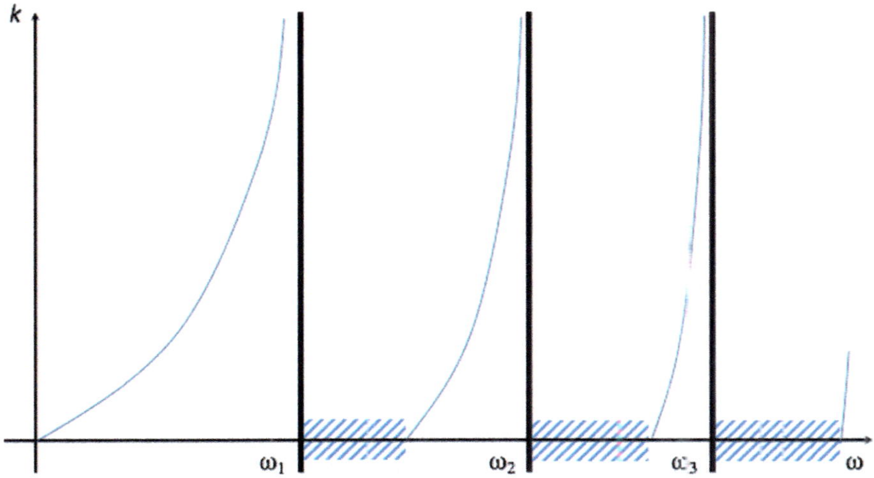

Fig. 6.1: Wave number versus angular frequency, inspired from Auriault and Bonnet (1985). Hatched areas: stopping bands. Solid lines: dispersion curves on passing bands

behaviours studied thereafter, which correspond to "decoupled resonance". If the recent advances provide a number of results devoted to particular cases, a general clear and comprehensive understanding is still challenging. Furthermore, academic and numerical studies investigate the features of the band-gaps of a great variety of 1D or 2D discrete lattice of rigid point mass connected by massless extension/compression springs. Meanwhile this approach enables to identify some basic possible mechanisms, the direct transposition of these studies to real materials is far from being straightforward. Indeed, replacing the discrete mass and stiffness distribution by a continuous 3D distribution can significantly modify the description, both quantitatively and qualitatively. The present paper aims at contributing to a synthetic picture of the inner resonance phenomena derived by using the asymptotic homogenization method (Sanchez-Palencia, 1980). The analysis is based on the comparative study of a few canonical realistic 3D continuous media. Investigating different materials

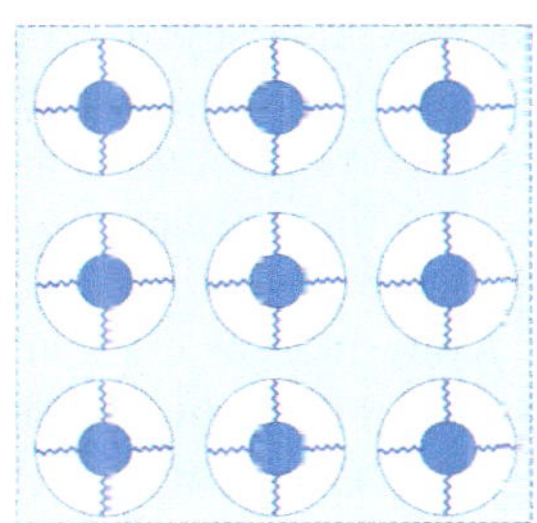

Fig. 6.2 The Maxwell - Rayleigh model of anomalous dispersion, inspired from Maugin (1995)

and different mechanisms enables to disclose the common principles and the specific features of the inner resonance situations and to point out their consequences on the effective behavior. Some general design rules enabling to reach the "co-dynamic" regime with a desired effect can also be established. In this view, we address:

- materials of different morphologies as elastic composite, reticulated media, permeable rigid media,
- undergoing phenomena governed either by momentum transfer or/and mass transfer, in which the inner resonance mechanisms can be highly or weakly dissipative, and is generally of mono-polar type but also of multi-polar type in the case of inner anti-resonance.

In these different cases, we investigate situations where the materials, conveniently represented by periodic media, experience a wave propagation with a long wavelength Λ much larger than the characteristic size ℓ of the REV, or of the period. Thus the condition of scale separation is satisfied and the homogenization method applies (Auriault et al, 2009; Sanchez-Palencia, 1980). The macroscopic characteristic length L of the wave field is defined by the reduced wavelength i.e. $L = \Lambda/2\pi = 1/k(\omega)$, where $k(\omega)$ is the wave number at the considered frequency ω. Denoting by ε the small scale ratio parameter, we have

$$\varepsilon = \frac{2\pi\ell}{\Lambda} = k(\omega)\ell = \frac{\ell}{L} \ll 1$$

This method enables to build up the equivalent macroscopic description by means of multiple scale expansions in ε-power. Recall that the macroscopic formulation - namely the differential operator and the effective parameters - is derived from the physics at the heterogeneity scale, in the condition of scale separation, *without any macroscopic prerequisites*. Compared to the self-consistent approaches, also known as "coherent potential approximation", where the structure of the macroscopic differential operator is postulated a priori, this procures some decisive advantages that will be exploited in the sequel. In particular it enables determining whether or not a macroscopic description exists and therefore providing the domains of validity of the macroscopic models. Throughout the study, we will show that:

- despite the different physical kinds of inner resonance, the phenomena present common features that result in a similar formalism of the macroscopic description,
- the inner resonance may be reached either by introducing large contrasts in the mechanical properties, or by a geometrical contrast in the morphology (e.g. including plates or beams within the microstructure),
- sparse co-dynamic regimes may exist in situation of inner anti-resonance,
- considering resonance with low or high dissipation, drastically changes the nature of the resonance effect,
- unconventional mass arises when the physics is governed by the momentum balance, while unconventional compressibility arises when physics is governed by the mass balance.

The paper is structured as follows. In Sect. 6.2, the specificity of inner-resonance media is presented in the general framework of the dynamic descriptions of heterogeneous linear media in comparison with situations where the contrast of physical properties is moderate. Section 6.3 deals with the case of inner resonance in elastic composites. Both situations of resonance and anti-resonance are analyzed. Section 6.4 is devoted to inner-resonance in poro-acoustics, considering the cases of resonance in media characterized by "parabolic" and "hyperbolic" dynamic equations. Section 6.5 addresses poro-elastic media in which momentum and mass balance are coupled, leading simultaneously to unconventional stiffness and mass. The main theoretical outcomes and practical perspectives are discussed in the conclusion.

6.2 Dynamic Descriptions of Heterogeneous Linear Elastic Media Without and With Inner Resonance

The "co-dynamics" regime where waves of long wavelength interfere with a local dynamic state departs from the dynamic regime usually considered in heterogeneous media. To highlight this specificity it is of interest to briefly review the dynamic descriptions derived by different approaches in situations where the contrast between physical properties is moderate. In every case, the heterogeneous media is assumed to present a Representative Elementary Volume (REV), assumption that is satisfied considering either statistically invariant media or Ω-periodic media (i.e. periodic media characterized by a unit cell Ω reproduced periodically). In this section we will focus on the elasto-dynamic case, but most of the comments also apply to linear phenomena governed by hyperbolic or parabolic dynamic equations (see e.g. Boutin, 1995; Auriault et al, 2005). The schematic diagram in Fig. 6 3, provides a classification of the possible descriptions, according to the frequency range, the periodic or non periodic nature of REV, and the contrast between physical parameters.

6.2.1 Long Wavelength Descriptions

In the long wavelength assumption, the homogenization method of periodic media (Auriault et al, 2009; Sanchez-Palencia, 1980) is a rigorous upscaling method. The macroscopic description of elastic composites is established on the assumption of scale separation and on multi-scale asymptotic expansions in powers of the scale ratio $\varepsilon = k(\omega)\ell = 2\pi\ell/\Lambda \ll 1$, where $k(\omega)$ denotes the wave number. By principle, the upscaled description is valid for frequencies lower than the diffraction frequency ω_d such that $k(\omega_d)\ell = 1$.

Consider a composite with a Ω-periodic elastic tensor $\mathbf{a}$ and density ρ. At the leading order, under an harmonic regime at frequency ω, the homogenization yields the classical elasto-dynamic formulation in which the effective elastic tensor $\boldsymbol{A}^0$ is

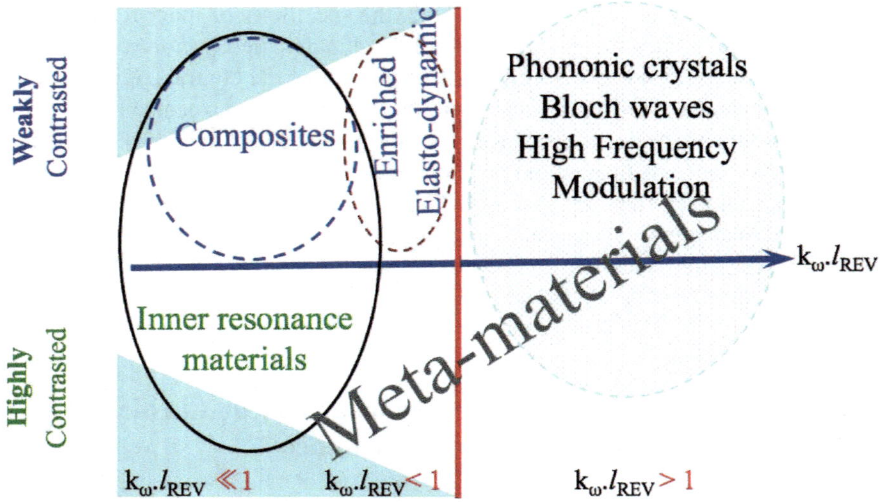

Fig. 6.3: Mapping of the dynamic descriptions, according to the normalized wavenumber $k(\omega)\ell$ and the contrast of the parameters. The descriptions apply to periodic media (white and dashed zones) but also to non periodic media (left grey triangles) characterized by a REV, at sufficiently low frequencies

the same as in statics, and the effective density $R^0 = \langle \rho \rangle$ is the scalar mean density of the composite. Physically, this description results from the fact that at the scale of the period Ω, each constituent is in a quasi-static regime and moves with the same rigid translation $\boldsymbol{U}^{(0)}$ at the leading order. This leads to the following representation, where $\boldsymbol{e}(\boldsymbol{U}^{(0)})$ stands for the macroscopic strain tensor

$$\mathrm{div}(\boldsymbol{A}^0 : \boldsymbol{e}(\boldsymbol{U}^{(0)})) + \omega^2 \langle \rho \rangle \boldsymbol{U}^{(0)} = 0 \tag{6.1}$$

Thus, at the first order of approximation, the wave propagation is dispersionless. For comparison with other descriptions one may also adopt the following alternative formulation where $\langle \boldsymbol{\sigma}^{(0)} \rangle$ is the mean stress on Ω and $\langle \mathbf{p}^{(0)} \rangle$ the mean moment density:

$$\mathrm{div}(\langle \boldsymbol{\sigma}^{(0)} \rangle) = i\omega \langle \boldsymbol{p}^{(0)} \rangle \quad \text{with} \quad \begin{pmatrix} \langle \boldsymbol{\sigma}^{(0)} \rangle \\ \langle \boldsymbol{p}^{(0)} \rangle \end{pmatrix} = \begin{pmatrix} \boldsymbol{A}^0 & 0 \\ 0 & i\omega \langle \rho \rangle \end{pmatrix} \begin{pmatrix} : \boldsymbol{e}(\boldsymbol{U}^{(0)}) \\ \cdot i\omega \boldsymbol{U}^{(0)} \end{pmatrix}$$

When the wavelength, is large but not very large compared to ℓ, the leading order model can be enhanced by considering the correctors associated to the high order terms. Doing so one obtains the following equation governing the ith corrector of the mean cell motion $\widetilde{\boldsymbol{U}}^{(i)} = O(\varepsilon^i \boldsymbol{U}^{(0)})$ (see Boutin and Auriault, 1993, for the developments and the analysis)

$$\mathrm{div}(\boldsymbol{A}^0 : \boldsymbol{e}(\widetilde{\boldsymbol{U}}^{(i)})) + \omega^2 \langle \rho \rangle \widetilde{\boldsymbol{U}}^{(i)} = -\mathrm{div}\left(\sum_{j=1}^{i} \boldsymbol{A}^j \underset{j+2}{\cdot} \nabla^j \boldsymbol{e}(\widetilde{\boldsymbol{U}}^{(i-j)}) \right) - \omega^2 \left(\sum_{j=1}^{i} \boldsymbol{R}^j \underset{j+1}{\cdot} \nabla^j \widetilde{\boldsymbol{U}}^{(i-j)} \right)$$

The detailed analysis shows that the higher order terms describe the Rayleigh scattering. In particular, the corrector of first order leads to a correction of polarization varying according to ω/ω_d, the second order induces a dispersion of wave celerity proportional to $(\omega/\omega_d)^2$, and the third order introduces a geometric attenuation increasing as $(\omega/\omega_d)^3$, where ω_d is the diffraction frequency.

The upscaled equivalent medium is a generalized continuum media, characterized by constitutive laws that are non-local in space. This relies on the higher gradients of strains, associated to the effective elastic and density tensors of higher orders, respectively $\boldsymbol{A}^j$ and $\boldsymbol{R}^j$. Furthermore, considering the total field $\boldsymbol{U} = \sum_{i=0}^{\infty} \widetilde{\boldsymbol{U}}^{(i)}$, the description reads

$$\mathrm{div}\left(\sum_{j=0}^{\infty} \boldsymbol{A}^j \underset{j+2}{\cdot} \nabla^j \boldsymbol{e}(\boldsymbol{U}) \right) + \omega^2 \left(\sum_{j=0}^{\infty} \boldsymbol{R}^j \underset{j+1}{\cdot} \nabla^j \boldsymbol{U} \right) = 0$$

Noting that the infinite sum of the terms with successive derivation can be considered as the expansion of a convolution product, the above description can be rewritten in the following condensed form expressed in terms of generalized stress $\boldsymbol{S}$ and momentum density $\boldsymbol{P}$:

$$\mathrm{div}(\boldsymbol{S}) = i\omega \boldsymbol{P} \quad \text{with} \quad \begin{pmatrix} \boldsymbol{S} \\ \boldsymbol{P} \end{pmatrix} = \begin{pmatrix} \widetilde{\boldsymbol{A}} & 0 \\ 0 & \widetilde{\boldsymbol{R}} \end{pmatrix} * \begin{pmatrix} : \boldsymbol{e}(\boldsymbol{U}) \\ \cdot i\omega \boldsymbol{U} \end{pmatrix}$$

where $*$ stand for the convolution product. The enhanced description introduces a dynamic effect at the period scale and improves the approximation of quasi-static regime made at the leading order. Nevertheless, by construction, the period experiences a regime of weak dynamics, and consequently the effects of the correctors are of weak magnitude, i.e. $|\widetilde{\boldsymbol{A}} - \boldsymbol{A}^0| \ll |\boldsymbol{A}^0|$, and $|\widetilde{\boldsymbol{R}} - \boldsymbol{R}^0| \ll \boldsymbol{R}^0$.

The leading order and higher order descriptions have been established by assuming implicitly that the contrast of the mechanical parameters of the constituents of the composite are moderate i.e $O(1)$ compared to the scale ratio $\varepsilon = k(\omega)\ell$, and also that the morphology of the constituent in Ω, do not presents geometrical contrast as in Subsect. 6.3.5 thereafter. Consequently, the natural way to deviate from these standard formulations is to remove these implicit assumptions as done in the next sections.

Remark finally that in the long wavelength range, periodic materials and non-periodic materials with REV show macroscopic behavior of the same nature (Auriault et al, 2009), however, the solution is not reducible to well-defined problems on a periodic cell in the case of a non-periodic medium.

6.2.2 Short Wavelength Descriptions

In the short wavelength range, i.e. $k(\omega_d)\ell \geq 1$, or $\omega \geq \omega_d$, the wavelength strongly interacts with the cell size. Then the description is of different nature for periodic materials and non-periodic materials with REV, while, in both cases, a full dynamic regime is observed at the local scale. Direct simulations based on multiscattering approach can be performed. However the numerical cost is in general extremely expansive. Theoretical formulations dedicated to periodic case, i.e. phononic crystals, or non periodic cases are also available.

Periodic media at high frequency are usually named phononic crystals. They are exactly described by the Floquet-Bloch theory (Brillouin, 1946), that provides through the Bloch wave decomposition (Allaire and Conca, 1998) the equation of dispersion $E(k,\omega) = 0$ which can be solved numerically. This approach describes Bragg scattering, with dispersion and occurrence of band-gaps. The latter corresponds to intervals of real frequency where the solutions of $E(k,\omega) = 0$ are purely imaginary wave numbers. A wide amount of literature is devoted to this approach and the specific behavior of phononic crystal is now largely studied from the first experimental evidence presented in Vasseur et al (1998). However, meanwhile the Floquet-Bloch theory gives a comprehensive description of wave fields with short wavelengths, the fact that such wave fields may also present large evolution lengths is not directly accessible by this approach. This question of large scale modulation of wave fields with short waves can be handled through a revisited asymptotic method that have been initiated by Daya et al (2002); Moustaghfir et al (2007) for periodic structures and followed by the works of Craster et al (2010); Boutin et al (2012, 2014) for 3D elastic composites. Compared to the usual homogenization method, the main change lies in the fact that the macroscopic variables are the *amplitudes of periodic eigenmodes* of the cell (or multi-cell) instead of being the *displacement field* itself. This induces significant differences in the nature of the up-scaled description. The approach by modulation enables simple calculations of high frequency wave fields based on a two-step procedure separating the periodic eigenmode of the cell (or multi-cell) and the large modulation scale using the derived modulation equation.

Non-periodic media with REV have been widely addressed by Willis (2012). In accordance with the statistic character of the REV, the method is based on the ensemble average (denoted $< . >$) instead of the volume average (denoted $\langle . \rangle$) used in periodic homogenization (note however that both averages coincide in periodic media). The developments yield a conservation equation expressed in terms of averaged stress $< \boldsymbol{\sigma} >$ and momentum density $< \mathbf{p} >$. These latter quantities are determined by non-local elasto-dynamic constitutive laws coupling the strain and the velocity. This formulation reads

$$\mathrm{div}_x(< \boldsymbol{\sigma} >) = i\omega < \mathbf{p} > \quad \text{with} \quad \begin{pmatrix} < \boldsymbol{\sigma} > \\ < \mathbf{p} > \end{pmatrix} = \begin{pmatrix} \mathscr{A}(\omega) & \mathscr{D}(\omega) \\ {}^t\mathscr{D}(\omega) & \mathscr{R}(\omega) \end{pmatrix} * \begin{pmatrix} : \boldsymbol{e}(\boldsymbol{U}) \\ \cdot i\omega \boldsymbol{U} \end{pmatrix}$$

The convolution operators $\mathscr{A}(\omega)$, $\mathscr{D}(\omega)$, ${}^t\mathscr{D}(\omega)$ and $\mathscr{R}(\omega)$ involved in the constitutive laws account for the dynamics at the REV scale. A main difficulty of this approach lies in the determination of these operators. Note that despite the different context of application and the more complex constitutive laws, this description presents formally some common features with the Rayleigh scattering description. Recent developments (Nassar et al, 2016a,b) deal with the transposition of this approach to periodic materials.

In the above descriptions, the alternative is that either the whole cell is in quasi-static or weakly dynamic regime, or the whole cell is in dynamic regime. Thus the specificity of the co-dynamic regime characterizing inner resonant media is not captured. In fact, the inner resonance situation mixes some of the assumptions of the low and high frequency regime which are usually incompatible. For this reason a specific description is required. Indeed, whatever the physics in consideration, a co-dynamics regime implies that in any cell one may distinguish

i) a part of the cell that acts as the long wavelength conveyor - and therefore undergoes a quasi-static regime -, and in the same time and same cell,
ii) an other part that experiences a dynamic state.

Thus, in presence of long waves, the REV responds with "partial" non equilibrium state at the local scale. This specific regime changes the fundamental assumption usually considered in continuum mechanics applied to heterogeneous material. As a consequence non conventional behaviors are obtained.

6.3 Inner Resonance in Elastic Composites

We investigate two types of inner resonant elastic media: bi-composites materials made of constituents having contrasted stiffness and/or densities; reticulated media made of a single material, for which the cell morphology presents a geometrical contrast (i.e. containing beams or plates).

6.3.1 Requirements for the Occurrence of Inner Resonance in Elastic Bi-Composites

Consider first a periodic bi-composite of period Ω and characteristic period size ℓ made of two homogeneous constituents namely the $\mathscr{C}$-constituent (denoted by index c) and the $\mathscr{R}$-constituent (denoted by index r), with uniform elastic tensors $\boldsymbol{a}_{\rm c}$ and $\boldsymbol{a}_{\rm r}$ and constant densities $\rho_{\rm c}$ and $\rho_{\rm r}$, occupying the domains $\Omega_{\rm c}$ and $\Omega_{\rm r}$ respectively (see Fig. 6.4). The interface between $\Omega_{\rm c}$ and $\Omega_{\rm r}$ is denoted Γ with unit normal $\boldsymbol{n}$, exterior to $\Omega_{\rm c}$.

As stated in the introduction, in the frequencies range of interest the two constituents experience distinct regimes at the cell scale. By convention, here and thereafter the $\mathscr{C}$-constituent will convey the long wavelength and then undergoes a local quasi-static regime, while the $\mathscr{R}$-constituent experiences a local dynamic regime. In other words, the wavelength $\Lambda_{\rm c}$ in $\mathscr{C}$-constituent is large with respect to ℓ, while the

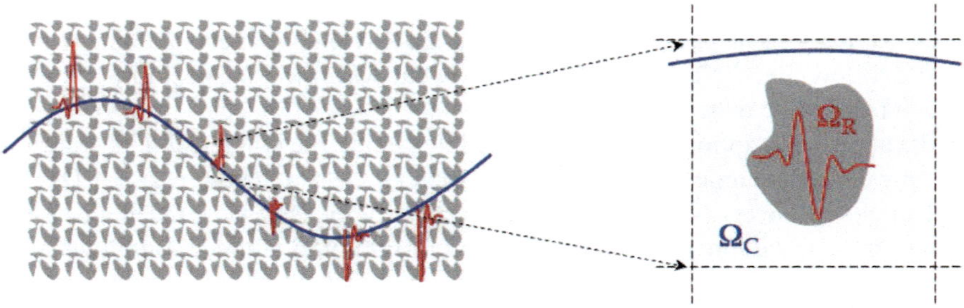

Fig. 6.4: Period Ω of a bi-composite elastic media. Illustration of the "co-dynamic" regime

wavelength Λ_r in $\mathscr{R}$-constituent is of the order of ℓ. To achieve this, the $\mathscr{C}$-constituent will be systematically considered as *connected*, while the $\mathscr{R}$-constituent can either be connected or dispersed. Hence, one has the following *a priori* estimates

$$L=\varepsilon^{-1}\ell=O(\frac{\Lambda_c}{2\pi})=\frac{1}{\omega}\sqrt{\frac{|\boldsymbol{a}_c|}{\rho_c}} \quad ; \quad \frac{\Lambda_r}{2\pi}=\frac{1}{\omega}\sqrt{\frac{|\boldsymbol{a}_r|}{\rho_r}}=O(\ell) \tag{6.2}$$

These requirements implies that a co-dynamic regime may be reached when

$$\frac{|\boldsymbol{a}_r|}{|\boldsymbol{a}_c|}\frac{\rho_c}{\rho_r}=O(\frac{\ell^2}{L^2})=O(\varepsilon^2) \tag{6.3}$$

Besides, the resonance of the $\mathscr{R}$-constituent will be enhanced if this constituent partially traps the energy carried by the surrounding $\mathscr{C}$-constituent. The energy storage is significant when the impedance of the $\mathscr{R}$-constituent is much smaller than that of the $\mathscr{C}$-constituent (so that the incoming energy cannot escape and stays trapped in $\mathscr{R}$-constituent). This leads to consider that

$$\frac{\sqrt{|\boldsymbol{a}_r|\rho_r}}{\sqrt{|\boldsymbol{a}_c|\rho_c}}=O(\varepsilon) \tag{6.4}$$

Combining relations (6.3)-(6.4), a "co-dynamic" regime is "naturally" expected when the $\mathscr{C}$-constituent is highly stiffer than the $\mathscr{R}$-constituent, while their contrast of density is moderate, i.e:

$$\frac{|\boldsymbol{a}_r|}{|\boldsymbol{a}_c|}=O(\varepsilon^2) \qquad \text{and} \qquad \frac{\rho_c}{\rho_r}=O(1) \tag{6.5}$$

This case will be addressed in Subsect. 6.3.2.

Aside from this canonical case that favors the resonance, the $\mathscr{R}$-constituent can "a priori" also be set in resonance in situations of moderate contrast of impedance, i.e. when

$$\frac{\sqrt{|\boldsymbol{a}_\mathrm{r}|\rho_\mathrm{r}}}{\sqrt{|\boldsymbol{a}_\mathrm{c}|\rho_\mathrm{c}}} = O(1) \tag{6.6}$$

This second case (6.3)-(6.6) that corresponds to a $\mathscr{C}$-constituent significantly stiffer and lighter than the $\mathscr{R}$-constituent, i.e.:

$$\frac{|\boldsymbol{a}_\mathrm{r}|}{|\boldsymbol{a}_\mathrm{c}|} = O(\varepsilon) \qquad \text{and} \qquad \frac{\rho_\mathrm{r}}{\rho_\mathrm{c}} = O(\varepsilon^{-1}) \tag{6.7}$$

deserves also to be studied. It will be investigated in Sect. 6.3.3.

Finally, for completeness, the case of moderate contrast of stiffness and highly denser $\mathscr{R}$-constituent, i.e.

$$\frac{|\boldsymbol{a}_\mathrm{r}|}{|\boldsymbol{a}_\mathrm{c}|} = O(1) \quad \text{and} \quad \frac{\rho_\mathrm{r}}{\rho_\mathrm{c}} = O(\varepsilon^{-2})$$

could also be considered. However, as the impedance of the $\mathscr{R}$-constituent would be significantly larger than that of the $\mathscr{C}$-constituent, it is expected that such media would behave as an usual composite with an apparent mass dominated by that of the denser constituent. Indeed, it will be shown in Appendix that this is actually the case, and that this situation is not compatible with the occurrence of a "co-dynamic" regime.

Note 6.1. By convention, in this paper we use the terms of high, significant or moderate contrast when the ratio of parameters are of the order of $O(\varepsilon^{\pm 2})$, $O(\varepsilon^{\pm 1})$ or $O(1)$.

6.3.2 Elastic Bi-Composites: High Contrast of Stiffness, Moderate Contrast of Density

The situation of high stiffness contrast and moderate density contrast (6.7) corresponds to that addressed in Auriault and Bonnet (1985); Auriault and Boutin (2012). In that case $O(\boldsymbol{a}_\mathrm{c}) \gg O(\boldsymbol{a}_\mathrm{r})$ and $\rho_\mathrm{c} = O(\rho_\mathrm{r})$ and the $\mathscr{C}$-constituent is connected. Independently of the fact that the $\mathscr{R}$-constituent can be connected or not, the stress is mainly carried by the $\mathscr{C}$-constituent. Thus the long wave is actually conveyed by the $\mathscr{C}$-constituent and estimates (6.2) clearly apply. For the sequel it is useful to recall the main results in this canonical case.

In harmonic regime at angular frequency ω, the medium satisfies the Navier equation in Ω_c and Ω_r, with the continuity of stress vector and of displacement on the interface Γ. The time dependance $\exp(+i\omega t)$ simplifies by linearity and is systematically omitted. Thus, the medium is governed by the following set of equations

$$\mathrm{div}(\boldsymbol{\sigma}) = -\omega^2 \rho \boldsymbol{u} \quad \text{in } \Omega \tag{6.8a}$$
$$\boldsymbol{\sigma} = \boldsymbol{a} : \boldsymbol{e}(\boldsymbol{u}) \quad \text{in } \Omega \tag{6.8b}$$
$$(\boldsymbol{\sigma}_c - \boldsymbol{\sigma}_r) \cdot \boldsymbol{n} = 0 \quad \text{on } \Gamma \tag{6.8c}$$
$$\boldsymbol{u}_c - \boldsymbol{u}_r = 0 \quad \text{on } \Gamma \tag{6.8d}$$
$$\boldsymbol{\sigma} \quad \text{and} \quad \boldsymbol{u} \quad \Omega - \text{periodic} \tag{6.8e}$$

where $\boldsymbol{\sigma}$ is the stress tensor, $\boldsymbol{e}(\boldsymbol{u})$ is the strain tensor, and $\boldsymbol{u}$ is the displacement. The elastic tensor $\mathbf{a}$ and density ρ take the values indexed by $_c$ and $_r$ in the domains Ω_c and Ω_r respectively and similarly for $\boldsymbol{\sigma}$ and $\boldsymbol{u}$. The homogenization process consists in introducing two space variables, $\boldsymbol{x}$ and $\boldsymbol{y} = \varepsilon^{-1}\boldsymbol{x}$, (hence in replacing the usual gradient by $\nabla_x + \varepsilon^{-1}\nabla_y$) and to look for the displacement and other physical quantities in the form of Ω-periodic asymptotic expansions

$$\boldsymbol{u} = \boldsymbol{u}^{(0)}(\boldsymbol{x},\boldsymbol{y}) + \varepsilon \boldsymbol{u}^{(1)}(\boldsymbol{x},\boldsymbol{y}) + \varepsilon^2 \boldsymbol{u}^{(2)}(\boldsymbol{x},\boldsymbol{y}) + \ldots, \quad \boldsymbol{u}^{(i)}(\boldsymbol{x},\boldsymbol{y})\, \Omega - \text{periodic in}\, \boldsymbol{y} \tag{6.9}$$

Further, to properly account for the contrast of stiffness in the $(\boldsymbol{x},\boldsymbol{y})$ formulation, a rescaling of the elastic tensors must be considered. Since the overall stiffness is dominated by that of the $\mathscr{C}$-constituent, $\boldsymbol{a}_c$ is taken as the reference stiffness tensor, and as such it is not rescaled. Now, by comparison with the reference tensor $\boldsymbol{a}_c$ and in accordance with the stiffness contrast $|\boldsymbol{a}_r|/|\boldsymbol{a}_c| = O(\varepsilon^2)$ the elastic tensor of the $\mathscr{R}$-constituent is consistently rescaled as $\varepsilon^2 \boldsymbol{a}_r$. Thus, using the reduced wavelength L as characteristic length, and considering in compliance with the continuity condition on Γ that the displacements in both constituents are of similar order of magnitude $\boldsymbol{u}_c = O(\boldsymbol{u}_r)$, the $(\boldsymbol{x},\boldsymbol{y})$ governing equations take the following scaled form

$$\mathrm{div}(\boldsymbol{a}_c : \boldsymbol{e}(\boldsymbol{u}_c)) = -\omega^2 \rho_c \boldsymbol{u}_c \quad \text{in } \Omega_c \tag{6.10a}$$
$$\mathrm{div}(\varepsilon^2 \boldsymbol{a}_r : \boldsymbol{e}(\boldsymbol{u}_r)) = -\omega^2 \rho_r \boldsymbol{u}_r \quad \text{in } \Omega_r \tag{6.10b}$$
$$(\boldsymbol{a}_c : \boldsymbol{e}(\boldsymbol{u}_c) - \varepsilon^2 \boldsymbol{a}_r : \boldsymbol{e}(\boldsymbol{u}_r)) \cdot \boldsymbol{n} = 0 \quad \text{onr } \Gamma \tag{6.10c}$$
$$\boldsymbol{u}_r - \boldsymbol{u}_c = 0 \quad \text{on } \Gamma \tag{6.10d}$$
$$\boldsymbol{u}_c \quad \text{and} \quad \boldsymbol{u}_r \quad \Omega - \text{periodic} \tag{6.10e}$$

6.3.2.1 Derivation of the Inner-Resonance Behavior by Homogenization

Introducing expansion (6.9) into the $(\boldsymbol{x},\boldsymbol{y})$-differential set (6.10) and equating the terms with powers of ε yield successive boundary value problems on the period.

- At the dominant order, the $\mathscr{C}$-constituent is governed by an elasto-*static* balance equation with *free* boundary condition on the interface Γ, namely:

$$\begin{cases} \mathrm{div}_y(\boldsymbol{a}_c : \boldsymbol{e}_y(\boldsymbol{u}_c^{(0)})) = 0 & \text{in } \Omega_c \\ (\boldsymbol{a}_c : \boldsymbol{e}_y(\boldsymbol{u}_c^{(0)})) \cdot \boldsymbol{n}_c = 0 & \text{on } \Gamma \\ \boldsymbol{u}_c^{(0)} & \Omega - \text{periodic} \end{cases} \tag{6.11}$$

Therefore, Ω_c experiences a periodic rigid motion at the leading order, and since Ω_c is connected, this rigid motion reduces to a translation. Thus

$$\boldsymbol{u}_c^{(0)} = \boldsymbol{U}^{(0)}(\boldsymbol{x}) \tag{6.12}$$

This is consistent with a local quasi static regime of the $\mathscr{C}$-constituent.

- The following order leads to the classic elastic local problem in the $\mathscr{C}$-constituent forced by the macroscopic strain $\boldsymbol{e}_x(\boldsymbol{U}^{(0)})$, with *free* boundary conditions on Γ:

$$\begin{cases} \mathrm{div}_y(\boldsymbol{a}_c : (\boldsymbol{e}_y(\boldsymbol{u}_c^{(1)}) + \boldsymbol{e}_x(\boldsymbol{U}^{(0)})) = 0 & \text{in } \Omega_c \\ \boldsymbol{a}_c : (\boldsymbol{e}_y(\boldsymbol{u}_c^{(1)}) + \boldsymbol{e}_x(\boldsymbol{U}^{(0)})) \cdot \boldsymbol{n} = 0 & \text{on } \Gamma \\ \boldsymbol{u}_c^{(1)} & \Omega - \text{periodic} \end{cases} \tag{6.13}$$

Hence, at this order, the $\mathscr{C}$-constituent behaves locally as if the $\mathscr{R}$-constituent were absent (or of zero stiffness). By linearity, the solution has the form below, where fields $\boldsymbol{\chi}^{pq}(\boldsymbol{y})$ are the real particular displacements solutions for unit macroscopic strains, i.e. $\boldsymbol{e}_x(\boldsymbol{U}^{(0)})_{ij} = (\delta_{ip}\delta_{jq} + \delta_{iq}\delta_{jp})/2$ (conveniently $\boldsymbol{\chi}(\boldsymbol{y})$ is the third order tensor built from the vectors $\boldsymbol{\chi}^{pq}(\boldsymbol{y}) = \chi_j^{pq}(\boldsymbol{y})\boldsymbol{E}^j$, where $\{\boldsymbol{E}^j, \mathrm{j} = 1,2,3\}$ denotes the unit orthogonal vectors of the reference frame

$$\begin{cases} \boldsymbol{u}_c^{(1)} = & \boldsymbol{\chi}(\boldsymbol{y}) : \boldsymbol{e}_x(\boldsymbol{U}^{(0)}) + \boldsymbol{U}_c^{(1)}(\boldsymbol{x}) \\ \boldsymbol{\sigma}_c^{(0)} = & (\boldsymbol{a}_c : \boldsymbol{e}_y(\boldsymbol{\chi}) + \boldsymbol{a}_c) : \boldsymbol{e}_x(\boldsymbol{U}^{(0)}) \end{cases} \tag{6.14}$$

- At the next order, the balance and boundary condition concerning both constituents write

$$\begin{cases} \mathrm{div}_y(\boldsymbol{\sigma}_c^{(1)}) + \mathrm{div}_x(\boldsymbol{\sigma}_c^{(0)}) = -\omega^2\rho_c\boldsymbol{U}^{(0)} & \text{in } \Omega_c \\ \boldsymbol{\sigma}_c^{(1)} \cdot \boldsymbol{n} = (\boldsymbol{a}_r : \boldsymbol{e}_y(\boldsymbol{u}_r^{(0)})) \cdot \boldsymbol{n} & \text{on } \Gamma \\ \mathrm{div}_y(\boldsymbol{a}_r : \boldsymbol{e}_y(\boldsymbol{u}_r^{(0)})) = -\omega^2\rho_r\boldsymbol{u}_r^{(0)} & \text{in } \Omega_r \\ \boldsymbol{\sigma}_c^{(1)} & \Omega - \text{periodic} \end{cases} \tag{6.15}$$

After integration over Ω_c, and Ω_r, using the divergence theorem, the Neumann boundary condition on Γ and the periodicity condition, it comes the macroscopic relation, independent on $\boldsymbol{y}$

$$\begin{aligned} \mathrm{div}_x(\boldsymbol{A}_{\mathscr{C}} : \boldsymbol{e}_x(\boldsymbol{U}^{(0)})) &= -\omega^2\langle\rho\boldsymbol{u}^{(0)}\rangle \\ &= -\omega^2\left(\rho_c\frac{|\Omega_c|}{|\Omega|}\boldsymbol{U}^{(0)} + \rho_r\frac{1}{|\Omega|}\int_{\Omega_c}\boldsymbol{u}_r^{(0)}\,d\Omega\right) \end{aligned} \tag{6.16}$$

Let us introduce the following notations

$$\langle . \rangle_i = \frac{1}{|\Omega|}\int_{\Omega_i} .\, d\Omega, \qquad i = c, r, \qquad \text{and} \qquad \langle . \rangle = \frac{1}{|\Omega|}\int_{\Omega} .\, d\Omega$$

The effective elasticity tensor reads

$$\boldsymbol{A}_{\mathscr{C}} = \langle \boldsymbol{a}_{\mathrm{c}} : \boldsymbol{e}_y(\boldsymbol{\chi}) + \boldsymbol{a}_{\mathrm{c}} \rangle_{\mathrm{c}} \tag{6.17}$$

and is therefore defined *independently* of the mechanical parameters of the soft $\mathscr{R}$-constituent (but depends on its geometry). It presents the properties of symmetry and positiveness classically established for elastic composites (Sanchez-Palencia, 1980). The notation $\boldsymbol{A}_{\mathscr{C}}$ for the effective elasticity tensor recalls that this tensor results from the constituent $\mathscr{C}$, obviously with a correction due to the presence of the $\mathscr{R}$-inclusions of negligible stiffness. Conversely, the effective inertia $\langle \rho \boldsymbol{u}^{(0)} \rangle$ on the right hand side of (6.16) results from both constituents.

- To close the description it remains to express the mean inertia of the $\mathscr{R}$-constituent. For this, we re-consider the balance equations in Ω_r (6.10b) together with the *Dirichlet* condition (6.10d) at the leading order on Γ.

$$\begin{cases} \mathrm{div}_y(\boldsymbol{a}_{\mathrm{r}} : \boldsymbol{e}_y(\boldsymbol{u}_{\mathrm{r}}^{(0)})) = -\omega^2 \rho_{\mathrm{r}} \boldsymbol{u}_{\mathrm{r}}^{(0)} \quad \text{in } \Omega_{\mathrm{r}} \\ \boldsymbol{u}_{\mathrm{r}}^{(0)} = \boldsymbol{U}^{(0)}(\boldsymbol{x}) \quad \text{on } \Gamma \\ \boldsymbol{u}_{\mathrm{r}}^{(0)} \quad \Omega - \text{periodic} \end{cases} \tag{6.18}$$

Consistently with the inner resonance regime, the $\mathscr{R}$-constituent at the leading order experiences a *dynamic* regime forced by the rigid translation $\boldsymbol{U}^{(0)}(\boldsymbol{x})$ imposed on its boundary. The fields $\boldsymbol{E}^p + \boldsymbol{\alpha}^p(\boldsymbol{y})$ are the real particular solutions for unit imposed displacements, i.e. $\boldsymbol{U}_i^{(0)} = \delta_{ip}$ (recall that $\{\boldsymbol{E}^p,\ p = 1,2,3\}$ are the vectors of the reference frame). By construction, the vectors $\boldsymbol{\alpha}^p$, $p = 1,2,3$ are the solutions of :

$$\begin{cases} \mathrm{div}_y(\boldsymbol{a}_{\mathrm{r}} : \boldsymbol{e}_y(\boldsymbol{\alpha}^p)) = -\omega^2 \rho_{\mathrm{r}} (\boldsymbol{E}^p + \boldsymbol{\alpha}^p) \quad \text{in } \Omega_{\mathrm{r}} \\ \boldsymbol{\alpha}^p = 0 \quad \text{on } \Gamma \\ \boldsymbol{\alpha}^p \quad \Omega - \text{periodic} \end{cases} \tag{6.19}$$

Denoting by $\boldsymbol{\alpha}$ the second order tensor build from the vectors $\boldsymbol{\alpha}^p = \alpha_j^p \boldsymbol{E}^j$, one derives by linearity that

$$\boldsymbol{u}_{\mathrm{r}}^{(0)} = \boldsymbol{U}^{(0)} + \boldsymbol{\alpha}(\boldsymbol{y}, \omega) \cdot \boldsymbol{U}^{(0)} \tag{6.20}$$

It is necessary for the next steps to identify the specific features of the tensor $\boldsymbol{\alpha}$. From (6.19) we deduce directly the low and high frequencies limits, that read

$$\boldsymbol{\alpha}^p \underset{\omega \to 0}{\to} 0 \quad \text{and} \quad \boldsymbol{\alpha}^p \underset{\omega \to \infty}{\to} -\boldsymbol{E}^p,$$

respectively. However, to thoroughly highlight the features of the frequency dependent fields $\boldsymbol{\alpha}^p(\boldsymbol{y}, \omega)$, let us consider the eigenvalue problem associated to (6.19)

$$\begin{cases} \mathrm{div}_y(\boldsymbol{a}_\mathrm{r} : \boldsymbol{e}_y(\boldsymbol{\phi})) = -\lambda \boldsymbol{\phi} & \text{in } \Omega_\mathrm{r} \\ \boldsymbol{\phi} = 0 & \text{on } \Gamma \\ \boldsymbol{\phi} & \Omega-\text{periodic} \end{cases} \tag{6.21}$$

The elastic tensor $\boldsymbol{a}_\mathrm{r}$ being symmetric and coercive, this problem presents a discrete series of eigen values $\lambda_J > 0$ and corresponding orthonormal eigenmodes $\boldsymbol{\phi}^J$ such that $\langle \boldsymbol{\phi}^J \cdot \boldsymbol{\phi}^K \rangle_\mathrm{r} = \delta_{JK}$. Then, using the modal decomposition of fields $\boldsymbol{\alpha}^p(y,\omega)$ the expression of tensors $\boldsymbol{\alpha}(y,\omega)$ and $\langle \boldsymbol{\alpha} \rangle_\mathrm{r}$ are obtained in the form

$$\boldsymbol{\alpha}(y,\omega) = \sum_{J=1}^{\infty} \frac{\boldsymbol{\phi}^J \otimes \langle \boldsymbol{\phi}^J \rangle_\mathrm{r}}{\frac{\omega_J^2}{\omega^2} - 1}; \quad \langle \boldsymbol{\alpha} \rangle_\mathrm{r} = \sum_{J=1}^{\infty} \frac{\langle \boldsymbol{\phi}^J \rangle_\mathrm{r} \otimes \langle \boldsymbol{\phi}^J \rangle_\mathrm{r}}{\frac{\omega_J^2}{\omega^2} - 1}; \quad \omega_J = \sqrt{\lambda_J/\rho_\mathrm{r}} \tag{6.22}$$

Expressions (6.22) show that $\langle \boldsymbol{\alpha} \rangle_\mathrm{r}$ is a second order real and symmetric tensor that depends on

i) the frequency,
ii) the elastic properties, density and geometry of the $\mathscr{R}$-constituent *only*.

This implies that $\langle \boldsymbol{\alpha} \rangle_\mathrm{r}$ is isotropic if the $\mathscr{R}$-constituent presents at least three orthogonal plans of symmetry. Moreover $\langle \boldsymbol{\alpha} \rangle_\mathrm{r}$ is not bounded, changes its sign at the eigenfrequencies $\omega = \omega_J$ of modes such that $\langle \boldsymbol{\phi}^J \rangle_\mathrm{r} \neq 0$, and presents continuous variations between $[-\infty, +\infty]$ in the interval of two consecutive poles $[\omega_J, \omega_{J+1}]$.

Note that the decomposition is realized on the subset of modes "activated " by the forced uniform motion $\boldsymbol{U}^{(0)}$ in the sense that only modes such that $\langle \boldsymbol{\phi}^J \rangle_\mathrm{r} \neq 0$ are involved in $\langle \boldsymbol{\alpha} \rangle_\mathrm{r}$. Modes $\boldsymbol{\phi}^{J^*}$ of zero mean value, i.e. $\langle \boldsymbol{\phi}^{J^*} \rangle_\mathrm{r} = 0$ (as antisymmetric modes in a symmetric domain) are not activated by the uniform motion of the $\mathscr{C}$-constituent. Consequently, the eigen frequencies ω_{J^*} of modes $\boldsymbol{\phi}^{J^*}$ are not poles of $\langle \boldsymbol{\alpha} \rangle_\mathrm{r}$. Thus, as the $\{\boldsymbol{\phi}^{J^*}\}$ are not involved in the local response, the macroscopic description ignores this subset of modes in the whole frequency range, including the vicinity of the ω_{J^*}. An example of such a strong reduction of modes involved in the computation of $< \boldsymbol{\alpha} >$ is the case of spheres. The eigenfrequencies of spheres are numerous (related to a double infinity of integer indices), but only one set of only two integers is involved in the computation of $< \boldsymbol{\alpha} >$ (Bonnet and Monchiet, 2017).

Finally, reporting the result (6.20) in (6.16), the effective behavior at the leading order reads

$$\begin{cases} \mathrm{div}_x(\boldsymbol{A}_\mathscr{C} : \boldsymbol{e}_x(\boldsymbol{U}^{(0)})) = & -\omega^2 \boldsymbol{\rho}(\omega) \cdot \boldsymbol{U}^{(0)}, \\ \boldsymbol{\rho}(\omega) = & \langle \rho \rangle \boldsymbol{I} + \rho_\mathrm{r} \langle \boldsymbol{\alpha}(\omega) \rangle_\mathrm{r} \end{cases} \tag{6.23}$$

This description combines conventional elasticity, characterized by the elastic tensor $\boldsymbol{A}_\mathscr{C}$ of the material made of the $\mathscr{C}$-constituent only, and by a non conventional mass density associated to the frequency dependent density $\boldsymbol{\rho}(\omega)$. The equivalent formulation

$$\mathrm{div}(\langle \boldsymbol{\sigma}^{(0)} \rangle) = i\omega \langle \boldsymbol{p}^{(0)} \rangle \tag{6.24}$$

with

$$\begin{pmatrix} \langle \boldsymbol{\sigma}^{(0)} \rangle \\ \langle \boldsymbol{p}^{(0)} \rangle \end{pmatrix} = \begin{pmatrix} \boldsymbol{A}_{\mathscr{C}} & 0 \\ 0 & i\omega(\langle \rho \rangle \boldsymbol{I} + \rho_{\mathrm{r}} \langle \boldsymbol{\alpha}(\omega) \rangle_{\mathrm{r}}) \end{pmatrix} \begin{pmatrix} : \boldsymbol{e}_x(\boldsymbol{U}^{(0)}) \\ \cdot i\omega \boldsymbol{U}^{(0)} \end{pmatrix}$$

highlights the difference with the Rayleigh scattering regime, as here the constitutive laws are local in space and frequency dependent, instead of being non-local in space and frequency independent.

6.3.2.2 Comments

The key point to obtain (6.24) is that the stress sustained by the $\mathscr{R}$-constituent is of one order smaller than the one prevailing in the $\mathscr{C}$-constituent. This is a direct consequence of the scale separation assumption as shown by the quite basic following arguments. Consider a period of face area S and denote symbolically $|S\sigma_{\mathrm{c}1}|$ and $|S\sigma_{\mathrm{c}2}|$ the ingoing and outgoing stress fluxes on opposite faces, and $|\Gamma\sigma_{\mathrm{r}}|$ the stress flux induced by the $\mathscr{R}$-constituent. By hypothesis a regime of long wavelength $L \gg \ell$ is considered, thus $(|S\sigma_{\mathrm{c}2}| - |S\sigma_{\mathrm{c}1}|)/|S\sigma_1| \approx \ell/L = O(\varepsilon)$. In addition, as the $\mathscr{C}$-constituent is in quasi static regime, its equilibrium at the scale of the cell implies that $|S\sigma_{\mathrm{c}2}| \approx |S\sigma_{\mathrm{c}1}| + |\Gamma\sigma_{\mathrm{r}}|$, and consequently $|\Gamma\sigma_{\mathrm{r}}|/|S\sigma_{\mathrm{c}1}| = O(\varepsilon)$. Thus, the stress external to the resonating domain is not significantly modified by the small stress induced by the resonating domain. In other words, the long-wavelength condition automatically ensures that at the leading order

i) the motion in the stiff $\mathscr{C}$-constituent is governed by ordinary ("low frenquency") elastic equations,
ii) the resonating domain acts as a source term in the balance equation within the cell.

As at low frequencies $\langle \boldsymbol{\alpha} \rangle_{\mathrm{r}} \to 0$, then $\boldsymbol{\rho}(\omega) \to \langle \rho \rangle \boldsymbol{I}$, and the description (6.24) tends to the usual elasto-dynamic formulation of porous composites. However, the description (6.24) can not be obtained by performing homogenization with the usual assumption of moderate stiffness contrast i.e. $|\boldsymbol{a}_{\mathrm{r}}|/|\boldsymbol{a}_{\mathrm{c}}| = O(1)$, and then taking the limit of the model by making $|\boldsymbol{a}_{\mathrm{r}}|/|\boldsymbol{a}_{\mathrm{c}}| \to 0$. Doing so, the same elastic tensor $\boldsymbol{A}_{\mathscr{C}}$ would be derived but the unconventional inertia would be missed. The reason is that, considering a moderate contrast leads to assume (at least implicitly) that the wavelengths in both constituents are of the same order, and throughout the homogenization process, to keep this physics at the limit $\varepsilon \to 0$. Thus, by construction such a description cannot capture the inner resonance effect, and this effect will be necessarily missed even if a contrast is introduced *a posteriori* in the macroscopic model. Conversely, when the contrast is introduced *ab initio*, the co-dynamic regime is explicitly formulated and preserved at the limit $\varepsilon \to 0$. The fact that the alternative approach is erroneous in dynamics (meanwhile correct in statics) highlights the fact that the homogenized model is only valid in the framework of the assumptions made at the microscopic scale. If one physical phenomena is disregarded it cannot be retrieved afterwards. Consequently, performing on the macroscopic model a passage to the limit on some parameters is only valid if it does not introduce an effect which is not accounted for in the initial assumptions.

The non conventional feature involved in description (6.24) lies in the fact that the effective density $\boldsymbol{\rho}(\omega)$

i) is of tensorial nature,
ii) is not bounded and depends on the frequency.

It changes of sign in the vicinity of the series of poles ω_J. This arises from the inner resonance of the $\mathscr{R}$-constituent domain that experiences a non uniform, frequency dependent motion. Hence, its effective inertia takes large positive or negative values in the vicinity of its eigen-frequencies.

As a consequence, the material shows a significant dispersion of the wave velocity with band-gaps when $\boldsymbol{\rho}(\omega)$ takes negative values. The specificity - and practical interest - of these band gaps is that they appear in a frequency range where the wavelengths that would propagate in the $\mathscr{C}$-constituent, in absence of the $\mathscr{R}$-constituent, are much larger than the period size. These band gaps are tunable with the $\mathscr{R}$-constituent properties and the geometry of its domain. For simple geometry, exact analytical expressions of the effective density are available. For periodic bi-stratified media made of isotropic materials, the effective density tensor is orthotropic with the classical mean value along the direction perpendicular to the layers, and takes the following value along the directions in the plane of the layers (Auriault and Bonnet, 1985):

$$\boldsymbol{\rho}(\omega) = \langle\rho\rangle + \rho_{\mathrm{r}}\frac{\ell_{\mathrm{r}}}{\ell}\left(\frac{\tan(\omega^*)}{\omega^*} - 1\right); \quad \omega^* = \omega\frac{\ell_{\mathrm{r}}}{2}\sqrt{\frac{\rho_{\mathrm{r}}}{\mu_{\mathrm{r}}}}$$

where ℓ and ℓ_{r} are respectively the thickness of the period and of the resonating layer (of shear modulus μ_{r}). One may refer to Auriault and Boutin (2012) for contrasted tri-stratified media, and to Bonnet and Monchiet (2015) for cylindrical or spherical composite resonant domains.

The description (6.24) also departs from usual homogenized models by the fact that the macroscopic kinematic descriptor of the media, $\boldsymbol{U}^{(0)}$ is not the mean average of the displacement over the cell, but is the uniform displacement of the $\mathscr{C}$-constituent. This is a signature of a non-local behavior as already pointed out by Lafarge and Nemati (2013) in the context of the acoustics of porous media.

In the time domain, due to the frequency dependence of $\boldsymbol{\rho}(\omega)$, a convolution product (denoted by $*$) appears in the inertial terms and the macroscopic equation reads

$$\mathrm{div}_x(\boldsymbol{A}_{\mathscr{C}} : \boldsymbol{e}_x(\boldsymbol{U}^{(0)})) = \langle\rho\rangle\frac{\partial^2\boldsymbol{U}^{(0)}}{\partial t^2} + \rho_{\mathrm{r}}\langle\boldsymbol{\alpha}\rangle_{\mathrm{r}} * \frac{\partial^2\boldsymbol{U}^{(0)}}{\partial t^2} \tag{6.25}$$

The modal decomposition of the local fields $\langle\boldsymbol{\alpha}\rangle_{\mathrm{r}}$ provides an alternative manner to account for the inner-resonance effects without introducing convolution products. It consists in introducing the instantaneous amplitudes of eigenfunctions as an infinite set of additional variables $\{b^J(x,t)\}$ which can be considered as the microscopic descriptors of the media. This leads to the following formulation, which is equivalent to (6.25):

$$
\left\{
\begin{aligned}
\operatorname{div}_x(\boldsymbol{A}_{\mathscr{C}} : \boldsymbol{e}_x(\boldsymbol{U}^{(0)})) &= \langle\rho\rangle \frac{\partial^2 \boldsymbol{U}^{(0)}}{\partial t^2} + \rho_{\mathrm{r}} \sum_{J=1}^{\infty} \frac{\partial^2 b^J}{\partial t^2} \langle \phi^J \rangle_{\mathrm{r}}, \\
b^J \omega_J^2 + \frac{\partial^2 b^J}{\partial t^2} &= \frac{\partial^2 \boldsymbol{U}^{(0)}}{\partial t^2} \cdot \langle \phi^J \rangle_{\mathrm{r}}
\end{aligned}
\right.
\tag{6.26}
$$

Thus, both the macro variables $\boldsymbol{U}^{(0)}(x,t)$ and the microscopic descriptors $b^J(x,t)$ can be determined in parallel by solving the coupled linear differential set governing their time/space evolution. Such a formulation is usual in the framework of generalized continua where the microstructural effects are "condensed" into additional variables governed by specific equations of evolution (Eringen, 1968; Maugin and Metrikine, 2010). The formulation in terms of microscopic descriptors may be convenient for computational methods as it

i) avoids the difficulties associated to the numerical treatment of convolution products, which requires to store the data of the preceding steps of calculation,
ii) does not require multi-scale mesh, as for example using the 'square' finite element method (FEM2).

In fact, local fields are "pre-integrated" owing to the calculation of the eigenmodes of the resonant domain. Furthermore, the relevant number of eigenmodes required for the calculations can be adapted according to the frequency spectrum content. Inner resonant media differ from "standard" generalized continua, mainly by the fact that the time evolution of each microscopic descriptor is ruled by differential equations of the second order, with a forcing term related to the macroscopic variables. Consequently, the descriptors can formally be determined as a function of the macro variables and their combination constitutes the kernel of the convolution product.

The above analysis can be extended to more complicated microstructures. Different cases have been investigated in Auriault and Boutin (2012). For instance, the regularizing effect of slight damping in case of viscoelastic $\mathscr{R}$-constituent can easily be studied in the same framework. One may also replace the homogeneous $\mathscr{R}$-constituent by a composite (or nonhomogeneous) inclusion. In this latter case, asymmetric morphology may induce also resonance in rotation.

Experimentally the co-dynamic regime has been actually evidenced in elastic composites, for instance in Liu et al (2000). In fact, a great variety of inner resonant materials can be contemplated by following the principles identified in the present simple case:

i) the contrast of properties imposes that the effective constitutive law is determined by the stiffer constituent, and that the resonant domain responds in a forced regime imposed by the motion of the stiff constituent,
ii) in turn the mean inertial force of the resonant domain acts as a source term in the macroscopic balance equations.

Such situations of "partial" non equilibrium local state would result in a non-conventional behavior characterized by frequency dependent inertia parameters.

Despite the local dynamics, the existence of the long wavelength enables to prevent the use of much complicated theoretical approaches based on multi-scattering formulation (that, furthermore, can only be handled numerically). In fact, one of the most practical interests of inner resonant materials lies in the fact that their constitutive parameters can easily be determined by two independent local problems, one related to the $\mathscr{C}$-constituent only, that corresponds to standard periodic homogenization problem in statics, and one related to the $\mathscr{R}$-constituent only, that corresponds to standard well-posed dynamic problem in a finite domain. This is a significant advantage for designing such materials with unconventional effective properties.

6.3.3 Elastic Bi-Composites: Significant Contrast of Stiffness and of Density

In this section we consider the second case defined in (6.7), i.e. $|\boldsymbol{a}_r|/|\boldsymbol{a}_c| = O(\varepsilon)$ and $\rho_r/\rho_c = O(\varepsilon^{-1})$. As previously the $\mathscr{C}$-constituent is connected so that the stress is mainly carried by the $\mathscr{C}$-constituent. However, due to the density contrast, the macroscopic wavelength in the media does not match the wavelength in the $\mathscr{C}$-constituent. Two estimates can be proposed - each corresponding to a different frequency range - according to the magnitude of the effective density.

Let us first investigate the situation where the effective density is dominated by the larger value ρ_r, in accordance with the physical intuition. In this situation, the overall stiffness is dominated by that of the $\mathscr{C}$-constituent while the density is dominated by that of the $\mathscr{R}$-constituent. As previously, to properly account for the contrast of parameters, their rescaling must be considered. The dominating stiffness $\boldsymbol{a}_c$ and density ρ_r are taken as reference values, and as such are not rescaled, while, compared with these reference values and in accordance with the contrasts $|\boldsymbol{a}_r|/|\boldsymbol{a}_c| = O(\varepsilon)$ and $\rho_r/\rho_c = O(\varepsilon^{-1})$ the elastic tensor of the $\mathscr{R}$-constituent and the density of the $\mathscr{C}$-constituent are consistently rescaled as $\varepsilon \boldsymbol{a}_r$ and $\varepsilon\rho_c$. Hence the scaled $(\boldsymbol{x},\boldsymbol{y})$ governing equations read in that case:

$$\left\{\begin{array}{rl}
\mathrm{div}(\boldsymbol{a}_c : \boldsymbol{e}(\boldsymbol{u}_c)) = -\omega^2 \varepsilon\rho_c \boldsymbol{u}_c & \text{in } \Omega_c \\
\mathrm{div}(\varepsilon\boldsymbol{a}_r : \boldsymbol{e}(\boldsymbol{u}_r)) = -\omega^2 \rho_r \boldsymbol{u}_r & \text{in } \Omega_r \\
(\boldsymbol{a}_c : \boldsymbol{e}(\boldsymbol{u}_c) - \varepsilon\boldsymbol{a}_r : \boldsymbol{e}(\boldsymbol{u}_r)) \cdot \boldsymbol{n} = 0 & \text{on } \Gamma \\
\boldsymbol{u}_r - \boldsymbol{u}_c = 0 & \text{on } \Gamma \\
\boldsymbol{u}_r \quad \text{and} \quad \boldsymbol{u}_c & \Omega - \text{periodic}
\end{array}\right. \tag{6.27}$$

As in the previous case, we successively obtain for the $\mathscr{C}$-constituent, at the leading order $\boldsymbol{u}_c^{(0)} = \boldsymbol{U}^{(0)}(\boldsymbol{x})$, and at the next order, that the local elastc-static set (6.13) is unchanged so that the solution (6.14) still apply. However the $\mathscr{R}$-constituent is here governed by the following *elasto-static* problem with a *Dirichlet* condition on its border:

$$\begin{cases} \mathrm{div}_y(\boldsymbol{a}_\mathrm{r} : \boldsymbol{e}_y(\boldsymbol{u}_\mathrm{r}^{(0)})) = 0 & \text{in } \Omega_\mathrm{r} \\ \boldsymbol{u}_\mathrm{r}^{(0)} = \boldsymbol{U}^{(0)}(\boldsymbol{x}) & \text{on } \Gamma \\ \boldsymbol{u}_\mathrm{r}^{(0)} & \Omega-\text{periodic} \end{cases} \tag{6.28}$$

The obvious solution is $\boldsymbol{u}_\mathrm{r}^{(0)} = \boldsymbol{U}^{(0)}(\boldsymbol{x})$. Then it is easy to see that the medium is described by a conventional elasto-dynamic formulation

$$\mathrm{div}_x(\boldsymbol{A}_{\mathscr{C}} : \boldsymbol{e}_x(\boldsymbol{U}^{(0)})) = -\omega^2 \langle \rho_\mathrm{r} \rangle_\mathrm{r} \cdot \boldsymbol{U}^{(0)}$$

where the effective elastic tensor $\boldsymbol{A}_{\mathscr{C}}$ is defined in (6.17) and the effective density is nothing but the mean density. It is clear that this situation does not corresponds to inner resonance as both constituents experience a quasi-static local regime. This implies that this conventional modeling applies at frequencies sufficiently low so that the local dynamic effects can be disregarded at the leading order.

6.3.3.1 Co-Dynamics Regime at Anti-Resonance Frequencies

Let us still consider that $|\boldsymbol{a}_\mathrm{r}|/|\boldsymbol{a}_\mathrm{c}| = O(\varepsilon)$ and $\rho_\mathrm{r}/\rho_\mathrm{c} = O(\varepsilon^{-1})$, but investigate now the less intuitive situation where the effective density is estimated of the order of the smaller value ρ_c. This assumption is supported by the fact that, as seen in the previous section, the effective density can indeed be very different from the mean density.

This assumption is formulated by considering that ρ_c is the reference value, and therefore is not rescaled, while the density of the $\mathscr{R}$-constituent, is consequently rescaled as $\varepsilon^{-1}\rho_\mathrm{r}$, while the elastic tensor of the $\mathscr{C}$-constituent is rescaled as previously as $\varepsilon \boldsymbol{a}_\mathrm{r}$. Hence the scaled $(\boldsymbol{x},\boldsymbol{y})$ governing equations read now

$$\begin{cases} \mathrm{div}(\boldsymbol{a}_\mathrm{c} : \boldsymbol{e}(\boldsymbol{u}_\mathrm{c})) = -\omega^2 \rho_\mathrm{c} \boldsymbol{u}_\mathrm{c} & \text{in } \Omega_\mathrm{c} \\ \mathrm{div}(\varepsilon \boldsymbol{a}_\mathrm{r} : \boldsymbol{e}(\boldsymbol{u}_\mathrm{r})) = -\omega^2 \varepsilon^{-1} \rho_\mathrm{r} \boldsymbol{u}_\mathrm{r} & \text{in } \Omega_\mathrm{r} \\ (\boldsymbol{a}_\mathrm{c} : \boldsymbol{e}(\boldsymbol{u}_\mathrm{c}) - \varepsilon \boldsymbol{a}_\mathrm{r} : \boldsymbol{e}(\boldsymbol{u}_\mathrm{r})).\boldsymbol{n} = 0 & \text{on } \Gamma \\ \boldsymbol{u}_\mathrm{r} - \boldsymbol{u}_\mathrm{c} = 0 & \text{on } \Gamma \\ \boldsymbol{u}_\mathrm{r} \quad \text{and} \quad \boldsymbol{u}_\mathrm{c} & \Omega-\text{periodic} \end{cases} \tag{6.29}$$

- At the dominant order, the $\mathscr{C}$-constituent and the $\mathscr{R}$-constituent are each governed by the same sets as previously, namely (6.11) and (6.18). Thus we have,

  $$\boldsymbol{u}_\mathrm{c}^{(0)} = \boldsymbol{U}^{(0)}(\boldsymbol{x}); \quad \boldsymbol{u}_\mathrm{r}^{(0)} = \boldsymbol{U}^{(0)} + \boldsymbol{\alpha}(\boldsymbol{y},\omega) \cdot \boldsymbol{U}^{(0)} \text{ and thus } \boldsymbol{\sigma}_\mathrm{r}^{(0)} = (\boldsymbol{a}_r : \boldsymbol{e}_y(\boldsymbol{\alpha})) \cdot \boldsymbol{U}^{(0)} \tag{6.30}$$

  where $\boldsymbol{\alpha}(\boldsymbol{y},\omega)$ is given by (6.22). Again this result is consistent with a local quasi static regime of the $\mathscr{C}$-constituent and a dynamic regime of the $\mathscr{R}$-constituent at the leading order.
- The following order for the $\mathscr{C}$-constituent leads to a specific local problem where the $\mathscr{C}$-constituent experiences a static regime while it undergoes the stresses

exerted by the $\mathscr{R}$-constituent in dynamic regime

$$\begin{cases} \operatorname{div}_y(\boldsymbol{a}_c : (\boldsymbol{e}_y(\boldsymbol{u}_c^{(1)}) + \boldsymbol{e}_x(\boldsymbol{U}^{(0)})) = 0 & \text{in } \Omega_c \\ (\boldsymbol{a}_c : (\boldsymbol{e}_y(\boldsymbol{u}_c^{(1)}) + \boldsymbol{e}_x(\boldsymbol{U}^{(0)})) \cdot \boldsymbol{n} = (\boldsymbol{a}_r : \boldsymbol{e}_y(\boldsymbol{u}_r^{(0)})) \cdot \boldsymbol{n} & \text{on } \Gamma \\ \boldsymbol{u}_c^{(1)} & \Omega - \text{periodic} \end{cases} \tag{6.31}$$

where $\boldsymbol{u}_r^{(0)}$ is given by its expression in (6.30) that involves the second order tensor $\boldsymbol{\alpha}$. The integration of the balance equation over Ω_c, combined with the use of the divergence theorem, the Neumann boundary condition on Γ and the periodicity condition, provides the necessary condition for the existence of a solution. This condition expresses the overall balance and implies that

$$\frac{1}{\Omega}\int_\Gamma (\boldsymbol{a}_r : \boldsymbol{e}_y(\boldsymbol{u}_r^{(0)})) \cdot \boldsymbol{n} = \frac{1}{\Omega}\int_\Gamma \boldsymbol{\sigma}_r^{(0)} \cdot \boldsymbol{n} ds = 0.$$

Now integrating the momentum balance

$$\operatorname{div}_y(\boldsymbol{a}_r : \boldsymbol{e}_y(\boldsymbol{u}_r^{(0)})) = -\omega^2 \rho_r \boldsymbol{u}_r^{(0)} \quad \text{in} \quad \Omega_r$$

and making use of the divergence theorem provides

$$\frac{1}{\Omega}\int_\Gamma \boldsymbol{\sigma}_r^{(0)} \cdot \boldsymbol{n} ds = -\omega^2 \langle \rho_r \boldsymbol{u}_r^{(0)} \rangle_r = 0.$$

According to the expression of $\boldsymbol{u}_r^{(0)}$ this requirement leads to

$$\omega^2 \rho_r \langle \boldsymbol{I} + \boldsymbol{\alpha} \rangle_r \cdot \boldsymbol{U}^{(0)} = 0$$

As $\omega \neq 0$, such a situation is *generally impossible except* for specific frequencies where

i) the tensor $\langle \boldsymbol{I} + \boldsymbol{\alpha} \rangle_r$ is singular and
ii) the polarization of the macroscopic motion $\boldsymbol{U}^{(0)}$ is constant and belongs to the kernel of $\langle \boldsymbol{I} + \boldsymbol{\alpha} \rangle_r$. Note that if $\langle \boldsymbol{\alpha} \rangle_r$ is isotropic, the singularity arises for the three directions of the space and then the orientation of $\boldsymbol{U}^{(0)}$ can be arbitrary.

Hence, in the following we will assume the isotropy of $\langle \boldsymbol{\alpha} \rangle_r$ (thus $\langle \boldsymbol{\alpha} \rangle_r$ can be replaced by a scalar) that enables less restrictive conditions.

To sum up, when $\langle \boldsymbol{\alpha} \rangle_r$ is isotropic, the set (6.31) can only have solutions in the vicinity of specific frequencies such that the effective inertia of the $\mathscr{R}$-constituent vanishes at the considered order i.e.

$$\langle \boldsymbol{I} + \boldsymbol{\alpha} \rangle_r = \langle \boldsymbol{\alpha} \rangle_r + \frac{|\Omega_r|}{|\Omega|}\boldsymbol{I} = O(\varepsilon) \tag{6.32}$$

The requirement (6.32) is consistent

i) with the a priori estimate of the macroscopic wavelength based on the lower density ρ_c and
ii) with the fact that the co-dynamic regime imposes that the stress flux induced by the $\mathscr{R}$-constituent is of one order smaller than the stress prevailing in the $\mathscr{C}$-constituent.

In fact meanwhile the stress tensor $\boldsymbol{\sigma}_r^{(0)}$ is actually of zero order, its mean value on Γ is of one order smaller. This can correspond to dissymmetric modes that may be seen as a dipolar source (instead of mono-polar source as in the previous case studied in Sect. 6.3.2). According to the above described properties of $\langle\boldsymbol{\alpha}\rangle_r$, and assuming its isotropy, this possibility occurs for a discrete series of frequency $\widetilde{\omega}_K$ localized in between each pole, i.e. $\omega_K < \widetilde{\omega}_K < \omega_{K+1}$. Such frequencies $\widetilde{\omega}_K$ will be designated thereafter as *anti-resonant frequencies*.

Now, focusing on the close vicinity of one of anti-resonant frequencies $\widetilde{\omega}_K$, the local elastodynamic problem (6.31) is forced by $\boldsymbol{e}_x(\boldsymbol{U}^{(0)})$, and $\boldsymbol{u}_r^{(0)} = (\boldsymbol{I}+\boldsymbol{\alpha})\cdot\boldsymbol{U}^{(0)}$. Accounting for the linearity, we can express the fields $\boldsymbol{u}_c^{(1)}$ and $\boldsymbol{\sigma}_c^{(0)}$ in the form

$$\begin{cases} \boldsymbol{u}_c^{(1)} = & \boldsymbol{\chi}(\boldsymbol{y}) : \boldsymbol{e}_x(\boldsymbol{U}^{(0)}) + \boldsymbol{\xi}(\boldsymbol{y},\widetilde{\omega}_K)\cdot\boldsymbol{U}^{(0)} + \boldsymbol{U}_c^{(1)}(\boldsymbol{x}) \\ \boldsymbol{\sigma}_c^{(0)} = & [\boldsymbol{a} : \boldsymbol{e}_y(\boldsymbol{\chi}) + \boldsymbol{a}] : \boldsymbol{e}_x(\boldsymbol{U}^{(0)}) + (\boldsymbol{a} : \boldsymbol{e}_y(\boldsymbol{\xi}))\cdot\boldsymbol{U}^{(0)} \end{cases} \tag{6.33}$$

where $\boldsymbol{\chi}(\boldsymbol{y})$ is the same real tensor as in the previous section, see (6.14). The second order tensor $\boldsymbol{\xi}(\boldsymbol{y},\widetilde{\omega}_K)$ is built from the particular fields $\boldsymbol{\xi}^p(\boldsymbol{y},\widetilde{\omega}_K)$. These fields are the frequency dependent solutions corresponding to unit values of the macroscopic motion i.e. $\boldsymbol{U}^{(0)} = \boldsymbol{E}^p$. From (6.31) the corresponding local problem reads

$$\begin{cases} \mathrm{div}_y(\boldsymbol{a}_c : (\boldsymbol{e}_y(\boldsymbol{\xi}^p)) = 0 \quad \text{in } \Omega_c \\ (\boldsymbol{a}_c : (\boldsymbol{e}_y(\boldsymbol{\xi}^p)\cdot\boldsymbol{n} = (\boldsymbol{a}_r : \boldsymbol{e}_y(\boldsymbol{\alpha}^p))\cdot\boldsymbol{n} \quad \text{on } \Gamma \\ \boldsymbol{\xi}^p \qquad \Omega-\text{periodic} \end{cases} \tag{6.34}$$

In other words, the first order fields $\boldsymbol{\xi}^p(\boldsymbol{y},\widetilde{\omega}_K)$ are the elasto-static displacements in the $\mathscr{C}$-constituent that balance the stresses $(\boldsymbol{a}_r : \boldsymbol{e}_y(\boldsymbol{\alpha}^p))\cdot\boldsymbol{n}$ exerted on its boundary by the $\mathscr{R}$-constituent, this latter undergoing a dynamic regime forced by the uniform motion $\boldsymbol{E}^p$ of the $\mathscr{C}$-constituent at the leading order.

- At the next order, the balance and boundary condition concerning both constituents write

$$\begin{cases} \mathrm{div}_y(\boldsymbol{\sigma}_c^{(1)}) + \mathrm{div}_x(\boldsymbol{\sigma}_c^{(0)}) = -\widetilde{\omega}_K^2\rho_c\boldsymbol{U}^{(0)} \quad \text{in } \Omega_c \\ \boldsymbol{\sigma}_c^{(1)}\cdot\boldsymbol{n} = \boldsymbol{\sigma}_r^{(1)}\cdot\boldsymbol{n} \\ \mathrm{div}_y(\boldsymbol{\sigma}_r^{(1)}) + \mathrm{div}_x(\boldsymbol{\sigma}_r^{(0)}) = -\widetilde{\omega}_K^2\rho_r\boldsymbol{u}_r^{(1)} \\ \boldsymbol{\sigma}_r^{(1)} \quad \text{and} \quad \boldsymbol{\sigma}_c^{(1)} \qquad \Omega-\text{periodic} \end{cases} \tag{6.35}$$

Integrating over Ω_c, and Ω_r and using the usual integral transformation, yield

$$\mathrm{div}_x(\langle\boldsymbol{\sigma}^{(0)}\rangle) = -\widetilde{\omega}_K^2(\langle\rho_c\rangle_c\boldsymbol{U}^{(0)} + \rho_r\langle\boldsymbol{u}_r^{(1)}\rangle_r) \tag{6.36}$$

From the expression of $\boldsymbol{\sigma}^{(0)}$ in both constituents, the mean stress term $\langle\boldsymbol{\sigma}^{(0)}\rangle$ reads:

$$\langle\boldsymbol{\sigma}^{(0)}\rangle = \boldsymbol{A}_{\mathscr{C}} : \boldsymbol{e}_x(\boldsymbol{U}^{(0)}) + \boldsymbol{D}_{\widetilde{\omega}_K} \cdot \boldsymbol{U}^{(0)}$$

The effective elasticity tensor $\boldsymbol{A}_{\mathscr{C}}$ is the same as in the previous case, see (6.17). The new elasto-inertial tensor $\boldsymbol{D}_{\widetilde{\omega}_K}$ is defined by

$$\boldsymbol{D}_{\widetilde{\omega}_K} = \langle\boldsymbol{a}_{\mathrm{c}} : \boldsymbol{e}_y(\boldsymbol{\xi})\rangle_{\mathrm{c}} + \langle\boldsymbol{a}_{\mathrm{r}} : \boldsymbol{e}_y(\boldsymbol{\alpha})\rangle_{\mathrm{r}}$$

As $\boldsymbol{\xi}$ and $\boldsymbol{\alpha}$ are of second rank, the tensor $\boldsymbol{D}_{\widetilde{\omega}_K}$ is of the third rank. Furthermore note that, as one considers $\boldsymbol{a}_{\mathrm{r}}$ constant then

$$\langle\boldsymbol{a}_{\mathrm{r}} : \boldsymbol{e}_y(\boldsymbol{\alpha})\rangle_{\mathrm{r}} = \boldsymbol{a}_{\mathrm{r}}\langle\boldsymbol{e}_y(\boldsymbol{\alpha})\rangle_{\mathrm{r}} = 0$$

because

$$\langle 2\boldsymbol{e}_y(\boldsymbol{\alpha})\rangle_{\mathrm{r}} = \int_{\Gamma} \boldsymbol{\alpha}\otimes\boldsymbol{n} + \boldsymbol{n}\otimes\boldsymbol{\alpha}\,\mathrm{d}s = 0$$

since $\boldsymbol{\alpha} = 0$ on Γ. Thus, in that case the last term of $\boldsymbol{D}_{\widetilde{\omega}_K}$ vanishes and

$$\boldsymbol{D}_{\widetilde{\omega}_K} = \langle\boldsymbol{a}_{\mathrm{c}} : \boldsymbol{e}_y(\boldsymbol{\xi})\rangle_{\mathrm{c}}$$

- To close the description it remains to express the inertial term $\rho_{\mathrm{r}}\langle\boldsymbol{u}_r^{(1)}\rangle_r$ that appears in (6.36). The field $\boldsymbol{u}_r^{(1)}$ is determined by the following balance equation in Ω_r together with the Dirichlet condition on Γ.

$$\begin{cases} \mathrm{div}_y(\boldsymbol{a}_{\mathrm{r}} : (\boldsymbol{e}_y(\boldsymbol{u}_{\mathrm{r}}^{(1)}) + \boldsymbol{e}_x(\boldsymbol{u}_{\mathrm{r}}^{(0)})) + \mathrm{div}_x(\boldsymbol{a}_{\mathrm{r}} : \boldsymbol{e}_y(\boldsymbol{u}_{\mathrm{r}}^{(0)}))) = -\widetilde{\omega}_K^2\rho_{\mathrm{r}}\boldsymbol{u}_{\mathrm{r}}^{(1)} & \text{in } \Omega_{\mathrm{r}} \\ \boldsymbol{u}_{\mathrm{r}}^{(1)} = \boldsymbol{u}_{\mathrm{c}}^{(1)} & \text{in } \Gamma \\ \boldsymbol{u}_{\mathrm{r}}^{(1)} \quad \Omega - \text{periodic} \end{cases} \tag{6.37}$$

According to the expressions of $\boldsymbol{u}_{\mathrm{c}}^{(1)}$ this local elastodynamic problem is forced by $\boldsymbol{e}_x(\boldsymbol{U}^{(0)})$, $\boldsymbol{U}^{(0)}$ and $\boldsymbol{U}^{(1)}$. Thus, by linearity we have:

$$\boldsymbol{u}_{\mathrm{r}}^{(1)} = \boldsymbol{\theta}_{\widetilde{\omega}_K}(\boldsymbol{y},\widetilde{\omega}_K) : \boldsymbol{e}_x(\boldsymbol{U}^{(0)}) + \boldsymbol{\zeta}_{\widetilde{\omega}_K}(\boldsymbol{y},\widetilde{\omega}_K)\cdot\boldsymbol{U}^{(0)} + \boldsymbol{U}^{(1)} + \boldsymbol{\alpha}(\boldsymbol{y},\widetilde{\omega}_K)\cdot\boldsymbol{U}^{(1)}$$

and as (at order 1)

$$\langle\boldsymbol{\alpha}\rangle_{\mathrm{r}}(\widetilde{\omega}_K) + \frac{|\Omega_{\mathrm{r}}|}{|\Omega|}\boldsymbol{I} = 0,$$

$$\langle\boldsymbol{u}_{\mathrm{r}}^{(1)}\rangle_{\mathrm{r}} = \langle\boldsymbol{\theta}_{\widetilde{\omega}_K}\rangle_{\mathrm{r}} : \boldsymbol{e}_x(\boldsymbol{U}^{(0)}) + \langle\boldsymbol{\zeta}_{\widetilde{\omega}_K}\rangle_{\mathrm{r}}.\boldsymbol{U}^{(0)}$$

where $\langle\boldsymbol{\theta}_{\widetilde{\omega}_K}\rangle_{\mathrm{r}}$ and $\langle\boldsymbol{\zeta}_{\widetilde{\omega}_K}\rangle_{\mathrm{r}}$ are tensors of rank three and two respectively. The field $\boldsymbol{u}_{\mathrm{r}}^{(1)}$ is the first corrector of $\boldsymbol{u}_{\mathrm{r}}^{(0)}$:
The contribution $\langle\boldsymbol{\theta}_{\widetilde{\omega}_K}\rangle_{\mathrm{r}} : \boldsymbol{e}_x(\boldsymbol{U}^{(0)})$ results

i) from the non uniform motion $\boldsymbol{\chi} : \boldsymbol{e}_x(\boldsymbol{U}^{(0)})$ imposed by the deformed stiff constituent on the boundary Γ, and
ii) from the volume forces accounting for the variation at the macroscopic scale of the local elasto-dynamic field $\boldsymbol{\alpha}$;

the contribution $\langle\boldsymbol{\zeta}_{\widetilde{\omega}_K}\rangle_{\mathrm{r}} \cdot \boldsymbol{U}^{(0)}$ results from the non uniform motion $\boldsymbol{\xi} \cdot \boldsymbol{U}^{(0)}$ imposed by the deformed stiff constituent on the boundary Γ.
Remark that as the fields constituting $\boldsymbol{\alpha}$, $\boldsymbol{\xi}$, and $\boldsymbol{\zeta}$ are determined at an anti-resonance frequency, one may reasonably infer that they remain bounded as well as the effective tensors in which they are involved. However, the proof would require a detailed analysis that is beyond the scope of this paper.

- Finally, focusing on the vicinity of the *anti-resonance frequencies* $\widetilde{\omega}_K$, the equivalent macroscopic behavior at the leading order reads

$$\begin{cases} \operatorname{div}_x\left(\boldsymbol{A}_{\mathscr{C}} : \boldsymbol{e}_x(\boldsymbol{U}^{(0)}) + \boldsymbol{D}_{\widetilde{\omega}_K} \cdot \boldsymbol{U}^{(0)}\right) = -\widetilde{\omega}_K^2\left(\widetilde{\boldsymbol{\rho}}_{\widetilde{\omega}_K} \cdot \boldsymbol{U}^{(0)} + \rho_{\mathrm{r}}\langle\boldsymbol{\theta}_{\widetilde{\omega}_K}\rangle_{\mathrm{r}} : \boldsymbol{e}_x(\boldsymbol{U}^{(0)})\right), \\ \widetilde{\boldsymbol{\rho}}_{\widetilde{\omega}_K} = \rho_{\mathrm{c}}\dfrac{|\Omega_{\mathrm{c}}|}{|\Omega|}\boldsymbol{I} + \rho_{\mathrm{r}}\langle\boldsymbol{\zeta}_{\widetilde{\omega}_K}\rangle_{\mathrm{r}} \end{cases} \tag{6.38}$$

The behaviour is finally expressed by using two real valued third order tensors, $\boldsymbol{D}_{\widetilde{\omega}_K}$ and $\langle\boldsymbol{\zeta}_{\widetilde{\omega}_K}\rangle_{\mathrm{r}}$, which are not null only if the material is not isotropic. This result is not inconsistent with the assumed isotropy of the second order tensor $\langle\boldsymbol{\alpha}\rangle_{\mathrm{r}}$, because the isotropy of $\langle\boldsymbol{\alpha}\rangle_{\mathrm{r}}$ is strictly related to the mechanical properties and geometry of the $\mathscr{R}$-constituent. At the contrary, the third order tensors are related to the mechanical properties and geometry of both constituents and can be anisotropic.

6.3.3.2 Comments

The description (6.38) is non conventional for the following reasons:

- It is only valid in the vicinity of the discrete spectrum of anti-resonance $\{\widetilde{\omega}_K\}$, and the effective parameters, other than the elastic tensor $\boldsymbol{A}_{\mathscr{C}}$, differ for each of these frequencies,
- The classic elastic constitutive law is complemented by an elasto-inertial term involving the $\{\widetilde{\omega}_K\}$-harmonic motion itself (not the strain). As the effective tensor $\boldsymbol{D}_{\widetilde{\omega}_K}$ is of third rank, it is necessarily anisotropic when non null, and it vanishes provided that the cell presents three orthogonal planes of symmetry,
- The effective density is of one order smaller that the mean density and the effective inertia involves a non-local term associated to the strain. The comment made on $\boldsymbol{D}_{\widetilde{\omega}_K}$ also applies to the third rank effective tensor $\langle\boldsymbol{\theta}_{\widetilde{\omega}_K}\rangle_{\mathrm{r}}$.

The description (6.38) can equivalently be expressed in terms of mean stress $\langle\boldsymbol{\sigma}^{(0)}\rangle$ and momentum density $\langle\mathbf{p}^{(0)}\rangle$ as follows:

$$\operatorname{div}_x(\langle\boldsymbol{\sigma}^{(0)}\rangle) = i\widetilde{\omega}_K\langle\mathbf{p}^{(0)}\rangle$$

with

$$\begin{pmatrix} \langle \boldsymbol{\sigma}^{(0)} \rangle \\ \langle \mathbf{p}^{(0)} \rangle \end{pmatrix} = \begin{pmatrix} \mathbf{A}_{\mathscr{C}} & \frac{\boldsymbol{D}_{\widetilde{\omega}_K}}{i\widetilde{\omega}_K} \\ \rho_r \imath \widetilde{\omega}_K \langle \boldsymbol{\theta}_{\widetilde{\omega}_K} \rangle_r & \widetilde{\boldsymbol{\rho}}_{\widetilde{\omega}_K} \end{pmatrix} \begin{pmatrix} : e_x(\boldsymbol{U}^{(0)}) \\ \cdot i\widetilde{\omega_K} \boldsymbol{U}^{(0)} \end{pmatrix}$$

It is worth mentioning that qualitatively, and physically, this formulation takes a form close to the one proposed in Willis (2012). However, a detailed comparison is far to be straightforward as the present equations are derived in the framework of largely contrasted periodic media (instead of random moderately contrasted media). The macroscopic descriptor $\boldsymbol{U}^{(0)}$ corresponds to the motion of the non-resonant domain of the cell (hence it differs form the mean value overall the cell), and the effective tensors are local and frequency dependent (instead of being non-local, i.e., convolution operators). It should be noticed that in the present case, the effective parameters can actually be determined from the knowledge of the microstructure.

Note finally that non-zero third rank coupling tensors yield a wave equation (with frequency dependent coefficients) involving classical space derivatives of the second order but also non-classical derivatives of the first order. These latter may induce a non-symmetry for waves propagating in opposite directions at the anti-resonance frequencies, i.e. corresponding to an effect of "acoustic diode".

6.3.4 Synthesis on the Resonant and Anti-Resonant Co-Dynamic Regimes

The main learning of the two above sub-sections can be summarized as follows:

- The occurrence of a co-dynamic regime requires strong contrasts of properties, and imposes that the stress flux induced by the resonating constituent is of one order smaller than the stress flux conveyed by the other constituent.
- The simplest case is that of a soft resonant domain in a much stiffer matrix, both having similar densities. In that case, the macroscopic constitutive law is of standard elastic type, and the inner resonance phenomenon emerges as a specific inertial source term in the stress balance. This mechanism presents a strong analogy with what is sometimes designed as mono-polar resonance. Consequently, this situation leads to an unconventional density, and a conventional elasticity. The description is valid on a wide frequency range, provided that the scale separation is respected.
- A more complicated case is that of a soft and dense resonant domain in a stiff and light matrix. Then a co-dynamic situation is possible only in the vicinity of the anti-resonance of the $\mathscr{R}$-constituent. This is a similar situation to what is sometimes designed as bi- or multi-polar resonance. Such a phenomenon leads to an unconventional elastic constitutive law and inertia, both including elasto-inertial terms. However the degrees of these additional differential operators are different from that of the conventional one. Consequently, due to the nature of the

description itself, it is not possible *in this context* to speak of equivalent media having possibly negative elasticity or density.

Compared to the co-dynamic regime at resonant frequencies, a co-dynamic regime at anti-resonant frequencies requires much more severe and specific conditions to be observed experimentally. In particular, in presence of slight damping, condition (6.32) of vanishing inertia of the $\mathscr{R}$-constituent at the leading order may be difficult to reach, specially as the resonant domain has a larger density than the one of the other constituent. The existence of anti-resonance lies on the combined interaction of, at least, two resonant modes. Thus, specific resonant systems with an unique degree of freedom (as a simple spring-mass system) cannot exhibit this type of behavior. Note also that the practical design of such materials governed by antiresonance is more complicated than materials governed by resonance, since, meanwhile the discrete anti-resonant frequencies can be determined from the knowledge of the resonant domain only, the effective properties result from coupling effects between the $\mathscr{R}$- and $\mathscr{C}$-constituents. These coupling effects require (even in the simple case where the coupling tensors $\boldsymbol{D}_{\widetilde{\omega}_K}$ and $\langle \boldsymbol{\theta}_{\widetilde{\omega}_K} \rangle_{\mathrm{r}}$ vanish) the resolution of the three successive local problems related to $\boldsymbol{\alpha}$, $\boldsymbol{\xi}$, and $\boldsymbol{\zeta}$. For this reason we focus in the sequel on the co-dynamic regime governed by resonance.

6.3.5 *Reticulated Media: Inner Resonance by Geometrical Contrast*

We illustrate in this subsection the possibility of realizing inner resonance media with a single material, by introducing geometrical contrasts in the morphology, such contrasts being achieved by the presence of beams or plates at the microscale. We follow here the work of Boutin et al (2010); Chesnais et al (2007) on reticulated structures experiencing global vibrations, where inner dynamic phenomena due to local resonance in bending have been evidenced.

Consider a periodic reticulated medium (Fig. 6.5-left), whose cubic period of side length ℓ simply consists of bars located on the edges, rigidly connected at their extremities (the cubic corners). For simplicity, we investigate the configuration where the bars are all identical, have a square section of side length a and are made of the same material of Young modulus E, Poisson's ratio ν and volumic mass ρ. Further, we consider a morphology that presents a strong geometrical contrast. This strong geometrical contrast is achieved by assuming that the bars are slender enough, i.e. $a \ll \ell$, to be described locally as Euler-Bernoulli beams. Hence the axial (compression) and transverse (bending) harmonic behaviours of the bars are described by the following set of equations (s stands here for beam axis coordinate, u the axial displacement, N the normal force, $S = a^2$ the beam section; v the transverse displacement, M the bending momentum, T the transverse shear force, and $I = a^4/12$ is the geometrical inertia of the beam section)

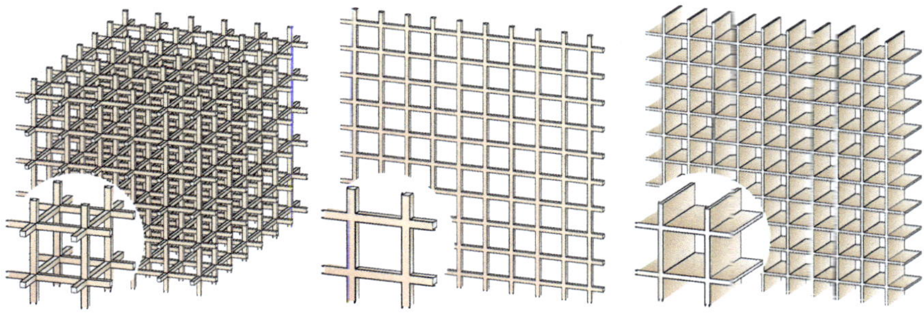

Fig. 6.5: Examples of reticulated elastic media. Left: 3D media made of a cubic array of identical beams; Center: 2D square array of beams ; Right: 2D square array of identical plates (courtesy of Logan Schwan)

$$\left\{\begin{array}{lll} \dfrac{\mathrm{d}N}{\mathrm{d}s} = -\omega^2 \rho S u \quad , & N = ES\dfrac{\mathrm{d}u}{\mathrm{d}s}, & \\ \dfrac{\mathrm{d}T}{\mathrm{d}s} = -\omega^2 \rho S v \quad , & T - \dfrac{\mathrm{d}M}{\mathrm{d}s} = 0 \quad , & M = -EI\dfrac{\mathrm{d}^2 v}{\mathrm{d}s^2} \end{array}\right. \tag{6.39}$$

and consequently the compression and bending wavelengths are respectively

$$\frac{\Lambda_c}{2\pi} = \frac{1}{\omega}\sqrt{\frac{E}{\rho}} \quad , \quad \frac{\Lambda_b}{2\pi} = \frac{1}{\sqrt{\omega}}\sqrt[4]{\frac{EI}{\rho S}}$$

Note that the validity of the beam model requires that the bending wavelength is significantly larger than the size of the section, i.e. $\Lambda_b \gg a$. Consequently, at a given frequency, the compression wavelength is much larger than the bending wavelength, since from the above expressions (for rectangular sections):

$$\frac{\Lambda_c}{\Lambda_b^2} = \frac{1}{2\pi}\sqrt{\frac{S}{I}} = \frac{\sqrt{3}}{\pi a} \qquad \text{hence} \qquad \frac{\Lambda_c}{\Lambda_b} = \frac{\Lambda_b\sqrt{3}}{\pi a} \gg 1$$

This corresponds to the fact that the beams present a much higher stiffness in compression than in bending.

In the considered reticulated medium, let us focus on the propagation of compressional waves in the direction $\boldsymbol{E}_1$ coinciding with one axis of the cubic period. In that case, we can distinguish the axial bars oriented along $\boldsymbol{E}_1$ that sustain a compression state and the transversal bars oriented at $\pi/2$ that undergo bending. Because of the compression/bending stiffness contrast, the stress is mainly carried by the axial bars that play the role of the conveying $\mathscr{C}$-constituent. Thus, the macroscopic wavelength is the compression wavelength, which is much larger than the period size ℓ in the assumption of scale separation. To reach a co-dynamics regime, the transversal bars should play the role of the resonating $\mathscr{R}$-constituent, i.e. they should be in dynamic bending regime. In other terms the bending wavelength should be of the order of magnitude of the period size ℓ. Consequently, the requirements for a co-dynamic regime are

$$L = O(\frac{\Lambda_c}{2\pi}) = \varepsilon^{-1}\ell; \qquad \frac{\Lambda_b}{2\pi} = O(\ell)$$

i.e., according to the expressions of Λ_c and Λ_b

$$O(\frac{1}{\omega^2}\frac{E}{\rho}) = \varepsilon^{-2}\ell^2 \; ; \quad O(\frac{1}{\omega^2}\frac{Ea^2}{12\rho}) = \ell^4$$

These relations impose that the inverse of the slenderness parameter of the beam is of the same order as the scale ratio:

$$O(\frac{a}{\ell}) = \varepsilon \ll 1$$

Note that with this geometry, the transverse force T_b in the bended beams and the axial force N_c in the compressed beam can be assessed as:

$$N_c = ES\frac{\mathrm{d}u}{\mathrm{d}s_c} = O(Ea^2\frac{u}{L}) \; ; \quad T_b = \frac{\mathrm{d}M}{\mathrm{d}s_r} = O(E\frac{a^4}{12}\frac{v}{\ell^3})$$

Thus, since the axial motion u of the compressed beams oriented along $\boldsymbol{E}_1$ is of the same order of magnitude as the transverse motion v of the bended beams oriented along $\boldsymbol{E}_2$ we have

$$\frac{T_b}{N_c} = O(\frac{a^2L}{12\ell^3}) = \varepsilon$$

Consequently the transverse force T_b in the bended beams is actually one order smaller than the axial force N_c in the compressed beam.

Such a reticulated medium can be analyzed by the asymptotic homogenization of periodic discrete media (Caillerie et al, 1989; Tollenaere and Caillerie, 1998). The detailed process is not reported here, and the reader can refer to Chesnais et al (2007, 2012); Hans and Boutin (2008) for a further insight. The principle consists in two steps: first the reduction of the media to an exact discrete representation, then the homogenization process itself.

The discretization consists in taking advantage of the integration of the beam constitutive laws in harmonic regime to express explicitly the forces and couples at the endpoints of a bar as functions of the kinematic variables (displacement and rotations) at these endpoints. Then, as the equilibrium of each beam is already achieved, the balance of the medium exactly reduces to the balance of forces and moments applied by the bars connected to a given node. These equations take the form of a finite difference set and provide the discrete description of the medium.

The homogenization process transforms these discrete descriptions into an equivalent continuum model. The nodal variables are identified to the values taken *at the discrete locations of the nodes* by continuous functions. These latter are looked for in the form of asymptotic expansions in powers of ε. Now, according to the scale separation assumption the nodal variables vary slowly from one node to the next.

Therefore, the finite differences can be re-expressed in terms of Taylor's series of the continuous functions, which introduces the macroscopic derivatives. Reporting the expansions in the discrete description of the medium yields a set of differential equations from which the macroscopic continuous description is derived.

In the present case this method leads successively to show that:

- at the leading order the normal forces in the axial bars is simply given by

$$ES\frac{\mathrm{d}U^{(0)}(x)}{\mathrm{d}x},$$

consistently with the fact that the transverse force in the transversal bar is of one order smaller,
- the transverse bars undergo a forced dynamic regime in bending resulting from the motion imposed at their extremities by the axial bars,
- the balance equation of normal forces accounts for the conventional inertia of the axial bar and the non-conventional inertia of the transversal bars.

Thus the derived 1D equation governing the compressional wave reads

$$\begin{cases} \frac{\mathrm{d}}{\mathrm{d}x}(E\frac{\mathrm{d}U^{(0)}}{\mathrm{d}x}) = -\omega^2\widetilde{\rho}(\omega)U^{(0)}, \\ \qquad \widetilde{\rho} = \rho\,(1+2\psi(\omega^*))\,; \quad \omega^* = \frac{\omega}{\omega_1} \text{ with } \omega_1 = \left(\frac{2}{\ell}\right)^2\sqrt{\frac{EI}{\rho S}} \end{cases} \tag{6.40}$$

The frequency dependent complex valued function $\psi(\omega^\star)$ accounts for the non uniform motion within the transverse bars. As for the composite, this motion can be expressed on the basis of the bending modes of the bar. Alternatively, an explicit expression of the motion can be derived by integrating the beam equation with the conditions of imposed motion and no rotation at its extremity Taking the average over the beam length yields:

$$\psi(\omega^*) = \frac{1}{\sqrt{\omega^*}}\frac{2}{\coth(\sqrt{\omega^*})+\cot(\sqrt{\omega^*})}$$

The function $\psi(\omega^\star)$ presents the same general features as the tensor $\boldsymbol{I}+\langle\boldsymbol{\alpha}\rangle_r$ involved in the bi-composite description: it is not bounded and changes of sign in the vicinity of the poles ω_i of ψ, that corresponds to the *odd bending* modes of the transversal bars, with clamped conditions at their extremities (the average displacement of even bending modes is null and consequently these modes do not participate to the effective density). Similarly to bi-composites, the negative effective density leads to dispersive waves and band-gaps in the frequency range where, in absence of inner resonance, the wavelength would be much larger than the period size. This effect results from the morphology instead of the interaction between contrasted constituents. Another significant difference between bi-composites, is that the role of conveying and resonating elements are not determined *a priori* but depends on the direction of the wave: they switch for compressional wave propagating in one or other of the two material orthogonal directions. Experimentally, inner resonance by bending has been reported in reticulated beams by Baravelli and Ruzzene (2013).

Many variants of this simple example based on bending/compression contrast are possible. For instance, considering the 2D arrays described in Fig. 6.5, a description similar to (6.40) applies, except that $\widetilde{\rho} = \rho\,(1+\psi(\omega^*))$ and, in the case of plates, E has to be changed in to $E/(1-\nu^2)$. With parallelepipedic cells the inner-resonant frequencies will be different in the three directions of propagation if the three families of orthogonal bars are different. One may also add localized masses to tune the bending resonance. Considering rhomboidal cell enables oblique orientations of inner-resonance effects, bars can also be replaced by plates, etc...

Despite the strong morphologic difference the same principles and consequences identified for bi-composites apply to reticulated media: the effective elasticity is conventional while the effective density is non-conventional. This relies on the fact that, in the momentum balance equation the "flux" is the stress, which is not influenced by the inner resonance effect, while the inner resonance effect is comprised in the source term of inertial nature. This leads to infer that, in the same framework, to obtain the reverse situation, and particularly non-conventional stiffness, one should consider phenomena governed by a balance equation such that the elastic effect appears as a source term instead of a flux. This is the case in acoustics of porous media where the flux of mass is balanced by the fluid compressibility. This topic is investigated in the next section.

6.4 Inner Resonance in Poro-Acoustics

The linear acoustics in homogeneous rigid porous media of porosity ϕ is classically described by the mass balance equation of the gas and the constitutive equation governing the gas flow, namely the dynamic Darcy's law (Auriault, 1980; Smeulders et al, 1992). This formulation applies in a frequency range such that the wavelength is significantly larger than the characteristic pore size. In harmonic regime at angular frequency ω, the set of differential equations reads (with the convention that each variable is multiplied by $\exp(+i\omega t)$):

$$\begin{cases} \operatorname{div}(\boldsymbol{v}) = -i\omega \dfrac{\phi}{\beta} p, \\ \boldsymbol{v} = -\dfrac{\mathcal{K}(\omega)}{\eta} \cdot \nabla p \end{cases} \tag{6.41}$$

where p is the acoustic pressure perturbation around the equilibrium pressure P^e, and $\boldsymbol{v}$ is the mean gas flux, i.e., following the usual convention, the flux averaged over the elementary representative volume (the flux averaged over the volume of the pores reads $\boldsymbol{v}/\phi$). By simplicity we assume that the thermal effects at the pores scale can be disregarded so that the wave propagation occurs in adiabatic regime. Thus the gas bulk modulus β (compressibility β^{-1}) reads $\beta = \gamma P^e$. This simplifying assumption can be removed by introducing a complex frequency dependent coefficient $\tilde{\gamma}(\omega)$ varying from the isothermal value $\tilde{\gamma}(0) = 1$ at low frequency to the adiabatic value γ at higher frequencies, see Lafarge et al (1997). The gas viscosity is η and $\mathscr{K}(\omega)$ is the frequency dependent dynamic permeability tensor. At low frequencies, the flow in the pores is driven by the viscosity, then

$$\mathscr{K}(\omega) \underset{\omega \to 0}{\to} \boldsymbol{\kappa}$$

where the real-valued intrinsic permeability $\boldsymbol{\kappa}$ is of the order of magnitude of $O((\text{pores size})^2)$. At high frequencies, the flow is driven by the inertia, then

$$\eta\phi\mathscr{K}^{-1}(\omega) \underset{\omega \to \infty}{\to} i\omega\rho^e\boldsymbol{\alpha}_\infty$$

where ρ^e is the gas density at equilibrium and $\boldsymbol{\alpha}_\infty$ is the real-valued high frequency limit of the tortuosity tensor whose order of magnitude is $O(1)$. The two regimes are delimited by the Biot transition frequency ω_b for which the low frequency viscous component within the balance of momentum equals the inertial component, so that

$$\omega_b = \frac{\eta\phi}{|\boldsymbol{\kappa}|\rho^e|\boldsymbol{\alpha}_\infty|}$$

Despite the differences in the conservative and dissipative phenomena we can highlight the formal similarity with the elastic case (6.8a)-(6.8b). The analogous variables and parameters in elastic/porous problems are: vectorial solid elastic motion/scalar gas pressure, i.e $\boldsymbol{u} \to -p$; tensorial elastic stress/vectorial gas flux, $\boldsymbol{\sigma} \to \boldsymbol{v}$; solid density /gas compressibility $\rho \to \phi\beta^{-1}$; fourth rank elastic tensor/second rank tensor related to permeability $\boldsymbol{a} \to i\omega\mathscr{K}/\eta$. The fact that the "force-type" and "kinematic-type" variables play an inverse role in the two cases, results from the different nature of the balance equation that expresses the balance of momentum (for elastic media) or the mass balance (for porous media).

Similarly to elastic composites, the description (6.41) can be recast in terms of mass flux $\mathbf{v}$ instead of stress and, instead of momentum density, condensation b (following the Lafarge and Nemati, 2013, formulation) as follows:

$$\text{div}_x(\boldsymbol{v}) = i\omega b \quad \text{with} \quad \begin{pmatrix} v \\ b \end{pmatrix} = - \begin{pmatrix} \frac{\mathscr{K}(\omega)}{\eta} & ,0 \\ 0 & ,\frac{\phi}{\beta} \end{pmatrix} \begin{pmatrix} \cdot\nabla_x p \\ p \end{pmatrix}$$

Combining the mass balance and the dynamic Darcy law, and eliminating the flux leads to the wave equation satisfied by the pressure:

$$\text{div}\left(-\frac{\mathscr{K}(\boldsymbol{\omega})}{\eta} \cdot \nabla p\right) + i\omega\frac{\phi}{\beta}p = 0$$

Then the acoustic wavelength is

$$\Lambda(\omega) = O\left(\frac{2\pi}{\omega}\sqrt{\frac{|i\omega\mathscr{K}(\omega)|\beta}{\eta\phi}}\right)$$

Consequently, for $\omega \ll \omega_b$ the wave is of diffusion type and

$$\Lambda \approx 2\pi\sqrt{\frac{|\boldsymbol{\kappa}|\beta}{\omega\eta\phi}}$$

while for $\omega \ll \omega_c$ the wave is propagative and

$$\Lambda \approx \frac{2\pi}{\omega}\sqrt{\frac{\beta}{|\boldsymbol{\alpha}_\infty|\rho^e}}$$

We investigate hereafter two types of inner resonant rigid porous media saturated by air. First, we analyze double porosity media, i.e. a periodic bi-composite porous medium with a large contrast in permeability (instead of elastic stiffness) of the constituents (Auriault and Boutin, 1994; Boutin et al, 1998; Hornung and Showalter, 1990; Venegas and Umnova, 2011). This case leads to a local dynamic state governed by parabolic equations. Second, porous media with embedded Helmholtz resonators are examined. In that case the required effect is introduced by a geometrical contrast coming from the resonator morphology. This configuration enables to reach a local dynamic state governed by hyperbolic equations.

6.4.1 Double Porosity Media: Inner Resonance by High Permeability Contrast

Consider a double porosity medium, i.e., a periodic bi-porous composite analogous to the elastic bi-composites studied in Subsect. 6.3.2. By simplicity we keep the same notations: the $\mathscr{C}$-constituent of porosity ϕ_c and the $\mathscr{R}$-constituent of porosity ϕ_r occupy respectively the domains Ω_c and Ω_r interfaced by Γ. They present uniform but highly contrasted permeability tensors $\mathscr{K}_c$ and $\mathscr{K}_r$ respectively, and are saturated by the *same gas* of identical viscosity, density and compressibility at equilibrium, η, ρ^e and γP^e, respectively, see Fig. 6.6. We keep also the convention that the $\mathscr{C}$-constituent conveys the long wavelength and then undergoes a local quasi-static regime, while the $\mathscr{R}$-constituent experiences a local dynamic regime. Therefore, the $\mathscr{C}$-constituent must be connected and much more permeable than the connected or dispersed $\mathscr{R}$-constituent, i.e. $O(\boldsymbol{\kappa}_c) \gg O(\boldsymbol{\kappa}_r)$.

In co-dynamics regime the wavelength Λ_c in $\mathscr{C}$-constituent is large with respect to ℓ. As a consequence of the permeability contrast, the mass flux is mainly carried by the $\mathscr{C}$-constituent. Thus the order of magnitude of the macroscopic length L is

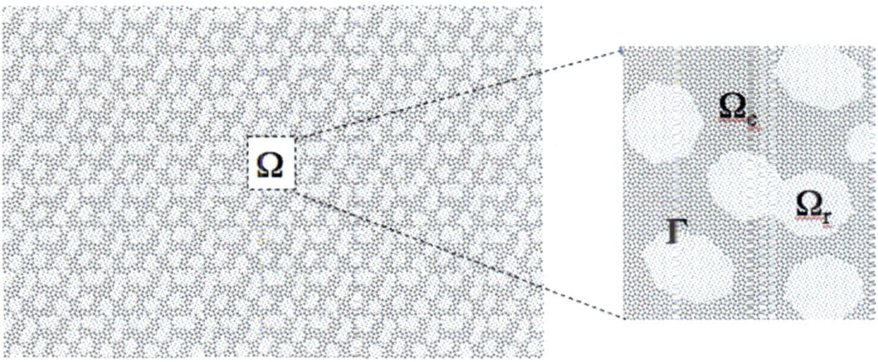

Fig. 6.6: Double porosity media

$$L = O(\frac{\Lambda_c}{2\pi}) = \frac{1}{\omega}\sqrt{\frac{\omega|\mathscr{K}_c|\beta}{\eta\phi_c}} \gg \ell \quad \text{i.e.} \quad \frac{\omega^2}{\beta} = \frac{\omega|\mathscr{K}_c|}{\eta\phi_c L^2}$$

while, the requirement of a dynamic regime in the $\mathscr{R}$-constituent at the period scale implies that

$$\frac{\Lambda_r}{2\pi} = \frac{1}{\omega}\sqrt{\frac{\omega|\mathscr{K}_r|\beta}{\eta\phi_r}} = O(\ell) \quad \text{i.e.} \quad \frac{\omega^2}{\beta} = \frac{\omega|\mathscr{K}_r|}{\eta\phi_r\ell^2} \tag{6.42}$$

Consequently a co-dynamics regime can be reached when the contrast of permeability tensors is:

$$\frac{|\mathscr{K}_r|}{|\mathscr{K}_c|}\frac{\phi_c}{\phi_r} = O(\frac{\ell^2}{L^2}) = \varepsilon^2 \ll 1 \tag{6.43}$$

Note that as the gas compressibility is identical in both constituents the condition (6.43) involves the permeability ratio only. Hence, in poro-acoustics there is no analogous of the case of significant density contrast encountered in elastic composites.

If the Darcy flow in both constituents is in inertial regime (i.e. $\omega > \omega_{bc}$ and $\omega > \omega_{br}$) the condition (6.43) simplifies into

$$\frac{|\boldsymbol{\alpha}_r|}{|\boldsymbol{\alpha}_c|} = \varepsilon^{-2} \gg 1$$

which is impossible to reach since the tortuosity $|\alpha_\infty|$ is a dimensionless factor O(1). However, (6.43) can be fulfilled provided that the Darcy flow in the low permeability $\mathscr{R}$-constituent is in viscous regime, i.e. $\omega \ll \omega_{br}$. Then, we will consider in the following that $\mathscr{K}_r = \boldsymbol{\kappa}_r(1 + O(\varepsilon))$, so that, at the leading order, we can approximate $\mathscr{K}_r$ by the real intrinsic permeability $\boldsymbol{\kappa}_r$. Consequently, the frequency range of interest as defined by (6.42) is

$$\omega = O(\omega_r), \qquad \omega_r = \frac{\beta |\boldsymbol{\kappa}_r|}{\eta \ell^2}$$

In compliance with the continuity condition on Γ, the pressure in both constituents are of the same order of magnitude $p_c = O(p_r)$. In accordance with the permeability contrast (6.43), the overall permeability is dominated by that of the $\mathscr{C}$-constituent which is taken as the reference value. Then, compared to this reference value, the permeability of the $\mathscr{R}$-constituent is rescaled as $\varepsilon^2 \mathscr{K}_r$. Thus, taking L as the characteristic length, the re-scaled governing equations accounting for the permeability contrast take the following form

$$\left\{\begin{array}{ll} \operatorname{div}(\dfrac{\mathscr{K}_c}{\eta} \cdot \boldsymbol{\nabla} p_c) = i\omega \dfrac{\phi_c}{\beta} p_c & \text{in } \Omega_c, \\ \operatorname{div}(\varepsilon^2 \dfrac{\mathscr{K}_r}{\eta} \cdot \boldsymbol{\nabla} p_r) = i\omega \dfrac{\phi_r}{\beta} p_r & \text{in } \Omega_r, \\ (\dfrac{\mathscr{K}_c}{\eta} \cdot \boldsymbol{\nabla} p_c - \varepsilon^2 \dfrac{\mathscr{K}_r}{\eta} \cdot \boldsymbol{\nabla} p_r) \cdot \boldsymbol{n} = 0 & \text{on } \Gamma, \\ p_r - p_c = 0 & \text{on } \Gamma, \\ p_r \quad \text{and} \quad p_c & \Omega - \text{periodic} \end{array}\right. \tag{6.44}$$

6.4.1.1 Homogenized Behavior

The homogenization process of this differential set is close to the one developed in elasticity.

- At the dominant order, the $\mathscr{C}$-constituent is governed by a *static* balance equation with *free* boundary condition. Therefore, consistently with a local quasi static regime of the $\mathscr{C}$-constituent, at the dominant order, the pressure is uniform,

$$p_c^{(0)} = P^{(0)}(\boldsymbol{x})$$

- The following order leads to a classic local problem of conduction in the $\mathscr{C}$-constituent with *zero flux* boundary conditions:

$$\left\{\begin{array}{ll} \operatorname{div}_y \left(\dfrac{\mathscr{K}_c}{\eta} \cdot (\boldsymbol{\nabla}_y p_c^{(1)} + \nabla_x P^{(0)}) \right) = 0 & \text{in } \Omega_c, \\ \dfrac{\mathscr{K}_c}{\eta} \cdot \left(\nabla_y p_c^{(1)} + \boldsymbol{\nabla}_x P^{(0)} \right) \cdot \boldsymbol{n} = 0 & \text{on } \Gamma, \\ p_c^{(1)} & \Omega - \text{periodic} \end{array}\right. \tag{6.45}$$

Hence, at this order, the $\mathscr{C}$-constituent behaves as if the $\mathscr{R}$-constituent were *impervious*. By linearity we have

$$\begin{cases} p_c^{(1)} = \boldsymbol{\theta}(\boldsymbol{y}) \cdot \boldsymbol{\nabla}_x P^{(0)} + P^{(1)}(\boldsymbol{x}), \\ \boldsymbol{v}_c^{(0)} = -\left(\dfrac{\mathscr{K}_c}{\eta} \cdot \boldsymbol{\nabla}_y \boldsymbol{\theta} + \dfrac{\mathscr{K}_c}{\eta}\right) . \boldsymbol{\nabla}_x P^{(0)} \end{cases} \tag{6.46}$$

where fields $\boldsymbol{\theta}^q(\boldsymbol{y})$ are the *real valued* particular solutions for unit pressure gradient, i.e. $\boldsymbol{\nabla}_x P_i^{(0)} = \delta_{iq}$.

- At the next order, the balance equation and the boundary conditions concerning the $\mathscr{C}$-constituent write

$$\begin{cases} \mathrm{div}_y(\boldsymbol{v}_c^{(1)}) + \mathrm{div}_x(\boldsymbol{v}_c^{(0)}) = -i\omega \dfrac{\phi_c}{\beta} P^{(0)} \quad \text{in } \Omega_c, \\ \boldsymbol{v}_c^{(1)} \cdot \boldsymbol{n} = -\dfrac{\boldsymbol{\kappa}_r}{\eta} \cdot \boldsymbol{\nabla}_y p_r^{(0)} \cdot \boldsymbol{n} \quad \text{on } \Gamma, \\ \boldsymbol{v}_c^{(1)} \quad \Omega - \text{periodic} \end{cases} \tag{6.47}$$

Integrating over Ω_c, and using the usual integral transformation and the Ω-periodicity yields

$$\mathrm{div}_x\left(-\frac{\boldsymbol{K}_{\mathscr{C}}}{\eta} \cdot \boldsymbol{\nabla}_x P^{(0)}\right) = -i\omega \frac{\phi_c}{\beta} \frac{|\Omega_c|}{|\Omega|} P^{(0)} - \frac{1}{|\Omega|} \int_\Gamma \left(\frac{\boldsymbol{\kappa}_r}{\eta} \cdot \boldsymbol{\nabla}_y p_r^{(0)}\right) \cdot \boldsymbol{n} \, d\Gamma \tag{6.48}$$

where the effective dynamic permeability $\boldsymbol{K}_c$ is *independent* of the (weak) permeability of the $\mathscr{R}$-constituent, which, at this order, appears as impervious

$$\boldsymbol{K}_{\mathscr{C}} = \langle \mathscr{K}_c \cdot \boldsymbol{\nabla}_y(\boldsymbol{\theta}) + \mathscr{K}_c \rangle_c$$

- To express the boundary flux term in (6.48), we focus now on the leading order problem that governs the response of the $\mathscr{R}$-constituent

$$\begin{cases} \mathrm{div}_y\left(\dfrac{\boldsymbol{\kappa}_r}{\eta} . \boldsymbol{\nabla}_y p_r^{(0)}\right) = i\omega \dfrac{\phi_r}{\beta} p_r^{(0)} \quad \text{in } \Omega_r, \\ p_r^{(0)} = P^{(0)}(\boldsymbol{x}) \quad \text{on } \Gamma, \\ p_r^{(0)} \quad \Omega - \text{periodic} \end{cases} \tag{6.49}$$

that, as expected, describes a local *dynamic* regime with *Dirichlet* condition. However, since $\boldsymbol{\kappa}_r$ is real valued, the dynamic phenomena is of *parabolic* type. Indeed, the mass transfer ruled by the viscous permeability is balanced by the gas compressibility. This induces a transient diffusion of pressure in the $\mathscr{R}$-constituent. By linearity, the pressure $p_r^{(0)}$ takes the form

$$p_r^{(0)} = P^{(0)} + \zeta(\boldsymbol{y}, \omega) P^{(0)}$$

the pressure field $1 + \zeta(\boldsymbol{y}, \omega)$ is the real particular solution for unit imposed pressure, i.e. $P^{(0)} = 1$, and is therefore the solution of:

$$\left\{\begin{array}{l} \mathrm{div}_y(\dfrac{\boldsymbol{\kappa}_r}{\eta}\cdot\boldsymbol{\nabla}_y\zeta) = i\omega\dfrac{\phi_r}{\beta}(1+\zeta) \quad \text{in } \Omega_\mathrm{r}, \\ \zeta = 0 \quad \text{on } \Gamma, \\ \zeta \qquad \Omega-\text{periodic} \end{array}\right. \tag{6.50}$$

Let us identify the features of the function ζ. From (6.50) the low and high frequencies limits read

$$\zeta \underset{\omega\to 0}{\rightarrow} 0 \quad \text{and} \quad \zeta \underset{\omega\to\infty}{\rightarrow} -1$$

The frequency dependence of field ζ is established by considering the eigen value problem associated to (6.50)

$$\left\{\begin{array}{l} \mathrm{div}_y(\dfrac{\boldsymbol{\kappa}_\mathrm{r}}{\eta}\cdot\boldsymbol{\nabla}_y\psi) = -\lambda\psi \quad \text{in } \Omega_\mathrm{r} \\ \psi = 0 \quad \text{on } \Gamma \\ \psi \qquad \Omega-\text{periodic} \end{array}\right. \tag{6.51}$$

The permeability tensor $\boldsymbol{\kappa}_\mathrm{r}$ being symmetric and coercive, the problem (6.51) presents a discrete series of positive eigenvalues and corresponding orthonormal eigenmodes $\{\lambda_J, \psi^J\}$ such that $\langle\psi^J.\psi^K\rangle_\mathrm{r} = \delta_{JK}$. Then, the expressions of ζ and $\langle\zeta\rangle_\mathrm{r}$ are derived by modal decomposition and read

$$\zeta(\boldsymbol{y},\omega) = \sum_{J=1}^{\infty}\frac{\boldsymbol{\psi}^J\otimes\langle\boldsymbol{\psi}^J\rangle_\mathrm{r}}{\dfrac{\omega_J}{i\omega}-1}; \qquad \langle\zeta\rangle_\mathrm{r} = \sum_{J=1}^{\infty}\frac{\langle\boldsymbol{\psi}^J\rangle_\mathrm{r}\otimes\langle\boldsymbol{\psi}^J\rangle_\mathrm{r}}{\dfrac{\omega_J}{i\omega}-1}; \qquad \omega_J = \frac{\lambda_J\beta}{\phi_\mathrm{r}}$$

Hence, $\langle\zeta\rangle_\mathrm{r}$ takes complex values and varies continuously without poles nor zero values (for $\omega > 0$), from $\langle\zeta(0)\rangle_\mathrm{r} = 0$ at $\omega = 0$ to $\langle\zeta(\infty)\rangle_\mathrm{r} = -1$ when $\omega \to \infty$.

- Finally, following the same procedure as for elastic media, we derive the macroscopic description:

$$\left\{\begin{array}{l} \mathrm{div}_x(\dfrac{\boldsymbol{K}_\mathscr{C}}{\eta}\cdot\boldsymbol{\nabla}_x P^{(0)}) = i\omega\dfrac{P^{(0)}}{B(\omega)}, \\ \dfrac{1}{B(\omega)} = \dfrac{1}{\beta}\left(\dfrac{\phi_\mathrm{c}|\Omega_\mathrm{c}| + \phi_\mathrm{r}|\Omega_\mathrm{r}|}{|\Omega|} + \langle\zeta\rangle_\mathrm{r}\right) \end{array}\right. \tag{6.52}$$

The description combines conventional dynamic Darcy flow, characterized by the complex permeability tensor $\boldsymbol{K}_\mathscr{C}$ of the material made of the $\mathscr{C}$-constituent and impervious $\mathscr{R}$-constituent, and by a non conventional compressibility associated to the frequency dependent density bulk modulus $B(\omega)$. The formal similarity with elastic composites (6.24), is obvious when the description is rewritten in the form of conservation equation, involving the mean mass flux $\langle\boldsymbol{v}^{(0)}\rangle$ and the mean condensation $\langle b^{(0)}\rangle$:

$$\mathrm{div}_x(\langle \mathbf{v}^{(0)}\rangle) = i\omega\langle b^{(0)}\rangle \quad \text{with} \quad \begin{pmatrix} \langle \boldsymbol{v}^{(0)}\rangle \\ \langle b^{(0)}\rangle \end{pmatrix} = -\begin{pmatrix} \dfrac{\boldsymbol{K}_{\mathscr{C}}}{\eta} & 0 \\ 0 & \dfrac{1}{B(\omega)} \end{pmatrix} \begin{pmatrix} \cdot\nabla_x P^{(0)} \\ P^{(0)} \end{pmatrix} \tag{6.53}$$

6.4.1.2 Comments and Generalization to Other Diffusion Phenomena

As suggested by the elasto-dynamic/poro-acoustic analogy, instead of obtaining a frequency dependent density, one obtains here a frequency dependent bulk modulus. However, the different hyperbolic or parabolic nature of the local problems leads to different physical effects, meanwhile the similarity in the formulations, see (6.23) and (6.52). As further similarities, note (i) that conversely to standard poro-acoustic models, the pressure $P^{(0)}$ (that plays the role of the displacement in elastic composites) is not the mean pressure of the cell, but that prevailing in the $\mathscr{C}$-constituent, and (ii) that the description (6.52) relies also on the fact that the flux pulsed by the $\mathscr{R}$-constituent is of one order smaller than that carried by the $\mathscr{C}$-constituent.

The effective bulk modulus $B(\omega)$, as $\langle\zeta(\omega)\rangle_r$, is complex valued and varies continuously without singularity according to the frequency. Due to the dissipative character of the "parabolic resonance", instead of the conservative character of hyperbolic resonance, there is no resonance-singularity for a series of specific frequencies, but a wide-band effect centered around the frequency of the fundamental mode

$$\omega_1 \approx \omega_{\mathrm{r}} = O(\frac{\beta|\boldsymbol{\kappa}_{\mathrm{r}}|}{\eta\ell^2})$$

Consequently no band-gap occurs, yet a significant increase of dissipation appears around ω_{r}. This latter frequency is tunable by playing either on the permeability value and/or on the size of the $\mathscr{R}$-constituent. Since the local problem (6.50) is scalar, the determination of the effective complex modulus can be achieved analytically for simple geometries. For example, if the $\mathscr{R}$-constituent is a sphere of diameter R made of an isotropic media of intrinsic permeability κ_{r} one has:

$$\langle\zeta(\omega)\rangle_{\mathrm{r}} = 1 + 3\left(\frac{1}{i\omega^*} - \frac{\coth(\sqrt{i\omega^*})}{\sqrt{i\omega^*}}\right) \quad ; \quad \omega^* = \frac{\omega}{\omega_{\mathrm{r}}} \quad \text{with} \quad \omega_{\mathrm{r}} = \frac{\beta\kappa_{\mathrm{r}}}{\eta R^2}$$

Assuming in addition that the flow within the $\mathscr{C}$-constituent is dominated by viscosity, then $\boldsymbol{K}$ is real. Therefore, transposed in the time domain, the macroscopic description becomes

$$\mathrm{div}_x(\frac{\boldsymbol{K}_{\mathscr{C}}}{\eta}\cdot\nabla_x P^{(0)}) = \frac{1}{\beta}\left(\frac{\phi_{\mathrm{c}}|\Omega_{\mathrm{c}}| + \phi_{\mathrm{r}}|\Omega_{\mathrm{r}}|}{|\Omega|}\frac{\partial P^{(0)}}{\partial t} + \langle\zeta\rangle_{\mathrm{r}} * \frac{\partial P^{(0)}}{\partial t}\right) \tag{6.54}$$

The convolution product introduces a memory effect. As in elastic composites, the convolution can be "replaced" by the introduction of an infinite set of microscopic descriptors of the medium $\{b^J(x,t)\}$ associated to the modal decomposition of ζ. The equivalent coupled linear differential set of (6.52) enabling the resolution in parallel of the time/space evolution of the macroscopic pressure $P^{(0)}(x,t)$ and of the microscopic descriptors $b^J(x,t)$ reads:

$$\begin{cases} \mathrm{div}_x(\dfrac{\boldsymbol{K}_{\mathscr{C}}}{\eta} \cdot \boldsymbol{\nabla}_x P^{(0)}) = \dfrac{1}{\beta} \left(\dfrac{\phi_\mathrm{c}|\Omega_\mathrm{c}| + \phi_\mathrm{r}|\Omega_\mathrm{r}|}{|\Omega|} \dfrac{\partial P^{(0)}}{\partial t} + \sum_{J=1}^{\infty} \dfrac{\partial b^J}{\partial t} \langle \psi^J \rangle_\mathrm{r} \right), \\ b^J \omega_J + \dfrac{\partial b^J}{\partial t} = \langle \dfrac{\partial P^{(0)}}{\partial t} . \psi^J \rangle_\mathrm{r} \end{cases} \tag{6.55}$$

The parabolic inner-resonance description (6.52) has been confirmed experimentally (Olny and Boutin, 2003): measurements realized on double porosity materials show the high absorption performance of such materials and the tunability of the frequency range of high dissipation.

In the poro-acoustic context a co-dynamic regime with anti-resonance cannot be reached for two reasons: firstly, as the gas is identical in the two pores networks, no compressibility contrast can be introduced, secondly, the complex term $\langle 1+\zeta \rangle_r$ never vanishes whatever the frequency is. Similar developments and outcomes can be obtained for other dissipative phenomena governed by parabolic equations. For instance, the same conclusions were established more than 30 years ago in Auriault (1983) dealing with double conductivity in thermal transfer. In that case the physical analogs of the pressure, the mass flux, the permeability and the compressibility become respectively the temperature, the heat flux, the conductivity and the specific volume heat capacity (in standard notation ρc_p). Then the effective double conductivity media present a complex frequency dependent specific volume heat capacity, associated to a non uniform temperature field in the low conducting $\mathscr{R}$-constituent. One may also mention molecular diffusion in double porosity media that leads to Non-Fickean diffusion process (Auriault and Lewandowska, 1995), or situations mixing diffusion and mass transfer.

6.4.2 Embedded Resonators in Porous Media: Inner Resonance by Geometrical Contrast

We consider in this section a bi-composite porous medium in which the $\mathscr{R}$-constituent is governed by an elasto-inertial dynamics of hyperbolic nature. To realize this, one may think about introducing in the pores matrix cavities significantly larger than the pores size. However, the resonance of sufficiently regular cavities occurs when the acoustic wavelength is of the order of magnitude of the size of the cavity. This is not compatible with a co-dynamic regime since the acoustic wavelengths in the matrix and in the resonating cavity would be of the same order. To overcome this difficulty, we can modify the resonance of the cavity by introducing a geometrical contrast in its morphology. This is performed by considering Helmholtz resonators, periodically embedded in the porous $\mathscr{C}$-constituent (Boutin, 2013), see Fig. 6.7.

6.4.2.1 Helmholtz Resonator

An Helmholtz resonator Ω_r is made of a 'chamber' with rigid and impervious wall connected to a constricted duct Ω' of length $\ell' = O(\ell)$ which is open toward the outer porous medium. This specific morphology presents a geometrical contrast (Fig. 6.7) between the section $|\mathrm{s}|$ of the duct and the one of the chamber whose order is $O(\ell^2)$. As a consequence, a particular dynamic regime occurs where the mass and stiffness effects are uncoupled since the inertia is localized in the duct in which the mass of the gas moves almost as a rigid body, while the stiffness is localized in the chamber where the gas is almost uniformly compressed. This situation corresponds to a spring-mass system and departs significantly from standard resonances where the mass and stiffness are distributed in the whole cavity.

The chamber spring k and duct mass m are straightforwardly given by $k = \beta\ |\mathrm{s}|^2/|\Omega_\mathrm{r}|$ and $m = \rho^e|\Omega'|$. Hence, the eigenfrequency $f_0 = \omega_0/2\pi$ of the resonator reads ($C^e = \sqrt{\beta/\rho^e}$ is the sound velocity in air):

$$\omega_0 = \sqrt{\frac{k}{m}} = \frac{C^e}{\delta}; \quad \delta = \frac{\sqrt{|\Omega_\mathrm{r}||\Omega'|}}{|\mathrm{s}|} = \ell O(\frac{\ell}{\sqrt{|\mathrm{s}|}})$$

Since $|\mathrm{s}|/\ell^2 \ll 1$, then $\delta \gg \ell$. Consequently, the resonance occurs at a frequency much lower than the diffraction frequency of the periodic medium, namely $\omega_0 = C^e/\delta \ll \omega_d = O(C^e/\ell)$. Note that this latter estimate of ω_d assumes that the flow regime in the porous matrix is dominated by inertia and thus the sound celerity is of the order of $C^e/\sqrt{\alpha_{C\infty}} = O(C^e)$. In other words, at the Helmholtz eigenfrequency, the reduced wavelength in the porous matrix is $\Lambda/(2\pi) = O(C^e/\omega_0) = O(\delta) \gg \ell$ hence, both requirements of scale separation and co-dynamics regime are fulfilled.

The classic undamped response (disregarding the viscous dissipation on the duct) of an Helmholtz resonator to an external harmonic perturbation of pressure P, is immediately derived from the following balance equation of the duct mass undergoing

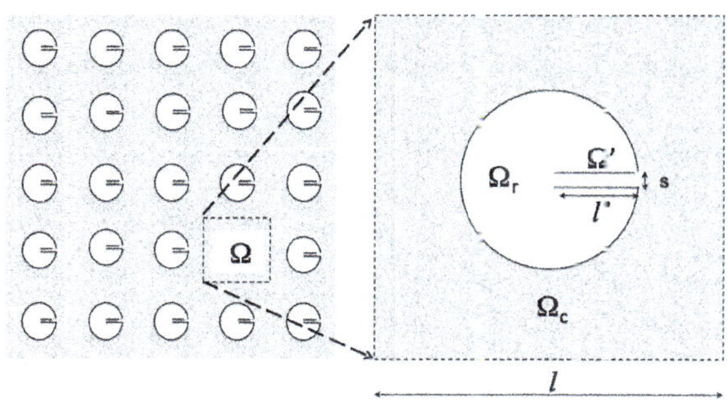

Fig. 6.7 Porous medium with a periodic array of embedded Helmholtz resonators

a rigid body motion u_r,

$$m\omega^2 u_r - k u_r = Ps$$

Thus, introducing the above mentioned expressions of k and m the outgoing flux induced by the external pressure P reads

$$q_r(P) = i\omega u_r|s| = -i\omega\frac{P}{\beta}\frac{|\Omega_r|}{1-(\omega/\omega_0)^2} \tag{6.56}$$

6.4.2.2 Homogenized Behavior

As previously, the $\mathscr{C}$-constituent of porosity ϕ_c is described by the classic poro-acoustic formulation. Now, to account for the geometrical contrast in the asymptotic expansions, we specify that

$$|s| = O(\varepsilon\ell^2)$$

The boundary condition at the interface Γ, is split into

i) an impervious condition on $\Gamma - s$ corresponding to the wall of the Helmholtz resonator, and
ii) the conditions of continuity of pressure and of flux on the aperture s.

This latter condition implies that $O(|v_r||s|) = O(|v_c||s|)$ and thus, accounting for the geometrical contrast, the flux pulsed by the resonator is of one order smaller than the flux carried through the $\mathscr{C}$-constituent, since $O(v_r||s/|v_c|\ell^2) = O(|s|/\ell^2) = \varepsilon \ll 1$. Consistently, q_r must be rescaled as $\varepsilon q_r(p_c)$. This analysis leads to the following rescaled differential set

$$\begin{cases} \mathrm{div}(\dfrac{\mathscr{K}_c}{\eta}\cdot\nabla p_c) = i\omega\dfrac{\phi_c}{\beta}p_c \quad \text{in } \Omega_c \\ (\dfrac{\mathscr{K}_c}{\eta}\cdot\nabla p_c)\cdot\boldsymbol{n}_c = 0 \quad \text{on } \Gamma - s \\ (\dfrac{\mathscr{K}_c}{\eta}\cdot\nabla p_c)\cdot\boldsymbol{n}_c = -\varepsilon v_r(p_c); \quad |s|v_r = q_r(p_c) \quad \text{on s} \\ \qquad p_c \qquad \Omega - \text{periodic} \end{cases} \tag{6.57}$$

When processing homogenization, since the flux brought by the resonator is of one order smaller than the one carried by the $\mathscr{C}$-constituent, the first and second problems in the $\mathscr{C}$-constituent remain unchanged compared to the double porosity case. The difference arises on the global mass balance, only, that takes now the form

$$\mathrm{div}_x(\frac{\boldsymbol{K}_{\mathscr{C}}}{\eta}\cdot\nabla_x P^{(0)}) = i\omega\frac{|\Omega_c|\phi_c}{|\Omega|}\frac{P^{(0)}}{\beta} - \frac{1}{|\Omega|}\int_s \boldsymbol{v}_c\cdot\boldsymbol{n}_c \,\mathrm{d}\Gamma$$

and, according to the flux continuity the last integral equals $-q_r(p_c)$. Consequently, reporting the expression (6.56) of the pulsed flux, the macroscopic description writes:

$$\begin{cases} \mathrm{div}_x\left(\dfrac{\boldsymbol{K}_{\mathscr{C}}}{\eta}\cdot\boldsymbol{\nabla}_x P^{(0)}\right) = i\omega\dfrac{P^{(0)}}{\widehat{B}(\omega)}, \\ \dfrac{1}{\widehat{B}(\omega)} = \dfrac{1}{\beta}\left(\dfrac{|\Omega_c|\phi_c}{|\Omega|} + \dfrac{|\Omega_r|}{|\Omega|}\dfrac{1}{1-(\omega/\omega_0)^2}\right) \end{cases} \tag{6.58}$$

The description involves the same conventional dynamic Darcy flow, than that obtained in the double porosity case (for the same geometry and parameters of the $\mathscr{C}$-constituent), but presents a non conventional frequency dependent compressibility. Formally, this is the same description as (6.53) except that the effective modulus $B(\omega)$ is now replaced by $\widehat{B}(\omega)$.

6.4.2.3 Comments

Around resonance, the resonator brings a negative contribution to the effective bulk modulus $\widehat{B}(\omega)$. Then $\widehat{B}(\omega)$ is negative in a broad "atypical band" $[\omega_0, \omega_0^*]$

$$\begin{cases} \widehat{B}(\omega_0) = 0 \quad ; \quad \widehat{B}(\omega_{0_\pm}^*) = \pm\infty, \\ \widehat{B}(\omega) \leq 0 \quad \text{for} \quad \omega_0 \leq \omega \leq \omega_0^* = \omega_0\sqrt{1+\dfrac{|\Omega_r|}{|\Omega_c|\phi_c}}, \\ \widehat{B}(\omega) \geq 0 \quad \text{for} \quad \omega \leq \omega_0 \quad \text{and} \quad \omega \geq \omega_0^* \end{cases} \tag{6.59}$$

Hence a broad band gap along with strongly dispersed waves occurs, and is tunable with the parameters of the porous matrix and the resonators. As for the elastic case, the local hyperbolic problem introduces an unbounded effective parameter that takes real positive or negative values. However, as the material is governed by a mass balance, the inner resonance effect induces an unconventional effective bulk modulus. This latter presents a single pole due to the single eigenfrequency of the resonator. Indeed the higher eigenfrequencies of the resonator are related to the modes of the chamber, which, as previously explained, do not respect the scale separation.

The "co-dynamics" condition is naturally fulfilled when $\omega_0 > \omega_{bc}$, i.e. when the flow in the $\mathscr{C}$-constituent is driven by inertia at the pore scale (much smaller than ℓ). From the expressions of ω_0 and ω_{bc} this situation arises when the $\mathscr{C}$-constituent is sufficiently permeable, as defined by:

$$\frac{\mathscr{K}_c \alpha_{c_\infty}}{\phi_c} > \frac{\eta}{\rho^e C^e}\delta$$

where $\mathscr{K}_c$ stands for the modulus of $\boldsymbol{\mathscr{K}}_c$. Conversely, when $\omega_0 < \omega_{bc}$, which means that the flow in the $\mathscr{C}$-constituent is in a viscous regime, then "co-dynamics" condition is rather restrictive. Indeed, in this situation the scale separation must also be satisfied, that is $\Lambda_c(\omega_0)/(2\pi) \gg \ell$. Hence, in this case, we have the two requirements:

$$\frac{\mathscr{K}_c \alpha_{c\infty}}{\phi_c} < \frac{\eta}{\rho^e C^e}\delta; \qquad \sqrt{\mathscr{K}_c P^e/(\omega_0 \eta \phi_c)} \gg \ell$$

that can be rewritten as:

$$\alpha_{c\infty}\frac{\ell}{\delta} \ll \frac{\mathscr{K}_c \alpha_{c\infty}}{\phi_c}\frac{\rho^e C^e}{\eta}\frac{1}{\ell} < \frac{\delta}{\ell}$$

and as $\alpha_{c\infty} = O(1)$, $\ell/\delta = \sqrt{|\Sigma|}/\ell = O(\sqrt{\varepsilon})$ then

$$O(\sqrt{\varepsilon}) \ll \frac{\mathscr{K}_c \alpha_{c\infty}}{\ell \phi_r}\frac{\rho^e C^e}{\eta} < O(\frac{1}{\sqrt{\varepsilon}})$$

and therefore we must necessarily have

$$\frac{\mathscr{K}_c \alpha_{c\infty}}{\ell \phi_r}\frac{\rho^e C^e}{\eta} = O(1)$$

Consequently, both viscous or inertial regimes in the $\mathscr{C}$-constituent lead to the conclusion that "co-dynamics" situations can be reached only when:

$$\frac{\mathscr{K}_c \alpha_{c\infty}}{\phi_r} \geq \frac{\eta}{\rho^e C^e}O(\ell) \quad \text{i.e.,} \quad \omega_0/\omega_{bc} \geq O(\ell/\delta)$$

which means that the Darcy flow in the porous $\mathscr{C}$-constituent may be in inertial or visco-inertial regime. Conversely, if the pores network is too resistive, i.e. when the Helmholtz resonance belongs to the almost purely viscous regime ($\omega_0/\omega_{bc} \ll 1$), then this outside resistance avoids the resonance phenomenon, and consequently the co-dynamics regime is not possible.

Several other morphological options can lead to inner resonance. Whatever the case, the resonator design should be such that the resonating effect induces a small flux (on average over the period) compared to the macroscopic flux. For instance, instead of having a single type of oscillator, the period may contain several different oscillators, or additional impervious inclusions (of the same magnitude as the size of the resonator). Instead of the spring-mass resonance of Helmholtz resonators, one may also consider quarter wavelength resonators. In a frequency range such that the length l_r of a resonating tube of small section s matches the quarter wavelength of the acoustic wave, a resonance occurs. Nevertheless, to ensure the scale separation necessary for the co-dynamic regime, the size of the resonating domain should be much smaller than the wavelength. This implies to rearrange the tube in a compact domain Ω_r of characteristic size ℓ, for instance by wrapping the tube. In that case the flux pulsed by the resonator reads

$$q_r(P) = i\omega P \frac{|s|}{\rho^e C^e}\tan(\frac{\omega l_r}{C^e})$$

and the compressibility can present several poles at the eigenmodes frequencies.

Negative bulk modulus has been actually observed experimentally, either in a tube connected to a 1D array of resonators (Fang et al, 2006) or in 3D materials made of a packing of resonators, each of them being made of an impervious spherical shell drilled and connected to a constricted tube (Boutin and Becot, 2015). Note also that in acoustics the membrane based meta-materials e.g. Naify et al (2012); Yang et al (2008) leads to unconventional densities, as they are governed by momentum balance.

To sum up it is clear that for *fluid* in porous media a non conventional bulk modulus can appear. However, these examples do not disclose the possibility of having non conventional stiffness of a *solid* material by inner resonance effect. One of the possible way to reach this situation is presented in the next section.

6.5 Inner Resonance in Poroelastic Media: Coupling Effect

Poro-elastic media are governed by the Biot model (Biot, 1956). This latter involves the two types of balance, namely of momentum and of mass, that couple the displacements and the pressure. For this reason, one may expect that an inner resonance phenomenon related to one type of balance, entails a non conventional parameter on the other type of balance also. This idea is illustrated below in the case of double porosity poro-elastic media.

6.5.1 Double Porosity Poro-Elastic Media - Problem Statement

We investigate the behavior of a fluid-saturated deformable elastic double porosity medium, see Fig. 6.8. The medium is periodic of period $\Omega = \Omega_\mathrm{b} \cup \Omega_\mathrm{f}$. Ω_b is the domain occupied by the fluid-saturated microporous poroelastic medium of porosity ϕ (instead of being rigid as studied in Sect. 6.4), and Ω_f is the pore space saturated by the same fluid and presenting the porosity $\phi_\mathrm{f} = |\Omega_\mathrm{f}|/|\Omega|$. For simplicity, we focus on a frequency range sufficiently low so that the flow regimes in the pores and the micropores are dominated by the viscosity. This means that the permeabilities of the two pores networks take the real values $\boldsymbol{\kappa}_\mathrm{b}/\eta$ and $\boldsymbol{\kappa}_\mathrm{f}/\eta$. Following the above analysis on rigid double porosity media, we assume that the permeabilities are highly contrasted, i.e., $\boldsymbol{\kappa}_\mathrm{b}/\boldsymbol{\kappa}_\mathrm{f} = O(\varepsilon^2)$. Such media were studied in Auriault and Boutin (1994) and we sumarize here below the approach based on reiterated homogenization presented in Boutin and Royer (2015).

The microporous poroelastic medium is governed by the classical Biot model (Biot, 1956). This latter consists in the momentum balance (6.60a), the poroelastic constitutive law (6.60b), the conservation of fluid mass (6.60c) and the rescaled Darcy's law (6.60d), which read in harmonic regime (to lighten the notations the variables in Ω_b are not indexed)

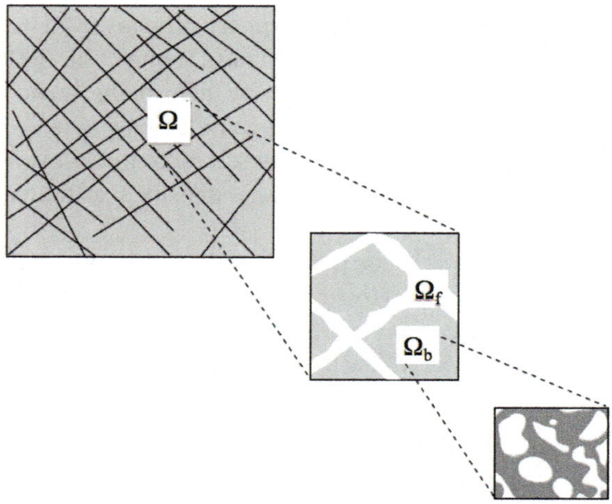

Fig. 6.8 A period Ω of double porosity media representative of the morphology of fissured microporous medium

$$\mathrm{div}(\boldsymbol{\Sigma}) = -\omega^2(1-\phi)\rho_s\boldsymbol{u} + i\omega\phi\rho_f\boldsymbol{v} \tag{6.60a}$$

$$\boldsymbol{\Sigma} = \boldsymbol{c} : \boldsymbol{e}(\boldsymbol{u}) - \boldsymbol{\alpha} p \tag{6.60b}$$

$$\mathrm{div}(\boldsymbol{q}) = i\omega\left(-\boldsymbol{\alpha} : \boldsymbol{e}(\boldsymbol{u}) - \frac{p}{M}\right) \tag{6.60c}$$

$$\boldsymbol{q} = \phi(\boldsymbol{v} - i\omega\boldsymbol{u}) = -\varepsilon^2\frac{\boldsymbol{\kappa}}{\eta}\cdot\nabla p \tag{6.60d}$$

where in addition to the notations already introduced previously:

- $\boldsymbol{\alpha}$ represents the symmetric and positive Biot coupling tensor. For isotropic media, this tensor is diagonal and all its components are scalar and identical. They comply with: $\phi \leq \alpha \leq 1$.
- $1/M$ is the Biot's bulk modulus:

$$\frac{1}{M} = \frac{\alpha - \phi}{K_s} + \frac{\phi}{\beta},$$

  where K_s and β are the bulk modulus of elastic material forming the microporous matrix and of the saturating fluid, respectively;
- $\boldsymbol{u}$ is the solid displacement, while $\boldsymbol{v}$ stands for the mean fluid velocity within the volume of the micropores. Darcy's flux is thus given by: $\boldsymbol{q} = \phi(\boldsymbol{v} - i\omega\boldsymbol{u})$;
- $\boldsymbol{\Sigma}$, $\boldsymbol{\sigma} = \boldsymbol{c} : \boldsymbol{e}(\boldsymbol{u})$ and p represent the tensor of total stress, the tensor of effective stress and the interstitial pressure, respectively; $\boldsymbol{c}$ is is the elastic tensor of the drained porous media;
- ρ_s and ρ_f are the density of the solid and of the fluid, respectively.

In the fluid domain Ω_f, the flow is governed by the Darcy law and the mass balance, that read:

$$q_f = \phi_f(v_f - i\omega u) = -\frac{\kappa_f}{\eta} \cdot \nabla p_f, \tag{6.61a}$$

$$\mathrm{div}(q_f) = -i\omega \frac{p_f}{\beta} \tag{6.61b}$$

On the interface, we express the continuity of normal stresses (6.62a), of pressures (6.62b), and of the fluid mass fluxes (6.62c) as:

$$\Sigma \cdot n = p_f \cdot n, \tag{6.62a}$$

$$p = p_f, \tag{6.62b}$$

$$q \cdot n = \phi(v - i\omega u) \cdot n = (v_f - i\omega u) \cdot n = q_f \cdot n \tag{6.62c}$$

6.5.2 Homogenized Behavior

We do not present the homogenization procedure; the detailed developments can be found in Boutin and Royer (2015). The leading order leads to an uniform solid motion $u^{(0)} = U(x)$ of the solid and to an uniform pressure in the fluid domain $p_f = P_f(x)$. Hence the Darcy law at the leading order reads

$$Q_f = \phi_f(V_f - i\omega U) = -\frac{\kappa_f}{\eta} \cdot \nabla P_f$$

In the microporous domain, the first order problem takes the form

$$\begin{cases} \mathbf{div}_y\left(c : (e_x(U) + e_y(u^{(1)})) - \alpha p^{(0)}\right) = 0 & \text{in } \Omega \\ \left(c : \left(e_x(U) + e_y(u^{(1)})\right) - \alpha p^{(0)}\right) \cdot n = -P_f n & \text{on } \Gamma \\ \mathrm{div}_y\left(\left(\frac{\kappa_b}{\eta}\right) \cdot \nabla_y p^{(0)}\right) = i\omega\left(\alpha : \left(e_x(U) + e_y(u^{(1)})\right) + \frac{1}{M} p^{(0)}\right) & \text{in } \Omega \\ p^{(0)} = P_f & \text{on } \Gamma \\ u^{(1)} \text{ and } p^{(0)}, & \widehat{\Omega}-\text{periodic} \end{cases} \tag{6.63}$$

This differential set coupling variables $u^{(1)}$ and $p^{(0)}$ describes a non stationary local regime forced by both the macro deformation of the solid matrix $e_x(U)$ and the pore pressure P_f. Due to the pressure/displacement coupling, the non-stationarity resulting from the transient pressure diffusion in the microporous Biot matrix concerns both variables $u^{(1)}$ and $p^{(0)}$. By linearity the local fields read:

$$\begin{cases} u^{(1)} = \tilde{u} + u_0^{(1)} = (\xi_0(y) + \tilde{\xi}(y,\omega)) : e_x(U) - (\zeta_0(y) + \tilde{\zeta}(y,\omega)) P_f, \\ p^{(0)} = \tilde{p} + P_f = \tilde{\theta}(y,\omega) : e_x(U) + (1 - \tilde{\varpi}(y,\omega)) P_f \end{cases} \tag{6.64}$$

In the above equations, the particular local fields are decomposed into their long term parts $(\boldsymbol{u}_0^{(1)}, P_{\mathrm{f}})$, (*i.e.* when $\omega = 0$, namely, $\boldsymbol{\xi}_0(\boldsymbol{y})$ and $\boldsymbol{\zeta}_0(\boldsymbol{y})$) and their transient parts $((\widetilde{\boldsymbol{u}}, \widetilde{p})$, namely, $\widetilde{\boldsymbol{\xi}}(\boldsymbol{y},\omega)$, $\widetilde{\boldsymbol{\zeta}}(\boldsymbol{y},\omega)$, $\widetilde{\boldsymbol{\theta}}(\boldsymbol{y},\omega)$ and $\widetilde{\boldsymbol{\varpi}}(\boldsymbol{y},\omega)$. Note that the micropore pressure is not uniform, except at both limits of zero frequency and of "infinite" frequency. In this latter case, a boundary layer arises at the pore/micropore interface Γ, which enables the sharp transition between both uniform pressure fields.

Finally, the macroscopic model is established from the momentum and mass balances of the local second order problems in both domains. Integrating these balance equations over each volume, and accounting for the boundary condition and periodicity leads to the balance equations governing the double porosity medium:

- As for the momentum balance one obtains

$$\mathbf{div}_x(\boldsymbol{S}) = -\omega^2(1-\phi_{\mathrm{f}})((1-\phi)\rho_s + \phi\rho_{\mathrm{f}})\boldsymbol{U}) + i\omega\phi_{\mathrm{f}}\rho_{\mathrm{f}}\boldsymbol{V}_{\mathrm{f}}$$

with

$$\boldsymbol{S} = \frac{1}{\Omega}\int_{\Omega_{\mathrm{b}}} \boldsymbol{\Sigma}\, d\Omega - \phi_{\mathrm{f}} P_{\mathrm{f}} \boldsymbol{I}$$

where $\boldsymbol{\Sigma} = \boldsymbol{c} : \left(\boldsymbol{e}_x(\boldsymbol{U}) + \boldsymbol{e}_y(\boldsymbol{u}^{(1)})\right) - \boldsymbol{\alpha} p^{(0)}$ is the total stress in the microporous domain and where $\boldsymbol{V}_{\mathrm{f}}$ is the macroscopic fluid velocity.
Recall that $\boldsymbol{u}^{(1)}$ and $p^{(0)}$ given by (6.64) are the frequency dependent solutions of the pressure/displacement localization problem 6.63. Hence, the constitutive law for the macroscopic total stress tensor reads:

$$\boldsymbol{S} = \boldsymbol{C}(\omega) : \boldsymbol{e}_x(\boldsymbol{U}) - \boldsymbol{A}(\omega) P_{\mathrm{f}} = \left(\boldsymbol{C}_0 + \widetilde{\boldsymbol{C}}(\omega)\right) : \boldsymbol{e}_x(\boldsymbol{U}) - \left(\boldsymbol{A}_0 + \widetilde{\boldsymbol{A}}(\omega)\right) P_{\mathrm{f}} \quad (6.65)$$

where the effective tensors are decomposed into their long term and transient contribution. The homogenized tensors $\boldsymbol{C}(\omega), \widetilde{\boldsymbol{C}}(\omega), \widetilde{\boldsymbol{A}}(\omega)$ are obtained by integration of the localization tensors appearing in Eq. 6.64.

- As for the mass balance

$$\phi_{\mathrm{f}} \mathrm{div}_x(\underline{V}_{\mathrm{f}} - i\omega \boldsymbol{U}) - i\omega \Bigg(\phi_{\mathrm{f}}(\mathrm{div}_x(\boldsymbol{U}) + \frac{P_{\mathrm{f}}}{\beta})$$
$$+ \frac{1}{|\Omega|}\int_{\Omega_{\mathrm{b}}} (\boldsymbol{\alpha} : (\boldsymbol{e}_{\mathrm{y}}(\boldsymbol{u}^{(1)}) + \boldsymbol{e}_{\mathrm{x}}(\boldsymbol{U})) - \mathrm{div}_{\mathrm{y}}(\boldsymbol{u}^{(1)}) + \frac{p^{(0)}}{M}) \Bigg) d\Omega \quad (6.66)$$

Replacing $\boldsymbol{u}^{(1)}$ and $p^{(0)}$ by their expressions (6.64), the macroscopic mass balance reads:

$$\phi_{\mathrm{f}} \mathrm{div}_x(\underline{V}_{\mathrm{f}} - i\omega \boldsymbol{U}) = i\omega \left(-\boldsymbol{A}(\omega) : \boldsymbol{e}_x(\boldsymbol{U}) - \frac{P_{\mathrm{f}}}{\mathcal{M}(\omega)} \right) \quad (6.67)$$

As above, the effective bulk stiffness $\mathcal{M}(\omega)$ can be decomposed into its long term and transient contributions.

To sum up the double porosity poro-elastic medium is described by the following set

$$\begin{cases} \mathrm{div}(\boldsymbol{\Sigma}) = -\omega^2\left((1-\phi_f)((1-\phi)\rho_s+\phi\rho_f)\boldsymbol{U}\right)+i\omega\phi_f\rho_f\boldsymbol{V}_f, \\ \boldsymbol{\Sigma} = \boldsymbol{C}(\omega):\boldsymbol{e}(\boldsymbol{U})-\boldsymbol{A}(\omega)P_f, \\ \mathrm{div}(\boldsymbol{Q}) = i\omega\left(-\boldsymbol{A}(\omega):\boldsymbol{e}(\boldsymbol{U})-\dfrac{P_f}{\mathscr{M}(\omega)}\right), \\ \boldsymbol{Q} = \phi(\boldsymbol{V}_f-i\omega\boldsymbol{U}) = -\dfrac{\boldsymbol{K}_f}{\eta}\cdot\nabla P_f \end{cases} \tag{6.68}$$

6.5.3 Comments

This description is similar to a Biot description except that the effective parameters $\boldsymbol{C}, \boldsymbol{A}, \mathscr{M}$ are now frequency dependent. Thus the macroscopic behavior is that of an equivalent porous medium with viscoelastic skeleton, saturated by a viscoelastic fluid (in compression). The permeability is that of the highly permeable fissure network, and the density is the mean density of the medium. Consistently with the fact that the inner resonance is associated to the mass balance, the bulk modulus $\mathscr{M}$ is non conventional. Furthermore, the non conventional natures of the effective elastic tensor $\boldsymbol{C}$, and of the coupling tensor $\boldsymbol{A}$, are directly inherited from the pressure/displacement coupling. In fact the feature of parabolic inner-resonance are also observed on $\boldsymbol{C}$ and $\boldsymbol{A}$. Note the classical nature of the permeability and of the density resulting from the fact that the inner resonance concerns the mass balance and not the momentum balance.

As an illustrating example, consider a periodic bi-stratified media whose periodic cell is made of an isotropic porous medium and a fluid layer of respective thickness $(1-\phi)\ell$ and $\phi\ell$. The effective stiffness modulus along the direction of the layers is complex valued and reads (see Boutin and Royer, 2015, for the expressions of the other effective coefficients):

$$C_{11}^{11} = (1-\phi)\frac{2\mu}{\lambda+2\mu}\left(2(\lambda+\mu)+\frac{2\mu}{\lambda+2\mu}\alpha^2 BF(\omega^*)\right);\ F(\omega^*) = 1-\frac{\tanh(\sqrt{i\omega^*})}{\sqrt{i\omega^*}}$$

where

$$\omega^* = \omega\left(\frac{(1-\phi)\ell}{2}\right)^2\frac{\eta}{\kappa_b B};\qquad \frac{1}{B} = \frac{1}{M}+\frac{1}{\lambda+2\mu}$$

and λ and μ are the Lamé coefficients of the porous medium. One may refer to Boutin and Venegas (2016) for other geometries.

As thermo-elasticity is analogous to poro-elasticity, the same kind of phenomenon would be observed in a thermoelastic matrix embedding a connected network of a highly conductive material. Note also that in poro-acoustics a coupling mechanism explains the complex component of the effective compressibility evoked in Sect. 6.4: at the pores scale, the presence of the solid walls implies a thermal transfer within the gas undergoing the uniform pressure. As a consequence of the non-thermal equilibrium, there is a phase shift between the mean temperature and the forced pressure. Hence, through the state equation, a complex valued effective compressibility is obtained.

6.6 Conclusions

The analysis of different physical systems shows that the inner resonance requires a highly contrasted microstructure. This can be reached either by considering composites with contrasted constituents or by specific morphologies with a "contrasted geometry". Whatever the physics in consideration, the common principle for a co-dynamics regime is that an internal dynamic phenomenon periodically or at least regularly distributed in the whole medium is activated by a locally quasi-static excitation. It results that at the local scale,

- a part of the period acts as the long wavelength conveyor and undergoes a quasi-static regime, while
- an other part experiences a dynamic state.

This implies that the resonating part responds in a forced regime imposed by the quasi-static part. Then, the effective constitutive law is determined by the static part, while the resonating part acts as a source term in the balance equations. The situation of "partial" non equilibrium local state results in a non conventional behavior characterized by frequency dependent constitutive parameters associated to the transient term. Hence, when the macroscopic governing equation expresses a momentum balance - respectively a mass balance - an atypical effective density - respectively bulk modulus - appears. However, if the physics involves coupled balance equations of both types, the inner resonance effect associated to a given type of balance is reflected in the effective parameters of the other type of balance as a consequence of the coupling. When the local dynamic problem is described by an equation of hyperbolic type, the effective parameter of the transient term presents poles and takes real positive or negative values that are not bonded in the vicinity of the eigenfrequency of the locally resonating domain. Thus band gaps and a strong dispersion exists at low frequencies (i.e. much lower that the diffraction frequency). Conversely, when the dynamic problem is described by an equation of parabolic type, the effective parameter varies continuously without poles, and takes complex values of significant phase in the frequency range of the "eigenfrequencies". Thus no band gaps arise but a strong dissipation and a dispersion appear at low frequencies. This points out the key effect of the local dissipation in the description of phenomena in materials involving co-dynamics state.

The possibility and the conditions for having a co-dynamic regime at anti-resonance have also been proven and described. Despite its theoretical interest, it seems however that the practical conditions to reach such a situation are not easy to implement. This likely may reduce the potential applications.

Note that as a scale separation exists, provided that the inner resonant domain are identical, their periodic or statistical invariant distribution would lead to a similar behavior. Compared to phononic crystals whose properties are strongly dependent on their geometry, the fact that a perfect periodicity is not a mandatory requirement is an interesting feature of inner-resonant materials in view of practical applications.

To conclude, the asymptotic approach enables to identify design rules in terms of morphology and/or specificity of the mechanical parameters of the constituents for the practical realization of microstructured materials, with tailored unconventional properties that are seemingly impossible to reach with classical materials.

Acknowledgements This work was performed within the framework of the Labex CeLyA of Université de Lyon, operated by the French National Research Agency (ANR-10-LABX-0060/ ANR-11-IDEX-0007).

Appendix: Elastic Bi-Composites: Moderate Stiffness Contrast and High Density Contrast

Consider now that $|\boldsymbol{a}_r|/|\boldsymbol{a}_c| = O(1)$ and $\rho_r/\rho_c = O(\varepsilon^{-2})$. If the effective density is dominated by the larger value ρ_r the scaled $(\boldsymbol{x},\boldsymbol{y})$ governing equations are:

$$\left\{\begin{array}{ll} \operatorname{div}(\boldsymbol{a}_c : \boldsymbol{e}(\boldsymbol{u}_c)) = -\omega^2\varepsilon^2\rho_c\boldsymbol{u}_c & \text{in } \Omega_c \\ \operatorname{div}(\boldsymbol{a}_r : \boldsymbol{e}(\boldsymbol{u}_r)) = -\omega^2\rho_r\boldsymbol{u}_r & \text{in } \Omega_r \\ (\boldsymbol{a}_c : \boldsymbol{e}(\boldsymbol{u}_c) - \boldsymbol{a}_r : \boldsymbol{e}(\boldsymbol{u}_r)) \cdot \boldsymbol{n} = 0 & \text{on } \Gamma \\ \boldsymbol{u}_r - \boldsymbol{u}_c = 0 & \text{on } \Gamma \\ \boldsymbol{u}_c \quad \text{and} \quad \boldsymbol{u}_r & \Omega-\text{periodic} \end{array}\right. \tag{6.69}$$

Using the same procedure as above it is easy to show that no inner resonance arises. The description is the classical elasto-dynamic formulation, identical to (6.1) (where the density of the $\mathscr{C}$-constituent can be neglected).

Let us now examine the possibility of a less intuitive situation where the effective density is dominated by the smaller value ρ_c. Then the scaled $(\boldsymbol{x},\boldsymbol{y})$ governing equations becomes

$$\left\{\begin{array}{ll} \operatorname{div}(\boldsymbol{a}_c : \boldsymbol{e}(\boldsymbol{u}_c)) = -\omega^2\rho_c\boldsymbol{u}_c & \text{in } \Omega_c \\ \operatorname{div}(\boldsymbol{a}_r : \boldsymbol{e}(\boldsymbol{u}_r)) = -\omega^2\varepsilon^{-2}\rho_r\boldsymbol{u}_r & \text{in } \Omega_r \\ (\boldsymbol{a}_c : \boldsymbol{e}(\boldsymbol{u}_c) - \boldsymbol{a}_r : \boldsymbol{e}(\boldsymbol{u}_r)).\boldsymbol{n} = 0 & \text{on } \Gamma \\ \boldsymbol{u}_r - \boldsymbol{u}_c = 0 & \text{on } \Gamma \\ \boldsymbol{u}_c \quad \text{and} \quad \boldsymbol{u}_r & \Omega-\text{periodic} \end{array}\right. \tag{6.70}$$

The local problem at the dominant order reads

$$\left\{\begin{array}{ll} \operatorname{div}(\boldsymbol{a}_{\mathrm{c}} : \boldsymbol{e}(\boldsymbol{u}_{\mathrm{c}}^{(0)})) = 0 & \text{in } \Omega_{\mathrm{c}} \\ \operatorname{div}(\boldsymbol{a}_{\mathrm{r}} : \boldsymbol{e}(\boldsymbol{u}_{\mathrm{r}}^{(0)})) = -\omega^2 \rho_{\mathrm{r}} \boldsymbol{u}_{\mathrm{r}}^{(0)} & \text{in } \Omega_{\mathrm{r}} \\ (\boldsymbol{a}_{\mathrm{c}} : \boldsymbol{e}(\boldsymbol{u}_{\mathrm{c}}^{(0)}) - \boldsymbol{a}_{\mathrm{r}} : \boldsymbol{e}(\boldsymbol{u}_{\mathrm{r}}^{(0)})) \cdot \boldsymbol{n} = 0 & \text{on } \Gamma \\ \boldsymbol{u}_{\mathrm{r}}^{(0)} - \boldsymbol{u}_{\mathrm{c}}^{(0)} = 0 & \text{on } \Gamma \\ \boldsymbol{u}_{\mathrm{c}}^{(0)} \quad \text{and} \quad \boldsymbol{u}_{\mathrm{r}}^{(0)} & \Omega - \text{periodic} \end{array}\right. \qquad (6.71)$$

This corresponds to an elasto-dynamics regime of the cell where both constituents are coupled. The solutions of this eigenvalue problem, are the periodic eigenmodes that occur at the discrete frequency spectrum $\{\omega_J\}$ of system described by (6.71). Hence, the whole field in the period varies at the local scale and therefore this situation does not correspond to a "co-dynamic" regime.

Note finally that the intermediary scaling of both densities, namely $\varepsilon^{-1}\rho_{\mathrm{r}}$ and $\varepsilon\rho_{\mathrm{c}}$ is not admissible since it would leads to an uniform motion at the leading order, and then to an unbalanced inertial term at the next order.

References

Allaire G, Conca C (1998) Bloch wave homogenization and spectral asymptotic analysis. Int J Eng Sci 77:153–208

Auriault JL (1980) Dynamic behavior of a porous medium saturated by a Newtonian fluid. Int J Eng Sci 18:775–785

Auriault JL (1983) Effective macroscopic description for heat-conduction in periodic composites. Int J Heat Mass Tran 26:861–869

Auriault JL (1994) Acoustics of heterogeneous media: Macroscopic behavior by homogenization. Current Topics in Acoust Res I:63–90

Auriault JL, Bonnet G (1985) Dynamique des composites élastiques périodiques. Arch Mech 37:269–284

Auriault JL, Boutin C (1994) Deformable porous media with double porosity. III Acoustics. Transport in Porous Media 14:143–162

Auriault JL, Boutin C (2012) Long wavelength inner-resonance cut-off frequencies in elastic composite materials. Int Jour Solid and Structures 49:3269–3281

Auriault JL, Lewandowska J (1995) Non-Gaussian Diffusion Modeling in Composite Porous Media by Homogenization: Tail effect. Transport in Porous Media 21:47–70

Auriault JL, Geindreau C, Boutin C (2005) Filtration law in porous media with poor scale separation. Transport in Porous Media 60(1):89–108

Auriault JL, Boutin C, Geindreau C (2009) Homogenization of Coupled Phenomena in Heterogenous Media. ISTE and Wiley

Ávila A, Griso G, Miara B (2005) Bandes phoniques interdites en élasticité linéarisée. C R Acad Sci Paris, Ser I 340:933–938

Babych NO, Kamotski IV, Smyshlyaev VP (2008) Homogenization of spectral problems in bounded domains with doubly high contrasts. Networks and Heterogeneous Media 3(3):413–436

Baravelli E, Ruzzene M (2013) Internally resonating lattices for bandgap generation and low-frequency vibration control. J Sound Vibr 332:6562–6579

Biot MA (1956) Theory of propagation of elastic waves in a fluid-saturated porous solid. II. Higher frequency range. J Acoust Soc Am 28:179–191

Bonnet G, Monchiet V (2015) Low frequency locally resonant metamaterials containing composite inclusions. J Acoust Soc Am 137(6):3263–3271
Bonnet G, Monchiet V (2017) Dynamic mass density of resonant metamaterials with homogeneous inclusions. J Acoust Soc Am 142(2):890–901
Boutin C (1995) Microstructural influence on heat conduction. Int J Heat Mass Transf 38(17):3181–3195
Boutin C (2013) Acoustics of porous media with inner resonators. J Acoust Soc Am 134(6):4717–4730
Boutin C, Auriault JL (1993) Rayleigh scattering in elastic composite materials. Int J Eng Sci 31:1669–1689
Boutin C, Becot FX (2015) Theory and experiments on poro-acoustics with inner resonators. Wave Motion 54:77–99
Boutin C, Royer P (2015) On models of double porosity poroelastic media. Geophys J Int 203:1694–1725
Boutin C, Venegas R (2016) Assessment of the effective parameters of dual porosity deformable media. Mechanics of Materials 102:26–46
Boutin C, Royer P, Auriault JL (1998) Acoustic absorption of porous surfacing with dual porosity. Int J Solids Struct 35:4709–4737
Boutin C, Hans S, Chesnais C (2010) Generalized beam and continua. Dynamics of reticulated structures. In: Maugin GA, Metrikine AV (eds) Mechanics of Generalized Continua, Springer, New York, pp 131–141
Boutin C, Rallu A, Hans S (2012) Large scale modulation of high frequency acoustics waves in periodic porous media. J Acoust Soc Am 134(6):3622–3636
Boutin C, Rallu A, Hans S (2014) Large scale modulation of high frequency waves in periodic elastic composites. J Mech Phys Solids 70:362–381
Brillouin L (1946) Wave Propagation in Periodic Structures. McGraw-Hill, New York
Caillerie D, Trompette P, Verna P (1989) Homogenisation of periodic trusses. In: IASS Symposium, 10 Years of Progress in Shell and Spatial Structures, Madrid
Chesnais C, Hans S, Boutin C (2007) Wave propagation and diffraction in discrete structures: Anisotropy and internal resonance. PAMM 7:1090,401–1090,402
Chesnais C, Boutin C, Hans S (2012) Effects of the local resonance on the wave propagation in periodic frame structures: generalized Newtonian mechanics. J Acoust Soc Am 132(4):2873–2886
Craster RV, Kaplunov J, Pichugin AV (2010) High-frequency homogenization for periodic media. Proc R Soc A 466:2341–2362
Daya EM, Braikat B, Damil N, Potier-Ferry M (2002) Continuum modeling for the modulated vibration modes of large repetitive structures. Comptes Rendus - Mec 330:333–338
Eringen AC (1968) Mechanics of micromorphic continua. In: Kröner E (ed) Mechanics of Generalized Continua - Proceedings of the IUTAM-Symposium on The Generalized Cosserat Continuum and the Continuum Theory of Dislocations with Applications, Springer, Freudenstadt and Stuttgart, pp 18–35
Fang N, Xi D, Xu J, Ambati M, Srituravanich W, Sun C, Zhang X (2006) Ultrasonic metamaterials with negative modulus. Nature materials 5:452–456
Hans S, Boutin C (2008) Dynamics of discrete framed structures: A unified homogenized description. J Mech Mater Struct 3:1709–1739
Hornung U, Showalter RR (1990) Diffusion models for fractured media. J Math Anal Appl 147:69–80
Krynkin A, Umnova O, Chong YBA, Taherzadeh S, Attenborough K (2011) Scattering by coupled resonating elements in air. J Phys D - Appl Phys 44:125,101
Lafarge D, Nemati N (2013) Nonlocal Maxwellian Theory of Sound Propagation in Fluid-Saturated Rigid-Framed Porous Media. Wave Motion 50:1016–1035
Lafarge D, Lemarinier P, Allard JF, Tarnow V (1997) Dynamic compressibility of air in porous structures at audible frequencies. J Acoust Soc Am 102:1995–2006

Liu Z, Zhang XX, Mao YW, Zhu YY, Yang ZY, Chan CT, Sheng P (2000) Locally resonant sonic materials. Science 289:1734–1736

Liu Z, Chan CT, Sheng P (2005) Analytic model of phononic crystals with local resonances. Phys Rev B 71:014,103–014,110

Ma G, Sheng P (2016) Acoustic metamaterials: From local resonances to broad horizons. Science Advance 2:e1501,595

Maugin GA (1995) On some generalization of Boussinesq and Korteweg de Vries systems. Proc Estonian Acad Sci Phys Math 44(1):40–55

Maugin GA, Metrikine AV (eds) (2010) Mechanics of Generalized Continua - One Hundred Years After the Cosserats, Advances in Mechanics and Mathematics, Springer, New York

Milton GW (2007) New metamaterials with macroscopic behavior outside that of continuum elastodynamics. New J Phys 9:359

Moustaghfir N, Daya EM, Braikat B, Damil N, Potier-Ferry M (2007) Evaluation of continuous modelings for the modulated vibration modes of long repetitive structures. Int J Solids Struct 44:7061–7072

Naify CJ, Chang CM, Mcknight G, Nutt SR (2012) Scaling of membrane-type locally resonant acoustic metamaterial arrays. J Acous Soc Am 132(4):2784–2792

Nassar H, He QC, Auffray N (2016a) A generalized theory of elastodynamic homogenization for periodic media. Int J Solids Struct 84:139–146

Nassar H, He QC, Auffray N (2016b) On asymptotic elastodynamic homogenization approaches for periodic media. J Mech Phys Solids 88:274–290

Olny X, Boutin C (2003) Acoustic wave propagation in double porosity media. J Acoust Soc Am 144:73–89

Sanchez-Palencia E (1980) Non-homogeneous media and vibration theory, Lecture Notes in Physics, vol 127. Springer, Berlin

Shanshan Y, Xiaoming Z, Gengkai H (2008) Experimental study on negative effective mass in a 1D mass-spring system. New J Phys 10:11

Sheng P, Zhang XX, Liu Z, Chan CT (2003) Locally resonant sonic materials. Physica B 338:201–205

Smeulders DMJ, Eggels RLGM, van Dongen MEH (1992) Dynamic permeability: reformulation of theory and new experimental and numerical data. J Fluid Mech 245:211–227

Smyshlyaev VP (2009) Propagation and localization of elastic waves in highly anisotropic periodic composites via two-scale homogenization. Mechanics of Materials 41(4):434–447

Tollenaere H, Caillerie D (1998) Continuous modeling of lattice structures by homogenization. Adv Eng Softw 29:699–705

Vasseur JO, Deymier PA, Prantziskonis G, Hong G (1998) Experimental evidence for the existence of absolute acoustic band gaps in two-dimensional periodic composite media. J Phys: Condens Mater 10:6051–6064

Venegas R, Umnova O (2011) Acoustical properties of double porosity granular materials". J Acoust Soc Am 130(5):2765–2776

Willis JR (2012) The construction of effective relation for waves in composites. C R Mecanique 340:181–192

Yang Z, Mei J, Yang M, Chan NH, Sheng P (2008) Membrane-type acoustic metamaterial with negative dynamic mass. Phys Rev Lett 101(20):204,301

Zhikov VV (2000) On an extension of the method of two-scale convergence and its applications. Sb Math 191:973–1014

Chapter 7
The Balance of Material Momentum Applied to Water Waves

Manfred Braun

Abstract The balance of material momentum is applied to the motion of an ideal, incompressible fluid with special emphasis on water waves. To this end, the fluid flow is represented by its material or Lagrangian description. A variational approach using Hamilton's principle is employed, with the incompressibility condition incorporated into the Lagrangian by means of a Lagrange multiplier. The balance of material momentum is obtained in its standard form known from nonlinear elasticity, however with the peculiarity that the dynamic Eshelby stress becomes hydrostatic and its divergence reduces to the (negative) gradient of an "Eshelby pressure" The balance is applied to Gerstner's nonlinear theory of water waves.

7.1 Introduction

Configurational Mechanics or *Mechanics in Material Space* is closely linked to the Lagrangian description of deformation. This way of describing the deformation is quite common in non-linear elasticity, where it offers several advantages over the Eulerian description. For instance, the boundary conditions at a stress-free surface are easily formulated at the undeformed body, while in the Eulerian description one would need the deformed surface, which is already most of the solution of the problem.

The use of material coordinates alone does not yet qualify an equation as "configurational". For instance, the Piola-Kirchhoff format of the balance of physical momentum employs material coordinates, but the physical momentum lives in the actual configuration. The balance of *material* momentum, in contrast, pertains to quantities that reside totally in the material space. It is, therefore, a legitimate equation

Manfred Braun
Chair of Mechanics and Robotics, University of Duisburg–Essen, Lotharstraße 1, 47057 Duisburg, Germany
e-mail: manfred.braun@uni-due.de

© Springer International Publishing AG, part of Springer Nature 2018

H. Altenbach et al. (eds.), *Generalized Models and Non-classical Approaches in Complex Materials 1*, Advanced Structured Materials 89,
https://doi.org/10.1007/978-3-319-72440-9_7

of configurational mechanics. This does not mean, of course, that the actual configuration has become obsolete. The vector of material momentum and the Eshelby stress tensor, although quantities of the material space, depend on the deformation, i.e., on the mapping from the material to the physical space.

Configurational mechanics as presented in the standard works of Maugin (1993, 2010); Kienzler and Herrmann (2000); Gurtin (2000) evolves from the theory of elasticity with all its generalizations. Fluids are mentioned only briefly by Gurtin (2000, page 45). This is related to the fact that fluid mechanics is the realm of Eulerian description. In some situations it would be well-nigh impossible to employ Lagrangian coordinates. Take, as an example, the flow around an airfoil. A Lagrangian grid convected by the flow is so heavily distorted in the vicinity of the airfoil that it cannot be used as a coordinate system. Also the boundary conditions pertain to the contour in the *actual* configuration. For such kind of problems the material description is not appropriate. A material reference is still needed though for the sole reason that the *material points* in the fluid flow must be given an identity.

There are, however, examples of fluid flow, where the Lagrangian description can be used. In a propagating surface wave the individual particles remain close to their initial positions. Even for rather steep waves a Lagrangian grid convected by the flow is not distorted too much and can well serve as a reference. It is quite interesting that the very first nonlinear theory of water waves proposed already in 1802 by Gerstner (1804) is based on a Lagrangian description. Gerstner's wave, although an exact solution of the equations of motion of an ideal fluid, has a severe drawback. It is rotational, and the vorticity is just opposite to the circulatory motion of the material particles. Apparently there is no simple way to overcome this deficiency. The later approximate theories of *irrotational* waves, notably Stokes' theory from 1847 (Stokes, 2009), use the Eulerian description.

Even though Gerstner's theory could be considered as disqualified by its strange vorticity behavior, it has kept its place in the textbooks on water waves, like Lamb (1932); Le Méhauté (1976); Rahman (1995), since it represents, up to now, the only closed-form solution of the nonlinear equations. In recent times, there is even a renaissance of the Lagrangian description of water waves, mainly initiated by Constantin (2001), that has revived also the interest in Gerstner's old theory (Clamond, 2007; Kalisch, 2004; Stuhlmeier, 2015) and some extensions (Constantin and Monismith, 2017). Clamond (2007) has pointed out the advantages of the Lagrangian description, especially for steep waves.

Once the Lagrangian description is used the way is paved for a configurational theory of water waves. It is certainly simpler than the configurational mechanics of elasticity due to the lack of a strain energy and might be considered as trivial. Nonetheless it is worthwhile to study the balance of material momentum in ideal fluids and apply it to Gerstner's wave.

Following this introduction about the motivation of the paper the balance of physical momentum of an incompressible fluid will be derived from a Lagrange function. There is one special point to be considered, namely that the reference configuration may be chosen such that the fluid cannot attain it without violating the incompressibility condition. Actually this is the case in Gerstner's theory. Incompressibility means only that the Jacobian of the deformation is time-independent, but not necessarily unity. This has to be implemented into the Lagrangian.

The third section devoted to the balance of *material* momentum represents the central part of the paper. Section 7.4 examines briefly the energy balance considered as another material conservation law. Gerstner's wave under the aspect of material momentum is presented in Sect. 7.5. As already mentioned the reference configuration in Gerstner's serves only to label the material points and cannot be attained by the fluid. In Sect. 7.6 it is shown how the reference configuration has to be changed such that it could be attained by the incompressible fluid. The paper closes with some concluding remarks.

Diverse notations are used throughout the paper, depending on the context. For instance, the deformation gradient and the velocity vector are usually denoted by $\boldsymbol{F}$ and $\boldsymbol{v}$, respectively. If it is to be emphasized that these quantities are derivatives of the mapping $\boldsymbol{\varphi}(\boldsymbol{\xi},t)$ they are written as $\boldsymbol{\varphi}_{\xi}$ and $\boldsymbol{\varphi}_{t}$. When, finally, it comes to detailed calculations the component representations $F^i{}_\alpha = x^i{}_{,\alpha}$ and $v^i = \dot{x}^i$ are preferred. In the example of Gerstner's wave, which is simply two-dimensional, coordinate pairs (ξ,η) and (x,y) are used rather than the coordinates with superscripts, ξ^α and x^i. Capitalized differential operators, like $\mathrm{D}/\mathrm{D}t$ or Grad, refer to the Lagrangian description with $\boldsymbol{\xi}$ and t as independent variables. Otherwise the used notations should be self-explanatory.

7.2 The Balance of Physical Momentum

The motion of the continuum is described as usual by a mapping

$$\boldsymbol{x} = \boldsymbol{\varphi}(\boldsymbol{\xi},t) \tag{7.1}$$

that assigns the position $\boldsymbol{x} \in \mathscr{B}(t)$ in the actual configuration at time t, of a material point specified by its position $\boldsymbol{\xi} \in \mathscr{B}_\mathrm{R}$ in the reference configuration. The governing equations will involve the derivatives

$$\boldsymbol{\varphi}_t = \frac{\mathrm{D}\boldsymbol{x}}{\mathrm{D}t} = \boldsymbol{v}, \quad \boldsymbol{\varphi}_\xi = \operatorname{Grad}\boldsymbol{x} = \boldsymbol{F}, \quad \text{and} \quad \boldsymbol{\varphi}_{\xi t} = \frac{\mathrm{D}\boldsymbol{F}}{\mathrm{D}t} = \operatorname{Grad}\boldsymbol{v}, \tag{7.2}$$

i.e., the velocity vector, the deformation gradient, and the material velocity gradient. The mass densities in the actual and the reference configurations are related, in general, by

$$\rho(\boldsymbol{\xi},t) = \frac{\rho_\mathrm{R}(\boldsymbol{\xi})}{J(\boldsymbol{\xi},t)}, \qquad J = \det\boldsymbol{F}. \tag{7.3}$$

The conservation of mass is reflected by the fact that the referential density ρ_R does not depend on time, though it may depend on the material point $\boldsymbol{\xi}$. In an *isochoric* motion also the actual density ρ at a fixed material point does not change with time. This means that, according to (7.3), the material time derivative of the Jacobian,

$$\frac{DJ}{Dt} = \frac{\partial J}{\partial \boldsymbol{F}} \cdot \frac{D\boldsymbol{F}}{Dt} = J\boldsymbol{F}^{-T} \cdot \frac{D\boldsymbol{F}}{Dt}, \tag{7.4}$$

vanishes. An *incompressible* material is characterized by the property that it can undergo only isochoric motions. These motions are restricted by the constraint

$$h(\boldsymbol{\varphi}_{\xi}, \boldsymbol{\varphi}_{\xi t}) \equiv \frac{DJ}{Dt} = 0 \tag{7.5}$$

that depends on the derivatives $\boldsymbol{\varphi}_{\xi}$ and $\boldsymbol{\varphi}_{\xi t}$ of the motion (7.1).

The kinetic energy density, per unit *reference* volume, is

$$\mathscr{T}(\boldsymbol{\varphi}_t; \boldsymbol{\xi}) = \frac{1}{2}\rho_R(\boldsymbol{\xi})\, v^2. \tag{7.6}$$

Since there is no strain energy in an incompressible fluid, the only potential energy is the external gravity potential. Its density, again per unit reference volume, is

$$\mathscr{U}(\boldsymbol{\varphi}; \boldsymbol{\xi}) = -\rho_R(\boldsymbol{\xi})\, \boldsymbol{g} \cdot \boldsymbol{x}. \tag{7.7}$$

If there were no constraint, the Lagrangian would be simply the difference $\mathscr{T} - \mathscr{U}$ of the kinetic and potential energy densities. To incorporate the incompressibility constraint the Lagrangian is enhanced by the constraint function (7.5) with a Lagrange multiplier λ. Thus the Lagrangian density becomes

$$\mathscr{L} = \mathscr{T} - \mathscr{U} + \lambda h = \frac{1}{2}\rho_R(\boldsymbol{\xi})\, \boldsymbol{v} \cdot \boldsymbol{v} + \rho_R(\boldsymbol{\xi})\, \boldsymbol{g} \cdot \boldsymbol{x} + \lambda J\boldsymbol{F}^{-T} \cdot \frac{D\boldsymbol{F}}{Dt}. \tag{7.8}$$

It is a function

$$\mathscr{L} = \mathscr{L}\left(\boldsymbol{\varphi}, \boldsymbol{\varphi}_t, \boldsymbol{\varphi}_{\xi}, \boldsymbol{\varphi}_{\xi t}, \lambda; \boldsymbol{\xi}\right) \tag{7.9}$$

that depends on the motion $\boldsymbol{\varphi}(\boldsymbol{\xi}, t)$ and its derivatives, on the Lagrange multiplier λ, and, via the non-uniform density, also explicitly on the material point $\boldsymbol{\xi}$.

Hamilton's principle states that the action integral

$$A = \int_{t_0}^{t_1} \int_{\mathscr{V}_R} \mathscr{L}\left(\boldsymbol{\varphi}, \boldsymbol{\varphi}_t, \boldsymbol{\varphi}_{\xi}, \boldsymbol{\varphi}_{\xi t}, \lambda; \boldsymbol{\xi}\right) dV_R dt \tag{7.10}$$

taken over a fixed volume $\mathscr{V}_R$ in the reference configuration $\mathscr{B}_R$ and a fixed time interval $[t_0, t_1]$, is stationary with respect to arbitrary variations $\delta\boldsymbol{\varphi}$ and $\delta\lambda$ of the motion and the Lagrange multiplier. Necessary conditions are the corresponding Euler-Lagrange equations

$$\frac{\delta \mathscr{L}}{\delta x^i} \equiv \frac{\partial \mathscr{L}}{\partial x^i} - D_t \frac{\partial \mathscr{L}}{\partial \dot{x}^i} - D_\alpha \frac{\partial \mathscr{L}}{\partial x^i{}_{,\alpha}} + D_t D_\alpha \frac{\partial \mathscr{L}}{\partial \dot{x}^i{}_{,\alpha}} \overset{!}{=} 0 \quad \text{and} \quad \frac{\delta \mathscr{L}}{\delta \lambda} \equiv \frac{\partial \mathscr{L}}{\partial \lambda} \overset{!}{=} 0, \tag{7.11}$$

where D_t and D_α denote the *total* derivatives with respect to the time t and the material coordinate ξ^α, respectively. The second, scalar equation just recovers the incompressibility constraint (7.5), while the first equation reveals the balance of physical momentum. Now the special form (7.8) of the Lagrangian has to be inserted. The two last terms are a little bit intricate to form. The details are worked out in the Appendix. Altogether one obtains the vectorial equation

$$\rho_{\mathrm{R}} \frac{\mathrm{D}\boldsymbol{v}}{\mathrm{D}t} = J\boldsymbol{F}^{-\mathrm{T}}\mathrm{Grad}\,\frac{\mathrm{D}\lambda}{\mathrm{D}t} + \rho_{\mathrm{R}}\boldsymbol{g}, \tag{7.12}$$

which is the Lagrangian form of the balance of physical momentum. It has to be complemented by the equations

$$\frac{\mathrm{D}\rho_{\mathrm{R}}}{\mathrm{D}t} = 0 \quad \text{and} \quad \frac{\mathrm{D}J}{\mathrm{D}t} = 0 \tag{7.13}$$

reflecting the balance of mass and the condition of incompressibility.

The Eulerian form of the balance of momentum is obtained by transforming the material gradient to the spatial gradient. Also the referential density is expressed via (7.3) in the actual density. Thus the Eulerian version of the balance reads

$$\rho \frac{\mathrm{D}\boldsymbol{v}}{\mathrm{D}t} = \mathrm{grad}\,\frac{\mathrm{D}\lambda}{\mathrm{D}t} + \rho\boldsymbol{g}. \tag{7.14}$$

From this equation one sees that the Lagrange multiplier is related to the hydrostatic pressure p by

$$\frac{\mathrm{D}\lambda}{\mathrm{D}t} = -p. \tag{7.15}$$

Otherwise the Eulerian form of the balance of momentum will not be used any further. It should be noted that, despite incompressibility, the density ρ may depend on the position $\boldsymbol{x}$, as in the case of a density stratified fluid.

If one had assumed the incompressibility condition in the form $J - 1 = 0$, rather than $\mathrm{D}J/\mathrm{D}t = 0$, the result would have been the same, except that the Lagrange multiplier of $(J-1)$ can be identified directly as $-p$. The balance of physical momentum, even in the Lagrange description, does not depend on whether the reference configuration can be attained by the fluid or not. For the present application it is important to allow reference configurations that violate the incompressibility condition and, therefore, cannot be attained by the fluid at any time. It is only the *actual* configuration that must obey the incompressibility condition.

7.3 The Balance of Material Momentum

There are several ways to derive the balance of material momentum of a continuum that admits a Lagrangian. The most straightforward procedure has been proposed by Gołębiewska-Herrmann (1981). She simply takes the total derivative of the Lagrangian with respect to the material coordinates and recasts the expression such that the functional derivative of the Lagrangian appears[1]. The remaining terms constitute the balance of material momentum.

The approach has to be modified here to allow for the constraint of incompressibility, which introduces a second derivative $\boldsymbol{\varphi}_{\xi t}$ into the Lagrangian (7.9). The Lagrangian depends on the material point $\boldsymbol{\xi}$ via all its arguments including possibly an explicit dependence. Its total derivative with respect to the material coordinate ξ^α is

$$\mathrm{D}_\alpha \mathscr{L} = \frac{\partial \mathscr{L}}{\partial x^i} x^i_{,\alpha} + \frac{\partial \mathscr{L}}{\partial \dot{x}^i} \dot{x}^i_{,\alpha} + \frac{\partial \mathscr{L}}{\partial x^i_{,\beta}} x^i_{,\beta\alpha} + \frac{\partial \mathscr{L}}{\partial \dot{x}^i_{,\beta}} \dot{x}^i_{,\beta\alpha} + \frac{\partial \mathscr{L}}{\partial \lambda} \lambda_{,\alpha} + \frac{\partial \mathscr{L}}{\partial \xi^\alpha}. \tag{7.16}$$

The second and third terms can be written, using the product rule, as

$$\begin{aligned} \frac{\partial \mathscr{L}}{\partial \dot{x}^i} \dot{x}^i_{,\alpha} &= \mathrm{D}_t \left(\frac{\partial \mathscr{L}}{\partial \dot{x}^i} x^i_{,\alpha} \right) - \mathrm{D}_t \frac{\partial \mathscr{L}}{\partial \dot{x}^i} x^i_{,\alpha}, \\ \frac{\partial \mathscr{L}}{\partial x^i_{,\beta}} x^i_{,\beta\alpha} &= \mathrm{D}_\beta \left(\frac{\partial \mathscr{L}}{\partial x^i_{,\beta}} x^i_{,\alpha} \right) - \mathrm{D}_\beta \frac{\partial \mathscr{L}}{\partial x^i_{,\beta}} x^i_{,\alpha}. \end{aligned} \tag{7.17}$$

Applying the product rule twice, the fourth term in (7.16) becomes

$$\begin{aligned} \frac{\partial \mathscr{L}}{\partial \dot{x}^i_{,\beta}} \dot{x}^i_{,\alpha\beta} &= \mathrm{D}_\beta \left(\frac{\partial \mathscr{L}}{\partial \dot{x}^i_{,\beta}} \dot{x}^i_{,\alpha} \right) - \mathrm{D}_\beta \frac{\partial \mathscr{L}}{\partial \dot{x}^i_{,\beta}} \dot{x}^i_{,\alpha} = \\ &= \mathrm{D}_\beta \left(\frac{\partial \mathscr{L}}{\partial \dot{x}^i_{,\beta}} \dot{x}^i_{,\alpha} \right) - \mathrm{D}_t \left(\mathrm{D}_\beta \frac{\partial \mathscr{L}}{\partial \dot{x}^i_{,\beta}} x^i_{,\alpha} \right) + \mathrm{D}_t \mathrm{D}_\beta \frac{\partial \mathscr{L}}{\partial \dot{x}^i_{,\beta}} x^i_{,\alpha}. \end{aligned} \tag{7.18}$$

Now these expressions are inserted into (7.16). The resulting terms are sorted such that first all terms with the factor $x^i_{,\beta}$ are collected, then the single term with $\lambda_{,\alpha}$, next those terms preceded by the operator D_t, then those preceded by the operator D_β, and finally the partial derivative with respect to ξ^α. Thus one arrives at the equation

$$\begin{aligned} \mathrm{D}_\alpha \mathscr{L} = \left(\frac{\partial \mathscr{L}}{\partial x^i} - \mathrm{D}_t \frac{\partial \mathscr{L}}{\partial \dot{x}^i} - \mathrm{D}_\beta \frac{\partial \mathscr{L}}{\partial x^i_{,\beta}} + \mathrm{D}_t \mathrm{D}_\beta \frac{\partial \mathscr{L}}{\partial \dot{x}^i_{,\beta}} \right) x^i_{,\alpha} + \frac{\partial \mathscr{L}}{\partial \lambda} \lambda_{,\alpha} + \\ + \mathrm{D}_t \left(\frac{\partial \mathscr{L}}{\partial \dot{x}^i} x^i_{,\alpha} - \mathrm{D}_\beta \frac{\partial \mathscr{L}}{\partial \dot{x}^i_{,\beta}} x^i_{,\alpha} \right) + \\ + \mathrm{D}_\beta \left(\frac{\partial \mathscr{L}}{\partial x^i_{,\beta}} x^i_{,\alpha} + \frac{\partial \mathscr{L}}{\partial \dot{x}^i_{,\beta}} \dot{x}^i_{,\alpha} \right) + \frac{\partial \mathscr{L}}{\partial \xi^\alpha}. \end{aligned} \tag{7.19}$$

On the right-hand side the two functional derivatives (7.11) appear, the second being simply a partial derivative. Moreover, the derivative $\mathrm{D}_\alpha \mathscr{L}$ on the left-hand side can be incorporated into the divergence term on the right-hand side. This leads to the final equation

[1] There is a misprint in Gołębiewska-Herrmann (1981): In the unnumbered equation preceding Eq. (5) a plus sign has to be inserted before the square brackets and the factor $\phi_{j,i}$ behind. The subsequent equations are correct.

$$0 = \frac{\delta \mathscr{L}}{\delta x^i} x^i{}_{,\alpha} + \frac{\delta \mathscr{L}}{\delta \lambda} \lambda_{,\alpha} + \mathrm{D}_t \left(\frac{\partial \mathscr{L}}{\partial \dot{x}^i} x^i{}_{,\alpha} - \mathrm{D}_\beta \frac{\partial \mathscr{L}}{\partial \dot{x}^i{}_{,\beta}} x^i{}_{,\alpha} \right) +$$
$$+ \mathrm{D}_\beta \left(\frac{\partial \mathscr{L}}{\partial x^i{}_{,\beta}} x^i{}_{,\alpha} + \frac{\partial \mathscr{L}}{\partial \dot{x}^i{}_{,\beta}} \dot{x}^i{}_{,\alpha} - \mathscr{L} \delta^\beta_\alpha \right) + \frac{\partial \mathscr{L}}{\partial \xi^\alpha} . \tag{7.20}$$

If the Euler–Lagrange equations (7.11) are fulfilled, the two functional derivatives in (7.20) vanish and the equation assumes the form of a balance law. In the homogeneous case, i.e., if the Lagrangian does not depend explicitly on the material coordinate ξ^α, it becomes a strict conservation law. Homogeneity means here that there is a symmetry with respect to translations in material space. The conservation law is, therefore, just the one predicted by Noether's theorem.

It would be straightforward to interpret the vector preceded by the time derivative D_t as the negative material momentum and the tensor preceded by the total derivative D_β as the dynamic Eshelby tensor. We shall postpone and even modify the introduction of these quantities for reasons that become clear after the next step.

Up to this point, the Lagrangian in (7.20) has not been specified except that it depends on the arguments listed in (7.9). Let us now take into account the special form (7.8) of the Lagrangian. Then the derivatives occurring in (7.20) are

$$\frac{\partial \mathscr{L}}{\partial \dot{x}^i} = \rho_{\mathrm{R}} \dot{x}_i , \quad \frac{\partial \mathscr{L}}{\partial x^i{}_{,\beta}} = \lambda J \left(\xi^\beta{}_{,i} \xi^\gamma{}_{,k} - \xi^\beta{}_{,k} \xi^\gamma{}_{,i} \right) \dot{x}^k{}_{,\gamma} , \quad \frac{\partial \mathscr{L}}{\partial \dot{x}^i{}_{,\beta}} = \lambda J \xi^\beta{}_{,i} . \tag{7.21}$$

The last of these derivatives appears twice in (7.20). In one case also its total derivative

$$\mathrm{D}_\beta \frac{\partial \mathscr{L}}{\partial \dot{x}^i{}_{,\beta}} = \lambda_{,\beta} J \xi^\beta{}_{,i} + \lambda J \left(\xi^\beta{}_{,i} \xi^\gamma{}_{,k} - \xi^\beta{}_{,k} \xi^\gamma{}_{,i} \right) x^k{}_{,\gamma\beta} = \lambda_{,\beta} J \xi^\beta{}_{,i} \tag{7.22}$$

is needed. This, in turn, is embedded into the expression

$$\mathrm{D}_t \left(\mathrm{D}_\beta \frac{\partial \mathscr{L}}{\partial \dot{x}^i{}_{,\beta}} x^i{}_{,\alpha} \right) = \mathrm{D}_t \left(\lambda_{,\beta} J \xi^\beta{}_{,i} x^i{}_{,\alpha} \right) = \mathrm{D}_t \left(\lambda_{,\alpha} J \right) = J \dot{\lambda}_{,\alpha} = -J p_{,\alpha} , \tag{7.23}$$

where the incompressibility condition (7.5) has been used. Also the time derivative of the Lagrange multiplier has been replaced by $-p$ according to (7.15).

The Lagrange multiplier λ itself hasn't any direct physical meaning, only its time derivative can be interpreted as the negative hydrostatic pressure. Due to the special structure of the Lagrangian the time derivative (7.23) has been reformulated such that there is no differentiation with respect to time anymore. Therefore it is appropriate to separate this expression from the other term whose time derivative has to be formed in (7.20). We *define* the vector of material momentum by

$$\mathfrak{p}_\alpha = -\frac{\partial \mathscr{L}}{\partial \dot{x}^i} x^i{}_{,\alpha} = -\rho_{\mathrm{R}} \dot{x}_i x^i{}_{,\alpha} \quad \text{or} \quad \boldsymbol{\mathfrak{p}} = -\rho_{\mathrm{R}} \boldsymbol{F}^{\mathsf{T}} \boldsymbol{v} . \tag{7.24}$$

This means that the material momentum is the same as in the case of a Lagrangian that does not depend on $\mathrm{D}\boldsymbol{F}/\mathrm{D}t$.

The term (7.23) that has been omitted so far must now be considered in the remaining parts of the equation. To this end it is written as

$$Jp_{,\alpha} = \mathrm{D}_\beta \left(Jp\delta_\alpha^\beta \right) - pJ_{,\alpha}\,. \tag{7.25}$$

The first term can be included in the divergence expression, while the second term that reflects some inhomogeneity is combined with the other inhomogeneity term $\partial\mathscr{L}/\partial\xi^\alpha$. Thus we define the dynamic Eshelby stress tensor by

$$\mathfrak{E}_\alpha{}^\beta = \frac{\partial\mathscr{L}}{\partial x^i{}_{,\beta}} x^i{}_{,\alpha} + \frac{\partial\mathscr{L}}{\partial \dot{x}^i{}_{,\beta}} \dot{x}^i{}_{,\alpha} + (Jp - \mathscr{L})\,\delta_\alpha^\beta\,. \tag{7.26}$$

Inserting the derivatives $(7.21)_{2,3}$ and the Lagrangian (7.8) several expressions cancel each other and the Eshelby tensor becomes

$$\mathfrak{E}_\alpha{}^\beta = -\left[\rho_\mathrm{R} \left(\frac{1}{2}v^2 + g_i x^i \right) - Jp \right] \delta_\alpha^\beta\,. \tag{7.27}$$

It is proportional to the identity tensor.

What remains is the inhomogeneity force

$$\mathfrak{f}_\alpha = \frac{\partial\mathscr{L}}{\partial\xi^\alpha} - pJ_{,\alpha} = \left(\frac{1}{2}v^2 + g_i x^i \right) \rho_{\mathrm{R},\alpha} - pJ_{,\alpha}\,. \tag{7.28}$$

If the reference density ρ_R is expressed via (7.3) in the actual density ρ, the inhomogeneity force can be decomposed into the two parts

$$\mathfrak{f}_\alpha = \left(\frac{1}{2}v^2 + g_i x^i \right) J\rho_{,\alpha} + \left[\rho \left(\frac{1}{2}v^2 + g_i x^i \right) - p \right] J_{,\alpha}\,. \tag{7.29}$$

The first part represents the actual inhomogeneity of the fluid due to a prescribed density stratification $\rho(\boldsymbol{\xi})$. The second part is not a property of the fluid, but arises from the choice of the reference configuration. So even if the fluid is homogeneous ($\rho = \text{const}$), the reference configuration may be chosen such that the Jacobian J depends on the material point giving rise to an inhomogeneity force.

We have now obtained the balance of material momentum in its standard form

$$\frac{\mathrm{D}\mathfrak{p}_\alpha}{\mathrm{D}t} = \mathfrak{E}_\alpha{}^\beta{}_{,\beta} + \mathfrak{f}_\alpha \quad \text{or} \quad \frac{\mathrm{D}\boldsymbol{\mathfrak{p}}}{\mathrm{D}t} = \mathrm{Div}\,\boldsymbol{\mathfrak{E}} + \boldsymbol{\mathfrak{f}} \tag{7.30}$$

with the material momentum vector $\boldsymbol{\mathfrak{p}}$, the dynamic Eshelby tensor $\boldsymbol{\mathfrak{E}}$, and the inhomogeneity force $\boldsymbol{\mathfrak{f}}$ according to (7.24), (7.27), and (7.29), respectively. The special structure of the Eshelby tensor (7.27) suggests to introduce a scalar "Eshelby pressure"

$$\mathfrak{q} = \rho_\mathrm{R} \left(\frac{1}{2}v^2 + \boldsymbol{g}\cdot\boldsymbol{x} \right) - Jp, \tag{7.31}$$

such that $\boldsymbol{\mathfrak{E}} = -\mathfrak{q}\boldsymbol{I}$. This simplifies the balance equation to

$$\frac{\mathrm{D}\mathfrak{p}}{\mathrm{D}t} = -\mathrm{Grad}\,\mathfrak{q} + \mathfrak{f}, \tag{7.32}$$

which resembles the Eulerian form of the balance of *physical* momentum. In the case of vanishing inhomogeneity force $\mathfrak{f}$, the balance becomes a strict conservation law. The pressure p in an incompressible material is an independent function $p(\boldsymbol{\xi},t)$ that is not determined by the motion $\boldsymbol{\varphi}(\boldsymbol{\xi},t)$ of the medium. The Eshelby pressure (7.31) depends on the motion and, additionally, on the pressure p. Therefore, it may be considered as an independent function $\mathfrak{q}(\boldsymbol{\xi},t)$ itself.

An alternate, though equivalent form of the balance of material momentum can be stated by modifying its ingredients as follows: Instead of the quantities $\mathfrak{p}$, $\mathfrak{q}$, $\mathfrak{f}$ we introduce

$$\bar{\mathfrak{p}} = \frac{1}{J}\mathfrak{p} = -\rho \boldsymbol{F}^{\mathsf{T}}\boldsymbol{v},\ \bar{\mathfrak{q}} = \frac{\mathfrak{q}}{J} = \rho\left(\frac{1}{2}v^2 + \boldsymbol{g}\cdot\boldsymbol{x}\right) - p,\ \bar{\mathfrak{f}} = \left(\frac{1}{2}v^2 + \boldsymbol{g}\cdot\boldsymbol{x}\right)\frac{\partial\rho}{\partial\boldsymbol{\xi}} \tag{7.33}$$

as modified material momentum, Eshelby pressure, and inhomogeneity force. Also these quantities satisfy the balance equation

$$\frac{\mathrm{D}\bar{\mathfrak{p}}}{\mathrm{D}t} = -\mathrm{Grad}\,\bar{\mathfrak{q}} + \bar{\mathfrak{f}} \tag{7.34}$$

that can be interpreted as the balance of material momentum. When comparing this new version to the original one, it might be considered a stylistic clash to define the quantities $\bar{\mathfrak{p}}$, $\bar{\mathfrak{q}}$, $\bar{\mathfrak{f}}$ in terms of the *actual* density ρ rather than the referential density ρ_{R}. It should be emphasized, however, that the actual density is considered here as a function $\rho(\boldsymbol{\xi})$, in the same way as the velocity vector is a function $\boldsymbol{v}(\boldsymbol{\xi},t)$, although both live in the actual configuration. An advantage of the alternate version is the fact that the formulas for the Eshelby pressure and the inhomogeneity force have become simpler and, above all, that we have eliminated the fictitious inhomogeneity that is brought in by the choice of the reference configuration. Unlike $\mathfrak{f}$, the inhomogeneity force $\bar{\mathfrak{f}}$ indicates only the true inhomogeneity of the fluid, independently of the chosen reference configuration.

The incompressibility condition (7.5) introduces a second derivative of the deformation, namely $\boldsymbol{\varphi}_{\xi t}$, into the Lagrangian. This puts the incompressible fluid close to a second-grade elastic material, where, of course, also the more important spatial second gradient $\boldsymbol{\varphi}_{\xi\xi}$ is included. Lazar and Anastassiadis (2007) have shown that there are different ways of defining material momentum and Eshelby stress of a second-grade material. The reason is that balance equations are always understood as first-order differential equations containing a time derivative and a divergence term. Higher-order derivatives must be included in either of them, and the assignment of mixed derivatives is not always unique.

Lazar and Anastassiadis (2007) emphasize that the inclusion of the time derivative of a *tensor* in the Eshelby stress makes it non-objective. Actually we have included a time-derivative in the Eshelby stress. But in our case it is only $\mathrm{D}\lambda/\mathrm{D}t = -p$, the derivative of a scalar, and this is always an objective time rate.

7.4 The Energy Balance

Although it is outside of the title problem, another material balance law should be mentioned, which is obtained in the same way as the balance of material momentum. Following again Gołębiewska-Herrmann (1981) we form the total derivative of the Lagrangian, now with respect to time t rather than with respect to the material coordinate ξ^α. Then, instead of (7.16), we get

$$\mathrm{D}_t\mathscr{L} = \frac{\partial\mathscr{L}}{\partial x^i}\dot{x}^i + \frac{\partial\mathscr{L}}{\partial\dot{x}^i}\ddot{x}^i + \frac{\partial\mathscr{L}}{\partial x^i{}_{,\alpha}}\dot{x}^i{}_{,\alpha} + \frac{\partial\mathscr{L}}{\partial\dot{x}^i{}_{,\alpha}}\ddot{x}^i{}_{,\alpha} + \frac{\partial\mathscr{L}}{\partial\lambda}\dot{\lambda} + \frac{\partial\mathscr{L}}{\partial t}. \tag{7.35}$$

Since there is no explicit dependence on time, the partial derivative $\partial\mathscr{L}/\partial t$ can be omitted. The subsequent steps are similar to those for the derivation of the material momentum balance. Again the product rule has to be applied to some terms, and the equation has to be recast such that the functional derivatives of the Lagrangian become visible. The resulting equation, the counterpart of (7.20), is here

$$\begin{aligned} 0 = {} & \frac{\delta\mathscr{L}}{\delta x^i}\dot{x}^i + \frac{\delta\mathscr{L}}{\delta\lambda}\dot{\lambda} + \\ & + \mathrm{D}_t\left(\frac{\partial\mathscr{L}}{\partial\dot{x}^i}\dot{x}^i + \frac{\partial\mathscr{L}}{\partial\dot{x}^i{}_{,\alpha}}\dot{x}^i{}_{,\alpha} - \mathscr{L}\right) + \mathrm{D}_\alpha\left(\frac{\partial\mathscr{L}}{\partial x^i{}_{,\alpha}}\dot{x}^i - \mathrm{D}_t\frac{\partial\mathscr{L}}{\partial\dot{x}^i{}_{,\alpha}}\dot{x}^i\right). \end{aligned} \tag{7.36}$$

If the functional derivatives (7.11) vanish, the equation assumes the form of a conservation law for the energy density. This is predicted by Noether's law applied to the symmetry of the Lagrangian with respect to a translation in time.

Taking into account the special form (7.8) of the Lagrangian the total energy density, or Hamiltonian, becomes

$$\mathscr{H} = \frac{\partial\mathscr{L}}{\partial\dot{x}^i}\dot{x}^i + \frac{\partial\mathscr{L}}{\partial\dot{x}^i{}_{,\alpha}}\dot{x}^i{}_{,\alpha} - \mathscr{L} = \rho_\mathrm{R}\left(\frac{1}{2}v^2 - \boldsymbol{g}\cdot\boldsymbol{x}\right). \tag{7.37}$$

The term $\lambda\dot{J}$ in the Lagrangian has canceled out, such that the Hamiltonian is simply the sum of kinetic and potential energy. Further we define the components of the energy flux or Poynting vector by

$$S^\alpha = -\left(\frac{\partial\mathscr{L}}{\partial x^i{}_{,\alpha}} - \mathrm{D}_t\frac{\partial\mathscr{L}}{\partial\dot{x}^i{}_{,\alpha}}\right)\dot{x}^i = \dot{\lambda}J\xi^\alpha{}_{,i}\dot{x}^i = pJV^\alpha, \tag{7.38}$$

where $\boldsymbol{V} = \partial\boldsymbol{\xi}(\boldsymbol{x},t)/\partial t$ denotes the velocity at which a material point has to move in reference space in order to keep its actual position $\boldsymbol{x}$ fixed. Now the conservation law assumes the form

$$\frac{\mathrm{D}\mathscr{H}}{\mathrm{D}t} = S^\alpha{}_{,\alpha} \quad \text{or} \quad \frac{\mathrm{D}\mathscr{H}}{\mathrm{D}t} = \mathrm{Div}\,\boldsymbol{S}. \tag{7.39}$$

Since the Lagrangian does not explicitly depend on time, there is no source term and the balance law is a strict conservation law. A source term would appear, if material parameters – in our case it could be only the density ρ_R – depend on time. Materials with time-dependent properties were studied, for instance, by Rousseau et al (2011).

7.5 Gerstner's Wave

To describe Gerstner's theory of water waves we use cartesian coordinates in both the reference and the actual configuration. Gerstner's theory is two-dimensional. To simplify the notation the index notation is abandoned, and the coordinates are denoted by (ξ,η) in the reference configuration and by (x,y) in the actual configuration, the latter with the y-axis pointing upward. In the reference configuration the fluid occupies the half-plane $\eta \leq 0$.

Following Gerstner we assume the motion $\boldsymbol{x} = \boldsymbol{\varphi}(\boldsymbol{\xi},t)$ to be described by

$$x = \xi + r(\eta)\sin\theta, \quad y = y_0 + \eta - r(\eta)\cos\theta \tag{7.40}$$

with the phase

$$\theta = \kappa\xi - \omega t. \tag{7.41}$$

The wave number κ and the frequency ω are assumed constant. The function $r(\eta)$ and the constant displacement y_0 have still to be specified.

The deformation gradient of the motion has the components

$$\boldsymbol{F} = \begin{bmatrix} 1+\kappa r\cos\theta & r'\sin\theta \\ \kappa r\sin\theta & 1-r'\cos\theta \end{bmatrix} \tag{7.42}$$

and leads to the Jacobian

$$J = \det\boldsymbol{F} = 1 - \kappa r r' - \left(r' - \kappa r\right)\cos\theta. \tag{7.43}$$

In general, this Jacobian would depend on time via the phase (7.41), thus violating the incompressibility condition. Therefore, the factor in front of $\cos\theta$ must vanish, which means that the function $r(\eta)$ has to satisfy the differential equation

$$r' = \kappa r. \tag{7.44}$$

Then the deformation gradient and the Jacobian become

$$\boldsymbol{F} = \begin{bmatrix} 1+\kappa r\cos\theta & \kappa r\sin\theta \\ \kappa r\sin\theta & 1-\kappa r\cos\theta \end{bmatrix} \quad \text{and} \quad J = 1 - (\kappa r)^2. \tag{7.45}$$

The general solution of the differential equation (7.44) is

$$r = a e^{\kappa\eta} \tag{7.46}$$

with an integration constant a that plays the role of a wave amplitude. Since the Jacobian must be positive in the whole fluid-filled area $\eta \leq 0$ of the reference configuration, the amplitude a and the wavenumber κ have to satisfy the inequality

$$\kappa a < 1, \tag{7.47}$$

which guarantees $J > 0$. If the wave amplitude is nonzero, the Jacobian will also be limited by $J < 1$. Therefore, the fluid cannot adopt the reference configuration at any time without violating the incompressibility constraint. For this reason, the more general condition of incompressibility $\mathrm{D}J/\mathrm{D}t = 0$ has been employed, rather than $J = 1$.

The ansatz (7.40) contains still a constant displacement y_0 that has to be specified. Gerstner himself and most of subsequent presentations take $y_0 = 0$, which means that crests and troughs have the elevations $y = \pm a$. However, since the troughs are wider than the peaks, the *mean* water level is then below the still-water level $y = 0$. To overcome this discrepancy one should choose the elevation y_0 such that the mean water level always coincides with the still-water level $y = 0$, as suggested by Clamond (2007). According to $(7.40)_2$ the mean water level is located at

$$\overline{y(\xi,0,t)} = \frac{\kappa}{2\pi} \int\limits_{\xi=0}^{2\pi/\kappa} y(\xi,0,t)\mathrm{d}x(\xi,0,t) = y_0 - \frac{1}{2}\kappa a^2. \tag{7.48}$$

Henceforth we shall take

$$y_0 = \frac{1}{2}\kappa a^2, \tag{7.49}$$

which means that the mean water level coincides with the still-water level $y = 0$.

The deformation at the fixed time $t = 0$ represented by the ansatz (7.40) with $r(\eta)$ and y_0 inserted from (7.46) and (7.49), respectively, is shown in Fig. 7.1. A square grid in the half-plane $\eta \leq 0$ of the reference configuration is transformed to the actual configuration. The parallels $\eta = \text{const}$ are mapped to trochoids. Below a depth of about half a wavelength, the material square grid is nearly undistorted. The deformed grid in Fig. 7.1 gives an impression of the mapping $\boldsymbol{\xi} \mapsto \boldsymbol{x} = \boldsymbol{\varphi}(\boldsymbol{\xi}, t_0)$ at a fixed time t_0. In contrast, a pathline tracks the motion $t \mapsto \boldsymbol{x} = \boldsymbol{\varphi}(\boldsymbol{\xi}_0, t)$ of a fixed material

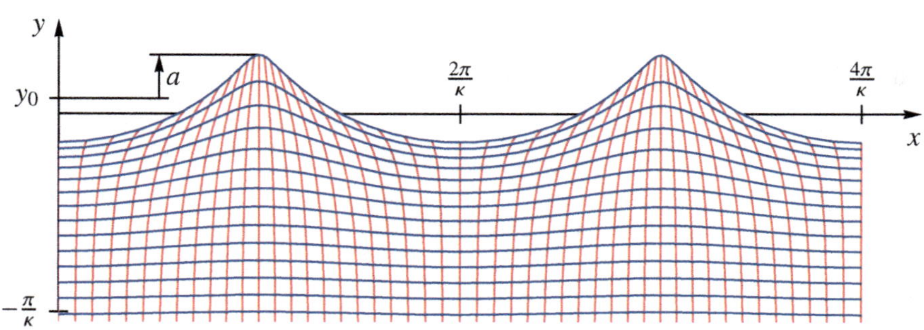

Fig. 7.1: Gerstner's wave: deformation at time $t = 0$, amplitude $a = 0.7/\kappa$

point ξ_0. In Gerstner's wave, any material point moves along a circular pathline of radius $r(\eta) = a\exp(\kappa\eta)$ with constant velocity ωr. If the wave propagates to the right, the circles are traversed clockwise. Figure 7.2 shows the pathlines of selected material points.

The actual density ρ of the fluid is taken constant in Gerstner's theory, although a generalization to stratified fluid would be possible. According to (7.3), the referential density

$$\rho_\mathrm{R} = J\rho = \left(1 - \kappa^2 r^2\right)\rho = \left(1 - \kappa^2 a^2 \mathrm{e}^{2\kappa\eta}\right)\rho \tag{7.50}$$

has its minimum value at the surface $\eta = 0$ and increases with depth tending asymptotically to $\rho_\mathrm{R} \to \rho$ as $\eta \to -\infty$. This "stratification" in the reference configuration is not a property of the fluid, it has rather been carried in by the choice of the reference configuration. As will be shown in the next section there exist also reference configurations with a constant reference density.

The ansatz (7.40) represents an isochoric motion, but it has not yet been shown that the governing equations are fulfilled. To this end we shall check the alternate form (7.34) of the balance of material momentum. It has the advantage that the inhomogeneity force $\bar{\mathfrak{f}}$ vanishes, since the actual density ρ is assumed constant. The balance of material momentum is reduced to the conservation law

$$\frac{\mathrm{D}\bar{\mathfrak{p}}}{\mathrm{D}t} = -\mathrm{Grad}\,\bar{\mathfrak{q}}. \tag{7.51}$$

Therefore, the time derivative of the material momentum has to satisfy the compatibility condition

$$\mathrm{Curl}\,\frac{\mathrm{D}\bar{\mathfrak{p}}}{\mathrm{D}t} = 0. \tag{7.52}$$

The vectors of velocity and material momentum have the components

$$\boldsymbol{v} = -\omega r \begin{bmatrix} \cos\theta \\ \sin\theta \end{bmatrix} \quad \text{and} \quad \bar{\mathfrak{p}} = -\rho \boldsymbol{F}^\mathsf{T}\boldsymbol{v} = \rho\omega r \begin{bmatrix} \cos\theta + \kappa r \\ \sin\theta \end{bmatrix}, \tag{7.53}$$

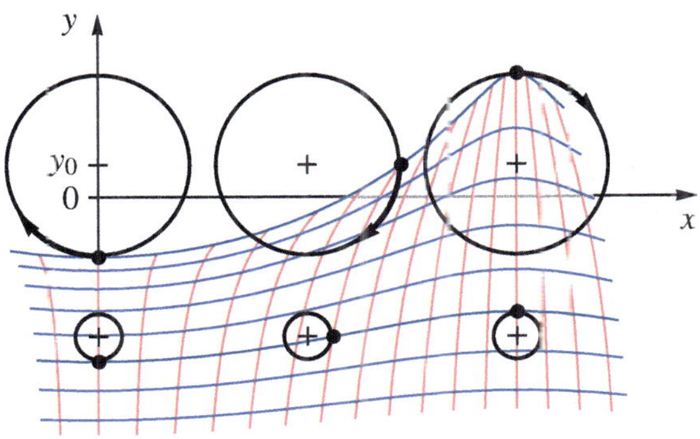

Fig. 7.2 Pathlines of a Gerstner wave propagating in positive x-direction. The initial positions of the material points at time $t = 0$ are marked by a dot. Also the deformed grid is shown for $t = 0$.

respectively, the material momentum depending on time via the phase θ. In two dimensions the compatibility condition (7.52) is reduced to a single equation, and one can easily verify that it is fulfilled by the material momentum vector $(7.53)_2$. This confirms that Gerstner's ansatz satisfies the balance of material momentum in its alternate form (7.34).

The Eshelby pressure is obtained explicitly by integrating the equation

$$\operatorname{Grad} \bar{\mathfrak{q}} = -\frac{\mathrm{D}\bar{\mathfrak{p}}}{\mathrm{D}t} = \rho\omega^2 r \begin{bmatrix} -\sin\theta \\ \cos\theta \end{bmatrix}, \tag{7.54}$$

resulting in

$$\bar{\mathfrak{q}} = \rho \frac{\omega^2 r}{\kappa} \cos\theta + C \tag{7.55}$$

with an integration constant C. According to $(7.33)_2$ the Eshelby pressure is related to the kinetic and potential energy densities and to the hydrostatic pressure p. Solving for the latter yields

$$p = \rho \left[\frac{1}{2}(\omega r)^2 - g(y_0 + \eta) + \left(g - \frac{\omega^2}{\kappa} \right) r \cos\theta \right] - C. \tag{7.56}$$

At the free surface $\eta = 0$ we have $r = a$ and the pressure becomes

$$p(\xi, 0, t) = \rho \left[\frac{1}{2}(\omega a)^2 - g y_0 + \left(g - \frac{\omega^2}{\kappa} \right) a \cos\theta \right] - C. \tag{7.57}$$

It depends, in general, on the phase θ of the wave. In principle, Gerstner's wave (7.40) with *arbitrary* values of the wave number κ, the frequency ω, and the amplitude a can propagate, provided the free surface is exposed to the pressure field (7.57) traveling at the phase speed $c = \omega/\kappa$.

Under realistic conditions, the pressure acting on the free water surface may have some modulation due to the interaction of the wind with the wavy water surface. This is a subject of its own and treated, for instance, in Janssen (2004). It is outside the scope of the present analysis. Here we assume, as Gerstner did, that the water surface is exposed to a constant pressure p_0. In this case, the variable term in (7.57) has to vanish, which leads to

$$\omega^2 = g\kappa. \tag{7.58}$$

It is the dispersion law of the linear theory of deep-water waves that holds here within a nonlinear theory. Below the water surface, the pressure is a function

$$p(\eta) = p_0 - \rho g \eta - \frac{1}{2}\rho\omega^2 \left(a^2 - r^2\right) = p_0 - \rho g \left[\eta + \frac{1}{2}\kappa a^2 \left(1 - \mathrm{e}^{2\kappa\eta}\right) \right]. \tag{7.59}$$

This means that not only the free surface $\eta = 0$, but all levels $\eta = \text{const}$ are isobaric surfaces. For $\eta \to -\infty$, where the wave motion is not perceptible anymore, the pressure tends to the hydrostatic pressure $p_0 - \rho g y$.

Gerstner's wave stands out for being the only known closed-form solution of the *nonlinear* equations. Its big deficiency is that the flow field is rotational. This should be pointed out in brief, even if it is not related to the balance of material momentum. The Eulerian velocity gradient of Gerstner's wave, expressed in the deformation gradient and its time-derivative, is

$$\operatorname{grad} \boldsymbol{v} = \frac{\mathrm{D}\boldsymbol{F}}{\mathrm{D}t}\boldsymbol{F}^{-1} = -\frac{\omega\kappa r}{J}\begin{bmatrix} -\sin\theta & \cos\theta + \kappa r \\ \cos\theta - \kappa r & \sin\theta \end{bmatrix}. \tag{7.60}$$

Its skew-symmetric part,

$$\boldsymbol{W} = \frac{1}{2}\left[\operatorname{grad} \boldsymbol{v} - (\operatorname{grad} \boldsymbol{v})^{\mathsf{T}}\right] = \frac{\omega(\kappa r)^2}{J}\begin{bmatrix} 0 & -1 \\ +1 & 0 \end{bmatrix}, \tag{7.61}$$

is the vorticity tensor. In three dimensions, vorticity is the corresponding axial vector. In our two dimensional flow, vorticity is the pseudo-scalar

$$w = W_{21} - W_{12} = \frac{2\omega(\kappa r)^2}{J} = \frac{2\omega(\kappa a \mathrm{e}^{\kappa\eta})^2}{1-(\kappa a \mathrm{e}^{\kappa\eta})^2}. \tag{7.62}$$

The vorticity of Gerstner's wave propagating to the right is positive, i.e., opposite to the clockwise rotation of the material points along their pathlines. It has its maximum at the water surface and decays exponentially with depth. Figure 7.3 shows the dependence of the vorticity on the vertical coordinate η according to (7.62), for different values of the dimensionless amplitude κa. For $\kappa a \to 1$, the vorticity at the surface tends to infinity.

A parallel shear flow of the same vorticity distribution would flow in the opposite direction of the wave motion. This flow was analyzed by Stokes already in 1847 (Stokes, 2009, page 223). In the same paper he presented an approximate theory of *irrotational* waves, where the pathlines of particles are not closed, leading to what today is called the Stokes drift.

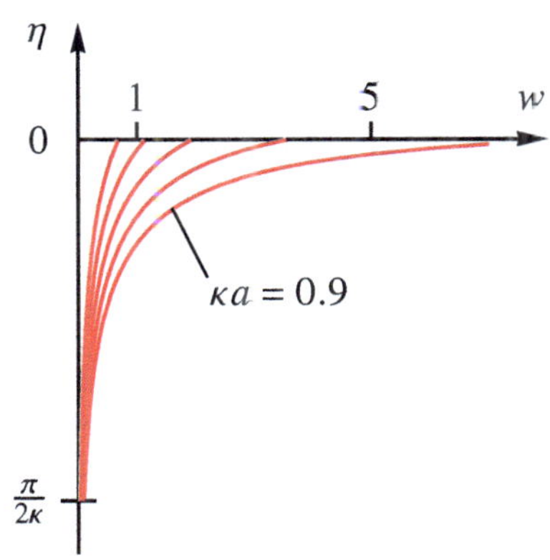

Fig. 7.3 Vorticity w as a function of the vertical Lagrangian coordinate η, for normalized amplitudes $\kappa a = 0.5,\ldots,0.9$

7.6 Change of Reference Configuration

Gerstner's representation (7.40) of the wave motion leads to a Jacobian $J(\eta) < 1$ that depends on the vertical material coordinate η. It should be possible to describe the same motion using a different reference configuration such that the Jacobian assumes the constant value $\hat{J} = 1$. This alternate reference configuration $\hat{\mathscr{B}}_\mathsf{R}$ can be generated from $\mathscr{B}_\mathsf{R}$ by an appropriate diffeomorphism

$$\boldsymbol{\Psi}\colon \mathscr{B}_\mathsf{R} \to \hat{\mathscr{B}}_\mathsf{R};\, \boldsymbol{\xi} \mapsto \boldsymbol{X} = \boldsymbol{\Psi}(\boldsymbol{\xi}), \quad \boldsymbol{K} = \frac{\partial \boldsymbol{X}}{\partial \boldsymbol{\xi}}. \tag{7.63}$$

The motion $\boldsymbol{x} = \boldsymbol{\varphi}(\boldsymbol{\xi},t)$ expressed in the new material coordinates $\boldsymbol{X}$ becomes

$$\boldsymbol{x} = \hat{\boldsymbol{\varphi}}(\boldsymbol{X},t) = \boldsymbol{\varphi}\left(\boldsymbol{\Psi}^{-1}(\boldsymbol{X}),t\right), \tag{7.64}$$

while the corresponding deformation gradient and Jacobian are transformed as

$$\hat{\boldsymbol{F}} = \boldsymbol{F}\boldsymbol{K}^{-1}, \quad \hat{J} = \frac{J}{\det \boldsymbol{K}}. \tag{7.65}$$

A constant Jacobian $\hat{J} = 1$ is achieved by any transformation $\boldsymbol{\Psi}$ whose Jacobian is $\det \boldsymbol{K} = J$.

Since, in the present case, the Jacobian J depends only on the vertical material coordinate η, it is reasonable to leave the horizontal coordinate unaltered and transform only the vertical coordinate:

$$X = \xi, \quad Y = \int_0^\eta J(\eta)\mathrm{d}\eta = \eta + \frac{1}{2}\kappa a^2\left(1 - \mathrm{e}^{2\kappa\eta}\right). \tag{7.66}$$

Then the differential of the diffeomorphism and its Jacobian are simply

$$\boldsymbol{K} = \begin{bmatrix} 1 & \\ & J(\eta) \end{bmatrix}, \quad \det \boldsymbol{K} = J(\eta). \tag{7.67}$$

The motion $\boldsymbol{x} = \hat{\boldsymbol{\varphi}}(\boldsymbol{X},t)$ has, according to $(7.65)_2$, the Jacobian $\hat{J} = 1$, which means that the new reference configuration $\hat{\mathscr{B}}_\mathsf{R}$ can be attained by the fluid. Up to now the motion is available only in the parametric form $\boldsymbol{X} = \boldsymbol{\Psi}(\boldsymbol{\xi})$, $\boldsymbol{x} = \boldsymbol{\varphi}(\boldsymbol{\xi},t)$, namely

$$\begin{aligned} X &= \xi, & x &= \xi + \kappa a \mathrm{e}^{\kappa\eta}\sin(\kappa\xi - \omega t), \\ Y &= \eta + \frac{1}{2}\kappa a^2\left(1 - \mathrm{e}^{2\kappa\eta}\right), & y &= \frac{1}{2}\kappa a^2 + \eta - \kappa a \mathrm{e}^{\kappa\eta}\cos(\kappa\xi - \omega t). \end{aligned} \tag{7.68}$$

To obtain the explicit representation $\boldsymbol{x} = \hat{\boldsymbol{\varphi}}(\boldsymbol{X},t)$ one has to invert the transformation (7.66), which is rather tedious. As shown by Clamond (2007) the inverse transformation can be expressed in terms of the Lambert W function, which is defined as the inverse of the function

$$W^{-1}(z) = z\mathrm{e}^{z}, \tag{7.69}$$

see Corless et al (1996) for details. Using the W function the inverse transformation of (7.66) becomes

$$\xi = X, \quad \eta = Y - \frac{1}{2\kappa}\left[\kappa^2 a^2 + W\left(-\kappa^2 a^2 \mathrm{e}^{2\kappa Y - \kappa^2 a^2}\right)\right]. \tag{7.70}$$

Now the unpleasant expression for η has to be inserted into $(7.68)_{2,4}$. Despite this "explicit" representation it is, for practical reasons, easier to invert the transformation numerically or to keep the parametric representation (7.68).

Figure 7.4 shows the two reference configurations and the actual configuration for a high-amplitude wave. A square grid is chosen in the alternate reference configuration $\hat{\mathscr{B}}_\mathsf{R}$, which is transformed by $\boldsymbol{\Psi}^{-1}$ to a grid of rectangles varying with η in the original reference configuration $\mathscr{B}_\mathsf{R}$. This, in turn, is mapped via $\boldsymbol{\varphi}$ to the actual configuration $\mathscr{B}(t)$. In Fig. 7.4, the three configurations are aligned such that the grids in the deep-water regime coincide. Therefore, the (ξ,η)-plane is raised by y_C. Unlike $\boldsymbol{\varphi}$ the mapping $\hat{\boldsymbol{\varphi}}$ is volume-preserving.

How does the change of the reference configuration influence the balance of material momentum? The material time derivative $\mathrm{D}/\mathrm{D}t$ is the same in either reference configuration, since the transformation $\boldsymbol{\Psi}$ is time-independent. The material momenta and the gradients of the Eshelby pressure in the two configurations are related by

$$\hat{\mathfrak{p}} = \boldsymbol{K}^\mathsf{T}\overline{\mathfrak{p}} \quad \text{and} \quad \mathrm{Grad}_{\boldsymbol{X}}\,\overline{\mathfrak{q}} = \boldsymbol{K}^\mathsf{T}\mathrm{Grad}_{\boldsymbol{\xi}}\,\overline{\mathfrak{q}}. \tag{7.71}$$

Therefore, changing the reference configuration from $\mathscr{B}_\mathsf{R}$ to $\hat{\mathscr{B}}_\mathsf{R}$ has the consequence that either side of the conservation law (7.51) is multiplied by $\boldsymbol{K}^\mathsf{T}$, where $\boldsymbol{K}$ is given by $(7.67)_1$. So essentially it is the same equation. Still of interest is the dependence of the pressure p on the vertical material coordinate. Using the new coordinate Y the function $p(\eta)$ according to (7.59) is transformed to

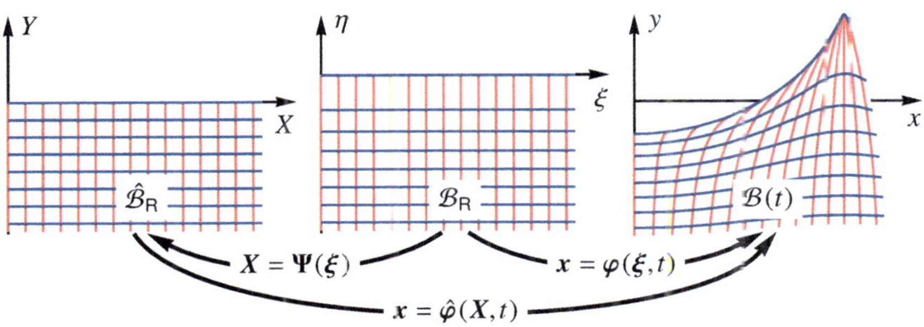

Fig. 7.4: Mappings between reference configurations $\hat{\mathscr{B}}_\mathsf{R}$, $\mathscr{B}_\mathsf{R}$ and actual configuration $\mathscr{B}(t)$; dimensionless amplitude $\kappa a = 0.9$

$$p(Y) = p_0 - \rho g Y. \tag{7.72}$$

In Gerstner's wave, the pressure at an arbitrary material point is equal to the hydrostatic pressure at that material point in the fluid at rest occupying the reference configuration $\hat{\mathscr{B}}_{\mathsf{R}}$. There is no change due to the motion, the levels of hydrostatic pressure are just mapped as material surfaces to the actual configuration.

7.7 Concluding Remarks

The balance of material momentum is applied to fluid motion, especially to the propagation of a surface gravity wave according to Gerstner's theory. In the Lagrangian description of fluid motion, the reference configuration is not necessarily some initial configuration. Even a configuration that cannot be attained by the fluid without violating the incompressibility condition is a legitimate reference configuration. Therefore, the incompressibility condition demands that the Jacobian of the deformation is only time-independent, but not necessarily unity. The Lagrangian, in which the incompressibility condition $\mathrm{D}J/\mathrm{D}t = 0$ is incorporated, exhibits a dependence on the mixed second derivative $\boldsymbol{\varphi}_{\boldsymbol{\xi}t}$ of the deformation.

The balance of material momentum of fluid motion is derived and the final result is presented in two versions. Formally they have the same appearance, resembling the balance of physical momentum in the actual configuration. The two versions differ, however, in the definition of material momentum, which is based either on the density ρ_{R} in the material configuration or on the density ρ in the actual configuration. In either case, the densities are considered as functions of the material coordinates. The density $\rho(\boldsymbol{\xi})$ has the advantage that it describes a property of the fluid only, while $\rho_{\mathsf{R}}(\boldsymbol{\xi})$ depends, in addition, on the choice of the reference configuration. For a reference configuration that complies with incompressibility, the densities are identical, $\rho_{\mathsf{R}} = \rho$, and the two versions of the balance of material momentum coincide. More as a side note, also the balance of energy in material space is derived.

Gerstner's theory of water waves relies on a Lagrangian description and is, therefore, amenable to configurational mechanics. It is shown that Gerstner's wave fulfills the balance of material momentum. The pressure at a material point depends only on the vertical material coordinate as in the hydrostatic case. The reference configuration used by Gerstner does not comply with the incompressibility condition and is, therefore, not attainable by the fluid. This application justifies the more general approach to incompressibility used throughout this paper.

Finally an alternate reference configuration is introduced which complies with the incompressibility constraint. This seems to be the most natural reference configuration. The Lagrangian description of the waves, however, becomes rather complicated and can be formulated only in terms of a not so well-known special function. Gerstner's simple and elegant representation works only with the original, distorted reference configuration that cannot be attained by the incompressible fluid.

The balance of material momentum for an ideal fluid, although much simpler than for an elastic body, is not so trivial as it might seem on the first sight. Still missing is an appropriate physical interpretation of the Eshelby pressure acting in the reference configuration of the fluid. In analogy to elasticity there should be a path integral that describes the configurational force acting on some inhomogeneity. Its significance is still an open question. Unfortunately, Gérard Maugin cannot be asked anymore for his advice. His comments, whether critical or encouraging, have always been of high value for the author.

Appendix: Derivatives of the Lagrangian

The derivatives of the Lagrangian (7.8) with respect to the motion $\boldsymbol{\varphi}$ and its derivatives are needed in Sects. 7.2 to 7.4. The derivatives with respect to $\boldsymbol{\varphi}$ and $\boldsymbol{\varphi}_t$ are immediately available as

$$\frac{\partial \mathscr{L}}{\partial x^i} = \rho_{\mathrm{R}} g_i \quad \text{and} \quad \frac{\partial \mathscr{L}}{\partial \dot{x}^i} = \rho_{\mathrm{R}} \dot{x}_i. \tag{7.73}$$

Additionally the Lagrangian depends on $\boldsymbol{\varphi}_\xi$ and $\boldsymbol{\varphi}_{\xi t}$ via the incompressibility term

$$\lambda h = \lambda \dot{J} = \lambda \frac{\partial J}{\partial x^j{}_{,\beta}} \dot{x}^j{}_{,\beta} \tag{7.74}$$

giving the derivatives

$$\frac{\partial \mathscr{L}}{\partial x^i{}_{,\alpha}} = \lambda \frac{\partial^2 J}{\partial x^i{}_{,\alpha} \partial x^j{}_{,\beta}} \dot{x}^j{}_{,\beta} \quad \text{and} \quad \frac{\partial \mathscr{L}}{\partial \dot{x}^i{}_{,\alpha}} = \lambda \frac{\partial J}{\partial x^i{}_{,\alpha}}. \tag{7.75}$$

These appear mainly in the combination

$$\frac{\delta \mathscr{L}}{\delta x^i{}_{,\alpha}} \equiv \frac{\partial \mathscr{L}}{\partial x^i{}_{,\alpha}} - \mathrm{D}_t \frac{\partial \mathscr{L}}{\partial \dot{x}^i{}_{,\alpha}} = -\dot{\lambda} \frac{\partial J}{\partial x^i{}_{,\alpha}}, \tag{7.76}$$

where the second derivatives of J have cancelled out. In (7.11) the total derivative

$$-\mathrm{D}_\alpha \left(\frac{\partial \mathscr{L}}{\partial x^i{}_{,\alpha}} - \mathrm{D}_t \frac{\partial \mathscr{L}}{\partial \dot{x}^i{}_{,\alpha}} \right) = \dot{\lambda}_{,\alpha} \frac{\partial J}{\partial x^i{}_{,\alpha}} + \dot{\lambda} \frac{\partial^2 J}{\partial x^i{}_{,\alpha} \partial x^j{}_{,\beta}} x^j{}_{,\alpha\beta} \tag{7.77}$$

is needed. At this point, the derivatives of the Jacobian with respect to the deformation gradient are written out explicitly as

$$\frac{\partial J}{\partial x^i{}_{,\alpha}} = J \xi^\alpha{}_{,i} \quad \text{and} \quad \frac{\partial^2 J}{\partial x^i{}_{,\alpha} \partial x^j{}_{,\beta}} = J \left(\xi^\alpha{}_{,i} \xi^\beta{}_{,j} - \xi^\alpha{}_{,j} \xi^\beta{}_{,i} \right), \tag{7.78}$$

whence the derivative (7.77) is reduced to

$$-\mathrm{D}_\alpha\left(\frac{\partial\mathscr{L}}{\partial x^i_{,\alpha}} - \mathrm{D}_t\frac{\partial\mathscr{L}}{\partial\dot{x}^i_{,\alpha}}\right) = \dot{\lambda}_{,\alpha}J\xi^\alpha{}_{,i}. \tag{7.79}$$

Collecting now the relevant derivatives of the Lagrangian the Euler–Lagrange equation $(7.11)_1$ becomes

$$\rho_\mathrm{R}g_i - \rho_\mathrm{R}\ddot{x}_i + \dot{\lambda}_{,\alpha}J\xi^\alpha{}_{,i} = 0. \tag{7.80}$$

Reformulated in index-free notation this is the balance of physical momentum (7.12).

References

Clamond D (2007) On the Lagrangian description of steady surface gravity waves. Journal of Fluid Mechanics 589:433–454

Constantin A (2001) On the deep water wave motion. Journal of Physics A: Mathematical and General 34(7):1405–1417

Constantin A, Monismith SG (2017) Gerstner waves in the presence of mean currents and rotation. Journal of Fluid Mechanics 820:511–528

Corless RM, Gonnet GH, Hare DEG, Jeffrey DJ, Knuth DE (1996) On the Lambert *W* function. Advances in Computational Mathematics 5(1):329–359

Gerstner F (1804) Theorie der Wellen samt einer daraus abgeleiteten Theorie der Deichprofile - Abh. Königl. Böhm. Ges. Wiss. Haase, Prag

Gołębiewska-Herrmann A (1981) On conservation laws of continuum mechanics. International Journal of Solids and Structures 17(1):1–9

Gurtin ME (2000) Configurational Forces as Basic Concepts of Continuum Physics. Springer, New York

Janssen P (2004) The Interaction of Ocean Waves and Wind. Cambridge University Press, Cambridge

Kalisch H (2004) Periodic traveling water waves with isobaric streamlines. Journal of Nonlinear Mathematical Physics 11(4):461–471

Kienzler R, Herrmann G (2000) Mechanics in Material Space - With Application to Defect and Fracture Mechanics. Springer, Berlin

Lamb H (1932) Hydrodynamics, 6th edn. Cambridge University Press, Cambridge

Lazar M, Anastassiadis C (2007) Lie point symmetries, conservation and balance laws in linear gradient elastodynamics. Journal of Elasticity 88(1):5–25

Le Méhauté B (1976) An Introduction to Hydrodynamics and Water Waves. Springer, New York

Maugin GA (1993) Material Inhomogeneities in Elasticity. Chapman & Hall, London

Maugin GA (2010) Configurational Forces: Thermomechanics, Physics, Mathematics, and Numerics. Chapman & Hall/CRC

Rahman M (1995) Water Waves. Clarendon Press, Oxford

Rousseau M, Maugin GA, Berezovski M (2011) Elements of study on dynamic materials. Archive of Applied Mechanics 81(7):925–942

Stokes GG (2009) On the Theory of Oscillatory Waves, Cambridge Library Collection - Mathematics, vol 1, Cambridge University Press, pp 197–229

Stuhlmeier R (2015) Gerstner's water wave and mass transport. Journal of Mathematical Fluid Mechanics 17(4):761–767

Chapter 8
Electromagnetic Fields in Meta-Media with Interfacial Surface Admittance

David C. Christie and Robin W. Tucker

Abstract We exploit Clemmow's complex plane-wave representation of electromagnetic fields to construct globally exact solutions of Maxwell's equations in a piecewise homogeneous dispersive conducting medium containing a plane interface that can sustain (possibly dissipative) field-induced surface electric currents. Families of solutions, parametrised by the complex rotation group $SO(3, \mathbb{C})$, are constructed from the roots of complex polynomials with coefficients determined by constitutive properties of the medium and a particular interface admittance tensor. Such solutions include coupled TE and TM-type surface polariton and Brewster modes and offer a means to analyse analytically their physical properties given the constitutive characteristics of bulk meta-materials containing fabricated meta-surface interfaces.

8.1 Introduction

This article is concerned with the behaviour of electromagnetic fields in regions of materials that possess rapid spatial variations in their constitutive properties in the vicinity of two-dimensional surfaces. In such regions certain components of these fields are also expected to exhibit enhanced spatial variations. An exact mathematical treatment of such systems is feasible if one idealises such regions as two-dimensional physical interfaces across which the material constitutive properties and electromagnetic fields become discontinuous. Since the fundamental structure of all materials is molecular and electronic one may consider these physical interfaces to be endowed with active or passive electromagnetic properties described classically by constitutive properties that are distinct from those of the media that they separate. Furthermore, with the recent rapid advances in meta-surface technology one may contemplate

David Christie · Robin Tucker
Physics department, Lancaster University, Lancaster LA1 4YB and The Cockcroft Institute of Accelerator Science, Keckwick Lane, Warington WA4 4AD, Unided Kingdom
e-mail: d.christie@lancaster.ac.uk, r.tucker@lancaster.ac.uk

© Springer International Publishing AG, part of Springer Nature 2018

H. Altenbach et al. (eds.), *Generalized Models and Non-classical Approaches in Complex Materials 1*, Advanced Structured Materials 89,
https://doi.org/10.1007/978-3-319-72440-9_8

systems with artificially fabricated meta-surfaces that offer novel possibilities for controlling the behaviour of electromagnetic fields in new meta-materials.

There exists a vast literature on recent technological developments in this field. In particular the search for an effective control of surface polariton excitations in various meta-structures remains an active area of current experimental endeavour (Raether, 1988; Pitarke et al, 2006; Sarid and Challener, 2010; Maier, 2007). The interesting properties of meta-surfaces with a tuneable surface impedance have been discussed in Zhu et al (2014); Nemilentsau et al (2016) and the recent discovery of the novel properties of graphene lends new impetus to exploring surface excitations in meta-structures involving this material (Vakil and Engheta, 2011; Gusynin et al, 2009; Sounas and Caloz, 2011, 2012).

Theoretical approaches to the behaviour of meta-surfaces often rely on simplified models that are bench-marked against various numerical computer codes while descriptions involving Greens' functions inevitably lead to various approximation schemes (Hanson, 2008). In this article we outline an approach based on an exact analysis of Maxwell's equations in media with interfaces with particular attention devoted to developing tools for finding solutions induced by the presence of a single planar-interface with an intrinsic frequency dependent complex rank (1,1) surface admittance tensor. Such a tensor can be employed to invoke surface currents from surface electromagnetic fields. The formulation of the entire theory benefits from the use of exterior differential forms in a 3-dimensional Euclidean space.

In Sect. 8.2 some mathematical preliminaries and notations are given for those unfamiliar with the geometrical language of differential forms. Further introductory information can be found in Benn and Tucker (1987); Burton (2003). It also introduces a complex extension of Rodrigues formula (Cheng and Gupta, 1989) for describing complex rotations that is used extensively in subsequent sections. Sections 8.3 and 8.4 use this language in the formulation of Maxwell's equations for fields in any regular domain of a medium free of interfaces. Section 8.5 introduces complex field structures based on Clemmow's plane-wave representation (Clemmow, 1966) and Sect. 8.6 indicates how a general family of solutions can be constructed in local regions (without interfaces) using elements from the complex rotation group $SO(3,\mathbb{C})$. In Sect. 8.7 interface conditions are given that must be satisfied in order that solutions can be constructed describing electromagnetic fields in a piecewise-homogeneous material system containing a planar-interface possessing a complex admittance tensor. In Sect. 8.8 these interface conditions are reduced to a set involving *complex rotated amplitudes*. This set facilitates their reduction in Sect. 8.9 to the solution of certain polynomials with complex coefficients. Furthermore, certain roots of these polynomials are then shown to lead to electromagnetic field solutions that can be identified with *surface polariton or bulk Brewster mode configurations* in particular materials possessing a complex interface admittance tensor. In particular when its components constitute a 2×2 Hermitian matrix the interface can sustain an anisotropic Ohmic surface current.

8.2 Mathematical Preliminaries

Our formulation is in terms of time or frequency dependent tensors on $\mathbb{R}^3$ endowed with the three-dimensional Euclidean metric

$$g = \sum_{a=1}^{3} e^a \otimes e^a,$$

and inverse metric

$$G = \sum_{a=1}^{3} X_a \otimes X_a$$

where $e^a(X_b) = \delta_a^b$ with $\delta_a^b = 1$ when $a = b = 1,2,3$ and zero otherwise.

For $\mathscr{U} \subset \mathbb{R}^3$, define $\Gamma T_s^r \mathscr{U}$ as the space of complex, time-dependent r (contravariant), s (covariant) type tensors on $\mathscr{U}$ and $\Gamma \Lambda^p \mathscr{U}$ as the space of complex, time-dependent p-forms on $\mathscr{U}$.

A 1-form $\gamma \in \Gamma \Lambda^1 \mathscr{U}$ which is the metric-dual of a vector field $X \in \Gamma T \mathscr{U}$ may be written $\gamma = \widetilde{X} \equiv g(X,-)$. Equivalently, one may write $X = \widetilde{\gamma} \equiv G(\gamma,-)$.[1]

For $\Phi \in \Gamma \Lambda^p \mathscr{U}$, $\Psi \in \Gamma \Lambda^q \mathscr{U}$ and $f \in \Gamma \Lambda^0 \mathscr{U}$, define the exterior derivative $\mathrm{d} : \Gamma \Lambda^p \mathscr{U} \to \Gamma \Lambda^{p+1} \mathscr{U}$ with the properties

$$\mathrm{d}f(X) = Xf, \qquad \mathrm{d}(\Phi \wedge \Psi) = \mathrm{d}\Phi \wedge \Psi + (-1)^p \Phi \wedge \mathrm{d}\Psi, \qquad \mathrm{d} \circ \mathrm{d} = 0 \tag{8.1}$$

and the linear interior contraction operator $\mathrm{i}_X : \Gamma \Lambda^p \mathscr{U} \to \Gamma \Lambda^{p-1} \mathscr{U}$ with the properties

$$\mathrm{i}_X \gamma = \gamma(X), \ \mathrm{i}_X f = 0, \ \mathrm{i}_X(\Phi \wedge \Psi) = \mathrm{i}_X \Phi \wedge \Psi + (-1)^p \Phi \wedge \mathrm{i}_X \Psi, \ \mathrm{i}_X \circ \mathrm{i}_X = 0. \tag{8.2}$$

In these relations, the exterior product $\Psi \wedge \Phi$ satisfies $\Psi \wedge \Phi = (-1)^{pq} \Phi \wedge \Psi$. Furthermore, associated with the metric g one may define the *linear* Hodge map $\# : \Gamma \Lambda^p \mathscr{U} \to \Gamma \Lambda^{3-p} \mathscr{U}$ with the properties[2]

$$\#(\Phi \wedge \gamma) = \mathrm{i}_{\widetilde{\gamma}} \# \Phi, \qquad \#1 = e^1 \wedge e^2 \wedge e^3, \qquad \# \circ \# = \mathbb{1} \tag{8.3}$$

where $\mathbb{1} \in T_1^1 \mathbb{R}^3$ is the unit (1,1) tensor on $\mathbb{R}^3$. From (8.2), one obtains the useful result that if Ψ is any decomposable p-form containing the 1-form γ, then

$$\mathrm{i}_{\widetilde{\gamma}} \# \Psi = 0. \tag{8.4}$$

---

[1] In a Cartesian coordinate system (x,y,z) for $\mathbb{R}^3$, a global co-frame $e^1 \equiv \mathrm{d}x$ $e^2 \equiv \mathrm{d}y$, $e^3 \equiv \mathrm{d}z$, and $\widetilde{\mathrm{d}x} = \partial_x$, $\widetilde{\mathrm{d}y} = \partial_y$ and $\widetilde{\mathrm{d}z} = \partial_z$. Similarly, $\widetilde{\partial_x} = \mathrm{d}x$, $\widetilde{\partial_y} = \mathrm{d}y$ and $\widetilde{\partial_z} = \mathrm{d}z$.

[2] A p-form is said to be *decomposable* if it can be written as the exterior product of p 1-forms. Then (8.3) is sufficient to define the Hodge map on any decomposable p-form since recursive application gives $\#(\gamma^1 \wedge \cdots \wedge \gamma^p) = \mathrm{i}_{\widetilde{\gamma^p}} \ldots \mathrm{i}_{\widetilde{\gamma^1}} \#1$ for $\gamma^1, \ldots, \gamma^p \in \Gamma \Lambda^1 \mathscr{U}$, and its action on a non-decomposable p-form follows by linearity. In a Cartesian coframe, $\#1 = \mathrm{d}x \wedge \mathrm{d}y \wedge \mathrm{d}z$, and (8.3) yields $\#\mathrm{d}x = \mathrm{d}y \wedge \mathrm{d}z$, $\#\mathrm{d}y = \mathrm{d}z \wedge \mathrm{d}x$ and $\#\mathrm{d}z = \mathrm{d}x \wedge \mathrm{d}y$.

For $\mathscr{U} \subset \mathbb{R}^3$ and any *unit norm real* vector field $N \in \Gamma T\mathscr{U}$ satisfying $g(N,N) = 1$, we define the normal and tangential projection operators $\mathfrak{n}_N, \mathfrak{t}_N$ and the *tangential Hodge map* $\#_N$ as

$$\mathfrak{n}_N : \Gamma\Lambda^p\mathscr{U} \to \Gamma\Lambda^p\mathscr{U}, \quad \xi \to \mathfrak{n}_N\xi \equiv \tilde{N} \wedge \mathrm{i}_N\xi \tag{8.5}$$

$$\mathfrak{t}_N : \Gamma\Lambda^p\mathscr{U} \to \Gamma\Lambda^p\mathscr{U}, \quad \xi \to \mathfrak{t}_N\xi \equiv \xi - \mathfrak{n}_N\xi = \mathrm{i}_N\left(\tilde{N} \wedge \xi\right) \tag{8.6}$$

where $p = 0,1,2,3$ and

$$\#_N : \Gamma\Lambda^p\mathscr{U} \to \Gamma\Lambda^{2-p}\mathscr{U}, \quad \xi \to \#_N\xi = (-1)^p\mathrm{i}_N\#\xi \equiv \#\left(\tilde{N} \wedge \xi\right), \tag{8.7}$$

where $p = 0,1,2$. Since N has unit norm, $\mathfrak{n}_N\tilde{N} = \tilde{N}$, $\mathfrak{t}_N\tilde{N} = 0$ and one has the operator relations:

$$\mathfrak{n}_N \circ \mathfrak{n}_N = \mathfrak{n}_N, \qquad \mathfrak{t}_N \circ \mathfrak{t}_N = \mathfrak{t}_N, \tag{8.8}$$

$$\mathfrak{n}_N \circ \mathfrak{t}_N = \mathfrak{t}_N \circ \mathfrak{n}_N = 0, \tag{8.9}$$

$$\mathfrak{n}_N \circ \# = \# \circ \mathfrak{t}_N, \qquad \mathfrak{t}_N \circ \# = \# \circ \mathfrak{n}_N \tag{8.10}$$

$$\mathfrak{n}_N \circ \#_N = \#_N \circ \mathfrak{n}_N = 0, \tag{8.11}$$

$$\mathfrak{t}_N \circ \#_N = \#_N \circ \mathfrak{t}_N = \#_N, \tag{8.12}$$

$$\#_N \circ \#_N = \mathfrak{t}_N \circ \eta, \tag{8.13}$$

$$\mathrm{i}_N \circ \mathfrak{n}_N = \mathrm{i}_N, \quad \mathrm{i}_N \circ \mathfrak{t}_N = \mathrm{i}_N \circ \#_N = 0, \tag{8.14}$$

$$\#_N \circ \mathrm{i}_N = \# \circ \mathfrak{n}_N = \mathfrak{t}_N \circ \#, \tag{8.15}$$

$$\# = \#_N \circ \mathrm{i}_N + \tilde{N} \wedge \#_N \circ \eta, \tag{8.16}$$

where $\eta(\Phi) \equiv (-1)^p(\Phi)$. Furthermore, for $\alpha, \beta \in \Gamma\Lambda^1\mathscr{U}$,

$$\mathfrak{n}_N(\alpha \wedge \beta) = \mathfrak{n}_N\alpha \wedge \beta + \alpha \wedge \mathfrak{n}_N\beta, \tag{8.17}$$

$$\mathfrak{t}_N(\alpha \wedge \beta) = \mathfrak{t}_N\alpha \wedge \mathfrak{t}_N\beta = \#_N\alpha \wedge \#_N\beta \tag{8.18}$$

$$\#_N(\alpha \wedge \beta) = \#_N(\#_N\alpha \wedge \#_N\beta) = G(\#_N\alpha, \beta) = -G(\alpha, \#_N\beta). \tag{8.19}$$

Let φ denote a *complex* angle and introduce the complex rotation operator[3]

$$R_N(\varphi) : \Gamma\Lambda^1\mathscr{U} \to \Gamma\Lambda^1\mathscr{U} : \quad \alpha \to R_N(\varphi)(\alpha) \equiv \mathfrak{n}_N\alpha + \cos\varphi\,\mathfrak{t}_N\alpha - \sin\varphi\,\#_N\alpha. \tag{8.20}$$

From (8.8)-(8.14), one has the relations

$$R_N(\varphi)\tilde{N} = \tilde{N} \tag{8.21}$$

$$\mathrm{i}_N \circ R_N(\varphi) = \mathrm{i}_N \tag{8.22}$$

$$R_N(\varphi) \circ \mathfrak{n}_N = \mathfrak{n}_N \circ R_N(\varphi) = \mathfrak{n}_N, \tag{8.23}$$

$$R_N(\varphi) \circ \mathfrak{t}_N = \mathfrak{t}_N \circ R_N(\varphi) \tag{8.24}$$

[3] This is a complex extension of the Rodrigues rotation formula.

i.e. the operator R_N commutes with the projectors $\mathfrak{t}_N$ and $\mathfrak{n}_N$. For a pair of complex angles φ_1 and φ_2, one has

$$R_N(\varphi_1+\varphi_2) = R_N(\varphi_1)\circ R_N(\varphi_2), \tag{8.25}$$

and hence

$$R_N(-\varphi)\circ R_N(\varphi) = R_N(0) = \mathbb{1}. \tag{8.26}$$

Identity (8.9) and definition (8.20) imply that

$$R_N\left(-\frac{\pi}{2}\right)\circ \mathfrak{t}_N = \#_N, \tag{8.27}$$

so that

$$R_N(\varphi)\circ\#_N = \#_N\circ R_N(\varphi) = R_N(\varphi-\frac{\pi}{2})\circ\mathfrak{t}_N = \mathfrak{t}_N\circ R_N(\varphi-\frac{\pi}{2}). \tag{8.28}$$

Using (8.18) and (8.19), it can be shown that

$$G(R_N(\varphi)(\alpha),R_N(\varphi)(\beta)) = G(\alpha,\beta), \tag{8.29}$$

$$\#_N(R_N(\varphi)(\alpha)\wedge R_N(\varphi)(\beta)) = \#_N(\alpha\wedge\beta) \tag{8.30}$$

i.e. the operator R_N is an isometry on the space of 1-forms on $\mathscr{U}$. Finally, since (8.16), (8.25), (8.28) and (8.30) give

$$\begin{aligned}
&(R_N(-\varphi)\circ\#)\,(R_N(\varphi)(\alpha)\wedge R_N(\varphi)(\beta))\\
&= R_N(-\varphi)\Big[(\mathrm{i}_N(\alpha))(\#_N\circ R_N(\varphi))(\beta)-(\mathrm{i}_N\beta)(\#_N\circ R_N(\varphi))(\alpha)\\
&\qquad +\#_N\,(R_N(\varphi)(\alpha)\wedge R_N(\varphi)(\beta))\,\widetilde{N}\Big]\\
&= R_N(-\varphi)\,\Big[(\mathrm{i}_N\alpha)R_N(\varphi)(\#_N\beta)-(\mathrm{i}_N\beta)R_N(\varphi)(\#_N\alpha)+\#_N\,(\alpha\wedge\beta)\,\widetilde{N}\Big]\\
&= (\mathrm{i}_N\alpha)\#_N\beta-(\mathrm{i}_N\beta)\#_N\alpha+\#_N\,(\alpha\wedge\beta)\,\widetilde{N}\\
&= \#\,(\alpha\wedge\beta),
\end{aligned}$$

one has the useful identity[4]

$$R_N(\varphi)\#(\alpha\wedge\beta) = \#(R_N(\varphi)\alpha\wedge R_N(\varphi)\beta). \tag{8.31}$$

8.3 General Maxwell Equations and their Fourier Transform

In terms of the electric field $\boldsymbol{e}\in\Gamma\Lambda^1\mathscr{U}$, electric displacement $\boldsymbol{d}\in\Gamma\Lambda^1\mathscr{U}$, magnetic flux density $\boldsymbol{b}\in\Gamma\Lambda^1\mathscr{U}$, magnetic field $\boldsymbol{h}\in\Gamma\Lambda^1\mathscr{U}$, *total* free current $\boldsymbol{j}\in\Gamma\Lambda^1\mathscr{U}$

[4] Analogous to the preservation of the real angle between two vectors under a Rodrigues' real rotation.

and *total* free charge density $\rho \in \Gamma\Lambda^0\mathscr{U}$, these time ($t$)-dependent fields satisfy the Maxwell system:

$$\#\mathrm{d}\boldsymbol{e} + \frac{\partial}{\partial t}\boldsymbol{b} = 0, \tag{8.32}$$

$$\#\mathrm{d}\boldsymbol{h} - \frac{\partial}{\partial t}\boldsymbol{d} - \boldsymbol{j} = 0, \tag{8.33}$$

$$\mathrm{d}\#\boldsymbol{b} = 0, \tag{8.34}$$

$$\#\mathrm{d}\#\boldsymbol{d} - \rho = 0, \tag{8.35}$$

provided

$$\#\mathrm{d}\#\boldsymbol{j} + \partial_t \rho = 0.$$

This system is assumed closed with the addition of (possibly nonlocal and nonlinear) constitutive relations correlating $\boldsymbol{e}, \boldsymbol{b}, \boldsymbol{d}, \boldsymbol{h}$ and $\boldsymbol{j}$ on $\mathscr{U}$.

If a t-parametrised (r,s)-type tensor $T \in \Gamma T^r_s\mathscr{U}$ in a t-independent tensor-basis can be related to an ω-parametrised tensor $\widehat{T} \in \widehat{\Gamma} T^r_s\mathscr{U}$ by the Fourier transform:

$$T(x,y,z,t) = \int_{-\infty}^{\infty} \widehat{T}(x,y,z,\omega)e^{-i\omega t}\mathrm{d}\omega, \tag{8.36}$$

then

$$\widehat{T}(x,y,z,\omega) = \frac{1}{2\pi}\int_{-\infty}^{\infty} T(x,y,z,t)e^{i\omega t}\mathrm{d}t \qquad \text{where } \omega \in \mathbb{R}. \tag{8.37}$$

Similarly, $\widehat{\Gamma}\Lambda^p\mathscr{U}$ is the space of complex, ω-dependent Fourier transformed p-forms on $\mathscr{U}$. The complex ω-domain Maxwell equations then satisfy

$$\#\mathrm{d}\widehat{\boldsymbol{e}} - i\omega\widehat{\boldsymbol{b}} = 0, \tag{8.38}$$

$$\#\mathrm{d}\widehat{\boldsymbol{h}} + i\omega\widehat{\boldsymbol{d}} - \widehat{\boldsymbol{j}} = 0, \tag{8.39}$$

$$\mathrm{d}\#\widehat{\boldsymbol{b}} = 0, \tag{8.40}$$

$$\#\mathrm{d}\#\widehat{\boldsymbol{d}} - \widehat{\rho} = 0 \tag{8.41}$$

together with the Fourier transformed constitutive relations on $\mathscr{U}$. Since

$$\mathrm{d}\#\widehat{\boldsymbol{b}} = -\frac{i}{\omega}\mathrm{d}\#\#\mathrm{d}\widehat{\boldsymbol{e}} = -\frac{i}{\omega}\mathrm{d}(\mathrm{d}\widehat{\boldsymbol{e}}) = 0,$$

(8.38) implies that equation (8.40) is satisfied automatically for $\omega \neq 0$.

8.4 Maxwell Equations in a Source-Free Domain of an Ohmic, Homogeneous, Isotropic, Dispersive, Linear Medium

If $\mathscr{U}$ contains a simple *linear* medium, the fields $\widehat{\boldsymbol{e}}$, $\widehat{\boldsymbol{d}}$, $\widehat{\boldsymbol{b}}$ and $\widehat{\boldsymbol{h}}$ are related by the following linear constitutive relations

$$\widehat{\boldsymbol{d}} = \widehat{\boldsymbol{\epsilon}}(\widehat{\boldsymbol{e}}), \qquad \widehat{\boldsymbol{b}} = \widehat{\boldsymbol{\mu}}(\widehat{\boldsymbol{h}}), \tag{8.42}$$

where $\widehat{\boldsymbol{\epsilon}}, \widehat{\boldsymbol{\mu}} \in \widehat{\Gamma T}_1^1\mathscr{U}$ are respectively complex dispersive permittivity and permeability tensors.

If $\mathscr{U}$ contains a stationary, isothermal, conducting medium, a contribution to the total current 1-form $\widehat{\boldsymbol{j}}$ may depend on the local electromagnetic fields in $\mathscr{U}$. The medium is said to be an Ohmic conductor if $\widehat{\boldsymbol{j}}$ contains a contribution $\widehat{\boldsymbol{\sigma}}(\widehat{\boldsymbol{e}})$ where $\widehat{\boldsymbol{\sigma}} \in \widehat{\Gamma T}_1^1\mathscr{U}$ denotes the Hermitian conductivity tensor of the medium. In an *isotropic, homogeneous* medium, one also has $\widehat{\boldsymbol{\sigma}} = \widehat{\sigma}\mathbb{1}$ where $\mathrm{d}\widehat{\sigma} = 0$. When $\widehat{\boldsymbol{\sigma}} \neq 0$, the *bulk* medium is dissipative. In the following, we assume all bulk regions without interfaces to be homogeneous and isotropic. Thus,

$$\widehat{\boldsymbol{\epsilon}} = \widehat{\epsilon}\mathbb{1}, \qquad \widehat{\boldsymbol{\mu}} = \widehat{\mu}\mathbb{1}, \tag{8.43}$$

where $\widehat{\epsilon}$, $\widehat{\mu}$ are frequency-dependent complex 0-forms satisfying $\mathrm{d}\widehat{\epsilon} = \mathrm{d}\widehat{\mu} = 0$ on $\mathscr{U}$. More generally, any form $\widehat{\alpha} \in \widehat{\Gamma\Lambda}^p\mathscr{U}$ is said to be *closed* on $\mathscr{U}$ if $\mathrm{d}\widehat{\alpha} = 0$ When $\widehat{\boldsymbol{\epsilon}} = \epsilon_0\boldsymbol{\kappa}_r$ or $\widehat{\boldsymbol{\mu}} = \mu_0\boldsymbol{\eta}_r$ where $\boldsymbol{\kappa}_r$ and $\boldsymbol{\eta}_r$ are (possibly complex) dimensionless (1,1) tensors *independent of frequency*, their Fourier transforms do not exist as smooth tensors. In these circumstances, we write $\boldsymbol{\epsilon} \equiv \epsilon_0\boldsymbol{\kappa}_r$, $\boldsymbol{\mu} \equiv \mu_0\boldsymbol{\eta}_r$ in terms of the permittivity ϵ_0 and the permeability μ_0 of free space.

We henceforth assume that the isotropic, homogeneous medium $\mathscr{U}$ is also *source-free*, so that $\widehat{\boldsymbol{j}} = \widehat{\sigma}\widehat{\boldsymbol{e}}$ and the Maxwell equations (8.38)-(8.39) may then be rewritten

$$\widehat{\boldsymbol{h}} = -\frac{i}{\omega\widehat{\mu}}\#\mathrm{d}\widehat{\boldsymbol{e}}, \qquad \widehat{\boldsymbol{e}} = \frac{i}{\omega\widehat{\epsilon'}}\#\mathrm{d}\widehat{\boldsymbol{h}}. \tag{8.44}$$

where

$$\widehat{\epsilon'} \equiv \widehat{\epsilon} + \frac{i\widehat{\sigma}}{\omega}.$$

Since

$$\widehat{\rho} = \widehat{\epsilon}\#\mathrm{d}\#\widehat{\boldsymbol{e}} = -\frac{\widehat{\epsilon}}{i\omega\widehat{\epsilon'}}\#\mathrm{d}\#(\#\mathrm{d}\widehat{\boldsymbol{h}}) = -\frac{\widehat{\epsilon}}{i\omega\widehat{\epsilon'}}\#\mathrm{d}(\mathrm{d}\widehat{\boldsymbol{h}}) = 0,$$

the remaining independent Maxwell equation (8.41) reduces to $\widehat{\rho} = 0$, which is consistent with $\#\mathrm{d}\#\widehat{\boldsymbol{j}} = 0$.

8.5 Electromagnetic Fields in a Source-Free Domain of an Ohmic, Homogeneous, Isotropic, Dispersive, Linear Medium

In the following, we explore particular real solutions to equations (8.44) from the particular *complex* forms

$$\widehat{\boldsymbol{e}} = \widehat{\mathscr{E}} e^{i\widehat{\chi}}, \qquad \widehat{\boldsymbol{h}} = \widehat{\mathscr{H}} e^{i\widehat{\chi}}, \tag{8.45}$$

where $\widehat{\mathscr{E}}$ is a closed complex electric polarisation 1-form on $\mathscr{U}$, $\widehat{\mathscr{H}}$ is a closed complex magnetic polarisation 1-form on $\mathscr{U}$ and $\widehat{\chi}$ is a complex 0-form. Substituting (8.45) into (8.44) and recalling that $\widehat{\mathscr{E}}$ and $\widehat{\mathscr{H}}$ are closed yields

$$\widehat{\mathscr{H}} = \frac{1}{\omega\widehat{\mu}}\#(\widehat{K}\wedge\widehat{\mathscr{E}}), \qquad \widehat{\mathscr{E}} = -\frac{1}{\omega\widehat{\epsilon'}}\#(\widehat{K}\wedge\widehat{\mathscr{H}}), \tag{8.46}$$

where $\widehat{K} = d\widehat{\chi}$ is a complex propagation 1-form. Using (8.3) and (8.4), it follows from (8.46) that the 1-forms $\widehat{\mathscr{E}}$, $\widehat{\mathscr{H}}$ and $\widehat{K}$ must form a mutually orthogonal triplet, i.e.

$$G(\widehat{\mathscr{E}},\widehat{\mathscr{H}}) = G(\widehat{K},\widehat{\mathscr{H}}) = G(\widehat{K},\widehat{\mathscr{E}}) = 0. \tag{8.47}$$

Eliminating $\widehat{\mathscr{H}}$ from (8.46) gives

$$\widehat{\mathscr{E}} = -\frac{1}{\omega^2\widehat{\epsilon'}\widehat{\mu}}\#(\widehat{K}\wedge\#(\widehat{K}\wedge\widehat{\mathscr{E}})) = \frac{G(\widehat{K},\widehat{K})}{\omega^2\widehat{\epsilon'}\widehat{\mu}}\widehat{\mathscr{E}}. \tag{8.48}$$

since (using (8.3))

$$-\#(\widehat{K}\wedge\#(\widehat{K}\wedge\widehat{\mathscr{E}})) = \#(\#(\widehat{K}\wedge\widehat{\mathscr{E}})\wedge\widehat{K}) = i_{\widetilde{\widehat{K}}}\#\#(\widehat{K}\wedge\widehat{\mathscr{E}}) = i_{\widetilde{\widehat{K}}}(\widehat{K}\wedge\widehat{\mathscr{E}}) = \left(i_{\widetilde{\widehat{K}}}\widehat{K}\right)\widehat{\mathscr{E}}$$

as

$$i_{\widetilde{\widehat{K}}}\widehat{\mathscr{E}} = G(\widehat{K},\widehat{\mathscr{E}}) = 0.$$

Therefore, for nontrivial electromagnetic fields, one requires

$$G(\widehat{K},\widehat{K}) = \omega^2\widehat{\epsilon'}\widehat{\mu} = \frac{\omega^2\widehat{\epsilon'}_r\widehat{\mu}_r}{c^2}, \tag{8.49}$$

where

$$\widehat{\epsilon'}_r = \frac{\widehat{\epsilon'}}{\epsilon_0}, \quad \widehat{\mu}_r = \frac{\widehat{\mu}}{\widehat{\mu}_0}$$

and

$$c = \sqrt{\mu_0\epsilon_0}$$

is the speed of light in vacuo.

In summary, a 0-form $\widehat{\chi}$ and 1-form $\widehat{\mathscr{E}}$ which satisfy (8.49) and (8.47) may be used to construct a complex Maxwell solution of the form (8.45), where the magnetic field amplitude

$$\widehat{\mathscr{H}} = \frac{1}{\omega\widehat{\mu}} \#(\mathrm{d}\widehat{\chi} \wedge \widehat{\mathscr{E}}), \tag{8.50}$$

provided $\mathrm{d}\widehat{\mathscr{E}} = \mathrm{d}\widehat{\mathscr{H}} = 0$. In the next section, we indicate the form of the scalar field $\widehat{\chi}$ which leads to a class of plane-fronted harmonic solutions.

8.6 Plane Wave Solutions in Terms of the Complex Rotation Group

If the Cartesian components of the vector field $\widetilde{\widehat{K}}$ were real constants (and $\widehat{\boldsymbol{j}} = 0$), the fields (8.45) would give rise to a family of plane harmonic electromagnetic waves with polarisation vectors orthogonal to the direction of propagation $\widetilde{\widehat{K}}$.

In a homogeneous medium, the only constraint on $\widehat{K}$ would be that it has modulus $\omega\sqrt{\widehat{\epsilon}\widehat{\mu}}$. Thus, all such plane wave solutions could be related to each other by an $SO(3,\mathbb{R})$ group action. In the presence of a meta-material with non-zero complex conductivity (leading to possible amplification and attenuation of electromagnetic waves) one expects that the group $SO(3,\mathbb{C})$ should play a role in relating solutions that propagate with attenuation (or amplification)[5]. In this section, we approach a representation of $SO(3,\mathbb{C})$ generated from a group action on differential 1-forms. This will then yield a construction of the fields (8.45) in terms of particular elements $\{\widehat{K}, \widehat{\mathscr{E}}, \widehat{\mathscr{H}}\}$ "rotated" by elements of $SO(3,\mathbb{C})$, satisfying the conditions (8.49) and (8.47), and maintaining the *relation* (8.50), thereby automatically satisfying the Maxwell system on $\mathscr{U}$.

Given a g-orthonormal triplet $\{N_1, N_2, N_3\}$ of vector fields on $\mathscr{U}$ and a triplet $\tau = \{\widehat{\psi}, \widehat{\theta}, \widehat{\phi}\}$ of complex functions of ω, use (8.20) to construct a three-complex-parameter linear map

$$\mathscr{R}_\tau : \widehat{\Gamma}\Lambda^1\mathscr{U} \to \widehat{\Gamma}\Lambda^1\mathscr{U}, \quad \widehat{\alpha} \to \mathscr{R}_\tau\widehat{\alpha} \equiv \left(R_{N_3}(\widehat{\psi}) \circ R_{N_2}(\widehat{\theta}) \circ R_{N_1}(\widehat{\phi})\right)(\widehat{\alpha}), \tag{8.51}$$

such that $\widehat{\psi}$, $\widehat{\theta}$, $\widehat{\phi}$ locally parametrise $SO(3,\mathbb{C})$. Repeated application of identity (8.29) gives

$$G(\mathscr{R}_\tau(\widehat{\alpha}), \mathscr{R}_\tau(\widehat{\beta})) = G(\widehat{\alpha}, \widehat{\beta}), \tag{8.52}$$

i.e., the operator $\mathscr{R}_\tau$ is an isometry on the space of 1-forms on $\mathscr{U}$.

A particularly simple solution satisfying the system (8.44) is

$$\widehat{\boldsymbol{e}}_{\tau_0} = \widehat{\mathscr{E}}_{\tau_0} e^{i\widehat{\chi}_{\tau_0}}, \qquad \widehat{\boldsymbol{h}}_{\tau_0} = \widehat{\mathscr{H}}_{\tau_0} e^{i\widehat{\chi}_{\tau_0}}, \tag{8.53}$$

where

[5] The elements of $SO(3,\mathbb{C})$ connected to the identity element are more familiar as elements of the Lorentz group $SO(3,1,\mathbb{R})$ after parametrisation in terms of real group coordinates

$$\widehat{K}_{\tau_0} = \mathrm{d}\widehat{\chi}_{\tau_0} = \omega\sqrt{\widehat{\epsilon'}\widehat{\mu}}\,\mathrm{d}x, \quad \widehat{\mathscr{E}}_{\tau_0} = \widehat{A}\mathrm{d}y, \quad \widehat{\mathscr{H}} = \widehat{A}\sqrt{\frac{\widehat{\epsilon'}}{\widehat{\mu}}}\mathrm{d}z \tag{8.54}$$

in a global Cartesian co-frame, where $\widehat{A}$ is a complex constant.

From this solution, one may use a generic element $\mathscr{R}_\tau$ of $SO(3,\mathbb{C})$ to generate an orbit of solutions:

$$\widehat{\boldsymbol{e}}_\tau = \mathscr{R}_\tau \widehat{\mathscr{E}}_{\tau_0} e^{i\widehat{\chi}_\tau}, \qquad \widehat{\boldsymbol{h}}_\tau = \mathscr{R}_\tau \widehat{\mathscr{H}_\tau} e^{i\widehat{\chi}_{\tau_0}}, \tag{8.55}$$

with

$$\widehat{K}_\tau \equiv \mathrm{d}\widehat{\chi}_\tau = \mathscr{R}_\tau \widehat{K}_{\tau_0} = \widehat{K}_{\tau,x}\mathrm{d}x + \widehat{K}_{\tau,y}\mathrm{d}y + \widehat{K}_{\tau,z}\mathrm{d}z \tag{8.56}$$

and

$$\widehat{\chi}_\tau \equiv \widehat{K}_{\tau,x}x + \widehat{K}_{\tau,y}y + \widehat{K}_{\tau,z}z = G\left(\widehat{K}_\tau,\ x\mathrm{d}x + y\mathrm{d}y + z\mathrm{d}z\right), \tag{8.57}$$

since (8.52) implies

$$G(\widehat{K}_\tau, \widehat{K}_\tau) = G(\widehat{K}_{\tau_0}, \widehat{K}_{\tau_0}) = \frac{\omega^2 \widehat{\epsilon'}_r \widehat{\mu}_r}{c^2}, \tag{8.58}$$

$$G(\widehat{K}_\tau, \widehat{\mathscr{E}}_\tau) = G(\widehat{K}_{\tau_0}, \widehat{\mathscr{E}}_{\tau_0}) = 0, \tag{8.59}$$

and repeated application of (8.31) yields

$$\widehat{\mathscr{H}_\tau} = \frac{1}{\omega\widehat{\mu}} \# \left(\widehat{K}_\tau \wedge \widehat{\mathscr{E}}_\tau\right) = \mathscr{R}_\tau \widehat{\mathscr{H}_{\tau_0}} \tag{8.60}$$

This approach can be extended to situations involving multiple regions with different material properties. The associated field components must then be matched at interfaces between such regions using junction conditions as described in the next section.

8.7 Interface Conditions for Media Containing Anisotropic, Homogeneous, Planar Interface Consitutive Relations

Let $f = 0$ be a particular smooth interface S belonging to a foliation of a region $\mathscr{M} \subset \mathbb{R}^3$ and assign a normal unit vector field

$$N = \frac{\widetilde{\mathrm{d}f}}{|\mathrm{d}f|} \neq 0,$$

where $|\mathrm{d}f|^2 \equiv G(\mathrm{d}f, \mathrm{d}f)$ with N oriented from a region I where $f \leq 0$ to the region II where $f \geq 0$.

If regions I and II in $\mathscr{M}$ contain material with different constitutive properties then the electromagnetic fields in these regions will in general exhibit discontinuities in certain of their components across the interface $f = 0$, corresponding to surface charge and current densities.

The interface conditions for general electric and magnetic 1-forms in the frequency domain are given in this language as:

$$S^* \mathrm{i}_N \widehat{\boldsymbol{b}}^{II} - S^* \mathrm{i}_N \widehat{\boldsymbol{b}}^{I} = 0 \tag{8.61}$$

$$S^* \mathrm{t}_N \widehat{\boldsymbol{e}}^{II} - S^* \mathrm{t}_N \widehat{\boldsymbol{e}}^{I} = 0 \tag{8.62}$$

$$S^* \mathrm{i}_N \widehat{\boldsymbol{d}}^{II} - S^* \mathrm{i}_N \widehat{\boldsymbol{d}}^{I} - \widehat{\rho}_S = 0 \tag{8.63}$$

$$S^* \mathrm{t}_N \widehat{\boldsymbol{h}}^{II} - S^* \mathrm{t}_N \widehat{\boldsymbol{h}}^{I} + \#_N \widehat{\boldsymbol{j}}_S = 0, \tag{8.64}$$

where for any p-form $\widehat{\gamma} \equiv \widehat{\Gamma} \Lambda^p \mathscr{M}$, $S^* \widehat{\gamma}$ denotes the pullback onto the interface S. Since $\#_N$ maps forms on any domain to their tangential parts with respect to N, they have a natural extension to maps on the pullbacks of such forms to any surface $S \subset \mathscr{M}$ with the local normal N. Furthermore, $\widehat{\rho}_S \in \widehat{\Gamma} \Lambda^0 S$ and $\widehat{\boldsymbol{j}}_S \in \widehat{\Gamma} \Lambda^1 S$ are the surface charge density 0-form and surface current density 1-form on S respectively. In general, the interface surface forms $\widehat{\rho}_S$ and $\widehat{\boldsymbol{j}}_S$ in (8.63)-(8.64) correspond to surface charge and current densities (including those produced by possible external sources of surface charge density, possible external sources of interface current density, and surface electromagnetic fields).

At this point, we assume that the surface charge density $\widehat{\rho}_S$ is determined solely by electromagnetic fields in the *bulk* media and is given by (8.63). Furthermore, we assume that $\widehat{\boldsymbol{j}}_S$ is determined solely by the fields satisfying the interface constitutive relation

$$\widehat{\boldsymbol{j}}_S = \frac{1}{Z_0} \widehat{\boldsymbol{\Sigma}} \left(S^* \mathrm{t}_N \widehat{\boldsymbol{e}}^{I} \right) \equiv \frac{1}{Z_0} \widehat{\boldsymbol{\Sigma}} \left(S^* \mathrm{t}_N \widehat{\boldsymbol{e}}^{II} \right) \tag{8.65}$$

where the rank (1,1) complex *surface admittance* tensor $\widehat{\boldsymbol{\Sigma}} \in \widehat{T}^1 S$ is defined to act only on the *tangential* components of the electric field on S and

$$Z_0 = \sqrt{\frac{\epsilon_0}{\mu_0}} = \frac{1}{c\mu_0} = c\epsilon_0$$

is the impedance of free space.

Assume that in a Cartesian coordinate system the surface S, $(z = 0)$, separates $\mathscr{M}$ into two semi-infinite volume regions $\mathscr{V}^I$ $(z < 0)$ and $\mathscr{V}^{II}$ $(z > 0)$ For notational simplicity, denote the unit normal vector field on S, $\partial_z|_S$, by ∂_z. We suppose that the domains $\mathscr{V}^I$ and $\mathscr{V}^{II}$ are filled with *isotropic, homogeneous* media with distinct complex permittivity, permeability and conductivity constitutive *scalars* $\{\widehat{\epsilon}^I, \widehat{\mu}^I, \widehat{\sigma}^I, \widehat{\epsilon}^{II}, \widehat{\mu}^{II}, \widehat{\sigma}^{II}\}$.

The electric and magnetic fields are given by

$$\widehat{\boldsymbol{e}}^L = \widehat{\mathscr{E}}^L e^{i\widehat{\chi}^L}, \qquad \widehat{\boldsymbol{h}}^L = \widehat{\mathscr{H}}^L e^{i\widehat{\chi}^L} \tag{8.66}$$

in terms of $\{\widehat{\chi}^L, \widehat{\mathscr{E}}^L, \widehat{\mathscr{H}^L}\}$ where $L \in \{I, II\}$. We refer to these fields as modes in $\mathscr{V} \equiv \mathscr{V}^I \cup \mathscr{V}^{II} \cup S$. The interface conditions (8.61)-(8.64) become

$$\widehat{\mu}^{II} \mathrm{i}_{\partial_z} \widehat{\mathscr{H}^{II}} e^{i\widehat{\chi}_0^{II}} - \widehat{\mu}^I \mathrm{i}_{\partial_z} \widehat{\mathscr{H}^I} e^{i\widehat{\chi}_0^I} = 0, \tag{8.67}$$

$$\mathrm{t}_{\partial_z} \widehat{\mathscr{E}}^{II} e^{i\widehat{\chi}_0^{II}} - \mathrm{t}_{\partial_z} \widehat{\mathscr{E}}^I e^{i\widehat{\chi}_0^I} = 0, \tag{8.68}$$

$$\mathrm{i}_{\partial_z} \widehat{\mathscr{E}}^{II} e^{i\widehat{\chi}_0^{II}} - \mathrm{i}_{\partial_z} \widehat{\mathscr{E}}^I e^{i\widehat{\chi}_0^I} = \widehat{\rho}_S, \tag{8.69}$$

$$\mathrm{t}_{\partial_z} \widehat{\mathscr{H}^{II}} e^{i\widehat{\chi}_0^{II}} - \mathrm{t}_{\partial_z} \widehat{\mathscr{H}^I} e^{i\widehat{\chi}_0^I} = -\#_N \widehat{\mathbf{j}}_S, \tag{8.70}$$

where

$$\widehat{\chi}_0^L = S^* \widehat{\chi}^L \tag{8.71}$$

depends only on x and y[6]. The interface conditions (8.67), (8.68) and (8.70) imply correlations amongst $\{\widehat{\chi}^L, \widehat{\mathscr{E}}^L, \widehat{\mathscr{H}^L}\}$ and $\widehat{\mathbf{j}}_S$. The remaining interface condition (8.69) defines the charge density $\widehat{\rho}_S \in \widehat{\Gamma\Lambda}^0 S$ that will arise on the interface. For (8.68) and (8.70) to be satisfied at every point on S we require

$$\widehat{\chi}_0^{II} = \widehat{\chi}_0^I \tag{8.72}$$

$$\widehat{\rho}_S = \widehat{\rho}_0 e^{i\widehat{\chi}_0^I} = \widehat{\rho}_0 e^{i\widehat{\chi}_0^{II}} \tag{8.73}$$

$$\widehat{\mathbf{j}}_S = \widehat{\mathscr{J}} e^{i\widehat{\chi}_0^I} = \widehat{\mathscr{J}} e^{i\widehat{\chi}_0^{II}} \tag{8.74}$$

where $\widehat{\rho}_0 \in \widehat{\Gamma\Lambda}^0 S$ and $\widehat{\mathscr{J}} \in \widehat{\Gamma\Lambda}^1 S$ denote the Fourier amplitudes of the surface charge 0-form and current density 1-form respectively.

From (8.56) and (8.56), it follows that (8.72) is equivalent to

$$\mathrm{t}_{\partial_z} \widehat{K}^{II} - \mathrm{t}_{\partial_z} \widehat{K}^I = 0. \tag{8.75}$$

The complex exponential terms now factor out of all the interface conditions which then reduce to

$$\widehat{\mu}^{II} \mathrm{i}_{\partial_z} \widehat{\mathscr{H}^{II}} - \widehat{\mu}^I \mathrm{i}_{\partial_z} \widehat{\mathscr{H}^I} = 0, \tag{8.76}$$

$$\mathrm{t}_{\partial_z} \widehat{\mathscr{E}}^{II} - \mathrm{t}_{\partial_z} \widehat{\mathscr{E}}^I = 0, \tag{8.77}$$

$$\mathrm{i}_{\partial_z} \widehat{\mathscr{E}}^{II} - \mathrm{i}_{\partial_z} \widehat{\mathscr{E}}^I = \widehat{\rho}_0, \tag{8.78}$$

$$\mathrm{t}_{\partial_z} \widehat{\mathscr{H}^{II}} - \mathrm{t}_{\partial_z} \widehat{\mathscr{H}^I} = -\#_N \widehat{\mathscr{J}} \tag{8.79}$$

with

$$\widehat{\mathscr{J}} = \frac{1}{Z_0} \widehat{\mathbf{\Sigma}} \left(\mathrm{t}_N \widehat{\mathscr{E}}^I \right) = \frac{1}{Z_0} \widehat{\mathbf{\Sigma}} \left(\mathrm{t}_N \widehat{\mathscr{E}}^{II} \right). \tag{8.80}$$

A *homogeneous* but *anisotropic* interface *admittance tensor* with components $(\widehat{\sigma}_{xx}, \widehat{\sigma}_{xy}, \widehat{\sigma}_{yx}, \widehat{\sigma}_{yy})$ depending only on ω may be written in terms of the induced, fixed-frame $(\mathrm{d}x, \mathrm{d}y)$ on S as

---

[6] For the 1-form $\widehat{\gamma} = \widehat{\gamma}_x \mathrm{d}x + \widehat{\gamma}_y \mathrm{d}y + \widehat{\gamma}_z \mathrm{d}z$ where $\widehat{\gamma}_x, \widehat{\gamma}_y, \widehat{\gamma}_z \in \widehat{\Gamma\Lambda}_0 \mathscr{U}$, $\mathrm{i}_{\partial_z} \widehat{\gamma} = \widehat{\gamma}_z$ and $\mathrm{t}_{\partial_z} \widehat{\gamma} = \widehat{\gamma}_x \mathrm{d}x + \widehat{\gamma}_y \mathrm{d}y$.

$$\widehat{\boldsymbol{\Sigma}} = \widehat{\sigma}_{xx}\mathrm{d}x \otimes \mathrm{i}_{\partial_x} + \widehat{\sigma}_{xy}\mathrm{d}x \otimes \mathrm{i}_{\partial_y} + \widehat{\sigma}_{yx}\mathrm{d}y \otimes \mathrm{i}_{\partial_x} + \widehat{\sigma}_{yy}\mathrm{d}y \otimes \mathrm{i}_{\partial_y}. \tag{8.81}$$

The junction condition (8.79) then becomes

$$\mathrm{t}_{\partial_z}\widehat{\mathscr{H}^{II}} - \mathrm{t}_{\partial_z}\widehat{\mathscr{H}^{I}} = -\frac{1}{Z_0}\left(\#_{\partial_z} \circ \widehat{\boldsymbol{\Sigma}} \circ \mathrm{t}_{\partial_z}\right)(\widehat{\mathscr{E}^L}), \tag{8.82}$$

where one can choose L to be either I or II. It will be useful in the following to rewrite (8.82) using (8.27) as

$$\mathrm{t}_{\partial_z}\widehat{\mathscr{H}^{II}} - \mathrm{t}_{\partial_z}\widehat{\mathscr{H}^{I}} = -\frac{1}{Z_0}\left(R_{\partial_z}\left(-\frac{\pi}{2}\right) \circ \widehat{\boldsymbol{\Sigma}} \circ \mathrm{t}_{\partial_z}\right)(\widehat{\mathscr{E}^L}). \tag{8.83}$$

8.8 Consequences of the Interface Conditions

The structure of the modes in $\mathscr{V} = \mathscr{V}^I \cup \mathscr{V}^{II} \cup S$ is given by (8.66) in terms of the specific 1-forms $\{\widehat{K}^I, \widehat{\mathscr{E}^I}, \widehat{\mathscr{H}^I}, \widehat{K}^{II}, \widehat{\mathscr{E}^{II}}, \widehat{\mathscr{H}^{II}}\}$ where $\widehat{K}^L = \mathrm{d}\widehat{\chi}^L$ for $L \in \{I, II\}$. In each region, $\widehat{\mathscr{E}^L}$ and $\widehat{K}^L$ must satisfy the (nonlinear) dispersion relation and orthogonality conditions (8.49) and (8.47), as well as the interface conditions (8.75), (8.76), (8.77) and (8.83) derived in Sect. 8.7, with $\widehat{\mathscr{H}^L}$ given by (8.50).

The construction of solutions parametrised by elements of $SO(3,\mathbb{C})$ acting on particular solutions in a domain free of interfaces developed in Sect. 8.6 is now extended to the parametrisation of solutions in region $\mathscr{M}$ containing a planar interface, in terms of the six complex angles $\{\widehat{\phi}^I, \widehat{\theta}^I, \widehat{\psi}^I, \widehat{\phi}^{II}, \widehat{\theta}^{II}, \widehat{\psi}^{II}\}$ using composite rotation operators $\mathscr{R}_{\tau^I}$ and $\mathscr{R}_{\tau^{II}}$ by writing

$$\widehat{\gamma}_{\tau^L} = \mathscr{R}_{\tau^L}\widehat{\gamma}^L_{\tau_0} \tag{8.84}$$

where

$$\begin{aligned} \widehat{\gamma}_{\tau^L} &\in \{\widehat{K}_{\tau^L}, \widehat{\mathscr{E}}_{\tau^L}, \widehat{\mathscr{H}}_{\tau^L}\}, \\ \widehat{\gamma}_{\tau^L_0} &\in \{\widehat{K}_{\tau^L_0}, \widehat{\mathscr{E}}_{\tau^L_0}, \widehat{\mathscr{H}}_{\tau^L_0}\}, \end{aligned} \tag{8.85}$$

$$\mathscr{R}_{\tau^L} = R_{\partial_z}(\widehat{\psi}^L) \circ R_{\partial_y}(\widehat{\theta}^L) \circ R_{\partial_x}(\widehat{\phi}^L), \tag{8.86}$$

$$\begin{aligned} \widehat{K}^L_{\tau_0} &= \widehat{K}^L_0\mathrm{d}x = \omega\sqrt{\widehat{\epsilon'}^L}\sqrt{\widehat{\mu}^L}\mathrm{d}x, \\ \widehat{\mathscr{E}}^L_{\tau_0} &= \widehat{A}^L\mathrm{d}y, \\ \widehat{\mathscr{H}}^L_{\tau_0} &= \frac{\widehat{A}^L\widehat{K}^L_0}{\omega\widehat{\mu}^L}\mathrm{d}z = \frac{\widehat{A}^L\sqrt{\widehat{\epsilon'}^L}}{\sqrt{\widehat{\mu}^L}}\mathrm{d}z, \end{aligned} \tag{8.87}$$

so that

$$\widehat{K}_{\tau^L} = \widehat{K}_0^L \left[\cos\widehat{\theta}^L \cos\widehat{\psi}^L \mathrm{d}x - \cos\widehat{\theta}^L \sin\widehat{\psi}^L \mathrm{d}y + \sin\widehat{\theta}^L \mathrm{d}z\right] \tag{8.88}$$

$$\begin{aligned}\widehat{\mathscr{E}}_{\tau^L} = \widehat{A}^L \Big[&\left(\cos\widehat{\phi}^L \sin\widehat{\psi}^L + \sin\widehat{\theta}^L \sin\widehat{\phi}^L \cos\widehat{\psi}^L\right) \mathrm{d}x \\ &+ \left(\cos\widehat{\phi}^L \cos\widehat{\psi}^L - \sin\widehat{\theta}^L \sin\widehat{\phi}^L \sin\widehat{\psi}^L\right) \mathrm{d}y - \cos\widehat{\theta}^L \sin\widehat{\phi}^L \mathrm{d}z\Big]\end{aligned} \tag{8.89}$$

$$\begin{aligned}\widehat{\mathscr{H}_{\tau^L}} = \frac{\widehat{A}^L \widehat{K}_0^L}{\omega \widehat{\mu}^L} \Big[&\left(\sin\widehat{\phi}^L \sin\widehat{\psi}^L - \sin\widehat{\theta}^L \cos\widehat{\phi}^L \cos\widehat{\psi}^L\right) \mathrm{d}x \\ &+ \left(\sin\widehat{\phi}^L \cos\widehat{\psi}^L + \sin\widehat{\theta}^L \cos\widehat{\phi}^L \sin\widehat{\psi}^L\right) \mathrm{d}y + \cos\widehat{\theta}^L \cos\widehat{\phi}^L \mathrm{d}z\Big].\end{aligned} \tag{8.90}$$

Such 1-forms already satisfy (8.49), (8.47) and (8.50), so one must only address the interface conditions (8.75), (8.76), (8.77) and (8.83) to obtain a suitably correlated solution in both regions. In terms of $\{\widehat{K}^L_{\tau^L}, \widehat{\mathscr{E}}^L_{\tau^L}, \widehat{\mathscr{H}}^L_{\tau^L}\}$, these interface conditions are

$$\mathrm{t}_{\partial_z}\widehat{K}_{\tau^{II}} - \mathrm{t}_{\partial_z}\widehat{K}_{\tau^I} = 0, \tag{8.91}$$

$$\widehat{\mu}^{II}\, \mathrm{i}_{\partial_z}\widehat{\mathscr{H}_{\tau^{II}}} - \widehat{\mu}^I\, \mathrm{i}_{\partial_z}\widehat{\mathscr{H}_{\tau^I}} = 0, \tag{8.92}$$

$$\mathrm{t}_{\partial_z}\widehat{\mathscr{E}}_{\tau^{II}} - \mathrm{t}_{\partial_z}\widehat{\mathscr{E}}_{\tau^I} = 0, \tag{8.93}$$

$$\mathrm{t}_{\partial_z}\widehat{\mathscr{H}_{\tau^{II}}} - \mathrm{t}_{\partial_z}\widehat{\mathscr{H}_{\tau^I}} + \frac{1}{Z_0}\left(R_{\partial_z}\left(-\frac{\pi}{2}\right) \circ \widehat{\boldsymbol{\Sigma}} \circ \mathrm{t}_N\right)(\widehat{\mathscr{E}}_{\tau^I}) = 0. \tag{8.94}$$

To decouple the system, it will prove expedient to work with a new set of 1-forms "rotated" with respect to ∂_z by the complex angle $-\widehat{\psi}^I$ using the operator (8.20). In terms of (8.84)-(8.85), define the *rotated forms*

$$\widehat{\gamma}_R^L \equiv R_{\partial_z}(-\widehat{\psi}^I)\widehat{\gamma}_{\tau^L} = \left(R_{\partial_z}(-\widehat{\psi}^I) \circ \mathscr{R}_{\tau^L}\right)(\widehat{\gamma}^L_{\tau_0}). \tag{8.95}$$

The identity (8.25) then gives

$$\begin{aligned}\widehat{\gamma}_R^I = \left(R_{\partial_z}(-\widehat{\psi}^I) \circ \mathscr{R}_{\tau^I}\right)(\widehat{\gamma}^I_{\tau_0}) &= \left(R_{\partial_z}(-\widehat{\psi}^I) \circ R_{\partial_z}(\widehat{\psi}^I) \circ R_{\partial_y}(\widehat{\theta}^I) \circ R_{\partial_x}(\widehat{\phi}^I)\right)(\widehat{\gamma}^I_{\tau_0}) \\ &= \left(R_{\partial_y}(\widehat{\theta}^I) \circ R_{\partial_x}(\widehat{\phi}^I)\right)(\widehat{\gamma}^I_{\tau_0})\end{aligned} \tag{8.96}$$

$$\begin{aligned}\widehat{\gamma}_R^{II} = \left(R_{\partial_z}(-\widehat{\psi}^I) \circ \mathscr{R}_{\tau^{II}}\right)(\widehat{\gamma}^{II}_{\tau_0}) &= \left(R_{\partial_z}(-\widehat{\psi}^I) \circ R_{\partial_z}(\widehat{\psi}^{II}) \circ R_{\partial_y}(\widehat{\theta}^{II}) \circ R_{\partial_x}(\widehat{\phi}^{II})\right)(\widehat{\gamma}^{II}_{\tau_0}) \\ &= \left(R_{\partial_z}(\widehat{\psi}^\Delta) \circ R_{\partial_y}(\widehat{\theta}^I) \circ R_{\partial_x}(\widehat{\phi}^I)\right)(\widehat{\gamma}^I_{\tau_0})\end{aligned} \tag{8.97}$$

where

$$\widehat{\psi}^\Delta \equiv \widehat{\psi}^{II} - \widehat{\psi}^I \tag{8.98}$$

and the system of rotated 1-forms in (8.96)-(8.97) now depends on the *five* complex angles $\{\widehat{\phi}^I, \widehat{\theta}^I, \widehat{\phi}^{II}, \widehat{\theta}^{II}, \widehat{\psi}^\Delta\}$ rather than the original six. Equations (8.88)-(8.90) may be replaced by the partially simplified formulae

$$\widehat{K}_R^I = \widehat{K}_0^I\left[\cos\widehat{\theta}^I\mathrm{d}x + \sin\widehat{\theta}^I\mathrm{d}z\right], \tag{8.99}$$

$$\widehat{\mathscr{E}}_R^I = \widehat{A}^I\left[\sin\widehat{\theta}^I\sin\widehat{\phi}^I\mathrm{d}x + \cos\widehat{\phi}^I\mathrm{d}y - \cos\widehat{\theta}^I\sin\widehat{\phi}^I\mathrm{d}z\right], \tag{8.100}$$

$$\widehat{\mathscr{H}}_R^I = \frac{\widehat{A}^I\widehat{K}_0^I}{\omega\widehat{\mu}^I}\left[-\sin\widehat{\theta}^I\cos\widehat{\phi}^I\mathrm{d}x + \sin\widehat{\phi}^I\mathrm{d}y + \cos\widehat{\theta}^I\cos\widehat{\phi}^I\mathrm{d}z\right], \tag{8.101}$$

$$\widehat{K}_R^{II} = \widehat{K}_0^{II}\left[\cos\widehat{\theta}^{II}\cos\widehat{\psi}^\Delta\mathrm{d}x - \cos\widehat{\theta}^{II}\sin\widehat{\psi}^\Delta\mathrm{d}y + \sin\widehat{\theta}^{II}\mathrm{d}z\right], \tag{8.102}$$

$$\begin{aligned}\widehat{\mathscr{E}}_R^{II} = \widehat{A}^{II}\Big[&\left(\cos\widehat{\phi}^{II}\sin\widehat{\psi}^\Delta + \sin\widehat{\theta}^{II}\sin\widehat{\phi}^{II}\cos\widehat{\psi}^\Delta\right)\mathrm{d}x\\ &+\left(\cos\widehat{\phi}^{II}\cos\widehat{\psi}^\Delta - \sin\widehat{\theta}^{II}\sin\widehat{\phi}^{II}\sin\widehat{\psi}^\Delta\right)\mathrm{d}y - \cos\widehat{\theta}^{II}\sin\widehat{\phi}^{II}\mathrm{d}z\Big],\end{aligned} \tag{8.103}$$

$$\begin{aligned}\widehat{\mathscr{H}}_R^{II} = \frac{\widehat{A}^{II}\widehat{K}_0^{II}}{\omega\widehat{\mu}^{II}}\Big[&\left(\sin\widehat{\phi}^{II}\sin\widehat{\psi}^\Delta - \sin\widehat{\theta}^{II}\cos\widehat{\phi}^{II}\cos\widehat{\psi}^\Delta\right)\mathrm{d}x\\ &+\left(\sin\widehat{\phi}^{II}\cos\widehat{\psi}^\Delta + \sin\widehat{\theta}^{II}\cos\widehat{\phi}^{II}\sin\widehat{\psi}^\Delta\right)\mathrm{d}y + \cos\widehat{\theta}^{II}\cos\widehat{\phi}^{II}\mathrm{d}z\Big].\end{aligned} \tag{8.104}$$

The interface conditions (8.91)-(8.94) may now be readily expressed in terms of the rotated 1-forms. Condition (8.92) is satisfied if

$$\widehat{\mu}^{II}\,\mathrm{i}_{\partial_z}\widehat{\mathscr{H}}_R^{II} - \widehat{\mu}^I\,\mathrm{i}_{\partial_z}\widehat{\mathscr{H}}_R^I = 0, \tag{8.105}$$

since identity (8.22) gives

$$\mathrm{i}_{\partial_z}\widehat{\mathscr{H}}_R^L = \mathrm{i}_{\partial_z}R_{\partial_z}(-\widehat{\psi}^I)(\widehat{\mathscr{H}}^L) = \mathrm{i}_{\partial_z}\widehat{\mathscr{H}}^L.$$

From the identities (8.24) and (8.25), the relation

$$\mathrm{t}_N(R_{\partial_z}(\varphi)(\widehat{\gamma}_{\tau^{II}})) - \mathrm{t}_N(R_{\partial_z}(\varphi)(\widehat{\gamma}_{\tau^I})) - R_{\partial_z}(\varphi)(\widehat{\beta}) = 0, \tag{8.106}$$

for any 1-form $\widehat{\beta}$ is readily converted by the action of $R_{\partial_z}(-\varphi)$ to the condition

$$\mathrm{t}_N\widehat{\gamma}_{\tau^{II}} - \mathrm{t}_N\widehat{\gamma}_{\tau^I} - \widehat{\beta} = 0. \tag{8.107}$$

Thus conditions (8.91) and (8.93) are satisfied by (8.106) with $\varphi = -\widehat{\psi}^I$ and $\widehat{\beta} = 0$ while condition (8.94) is satisfied with $\varphi = (\pi/2) - \widehat{\psi}^I$ and $\widehat{\beta} = \frac{1}{Z_0}\left(R_N(-\frac{\pi}{2})\circ\mathbf{\Sigma}\circ\mathrm{t}_N\right)\left(\widehat{\mathscr{E}}^L\right)$ thereby replacing these with the new interface conditions:

$$\begin{aligned}0 &= \mathrm{t}_{\partial_z}\left(R_{\partial_z}(\varphi)(\widehat{\gamma}_{\tau^{II}})\right) - \mathrm{t}_{\partial_z}\left(R_{\partial_z}(\varphi)(\widehat{\gamma}_{\tau^{II}})\right) - R_{\partial_z}(\varphi)(\widehat{\beta})\\ &= R_{\partial_z}(\varphi)\left(\mathrm{t}_{\partial_z}(\widehat{\gamma}_{\tau^{II}})\right) - R_{\partial_z}(\varphi)\left(\mathrm{t}_{\partial_z}(\widehat{\gamma}_{\tau^{II}})\right) - R_{\partial_z}(\varphi)(\widehat{\beta})\\ &= R_{\partial_z}(-\varphi)\left(R_{\partial_z}(\varphi)\left(\mathrm{t}_{\partial_z}(\widehat{\gamma}_{\tau^{II}})\right) - R_{\partial_z}(\varphi)\left(\mathrm{t}_{\partial_z}(\widehat{\gamma}_{\tau^{II}})\right) - R_{\partial_z}(\varphi)(\widehat{\beta})\right)\\ &= \mathrm{t}_{\partial_z}\widehat{\gamma}_{\tau^{II}} - \mathrm{t}_{\partial_z}\widehat{\gamma}_{\tau^{II}} - \widehat{\beta}.\end{aligned}$$

Hence, taking $\varphi = -\widehat{\psi}^I$, $\widehat{\beta} = 0$ in (8.75), (8.77) and

$$\widehat{\varphi} = \frac{\pi}{2} - \widehat{\psi}^I, \widehat{\beta} = \frac{1}{Z_0}\left(R_{\partial_z}\left(-\frac{\pi}{2}\right) \circ \widehat{\boldsymbol{\Sigma}} \circ \mathrm{t}_{\partial_z}\right)\left(\widehat{\mathscr{E}}^L\right)$$

in (8.83) gives the new conditions

$$\mathrm{t}_{\partial_z}\widehat{K}_R^{II} - \mathrm{t}_{\partial_z}\widehat{K}_R^{I} = 0, \tag{8.108}$$

$$\mathrm{t}_{\partial_z}\widehat{\mathscr{E}}_R^{II} - \mathrm{t}_{\partial_z}\widehat{\mathscr{E}}_R^{I} = 0, \tag{8.109}$$

$$\left[\mathrm{t}_{\partial_z} \circ R_{\partial_z}\left(\frac{\pi}{2}\right)\right]\left(\widehat{\mathscr{H}}_R^{II} - \widehat{\mathscr{H}}_R^{I}\right) + \frac{1}{Z_0}\left[R_{\partial_z}\left(\frac{\pi}{2} - \widehat{\psi}^I\right) \circ R_{\partial_z}\left(-\frac{\pi}{2}\right) \circ \widehat{\boldsymbol{\Sigma}} \circ \mathrm{t}_{\partial_z}\right]\left(\widehat{\mathscr{E}}^I\right) = 0. \tag{8.110}$$

Using identities (8.25), (8.24) and definition (8.95), (8.110) can be rewritten:

$$\left(\mathrm{t}_{\partial_z} \circ R_{\partial_z}\left(\frac{\pi}{2}\right)\right)\left(\widehat{\mathscr{H}}_R^{II} - \widehat{\mathscr{H}}_R^{I}\right) = -\frac{1}{Z_0}\left(R_{\partial_z}\left(-\widehat{\psi}^I\right) \circ \widehat{\boldsymbol{\Sigma}} \circ \mathrm{t}_{\partial_z}\right)\left(\widehat{\mathscr{E}}^I\right). \tag{8.111}$$

To further simplify (8.111), introduce the rank (1,1) tensor $\widehat{\boldsymbol{\Lambda}} : \widehat{\Gamma}\Lambda^1 S \to \widehat{\Gamma}\Lambda^1 S$ where

$$\begin{aligned}\widehat{\boldsymbol{\Lambda}} &= R_{\partial_z}\left(-\widehat{\psi}^I\right) \circ \widehat{\boldsymbol{\Sigma}} \circ R_{\partial_z}\left(\widehat{\psi}^I\right) \circ \mathrm{t}_{\partial_z} \\ &= \widehat{\lambda}_{xx}\mathrm{d}x \otimes \mathrm{i}_{\partial_x} + \widehat{\lambda}_{xy}\mathrm{d}x \otimes \mathrm{i}_{\partial_y} + \widehat{\lambda}_{yx}\mathrm{d}y \otimes \mathrm{i}_{\partial_x} + \widehat{\lambda}_{yy}\mathrm{d}y \otimes \mathrm{i}_{\partial_y}.\end{aligned} \tag{8.112}$$

After some algebra, it can be shown that the components of the tensor $\widehat{\boldsymbol{\Lambda}}$ are given by

$$\widehat{\lambda}_{xx} = \frac{1}{2}\left[(\widehat{\sigma}_{xx} + \widehat{\sigma}_{yy}) + (\widehat{\sigma}_{xx} - \widehat{\sigma}_{yy})\cos 2\widehat{\psi}^I - (\widehat{\sigma}_{xy} + \widehat{\sigma}_{yx})\sin 2\widehat{\psi}^I\right], \tag{8.113}$$

$$\widehat{\lambda}_{xy} = \frac{1}{2}\left[(\widehat{\sigma}_{xy} - \widehat{\sigma}_{yx}) + (\widehat{\sigma}_{xy} + \widehat{\sigma}_{yx})\cos 2\widehat{\psi}^I + (\widehat{\sigma}_{xx} - \widehat{\sigma}_{yy})\sin 2\widehat{\psi}^I\right], \tag{8.114}$$

$$\widehat{\lambda}_{yx} = \frac{1}{2}\left[-(\widehat{\sigma}_{xy} - \widehat{\sigma}_{yx}) + (\widehat{\sigma}_{xy} + \widehat{\sigma}_{yx})\cos 2\widehat{\psi}^I + (\widehat{\sigma}_{xx} - \widehat{\sigma}_{yy})\sin 2\widehat{\psi}^I\right], \tag{8.115}$$

$$\widehat{\lambda}_{yy} = \frac{1}{2}\left[(\widehat{\sigma}_{xx} + \widehat{\sigma}_{yy}) - (\widehat{\sigma}_{xx} - \widehat{\sigma}_{yy})\cos 2\widehat{\psi}^I + (\widehat{\sigma}_{xy} + \widehat{\sigma}_{yx})\sin 2\widehat{\psi}^I\right]. \tag{8.116}$$

Condition (8.111) can thus be rewritten in terms of the tensor $\widehat{\boldsymbol{\Lambda}}$ as

$$\left(\mathrm{t}_{\partial_z} \circ R_{\partial_z}\left(\frac{\pi}{2}\right)\right)\left(\widehat{\mathscr{H}}_R^{II} - \widehat{\mathscr{H}}_R^{I}\right) = -\frac{1}{Z_0}\widehat{\boldsymbol{\Lambda}}\left(\widehat{\mathscr{E}}_R^{I}\right). \tag{8.117}$$

Having written the interface conditions (8.91)-(8.94) in terms of the rotated 1-forms $\widehat{\gamma}_R^L$, as (8.105), (8.108)-(8.110) they may be further decoupled. Substituting (8.99) and (8.102) into (8.108) gives

$$\left(\widehat{K}_0^{II}\cos\widehat{\theta}^{II}\cos\widehat{\psi}^{\Delta} - \widehat{K}_0^{I}\cos\widehat{\theta}^{I}\right)\mathrm{d}x + \widehat{K}_0^{II}\cos\widehat{\theta}^{II}\sin\widehat{\psi}^{\Delta}\mathrm{d}y = 0, \tag{8.118}$$

One solution of (8.118) is $\cos\widehat{\theta}^I = \cos\widehat{\theta}^{II} = 0$. This implies from equation (8.88) that the tangential components of the wave vector $\widehat{K}^L$ are zero in both regions and the fields can only vary with z, thereby excluding any solutions propagating along the interface. In the following, we instead restrict to solutions with $\widehat{K}_0^I\cos\widehat{\theta}^I \neq 0$ and $\widehat{K}_0^{II}\cos\widehat{\theta}^{II} \neq 0$, which include both single-interface surface polariton and Brewster modes (see below). In this case, the dy component of (8.118) requires $\sin\widehat{\psi}^\Delta = 0$ and therefore

$$\widehat{\psi}^\Delta = m\pi, \tag{8.119}$$

where $m \in \mathbb{Z}$. Thus $\cos\widehat{\psi}^\Delta = (-1)^m$, $\sin\widehat{\psi}^\Delta = 0$, and (8.99)-(8 104) can be written in terms of the scalings $\{\widehat{A}^L\}$ and *four* complex angles $\{\widehat{\phi}^I, \widehat{\theta}^I, \widehat{\phi}^{II}, \widehat{\theta}^{II}\}$:

$$\widehat{K}_R^L = \widehat{K}_0^L\left[\left(\zeta^L\right)^m\cos\widehat{\theta}^L\mathrm{d}x + \sin\widehat{\theta}^L\mathrm{d}z\right], \tag{8.120}$$

$$\widehat{\mathscr{E}}_R^L = \widehat{A}^L\left[\left(\zeta^L\right)^m\left(\sin\widehat{\theta}^L\sin\widehat{\phi}^L\mathrm{d}x + \cos\widehat{\phi}^L\mathrm{d}y\right) - \cos\widehat{\theta}^L\sin\widehat{\phi}^L\mathrm{d}z\right], \tag{8.121}$$

$$\widehat{\mathscr{H}}_R^L = \frac{\widehat{A}^L\widehat{K}_0^L}{\omega\widehat{\mu}^L}\left[\left(\zeta^L\right)^m\left(-\sin\widehat{\theta}^L\cos\widehat{\phi}^L\mathrm{d}x + \sin\widehat{\phi}^L\mathrm{d}y\right) + \cos\widehat{\theta}^L\cos\widehat{\phi}^L\mathrm{d}z\right]. \tag{8.122}$$

where the constant ζ^L is defined in each region as:

$$\zeta^I = 1 \qquad \zeta^{II} = -1. \tag{8.123}$$

The interface condition (8.108) (and hence (8.118)) gives (8.124) below, and (8.105) gives (8.125). Furthermore, the components of (8.109) yield (8.126) and (8.127) and the components of (8.117) give (8.128) and (8.129). The complete set of interface conditions therefore becomes

$$\widehat{K}_0^{II}\cos\widehat{\theta}^{II} = (-1)^m\widehat{K}_0^I\cos\widehat{\theta}^I, \tag{8.124}$$

$$\widehat{K}_0^{II}\widehat{A}^{II}\cos\widehat{\theta}^{II}\cos\widehat{\phi}^{II} = \widehat{K}_0^I\widehat{A}^I\cos\widehat{\theta}^I\cos\widehat{\phi}^I, \tag{8.125}$$

$$\widehat{A}^{II}\sin\widehat{\theta}^{II}\sin\widehat{\phi}^{II} = (-1)^m\widehat{A}^I\sin\widehat{\theta}^I\sin\widehat{\phi}^I, \tag{8.126}$$

$$\widehat{A}^{II}\cos\widehat{\phi}^{II} = (-1)^m\widehat{A}^I\cos\widehat{\phi}^I. \tag{8.127}$$

$$(-1)^m\frac{c\widehat{K}_0^{II}}{\omega\widehat{\mu}_r^{II}}\widehat{A}^{II}\sin\widehat{\phi}^{II} - \frac{c\widehat{K}_0^I}{\omega\widehat{\mu}_r^I}\widehat{A}^I\sin\widehat{\phi}^I = -\widehat{\lambda}_{xx}\widehat{A}^I\sin\widehat{\theta}^I\sin\widehat{\phi}^I - \widehat{\lambda}_{xy}\widehat{A}^I\cos\widehat{\phi}^I, \tag{8.128}$$

$$\begin{aligned}(-1)^m\frac{c\widehat{K}_0^{II}}{\omega\widehat{\mu}_r^{II}}\widehat{A}^{II}\sin\widehat{\theta}^{II}\cos\widehat{\phi}^{II} - \frac{c\widehat{K}_0^I}{\omega\widehat{\mu}_r^I}\widehat{A}^I\sin\widehat{\theta}^I\cos\widehat{\phi}^I \\ = -\widehat{\lambda}_{yx}\widehat{A}^I\sin\widehat{\theta}^I\sin\widehat{\phi}^I - \widehat{\lambda}_{yy}\widehat{A}^I\cos\widehat{\phi}^I.\end{aligned} \tag{8.129}$$

with

$$\widehat{\mu}_r^L = \frac{\widehat{\mu}^L}{\mu_0} = \frac{\widehat{\mu}^L c}{Z_0}.$$

We observe that (8.125) automatically follows if (8.124) and (8.127) are satisfied.

8.9 Solving the Interface Conditions

We now show that solving the interface conditions (8.124)-(8.129) reduces to finding solutions to a complex polynomial equation. Introduce the complex quantities

$$\widehat{\Psi} \equiv \widehat{A}^I \sin\widehat{\theta}^I \sin\widehat{\phi}^I, \tag{8.130}$$

$$\widehat{\Phi} \equiv \widehat{A}^I \cos\widehat{\phi}^I, \tag{8.131}$$

$$\widehat{Q} \equiv \frac{c\widehat{K}_0^I}{\omega}\cos\widehat{\theta}^I \neq 0, \tag{8.132}$$

$$\widehat{\alpha} \equiv \frac{c}{2\omega}\left(\widehat{K}_0{}^I \sin\widehat{\theta}^I + \widehat{K}_0{}^{II}\sin\widehat{\theta}^{II}\right), \tag{8.133}$$

$$\widehat{\beta} \equiv \frac{c}{2\omega}\left(\widehat{K}_0{}^I \sin\widehat{\theta}^I - \widehat{K}_0{}^{II}\sin\widehat{\theta}^{II}\right). \tag{8.134}$$

Equation (8.108) is then satisfied with

$$\cos\widehat{\theta}^L = \left(\zeta^L\right)^m \frac{\omega\widehat{Q}}{c\widehat{K}_0^L} \tag{8.135}$$

and (8.133) and (8.134) yield

$$\sin\widehat{\theta}^L = \frac{\omega}{c\widehat{K}_0^L}\left(\widehat{\alpha} + \zeta^L\widehat{\beta}\right). \tag{8.136}$$

Equations (8.133)-(8.134) give

$$\begin{aligned}
\widehat{\alpha}\widehat{\beta} &= \frac{c^2}{4\omega^2}\left(\widehat{K}_0{}^{I\,2}\sin^2\widehat{\theta}^I - \widehat{K}_0{}^{II\,2}\sin^2\widehat{\theta}^{II}\right)\\
&= \frac{c^2}{4\omega^2}\left(\widehat{K}_0{}^{I\,2} - \widehat{K}_0{}^{II\,2} - \widehat{K}_0{}^{I\,2}\cos^2\widehat{\theta}^I + \widehat{K}_0{}^{II\,2}\cos^2\widehat{\theta}^{II}\right)\\
&= \frac{c^2}{4\omega^2}\left(\widehat{K}_0{}^{I\,2} - \widehat{K}_0{}^{II\,2}\right) = \frac{1}{4}\left(\widehat{\epsilon'}^I_r\widehat{\mu}^I_r - \widehat{\epsilon'}^{II}_r\widehat{\mu}^{II}_r\right),
\end{aligned}$$

since $\widehat{K}_0{}^{I\,2}\cos^2\widehat{\theta}^I = \widehat{K}_0{}^{II\,2}\cos^2\widehat{\theta}^{II}$ from (8.124). Hence,

$$\widehat{\beta} = \frac{\widehat{\nu}}{\widehat{\alpha}} \tag{8.137}$$

where

$$\widehat{\nu} \equiv \frac{1}{4}\left(\widehat{\epsilon'}^I_r\widehat{\mu}^I_r - \widehat{\epsilon'}^{II}_r\widehat{\mu}^{II}_r\right) \tag{8.138}$$

depends only on the constitutive properties of the two bulk regions. Equation (8.136) is now written in terms of $\widehat{\alpha}$ as

$$\sin\widehat{\theta}^L = \frac{\omega}{c\widehat{K}_0^L}\left(\widehat{\alpha}+\zeta^L\frac{\widehat{v}}{\widehat{\alpha}}\right). \tag{8.139}$$

Definition (8.130) and junction condition (8.126) imply

$$A^L\sin\widehat{\phi}^L = \left(\zeta^L\right)^m\frac{\widehat{\Psi}}{\sin\widehat{\theta}^L} = \left(\zeta^L\right)^m\frac{c\widehat{K}_0^L}{\omega}\frac{\widehat{\Psi}}{\widehat{\alpha}+\zeta^L\frac{\widehat{v}}{\widehat{\alpha}}}, \tag{8.140}$$

while (8.131) and junction condition (8.127) yield

$$A^L\cos\widehat{\phi}^L = \left(\zeta^L\right)^m\widehat{\Phi}. \tag{8.141}$$

Substituting equations (8.135) and (8.139) into the complex identity $\cos^2\widehat{\theta}^L + \sin^2\widehat{\theta}^L = 1$ for either $L = I$ or $L = II$ and solving for $\widehat{Q}$ gives

$$\widehat{Q} = \pm\sqrt{\frac{\widehat{\epsilon'}_r^I\widehat{\mu}_r^I}{2}+\frac{\widehat{\epsilon'}_r^{II}\widehat{\mu}_r^{II}}{2}-\widehat{\alpha}^2-\frac{\widehat{v}^2}{\widehat{\alpha}^2}}. \tag{8.142}$$

In summary, the trigonometric functions of $\widehat{\theta}^L$ and $\widehat{\phi}^L$ are expressed in terms of $\widehat{\alpha}$, $\widehat{\Phi}$, $\widehat{\Psi}$ and the constitutive properties of $\mathscr{V}^I$ and $\mathscr{V}^{II}$ by the relations

$$\begin{aligned}
\sin\widehat{\theta}^L &= \frac{\omega}{c\widehat{K}_0^L}\left(\widehat{\alpha}+\zeta^L\frac{\widehat{v}}{\widehat{\alpha}}\right), & \cos\widehat{\theta}^L &= \left(\zeta^L\right)^m\frac{\omega\widehat{Q}}{c\widehat{K}_0^L},\\
\widehat{A}^L\sin\widehat{\phi}^L &= \left(\zeta^L\right)^m\frac{c\widehat{K}_0^L}{\omega}\frac{\widehat{\Psi}}{\widehat{\alpha}+\zeta^L\frac{\widehat{v}}{\widehat{\alpha}}}, & \widehat{A}^L\cos\widehat{\phi}^L &= \left(\zeta^L\right)^m\widehat{\Phi},
\end{aligned} \tag{8.143}$$

where ζ^L is defined by (8.123) and $\widehat{Q}$ is given by (8.142). It may be readily confirmed that the interface conditions (8.124)-(8.127) are satified by the equations (8.143). We now turn to the final two conditions (8.128) and (8.129).

Substituting (8.143) into (8.128) and using (8.49) gives

$$\widehat{X}(\widehat{\alpha})\widehat{\Psi} - \widehat{\lambda}_{xy}\widehat{\Phi} = 0, \tag{8.144}$$

where

$$\widehat{X}(\widehat{\alpha}) \equiv \frac{\widehat{\alpha}\left(\widehat{\epsilon'}_r^I-\widehat{\epsilon'}_r^{II}\right)-\frac{\widehat{v}}{\widehat{\alpha}}\left(\widehat{\epsilon'}_r^I+\widehat{\epsilon'}_r^{II}\right)-\widehat{\lambda}_{xx}\left(\widehat{\alpha}^2-\frac{\widehat{v}^2}{\widehat{\alpha}^2}\right)}{\widehat{\alpha}^2-\frac{\widehat{v}^2}{\widehat{\alpha}^2}}. \tag{8.145}$$

Substituting (8.143) into (8.129) then implies

$$\widehat{Y}(\widehat{\alpha})\widehat{\Phi} - \widehat{\lambda}_{yx}\widehat{\Psi} = 0, \tag{8.146}$$

where

$$\widehat{Y}(\alpha) \equiv \left(\frac{1}{\widehat{\mu}_r^I} - \frac{1}{\widehat{\mu}_r^{II}}\right)\widehat{\alpha} + \left(\frac{1}{\widehat{\mu}_r^I} + \frac{1}{\widehat{\mu}_r^{II}}\right)\frac{\widehat{\nu}}{\widehat{\alpha}} - \widehat{\lambda}_{yy}. \tag{8.147}$$

Equations (8.144) and (8.146) admit different classes of solution for $\widehat{\Phi}$ and $\widehat{\Psi}$ depending on the values of the constitutive functions and their dependence on frequency.

For solutions with $\widehat{\Phi} = 0$ and $\widehat{\Psi} \neq 0$, these equations degenerate to the set

$$\widehat{X}(\widehat{\alpha}) = 0, \qquad \widehat{\lambda}_{yx} = 0 \tag{8.148}$$

where the first equation is a complex *quartic* equation for $\widehat{\alpha}$. We will call the fields determined by such solutions "type-$\widehat{\Psi}$ modes".

When $\widehat{\Psi} = 0$ and $\widehat{\Phi} \neq 0$ one obtains

$$\widehat{Y}(\widehat{\alpha}) = 0, \qquad \widehat{\lambda}_{xy} = 0 \tag{8.149}$$

where the first equation is a complex *quadratic* equation for $\widehat{\alpha}$. We will call the fields determined by such solutions "type-$\widehat{\Phi}$ modes".

In the general case (with both $\widehat{\Phi}$ and $\widehat{\Psi}$ non-zero), one obtains coupled type-$\widehat{\Phi}$-$\widehat{\Psi}$ modes from solutions $\widehat{\alpha}$ satisfying the determinantal condition

$$\widehat{X}(\widehat{\alpha})\widehat{Y}(\widehat{\alpha}) - \widehat{\lambda}_{xy}\widehat{\lambda}_{yx} = 0 \tag{8.150}$$

i.e. the *degree six complex polynomial*

$$\begin{aligned} 0 = &\left(\widehat{\alpha}^3\left(\widehat{\epsilon'}^I_r - \widehat{\epsilon'}^{II}_r\right) - \widehat{\nu}\widehat{\alpha}\left(\widehat{\epsilon'}^I_r + \widehat{\epsilon'}^{II}_r\right) - \widehat{\lambda}_{xx}\left(\widehat{\alpha}^4 - \widehat{\nu}^2\right)\right) \\ &\times\left(\left(\frac{1}{\widehat{\mu}_r^I} - \frac{1}{\widehat{\mu}_r^{II}}\right)\widehat{\alpha}^2 + \left(\frac{1}{\widehat{\mu}_r^I} + \frac{1}{\widehat{\mu}_r^{II}}\right)\widehat{\nu} - \widehat{\lambda}_{yy}\widehat{\alpha}\right) - \widehat{\lambda}_{xy}\widehat{\lambda}_{yx}\widehat{\alpha}\left(\widehat{\alpha}^4 - \widehat{\nu}^2\right). \end{aligned} \tag{8.151}$$

Then for each root $\widehat{\alpha}$,

$$\frac{\widehat{\Psi}}{\widehat{\Phi}} = \frac{1}{\widehat{\lambda}_{yx}}\left\{\frac{1}{\widehat{\mu}_r^I}\left(\widehat{\alpha} + \frac{\widehat{\nu}}{\widehat{\alpha}}\right) - \frac{1}{\widehat{\mu}_r^{II}}\left(\widehat{\alpha} - \frac{\widehat{\nu}}{\widehat{\alpha}}\right) - \widehat{\lambda}_{yy}\right\}. \tag{8.152}$$

and the rotated field 1-forms are given by substituting (8.143) into (8.120)-(8.122), for each root $\widehat{\alpha}$ of (8.151), using (8.49) to rewrite $\widehat{K}_0^2$:

$$\widehat{K}_R^L = \frac{\omega}{c}\left\{\widehat{Q}\mathrm{d}x + \left(\widehat{\alpha} + \frac{\zeta^L\widehat{\nu}}{\widehat{\alpha}}\right)\mathrm{d}z\right\}, \tag{8.153}$$

$$\widehat{\mathscr{E}}_R^L = \widehat{\Psi}\mathrm{d}x + \widehat{\Phi}\mathrm{d}y - \frac{\widehat{Q}\widehat{\Psi}}{\widehat{\alpha} + \frac{\zeta^L\widehat{\nu}}{\widehat{\alpha}}}\mathrm{d}z, \tag{8.154}$$

$$\widehat{\mathscr{H}}_R^L = \frac{1}{Z_0}\left\{-\left(\widehat{\alpha} + \frac{\zeta^L\widehat{\nu}}{\widehat{\alpha}}\right)\frac{\widehat{\Phi}}{\widehat{\mu}_r^L}\mathrm{d}x + \frac{\widehat{\epsilon'}_r\widehat{\Psi}}{\widehat{\alpha} + \frac{\zeta^L\widehat{\nu}}{\widehat{\alpha}}}\mathrm{d}y + \frac{\widehat{Q}\widehat{\Phi}}{\widehat{\mu}_r^L}\mathrm{d}z\right\}, \tag{8.155}$$

where $\widehat{\nu}$ is given by (8.138) and $\widehat{Q}$ is given by (8.142). Finally,

$$\{\widehat{K}^L, \widehat{\mathscr{E}}^L, \widehat{\mathscr{H}}^L\} = R_{\partial_z}(\widehat{\psi}^I)\{\widehat{K}_R^L, \widehat{\mathscr{E}}_R^L, \widehat{\mathscr{H}}_R^L\},$$

where

$$\widehat{K}^L = \frac{\omega}{c}\left\{\widehat{Q}\left(\cos\widehat{\psi}^I\mathrm{d}x - \sin\widehat{\psi}^I\mathrm{d}y\right) - \left(\widehat{\alpha} + \frac{\zeta^L\widehat{\nu}}{\widehat{\alpha}}\right)\mathrm{d}z\right\}, \tag{8.156}$$

$$\widehat{\mathscr{E}}^L = \left(\widehat{\Psi}\cos\widehat{\psi}^I + \widehat{\Phi}\sin\widehat{\psi}^I\right)\mathrm{d}x + \left(\widehat{\Phi}\cos\widehat{\psi}^I - \widehat{\Psi}\sin\widehat{\psi}^I\right)\mathrm{d}y + \frac{\widehat{Q}\widehat{\Psi}}{\widehat{\alpha} + \frac{\zeta^L\widehat{\nu}}{\widehat{\alpha}}}\mathrm{d}z, \tag{8.157}$$

$$\begin{aligned}\widehat{\mathscr{H}}^L = \frac{1}{Z_0}\Bigg\{&\left(\frac{\widehat{\epsilon'}_r\widehat{\Psi}}{\widehat{\alpha} + \frac{\zeta^L\widehat{\nu}}{\widehat{\alpha}}}\sin\widehat{\psi}^I - \left(\widehat{\alpha} + \frac{\zeta^L\widehat{\nu}}{\widehat{\alpha}}\right)\frac{\widehat{\Phi}}{\widehat{\mu}_r^L}\cos\widehat{\psi}^I\right)\mathrm{d}x,\\ &+\left(\left(\widehat{\alpha} + \frac{\zeta^L\widehat{\nu}}{\widehat{\alpha}}\right)\frac{\widehat{\Phi}}{\widehat{\mu}_r^L}\sin\widehat{\psi}^I + \frac{\widehat{\epsilon'}_r\widehat{\Psi}}{\widehat{\alpha} + \frac{\zeta^L\widehat{\nu}}{\widehat{\alpha}}}\cos\widehat{\psi}^I\right)\mathrm{d}y + \frac{\widehat{Q}\widehat{\Phi}}{\widehat{\mu}_r^L}\mathrm{d}z\Bigg\}.\end{aligned} \tag{8.158}$$

The complex parameter $\widehat{\psi}^I$ remains an arbitrary complex angle. Thus (8.156)-(8.158) constitute a family of solutions parametrised by $\widehat{\psi}^I$ along with the bulk and surface constitutive scalars.

For each m, $\widehat{\psi}^I$ and root $\widehat{\alpha}$, the final electric and magnetic field configurations are given by

$$\widehat{\boldsymbol{e}}^L = \widehat{\mathscr{E}}^L e^{i\widehat{\chi}^L}, \qquad \widehat{\boldsymbol{h}}^L = \widehat{\mathscr{H}}^L e^{i\widehat{\chi}^L},$$

where

$$\widehat{\chi}^L = \frac{\omega}{c}\left\{\widehat{Q}\left(\cos\widehat{\psi}^I x - \sin\widehat{\psi}^I y\right) + \left(\widehat{\alpha} + \frac{\zeta^L\widehat{\nu}}{\widehat{\alpha}}\right)z\right\}. \tag{8.159}$$

8.10 Conclusion

We have developed a strategy for analysing a large class of solutions to Maxwell's equations for the electromagnetic fields in piecewise homogeneous material media containing a plane interface. The constitutive properties on either side of the interface have been assumed dispersive but isotropic while the interface has been endowed with a complex homogeneous anisotropic admittance tensor relating surface currents to electric fields in the interface. Such a model accommodates as a special case both active and passive (including Ohmic) interface electromagnetic characteristics. The analysis yields a family of solutions to this problem characterised by both constitutive properties of the system and a number of arbitrary, complex (frequency-dependent) constants. Different choices of these constants and constitutive scalars determine the physical characteristics of the solutions.

For a planar interfaces $f = 0$ with surface normal unit vector $N = \mathrm{d}f/|\mathrm{d}f|$, bounded solutions with $\widehat{K}^I$ and $\widehat{K}^{II}$ complex and

$$\Im(\mathrm{i}_N(\widehat{K}^L)) \neq 0,$$

exhibit mode attenuation in directions where $|f| \leftrightarrow \infty$ and are referred to as *surface polariton* modes. If the bulk constitutive permittivities or permeabilities are complex, with bulk conductivity non-zero or the interface possesses surface admittance, the *tangential* components of $\widehat{K}^I$ (or $\widehat{K}^{II}$) may also become complex. In these circumstances, the physically acceptable plane-fronted polariton modes will propagate (in half-spaces) with attenuation in directions orthogonal to N. When $\widehat{K}^I$ and $\widehat{K}^{II}$ are real 1-forms with

$$\mathrm{sign}\,(\mathrm{i}_N\widehat{K}^I) = \mathrm{sign}\,(\mathrm{i}_N\widehat{K}^{II}),$$

one speaks of plane-fronted *Brewster* modes.

Particular solutions may be classified further by introducing the notion of a *real* plane of propagation at any point as the span of the *real* vector fields N and $\widetilde{\widehat{\Upsilon}}$ where

$$\widehat{\Upsilon} \equiv \Re(\mathfrak{t}_N\widehat{K}^I) = \Re(\mathfrak{t}_N\widehat{K}^{II}). \tag{8.160}$$

Such solutions are said to generate TE-type modes in the domain L if

$$\mathrm{i}_{\widetilde{\widehat{\mathscr{E}}}^L}\left(\widetilde{N} \wedge \widehat{\Upsilon}\right) = 0, \tag{8.161}$$

and TM-type modes in the domain L if

$$\mathrm{i}_{\widetilde{\widehat{\mathscr{H}}}^L}\left(\widetilde{N} \wedge \widehat{\Upsilon}\right) = 0. \tag{8.162}$$

By definition the complex propagation vectors for surface polaritons without attenuation in any plane orthogonal to N can be written in the form

$$\widetilde{\widehat{K}}^L_{SP} = \widetilde{\widehat{\Upsilon}}_{SP} + i\widehat{K}^L_N N \tag{8.163}$$

for some $\widehat{K}^L_N \in \mathbb{R}$ and the real propagation vectors for Brewster modes as $\widehat{K}^L_B$. Since

$$\mathrm{i}_{\widetilde{\widehat{\mathscr{E}}}^L}\widehat{K}^L = \mathrm{i}_{\widetilde{\widehat{\mathscr{H}}}^L}\widehat{K}^L = 0 \tag{8.164}$$

by construction, it follows that TE-type solutions in this class also satisfy

$$0 = \mathrm{i}_{\widetilde{\mathscr{E}}^L}\left(\widetilde{N} \wedge \widehat{\Upsilon}\right) = \mathrm{i}_{\widetilde{\mathscr{E}}^L}\left(\widetilde{N} \wedge \widehat{K}^L\right) = \mathrm{i}_N \widehat{\mathscr{E}}^L. \tag{8.165}$$

From (8.157), with $N = \partial_z$ and $\widehat{Q} \neq 0$, this implies that

$$\widehat{\Psi} = 0. \tag{8.166}$$

Similarly, (8.164) implies that TM-type solutions in this class also satisfy

$$0 = \mathrm{i}_{\widetilde{\mathscr{H}}^L}\left(\widetilde{N} \wedge \widehat{\Upsilon}\right) = \mathrm{i}_{\widetilde{\mathscr{H}}^L}\left(\widetilde{N} \wedge \widehat{K}^L\right) = \mathrm{i}_N \widehat{\mathscr{H}}^L. \tag{8.167}$$

From (8.158), this implies that

$$\widehat{\Phi} = 0. \tag{8.168}$$

Thus type-Φ modes in this class correspond to TE polarised fields in the presence of a planar interface, and type-Ψ modes to TM polarised fields[7].

One may observe in this way how, for example, the presence of a non-zero surface conductivity can dramatically change the standard mode structure of surface polariton and Brewster electromagnetic field configurations. These and other effects will be presented elsewhere (Christie and Tucker, 2018).

The systematic strategy outlined in the paper for finding analytic expressions describing the electromagnetic fields in a dispersive medium containing a planar meta-interface can be generalised to accommodate more intricate interface conditions where surface currents are induced by electric and/or magnetic fields that are normal and/or transverse to the planar interface. Such conditions have been contemplated in Epstein and Eleftheriades (2016) in efforts to construct a "tunable meta-surface".

Our geometric formulation in terms of differential forms offers a consistent and compact way to approach more challenging problems involving media with inhomogeneous constitutive components and curved interfaces. Such problems will be discussed more fully elsewhere.

Acknowledgements The authors are grateful to J. Gratus, T. Walton, colleagues in the Cockcroft Institute and the Alpha-X collaboration for useful discussions. They are also grateful to STFC (ST/G008248/1) and EPSRC (EP/J018171/1) for supporting this research.

References

Benn IM, Tucker RW (1987) An Introduction to Spinors and Geometry With Applications in Physics. Adam Hilger

[7] More generally, if a region $\mathscr{U}$ contains a real unit vector field N we may *define* field components $\{\widehat{\mathscr{E}}, \widehat{\mathscr{H}}\} \in \widehat{\Gamma}\Lambda^1\mathscr{U}$ to be of type TE with respect to N if they satisfy $\mathrm{i}_N\widehat{\mathscr{E}} = 0$ and of type TM with respect to N if they satisfy $\mathrm{i}_N\widehat{\mathscr{H}} = 0$.

Burton DA (2003) A primer on exterior differential calculus. Theoretical and Applied Mechanics 30(2):85–162
Cheng H, Gupta KC (1989) An historical note on finite rotations. Journal of Applied Mechanics 56(1):139–145
Christie DC, Tucker RW (2018) Electromagnetic polariton and Brewster mode dynamics in media with interfacial surface-admittance. Article in Preparation
Clemmow PC (1966) The Plane Wave Spectrum Representation of Electromagnetic Fields. Pergamon Press, Oxford, New York
Epstein A, Eleftheriades GV (2016) Synthesis of passive lossless metasurfaces using auxiliary fields for reflectionless beam splitting and perfect reflection. Physical Review Letters 117(25):256,103
Gusynin VP, Sharapov SG, Carbotte JP (2009) On the universal ac optical background in graphene. New Journal of Physics 11(9):095,013
Hanson GW (2008) Dyadic Green's functions and guided surface waves for a surface conductivity model of graphene. Journal of Applied Physics 103(6):064,302
Maier SA (2007) Plasmonics: Fundamentals and Applications. Springer
Nemilentsau A, Low T, Hanson G (2016) Anisotropic 2D materials for tunable hyperbolic plasmonics. Physical Review Letters 116(6):066,804
Pitarke JM, Silkin VM, Chulkov EV, Echenique PM (2006) Theory of surface plasmons and surface-plasmon polaritons. Reports on Progress in Physics 70(1):1
Raether H (1988) Surface Plasmons on Smooth and Rough Surfaces and on Gratings, Springer Tracts in Modern Physics, vol 111. Springer
Sarid D, Challener W (2010) Modern Introduction to Surface Plasmons: Theory, Mathematica Modeling, and Applications. Cambridge University Press
Sounas DL, Caloz C (2011) Graphene-based non-reciprocal metasurface. In: Proceedings of the 5th European Conference on Antennas and Propagation (EUCAP), IEEE, pp 2419–2422
Sounas DL, Caloz C (2012) Gyrotropy and nonreciprocity of graphene for microwave applications. IEEE Transactions on Microwave Theory and Techniques 60(4):901–914
Vakil A, Engheta N (2011) Transformation optics using graphene. Science 332(6035):1291–1294
Zhu BO, Chen K, Jia N, Sun L, Zhao J, Jiang T, Feng Y (2014) Dynamic control of electromagnetic wave propagation with the equivalent principle inspired tunable metasurface. Scientific Reports 4

Chapter 9
Evolution Equations for Defects in Finite Elasto-Plasticity

Sanda Cleja-Ţigoiu

Abstract The paper deals with continuous models of elasto-plastic materials with microstructural defects such as dislocations and disclinations. The basic assumptions concern the existence of plastic distortion and so-called plastic connection with metric property and the existence of the free energy function. This is dependent on the Cauchy-Green strain tensor, and its gradient with respect to the plastically deformed anholonomic configuration, and on the dislocation and disclination densities. The defect densities are defined in terms of the incompatibility of the plastic distortion and non-integrability of the plastic connection. The evolution of plastic distortion and disclination tensor has been postulated under the appropriate viscoplastic and dissipative type equations, which are compatible with the principle of the free energy imbalance. The associated small distortion model is provided The present model and the previous ones have been also compared.

9.1 Introduction

The paper deals with defects in crystalline materials, when the differential geometry description is used in order to characterize lattice defects existing at the micro structural level, see Kröner (1990). Here we restrict ourselves to dislocations and disclinations and we make reference to different continuous descriptions which are close to the background of our finite elasto-plastic model.

Sanda Cleja-Ţigoiu
University of Bucharest, Faculty of Mathematics and Computer Science, str. Academiei 14, 010014-Bucharest, Romania
e-mail: tigoiu@fmi.unibuc.ro

© Springer International Publishing AG, part of Springer Nature 2018
H. Altenbach et al. (eds.), *Generalized Models and Non-classical Approaches in Complex Materials 1*, Advanced Structured Materials 89,
https://doi.org/10.1007/978-3-319-72440-9_9

9.1.1 Defects in Linear eElasticity

The elastic model for the defects such as dislocations and disclinations have been described by the solutions of the linear theory of elasticity having the displacement fields discontinuous along cut-off surfaces, see de Wit (1973a,b); Teodosiu (1982). The problems formulated by de Wit (1973c) and Kossecka and de Wit (1977) concern the finding of the elastic basic fields (i.e. strain and bent twist tensors) and the stress, when the plastic fields, namely the defect densities, $\boldsymbol{\alpha}$ for dislocations and $\boldsymbol{\omega}$ for disclinations, have been prescribed (without specifying the nature of these defects). The incompatibilities in linear elasticity were reviewed by de Wit (1970, 1981); Kossecka and de Wit (1977), see also Fressengeas et al (2011). Traditionally the dislocations are determined by Burgers vector, $\boldsymbol{b}$, which is equal to the translational displacement, and the disclinations are characterized by Frank vector, $\boldsymbol{\Omega}$, which is equal to the rotational displacement, see de Wit (1973a,c); Kossecka and de Wit (1977). The internal mechanical state of solids with defects leads Kröner (1992), to solve the elastic problems with given incompatibilities.

9.1.2 Defects in Non-Linear Elasticity

Yavari and Goriely (2013) considered the cases of a single wedge disclination and a parallel cylindrically-symmetric distribution of wedge disclination, respectively. They solved the problem of existence of the residual stress: find the stress distribution in an Neo-Hookean material, which is stress free, namely zero traction is applied on the outer radius of the cylinder, when the wedge disclination densities, as those mentioned above, have been given. First Yavari and Goriely (2013) construct a Riemannian material manifold which is metric compatible having zero torsion and non-zero curvature, for the given disclination density identified with the curvature tensor. The manifold corresponds to Volterra's geometrical description of wedge disclination in a cylindrical body: cut, following by removing or inserting material, and weld operations. As an example of Riemann-Cartan geometry Yavari and Goriely (2012) built the material manifold which is dependent on the distribution of dislocations, namely for a screw and a cylindrically-symmetric distribution of parallel screw dislocations, respectively. The material manifold has torsion and vanishing non-metricity, and corresponds to Volterra's description of screw dislocation: cut a half axial plane of the cylinder followed by the displacement with Burgers vector along the symmetry axis, and weld procedure with removing the axis. The authors found the residual stress for a generalized Neo-Hookean elastic body induced by several distributed dislocations, including also the radially-symmetric distributed edge dislocations.

9.1.3 Defects in Nonlocal Elasticity

An improvement has been obtained using the nonlocal elasticity instead of the classical one, namely the stress and strain energy singularities, which are present in classical elasticity, have been eliminated. The nonlocal elasticity is given by an integral type constitutive equations, which are characterized by the so-called nonlocal kernel. The approaches to nonlocal elasticity proposed and discussed in Eringen (2002) replace the integral operators with a special class of kernels by certain differential operators. The integral representation of the stress is given in terms of the Hookean stress. The displacement fields are identical with the classical forms obtained by integrating the stress-strain relations of linear elasticity. Solutions for screw dislocation, edge dislocation and wedge disclination have been described and analyzed within the nonlocal elasticity, with Gaussian kernel, and for a special class of kernels, which are Green functions of the Helmholtz equation in Eringen (2002) and for bi-Helmholtz equation in Lazar et al (2006). Lazar and Maugin (2004a,b), developed first a constitutive framework of gradient micropolar isotropic elasticity, which was connected to the nonlocal micropolar elasticity given by Eringen (2002). Second, Lazar and Maugin examined the mentioned defects in gradient micropolar elasticity. The micropolar distortion and bent twist tensors satisfy the appropriate inhomogeneous Helmholtz equations, with the inhomogeneities identified with classical elastic expressions for the stress and couple stress tensor. The authors did not derive the associate boundary conditions because they considered an infinite extended medium. Only in a small region in the vicinity of $r = 0$, the stress calculated in nonlocal elasticity of Helmholtz or bi-Helmholtz type is different, both of them being zero at $r = 0$. Eringen's results were recovered too.

9.1.4 Elasto-Plastic Models for Defects

Continuum models of these defects involve the couple stresses within the micropolar materials and Cosserat continuum, see Clayton et al (2006); Fressengeas et al (2011). In the models developed by Arsenlis and Parks (1999); Gurtin (2002) the Burgers vector has been defined by the geometrically necessary dislocation (GND) tensor $\boldsymbol{G} = \boldsymbol{F}^p \mathrm{curl} \boldsymbol{F}^p$ in the lattice space. The GND density tensor is decomposed in the appropriate edge and screw dislocations. Clayton et al (2006) introduced the geometrically necessary defect density tensors in the deformed configuration, $\boldsymbol{\alpha}$ and $\boldsymbol{\theta}$, accounting for the incompatibilities induced by the torsion and curvature tensor, which are associated with the connection $\boldsymbol{\Gamma}$. The connection coefficients were defined as in Minagawa (1979) in terms of non-Riemannian's type connection (using $(\boldsymbol{F}^{\mathscr{L}})^{-1}$) and a third order tensor field $\boldsymbol{Q}$ with assigned skew-symmetry. We recall the multiplicative decomposition $\boldsymbol{F} = \boldsymbol{F}^{\mathscr{L}} \boldsymbol{F}^p$ assumed by Clayton et al (2006), where the lattice part $\boldsymbol{F}^{\mathscr{L}}$ is given by $\boldsymbol{F}^{\mathscr{L}} = \boldsymbol{F}^e \boldsymbol{F}^i$, $\boldsymbol{F}^e$ is the elastic part, $\boldsymbol{F}^i$ is the residual part due to the micro-heterogeneity in the presence in the of lattice defects. The free energy function in the intermediate configuration is dependent on elastic Cauchy-

Green strain tensor (expressed in terms of $\boldsymbol{F}^e$), defect density tensors pulled back to the intermediate configuration by $\boldsymbol{F}^e$, i.e. $\widetilde{\boldsymbol{\alpha}}$ and $\widetilde{\boldsymbol{\theta}}$, the symmetric and positive definite part of $\boldsymbol{F}^i$, and so on.

9.1.5 Aim of this Paper

Fressengeas et al (2011) proposed a *field defect* (dislocation and disclination), restricted to small strains. The non-symmetric Cauchy stress, $\boldsymbol{T}$, and couple- stress tensor, $\boldsymbol{m}$, are described in terms of elastic strain and bent-twist, $\boldsymbol{\varepsilon}^e$ and $\boldsymbol{\kappa}^e$, with macro forces, $\boldsymbol{T}$ and $\boldsymbol{m}$, satisfying the balance equation formulated by Fleck et al (1994). The evolution equations for basic plastic fields, $\boldsymbol{\varepsilon}^p, \boldsymbol{\kappa}^p$, are dependent on the density of dislocations and disclinations, $\boldsymbol{\alpha}$ and $\boldsymbol{\theta}$, and on the macro forces. The dislocation density $\boldsymbol{\theta}$ generates a Frank vector $\boldsymbol{\Omega}$.

In this paper we propose a model for structural defects such as dislocations and disclinations, which can be viewed as an improvement of the models provided by Cleja-Ţigoiu (2014); Cleja-Ţigoiu et al (2016), within the constitutive framework developed by Cleja-Ţigoiu (2007, 2010). The key point is related to the expression of the free energy density, this time also dependent on the gradient of the Cauchy-Green elastic strain with respect to the plastically deformed configuration or the so-called configuration with torsion. The basic assumptions concern the existence of plastic distortion and so-called plastic connection with metric property and the existence of the free energy function. This function is dependent on the Cauchy-Green strain tensor and its gradient with respect to the plastically deformed anholonomic configuration, and on dislocation and disclination densities. The defect densities are defined in terms of the incompatibility of the plastic distortion and non-integrability of the plastic connection, respectively. The free energy imbalance principle is postulated in a similar form with those presented by Cleja-Ţigoiu (2007, 2010), following the ideas given by Gurtin (2002); Gurtin et al (2010). The balance equation for micro forces have been considered in the form provided by Cleja-Ţigoiu (2007, 2017). The constitutive and evolution equations for plastic distortion and disclination tensor are derived to be compatible with the free energy imbalance principle. The evolution of plastic distortion and disclination tensor has been postulated under the appropriate viscoplastic and dissipative type equations. The associated small distortion model is also provided. The proposed model is compared with the previous models discussed in Cleja-Ţigoiu et al (2016); Cleja-Tigoiu and Maugin (2000).

9.1.6 List of Notations

Further the following notations will be used:
$\mathscr{E}$ - the three dimensional Euclidean space, with the vector space of translations $\mathscr{V}$;
Lin - the set of the linear mappings from $\mathscr{V}$ to $\mathscr{V}$, i.e the set of second order tensor,

$Sym_2, Skw_2 \subset Lin$ are the sets of all symmetric and skew-symmetric second order tensors, respectively; $\boldsymbol{u}\cdot\boldsymbol{v}, \boldsymbol{u}\times\boldsymbol{v}, \boldsymbol{u}\otimes\boldsymbol{v}$ denote scalar, cross and tensorial products of vectors; $(\boldsymbol{u},\boldsymbol{v},\boldsymbol{z}) := (\boldsymbol{u}\times\boldsymbol{v})\cdot\boldsymbol{z}$ is the mixed product of the vectors from $\mathscr{V}$. $\boldsymbol{a}\otimes\boldsymbol{b}$ and $\boldsymbol{a}\otimes\boldsymbol{b}\otimes\boldsymbol{c}$ are defined to be a second order tensor and a third order tensor by $(\boldsymbol{a}\otimes\boldsymbol{b})\boldsymbol{u} = \boldsymbol{a}(\boldsymbol{b}\cdot\boldsymbol{u}), (\boldsymbol{a}\otimes\boldsymbol{b}\otimes\boldsymbol{c})\boldsymbol{u} = (\boldsymbol{a}\otimes\boldsymbol{b})(\boldsymbol{c}\cdot\boldsymbol{u})$, for all vectors $\boldsymbol{u}$.

For $\boldsymbol{A}\in Lin-$ a second order tensor, we introduce:
the notations $\{\boldsymbol{A}\}^S, \{\boldsymbol{A}\}^a$ for the symmetric and skew-symmetric parts of the tensor; definition of the trace: $\mathrm{tr}\boldsymbol{A}((\boldsymbol{u}\times\boldsymbol{v})\cdot\boldsymbol{z}) = (\boldsymbol{A}\boldsymbol{u},\boldsymbol{v},\boldsymbol{z}) + (\boldsymbol{u},\boldsymbol{A}\boldsymbol{v},\boldsymbol{z}) + (\boldsymbol{u},\boldsymbol{v},\boldsymbol{A}\boldsymbol{z})$. $\boldsymbol{I}$ is the identity tensor in *Lin*, $\boldsymbol{A}^T$ denotes the transpose of $\boldsymbol{A}\in Lin$, $\partial_{\boldsymbol{A}}\phi(x)$ denotes the partial differential of the function ϕ with respect to the field $\boldsymbol{A}$.

Let $\chi : \mathscr{B}\times\boldsymbol{R}\to\mathscr{V}$ defines the motion of the body $\mathscr{B}$. The deformation gradient and its gradient are expressed in coordinate systems by

$$\boldsymbol{F}(\boldsymbol{X},t) = \nabla\chi(\boldsymbol{X},t) = \frac{\partial x^i}{\partial X^j}\boldsymbol{g}_i\otimes\boldsymbol{G}^j, \quad \nabla\boldsymbol{F}(\boldsymbol{X},t) = \frac{\partial^2 x^i}{\partial X^j\partial X^k}\boldsymbol{g}_i\otimes\boldsymbol{G}^j\otimes\boldsymbol{G}^k. \quad (9.1)$$

Here $\{\boldsymbol{g}_i\}_{i=1,2,3}$ and $\{\boldsymbol{G}_i\}_{i=1,2,3}$ are local bases in the actual and reference configurations, respectively.

In what follows the anholonomic basis vectors, in the so-called plastically deformed configuration or the configuration with torsion, generically denoted by $\mathscr{K}$, are related with *the crystal* and defined by $\boldsymbol{e}_j = \boldsymbol{F}^p\boldsymbol{G}_j$. Let $\{\boldsymbol{G}^i\}_{i=1,2,3}$ be the reciprocal basis in the reference configuration. The plastic connection is represented by its coefficients in a component representation given by

$$\overset{(\mathrm{p})}{\boldsymbol{\Gamma}} = \Gamma^{\alpha}_{\beta\gamma}\boldsymbol{G}_\alpha\otimes\boldsymbol{G}^\beta\otimes\boldsymbol{G}^\gamma. \quad (9.2)$$

The differential of smooth field $\boldsymbol{A}$, with respect to the anholonomic configuration $\mathscr{K}$, obeys the rule

$$\nabla_{\mathscr{K}}\boldsymbol{A} = (\nabla\boldsymbol{A})(\boldsymbol{F}^p)^{-1}. \quad (9.3)$$

curl of a second order tensor field $\boldsymbol{A}$ is defined by the second order tensor field

$$\begin{aligned}(curl\boldsymbol{A})(\boldsymbol{u}\times\boldsymbol{v}) &:= (\nabla\boldsymbol{A}(\boldsymbol{u}))\boldsymbol{v} - (\nabla\boldsymbol{A}(\boldsymbol{v}))\boldsymbol{u} \quad \forall\boldsymbol{u},\boldsymbol{v}\in\mathscr{V},\\ (curl\boldsymbol{A})_{pi} &= \epsilon_{ijk}\frac{\partial A_{pk}}{\partial x^j}\end{aligned} \quad (9.4)$$

are the components of *curl* $\boldsymbol{A}$ given in a Cartesian basis.

The transpose of the third order tensor tensor field is defined by

$$\mathscr{A}^T(\boldsymbol{u}) := (\mathscr{A}\boldsymbol{u})^T, \quad \forall\boldsymbol{u}\in\mathscr{V}. \quad (9.5)$$

The third order tensors, denoted by $\mathscr{A}$, are linear mapping defined as element of the set $Lin\{\mathscr{V}, Lin\}$, which are represented in a Cartesian basis $\{\boldsymbol{e}_i\}_{i=1,2,3}$ as

$$\mathscr{A} := \mathscr{A}_{ijk}\boldsymbol{e}_i\otimes\boldsymbol{e}_j\otimes\boldsymbol{e}_k. \quad (9.6)$$

The scalar product of second order tensors $\boldsymbol{A}$ and $\boldsymbol{B}$, and of the third order tensors $\mathscr{A}$ and $\mathscr{B}$, are defined in terms of their Cartesian components by

$$\begin{aligned} \boldsymbol{A}\cdot\boldsymbol{B} &= A_{ij}B_{ij}, \quad \forall \boldsymbol{A},\boldsymbol{B} \in Lin, \\ \mathscr{A}\cdot\mathscr{B} &= \mathscr{A}_{ijk}\mathscr{B}_{ijk}, \quad \forall \mathscr{A},\mathscr{B} \in Lin\{\mathscr{V}, Lin\}. \end{aligned} \tag{9.7}$$

For any $\boldsymbol{\Lambda}_1, \boldsymbol{\Lambda}_2 \in Lin$ we define a third order tensor associated with them, denoted $\boldsymbol{\Lambda}_1 \times \boldsymbol{\Lambda}_2$, by

$$((\boldsymbol{\Lambda}_1 \times \boldsymbol{\Lambda}_2)\boldsymbol{u})\boldsymbol{v} = (\boldsymbol{\Lambda}_1\boldsymbol{u}) \times (\boldsymbol{\Lambda}_2\boldsymbol{v}), \quad \forall \quad \boldsymbol{u},\boldsymbol{v}. \tag{9.8}$$

The notations $Sym\mathscr{A}$ and $Skw\mathscr{A}$ are introduced for all $\mathscr{A} \in Lin\{\mathscr{V}, Lin,\}$ by

$$\begin{aligned} &Sym\mathscr{A} \in Lin\{\mathscr{V}, Lin\}, \quad Sym(\mathscr{A}) = \mathscr{A} + \mathscr{A}^T, \\ &Skw\mathscr{A} \in Lin\{\mathscr{V}, Lin\}, \quad ((Skw\mathscr{A})\boldsymbol{u})\boldsymbol{v} := (\mathscr{A}\boldsymbol{u})\boldsymbol{v} - (\mathscr{A}\boldsymbol{v})\boldsymbol{u} \quad \forall \boldsymbol{u} \in \mathscr{V}. \end{aligned} \tag{9.9}$$

The following identity holds

$$\mathscr{A}\cdot Sym(\boldsymbol{C}\mathscr{B}) = (\boldsymbol{C}(Sym\mathscr{A}))\cdot\mathscr{B} \,\forall\, \boldsymbol{C} \in Sym_2,\ \mathscr{A},\mathscr{B} \in Lin\{\mathscr{V}, Lin\}. \tag{9.10}$$

The third order tensor, denoted by $\mathscr{A}[\boldsymbol{F}_1,\boldsymbol{F}_2]$, is associated to the set of tensors $\mathscr{A} \in Lin(\mathscr{V}, Lin)$, and $\boldsymbol{F}_1,\boldsymbol{F}_2 \in Lin$, and is defined by

$$((\mathscr{A}[\boldsymbol{F}_1,\boldsymbol{F}_2])\boldsymbol{u})\boldsymbol{v} = (\mathscr{A}(\boldsymbol{F}_1\boldsymbol{u}))\boldsymbol{F}_2\boldsymbol{v}, \quad \forall\, \boldsymbol{u},\boldsymbol{v} \in \mathscr{V}. \tag{9.11}$$

or by its Cartesian components $(\mathscr{A}[\boldsymbol{F}_1,\boldsymbol{F}_2])_{ilq} = \mathscr{A}_{ijk}(\boldsymbol{F}_1)_{jl}(\boldsymbol{F}_2)_{kq}$. Two types of second order tensors that can be associated with any pair of third order tensors, $\mathscr{A},\mathscr{B}$, following the rules written below

$$\begin{aligned} &(\mathscr{A} \odot \mathscr{B})\cdot\boldsymbol{L} = \mathscr{A}[\boldsymbol{I},\boldsymbol{L}]\cdot\mathscr{B} = \mathscr{A}_{isk}L_{sn}\mathscr{B}_{ink}, \\ &(\mathscr{A}\ _r\!\odot \mathscr{B})\cdot\boldsymbol{L} = \mathscr{A}\cdot(\boldsymbol{L}\mathscr{B}) = \mathscr{A}_{ijk}L_{in}\mathscr{B}_{njk}, \quad \boldsymbol{L} \in Lin. \end{aligned} \tag{9.12}$$

9.2 Elasto-Plastic Materials with Lattice Defects

We recall our basic relationships which characterize the elasto-plastic material from the geometrical point of view. The motion of the body, χ, induces a second order deformation which is defined by $(\boldsymbol{F}, \boldsymbol{\Gamma} := (\boldsymbol{F})^{-1}(\nabla\boldsymbol{F}))$, $\boldsymbol{\Gamma}$ is called the motion connection.

The *multiplicative decomposition* of the deformation gradient $\boldsymbol{F}$ into its elastic and plastic components, $\boldsymbol{F}^e$ and $\boldsymbol{F}^p$, called distortions, namely

$$\boldsymbol{F} = \boldsymbol{F}^e\boldsymbol{F}^p \tag{9.13}$$

is considered.

Definition 9.1. The Cauchy-Green elastic strain tensor, $\boldsymbol{C}^e$, and Cauchy-Green plastic strain tensor, $\boldsymbol{C}^p$, are expressed by

$$\begin{aligned} &\boldsymbol{C}^e = (\boldsymbol{F}^e)^T \boldsymbol{F}^e, \quad \Longrightarrow \quad \boldsymbol{C}^e = (\boldsymbol{F}^p)^{-T} \boldsymbol{C} (\boldsymbol{F}^p)^{-1}, \quad \text{where} \quad \boldsymbol{C} = \boldsymbol{F}^T \boldsymbol{F}, \\ &\boldsymbol{C}^p = (\boldsymbol{F}^p)^T \boldsymbol{F}^p. \end{aligned} \tag{9.14}$$

The gradient of $\boldsymbol{C}^e$ with respect to the configuration $\mathscr{K}$ can be expressed by

$$\nabla_{\mathscr{K}} \boldsymbol{C}^e = (\boldsymbol{F}^p)^{-T} \big(\nabla \boldsymbol{C} - Sym\{\boldsymbol{C} \overset{(p)}{\mathscr{A}}\}\big) [(\boldsymbol{F}^p)^{-1}, (\boldsymbol{F}^p)^{-1}] \tag{9.15}$$

As a direct consequence of the multiplicative decomposition formula (9.19) the following relationships hold

$$\boldsymbol{L} = \boldsymbol{L}^e + \boldsymbol{F}^e \boldsymbol{L}^p (\boldsymbol{F}^e)^{-1}, \quad \text{where} \quad \boldsymbol{L}^e = \dot{\boldsymbol{F}}^e (\boldsymbol{F}^e)^{-1}, \quad \boldsymbol{L}^p = \dot{\boldsymbol{F}}^p (\boldsymbol{F}^p)^{-1}. \tag{9.16}$$

The plastic rate tensors with respect to the plastically deformed and reference configurations, respectively, and denoted by $\boldsymbol{L}^p$ and $\underline{\boldsymbol{l}}^p$, are related by

$$\boldsymbol{L}^p = \boldsymbol{F}^p \underline{\boldsymbol{l}}^p (\boldsymbol{F}^p)^{-1}, \quad \text{where} \quad \underline{\boldsymbol{l}}^p = (\boldsymbol{F}^p)^{-1} \dot{\boldsymbol{F}}^p. \tag{9.17}$$

The following time derivatives can be computed

$$\begin{aligned} &\frac{d}{dt} (\boldsymbol{C}^p)^{-1} = -\underline{\boldsymbol{l}}^p (\boldsymbol{C}^p)^{-1} - (\boldsymbol{C}^p)^{-1} (\underline{\boldsymbol{l}}^p)^T, \\ &\dot{\boldsymbol{C}} = 2\boldsymbol{F}^T \boldsymbol{D} \boldsymbol{F}, \quad \text{where} \quad \boldsymbol{D} = \{\boldsymbol{L}\}^S, \\ &\nabla \dot{\boldsymbol{C}} = (\dot{\boldsymbol{C}} \boldsymbol{\Gamma})^T + \boldsymbol{F}^T (\nabla_\chi \boldsymbol{D}) [\boldsymbol{F}, \boldsymbol{F}] + \dot{\boldsymbol{C}} \boldsymbol{\Gamma}, \quad \text{where} \quad \boldsymbol{\Gamma} = \boldsymbol{F}^{-1} \nabla \boldsymbol{F}, \\ &\frac{d}{dt} \{(\boldsymbol{C}^p)^{-1} (\boldsymbol{\Lambda} \times \boldsymbol{I})\} = -(\underline{\boldsymbol{l}}^p (\boldsymbol{C}^p)^{-1} + (\boldsymbol{C}^p)^{-1} (\underline{\boldsymbol{l}}^p)^T)(\boldsymbol{\Lambda} \times \boldsymbol{I}) + \\ &\qquad\qquad + (\boldsymbol{C}^p)^{-1} (\dot{\boldsymbol{\Lambda}} \times \boldsymbol{I}). \end{aligned} \tag{9.18}$$

Here $\boldsymbol{\Lambda}$ is a second order tensor.

Proposition 9.1. *Under the hypothesis of the multiplicative decomposition of the deformation gradient* $\boldsymbol{F}$, *postulated in (9.19), we get the composition rule of the motion connection*

$$\begin{aligned} &\boldsymbol{\Gamma} = (\boldsymbol{F}^p)^{-1} \overset{(e)}{\mathscr{A}}_{\mathscr{K}} [\boldsymbol{F}^p, \boldsymbol{F}^p] + \overset{(p)}{\mathscr{A}}, \\ &\textit{where} \quad \overset{(e)}{\mathscr{A}}_{\mathscr{K}} := (\boldsymbol{F}^e)^{-1} \nabla_{\mathscr{K}} \boldsymbol{F}^e, \quad \overset{(p)}{\mathscr{A}} := (\boldsymbol{F}^p)^{-1} \nabla \boldsymbol{F}^p. \end{aligned} \tag{9.19}$$

$\overset{(e)}{\mathscr{A}}_{\mathscr{K}}$ and $\overset{(p)}{\mathscr{A}}$ define Bilby's type connections (see Bilby, 1960) with respect to the so-called configuration with torsion, and initial one, respectively.

Based on the definitions for elastic strain $\boldsymbol{C}^e$ and its gradient $\nabla_{\mathscr{K}}\boldsymbol{C}^e$ the following relation holds

$$\nabla_{\mathscr{K}}\boldsymbol{C}^e = (\boldsymbol{C}^e \overset{(e)}{\mathscr{A}}_{\mathscr{K}})^T + \boldsymbol{C}^e(\overset{(e)}{\mathscr{A}}_{\mathscr{K}}), \tag{9.20}$$

see (9.19) together with (9.14).

The formula (9.20) states that the so-called elastic Bilby's connection $\overset{(e)}{\mathscr{A}}_{\mathscr{K}}$ has metric property in $\mathscr{K}$ with respect to the elastic metric tensor $\boldsymbol{C}^e$, using a definition given by Schouten (1954), see also Yavari and Goriely (2012).

9.2.1 Plastic Connection with Metric Property

We accepted the existence of the plastic connection with metric property, see Cleja-Ţigoiu (2010).

Definition 9.2. The connection $\overset{(p)}{\boldsymbol{\Gamma}}$ has metric property if and only if the following relationship holds

$$\nabla\boldsymbol{C}^p = (\boldsymbol{C}^p \overset{(p)}{\boldsymbol{\Gamma}})^T + \boldsymbol{C}^p \overset{(p)}{\boldsymbol{\Gamma}}. \tag{9.21}$$

The relationship (9.21) is similarly to (9.20).

Proposition 9.2. *The plastic connection, which is metric compatible with respect to the metric tensor $\boldsymbol{C}^p$, is represented under the form*

$$\overset{(p)}{\boldsymbol{\Gamma}} = \overset{(p)}{\mathscr{A}} + (\boldsymbol{C}^p)^{-1}(\boldsymbol{\Lambda} \times \boldsymbol{I}), \tag{9.22}$$

where the third order tensor $\boldsymbol{\Lambda} \times \boldsymbol{I}$ is generated by the second order (covariant) tensor $\boldsymbol{\Lambda}$, which is called disclination tensor.

Proof. The proof can be found in Cleja-Ţigoiu (2010).

Definition 9.3. The (Cartan) torsion associated with the plastic connection (9.22) is calculated in a given coordinate system by the skew-symmetric part of the connection, as it follows

$$(\boldsymbol{S}^p\boldsymbol{v})\boldsymbol{u} \equiv (\overset{(p)}{\boldsymbol{\Gamma}}\,\boldsymbol{v})\boldsymbol{u} - (\overset{(p)}{\boldsymbol{\Gamma}}\,\boldsymbol{u})\boldsymbol{v} \equiv ((Skw\boldsymbol{\Gamma})\boldsymbol{u})\boldsymbol{v}. \tag{9.23}$$

Let us remark that the Cartan torsion (9.23) can be expressed as a consequence of the formula (9.22) by

$$\boldsymbol{S}^p = Skw\,\overset{(p)}{\mathscr{A}} + Skw((\boldsymbol{C}^p)^{-1}(\overset{(p)}{\boldsymbol{\Lambda}} \times \boldsymbol{I})). \tag{9.24}$$

Proposition 9.3. *The second order torsion tensor $\mathscr{N}^p$, which is associated with Cartan torsion (9.24), is expressed by*

$$\begin{aligned} &\mathscr{N}^p = (\boldsymbol{F}^p)^{-1} curl \boldsymbol{F}^p + (\boldsymbol{C}^p)^{-1}\left((tr\,\boldsymbol{\Lambda})\boldsymbol{I} - (\boldsymbol{\Lambda})^T\right), \\ &where \quad (\boldsymbol{S}^p\boldsymbol{u})\boldsymbol{v} = \mathscr{N}^p(\boldsymbol{u}\times\boldsymbol{v}). \end{aligned} \tag{9.25}$$

Proof. The formula (9.31) follows directely from the mentioned calculus rules:

1. $\forall \mathscr{N} \in Lin(\mathscr{V}, Lin)$ there exists

$$\begin{aligned} \boldsymbol{\Omega}_{<\mathscr{N}>} \in Skw_2, \quad &\text{such that} \quad Skw\mathscr{N} = \boldsymbol{\Omega}_{<\mathscr{N}>}(\boldsymbol{I}\times\boldsymbol{I}), \\ \text{which means} \quad &((Skw\mathscr{N})\boldsymbol{u})\boldsymbol{v} = \boldsymbol{\Omega}_{<\mathscr{N}>}(\boldsymbol{u}\times\boldsymbol{v}),\ \forall\,\boldsymbol{u},\boldsymbol{v}\in\mathscr{V}. \end{aligned} \tag{9.26}$$

2. The following component representations hold

$$\begin{aligned} (Skw\mathscr{N})_{ijk} &= (\boldsymbol{\Omega}_{<\mathscr{N}>})_{im}\epsilon_{mkj}, \\ 2(\boldsymbol{\Omega}_{<\mathscr{N}>})_{im} &= (Skw\mathscr{N})_{ijk}\epsilon_{mkj}. \end{aligned} \tag{9.27}$$

   Here ϵ_{mkj} denotes components of Ricci's permutation tensor, namely $\boldsymbol{\epsilon}$.
3. The following results can be proved

$$\begin{aligned} &Skw \overset{(p)}{\mathscr{A}} = curl\boldsymbol{F}^p(\boldsymbol{I}\times\boldsymbol{I}), \quad \text{that is} \\ &((Skw \overset{(p)}{\mathscr{A}})\boldsymbol{u})\boldsymbol{v} = (curl\boldsymbol{F}^p)(\boldsymbol{u}\times\boldsymbol{v}),\ \forall\,\boldsymbol{u},\boldsymbol{v}\in\mathscr{V}, \\ &Skw(\boldsymbol{\Lambda}\times\boldsymbol{I}) = (tr\boldsymbol{\Lambda}\boldsymbol{I} - \boldsymbol{\Lambda}^T)(\boldsymbol{I}\times\boldsymbol{I}), \quad \text{namely} \\ &(Skw(\boldsymbol{\Lambda}\times\boldsymbol{I})\boldsymbol{u})\boldsymbol{v} = (tr\boldsymbol{\Lambda}\boldsymbol{I} - \boldsymbol{\Lambda}^T)(\boldsymbol{u}\times\boldsymbol{v}). \end{aligned} \tag{9.28}$$

9.2.2 Measure of Defects

The disclination tensor with respect to the configuration with torsion will be denoted by $\widetilde{\boldsymbol{\Lambda}}$, see Cleja-Ţigoiu (2010), and it will be introduced here through

$$\widetilde{\boldsymbol{\Lambda}} = \boldsymbol{F}^p\boldsymbol{\Lambda}(\boldsymbol{F}^p)^{-1}. \tag{9.29}$$

We define *Burgers* and *Frank vectors* in terms of the plastic distortion $\boldsymbol{F}^p$ and disclination tensor $\widetilde{\boldsymbol{\Lambda}}$. Both vectors are associated with a circuit C_0. Let $\mathscr{A}_0$ be a surface with normal $\boldsymbol{N}$, which is surrounded by C_0 in the reference configuration.

Definition 9.4. The Frank vector associated with a circuit C_0 is defined by

$$\boldsymbol{\Omega}_{\mathscr{K}} = \int_{C_{\mathscr{K}}} \widetilde{\boldsymbol{\Lambda}}\, d\boldsymbol{x}_{\mathscr{K}} = \int_{C_0} \widetilde{\boldsymbol{\Lambda}}\boldsymbol{F}^p\, d\boldsymbol{X} = \int_{\mathscr{A}_0} curl(\boldsymbol{F}^p\boldsymbol{\Lambda})\boldsymbol{N} dA. \tag{9.30}$$

Definition 9.5. The *disclination density tensor* with respect to the reference configuration is defined by

$$\boldsymbol{\omega} = curl(\boldsymbol{F}^p \boldsymbol{\Lambda}). \tag{9.31}$$

The Burgers vector is defined in terms of the plastic distortion $\boldsymbol{F}^p$.

Definition 9.6. The Burgers vector associated with the circuit C_0 is defined by

$$\boldsymbol{b}_{\mathscr{K}} = \int_{C_0} \boldsymbol{F}^p \, d\boldsymbol{X} = \int_{\mathscr{A}_0} (curl \boldsymbol{F}^p) \boldsymbol{N} dA. \tag{9.32}$$

The *dislocation density tensor* $\boldsymbol{\alpha}$ is expressed by

$$\boldsymbol{\alpha} := (\boldsymbol{F}^p)^{-1}(curl\, \boldsymbol{F}^p), \quad \text{or} \quad \boldsymbol{\alpha}(\boldsymbol{I} \times \boldsymbol{I}) = Skw(\boldsymbol{S}^p). \tag{9.33}$$

$\boldsymbol{\alpha}$ is a *measure of the incompatibility* of the plastic distortion $\boldsymbol{F}^p$, and its expression is involved in (9.25).

Note 9.1. Starting from the definition of the Cartan torsion $\boldsymbol{S}^p$, via the second order torsion tensor $\mathscr{N}^p$ expressed by (9.25), and using the definitions of the defect densities, (9.31) and (9.33), we can say that $\boldsymbol{S}^p$ is a *measure of the coupling between continuously distributed dislocations and disclinations.*

9.3 Free Energy Imbalance Principle Formulated in $\mathscr{K}$

The local free energy imbalance is formulated with respect to the configuration with torsion $\mathscr{K}$, since the defects are relevant at the level of the lattice microstructure. First we introduce the expression of the free energy density postulated with respect to the configuration with torsion.

9.3.1 Free Energy Function

We assume that the free energy density in $\mathscr{K}$ is dependent on the second order elastic deformation in terms of $(\boldsymbol{C}^e, \nabla_{\mathscr{K}} \boldsymbol{C}^e)$, and it is also influenced by the state of defects, i.e. $\boldsymbol{S}^p_{\mathscr{K}}$, $\boldsymbol{\Lambda}$ and $\nabla_{\mathscr{K}} \boldsymbol{\Lambda}$. The Cartan torsion $\boldsymbol{S}^p$ pushed away to the plastically deformed configuration is related to $\boldsymbol{S}^p_{\mathscr{K}}$, in terms of the plastic distortion as it follows

$$\boldsymbol{S}^p_{\mathscr{K}} = -\boldsymbol{F}^p \boldsymbol{S}^p[(\boldsymbol{F}^p)^{-1}, (\boldsymbol{F}^p)^{-1}]. \tag{9.34}$$

Axiom 9.1 *The free energy density with respect to the plastically deformed configuration,* $\mathscr{K}$, *is postulated to be a function of the second order elastic deformation, which is also dependent on the defects, and it is given by*

$$\psi = \psi_{\mathscr{K}}(\boldsymbol{C}^e - \boldsymbol{I}, \nabla_{\mathscr{K}}\boldsymbol{C}^e, \boldsymbol{S}^p_{\mathscr{K}}, \widetilde{\boldsymbol{\Lambda}}, \nabla_{\mathscr{K}}\widetilde{\boldsymbol{\Lambda}}). \tag{9.35}$$

Now we compare the free energy density function postulated in the paper Cleja-Ţigoiu (2014) and the expression (9.35) considered herein. We recall here the basic relationships between $\boldsymbol{\Gamma}$, $\overset{(p)}{\boldsymbol{\Gamma}}$ and the so-called elastic connection in $\mathscr{K}$, $\overset{(e)}{\boldsymbol{\Gamma}}_{\mathscr{K}}$

$$\begin{aligned} \boldsymbol{\Gamma} &= \overset{(p)}{\boldsymbol{\Gamma}} + (\boldsymbol{F}^p)^{-1}\overset{(e)}{\boldsymbol{\Gamma}}_{\mathscr{K}}[\boldsymbol{F}^p, \boldsymbol{F}^p], \\ \overset{(e)}{\boldsymbol{\Gamma}}_{\mathscr{K}} &= \overset{(e)}{\mathscr{A}}_{\mathscr{K}} - \boldsymbol{\Lambda}_{\mathscr{K}} \times \boldsymbol{I}, \quad \text{where} \quad \boldsymbol{\Lambda}_{\mathscr{K}} = \frac{1}{det\boldsymbol{F}^p}\widetilde{\boldsymbol{\Lambda}}, \end{aligned} \tag{9.36}$$

that can be found in a detailed presentation in Cleja-Ţigoiu (2007) and Cleja-Tigoiu and Maugin (2000).

The torsion of the elastic type connection $\overset{(e)}{\boldsymbol{\Gamma}}_{\mathscr{K}}$ is defined by

$$\boldsymbol{S}^e_{\mathscr{K}} = Skw\overset{(e)}{\boldsymbol{\Gamma}}_{\mathscr{K}} \tag{9.37}$$

and the relationship between torsions of the appropriate connections

$$\boldsymbol{S}^e_{\mathscr{K}} = -\boldsymbol{S}^p_{\mathscr{K}} \tag{9.38}$$

can be proved as a direct consequences of the formulae (9.36), (9.37) together with (9.29) and (9.37).

Proposition 9.4. *The constitutive representation for the free energy density as dependent on the second order elastic deformation in terms of* $(\boldsymbol{C}^e, \overset{(e)}{\mathscr{A}}_{\mathscr{K}})$ *and on the defects through* $(\boldsymbol{S}^e_{\mathscr{K}}, \widetilde{\boldsymbol{\Lambda}})$*, namely*

$$\psi = \psi_{\mathscr{K}}(\boldsymbol{C}^e, \overset{(e)}{\mathscr{A}}, \boldsymbol{S}^e_{\mathscr{K}}, \widetilde{\boldsymbol{\Lambda}}), \tag{9.39}$$

which has been postulated in Cleja-Ţigoiu (2014), can be viewed as a function of arguments given by (9.35) if the dependence on $\nabla\boldsymbol{\Lambda}$ *is ignored.*

Proof. In order to justify the statement, first we recall the following theorem referring to the compatible connection and which is written in component representations.

Theorem 9.1. *The plastic Bilby connection allows the following representation*

$$\begin{aligned} \overset{(p)}{\mathscr{A}} &= \boldsymbol{\gamma}^p + \boldsymbol{W}^p, \\ ((\boldsymbol{\gamma}^p\boldsymbol{u})\boldsymbol{v})\cdot\boldsymbol{z} &= \frac{1}{2}(\boldsymbol{C}^p)^{-1}\big[((\nabla\boldsymbol{C}^p)\boldsymbol{u})\boldsymbol{v}\cdot\boldsymbol{z} + ((\nabla\boldsymbol{C}^p)\boldsymbol{v})\boldsymbol{u}\cdot\boldsymbol{z} - ((\nabla\boldsymbol{C}^p)\boldsymbol{z})\boldsymbol{u}\cdot\boldsymbol{v}\big], \\ (\boldsymbol{W}^p\boldsymbol{u})\boldsymbol{v} &= \frac{1}{2}((\boldsymbol{S}^p)\boldsymbol{u})\boldsymbol{v} - \frac{1}{2}(\boldsymbol{C}^p)^{-1}\big((\boldsymbol{C}^p\boldsymbol{S}^p\boldsymbol{u})^T\boldsymbol{v} + (\boldsymbol{C}^p\boldsymbol{S}^p\boldsymbol{v})^T\boldsymbol{u}\big), \end{aligned} \tag{9.40}$$

defined for all $\boldsymbol{u},\boldsymbol{v},\boldsymbol{z} \in \mathscr{V}$ *or in component representation the Levi-Civita (plastic) connection is given by*

$$(\boldsymbol{\gamma}^p)^s_{jk} = \frac{1}{2}G^{si}\big(\frac{\partial G_{ik}}{\partial X^j} + \frac{\partial G_{ij}}{\partial X^k} - \frac{\partial G_{jk}}{\partial X^i}\big),$$

and the (plastic) contorsion is expressed in terms of the torsion components as

$$(\boldsymbol{W}^p)^i_{jk} = \frac{1}{2}(\boldsymbol{S}^p)^i_{jk} - \frac{1}{2}G^{is}\big(G_{jm}(\boldsymbol{S}^p)^m_{sk} + G_{km}(\boldsymbol{S}^p)^m_{sj}\big).$$

Here the components of the plastic metric tensors $\boldsymbol{C}^p$ *and* $(\boldsymbol{C}^p)^{-1}$ *are denoted by* G_{ij} *and* G^{ij}*, respectively. We mention here the symmetry properties of the fields defined above*

$$\begin{aligned} &((\boldsymbol{\gamma}^p)\boldsymbol{u})\boldsymbol{v} = (\boldsymbol{\gamma}^p\boldsymbol{v})\boldsymbol{u} \quad \forall\, \boldsymbol{u},\boldsymbol{v} \in \mathscr{V}, \\ &\boldsymbol{W}^p\boldsymbol{u} \in Skew_2 \quad \forall\, \boldsymbol{u} \in \mathscr{V}. \end{aligned} \tag{9.41}$$

The proof can be found, for instance, in Schouten (1954), see also Yavari and Goriely (2012). In addition, we apply the formulae (9.40) to the connection $\overset{(e)}{\mathscr{A}}_{\mathscr{K}}$, with respect to the anholonomic configuration $\mathscr{K}$. This means that gradient ∇ is replaced by $\nabla_{\mathscr{K}}$. Consequently, having in mind the decomposition (9.40) the presence of the fields $(\boldsymbol{C}^e, \nabla_{\mathscr{K}}\boldsymbol{C}^e, \boldsymbol{S}^e_{\mathscr{K}})$ in the formula (9.35) can be justified. Thus the presence of the elastic torsion written with respect to the configuration $\mathscr{K}$ has been replaced by the plastic torsion with respect to the same configuration via the relationship (9.38).

Note 9.2. The free energy density is influenced by the dislocation density $\boldsymbol{\alpha}_{\mathscr{K}}$, which is defined by

$$\begin{aligned} &\boldsymbol{\alpha}_{\mathscr{K}}(\tilde{\boldsymbol{u}} \times \tilde{\boldsymbol{v}}) = (Skw(\overset{(e)}{\mathscr{A}}_{\mathscr{K}})\tilde{\boldsymbol{u}})\tilde{\boldsymbol{v}}, \quad \text{with the property} \\ &\boldsymbol{\alpha}_{\mathscr{K}}(\tilde{\boldsymbol{u}} \times \tilde{\boldsymbol{v}}) = -\frac{1}{det\boldsymbol{F}^p}\boldsymbol{F}^p\boldsymbol{\alpha}(\boldsymbol{F}^p)^{-1}(\boldsymbol{u} \times \boldsymbol{v}), \quad (\boldsymbol{F}^p)^{-1}(\tilde{\boldsymbol{u}}) = \boldsymbol{u}, (\boldsymbol{F}^p)^{-1}(\tilde{\boldsymbol{v}}) = \boldsymbol{v}, \end{aligned} \tag{9.42}$$

and by the disclination tensor $\widetilde{\boldsymbol{\Lambda}}$, both tensors being defined with respect to the configuration with torsion $\mathscr{K}$.

We mention that the elastic strain field

$$\boldsymbol{C}^e - \boldsymbol{I} = (\boldsymbol{F}^p)^{-T}(\boldsymbol{C} - \boldsymbol{C}^p)(\boldsymbol{F}^p)^{-1}, \tag{9.43}$$

its gradient formula written in (9.15) together with (9.34), and (9.29) suggest that the free energy density can be rewritten with respect to the reference configuration. When the fields were pulled back to the reference configuration by $(\boldsymbol{F}^p)^{-1}$, the function ψ can be written under the form

$$\psi = \psi(\boldsymbol{C} - \boldsymbol{C}^p, \nabla\boldsymbol{C} - Sym\{\boldsymbol{C}\overset{(p)}{\mathscr{A}}\}, \boldsymbol{S}^p, \boldsymbol{\Lambda}, \nabla\boldsymbol{\Lambda}). \tag{9.44}$$

As $\boldsymbol{S}^p$ is considered to be a measure of dislocation-disclination interplay we introduce a more general representation, which contains separately the influence of dislocation and disclination defects, given by

$$\begin{aligned}\psi = \psi^e(\boldsymbol{C}-\boldsymbol{C}^p, \nabla\boldsymbol{C} - Sym\{\boldsymbol{C} \overset{(p)}{\mathscr{A}}\})+ \\ + \underline{\psi}(Skw(\overset{(p)}{\mathscr{A}}), (\boldsymbol{C}^p)^{-1}Skw(\boldsymbol{\Lambda}\times\boldsymbol{I}), \boldsymbol{\Lambda}, \nabla\boldsymbol{\Lambda}).\end{aligned} \tag{9.45}$$

The time derivative of the free density function (9.45) is computed by

$$\begin{aligned}\dot{\psi} = \partial_{\boldsymbol{C}^e}\psi^e \cdot (\dot{\boldsymbol{C}} - \dot{\boldsymbol{C}}^p)+ \\ +\partial_{\nabla\boldsymbol{C}^e}\psi^e \cdot [\nabla\dot{\boldsymbol{C}} - Sym\{\boldsymbol{C}\frac{d}{dt}(\overset{(p)}{\mathscr{A}})\} - Sym\{\dot{\boldsymbol{C}} \overset{(p)}{\mathscr{A}}\}] + \dot{\underline{\psi}}.\end{aligned} \tag{9.46}$$

The derivatives of the mentioned fields will be replaced by their appropriate expressions.

9.3.2 Free Energy Imbalance Principle

The local free energy imbalance states the internal power expended during the elasto-plastic process is equal or greater than the time derivative of the free energy density.

Axiom 9.2 *The elasto-plastic constitutive description of the material is restricted to satisfy in $\mathscr{K}$ the free energy imbalance principle*

$$(\mathscr{P}_{int})_{\mathscr{K}} - \dot{\psi}_{\mathscr{K}} \geq 0, \tag{9.47}$$

for any virtual (isothermal) processes.

The expression of the internal power is the result of the superposed elastic, plastic and defect effects and will be written here in a slightly modified version of the corresponding expression postulated in Cleja-Ţigoiu (2010).

Axiom 9.3 *The internal power in the configuration with torsion is postulated to be given by the expression*

$$\begin{aligned}(\mathscr{P}_{int})_{\mathscr{K}} = \frac{1}{\rho}(\boldsymbol{T}^s)\cdot\boldsymbol{L}^e + \frac{1}{\tilde{\tilde{\rho}}}\boldsymbol{\Upsilon}^p\cdot\boldsymbol{L}^p + \frac{1}{\tilde{\tilde{\rho}}}\boldsymbol{\mu}^p\cdot\nabla_{\mathscr{K}}\boldsymbol{L}^p + \\ +\frac{1}{\tilde{\tilde{\rho}}}\boldsymbol{\mu}_{\mathscr{K}}\cdot((\boldsymbol{F}^e)^{-1}(\nabla_\chi\boldsymbol{L})[\boldsymbol{F}^e,\boldsymbol{F}^e] - \nabla_{\mathscr{K}}\boldsymbol{L}^p)+ \\ +\frac{1}{\tilde{\tilde{\rho}}}\boldsymbol{\Upsilon}^\lambda\cdot\frac{D}{Dt}\widetilde{\boldsymbol{\Lambda}} + \frac{1}{\tilde{\tilde{\rho}}}\boldsymbol{\mu}^\lambda\cdot\nabla_{\mathscr{K}}\frac{D}{Dt}\widetilde{\boldsymbol{\Lambda}}.\end{aligned} \tag{9.48}$$

Note 9.3. The gradient of the plastic rate $\boldsymbol{L}^p = \dot{\boldsymbol{F}}^p(\boldsymbol{F}^p)^{-1}$ with respect to the plastically deformed configuration is related to the time derivative of $\overset{(p)}{\mathscr{A}} = (\boldsymbol{F}^p)^{-1}\nabla\boldsymbol{F}^p$, and moreover $\nabla \underline{l}^p$ is involved in the expression of the previous one, as can be seen from the following formulae:

$$\frac{d}{dt}(\overset{(\mathrm{p})}{\mathscr{A}}) = (\boldsymbol{F}^p)^{-1}(\nabla_{\mathscr{K}}\boldsymbol{L}^p)[\boldsymbol{F}^p,\boldsymbol{F}^p] = \nabla \underline{l}^p - \underline{l}^p\, \overset{(\mathrm{p})}{\mathscr{A}} + \overset{(\mathrm{p})}{\mathscr{A}}\,[\boldsymbol{I},\underline{l}^p]. \tag{9.49}$$

We represent now the stresses and stress momenta as associated measures with respect to the reference configuration by pulled back procedure, see Cleja-Ţigoiu et al (2016). For instance the Mandel stress tensors, associated with plastic and disclination behaviour, are introduced with respect to the reference configuration by

$$\frac{1}{\rho_0}\boldsymbol{\Sigma}_0^p = \frac{1}{\tilde{\tilde{\rho}}}(\boldsymbol{F}^p)^T\boldsymbol{\Upsilon}^p(\boldsymbol{F}^p)^{-T}, \quad \frac{1}{\rho_0}\boldsymbol{\Sigma}_0^\lambda = \frac{1}{\tilde{\tilde{\rho}}}(\boldsymbol{F}^p)^T\boldsymbol{\Upsilon}^\lambda(\boldsymbol{F}^p)^{-T}, \tag{9.50}$$

while the appropriate micro stress momenta with respect to the reference configuration are given by

$$\begin{aligned}
\frac{1}{\rho_0}\boldsymbol{\mu}_0 &= (\boldsymbol{F}^p)^T\frac{1}{\tilde{\tilde{\rho}}}\boldsymbol{\mu}_{\mathscr{K}}[(\boldsymbol{F}^p)^{-T},(\boldsymbol{F}^p)^{-T}],\\
\frac{1}{\rho_0}\boldsymbol{\mu}_0^p &= (\boldsymbol{F}^p)^T\frac{1}{\tilde{\tilde{\rho}}}\boldsymbol{\mu}^p[(\boldsymbol{F}^p)^{-T},(\boldsymbol{F}^p)^{-T}],\\
\frac{1}{\rho_0}\boldsymbol{\mu}_0^\lambda &= (\boldsymbol{F}^p)^T\frac{1}{\tilde{\tilde{\rho}}}\boldsymbol{\mu}^\lambda[(\boldsymbol{F}^p)^{-T},(\boldsymbol{F}^p)^{-T}].
\end{aligned} \tag{9.51}$$

Proposition 9.5. *The internal power postulated by (9.48) is reformulated in terms of the stresses and stress momenta associated with the reference configuration, (9.50) and (9.51), under the form*

$$\begin{aligned}
&\frac{1}{\rho}(\boldsymbol{T}^s)\cdot(\boldsymbol{L}-\boldsymbol{F}\underline{l}^p\boldsymbol{F}^{-1}) - 2\partial_{\boldsymbol{C}^e}\psi^e\cdot\boldsymbol{F}^T\boldsymbol{D}\boldsymbol{F} + \frac{1}{\rho_0}\boldsymbol{\Sigma}_0^p\cdot\underline{l}^p + \frac{1}{\rho_0}\boldsymbol{\mu}_0^p\cdot\frac{d}{dt}(\overset{(\mathrm{p})}{\mathscr{A}})+\\
&+\frac{1}{\tilde{\tilde{\rho}}_0}\boldsymbol{\mu}_0\cdot((\boldsymbol{F}^{-1}(\nabla_\chi\boldsymbol{L})[\boldsymbol{F},\boldsymbol{F}] - \frac{d}{dt}(\overset{(\mathrm{p})}{\mathscr{A}})) + \partial_{\boldsymbol{C}^e}\psi^e\cdot[\boldsymbol{C}^p\underline{l}^p + (\underline{l}^p)^T\boldsymbol{C}^p]-\\
&-\partial_{\nabla\boldsymbol{C}^e}\psi^e\cdot\{Sym(\dot{\boldsymbol{C}}\boldsymbol{\Gamma}) + \boldsymbol{F}^T(\nabla_\chi\boldsymbol{D})[\boldsymbol{F},\boldsymbol{F}]\}+\\
&+\partial_{\nabla\boldsymbol{C}^e}\psi^e\cdot[Sym\{\boldsymbol{C}\frac{d}{dt}(\overset{(\mathrm{p})}{\mathscr{A}})\} + Sym\{\dot{\boldsymbol{C}}\,\overset{(\mathrm{p})}{\mathscr{A}}\}]+\\
&+\frac{1}{\rho_0}\boldsymbol{\Sigma}_0^\lambda\cdot\dot{\boldsymbol{\Lambda}} + \frac{1}{\rho_0}\boldsymbol{\mu}_0^\lambda\cdot\{\overset{(\mathrm{p})}{\mathscr{A}}\,[\boldsymbol{I},\dot{\boldsymbol{\Lambda}}] + \nabla\dot{\boldsymbol{\Lambda}} - \dot{\boldsymbol{\Lambda}}\,\overset{(\mathrm{p})}{\mathscr{A}}\} - \underline{\dot{\psi}} \geq 0.
\end{aligned} \tag{9.52}$$

In order to derive the consequences that follow from the dissipation inequality (9.52) we introduce certain identities involving the operators defined by (9.12).

The following identities written for any $\mathscr{A}$ and $\mathscr{B} \in Lin(\mathscr{V}, Lin)$ are direct consequences of the given definitions

$$
\begin{aligned}
&(\mathscr{A} \odot \mathscr{B})^T = \mathscr{B} \odot \mathscr{A}, \quad (\mathscr{A}\ _r\odot \mathscr{B})^T = \mathscr{B}\ _r\odot \mathscr{A},\\
&\{Sym(\mathscr{A} \odot \mathscr{B})\}^s = \frac{1}{2}(\mathscr{A} \odot \mathscr{B} + \mathscr{A}^T \odot \mathscr{B} + \mathscr{B} \odot \mathscr{A} + \mathscr{B} \odot \mathscr{A}^T),\\
&\{Sym(\mathscr{A}\ _r\odot \mathscr{B})\}^s = \frac{1}{2}(\mathscr{A}\ _r\odot \mathscr{B} + \mathscr{A}^T\ _r\odot \mathscr{B} + \mathscr{B}\ _r\odot \mathscr{A} + \mathscr{B}\ _r\odot \mathscr{A}).
\end{aligned}
\tag{9.53}
$$

$$
\begin{aligned}
&(Skw\mathscr{A}) \odot (Skw\mathscr{B}) = (\boldsymbol{\Omega}_{<\mathscr{A}>} \cdot \boldsymbol{\Omega}_{<\mathscr{B}>})\boldsymbol{I} + (\boldsymbol{\Omega}_{<\mathscr{B}>})^T \boldsymbol{\Omega}_{<\mathscr{A}>},\\
&(Skw\mathscr{A})\ _r\odot (Skw\mathscr{B}) = 2\boldsymbol{\Omega}_{<\mathscr{A}>}(\boldsymbol{\Omega}_{<\mathscr{B}>})^T,
\end{aligned}
\tag{9.54}
$$

Moreover

$$
\mathscr{A} \cdot \mathscr{B} = 2\boldsymbol{\Omega}_{<\mathscr{A}>} \cdot \boldsymbol{\Omega}_{<\mathscr{B}>}. \tag{9.55}
$$

9.4 Constitutive Restrictions Imposed by the Imbalance Free Energy Principle

First we derive the elastic type constitutive equations, starting from the supposition that no variation of the irreversible behaviour can occur.

9.4.1 Elastic Type Constitutive Equations

Proposition 9.6. *We suppose that $\mathbf{L}^p = 0$ or $\underline{\mathbf{l}}^p = 0$ (then $\mathbf{L}^e = \mathbf{L}$) and $\dot{\mathbf{\Lambda}} = 0$. Thus the imbalance free energy relation is reduced to the following inequality*

$$
\begin{aligned}
&(\frac{1}{\rho}(\boldsymbol{T}^s) - 2\boldsymbol{F}\partial_{\boldsymbol{C}^e}\psi^e\boldsymbol{F}^T)\cdot\boldsymbol{D} + \frac{1}{\rho_0}\boldsymbol{\mu}_0\cdot\boldsymbol{F}^{-1}(\nabla_\chi\mathbf{L})[\boldsymbol{F},\boldsymbol{F}]) -\\
&-\partial_{\nabla\boldsymbol{C}^e}\psi^e\cdot\{Sym(\dot{\boldsymbol{C}}\boldsymbol{\Gamma}) + \boldsymbol{F}^T\nabla_\chi\boldsymbol{D}[\boldsymbol{F},\boldsymbol{F}]\} + \partial_{\nabla\boldsymbol{C}^e}\psi^e\cdot Sym\{\dot{\boldsymbol{C}}\overset{(p)}{\mathscr{A}}\} \geq 0,
\end{aligned}
\tag{9.56}
$$

which holds for any $\mathbf{L}$ and $\nabla_\chi\mathbf{L}$.

Theorem 9.2. *The elastic free energy, denoted by ψ^e is potential for the macro stress and macro stress momentum, respectively, related to the reference configuration, namely*

$$
\begin{aligned}
&\frac{1}{2\rho}\boldsymbol{\pi}_0 = \partial_{\boldsymbol{C}^e}\psi^e + \{Sym(\partial_{\nabla\boldsymbol{C}^e}\psi^e)\ _r\odot(\boldsymbol{\Gamma} - \overset{(p)}{\mathscr{A}})\}^S,\\
&\frac{1}{\rho_0}\boldsymbol{\mu}_0 = Sym(\boldsymbol{C}\partial_{\nabla\boldsymbol{C}^e}\psi^e),
\end{aligned}
\tag{9.57}
$$

were $\boldsymbol{\Gamma} - \overset{(p)}{\mathscr{A}} \equiv (\boldsymbol{F}^p)^{-1}\overset{(e)}{\mathscr{A}}[\boldsymbol{F}^p,\boldsymbol{F}^p]$.

Proof. In order to compare the terms written in (9.56) we use the rule (9.12)

$$\partial_{\nabla C^e}\psi^e \cdot Sym\big(\dot{C}(\Gamma - \overset{(p)}{\mathscr{A}})\big) = \big((Sym\partial_{\nabla C^e}\psi^e)\,_r\odot(\Gamma - \overset{(p)}{\mathscr{A}})\big)\cdot\dot{C}, \tag{9.58}$$

and we pass from D to $\dot{C}$ via the relation $(9.18)_2$. The inequality (9.56) is written finally under the form

$$\begin{aligned}&\{\frac{1}{2\rho}\pi_0 - \partial_{C^e}\psi^e)\cdot\dot{C} - \big((Sym\partial_{\nabla C^e}\psi^e)_r\odot(\Gamma - \overset{(p)}{\mathscr{A}})\big)\}\cdot\dot{C}+\\&+\frac{1}{\rho_0}F^{-T}(\mu_0 - Sym(C\partial_{\nabla C^e}\psi^e))[F^T,F^T]\cdot\nabla_\chi L \geq 0,\end{aligned} \tag{9.59}$$

which holds for any L and $\nabla_\chi L$. In (9.59) the expression of the Piola-Kirchhoff stress tensor with respect to the reference configuration, π_0, has been introduced

$$\frac{1}{\rho_0}\pi_0 = \frac{1}{\rho}F^{-1}TF^{-T}, \tag{9.60}$$

Note 9.4. The non-symmetric Cauchy stress, T, and couple-stress tensor, m, satisfy the balance equations formulated by Fleck et al (1994), see also Cleja-Ţigoiu and Ţigoiu (2011). In this model $\mu_0 z \in Sym_2$ and consequently T^a is vanishing. The (equilibrium) balance equations for macro forces, say (T,μ) in the actual configuration is reduced to the classical one, $divT = 0$, if the mass density of the body and couple forces are neglected.

9.4.2 Dissipation Inequality

In order to derive the restrictions imposed by the free energy imbalance related to the plastic behaviour, we return to the inequality (9.52).

Theorem 9.3. *The reduced dissipation inequality is derived under the form*

$$\begin{aligned}&\{\frac{1}{\rho_0}(\mu_0^p - \mu_0) + CSym(\partial_{\nabla C^e}\psi^e) - Skw(\partial_{\mathscr{Z}_1}\underline{\psi})\}\cdot\frac{d}{dt}\overset{(p)}{\mathscr{A}}+\\&+\{\frac{1}{\rho_0}\Sigma_0^p + 2C^p\partial_{C^e}\psi^e + \partial_{\mathscr{Z}_2}\underline{\psi}\,_r\odot(C^p)^{-1}Skw(\Lambda\times I)+\\&+Skw(\Lambda\times I)\,_r\odot(C^p)^{-1}\partial_{\mathscr{Z}_2}\underline{\psi}\}\cdot\underline{l}^p + (\frac{1}{\rho_0}\mu_0^\lambda - \partial_{\nabla\Lambda}\underline{\psi})\cdot\nabla\dot{\Lambda}+\\&+(\frac{1}{\rho_0}\Sigma_0^\lambda - \partial_\Lambda\underline{\psi} + \overset{(p)}{\mathscr{A}}\odot\frac{1}{\rho_0}\mu_0^\lambda - \frac{1}{\rho_0}\mu_0^\lambda\,_r\odot\overset{(p)}{\mathscr{A}})\cdot\dot{\Lambda}+\\&+\{\epsilon\cdot((C^p)^{-1}\partial_{\mathscr{Z}_2}\underline{\psi})I - (\epsilon_r\odot((C^p)^{-1}\partial_{\mathscr{Z}_2}\underline{\psi}))\}\cdot\dot{\Lambda}\geq 0,\end{aligned} \tag{9.61}$$

when we put into evidence the terms which contain the rates of the appropriate fields and their gradients, namely $\underline{l}^p$, $\dfrac{d}{dt}(\overset{(p)}{\mathscr{A}})$, $\dot{\boldsymbol{\Lambda}}$ *and* $\nabla\dot{\boldsymbol{\Lambda}}$. *Here* $\boldsymbol{\epsilon}$ *denotes Ricci's permutation tensor.*

Proof. First we introduce the expression for the time derivative of non-elastic free energy function $\underline{\psi}$, in which we use the notations mentioned below

$$\underline{\psi} = \underline{\psi}(Skw(\overset{(p)}{\mathscr{A}}), (\mathbf{C}^p)^{-1}Skw(\boldsymbol{\Lambda}\times\boldsymbol{I}), \boldsymbol{\Lambda}, \nabla\boldsymbol{\Lambda}) \equiv \underline{\psi}(Skw(\mathscr{Z}_1, \mathscr{Z}_2, \boldsymbol{\Lambda}, \nabla\boldsymbol{\Lambda}), \tag{9.62}$$
$$\text{where} \quad \mathscr{Z}_1 = Skw(\overset{(p)}{\mathscr{A}}), \quad \mathscr{Z}_2 = (\mathbf{C}^p)^{-1}Skw(\boldsymbol{\Lambda}\times\boldsymbol{I}).$$

The time derivative of the non elastic free energy function (9.62) is expressed as

$$\begin{aligned}\dot{\underline{\psi}} &= \partial_{\mathscr{Z}_1}\underline{\psi}\cdot Skw(\frac{d}{dt}(\overset{(p)}{\mathscr{A}}) + \partial_{\mathscr{Z}_2}\underline{\psi}\cdot Skw(\frac{d}{dt}((\mathbf{C}^p)^{-1}(\boldsymbol{\Lambda}\times\boldsymbol{I})) + \\ &+\partial_{\boldsymbol{\Lambda}}\underline{\psi}\cdot\dot{\boldsymbol{\Lambda}} + \partial_{\nabla\boldsymbol{\Lambda}}\underline{\psi}\cdot\nabla\dot{\boldsymbol{\Lambda}}.\end{aligned} \tag{9.63}$$

Using the time derivative formula written in $(9.18)_4$, $(9.7)_4$ and the rules (9.12) we obtain

$$\begin{aligned}\partial_{\mathscr{Z}_2}\underline{\psi}\cdot Skw\frac{d}{dt}((\mathbf{C}^p)^{-1}(\boldsymbol{\Lambda}\times\boldsymbol{I})) &= -(\partial_{\mathscr{Z}_2}\underline{\psi}_r \odot Skw(\mathbf{C}^p)^{-1}(\boldsymbol{\Lambda}\times\boldsymbol{I})) \; \underline{l}^p - \\ &-((\mathbf{C}^p)^{-1}\partial_{\mathscr{Z}_2}\underline{\psi}_r \odot Skw(\boldsymbol{\Lambda}\times\boldsymbol{I}))\cdot(\underline{l}^p)^T - \\ &-\boldsymbol{\epsilon}\cdot((\mathbf{C}^p)^{-1}\partial_{\mathscr{Z}_2}\underline{\psi})\boldsymbol{I}\cdot\dot{\boldsymbol{\Lambda}} + (\boldsymbol{\epsilon}_r\odot((\mathbf{C}^p)^{-1}\partial_{\mathscr{Z}_2}\underline{\psi}))\cdot\dot{\boldsymbol{\Lambda}}.\end{aligned} \tag{9.64}$$

The reduced dissipation inequality is derived from (9.52) together with the elastic type constitutive relation $(9.57)_1$, were $\dot{\underline{\psi}}$ is given by the formulae (9.63) together with (9.64).

9.5 Viscoplastic Type Evolution Equations for Plastic Distortion and Disclination Tensor

Hypotheses. The energetic type constitutive equations will be defined for micro momenta related to the plastic and disclination mechanism, namely

$$\begin{aligned}\frac{1}{\rho_0}\boldsymbol{\mu}_0^p &= \frac{1}{\rho_0}\boldsymbol{\mu}_0 - \mathbf{C}Sym(\partial_{\nabla\mathbf{C}^e}\psi^e) + Skw(\partial_{\mathscr{Z}_1}\underline{\psi}), \\ \frac{1}{\rho_0}\boldsymbol{\mu}_0^\lambda &= \partial_{\nabla\boldsymbol{\Lambda}}\underline{\psi}.\end{aligned} \tag{9.65}$$

Theorem 9.4. *Under the hypotheses formulated by (9.65) the reduced dissipation inequality can be written under the form*

$$
\begin{aligned}
&\{\frac{1}{\rho_0}\boldsymbol{\Sigma}_0^p + 2Skw\partial_{\mathscr{X}_2}\underline{\psi}\,{}_r\odot(\boldsymbol{C}^p)^{-1}(\boldsymbol{\Lambda}\times\boldsymbol{I}) + \\
&+(\boldsymbol{\Lambda}\times\boldsymbol{I})\,{}_r\odot\big((\boldsymbol{C}^p)^{-1}Skw(\partial_{\mathscr{X}_2}\underline{\psi})\big) + 2\boldsymbol{C}^p\partial_{\boldsymbol{C}^e}\psi^e\}\cdot\underline{\boldsymbol{l}}^p + \\
&+(\frac{1}{\rho_0}\boldsymbol{\Sigma}_0^\lambda - \partial_{\boldsymbol{\Lambda}}\underline{\psi} + \overset{(\mathrm{p})}{\mathscr{A}}\odot\frac{1}{\rho_0}\boldsymbol{\mu}_0^\lambda - \frac{1}{\rho_0}\boldsymbol{\mu}_0^\lambda{}_r\odot\overset{(\mathrm{p})}{\mathscr{A}})\cdot\dot{\boldsymbol{\Lambda}} + \\
&+\{\boldsymbol{\epsilon}\cdot\big((\boldsymbol{C}^p)^{-1}\partial_{\mathscr{X}_2}\underline{\psi}\big)\boldsymbol{I} - \big(\boldsymbol{\epsilon}_r\odot((\boldsymbol{C}^p)^{-1}\partial_{\mathscr{X}_2}\underline{\psi})\big)\}\cdot\dot{\boldsymbol{\Lambda}} \geq 0
\end{aligned}
\tag{9.66}
$$

Proof. As a direct consequence of (9.61) together with (9.65) the inequality (9.66) follows at once. The variables $\mathscr{X}_1$ and $\mathscr{X}_2$ have been defined in (9.62). Here Mandel's type stress tensor, $\frac{1}{\rho_0}\boldsymbol{\Sigma}_0^p$, appears to be power conjugate to the rate of plastic distortion $\underline{\boldsymbol{l}}^p$. For physical meaning and properties of Mandel and Eshelby stress tensors see for instance Maugin (1994); Cleja-Tigoiu and Maugin (2000).

Axiom 9.4 *The evolution equations for plastic distortion and disclination tensor are supposed to be given by*

$$
\begin{aligned}
\xi_1\,\underline{\boldsymbol{l}}^p &= \frac{1}{\rho_0}\boldsymbol{\Sigma}_0^p + Skw(\partial_{\mathscr{X}_2}\underline{\psi})\,{}_r\odot\big((\boldsymbol{C}^p)^{-1}Skw(\boldsymbol{\Lambda}\times\boldsymbol{I})\big) + \\
&\quad + Skw(\boldsymbol{\Lambda}\times\boldsymbol{I})\,{}_r\odot\big((\boldsymbol{C}^p)^{-1}Skw(\partial_{\mathscr{X}_2}\underline{\psi})\big) + 2\boldsymbol{C}^p\partial_{\boldsymbol{C}^e}\psi^e, \\
\xi_2\,\dot{\boldsymbol{\Lambda}} &= \frac{1}{\rho_0}\boldsymbol{\Sigma}_0^\lambda - \partial_{\boldsymbol{\Lambda}}\underline{\psi} + \overset{(\mathrm{p})}{\mathscr{A}}\odot\frac{1}{\rho_0}\boldsymbol{\mu}_0^\lambda - \frac{1}{\rho_0}\boldsymbol{\mu}_0^\lambda\,{}_r\odot\overset{(\mathrm{p})}{\mathscr{A}} + \\
&\quad + \boldsymbol{\epsilon}\cdot\big((\boldsymbol{C}^p)^{-1}\partial_{\mathscr{X}_2}\underline{\psi}\big)\boldsymbol{I} - \boldsymbol{\epsilon}\odot\big((\boldsymbol{C}^p)^{-1}\partial_{\mathscr{X}_2}\underline{\psi}\big).
\end{aligned}
\tag{9.67}
$$

As a direct consequence of (9.67) the dissipation inequality (9.66) becomes

$$
\xi_1\,\underline{\boldsymbol{l}}^p\cdot\underline{\boldsymbol{l}}^p + \xi_2\,\dot{\boldsymbol{\Lambda}}\cdot\dot{\boldsymbol{\Lambda}} \geq 0. \tag{9.68}
$$

The last inequality holds for any non-negative constitutive functions ξ_1, ξ_2.

Concerning the expression of the Mandel type stress tensors, $\frac{1}{\rho_0}\boldsymbol{\Sigma}^p$ and $\frac{1}{\rho_0}\boldsymbol{\Sigma}^\lambda$ defined by (9.50), we shall use the balance equations for micro forces provided in the paper by Cleja-Ţigoiu (2017). We recall the micro balance equation for the plastic mechanism

$$
\frac{1}{\overline{\overline{\rho}}}(\boldsymbol{\Upsilon}^p - \boldsymbol{\Sigma}_{\mathscr{K}}) = \mathrm{div}\,\big(\frac{1}{\overline{\overline{\rho}}}(\boldsymbol{\mu}^p - \boldsymbol{\mu}_{\mathscr{K}})(\boldsymbol{F}^p)^{-T}\big) + \boldsymbol{B}^p, \tag{9.69}
$$

and the appropriate micro balance equation related to the disclination mechanism

$$
\frac{1}{\overline{\overline{\rho}}}\boldsymbol{\Upsilon}^\lambda = \mathrm{div}\big(\frac{1}{\overline{\overline{\rho}}}\boldsymbol{\mu}^\lambda\,(\boldsymbol{F}^p)^{-T}\big) + \boldsymbol{B}^\lambda. \tag{9.70}
$$

Here $\tilde{\rho}\boldsymbol{B}^p$ and $\tilde{\rho}\boldsymbol{B}^\lambda$ are mass densities of couple body forces.

Definition 9.7. The Piola-Kirchhoff macroscopic stress tensor, $\boldsymbol{\pi}_0$, the macro stress tensor, $\boldsymbol{\Sigma}_{\mathscr{K}}$, and the Cauchy stress $\boldsymbol{T}$ are related by the following relationships

$$\frac{1}{\tilde{\rho}}\boldsymbol{\Sigma}_{\mathscr{K}} = \frac{1}{\rho_0}(\boldsymbol{F}^p)^{-T}\boldsymbol{C}\boldsymbol{\pi}_0(\boldsymbol{F}^p)^T = \frac{1}{\rho}(\boldsymbol{F}^e)^{-T}\boldsymbol{T}(\boldsymbol{F}^e)^T. \tag{9.71}$$

Note 9.5. Clayton et al (2006) assumed that the geometrically necessary density tensors $\tilde{\boldsymbol{\alpha}}$ and $\tilde{\boldsymbol{\theta}}$ do not contribute to the free energy dissipation, namely

$$\tilde{\sigma} = \frac{\partial \tilde{\psi}}{\partial \tilde{\boldsymbol{\alpha}}}, \quad \tilde{\boldsymbol{\mu}} = \frac{\partial \tilde{\psi}}{\partial \tilde{\boldsymbol{\theta}}},$$

and the microforces with respect to the intermediate configuration satisfy the Fleck et al (1994) type balance equations.

9.5.1 Quadratic Free Energy

We restrict ourself to the case of the free energy function which is quadratic with respect to above mentioned variables, given by

$$\begin{aligned}
\psi &= \psi^e + \underline{\psi}, \\
\psi^e &= \frac{1}{8}\mathscr{E}(\boldsymbol{C}-\boldsymbol{C}^p)\cdot(\boldsymbol{C}-\boldsymbol{C}^p) + \frac{1}{4}\beta_1(\nabla\boldsymbol{C}- \\
&\quad -Sym\{\boldsymbol{C}\overset{(p)}{\mathscr{A}}\})\cdot(\nabla\boldsymbol{C} - Sym\{\boldsymbol{C}\overset{(p)}{\mathscr{A}}\}), \\
\underline{\psi} &= \frac{1}{4}\beta_2\{Skw\overset{(p)}{\mathscr{A}} + \tilde{\beta}(\boldsymbol{C}^p)^{-1}Skw(\boldsymbol{\Lambda}\times\boldsymbol{I})\}\cdot\{Skw\overset{(p)}{\mathscr{A}} + \\
&\quad +\tilde{\beta}(\boldsymbol{C}^p)^{-1}Skw(\boldsymbol{\Lambda}\times\boldsymbol{I})\} + \frac{1}{2}\beta_3\boldsymbol{\Lambda}\cdot\boldsymbol{\Lambda} + \frac{1}{2}\beta_4\nabla\boldsymbol{\Lambda}\cdot\nabla\boldsymbol{\Lambda}.
\end{aligned} \tag{9.72}$$

Note 9.6. If $\tilde{\beta} = 1$ then the non-elastic part of the free energy postulated here coincides with those introduced by Cleja-Ţigoiu et al (2016).

The appropriate partial derivatives of the free energy function (9.72) determine the elastic type constitutive equations (9.57), the energetic representation for the micro stress momenta (9.65) as well as the evolution equations for plastic distortion and disclination tensor (9.67).

We do not provide here the particular constitutive model associated with the free energy function (9.72) for the conciseness of the exposure. We pass directly to the case of small distortions, that follows directly from the finite deformation model.

9.5.2 Elasto-Plastic Model for Dislocations and Disclinations in the Case of Small Distortions

The case of small elastic and plastic distortions is defined by the following conditions

$$\begin{array}{l} \boldsymbol{F}^e = \boldsymbol{I} + \boldsymbol{H}^e, \quad \boldsymbol{F}^p = \boldsymbol{I} + \boldsymbol{H}^p, \quad \boldsymbol{F} = \boldsymbol{I} + \boldsymbol{H}, \\ \boldsymbol{H} = \nabla \boldsymbol{u}, \quad \boldsymbol{H} = \boldsymbol{H}^e + \boldsymbol{H}^e, \quad \text{for} \quad \| \boldsymbol{H}^e \| \ll 1, \| \boldsymbol{H}^p \| \ll 1, \end{array} \tag{9.73}$$

The following approximated formulae can be put into evidence

$$\boldsymbol{C} = \boldsymbol{I} + 2\boldsymbol{\varepsilon}, \quad \boldsymbol{\varepsilon} = \frac{1}{2}(\nabla \boldsymbol{u} + \nabla \boldsymbol{u}^T), \quad \boldsymbol{C}^p = \boldsymbol{I} + \boldsymbol{\varepsilon}^p, \tag{9.74}$$

where

$$\begin{array}{l} \boldsymbol{\varepsilon}^p = \frac{1}{2}(\boldsymbol{H}^p + (\boldsymbol{H}^p)^T) \\ \frac{1}{2}\beta_1 \big(\nabla \boldsymbol{C} - Sym(\{\boldsymbol{C} \overset{(\mathrm{p})}{\mathscr{A}}\})\big) = \nabla \boldsymbol{\varepsilon} - \nabla \boldsymbol{\varepsilon}^p, \\ \overset{(\mathrm{p})}{\mathscr{A}} = \nabla \boldsymbol{H}^p, \quad \boldsymbol{\Gamma} = \nabla \boldsymbol{H}, \quad \mathscr{Z}_1 = Skw \nabla \boldsymbol{H}^p, \quad \mathscr{Z}_2 = Skw(\boldsymbol{\Lambda} \times \boldsymbol{I}). \end{array}$$

The elastic type constitutive equations, namely the formulae (9.57), can be represented under the form

$$\begin{array}{rl} \frac{1}{\rho_0} \boldsymbol{\pi}_0 = & \mathscr{E}(\boldsymbol{\varepsilon} - \boldsymbol{\varepsilon}^p) + \frac{1}{2}\beta_1 \{(\nabla \boldsymbol{\varepsilon} - \nabla \boldsymbol{\varepsilon}^p)\,{}_r\odot (\nabla \boldsymbol{H} - \nabla \boldsymbol{H}^p) \\ & + (\nabla \boldsymbol{H} - \nabla \boldsymbol{H}^p)\,{}_r\odot (\nabla \boldsymbol{\varepsilon} - \nabla \boldsymbol{\varepsilon}^p)\}, \\ \frac{1}{\rho_0} \boldsymbol{\mu}_0 = & \beta_1 (\nabla \boldsymbol{\varepsilon} - \nabla \boldsymbol{\varepsilon}^p). \end{array} \tag{9.75}$$

The energetic expressions for the plastic and disclination momenta are given by

$$\begin{array}{l} \frac{1}{\rho_0} \boldsymbol{\mu}^p = \beta_2 \big(Skw \nabla \boldsymbol{H}^p + \tilde{\beta} Skw(\boldsymbol{\Lambda} \times \boldsymbol{I})\big), \\ \frac{1}{\rho_0} \boldsymbol{\mu}^\lambda = \beta_4 \nabla \boldsymbol{\Lambda}. \end{array} \tag{9.76}$$

The evolution equations for plastic distortion and disclination tensor become

$$\begin{array}{rl} \xi_1 \underline{\boldsymbol{l}}^p = & \frac{1}{\rho_0} \boldsymbol{\Sigma}_0^p + \mathscr{E}(\boldsymbol{\varepsilon} - \boldsymbol{\varepsilon}^p) + \\ & + \beta_2 Skw(\boldsymbol{\Lambda} \times \boldsymbol{I})\,{}_r\odot \big(Skw \nabla \boldsymbol{H}^p + \tilde{\beta} Skw(\boldsymbol{\Lambda} \times \boldsymbol{I})\big) + \\ & + \beta_2 \big(Skw \nabla \boldsymbol{H}^p + \tilde{\beta} Skw(\boldsymbol{\Lambda} \times \boldsymbol{I})\big)\,{}_r\odot Skw(\boldsymbol{\Lambda} \times \boldsymbol{I}), \end{array} \tag{9.77}$$

$$\begin{aligned}
\xi_2\,\dot{\boldsymbol{\Lambda}} &= \frac{1}{\rho_0}\boldsymbol{\Sigma}_0^\lambda + \beta_4\,\nabla\boldsymbol{H}^p \odot \nabla\boldsymbol{\Lambda} - \beta_4 \nabla\boldsymbol{\Lambda}\,{}_r\!\odot \nabla\boldsymbol{H}^p - \beta_3\boldsymbol{\Lambda} + \\
&+ \frac{1}{2}\beta_2 \in \cdot\big(Skw\nabla\boldsymbol{H}^p + \tilde{\beta} Skw(\boldsymbol{\Lambda}\times\boldsymbol{I})\boldsymbol{I} - \\
&- \frac{1}{2}\beta_2 \in {}_r\!\odot \big(Skw\nabla\boldsymbol{H}^p + \tilde{\beta} Skw(\boldsymbol{\Lambda}\times\boldsymbol{I})\big).
\end{aligned} \tag{9.78}$$

Using the identities (9.54) and (9.55), together with the property $\Omega_{<\in>} = -\boldsymbol{I}$ the evolution equation for the disclination tensor $\boldsymbol{\Lambda}$ will be given by

$$\begin{aligned}
\xi_2\,\dot{\boldsymbol{\Lambda}} &= \frac{1}{\rho_0}\boldsymbol{\Sigma}_0^\lambda + \beta_4\,\nabla\boldsymbol{H}^p \odot \nabla\boldsymbol{\Lambda} - \beta_4 \nabla\boldsymbol{\Lambda}\,{}_r\!\odot \nabla\boldsymbol{H}^p - \beta_3\boldsymbol{\Lambda} - \\
&-2\beta_2\, tr\,(curl\boldsymbol{H}^p)\boldsymbol{I} - 4\beta_2\,\tilde{\beta} tr(\boldsymbol{\Lambda})\boldsymbol{I} + \beta_2\,(curl\boldsymbol{H}^p)^T + \beta_2\tilde{\beta}\big(tr(\boldsymbol{\Lambda})\boldsymbol{I} - \boldsymbol{\Lambda})\big)
\end{aligned} \tag{9.79}$$

Proposition 9.7. *In the case of small distortions the evolution equations for the plastic distortion* $\boldsymbol{H}^p$ *and* $\boldsymbol{\Lambda}$ *are given by*

$$\begin{aligned}
\xi_1\,\dot{\boldsymbol{H}}^p &= \frac{1}{\rho_0}\boldsymbol{\Sigma}_0^p + \mathscr{E}(\boldsymbol{\varepsilon} - \boldsymbol{\varepsilon}^p) + \beta_2\,\big((tr\boldsymbol{\Lambda})\boldsymbol{I} - \boldsymbol{\Lambda}^T\big)(curl\boldsymbol{H}^p)^T + \\
&+\beta_2\,(curl\boldsymbol{H}^p)\big((tr\boldsymbol{\Lambda})\boldsymbol{I} - \boldsymbol{\Lambda}\big) + 2\beta_2\tilde{\beta}\,\big((tr\boldsymbol{\Lambda})\boldsymbol{I} - \boldsymbol{\Lambda}^T\big)\big((tr\boldsymbol{\Lambda})\boldsymbol{I} - \boldsymbol{\Lambda}\big), \\
\xi_2\,\dot{\boldsymbol{\Lambda}} &= \frac{1}{\rho_0}\boldsymbol{\Sigma}_0^\lambda + \beta_4\,\nabla\boldsymbol{H}^p \odot \nabla\boldsymbol{\Lambda} - \beta_4 \nabla\boldsymbol{\Lambda}\,{}_r\!\odot \nabla\boldsymbol{H}^p - \beta_3\boldsymbol{\Lambda} - \\
&-2\beta_2\, tr\,(curl\boldsymbol{H}^p)\boldsymbol{I} - 4\beta_2\,\tilde{\beta} tr(\boldsymbol{\Lambda})\boldsymbol{I} + \\
&+\beta_2\,(curl\boldsymbol{H}^p)^T + \beta_2\tilde{\beta}\big(tr(\boldsymbol{\Lambda})\boldsymbol{I} - \boldsymbol{\Lambda}\big).
\end{aligned} \tag{9.80}$$

The appropriate Mandel stress tensors are approximated by

$$\boldsymbol{\Sigma}_0^p = \boldsymbol{\Upsilon}^p, \quad \boldsymbol{\Sigma}_0^\lambda = \boldsymbol{\Upsilon}^\lambda, \quad \boldsymbol{\Sigma}_{\mathscr{K}} = \boldsymbol{\pi}_0 = \boldsymbol{T}, \quad \tilde{\rho} = \rho_0, \tag{9.81}$$

and are characterized by the micro balance equations (9.69) and (9.70) reduced to the following ones

$$\frac{1}{\rho_0}\big(\boldsymbol{\Upsilon}^p - \boldsymbol{\pi}_0\big) = div\,\frac{1}{\rho_0}(\boldsymbol{\mu}_0^p - \mu), \tag{9.82}$$

where

$$\begin{aligned}
&div\,(\frac{1}{\rho_0}\boldsymbol{\mu}_0^p) = -\beta_2 curl\big(curl\boldsymbol{H}^p\big) - \beta_2\tilde{\beta} \in (\nabla tr\boldsymbol{\Lambda}) + \beta_2\tilde{\beta}\; curl(\boldsymbol{\Lambda}^T), \\
&div\,(\frac{1}{\rho_0}\boldsymbol{\mu}_0) = \beta_1(\Delta\boldsymbol{\varepsilon} - \Delta\boldsymbol{\varepsilon}^p).
\end{aligned}$$

and

$$\frac{1}{\rho_0}\boldsymbol{\Upsilon}^\lambda = \beta_4\,\Delta\boldsymbol{\Lambda}, \tag{9.83}$$

when the mass density of couple body forces were neglected.

Theorem 9.5. *Finally, the evolution equations for the unknowns $\boldsymbol{H}^p$ and $\boldsymbol{\Lambda}$ are given by*

$$
\begin{aligned}
\xi_1 \dot{\boldsymbol{H}}^p &= \frac{1}{\rho_0}\boldsymbol{T} + \mathscr{E}(\boldsymbol{\varepsilon} - \boldsymbol{\varepsilon}^p) - \beta_1(\Delta\boldsymbol{\varepsilon} - \Delta\boldsymbol{\varepsilon}^p) + \\
&-\beta_2 curl\big(curl\boldsymbol{H}^p\big) - \beta_2\tilde{\beta} \in (\nabla tr\boldsymbol{\Lambda}) + \beta_2\tilde{\beta}\, curl(\boldsymbol{\Lambda}^T) + \\
&+\beta_2\left((tr\boldsymbol{\Lambda})\boldsymbol{I} - \boldsymbol{\Lambda}^T\right)(curl\boldsymbol{H}^p)^T + \beta_2\,(curl\boldsymbol{H}^p)\left((tr\boldsymbol{\Lambda})\boldsymbol{I} - \boldsymbol{\Lambda}\right) + \\
&+2\beta_2\tilde{\beta}\left((tr\boldsymbol{\Lambda})\boldsymbol{I} - \boldsymbol{\Lambda}^T\right)\left((tr\boldsymbol{\Lambda})\boldsymbol{I} - \boldsymbol{\Lambda}\right), \\
\xi_2 \dot{\boldsymbol{\Lambda}} &= \beta_4\,\Delta\boldsymbol{\Lambda} + \beta_4\,\nabla\boldsymbol{H}^p \odot \nabla\boldsymbol{\Lambda} - \beta_4\nabla\boldsymbol{\Lambda}\,{}_r\!\odot \nabla\boldsymbol{H}^p - \beta_3\boldsymbol{\Lambda} - \\
&-2\beta_2\, tr\,(curl\boldsymbol{H}^p)\boldsymbol{I} - 4\beta_2\,\tilde{\beta} tr(\boldsymbol{\Lambda})\boldsymbol{I} + \\
&+\beta_2\,(curl\boldsymbol{H}^p)^T + \beta_2\tilde{\beta}\left(tr(\boldsymbol{\Lambda})\boldsymbol{I} - \boldsymbol{\Lambda}\right).
\end{aligned}
\tag{9.84}
$$

Here $\boldsymbol{T}$ is given by the relationship (9.75) when the hypothesis of small distortions is accepted.

9.6 Conclusions

The proposed model of structural defects such as dislocations and disclinations appears to be a continuation of the previous ones proposed by Cleja-Ţigoiu (2014); Cleja-Ţigoiu et al (2016).

- The postulated free energy functions contain somehow the same variables describing the defects, excepting the gradient of the disclination tensor which is not involved in Cleja-Ţigoiu (2014).

The elastic constitutive functions have been essentially changed, as follows

- The elastic response is characterized here by the formulae (9.57). The Piola-Kirchhoff stress tensor is expressed in terms of the partial derivatives $\partial_{\boldsymbol{C}^e}\psi, \partial_{\nabla\boldsymbol{C}^e}\psi$, as well as Bilby's elastic connection, $\overset{(e)}{\mathscr{A}}$, while
$$\frac{1}{\rho_0}\boldsymbol{\pi} = \partial_{\boldsymbol{C}^e}\psi$$
in Cleja-Ţigoiu (2014).
- The macro stress momentum with respect to the reference configuration is given in terms of $\partial_{\nabla\boldsymbol{C}^e}\psi$ via the relation $(9.57)_2$, while in Cleja-Ţigoiu (2014) the macro stress momentum is not a third order symmetric tensor and it depends on $\partial_{\mathscr{A}^e}\psi$, and $\partial_{\boldsymbol{S}^e}\psi$. We used the notations
$$\mathscr{A}^e \equiv \overset{(e)}{\mathscr{A}}_{\mathscr{K}}$$
and
$$\boldsymbol{S}^e \equiv \boldsymbol{S}^e_{\mathscr{K}}.$$

The evolution equations for plastic distortion, $\boldsymbol{F}^p$, and disclination tensor, $\widetilde{\boldsymbol{\Lambda}}$, were provided to be compatible with the reduced dissipation inequality. We derived also peculiar evolution equations for $\boldsymbol{H}^p$ and $\widetilde{\boldsymbol{\Lambda}}$ within the small distortions framework.

The evolution equations provided here for $\underline{\beta} = 1$ are similar with those derived in Cleja-Ţigoiu et al (2016) for the small strains, apart from the terms induced by the elastic effect, namely the first three terms involved in right-hand side of the evolution equation $(9.84)_1$. In Cleja-Ţigoiu (2014) the micro stress associated with the disclination mechanism remained undefined, and the disclination tensor, $\widetilde{\boldsymbol{\Lambda}}$, was viewed as internal variable, see Maugin (2006). The presence of the gradient $\nabla_{\mathscr{K}}\widetilde{\boldsymbol{\Lambda}}$ in the free energy function allowed us to define the micro stress $\boldsymbol{\Upsilon}^\lambda$, introduced in (9.48) via (9.70). Thus both evolution equations are viscoplastic and diffusion type.

Acknowledgements This work was supported by a grant of Ministery of Research and Innovation, CNCS-UEFISCDI, project number PN-III-P4-ID-PCE-2016-0083 within PNCDI III.

References

Arsenlis A, Parks DM (1999) Crystallographic aspects of geometrically-necessary and statistically-stored dislocation density. Acta Materialia 47(5):1597 – 1611

Bilby BA (1960) Continuous distribution of dislocations. In: Sneddon IN, Hill R (eds) Progress in Solid Mechanics, North-Holland, Amsterdam, pp 329–398

Clayton JD, McDowell DL, Bammann DJ (2006) Modeling dislocations and disclinations with finite micropolar elastoplasticity. International Journal of Plasticity 22(2):210 – 256

Cleja-Ţigoiu S, Ţigoiu V (2011) Strain gradient effect in finite elasto-plastic damaged materials. International Journal of Damage Mechanics 20(4):484–514

Cleja-Ţigoiu S, Paşcan R, Ţigoiu V (2016) Interplay between continuous dislocations and disclinations in elasto-plasticity. International Journal of Plasticity 79:68 – 110

Cleja-Ţigoiu S (2007) Material forces in finite elasto-plasticity with continuously distributed dislocations. International Journal of Fracture 147(1):67–81

Cleja-Ţigoiu S (2010) Elasto-plastic materials with lattice defects modeled by second order deformations with non-zero curvature. International Journal of Fracture 166(1):61–75

Cleja-Ţigoiu S (2014) Dislocations and disclinations: continuously distributed defects in elasto-plastic crystalline materials. Archive of Applied Mechanics 84(9):1293–1306

Cleja-Ţigoiu S (2017) Finite elasto-plastic models for lattice defects in crystalline materials. In: dell'Isola F, Sofonea M, Steigmann D (eds) Mathematical Modelling in Solid Mechanics, Springer, Singapore, Advanced Structured Materials, vol 69, pp 43–57

Cleja-Tigoiu S, Maugin GA (2000) Eshelby's stress tensors in finite elastoplasticity. Acta Mechanica 139(1):231–249

de Wit R (1970) Linear theory of static disclinations. In: Simmons JA, de Wit R, Bullough R (eds) Fundamental Aspects of Dislocation Theory, Nat. Bur. Stand. (U.S.) Spec. Publ., vol 317, I, pp 651–673

de Wit R (1973a) Theory of disclinations: II. Continuous and discrete disclinations in anisotropic elasticity. J Res Nat Bur Stand - A Phys Chem 77A(1):49 – 100

de Wit R (1973b) Theory of disclinations: III. Continuous and discrete disclinations in isotropic elasticity. J Res Nat Bur Stand - A Phys Chem 77A(3):359 – 368

de Wit R (1973c) Theory of disclinations: IV. Straight disclinations. J Res Nat Bur Stand - A Phys Chem 77A(5):607 – 658

de Wit R (1981) A view of the relation between the continuum theory of lattice defects and non-euclidean geometry in the linear approximation. International Journal of Engineering Science 19(12):1475 – 1506
Eringen AC (2002) Nonlocal Continuum Field Theories. Springer, New York
Fleck NA, Muller GM, Ashby MF, Hutchinson JW (1994) Strain gradient plasticity: Theory and experiment. Acta Metallurgica et Materialia 42(2):475 – 487
Fressengeas C, Taupin V, Capolungo L (2011) An elasto-plastic theory of dislocation and disclination fields. International Journal of Solids and Structures 48(25):3499 – 3509
Gurtin ME (2002) A gradient theory of single-crystal viscoplasticity that accounts for geometrically necessary dislocations. Journal of the Mechanics and Physics of Solids 50(1):5 – 32
Gurtin ME, Fried E, Anand L (2010) The Mechanics and Thermodynamics of Continua. University Press, Cambridge
Kossecka E, de Wit R (1977) Disclination kinematics. Archives of Mechanics 29:633 – 650
Kröner E (1990) The differential geometry of elementary point and line defects in Bravais crystals. International Journal of Theoretical Physics 29(11):1219–1237
Kröner E (1992) The internal mechanical state of solids with defects. International Journal of Solids and Structures 29(14):1849–1857
Lazar M, Maugin GA (2004a) Defects in gradient micropolar elasticity. I: screw dislocation. Journal of the Mechanics and Physics of Solids 52(10):2263 – 2284
Lazar M, Maugin GA (2004b) Defects in gradient micropolar elasticity. II: edge dislocation and wedge disclination. Journal of the Mechanics and Physics of Solids 52(10):2285 – 2307
Lazar M, Maugin GA, Aifantis EC (2006) On a theory of nonlocal elasticity of bi-Helmholtz type and some applications. International Journal of Solids and Structures 43(6):1404 – 1421
Maugin GA (1994) Eshelby stress in elastoplasticity and ductile fracture. International Journal of Plasticity 10(4):393 – 408
Maugin GA (2006) Internal variables and dissipative structures. J Non-Equilib Thermodyn 15:173 – 192
Minagawa S (1979) A non-Riemannian geometrica theory of imperfections in a Cosserat continuum. Arch Mech 31(6):783 – 792
Schouten J (1954) Ricci Calculus. Springer, Berlink
Teodosiu C (1982) Elastic Models of Crystal Defects. Springer, Berlin, Heidelberg, New York
Yavari A, Goriely A (2012) Riemann–Cartan geometry of nonlinear dislocation mechanics. Archive for Rational Mechanics and Analysis 205(1):59–118
Yavari A, Goriely A (2013) Riemann-Cartan geometry of nonlinear disclination mechanics. Mathematics and Mechanics of Solids 18(1):91–102

Chapter 10
Viscoelastic effective properties for composites with rectangular cross-section fibers using the asymptotic homogenization method

Oscar L. Cruz-González, Reinaldo Rodríguez-Ramos, José A. Otero, Julián Bravo-Castillero, Raúl Guinovart-Díaz, Raúl Martínez-Rosado, Federico J. Sabina, Serge Dumont, Frederic Lebon, and Igor Sevostianov

Abstract The present work deals with the estimation of the linear viscoelastic effective properties for composites with periodic structure and rectangular cross-section fibers, using the two-scale asymptotic homogenization method (AHM). As a particular case, the effective properties for a layered medium with transversely isotropic

---

Oscar L. Cruz-González
Facultad de Ciencias Técnicas, Departamento de Matemática, Universidad de Matanzas, Varadero road, Km. 2 1/2, Matanzas, Cuba
e-mail: oscar.lcg93@gmail.com

Reinaldo Rodríguez-Ramos · Julián Bravo-Castillero · Raúl Guinovart-Díaz
Facultad de Matemática y Computación, Universidad de La Habana, San Lázaro y L, Vedado, La Habana, CP 10400, Cuba
e-mail: reinaldo@matcom.uh.cu, jbravo@matcom.uh.cu, guino@matcom.uh.cu

José A. Otero · Raúl Martínez-Rosado
Tecnologico de Monterrey, Escuela de Ingenieria y Ciencias, Atizapan de Zaragoza, Estado de México, México,
e-mail: j.a.otero@itesm.mx, rrosado@itesm.mx

Federico J. Sabina
Instituto de Investigaciones en Matemáticas Aplicadas y en Sistemas, Universidad Nacional Autónoma de México, Apartado Postal 20-126, 01000 CDMX, México
e-mail: fjs@mym.iimas.unam.mx

Serge Dumont
Université de Nîmes, Institut de Mathématiques Alexander Grothendieck, CNRS UMR 5149, CC.051, Pl. E. Bataillon, 34 095 Montpellier Cedex 5, France
e-mail: serge.dumont@unimes.fr

Frederic Lebon
Aix-Marseille Univ., CNRS, Centrale Marseille, LMA, 4 Impasse Nikola Tesla, CS 40006, 13453 Marseille Cedex 13, France
e-mail: lebon@lma.cnrs-mrs.fr

Igor Sevostianov
Department of Mechanical and Aerospace Engineering, New Mexico State University, Las Cruces, NM 88003, USA
e-mail: igor@nmsu.edu

© Springer International Publishing AG, part of Springer Nature 2018

H. Altenbach et al. (eds.), *Generalized Models and Non-classical Approaches in Complex Materials 1*, Advanced Structured Materials 89,
https://doi.org/10.1007/978-3-319-72440-9_10

properties are obtained. Two times the homogenization method, in different directions, according to the geometrical configuration of the composite material is applied for deriving the analytical expressions of the viscoelastic effective properties for a composite material with rectangular cross-section fibers, periodically distributed along one axis. In addition to that, models with different creep kernels, in particular, the Rabotnov's kernel are analyzed. Finally, the numerical computation of the effective viscoelastic properties is developed for the analysis of the results. Moreover, a numerical algorithm using FEM is developed in the present work. Comparisons with other approaches are given as a validation of the present model.

10.1 Introduction

Many materials like metals, polymers, and ceramics especially at high temperature exhibit time dependent behavior. For linear non ageing viscoelastic materials, the convolution product in the Stieltjes space is often used for modeling of the time dependent constitutive law. The overall behavior of viscoelastic composite materials is investigated by many researchers based on homogenization techniques. Classically, the investigation of effective properties of non ageing linear viscoelastic composites are mainly based on the correspondence principle and Laplace transform. This approach changes the convolution constitutive law describing the linear non ageing viscoelastic behavior into a fictitious linear elastic one in the Laplace domain. Linear homogenization method can then be used to drive the effective properties in the frequency domain. The time dependent effective properties can be obtained by performing numerical inversion of their Laplace transform. Based on this methodological approach, many researchers investigated the overall non ageing behavior of linear viscoelastic composites.

Important results obtained for elastic materials can be translated to linear non ageing viscoelastic materials using the correspondence principle (Hashin, 1965, 1970b; Christensen, 1969; Schapery, 1967; Wang and Weng, 1992; Kachanov, 1992; Lahellec and Suquet, 2007; Dormieux et al, 2006). It can be shown that in the Laplace-Carson (LC) transform space, the writing of the constitutive behavior is the same as in elasticity, the stiffness tensors being nevertheless functions of the LC transform variable (denoted thereafter LC variable). However, assuming that the solution in LC space is obtained, there are still considerable difficulties in obtaining the inverse LC transform to find the corresponding results in time space (Lévesque et al, 2007; Le et al, 2007).

Homogenization of viscoelastic composites can be performed analytically by solving the homogenization equations in Laplace-Carson space with the so-called correspondence principle (Hashin, 1966, 1970a; Laws and McLaughlin, 1978; Beurthey and Zaoui, 2000; Lévesque et al, 2007). The time domain solution is usually obtained with inversion algorithms,such as the collocation method (Schapery, 1964; Lévesque et al, 2007). The recently developed method of Lévesque et al (2007) is quite accurate, leads to thermodynamically admissible materials but requires a moderate computation time. Another approach consists of a time-integration approach, relying on variational principles (Lahellec and Suquet, 2007). While it avoids Laplace-Carson transforms and solves the viscoelastic problem directly in the time domain, its numerical implementation is challenging. Finally, another technique relies on a direct quasi-elastic approximation in the Laplace-Carson space (Brenner et al, 2002). This method is computationally-efficient but at the expense of accuracy. Finally, Ricaud and Masson (2009) have shown that the Laplace-Carson scheme, making use of the Prony series approximation, is equivalent to a time-integration scheme of the internal variable formulation, establishing a link between the two approaches.

This work is devoted to Gerard A. Maugin, who apported significant contributions in the micro-mechanic area. In particular, the authors had the pleasant opportunity to collaborate with him in the piezoelectric composites area (Berger et al, 2003, 2006; Otero et al, 2003). In the present contribution, making use of the two-scale asymptotic homogenization method (AHM) and correspondence principle, an equilibrium viscoelastic heterogeneous problem is solved similar to the elastic case, but in Laplace-Carson space. The overall behavior of composite non-ageing constituents is investigated with the estimation of the linear viscoelastic effective properties. As a particular case, the explicit formulae to predict the effective relaxation modulus for a two-layered medium with transversely isotropic properties are obtained. The numerical algorithm proposed by Hollenbeck (1998), to invert the Laplace transform, is using to calculate the properties to the homogenized composites in the time domain. Different creep kernels are analyzed, in particular, a time-dependent function considerated by Dischinger's model (see Maghous and Creus, 2003) and the fraction-exponential function or Rabotnov's kernel (see Sevostianov et al, 2015, 2016) with the aim to make several comparison and to validate the numerical results. A model of finite element method is implemented and comparisons with the obtained results using finite element method is also realized. An alternative approach, using double homogenization scheme, is considered to estimate the linear viscoelastic effective properties for composites with periodic structure and rectangular cross-section fibers. This approach is inspired in the geometric design of the material.

10.2 Statement of the Viscoelastic Heterogeneous Problem

A heterogeneous material with periodic structure exhibiting a linear viscoelastic behavior is considered. It occupies a region Σ in $\mathbb{R}^3$. Cartesian coordinate system $\boldsymbol{x}\,(x_i)$ is used where the stress tensor, external force field, surface force field, displacement field and outer unit normal to the boundary $\partial\Sigma$ of Σ (see Fig. 10.1) are denoted by $\boldsymbol{\sigma}\,(\sigma_{ij})$, $\boldsymbol{f}\,(f_i)$, $\boldsymbol{s}^0\,(s_i^0)$, $\boldsymbol{u}\,(u_i)$ and $\boldsymbol{n}\,(n_i)$, respectively. The equilibrium equation under the action of external force field is written as (see Persson et al, 1993)

$$\operatorname{div} \boldsymbol{\sigma}(\boldsymbol{x},\mathrm{t}) + \boldsymbol{f}(\boldsymbol{x}) = \boldsymbol{0}, \quad \text{in} \quad \Sigma. \tag{10.1}$$

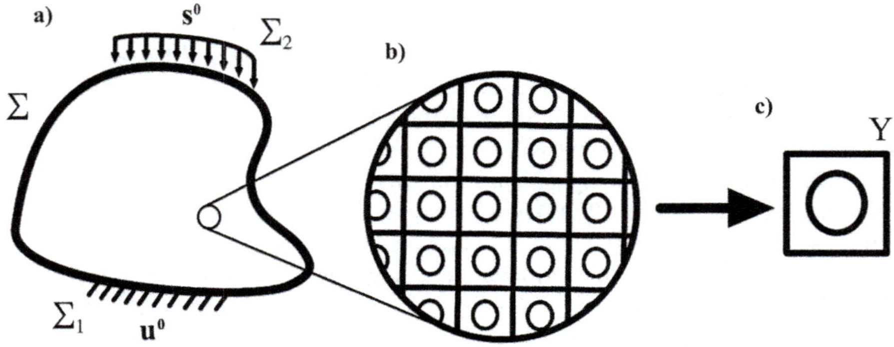

Fig. 10.1: Scales of the heterogeneous structure. a) Macroscopic heterogeneous structure, b) periodic microstructure and c) periodic cell.

The corresponding boundary conditions associated to (10.1) are

$$\boldsymbol{u}(\boldsymbol{x},t) = \boldsymbol{u}^0, \quad \text{on} \quad \Sigma_1, \tag{10.2}$$

$$\boldsymbol{\sigma}(\boldsymbol{x},t)\cdot\boldsymbol{n} = \boldsymbol{s}^0, \quad \text{on} \quad \Sigma_2. \tag{10.3}$$

The initial condition for Eq. (10.1) is

$$\boldsymbol{u}(\boldsymbol{x},t) = \boldsymbol{0}, \quad \text{in} \quad \Sigma \times \{0\}, \tag{10.4}$$

where $\Sigma_1 \cup \Sigma_2 = \partial\Sigma$ and $\Sigma_1 \cap \Sigma_2 = \varnothing$, (see Fig 10.1).

The stress and strain fields are linearly related to the constitutive law (see viscoelastic theory in Christensen, 1971; Pipkin, 1986)

$$\boldsymbol{\sigma}(\boldsymbol{x},t) = \int_0^t \boldsymbol{R}(\boldsymbol{x},t-\tau) : \frac{\partial \boldsymbol{\varepsilon}\big(\boldsymbol{u}(\boldsymbol{x},\tau)\big)}{\partial \tau} d\tau, \tag{10.5}$$

where $\boldsymbol{R}$ (R_{ijkl}) and $\boldsymbol{\varepsilon}$ (ε_{kl}) denote a fourth rank tensor (the creep kernel or relaxation modulus) and the Cauchy strain tensors, respectively.

The Eq. (10.5) can be expressed in a simplified form

$$\boldsymbol{\sigma}(\boldsymbol{x},t) = \boldsymbol{R}(\boldsymbol{x},t) \circ \boldsymbol{\varepsilon}\big(\boldsymbol{u}(\boldsymbol{x},t)\big), \tag{10.6}$$

where $\circ$ is indicating the convolution integral (Zhang and Ostoja-Starzewski, 2015).

The following relationship is satisfied for small displacements

$$\varepsilon_{kl}\big(\boldsymbol{u}(\boldsymbol{x},t)\big) = \frac{1}{2}\left(\frac{\partial u_k(\boldsymbol{x},t)}{\partial x_l} + \frac{\partial u_l(\boldsymbol{x},t)}{\partial x_k}\right). \tag{10.7}$$

Replacing (10.6) into (10.1) and using (10.2) - (10.4), the mathematical statement for the equilibrium viscoelastic heterogeneous problem is obtained

$$-\text{div}\left(\boldsymbol{R}(\boldsymbol{x},\text{t})\circ\boldsymbol{\varepsilon}\left(\boldsymbol{u}(\boldsymbol{x},\text{t})\right)\right)=\boldsymbol{f}(\boldsymbol{x}), \tag{10.8}$$

$$\boldsymbol{u}(\boldsymbol{x},t)=\boldsymbol{u}^0,\quad \forall\boldsymbol{x}\in\Sigma_1\wedge\forall t\in\mathbb{R}, \tag{10.9}$$

$$\boldsymbol{R}(\boldsymbol{x},t)\circ\boldsymbol{\varepsilon}\left(\boldsymbol{u}(\boldsymbol{x},t)\right)\cdot\boldsymbol{n}=\boldsymbol{s}^0,\quad \forall\boldsymbol{x}\in\Sigma_2\wedge\forall t\in\mathbb{R}, \tag{10.10}$$

$$\boldsymbol{u}(\boldsymbol{x},0)=\boldsymbol{0},\quad \forall\boldsymbol{x}\in\Sigma. \tag{10.11}$$

The following *additional conditions* are considered:

1. $\boldsymbol{x}$ (x_i) is called the global coordinate. Also, is introduced the local or fast scale coordinate $\boldsymbol{y}$ (y_i), where $\boldsymbol{y}=\xi^{-1}\boldsymbol{x}$. The parameter ξ is the fine mesh size of the cell structure (see Persson et al, 1993).
2. The relaxation modulus fulfills the memory principle $\boldsymbol{R}(\boldsymbol{x},t)\to 0$ as $t\to\infty$ (see Sevostianov et al, 2016). In particular, we have $\boldsymbol{R}^{\xi}(\boldsymbol{x},t)=\boldsymbol{R}(\boldsymbol{x}/\xi,t)=\boldsymbol{R}(\boldsymbol{y},t)$ is $Y-$periodic related to the fast variable $\boldsymbol{y}$.
3. $\boldsymbol{R}(\boldsymbol{y},t)\in C^{\infty}(\mathbb{R}^3\times\mathbb{R})$.
4. $\epsilon\alpha,\beta,t_0$ such that $0\le\alpha\le\boldsymbol{R}(\boldsymbol{y},t_0)\le\beta\le\infty$ $\forall\boldsymbol{y}\in\mathbb{R}^3$ $(\xi\to 0)$.
5. $\boldsymbol{f}(\boldsymbol{x})\in C^{\infty}(\Sigma)$.

The non-aging linear viscoelastic problem corresponds to elastic problems thanks to the Laplace-Carson transform (Lavergne et al, 2016). The transformed of a function $\boldsymbol{g}(\boldsymbol{x},t)$ is defined by

$$L_C[\boldsymbol{g}(\boldsymbol{x},t)]=\widehat{\boldsymbol{g}}(\boldsymbol{x},p)=p\int_0^{\infty}e^{-pt}\boldsymbol{g}(\boldsymbol{x},t)dt.$$

From now on, the functions with the symbol $\widehat{()}$ depending on the parameter p denotes the Laplace-Carson space.

Considering the convolution theorem (see Sokolnikoff and Redheffer, 1968), the equilibrium viscoelastic heterogeneous problem (10.8)-(10.11) becomes,

$$P\,\widehat{\boldsymbol{u}}(\boldsymbol{x},p)=\boldsymbol{f}(\boldsymbol{x}), \tag{10.12}$$

$$\widehat{\boldsymbol{u}}(\boldsymbol{x},p)=\boldsymbol{u}^0,\quad \forall\boldsymbol{x}\in\Sigma_1\wedge\forall p\in[0,\infty], \tag{10.13}$$

$$\widehat{\boldsymbol{R}}(\boldsymbol{x},p):\boldsymbol{\varepsilon}\left(\widehat{\boldsymbol{u}}(\boldsymbol{x},p)\right)\cdot\boldsymbol{n}=\boldsymbol{s}^0,\quad \forall\boldsymbol{x}\in\Sigma_2\wedge\forall p\in[0,\infty], \tag{10.14}$$

$$\widehat{\boldsymbol{u}}(\boldsymbol{x},0)=\boldsymbol{0},\quad \forall\boldsymbol{x}\in\Sigma, \tag{10.15}$$

where

$$P\,\widehat{\boldsymbol{u}}(\boldsymbol{x},p)=-\text{div}\left(\widehat{\boldsymbol{R}}(\boldsymbol{x},\text{p}):\boldsymbol{\varepsilon}\left(\widehat{\boldsymbol{u}}(\boldsymbol{x},\text{p})\right)\right).$$

10.3 Two-Scale Asymptotic Homogenization Method to Solve the Heterogeneous Problem

In this section the two-scales homogenization technique is used to obtain the basic equations and the effective characteristic of the composite. *A formal asymptotic solution* (see definition in Bakhvalov and Panasenko, 1989) for finding the solution of the problem is proposed. The asymptotic according to Bakhvalov and Panasenko (1989) for solving (10.12)-(10.15) is given in the form

$$\widehat{\boldsymbol{u}}(\boldsymbol{x},\xi,p)=\sum_{a=0}^{\infty}\xi^{a}\,\widehat{\boldsymbol{u}}^{(a)}(\boldsymbol{x},\boldsymbol{y},p), \tag{10.16}$$

where $\widehat{\boldsymbol{u}}^{(a)}$ ($\widehat{u}_i^{(a)}$) is $Y-$periodic related to the variable $\boldsymbol{y}$ $\forall a$, $\forall \boldsymbol{x}\in\Sigma$, $\forall p\in[0,\infty]$ and $\widehat{\boldsymbol{u}}^{(a)}(\boldsymbol{x},\boldsymbol{y},p)\in C^{\infty}(\Sigma\times\mathbb{R}^3\times[0,\infty])$.

Besides, the expressions e_{klx} and e_{kly} are defined as follows (see Persson et al, 1993)

$$\varepsilon_{klx}\big(\widehat{\boldsymbol{\Phi}}(\boldsymbol{x},p)\big)=\frac{1}{2}\Big(\frac{\partial\widehat{\Phi}_k(\boldsymbol{x},p)}{\partial x_l}+\frac{\partial\widehat{\Phi}_l(\boldsymbol{x},p)}{\partial x_k}\Big), \tag{10.17}$$

$$\varepsilon_{kly}\big(\widehat{\boldsymbol{\Phi}}(\boldsymbol{y},p)\big)=\frac{1}{2}\Big(\frac{\partial\widehat{\Phi}_k(\boldsymbol{y},p)}{\partial y_l}+\frac{\partial\widehat{\Phi}_l(\boldsymbol{y},p)}{\partial y_k}\Big). \tag{10.18}$$

According to the chain rule and from (10.7), (10.16) - (1.18) can be obtained

$$\varepsilon_{kl}\big(\widehat{\boldsymbol{u}}^{(a)}(\boldsymbol{x},\boldsymbol{x}/\xi,p)\big)=\varepsilon_{klx}\big(\widehat{\boldsymbol{u}}^{(a)}(\boldsymbol{x},\boldsymbol{y},p)\big)+\xi^{-1}\,\varepsilon_{kly}\big(\widehat{\boldsymbol{u}}^{(a)}(\boldsymbol{x},\boldsymbol{y},p)\big). \tag{10.19}$$

Now, the objective is to find the expression of the coefficients such that the following equality is satisfied (Bakhvalov and Panasenko, 1989)

$$P^{(\xi)}\,\widehat{\boldsymbol{u}}(\boldsymbol{x},\xi,p)-\boldsymbol{f}(\boldsymbol{x})=\mathbf{O}(\xi). \tag{10.20}$$

In order to guarantee (10.20), the following operator is defined

$$L_{\alpha\beta}(\bullet):=-\frac{\partial}{\partial\alpha_j}\left(\widehat{R}_{ijkl}(\boldsymbol{y},p)\varepsilon_{kl\beta}(\bullet)\right), \tag{10.21}$$

for $\alpha,\beta=\boldsymbol{x},\boldsymbol{y}$ indistinctly and taken into account the rule for derivation

$$\frac{\partial(\bullet)}{\partial x_j}\equiv\frac{\partial(\bullet)}{\partial x_j}+\frac{1}{\xi}\frac{\partial(\bullet)}{\partial y_j}. \tag{10.22}$$

Then, replacing (10.16), (10.19) - (10.22) into (10.20), applying some simplifications and grouping in powers of ϵ, the following sequence of problems are obtained

$$\xi^{-2}\to L_{yy}\widehat{\boldsymbol{u}}^{(0)}(\boldsymbol{x},\boldsymbol{y},p)=\mathbf{0}, \tag{10.23}$$

$$\xi^{-1}\to L_{xy}\widehat{\boldsymbol{u}}^{(0)}(\boldsymbol{x},\boldsymbol{y},p)+L_{yx}\widehat{\boldsymbol{u}}^{(0)}(\boldsymbol{x},\boldsymbol{y},p)+L_{yy}\widehat{\boldsymbol{u}}^{(1)}(\boldsymbol{x},\boldsymbol{y},p)=\mathbf{0}, \tag{10.24}$$

$$\begin{aligned}\xi^{0}\to\ & L_{xx}\widehat{\boldsymbol{u}}^{(0)}(\boldsymbol{x},\boldsymbol{y},p)+L_{xy}\widehat{\boldsymbol{u}}^{(1)}(\boldsymbol{x},\boldsymbol{y},p)+L_{yx}\widehat{\boldsymbol{u}}^{(1)}(\boldsymbol{x},\boldsymbol{y},p)\\ &+L_{yy}\widehat{\boldsymbol{u}}^{(2)}(\boldsymbol{x},\boldsymbol{y},p)-\boldsymbol{f}(\boldsymbol{x})=\mathbf{0}.\end{aligned} \tag{10.25}$$

The problems (10.23) - (10.25) can be solved in a recursive form.

Considering the asymptotic (10.16) as an approximation of the exact solution of the original problem (10.12)-(10.15) and replacing (10.16) into the boundary conditions (10.13) and (10.14)

$$\widehat{\boldsymbol{u}}^{(0)}(\boldsymbol{x},\boldsymbol{y},p) = \boldsymbol{u}^0, \quad \forall \boldsymbol{x} \in \Sigma_1 \wedge \forall p \in [0,\infty], \tag{10.26}$$

$$\widehat{\boldsymbol{R}}(\boldsymbol{x},p) : \boldsymbol{\varepsilon}(\widehat{\boldsymbol{u}}^{(0)}(\boldsymbol{x},\boldsymbol{y},p))\cdot\boldsymbol{n} = \boldsymbol{s}^0, \quad \forall \boldsymbol{x} \in \Sigma_2 \wedge \forall p \in [0,\infty]. \tag{10.27}$$

Besides, the remaining terms are assigned by

$$\widehat{\boldsymbol{u}}^{(a)}(\boldsymbol{x},\boldsymbol{y},p) = \boldsymbol{0}, \quad \forall \boldsymbol{x} \in \Sigma_1 \wedge \forall p \in [0,\infty] \wedge \forall a > 0, \tag{10.28}$$

$$\widehat{\boldsymbol{R}}(\boldsymbol{x},p) : \boldsymbol{\varepsilon}(\widehat{\boldsymbol{u}}^{(a)}(\boldsymbol{x},\boldsymbol{y},p))\cdot\boldsymbol{n} = \boldsymbol{0}, \quad \forall \boldsymbol{x} \in \Sigma_2 \wedge \forall p \in [0,\infty] \wedge \forall a > 0 \tag{10.29}$$

The conditions (10.26)-(10.29) are justified by *construction* because of the formal asymptotic solution is considered.

In order to solve (10.23)-(10.25) the following lemma is required. The proof of this lemma is given in Sect. 4.3 of Persson et al (1993).

Lemma 10.1. *Let* $\boldsymbol{F}$ (F_i) *be square integrable function over* Y *and consider the boundary value problem*

$$L_{yy}\boldsymbol{\Phi} = \boldsymbol{F},$$

where $\boldsymbol{\Phi}$ *is* Y*-periodic. Then the following conditions hold,*

(i) A Y*-periodic solution* $\boldsymbol{\Phi}$ *exists if and only if* $\langle \boldsymbol{F} \rangle = 0$.
(ii) If a Y*-periodic solution* $\boldsymbol{\Phi}$ *exists, then it is unique up to a constant vector* $\boldsymbol{c}$.

The notation $\langle \bullet \rangle$ defines the average over the Y-cell, i.e.,

$$\langle \boldsymbol{F} \rangle := \frac{1}{|Y|}\int_Y \boldsymbol{F} dy,$$

where $|Y|$ is the measure of Y. Subsequently, the main results for each power of ϵ are summarized.

10.3.1 Contribution of the Level ξ^{-2} Problem

The problem (10.23) and (10.26)-(10.27) is stated as follows

$$L_{yy}\widehat{\boldsymbol{u}}^{(0)}(\boldsymbol{x},\boldsymbol{y},p) = \boldsymbol{0}, \tag{10.30}$$

$$\widehat{\boldsymbol{u}}^{(0)}(\boldsymbol{x},\boldsymbol{y},p) = \boldsymbol{u}^0, \quad \forall \boldsymbol{x} \in \Sigma_1 \wedge \forall p \in [0,\infty], \tag{10.31}$$

$$\widehat{\boldsymbol{R}}(\boldsymbol{x},p) : \boldsymbol{\varepsilon}(\widehat{\boldsymbol{u}}^{(0)}(\boldsymbol{x},\boldsymbol{y},p))\cdot\boldsymbol{n} = \boldsymbol{s}^0, \quad \forall \boldsymbol{x} \in \Sigma_2 \wedge \forall p \in [0,\infty], \tag{10.32}$$

$$\widehat{\boldsymbol{u}}^{(0)}(\boldsymbol{x},\boldsymbol{y},0) = \boldsymbol{0}, \qquad \forall \boldsymbol{x} \in \Sigma. \tag{10.33}$$

Equation (10.30) has the trivial solution $\widehat{\boldsymbol{u}}^{(0)}(\boldsymbol{x},\boldsymbol{y},p) \equiv \boldsymbol{0}$. Thus, Lemma 10.1 indicates that $\widehat{\boldsymbol{u}}^{(0)}(\boldsymbol{x},\boldsymbol{y},p)$ is a solution of (10.30) if and only if it is a constant with respect to the variable $\boldsymbol{y}$. It implies that,

$$\widehat{\boldsymbol{u}}^{(0)}(\boldsymbol{x},\boldsymbol{y},p) = \widehat{\boldsymbol{v}}(\boldsymbol{x},p), \tag{10.34}$$

where $\widehat{\boldsymbol{v}}(\boldsymbol{x},t)$ is a infinitely differentiable function (see Persson et al, 1993).

Now, from (10.31) - (10.33) are obtained

$$\widehat{\boldsymbol{u}}^{(0)}(\boldsymbol{x},\boldsymbol{y},p) = \widehat{\boldsymbol{v}}(\boldsymbol{x},p) = \boldsymbol{u}^0, \quad \forall \boldsymbol{x} \in \Sigma_1 \wedge \forall p \in [0,\infty], \tag{10.35}$$

$$\widehat{\boldsymbol{R}}(\boldsymbol{x},p) : \boldsymbol{\varepsilon}(\widehat{\boldsymbol{v}}(\boldsymbol{x},p))\cdot\boldsymbol{n} = \boldsymbol{s}^0, \ \forall \boldsymbol{x} \in \Sigma_2 \wedge \forall p \in [0,\infty], \tag{10.36}$$

$$\widehat{\boldsymbol{u}}^{(0)}(\boldsymbol{x},\boldsymbol{y},0) = \widehat{\boldsymbol{v}}(\boldsymbol{x},0) = \boldsymbol{0}, \quad \forall \boldsymbol{x} \in \Sigma. \tag{10.37}$$

10.3.2 Contribution of the Level ξ^{-1} Problem

According to (10.34), the first term of (10.24) is zero, $L_{xy}\widehat{\boldsymbol{u}}^{(0)}(\boldsymbol{x},\boldsymbol{y},p) = L_{xy}\widehat{\boldsymbol{v}}(\boldsymbol{x},p) = \boldsymbol{0}$. In this case, the problems (10.24) and (10.28)-(10.29) are reduced to

$$L_{yy}\widehat{\boldsymbol{u}}^{(1)}(\boldsymbol{x},\boldsymbol{y},p) = -L_{yx}\widehat{\boldsymbol{u}}^{(0)}(\boldsymbol{x},\boldsymbol{y},p), \tag{10.38}$$

$$\widehat{\boldsymbol{u}}^{(1)}(\boldsymbol{x},\boldsymbol{y},p) = \boldsymbol{0}, \quad \forall \boldsymbol{x} \in \Sigma_1 \wedge \forall p \in [0,\infty], \tag{10.39}$$

$$\widehat{\boldsymbol{R}}(\boldsymbol{x},p) : \boldsymbol{\varepsilon}(\widehat{\boldsymbol{u}}^{(1)}(\boldsymbol{x},\boldsymbol{y},p))\cdot\boldsymbol{n} = \boldsymbol{0}, \quad \forall \boldsymbol{x} \in \Sigma_2 \wedge \forall p \in [0,\infty], \tag{10.40}$$

$$\widehat{\boldsymbol{u}}^{(1)}(\boldsymbol{x},\boldsymbol{y},0) = \boldsymbol{0}, \qquad \forall \boldsymbol{x} \in \Sigma. \tag{10.41}$$

Applying the Lemma 10.1 on (10.38), having into account (10.34), the divergence theorem and the $Y-$periodicity condition of $\widehat{\boldsymbol{R}}(\boldsymbol{y},p)$ (see the *additional condition* (2)), the following result is satisfied

$$\left\langle -L_{yx}\widehat{\boldsymbol{u}}^{(0)}(\boldsymbol{x},\boldsymbol{y},p)\right\rangle = 0.$$

Consequently, the existence of one solution for the problem (10.38) is guaranteed. Now, using separation of variables and the condition (ii) of Lemma 10.1, a general solution of (10.38) can be given by

$$\widehat{\boldsymbol{u}}^{(1)}(\boldsymbol{x},\boldsymbol{y},p) = \widehat{\boldsymbol{N}}^{rs}(\boldsymbol{y},p)\varepsilon_{rsx}(\widehat{\boldsymbol{v}}(\boldsymbol{x},p)) + \widehat{\boldsymbol{w}}(\boldsymbol{x},p), \tag{10.42}$$

where $\widehat{\boldsymbol{N}}^{rs}$ ($\widehat{N}_i^{rs}$) is called *the local function* and $\widehat{\boldsymbol{w}}$ ($\widehat{w}_i$) is infinitely differentiable function.

Finally, replacing (10.34) and (10.42) into (10.38) and after some simplifications the *cell problem* is obtained

$$-\frac{\partial}{\partial y_j}\left(\widehat{R}_{ijkl}(\boldsymbol{y},p)\varepsilon_{kly}(\widehat{\boldsymbol{N}}^{rs}(\boldsymbol{y},p))\right)=\frac{\partial}{\partial y_j}\left(\widehat{R}_{ijrs}(\boldsymbol{y},p)\right), \tag{10.43}$$

where, employing the Lemma 10.1 on (10.43), $\widehat{\boldsymbol{N}}^{rs}$ is Y-periodic function.

Having into account (10.35) - (10.37), (10.39) - (10.42), the boundary conditions and initial condition to the *cell problem* are written in term of the local function

$$\widehat{\boldsymbol{N}}^{rs}(\boldsymbol{y},p)=\mathbf{0},\quad \forall \boldsymbol{y}\in Y\ \wedge\ \forall p\in[0,\infty], \tag{10.44}$$

$$\widehat{\boldsymbol{R}}(\boldsymbol{x},p):\boldsymbol{\varepsilon}(\widehat{\boldsymbol{N}}^{rs}(\boldsymbol{y},p))\cdot\boldsymbol{n}=\mathbf{0},\quad \forall \boldsymbol{y}\in Y\ \wedge\ \forall p\in[0,\infty], \tag{10.45}$$

$$\widehat{\boldsymbol{N}}^{rs}(\boldsymbol{y},0)=\mathbf{0},\qquad \forall \boldsymbol{y}\in\mathbb{R}^3. \tag{10.46}$$

10.3.3 Contribution of the Level ξ^0 Problem

The problem (10.25) and (10.28)-(10.29) are given as follows

$$L_{yy}\widehat{\boldsymbol{u}}^{(2)}(\boldsymbol{x},\boldsymbol{y},p)=\boldsymbol{f}(\boldsymbol{x})-L_{xx}\widehat{\boldsymbol{u}}^{(0)}(\boldsymbol{x},\boldsymbol{y},p)-L_{xy}\widehat{\boldsymbol{u}}^{(1)}(\boldsymbol{x},\boldsymbol{y},p)-L_{yx}\widehat{\boldsymbol{u}}^{(1)}(\boldsymbol{x},\boldsymbol{y},p), \tag{10.47}$$

$$\widehat{\boldsymbol{u}}^{(2)}(\boldsymbol{x},\boldsymbol{y},p)=\mathbf{0},\quad \forall \boldsymbol{x}\in\Sigma_1\ \wedge\ \forall p\in[0,\infty], \tag{10.48}$$

$$\widehat{\boldsymbol{R}}(\boldsymbol{x},p):\boldsymbol{\varepsilon}(\widehat{\boldsymbol{u}}^{(2)}(\boldsymbol{x},\boldsymbol{y},p))\cdot\boldsymbol{n}=\mathbf{0},\quad \forall \boldsymbol{x}\in\Sigma_2\ \wedge\ \forall p\in[0,\infty], \tag{10.49}$$

$$\widehat{\boldsymbol{u}}^{(2)}(\boldsymbol{x},\boldsymbol{y},0)=\mathbf{0},\qquad \forall \boldsymbol{x}\in\Sigma. \tag{10.50}$$

The Lemma 10.1 guarantees the existence of one Y-periodic solution of the problem (10.47), if and only if

$$\left\langle \boldsymbol{f}(\boldsymbol{x})-L_{xx}\widehat{\boldsymbol{u}}^{(0)}(\boldsymbol{x},\boldsymbol{y},p)-L_{xy}\widehat{\boldsymbol{u}}^{(1)}(\boldsymbol{x},\boldsymbol{y},p)-L_{yx}\widehat{\boldsymbol{u}}^{(1)}(\boldsymbol{x},\boldsymbol{y},p)\right\rangle=0. \tag{10.51}$$

The functions $\widehat{\boldsymbol{R}}(\boldsymbol{y},p)$ and $\widehat{\boldsymbol{N}}^{rs}(\boldsymbol{y},p)$ are $Y-$periodic and the function $\widehat{\boldsymbol{v}}(\boldsymbol{x},p)$ is independent of $\boldsymbol{y}$. Therefore, using (10.42) and the divergence theorem can be proved

$$\left\langle L_{yx}\widehat{\boldsymbol{u}}^{(1)}(\boldsymbol{x},\boldsymbol{y},p)\right\rangle=0.$$

Then, from (10.51), the *homogenized equation* is obtained and it can be written in the form

$$-\widehat{R}^{(e)}_{ijrs}(p)\frac{\partial}{\partial x_j}\varepsilon_{rsx}(\widehat{\boldsymbol{v}}(\boldsymbol{x},p))=f_i(\boldsymbol{x}), \tag{10.52}$$

where

$$\widehat{R}^{(e)}_{ijrs}(p)=\left\langle \widehat{R}_{ijrs}(\boldsymbol{y},p)+\widehat{R}_{ijkl}(\boldsymbol{y},p)\varepsilon_{kly}(\widehat{\boldsymbol{N}}^{rs}(\boldsymbol{y},p))\right\rangle, \tag{10.53}$$

are *the homogenized coefficients* or *the effective coefficients*.

To obtain a well posed problem in (10.52), the boundary conditions and initial condition are needed for $\widehat{\boldsymbol{v}}(\boldsymbol{x},p)$. From (10.35) and (10.36) the boundary conditions for the Eq. (10.52) are given in the form

$$\widehat{\boldsymbol{v}}(\boldsymbol{x},p) = \boldsymbol{u}^0, \quad \forall \boldsymbol{x} \in \Sigma_1 \wedge \forall p \in [0,\infty], \tag{10.54}$$

$$\widehat{\boldsymbol{R}}(\boldsymbol{x},p) : \boldsymbol{\varepsilon}(\widehat{\boldsymbol{v}}(\boldsymbol{x},p)) \cdot \boldsymbol{n} = \boldsymbol{s}^0, \quad \forall \boldsymbol{x} \in \Sigma_2 \wedge \forall p \in [0,\infty], \tag{10.55}$$

and from (10.37) the initial condition is

$$\widehat{\boldsymbol{v}}(\boldsymbol{x},0) = \boldsymbol{0}, \ \forall \boldsymbol{x} \in \Sigma. \tag{10.56}$$

10.4 Two Phase Viscoelastic Composite

Now, we study a layered medium (see Fig 10.2). It is a composite formed by cells that are periodically distributed along one axis and each cell is made of a finite number of layers. In particular, we take x_3 the axis that describe the periodicity and it is perpendicular to the layers. The relaxation modulus $\widehat{\boldsymbol{R}}(\boldsymbol{x},p)$ is periodic function of the coordinate x_3 and it does not depend on x_1 and x_2.

Expanding the cell problem (10.43) for the index j, the first two terms are vanished because the fast variable depends only on y_3, ($y_3 = \xi^{-1}x_3$), and therefore

$$-\frac{\partial}{\partial y_3}\left(\widehat{R}_{i3kl}(\boldsymbol{y},p)\varepsilon_{kly}(\widehat{\boldsymbol{N}}^{rs}(\boldsymbol{y},p)) + \widehat{R}_{i3rs}(\boldsymbol{y},p)\right) = 0. \tag{10.57}$$

The symmetry properties of the relaxation viscoelastic modulus are considered

$$R_{ijkl}(\boldsymbol{x},t) = R_{jikl}(\boldsymbol{x},t) = R_{ijlk}(\boldsymbol{x},t) = R_{klij}(\boldsymbol{x},t),$$

where the last equality is consistent with the reciprocity principle stated by Onsager (see Maghous and Creus, 2003). Expanding (10.57) for the index k and l, this equation

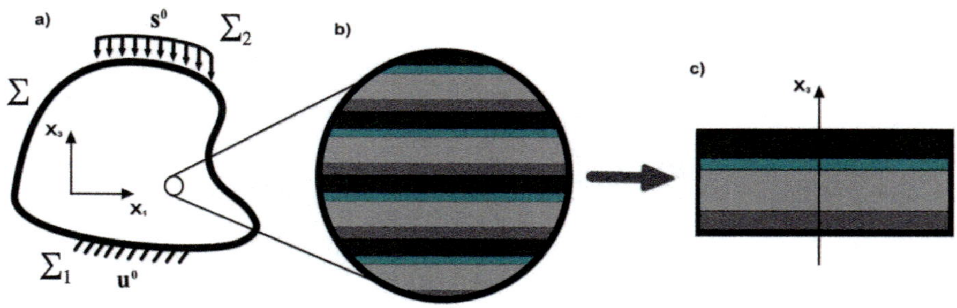

Fig. 10.2: Scales of a layered structure. a) Macroscopic structure, b) periodic microstructure, c) unit cell.

becomes in the simplified form

$$-\frac{\partial}{\partial y_3}\left(\widehat{R}_{i3k3}(\mathbf{y},p)\frac{\partial \widehat{N}_k^{rs}(\mathbf{y},p)}{\partial y_3}+\widehat{R}_{i3rs}(\mathbf{y},p)\right)=0. \tag{10.58}$$

From (10.58) is obtained

$$\widehat{R}_{i3k3}(\mathbf{y},p)\frac{\partial \widehat{N}_k^{rs}(\mathbf{y},p)}{\partial y_3}+\widehat{R}_{i3rs}(\mathbf{y},p)=\widehat{A}_{i3rs}(p). \tag{10.59}$$

and therefore,

$$\frac{\partial \widehat{N}_k^{rs}(\mathbf{y},p)}{\partial y_3}=\widehat{R}_{i3k3}^{-1}(\mathbf{y},p)\left(\widehat{A}_{i3rs}(p)-\widehat{R}_{i3rs}(\mathbf{y},p)\right). \tag{10.60}$$

As $\widehat{\boldsymbol{N}}^{rs}$ is $Y-$periodic function, applying the average operator to both sides of (10.60) and using the divergence theorem we obtain

$$\left\langle\frac{\partial \widehat{N}_k^{rs}(\mathbf{y},p)}{\partial y_3}\right\rangle=0, \tag{10.61}$$

and consequently

$$\widehat{A}_{i3rs}(p)=\left\langle\widehat{R}_{i3q3}^{-1}(\mathbf{y},p)\right\rangle^{-1}\left\langle\widehat{R}_{p3q3}^{-1}(\mathbf{y},p)\widehat{R}_{p3rs}(\mathbf{y},p)\right\rangle. \tag{10.62}$$

Replacing (10.62) into (10.60) and after some algebraic manipulations

$$\begin{aligned}\frac{\partial \widehat{N}_k^{rs}(\mathbf{y},p)}{\partial y_3}=&\ \widehat{R}_{l3k3}^{-1}(\mathbf{y},p)\left\langle\widehat{R}_{l3q3}^{-1}(\mathbf{y},p)\right\rangle^{-1}\left\langle\widehat{R}_{p3q3}^{-1}(\mathbf{y},p)\widehat{R}_{p3rs}(\mathbf{y},p)\right\rangle\\ &-\widehat{R}_{l3k3}^{-1}(\mathbf{y},p)\widehat{R}_{l3rs}(\mathbf{y},p).\end{aligned} \tag{10.63}$$

Using the formula of the effective coefficients (10.53)

$$\begin{aligned}\widehat{R}_{ijrs}^{(e)}(p)&=\left\langle\widehat{R}_{ijrs}(\mathbf{y},p)+\widehat{R}_{ijkl}(\mathbf{y},p)\varepsilon_{kly}\left(\widehat{\boldsymbol{N}}^{rs}(\mathbf{y},p)\right)\right\rangle\\ &=\left\langle\widehat{R}_{ijrs}(\mathbf{y},p)+\widehat{R}_{ijk3}(\mathbf{y},p)\left(\frac{\partial \widehat{N}_k^{rs}(\mathbf{y},p)}{\partial y_3}\right)\right\rangle.\end{aligned} \tag{10.64}$$

Replacing (10.63) into (10.64), the expression of the effective coefficients is given in the final form

$$\begin{aligned}\widehat{R}_{ijrs}^{(e)}(p)=&\left\langle\widehat{R}_{ijrs}(\mathbf{y},p)\right\rangle+\left\langle\widehat{R}_{ijk3}(\mathbf{y},p)\widehat{R}_{l3k3}^{-1}(\mathbf{y},p)\right\rangle\left\langle\widehat{R}_{l3q3}^{-1}(\mathbf{y},p)\right\rangle^{-1}\\ &\cdot\left\langle\widehat{R}_{p3q3}^{-1}(\mathbf{y},p)\widehat{R}_{p3rs}(\mathbf{y},p)\right\rangle-\left\langle\widehat{R}_{ijk3}(\mathbf{y},p)\widehat{R}_{l3k3}^{-1}(\mathbf{y},p)\widehat{R}_{l3rs}(\mathbf{y},p)\right\rangle.\end{aligned} \tag{10.65}$$

The Eq. (10.65) represents the general expression for the effective coefficients in Laplace-Carson space (see Pobedria, 1984).

In case of a two-layered medium where each constituent has isotropic characteristic, the expression of the effective coefficients derived from (10.57) are the following

$$\widehat{R}^{(e)}_{1111}(p) = \widehat{R}^{(e)}_{2222}(p)$$
$$= \langle \lambda + 2\mu \rangle + \left\langle \frac{1}{\lambda + 2\mu} \right\rangle^{-1} \left\langle \frac{\lambda}{\lambda + 2\mu} \right\rangle^{2} - \left\langle \frac{\lambda^2}{\lambda + 2\mu} \right\rangle, \tag{10.66}$$

$$\widehat{R}^{(e)}_{3333}(p) = \left\langle \frac{1}{\lambda + 2\mu} \right\rangle^{-1}, \tag{10.67}$$

$$\widehat{R}^{(e)}_{1122}(p) = \widehat{R}^{(e)}_{2211}(p) = \langle \lambda \rangle + \left\langle \frac{1}{\lambda + 2\mu} \right\rangle^{-1} \left\langle \frac{\lambda}{\lambda + 2\mu} \right\rangle^{2} - \left\langle \frac{\lambda^2}{\lambda + 2\mu} \right\rangle, \tag{10.68}$$

$$\widehat{R}^{(e)}_{1133}(p) = \widehat{R}^{(e)}_{3311}(p) = \widehat{R}^{(e)}_{3322}(p) = \widehat{R}^{(e)}_{2233}(p) = \left\langle \frac{\lambda}{\lambda + 2\mu} \right\rangle \left\langle \frac{1}{\lambda + 2\mu} \right\rangle^{-1}, \tag{10.69}$$

$$\widehat{R}^{(e)}_{1313}(p) = \widehat{R}^{(e)}_{2323}(p) = \left\langle \frac{1}{\mu} \right\rangle^{-1}, \tag{10.70}$$

$$\widehat{R}^{(e)}_{1212}(p) = \langle \mu \rangle. \tag{10.71}$$

Notice that the global behavior is transversely isotropic for this type of laminate composite with isotropic constituents where their properties are denoted by the Lame's constants λ and μ.

10.5 Numerical Results

The above described method allows to find for two-layered viscoelastic medium with isotropic constituents, the value of the effective properties. In the literature, there are different creep kernels that describe the viscoelastic properties of a material. As examples, two different models are analyzed.

10.5.1 Model I

The first case is a two-layered medium (see Fig 10.3) structured by a layer 1 with linear elastic behavior and a layer 2 with viscoelastic behavior. This last layer is described by Dischinger's model (see Maghous and Creus, 2003). The Dischinger's model considers a time-dependent function, given in the form

$$\varphi(\tau, t) = \exp(-\alpha t) - \exp(-\alpha \tau). \tag{10.72}$$

The correspondent Lame's constants for the elastic layer 1 are taken by the relations

$$\mu_1 = \mu_{2,0}, \quad \lambda_1 = K - \frac{2}{3}\,\mu_{2,0}, \tag{10.73}$$

and the relaxation functions for the viscoelastic layer 2 are defined for $\tau \leq t$ by

$$\mu_2(\tau,t) = \mu_{2,0}\,\exp\left(\frac{\mu_{2,0}}{\alpha\beta}\,\varphi(\tau,t)\right), \qquad \lambda_2(\tau,t) = K - \frac{2}{3}\,\mu_2(\tau,t), \tag{10.74}$$

where K is the bulk elastic modulus and $\mu_{2,0}$ represents a constant. Besides, α and β are parameters of model. In Table 10.1 can be found the values of the parameters related to Dischinger's model. The results are obtained for $\tau = 0$, a medium with non-ageing.

The computation of the effective coefficients are done using the formulae (10.66)-(10.68) and (10.70). The behavior of the macroscopic properties are displayed in Fig. 10.4. This result by AHM is compared with Maghous and Creus (2003). Besides, in Fig. 10.4 are added the Voigt upper bound (VUB) and Reuss lower bound (RLB) to verify that the overall properties of the composite are between them. Since the composite has two phases and piecewise constant properties, the bounds can be calculated as follows

$$\widehat{R}^{(\text{Voigt})}_{ijkl}(p) = \eta_1 \widehat{R}^{(1)}_{ijkl}(p) + \eta_2 \widehat{R}^{(2)}_{ijkl}(p), \tag{10.75}$$

$$\widehat{R}^{(\text{Reuss})}_{ijkl}(p) = \left(\frac{\eta_1}{\widehat{R}^{(1)}_{ijkl}(p)} + \frac{\eta_2}{\widehat{R}^{(2)}_{ijkl}(p)}\right)^{-1} \tag{10.76}$$

where η_1 and η_2 are the volume fractions of layer 1 and layer 2 respectively. The superscripts (1), (2) in the kernels are indicating the corresponding material in each

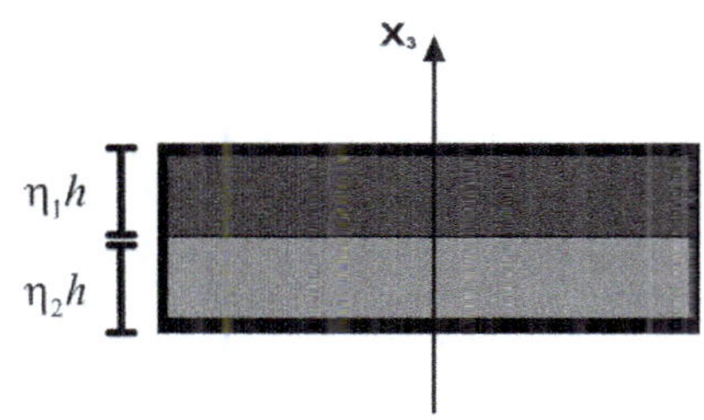

Fig. 10.3 Cell of two-layered medium.

Table 10.1: Parameters of Dischinger's model.

K (GPa)	μ_{20} (GPa)	$\alpha \cdot \beta$ (GPa)	α (days^{-1})	η_1 (dimensionless)	η_2 (dimensionless)
10 000	8571	35667	0.026	0.5	0.5

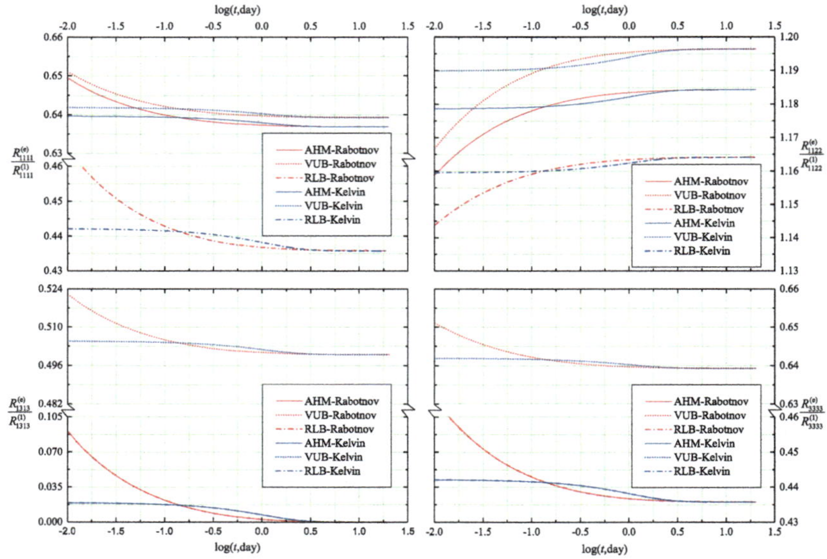

Fig. 10.4: Comparison between AHM, FEM, Maghous and Creus model and the Voigt and Reuss bounds. The results have been normalized with respect to the first phase and are obtained for $\tau = 0$.

layer. In the computational process the numerical implementation of the effective coefficients is necessary. The function INVLAP2, developed in Matlab and the explanation given in Hollenbeck (1998), is adapted to guarantee the necessary technical support for the numerical inversion of Laplace-Carson transform.

As another validation of the present algorithm, a finite element method to compute homogenized coefficients, using Eqs. (10.58) and (10.61) written in the physical space, is implemented. For that purpose, the time interval $[0,T]$ is firstly discretized $t_i = k\Delta t$, $(k = 0,\ldots,N_T)$, and the time integral is approximated using a midpoint quadrature rule. As a consequence, the kernel of the convolution is evaluated on a staggered grid. Then, we obtain a sequence of Poisson's equations that are discretized using a standard finite element method (for more details, see for example Chen et al, 2010; Dumont and Duval, 2013).

10.5.2 Model II

In the previous model, the proposed creep kernel is a composed exponential function. The main problem is that, simplest kernels as exponential function or their linear combinations, does not always describe correctly the viscoelastic behavior for determined materials (see Sevostianov et al, 2016). The use of fraction-exponential functions (see Rabotnov, 1948, 2014; Blair and Coppen, 1939, 1943)

$$\ni_\alpha(\beta,t) = t^{-\alpha}\sum_{n=0}^{\infty}\frac{(-\beta)^n t^{n(1-\alpha)}}{\Gamma[(n+1)(1-\alpha)]}, \tag{10.77}$$

with $0 \leq \alpha < 1$ and $0 < \beta$, as kernel of viscoelastic operators, simplify and solve this situation. Furthermore, they allow, analytically, to obtain results with the use of Laplace transform and at the same time, offer an excellent concordance with the experimental data (see Sevostianov et al, 2016).

The algebraic properties of the Rabotnov's kernel are very well developed in Rabotnov (1977). One of the most important aspect described in the theory of this function is the analytical treatment of its Laplace transform

$$L\left[\ni_\alpha(\beta,t)\right] \equiv \int_0^{\infty}\ni_\alpha(\beta,t)e^{-pt} = \frac{1}{p^{1-\alpha}+\beta}. \tag{10.78}$$

Now, consider a two-layered medium with the structure: the layer 1 exhibits elastic properties and the layer 2 viscoelastic properties. The following data, obtained from Sevostianov et al (2016), has been selected to guarantee the numerical computation. The corresponding Lame's constants for the layer 1 are given as follows

$$\mu_1 = 8.571\text{GPa}, \qquad \lambda_1 = K_1 - \frac{2}{3}\mu_1\text{GPa}. \tag{10.79}$$

Taking the relaxation modulus of layer 2 as a fraction-exponential function or Rabotnov's kernel and applying the Laplace-Carson transform, having in mind the expression (10.71), the following analytic formulae are obtained,

$$\widehat{\mu}_2(p) = p\cdot\mu_{20}\left(1+\frac{\gamma_2}{p^{(1-\alpha_2)}+\beta_2}\right), \qquad \widehat{\lambda}_2(p) = K_2 - \frac{2}{3}\widehat{u}_2(p), \tag{10.80}$$

where μ_{20} is the instantaneous shear modulus and $\mu_{20},\alpha_2,\beta_2$ and γ_2 are parameters of the material given experimentally. The material parameters are shown in Table 10.2.

The obtained effective coefficients (10.66)-(10.68) and (10.70), for this model, are displayed in Fig. 10.5. The expression (10.77) is reduced to exponential function for $\alpha = 0$. In this case, it describes the properties of standard viscoelastic material or Kelvin model (see Sevostianov et al, 2015). Therefore, the numerical results of the effective coefficients by AHM are compared with the corresponding coefficients calculated by Kelvin's model (AHM-Kelvin). Besides, the Voigt upper bound (VUB) and Reuss lower bound (RLB) are added to the comparison (see Fig. 10.5) as a validation of the results obtained by the present model (AHM).

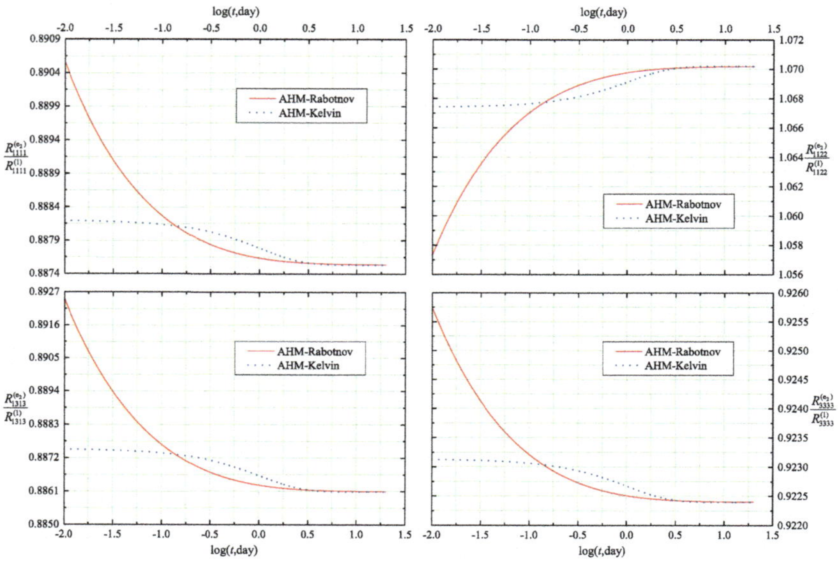

Fig. 10.5: Computation of the effective coefficients using Rabotnov's kernel. Comparison with Kelvin's model, Voigt and Reuss bounds. The coefficients have been normalized with respect to the first phase.

10.5.3 Viscoelastic Effective Constants for Composites with Rectangular Cross-Section Fibers: Double Homogenization

The purpose now is to obtain, using the asymptotic homogenization method, the viscoelastic effective properties for a composite material with rectangular cross-

Table 10.2: Material parameters.

Layer 1	K_1 (GPa)	η_1 (-)	-	-	-	-
	10.0	0.5	-	-	-	-
Layer 2	K_2 (GPa)	η_2 (-)	α_2 (-)	β_2 (day$^{1-\alpha}$)	γ_2 (day$^{1-\alpha}$)	μ_{20} (MPa)
	5.97	0.5	0.47	0.98	49.6	1.7

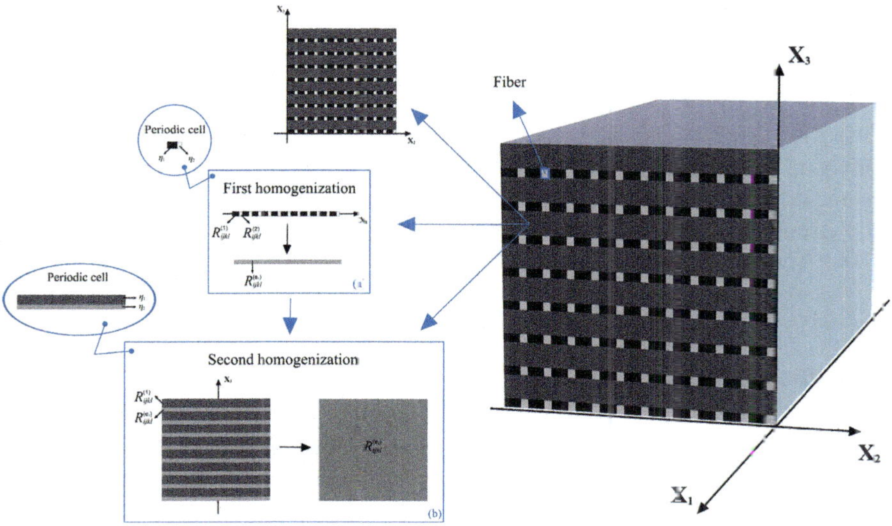

Fig. 10.6: Composites material with rectangular cross-section fibers. (a) First homogenization structure. (b) Second homogenization structure.

section fibers, periodically distributed (see Fig 10.6) along one axis with the aim of the aforementioned results applied to layered composites.

The basic idea is applied two times the homogenization method, in different directions, according to the geometrical configuration of the composite material. Therefore, we can divided the problem into two homogenization stages.

1. In a first moment, the composite is homogenized in the direction x_2. This structure can be analyzed as a two-layered medium with transversely isotropic properties after the homogenization process (see Fig. 10.6 (a)). In Sec. 10.4 is applied the homogenization method on this type of material in the direction x_3, obtaining the effective coefficients using (10.66)-(10.71). Now, as the problem is formulated in the direction x_2, it is affected the disposition of the subscripts in the formulae of the effective coefficients. In order to solve this issue, only it is necessary to change the subscripts $(3 \to 2)$ and $(2 \to 3)$ in all the formulae. Moreover, the superscript (e_1) is added indicating the first homogenization step. For example, (10.71) is transformed as follows $\widehat{R}^{(\mathrm{e}_1)}_{1313}(p) = \langle \mu \rangle$. Clearly, for this two-layered medium $\langle f \rangle = \eta_1 f^{(1)} + \eta_2 f^{(2)}$, where the superscripts (1), (2) are indicating the corresponding layer (see Fig. 10.6 (a)).
2. Finally, the effective coefficients are calculated in the direction x_3. The structure can be analyzed as a two-layered medium (see Fig 10.6 (b)). In this case, the formulae obtained are just the same that (10.66)-(10.71). Besides, the notation (e_2) is added as superscript denoting the second homogenization step. For example, the coefficient $R^{(\mathrm{e}_2)}_{1212}(t)$ can be obtained from (10.71) as follows, $\widehat{R}^{(\mathrm{e}_2)}_{1212}(p) = \langle \mu \rangle$. Now, is necessary to take into account that $\langle f \rangle = \eta_1 f^{(1)} + \eta_2 f^{(\mathrm{e}_1)}$, where the

superscript (1) indicates the elastic property of the layer 1 and the superscript (e_1) denotes the effective viscoelastic property of the layer 2. This last properties are calculated in the first homogenization step, see Fig. 10.6 (b).

In the Fig. 10.7 is displayed the computation of the effective coefficients for a composite material reinforced with rectangular cross-section fibers. A fraction-exponential function (see Model II in Sect. 10.5) is assigned as relaxation modulus for the viscoelastic layer 2. The volume fractions are taken as follow $\eta_1 = 0.7$ and $\eta_2 = 0.3$. The remaining data has been taken from Table 10.2.

10.6 Conclusions

In this work, the two-scale asymptotic homogenization method is applied to calculate the linear viscoelastic effective properties for layered composites where the distribution of the layers are perpendicular to one preferential axis. The local problems and the analytic expressions of the effective coefficients are derived. Based on this result, two times, the homogenization method is applied in different directions, according to the geometrical configuration of the composite material for deriving the analytical

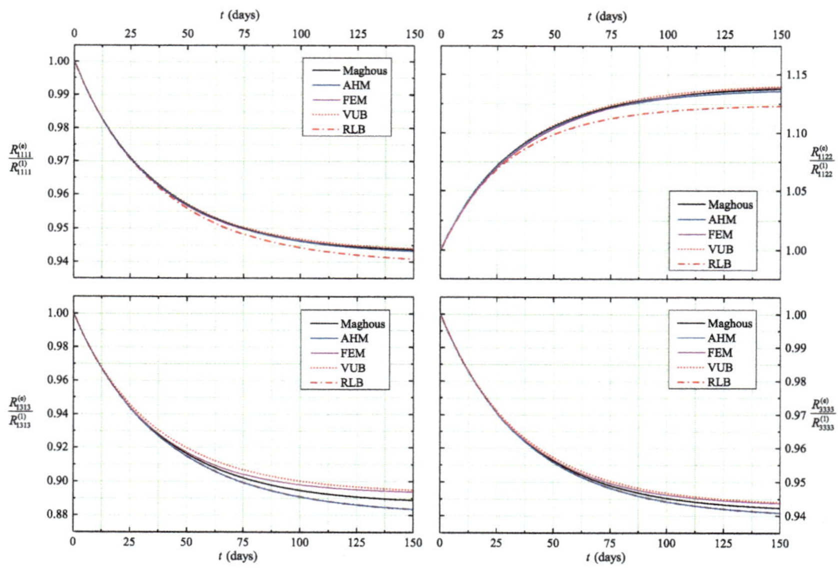

Fig. 10.7: Effective coefficients for rectangular cross-section fibers, using double homogenization, with Rabotnov's kernel. Comparison with the Kelvin's model. The results have been normalized with respect to the first phase.

expressions of the viscoelastic effective properties for composites with rectangular cross-section fibers, periodically distributed along one axis. In addition, models with different creep kernels are studied, in particular, the Rabotnov's kernel is analyzed. A numerical computation where the inverse of Laplace-Carson is implemented numerically for the computation of the effective viscoelastic properties. An algorithm using FEM is developed in the present work and comparisons with other approaches are given as a validation of the present model.

Acknowledgements The authors would like to be grateful to University of Matanzas and Tecnologico de Monterrey, Escuela de Ingenieria y Ciencias, Atizapan de Zaragoza, Estado de Mexico, for its support. France-Cuba project "Partenariat Hubert Curien franco-cubain Carlos J. Finlay" 2017-2018 and Proyecto Nacional de Ciencias Básicas 2017-2019 are gratefully acknowledged. Thanks to Departamento de Matemáticas y Mecánica IIMAS-UNAM and FENOMEC for their support and Ramiro Chávez Tovar and Ana Pérez Arteaga for computational assistance.

References

Bakhvalov N, Panasenko GP (1989) Homogenisation: Averaging Processes in Periodic Media. Kluwer, Dordrecht

Berger H, Gabbert U, Köppe H, Rodriguez-Ramos R, Bravo-Castillero J, Guinovart-Diaz R, Otero JA, Maugin GA (2003) Finite element and asymptotic homogenization methods applied to smart composite materials. Computational Mechanics 33(1):61–67

Berger H, Kari S, Gabbert U, Rodriguez-Ramos R, Bravo-Castillero J, Guinovart-Diaz R, Sabina FJ, Maugin GA (2006) Unit cell models of piezoelectric fiber composites for numerical and analytical calculation of effective properties. Smart Materials and Structures 15(2):451–458

Beurthey S, Zaoui A (2000) Structural morphology and relaxation spectra of viscoelastic heterogeneous materials. European Journal of Mechanics - A/Solids 19(1):1–16

Blair GWS, Coppen FMV (1939) The subjective judgment of the elastic and plastic properties of soft bodies; the "differential thresholds" for viscosities and compression moduli. Proceedings of the Royal Society of London Series B, Biological Sciences 128(850):109–125

Blair GWS, Coppen FMV (1943) The estimation of firmness in soft materials. The American Journal of Psychology 56(2):234–246

Brenner R, Masson R, Castelnau O, Zaoui A (2002) A quasi-elastic affine formulation for the homogenised behaviour of nonlinear viscoelastic polycrystals and composites. European Journal of Mechanics - A/Solids 21(6):943–960

Chen M, Dumont S, Dupaigne L, Goubet O (2010) Decay of solutions to a water wave model with a nonlocal viscous dispersive term. Discrete and Continuous Dynamical Systems 27(4):1473–1492

Christensen RM (1969) Viscoelastic properties of heterogeneous media. Journal of the Mechanics and Physics of Solids 17(1):23–41

Christensen RM (1971) Theory of Viscoelasticity. Academic Press, New York

Dormieux L, Kondo D, Ulm FJ (2006) Microporomechanics. John Wiley & Sons, Chichester

Dumont S, Duval JB (2013) Numerical investigation of asymptotical properties of solutions to models for waterways with non local viscosity. Int J Num Anal Modeling 10(2):333–349

Hashin Z (1965) Viscoelastic behavior of heterogeneous media. Trans ASME J Appl Mech 32:630–636

Hashin Z (1966) Viscoelastic fibre reinforced materials. AIAA Journal 4:1411–1417

Hashin Z (1970a) Complex moduli of viscoelastic composites - I. General theory and application to particulate composites. Int J Solids Struct 6:539–552

Hashin Z (1970b) Complex moduli of viscoelastic composites - II. Fibre reinforced materials. Int J Solids Struct 6:797–807

Hollenbeck KJ (1998) Invlap.m: a Matlab function for numerical inversion of Laplace transforms by the Hoog algorithm URL `http://www.mathworks.com`

Kachanov M (1992) Effective elastic properties of cracked solids: critical review of some basic concepts. Appl Mech Rev 45(8):304–335

Lahellec N, Suquet P (2007) Effective behavior of linear viscoelastic composites: A time-integration approach. International Journal of Solids and Structures 44(2):507 – 529

Lavergne F, Sab K, Sanahuja J, Bornert M, Toulemonde C (2016) Homogenization schemes for aging linear viscoelastic matrix-inclusion composite materials with elongated inclusions. International Journal of Solids and Structures 80:545 – 560

Laws N, McLaughlin R (1978) Self-consistent estimates for the viscoelastic creep compliances of composite materials. Proceedings of the Royal Society of London Series A, Mathematical and Physical Sciences 359(1697):251–273

Le QV, Meftah F, He QC, Le Pape Y (2007) Creep and relaxation functions of a heterogeneous viscoelastic porous medium using the Mori-Tanaka homogenization scheme and a discrete microscopic retardation spectrum. Mechanics of Time-Dependent Materials 11(3):309–331

Lévesque M, Gilchrist MD, Bouleau N, Derrien K, Baptiste D (2007) Numerical inversion of the Laplace–Carson transform applied to homogenization of randomly reinforced linear viscoelastic media. Computational Mechanics 40(4):771–789

Maghous S, Creus GJ (2003) Periodic homogenization in thermoviscoelasticity: case of multilayered media with ageing. International Journal of Solids and Structures 40(4):851 – 870

Otero JA, Bravo-Castillero J, Guinovart-Díaz R, Rodríguez-Ramos R, Maugin GA (2003) Analytical expressions of effective constants for a piezoelectric composite reinforced with square cross-section fibers. Arch Mech 55:357 – 371

Persson LE, Persson L, Svanstedt N, Wyller J (1993) The Homogenization Method. An Introduction. Student litteratur, Lund

Pipkin AC (1986) Lectures on Viscoelastic Theory. Springer, New York, Berlin, Heidelberg

Pobedria BE (1984) Mechanics of Composite Materials (in Russ.). Moscow State University Press, Moscow

Rabotnov YN (1948) Equilibrium of an elastic medium with after-effect (in Russ.). Prikladnaya Matematika i Mekhanika (J Appl Math Mech) 12(1):53 – 62

Rabotnov YN (1977) Elements of Hereditary Solid Mechanics. Mir, Moscow

Rabotnov YN (2014) Equilibrium of an elastic medium with after-effect. Fractional Calculus and Applied Analysis 17(3):684 – 696

Ricaud JM, Masson R (2009) Effective properties of linear viscoelastic heterogeneous media: Internal variables formulation and extension to ageing behaviours. International Journal of Solids and Structures 46(7):1599 – 1606

Schapery RA (1964) Application of thermodynamics to thermomechanical, fracture, and birefringent phenomena in viscoelastic media. Journal of Applied Physics 35(5):1451–1465

Schapery RA (1967) Stress analysis of viscoelastic composite materials. Journal of Composite Materials 1(3):228–267

Sevostianov I, Levin V, Radi E (2015) Effective properties of linear viscoelastic microcracked materials: Application of Maxwell homogenization scheme. Mechanics of Materials 84:28 – 43

Sevostianov I, Levin V, Radi E (2016) Effective viscoelastic properties of short-fiber reinforced composites. International Journal of Engineering Science 100:61 – 73

Sokolnikoff IS, Redheffer RM (1968) Mathematics of Physics and Modern Engineering. McGraw-Hill Book Company, Inc, New York, Toronto, London

Wang YM, Weng GJ (1992) The influence of inclusion shape on the overall viscoelastic behavior of composites. Trans ASME J Appl Mech 59(3):510 – 518

Zhang J, Ostoja-Starzewski M (2015) Mesoscale bounds in viscoelasticity of random composites. Mechanics Research Communications 68:98 – 104

Chapter 11
A Single Crystal Beam Bent in Double Slip

Xiangyu Cui and Khanh Chau Le

Abstract A theory of plastic bending of single crystal beam with two active slip systems accounting for continuously distributed excess dislocations is proposed. If the resistance to dislocation motion is negligibly small, then excess dislocations pile up against the beam's neutral line, leaving two small layers near the lateral faces dislocation-free. The threshold value at the onset of plastic yielding, the dislocation density, as well as the moment-curvature curve are found. If the energy dissipation is taken into account, excess dislocations at the beginning of plastic yielding occupy two thin layers, leaving the zone near the middle line as well as two layers near the beam's lateral faces dislocation-free. The threshold bending moment at the dislocation nucleation and the hardening rate are higher than those in the case of zero dissipation.

11.1 Introduction

When a single crystal beam is plastically bent, excess dislocations occur on certain active slip planes to reduce its energy. As a rule, the number of dislocations is huge, so, instead of a difficult and time-consuming task of describing individual dislocations (Cleveringa et al, 1999; Motz et al, 2008), the use of continuum approach to find the dislocation densities in the bent state turns out more effective. The first attempt of taking into account continuously distributed excess dislocations in the plastically bent beam was made by Nye (1953) who expressed the curvature of the beam caused by dislocations in terms of the dislocation density tensor bearing now his name. Read

Xiangyu Cui
Computational Engineering, Ruhr University Bochum, D-44780 Bochum, Germany
e-mail: 15120080178@163.com

Khanh Chau Le
Lehrstuhl für Mechanik - Materialtheorie, Ruhr-Universität Bochum, D-44780 Bochum, Germany
e-mail: chau.le@rub.de

© Springer International Publishing AG, part of Springer Nature 2018

H. Altenbach et al. (eds.), *Generalized Models and Non-classical Approaches in Complex Materials 1*, Advanced Structured Materials 89,
https://doi.org/10.1007/978-3-319-72440-9_11

(1957) and Bilby et al (1958) have extended this result to the case when the stress due to excess dislocations does not vanish (see also Ashby, 1970; Evans, 1995; Mura, 1987; Wang et al, 2003). The first quantitative prediction of the dislocation density based on the asymptotically exact solution for single crystal beam bent in single slip has been made in Le and Nguyen (2013). The analyzed model has been the simplest one within the continuum dislocation theory (CDT) proposed by Berdichevsky (2006a,b) and developed further in Berdichevsky and Le (2007); Kaluza and Le (2011); Kochmann and Le (2008, 2009b,a); Le and Sembiring (2008a,b, 2009) (see also the alternative approaches in Gurtin (2002); Gurtin et al (2007); Sandfeld et al (2010)). The information about dislocation distribution obtained in Le and Nguyen (2013) has been used to explain the formation of small angle tilt boundaries and predict the number of polygons in a polygonized beam after annealing (see also Le and Nguyen, 2010, 2012).

The real bending of single crystal beam is often more involved because two or more slip systems could be activated during the plastic bending, depending on the orientation of crystal with respect to the beam axis affecting the Schmid factors. For instance, if the fcc single crystal beam having the axis parallel to $[\underline{1}10]$-direction is bent about the $[110]$ direction, then there are two symmetric active slip systems oriented at angles $\pm 54.7^\circ$ with respect to the beam axis (see, e.g., Kysar et al, 2010). Simulations of such bending problems require more elaborate models because of the interaction between edge dislocations of different groups (Cleveringa et al, 1999). This paper aims at constructing an approximate one-dimensional theory of bending of single crystal beams having two active slip systems and, thus, extending the results obtained in Le and Nguyen (2013) to double slip. We incorporate the interactions of excess dislocations in a material model similar to that proposed by Le and Sembiring (2009). We then use the variational-asymptotic method developed in Berdichevsky (1983); Le (1999) to derive the one-dimensional approximate theories from the exact three-dimensional problems in two cases:

i) the resistance to dislocation motion (and, consequently, the energy dissipation) is negligibly small so that the displacements and plastic slips can be found by minimizing the energy functional,
ii) the rate-independent dissipation is taken into account.

The obtained one-dimensional variational problems admit analytical solutions representing the smooth minimizers for the symmetric active slip systems. It is established that there exist energetic as well as dissipative thresholds for the dislocation nucleation which depend on the thickness of the beam. However, the distribution of excess dislocations are quite different in case i) and ii). In case of negligibly small resistance to dislocation motion the excess dislocations pile up against the beam's neutral line, leaving its lateral faces dislocation-free. In contrary, in the case of non-zero dissipation, dislocations cannot reach the zone near the neutral line at the beginning of plastic yielding because the driving force cannot overcome Peierls' barrier. At the later stage of plastic bending dislocations may reach the neutral line, so they pile up against it. Thus, in case ii) excess dislocations either occupy two thin layers, leaving the zone near the middle line as well as two plastically deformed layers near the beam's lateral faces dislocation-free, in the early stage of plastic yielding, or pile up against the neutral line in the later stage of plastic bending. The thickness of dislocation layer depends on the thickness of the beam causing the typical size effect of the gradient theory. Based on this analytical solution the deflection of the beam, the dislocation density, and the moment-curvature curve are analyzed in terms of the bending moment for different loading/unloading processes of both case i) and ii).

The paper is organized as follows. In the next section the 3-D models of single crystal beam with continuously distributed dislocations bent in double slip are formulated. Sections 11.3 and 11.4 derive the one-dimensional equations and solve them for the case of zero dissipation. Sections 11.5 and 11.6 consider the same problems but with dissipation. Finally, Sect. 11.7 concludes the paper.

11.2 3-D Models of Crystal Beam Bent in Double Slip

Consider a single crystal beam bent by a moment M. In the undeformed state the beam occupies the domain $\mathscr{V}$ of the three-dimensional euclidean space, with the cartesian coordinates $(x,y,z) \in (0,L)\times(-h/2,h/2)\times(0,a)$ where L, h, and a are the length, height, and width of the beam, respectively. We will assume that $h \ll a \ll L$ and that the beam is bent under the plane strain condition. For definiteness, let the beam be clamped at $x=0$ and subjected to a linearly distributed traction $-\tau y$ having the resultant moment M at $x=L$. Besides, the upper and lower boundaries of the beam at $y=h/2$ and $y=-h/2$ are traction free. If the applied moment is small, it is natural to expect that the beam deforms elastically and the stress distribution is linear over the thickness according to the elementary beam theory. However, if the applied moment exceeds some critical value, edge dislocations with dislocation lines parallel to the z-axis may appear to reduce energy of the bent beam. We assume that, at the onset of yielding and during the plastic deformations, the crystal admits two active slip systems with slip planes inclined at angles φ_1 and φ_2 to the plane $y=0$ as shown in Fig. 11.1.

Under the plane strain state condition there are only two non-zero components of the displacements that do not depend on z, $u_x(x,y)$ and $u_y(x,y)$. Consequently, the non-zero components of the total strain tensor are

$$\varepsilon_{xx}=u_{x,x}, \quad \varepsilon_{yy}=u_{y,y}, \quad \varepsilon_{xy}=\varepsilon_{yx}=\frac{1}{2}(u_{x,y}+u_{y,x}).$$

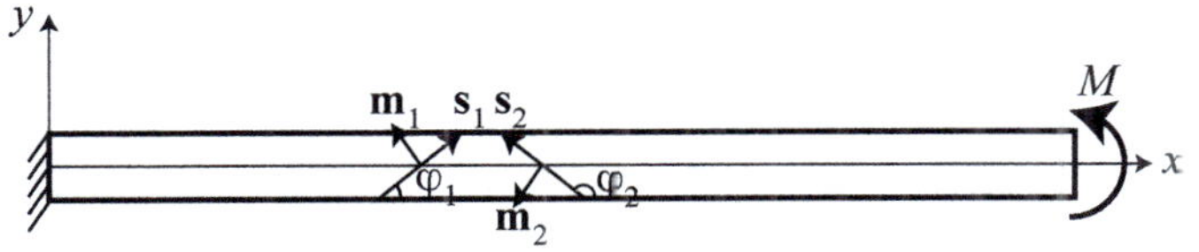

Fig. 11.1: Single crystal beam bent in double slip by a moment M.

Throughout the paper the comma before an index is used to denote the partial derivative with respect to the corresponding coordinate. Since there are two active slip systems, the plastic distortion tensor is given by

$$\beta_{ij} = \beta_1(x,y)s_i^1 m_j^1 + \beta_2(x,y)s_i^2 m_j^2,$$

with

$$\begin{aligned} \boldsymbol{s}^1 &= (\cos\varphi_1, \sin\varphi_1, 0), & \boldsymbol{m}^1 &= (-\sin\varphi_1, \cos\varphi_1, 0), \\ \boldsymbol{s}^2 &= (\cos\varphi_2, \sin\varphi_2, 0), & \boldsymbol{m}^2 &= (-\sin\varphi_2, \cos\varphi_2, 0) \end{aligned}$$

being the unit vectors denoting the slip direction and the normal to the slip planes of the corresponding slip system. We call $\beta_1(x,y)$ and $\beta_2(x,y)$ the plastic slips. Thus, the finding of plastic distortion tensor reduces to determining the functions $\beta_1(x,y)$ and $\beta_2(x,y)$. Non-zero components of the plastic strain tensor, $\varepsilon_{ij}^p = \frac{1}{2}(\beta_{ij} + \beta_{ji})$, read

$$\begin{aligned} \varepsilon_{xx}^p &= -\frac{1}{2}\beta_1 \sin 2\varphi_1 - \frac{1}{2}\beta_2 \sin 2\varphi_2, \\ \varepsilon_{yy}^p &= \frac{1}{2}\beta_1 \sin 2\varphi_1 + \frac{1}{2}\beta_2 \sin 2\varphi_2, \\ \varepsilon_{xy}^p &= \varepsilon_{yx}^p = \frac{1}{2}\beta_1 \cos 2\varphi_1 + \frac{1}{2}\beta_2 \cos 2\varphi_2. \end{aligned}$$

Accordingly, the non-zero components of the elastic strain tensor, $\varepsilon_{ij}^e = \varepsilon_{ij} - \varepsilon_{ij}^p$, are

$$\begin{aligned} \varepsilon_{xx}^e &= u_{x,x} + \frac{1}{2}\beta_1 \sin 2\varphi_1 + \frac{1}{2}\beta_2 \sin 2\varphi_2, \\ \varepsilon_{yy}^e &= u_{y,y} - \frac{1}{2}\beta_1 \sin 2\varphi_1 - \frac{1}{2}\beta_2 \sin 2\varphi_2, \\ \varepsilon_{xy}^e &= \varepsilon_{yx}^e = \frac{1}{2}\left(u_{x,y} + u_{y,x} - \beta_1 \cos 2\varphi_1 - \beta_2 \cos 2\varphi_2\right). \end{aligned}$$

The distribution of excess dislocations associated with these two active slip systems is described by the dislocation density tensor (introduced by Nye (1953); Bilby (1955), and Kröner (1955)), $\alpha_{ij} = \varepsilon_{jkl}\beta_{il,k}$, whose non-zero components read

$$\begin{aligned} \alpha_{xz} &= \beta_{1,x}\cos^2\varphi_1 + \beta_{1,y}\cos\varphi_1 \sin\varphi_1 + \beta_{2,x}\cos^2\varphi_2 + \beta_{2,y}\cos\varphi_2 \sin\varphi_2, \\ \alpha_{yz} &= \beta_{1,x}\cos\varphi_1 \sin\varphi_1 + \beta_{1,y}\sin^2\varphi_1 + \beta_{2,x}\cos\varphi_2 \sin\varphi_2 + \beta_{2,y}\sin^2\varphi_2. \end{aligned}$$

Thus, we see that there are two different groups of excess edge dislocations associated with the slip systems 1 and 2, whose resultant Burgers' vectors show in the directions $\boldsymbol{s}^1$ and $\boldsymbol{s}^2$, respectively. These give two scalar dislocation densities

$$\rho_1 = \frac{1}{b}|\beta_{1,x}\cos\varphi_1 + \beta_{1,y}\sin\varphi_1|, \quad \rho_2 = \frac{1}{b}|\beta_{2,x}\cos\varphi_2 + \beta_{2,y}\sin\varphi_2|, \qquad (11.1)$$

with b being the magnitude of the Burgers vector.

Let us propose the free energy density of the single crystal deforming in double slip in the form (see Le and Sembiring, 2009)

$$\phi = \frac{1}{2}\lambda(\varepsilon^e_{kk})^2 + \mu\varepsilon^e_{ij}\varepsilon^e_{ij} + \mu k \ln\frac{1}{1-\frac{\rho_1}{\rho_s}} + \mu k \ln\frac{1}{1-\frac{\rho_2}{\rho_s}} + \mu\chi\frac{\rho_1\rho_2}{\rho_s^2}, \tag{11.2}$$

with λ and μ being the Lamé constants, ρ_s the saturated dislocation density, and k and χ the constants of the material. The first two terms in (11.2) represent the contribution to the energy of crystal by the elastic strain. The third and fourth terms are the contributions to the energy from the two separate groups of excess dislocations. The logarithmic energy term (Berdichevsky, 2006b) stems from two facts: i) energy of the dislocation network for small dislocation densities is the sum of energy of non-interacting dislocations, and ii) there exists a saturated dislocation density which characterizes the closest packing of dislocations admissible in the discrete crystal lattice. The logarithmic term ensures a linear increase of the energy for small dislocation density ρ (Ortiz and Repetto, 1999) and tends to infinity as ρ approaches the saturated dislocation density ρ_s hence providing an energetic barrier against over-saturation. Finally, the last term takes the interaction of dislocations between these two groups into account. We restrict ourselves to the range of small up to moderate dislocation densities for which the logarithmic terms can be approximated by the Taylor series

$$\ln\frac{1}{1-\frac{\rho}{\rho_s}} \cong \frac{\rho}{\rho_s} + \frac{1}{2}\frac{\rho^2}{\rho_s^2}$$

and shall use further only this approximation.

With (11.1) and (11.2) the total energy functional of the bent beam becomes

$$\begin{aligned}
I = a\int_0^L\int_{-h/2}^{h/2} \Big[&\frac{1}{2}\lambda(u_{x,x}+u_{y,y})^2 + \mu(u_{x,x}+\frac{1}{2}\beta_1\sin 2\varphi_1 + \frac{1}{2}\beta_2\sin 2\varphi_2)^2 \\
&+ \mu(u_{y,y}-\frac{1}{2}\beta_1\sin 2\varphi_1 - \frac{1}{2}\beta_2\sin 2\varphi_2)^2 + \frac{1}{2}\mu(u_{x,y}+u_{y,x}-\beta_1\cos 2\varphi_1 \\
&- \beta_2\cos 2\varphi_2)^2 + \mu k\frac{|\beta_{1,x}\cos\varphi_1+\beta_{1,y}\sin\varphi_1|}{b\rho_s} + \frac{1}{2}\mu k\frac{(\beta_{1,x}\cos\varphi_1+\beta_{1,y}\sin\varphi_1)^2}{b^2\rho_s^2} \\
&+ \mu k\frac{|\beta_{2,x}\cos\varphi_2+\beta_{2,y}\sin\varphi_2|}{b\rho_s} + \frac{1}{2}\mu k\frac{(\beta_{2,x}\cos\varphi_2+\beta_{2,y}\sin\varphi_2)^2}{b^2\rho_s^2} \\
&+ k\mu\chi\frac{|\beta_{1,x}\cos\varphi_1+\beta_{1,y}\sin\varphi_1||\beta_{2,x}\cos\varphi_2+\beta_{2,y}\sin\varphi_2|}{b^2\rho_s^2}\Big]\,dx\,dy \\
&+ a\int_{-h/2}^{h/2}\tau y u_x|_{x=L}\,dy.
\end{aligned} \tag{11.3}$$

The last term in (11.3) is the work done by the linearly distributed traction $-\tau y$ acting at the boundary $x=L$. If the dissipation caused by the dislocation motion is negligible, then the true displacements u_x, u_y and plastic slips β_1, β_2 in the *final* state of bending minimize energy functional (11.3) among all admissible displacements and plastic slips satisfying the kinematic boundary condition. By the final state of bending we means the equilibrium state which is established at fixed loading condition after dislocation nucleation and after the movement of dislocations toward their equilibrium positions is finished. Thus, the whole process of dislocation nucleation and dislocation dynamics cannot be described by this theory. The bending moment $M = \tau a h^3/12$ is regarded as a control parameter, so one can study the evolution of the dislocation network in terms of M.

If the dissipation due to dislocation motion cannot be neglected, energy minimization should be replaced by the variational equation (Sedov, 1965)

$$\delta I + a \int_0^L \int_{-h/2}^{h/2} \left(\frac{\partial D}{\partial \dot{\beta}_1}\delta\beta_1 + \frac{\partial D}{\partial \dot{\beta}_2}\delta\beta_2\right) dxdy = 0. \tag{11.4}$$

The last integral in this equation describes the energy dissipation due to dislocation motion, where the dissipation function $D(\dot{\beta}_1, \dot{\beta}_2)$ is assumed to depend only on the plastic slip rates. We shall consider the simplest rate-independent theory (Maugin, 1990, 1992) for which

$$D(\dot{\beta}_1, \dot{\beta}_2) = K(|\dot{\beta}_1| + |\dot{\beta}_2|),$$

with K being the critical resolved shear stress. Then, provided the signs of $\dot{\beta}_1$ and $\dot{\beta}_2$ do not change during the evolution of β_1 and β_2, the variational equation (11.4) reduces to minimizing the following "relaxed energy" functional

$$\begin{aligned}
I_d = a \int_0^L \int_{-h/2}^{h/2} \Big[& \frac{1}{2}\lambda(u_{x,x}+u_{y,y})^2 + \mu(u_{x,x} + \frac{1}{2}\beta_1 \sin 2\varphi_1 + \frac{1}{2}\beta_2 \sin 2\varphi_2)^2 \\
& + \mu(u_{y,y} - \frac{1}{2}\beta_1 \sin 2\varphi_1 - \frac{1}{2}\beta_2 \sin 2\varphi_2)^2 + \frac{1}{2}\mu(u_{x,y} + u_{y,x} - \beta_1 \cos 2\varphi_1 \\
& - \beta_2 \cos 2\varphi_2)^2 + \mu k \frac{|\beta_{1,x}\cos\varphi_1 + \beta_{1,y}\sin\varphi_1|}{b\rho_s} + \frac{1}{2}\mu k \frac{(\beta_{1,x}\cos\varphi_1 + \beta_{1,y}\sin\varphi_1)^2}{b^2\rho_s^2} \\
& + \mu k \frac{|\beta_{2,x}\cos\varphi_2 + \beta_{2,y}\sin\varphi_2|}{b\rho_s} + \frac{1}{2}\mu k \frac{(\beta_{2,x}\cos\varphi_2 + \beta_{2,y}\sin\varphi_2)^2}{b^2\rho_s^2} \\
& + k\mu\chi \frac{|\beta_{1,x}\cos\varphi_1 + \beta_{1,y}\sin\varphi_1||\beta_{2,x}\cos\varphi_2 + \beta_{2,y}\sin\varphi_2|}{b^2\rho_s^2} \\
& + K \operatorname{sign}(\dot{\beta}_1)\beta_1 + K \operatorname{sign}(\dot{\beta}_2)\beta_2 \Big] dxdy + a \int_{-h/2}^{h/2} \tau y u_x|_{x=L} dy.
\end{aligned} \tag{11.5}$$

So, the true displacement and plastic slip fields in the *final* equilibrium state of plastic bending minimize the "relaxed" energy functional I_d among all admissible displacements and plastic slips. Finally, if $\dot{\beta}_1 = 0$ and $\dot{\beta}_2 = 0$, then the plastic slips are frozen, while the displacements should be found by minimizing (11.5) with the frozen plastic slips.

11.3 Energy Minimization

We first focus on the minimization problem (11.3). It is convenient to introduce the following dimensionless variables and quantities

$$\underline{x} = b\rho_s x, \quad \underline{y} = b\rho_s y, \quad \underline{h} = b\rho_s h, \quad \underline{L} = b\rho_s L, \quad \underline{\tau} = \frac{\tau}{\mu b \rho_s},$$
$$\underline{u}_x = b\rho_s u_x, \quad \underline{u}_y = b\rho_s u_y, \quad \underline{I} = \frac{I(b\rho_s)^2}{\mu a}, \quad \gamma = \frac{\lambda}{\mu}, \tag{11.6}$$

which simplify the minimization problem. Now the energy functional can be rewritten in the dimensionless form as follows

$$\begin{aligned} I = \int_0^L \int_{-h/2}^{h/2} \Big[& \frac{1}{2}\gamma(u_{x,x} + u_{y,y})^2 + (u_{x,x} + \frac{1}{2}\beta_1 \sin 2\varphi_1 + \frac{1}{2}\beta_2 \sin 2\varphi_2)^2 \\ & + (u_{y,y} - \frac{1}{2}\beta_1 \sin 2\varphi_1 - \frac{1}{2}\beta_2 \sin 2\varphi_2)^2 + \frac{1}{2}(u_{x,y} + u_{y,x} - \beta_1 \cos 2\varphi_1 \\ & - \beta_2 \cos 2\varphi_2)^2 + k|\beta_{1,x} \cos\varphi_1 + \beta_{1,y} \sin\varphi_1| + \frac{1}{2}k(\beta_{1,x} \cos\varphi_1 + \beta_{1,y} \sin\varphi_1)^2 \\ & + k|\beta_{2,x} \cos\varphi_2 + \beta_{2,y} \sin\varphi_2| + \frac{1}{2}k(\beta_{2,x} \cos\varphi_2 + \beta_{2,y} \sin\varphi_2)^2 + k\chi|\beta_{1,x} \cos\varphi_1 \\ & + \beta_{1,y} \sin\varphi_1||\beta_{2,x} \cos\varphi_2 + \beta_{2,y} \sin\varphi_2| \Big] dxdy + \int_{-h/2}^{h/2} \tau y u_x|_{x=L} dy. \end{aligned} \tag{11.7}$$

The bar over the dimensionless quantities is dropped for short because in the subsequent analysis we shall deal only with them.

Energy functional (11.7) contains a small parameter h, so it can be reduced to 1-D energy functional by the variational asymptotic method (see Berdichevsky, 1983; Le, 1999). For this purpose let us introduce the rescaled coordinate $\zeta = y/h$, $\zeta \in (-1/2, 1/2)$ to fix the domain over the thickness in the passage to the limit $h \to 0$. With this new variable ζ the small parameter h enters the functional explicitly through the formulas

$$u_{i,y} = \frac{1}{h} u_{i,\zeta}, \quad \beta_{\alpha,y} = \frac{1}{h}\beta_{\alpha,\zeta}, \quad \alpha = 1, 2.$$

Since the boundary condition at $x = L$ does not influence the inner asymptotic distributions of the displacements and the plastic slips over the thickness, we put

$\tau = 0$ for a while. At the first step of the variational-asymptotic procedure we keep the asymptotic principal terms in (11.7) to obtain

$$\begin{aligned} I_0 = h \int_0^L \int_{1/2}^{1/2} \Big[& \frac{1}{2h^2}\gamma(u_{y,\zeta})^2 + \frac{1}{h^2}(u_{y,\zeta})^2 + \frac{1}{2h^2}(u_{x,\zeta})^2 + \frac{k}{h}|\beta_{1,\zeta}\sin\varphi_1| \\ & + \frac{k}{2h^2}(\beta_{1,\zeta}\sin\varphi_1)^2 + \frac{k}{h}|\beta_{2,\zeta}\sin\varphi_2| + \frac{k}{2h^2}(\beta_{2,\zeta}\sin\varphi_2)^2 \\ & + \frac{\chi k}{h^2}|\beta_{1,\zeta}\sin\varphi_1||\beta_{2,\zeta}\sin\varphi_2| \Big] dx d\zeta. \end{aligned}$$

Functional I_0 is positive definite, so its minimum must be zero and is achieved at $u_{x,\zeta} = u_{y,\zeta} = \beta_{1,\zeta} = \beta_{2,\zeta} = 0$. For the bending states which we are interested in let us set at this step $u_x = 0$ and $\beta_1 = \beta_2 = 0$ to obtain

$$u_x = 0, \quad u_y = v(x), \quad \beta_1 = \beta_2 = 0.$$

At the second step, we fix $v(x)$ and seek the minimizer in the form

$$u_x = u_x'(x,\zeta), \quad u_y = v(x) + u_y'(x,\zeta), \quad \beta_1 = \beta_1'(x,\zeta), \quad \beta_2 = \beta_2'(x,\zeta),$$

with $u_x'(x,\zeta), u_y'(x,\zeta), \beta_1'(x,\zeta), \beta_2'(x,\zeta)$ being the correction terms which are small compared to $v(x)$. Substituting these formulas into (11.7) and then keeping the asymptotically leading terms containing $u_x'(x,\zeta)$, $u_y'(x,\zeta)$, and $\beta_\alpha'(x,\zeta)$, we obtain

$$\begin{aligned} I_1 = h \int_0^L \int_{1/2}^{1/2} \Big[& \frac{1}{2h^2}\gamma(u'_{y,\zeta})^2 + \frac{1}{h^2}(u'_{y,\zeta})^2 + \frac{1}{2}(\frac{1}{h}u'_{x,\zeta} + v_{,x})^2 + \frac{k}{h}|\beta'_{1,\zeta}\sin\varphi_1| \\ & + \frac{k}{2h^2}(\beta'_{1,\zeta}\sin\varphi_1)^2 + \frac{k}{h}|\beta'_{2,\zeta}\sin\varphi_2| + \frac{k}{2h^2}(\beta'_{2,\zeta}\sin\varphi_2)^2 \\ & + \frac{\chi k}{h^2}|\beta'_{1,\zeta}\sin\varphi_1||\beta'_{2,\zeta}\sin\varphi_2| \Big] dx d\zeta \end{aligned}$$

Since functional I_1 is also positive definite, its minimum is again zero and is achieved at

$$u'_{y,\zeta} = 0, \quad u'_{x,\zeta} = -hv_{,x}, \quad \beta'_{1,\zeta} = 0, \quad \beta'_{2,\zeta} = 0.$$

This implies

$$u_y' = 0, \quad u_x' = -h\zeta v_{,x}, \quad \beta_1' = 0, \quad \beta_2' = 0.$$

At the third step of the variational-asymptotic procedure, we look for the minimizer in the form

$$u_x = -h\zeta v_{,x} + u''_x(x,\zeta), \quad u_y = v(x) + u''_y(x,\zeta), \tag{11.8}$$
$$\beta_1 = \beta''_1(x,\zeta), \quad \beta_2 = \beta''_2(x,\zeta),$$

with $u''_x(x,\zeta)$, $u''_y(x,\zeta)$, $\beta''_1(x,\zeta)$, and $\beta''_2(x,\zeta)$ being the correction terms which are assumed small compared with those found at the second step. By redefining $v(x)$ if required, we may put the following constraints on these correction terms

$$\langle u''_y \rangle = 0, \quad \langle \beta''_1 \rangle = 0, \quad \langle \beta''_2 \rangle = 0, \quad \text{where} \quad \langle \cdot \rangle = \int_{-1/2}^{1/2} \cdot \, d\zeta. \tag{11.9}$$

Constraint $\langle u''_y \rangle = 0$ means that $v(x)$ is the average deflection (over the thickness) of the beam. Substituting (11.8) into (11.7), then keeping the asymptotically leading terms containing $u''_x(x,\zeta)$, $u''_y(x,\zeta)$, $\beta''_1(x,\zeta)$, and $\beta''_2(x,\zeta)$ to get

$$\begin{aligned} I_2 = h\int_0^L \int_{-1/2}^{1/2} \Big[& \frac{1}{2}\gamma(-h\zeta v_{,xx} + \frac{1}{h}u''_{y,\zeta})^2 + (-h\zeta v_{,xx} + \frac{\beta''_1}{2}\sin 2\varphi_1 + \frac{\beta''_2}{2}\sin 2\varphi_2)^2 \\ & + (\frac{1}{h}u''_{y,\zeta} - \frac{1}{2}\beta''_1 \sin 2\varphi_1 - \frac{1}{2}\beta''_2 \sin 2\varphi_2)^2 + \frac{1}{2}(\frac{1}{h}u''_{x,\zeta} - \beta''_1 \cos 2\varphi_1 - \beta''_2 \cos 2\varphi_2)^2 \\ & + \frac{k}{h}|\beta''_{1,\zeta} \sin\varphi_1| + \frac{k}{2h^2}(\beta''_{1,\zeta} \sin\varphi_1)^2 + \frac{k}{h}|\beta''_{2,\zeta} \sin\varphi_2| + \frac{k}{2h^2}(\beta''_{2,\zeta} \sin\varphi_2)^2 \\ & + \frac{\chi k}{h^2}|\beta''_{1,\zeta} \sin\varphi_1||\beta''_{2,\zeta} \sin\varphi_2| \Big] dx d\zeta. \end{aligned}$$

Functional I_2 can be reduced to a functional depending only on β''_1 and β''_2. Indeed, fixing first β''_1 and β''_2 and varying this functional with respect to u''_x and u''_y, then using the natural boundary conditions at $\zeta = \pm 1/2$, we obtain,

$$\begin{aligned} \frac{1}{h}(\gamma+2)u''_{y,\zeta} &= \gamma h\zeta v_{,xx} + \beta''_1 \sin 2\varphi_1 + \beta''_2 \sin 2\varphi_2, \\ \frac{1}{h}u''_{x,\zeta} &= \beta''_1 \cos 2\varphi_1 + \beta''_2 \cos 2\varphi_2. \end{aligned} \tag{11.10}$$

After finding u''_x and u''_y according to these equations, we substitute (11.10) into (11.7) and change ζ back to y. Since the functional does not contain $\beta''_{1,x}$ and $\beta''_{2,x}$, the thickness problem reduces to minimizing the following functional with respect to $\beta''_1(x,y)$ and $\beta''_2(x,y)$

$$
\begin{aligned}
I_3 = \int_{-h/2}^{h/2} \Big[& \frac{\kappa}{2}(-2v_{,xx}y + \beta_1'' \sin 2\varphi_1 + \beta_2'' \sin 2\varphi_2)^2 + k|\beta_{1,y}'' \sin\varphi_1| \\
& + \frac{k}{2}(\beta_{1,y}'' \sin\varphi_1)^2 + k|\beta_{2,y}'' \sin\varphi_2| + \frac{k}{2}(\beta_{2,y}'' \sin\varphi_2)^2 \\
& + \chi k|\beta_{1,y}'' \sin\varphi_1||\beta_{2,y}'' \sin\varphi_2| \Big] dy,
\end{aligned}
$$

where

$$
\kappa = \frac{1}{2(1-\nu)}.
$$

Solution of this minimization problem enables one to find $\beta_1''(x,y)$ and $\beta_2''(x,y)$ required for the reduction to 1-D theory. The analytical solution of the thickness problem can be found in the special case of symmetric slip systems with the angles $\varphi_1 = \varphi \in (0, \pi/2)$, $\varphi_2 = \pi - \varphi$. Obviously, $\beta_1'' = \beta = -\beta_2''$ in this case due to the symmetry of the problem. Thus, functional I_3 simplifies to

$$
I_3 = \int_{-h/2}^{h/2} [\frac{\kappa}{2}(-2v_{,xx}y + 2\beta \sin 2\varphi)^2 + 2k|\beta_{,y} \sin\varphi| + k(1+\chi)(\beta_{,y} \sin\varphi)^2]\, dy. \tag{11.11}
$$

Functional (11.11) is similar to the functional obtained in Le and Nguyen (2013), so we present the solution of the thickness problem without detailed derivation. The minimizer $\beta(x,y)$ is

$$
\beta(x,y) = \begin{cases} \beta_m(x) & \text{for } y \in (-h/2, -l(x)/2), \\ \beta_0(x,y) & \text{for } y \in (-l(x)/2, 0), \\ -\beta(x,-y) & \text{for } y \in (0, h/2). \end{cases} \tag{11.12}
$$

Function $\beta_0(x,y)$ is found to be

$$
\beta_0(x,y) = \frac{v_{,xx}}{\sin 2\varphi} \frac{1}{\eta} \left(\eta y - \frac{1}{\cosh \frac{\eta l}{2}} \sinh \eta y \right),
$$

where

$$
\eta = 2\sqrt{\frac{2\kappa}{k(1+\chi)}} \cos\varphi. \tag{11.13}
$$

Function $\beta_m(x)$ is given by

$$
\beta_m(x) = \frac{v_{,xx}}{\sin 2\varphi} \frac{1}{\eta} \left(-\frac{\eta l}{2} + \tanh \frac{\eta l}{2} \right).
$$

Finally, the length $l(x)$ should be found by solving the following transcendental equation

$$\kappa v_{,xx}[\frac{1}{4}(h^2-l^2)+\frac{1}{\eta}(-\frac{\eta l}{2}+\tanh\frac{\eta l}{2})(h-l)]\sin 2\varphi - k\,\text{sign}(\beta_{0,y})\sin\varphi = 0. \tag{11.14}$$

If the length l is small, then the sign of $v_{,xx}$ coincides with the sign of $\beta_{0,y}$ evaluated right from the point $y=-l/2$. For definiteness let $\beta_{0,y}(-l/2+0)>0$ so that $v_{,xx}>0$. Note that, if the curvature of the beam is constant, then $l(x)$ does not depend on x and remains constant over the whole length of the beam. In this case (11.14) can be regarded as the equation for $v_{,xx}$ once l is known. By integrating (11.10) in our special case and taking the constraints (11.9) into account, we obtain the displacements u_x and u_y in the form

$$u_x = -v_{,x}y,$$

$$u_y = v(x) + \frac{\gamma}{\gamma+2}\frac{1}{2}(y^2-\frac{h^2}{12})v_{,xx} + \frac{2}{\gamma+2}\left(\int_0^y \beta(x,\xi)\,d\xi - \Lambda\right)\sin 2\varphi,$$

where

$$\Lambda = \langle \int_0^y \beta(x,\xi)d\xi \rangle.$$

Having found the solution of the thickness problem, let us now substitute the above formulas for the displacements together with the expression of β into the energy functional (11.7). Keeping the asymptotically principal terms and integrating over the thickness we obtain the one-dimensional functional

$$J[v(x)] = \int_0^L \Phi(v_{,xx})\,dx - Mv_{,x}|_{x=L}, \tag{11.15}$$

where the bending energy density reads

$$\Phi(\omega) = c_1\omega^2 + c_2\omega.$$

In these formulas we use ω to denote the curvature $v_{,xx}$, $M=\tau h^2/12$ is the resultant moment, and

$$c_1 = 2\int_{-h/2}^{-l/2} \frac{\kappa}{2}(2q_0-2y)^2\,dy + 2\int_{-l/2}^{0}\left[\frac{\kappa}{2}(2q-2y)^2 + \frac{k(1+\chi)}{4\cos^2\varphi}(q_{,y})^2\right]dy,$$

$$c_2 = 2\int_{-l/2}^{0} 2kq_{,y}\frac{\sin\varphi}{\sin 2\varphi}dy = -\frac{2k}{\cos\varphi}q_0, \tag{11.16}$$

with

$$q(y) = \frac{1}{\eta}\left(\eta y - \frac{1}{\cosh\frac{\eta l}{2}}\sinh\eta y\right), \tag{11.17}$$

and $q_0 = q(-l/2)$. Note that, as the coefficients c_1 and c_2 depends on the curvature ω through l, the energy density is not quadratic with respect to ω. Besides, if $q(y) = 0$, then $c_1 = \kappa h^3/6$ and $c_2 = 0$, so the obtained functional reduces to the classical 1-D functional of the elastic beam as expected (Le, 1999).

Varying functional (11.15) with respect to the deflection v, we obtain the differential equation of bending

$$m_{,xx} = 0, \quad m = \frac{\partial\Phi}{\partial\omega}, \tag{11.18}$$

subject to the boundary conditions

$$\begin{cases} v(0) = 0, \quad v_{,x}(0) = 0, \\ m(L) - M = 0, \quad m_{,x}(L) = 0. \end{cases} \tag{11.19}$$

Equation (11.18), together with the conditions $(11.19)_2$, implies that

$$m(\omega) = 2c_1\omega + \frac{dc_1}{d\omega}\omega^2 + c_2 + \frac{dc_2}{d\omega}\omega = M. \tag{11.20}$$

Since the bending moment m is independent of x, the curvature must also be constant over the length of the beam. Consequently, l and β_0 are independent of x and the upper and lower layers of the beam are dislocation-free. Together with (11.14), this equation determines the moment-curvature curve during the plastic bending. To plot this curve let us compute the derivatives of c_1 and c_2 with respect to ω

$$\frac{dc_1}{d\omega} = \frac{dc_1}{dl}\frac{dl}{d\omega}, \quad \frac{dc_2}{d\omega} = \frac{dc_2}{dl}\frac{dl}{d\omega}.$$

From (11.14) we find that

$$\frac{d\omega}{dl} = \frac{4k\eta\tanh\frac{l\eta}{2}\left(\eta\,(h-l)\tanh\frac{l\eta}{2} + 2\right)}{\kappa\cos\varphi(h-l)^2\left(\eta\,(h-l) + 4\tanh\frac{l\eta}{2}\right)^2}.$$

This formula, together with (11.16), enables one to determine $dc_1/d\omega$ and $dc_2/d\omega$, required for plotting the moment-curvature curve.

The threshold value of curvature at which dislocations begin to nucleate is calculated by letting l go to zero in (11.14). This yields

$$\omega_{en} = \frac{2k}{\kappa h^2\cos\varphi}.$$

The threshold value of moment can be computed from (11.20). Taking into account that, at $l = 0$, $c_1 = \kappa h^3/5$, $c_2 = 0$, while $dc_1/d\omega = dc_2/d\omega = 0$, we get

$$M_{en} = \frac{2kh}{3\cos\varphi}.$$

Returning to the original physical quantities according to (11.6) we have

$$M_{en} = \frac{\tau_{en} h^3 a}{12} = \frac{2\mu kah}{3b\rho_s \cos\varphi},$$

which exhibits explicitly the size effect. Besides, M_{en} goes to infinity as the shear modulus μ goes to infinity, so this coincides well with the result obtained by Ashby (1970). Thus, if $M < M_{en}$, then $\beta = 0$, so no dislocation is nucleated and we have purely elastic solution. The plastic yielding begins at $M = M_{en}$.

Combining the elastic and plastic responses together, we get the moment-curvature relation in the following form

$$M = \begin{cases} \dfrac{\kappa h^3}{3}\omega & \text{for } M < M_{en}, \\ 2c_1\omega + \dfrac{dc_1}{d\omega}\omega^2 + c_2 + \dfrac{dc_2}{d\omega}\omega & \text{for } M > M_{en}. \end{cases}$$

Knowing the curvature ω from M, we integrate the equation $v_{,xx} = \omega$ and use the boundary conditions $(11.19)_1$ to obtain the deflection of the beam

$$v(x) = \frac{1}{2}\omega x^2.$$

11.4 Numerical Simulations

In order to simulate the minimizer numerically, we choose $h = 0.1$, $L = 10$, $\nu = 0.25$, $k = 10^{-4}$, and $\chi = 0.8$. Figure 11.2 shows the plot of energetic threshold of the bending moment M_{en} as function of the angle φ. It can be seen that the threshold moment becomes infinite as φ goes to $\pi/2$ and has a local minimum at $\varphi = 0$. It must be emphasized that this result is obtained under the assumption that the crystal admits two active symmetric slip systems. So, if $\varphi = \pi/2$ the slip systems merge into one that remains inactive during the whole process of bending because of its vanishing resolved shear stress (cf. with Le and Nguyen, 2013). However, if the slip systems are oriented unsymmetrically, then the one with largest Schmid factor will be activated at a certain finite bending moment.

For $M > M_{en}$ the plastic slip becomes non-zero. Figure 11.3 show the plots of plastic slip $\beta(y)$ for $M = 0.0001$ at two different orientations of slip systems. The plastic strain vanishes on the middle line of the beam as expected and reaches its

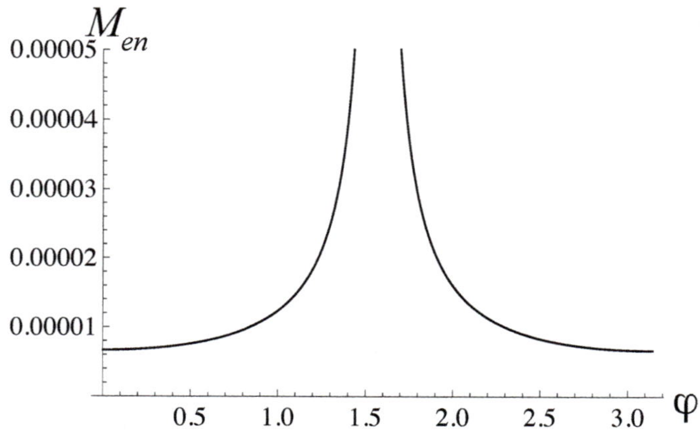

Fig. 11.2: Function $M_{en}(\varphi)$

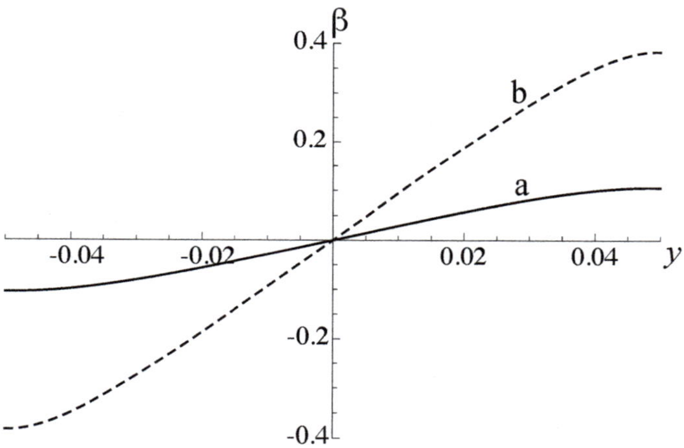

Fig. 11.3: Function $\beta(y)$ for $M = 0.0001$. a) $\varphi = \pi/3$, b) $\varphi = \pi/6$

maximum and minimum at the traction-free faces. There are two small boundary layers near these faces where the plastic slip remains constant.

On Fig. 11.4, where the dimensionless dislocation density $\rho(y) = \beta_{,y} \sin\varphi$ is plotted for $M = 0.0001$ and for two different angles, it is seen that the excess edge dislocations of the same sign of each group are concentrated in the middle of the beam thickness, with maximum dislocation density achieved at $y = 0$. Although no obstacle exists on the middle line, the repulsive forces between dislocations of the same sign prevent them from colliding. Thus, the high concentration of dislocations can be regarded as the dislocation pile-up against the middle line. The dislocation free zones are $y \in (-h/2, -l/2)$ and $y \in (l/2, h/2)$.

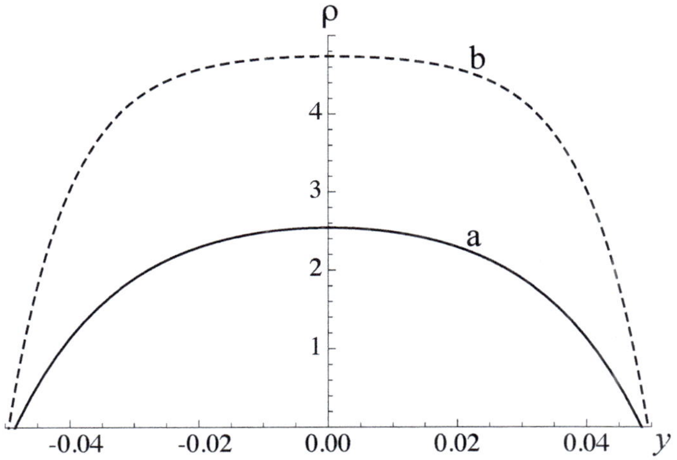

Fig. 11.4: Dislocation density for $M = 0.0001$. a) $\varphi = \pi/3$ ($l \approx 0.097$), b) $\varphi = \pi/6$ ($l \approx 0.099$)

The dimensionless deflection of the beam, $v(x)$, is shown in Fig. 11.5 for $M = 0.0001$ and (a) $\varphi = \pi/3$, (b) $\varphi = \pi/6$. Since the curvature of the beam is constant, the thickness of the dislocation zone l does not depend on x.

On Fig. 11.6 we show the moment-curvature curve for $\varphi = \pi/3$ and $\varphi = \pi/6$. Up to the threshold moment M_{en} (corresponding to point A on this figure) the moment-curvature curve is a straight line corresponding to the linear elastic beam theory. Then the curve becomes non-linear and increasing as M increases and l increases

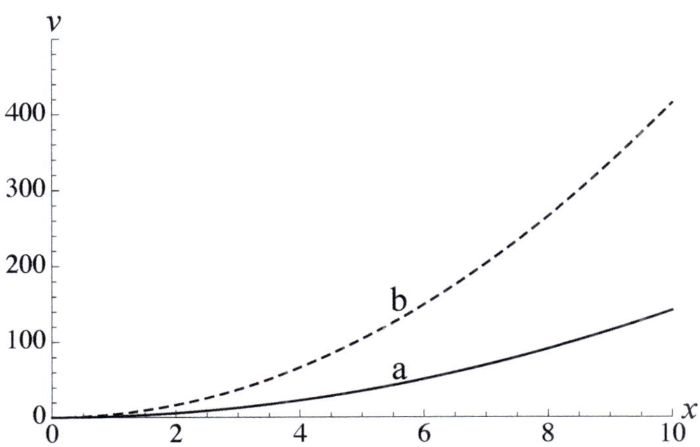

Fig. 11.5: Deflection of the beam

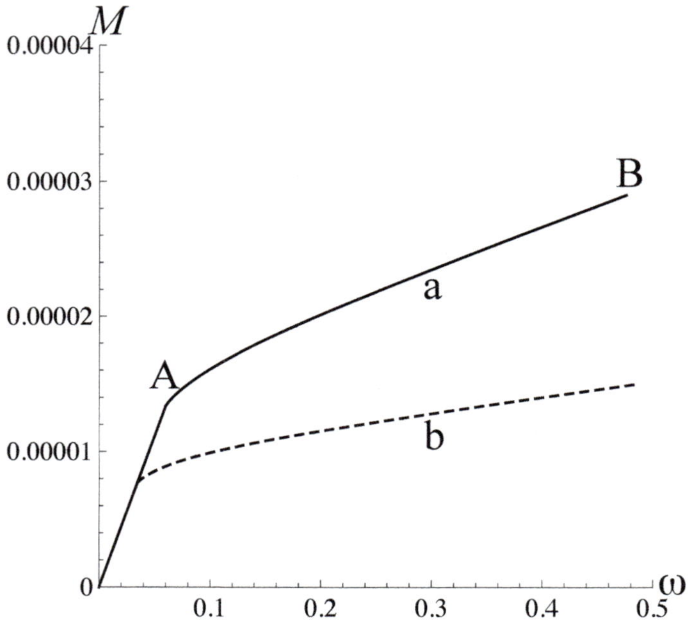

Fig. 11.6: Moment-curvature curve. a) $\varphi = \pi/3$, b) $\varphi = \pi/6$

from zero to $h/2$. This nonlinear portion describes the work hardening due to the dislocation pile-up against the middle line of the beam. If the bending moment is increased from zero up to the moment M_B corresponding to B (the loading case) we follow the moment-curvature curve from O through A to B. Now, if we unload the beam by decreasing the bending moment from M_B to zero, the curvature and moment follow the same path BAO (in the inverse direction), and at the end of the unloading process no residual curvature of the beam is observed.

11.5 Non-Zero Dissipation

If the dissipation cannot be neglected, the problems reduce to minimizing the relaxed energy functionals (11.5). The signs of the dissipation terms in this functional depend on whether $\dot{\beta}_\alpha > 0$ or $\dot{\beta}_\alpha < 0$, $\alpha = 1,2$. However, for the beam bending it is easy to see that both cases occur simultaneously during the plastic deformations. Indeed, from the elementary beam theory (and also from the previous simulations) we know that, as the bending moment is increased (loading), $\dot{\beta}_1 > 0$ for $y > 0$ and $\dot{\beta}_1 < 0$ for $y < 0$, while $\dot{\beta}_2 < 0$ for $y > 0$ and $\dot{\beta}_2 > 0$ for $y < 0$. In contrary, if the bending moment is decreased (unloading or loading in the opposite direction), β_1 is either frozen or $\dot{\beta}_1 > 0$ for $y < 0$ and $\dot{\beta}_1 < 0$ for $y > 0$ while the opposite happens to β_2.

Since these functionals differs from each other in the loading and unloading case, the case study must be done separately.

The case study becomes simpler for the symmetric slip systems with $\varphi_1 = \varphi \in (0, \pi/2)$, $\varphi_2 = \pi - \varphi$. Due to the symmetry of the problem, we have $\beta_1 = \beta = -\beta_2$. Therefore $K\,\mathrm{sign}(\dot{\beta}_1)\beta_1 = K\,\mathrm{sign}(\dot{\beta}_2)\beta_2$. According to the above property, both terms can be replaced by $K\,\mathrm{sign}\, y\,\beta$. We use again the dimensionless quantities (11.6) and write the functional (11.5) in this special case in the form

$$I_d = \int_0^L \int_{-h/2}^{h/2} \Big[\frac{1}{2}\gamma(u_{x,x}+u_{y,y})^2 + (u_{x,x}+\beta\sin 2\varphi)^2 + (u_{y,y}-\beta\sin 2\varphi)^2 + \frac{1}{2}(u_{x,y}+u_{y,x})^2 + k|\beta_{,x}\cos\varphi+\beta_{,y}\sin\varphi| + \frac{1}{2}k(\beta_{,x}\cos\varphi+\beta_{,y}\sin\varphi)^2 + k|\beta_{,x}\cos\varphi-\beta_{,y}\sin\varphi| + \frac{1}{2}k(\beta_{,x}\cos\varphi-\beta_{,y}\sin\varphi)^2 + k\chi|\beta_{,x}\cos\varphi+\beta_{,y}\sin\varphi|\cdot|\beta_{,x}\cos\varphi-\beta_{,y}\sin\varphi| + 2\epsilon\,\mathrm{sign}\, y\,\beta\Big]\,dx\,dy + \int_{-h/2}^{h/2} \tau y u_x|_{x=L}\,dy,$$

where $\epsilon = K/\mu$. As compared to the previous functional (11.7) the additional term $2\epsilon\,\mathrm{sign}\, y\,\beta$ does not belong to the asymptotically principal terms. Therefore, up to the second step of the variational-asymptotic procedure this term does not have any influence on the inner asymptotic expansion. At the third step, we are looking for the minimize in the form (11.8) such that the constraints (11.9) are fulfilled. Fixing β'' and minimizing the relaxed energy with respect to u''_x and u''_y we find them in the form (11.10). Then, the functional reduces to

$$I_d = \int_{-h/2}^{h/2} \Big[\frac{\kappa}{2}(-2v_{,xx}y + 2\beta''\sin 2\varphi)^2 + 2k|\beta''_{,y}\sin\varphi| + k(\beta''_{,y}\sin\varphi)^2 + \chi k(\beta''_{,y}\sin\varphi)^2 + 2\epsilon\,\mathrm{sign}\, y\,\beta''\Big]\,dy$$

Now the term $2\epsilon\,\mathrm{sign}\, y\,\beta''$ should be kept because it has the same order as the cross term $-4\kappa v_{,xx} y \beta'' \sin 2\varphi$. The presence of this term changes radically the behavior of the minimizer. Since at the beginning of plastic yielding the resolve shear stress near the beam axis is always smaller that the critical resolved shear stress K, dislocations cannot reach this neutral axis[1]. Thus, near the beam axis the plastic slip must vanish identically at this stage. Therefore in the first stage of plastic bending we look for the minimizer in the form

[1] Our paper (Le and Nguyen, 2013) missed this early stage of plastic yielding.

$$\beta''(y) = \begin{cases} \beta_m & \text{for } y \in (-h/2, -l_1/2), \\ \beta_1(y) & \text{for } y \in (-l_1/2, -l_2/2), \\ 0 & \text{for } y \in (-l_2/2, 0), \\ -\beta(-y) & \text{for } y \in (0, h/2). \end{cases} \tag{11.21}$$

According to (11.21), the layer $y \in (-l_2, l_2)$ deforms elastically, and this agrees well with the classical beam theory for elasto-plastic materials.

With the solution Ansatz (11.21) being plugged into functional (11.21), we get the variational (thickness) problem that can be solved in a similar way as that presented in Sect. 11.3. For function $\beta_1(y)$ we find

$$\beta_1(y) = \frac{v_{,xx}}{\sin 2\varphi} y + \frac{\epsilon}{2\kappa \sin^2 2\varphi} + C_1 \cosh \eta y + C_2 \sinh \eta y, \tag{11.22}$$

where

$$C_1 = \Big(\frac{l_1 \omega}{2 \sin 2\varphi} - \frac{\epsilon}{2\kappa \sin^2 2\varphi} + \beta_m\Big) \cosh \frac{\eta l_1}{2} - \frac{\omega}{\eta \sin 2\varphi} \sinh \frac{\eta l_1}{2}, \tag{11.23}$$

$$C_2 = -\frac{\omega}{\eta \sin 2\varphi} \cosh \frac{\eta l_1}{2} + \Big(\frac{l_1 \omega}{2 \sin 2\varphi} - \frac{\epsilon}{2\kappa \sin^2 2\varphi} + \beta_m\Big) \sinh \frac{\eta l_1}{2}. \tag{11.24}$$

Here, $\omega = v_{,xx}$ denotes as before the curvature of the deformed beam axis, while η is given by (11.13). One may find the unknown quantities β_m, l_1, and l_2 in terms of the given curvature ω, or alternatively, use β_m as the parameter and express ω, l_1, and l_2 through it. We find the latter way more convenient for numerical simulations. In this case the curvature ω is given by

$$\omega = \frac{2k \sin \varphi}{\kappa(h^2 - l_1^2) \sin 2\varphi} + \frac{2\epsilon - 4\kappa \beta_m \sin^2 2\varphi}{\kappa(h + l_1) \sin 2\varphi}, \tag{11.25}$$

while l_1 and l_2 must be found from the equations

$$\beta_1(-l_2/2) = 0, \quad \beta_{1,y}(-l_1/2) = 0. \tag{11.26}$$

If the bending moment and therefore the curvature increases further, it turns out that equations (11.26) have no longer solutions. This means that in the later stage of plastic deformation dislocations may reach the neutral line. In this case we use the same solution Ansatz as (11.12). For function $\beta_0(y)$ we find

$$\begin{aligned} \beta_0(y) = {} & \frac{v_{,xx}}{\sin 2\varphi} \frac{1}{\eta} \left(\eta y - \frac{1}{\cosh \frac{\eta l}{2}} \sinh \eta y \right) \\ & + \frac{\epsilon}{2\kappa \sin^2 2\varphi} \left(1 - \cosh \eta y - \tanh \frac{\eta l}{2} \sinh \eta y \right). \end{aligned} \tag{11.27}$$

Computing $\beta_0(y)$ at $y = -l/2$ we get

$$\beta_m = \frac{v_{,xx}}{\sin 2\varphi \eta}\left(-\frac{\eta l}{2} + \tanh\frac{\eta l}{2}\right) + \frac{\epsilon}{2\kappa \sin^2 2\varphi}\left(1 - \frac{1}{\cosh\frac{\eta l}{2}}\right)$$

The transcendental equation for l reads

$$\kappa v_{,xx}[\frac{1}{4}(h^2 - l^2) + \frac{1}{\eta}(-\frac{\eta l}{2} + \tanh\frac{\eta l}{2})(h-l)]\sin 2\varphi$$
$$-\frac{\epsilon}{2}\frac{1}{\cosh\frac{\eta l}{2}}(h-l) - k\,\mathrm{sign}(\beta_{1,y})\sin\varphi = 0.$$

Having found the plastic slip, we can now compute the dimensionless bending moment. In the first stage of plastic yielding it is convenient to compute it according to

$$m = \int_{-h/2}^{h/2} 4\kappa(\omega y - \beta''(y)\sin 2\varphi) y\,dy. \tag{11.28}$$

Substituting (11.21) into this formula and making use of (11.22) and (11.23), we obtain

$$\begin{aligned} m =& \frac{1}{3}\kappa\omega h^3 + \kappa \sin 2\varphi \beta_m (h^2 - l_1^2) + \frac{1}{3}\kappa\omega(l_1^3 - l_2^3) - \frac{\epsilon(l_1^2 - l_2^2)}{2\sin 2\varphi} \\ &+ \frac{4\kappa\sin 2\varphi}{\eta^2}[(2C_1 + C_2 l_1\eta)\cosh\frac{\eta l_1}{2} - (2C_1 + C_2 l_2 \eta)\cosh\frac{\eta l_2}{2} \\ &- (2C_2 + C_1 l_1 \eta)\sinh\frac{\eta l_1}{2} + (2C_2 + C_1 l_2\eta)\sinh\frac{\eta l_2}{2}]. \end{aligned} \tag{11.29}$$

Note that when $\beta''(y) \equiv 0$, the bending moment $m = \frac{1}{3}\kappa\omega h^3$ is exactly that of the purely elastic solution. In the second stage of bending we compute the bending energy density

$$\Phi(\omega) = c_1\omega^2 + c_2\omega,$$

with the coefficient c_1 and c_2 being given by

$$c_1 = 2\int_{-h/2}^{-l/2} \frac{\kappa}{2}(2q_0 - 2y)^2\,dy + 2\int_{-l/2}^{0}\left[\frac{\kappa}{2}(2q - 2y)^2 + \frac{k(1+\chi)}{4\cos^2\varphi}(q_{,y})^2\right]dy,$$

$$c_2 = 2\int_{-l/2}^{0}\left[\frac{k}{\cos\varphi}q_{,y} + \frac{\epsilon}{\kappa \sin 2\varphi}(\frac{k(1+\chi)}{4\cos^2\varphi}q_{,y}p_{,y} + \kappa(p(y) - 1)(2q(y)\right.$$
$$\left.- 2y))\right]dy + 2\int_{-h/2}^{-l/2}\left[\frac{\epsilon}{\sin 2\varphi}(p_0 - 1)(2q_0 - 2y)\right]dy,$$

with

$$q(y) = \frac{1}{\eta}(\eta y - \frac{\sinh \eta y}{\cosh \frac{\eta l}{2}}), \quad p(y) = 1 - \cosh \eta y - \tanh \frac{\eta l}{2} \sinh \eta y,$$

and $p_0 = p(-l/2)$, $q_0 = q(-l/2)$. Then the bending moment can be computed from this bending energy in accordance with (11.20).

Consider now the following loading/unloading process. We first bend the beam slowly and successively by increasing the moment until some maximum value $M_* > M_d$ such that at the end of the loading process the plastic slip becomes β_*. Then we unload the beam by reducing the bending moment back to zero. It is reasonable to assume that dislocations are frozen during this unloading process with $\beta = \beta_*$ and $\dot{\beta} = 0$. In this case the bending moment becomes

$$m = \int_{-h/2}^{h/2} 4\kappa(\omega y - \beta_*''(y)\sin 2\varphi) y\, dy, \tag{11.30}$$

where $\beta_*''(y)$ is the plastic slip found at the end of the loading process. Thus, during the unloading process the moment-curvature curve becomes a straight line.

11.6 Numerical Simulation

In order to simulate the minimizer numerically, we choose the same numerical values for h, L, ν, k, χ as in Sect. 11.4, and $\varphi = \pi/6$. For $\epsilon = K/\mu$ we choose the value $\epsilon = 0.001$. In order to find the dissipative threshold for the bending moment we need to study equations (11.26) in the limit $\beta_m \to 0$. It turns out that in this limit $l_2 \to l_1$ which is given by

$$l_1 = \frac{\epsilon}{\kappa \omega \sin 2\varphi}.$$

Besides, as β_m goes to zero,

$$\omega = \frac{4k \sin \varphi + 2\epsilon(h - l_1)}{(h^2 - l_1^2)\sin 2\varphi}.$$

Solving these equations, we find the dissipative threshold of dislocation nucleation to be

$$\omega_d = \frac{\epsilon}{\kappa l_1 \sin 2\varphi}, \tag{11.31}$$

where

$$l_1 = h + 2\frac{k}{\epsilon}\sin\varphi - \sqrt{(h + 2\frac{k}{\epsilon}\sin\varphi)^2 - h^2}. \tag{11.32}$$

As this threshold value depends on h, we have clearly the size effect. For the above chosen parameters $l_1 \approx 0.01896$, while $\omega_d \approx 0.09136$. The corresponding threshold moment is

$$M_d = \frac{1}{3}\kappa\omega_d h^3 \approx 0.0000203. \tag{11.33}$$

Figure 11.7 shows the evolution of the plastic slip in the first stage of plastic yielding. Thus, as the magnitude of β_m increases, the plastic slip increases too. The elastic zone in the middle decreases, and at $\beta_m = -0.0002196$, the length l_2 vanishes. After this equations (11.26) have no longer solutions. The plastic slip at the later stage of plastic bending (computed according to (11.27)) looks similar to that in Sect. 11.4.

On Fig. 11.8, where the dimensionless dislocation density $\rho(y) = \beta_{,y}\sin\varphi$ is plotted for three different β_m at the first stage of plastic yielding, it is seen that the excess edge dislocations of the same sign are concentrated in two layers, leaving the central layers $(-l_1/2, l_1/2)$ as well as the layers near the free faces dislocation-free. The dislocation density distributions at the later stage of plastic bending (computed according to (11.27)) look similar to those in Sect. 11.4.

On Fig. 11.9 we show the moment-curvature curve for the chosen parameters. Up to the threshold moment M_d (corresponding to point A on this figure) the moment-curvature curve is a straight line corresponding to the linear elastic beam theory. Then the curve becomes non-linear and, in the first stage of plastic yielding, monotone increasing as ω increases. At the beginning of the later stage of plastic bending we can observe a small softening behavior of the moment-curvature curve. The further monotonously increasing portion of the moment-curvature curve describes the work hardening due to the dislocation pile-up against the middle line of the beam. If the curvature is increased from zero up to the value ω_B corresponding to B (the loading

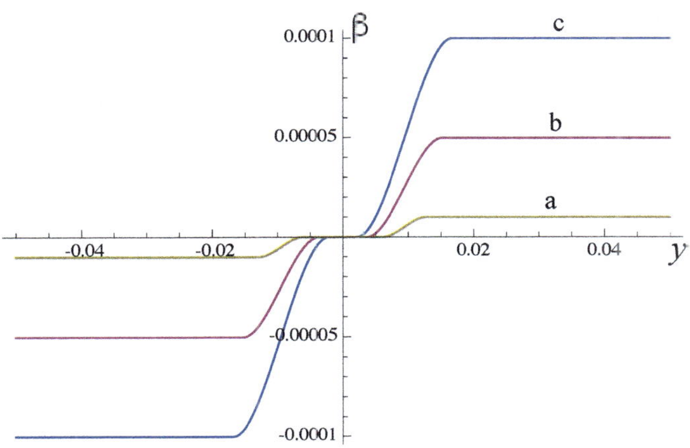

Fig. 11.7: Function $\beta(y)$. a) $\beta_m = -10^{-5}$ ($\omega = 0.09207$), b) $\beta_m = -5 \times 10^{-5}$ ($\omega = 0.094$), c) $\beta_m = -10^{-4}$ ($\omega = 0.096$)

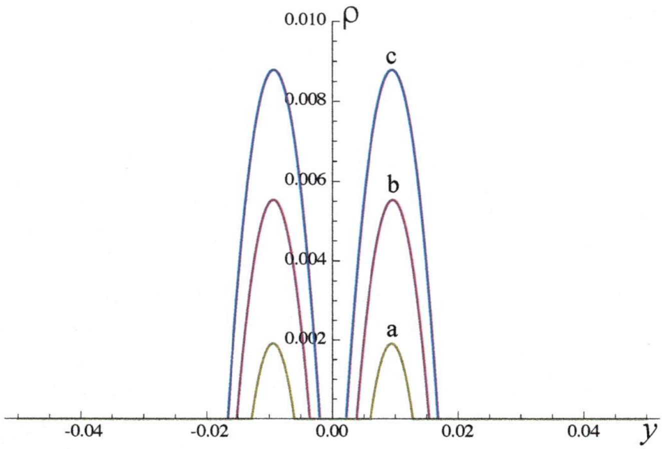

Fig. 11.8: Dislocation density. a) $\beta_m = -10^{-5}$ ($\omega = 0.09207$), b) $\beta_m = -5 \times 10^{-5}$ ($\omega = 0.094$), c) $\beta_m = -10^{-4}$ ($\omega = 0.096$)

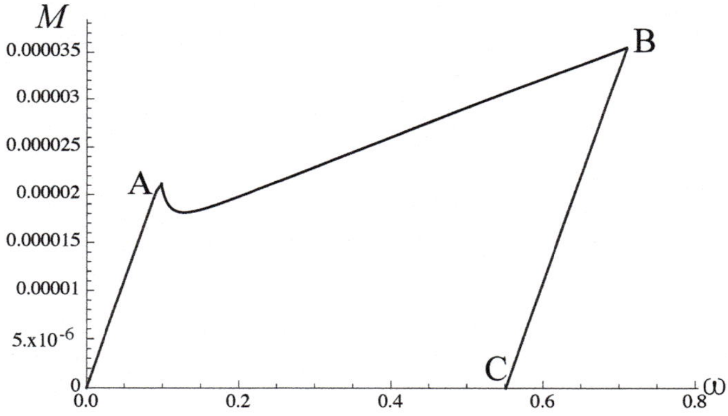

Fig. 11.9: Moment-curvature curve

case) we follow the moment-curvature curve from O through A to B. If the beam is then unloaded, the moment-curvature curve becomes a straight line BC with the same slope like that of OA as shown in Fig. 11.9.

11.7 Discussion and Outlook

In this paper the one-dimensional theory of bending of a single crystal beam with two active slip systems has been developed. The one-dimensional energy of bending has been found after the solution of the thickness problem. The latter has been solved for the symmetric slip systems. The threshold bending moment depends on the beam thickness and, thus, exhibits the size effect. We have found also the dislocation distributions and the moment-curvature curves at loading and unloading. It would be interesting to verify the obtained results with the measurement of excess dislocation density in a bent beam by the method of orientation imaging microscopy developed recently in Kysar et al (2010). The specimen could be for instance the fcc single crystal beam having the axis parallel to [$\underline{1}$10]-direction and bent about the [110] direction. For the unsymmetrically oriented slip systems the thickness problem can only be solved numerically. This will be addressed in our forthcoming paper.

Acknowledgements The financial support by the German Science Foundation (DFG) through the research project LE 1216/4-2 is gratefully acknowledged.

References

Ashby MF (1970) The deformation of plastically non-homogeneous materials. The Philosophical Magazine: A Journal of Theoretical Experimental and Applied Physics 21(170):399–424

Berdichevsky VL (1983) Variational Principles of Continuum Mechanics (in Russ.). Nauka, Moscow

Berdichevsky VL (2006a) Continuum theory of dislocations revisited. Continuum Mechanics and Thermodynamics 18(3):195–222

Berdichevsky VL (2006b) On thermodynamics of crystal plasticity. Scripta Materialia 54(5):711–716

Berdichevsky VL, Le KC (2007) Dislocation nucleation and work hardening in anti-plane constrained shear. Continuum Mechanics and Thermodynamics 18(7):455–467

Bilby BA (1955) Types of dislocation source. In: Defects in crystalline solids. Report of 1954 Bristol conference, The Physical Society, London, pp 123–133

Bilby BA, Gardner LRT, Smith E (1958) The relation between dislocation density and stress. Acta Metallurgica 6(1):29–33

Cleveringa HHM, van der Giessen E, Needleman A (1999) A discrete dislocation analysis of bending. International Journal of Plasticity 15(8):837–868

Evans JT (1995) Bending produced by dislocations in a beam. International Journal of Engineering Science 33:1321–1329

Gurtin ME (2002) A gradient theory of single-crystal viscoplasticity that accounts for geometrically necessary dislocations. Journal of the Mechanics and Physics of Solids 50(1):5–32

Gurtin ME, Anand L, Lele SP (2007) Gradient single-crystal plasticity with free energy dependent on dislocation densities. Journal of the Mechanics and Physics of Solids 55(9):1853–1878

Kaluza M, Le KC (2011) On torsion of a single crystal rod. International Journal of Plasticity 27(3):460–469

Kochmann DM, Le KC (2008) Dislocation pile-ups in bicrystals within continuum dislocation theory. International Journal of Plasticity 24(11):2125–2147

Kochmann DM, Le KC (2009a) A continuum model for initiation and evolution of deformation twinning. Journal of the Mechanics and Physics of Solids 57(6):987–1002

Kochmann DM, Le KC (2009b) Plastic deformation of bicrystals within continuum dislocation theory. Mathematics and Mechanics of Solids 14(6):540–563

Kröner E (1955) Der fundamentale Zusammenhang zwischen Versetzungsdichte und Spannungsfunktionen. Zeitschrift für Physik 142(4):463–475

Kysar JW, Saito Y, Oztop MS, Lee D, Huh WT (2010) Experimental lower bounds on geometrically necessary dislocation density. International Journal of Plasticity 26(8):1097–1123

Le KC (1999) Vibrations of Shells and Rods. Springer, Berlin

Le KC, Nguyen BD (2012) Polygonization: Theory and comparison with experiments. International Journal of Engineering Science 59:211–218

Le KC, Nguyen BD (2013) On bending of single crystal beam with continuously distributed dislocations. International Journal of Plasticity 48:152–167

Le KC, Nguyen QS (2010) Polygonization as low energy dislocation structure. Continuum Mechanics and Thermodynamics 22(4):291–298

Le KC, Sembiring P (2008a) Analytical solution of plane constrained shear problem for single crystals within continuum dislocation theory. Archive of Applied Mechanics 78(8):587–597

Le KC, Sembiring P (2008b) Plane constrained shear of single crystal strip with two active slip systems. Journal of the Mechanics and Physics of Solids 56(8):2541–2554

Le KC, Sembiring P (2009) Plane constrained uniaxial extension of a single crystal strip. International Journal of Plasticity 25(10):1950–1969

Maugin GA (1990) Infernal variables and dissipative structures. Journal of Non-Equilibrium Thermodynamics 15(2):173–192

Maugin GA (1992) The Thermomechanics of Plasticity and Fracture, Cambridge Texts in Applied Mathematics, vol 7. Cambridge University Press, Cambridge

Motz C, Weygand D, Senger J, Gumbsch P (2008) Micro-bending tests: A comparison between three-dimensional discrete dislocation dynamics simulations and experiments. Acta Materialia 56(9):1942–1955

Mura T (1987) Micromechanics of Defects in Solids. Kluwer Academic Publishers, Oxford

Nye JF (1953) Some geometrical relations in dislocated crystals. Acta Metallurgica 1(2):153–162

Ortiz M, Repetto EA (1999) Nonconvex energy minimization and dislocation structures in ductile single crystals. Journal of the Mechanics and Physics of Solids 47(2):397–462

Read WT (1957) Dislocation theory of plastic bending. Acta Metallurgica 5(2):83–88

Sandfeld S, Hochrainer T, Gumbsch P, Zaiser M (2010) Numerical implementation of a 3D continuum theory of dislocation dynamics and application to micro-bending. Philosophical Magazine 90(27–28):3697–3728

Sedov LI (1965) Mathematical methods of constructing models of continuum media (in Russ.). Usp Matem Nauk 20:123–182

Wang W, Huang Y, Hsia KJ, Hu KX, Chandra A (2003) A study of microbend test by strain gradient plasticity. International Journal of Plasticity 19(3):365–382

Chapter 12
Acoustic Metamaterials Based on Local Resonances: Homogenization, Optimization and Applications

Fabio di Cosmo, Marco Laudato, and Mario Spagnuolo

Abstract The aim of this review is to give an overview of techniques and methods used in the modeling of acoustic and elastic metamaterials. Acoustic and elastic metamaterials are man-made materials which present exotic properties capable to modify and drive wave propagation. In particular in this work we will focus on locally resonant microstructures. Such metamaterials are based on local resonances of the internal structure, the dimensions of which are much smaller than the wavelengths of the waves under analysis. We will consider the seminal papers in the fields to grasp the most important ideas used to develop locally resonant metamaterials, such as homogenization techniques and optimization topology. Finally, we will discuss some interesting application to clarify the aforementioned methods.

12.1 Introduction

The term acoustic (elastic) metamaterials refers to all those artificial materials which have been conceived to manipulate the propagation of acoustic (elastic) waves.

Fabio di Cosmo
International Research Center on Mathematics and Mechanics of Complex System (M&MOCS), Universitá degli Studi dell'Aquila, Via Giovanni Gronchi 18 - Zona industriale di Pile, 67100, L'Aquila, Italy
e-mail: fabio.dicosmo.memocs@gmail.com

Marco Laudato
Dipartimento di Ingegneria e Scienze dell'Informazione e Matematica, Universitá degli Studi dell'Aquila, Via Vetoio (Coppito 1), 67100 Coppito & International Research Center on Mathematics and Mechanics of Complex System (M&MOCS), Universitá degli Studi dell'Aquila, Via Giovanni Gronchi 18 - Zona industriale di Pile, 67100, L'Aquila, Italy
e-mail: laudato.memocs@gmail.com

Mario Spagnuolo
CNRS, LSPM UPR3407, Université Paris 13, Sorbonne Paris Cité, 93430 Villetaneuse, France
e-mail: mario.spagnuolo@lspm.cnrs.fr

© Springer International Publishing AG, part of Springer Nature 2018
H. Altenbach et al. (eds.), *Generalized Models and Non-classical Approaches in Complex Materials 1*, Advanced Structured Materials 89,
https://doi.org/10.1007/978-3-319-72440-9_12

Usually they are composed by a periodic repetition of a fundamental cell, whose dimensions are much smaller than the characteristic wavelength of the phenomenon under investigation. Therefore, the macroscopic behaviour of the material can be very exotic, since it is given by the superposition of the responses produced by many cells.

Metamaterials of this kind were firstly introduced in electromagnetism and optics. Indeed, researches in these fields were motivated by some pioneering papers by Veselago (1968, 1967), which at the end of the 60's studied wave propagation in materials with magnetic permeability and electric susceptibility both negative. His analysis gave as a result the introduction of materials with negative refractive index, with extremely interesting applications related to manipulation of electromagnetic waves, for instance flat lenses. Interestingly, Pendry et al (1996) proposed a metamaterial made up of an array of thin wires. This microstructured continuum behaves like a low-density plasma with an effective dielectric constant which can become negative below the effective plasma frequency. The successes in the realm of electromagnetism and optics were rapidly extended to acoustics wave propagation. The milestone in this field was the work made by Liu et al (2000). As we will explicitly discuss in the following, Liu et al. considered a metamaterial formed by an array of rubber-coated lead spheres embedded in an epoxy matrix was designed, producing spectral band gaps in the dispersion diagram. Since then more and more researchers joined this activity and the efficient interchange of ideas in both fields, optics and acoustics, has produced a very rapid increase of new results and new applications.

As well-explained in dell'Isola et al (2016a) metamaterials together with the recent technological developments in manufacturing techniques have led to a drastic change of perspective in material engineering: the possibility of realizing microstructures with a desired macroscopic behaviour has reestablished the fundamental role of mathematical modelling and theoretical analysis in the production process. Interesting examples of architectured microstructure with a desired macroscopic behaviour can be found in Turco et al (2016); dell'Isola et al (2016b); Rahali et al (2015); Placidi et al (2016a); Giorgio (2016); Boutin et al (2017): pantographic sheets, indeed, are metamaterials which are described, at a larger scale, as second-gradient continua. Higher-gradient continua were already studied in the 19th century by Piola (1825); the possibility of defining microstructures modelled according to this theory has been considered in many papers at the beginning of this century (Pideri and Seppecher, 1997; Placidi et al, 2016b; Alibert and Della Corte, 2015; Rinaldi and Placidi, 2014; Auffray et al, 2015; Placidi, 2016, 2015; Cuomo et al, 2016; Battista et al, 2016; Yang and Misra, 2012; Andreaus et al, 2016; Barchiesi and Placidi, 2017); 3D-printing is opening new horizons from this point of view, since the accuracy of this technology allows for the creation of almost any kind of structure. Some analysis of wave propagation in higher-gradient continua has been considered in some papers (Placidi et al, 2014; dell'Isola et al, 2015; Madeo et al, 2014; Placidi et al, 2008; Berezovski et al, 2016; Alibert et al, 2003), but more investigations are necessary and we will not consider this topic in the rest of this review.

As already explained, the research in acoustic and elastic metamaterials is very active. There are many approaches to the problems and many different applications.

First of all there are metamaterials with negative effective parameters (Milton and Willis, 2007; Ding et al, 2007; Chan et al, 2006; Wu et al, 2011; Nemat-Nasser and Srivastava, 2011). These materials have been largely study in order to produce bandgaps in the dispersion diagram and are of great interest for acoustic insulation and elastic vibrations absorption.

Another trend in metamaterials is the one devoted to image focusing below the diffraction limit (Deng et al, 2009; Zhang et al, 2009; Ambati et al, 2007; Ao and Chan, 2008; Jia et al, 2010; Liu et al, 2009). From this point of view we can consider two family of structures: hyperlenses, which are able to transform evanescent waves that are produced close to an object to image into propagating ones which can be detected at larger distance, and superlenses that amplify these evanescent waves thanks to the superposition of Fabry-Perot resonance modes. Metafluids (Norris, 2009), which are fluid metamaterials with anisotropic tensorial mass density, have found interesting applications in this field: they have been produced, for instance, by inserting in a fluid an array of very small plaquettes with internal sphere of a different material, in order to affect both the isotropy and the effective bulk modulus (see Chap.7 in Craster and Guenneau, 2012).

Eventually there is the branch called transformation acoustics (Chen and Chan, 2007, 2010; Hu et al, 2011; Chang et al, 2011, 2010; Zhang et al, 2011; Brun et al, 2009; Hu et al, 2009; Torrent and Sánchez-Dehesa, 2008; Gokhale et al, 2012; Norris and Parnell, 2012), under which we are collecting all the techniques for wave manipulation which are derived from the transformation properties of acoustics and elastic equations. Indeed, when the equations which describe a certain phenomenon are invariant in form with respect to a certain class of coordinate transformations, one can think of the result of this transformation as an effect due to the presence of a new metamaterial. Therefore a certain mode of propagation obtained after a change of coordinates can be produced by the insertion of a metamaterial with some desired properties. These techniques have been widely analysed in connection to cloaking (Milton et al, 2006; Milton and Nicorovici, 2006; Milton et al, 2017; Norris and Shuvalov, 2011) and underwater stealthing.

In this paper we will touch this topics only superficially. Instead we will try to focus on some of the several phases which are required in the production of a metamaterial. This will give us the opportunity of surveying the principal methods used in this very active research field. We will start from the theoretical analysis: first of all the modelling of a single unit of the microstructure, then the macroscopic behaviour which is obtained by means of suitable homogenization techniques. A second fundamental step is the optimization of the structure in order to obtain the desired result. Eventually there are the applications and all the problems related to manufacturing; in particular we will consider applications to noise and vibration isolation. Many examples will support the theoretical discussion in order to facilitate the reading of the work.

According to the previous scheme, the paper will be divided into six sections. After this short introduction, in Sect. 12.2 we will present some of the locally resonant microstructures which have been designed in order to produce negative effective parameters. Section 12.3, instead, will be devoted to a survey of the principal homogenization techniques which have been adopted in literature for the macroscopic description of acoustic metamaterials. In Sect. 12.4 we will deal with topology optimization: how to obtain an optimal unit cell according to a certain desired behaviour. Section 12.5 will contain a short review of applications mainly related to isolation. Eventually some conclusions are derived.

12.2 Locally Resonant Microstructures

Acoustic and elastic metamaterials are man-made materials which present exotic properties capable to modify and drive wave propagation. Usually, these metamaterials are realized by periodically reproducing a fundamental cell or by considering heterogeneous assembly of different materials. Indeed, there are different phenomena which are able to affect wave propagation in elastic media depending on the wavelength characterizing the propagation phenomenon and the size characterizing the internal structure of the metamaterial. In particular one may distinguish between metamaterials which are based on diffraction effects, where the wavelengths of interest are of the same order of magnitude as the size of the internal structure, and metamaterials which are based on local resonances of the internal structure, the dimensions of which are much smaller than the wavelengths of the waves under analysis. In this paper we will focus on this second family of acoustic metamaterials, on their mathematical modelling and their subsequent applications.

The research field on acoustic metamaterials has fruitfully exploited its analogies with electromagnetic and optical metamaterials which were theoretically predicted in Veselago (1967, 1968, 2002) and concretely realized in Pendry et al (1996). Indeed, few years later Liu et al (2000) produced the first example of locally resonant acoustic metamaterial. This metamaterial was realized as an array of lead spheres coated by a soft rubber cloak and embedded in an epoxy matrix. Such a metamaterial presented some bandgaps in the dispersion diagram the feature of which were dependent on the parameters of the resonating unit cell. Since then more and more researchers joined this activity and the efficient interchange of ideas in both fields has produced a very rapid increase of new results and new applications.

The usual description of wave propagation in homogeneous linear elastic Cauchy continua is obtained by means of the equation

$$\nabla \cdot \tilde{\sigma} = \rho \ddot{\boldsymbol{u}}, \tag{12.1}$$

where $\boldsymbol{u}(\boldsymbol{x},t)$ is the displacement. $\tilde{\sigma}$ is the stress tensor which in linear elasticity is written as follows:

$$\tilde{\sigma} = \mathbb{C} : \left(\nabla \boldsymbol{u} + (\nabla \boldsymbol{u})^T \right). \tag{12.2}$$

$\mathbb{C}$ is the fourth-rank elasticity tensor which satisfies the following properties:

$$\mathbb{C}_{ijkl} = \mathbb{C}_{jikl} = \mathbb{C}_{ijlk}.$$

The equation governing acoustic propagation, on the other hand, involves a smaller number of fields and parameters. It is written as

$$\nabla^2 P - \frac{\rho}{K}\frac{\partial^2}{\partial t^2}P = 0, \tag{12.3}$$

where P is the pressure field, ρ the mass density and K the bulk modulus.

If one deals with microstructured continua the previous description is not satisfactory since all the information about the internal structure is lost. However many techniques have been developed which allow to replace the initial heterogeneous medium with an effective homogeneous one with exotic features. These exceptional properties are responsible for the manipulation of the wave propagation. These techniques are collected under the name of homogenization techniques and they will be wider illustrated in the next section.

Before passing to a more systematic investigations of homogenization procedures, in the rest of this section we will concentrate on the microstructure. Indeed, many discrete models involving systems of springs and other mechanical components, have been introduced in order to explain why local resonances can give rise to unusual effective properties. One of the major contributions to this activity has been given by Milton and Willis (2007).

They firstly understood that negative mass can be originated by local resonances in a unit cell. They proposed a very simple model illustrated in Fig. 12.1 a small mass M_1, described as a point particle, connected to a hollow sphere of mass M_2 by means of two springs with elastic constant k. Newton's Second Law of Dynamics implies that

$$M_1\ddot{x} = 2k(X - x), \tag{12.4}$$
$$M_2\ddot{X} = 2k(x - X), \tag{12.5}$$

where X is the horizontal displacement of the barycenter of the mass M_2, whereas x is the horizontal displacement of the material point M_1.

If we look for harmonic solutions, i.e.

$$x = \overline{x}e^{i\omega t}, \qquad X = \overline{X}e^{i\omega t} \tag{12.6}$$

Equation (12.4) can be rewritten as follows

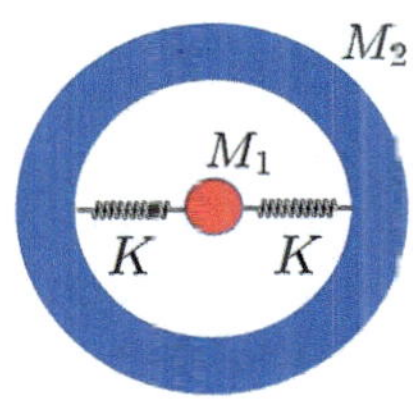

Fig. 12.1 A mass-spring system exhibiting negative effective mass. The periodic arrangement of this basic unit cell constitutes a negative mass acoustic metamaterial.

$$M_1\omega^2\overline{x} = 2k(\overline{x} - \overline{X})\,. \tag{12.7}$$

Therefore we get

$$\frac{\overline{x}}{\overline{X}} = \frac{\omega_1^2}{\omega_1^2 - \omega} \tag{12.8}$$

with

$$\omega_1 = \sqrt{2\frac{k}{M_1}}. \tag{12.9}$$

ω_1 is the resonance frequency of the whole system in Fig. 12.1. For frequencies close to this value the oscillation of the internal mass is amplified, resulting in a out-of-phase motion with respect to the motion of the external hollow sphere. If one defines the effective mass μ of the cell as the ratio between the total momentum of the cell P and the velocity of the external hollow sphere $\dot{X}$, one gets that

$$\mu = M_1\frac{\omega_1^2}{\omega_1^2 - \omega^2} + M_2\,. \tag{12.10}$$

It is worth noticing that the global effect of the resonance is that the effective mass can assume also negative values. Periodic arrays of these resonant cells have been used as acoustic insulators because, when the mass of a single unit becomes negative, wave propagation is forbidden: the dispersion diagram consequently presents some bandgaps whose features can be manipulated by changing the parameters of the single unit cell (Yao et al, 2008; Huang and Sun, 2009).

Similar mechanisms have been introduced to obtain fundamental units characterized by a frequency dependent bulk modulus (Ding et al, 2007; Li and Chan, 2004; Fang et al, 2006; Zhou and Hu, 2007; Liu et al, 2011a). Close to the resonant frequency the effective bulk modulus becomes negative. In this case the resonance must involve compressional-extensional motions, a phenomenon which occurs, for instance, in Helmholtz resonators.

One of the main objective of current research in acoustic metamaterials is finding structures with a wide range of frequencies in which both parameters, mass density and bulk modulus, are negative. Indeed, doubly negative acoustic metamaterials (Li and Chan, 2004; Fang et al, 2006; Ding et al, 2007; Nemat-Nasser and Srivastava, 2011) would be the analogue of right-handed electromagnetic metamaterials introduced by Veselago. The main feature of these metamaterials is their negative refractive index: for these systems group and phase velocity have opposite directions, giving rise to other exotic phenomena like negative Doppler effect and negative Cherenkov radiation. From the point of view of applications, doubly negative metamaterials could be efficiently used to manufacture flat lenses (Pendry, 2000) or to design acoustic cloaks (Milton et al, 2006; Milton and Nicorovici, 2006; Milton et al, 2017; Norris and Shuvalov, 2011).

Locally resonant microstructures are also studied for sub-wavelength imaging resolution (Deng et al, 2009; Zhang et al, 2009; Ambati et al, 2007; Ao and Chan, 2008; Jia et al, 2010; Liu et al, 2009). Indeed, by means of locally resonant acoustic metamaterials, it is possible to overcome the diffraction limit and use evanescent waves for better imaging a source. For instance superlenses use local Fabry-Perot resonant modes to amplify the evanescent waves which are produced in the vicinity of the imaging object.

The step to elastic metamaterials is now very short. Ways of reasoning similar to the ones above illustrated can be extended to elasticity: the main difference is that the number of parameters which appear in the equations of motion is higher. For instance homogeneous isotropic linear elastic materials are described by three parameters, the mass density, ρ, and the two Lamé's coefficients, λ, μ. In this case systems made up of a hard core coated by a elastic layer, behave like local resonators giving rise to frequency-dependent effective elastic parameters (Wu et al, 2011; Hu et al, 2011; Liu et al, 2011b; Zhou et al, 2008; Zhou and Hu, 2009). The out-of-phase motion of the internal core with respect to the external cloak creates resonances which affect either mass density or Lamé coefficients, depending on the kind of motion they produce on the unit cell. The principal applications of this family of metamaterials are related to absorption of elastic vibrations: new results in this direction could be fundamental in civil engineering to efficiently protect buildings from earthquakes, or for vibration attenuation.

In this section we have only outlined some of the macroscopic consequences deriving from locally resonant microstructures. Our aim was not to provide a full list of the microstructures which have been proposed in literature during the last decades. We have only presented some justifications for the use of local resonances to produce metamaterials with extra-ordinary properties. In the next section we will illustrate with more details some homogenization techniques which have been used in modelling this kind of metamaterials. This will help us to better understand how unit cells have to be designed in order to obtain a certain desired behaviour.

12.3 A Survey of Homogenization Techniques

As already stated, many homogenization techniques exist in literature. For instance it has to be remarked that in statistical mechanics some rigorous results are obtained which prove rigorously how one can get fluid continua from discrete microscopic systems (see the exhaustive reviews presented in Esposito and Pulvirenti (2004); Pulvirenti (1996)). Actually while one can obtain relevant and rigorous results in homogenizing microscopically inhomogeneous systems, also including in the picture thermal phenomena for systems behaving macroscopically as fluids (see the e.g. Caprino et al, 1993; De Masi and Olla, 2015; Carinci et al, 2014a; De Masi et al, 2009; Carinci et al, 2014b) and also for different class of biological systems (see e.g. De Masi et al, 2007, 2015), it is an open problem the determination, via statistical mechanics and rigorous reasonings, of micro properties leading at macro level, to solid behavior.

Therefore one has to expect that heuristic methods (Nadler et al, 2006; Steigmann and dell'Isola, 2015; Steigmann, 2008), mixed micro-macro approaches (Javili et al, 2013; Misra and Poorsolhjouy, 2015; Nadler et al, 2006), numerical identifications

(Saeb et al, 2016) and ad hoc phenomenological continuum models (Bertram and Glüge, 2016; Bevill et al, 2006; Eringen, 1976; Ganghoffer, 2016; Javili et al, 2013; Misra and Poorsolhjouy, 2015; Nadler et al, 2006; Steigmann and dell'Isola, 2015; Steigmann, 2008) still play a relevant role in the efforts for describing structured media.

In this section we will present some of these methods directly applied to acoustic/elastic metamaterials.

12.3.1 Periodic Homogenization

In this paragraph we will illustrate the method of periodic homogenization in a concrete example coming from elasticity. For other details on this method applied to acoustic/elastic metamaterials see for instance the references (Smith, 2011; Andrianov et al, 2008; Otero et al, 2005; Craster et al, 2010).

Let us consider the equation describing anti-plane shear waves in an elastic medium. If U represents the out of plane component of the displacement field $\boldsymbol{u}$, the above mentioned equation can be written as follows:

$$\nabla \cdot (\mu \nabla U) + \omega^2 \rho U = 0, \tag{12.11}$$

where μ is the shear modulus of the homogeneous elastic medium under analysis. Since we are dealing with a microstructured continuum, we will introduce a new variable, $\boldsymbol{y} = \boldsymbol{x}/\epsilon$, where ϵ denotes the length scale of the microstructure. In other words, this new variable describes the fast degrees of freedom which characterize the microscale dynamics of the system, whereas $\boldsymbol{x}$ describes its macroscopic behaviour. The previous field U will now depend on two variables, i.e. $U_\epsilon(\boldsymbol{x}) = U(\boldsymbol{x}, \boldsymbol{x}/\epsilon)$, and the system will be supposed to be periodic, with period h, in the new variable $\boldsymbol{y}$. This new field obeys the rescaled equation

$$\nabla_\epsilon \cdot (\mu \nabla_\epsilon U_\epsilon) + \rho \omega^2 U_\epsilon = 0, \tag{12.12}$$

where $\nabla_\epsilon = \nabla_x + \frac{1}{\epsilon} \nabla_y$. The final step of this homogenization technique is to obtain the field solution of Eq. (12.12) in the limit $\epsilon \to 0$. For this reason one looks for a solution U_ϵ in the form of a power series in the parameter ϵ:

$$U_\epsilon(\boldsymbol{x}) = \sum_{j \in \mathbb{N}} \epsilon^j U^{(j)}(\boldsymbol{x}, \boldsymbol{y}).$$

By replacing the ansatz in Eq. (12.12) one obtains the equations for the limiting field $U^{(0)}(\boldsymbol{x}, \boldsymbol{y})$. To retain only the dependence on the macroscopic variable $\boldsymbol{x}$ one can perform an integration over the unit cell of the periodic microstructure.

An interesting application of this homogenization procedure can be found in Chap.1 of Craster and Guenneau (2012). Guenneau et al (2007) have proved that if one consider a microstructure made up of square unit cells containing two semi-cylinders with a very small separation distance, the equation of the motion for the homogenized medium preserves the same form as the initial one (Eq. 12 11) but the shear modulus becomes anisotropic. This kind of structures have been largely studied in relationship with transformation acoustics.

12.3.2 Dynamic Homogenization and Willis-Type Constitutive Relations

This method focuses on the dynamical behaviour of the homogeneous effective medium, providing, in particular, a characterization of those properties which affect waves propagation (for more details see, for instance, Nemat-Nasser et al, 2011; Srivastava and Nemat-Nasser, 2012; Nemat-Nasser, 2015; Sieck et al, 2015). The quantities which are obtained by averaging techniques will, in the end, depend on the frequencies, resulting in an effective dispersive medium.

The main assumption in this method is the existence of a microstructure which can be either random and characterized by a probability density $\mathscr{P}(a)$ (a is the parameter representing the distribution of the microstructure) or periodic. Its validity, in both cases, is restricted to wavelengths much longer than the size of the microstructure. The presence of the probability distribution allows to define averages of the field as follows

$$\langle f(\boldsymbol{x},t)\rangle = \int_A \mathscr{P}(a) f(\boldsymbol{x},t,a)\,. \tag{12.13}$$

Analogously, in case of periodic structure one can consider different realizations of the fundamental cell. Since the structure is periodic, there will be a vector $\boldsymbol{y}$ which connects this realization with the cell considered from the beginning. If one considers equally probable all these realizations it is possible to compute averages as follows:

$$\langle f(\boldsymbol{x},t)\rangle = \int_Y f(\boldsymbol{x}+\boldsymbol{y},t)d\boldsymbol{y}\,, \tag{12.14}$$

where Y denotes the initial realization of the fundamental cell. In the rest of the paragraph we will limit to the case of periodic microstructures which permits to understand more easily the dispersive behaviour of the effective dynamical medium resulting from the average procedure.

Let us consider again (12.1),

$$\nabla\cdot\tilde{\sigma} = \rho\ddot{\boldsymbol{u}}\,. \tag{12.15}$$

with $\tilde{\sigma} = \mathbb{C} : \tilde{\epsilon}$.

The periodic hypothesis allows to introduce Bloch-Floquet solutions, i.e.

$$\boldsymbol{u}(\boldsymbol{x},t) = \hat{\boldsymbol{u}}(\boldsymbol{x})e^{i(\boldsymbol{k}\cdot\boldsymbol{x}-\omega t)}\,, \tag{12.16}$$

where the function $\hat{\boldsymbol{u}}(\boldsymbol{x})$ is periodic. The effective averaged field is

$$\boldsymbol{u}_{eff}(\boldsymbol{x},t) = \langle \hat{\boldsymbol{u}} \rangle \, e^{i(\boldsymbol{k}\cdot\boldsymbol{x}-\omega t)} .$$

Let us now consider a comparison homogenized medium, the properties of which will be denoted by a superscript 0, e.g. $\mathbb{C}^0$ will denote its elasticity tensor. The constitutive relations of linear elasticity can be rewritten as follows:

$$\tilde{\sigma} = \mathbb{C}^0 : \tilde{\epsilon} + \tilde{\Sigma}, \qquad p = \rho^0 \dot{u} + \Pi . \tag{12.17}$$

Therefore the equation of motion becomes

$$\nabla \cdot (\mathbb{C}^0 : \nabla u) + \nabla \cdot \tilde{\Sigma} - \dot{\Pi} = \rho^0 \ddot{\boldsymbol{u}} , \tag{12.18}$$

with appropriate boundary conditions. The solution of this equation can be written as follows:

$$\boldsymbol{u}(\boldsymbol{x},t) = \boldsymbol{u}^0(\boldsymbol{x},t) + G^0 * (\nabla \cdot \tilde{\Sigma} - \dot{\Pi}) , \tag{12.19}$$

where $*$ denotes the operation of convolution, both in space and time, $\boldsymbol{u}^0$ is solution of the equation $\nabla \cdot (\mathbb{C}^0 : \nabla \boldsymbol{u}^0) = \rho^0 \ddot{\boldsymbol{u}}^0$ with the same boundary conditions as Eq. (12.18) and G^0 is the Green function of the same PDE, satisfying homogeneous boundary conditions.

After integration by part in the convolution integral one obtains a linear relation between the field u and the fields $\tilde{\Sigma}$ and Π:

$$\boldsymbol{u} = \boldsymbol{u}^0 - S^0 * \tilde{\Sigma} - M^0 * \Pi ,$$

where S^0 and M^0 are integral operator deriving from the Green function G^0.

Consequently one can write

$$\tilde{\epsilon} = \tilde{\epsilon}^0 - S^0_x * \tilde{\Sigma} - M^0_x * \Pi, \qquad \boldsymbol{p} = \boldsymbol{p}^0 - S^0_t * \tilde{\Sigma} - M^0_t * \Pi . \tag{12.20}$$

On the other hand these expressions can be read as a set of relations between the fields $(\tilde{\Sigma}, \Pi)$ and the fields $\tilde{\epsilon}, \dot{\boldsymbol{u}}$. Therefore after averaging over the unit cell Y we get the final relations:

$$\langle \tilde{\sigma} \rangle = T_{11} \langle \tilde{\epsilon} \rangle + T_{12} \langle \dot{\boldsymbol{u}} \rangle , \qquad \langle \Pi \rangle = T_{21} \langle \tilde{\epsilon} \rangle + T_{22} \langle \dot{\boldsymbol{u}} \rangle \tag{12.21}$$

Replacing this expression in the constitutive relations one obtains the constitutive equations for the effective medium which are of the Willis-type, that is

$$\langle \tilde{\sigma} \rangle = \mathbb{C}_{eff} \langle \tilde{\epsilon} \rangle + \mathbb{S}_{eff} \langle \dot{\boldsymbol{u}} \rangle , \qquad \langle p \rangle = \mathbb{S}_{eff} \langle \tilde{\epsilon} \rangle + \rho_{eff} \langle \dot{\boldsymbol{u}} \rangle . \tag{12.22}$$

It has been proved in Norris et al (2012); Srivastava (2015) that this constitutive relations are closed under this homogenization scheme, that is a Willis type microstructure remains of the same kind after homogenization. Furthermore self-adjointness of the problem is also preserved.

Let us remark that the effective tensor $\mathbb{S}_{eff}$ is not univoquely determined since we can modify the integral operators involved in the homogenization scheme without altering the procedure. However, as observed by Willis, if one introduces an eigenstrain η in the constitutive relations the effective constitutive relations are unique (Willis, 2011).

12.3.3 Homogenization from Scattering Properties.

Another largely applied method which gives effective homogeneous media describing the properties of acoustic metamaterials is based on the comparison between the scattering properties of the internal microstructure and the ones of the corresponding homogenized medium. These methods have been largely used for acoustic transparency. Indeed one can design the microstructure in such a way that the effective parameters coincide with the parameters of the external host medium (Cummer et al, 2008; Cummer and Schurig, 2007; Alù and Engheta, 2005; Zhou and Hu, 2006; Zhou et al, 2008; Zhou and Hu, 2007; Guild et al, 2011).

Let us consider a metamaterial formed by a large number of subunits which we will consider as single scatterers. When an incident field impinges on a rotationally symmetric cluster of scatterers, its response is described by a scattered field which is linearly related with the incident wave, i.e.

$$\psi_{sc} = T_{cls}\psi_{in} .$$

The matrix T_{cls} depends only on the properties of the cluster and the frequency of the incident wave.

The idea at the base of this homogenization scheme is to replace the original metamaterial with a homogenized medium whose properties are obtained by the following equation:

$$\lim_{\omega\to 0} T_{cls} = \lim_{\omega\to 0} T_{eff} . \tag{12.23}$$

In other words one is supposing that in the quasi-static limit, that is $\omega \to 0$, the cluster behaves like an homogeneous medium characterized by some effective parameters.

Because of the rotational symmetry, the incident wave and the scattered field can be expanded in a multipolar series, and the scattering matrix actually connects the coefficient in the series. If one considers a bidimensional example one can write:

$$\psi_{in}(r,\theta) = \sum_{q\in\mathbb{N}} A_q j_q(kr)e^{iq\theta}, \tag{12.24}$$

$$\psi_{sc}(r,\theta) = \sum_{q\in\mathbb{N}} B_q j_q(kr)e^{iq\theta}, \tag{12.25}$$

$$B_q = \sum_{s\in\mathbb{N}} T_{qs}A_s, \tag{12.26}$$

where the symbol $j_q(kr)$ indicates one of the Bessel or Hankel functions (depending on the problem under analysis) and k is the background wave-vector.

Let us continue with this bidimensional example to illustrate how this homogenization scheme works. In the circular case, for a homogeneous and isotropic elastic medium, the scattering matrix T_{qs} becomes diagonal $T = T_q \delta_{qs}$. Furthermore all the T_q vanish in the limit $\omega \to 0$, but according to different trends. For instance, in the electromagnetic case the dominant terms in this quasi-static limit are the coefficient T_0 and T_1 both vanishing as ω^2.

In general for a homogeneous scatterer of radius R_a it can be shown (see Waterman, 1969, for instance) that the dominant terms in the scattering matrix have the form

$$T_q \approx \frac{i\pi k^2 R_a^2}{4} \Gamma_q^a , \tag{12.27}$$

where the Γ_q^a are coefficients which depend only on the field considered. Therefore for the effective medium we will have:

$$T_q^{eff} \approx \frac{i\pi k^2 R_{eff}^2}{4} \Gamma_q^{eff} .$$

On the other hand, if we neglect multiple scattering, the scattering coefficients of the cluster are the sum of the scattering coefficients of any single scatterer. Therefore if the medium is composed of N homogeneous scatterer of radius R_a the dominant terms of the scattering matrix for the cluster are written as follows:

$$T_q^{cls} \approx N \frac{i\pi k^2 R_a^2}{4} \Gamma_q^a .$$

Equating the two expressions one can find the effective parameters describing the homogenized medium:

$$\Gamma_q^{eff} = f \Gamma_q^a$$

where $f = NR_a^2/R_{eff}^2$ is the filling factor relative to the metamaterial. Some interesting applications of this general procedure to obtain homogenized plates is presented in Torrent et al (2014). In this case it is interesting to notice that, differently from the electromagnetic case, the properties of the homogenized medium are determined only by the coefficients T_0 and T_2.

After this short review of some homogenization techniques used for acoustic/elastic metamaterials, in the next section we will present another aspect which is fundamental in the design and modelling of metamaterials: optimization of the structures.

12.4 Topology Optimization

One of the key feature that has driven the attention of researcher on Locally Resonant Sonic Materials is the possibility, in principle, of controlling wave-propagation phenomena. Moreover, an important advantage from the realizability point of view is that, unlike other absorptive materials like for instance phononic crystals, they do not require in general any particular relation between the characteristic length of the microstructure and the wavelength of the acoustic waves (Deymier, 2013).

However, the modeling of such materials is not straightforward and a systematical method of design is not yet available. Here modeling means the way to design the local resonators in order to obtain an aimed band-gap in the dispersion relation of the material. Although several proposals have been done in the last years, a satisfying model to forecast the frequencies at which attenuation is maximized still miss. For instance one can mention the heuristic model to forecast resonance frequencies by Hirsekorn (2004), where the main problem is that, due to the action of elastic waves, resonance frequencies alone are not enough to determine the damping band.

Despite the difficulties of such a study, the advantages of a complete control of Locally Resonant Sonic Materials are formidable and one can mention applications to several area of engineering, physics, and biology. Therefore, it is not unexpected that several research groups are intensely working on this topic.

A promising idea which has given interesting and fruitful results is the so-called topology optimization procedure (Bendsøe and Kikuchi, 1988; Bendsøe and Sigmund, 2004). We will briefly show in the rest of the section the main idea behind the topology optimization by considering specific relevant examples from the literature. In each example, which deals with Locally Resonant Sonic Materials with different microstructures, we will discuss how the topology optimization procedure is realized and the particular advantages will be examined case by case.

12.4.1 Topology Optimization for Local Resonant Sonic Materials

In Liu et al (2000) introduced for the first time the concept of Local Resonant Sonic Materials. The microstructure consists of a epoxy resin matrix with inclusions of lead covered with silicon rubber. An incident wave induces a near eigen-frequency high-amplitude vibration of the inclusions which consequently prevents the passage of the wave through the Local Resonant Sonic Materials.

The first steps to produce Local Resonant Sonic Materials with satisfying attenuation properties were driven by a trial and error process in the design of the microstructure. Of course, these methods were not very reliable and they were high-time consuming from the computational point of view. Some efforts were done trying to forecast the damping frequency band (see for instance Hirsekorn et al, 2006) or to enlarge it (Gu et al, 2008), but the first systematic method for the design of Local Resonant Sonic Materials was presented in 2004 by Matsuki et al (2014).

They have proposed a systematic method to obtain the optimal unit cell, whose repetition gives rise at macroscopic level to the desired frequency band at which attenuation is maximized. The main tool is the structural topology optimization method, which essentially deals with the density distributions of the various components of the material. One of the main problems which usually arises dealing with topology optimization methods is that the optimal configuration may contain non-physical materials, i.e. it may include areas where the material density has an intermediate values between the densities of the actual materials. This problem was brilliantly overcome by Allaire et al (2004) and Wang et al (2003), independently, by proposing the so-called set-based structural optimization method (we refer to Yamada et al, 2010, for an application to topology optimization method).

The optimization problem is set as follows: one considers a single cell, harmonically loaded with frequency ω to the left side Γ_{in}. The cell is divided in three different areas: The design area Ω_{design} (around the inclusion), the coat area Ω_{coat}, and the kernel area Ω_{kernel}. The densities of the materials in Ω_{design} are defined as:

$$\begin{aligned} \rho_{\text{design}} &= \rho_{\text{epoxy}} + \chi\left(\rho_{\text{silicone}} - \rho_{\text{epoxy}}\right), \\ \lambda_{\text{design}} &= \lambda_{\text{epoxy}} + \chi\left(\lambda_{\text{silicone}} - \lambda_{\text{epoxy}}\right), \\ \mu_{\text{design}} &= \mu_{\text{epoxy}} + \chi\left(\mu_{\text{silicone}} - \mu_{\text{epoxy}}\right), \end{aligned} \tag{12.28}$$

where the characteristic function $\chi(\boldsymbol{x})$ describes the extension of the Ω_{silicone} area:

$$\begin{cases} \chi(\boldsymbol{x}) = 1, & \text{if } \boldsymbol{x} \in \Omega_{\text{silicone}}, \\ \chi(\boldsymbol{x}) = 0, & \text{if } \boldsymbol{x} \in \Omega_{\text{design}}/\Omega_{\text{silicone}}. \end{cases} \tag{12.29}$$

The optimization problem is stated by requiring the minimization with respect to the characteristic function χ of the following functional (Matsuki et al, 2014)

$$\inf_{\chi} \int_{\Gamma_{\text{out}}} u(\omega) \cdot u^*(\omega)\, d\Gamma, \tag{12.30}$$

being Γ_{out} the right-side of the cell, u the displacement field, and * the complex conjugate.

In conclusion, by means of a topology optimization procedure, based on a set-based approach, this method is able to give an engineering meaningful optimal configuration of the cell, providing an acoustic damping band around the desired frequency ω.

12.4.2 Topology Optimization for Hyperbolic Elastic Metamaterials

Hyperbolic elastic materials are metamaterials in which the wave propagation is allowed only in the radial direction while it is forbidden in the angular direction. They were introduced for the first time, for electromagnetic waves, independently, by

Salandrino and Engheta (2006) and Jacob et al (2006), but several examples were developed in the subsequent years (see for instance Liu et al, 2007; Lee et al, 2007) with several interesting applications, see for instance the results in high-resolution lithography (Xiong et al, 2009) and hyperbolic lenses (Kildishev and Narimanov, 2007).

Only in a second moment researchers, driven by possible applications to sub-wavelength imaging for medical purposes, considered elastic/acoustic hyperbolic metamaterials. The typical microstructure (see for instance the one proposed by in Oh et al (2015)) of a hyperbolic elastic metamaterial is a small plate (thickness $\sim$ 1mm) in aluminium with an internal corrugated structure which is able to slow the speed of the incident wave along the angular direction. The dispersion curve of such a metamaterial shows that waves can propagate along the angular direction until a certain frequency threshold $f < f_\theta$ is reached, while they can propagate in the radial direction at frequencies below f_r, with $f_r > f_\theta$. Therefore, the frequency band for which this system works as a hyperbolic metamaterial is $f_\theta < f < f_r$ One of the main challenges in this topic is to broaden this band as much as possible.

An interesting approach to this problem was proposed by Oh et al (2015) and it deals with a topology optimization procedure. The key role in the damping effect is played by the corrugated part in the microstructure and the optimization procedure looks for its optimal design. In this system, one has to consider that the physical phenomena in the radial and angular directions are different. Moreover, and this is the major feature which distinguishes this problem from all the similar optimization problems, the frequency band for which the system works as a true hyperbolic materials depends on both internal resonance effects and on the periodicity of the microstructure. To deal with this situation is not straightforward and one has to consider two independent finite element models which are able to describe the physical phenomena associated to the propagation in the two orthogonal directions at the same time. This is the so-called multi-model based finite element analysis.

The optimization problem is formulated by requiring that following functional has to be minimized

$$\psi(x_1,\ldots,x_N) = a\frac{\omega_{\text{Bragg}}}{\omega^i_{\text{Bragg}}} + b\frac{\omega^i_{\text{shear}}}{\omega_{\text{shear}}} + c\sum_{e=1}^{N}(1-x_e)x_e, \tag{12.31}$$

where $0 < x_e \leq 1$ is the element density design variable associated to the e-th cell, N is the total number of design variables, ω_{Bragg} is the Bragg frequency of the system, a, b, c are weights, and the index i is relative to the initial guess of the corrugated shape which is known and that for sure gives rise to truly-hyperbolic metamaterial. By means of this topology optimization problem, Oh et al (2015) were able to show that the optimal internal corrugated structure, which maximally broaden the working frequency band, is characterized by an hinge and a reinforced rectangular part.

12.4.3 Topology Optimization for Hyperelastic Plates

In the recent years the peculiar properties of hyperelastic materials have been deeply studied. The main reason is the growing interest in applications like bio-mimetic actuators and sensors, energy producing devices (Czech et al, 2010), and microjets (Oates and Liu, 2009) for which hyperelastic materials (Park et al, 2010) seems to be suitable.

One of the most interesting phenomenon which has been investigated is the nonlinear response of elastic structure to resonant excitations (Sathyamoorthy, 1997; Lacarbonara, 2013). In particular, when the natural frequencies associated to two (or more) oscillation modes are (nearly) commensurable, it is possible to observe the effects of transfer of energy between the modes due to resonances of the internal structure. For instance, internal resonance will happen if a sufficient huge external excitation is applied to a system described in terms of quadratic non-linearities which exhibits a 1:2 frequency ratio of two linear modes.

There are essentially two different possibilities to obtain quadratic nonlinearities in a considered structure: material nonlinearities, which are the ones related to a non linear stress-strain relation, and geometric nonlinearities, which are related to a large deformation regime.

In order to obtain internal resonance in hyperelastic materials, one has to look for the particular design of the microstructure that is able to produce the quadratic nonlinearities which allow 1:2 internal resonance. Among the many proposals that have been done in the last year, one of the most successful is the one presented in the recent work of Tripathi and Bajaj (2016) dealing with a topology optimization procedure.

In their model, the resonator whose design has to be optimized to obtain internal resonance, is an isotropic rectangular thin hyperelastic plate described as a two-parameter incompressible Mooney-Rivlin material. It is modeled by means of four node thin-plate elements which can rotate around the two axis of the plate and move on the transverse plane. Initially, they consider energy transfer only between two oscillation modes. In particular, an external resonant perturbation excites the second mode that in turn, once the optimal design is reached, excites the internal resonators with a 1:2 frequency ratio.

They performed the so-called method of moving asymptotes topology optimization (for a complete introduction, see for instance Svanberg, 1987). In this method, one divides the design region in N cells and a design variable x_i, the index i referring to the i-th cell, such that $0 < x_i < 1$ is defined. The density and the Young modulus of the i-th cell are then expressed as

$$\rho_i = \rho_{\min} + x_i^n \rho_0, \qquad E_i = E_{\min} + x_i^m E_0, \tag{12.32}$$

where the constants $\rho_{\min}$ and $E_{\min}$ are needed to avoid singularities during the finite element procedure and ρ_0 and E_0 are the density and Young's modulus of the material, respectively. Again, it is crucial to avoid gray-scales zones, i.e. the case in which the optimization procedure yields non-physical densities. This is the reason of the exponents n and m in (12.32) that, to prevent intermediate densities and Young's modulus, are usually defined higher than 3 (in Tripathi and Bajaj, 2016, they are fixed to 6 and 12, respectively). The topology optimization problem is the minimization of the following functional

$$c(\omega) = \left(\xi - \frac{\omega_2}{\omega_1}\right) + C_k \sum_{i=1}^{N} (x_i^n)(1 - x_i^m), \tag{12.33}$$

where ω_1 and ω_2 are the frequencies of the two modes and C_k is a fixed coupling constant. The optimization method of moving asymptotes topology essentially solves a series of iterative convex approximation of the functional, starting from an initial solution. In the first step, by means of the initial solutions we obtain the first gradient of the objective function. By means of the gradient, it is possible to define an approximate convex form of the objective function that can be solved by using usual methods. In the second step, the so-obtained solution will be now considered as a starting objective function and the process will be repeated until a satisfactory approximation is reached.

Despite the fact that topology optimization is not very developed for nonlinear dynamics, by means of the moving asymptotes topology method, in Tripathi and Bajaj (2016) a method to optimize the design of the microstructure to obtain internal resonators has been shown, providing us with another example of the usefulness of topology optimization method.

12.5 Principal Applications: Phononic Crystals

An important kind of acoustic metamaterial is the one including particular artificial periodic composites known as phononic crystals (PC) or sonic crystals, consisting in a periodic array of acoustic scatterers (see for instance Fig. 12.2 where a square unit cell is represented) embedded in a host material (Kushwaha et al, 1993; Sigalas et al, 2005; Liu et al, 2000). The fundamental feature of PC is that they may induce band gaps. The existence of band gaps in PC is of great interest in the area of noise and vibration isolation (Sánchez-Dehesa et al, 2011; Yu et al, 2008). A phononic band gap is a frequency range in which all the propagating Bloch waves (not depending on their wave vector and polarization) are forbidden.

The problem of wave propagation in two-dimensional acoustic metamaterials is investigated in Wang et al (2015), referring to locally resonant viscoelastic phononic crystals of two different kind associated to quasi-longitudinal and quasi-shear waves. A variational formulation of the complex band structure for in-plane polarized waves is considered: this kind of approach, in fact, allows to simply develop a finite element model. For linearly viscoelastic materials load and deformations are linearly related and the general relation between the components of the stress and strain tensors is given by

$$T_\alpha(\boldsymbol{r},t) = \psi_{\alpha\beta}(t) * S_\beta(\boldsymbol{r},t), \tag{12.34}$$

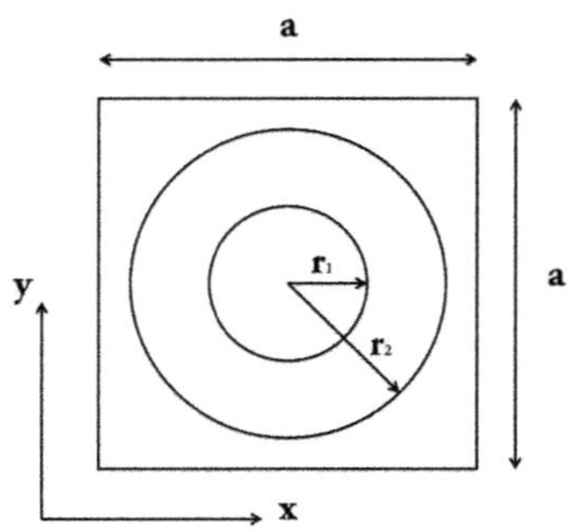

Fig. 12.2 Unit cell of the square-lattice metamaterial. The lattice constant is a.

where $*$ indicates the time convolution, $\psi_{\alpha\beta}$ are the 21 independent components of the relaxation function with indices $\alpha, \beta = 1, \ldots, 6$ and

$$\boldsymbol{T} = [\sigma_{xx}, \sigma_{yy}, \sigma_{zz}, \sigma_{yz}, \sigma_{xz}, \sigma_{xy}]^T, \qquad \boldsymbol{S} = [\varepsilon_{xx}, \varepsilon_{yy}, \varepsilon_{zz}, \varepsilon_{yz}, \varepsilon_{xz}, \varepsilon_{xy}]^T. \tag{12.35}$$

In the case of a time-harmonic plane wave $u_j(\boldsymbol{r}, t)$, the stress and strain vectors can be written as

$$T_\alpha(\boldsymbol{r}, t) = \tilde{T}_\alpha e^{i\omega t}, \quad S_\alpha(\boldsymbol{r}, t) = \tilde{S}_\alpha e^{i\omega t}, \tag{12.36}$$

and the constitutive relation becomes

$$\tilde{T}_\alpha = C^{v\alpha\beta}(i\omega) \cdot \tilde{S}^\beta \tag{12.37}$$

The complex stiffness matrix C^v, whose components are

$$C^v_{\alpha\beta}(i\omega) = \int_{-\infty}^{+\infty} \psi_{\alpha\beta}(t) e^{-i\omega t} dt, \tag{12.38}$$

can be expressed as

$$C^v = C + i\omega\eta, \tag{12.39}$$

where we call η the viscosity matrix.

The governing equation in absence of body forces is then (ρ is the mass density)

$$\nabla_{j\alpha}\tilde{T}_\alpha(\boldsymbol{r}) = -\rho\omega^2\tilde{u}_j. \tag{12.40}$$

If the elastic waves propagate in the transverse plane with displacements independent of the z-coordinate, they can be decomposed into mixed and shear modes and the wave equation for mixed modes is

$$\nabla \cdot (\mu^* \nabla \tilde{y}_j) + \nabla \cdot \left(\mu^* \frac{\partial}{\partial x_j} \tilde{\boldsymbol{u}}\right) + \frac{\partial}{\partial x_j}(\lambda^* \nabla \cdot \tilde{\boldsymbol{u}}) = -\rho\omega^2\tilde{u}_j, \tag{12.41}$$

where $\lambda^* = C_{12}^v(\boldsymbol{r}, i\omega)$ and $\mu^* = C_{44}^v(\boldsymbol{r}, i\omega)$ are the complex Lamè constant and shear modulus. According to the Bloch's theorem by substituting this expression in (12.41) we obtain:

$$(\nabla - i\boldsymbol{k}) \cdot \left[\mu^* (\nabla - i\boldsymbol{k}) \tilde{u}_{jk}\right] + (\nabla - i\boldsymbol{k}) \cdot \left[\mu^* (\nabla - i\boldsymbol{k})_j \tilde{\boldsymbol{u}}_{\boldsymbol{k}}\right] \\ + (\nabla - i\boldsymbol{k})_j \left[\lambda^* (\nabla - i\boldsymbol{k}) \cdot \tilde{\boldsymbol{u}}_{\boldsymbol{k}}\right] = -\rho\omega^2 \tilde{u}_{jk} . \tag{12.42}$$

After the presentation of the theoretical aim, the authors describe an algorithm for calculating the complex band structure (see Fig. 2 in Wang et al, 2015) for in-plane waves and the modal distribution of the displacement components (see Fig. 3 in Wang et al, 2015) based on a FEM implementation. Propagating and evanescent Bloch waves (see Fig. 5 in Wang et al, 2015) are calculated and analyzed in order to understand the physical meaning of the associated complex bands, also in presence of viscosity.

A similar problem is studied in Xiao et al (2012), where local resonators are also considered. The plane wave expansion method to study the flexural wave propagation in locally resonant beams with multiple periodic arrays of attached spring-mass resonators is considered. In this case it is possible to quantify the wave attenuation performance of band gaps by calculating the complex Bloch wave vectors. The interesting point consists in the fact that a locally resonant beam with multiple arrays of damped resonators can achieve much broader band gaps than a locally resonant beam with only a single array of resonators.

For the calculation of elastic wave band structures associated to the system in Fig. 12.3 the authors use the so-called plane wave expansion method (PWE) to deal with the LR beams containing multiple arrays of lumped resonant elements (i.e., spring-mass resonators). The governing equation for the time-harmonic flexural vibration of the LR beam system can be written, by referring to the Euler-Bernoulli beam theory, as

$$EI\frac{\partial^4 w(x)}{\partial x^4} - \omega^2 \rho A w(x) = \sum_{j=1}^{N} \sum_{n=-\infty}^{+\infty} f_j(x_j + na)\delta\left[x - (x_j + na)\right], \tag{12.43}$$

$$f_j(x_j + na) = -k_j\left[w(x_j + na) - u_j(x_j + na)\right], \quad j = 1, \dots, N, \tag{12.44}$$

$$-\omega^2 m_j u_j(x_j + na) = -f_j(x_j + na), \quad j = 1, \dots, N \tag{12.45}$$

with E Young's modulus, I the cross-sectional inertia moment, A the cross-sectional area, while $w(x)$ is the transverse displacement of the beam, $f_j(x_j + na)$ the force

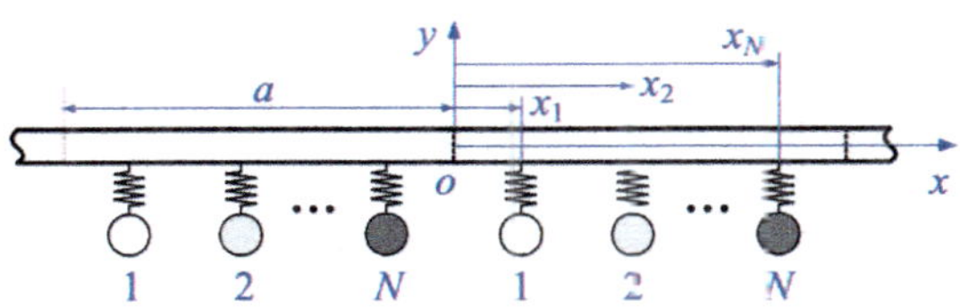

Fig. 12.3 Diagram of an infinite LR beam with multiple periodic arrays of attached spring-mass resonators.

applied to the beam by the resonator located at $x_j + na$ and $u_j(x_j + na)$ is the displacement of the mass of the resonator.

For the periodicity of the LR beam system, the displacement of the beam can be written as

$$w(x) = \sum_{m=-\infty}^{+\infty} W_m \, e^{-i(q+2m\pi/a)x} \tag{12.46}$$

and similarly can be done for the others terms present in Eq. (12.43). By substituting these quantities in Eq. (12.43) the following equations are obtained:

$$\begin{aligned} &EIa(q+2m\pi/a)^4 W_m - \omega^2 \rho A a W_m = \sum_{j=1}^{N} \left\{ -k_j \left[w(x_j) - u_j \right] e^{i(q+2m\pi/a)x_j} \right\} \\ &-\omega^2 m_j u_j = j_j \left[w(x_j) - u_j \right], \quad j = 1, \ldots, N. \end{aligned} \tag{12.47}$$

By solving the previous equations the predicted values of the Bloch vectors are obtained (see Figs. 2 and 3 in Xiao et al (2012)).

Flexural wave propagation is also studied in Xiao et al (2013), by using an exact analytical approach based on a combination of the spectral element method and periodic structure theory. The authors calculate the propagation constants and thus the complex band structures in the case of infinite and finite periodic structures are considered. The locally resonant beam structure is pictorially described in Fig. 12.4. As a result of numerical simulations, the complex band structures of a LR beam (see Fig. 3 in Xiao et al, 2013) and the relationship between the lattice constant L on the band gap behavior of a LR beam with fixed absorber parameters is obtained (see Fig.

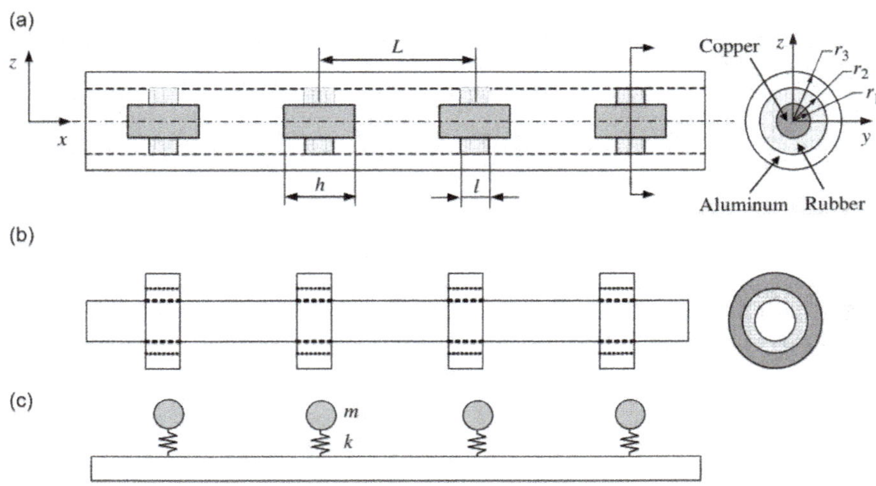

Fig. 12.4: Locally resonant (LR) beams with periodically attached dynamic vibration absorbers (DVAs): (a) and (b) are schematics of two realistic models, and (c) is the simplified physical model.

10-11 in Xiao et al, 2013). By changing the key physical parameters it is possible to show how to obtain different basic unit cell, as depicted in Fig. 12.5.

A numerical study and relative comparison with experiments of in-plane and out-of-plane guided waves in a thin plate with local resonators is presented in Zhu et al (2011). The authors achieve a new metamaterial plate design for a low-frequency bandgap in both in-plane and out-of-plane guided waves. A different point of view is considered in Zhu et al (2014), where chiral lattice-based EMM beam with multiple embedded local resonators is suggested to achieve broadband vibration suppression. Also in this case, the authors verify their principal hypothesis via comparison between numerical simulations and experimental measures.

In Chen et al (2011) the behaviour of wave propagation in a sandwich beam with internal resonators (see Fig. 12.6) is studied analytically and experimentally. The authors find that, near the local resonance frequency of the resonator, harmonic waves cannot propagate without attenuation in amplitude and that the extent of the bandgap can be selected by altering the local resonance frequency of the resonator. The magnitude of the internal mass has a greater influence on the size of the bandgap. Sandwich beams with internal resonators are also studied in Sharma and Sun (2016). Using FEM numerical tools the effectiveness of various local resonator frequencies under a given impact load is compared to the behavior of an equivalent mass beam. It is shown that addition of appropriately chosen local resonators into the sandwich beam is an effective method of improving its flexural bending behavior under impact loads.

In Thompson (2008) a damped mass–spring absorber system is considered attached continuously along the beam length. The aim of this study is to attenuate

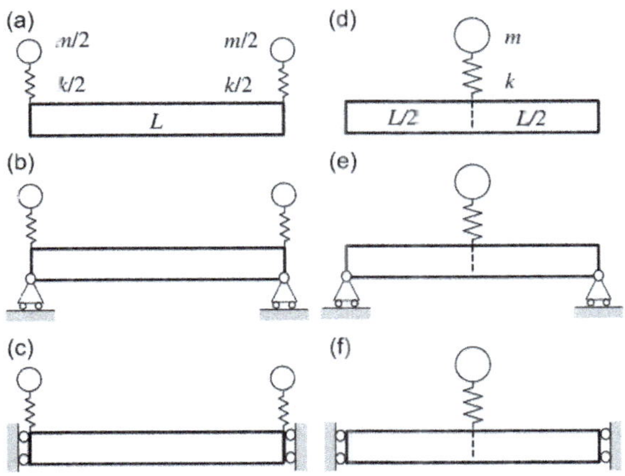

Fig. 12.5 Unit-cell models: (a) and (d) represent two choices of a single symmetric unit cell; (b), (c), (e) and (f) depict the unit cells with pinned-pinned or sliding–sliding boundary conditions.

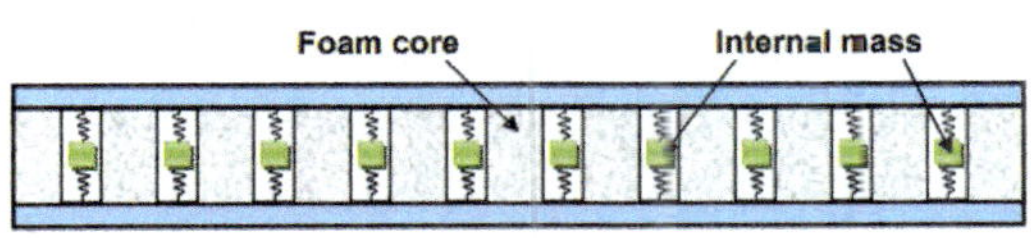

Fig. 12.6 Sandwich beam with internal resonators.

structural waves in beams. As particular application the authors give the reduction of noise from a railway track, which requires the attenuation of structural waves along the rail to be increased over a frequency band of two or more octaves.

The study in Wang (2014) presents a cell of elastic metamaterials in an effort to provide a model for generating negative mass and/or negative modulus. The authors present a model consisting of a series of properly arranged rigid bodies and linear springs. By introducing both translational and rotational motions in the representative cell, negative mass and negative modulus can be obtained in a controlled manner. Finally, a review on the recent advances in the microstructural designs of LR based EMM plates can be found in Zhu et al (2015).

12.6 Conclusions

In this review we have presented the actual development of the studies on acoustic metamaterials based on local resonances. We have explored the homogenization procedures developed for this kind of materials and a particular attention has been paid to the topological optimization processes. Finally, some applications have been presented. In particular the interesting case of phononic crystal has been treated. Some interesting applications for the methods described in this review could result in the development of brand-new kind of acoustic metamaterials. Indeed, the fast advancement of 3D printing technologies, allowing the realization of relatively "small structural architectures", has pushed the interest of several research groups into this field. History of science has many paradigmatic examples in which it is clear how technological advancement leads to novel theoretical progresses and opens new scientific perspectives (Dell'Isola et al, 2016; Russo and Levy, 2013; Eugster and dell'Isola, 2017; dell'Isola et al, 2016).

This review would answer to the need of understanding the main techniques and results in the modeling of acoustic metamaterials. The importance of such activity should not be underestimated since in the history of science it happens frequently that some results are lost and then rediscovered in subsequent works.

References

Alibert J, Della Corte A (2015) Second-gradient continua as homogenized limit of pantographic microstructured plates: a rigorous proof. ZAMP 66(5):2855–2870

Alibert JJ, Seppecher P, dell'Isola F (2003) Truss modular beams with deformation energy depending on higher displacement gradients. Mathematics and Mechanics of Solids 8(1):51–73

Allaire G, Jouve F, Toader AM (2004) Structural optimization using sensitivity analysis and a level-set method. Journal of Computational Physics 194(1):363–393

Alù A, Engheta N (2005) Achieving transparency with plasmonic and metamaterial coatings. Physical Review E 72(1):016,623

Ambati M, Fang N, Sun C, Zhang X (2007) Surface resonant states and superlensing in acoustic metamaterials. Physical Review B 75(19):195,447

Andreaus U, dell'Isola F, Giorgio I, Placidi L, Lekszycki T, Rizzi N (2016) Numerical simulations of classical problems in two-dimensional (non) linear second gradient elasticity. International Journal of Engineering Science 108:34–50

Andrianov I, Bolshakov V, Danishevs'kyy V, Weichert D (2008) Higher order asymptotic homogenization and wave propagation in periodic composite materials. Proc R Soc London A: Mathematical, Physical and Engineering Sciences 464(2093):1181–1201

Ao X, Chan C (2008) Far-field image magnification for acoustic waves using anisotropic acoustic metamaterials. Physical Review E 77(2):025,601

Auffray N, dell'Isola F, Eremeyev V, Madeo A, Rosi G (2015) Analytical continuum mechanics à la Hamilton–Piola least action principle for second gradient continua and capillary fluids Mathematics and Mechanics of Solids 20(4):375–417

Barchiesi E, Placidi L (2017) A review on models for the 3d statics and 2d dynamics of pantographic fabrics. In: Wave Dynamics and Composite Mechanics for Microstructured Materials and Metamaterials, Springer, pp 239–258

Battista A, Rosa L, dell'Erba R, Greco L (2016) Numerical investigation of a particle system compared with first and second gradient continua: Deformation and fracture phenomena. Mathematics and Mechanics of Solids 22:2120–2134

Bendsøe MP, Kikuchi N (1988) Generating optimal topologies in structural design using a homogenization method. Computer Methods in Applied Mechanics and Engineering 71(2):197–224

Bendsøe MP, Sigmund O (2004) Topology optimization by distribution of isotropic material. In: Topology Optimization, Springer, pp 1–69

Berezovski A, Giorgio I, Corte AD (2016) Interfaces in micromorphic materials: wave transmission and reflection with numerical simulations. Mathematics and Mechanics of Solids 21(1):37–51

Bertram A, Glüge R (2016) Gradient materials with internal constraints. Mathematics and Mechanics of Complex Systems 4(1):1–15

Bevill G, Eswaran SK, Gupta A, Papadopoulos P, Keaveny TM (2006) Influence of bone volume fraction and architecture on computed large-deformation failure mechanisms in human trabecular bone. Bone 39(6):1218–1225

Boutin C, Giorgio I, Placidi L, et al (2017) Linear pantographic sheets: Asymptotic micro-macro models identification. Mathematics and Mechanics of Complex Systems 5(2):127–162

Brun M, Guenneau S, Movchan A (2009) Achieving control of in-plane elastic waves. Applied Physics Letters 94(6):061,903

Caprino S, Esposito R, Marra R, Pulvirenti M (1993) Hydrodynamic limits of the vlasov equation. Communications in Partial Differential Equations 18(5):805–820

Carinci G, De Masi A, Giardinà C, Presutti E (2014a) Hydrodynamic limit in a particle system with topological interactions. Arabian Journal of Mathematics 3(4):381–417

Carinci G, De Masi A, Giardinà C, Presutti E (2014b) Super-hydrodynamic limit in interacting particle systems. Journal of Statistical Physics 155(5):867–887

Chan C, Li J, Fung K (2006) On extending the concept of double negativity to acoustic waves. Journal of Zhejiang University-SCIENCE A 7(1):24–28

Chang Z, Hu J, Hu G (2010) Transformation method and wave control. Acta Mechanica Sinica 26(6):889–898

Chang Z, Hu J, Hu G, Tao R, Wang Y (2011) Controlling elastic waves with isotropic materials. Applied Physics Letters 98(12):121,904

Chen H, Chan C (2007) Acoustic cloaking in three dimensions using acoustic metamaterials. Applied Physics Letters 91(18):183,518

Chen H, Chan C (2010) Acoustic cloaking and transformation acoustics. Journal of Physics D: Applied Physics 43(11):113,001

Chen J, Sharma B, Sun C (2011) Dynamic behaviour of sandwich structure containing spring-mass resonators. Composite Structures 93(8):2120–2125

Craster R, Guenneau S (2012) Acoustic metamaterials: Negative refraction, imaging, lensing and cloaking, vol 166. Springer Science & Business Media

Craster RV, Kaplunov J, Pichugin AV (2010) High-frequency homogenization for periodic media. Proc Royal Soc A 466:2341–2362

Cummer S, Schurig D (2007) One path to acoustic cloaking. New Journal of Physics 9(3):45
Cummer S, Popa B, Schurig D, Smith D, Pendry J, Rahm M, Starr A (2008) Scattering theory derivation of a 3d acoustic cloaking shell. Physical Review Letters 100(2):024,301
Cuomo M, dell'Isola F, Greco L, Rizzi N (2016) First versus second gradient energies for planar sheets with two families of inextensible fibres: Investigation on deformation boundary layers, discontinuities and geometrical instabilities. Composites Part B: Engineering 115:423–448
Czech B, van Kessel R, Bauer P, Ferreira JA, Wattez A (2010) Energy harvesting using dielectric elastomers. Power Electronics and Motion Control Conference (EPE/PEMC), 2010 14th International pp S4–18
De Masi A, Olla S (2015) Quasi-static hydrodynamic limits. Journal of Statistical Physics 161(5):1037–1058
De Masi A, Luckhaus S, Presutti E (2007) Two scales hydrodynamic limit for a model of malignant tumor cells. Annales de l'Institut Henri Poincare (B) Probability and Statistics 43(3):257–297
De Masi A, Merola I, Presutti E, Vignaud Y (2009) Coexistence of ordered and disordered phases in potts models in the continuum. Journal of Statistical Physics 134(2):243–306
De Masi A, Galves A, Löcherbach E, Presutti E (2015) Hydrodynamic limit for interacting neurons. Journal of Statistical Physics 158(4):866–902
dell'Isola F, Giorgio I, Andreaus U (2015) Elastic pantographic 2d lattices: a numerical analysis on static response and wave propagation. Proc Estonian Academy of Sciences 64(3):219–225
dell'Isola F, Bucci S, Battista A (2016a) Against the fragmentation of knowledge: The power of multidisciplinary research for the design of metamaterials. In: Advanced Methods of Continuum Mechanics for Materials and Structures, Springer, pp 523–545
dell'Isola F, Cuomo M, Greco L, Della Corte A (2016b) Bias extension test for pantographic sheets: numerical simulations based on second gradient shear energies. Journal of Engineering Mathematics pp 1–31
Dell'Isola F, Della Corte A, Esposito R, Russo L (2016) Some cases of unrecognized transmission of scientific knowledge: from antiquity to Gabrio Piola's peridynamics and generalized continuum theories. In: Altenbach H, Forest S (eds) Generalized Continua as Models for Classical and Advanced Materials, Advanced Structured Materials, vol 42, Springer, Cham, pp 77–128
dell'Isola F, Della Corte A, Giorgio I (2016) Higher-gradient continua: The legacy of piola, mindlin, sedov and toupin and some future research perspectives. Mathematics and Mechanics of Solids 22(4):852–872
Deng K, Ding Y, He Z, Zhao H, Shi J, Liu Z (2009) Theoretical study of subwavelength imaging by acoustic metamaterial slabs. Journal of Applied Physics 105(12):124,909
Deymier P (2013) Acoustic Metamaterials and Phononic Crystals. Springer Series in Solid-State Sciences, Springer Berlin Heidelberg
Ding Y, Liu Z, Qiu C, Shi J (2007) Metamaterial with simultaneously negative bulk modulus and mass density. Physical Review Letters 99(9):093,904
Eringen A (1976) Continuum Physics, vol 4: Polar and Nonlocal Field Theories. Academic Press, Inc., New York
Esposito R, Pulvirenti M (2004) From particles to fluids. Handbook of Mathematical Fluid Dynamics 3:1–82
Eugster SR, dell'Isola F (2017) Exegesis of the Introduction and Sect. I from "Fundamentals of the Mechanics of Continua" by E. Hellinger. ZAMM 97(4):477–506
Fang N, Xi D, Xu J, Ambati M, Srituravanich W, Sun C, Zhang X (2006) Ultrasonic metamaterials with negative modulus. Nature Materials 5(6):452–456
Ganghoffer J (2016) Spatial and material stress tensors in continuum mechanics of growing solid bodies. Mathematics and Mechanics of Complex Systems 3(4):341–363
Giorgio I (2016) Numerical identification procedure between a micro-cauchy model and a macro-second gradient model for planar pantographic structures. ZAMP 67(4)(95)
Gokhale N, Cipolla J, Norris A (2012) Special transformations for pentamode acoustic cloaking. The Journal of the Acoustical Society of America 132(4):2932–2941
Gu YW, Luo XD, Ma HR (2008) Optimization of the local resonant sonic material by tuning the shape of the resonator. Journal of Physics D: Applied Physics 41(20):205,402

Guenneau S, Movchan A, Pétursson G, Ramakrishna S (2007) Acoustic metamaterials for sound focusing and confinement. New Journal of Physics 9(11):399
Guild M, Alu A, Haberman M (2011) Cancellation of acoustic scattering from an elastic sphere. The Journal of the Acoustical Society of America 129(3):1355–1365
Hirsekorn M (2004) Small-size sonic crystals with strong attenuation bands in the audible frequency range. Applied Physics Letters 84(17):3364–3366
Hirsekorn M, Delsanto PP, Leung AC, Matic P (2006) Elastic wave propagation in locally resonant sonic material: Comparison between local interaction simulation approach and modal analysis. Journal of Applied Physics 99(12):124,912
Hu J, Zhou X, Hu G (2009) Design method for electromagnetic cloak with arbitrary shapes based on Laplace's equation. Optics Express 17(3):1308–1320
Hu J, Chang Z, Hu G (2011) Approximate method for controlling solid elastic waves by transformation media. Physical Review B 84(20):201,101
Huang H, Sun C (2009) Wave attenuation mechanism in an acoustic metamaterial with negative effective mass density. New Journal of Physics 11(1):013,003
Jacob Z, Alekseyev LV, Narimanov E (2006) Optical hyperlens: far-field imaging beyond the diffraction limit. Optics Express 14(18):8247–8256
Javili A, McBride A, Mergheim J, Steinmann P, Schmidt U (2013) Micro-to-macro transitions for continua with surface structure at the microscale. IJSS 50(16):2561–2572
Jia H, Ke M, Hao R, Ye Y, Liu F, Liu Z (2010) Subwavelength imaging by a simple planar acoustic superlens. Applied Physics Letters 97(17):173,507
Kildishev AV, Narimanov EE (2007) Impedance-matched hyperlens. Optics Letters 32(23):3432–3434
Kushwaha MS, Halevi P, Dobrzynski L, Djafari-Rouhani B (1993) Acoustic band structure of periodic elastic composites. Physical Review Letters 71(13):2022
Lacarbonara W (2013) Nonlinear Structural Mechanics: Theory, Dynamical Phenomena and Modeling. Springer US
Lee H, Liu Z, Xiong Y, Sun C, Zhang X (2007) Development of optical hyperlens for imaging below the diffraction limit. Optics Express 15(24):15,886–15,891
Li J, Chan C (2004) Double-negative acoustic metamaterial. Physical Review E 70(5):055,602
Liu F, Cai F, Peng S, Hao R, Ke M, Liu Z (2009) Parallel acoustic near-field microscope: A steel slab with a periodic array of slits. Physical Review E 80(2):026,603
Liu X, Hu G, Huang G, Sun C (2011a) An elastic metamaterial with simultaneously negative mass density and bulk modulus. Applied Physics Letters 98(25):251,907
Liu X, Hu G, Sun C, Huang G (2011b) Wave propagation characterization and design of two-dimensional elastic chiral metacomposite. Journal of Sound and Vibration 330(11):2536–2553
Liu Z, Zhang X, Mao Y, Zhu Y, Yang Z, Chan C, Sheng P (2000) Locally resonant sonic materials Science 289(5485):1734–1736
Liu Z, Lee H, Xiong Y, Sun C, Zhang X (2007) Far-field optical hyperlens magnifying sub-diffraction-limited objects. science 315(5819):1686–1686
Madeo A, Della Corte A, Greco L, Neff P (2014) Wave propagation in pantographic 2d lattices with internal discontinuities. arXiv preprint arXiv:14123926
Matsuki T, Yamada T, Izui K, Nishiwaki S (2014) Topology optimization for locally resonant sonic materials. Applied Physics Letters 104(19):191,905
Milton G, Nicorovici N (2006) On the cloaking effects associated with anomalous localized resonance. Proc Royal Soc A 462(2074):3027–3059
Milton G, Willis J (2007) On modifications of newton's second law and linear continuum elastodynamics. Proc Royal Soc A 463:855–880
Milton G, Briane M, Willis J (2006) On cloaking for elasticity and physical equations with a transformation invariant form. New Journal of Physics 8(10):248
Milton G, Briane M, Harutyunyan D (2017) On the possible effective elasticity tensors of 2-dimensional and 3-dimensional printed materials. Mathematics and Mechanics of Complex Systems 5(1):41–94

Misra A, Poorsolhjouy P (2015) Identification of higher-order elastic constants for grain assemblies based upon granular micromechanics. Mathematics and Mechanics of Complex Systems 3(3):285–308

Nadler B, Papadopoulos P, Steigmann D (2006) Multiscale constitutive modeling and numerical simulation of fabric material. IJSS 43(2):206–221

Nemat-Nasser J Sand Willis, Srivas tava A, Amirkhizi A (2011) Homogenization of periodic elastic composites and locally resonant sonic materials. Phy Rev B 83(10):104,103

Nemat-Nasser S (2015) Anti-plane shear waves in periodic elastic composites: band structure and anomalous wave refraction. Proc R Soc London A: Mathematical, Physical and Engineering Sciences 471(2180):20150,152

Nemat-Nasser S, Srivastava A (2011) Negative effective dynamic mass-density and stiffness: Micro-architecture and phononic transport in periodic composites. AIP Advances 1(4):041,502

Norris A (2009) Acoustic metafluids. The Journal of the Acoustical Society of America 125(2):839–849

Norris A, Parnell W (2012) Hyperelastic cloaking theory: transformation elasticity with pre-stressed solids. Proc Royal Soc A 468(2146):2881–2903

Norris A, Shuvalov A (2011) Elastic cloaking theory. Wave Motion 48(6):525–538

Norris A, Shuvalov A, Kutsenko A (2012) Analytical formulation of three-dimensional dynamic homogenization for periodic elastic systems. Proc Royal Soc A 468(2142):1629–1651

Oates WS, Liu F (2009) Piezohydraulic actuator development for microjet flow control. Journal of Mechanical Design 131(9):091,001

Oh JH, Ahn YK, Kim YY (2015) Maximization of operating frequency ranges of hyperbolic elastic metamaterials by topology optimization. Structural and Multidisciplinary Optimization 52(6):1023–1040

Otero JA, Rodriguez-Ramos R, Monsivais G, Perez-Alvarez R (2005) Dynamical behavior of a layered piezocomposite using the asymptotic homogenization method. Mechanics of Materials 37(1):33–44

Park YL, Majidi C, Kramer R, Bérard P, Wood RJ (2010) Hyperelastic pressure sensing with a liquid-embedded elastomer. Journal of Micromechanics and Microengineering 20(12):125,029

Pendry J (2000) Negative refraction makes a perfect lens. Physical Review Tetters 85(18):3966

Pendry J, Holden A, Stewart W, Youngs I (1996) Extremely low frequency plasmons in metallic mesostructures. Physical Review Letters 76(25):4773

Pideri C, Seppecher P (1997) A second gradient material resulting from the homogenization of an heterogeneous linear elastic medium. Continuum Mechanics and Thermodynamics 9(5):241–257

Piola G (1825) Sull'applicazione de'principj della meccanica analitica del Lagrange ai principali problemi. Memoria di Gabrio Piola presentata al concorso del premio e coronata dall'IR Istituto di Scienze, ecc. nella solennità del giorno 4 ottobre 1824. dall'Imp. Regia stamperia

Placidi L (2015) A variational approach for a nonlinear 1-dimensional second gradient continuum damage model. Continuum Mechanics and Thermodynamics 27(4-5):623

Placidi L (2016) A variational approach for a nonlinear one-dimensional damage-elasto-plastic second-gradient continuum model. Continuum Mechanics and Thermodynamics 28(1-2):119–137

Placidi L, dell'Isola F, Ianiro N, Sciarra G (2008) Variational formulation of pre-stressed solid–fluid mixture theory, with an application to wave phenomena. European Journal of Mechanics-A/Solids 27(4):582–606

Placidi L, Rosi G, Giorgio I, Madeo A (2014) Reflection and transmission of plane waves at surfaces carrying material properties and embedded in second-gradient materials. Mathematics and Mechanics of Solids 19(5):555–578

Placidi L, Andreaus U, Giorgio I (2016a) Identification of two-dimensional pantographic structure via a linear d4 orthotropic second gradient elastic model. Journal of Engineering Mathematics pp 1–21

Placidi L, Greco L, Bucci S, Turco E, Rizzi N (2016b) A second gradient formulation for a 2d fabric sheet with inextensible fibres. ZAMP 67(5)(114)

Pulvirenti M (1996) Kinetic limits for stochastic particle systems. Lecture Notes in Mathematics, Springer

Rahali Y, Giorgio I, Ganghoffer J, Dell'Isola F (2015) Homogenization à la Piola produces second gradient continuum models for linear pantographic lattices. International Journal of Engineering Science 97:148–172

Rinaldi A, Placidi L (2014) A microscale second gradient approximation of the damage parameter of quasi-brittle heterogeneous lattices. ZAMM 94(10):862–877

Russo L, Levy S (2013) The Forgotten Revolution: How Science Was Born in 300 BC and Why it Had to Be Reborn. Springer Berlin Heidelberg

Saeb S, Steinmann P, Javili A (2016) Aspects of computational homogenization at finite deformations: A unifying review from Reuss' to Voigt's bound. Applied Mechanics Reviews 68(5):050,801

Salandrino A, Engheta N (2006) Far-field subdiffraction optical microscopy using metamaterial crystals: Theory and simulations. Physical Review B 74(7):075,103

Sánchez-Dehesa J, Garcia-Chocano VM, Torrent D, Cervera F, Cabrera S, Simon F (2011) Noise control by sonic crystal barriers made of recycled materials. The Journal of the Acoustical Society of America 129(3):1173–1183

Sathyamoorthy M (1997) Nonlinear Analysis of Structures, Mechanical and Aerospace Engineering Series, vol 8. CRC Press

Sharma B, Sun C (2016) Impact load mitigation in sandwich beams using local resonators. Journal of Sandwich Structures & Materials 18(1):50–64

Sieck C, Alù A, Haberman M (2015) Dynamic homogenization of acoustic metamaterials with coupled field response. Physics Procedia 70:275–278

Sigalas M, Kushwaha MS, Economou EN, Kafesaki M, Psarobas IE, Steurer W (2005) Classical vibrational modes in phononic lattices: theory and experiment. Zeitschrift für Kristallographie-Crystalline Materials 220(9-10):765–809

Smith J (2011) Application of the method of asymptotic homogenization to an acoustic metafluid. Proc R Soc London A: Mathematical, Physical and Engineering Sciences 467(2135):3318–3331

Srivastava A (2015) Elastic metamaterials and dynamic homogenization: a review. International Journal of Smart and Nano Materials 6(1):41–60

Srivastava A, Nemat-Nasser S (2012) Overall dynamic properties of three-dimensional periodic elastic composites. Proc R Soc London A: Mathematical, Physical and Engineering Sciences 468(2137):269–287

Steigmann D (2008) Two-dimensional models for the combined bending and stretching of plates and shells based on three-dimensional linear elasticity. IJ Engng Sci 46(7):654–676

Steigmann D, dell'Isola F (2015) Mechanical response of fabric sheets to three-dimensional bending, twisting, and stretching. Acta Mechanica Sinica 31(3):372–382

Svanberg K (1987) The method of moving asymptotes' new method for structural optimization. International Journal for Numerical Methods in Engineering 24(2):359–373

Thompson DJ (2008) A continuous damped vibration absorber to reduce broad-band wave propagation in beams. Journal of Sound and Vibration 311(3):824–842

Torrent D, Sánchez-Dehesa J (2008) Acoustic cloaking in two dimensions: a feasible approach. New Journal of Physics 10(6):063,015

Torrent D, Pennec Y, Djafari-Rouhani B (2014) Effective medium theory for elastic metamaterials in thin elastic plates. Physical Review B 90(10):104–110

Tripathi A, Bajaj AK (2016) Topology optimization and internal resonances in transverse vibrations of hyperelastic plates. IJSS 81:311–328

Turco E, dell'Isola F, Cazzani A, Rizzi N (2016) Hencky-type discrete model for pantographic structures: numerical comparison with second gradient continuum models. ZAMP 67

Veselago V (1967) Properties of materials having simultaneously negative values of the dielectric and magnetic susceptibilities. Soviet Physics Solid State USSR 8:2854–2856

Veselago V (1968) The electrodynamics of substances with simultaneously negative values of ϵ and μ. Soviet Physics Uspekhi 10(4):509

Veselago VG (2002) Electrodynamics of media with simultaneously negative electric permittivity and magnetic permeability. In: Advances in Electromagnetics of Complex Media and Metamaterials, Springer, pp 83–97
Wang MY, Wang X, Guo D (2003) A level set method for structural topology optimization. Computer Methods in Applied Mechanics and Engineering 192(1):227–246
Wang X (2014) Dynamic behaviour of a metamaterial system with negative mass and modulus. IJSS 51(7):1534–1541
Wang YF, Wang YS, Laude V (2015) Wave propagation in two-dimensional viscoelastic metamaterials. Physical Review B 92(10):104,110
Waterman PC (1969) New formulation of acoustic scattering. The Journal of the Acoustical Society of America 45(6):1417–1429
Willis J (2011) Effective constitutive relations for waves in composites and metamaterials. Proc Royal Soc A 467(2131):1865–1879
Wu Y, Lai Y, Zhang Z (2011) Elastic metamaterials with simultaneously negative effective shear modulus and mass density. Physical Review Letters 107(10):105,506
Xiao Y, Wen J, Wen X (2012) Broadband locally resonant beams containing multiple periodic arrays of attached resonators. Physics Letters A 376(16):1384–1390
Xiao Y, Wen J, Yu D, Wen X (2013) Flexural wave propagation in beams with periodically attached vibration absorbers: Band-gap behavior and band formation mechanisms. Journal of Sound and Vibration 332(4):867–893
Xiong Y, Liu Z, Zhang X (2009) A simple design of flat hyperlens for lithography and imaging with half-pitch resolution down to 20 nm. Applied Physics Letters 94(20):203,108
Yamada T, Izui K, Nishiwaki S, Takezawa A (2010) A topology optimization method based on the level set method incorporating a fictitious interface energy. Computer Methods in Applied Mechanics and Engineering 199(45):2876–2891
Yang Y, Misra A (2012) Micromechanics based second gradient continuum theory for shear band modeling in cohesive granular materials following damage elasticity. IJSS 49(18):2500–2514
Yao S, Zhou X, Hu G (2008) Experimental study on negative effective mass in a 1D mass-spring system. New Journal of Physics 10(4):043,020
Yu D, Wen J, Zhao H, Liu Y, Wen X (2008) Vibration reduction by using the idea of phononic crystals in a pipe-conveying fluid. Journal of Sound and vibration 318(1):193–205
Zhang S, Yin L, Fang N (2009) Focusing ultrasound with an acoustic metamaterial network. Physical Review Letters 102(19):194,301
Zhang S, Xia C, Fang N (2011) Broadband acoustic cloak for ultrasound waves. Physical Review Letters 106(2):024,301
Zhou X, Hu G (2006) Design for electromagnetic wave transparency with metamaterials. Physical Review E 74(2):026,607
Zhou X, Hu G (2007) Acoustic wave transparency for a multilayered sphere with acoustic metamaterials. Physical Review E 75(4):046,606
Zhou X, Hu G (2009) Analytic model of elastic metamaterials with local resonances. Physical Review B 79(19):195,109
Zhou X, Hu G, Lu T (2008) Elastic wave transparency of a solid sphere coated with metamaterials. Physical Review B 77(2):024,101
Zhu R, Huang G, Huang H, Sun C (2011) Experimental and numerical study of guided wave propagation in a thin metamaterial plate. Physics Letters A 375(30):2863–2867
Zhu R, Liu X, Hu G, Sun C, Huang G (2014) A chiral elastic metamaterial beam for broadband vibration suppression. Journal of Sound and Vibration 333(10):2759–2773
Zhu R, Liu X, Hu G, Yuan F, Huang G (2015) Microstructural designs of plate-type elastic metamaterial and their potential applications: a review. International Journal of Smart and Nano Materials 6(1):14–40

Chapter 13
On Nonlinear Waves in Media with Complex Properties

Jüri Engelbrecht, Andrus Salupere, Arkadi Berezovski, Tanel Peets, and Kert Tamm

Abstract In nonlinear theories the axiom of equipresence requires all the effects of the same order to be taken account. In this paper the mathematical modelling of deformation waves in media is analysed involving nonlinear and dispersive effects together with accompanying phenomena caused by thermal or electrical fields. The modelling is based on principles of generalized continuum mechanics developed by G.A. Maugin. The analysis demonstrates the richness of models in describing the physical effects in media with complex properties.

13.1 Introduction

The legacy of G.A. Maugin is huge and has an imprint on many studies on continuum mechanics in the second half of the 20th century. His studies have cast light on many fundamental problems of continua like the principle of virtual power, generalized continuum mechanics, the concepts of internal variables and configurational forces, propagation of waves and fronts – just to name a few (dell'Isola et al, 2014). His sparkling ability to inspire his colleagues to collaborate and find new problems in the field of fundamental understanding of the behaviour of materials has been realized in numerous joint publications. In this paper the attention is paid to nonlinear wave propagation. G.A. Maugin himself has studied waves in elastic crystals (Maugin, 1999), numerical methods used for the analysis of waves and fronts (Berezovski et al, 2008) and published several overviews on waves (Maugin, 2011; Christov et al, 2007).The cooperation with colleagues in Tallinn has resulted in describing complexities of soliton theory (Salupere et al, 1994, 2001), in nerve pulse analysis

Jüri Engelbrecht · Andrus Salupere · Arkadi Berezovski · Tanel Peets · Kert Tamm
Department of Cybernetics, School of Science, Tallinn University of Technology, Akadeemia tee 21, 12618, Tallinn, Estonia
e-mail: je@ioc.ee, salupere@ioc.ee, berezovski.arkadi@gmail.com, tanelp@ioc.ee, kert@ioc.ee

© Springer International Publishing AG, part of Springer Nature 2018
H. Altenbach et al. (eds.), *Generalized Models and Non-classical Approaches in Complex Materials 1*, Advanced Structured Materials 89,
https://doi.org/10.1007/978-3-319-72440-9_13

(Maugin and Engelbrecht, 1994), in elaborating the concept of internal variables (Berezovski et al, 2011a,b), etc. Here we shall present some fundamental ideas from this cooperation and novel results developed recently. The basic problem is how to describe real properties of materials and how these are reflected in wave propagation. The importance of such an analysis is pointed out also by Maugin (2015).

In what follows, the problems in deriving the governing equations of nonlinear wave motion for describing complicated properties of media (materials) and the corresponding mathematical models are presented in Sect. 13.2. The physical effects resulting from these governing equations are analysed in Sect. 13.3. Finally, in Sect. 13.4 discussion is given together with ideas for the further research.

13.2 The Governing Equations

The governing equations for describing wave motion are based on the balance of momentum. Besides classical linear wave equations, the Boussinesq-type models are richer because they account also nonlinear and dispersive effects (Christov et al, 2007). Like classical wave equations in the 1D setting, these equations have bi-directional solutions. Another class of models describing nonlinear waves are evolution equations like the Korteweg-de Vries (KdV) equation and its modifications. Evolution equations describe uni-directional propagation and are usually derived from complicated systems by the reductive perturbation method using the moving frame of reference. Characteristically for both types of equations, the modelling of nonlinear and dispersive effects permits to describe many interesting physical phenomena. Below some results of modelling are briefly described. The 1D set-ups are used in order to reach transparent models where it would be easy to trace the influence of individual terms in models.

A typical form of a Boussinesq-type equation in terms of a displacement u is (Christov et al, 2007):

$$u_{tt} - c_0^2 u_{xx} - [F(u)]_{xx} = (\beta_1 u_{tt} - \beta_2 u_{xx})_{xx}, \tag{13.1}$$

where c_0 is the wave velocity, F_u is a polynomial and β_1, β_2 are physical constants. As usual, x denotes space and t denotes time. Here and further, indices x and t denote the differentiation with respect to the indicated variable. This equation can be found as in solid mechanics as well as in fluids where it was derived originally. It must be noted that the r.h.s. of Eq. (13.1) has often an order of $\mathcal{O}(\varepsilon)$ where ε is a small parameter.

The general form of a KdV-type evolution equation in terms of $v \sim u_x$ (or $v \sim u_t$) is (Salupere et al, 2001)

$$v_\tau + [P(v)]_\xi + D(v) = 0, \tag{13.2}$$

where τ is a scaled coordinate, $\xi = c_0 t - x$ is the moving frame coordinate, $P(v)$ is a polynomial and $D(v)$ is a dispersion operator including the odd derivatives of v with respect to ξ only.

13.2.1 Boussinesq-Type Models

In modelling of microstructured solids, it is possible to distinguish macro- and microstructure that must be taken into account in modelling the wave motion. Based on the Mindlin (1964) micromorphic theory, the governing equations can be derived for both coupled structures. The existence of the microstructure leads to dispersive effects while nonlinearity is of the physical character. The free energy W is assumed to have a form:

$$W = \frac{\rho_0 c^2}{2} u_x^2 + A_1 \varphi u_x + \frac{1}{2} B\varphi^2 + C\varphi_x^2 + \frac{1}{6} N u_x^3 + \frac{1}{6} M\varphi_x^3, \tag{13.3}$$

where ρ_0 and c are the density and the sound velocity of the macrostructure, u is the macrodisplacement, φ is the microdeformation in the sense of Mindlin (1964), and A, B, C, N, M are material parameters. The kinetic energy K is

$$K = \frac{\rho_0}{2} u_t^2 + \frac{I}{2} \varphi_t^2, \tag{13.4}$$

where I is the measure of microstructure inertia.

Then the governing equation in terms of the macrodisplacement u is (Engelbrecht et al, 2005; Berezovski et al, 2013):

$$u_{tt} - (c^2 - c_A^2) u_{xx} - \frac{1}{2} k_1 (u_x^2)_x = p^2 c_A^2 \left(u_{tt} - c_1^2 u_{xx} \right)_{xx} - \frac{1}{2} k_2 (u_{xx})_{xx}, \tag{13.5}$$

where $c_A^2 = A^2/\rho_0 B$, $c_1^2 = C/I$, $p^2 = I/B$ and k_1, k_2 are the coefficients of nonlinearities.

It must be stressed that Eq. (13.5) reflects the following: (i) the nonlinearities are of the deformation-type as usually in solid mechanics; (ii) microinertia of the microstructure is taken into account; (iii) the second wave operator at the r.h.s has a parameter p^2 which is usually small and therefore Eq. (13.5) is a hierarchical equation. In addition, the velocity of the wave operator for the macrostructure at the l.h.s is influenced by the properties of the microstructure. In such a way, the governing equation reflects real properties of the microstructured material.

In case of a multiscale (the scale in the scale) microstructure, the governing equation involves two wave operators reflecting the properties of microstructures (Engelbrecht et al, 2007). In the linear case this equation is:

$$u_{tt} - (c^2 - c_{A1}^2)u_{xx} = p_1^2 c_{A1}^2 \left[u_{tt} - \left(c_1^2 - c_{A2}^2\right) u_{xx}\right]_{xx} \\ - p_1^2 c_{A1}^2 p_2^2 c_{A2}^2 \left(u_{tt} - c_2^2 u_{xx}\right)_{xxxx}, \tag{13.6}$$

where indices 1 and 2 denote the microstructures. The smaller scales bring in higher order dispersive terms. Like the macrostructure, the level 1 microstructure is also influenced by the level 2 microstructure. Two wave operators at the r.h.s. of Eq. (13.6) indicate the hierarchical structure the governing equation. If internal variables are considered to include nonlinearities in the microscale then the structure of governing equations becomes even more complicated compared to Eq. (13.3) as shown by Berezovski (2015).

The asymptotic analysis demonstrates also the hierarchies for waves in Cosserat media and ferroelectrics, analysed by Maugin (1999). In the linear case, the governing equations are similar to those for microstructured solids (Salupere and Engelbrecht, 2014).

In biomechanics, the character of nonlinearities can be different from what is typical in solid mechanics. Based on experiments, it has been shown that in biomembranes where the microstructure is built up by lipid molecules, the nonlinearity of mechanical waves can be accounted in the effective velocity (Heimburg and Jackson, 2005)

$$c_e^2 = c_0^2 + pu + qu^2, \tag{13.7}$$

where c_0 is the velocity in the unperturbed state and u is the density change along the axis of the biomembrane, while p, q are material coefficients. Substituting c_e^2 into the balance of momentum and the adding dispersive terms, the governing equation for longitudinal waves in biomembranes takes the form

$$u_{tt} = \left[\left(c_0^2 + pu + qu^2\right) u_x\right]_x - h_1 u_{xxxx} + h_2 u_{xxtt}, \tag{13.8}$$

where h_1, h_2 are constants. This equation was proposed by Heimburg and Jackson (2005) with $h_2 = 0$ and later improved by Engelbrecht et al (2015). This improvement with $h_2 \neq 0$ is important because it accounts for the microstructure of the biomembrane made of lipids and removes the discrepancy that at higher frequencies the velocities are unbounded. This is a physically admissible situation as stressed by Maugin (1999). It must be stressed that as noted above, in Eq. (13.5) the nonlinearities are of the deformation-type, then in Eq. (13.8) they are of the displacement-type.

13.2.2 Evolution-Type (KdV-Type) Models

These one-wave asymptotical models have gained wide attention because of the iconic status of several nonlinear evolution equations like the KdV equation, Schrödinger equation which permit in some cases also analytical solutions (for example, Maugin, 1999, 2011; Ablowitz, 2011). The classical KdV equation combines the quadratic nonlinearity and cubic dispersion:

$$v_\tau + vv_\xi + dv_{\xi\xi\xi} = 0, \tag{13.9}$$

where d is the dispersion parameter. The numerical analysis of the KdV equation has revealed many details including the behaviour of multi-recurrence of solitons forming from a harmonic excitation (Salupere et al, 2002), the explanation of the importance of hidden solitons (Salupere et al, 2003; Engelbrecht and Salupere, 2005) and the influence of an additional force (Engelbrecht and Salupere, 2005). The modifications of the KdV equations involve more physical effects. For example, for martensitic-austenitic shape-memory alloys the governing equation takes the form (Salupere et al, 2001):

$$v_\tau + [P(v)]_\xi + dv_{\xi\xi\xi} + bv_{5\xi} = 0, \tag{13.10}$$

$$P(v) = -\frac{1}{2}v^2 + \frac{1}{4}v^4, \tag{13.11}$$

where d and b are the third- and the fifth-order dispersion parameters, respectively. The quartic potential (13.11) corresponds to the two-well energy distribution which has a direct influence on nonlinear effects. Equations (13.10), (13.11) are able to describe several solitonic structures (Ilison and Salupere, 2006).

It is also possible to derive an evolution equation from the bi-directional model (13.5). However, in this case the result is a modified KdV equation (Randrüüt and Braun, 2010)

$$v_\tau + a_1 vv_\xi + d_1 v_{\xi\xi\xi} + a_2 (v_\xi^2)_{\xi\xi} = 0, \tag{13.12}$$

where a_1 describes the nonlinearity of the macrostructure, $a_2 \sim \mathcal{O}(\varepsilon)$ - the nonlinearity of the microstructure and d_1 denotes the joint influence of dispersive terms (cf. Eq. (13.5)). It means that both of the effects – inertia of the microstructure (term u_{ttxx} in Eq. (13.5)) and elasticity of the microstructure (term u_{xxxx} in Eq. (13.5)) are involved in the dispersive term in Eq. (13.12), reflected by the sign of d_1 (Randrüüt and Braun, 2010). More detailed analysis of nonlinearities in the microscale demonstrates that also Benjamin-Bona-Mahoney or Camassa-Holm equations can be derived (Berezovski, 2015).

Like for the Boussinesq-type equations, the evolution equations may also have a hierarchical character reflecting the scale effects. This is the case of granular materials when the evolution equation can be written in the form (Giovine and Oliveri, 1995):

$$v_\tau + vv_\varsigma + \alpha_1 v_{\xi\xi\xi} + \beta\left(v_\tau + vv_\xi + \alpha_2 v_{\xi\xi\xi}\right)_{\xi\xi} = 0, \tag{13.13}$$

where α_1 and α_2 are macro- and microlevel nonlinearities and β is a parameter involving the ratio of the grain size and the wavelength. The solutions of Eq. (13.13) involve beside single solitons also soliton ensembles. This is a typical example of two concurrent dispersive effects (Ilison and Salupere, 2009).

13.2.3 Coupled Fields

Several physical situations need accounting for coupled fields. For example, in mechanics of solids, the presence of heat sources lead to coupling of deformation fields and temperature fields. For microstructured materials the processes in macro- and microstructures are influenced by both fields. Besides the deformations of macro- and microstructures, the temperature fields can also be divided: macrotemperature and microtemperature (fluctuation of temperature in microstructural elements). The corresponding governing equations can be derived by using the concept of internal variables (Berezovski et al, 2011a,b). However, due to the complicated structure of these equations, it is impossible to derive a single governing equation like it is done for elastic waves in microstructured solids (see above). In this coupled case the governing system of equations is (Berezovski et al, 2014):
balance of linear momentum:

$$u_{tt} - c_o^2 u_{xx} = m_1 \theta_x + m_2 \varphi_x + m_3 \alpha_{xx}; \tag{13.14}$$

balance of energy:

$$\theta_t = n(k\theta_x)_x + m_4 u_{xt} + r_1 \alpha_t^2; \tag{13.15}$$

governing equation for microtemperature:

$$\alpha_{tt} - c_d^2\, \alpha_{xx} = m_5 u_{xx} - r_2 \alpha_t; \tag{13.16}$$

governing equation for microdeformation:

$$\varphi_{tt} - c_t^2 \varphi_{xx} = -m_2 u_x - m_3 \varphi, \tag{13.17}$$

where u is the macrodisplacement, φ - the microdeformation, θ - the macrotemperature, α - the microtemperature; c_0, c_d, c_t denote velocities, k is the thermal conductivity and $m_1, m_2, m_3, m_4, m_5, r_1, r_2, n$ are coefficients. If conditions $\theta = const, \alpha = const$ are satisfied then Eqs. (13.14) and (13.16) can be reduced to the linear form of Eq. (13.5). The full system of Eqs. (13.14) – (13.17) includes three hyperbolic equations (Eqs. (13.14), (13.16), (13.17) and one parabolic equation (Eq. (13.15)). The coupling of physical effects is complicated – microdeformation and microtemperature are not coupled but both are coupled to the balance of linear momentum while macrostructure is affected by the macrodisplacement (like in the usual theory of thermoelasticity) and microtemperature.

In biophysics, a theoretical model for nerve signal propagation including all the physical effects is still a challenge calling "to frame a theory that incorporates all observed phenomena in one coherent and predictive theory of nerve signal propagation" (Andersen et al, 2009). The phenomena are following: the action potential (the electrical pulse) in a nerve fibre which carries the signal, generates also mechanical waves in the axoplasm within a fibre and in the surrounding biomembrane. The longitudinal wave in the biomembrane leads to the transverse displacement which is measurable. Leaving aside the detailed description on the origin of physical effects and corresponding models, a possible mathematical model uniting all the processes into one system has recently been proposed in the following form (Engelbrecht et al, 2016).

First, the action potential can be modelled by the simplified FitzHugh-Nagumo (FHN) equation governing the propagation of an electrical pulse z (Nagumo et al, 1962):

$$z_{txx} = z_{tt} + \mu(1 - a_1 z + a_2 z^2) z_t + z, \tag{13.18}$$

where a_1, a_2, μ are parameters and z is the scaled voltage.

Second, the pressure wave in axoplasm may be governed by a 1D Navier-Stokes model

$$\rho(V_t + VV_x) = -\overline{p}_x + \mu_V V_{xx} + F_1(z), \tag{13.19}$$

where V is the velocity, ρ – the density, $\overline{p}$ – the pressure and μ_V – the viscosity. The force acting from the action potential is denoted by $F_1(z)$.

Third, the longitudinal wave in the biomembrane is modelled by Eq. (13.8)

$$u_{tt} = \left[\left(c_0^2 + pu + qu^2\right) u_x\right]_x - h_1 u_{xxxx} + h_2 u_{xxtt} + F_2(z,V), \tag{13.20}$$

where $F_2(z,V)$ is a force from other waves. The system of equations (13.18), (13.19), (13.20) is solved for the initial condition

$$z\big|_{t=0} = f(x) \tag{13.21}$$

and the transverse wave (horizontal displacement w of the biomembrane) is calculated by

$$w = -kru_x, \tag{13.22}$$

like in rods (Porubov, 2003). All the governing equations are nonlinear and demonstrate explicitly the complexity of the process. The nature of forces $F_1(z), F_2(z,V)$ must be determined by experiments.

13.3 Physical Effects

The model equations described in Sect. 13.2 give an idea about how to account for complicated physical effects reflecting the properties of nonlinear media. In this Section, the most typical effects are described which have resulted from recent studies (many in cooperation with G.A. Maugin). As typical for the complex world, the interactions of effects lead to new phenomena.

Most of the mathematical models described above are the soliton-bearing systems. The nonlinear Boussinesq-type model like Eq. (13.5) demonstrates the emergence of soliton trains. Note that here the nonlinearity is of the deformation type. An initial condition produces left- and right-propagating trains of deformation solitons (Berezovski et al, 2013) where, as expected, the higher the amplitude, the faster the soliton. The interaction of solitons governed by non-integrable equation (13.5), however, is not fully elastic and produces some radiation explained already by Maugin (1999). Due to the nonlinearity at the microlevel, the emerging solitons are not fully symmetric (Salupere et al, 2008). By solving the corresponding evolution equation (13.12), the same effect is demonstrated (Randrüüt and Braun, 2010).

Another Boussinesq-type equation (13.8) involves displacement-type nonlinearities. It possesses a soliton solution and from an initial input, the soliton trains can be formed. Given the signs of the coefficients from experiments $(p < 0)$ contrary to the previous case, the soliton trains have an interesting property – the smaller the amplitude, the faster is the soliton. The analysed improved model (13.8) demonstrates clearly that the existence of the inertial term $h_2 u_{xxtt}$ leads to a narrower pulse which is important in determining its value from experiments by measuring the width of the pulse. The full analysis of Eq. (13.8) is given by Engelbrecht et al (2017) together with the demonstration of the existence of periodical waves (cf cnoidal waves for the KdV equation) governed by this equation. Like in the previous case, the interaction of solitons is not fully elastic resulting in some radiation during interactions.

The existence of solitary solutions or the emergence of regular soliton trains are like benchmarks of solitonics. However, due to complicated physics, the governing equations are different from well-studied classical models and interest should also be focused to the complicated solitonic structures. Such structures may emerge in phase memory alloys (Eq. (13.10)), in granular media (Eq. (13.13)) and forced KdV models. In order to understand properly the mechanisms of emerging solitonic structures, one should determine the number of possible emerging solitons. This depends on the energy sharing and redistribution between solitons. In general terms, starting from the seminal paper by Zabusky and Kruskal (2014) this problem has been analysed using various estimations (see references in Salupere et al, 2014). A detailed analysis of interaction of solitons shows that besides visible solitons there exist also hidden (or virtual) solitons (Salupere et al, 1996; Christov, 2012). The hidden solitons can be detected from the changes they cause in trajectories of other solitons during interactions and can be visible during the short time intervals due to the fluctuations of the reference level. What is important, is that these hidden solitons may serve as "energy pockets" which may become visible if an external force acts in a system (Engelbrecht and Salupere, 2005). This effect has been analysed for the KdV equation with the periodic external force (Engelbrecht and Salupere, 2005). Depending on the strength of the force, several features were established: weak, moderate, strong and dominating external fields. In the case of the weak field all hidden and smaller visible solitons are suppressed; in the case of the moderate field the resulting solitons include all visible and some hidden solitons; in the case of the strong field the number of emerging solitons is higher than in the corresponding conservative case; and in the dominating field no soliton complexes but wave packets are formed. If the external force has a polynomial character with one maximum and one minimum then a single soliton may be suppressed or amplified depending on its amplitude (Engelbrecht and Khamidullin, 1988). This phenomenon could explain the possible amplification of the precursors to seismic waves generated by earth-quakes. The hierarchical KdV equation (13.13) governs beside a single soliton several types of soliton complexes: a KdV soliton ensemble with or without a weak tail; a soliton with a strong tail; a solitary wave with a tail and wave packets (Ilison and Salupere, 2009; Salupere et al, 2014). Here the hidden solitons play a role in the emergence of soliton complexes.

It is obvious that the soliton "menagerie" is rich and above only a part of phenomena was described related to microstructured solids. For a more detailed review the reader is referred to Maugin (2011).

Besides solitons and soliton complexes, the microstructured solids reveal other interesting phenomena. In the case of multiscaled hierarchical microstructures (see Eq. (13.6)) the effect of the negative group velocity may appear (Peets et al, 2013). This phenomenon is related to the coupling effects between the two scales. In terms of dispersion analysis, this is a case when two optical branches of dispersion curves are very close to each other at certain frequencies. As far as the optical modes represent non-propagating oscillations, such a situation can be considered as a pre-resonant one.

The processes in thermoelastic microstructured solids are described by Eqs. (13.14) – (13.17). The numerical simulation shows that even in the absence of effects of the microdeformation, the wave propagation process is strongly influenced by the microtemperature (Berezovski and Engelbrecht, 2013). Namely although the leading terms in the balance of energy (13.15) reflect the parabolicity as expected, the macrotemperature is affected by the microtemperature changes (hyperbolic equation (13.17)) and demonstrates the wave-like behaviour. This result casts surely more light on the behaviour of microstructured solids.

The joint model of a nerve signal propagation (Eqs. (13.18) – (13.20)) is an attempt to explain this fascinating process by including all the possible waves into an ensemble where the nonlinearities play a decisive role. The waves in the ensemble interact with each other through the coupling forces. Certainly, the description of the electrical signal is here simplified because the FHN model takes into account only one (generalized) ionic current. This current plays a crucial role in the energy balance of the electrical pulse dictating its asymmetric shape. It would better to account for specified currents of Na and K ions but the more complicated models like the Hodgkin-Huxley model (Hodgkin and Huxley, 1952) taking these ionic currents into account need many more physical parameters. So at this stage we limit ourselves to the simple FHN equation (13.18). The pressure wave in the axoplasm is described by the classical Navier-Stokes equation. Finally, the longitudinal waves in the surrounding biomembrane are described by a recently derived equation (13.20). To make this model work, two important physical phenomena must be properly understood:

i) the mechanisms of opening the ion channels;
ii) the nature of coupling forces.

It means that in Eq. (13.18) the parameters a_1 and a_2 should be carefully estimated and the forces $F_1(z)$ and $F_2(z,V)$ determined. This work is in progress. A special challenge is to understand the synchronization of velocities and the possible phase-shifts between the waves in an ensemble.

13.4 Discussion

What has been described above, is the description of complexity in wave motion. Indeed, the main features of complex systems are (Érdi, 2008):

i) complex systems are comprised of many different constituents which are connected in multiple ways;
ii) complex systems produce global effects including emergent structures generated by local interactions;
iii)complex systems are typically nonlinear;
iv) emergent structures may occur far from the equilibrium.

The need for the inevitable introduction of complexity in the mechanics of real materials has been suggested also by Maugin (2015). The list of effects in nonlinear wave motion includes many fundamental phenomena such as the balance of nonlinear and dispersive effects, scale effects and hierarchies, coupling of different fields, etc. As a result, special wave structures could emerge and the interaction of waves may lead to amplification, instability and energy redistribution. The coupling of several fields like in thermoelasticity and biophysics leads to completely novel physical effects which can explain the behaviour of materials or systems in a more informative way. In general terms, the corresponding mathematical models are non-integrable (Maugin, 2011) and that is why numerical methods are used in the analysis. Most of the results described above are obtained either by using the finite volume (Berezovski et al, 2008) or the pseudospectral (Salupere, 2009) methods. A special attention is paid to the accuracy and convergence of numerical simulations.

The analysis of complexity of wave motion demonstrates clearly that the mechanical behaviour (stresses, velocities, deformation, temperature) of continua depends on the interactions of constituents and fields. From another point of view, the waves are the carriers of information and energy reflecting so the interaction processes. By measuring the physical properties of waves (amplitudes, velocities, spectra, shapes), this information can be used for the determining the properties of fields or internal structures, i.e. for solving the inverse problems (Janno and Engelbrecht, 2011).

There are many unsolved problems in the complexity of wave motion. One could ask about the soliton management, soliton tunability (generation of solitons with predetermined amplitudes or spectral densities), soliton turbulence (self-organization into spatially localized solitonic structures), etc. An interesting question is whether intuitively well understood microtemperatures in microstructured materials can be measured. Metamaterials and nanomaterials need more attention because of their properties which must be reflected also in wave motion.

The impact of G.A. Maugin in generating novel ideas is enormous.

Acknowledgements This research was supported by the EU through the European Regional Development Fund (project TK 124), by the Estonian Research Council (Project IUT 33-7, IUT 33-24, PUT 434) and by the French-Estonian Parrot programme. J. Engelbrecht, A. Salupere and A. Berezovski are indebted to G.A. Maugin for supporting their visits to the University of Marie and Pierre Curie.

References

Ablowitz MJ (2011) Nonlinear Dispersive Waves. Asymptotic Analysis and Solitons. Cambridge University Press, Cambridge

Andersen SSL, Jackson AD, Heimburg T (2009) Towards a thermodynamic theory of nerve pulse propagation. Progr Neurobiol 88:104–113

Berezovski A (2015) Nonlinear dispersive wave equations for microstructured solids. Proc Estonian Acad Sci 64(3):203–211

Berezovski A, Engelbrecht J (2013) Thermoelastic waves in microstructured solids: Dual internal variables approach. Journal of Coupled Systems and Multiscale Dynamics 1:112–119

Berezovski A, Engelbrecht J, Maugin GA (2008) Numerical Simulation of Waves and Fronts in Inhomogeneous Solids. World Scientific, Singapore

Berezovski A, Engelbrecht J, Maugin GA (2011a) Generalized thermomechanics with dual internal variables. Archive of Applied Mechanics 81(2):229–240

Berezovski A, Engelbrecht J, Maugin GA (2011b) Thermoelasticity with dual internal variables. Journal of Thermal Stresses 34(5–6):413–430

Berezovski A, Engelbrecht J, Salupere A, Tamm K, Peets T, Berezovski M (2013) Dispersive waves in microstructured solids. International Journal of Solids and Structures 50(11):1981–1990

Berezovski A, Engelbrecht J, Ván P (2014) Weakly nonlocal thermoelasticity for microstructured solids: microdeformation and microtemperature. Archive of Applied Mechanics 84(9-11):1249–1261

Christov CI (2012) Hidden solitons in the Zabusky-Kruskal experiment: analysis using the periodic, inverse scattering transform. Math Comput Simul 82(6):1069–1078

Christov CI, Maugin GA, Porubov AV (2007) On boussinesq's paradigm in nonlinear wave propagation. Comptes Rendus Mécanique 335(9):521–535

dell'Isola F, Pouget J, Rousseau M (2014) Gérard A. Maugin: engineering scientist: Celebrating his 70th anniversary. Math Comput Simul 84:1221–1227

Engelbrecht J, Khamidullin Y (1988) On the possible amplification of nonlinear seismic waves. Phys Earth Planet Int 50:39–45

Engelbrecht J, Salupere A (2005) On the problem of periodicity and hidden solitons for the KdV model. Phys Earth Planet Int 15:015,114

Engelbrecht J, Berezovski A, Pastrone F, Braun M (2005) Waves in microstructured materials and dispersion. Phil Mag 85(33–35):4127–4141

Engelbrecht J, Pastrone F, Braun M, Berezovski A (2007) Hierarchies of waves in nonclassical materials. In: Delsanto PP (ed) Universality of Nonclassical Nonlinearity: Applications to Non-Destructive Evaluation and Ultrasonics, Springer, New York, pp 29–47

Engelbrecht J, Tamm K, Peets T (2015) On mathemathical modelling of solitary pulses in cylindrical biomembranes. Biomech Model Mechanobiol 14:159–167

Engelbrecht J, Peets T, Tamm K, Laasmaa M, Vendelin M (2016) On modelling of physical effects accompanying the propagation of action potentials in nerve fibres. arXiv p 1601.01867

Engelbrecht J, Tamm K, Peets T (2017) On solutions of a Boussinesq-type equation with displacement-dependent nonlinearities: the case of biomembranes. Phil Mag 97(12):967–987

Giovine P, Oliveri F (1995) Dynamics and wave propagation in dilatant granular materials. Meccanica 30:341–357

Heimburg T, Jackson AD (2005) On soliton propagation in biomembranes and nerves. Proc Natl Acad Sci USA 102:9790–9795

Hodgkin AL, Huxley AF (1952) A quantitative description of membrane current and its application to conduction and excitation in nerve. J Physiol 117:500–544

Ilison O, Salupere A (2006) On the propagation of solitary pulses in microstructured materials. Chaos, Solitons, Fractals 29(1):202–214

Ilison O, Salupere A (2009) Propagation of sech2-type solitary waves in hierarchical KdV-type systems. Math Comput Simul 79:3314–3327

Janno J, Engelbrecht J (2011) Microstructured Materials: Inverse Problems. Springer, Heidelberg

Maugin GA (1999) Nonlinear Waves in Elastic Crystals. Oxford Univ. Press, Oxford et al.
Maugin GA (2011) Solitons in elastic solids (1938–2010). Mech Res Comm 38:341–349
Maugin GA (2015) Some remarks on generalized continuum mechanics. Math Mech Solids 20(3):280–291
Maugin GA, Engelbrecht J (1994) A thermodynamical viewpoint on nerve pulse dynamics. J Non-Equil Thermodyn 19:9–23
Mindlin RD (1964) Micro-structure in linear elasticity. Arch Rat Mech Anal 16:51–78
Nagumo J, Arimoto S, Yoshizawa S (1962) An active pulse transmission line simulating nerve axon. Proc IRE 50:2061–2070
Peets T, Kartofelev D, Tamm K, Engelbrecht J (2013) Waves in microstructured solids and negative group velocity. Proc IRE 103:16,001
Porubov AV (2003) Amplification of Nonlinear Strain Waves in Solids. World Scientific, Singapore
Randrüüt M, Braun M (2010) On one-dimensional solitary waves in microstructured solids. Wave Motion 47:217–230
Salupere A (2009) The pseudospectral method and discrete spectral analysis. In: Quak E, Soomere T (eds) Applied Wave Mathematics, Springer, Heidelberg, pp 301–33
Salupere A, Engelbrecht J (2014) Scaling and hierarchies of wave motion in solids. ZAMM 94(9):775–783
Salupere A, Maugin GA, Engelbrecht J (1994) KdV soliton detection from a harmonic input. Phys Lett A 192:5–8
Salupere A, Maugin GA, Engelbrecht J, Kalda J (1996) On the KdV soliton formation and discrete spectral analysis. Wave Motion 23(1):49–66
Salupere A, Engelbrecht J, Maugin GA (2001) Solitonic structures in KdV-based higher order systems. Wave Motion 34:51–61
Salupere A, Peterson P, Engelbrecht J (2002) Long-time behaviour of soliton ensembles. Part I – emergence of ensembles. Chaos, Solitons, Fractals 14(9):1413–1424
Salupere A, Peterson P, Engelbrecht J (2003) Long-time behaviour of soliton ensembles. Part I – periodical patterns of trajectories. Chaos, Solitons, Fractals 15(1):29–40
Salupere A, Tamm K, Engelbrecht J (2008) Numerical simulation of interaction of solitary deformation waves in microstructured solids. Int J Non-Lin Mech 43:201–208
Salupere A, Lints M, Engelbrecht J (2014) On solitons in media modelled by the hierarchical KdV equation. Arch Appl Mech 84(9–11):1583–1593
Zabusky NJ, Kruskal MD (2014) Interaction of solitons in a collisionless plasma and the recurrence of inititial states. Phys Rev Lett 15:240–243

Chapter 14
The Dual Approach to Smooth Defects

Marcelo Epstein

Abstract Within the theory of continuous distributions of defects in materials, this paper advocates a point of view based on the geometry of differential forms. Applications to smectic liquid crystals and to multi-walled nanotube composites serve to show how this dual mathematical approach fits perfectly with the intended physical reality. Moreover, the weak formulation of the theory in terms of de Rham currents delivers the description of discrete isolated dislocations as a generalization of the smooth theory.

14.1 Dedication

It is with not a small amount of sad yearning that I recollect Gérard's gift in celebration of our almost common seventieth birthday, a gift that he delivered in the form of a lecture in Calgary in August of 2015. It was the last time that I saw him and had a chance to witness the amazing scope of his knowledge, the power of his intellect, his keen sense of humour and the depth of his friendship. The topic of his talk on that occasion was "Geometry and Continuum Mechanics". In this tribute to his memory, therefore, it seems fitting that I should write a summary of some ideas within this field, which he knew to be my aesthetic ideal and scientific paradigm.

14.2 Summary of the Direct Approach

In the mental process of passing to the continuous limit of a crystal structure (such as a Bravais lattice), the most natural correspondence is obtained by smoothly assigning

---

Marcelo Epstein
University of Calgary, Canada,
e-mail: mepstein@ucalgary.ca

© Springer International Publishing AG, part of Springer Nature 2018

H. Altenbach et al. (eds.), *Generalized Models and Non-classical Approaches in Complex Materials 1*, Advanced Structured Materials 89,
https://doi.org/10.1007/978-3-319-72440-9_14

to each point X of a material body $\mathscr{B}$ a *frame*, that is, a basis $\boldsymbol{E}_\alpha$ $(\alpha = 1,2,3)$, of the *tangent space* $T_X\mathscr{B}$ at X. Smoothness means that for every point $X \in \mathscr{B}$ a neighbourhood $\mathscr{U}(X)$ can be found such that $X \in \mathscr{U}(X)$ and such that, for any system of coordinates X^I $(I = 1,2,3)$ in $\mathscr{U}(X)$, the components E^I_α of $\boldsymbol{E}_\alpha$ in the natural coordinate basis $\boldsymbol{e}_I = \frac{\partial}{\partial X^I}$ are smooth functions of X^I. Thus, locally at least, we obtain a *material parallelism*. A vector $\boldsymbol{v} \in T_Y\mathscr{B}$ at a point $Y \in \mathscr{U}(X)$ is said to be *materially parallel* to a given vector $\boldsymbol{u} \in T_X\mathscr{B}$ if their components in the respective material bases, $\boldsymbol{E}_\alpha(Y)$ and $\boldsymbol{E}_\alpha(X)$, are respectively equal.

Thus constructed, a *continuous lattice* is said to be (locally) *defect-free* if, for each $\mathscr{U}(X)$ above, a (reference) configuration of the body $\mathscr{B}$ exists such that the material parallelism coincides with the standard Euclidean parallelism in $\mathbb{R}^3$. In plain language, the lattice is (locally) defect-free if it can be 'straightened' in a neighbourhood of each point. A classical theorem (see, e.g., Bishop and Goldberg, 1980) in differential geometry establishes that a necessary and sufficient condition for this to be the case is that the *Lie bracket* (or *commutator*) of every pair of lattice base vectors must vanish identically in $\mathscr{U}(X)$, that is

$$[\boldsymbol{E}_\alpha, \boldsymbol{E}_\beta] = 0 \quad \forall \alpha, \beta = 1,2,3. \tag{14.1}$$

In components, using the summation convention, this condition reads

$$E^I_\alpha \frac{\partial E^J_\beta}{\partial X^I} - E^I_\beta \frac{\partial E^J_\alpha}{\partial X^I} = 0. \tag{14.2}$$

The description summarized above will be referred to as the *structural* description of the direct approach. The direct approach can also be motivated from a more sophisticated point of view, epitomized in Noll (1967), which we will call the *constitutive* description. Since from an epistemological standpoint Continuum Mechanics must stand alone, rather than justify its legitimacy on the basis of another physical model, one must conclude that any material geometry, such as the one exhibited above, must be "the *outcome*, not the first assumption, of the physical theory" (Truesdell and Noll, 1965). Accordingly, a notion of *material isomorphism* is established exclusively on the basis of a *constitutive law* of the body. In the elastic case, for a *simple* or *first-grade* material, the constitutive law is given by specifying the *Cauchy stress tensor* $\boldsymbol{t}$ as a function $\boldsymbol{t} = \boldsymbol{t}(\boldsymbol{F}, X)$ of the local value $\boldsymbol{F}$ of the *deformation gradient* at each point $X \in \mathscr{B}$. The deformation gradient is, of course, a non-singular two-point tensor. In any referential and spatial coordinate systems it is represented by a nonsingular matrix. Two material points $X, Y \in \mathscr{B}$ are *materially isomorphic* if there exists a linear map $\boldsymbol{P}(X,Y) : T_X\mathscr{B} \to T_Y\mathscr{B}$ such that

$$\boldsymbol{t}(\boldsymbol{F}, Y) = \boldsymbol{t}(\boldsymbol{F}P(X,Y), X) \quad \forall \boldsymbol{F}. \tag{14.3}$$

Physically, two points are materially isomorphic if they are made of the same material. Indeed, the material isomorphism map $\boldsymbol{P}$ only changes the local reference configuration induced at each point by a given global reference configuration. For this reason, a material isomorphism can also be described as a *material transplant*.

A body is said to be *materially uniform* if all its points are mutually materially isomorphic. We assume that a field of material isomorphisms in a uniform body can be chosen smoothly neighbourhood by neighbourhood. To relate the two points of view (structural and constitutive), it suffices to adopt arbitrarily a basis $\boldsymbol{E}_\alpha$ at any one point X and, by application of the transplant maps assumed to exist in a uniform body, to obtain the corresponding basis at all other points of $\mathscr{U}(X)$. In the terminology of the constitutive description, a (locally) defect-free body is called (locally) *homogeneous*.

Two significant differences between the two descriptions of the direct approach must be mentioned. The first discrepancy arises from the fact that, having already utilized the underlying constitutive law, the constitutive description must forego any possibility of including the contribution of the presence of defects to the constitutive response, beyond the contribution encoded in the lack of homogeneity. Put differently, we cannot include in the constitutive law an a posteriori dependence on the non-vanishing commutators, since they were obtained by an a priori specified law. The second significant difference between the two descriptions arises from the fact that the material at hand may have a continuous *symmetry group*, such as is the case in commonly found isotropic or transversely isotropic materials. A *material symmetry* is nothing but a material automorphism, that is, a material isomorphism of a point with itself. In a uniform body it is not difficult to show that the symmetry groups at all points are mutually conjugate. Thus, a smooth extra degree of freedom becomes available in the choice of material isomorphisms and, consequently, in the determination of the local material bases $\boldsymbol{E}_\alpha$. As a result of this extra degree of freedom, the homogeneity condition (14.2) does not necessarily hold for every possible choice of material parallelism in a defect-free neighbourhood. These differences should not be counted as deficiencies of either description.

14.3 The Dual Perspective

Given a finite-dimensional vector real space V its *dual space* V^* is defined as the vector space consisting of all real-valued linear functions $\omega : V \to \mathscr{R}$, with the obvious definitions of vector addition and multiplication by a scalar. An element of the dual space V^* is also known as a *covector*. Being of the same dimension, the spaces V and V^* are automatically isomorphic to each other, but no canonical isomorphism between them can be established. On the other hand, once a basis $\boldsymbol{E}_\alpha$ is specified in V, a *dual basis* $\boldsymbol{E}^\alpha$ is obtained uniquely in V^* according to the prescription

$$\langle \boldsymbol{E}^\alpha, \boldsymbol{E}_\beta \rangle = \delta^\alpha_\beta, \tag{14.4}$$

where δ^α_β is the Kronecker symbol and where $\langle \omega, \boldsymbol{v} \rangle$ denotes the evaluation of the linear functional $\omega \in V^*$ on the vector $\boldsymbol{v} \in V$.

It might appear, therefore, that the direct formulation summarized in Sect. 14.2 can be replicated almost word for word by moving from the tangent bundle $T\mathscr{B}$ to the

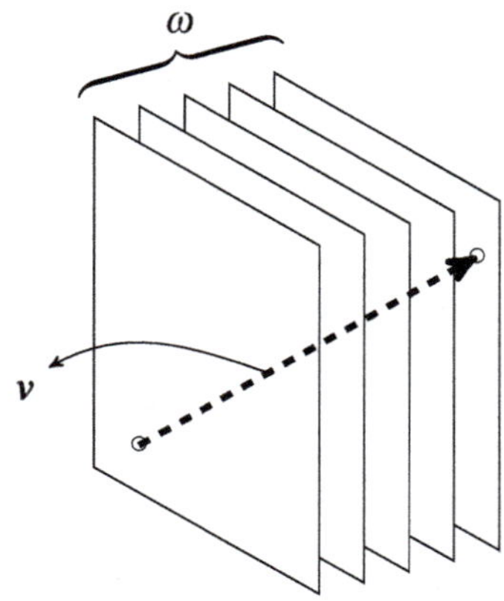

Fig. 14.1 A physicist's view of a covector ω and its evaluation on a vector $\boldsymbol{v}$

cotangent bundle $T^*\mathscr{B}$, which would amount to a perhaps unnecessary abstraction with little or no added benefit. Our intention in this note is precisely to demonstrate that this is not the case and that definite advantages may arise from adopting the dual point of view. Before doing so, however, it may prove useful to reveal the meaning of the homogeneity condition in the geometric language of the dual formulation.

Even from a purely aesthetical point of view, the elegance of the dual approach can be traced back to the fact that a *differential form*, that is, a field of covectors on a manifold, is endowed with an analytic structure known as *exterior differentiation*, which is not the case for vector fields. It is true that every vector space gives rise to an algebraic construct known as the *exterior algebra* (or *Grassmann algebra*) of multi-vectors of various orders. But it is only in a bundle of multi-covectors that an analytic construction can be carried out. The exterior derivative operator d assigns to a *differential k-form* ω (that is, to a field of multi-covectors of order k) a differential $(k+1)$-form $d\omega$. In this language, to be further elucidated in the next section, the homogeneity condition (14.2) is written as

$$d\boldsymbol{E}^{\alpha} = 0. \tag{14.5}$$

14.4 A Brief Review of Differential Forms

14.4.1 Pictorial Representation of Covectors and 1-Forms

We borrow from Misner et al (1973) a surprisingly useful representation of a covector ω as a stack of parallel planes arranged at a given density. Each plane in the stack is defined by all the vectors $\boldsymbol{v}$ such that $\langle \omega, \boldsymbol{v} \rangle =$ constant. Doubling the covector is tantamount to doubling the density without otherwise affecting the planes themselves. Within this imagery, the evaluation of a covector on a vector is obtained by 'counting how many planes' are pierced by the vector, as suggested in Fig. 14.1.

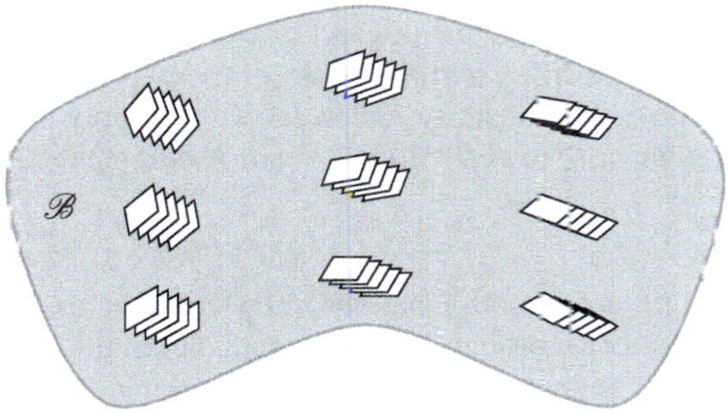

Fig. 14.2 Pictorial representation of a differential 1-form on $\mathscr{B}$

A differential 1-form is a smooth assignment of a co-vector at each point of $\mathscr{B}$. Following our previous depiction, a differential 1-form can be understood as the specification of a plane stack at each point of $\mathscr{B}$, as shown schematically in Fig. 14.2. Let us just point out for now that this picture looks very much like a field of Bravais planes in the continuous limit.

In a coordinate system X^I the dual basis of the natural basis $\boldsymbol{e}_I = \frac{\partial}{\partial X^I}$ is denoted by dX^I. A differential 1-form ω is given by

$$\omega = \omega_I \, dx^I, \tag{14.6}$$

where $\omega_I = \omega_I(X^1, X^2, X^3)$ are smooth functions. The evaluation of ω on the vector field $\boldsymbol{v}$ with components $v^I(X^1, X^2, X^3)$ is

$$\langle \omega(X), \boldsymbol{v}(X) \rangle = \omega_I \, v^I. \tag{14.7}$$

14.4.2 Exterior Algebra

Given a vector space V, we declare *2-covectors* to be the same as skew-symmetric second-order covariant tensors. Two 1-covectors, $\omega, \sigma \in V^*$, give rise to the 2-covector $\omega \wedge \sigma$ defined as

$$\omega \wedge \sigma = \frac{1}{2} \left(\omega \otimes \sigma - \sigma \otimes \omega \right), \tag{14.8}$$

where $\otimes$ denotes the tensor product. The 2-covector $\omega \wedge \sigma$ is called the *exterior product* (or *wedge product*) of ω with σ, and the symbol $\wedge$ is also known as the *wedge* operator.

By definition, the exterior product is skew-symmetric. In particular the product of a 1-covector by itself vanishes. In three dimensions, therefore, the exterior products $\boldsymbol{e}^I \wedge \boldsymbol{e}^J$ constitute essentially only 3 distinct non-zero values. These products can be regarded as a basis for all possible products of 1-covectors. In other words, any skew-symmetric tensor ρ can be expressed as a linear combination $\rho = \rho_{IJ} \, \boldsymbol{e}^I \wedge \boldsymbol{e}^J$, whether or not it arises from the product of two covectors. A 2-covector that happens to be expressible as the exterior product of two 1-covectors is said to be *decomposable*.

Given a coordinate system in $\mathscr{B}$, we can apply these ideas in a pointwise manner thus giving rise to *differential 2-forms*. A differential 2-form ρ is given in components by

$$\rho(X) = \rho_{IJ}\, dX^I \wedge dX^J, \tag{14.9}$$

where ρ_{IJ} are smooth functions of the coordinates.

These concepts can also be extended to multi-covectors of any order. The order, however, cannot non-trivially exceed the dimension of the underlying vector space. A 3-covector is a completely skew-symmetric covariant tensor of order 3. The exterior products $\boldsymbol{e}^I \wedge \boldsymbol{e}^J \wedge \boldsymbol{e}^K$ constitute a basis for all such tensors. By skew-symmetry, this basis boils down to the single 3-covector $\boldsymbol{e}^1 \wedge \boldsymbol{e}^2 \wedge \boldsymbol{e}^3$. In 3 dimensions we cannot go any further non-trivially.

14.4.3 The Exterior Derivative

It is convenient to regard scalar functions $f : \mathscr{B} \to \mathbb{R}$ as 0-forms. The differential of a 0-form is

$$df = \frac{\partial f}{\partial X^I}\, dX^I = f_{,I}\, dx^I. \tag{14.10}$$

The notation suggests that the differential of a smooth scalar field is a differential 1-form. In the same spirit, we define the *exterior derivative* or *exterior differential* of a 1-form $\omega = \omega_I\, dx^I$ as the 2-form

$$d\omega = \frac{\partial \omega_I}{\partial X^J}\, dX^J \wedge dX^I = \omega_{I,J}\, dX^J \wedge dX^I. \tag{14.11}$$

Similarly, the exterior derivative of a 2-form $\Omega = \Omega_{IJ}\, dX^I \wedge dX^J$ is the 3-form

$$d\Omega = \Omega_{IJ,K}\;\, dX^K \wedge dX^I \wedge dX^J. \tag{14.12}$$

An important property of the exterior differential is that a repeated application of the operator d to any differential form ω annihilates the result, namely,

$$d^2\omega = d(d\omega) = 0, \tag{14.13}$$

The reason for this result stems from the inherent symmetry of second partial derivatives combined with the inherent skew-symmetry of the exterior derivative. For example, if f is a scalar function, we obtain

$$d^2 f = d(df) = d(f_{,I}\, dX^I) = f_{,IJ}\, dX^J \wedge dX^I = 0. \tag{14.14}$$

The converse result, known as *Poincaré's lemma*, is a keystone in the theory of forms and its interpretation in various physical contexts.

Lemma 14.1. *(Poincaré's lemma) Let ω be a p-form (with $p \geq 1$) such that $d\omega = 0$ on a coordinate ball. Then, there exists a $(p-1)$-form f such that $\omega = df$ on that ball.*

A p-form ω with the property $d\omega = 0$ is said to be *closed*. A p-form obtained as the differential of a $(p-1)$-form is said to be *exact*. Every exact form is closed, but (according to Poincaré's lemma) a closed form is only guaranteed to be *locally exact*. If the domain of definition $\mathscr{B}$ is topologically *contractible*, then every closed p-form on $\mathscr{B}$ is exact. Notice that in n dimensions every n-form is automatically closed.

14.4.4 Integration

Differential p-forms are essentially objects that can be integrated over p-dimensional domains of integration. If $\mathscr{M}$ is a p-dimensional *oriented manifold with boundary* and ω is a p-form with *compact support* in $\mathscr{M}$, the integral of ω over $\mathscr{M}$ is denoted as $\int_{\mathscr{M}} \omega$. If the support of ω happens to be contained in the domain $\mathscr{U} \subset \mathscr{M}$ of coordinate chart ϕ, the integral is defined by working in the chart

$$\int_{\mathscr{M}} \omega = \int_{\phi(\mathscr{U})} f \, dX^1 ... dX^p, \tag{14.15}$$

where $\omega = f \, dX^1 \wedge ... \wedge dX^p$. The right-hand side is understood as the ordinary Riemann integral in $\mathbb{R}^p$. The more general case where the support is not contained in a single chart is handled technically by means of *partitions of unity* (see Chern et al, 2000).

The fundamental result in the theory of integration is contained in the following theorem. It generalizes and encompasses all the results of this kind covered by classical vector calculus.

Theorem 14.1. *(Stokes' formula): Let ω be a compactly supported $(p-1)$-form in a p-dimensional oriented manifold $\mathscr{M}$ with boundary. Then*

$$\int_{\mathscr{M}} d\omega = \int_{\partial \mathscr{M}} \omega, \tag{14.16}$$

where $\partial \mathscr{M}$ is the boundary of $\mathscr{M}$. On the right-hand side, ω is understood as the restriction of the given $(p-1)$-form to $\partial \mathscr{M}$. Moreover, the orientation of the boundary is assumed to be consistent with the orientation of $\mathscr{M}$.

14.5 An Application to Smectics

Quoting from Chen et al (2009),

> Smectic liquid crystals consist of rod-shaped molecules that spontaneously form both directional (nematic) order and a one-dimensional density wave, commonly described as a layered system; the spacing between the layers is approximately the rod length, a. In two dimensions, we can picture the layers as a set of nearly equally spaced curves lying in the plane. The ground states are characterized by both equal spacing between these curves and vanishing curvature.

Without pretending to treat this subject in depth, it is clear from the description above that defects are worthy of study in molecular structures that are not necessarily representable by Bravais lattices. In the continuous limit, we can conjecture that a smectic liquid crystal of the kind described in Chen et al (2009) is faithfully represented by a differential 1-form. Although the objective in Chen et al (2009) is the study and classification of isolated defects, the presence of smooth distributions of defects is also important. Notice that the direct approach would be at pains to describe the physical situation in a natural way. In the dual formulation, all we have to do is consider a single 1-form instead of the three required for a fully-fledged Bravais structure.

Let, therefore, ω be the 1-form representing the field of local stacks of layers of a smectic. We will call this the *layering form* (Epstein and Segev, 2014a,b). What is the meaning of the condition that the layering form be closed, that is, $d\omega = 0$? According to Poincaré's Lemma 14.1, for each point $X \in \mathscr{B}$ this condition implies the existence of a scalar function $f : \mathscr{U}(X) \to \mathbb{R}$ such that $\omega = df$. This is equivalent to stating that in the neighbourhood $\mathscr{U}(X)$ the stack of surfaces $f =$ constant coincides at each point with the local stack of the layering form in the sense that it is both tangent to it and has exactly the same local density. In physical terms, the smectic is defect-free, as schematically represented in Fig. 14.3.

On the other hand, if $d\omega$ does not vanish identically over $\mathscr{U}(X)$, we conclude that the smectic contains a continuous distribution of defects. It is, therefore, appropriate to call the 2-form $D = d\omega$ the *defect density*. We will now try to extract the physical meaning of the fact that the defect density D, as just defined, is itself an exact differential form.

In a smectic body $\mathscr{B}$, with a given layering 1-form ω, let γ be a closed curve in $\mathscr{B}$ and let $\mathscr{C}$ be an oriented 2-dimensional submanifold with boundary (a surface) such that $\gamma = \partial\mathscr{C}$. We may think of γ as a *Burgers' circuit*, by analogy with the

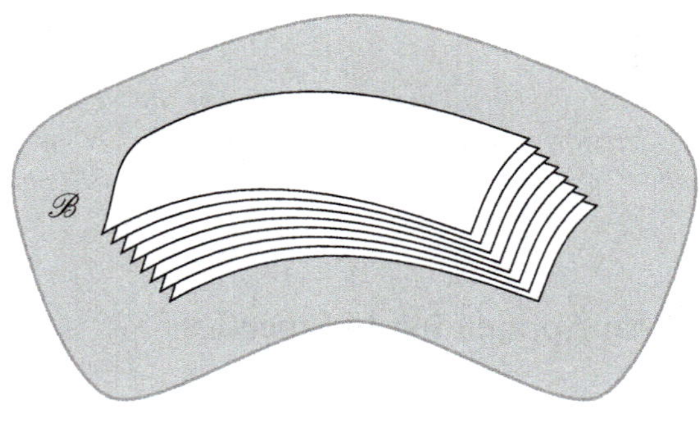

Fig. 14.3 Pictorial representation of a defect-free smectic ($d\omega = 0$)

theory of crystal dislocations. The surface $\mathscr{C}$ encounters defects (smoothly distributed mismatches in slope and/or density between neighbouring stacks). The net amount of defects encountered is given by the integral of the defect density $D = d\omega$ over $\mathscr{C}$. This net amount may vanish even if there are defects on $\mathscr{C}$ which end up canceling out mutually. Whatever the case, according to Stokes' Theorem 14.1, we must have

$$\int_{\mathscr{C}} D = \int_{\gamma} \omega, \tag{14.17}$$

Consider a different surface $\mathscr{C}'$ with *the same* boundary curve γ, as shown in Fig. 14.4. We conclude from Eq. (14.17) that the net amount of defects is the same as before. In other words, the net amount of defects is the same on all surfaces sharing a common boundary. Thus, a Burgers circuit γ can be said to embrace a fixed net amount of defects. This generalization of the concept of Burgers circuit (and, in this case, the associated Burgers *scalar*) is a direct elementary consequence of the geometrical setting, rather than of a clever physical insight.

A fundamental result, with a clear intuitive flavour, states that the boundary of the boundary of a manifold with boundary vanishes, namely,

$$\partial\partial\mathscr{M} = \emptyset. \tag{14.18}$$

This equation should be compared with its formal analogue (14.13). These two statements are intimately related via Stokes' theorem. To elucidate the physical significance of these formulas, consider the integration of the defect density $D = d\omega$ over the boundary $\partial\mathscr{R}$ of a three-dimensional domain in $\mathscr{B}$. Applying Stokes' theorem, we obtain

$$\int_{\partial\mathscr{R}} D = \int_{\partial\partial\mathscr{R}} \omega = 0. \tag{14.19}$$

This result can be interpreted physically as the fact that there can be no isolated sources of defects. It was obtained by exploiting the vanishing of the boundary of a boundary, that is, $\partial^2 = 0$. It could also have been independently obtained from $d^2 = 0$. Indeed, recalling that, by its very definition, D is an exact form, we could have claimed that

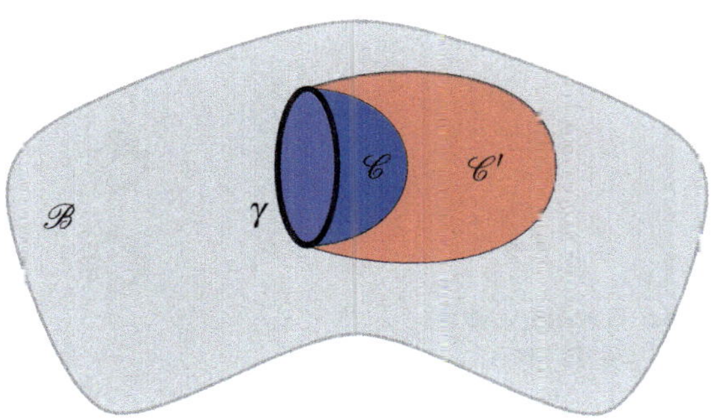

Fig. 14.4 A Burgers circuit γ and two associated surfaces $\mathscr{C}$ and $\mathscr{C}'$

$$\int_{\partial\mathscr{R}} D = \int_{\mathscr{R}} dD = \int_{\mathscr{R}} d(d\omega) = 0. \tag{14.20}$$

Assume, finally, that the defect density vanishes identically everywhere except around a very thin, wire-like, domain traversing the body. Enclosing this domain with a slightly thicker tubular neighbourhood, and focusing attention on the portion comprised between two cross sections, we conclude that the net defect density in any cross section of this tube is constant. We have thus essentially recovered Frank's rule (Frank, 1951) for line dislocations. A similar result can be obtained for branching dislocation lines. We again emphasize that, in this approach, the physical results emerge naturally from the geometric setting.

14.6 An Application to Nanotubes

Because of the perfect match between the physical reality of smooth defects in a smectic liquid crystal and the geometry of differential forms, it is difficult to imagine a better alternative procedure. An even closer agreement is encountered when trying to tackle the more difficult problem of smooth defects in nanotube composites, as proposed in Epstein (2016).

Carbon nanotubes may consist of several coaxial cylinders of carbon atoms. When embedded in large numbers within a matrix, therefore, our picture of local plane stacks breaks down as a geometric descriptor of the resulting composite. Instead, we have a situation schematically depicted in Fig. 14.5, whose mathematical expression we will consider next.

As a collection of coaxial cylinders, a single nanotube provides us with a one-parameter family of atomic plane stacks tangent to the cylindrical surfaces. In other words, instead of having a single layering 1-form as was the case in the smectic, we are now confronted with a one-parameter family of layering forms, namely

$$\omega(\lambda) = \omega_I(X^1, X^2, X^3; \lambda)\, dX^I, \tag{14.21}$$

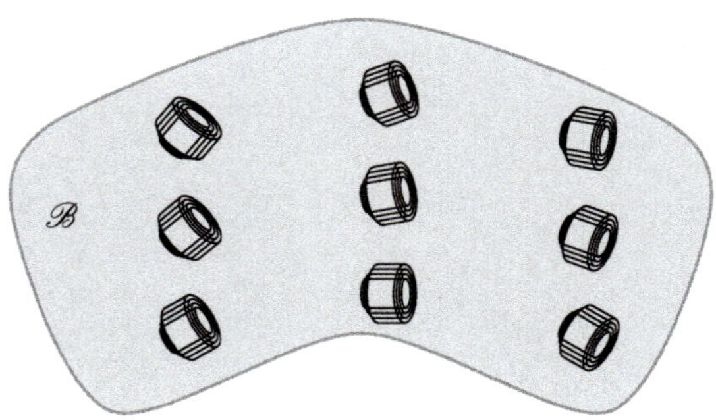

Fig. 14.5 A multi-walled nanotube composite

where λ is the parameter. A more general case can be imagined in which the family depends on more than one parameter, as would be the case for a composite with embedded multi-walled nanospheres.

When looking for defects, we are not referring to imperfections within each elementary cylinder. What we have in mind is a possible lack of alignment that transcends the individual elements. In a defect-free composite, we look for the existence of a family of integral surfaces, such as those depicted in Fig. 14.3, that are tangential to the elementary cylinders and have their prescribed stack density. Put differently, we will say that the composite is locally defect-free if each point $X \in \mathscr{B}$ belongs to a neighbourhood $\mathscr{U}(X)$ such that a scalar field $\lambda : \mathscr{U}(X) \to \mathbb{R}$ exists that renders the layering form (14.21) closed, namely, by the chain rule,

$$d\omega = \left(\omega_{I,J} + \omega_{I,\lambda}\lambda_{,J}\right) dX^J \wedge dX^I = 0. \tag{14.22}$$

We obtain thus a system of 3 quasi-linear partial differential equations for the function λ, namely,

$$\epsilon^{KIJ}\left(\omega_{I,J} + \omega_{I,\lambda}\lambda_{,J}\right) = 0, \quad (K = 1,2,3), \tag{14.23}$$

where ϵ^{KIJ} is the alternating symbol, equal to +1 (-1) for each even (odd) permutation of the values of the indices K, I, J) and vanishing whenever the values of any two indices are equal.

For a given one-parameter form $\omega(\lambda)$, the system (14.23) is overdetermined, so that there may be no functions $\lambda = \lambda(X^1, X^2, X^3)$ that satisfy (14.23). In this case, we say that the nanotube composite is defective. A necessary condition for this system to have a solution is of a purely algebraic nature. Indeed, if we attempt to read off the three partial derivatives $\lambda_{,I}$, we find that they are connected linearly and that the coefficient matrix of this linear system is

$$A^{IJ} = \epsilon^{IKJ}\,\omega_{K,\lambda}. \tag{14.24}$$

This is a singular matrix. It is not difficult to check that the rank consistency condition of the system (14.23) is

$$\epsilon^{IKJ}\,\omega_{I,\lambda}\,\omega_{K,J} = 0. \tag{14.25}$$

If this necessary algebraic condition yields no solutions for $\lambda(X^1, X^2, X^3)$, the structure is defective. If it has a solution, we still have to check whether the differential equations are satisfied. Finally, if the algebraic condition is satisfied identically, it can be shown (Epstein, 2016) that the system of PDEs has a solution. In this way, it is possible to construct configurations of nanotube composites that are defect-free, in the sense that there are atomic integral surfaces that transcend the individual strands and provide the setting for enhanced strength and conduction properties.

14.7 A Volterra Dislocation

Historically, the study of material defects is traceable to the pioneering work of Volterra (1907), who calculated the stress fields in multiply-connected elastic bodies subjected to a discontinuous displacement field. Although he did not use the modern terminology, this work established the basis for the study of isolated *dislocations* in solids. The passage to the continuous case was achieved later, using the Volterra paradigm and the crystal lattice model as its inspiration (see, e.g., Bilby, 1960). Can this process be reversed? In other words, using a continuous model ab initio, is it possible to recover rigorously an isolated dislocation, through a well-established mathematical process without any further assumptions? The positive answer to this question is given in Epstein and Segev (2014b,a), thus embracing both theories under a single umbrella. The dual approach is, once again, the secret of success. Indeed, although a theory of singular vector fields à la Schwartz distributions does not exist, the theory of singular differential forms has been well established some decades ago by de Rham (1984).

Recall that a *Schwartz distribution* (or *generalized function*) is a continuous linear functional on a space of C^∞ real-valued functions with compact support (*test functions*). The classical example is the δ function, which assigns to every test function its value at the origin. Similarly a de Rham *r-current* T on a manifold $\mathscr{M}$ is a continuous linear functional on the space of C^∞ r-forms with compact support in $\mathscr{M}$. By continuity in this context we mean that $T[\omega]_i \to 0$ as $i \to \infty$ whenever the components of the sequence of r-forms ω_i and all their derivatives tend to zero uniformly in a coordinate domain that includes the supports of the sequence.

As pointed out by de Rham himself in the introduction to his book (de Rham, 1984), the concept of a current is so general that it includes as special cases both differential forms and chains. In other words, both the integrands and the domains of integration can be regarded as currents! Indeed, let ω be a p-form on the n-dimensional manifold $\mathscr{M}$. We can assign to it a unique $(n-p)$-form T_ω by the prescription

$$T_\omega[\phi] = \int_{\mathscr{M}} \omega \wedge \phi, \tag{14.26}$$

for all $(n-p)$-forms ϕ with compact support in $\mathscr{M}$. On the other hand, let $\mathscr{S}$ be an oriented p-dimensional submanifold of $\mathscr{M}$ with boundary. We can uniquely associate with it the p-current T_S defined as

$$T_S[\phi] = \int_{\mathscr{S}} \phi. \tag{14.27}$$

for all p-forms ϕ with compact support in $\mathscr{M}$. Note how this p-current resembles the δ function in the sense that it perceives only the values of the test function restricted to $\mathscr{S}$.

By analogy with the exterior derivative of a differential form, the *boundary* of a p-current T is the $(p-1)$-current ∂T defined by

$$\partial T[\phi] = T[d\phi], \tag{14.28}$$

for all $(p-1)$-forms ϕ with compact support in $\mathscr{M}$. A p-current T is *closed* if $\partial T = 0$. The boundary of the current associated with a submanifold, turns elegantly to be equal to the current associated with the boundary of the submanifold. Indeed,

$$\partial T_S[\phi] = T_S[d\phi] = \int_{\mathscr{S}} d\phi = \int_{\partial S} \phi = T_{\partial \mathscr{S}}[\phi] \tag{14.29}$$

where Stokes' theorem has been invoked.

Translating the treatment of the smooth case to the language of currents, we stipulate that a (possibly singular) layering is expressed as a *layering* $(n-1)$-*current*. In 3 dimensions, therefore, we speak of a layering 2-current T. The *dislocation current* is given by $D = \partial T$. The layering is defect-free if $D = 0$. As an example, consider a body consisting of the open cube $\mathscr{B} = (-1,1)^3 \subset \mathbb{R}^3$ shown in Fig. 14.6. With the intention of emulating the addition or removal of the atoms contained within a half plane, let $\mathscr{S}$ denote the intersection of $\mathscr{B}$ with with the oriented lower half plane $\{\{x,y,z\} \in \mathbb{R}^3 \mid x = 0, z \leq 0\}$. The 2-current $T_{\mathscr{S}}$, as defined in Eq. (14.27), can be seen as the prescription of a non-smooth layering, since it is the distributional generalization of a smooth one-form. The dislocation current is given by the boundary current ∂T. It is not difficult to check that, as a consequence of the fact that the currents act on smooth forms with compact support in $\mathscr{B}$,

$$\partial T_{\mathscr{S}} = T_L, \tag{14.30}$$

where L is the open interval shown with thick dashes in Fig. 14.6. This is precisely the dislocation line of the textbook description of a Volterra edge dislocation.

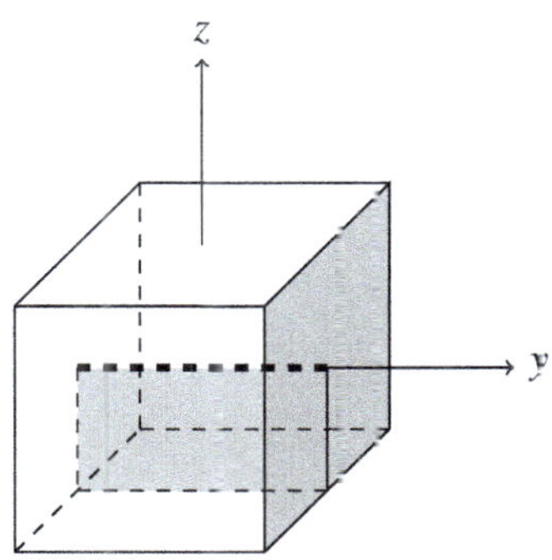

Fig. 14.6 An edge dislocation

References

Bilby DA (1960) Continuous distributions of dislocations. In: Sneddon IN, Hill R (eds) Progress in Solid Mechanics, North-Holland, vol 1, pp 329–398
Bishop RL, Goldberg SI (1980) Tensor Analysis on Manifolds. Dover Books on Mathematics, Dover
Chen BG, Alexander GP, Kamien RD (2009) Symmetry breaking in smectics and surface models of their singularities. Proceedings of the National Academy of Sciences of the United States of America 106(37):15,577–15,582
Chern SS, Chen WH, Lam KS (2000) Lectures on Differential Geometry. World Scientific
de Rham G (1984) Differentiable Manifolds. Springer, Berlin et al.
Epstein M (2016) Modeling continuous distributions of dislocations in multi-walled nanotube composites. Zeitschrift für angewandte Mathematik und Physik 67(6):141
Epstein M, Segev R (2014a) Geometric aspects of singular dislocations. Mathematics and Mechanics of Solids 19(4):337–349
Epstein M, Segev R (2014b) Geometric theory of smooth and singular defects. International Journal of Non-Linear Mechanics 66:105–110
Frank FC (1951) Crystal dislocations - Elementary concepts and definitions. Phil Mag 42:809–819
Misner W, Thorne KS, Wheeler JA (1973) Gravitation. Freeman, San Francisco
Noll W (1967) Materially uniform simple bodies with inhomogeneities. Archive for Rational Mechanics and Analysis 27:1–32
Truesdell C, Noll W (1965) The non-linear field theories of mechanics. In: Flügge S (ed) Handbuch der Physik, Springer, Heidelberg, vol III/3
Volterra V (1907) Sur l'équilibre des corps élastiques multiplement connexes. Annales scientifiques de l'École Normale Supérieure 24:401–517

Chapter 15
A Note on Reduced Strain Gradient Elasticity

Victor A. Eremeyev and Francesco dell'Isola

Abstract We discuss the particular class of strain-gradient elastic material models which we called the reduced or degenerated strain-gradient elasticity. For this class the strain energy density depends on functions which have different differential properties in different spatial directions. As an example of such media we consider the continual models of pantographic beam lattices and smectic and columnar liquid crystals.

15.1 Introduction

The Mechanics of Generalized Continua was one of the permanent interests of Prof. Gérard A. Maugin who is one of the main founder and contributor in the field, see, e.g., his recent original books and papers Maugin (2010, 2011, 2013, 2016, 2017) devoted to developments in the field. Among various generalized models of continuum the strain gradient elasticity found recently various applications in the mechanics of nano-sized solids and fluids (Forest et al, 2011; Cordero et al, 2016), for modelling of strain localization phenomena (Aifantis, 1992), as model of heterogeneous granular solids (Misra and Chang, 1993), as model of fabrics and composites (Harrison, 2016; dell'Isola and Steigmann, 2015; dell'Isola et al, 2016b; Soubestre and Boutin, 2012; d'Agostino et al, 2015; Rahali et al, 2015; Placidi et al, 2016), see also dell'Isola et al (2017); Aifantis (2014); Bertram and Glüge (2016);

Victor A. Eremeyev
Gdańsk University of Technology, ul. Gabriela Narutowicza 11/12, 80-233 Gdańsk, Poland
e-mail: eremeyev.victor@gmail.com

Francesco dell'Isola
Università di Roma "La Sapienza" & International Research Center on Mathematics and Mechanics of Complex System (M&MOCS), Universitá degli Studi dell'Aquila, Via Giovanni Gronchi 18 - Zona industriale di Pile, 67100, L'Aquila, Italy
e-mail: francescodellisola@uniroma1.it

© Springer International Publishing AG, part of Springer Nature 2018
H. Altenbach et al. (eds.), *Generalized Models and Non-classical Approaches in Complex Materials 1*, Advanced Structured Materials 89,
https://doi.org/10.1007/978-3-319-72440-9_15

Engelbrecht and Berezovski (2015); Chatzigeorgiou et al (2017). In this context, it is also worth mentioning continuum bodies, characterized at micro scale, by interacting spins in a crystalline lattice (see e.g. Grimmett (2016); de Masi et al (2008, 2009) for some relevant results using the Potts model). As some other generalized models it can predict size-effect since it includes the characteristic length observed at the nanoscale and can describe dispersion for waves propagation in microstructured solids. The typical form of the strain energy density $\mathscr{W}$ under infinitesimal deformations can be represented as follows

$$\mathscr{W} = \mathscr{W}(\boldsymbol{e}, \nabla \boldsymbol{e}) \tag{15.1}$$

or

$$\mathscr{W} = \mathscr{W}(\boldsymbol{e}, \nabla\nabla \boldsymbol{u}), \tag{15.2}$$

where

$$\boldsymbol{e} = \frac{1}{2}(\nabla \boldsymbol{u} + (\nabla \boldsymbol{u})^T)$$

is the strain tensor, $\boldsymbol{u}$ is the vector of displacements, and ∇ is the three-dimensional (3D) nabla-operator (Lebedev et al, 2010; Simmonds, 1994). In what follows for brevity we use form (15.2). The model was presented in the original works by Toupin (1962); Mindlin (1964); Mindlin and Eshel (1968) and in simplified form by Aifantis (1992, 2003); Askes and Aifantis (2011); Askes and Gitman (2017).

The characteristic feature of (15.1) and (15.2) is that $\mathscr{W}$ depends on a set of *all* second derivatives of $\boldsymbol{u}$. Considering small deformations we can transform $\mathscr{W}$ into a quadratic form

$$\mathscr{W} = \frac{1}{2}\boldsymbol{e} : \boldsymbol{C} : \boldsymbol{e} + \frac{1}{2}\nabla\nabla \boldsymbol{u} \,\vdots\, \boldsymbol{D} \,\vdots\, \nabla\nabla \boldsymbol{u}, \tag{15.3}$$

where $\boldsymbol{C}$ and $\boldsymbol{D}$ are forth- and six-order tensors of elastic moduli, respectively. The standard assumption is that $\boldsymbol{C}$ and $\boldsymbol{D}$ are not singular. Moreover, $\boldsymbol{C}$ and $\boldsymbol{D}$ are assumed to be positive definite, see, e.g., Healey and Krömer (2009); Mareno and Healey (2006), that is

$$\boldsymbol{e} : \boldsymbol{C} : \boldsymbol{e} \geq c\, \boldsymbol{e} : \boldsymbol{e}, \tag{15.4}$$

$$\nabla\nabla \boldsymbol{u} \,\vdots\, \boldsymbol{D} \,\vdots\, \nabla\nabla \boldsymbol{u} \geq d\, \nabla\nabla \boldsymbol{u} \,\vdots\, \nabla\nabla \boldsymbol{u} \tag{15.5}$$

with $c > 0$, $d > 0$. Here ":" and "$\vdots$" stand for scalar (inner) products in the spaces of second- and third-order tensors, respectively.

So, all spatial directions are equivalent in the sense of dependence on all second derivatives.

In the follows we consider another type of constitutive equations. We keep requirement (15.4) and relax (15.5) assuming that $\boldsymbol{D}$ can be degenerated, that is

$$\boldsymbol{e} : \boldsymbol{C} : \boldsymbol{e} \geq c\, \boldsymbol{e} : \boldsymbol{e}, \quad \nabla\nabla \boldsymbol{u} \,\vdots\, \boldsymbol{D} \,\vdots\, \nabla\nabla \boldsymbol{u} \geq 0. \tag{15.6}$$

We call such model the *reduced or degenerated strain gradient elasticity*. In what follows we discuss few examples of such models.

15.2 Reduced Strain Gradient Elasticity. Examples

15.2.1 Structural Mechanics

First, let us recall that such inequality in spatial directions is well-known in the structural mechanics. For example, the strain energy density of an extensible beam under tension and in-plane bending is given by

$$\mathscr{W} = \frac{1}{2}\mathbb{K}_e(u')^2 + \mathbb{K}_b(w'')^2, \tag{15.7}$$

where $\mathbb{K}_e > 0$ and $\mathbb{K}_b > 0$ are the extensional and bending stiffness parameters, $u = u(x)$ and $w = w(x)$ are longitudinal and transverse displacements, respectively, x is the coordinate along the beam axis, and the prime stands for derivative with respect to x.

The same situation can be observed in the Kirchhof-Love theory of plates and shells (Timoshenko and Woinowsky-Krieger, 1985; Lebedev et al, 2010). For example, the strain energy density of an elastic homogeneous plate of thickness h is given by

$$\begin{aligned}\mathscr{W} = &\frac{Eh}{2(1+\nu)}\left[u_{\alpha,\beta}u_{\alpha,\beta} + \frac{\nu}{1-\nu}\left(u_{\alpha,\alpha}\right)^2\right] \\ &+ \frac{Eh^3}{24(1+\nu)}\left[w_{,\alpha\beta}w_{,\alpha\beta} + \frac{\nu}{1-\nu}\left(w_{,\alpha\alpha}\right)^2\right],\end{aligned} \tag{15.8}$$

where E and ν are Young's modulus and Poisson's ratio, respectively, u_α and w are the in-plane and transverse displacements, x_α is a Cartesian coordinates, $\alpha = 1,2$. Hereinafter indices after comma denote derivatives with respect to x_α , so $w_{,\alpha}$ is the partial derivative of w with respect to x_α,

$$w_{,\alpha} = \frac{\partial w}{\partial x_\alpha}.$$

So, considering beams, plates and shells as 1D and 2D continua embedded into 3D physical space we see a difference in order of used derivatives in different directions. Obviously, for beams the preferable direction is tangent to the beam axis whereas for plates and shells it is normal to the middle surface. Nevertheless, here the mentioned difference concerns different components of displacements. For example, (15.8) includes *all* derivatives of the deflection w. In what follows we demonstrate another situation.

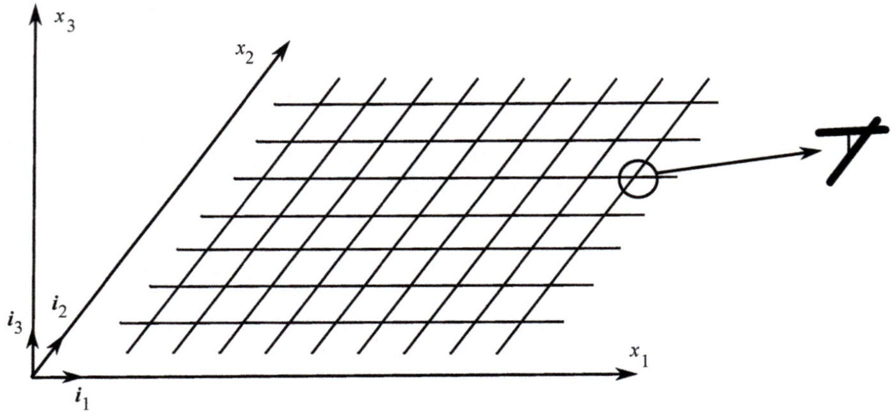

Fig. 15.1: Structure of a pantographic beam lattice.

15.2.2 Continual Models for Pantographic Beam Lattices

Recently, a new class of composites materials called pantographic beam lattices was introduced, see dell'Isola and Steigmann (2015); dell'Isola et al (2016a); Giorgio et al (2017); Placidi et al (2016). A scheme of a pantographic lattice is given in Fig. 15.1. It consists of two families of orthogonal long flexible beams connected by pivots. For in-plane deformations using heuristic homogenization the model of a pantographic lattice is reduced to the two-dimensional strain gradient continuum. Introducing the vector displacements as follows

$$\boldsymbol{u} = u_1 \boldsymbol{i}_1 + u_2 \boldsymbol{i}_2, \quad u_\alpha = u_\alpha(x_1, x_2),$$

we obtain the strain energy density in the form, see dell'Isola et al (2016a); Boutin et al (2017); Placidi et al (2017) for details,

$$\mathscr{W} = \frac{\mathbb{K}_e^{(1)}}{2} u_{1,1}^2 + \frac{\mathbb{K}_e^{(2)}}{2} u_{2,2}^2 + \frac{\mathbb{K}_p}{2} (u_{1,2} + u_{2,1})^2 + \frac{\mathbb{K}_b^{(2)}}{2} u_{1,22}^2 + \frac{\mathbb{K}_b^{(1)}}{2} u_{2,11}^2, \tag{15.9}$$

where the stiffness parameters $\mathbb{K}_e^{(\alpha)} > 0$ and $\mathbb{K}_b^{(\alpha)} > 0$ are related to the extensional and bending stiffnesses of the beams at the interpivot scale, respectively, while the coefficient $\mathbb{K}_p \geq 0$ relates to shear stiffness of the pivots. Here we also neglect torsional deformations of the beams. In particular, if the pivots are perfect then $\mathbb{K}_p$ vanishes and the strain energy transforms to

$$\mathscr{W} = \frac{\mathbb{K}_e^{(1)}}{2} u_{1,1}^2 + \frac{\mathbb{K}_e^{(2)}}{2} u_{2,2}^2 + \frac{\mathbb{K}_b^{(2)}}{2} u_{1,22}^2 + \frac{\mathbb{K}_b^{(1)}}{2} u_{2,11}^2. \tag{15.10}$$

Both energies are significantly degenerated since (15.9) and (15.10) include only few second derivatives. In fact, there are only $u_{2,11}$ and $u_{1,22}$. As it was shown in (dell'Isola et al, 2016a; Eremeyev et al, 2017) for (15.10) there exist nontrivial static solutions with zero energy which are different from rigid body motions.

Neglecting the torsional deformations of the beams and pivots we may also write the strain energy density for a pantographic beam lattice for three-dimensional deformations. Here $\boldsymbol{u} = u_k(x_1,x_2,x_3)\boldsymbol{i}_k$, $k = 1,2,3$, and vectors $\boldsymbol{i}_1$ and $\boldsymbol{i}_2$ are directed along the corresponding beams. Under certain assumptions on the beam cross-sections we obtain that

$$\mathscr{W} = \frac{\mathbb{K}_e^{(1)}}{2}u_{1,1}^2 + \frac{\mathbb{K}_e^{(2)}}{2}u_{2,2}^2 + \frac{\mathbb{K}_b^{(2)}}{2}\left(u_{1,22}^2 + u_{3,22}^2\right) + \frac{\mathbb{K}_b^{(1)}}{2}\left(u_{2,11}^2 + u_{3,11}^2\right). \qquad (15.11)$$

The form (15.11) corresponds to the perfect pivots that is without stored energy as in (15.10). In other words this 2D material inherits the properties of extensible beams, so its strain energy density is a "sum" of energies of beams undergoing bending and stretching deformations without torsion.

The advantages in 3D printing gives us also the possibility to produce three-dimensional pantographic beam lattices such as shown in Fig. 15.2. Neglecting the energies of beams due to torsion as well as energies due to torsion and bending in pivots we introduce the extension of the 2D model (15.11) to the 3D case as follows

$$\begin{aligned}\mathscr{W} =& \frac{\mathbb{K}_e^{(1)}}{2}u_{1,1}^2 + \frac{\mathbb{K}_e^{(2)}}{2}u_{2,2}^2 + \frac{\mathbb{K}_b^{(2)}}{2}\left(u_{1,22}^2 + u_{3,22}^2\right) + \frac{\mathbb{K}_b^{(1)}}{2}\left(u_{2,11}^2 + u_{3,11}^2\right)\\ &+ \frac{\mathbb{K}_e^{(3)}}{2}u_{3,3}^2. \end{aligned} \qquad (15.12)$$

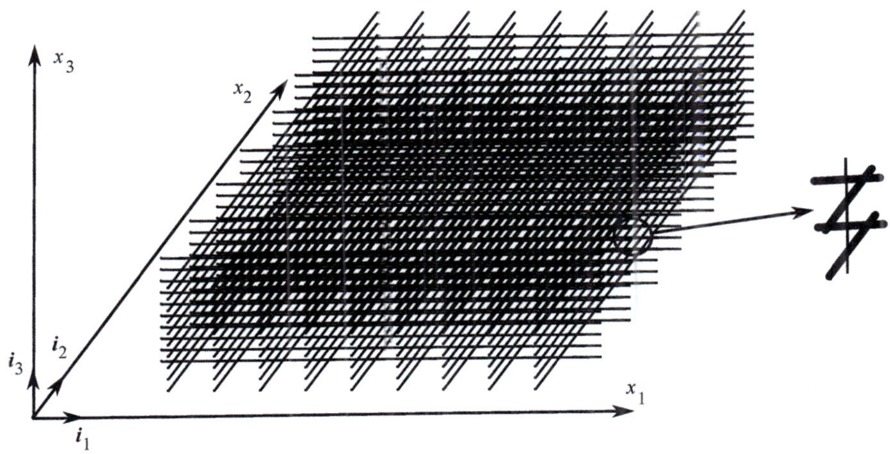

Fig. 15.2: Three-dimensional pantographic beam lattice.

Here $\mathbb{K}_e^{(3)}$ is the stiffness related to pivots undergoing tension/compression.

Comparing (15.12) with the general constitutive equation (15.3) we see that $\boldsymbol{D}$ is rather degenerated. Thus, the presented above models belong to the class of constitutive relations of reduced strain gradient elasticity. The pantographic beam structures are not unique example of such materials, other examples one can find in the hydrostatics of liquid crystals.

15.2.3 Smectics and Columnar Liquid Crystals

Smectics are the particular class of liquid crystals with layered structure, see Chandrasekhar (1977); de Gennes and Prost (1993); Oswald and Pieranski (2006). Each layer is a two-dimensional liquid. Nevertheless smectics demonstrate existence of non-hydrostatic stresses and resistance to bending. The schematic structure of smectic A is given in Fig. 15.3 (a) where ordered arrangements of molecules in each layer is shown. For small distortion of the initially plane layers of a smectic the main kinematical descriptor is the displacement field $\boldsymbol{u} = u(x,y,z)\boldsymbol{i}_3$. So, the kinematics of smectics is determined by one scalar function u.

The strain energy density of smectics A is given

$$\mathscr{W} = \frac{1}{2}\mathbb{B}(u_{,z})^2 + \frac{1}{2}\mathbb{K}\left(u_{,xx} + u_{,yy}\right)^2, \tag{15.13}$$

where $\mathbb{B}$ and $\mathbb{K}$ are elastic moduli describing the longitudinal stiffness along z-direction and the bending stiffness of the smectic layers, respectively.

Another type of liquid crystals are so-called columnar phases which structure is shown in Fig. 15.3 (b). From the physical point of view these liquid crystals look as set of flexible deformable tubes which can easily slide along each other. For small deformations of this class of liquid crystals we introduce the vector of displacements

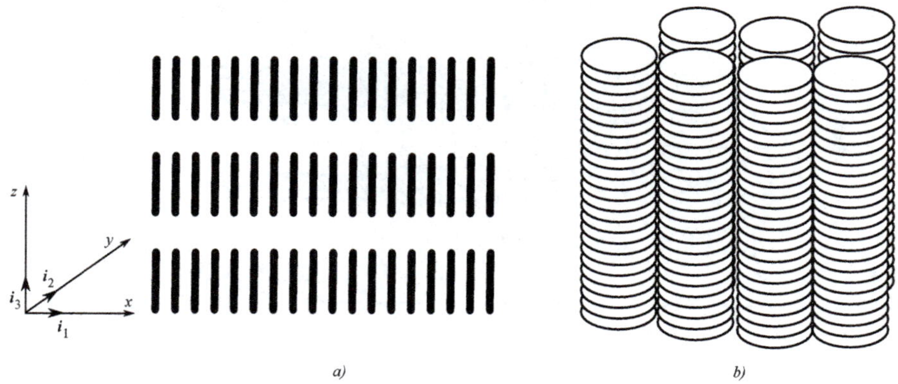

Fig. 15.3: Structures of liquid crystals. a) smectic A, b) a columnar.

$\boldsymbol{u} = u_1\boldsymbol{i}_1 + u_2\boldsymbol{i}_2 + u_3\boldsymbol{i}_3$, $u_k = u_k(x,y,z)$. The strain energy density is now given by de Gennes and Prost (1993) as follows

$$\mathscr{W} = \frac{1}{2}\mathbb{B}(u_{1,x}+u_{2,y})^2 + \frac{1}{2}\mathbb{K}_1\left[(u_{1,x}-u_{2,y})^2 + (u_{1,y}+u_{2,x})^2\right] + \frac{1}{2}\mathbb{K}_2\left[u_{1,zz}^2 + u_{2,zz}^2\right], \tag{15.14}$$

where $\mathbb{B}$, $\mathbb{K}_1$ and $\mathbb{K}_2$ are elastic moduli.

The similarity between strain energies (15.13) and (15.14) of these liquid crystals and the form of strain energy densities of pantographic beam lattices (15.12) is obvious. Indeed, all these forms contains incomplete set of second spatial derivatives, so they belong to the class of materials described using the reduced strain gradient elasticity.

15.2.4 Other Spatially Non-Symmetric Models

Almost all models known in structural and continuum mechanics are symmetric in spatial directions in mentioned above sense. Nevertheless, non-symmetric models are also known. Let us mention few equations of such kind.

The Kadomtsev-Petwiashvili equation is a two-dimensional analog of the famous Korteweg-de-Vries equation, see Kadomtsev and Petviashvili (1970) It takes the form

$$(u_{,t} + 6uu_{,x} + u_{,xxx})_{,x} + 3\lambda u_{,yy} = 0, \tag{15.15}$$

where $u = u(x,y,t)$, x and y are Cartesian coordinates and $\lambda = \pm 1$. Eq. (15.15) cames originally from acoustics and found further applications in the nonlinear theory of dispersive waves, theory of solitons, theory of ferromagnetics, string theory, see, e.g., Ablowitz and Segur (1981); Ablowitz and Clarkson (1991); Maugin (1999) The Kadomtsev-Petviashvili equation describes also the evolution of the localized structures over a large-time scale of a quasi-continuum model deduced from a non-linear lattice model, see Pouget (2005).

The next example is the PDE arose in the analysis of gas centrifuges, nowadays known as Onsager's pancake equation (Wood and Morton, 1980; Eastham and Peterson, 2004). It is given by

$$\left[e^x\left(e^x u_{,xx}\right)_{,xx}\right]_{,xx} + bu_{,yy} = f(x,y), \tag{15.16}$$

where $u = u(x,y)$, b and f are given.

Both equations (15.15) and (15.16) were derived using some approximation techniques in which the preferable spatial direction was chosen a priori.

15.3 Conclusions

We briefly discussed a new class of strain gradient elasticity models called the reduced or degenerated strain gradient elasticity. Among such continua there are continual models of pantographic beam lattices and smectic and columnar liquid crystals. The common peculiarity for both media is a layered structure and sensitivity to bending only in certain directions. From the mathematical point of view the corresponding systems of partial differential equations and natural boundary conditions requires special analysis, such as given in Eremeyev et al (2017) within the framework of anisotropic Sobolev spaces. Let us note that the considered model is obviously anisotropic with a special type of anisotropy. Usually, the anisotropic behavior of solids is determines by symmetries of the tensors of elastic moduli. Here, in addition to this symmetries we meet strong dependence of the material properties on the spatial directions since the strain energy density may be independent on some components of strain gradient tensor. In particular, for pantographic beam lattice as for smectic and columnar liquid crystals there is a preferable direction normal to the inner layers. Here we are restricted ourselves by infinitesimal deformations, but the models can be easily extended for finite deformations. So, we can easily consider the reduced strain gradient elasticity for large deformations. Another common peculiarity observed for pantographic beam lattices and liquid crystals is the existence of non-trivial solutions with zero energy. The further analysis of such constitutive equations can be performed using the material symmetry group defined in (Bertram, 2016) as was performed for shells in (Eremeyev and Pietraszkiewicz, 2006) and for micropolar media in (Eremeyev and Pietraszkiewicz, 2012, 2016).

References

Ablowitz MA, Clarkson PA (1991) Solitons, Nonlinear Evolution Equations and Inverse Scattering, London Mathematical Society lecture note series, vol 149. Cambridge University Press, Cambridge

Ablowitz MJ, Segur H (1981) Solitons and the inverse scattering transform. SIAM, Philadelphia

Aifantis EC (1992) On the role of gradients in the localization of deformation and fracture. Int J Engng Sci 30(10):1279 – 1299

Aifantis EC (2003) Update on a class of gradient theories. Mech Materials 35(3):259–280

Aifantis EC (2014) Gradient material mechanics: perspectives and prospects. Acta Mech 225(4-5):999–1012

Askes H, Aifantis EC (2011) Gradient elasticity in statics and dynamics: An overview of formulations, length scale identification procedures, finite element implementations and new results. Int J Solids Struct 48(13):1962–1990

Askes H, Gitman I (2017) Reducible and irreducible forms of stabilised gradient elasticity in dynamics. Math Mech Complex Systems 5(1):1–17

Bertram A (2016) Compendium on Gradient Materials . OvGU, Magdeburg

Bertram A, Glüge R (2016) Gradient materials with internal constraints. Math Mech Complex Systems 4(1):1–15

Boutin C, dell'Isola F, Giorgio I, Placidi L (2017) Linear pantographic sheets: Asymptotic micro-macro models identification. Math Mech Complex Systems 5(2):127–162

Chandrasekhar S (1977) Liquid Crystals. Cambridge University Press, Cambridge, UK
Chatzigeorgiou G, Meraghni F, Javili A (2017) Generalized interfacial energy and size effects in composites. J Mech Phys Solids 106:257–282
Cordero NM, Forest S, Busso EP (2016) Second strain gradient elasticity of nano-objects. J Mech Phys Solids 97:92–124
d'Agostino MV, Giorgio I, Greco L, Madeo A, Boisse P (2015) Continuum and discrete models for structures including (quasi-) inextensible elasticae with a view to the design and modeling of composite reinforcements. Int J Solids Struct 59:1–17
dell'Isola F, Steigmann D (2015) A two-dimensional gradient-elasticity theory for woven fabrics. J Elast 118(1):113–125
dell'Isola F, Giorgio I, Pawlikowski M, Rizzi N (2016a) Large deformations of planar extensible beams and pantographic lattices: Heuristic homogenisation, experimental and numerical examples of equilibrium. Proc Roy Soc London A 472(2185):20150,790
dell'Isola F, Steigmann D, della Corte A (2016b) Synthesis of fibrous complex structures: Designing microstructure to deliver targeted macroscale response. Appl Mech Rev 67(6):060,804–060,804–21
dell'Isola F, Della Corte A, Giorgio I (2017) Higher-gradient continua: The legacy of Piola, Mindlin, Sedov and Toupin and some future research perspectives. Math Mech Solids 22(4):852–872
Eastham JF, Peterson JS (2004) The finite element method in anisotropic Sobolev spaces. Computers & Mathematics with Applications 47(10):1775–1786
Engelbrecht J, Berezovski A (2015) Reflections on mathematical models of deformation waves in elastic microstructured solids. Math Mech Complex Systems 3(1):43–82
Eremeyev VA, Pietraszkiewicz W (2006) Local symmetry group in the general theory of elastic shells. J Elast 85(2):125–152
Eremeyev VA, Pietraszkiewicz W (2012) Material symmetry group of the non-linear polar-elastic continuum. Int J Solids Struct 49(14):1993–2005
Eremeyev VA, Pietraszkiewicz W (2016) Material symmetry group and constitutive equations of micropolar anisotropic elastic solids. Math Mech Solids 21(2):210–221
Eremeyev VA, dell'Isola F, Boutin C, Steigmann D (2017) Linear pantographic sheets: existence and uniqueness of weak solutions. J Elast DOI 10.1007/s10659-017-9660-3
Forest S, Cordero N, Busso EP (2011) First vs. second gradient of strain theory for capillarity effects in an elastic fluid at small length scales. Comput Materials Sci 50(4):1299–1304
de Gennes G P, Prost J (1993) The Physics of Liquid Crystals, 2nd edn. Clarendon Press, Oxford
Giorgio I, Rizzi N, Turco E (2017) Continuum modelling of pantographic sheets for out-of-plane bifurcation and vibrational analysis. Proc Roy Soc A 473(2207):21 pages DOI 10.1098/rspa.2017.0636
Grimmett G (2016) Correlation inequalities for the Potts model. Math Mech Complex Systems 4(3):327–334
Harrison P (2016) Modelling the forming mechanics of engineering fabrics using a mutually constrained pantographic beam and membrane mesh. Composites A 81:145–157
Healey TJ, Krömer S (2009) Injective weak solutions in second-gradient nonlinear elasticity. ESAIM: Control, Optimisation and Calculus of Variations 15(4):863–871
Kadomtsev BB, Petviashvili VI (1970) On the stability of solitary waves in weakly dispersing media. Sov Phys Doklady 15(6):539–541
Lebedev LP, Cloud MJ, Eremeyev VA (2010) Tensor Analysis with Applications in Mechanics. World Scientific, New Jersey
Mareno A, Healey TJ (2006) Global continuation in second-gradient nonlinear elasticity. SIAM J Math Analysis 38(1):103–115
de Masi A, Merola I, Presutti E, Vignaud Y (2008) Potts models in the continuum. uniqueness and exponential decay in the restricted ensembles. J Stat Phys 133(2):281–345
de Masi A, Merola I, Presutti E, Vignaud Y (2009) Coexistence of ordered and disordered phases in Potts models in the continuum. J Stat Phys 134(2):243–306
Maugin GA (1999) Nonlinear Waves in Elastic Crystals. Oxford University Press, Oxford

Maugin GA (2010) Generalized continuum mechanics: what do we mean by that? In: Maugin GA, Metrikine AV (eds) Mechanics of Generalized Continua. One Hundred Years after the Cosserats, Springer, pp 3–13

Maugin GA (2011) A historical perspective of generalized continuum mechanics. In: Altenbach H, Erofeev VI, Maugin GA (eds) Mechanics of Generalized Continua. From the Micromechanical Basics to Engineering Applications, Springer, Berlin, pp 3–19

Maugin GA (2013) Generalized Continuum Mechanics: Various Paths, Springer, Dordrecht, pp 223–241

Maugin GA (2016) Continuum Mechanics Through Ages. From the Renaissance to the Twentieth Century. Springer, Cham

Maugin GA (2017) Non-Classical Continuum Mechanics: A Dictionary. Springer, Singapore

Mindlin RD (1964) Micro-structure in linear elasticity. Arch Ration Mech Analysis 16(1):51–78

Mindlin RD, Eshel NN (1968) On first strain-gradient theories in linear elasticity. Int J Solids Struct 4(1):109–124

Misra A, Chang CS (1993) Effective elastic moduli of heterogeneous granular solids. Int J Solids Struct 30:2547–2566

Oswald P, Pieranski P (2006) Smectic and Columnar Liquid Crystals: Concepts and Physical Properties Illustrated by Experiments. The Liquid Crystals Book Series (eds GW Gray, JW Goodby, and A Fukuda), Taylor & Francis, Boca Raton

Placidi L, Barchiesi E, Turco E, Rizzi NL (2016) A review on 2D models for the description of pantographic fabrics. ZAMP 67(5):121

Placidi L, Andreaus U, Giorgio I (2017) Identification of two-dimensional pantographic structure via a linear D4 orthotropic second gradient elastic model. J Engng Math 103(1):1–21

Pouget J (2005) Non-linear lattice models: complex dynamics, pattern formation and aspects of chaos. Phil Magazine 85(33–35):4067–4094

Rahali Y, Giorgio I, Ganghoffer JF, dell'Isola F (2015) Homogenization à la Piola produces second gradient continuum models for linear pantographic lattices. Int J Engng Sci 97:148–172

Simmonds JG (1994) A Brief on Tensor Analysis, 2nd edn. Springer, New Yourk

Soubestre J, Boutin C (2012) Non-local dynamic behavior of linear fiber reinforced materials. Mech Materials 55:16–32

Timoshenko SP, Woinowsky-Krieger S (1985) Theory of Plates and Shells. McGraw Hill, New York

Toupin RA (1962) Elastic materials with couple-stresses. Arch Ration Mech Analysis 11(1):385–414

Wood HG, Morton JB (1980) Onsager's pancake approximation for the fluid dynamics of a gas centrifuge. J Fluid Mech 101(1):1–31

Chapter 16
Use and Abuse of the Method of Virtual Power in Generalized Continuum Mechanics and Thermodynamics

Samuel Forest

Abstract The method of virtual power, put forward by Paul Germain and celebrated by Gérard A. Maugin, is *used* (and *abused*) in the present work in combination with continuum thermodynamics concepts in order to develop generalized continuum, phase field, higher order temperature and diffusion theories. The systematic and effective character of the method is illustrated in the case of gradient and micromorphic plasticity models. It is then tentatively applied to the introduction of temperature and concentration gradient effects in diffusion theories leading to generalized heat and mass diffusion equations.

16.1 Introduction

There are essentially two equivalent ways to mathematically represent forces acting on continuum mechanical media: the introduction of forces and couples in the balance of moment and moment of momentum equations, on the one hand, and the method of virtual power, on the other hand, which is a variational statement of the dynamics of bodies (Germain, 1973a). The axiomatic and systematic character of the latter has been put forward by P. Germain and illustrated in the case of first and second gradient continuum theories. It has the merit of clearly separating universal balance laws from peculiar constitutive laws in contrast to Lagrangian/Hamiltonian mechanics which was used by Mindlin to first propose a consistent strain gradient elasticity theory with suitable boundary conditions (Mindlin, 1965). The derivation of complex boundary conditions as they arise in strain gradient and also plate theories probably is the most striking example where the method of virtual power is undoubtedly more effective than the usual procedure. This is due to the fact that the complex form of contact forces, in contrast to internal forces, can hardly be anticipated in that case,

Samuel Forest
Mines ParisTech, Centre des Matériaux, CNRS, UMR 7633, France
e-mail: samuel.forest@ensmp.fr

© Springer International Publishing AG, part of Springer Nature 2018

H. Altenbach et al. (eds.), *Generalized Models and Non-classical Approaches in Complex Materials 1*, Advanced Structured Materials 89,
https://doi.org/10.1007/978-3-319-72440-9_16

as discussed in dell'Isola and Seppecher (1995); dell'Isola et al (2012). Following Germain, Maugin has promoted the method of virtual power and extended it to non–mechanical fields in order to construct complex and coupled continuum theories (Maugin, 1980). The latter contribution deals with the coupling of mechanics with electromagnetism. Maugin also introduced the consideration of singular surfaces and interfaces for mechanical and thermodynamic field variables (Daher and Maugin, 1986). Maugin's passion for the method of virtual power in continuum physics led him recently to write a review paper on the merits of the approach and the many fields of application, namely mechanics of one, two and three–dimensional continuum mechanics, including beam, plate and shell theories, and the coupling with electromagnetic fields (Maugin, 2013).

The mechanics of generalized continua is undoubtedly the privileged domain of application of the principle of virtual power because it enables the systematic introduction of enriched kinematics of the material point with the conjugate forces. This is illustrated by the application of this method to the theory of micromorphic media by Germain (1973b). As claimed by Germain, the derived balance laws and boundary conditions were not new but the level of generality was increased by the systematic nature of the method and the definition by Germain of general micromorphic continua with additional degrees of freedom represented by tensors of increasing order. The method is illustrated in several recent books edited by G.A. Maugin and collecting various generalized continuum theories (Maugin and Metrikine, 2010; Altenbach et al, 2011).

The combination of the method of virtual power and the concepts of continuum thermodynamics, again following Germain's incentive (Germain et al, 1983), leads to a complete framework for the development of continuum theories including consistent constitutive equations. Within this framework, Maugin himself significantly developed the thermomechanics of continua (Maugin, 1992, 1999). He recently also applied the method to coupled diffusion theory (Maugin, 2006).

The objective of the present work is first to extend generalized continuum approaches like strain gradient and micromorphic models to plasticity and damage by means of the method of virtual power and continuum thermodynamics. This shows the systematic *use* and the merits of the method. The second part deals with the coupling of the mechanics with thermodynamic fields like temperature, phase field and concentration. The attention is drawn on the introduction of the gradient of these variables into thermodynamic potentials. For that purpose, the method of virtual power is used again in non-conventional situations, which may represent an *abuse* of the method in the sense that its application may be regarded as formal. It however delivers new types of balance and evolution equations that can be compared to more standard formulations. In particular, the two last sections deal with the application of the method of virtual power to the construction of phase field and higher order diffusion models. These formulations differ from the classical ones based on variational derivatives that do not take enough care of boundary conditions. The seminal contribution by Gurtin (1996) proposes a construction of the phase field continuum theory using additional balance laws for so–called microstress and microforce tensors. In the present work, the method of virtual power is used instead and applied to various situations.

An intrinsic notation system is used in this chapter whereby tensors of order 1, 2 and 4 are respectively denoted by $\underline{a}$, $\underset{\sim}{A}$ and $\underset{\approx}{A}$. Simple and double contractions read:

$$\underline{f} \cdot \underline{v} = f_i v_i, \quad \underset{\sim}{A} : \underset{\sim}{B} = A_{ij} B_{ij} \tag{16.1}$$

The gradient and divergence operators are written as

$$\nabla \underline{u} = u_{i,j} \underline{e}_i \otimes \underline{e}_j, \quad \underset{\sim}{\sigma} \cdot \nabla = \sigma_{ij,j} \underline{e}_i \tag{16.2}$$

in a Cartesian orthonormal basis $(\underline{e}_1, \underline{e}_2, \underline{e}_3)$.

All theories are presented within the small deformation framework for the sake of brevity. The readers are referred to Forest (2016) for the finite deformation formulations of some of them.

16.2 Micromorphic and Gradient Plasticity

The method of virtual power certainly is an powerful tool to construct generalized continuum mechanics theories ranging from strain gradient to micromorphic continua. Higher order or generalized stresses are introduced via the power density of internal forces and lead to additional boundary conditions for the partial differential equations to be solved. More recently, reduced models have been developed that concentrate the gradient effects on plastic or damage variables instead of the full kinematics (Aifantis, 1984; Frémond and Nedjar, 1996; Gurtin, 2003; Forest, 2009). A generic example is provided in this section dealing with micromorphic and strain gradient plasticity. It can be readily extended to microdamage and gradient damage models (Aslan et al, 2011). The present theory is limited to the quasi-static problem but dynamic contributions of the micromorphic variables can be included as proposed in Saanouni and Hamed (2013).

16.2.1 The Micromorphic Approach to Gradient Plasticity

Within the framework of thermomechanics with additional degrees of freedom (Maugin, 1990), the displacement degrees of freedom, $\underline{u}$, of the material point are complemented by a micromorphic degree of freedom, called here plastic microstrain, p_χ. Within a first gradient theory, the model variables are the strain tensor, $\underset{\sim}{\varepsilon}$, temperature T, internal variables α, the plastic microstrain and its gradient, $p_\chi, \nabla p_\chi$.

The virtual power of internal forces in a subdomain $\mathscr{D}$ of the body $\mathscr{B}$ is a linear form with respect to the degrees of freedom ant their gradients:

$$\mathscr{P}^{(i)}(\underline{v}^\star, \dot{p}_\chi^\star) = -\int_{\mathscr{D}} p^{(i)}(\underline{v}^\star, \dot{p}_\chi^\star)\, dV$$

$$p^{(i)}(\underline{v}^\star, \dot{p}_\chi^\star) = \underset{\sim}{\sigma} : \nabla \underline{v}^\star + a\dot{\phi}_\chi^\star + \underline{b} \cdot \nabla \dot{p}_\chi^\star \tag{16.3}$$

where $\underline{v}^\star, \dot{p}_\chi^\star$ are virtual velocity and plastic microstrain rate variables. The dual quantities are generalized stresses. The Cauchy stress is $\underset{\sim}{\sigma}$ and a and $\underline{b}$ are generalized stresses associated with the micromorphic variable and its first gradient. Similarly, the power of contact forces must be extended as follows:

$$\mathscr{P}^{(c)}(\underline{v}^\star, \dot{p}_\chi^\star) = \int_{\mathscr{D}} p^{(c)}(\underline{v}^\star, \dot{p}_\chi^\star)\, dV, \quad p^{(c)}(\underline{v}^\star, \dot{p}_\chi^\star) = \underline{t} \cdot \underline{v}^\star + a^c \dot{p}_\chi^\star \tag{16.4}$$

where $\underline{t}$ is the traction vector and a^c a generalized traction. For conciseness, we do not extend the power of forces acting at a distance and keep the classical form:

$$\mathscr{P}^{(e)}(\underline{v}^\star, \dot{p}_\chi^\star) = \int_{\mathscr{D}} p^{(e)}(\underline{v}^\star, \dot{p}_\chi^\star)\, dV, \quad p^{(e)}(\underline{v}^\star, \dot{p}_\chi^\star) = \rho \underline{f} \cdot \underline{v}^\star \tag{16.5}$$

where $\rho \underline{f}$ accounts for given simple body forces. Following Germain (1973a), given body couples and double forces working with the gradient of the velocity field could also be introduced in the theory. The generalized principle of virtual power with respect to the velocity and micromorphic variable fields, is presented here in the static case only:

$$\mathscr{P}^{(i)}(\underline{v}^\star, \dot{p}_\chi^\star) + \mathscr{P}^{(e)}(\underline{v}^\star, \dot{p}_\chi^\star) + \mathscr{P}^{(c)}(\underline{v}^\star, \dot{p}_\chi^\star) = 0, \quad \forall \mathscr{D} \subset \mathscr{B}, \quad \forall \underline{v}^\star, \dot{p}_\chi^\star \tag{16.6}$$

The method of virtual power according to Maugin (1980) is used then to derive the standard local balance of momentum equation:

$$\operatorname{div} \underset{\sim}{\sigma} + \rho \underline{f} = 0, \quad \forall \underline{x} \in \mathscr{B} \tag{16.7}$$

and the generalized balance of micromorphic momentum equation:

$$\operatorname{div} \underline{b} - a = 0, \quad \forall \underline{x} \in \mathscr{B} \tag{16.8}$$

The method also delivers the associated boundary conditions for the simple and generalized tractions:

$$\underline{t} = \underset{\sim}{\sigma} \cdot \underline{n}, \quad a^c = \underline{b} \cdot \underline{n}, \quad \forall \underline{x} \in \partial \mathscr{D} \tag{16.9}$$

The local balance of energy is also enhanced by the generalized micromorphic power already included in the power of internal forces (16.3):

$$\rho \dot{\epsilon} = p^{(i)} - \operatorname{div} \underline{q} \tag{16.10}$$

where ϵ is the specific internal energy and $\underline{q}$ the heat flux vector. The entropy principle takes the usual local form:

$$-\rho(\dot{\psi}+\eta\dot{T})+p^{(i)}-\frac{\underline{q}}{T}.\nabla T \geq 0 \tag{16.11}$$

where it is assumed that the entropy production vector is still equal to the heat vector divided by temperature, as in classical thermomechanics according to Coleman and Noll (1963). Again, the enhancement of the theory goes through the enriched power density of internal forces (16.3). The entropy principle is explcited according to classical continuum thermodynamics to derive the state laws. At this stage it is necessary to be more specific on the dependence of the state functions $\psi,\eta,\underset{\sim}{\sigma},a,\underline{b}$ on state variables and to distinguish between dissipative and non–dissipative mechanisms. The introduction of dissipative mechanisms may require an increase in the number of state variables. The presentation is limited here to non–dissipative contributions of generalized stresses (see Forest, 2009; Aslan and Forest, 2011) for more sophisticated cases including dissipative contributions). Dissipative events are assumed here to enter the model only via the classical mechanical part. Total strain is split into elastic and plastic parts:

$$\underset{\sim}{\varepsilon}=\underset{\sim}{\varepsilon}^{e}+\underset{\sim}{\varepsilon}^{p} \tag{16.12}$$

The constitutive functional are assumed to depend on the following set of state variables:

$$(\underset{\sim}{\varepsilon}^{e},T,\alpha,p_{\chi},\nabla p_{\chi})$$

The entropy inequality (16.11) can be expanded as:

$$(\underset{\sim}{\sigma}-\rho\frac{\partial\psi}{\partial\underset{\sim}{\varepsilon}^{e}}):\dot{\underset{\sim}{\varepsilon}}^{e}+\rho(\eta+\frac{\partial\psi}{\partial T})\dot{T}+(a-\rho\frac{\partial\psi}{\partial p_{\chi}})\dot{p}_{\chi}+(\underline{b}-\rho\frac{\partial\psi}{\partial\nabla p_{\chi}})\cdot\nabla\dot{p}_{\chi}$$

$$+\underset{\sim}{\sigma}:\dot{\underset{\sim}{\varepsilon}}^{p}-\rho\frac{\partial\psi}{\partial\alpha}\dot{\alpha}-\frac{\underline{q}}{T}\cdot\nabla T\geq 0 \tag{16.13}$$

Assuming that no dissipation is associated with the four first terms of the previous inequality, the following state laws are found

$$\underset{\sim}{\sigma}=\rho\frac{\partial\psi}{\partial\underset{\sim}{\varepsilon}^{e}},\quad \eta=-\frac{\partial\psi}{\partial T},\quad R=-\rho\frac{\partial\psi}{\partial\alpha}\quad a=\rho\frac{\partial\psi}{\partial p_{\chi}},\quad \underline{b}=\rho\frac{\partial\psi}{\partial\nabla p_{\chi}} \tag{16.14}$$

and the residual dissipation is

$$D^{res}=\underset{\sim}{\sigma}:\dot{\underset{\sim}{\varepsilon}}^{p}+R\dot{\alpha}-\frac{\underline{q}}{T}.\nabla T\geq 0 \tag{16.15}$$

where R is the thermodynamic force associated with the internal variable α. The existence of a convex dissipation potential, $\Omega(\underset{\sim}{\sigma},R)$ depending on the thermodynamic forces can then be assumed from which the evolution rules for internal variables are derived, that identically fulfil the entropy inequality, as usually done in classical continuum thermomechanics (Germain et al, 1983):

$$\dot{\underset{\sim}{\varepsilon}}^p = \frac{\partial \Omega}{\partial \underset{\sim}{\sigma}}, \quad \dot{\alpha} = \frac{\partial \Omega}{\partial R} \tag{16.16}$$

After presenting the general approach, we readily give the most simple example which provides a direct connection to several existing generalized continuum models. The free energy density function ψ is chosen as a function of the generalized relative strain variable e defined as:

$$e = p - p_\chi \tag{16.17}$$

where p is the cumulative plastic strain, thus introducing a coupling between macro and micromorphic plastic variables. Assuming isotropic material behaviour for brevity, the additional contributions to the free energy are taken as quadratic functions of e and ∇p_χ:

$$\psi(\underset{\sim}{\varepsilon}^e, T, \alpha, p_\chi, \nabla p_\chi) = \psi^{(1)}(\underset{\sim}{\varepsilon}^e, T, \alpha) + \frac{1}{2} H_\chi (p - p_\chi)^2 + \frac{1}{2} A \nabla p_\chi \cdot \nabla p_\chi \tag{16.18}$$

where H_χ and A are the additional moduli introduced by the micrmorphic model. The function $\psi^{(1)}(\underset{\sim}{\varepsilon}^e, T, \alpha)$ refers to any constitutive function in a classical continuum thermomechanical model with internal variables. After inserting the state laws (16.14)

$$a = \rho \frac{\partial \psi}{\partial p_\chi} = -H_\chi (p - p_\chi), \quad \underline{b} = \rho \frac{\partial \psi}{\partial \nabla p_\chi} = A \nabla p_\chi \tag{16.19}$$

into the additional balance equation (16.8), the following partial differential equation for p_χ is obtained, at least for a homogeneous material under isothermal conditions:

$$p = p_\chi - \frac{A}{H_\chi} \Delta p_\chi \tag{16.20}$$

where Δ is the Laplace operator. It involves a characteristic length scale defined by:

$$l_c^2 = \frac{A}{H_\chi} \tag{16.21}$$

The additional material parameters H_χ and A are assumed to be positive in this work. This does not exclude a softening material behaviour that can be induced by the proper evolution of the internal variables.

Let us now choose a yield function in the form

$$f(\underset{\sim}{\sigma}, R) = \sigma_{eq} - \sigma_Y - R \tag{16.22}$$

where σ_{eq} is an equivalent stress measure, σ_Y the initial yield stress and $R(p)$ is here the hardening (or softening) function. It follows from the state law (16.14) and from the balance equation (16.20) that

$$R = \frac{\partial \psi}{\partial p} = R(p) + H_\chi (p - p_\chi) = R(p) - A \Delta p_\chi \tag{16.23}$$

which shows the enhanced hardening due to the plastic microstrain. Under plastic loading condition,

$$\sigma_{eq} = \sigma_Y + R(p) - A\Delta p_\chi \tag{16.24}$$

which is reminiscent of Aifantis celebrated strain gradient plasticity model (Aifantis, 1984). The equivalence with Aifantis model is obtained for $H_\chi = \infty$ which enforces the internal constraint: $p_\chi \simeq p$.

16.2.2 Direct Construction of Gradient Plasticity Theory

The method of virtual power can also be used directly to construct the strain gradient plasticity model without resorting to the micromorphic model (Forest and Bertram, 2011). The enriched power density of internal forces and of contact forces are introduced as

$$p^{(i)} = \underset{\sim}{\sigma} : \underset{\sim}{\dot{\varepsilon}} + a\dot{p} + \underline{b} \cdot \nabla \dot{p}, \quad p^{(c)} = \underline{t} \cdot \underline{\dot{u}} + a^c \dot{p} \tag{16.25}$$

where a and $\underline{b}$ are generalized stresses acting on the virtual plastic field $\dot{p}$ and its gradient, respectively. The usual traction vector is $\underline{t}$ and a^c denotes the generalized traction. Such generalized stresses are called micro–forces by Gurtin (2003). A generalized principle of virtual power is stated with respect to the virtual fields of displacements and the p–variable. The application of this principle results in the same balance equations and boundary conditions as in Eq. (16.7) to (16.9).

The constitutive functions now depend on cumulative plastic strain and its gradient. The Clausius–Duhem inequality then becomes:

$$(\underset{\sim}{\sigma} - \rho\frac{\partial \psi}{\partial \underset{\sim}{\varepsilon}^e}) : \underset{\sim}{\dot{\varepsilon}}^e + (a - \rho\frac{\partial \psi}{\partial p})\dot{p} + (\underline{b} - \rho\frac{\partial \psi}{\partial \nabla p}) \cdot \nabla \dot{p} + \underset{\sim}{\sigma} : \underset{\sim}{\dot{\varepsilon}}^p \geq 0 \tag{16.26}$$

At this stage, the following state laws are adopted

$$\underset{\sim}{\sigma} = \rho\frac{\partial \psi}{\partial \underset{\sim}{\varepsilon}^e}, \quad a = \rho\frac{\partial \psi}{\partial p} + R, \quad \underline{b} = \rho\frac{\partial \psi}{\partial \nabla p} \tag{16.27}$$

thus assuming that no dissipation is associated with the generalized stress $\underline{b}$. This is the most simple assumption that is sufficient for deriving Aifantis model, in particular. The residual dissipation is then

$$\underset{\sim}{\sigma} : \underset{\sim}{\dot{\varepsilon}}^p + R\dot{p} \geq 0 \tag{16.28}$$

A simple quadratic free energy potential is chosen

$$\rho\psi(\underset{\sim}{\varepsilon}^e, p, \nabla p) = \frac{1}{2}\underset{\sim}{\varepsilon}^e : \underset{\approx}{C} : \underset{\sim}{\varepsilon}^e + \frac{1}{2}Hp^2 + \frac{1}{2}A\nabla p \cdot \nabla p \tag{16.29}$$

from which the state laws are derived:

$$\underset{\sim}{\sigma} = \underset{\approx}{C} : \underset{\sim}{\varepsilon}^e, \quad R = -Hp + a, \quad \underline{b} = A\nabla p \tag{16.30}$$

where $\underset{\approx}{C}$ is the four–rank tensor of the elastic moduli, H is the usual hardening modulus and A is an additional material parameter (unit MPa·mm^2). The yield function is taken as

$$f(\underset{\sim}{\sigma}, R) = \sigma_{eq} - \sigma_Y + R \tag{16.31}$$

Under plastic loading, this gives

$$\sigma_{eq} = \sigma_Y - R = \sigma_Y + Hp - a = \sigma_Y + Hp - \operatorname{div} \underline{b} = \sigma_Y + Hp - A\nabla^2 p \tag{16.32}$$

which is Aifantis celebrated equation, to be compared with Eq. (16.24). The plasticity flow and evolution rules are

$$\dot{\underset{\sim}{\varepsilon}}^p = \lambda \frac{\partial f}{\partial \underset{\sim}{\sigma}}, \quad \dot{p} = \lambda \frac{\partial f}{\partial R} = \lambda \tag{16.33}$$

in the rate–independent case, λ being the plastic multiplier.

The enhanced power of internal forces has been used also by Gurtin and Anand (2009) for gradient plasticity and by Frémond and Nedjar (1996) for gradient damage. However, there exist alternative formulations avoiding the modification of $p^{(i)}$, see Nguyen (2010b, 2016).

16.3 Gradient of Entropy or Temperature Models

It was shown in the previous section that the free energy function can depend on the gradient of strain or on any internal variable like plastic strain and damage. This requires an amendment of the principle of virtual power with the introduction of generalized stress tensors. A similar question arises in the case of heat transfer: can the free energy function depend on the temperature gradient? In the standard continuum thermodynamics framework, the answer is no (Coleman and Mizel, 1963). However, this impossibility can be overcome by suitable enhancement of the theoretical framework. Several tracks have been proposed in the literature for that purpose which are reviewed in Liu et al (2017) with special consideration of the consequences on the heat equations.

In this part, the mechanical contributions are omitted for the sake of conciseness.

16.3.1 A Principle of Virtual Power for Entropy

One track is presented here based on the formulation of a principle of virtual power for entropy (or temperature) and its gradient following (Forest and Amestoy, 2008).

It is postulated that the entropy rate and its gradient contribute to the power of internal and external forces, in the form:

$$\mathscr{P}^{(i)}(\dot{\eta}^\star) = -\int_{\mathscr{D}} (a_\eta \dot{\eta}^\star + \underline{b}_\eta \cdot \nabla \dot{\eta}^\star)\, dV \tag{16.34}$$

$$\mathscr{P}^{(e)}(\dot{\eta}^\star) = \int_{\mathscr{D}} (a^p_\eta \dot{\eta}^\star + \underline{b}^p_\eta \cdot \nabla \dot{\eta}^\star)\, dV + \int_{\partial\mathscr{D}} a^c_\eta \dot{\eta}^\star\, dS \tag{16.35}$$

in addition to the purely mechanical parts not recalled here, where $\dot{\eta}^\star$ is a field of virtual rate of change of entropy density. These power densities involve internal scalar and vector generalized stresses a_η, $\underline{b}_\eta$, on the one hand, and prescribed external scalar and vector volume microforces a^p_η, $\underline{b}^p_\eta$ and a generalized surface traction a^c_η, on the other hand.

The exploitation of the principle of virtual power with respect to the virtual field $\dot{\eta}^\star$ results in the following independent variational equation:

$$\mathscr{P}^{(i)}(\dot{\eta}^\star) + \mathscr{P}^{(e)}(\dot{\eta}^\star) = 0, \quad \forall \mathscr{D} \subset \mathscr{B}, \quad \forall \dot{\eta}^\star \tag{16.36}$$

provided that no "microinertia" effects are attached to variable η (Svendsen, 1999). Such terms involving the second derivative of temperature or entropy can be introduced following (Liu et al, 2017). It is used to derive a balance equation associated with variable η and the associated boundary condition:

$$\operatorname{div}(\underline{b}_\eta - \underline{b}^p_\eta) - a_\eta + a^p_\eta = 0, \quad \forall \underline{x} \in \mathscr{D}, \quad a^c_\eta = (\underline{b}_\eta - \underline{b}^p_\eta) \cdot \underline{n}, \quad \forall \underline{x} \in \partial\mathscr{D} \tag{16.37}$$

It is essential that the power of external forces $(16.35)_2$ contributes to the global balance of energy:

$$\dot{\mathscr{E}} = \int_{\mathscr{D}} \rho \dot{\varepsilon}\, dV = \mathscr{P}^{(e)}(\dot{\eta}) - \int_{\partial\mathscr{D}} \underline{q} \cdot \underline{n}\, dS \tag{16.38}$$

An alternative form is obtained after taking the generalized principle of virtual power into account:

$$\dot{\mathscr{E}} = -\mathscr{P}^{(i)}(\dot{\eta}) - \int_{\partial\mathscr{D}} \underline{q} \cdot \underline{n}\, dS \tag{16.39}$$

The local form of energy balance follows:

$$\rho \dot{\varepsilon} = a_\eta \dot{\eta} + \underline{b}_\eta \cdot \nabla \dot{\eta} - \operatorname{div} \underline{q} \tag{16.40}$$

The entropy principle is assumed to keep its classical global form, for any material subdomain $\mathscr{D} \subset \mathscr{B}$:

$$\frac{d}{dt} \int_{\mathscr{D}} \rho \eta\, dV \geq -\int_{\partial\mathscr{D}} \frac{\underline{q}}{T}\, dS \tag{16.41}$$

The additional contributions appear in the generalized Clausius–Duhem inequality:

$$\rho(T\dot{\eta} - \dot{\epsilon}) + a_\eta \dot{\eta} + \underline{b}_\eta \cdot \nabla\dot{\eta} - \frac{\underline{q}}{T} \cdot \nabla T \geq 0 \tag{16.42}$$

The constitutive functions of solids of this kind depend on the set of state variables $(\eta, \nabla\eta)$. After inserting these dependencies, the Clausius–Duhem inequality (16.42) becomes:

$$\rho(T - \frac{\partial \epsilon}{\partial \eta} + \frac{a_\eta}{\rho})\dot{\eta} + (\underline{b}_\eta - \rho\frac{\partial \epsilon}{\partial \nabla\eta}) \cdot \nabla\dot{\eta} - \frac{\underline{q}}{T} \cdot \nabla T \geq 0 \tag{16.43}$$

from which the state laws are derived,

$$T = \frac{\partial \epsilon}{\partial \eta} - \frac{a_\eta}{\rho}, \quad \underline{b}_\eta = \rho\frac{\partial \epsilon}{\partial \nabla\eta} \tag{16.44}$$

Accordingly, the temperature is found to be equal to the partial derivative of the internal energy function with respect to entropy complemented by a contribution of the internal scalar microstress a_η. The internal generalized stress vector $\underline{b}_\eta$ is nothing but the partial derivative of internal energy with respect to the entropy gradient. It is called the hypertemperature vector (Forest and Amestoy, 2008). The residual dissipation reduces to:

$$-\frac{\underline{q}}{T} \cdot \nabla T \geq 0 \tag{16.45}$$

First consequences of the additional or modified state laws established previously are investigated in the simplest case, namely that of the rigid heat conductor. After inserting the state laws (16.44) into the local energy balance (16.40), the following usual form of the heat equation is recovered:

$$\rho T \dot{\eta} = -\mathrm{div}\, \underline{q} \tag{16.46}$$

In the sequel, a specific constitutive function for internal energy is chosen for the illustration of the modifications brought in the governing equations of thermal conduction by the introduction of entropy gradient effects. A specific constitutive relation will also be needed for the heat flux vector in order to obtain an explicit partial differential equation for entropy. In the present theory, there is no need for departing from the classical Fourier law of heat conduction:

$$\underline{q} = -\kappa \nabla T \tag{16.47}$$

written here for isotropic materials for simplicity. So in the present theory, heat still flows from hot to cold and there is no up–hill heat diffusion.

The explicit heat equation associated with the gradient of entropy theory is now derived for isotropic materials. We take $a_\eta^p = 0$ and $\underline{b}_\eta^p = 0$ for the sake of brevity. Material homogeneity is also assumed for simplicity. The mass density ρ is constant and homogeneous.

The expression of the internal energy density function in a rigid heat conducting body is linearised around the reference entropy value η_0. According to the gradient of entropy model, it contains quadratic terms in the entropy and entropy gradient:

$$\rho \epsilon(\eta, \nabla \eta) = \rho \eta T_0 + \frac{\rho^2(\eta - \eta_0)^2}{4\beta} + \frac{1}{2} A_\eta \nabla \eta \cdot \nabla \eta \tag{16.48}$$

where β and A_η are (strictly positive) material parameters. The expression of the generalized stresses are derived from the additional state and balance laws:

$$\underline{b}_\eta = \rho \frac{\partial \epsilon}{\partial \nabla \eta} = A_\eta \nabla \eta, \quad a_\eta = \mathrm{div}\, \underline{b}_\eta = A_\eta \Delta \eta \tag{16.49}$$

where Δ is the Laplace operator. Compared to classical linearised thermal diffusion, the temperature function is modified as follows:

$$T = \frac{\partial \epsilon}{\partial \eta} - \frac{a_\eta}{\rho} = T_0 + \frac{\rho(\eta - \eta_0)}{2\beta} - \frac{A_\eta}{\rho} \Delta \eta \tag{16.50}$$

Keeping the usual form (16.47) of Fourier heat conduction law, the heat equation can now be derived as

$$\rho T_0 \dot{\eta} = \frac{\rho \kappa}{2\beta} (\Delta \eta - l_\eta^2 \Delta^2 \eta), \quad \text{with} \quad l_\eta^2 = \frac{2\beta A_\eta}{\rho^2} \tag{16.51}$$

This enhanced heat equation has the structure of the Cahn–Hilliard equation in mass transport theory as derived in Cahn and Hilliard (1958); Gurtin (1996). It involves a characteristic (positive) length l_η related to material parameters. The classical heat equation is retrieved for a vanishing intrinsic length scale $l_\eta = 0$, or equivalently $A_\eta = 0$, in the absence of prescribed external microforces

16.3.2 *Gradient of Entropy or Gradient of Temperature?*

The initial question raised in this part was the introduction of the temperature gradient in the free energy density function. The previous theory was developed for the entropy gradient instead. This is due to the fact that the first function of state in thermodynamics is the internal energy density which is a function of entropy and strain in general. The question of the introduction of a gradient term must therefore be asked first at this stage. A similar construction as before is possible for a gradient of temperature theory. It was shown in Forest and Amestoy (2008) to deliver a distinct theory from the gradient of entropy one. The gradient of temperature and gradient of entropy models are not dual contrary to the usual case where the entropy based and temperature based theories are dual in the absence of gradients. The same situation is encountered in mechanics where the strain gradient theory and the recently proposed stress gradient model are no dual theories (Forest and Sab 2012, 2017)

Alternative formulations of the gradient of temperature models were proposed in Ireman and Nguyen (2004); Nguyen (2010a,b) without resorting to additional power contributions but, instead, by modifying the definition of the free energy function and extending the concept of standard generalized materials to the gradient case.

They lead to heat equations that differ from Eq. (16.51) and that possibly include the possibility of heat wave propagation.

The method of virtual power can be used to propose new theories of heat conductors including microtemperature or microentropy concepts (Forest and Aifantis, 2010). The predicted effects are similar to those deduced from double temperature models where each material point is characterized by two distinct temperatures, similarly to mixture theory for fluids.

16.4 The Method of Virtual Power Applied to Phase Field Modelling

The phase field variable ϕ usually is an order parameter with continuous values ranging from 0 to 1. It very often serves as a phase indicator in combination with a concentration field, c, of some solute species in solid body (Finel et al, 2010). Virtual fields of order parameter $\phi^\star$ are considered with suitable regularity[1]. The virtual power of internal generalized forces is defined by the integral over the volume $\mathscr{D} \subset \mathscr{B}$ of a power density, which is assumed *a priori* to be a linear form represented by the generalized stress measures a and $\underline{b}$ (Ammar et al, 2009):

$$\begin{aligned}\mathscr{P}^{(i)}(\phi^\star,\mathscr{D}) &= -\int_{\mathscr{D}} (a\phi^\star + \underline{b}\cdot\nabla\phi^\star)\,dV \\ &= -\int_{\mathscr{D}} (a - \nabla\cdot\underline{b})\phi^\star\,dV - \int_{\partial\mathscr{D}} (\underline{b}\cdot\underline{n})\phi^\star dS \end{aligned} \tag{16.52}$$

The next step is to introduce the virtual power of external forces applied to the considered body. It can be split into a virtual power density of long range volume forces, which can include, in general, a volume density of scalar external generalized forces γ and vector external generalized force $\underline{\gamma}$:

$$\begin{aligned}\mathscr{P}^{(e)}(\phi^\star,\mathscr{D}) &= \int_{\mathscr{D}} (\gamma\phi^\star + \underline{\gamma}\cdot\nabla\phi^\star)\,dV \\ &= \int_{\mathscr{D}} (\gamma - \nabla\cdot\underline{\gamma})\phi^\star\,dV + \int_{\partial\mathscr{D}} (\underline{\gamma}\cdot\underline{n})\phi^\star\,dS \end{aligned} \tag{16.53}$$

and a virtual power density of generalized contact forces, represented by a surface density a^c of generalized traction:

$$\mathscr{P}^{(c)}(\phi^\star,\mathscr{D}) = \int_{\partial\mathscr{D}} a^c\phi^\star\,dS \tag{16.54}$$

---

[1] In fact, as in distribution theory, it is sufficient to take them as differentiable at any order with compact support.

A possible power of inertial microforces is not envisaged here. According to the principle of virtual power, the total virtual power of all forces vanishes on any subdomain $\mathscr{D} \subset \mathscr{B}$ and for any virtual order parameter field $\phi^\star$:

$$\mathscr{P}^{(i)}(\phi^\star,\mathscr{D})+\mathscr{P}^{(c)}(\phi^\star,\mathscr{D})+\mathscr{P}^{(e)}(\phi^\star,\mathscr{D})=0, \quad \forall\phi^\star,\forall\mathscr{D}\subset\mathscr{B} \tag{16.55}$$

$$\int_{\mathscr{D}}(-a+\nabla\cdot\underline{b}+\gamma-\nabla\cdot\underline{\gamma})\phi^\star\,dV+\int_{\partial\mathscr{D}}(a^c-\underline{b}\cdot\underline{n}+\underline{\gamma}\cdot\underline{n})\phi^\star\,dS=0 \tag{16.56}$$

This identity can be satisfied for any field $\phi^\star$ and $\forall\mathscr{D}$ if and only if:

$$a-\nabla\cdot(\underline{b}-\underline{\gamma})-\gamma=0 \quad \text{in} \quad \mathscr{B}, \qquad a^c=(\underline{b}-\underline{\gamma})\cdot\underline{n} \quad \text{on} \quad \partial\mathscr{B} \tag{16.57}$$

Equation $(16.57)_1$ expresses the general form of balance of generalized stresses. It is identical with Gurtin's balance of microforces (Gurtin, 1996), except the external microforce contribution γ that may exist in general. In the sequel, however, it is assumed that $\gamma=0$ and $\underline{\gamma}=0$ for the sake of brevity. The equation $(16.57)_2$ represents the boundary condition for the generalized traction vector.

State Laws and Dissipation Potential

According to the first principle of thermodynamics, the time variation of the total energy in a material subdomain is equal to the power of external forces acting on it. In the absence of inertial forces, the total energy is reduced to the internal energy with density e. Then, the energy balance is stated as:

$$\int_{\mathscr{D}}\dot{\epsilon}\,dV = \mathscr{P}^{(e)}+\mathscr{P}^{(c)} = -\mathscr{P}^{(i)} = \int_{\mathscr{D}} a\dot{\phi}+\underline{b}\cdot\nabla\dot{\phi}\,dV \tag{16.58}$$

This identity is valid for any subdomain $\mathscr{D}\subset\mathscr{B}$. The local form of the energy balance is obtained:

$$\dot{\epsilon}=a\dot{\varphi}+\underline{b}\cdot\nabla\dot{\phi}=\nabla\cdot(\dot{\phi}\,\underline{b}) \tag{16.59}$$

The entropy principle is formulated as follows:

$$\int_{\mathscr{D}}\dot{\eta}\,dV \geq -\int_{\partial\mathscr{D}}\underline{\Phi}\cdot\underline{n}\,dS \quad \text{and} \quad \underline{\Phi}=-\mu\frac{\underline{J}}{T} \tag{16.60}$$

where η is the entropy density, $\underline{\Phi}$ the entropy flux, $\underline{J}$ the diffusion flux and μ the diffusion potential (Villani et al, 2014). Using the equation of local conservation of mass:

$$\dot{c}=-\nabla\cdot\underline{J} \tag{16.61}$$

the following local form of the entropy inequality is obtained:

$$T\dot{\eta} - \nabla \cdot (\mu \underline{J}) \geq 0 \tag{16.62}$$

Combining the equation of the free energy density $\psi = \dot{\epsilon} - T\dot{\eta}$ in the isothermal case with Eqs. (16.59)–(16.62), leads to the Clausius-Duhem inequality:

$$-\dot{\psi} + a\dot{\phi} + \underline{b} \cdot \nabla\dot{\phi} - \underline{J} \cdot \nabla\mu - \mu\nabla \cdot \underline{J} \geq 0 \tag{16.63}$$

The free energy density is assumed to be a function of concentration c, order parameter ϕ, as well as its gradient $\nabla\phi$. The Clausius-Duhem inequality can then be written as follows:

$$\left(\mu - \frac{\partial \psi}{\partial c}\right)\dot{c} + \left(a - \frac{\partial \psi}{\partial \phi}\right)\dot{\phi} + \left(\underline{b} - \frac{\partial \psi}{\partial \nabla\phi}\right) \cdot \nabla\dot{\phi} - \underline{J} \cdot \nabla\mu \geq 0 \tag{16.64}$$

For every admissible process and for any given (c, ϕ,$\nabla\phi$), the inequality (16.64) must hold for arbitrary values of $\dot{c}$, $\dot{\phi}$ and $\nabla\dot{\phi}$. The microstress $\underline{b}\,(c,\phi,\nabla\phi)$ and the diffusion potential $\mu(c,\phi,\nabla\phi)$ are assumed independent of $\nabla\dot{\phi}$ and $\dot{c}$. The following state laws are deduced:

$$\mu = \frac{\partial \psi}{\partial c}, \quad \underline{b} = \frac{\partial f}{\partial \nabla\phi} = A\nabla\phi \tag{16.65}$$

the latter equation being valid in the case of a quadratic potential w.r.t. $\nabla\phi$. The Clausius-Duhem inequality then reduces to the residual dissipation:

$$D = -\underline{J} \cdot \nabla\mu + a^{dis}\dot{\phi} \geq 0 \quad \text{with} \quad a^{dis} = a - \frac{\partial \psi}{\partial \phi} \tag{16.66}$$

where a^{dis} is the chemical force associated with the dissipative processes, as introduced in Gurtin (1996).

In order to define the complementary laws related to the dissipative processes, the existence of a dissipation potential function $\Omega(\nabla\mu, \pi_{dis})$ is assumed. The retained specific form is the following:

$$\Omega(\nabla\mu, a^{dis}) = \frac{1}{2}L(\phi)\nabla\mu \cdot \nabla\mu + \frac{1}{2\beta}a^{dis2} \tag{16.67}$$

where $L(\phi)$ and β are material parameters or functions.

The complementary evolution laws derive from the dissipation potential:

$$\dot{\phi} = \frac{\partial \Omega}{\partial a^{dis}} = \frac{1}{\beta}a^{dis}, \quad \underline{J} = -\frac{\partial \Omega}{\partial \nabla\mu} = -L(\phi)\nabla\mu \tag{16.68}$$

The convexity of the dissipation potential ensures the positivity of dissipation.

Combining Eqs. (16.66) and (16.68), one gets:

$$a = \beta\,\dot{\phi} + \frac{\partial \psi}{\partial \phi} \tag{16.69}$$

The substitution of the two state laws and the complementary laws, into the balance equations for mass concentration and generalized stresses respectively leads to the evolution equations for concentration and order parameter:

$$\dot{c} = -\nabla \cdot (-L(\varphi)\nabla \mu) = -\nabla \cdot \left(-L(\phi)\nabla \frac{\partial \psi}{\partial c}\right) \tag{16.70}$$

$$a - \nabla \cdot \underline{b} = \beta \dot{\phi} - \alpha \Delta \phi + \frac{\partial \psi}{\partial \phi} = 0 \tag{16.71}$$

The usual diffusion and Cahn–Allen / Ginzburg–Landau equations are thus retrieved (Finel et al, 2010).

In the previous theory, the free energy density depends on the gradient of the order parameter but not on the concentration gradient. This possibility is investigated in the next section.

16.5 On the Construction of the Cahn–Hilliard Diffusion Theory

The partial differential equation for the concentration field according to Cahn and Hilliard (1958) is the following

$$\dot{c} = \kappa \Delta (\frac{\partial \psi}{\partial c} - A \Delta c) \tag{16.72}$$

with the special case: $\dot{c} = \beta \nabla^2 c - \chi \nabla^4 c$ when the free energy potential is quadratic w.r.t. c and ∇c. We present in this section three distinct derivations of these field equations from first principles. The first one was initially proposed by Cahn and Hilliard. The second one is based on the principle of virtual power with an additional generalized balance equation, following Gurtin (1996). In the last subsection, a variational formulation is proposed considering a second gradient theory of diffusion.

16.5.1 Usual Presentation Based on the Variational Derivative

A homogeneous material system is considered with composition c of a given solute element. The free energy density $\psi(c, \nabla c)$ is assumed to depend not only on concentration but also on its gradient, as initially proposed by Cahn and Hilliard (1958):

$$F = \int_{\mathscr{B}} \psi \, dV = \int_{\mathscr{B}} \left(\psi_0(c) + \kappa_1 \nabla^2 c + \kappa_2 (\nabla c) \cdot (\nabla c) \right) dV$$
$$= \int_{\mathscr{B}} \left(\psi_0(c) + \frac{1}{2} A (\nabla c) \cdot (\nabla c) \right) dV \tag{16.73}$$

in the isotropic case. The second expression holds assuming $(\nabla c) \cdot \underline{n} = 0$ on $\partial \mathscr{B}$. The potential $\psi_0(c)$ is the usual potential depending on concentration in classical diffusion theory. The variational derivative of the total free energy of the body is defined as

$$\delta F = \int_{\mathscr{B}} \delta \psi \, dV = \int_{\mathscr{B}} \left(\frac{\partial \psi}{\partial c} \delta c + \frac{\partial \psi}{\partial \nabla c} \delta \nabla c \right) dV \tag{16.74}$$

Note that $\nabla \cdot (\delta c \nabla c) = \nabla c \cdot \delta \nabla c + \delta c \Delta c$ so that

$$\int_{\mathscr{B}} \nabla \cdot (\delta c \nabla c) \, dV = \int_{\partial \mathscr{B}} \delta c \nabla c \cdot \underline{n} \, dS = 0, \quad \text{if} \quad \nabla c \cdot \underline{n} = 0 \text{ on } \partial \mathscr{B}$$
$$= \int_{\mathscr{B}} \nabla c \cdot \delta \nabla c \, dV + \int_{\mathscr{B}} \delta c \Delta c \, dV \tag{16.75}$$

Hence

$$\delta F = \int_{\mathscr{B}} \left(\frac{\partial \psi_0}{\partial c} - A \Delta c \right) \delta c \, dV \tag{16.76}$$

The variational derivative of the free energy function then is

$$\frac{\delta \psi}{\delta c} := \frac{\partial \psi_0}{\partial c} - A \Delta c \tag{16.77}$$

which makes sense only in the absence diffusion flux $\nabla c \cdot \underline{n}$ on the boundary of the body or in periodic systems. The balance of mass and Fick's law for the mass flux write

$$\dot{c} = -\nabla \cdot \underline{J}, \quad \underline{J} = -\kappa \nabla \mu \tag{16.78}$$

The diffusion potential μ is defined as the variational derivative of the free energy density function:

$$\mu := \frac{\delta \psi}{\delta c} \tag{16.79}$$

The combination of mass balance, Fick's law and constitutive potential ψ leads to the Cahn-Hilliard equation:

$$\dot{c} = \nabla \cdot \left(\kappa \nabla \frac{\delta \psi}{\delta c} \right) = \kappa \Delta \frac{\delta \psi}{\delta c} = \kappa \Delta \left(\frac{\partial \psi_0}{\partial c} - A \Delta c \right) \tag{16.80}$$

This is the general form for a conserved quantity in contrast to non–conserved phase field or order parameter considered in the previous section.

16.5.2 Method of Virtual Power with Additional Balance Equation

The existence of a principle of virtual power w.r.t. to the concentration field and its gradient is assumed. The power density of generalized internal and contact forces is

$$p^{(i)} = a\dot{c} + \underline{b} \cdot \nabla \dot{c}, \quad p^{(c)} = a^c \dot{c} \tag{16.81}$$

This leads to a field equation for the balance of generalized forces a and $\underline{b}$, and associated boundary conditions:

$$a = \operatorname{div} \underline{b}\,, \quad \underline{b} \cdot \underline{n} = a^c \tag{16.82}$$

in addition to the balance of mass $\dot{c} = -\operatorname{div} \underline{J}$. The first and second principles (isothermal case) take the form

$$\dot{\epsilon} = p^{(i)}, \quad -\dot{\psi} + p^{(i)} - \operatorname{div}(\mu \underline{J}) \geq 0 \tag{16.83}$$

It should be noted that in the absence of concentration gradient in the free energy potential, the generalized forces a and $\underline{b}$ identically vanish and the classical diffusion theory is retrieved, as it should. The Clausius-Duhem inequality is

$$(a + \mu - \frac{\partial \psi}{\partial c})\dot{c} + (\underline{b} - \frac{\partial \psi}{\partial \nabla c}) \cdot \nabla \dot{c} - \underline{J} \cdot \nabla \mu \geq 0 \tag{16.84}$$

leading to the state laws

$$\mu = \frac{\partial \psi}{\partial c} - a, \qquad \underline{b} = \frac{\partial \psi}{\partial \nabla c} \tag{16.85}$$

Fick's law (16.78) and the potential (16.73) are used again. The Cahn–Hilliard equations are now obtained by combining the balance and constitutive equations in the following way:

$$\mu = \frac{\partial \psi}{\partial c} - a = \frac{\partial \psi}{\partial c} - \operatorname{div} \underline{b} = \frac{\partial \psi}{\partial c} - A \Delta c, \tag{16.86}$$

$$\dot{c} = -\operatorname{div} \underline{J} = \kappa \Delta \mu = \kappa \Delta (\frac{\partial \psi}{\partial c} - A \Delta c) \tag{16.87}$$

The Cahn–Hilliard model can also be derived from a more general theory introducing micromorphic concentration or microconcentration variables c_χ and its gradient in addition to the concentration as done in Sect. 16.3 for the entropy or temperature. The internal constraint $c \equiv c_\chi$ leads to the previous equations (Forest, 2003). The microconcentration can be regarded as a description of the variance of the composition inside the volume element.

16.5.3 Second Gradient Diffusion Theory

The Cahn–Hilliard equations can also be interpreted as emerging from a second gradient theory of diffusion. For that purpose, let us first recall the

16.5.3.1 Variational Formulation of Classical Diffusion

The variational formulation of the field and boundary equations of classical diffusion, namely

$$-\nabla \cdot \underline{J} + \gamma = \dot{c} \quad \text{on} \quad \mathscr{B}, \qquad j = \underline{J} \cdot \underline{n} \quad \text{on} \quad \partial\mathscr{B} \tag{16.88}$$

can be written in the form $\forall c^\star, \forall \mathscr{D} \subset \mathscr{B}$

$$\mathscr{I}^{(i)}(c^\star, \mathscr{D}) + \mathscr{I}^{(c)}(c^\star, \mathscr{D}) + \mathscr{I}^{(e)}(c^\star, \mathscr{D}) + \mathscr{I}^{(a)}(c^\star, \mathscr{D}) = 0 \tag{16.89}$$

for test compositions $c^\star$ and with the following contributions

$$\mathscr{I}^{(i)}(c^\star, \mathscr{D}) = \int_{\mathscr{D}} \underline{J} \cdot \nabla c^\star \, dV, \quad \mathscr{I}^{(c)}(c^\star, \mathscr{D}) = -\int_{\partial\mathscr{D}} j c^\star \, dS \tag{16.90}$$

$$\mathscr{I}^{(e)}(c^\star, \mathscr{D}) = \int_{\mathscr{D}} \gamma c^\star \, dV, \quad \mathscr{I}^{(a)}(c^\star, \mathscr{D}) = -\int_{\mathscr{D}} \dot{c} c^\star \, dV \tag{16.91}$$

In other words,

$$\int_{\mathscr{D}} (-\nabla \cdot \underline{J} + \gamma - \dot{c}) c^\star \, dV + \int_{\partial\mathscr{D}} (\underline{J} \cdot \underline{n} - j) c^\star \, dS = 0 \tag{16.92}$$

16.5.3.2 Variational Formulation of Second Gradient Diffusion

The internal contribution (16.90) is extended by introducing the second gradient of the test functions and a generalized second order flux tensor:

$$\mathscr{I}^{(i)}(c^\star, \mathscr{D}) = \int_{\mathscr{D}} \left(\underline{J} \cdot \nabla c^\star + \underset{\sim}{K} : \nabla\nabla c^\star \right) dV \tag{16.93}$$

Two integration by parts are necessary to obtain the form of the generalized surface flux:

$$\begin{aligned}
\mathscr{I}^{(i)}(c^\star, \mathscr{D}) &= \int_{\mathscr{D}} \left((J_i c^\star)_{,i} + (K_{ij} c^\star_{,i})_{,j} - J_{i,i} c^\star - K_{ij,j} c^\star_{,i} \right) dV \\
&= \int_{\mathscr{D}} \left((J_i c^\star)_{,i} + (K_{ij} c^\star_{,i})_{,j} - J_{i,i} c^\star - (K_{ij,j} c^\star)_{,i} + K_{ij,ij} c^\star \right) dV \\
&= \int_{\partial \mathscr{D}} \left(J_i c^\star + K_{ij} c^\star_{,i} - K_{ij,j} c^\star \right) n_i \, dS - \int_{\mathscr{D}} (J_{i,i} - K_{ij,ij}) c^\star \, dV \\
&= \int_{\partial \mathscr{D}} \left(\underline{J} - \underset{\sim}{K} \cdot \nabla \right) \cdot \underline{n} \, c^\star \, dS + \int_{\partial V} (\underset{\sim}{K} \cdot \underline{n}) \cdot \nabla c^\star \, dS - \int_{V} ((\underline{J} - \underset{\sim}{K} \cdot \nabla) \cdot \nabla) c^\star \, dV
\end{aligned}$$

At this stage, tangent and normal derivatives, $\underline{D}_t$ and D_n of the concentration field on surfaces must be distinguished:

$$\nabla c = \underline{D}_t c + (D_n c) \underline{n}, \quad D_n c = \nabla c \cdot \underline{n} \tag{16.94}$$

It follows that

$$\int_{\partial \mathscr{D}} (\underset{\sim}{K} \cdot \underline{n}) \cdot \nabla c^\star \, dS = \int_{\partial \mathscr{D}} (\underset{\sim}{K} \cdot \underline{n}) \cdot \underline{D}_t c^\star \, dS + \int_{\partial \mathscr{D}} (\underline{n} \cdot \underset{\sim}{K} \cdot \underline{n}) D_r c^\star \, dS \tag{16.95}$$

An integration by parts of the integral involving the tangent derivative is possible:

$$\int_{\partial \mathscr{D}} (\underset{\sim}{K} \cdot \underline{n}) \cdot \underline{D}_t c^\star \, dS = \int_{\partial \mathscr{D}} \underline{D}_t \cdot (\underset{\sim}{K} \cdot \underline{n} \, c^\star) \, dS - \int_{\partial \mathscr{D}} (\underline{D}_t \cdot (\underset{\sim}{K} \cdot \underline{n})) c^\star \, dS \tag{16.96}$$

The divergence theorem for surfaces[2] can be applied to the first term of the right-hand side to get the final expression of

---

[2] The divergence theorem for a closed and smooth surface ∂V (no edge) is

$$\int_{\partial V} \underline{D}_t \cdot \underline{q} \, dS = \int_{\partial V} 2\mathscr{C} \, \underline{q} \cdot \underline{n} \, dS, \quad \text{with} \quad 2\mathscr{C} = \underline{D}_t \cdot \underline{n}$$

where $\mathscr{C}$ is the mean local curvature.

Proof. Evaluate

$$\begin{aligned}
\underline{\mathrm{curl}}\,(\underline{n} \times \underline{q}) &= \epsilon_{ijk} \epsilon_{jlm} (n_l q_m)_{,k} \underline{e}_i = -\epsilon_{jik} \epsilon_{jlm} (n_{l,k} q_m + n_l q_{m,k}) \\
&= -(\delta_{il} \delta_{kn} - \delta_{im} \delta_{kl}) (n_{l,k} q_m + n_l q_{m,k}) = -(n_{i,k} q_k - q_i n_{k,k} + n_i q_{k,k} - n_k q_{i,k}) \\
\underline{n} \cdot \underline{\mathrm{curl}}\,(\underline{n} \times \underline{q}) &= -(q_k n_i n_{i,k} - q_i n_i n_{k,k} + q_{k,k} - q_{i,k} n_i n_k) = q_i n_i n_{k,k} - q_{k,k} + q_{i,k} n_i n_k \\
&= q_i n_i n_{k,k} - D_{ti} q_i
\end{aligned}$$

since $n_i n_{i,k} = 0$ ($\|\underline{n}\| = 1$). Hence

$$\int_{\partial V} \underline{D}_t \cdot \underline{q} \, dS = \int_{\partial V} \underline{q} . \underline{n} \, (\mathrm{div}\, \underline{n}) \, dS - \int_{\partial V} \underline{n} \cdot \underline{\mathrm{curl}}\,(\underline{n} \times \underline{q}) dS$$

$$\mathscr{I}^{(i)}(c^\star,\mathscr{D}) = -\int_{\mathscr{D}} \left((\underline{J} - \underset{\sim}{K}\cdot\nabla)\cdot\nabla \right) c^\star \, dV$$

$$+\int_{\partial\mathscr{D}} \left((\underline{J} - \underset{\sim}{K}\cdot\nabla).\underline{n} + (\underline{n}\cdot\underset{\sim}{K}\cdot\underline{n})2\mathscr{C} - \underline{D}_t\cdot(\underset{\sim}{K}\cdot\underline{n}) \right) c^\star \, dS + \int_{\partial\mathscr{D}} \left(\underline{n}\cdot\underset{\sim}{K}\cdot\underline{n}\right) D_n c^\star \, dS \tag{16.97}$$

where $\mathscr{C}$ is the mean surface curvature. The form of the surface contribution can be deduced from the previous calculation:

$$\mathscr{I}^{(c)}(c^\star,\mathscr{D}) = \int_{\partial\mathscr{D}} (jc^\star + kD_n c^\star)\, dS \tag{16.98}$$

where j and k are generalized prescribed mass flux surface densities. External contributions take the form:

$$\mathscr{I}^{(e)}(c^\star,\mathscr{D}) = \int_{\mathscr{D}} (\gamma c^\star + \underline{\gamma}\cdot\nabla c^\star + \underset{\sim}{\gamma} : \nabla\nabla c^\star)\, dV = \int_V \left(\gamma - \underline{\gamma}\cdot\nabla + \underset{\sim}{\gamma}\cdot\nabla\cdot\nabla\right) c^\star \, dV$$

$$+\int_{\partial V} \left(\underline{\gamma}\cdot\underline{n} - (\underset{\sim}{\gamma}\cdot\nabla)\cdot\underline{n} - \underline{D}_t\cdot(\underset{\sim}{\gamma}\cdot\underline{n}) + 2\mathscr{C}\underline{n}\cdot\underset{\sim}{\gamma}\cdot\underline{n}\right) c^\star \, dS + \int_{\partial V} \underset{\sim}{\gamma} : (\underline{n}\otimes\underline{n}) D_n c^\star \, dS \tag{16.99}$$

The last contribution $\mathscr{I}^{(a)}(c^\star,\mathscr{D})$ keeps the classical form (16.91).

The variational principle (16.89) is invoked again to derive the field equations:

$$\dot{c} = -\underline{J}^{eff}\cdot\nabla + \gamma \tag{16.100}$$

with the effective[3] tensor diffusion flux defined as

$$\underline{J}^{eff} = \underline{J} + \underline{\gamma} - (\underset{\sim}{K} + \underset{\sim}{\gamma})\cdot\nabla \tag{16.101}$$

The associated boundary conditions for simple and double flux are

$$\begin{aligned} j &= \underline{J}^{eff}\cdot\underline{n} + 2R(\underset{\sim}{K} + \underset{\sim}{\gamma}) : (\underline{n}\otimes\underline{n}) - ((\underset{\sim}{K} + \underset{\sim}{\gamma})\cdot\underline{n})\cdot\underline{D}_t, \\ k &= (\underset{\sim}{K} + \underset{\sim}{\gamma}) : (\underline{n}\otimes\underline{n}) \end{aligned} \tag{16.102}$$

The energy and entropy principles (isothermal case) take the following form

$$\dot{E} = \mathscr{I}^{(e)}, \quad \dot{\epsilon} = \gamma\dot{c} + \underline{\gamma}\cdot\nabla\dot{c} + \underset{\sim}{\gamma} : \nabla\nabla\dot{c}, \quad T\dot{S} \geq \int_{\partial V} T\underline{\Phi}\cdot\underline{n}\, dV \tag{16.103}$$

The last term vanishes due to Stokes theorem: $\int_S (\underline{\text{curl}}\ \underline{A})\cdot\underline{n}\ dS = -\oint_\Gamma \underline{A}\cdot\underline{l}\ dl$ and $\Gamma = \emptyset$ for a closed surface. Finally, note that

$$n_{k,l} = D_{tl}n_k + n_{k,j}n_k n_l \Longrightarrow n_{k,k} = D_{tk}n_k, \quad \text{div}\underline{n} = \underline{D}_t\cdot\underline{n} \qquad \square$$

[3] This is similar to Germain's effective stress whose divergence arises in the strain gradient balance of momentum equation.

In contrast to the formulation (16.83), there is no internal energy contribution due to composition rate and its gradient (only the source terms $\gamma, \underline{\gamma}, \underset{\sim}{\gamma}$), as in the classical case. The new contributions are included in the extended entropy flux (Maugin and Muschik, 1994):

$$T\underline{\Phi} = \mu \underline{J}^{eff} + T\underline{\Phi}^{extra} \tag{16.104}$$

The Clausius–Duhem inequality (for vanishing $\gamma, \underline{\gamma}, \underset{\sim}{\gamma} \equiv 0$) then reads

$$-\dot{\psi} - \nabla \cdot (T\underline{\Phi}^{extra}) - \nabla \cdot (\mu \underline{J}^{eff}) \geq 0$$

$$-\left(\frac{\partial \psi}{\partial c} - \nabla \cdot \frac{\partial \psi}{\partial \nabla c} - \mu\right)\dot{c} - \nabla \cdot \left(T\underline{\Phi}^{extra} + \frac{\partial \psi}{\partial \nabla c}\dot{c}\right) - \underline{J}^{eff} \cdot \nabla \mu \geq 0$$

The chemical potential is now defined as

$$\mu = \frac{\partial \psi}{\partial c} - \nabla \cdot \frac{\partial \psi}{\partial \nabla c} \tag{16.105}$$

The following constitutive choices are made ensuring positivity of dissipation:

- extra–entropy flux

$$T\underline{\Phi}^{extra} = -\frac{\partial \psi}{\partial \nabla c}\dot{c} \tag{16.106}$$

- generalized Fick law

$$\underline{J}^{eff} = -\underset{\sim}{\kappa} \cdot \nabla \mu \tag{16.107}$$

- corresponding constitutive equations for $\underline{J}$ and $\underset{\sim}{K}$

$$\underline{J}^{eff} = \underline{J} - \underset{\sim}{K} \cdot \nabla = -\underset{\sim}{\kappa} \cdot \left(\nabla\left(\frac{\partial \psi}{\partial c} - \nabla \cdot \frac{\partial \psi}{\partial \nabla c}\right)\right) \tag{16.108}$$

  with

$$\underline{J} = -\underset{\sim}{\kappa} \cdot \frac{\partial \psi}{\partial c}, \quad \underset{\sim}{K} = -\left(\nabla \cdot \frac{\partial \psi}{\partial \nabla c}\right)\underset{\sim}{\kappa} \tag{16.109}$$

  (for uniform diffusion coefficients $\underset{\sim}{\kappa}$).

16.6 Conclusions

The development of continuum theories involving the gradient of field variables requires the introduction of suited (conjugate) generalized forces and stresses fulfilling higher order or additional balance equations. Such models can be constructed in an efficient and rigorous manner by applying the method of virtual power. It has been illustrated in the case of gradient and micromorphic plasticity and can be extended to damage and other internal variable theories in a straightforward way. The application to nonmechanical fields like temperature, microtemperature, concentration or microconcentration was shown to be possible even though it remains rather formal.

It provides new higher order diffusion equations that can be compared to existing ones derived from different concepts. For instance, micromechanically motivated higher-order continuum formulation of linear thermal conduction was proposed in Temizer and Wriggers (2010) based on a second gradient of temperature model very close to the formulation proposed in Sect. 16.5.3. In particular the higher order boundary conditions are the same. These boundary conditions, involving surface curvature effects, differ from the ones derived by means of an additional balance laws in Subsect. 16.5.2. Two different views of the Cahn–Hilliard model were presented, one based on an additional balance equation, the second one based on second gradient diffusion theory, all derived using the method of virtual power. The different interplay between generalized forces and the use of extra–entropy flux leads to different boundary conditions. This combination of kinematic and constitutive choices was discussed many times by G.A. Maugin who tried in each situation to reduce the arbitrariness in the choice of generalized forces and extra–entropy flux vectors.

References

Aifantis E (1984) On the microstructural origin of certain inelastic models. Journal of Engineering Materials and Technology 106:326–330

Altenbach H, Maugin GA, Erofeev V (2011) Mechanics of Generalized Continua, Advanced Structured Materials, vol 7. Springer, Heidelberg

Ammar K, Appolaire B, Cailletaud G, Feyel F, Forest S (2009) Finite element formulation of a phase field model based on the concept of generalized stresses. Computational Materials Science 45:800–805

Aslan O, Forest S (2011) The micromorphic versus phase field approach to gradient plasticity and damage with application to cracking in metal single crystals. In: de Borst R, Ramm E (eds) Multiscale Methods in Computational Mechanics, Lecture Notes in Applied and Computational Mechanics 55, Springer, pp 135–154

Aslan O, Cordero NM, Gaubert A, Forest S (2011) Micromorphic approach to single crystal plasticity and damage. International Journal of Engineering Science 49:1311–1325

Cahn J, Hilliard J (1958) Free energy of a nonuniform system. I. Interfacial free energy. The Journal of Chemical Physics 28:258–267

Coleman B, Mizel J (1963) Thermodynamics and departures from Fourier's law of heat conduction. Arch Rational Mech and Anal 13:245–261

Coleman B, Noll W (1963) The thermodynamics of elastic materials with heat conduction and viscosity. Arch Rational Mech and Anal 13:167–178

Daher N, Maugin GA (1986) The method of virtual power in continuum mechanics application to media presenting singular surfaces and interfaces. Acta Mechanica 60:217–240

dell'Isola F, Seppecher P (1995) The relationship between edge contact forces, double forces and intersticial working allowed by the principle of virtual power. CR Acad Sci Paris IIb 321:303–308

dell'Isola F, Seppecher P, Madeo A (2012) How contact interactions may depend on the shape of Cauchy cuts in N-th gradient continua: approach "à la D'Alembert". Zeitschrift für Angewandte Mathematik und Physik 63:1119–1141

Finel A, Le Bouar Y, Gaubert A, Salman U (2010) Phase field methods: Microstructures, mechanical properties and complexity. Comptes Rendus Physique 11:245–256

Forest S (2008) The micromorphic approach to plasticity and diffusion. In: Jeulin D, Forest S (eds) Continuum Models and Discrete Systems 11, Proceedings of the international conference CMDS11, Les Presses de l'Ecole des Mines de Paris, Paris, France, pp 105–112

Forest S (2009) The micromorphic approach for gradient elasticity, viscoplasticity and damage. ASCE Journal of Engineering Mechanics 135:117–131

Forest S (2016) Nonlinear regularisation operators as derived from the micromorphic approach to gradient elasticity, viscoplasticity and damage. Proceedings of the Royal Society of London A: Mathematical, Physical and Engineering Sciences 472(2188)

Forest S, Aifantis EC (2010) Some links between recent gradient thermo-elasto-plasticity theories and the thermomechanics of generalized continua. International Journal of Solids and Structures 47:3367–3376

Forest S, Amestoy M (2008) Hypertemperature in thermoelastic solids. Comptes Rendus Mécanique 336:347–353

Forest S, Bertram A (2011) Formulations of strain gradient plasticity. In: Altenbach H, Maugin GA, Erofeev V (eds) Mechanics of Generalized Continua, Advanced Structured Materials vol. 7, Springer, pp 137–150

Forest S, Sab K (2012) Continuum stress gradient theory. Mechanics Research Communications 40:16–25

Forest S, Sab K (2017) Finite-deformation second-order micromorphic theory and its relations to strain and stress gradient models. Mathematics and Mechanics of Solids

Frémond M, Nedjar B (1996) Damage, gradient of damage and principle of virtual power. Int J Solids Structures 33:1083–1103

Germain P (1973a) La méthode des puissances virtuelles en mécanique des milieux continus, première partie : théorie du second gradient. J de Mécanique 12:235–274

Germain P (1973b) The method of virtual power in continuum mechanics. Part 2: Microstructure. SIAM J Appl Math 25:556–575

Germain P, Nguyen QS, Suquet P (1983) Continuum thermodynamics. Journal of Applied Mechanics 50:1010–1020

Gurtin M (1996) Generalized Ginzburg–Landau and Cahn–Hilliard equations based on a microforce balance. Physica D 92:178–192

Gurtin M (2003) On a framework for small–deformation viscoplasticity: free energy, microforces, strain gradients. International Journal of Plasticity 19:47–90

Gurtin M, Anand L (2009) Thermodynamics applied to gradient theories involving the accumulated plastic strain: The theories of Aifantis and Fleck & Hutchinson and their generalization. Journal of the Mechanics and Physics of Solids 57:405–421

Ireman P, Nguyen QS (2004) Using the gradients of temperature and internal parameters in continuum mechanics. CR Mécanique 332:249–255

Liu W, Saanouni K, Forest S, Hu P (2017) The micromorphic approach to generalized heat equations. Journal of Non-Equilibrium Thermodynamics 42(4):327–358

Maugin GA (1980) The method of virtual power in continuum mechanics: Application to coupled fields. Acta Mechanica 35:1–70

Maugin GA (1990) Internal variables and dissipative structures. J Non–Equilib Thermodyn 15:173–192

Maugin GA (1992) Thermomechanics of Plasticity and Fracture. Cambridge University Press

Maugin GA (1999) Thermomechanics of Nonlinear Irreversible Behaviors. World Scientific

Maugin GA (2006) On the thermomechanics of continuous media with diffusion and/or weak nonlocality. Archives of Applied Mechanics 75:723–738

Maugin GA (2013) The principle of virtual power: from eliminating metaphysical forces to providing an efficient modelling tool. Continuum Mechanics and Thermodynamics 25:127–146

Maugin GA, Metrikine AV (eds) (2010) Mechanics of Generalized Continua - One Hundred Years After the Cosserats, Advances in Mechanics and Mathematics, vol 21. Springer, New York

Maugin GA, Muschik W (1994) Thermodynamics with internal variables, Part I. General concepts. J Non-Equilib Thermodyn 19:217–249

Mindlin R (1965) Second gradient of strain and surface–tension in linear elasticity. Int J Solids Structures 1:417–438

Nguyen QS (2010a) Gradient thermodynamics and heat equations. Comptes Rendus Mécanique 338:321–326

Nguyen QS (2010b) On standard dissipative gradient models. Annals of Solid and Structural Mechanics 1(2):79–86

Nguyen QS (2016) Quasi-static responses and variational principles in gradient plasticity. Journal of the Mechanics and Physics of Solids 97:156 – 167

Saanouni K, Hamed M (2013) Micromorphic approach for finite gradient-elastoplasticity fully coupled with ductile damage: Formulation and computational aspects. International Journal of Solids and Structures 50:2289–2309

Svendsen B (1999) On the thermodynamics of thermoelastic materials with additional scalar degrees of freedom. Continuum Mechanics and Thermodynamics 4:247–262

Temizer I, Wriggers P (2010) A micromechanically motivated higher–order continuum formulation of linear thermal conduction. ZAMM 90:768–782

Villani A, Busso E, Ammar K, Forest S, Geers M (2014) A fully coupled diffusional-mechanical formulation: numerical implementation, analytical validation, and effects of plasticity on equilibrium. Archive of Applied Mechanics 84:1647–1664

Chapter 17
Forbidden Strains and Stresses in Mechanochemistry of Chemical Reaction Fronts

Alexander B. Freidin and Leah L. Sharipova

Abstract The influence of stresses and strains on a chemical reaction rate and a chemical reaction front velocity is studied basing on the concept of the chemical affinity tensor. The notion of forbidden zones formed by strains or stresses at which the reaction cannot go is discussed. Examples of forbidden zones are constructed.

17.1 Introduction

The influence of a stress-strain state on a chemical reaction rate is still remaining questionable in spite of the fact that modern engineering challenges manifest the necessity of a detailed description of this influence, particulary due to increasing use of micro- and nano- structural elements in microsystem technologies combining thermomechanical and chemical actions.

It should be noted that coupling of stresses, diffusion and chemical reaction has been intensively discussed in the context of silicon oxidation. In a number of papers, the Deal-Grove model (Deal and Grove, 1965) has been modified by considering the classical diffusion equation with the stress-dependent diffusion coefficient and the stress-dependent reaction rate parameter. The choice of stress characteristics, which affect the reaction and the diffusion parameters, was made intuitively, e.g. Kao et al (1988); Rafferty (1990); Sutardja and Oldham (2005). Additionally, a concentration-dependent volumetric expansion was introduced, which led to the total stress-diffusion coupling, e.g. Rao and Hughes (2000); Rao et al (2000). The influence of stresses on diffusion via additional terms in the diffusion equation has been also considered, e.g. Knyazeva (2003); Toribio et al (2011).

Alexander B. Freidin · Leah L. Sharipova
Institute for Problems in Mechanical Engineering of the Russian Academy of Sciences, Bolshoy pr., 61, V.O., St. Petersburg, 199178 & Peter the Great St.Petersburg Polytechnic University, Polytechnicheskaya st., 29, St.Petersburg, 195251, Russia
e-mail: alexander.freidin@gmail.com, sleah07@gmail.com

© Springer International Publishing AG, part of Springer Nature 2018
H. Altenbach et al. (eds.), *Generalized Models and Non-classical Approaches in Complex Materials 1*, Advanced Structured Materials 89,
https://doi.org/10.1007/978-3-319-72440-9_17

Another group of models includes the influence of stresses on the diffusion flux and the chemical reaction rate via a scalar chemical potential, which depends on the concentration and the stresses, and the gradient of which governs the flux of the reactant, e.g. Loeffel and Anand (2011); Cui et al (2012); Brassart and Suo (2013); Levitas and Attariani (2014); Bower et al (2015), and it was noted that the velocity of the reaction front can be controlled by the reaction rate at the reaction front rather than by the diffusivity of the reactant (see e.g. Jia and Li, 2015, and reference therein).

In the present paper we develop the thermodynamically consistent approach to mechanochemistry of reaction fronts by which the influence of stresses and strains on the chemical reaction rate and the chemical reaction front velocity is studied basing on the concept of the chemical affinity tensor (Freidin, 2014; Freidin et al, 2014; Freidin, 2015; Freidin et al, 2016a,b). We focus on locking effects – arresting of the reaction front propagation by mechanical stresses.

Recall that in the classical physical chemistry the chemical affinity appears as a multiplier A in the expression of the entropy production $P[S]$ due to a chemical reaction (Prigogine and Defay, 1988):

$$TP[S] = A\omega, \tag{17.1}$$

where ω is the reaction rate, T is the temperature. In the case of a reaction

$$n_1B_1 + n_2B_2 + ... + n_lB_l \to n_{l+1}B_{l+1} + n_{l+2}B_{l+2} + ... + n_NB_N,$$

where B_k are chemical formulae of reacting constituents, n_k are the stoichiometric coefficients ($k = 1,\ldots,N$), the chemical affinity is defined as

$$A = -\sum_1^N n_k M_k \mu_k,$$

where μ_i and M_i ($i = 1,\ldots,4$) are chemical potentials per unit mass and molar masses of the reaction constituents, the stoichiometric coefficient n_k is substituted into the sum with the sign "+" if the kth constituent is produced in the reaction and with the sign "−" if the kth constituent is consumed.

The structure of the entropy production expression indicates that a kinetic equation can be formulated in a form of the dependence of the reaction rate on the affinity: $\omega = \omega(A)$. Chemical equilibrium corresponds to the case $\omega = 0$ and $A = 0$, and a direct reaction is possible only if $A > 0$. So called false equilibrium $A \neq 0$, $\omega = 0$, is not discussed in the present paper.

The reaction

$$n_-B_- + n_*B_* \to n_+B_+ \tag{17.2}$$

is considered further where B_- and B_+ are solid constituents, and B_* is a diffusive ("gaseous") component. The reaction is sustained by the diffusion of B_* through the material B_+ and is localized at the reaction front Γ that divides domains v_- and v_+ occupied by the solid constituents B_- and B_+, respectively, and the diffusive constituent is fully consumed at the reaction front (Fig. 17.1). Silicon oxidation

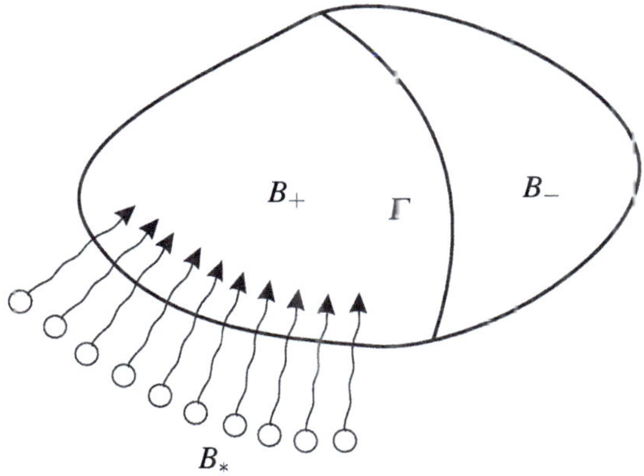

Fig. 17.1 Localized chemical reaction in a solid sustained by the diffusion

$$Si + O_2 \to SiO_2$$

performs a reaction of this type.

For the reaction (17.2) the chemical affinity equals

$$A = n_- M_- \mu_- + n_* M_* \mu_* - n_+ M_+ \mu_+,$$

and the kinetic equation can be taken in the form

$$\omega = \overrightarrow{\omega} \left\{ 1 - \exp\left(-\frac{A}{RT} \right) \right\}, \tag{17.3}$$

where $\overrightarrow{\omega} = k_* c$ is the direct reaction partial rate, which, in the case of a reaction between gaseous and solid constituents, is proportional to the concentrations of reacting solid constituent, k_* is the kinetic coefficient (reaction rate parameter), $c = \rho_* / M_*$ is the molar concentration of the diffusive constituent (Glansdorff and Prigogine, 1971). Then it follows from the mass balance that the front velocity equals

$$W = \frac{n_- M_-}{\rho_-} k_* c \left\{ 1 - \exp\left(-\frac{A}{RT} \right) \right\},$$

where ρ_- is the mass density of the constituent B_+.

In the linear thermodynamic approach $\dfrac{A}{RT} \ll 1$, then

$$\omega = k_* c \frac{A}{RT}, \quad W = \frac{n_- M_-}{\rho_-} k_* c \frac{A}{RT}.$$

The above considerations were based on a scalar chemical potential and was not specified for deformable constituents. In the last decades of the twentieth century it was realized that in the case of solid phases the chemical potential was to be a tensor (see Grinfeld, 1991, and reference therein), and in the case of a moving interface, the jump of the normal component of the chemical potential tenor acts as a configurational (driving, thermodynamical) force (see Maugin, 1993; Wilmanski, 1998; Gurtin, 2000; Kienzler and Herrmann, 2000; Abeyaratne and Knowles, 2006; Maugin, 2010, and reference therein). Tensorial nature of the chemical potential also have been pointed out in Rusanov (2005, 2006).

For the case of a chemical reaction front these results were generalised in Freidin (2014) (see also Appendix in Freidin, 2014, 2015). It was derived from mass, linear momentum and energy balances and the entropy inequality written down for the case of the reaction between a diffusive constituent and solids of arbitrary rheology that the chemical affinity is a tensor. For nonlinear elastic solids the chemical affinity tensor was derived in Freidin (2009). Tensorial nature of the chemical affinity reflects the fact that the reaction takes place at oriented surface elements that pass through a point, as opposed to a reaction just at a point, and the reaction rate depends on the orientation of the surface element with respect to stresses. It was shown that the normal component A_{nn} of the chemical affinity tensor $\boldsymbol{A}$ arises in the expression of the entropy production (17.1) instead of the scalar affinity A.

The use of A_{nn} in the kinetic equation (17.3) allows to describe the influence of stresses and strains on the reaction rate by the thermodynamically sound way, as they affect the chemical affinity tensor. The approach based on the chemical affinity tensor was applied to a number of boundary value problems for linear elastic solids undergoing chemical reactions (Freidin et al, 2014, 2016a,b). It was demonstrated that stresses can accelerate, retard and even block the chemical reaction. In Freidin et al (2016b) the notion of a forbidden zone was introduced as a domain formed by the strains or the stresses at which the reaction cannot go at any admissible concentrations of the diffusive constituent. The aim of the present paper is further development of the forbidden zones notion and their construction.

17.2 Chemical Affinity in the Case of Small Strains

If to neglect the input of pressure of a diffusive constituent into total stresses in comparison with the stresses produced by a transformation strain and external loading, then the normal component of the affinity tensor takes the following form in the case of small strains Freidin et al (2014); Freidin (2015):

$$A_{nn} = \frac{n_- M_-}{\rho_-} \left(w_- - g^3 w_+ + \boldsymbol{\sigma}_\pm : [\![\boldsymbol{\varepsilon}]\!] \right) + n_* \left(\eta_* + RT \ln \frac{c}{c_*} \right),$$

where $w_\pm$ are the volume densities of the Helmholtz free energy of solid constituents, $\boldsymbol{\varepsilon}_\pm$ are the deformations at the reaction front, the stress tensors $\boldsymbol{\sigma}_\pm$ can be taken at any side of the reaction front, the parameter g is determined by the volume part of the transformation strain $\boldsymbol{\varepsilon}^{tr}$: if $\boldsymbol{\varepsilon}^{tr} = \varepsilon^{tr}\boldsymbol{I}$, then $g = 1 + \varepsilon^{tr}$; it is accepted that the chemical potential μ_* of the diffusive constituent is defined as

$$M_* \mu_* = \eta_* + RT \ln \frac{c}{c_*},$$

where c is the molar concentration of the diffusive constituent, c_* and η_* are the reference concentration and the chemical potential of B_*.

The reference concentration c_* can be taken as the concentration at the part $\Omega_* \subset \Omega$ of the outer surface Ω through which the diffusive constituent is supplied, or as the constituent B_* solubility in the transformed material B_-. Then two types of the boundary condition can be state at Ω_* for the diffusion problem: which prescribe the concentration at the outer surface or the flux through the outer surface:

$$c|_{\Omega_*} = c_*, \tag{17.4}$$

$$D\frac{\partial c}{\partial n}\bigg|_{\Omega_*} - \alpha(c_* - c|_{\Omega_*}) = 0 \tag{17.5}$$

where α is the surface mass transfer coefficient, the normal $\boldsymbol{n}$ to the reaction front is directed from "−" to "+".

Stresses are assumed to be related with elastic strains by Hooke's law:

$$\boldsymbol{\sigma}_- = \boldsymbol{C}_- : (\boldsymbol{\varepsilon}_- - \boldsymbol{\varepsilon}_-^*), \quad \boldsymbol{\sigma}_+ = \boldsymbol{C}_+ : (\boldsymbol{\varepsilon}_+ - \boldsymbol{\varepsilon}_+^*), \tag{17.6}$$

where $\boldsymbol{\varepsilon}_-^* = \boldsymbol{\varepsilon}_-^{in}$ is the inelastic (viscous, plastic) strain in the material B_-, $\boldsymbol{\varepsilon}_+^* = \boldsymbol{\varepsilon}^{ch} + \boldsymbol{\varepsilon}_+^{in}$ is the sum of the transformation strain $\boldsymbol{\varepsilon}^{ch}$ and inelastic strain $\boldsymbol{\varepsilon}_+^{in}$ in the material B_+, $\boldsymbol{C}_\pm$ are the stiffness tensors of the materials $B_\pm$.

From (17.6) it follows that the volume densities of the Helmholtz free energies of the materials $B_\mp$ are

$$w_- = \eta_-(T) + \frac{1}{2}(\boldsymbol{\varepsilon}_- - \boldsymbol{\varepsilon}_-^*) : \boldsymbol{C}_- : (\boldsymbol{\varepsilon}_- - \boldsymbol{\varepsilon}_-^*),$$

$$w_+ = \eta_+(T) + \frac{1}{2}(\boldsymbol{\varepsilon}_+ - \boldsymbol{\varepsilon}_+^*) : \boldsymbol{C}_+ : (\boldsymbol{\varepsilon}_+ - \boldsymbol{\varepsilon}_+^*),$$

where $\eta_\pm(T)$ are the free energy densities of $B_\pm$ in stress-free states (chemical energies). Then the normal component of the chemical affinity tensor takes the form:

$$A_{nn} = \frac{n_- M_-}{\rho_-}\left\{\gamma(T) + \frac{1}{2}\boldsymbol{\sigma}_- : (\boldsymbol{\varepsilon}_- - \boldsymbol{\varepsilon}_-^*) - \frac{1}{2}\boldsymbol{\sigma}_+ : (\boldsymbol{\varepsilon}_+ - \boldsymbol{\varepsilon}_+^*) + \boldsymbol{\sigma}_- : (\boldsymbol{\varepsilon}_- - \boldsymbol{\varepsilon}_-)\right\} + n_* RT \ln\frac{c}{c_*}, \tag{17.7}$$

where the temperature dependent parameter γ is determined by the chemical energies of the reaction constituents:

$$\gamma = \eta_- - g\eta_+ + \frac{\rho_-}{n_- M_-} n_* \eta_*.$$

Note that if the role of stresses is neglected than the reaction is allowed only if the chemical energies are such that $\gamma > 0$, and the stresses affect the chemical affinity and, thus, the reaction rate, only if the input of the strain energies of the constituents is compatible with the input of the chemical energies.

The normal component of the affinity tensor can be expressed via strains or stresses at one sides of the reaction front, as it was done earlier for the interfaces in the case of phase transformations (see, e.g., Morozov and Freidin, 1998; Freidin, 2007). Indeed, if the contacting materials have constitutive equations (17.6), then from the continuity of displacements and traction across the reaction front it follows that the jumps of strains and stresses can be calculated as (Kunin, 1983):

$$[\![\boldsymbol{\varepsilon}]\!] = -\boldsymbol{K}_{\mp}(\boldsymbol{n}):\boldsymbol{q}_{\pm}, \quad [\![\boldsymbol{\sigma}]\!] = \boldsymbol{S}_{\mp}(\boldsymbol{n}):\boldsymbol{m}_{\pm}, \tag{17.8}$$

where the tensors $\boldsymbol{q}$ and $\boldsymbol{m}$ are linear transformations of strains and stresses with coefficients equal to the jumps of the stiffness tensor in combination with inelastic strains:

$$\begin{gathered}
\boldsymbol{q}_{\pm} = [\![\boldsymbol{C}]\!]:\boldsymbol{\varepsilon}_{\pm} - [\![\boldsymbol{C}:\boldsymbol{\varepsilon}^*]\!], \quad \boldsymbol{m}_{\pm} = [\![\boldsymbol{B}]\!]:\boldsymbol{\sigma}_{\pm} + [\![\boldsymbol{\varepsilon}^*]\!], \\
\boldsymbol{K}_{\mp}(\boldsymbol{n}) = \{\boldsymbol{n}\otimes\boldsymbol{G}_{\mp}(\boldsymbol{n})\otimes\boldsymbol{n}\}^s, \; \boldsymbol{G}_{\mp}(\boldsymbol{n}) = (\boldsymbol{n}\cdot\boldsymbol{C}_{\mp}\cdot\boldsymbol{n})^{-1}, \\
\boldsymbol{S}_{\mp}(\boldsymbol{n}) = \boldsymbol{C}_{\mp}:\boldsymbol{K}_{\mp}(\boldsymbol{n}):\boldsymbol{C}_{\mp} - \boldsymbol{C}_{\mp}, \; \boldsymbol{B}_{\pm} = \boldsymbol{C}_{\pm}^{-1}.
\end{gathered}$$

Tensors $\boldsymbol{G}_{\mp}(\boldsymbol{n})$ are Fourier transformations of Green tensors for the determination of displacements in elastic medium B_- or B_+, that equal to the acoustic tensors, s denote the symmetrization with respect to the indexes in the first and second pair: $K_{ijkl} = n_{(i}G_{j)(k}n_{l)}$, further in the expressions the upper and lower indexes and the signs "+" and "−" correspond with each other.

From (17.7) and (17.8) it follows the formulae expressing the normal component of the chemical affinity tensor via strains and stresses on a side of the transformation front:

$$\begin{aligned}
A_{nn} &= \frac{n_- M_-}{\rho_-}\left\{\varphi_\varepsilon(\boldsymbol{\varepsilon}_-) + \frac{1}{2}\boldsymbol{q}_-:\boldsymbol{K}_+(\boldsymbol{n}):\boldsymbol{q}_-\right\} + n_* RT\ln\frac{c}{c_*} \\
&= \frac{n_- M_-}{\rho_-}\left\{\varphi_\varepsilon(\boldsymbol{\varepsilon}_+) - \frac{1}{2}\boldsymbol{q}_+:\boldsymbol{K}_-(\boldsymbol{n}):\boldsymbol{q}_+\right\} + n_* RT\ln\frac{c}{c_*} \\
&= \frac{n_- M_-}{\rho_-}\left\{\varphi_\sigma(\boldsymbol{\sigma}_-) + \frac{1}{2}\boldsymbol{m}_-:\boldsymbol{S}_+(\boldsymbol{n}):\boldsymbol{m}_-\right\} + n_* RT\ln\frac{c}{c_*} \\
&= \frac{n_- M_-}{\rho_-}\left\{\varphi_\sigma(\boldsymbol{\sigma}_-) - \frac{1}{2}\boldsymbol{m}_+:\boldsymbol{S}_-(\boldsymbol{n}):\boldsymbol{m}_+\right\} + n_* RT\ln\frac{c}{c_*},
\end{aligned} \tag{17.9}$$

where $\varphi_\varepsilon(\boldsymbol{\varepsilon})$ and $\varphi_\sigma(\boldsymbol{\sigma})$ are defined by the differences of the volume densities of the Helmholtz and Gibbs free energies, respectively,

$$\begin{gathered}
\varphi_\varepsilon(\boldsymbol{\varepsilon}) = w_-(\boldsymbol{\varepsilon}) - g^3 w_+(\boldsymbol{\varepsilon}) + \eta_*, \\
\varphi_\sigma(\boldsymbol{\sigma}) = \psi_-(\boldsymbol{\sigma}) - g^3 \psi_+(\boldsymbol{\sigma}) + \eta_*, \\
\psi_{\mp}(\boldsymbol{\sigma}) = w_{\mp}(\boldsymbol{\varepsilon}_{\mp}(\boldsymbol{\sigma})) - \boldsymbol{\sigma}:\boldsymbol{\varepsilon}_{\mp}(\boldsymbol{\sigma}), \quad \boldsymbol{\varepsilon}_{\mp}(\boldsymbol{\sigma}) = \boldsymbol{B}_{\mp}:\boldsymbol{\sigma} + \boldsymbol{\varepsilon}^*_{\mp}
\end{gathered}$$

and, thus,

$$\varphi_\varepsilon(\boldsymbol{\varepsilon}_\mp) = \gamma - \frac{1}{2}[\![\boldsymbol{\varepsilon}^* : \boldsymbol{C} : \boldsymbol{\varepsilon}^*]\!] - \frac{1}{2}\boldsymbol{\varepsilon}_\mp : [\![\boldsymbol{C}]\!] : \boldsymbol{\varepsilon}_\mp + \boldsymbol{\varepsilon}_\mp : [\![\boldsymbol{C} : \boldsymbol{\varepsilon}^*]\!], \tag{17.10}$$

$$\varphi_\sigma(\boldsymbol{\sigma}_\mp) = \gamma + \frac{1}{2}\boldsymbol{\sigma}_\mp : [\![\boldsymbol{B}]\!] : \boldsymbol{\sigma}_\mp + \boldsymbol{\sigma}_\mp : [\![\boldsymbol{\varepsilon}^*]\!]. \tag{17.11}$$

Note that $\varphi_\varepsilon(\boldsymbol{\varepsilon})$ and $\varphi_\sigma(\boldsymbol{\sigma})$ can be also expressed via $\boldsymbol{q}$ and $\boldsymbol{m}$:

$$\varphi_\varepsilon(\boldsymbol{\varepsilon}) = \gamma_* - \frac{1}{2}\boldsymbol{q} : [\![\boldsymbol{C}]\!]^{-1} : \boldsymbol{q} \equiv \varphi_q(\boldsymbol{q}), \quad \varphi_\sigma(\boldsymbol{\sigma}) = \gamma_* + \frac{1}{2}\boldsymbol{m} : [\![\boldsymbol{B}]\!]^{-1} : \boldsymbol{m} \equiv \varphi_m(\boldsymbol{m}). \tag{17.12}$$

If the reverse $[\![\boldsymbol{C}]\!]^{-1}$ exists then the strains and stresses in A_{rn} can be expressed via $\boldsymbol{q}$ and $\boldsymbol{m}$. Then the normal component A_{nn} takes the form of quadratic forms of $\boldsymbol{q}$ and $\boldsymbol{m}$:

$$\begin{aligned} A_{nn} &= \frac{n_- M_-}{\rho_-}\left\{\gamma_* - \frac{1}{2}\boldsymbol{q}_- : \left([\![\boldsymbol{C}]\!]^{-1} - \boldsymbol{K}_+(\boldsymbol{n})\right) : \boldsymbol{q}_-\right\} + n_* RT \ln\frac{c}{c_*} \\ &= \frac{n_- M_-}{\rho_-}\left\{\gamma_* - \frac{1}{2}\boldsymbol{q}_+ : \left([\![\boldsymbol{C}]\!]^{-1} + \boldsymbol{K}_-(\boldsymbol{n})\right) : \boldsymbol{q}_+\right\} + n_* RT \ln\frac{c}{c_*} \\ &= \frac{n_- M_-}{\rho_-}\left\{\gamma_* + \frac{1}{2}\boldsymbol{m}_- : \left([\![\boldsymbol{B}]\!]^{-1} + \boldsymbol{S}_+(\boldsymbol{n})\right) : \boldsymbol{m}_-\right\} + n_* RT \ln\frac{c}{c_*} \\ &= \frac{n_- M_-}{\rho_-}\left\{\gamma_* + \frac{1}{2}\boldsymbol{m}_+ : \left([\![\boldsymbol{B}]\!]^{-1} - \boldsymbol{S}_-(\boldsymbol{n})\right) : \boldsymbol{m}_+\right\} + n_* RT \ln\frac{c}{c_*}, \end{aligned} \tag{17.13}$$

$$\gamma_* = \gamma - \frac{1}{2}[\![\boldsymbol{\varepsilon}^*]\!] : [\![\boldsymbol{B}]\!]^{-1} : [\![\boldsymbol{\varepsilon}^*]\!].$$

The representations (17.9)–(17.13) define how A_{nn} and, thus, the reaction rate and front velocity, depend on strains and stresses at the front and the front surface element orientation.

17.3 Forbidden Zones

The direct reaction is possible only if $A_{nn} > 0$, i.e. only if

$$\varphi_\varepsilon(\boldsymbol{\varepsilon}_-) + \frac{1}{2}\boldsymbol{q}_- : \boldsymbol{K}_+(\boldsymbol{n}) : \boldsymbol{q}_- \geq \frac{n_* \rho_-}{n_- M_-} RT \ln\frac{c_*}{c} \tag{17.14}$$

and/or

$$\varphi_\varepsilon(\boldsymbol{\varepsilon}_+) - \frac{1}{2}\boldsymbol{q}_+ : \boldsymbol{K}_-(\boldsymbol{n}) : \boldsymbol{q}_+ \geq \frac{n_* \rho_-}{n_- M_-} RT \ln\frac{c_*}{c}. \tag{17.15}$$

Since $c < c_*$ at the direct reaction front, $\ln(c_*/c) > 0$, and the following inequalities are to be fulfilled in the case of the propagating front:

$$\varphi_\varepsilon(\boldsymbol{\varepsilon}_-) + \frac{1}{2}\mathcal{K}^+_{\max}(\boldsymbol{q}_-) \geq 0, \quad \text{and/or} \quad \varphi_\varepsilon(\boldsymbol{\varepsilon}_+) - \frac{1}{2}\mathcal{K}^-_{\min}(\boldsymbol{q}_+) \geq 0, \tag{17.16}$$

where

$$\mathcal{K}^\pm_{\max}(\boldsymbol{q}_\mp) = \max_{\boldsymbol{n}} \boldsymbol{q}_\mp : \boldsymbol{K}_\pm(\boldsymbol{n}) : \boldsymbol{q}_\mp, \quad \mathcal{K}^\pm_{\min}(\boldsymbol{q}_\mp) = \min_{\boldsymbol{n}} \boldsymbol{q}_\mp : \boldsymbol{K}_\pm(\boldsymbol{n}) : \boldsymbol{q}_\mp$$

If

$$\varphi_\varepsilon(\boldsymbol{\varepsilon}_-) + \frac{1}{2}\mathcal{K}^+_{\max}(\boldsymbol{q}_-) \leq 0 \tag{17.17}$$

and/or

$$\varphi_\varepsilon(\boldsymbol{\varepsilon}_+) - \frac{1}{2}\mathcal{K}^-_{\min}(\boldsymbol{q}_+) \leq 0 \tag{17.18}$$

then the direct reaction cannot go at any area elements and can not be started-up by any diffusion supply of the diffusive constituent.

By (17.13), the inequalities (17.17) and (17.18) can be written via tensor $\boldsymbol{q}$: if

$$\varphi_q(\boldsymbol{q}_-) + \frac{1}{2}\mathcal{K}^+_{\max}(\boldsymbol{q}_-) \leq 0, \quad \text{and/or} \quad \varphi_q(\boldsymbol{q}_+) - \frac{1}{2}\mathcal{K}^-_{\min}(\boldsymbol{q}_+) \leq 0 \tag{17.19}$$

then the reaction cannot go.

The reverse reaction can go only if $A_{nn} \leq 0$. Since the reverse reaction is accompanied by the extraction and diffusive withdrawal of the constituent B_*, the concentration at the reaction front is greater than the reference concentration c_*, $c > c_*$. Then $\ln(c/c_*) > 0$ at the reverse reaction front and

$$A_{nn} \leq \frac{n_- M_-}{\rho_-}\left\{\varphi_\varepsilon(\boldsymbol{\varepsilon}_\mp) \pm \frac{1}{2}\boldsymbol{q}_\mp : \boldsymbol{K}_\pm(\boldsymbol{n}) : \boldsymbol{q}_\mp\right\}. \tag{17.20}$$

Then if the reverse reaction front propagates then

$$\varphi_\varepsilon(\boldsymbol{\varepsilon}_\mp) \pm \frac{1}{2}\boldsymbol{q}_\mp : \boldsymbol{K}_\pm(\boldsymbol{n}) : \boldsymbol{q}_\mp = \varphi_\sigma(\boldsymbol{\sigma}_\mp) \pm \frac{1}{2}\boldsymbol{m}_\mp : \boldsymbol{S}_\pm(\boldsymbol{n}) : \boldsymbol{m}_\mp \leq 0. \tag{17.21}$$

Thus, the reverse reaction cannot go if

$$\varphi_\varepsilon(\boldsymbol{\varepsilon}_-) + \frac{1}{2}\mathcal{K}^+_{\min}(\boldsymbol{q}_-) \geq 0 \quad \text{and/or} \quad \varphi_\varepsilon(\boldsymbol{\varepsilon}_+) - \frac{1}{2}\mathcal{K}^-_{\max}(\boldsymbol{q}_+) \geq 0. \tag{17.22}$$

Obviously, conditions (17.22) can be reformulated in terms of stresses as well as via tensors $\boldsymbol{q}$ and $\boldsymbol{m}$.

Restrictive inequalities (17.17)–(17.19), (17.22) form domains in the strain space or q−space which further will be referred as *forbidden zones*. The boundaries of the domains (17.17)–(17.19) were constructed earlier in the context of the phase transition zones construction formed by the strains which can exist at equilibrium interfaces (Morozov and Freidin, 1998; Freidin et al, 2002; Freidin and Sharipova,

2006; Freidin, 2007). The difference is only in the definition of the parameter γ and the sign of the inequalities.

If materials B_+ and B_- are elastic and isotropic then

$$\boldsymbol{C}_\pm = \lambda_\pm \boldsymbol{I}\otimes\boldsymbol{I} + 2\mu_\pm^4\boldsymbol{I}, \quad \boldsymbol{\varepsilon}_+^* = (\vartheta^{ch}/3)\boldsymbol{I}, \quad \boldsymbol{\varepsilon}_-^* = 0,$$
$$\boldsymbol{K}_\pm(\boldsymbol{n}) = \frac{1}{\mu_\pm}(\boldsymbol{n}\otimes\boldsymbol{I}\otimes\boldsymbol{n})^s - \frac{1}{2\mu_\pm(1-\nu_\pm)}\boldsymbol{n}\otimes\boldsymbol{n}\otimes\boldsymbol{n}\otimes\boldsymbol{n},$$

where $\boldsymbol{I}$ and $^4\boldsymbol{I}$ are the identity tensors of the second and fourth rank, respectively; λ and μ are the Lamé's constants; ν is the Poisson's ratio. The maximal and minimal values of the quadratic form

$$\mathscr{K}(\boldsymbol{n}|\boldsymbol{q}) = \boldsymbol{q} : \boldsymbol{K}(\boldsymbol{n}) : \boldsymbol{n}$$

in the case of isotropic material are given by the following relationships (see, e.g., Freidin, 2007). Denote the minimal, maximal and intermediate eigenvalues of the tensor $\boldsymbol{q}$ by $q_{\min}$, $q_{\max}$ and q_{mid}, and the minimal and maximal absolute values of the eigenvalues of $\boldsymbol{q}$ by $|q|_{\min}$ and $|q|_{\max}$, respectively. Let $\boldsymbol{e}_{\min}$, $\boldsymbol{e}_{\max}$ and $\boldsymbol{e}_{\mathrm{mid}}$, $\boldsymbol{e}_{|q|\min}$ and $\boldsymbol{e}_{|q|\max}$ be the corresponding eigenvectors of $\boldsymbol{q}$, and $n_{\min}$, $n_{\max}$ and n_{mid} be the components of the normal $\boldsymbol{n}$ in the basis $\boldsymbol{e}_{\min}, \boldsymbol{e}_{\max}, \boldsymbol{e}_{\mathrm{mid}}$. Then if

$$q_{\min}q_{\max} < 0 \quad \text{or} \quad \begin{cases} q_{\min}q_{\max} > 0 \\ (1-\nu)|q|_{\min} < \nu|q|_{\max} \end{cases} \tag{17.23}$$

then the normal lies in the plane of maximal and minimal eigenvalues of the tensor $\boldsymbol{q}$, $n_{\mathrm{mid}} = 0$, and

$$n_{\max}^2 = \frac{(1-\nu)q_{\max} - \nu q_{\min}}{q_{\max} - q_{\min}}, \quad n_{\min}^2 = \frac{\nu q_{\max} - (1-\nu)q_{\min}}{q_{\max} - q_{\min}}, \tag{17.24}$$

$$\mathscr{K}_{\max}(\boldsymbol{q}) = \frac{1-\nu}{2\mu}(q_{\max}^2 + q_{\min}^2) - \frac{\nu}{\mu}q_{\max}q_{\min}. \tag{17.25}$$

otherwise

$$\boldsymbol{n} = \boldsymbol{e}_{|q|\max}, \tag{17.26}$$
$$\mathscr{K}_{\max}(\boldsymbol{q}) = \frac{1-2\nu}{2\mu_-(1-\nu)}|q|_{\max}^2. \tag{17.27}$$

The minimal value of $\boldsymbol{q} : \boldsymbol{K}(\boldsymbol{n}) : \boldsymbol{q}$ and the corresponding normal are given by

$$\boldsymbol{n} = \boldsymbol{e}_{|q|\min}, \qquad \mathscr{K}_{\min}(\boldsymbol{q}) = \frac{1-2\nu_+}{2\mu_+(1-\nu_+)}|q|_{\min}^2. \tag{17.28}$$

The cross-sections of forbidden zones by the plane $\varepsilon_1 = \varepsilon_2$ in strain space and in q-space are represented in Fig. 17.2 for $\gamma > 0$ and in Fig. 17.3 for $\gamma < 0$. The lines

$\omega_- = \omega'_- \cup \omega''_-$ correspond to the equation

$$\varphi_q(\boldsymbol{q}_-) + \frac{1}{2}\mathscr{K}^+_{\max}(\boldsymbol{q}_-) = 0$$

with $\mathscr{K}^+_{\max}(\boldsymbol{q}_-)$ and the normal $\boldsymbol{n}$ given by (17.25) and (17.24) (the solid parts ω'') or given by (17.27) and (17.26) (the dotted parts ω') with $\nu = \nu_+$ and $\boldsymbol{q} = \boldsymbol{q}_-$.

Analogously, the lines $\omega_+ = \omega'_+ \cup \omega''_+$ correspond to the equation

$$\varphi_q(\boldsymbol{q}_+) - \frac{1}{2}\mathscr{K}^-_{\max}(\boldsymbol{q}_+) = 0.$$

The lines β_+ and β_- correspond to the equations

$$\varphi_q(\boldsymbol{q}_+) - \frac{1}{2}\mathscr{K}^-_{\min}(\boldsymbol{q}_+) = 0 \quad \text{and} \quad \varphi_q(\boldsymbol{q}_-) - \frac{1}{2}\mathscr{K}^+_{\min}(\boldsymbol{q}_-) = 0$$

with $\mathscr{K}^\mp_{\min}(\boldsymbol{q}_\pm)$ and the normal given by (17.28). The lines ξ correspond to

$$\varphi_\varepsilon(\boldsymbol{\varepsilon}) = \varphi_q(\boldsymbol{q}) = 0.$$

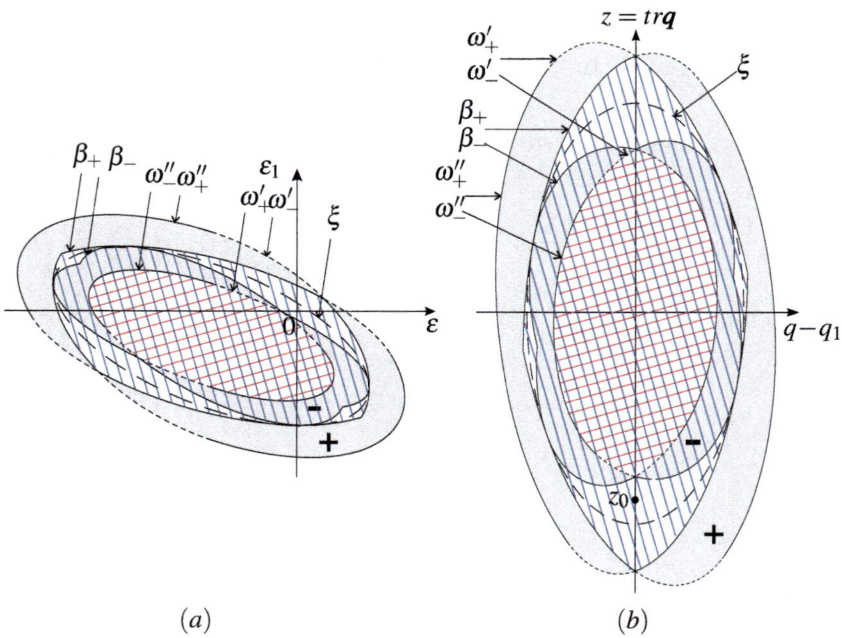

Fig. 17.2: The cross-sections of the forbidden zones by the planes (a) $\varepsilon_2 = \varepsilon_3 = \varepsilon$ and (b) $q_2 = q_3 = q$ at $\gamma = 0.03$, $K_+ = 39$, $K_- = 78$, $\mu_+ = 15$, $\mu_- = 30$, $\vartheta^{tr} = 0.015$, the point z_0 corresponds to $\varepsilon_1 = \varepsilon_2 = \varepsilon_3 = 0$.

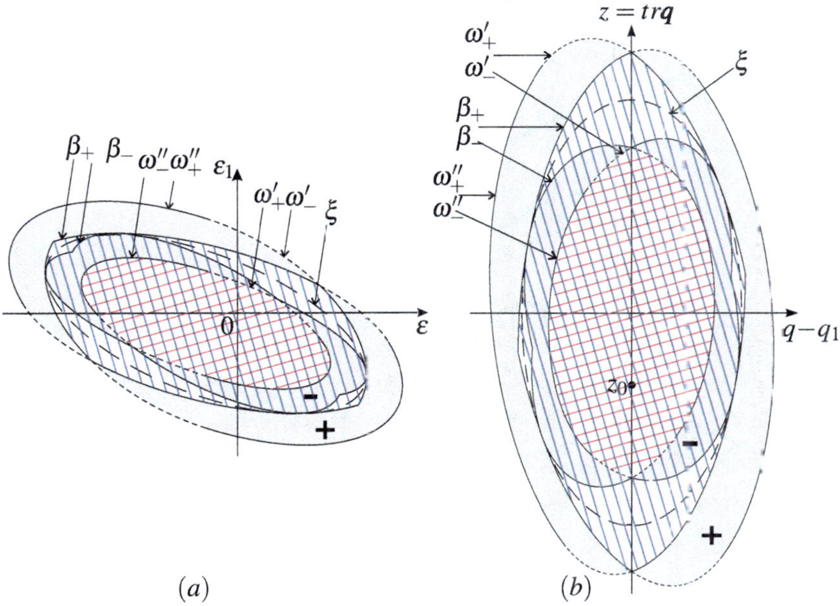

Fig. 17.3: The cross-sections of the forgbiden zones by the planes (a) $\varepsilon_2 = \varepsilon_3 = \varepsilon$ and (b) $q_2 = q_3 = q$ at $\gamma = -3$, $K_+ = 39$, $K_- = 78$, $\mu_+ = 15$, $\mu_- = 30$, $\vartheta^{tr} = 0.015$. the point z_0 corresponds to $\varepsilon_1 = \varepsilon_2 = \varepsilon_3 = 0$.

The forbidden zone is inside the line ω_- for the strains at the side "−" of the reaction front (red hatching) and inside the line β_+ for the strains at the side "+" of the front (blue hatching). One can see that reaction cannot go at $\gamma < 0$ without straining but can start at proper loading.

The lines ω which bound the forbidden zones for strain "−" are shown for various parameters γ in Fig. 17.4. One can see that if γ increases than the forbidden zone decreases and the reaction can be blocked only by strains from a relatively small domain. The forbidden zone disappears at further increase of gamma If γ decreases then the both effects become essential: from some value of γ the reaction cannot go without additional loading, at other strains the direct reaction in blocked

As final remarks, note that the forbidden zones reflect the influence not only of the values of strains and stresses, but also the influence of the strain state. The size and the position of the forbidden zone depend also on the strain $\boldsymbol{\varepsilon}^*$ which may be considered as a constant transformation strain if the solid constituents are elastic, but may depend on time in the case of viscoelastic or viscoelastoplastic constituents. In this case forbidden zones will evolve with time and, particularly, an arrested reaction can be unlocked.

Acknowledgements Gérard Maugin was one of those who determined the face of modern mechanics, and Alexander Freidin was privileged to communicate with him and to have a very supporting and encouraging discussions. Leah Sharipova spent six months in Gérard Maugin's laboratory

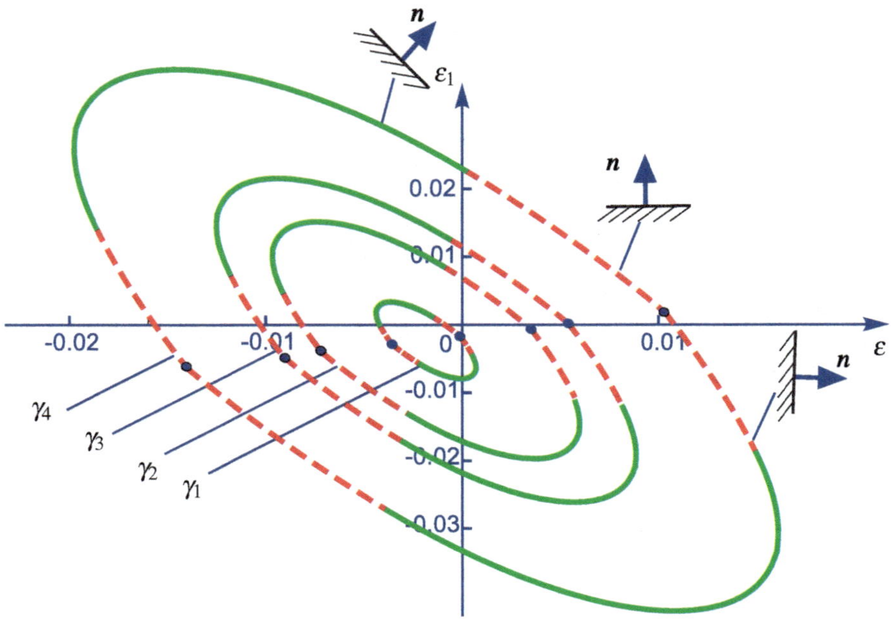

Fig. 17.4: Locking strains $\boldsymbol{\varepsilon}_-$ for direct reactions. Cross-sections $\varepsilon_2 = \varepsilon_3 \equiv \varepsilon$ of forbidden zones at various γ: $\gamma_1 = 0.008 > \gamma_2 = 0.001 > \gamma_3 = 0 > \gamma_4 = -0.03$.

where she learned important lessons of a deep scientific research. She will always remember his paternal relation. The subject of the present paper reflects the discussions of those times and now, with feeling deep sorrow, we dedicate this paper to memory of Gérard Maugin.

This work was supported by the Russian Foundation for Basic Research (Grants No 16-01-00815, No 17-51-12055).

References

Abeyaratne R, Knowles J (2006) Evolution of Phase Transitions. A Continuum Theory. Cambridge University Press, Cambridge

Bower AF, Guduru PR, Chason E (2015) Analytical solutions for composition and stress in spherical elastic-plastic lithium-ion electrode particles containing a propagating phase boundary. International Journal of Solids and Structures 69-70:328 – 342

Brassart L, Suo Z (2013) Reactive flow in solids. Journal of the Mechanics and Physics of Solids 61(1):61 – 77

Cui Z, Gao F, Qu J (2012) A finite deformation stress-dependent chemical potential and its applications to lithium ion batteries. Journal of the Mechanics and Physics of Solids 60(7):1280 – 1295

Deal BE, Grove AS (1965) General relationship for the thermal oxidation of silicon. Journal of Applied Physics 36(12):3770–3778

Freidin A (2014) Chemical affinity tensor and stress-assist chemical reactions front propagation in solids. In: ASME 2013 International Mechanical Engineering Congress and Exposition, November 13-21, 2013, San Diego, California, USA

Freidin A, Morozov N, Petrenko S, Vilchevskaya E (2016a) Chemical reactions in spherically symmetric problems of mechanochemistry. Acta Mechanica 227(1):43–56

Freidin AB (2007) On new phase inclusions in elastic solids. ZAMM 87(2):102–116

Freidin AB (2009) On chemical reaction fronts in nonlinear elastic solids. In: Indeitsev D, Krivtsov AM (eds) Proceedings of XXXVII International Summer School-Conference Advanced Problems in Mechanics, pp 231–237

Freidin AB (2015) On the chemical affinity tensor for chemical reactions in deformable materials. Mechanics of Solids 50(3):260–285

Freidin AB, Sharipova LL (2006) On a model of heterogenous deformation of elastic bodies by the mechanism of multiple appearance of new phase layers. Meccanica 41(3):321–339

Freidin AB, Vilchevskaya EN, Sharipova LL (2002) Two-phase deformations within the framework of phase transition zones. Theoretical and Applied Mechanics 28-29:149–172

Freidin AB, Vilchevskaya EN, Korolev IK (2014) Stress-assist chemical reactions front propagation in deformable solids. International Journal of Engineering Science 83:57 – 75

Freidin AB, Korolev IK, Aleshchenko SP, Vilchevskaya EN (2016b) Chemical affinity tensor and chemical reaction front propagation: theory and fe-simulations. International Journal of Fracture 202(2):245–259

Glansdorff P, Prigogine I (1971) Thermodynamic Theory of Stability and Fluctuation. Wiley-Interscience, New York

Grinfeld M (1991) Thermodynamic Methods in the Theory of Heterogeneous Systems. Longman, New York

Gurtin M (2000) Configurational Forces as Basic Concepts of Continuum Physics. Springer, New York

Jia Z, Li T (2015) Stress-modulated driving force for lithiation reaction in hollow nano-anodes. Journal of Power Sources 275:866 – 876

Kao DB, McVittie JP, Nix WD, Saraswat KC (1988) Two-dimensional thermal-oxidation of silicon – II. modeling stress effects in wet oxides. IEEE Transactions On Electron Devices 35(1):25 – 37

Kienzler R, Herrmann G (2000) Mechanics in Material Space with Application to Defect and Fracture Mechanics. Springer, Berlin

Knyazeva AG (2003) Cross effects in solid media with diffusion. Journal of Applied Mechanics and Technical Physics 44(3):373–384

Kunin I (1983) Elastic Media with Microstructure. Springer, Berlin

Levitas VI, Attariani H (2014) Anisotropic compositional expansion in elastoplastic materials and corresponding chemical potential: Large-strain formulation and application to amorphous lithiated silicon. Journal of the Mechanics and Physics of Solids 69:84 – 111

Loeffel K, Anand L (2011) A chemo-thermo-mechanically coupled theory for elastic-viscoplastic deformation, diffusion, and volumetric swelling due to a chemical reaction. International Journal of Plasticity 27(9):1409 – 1431

Maugin G (1993) Material Inhomogeneities in Elasticity. Chapman & Hall, London

Maugin G (2010) Configurational Forces. Thermomechanics, Physics, Mathematics, and Numerics. CRC, Boca Raton

Morozov NF, Freidin AA (1998) Phase transition zones and phase transformations of elastic solids under different stress states. Proc Steklov Mathe Inst 223:220 – 232

Prigogine I, Defay R (1988) Chemical Thermodynamics. Longmans, Green, London

Rafferty CS (1990) Stress effects in silicon oxidation - simulation and experiments. PhD thesis, Stanford University

Rao VS, Hughes TJR (2000) On modelling thermal oxidation of Silicon I: theory. International Journal for Numerical Methods in Engineering 47(1-3):341–358

Rao VS, Hughes TJR, Garikipati K (2000) On modelling thermal oxidation of Silicon II: numerical aspects. International Journal for Numerical Methods in Engineering 47(1-3):359–377

Rusanov AI (2005) Surface thermodynamics revisited. Surface Science Reports 58(5):111 – 239

Rusanov AI (2006) Thermodynamic Foundations of Mechanochemistry. Nauka, St. Petersburg
Sutardja P, Oldham WG (2005) Modeling of stress effects in silicon oxidation. IEEE Transactions on Electron Devices 36(11):2415 – 2421
Toribio J, Kharin V, Lorenzo M, Vergara D (2011) Role of drawing-induced residual stresses and strains in the hydrogen embrittlement susceptibility of prestressing steels. Corrosion Science 53(10):3346 – 3355
Wilmanski K (1998) Thermomechanics of Continua. Springer, Berlin

Chapter 18
Generalized Debye Series Theory for Acoustic Scattering: Some Applications

Alain Gérard

Abstract The problem of propagation of acoustic waves in the presence of a submerged elastic object of cylindrical or spherical shape is formulated here in a unified fashion. With this general formalism is associated a synthetic procedure of resolution of the continuity conditions for the fields at the interfaces. This procedure, based on the operator concept, leads in a simple and direct fashion to the generalized Debye series for the exact solution of wave propagation in the case of separable geometries with the acoustic source either inside or ouside to the scatterer. For cylindrical and spherical geometries, this result, which we term the "Generalized Debye Series Theory" (GDST), is exploited for various applications and appears as a complementary contribution to the Resonance Scattering Theory (RST) as established by H. Überall et al.

18.1 Introduction

Determining the exact solution of the propagation of elastic waves in the presence of a cylindrical or spherical object imbedded in another medium (fluid or elastic) is a problem that has been solved mathematically since a long time (Sezawa, 1927). However, only the last thirty years the power of electronic computers has permitted to start exploiting this solution, with numerous applications in seismology, nondestructive testing, underwater acoustics etc. This solution, which stems from satisfying the boundary (continuity) conditions, is usually obtained by solving Cramer's system of algebraic equations and appears in the form of a ratio of two determinants whose elements are linear combinations of Bessel functions and their derivatives. Apart from the complexity of this result, whose exploitation necessitates the use of a calculator, such a method masks the physical interpretation of the solution.

Alain Gérard
Bordeaux University, I2M-MPI, CNRS, UMR5295, 33600 Pessac, France
e-mail: ajr.gerard@gmail.com

© Springer International Publishing AG, part of Springer Nature 2018
H. Altenbach et al. (eds.), *Generalized Models and Non-classical Approaches in Complex Materials 1*, Advanced Structured Materials 89,
https://doi.org/10.1007/978-3-319-72440-9_18

Consequently, one of the major problems posed by the analysis of acoustic scattering consists in the difficulty of reaching, in all cases, at a physical interpretation, which constitutes the most interesting aspect of the phenomenon. The study of scattering often takes its starting point from a modal series of the normal modes of vibration. It is clear that the comprehension of a physical problem is intimately linked to the modeling of the problem. The modal series constitutes a typical example where the mathematical model does provide a global image of the physical situation under consideration, but where at the same time it hides most of the fine details that could be observed in the scattering process. Thus, for example, the modal series at first sight does not exhibit the multiple reflections and refractions that take place inside the scatterer.

It may be overly ambitions, in general, to expect that an exact mathematical solution should immediately reflect all the elementary mechanisms that one would like to understand, but it appears desirable to single out, as soon as possible, the elements of the analysis that correspond to the type of interpretation that is being looked for. However, numerous studies on the scattering of electromagnetic waves by spheres and cylinders have shown to permit appropriate techniques for this type of scattering problem for scalar fields.

Accordingly, since the work of Debye (1908), and of van-der-Pol and Bremmer (1937), on cylindrical and spherical objects, we know that for a scalar field the exact solution for the propagation of waves exterior to an obstacle can be expanded in a series of an elementary[1] waves (the Debye series) expressed by reflection and refraction coefficients. In particular, the Debye series has led to advances such as the understanding of the rainbow phenomenon, which is created by multiple reflections and refractions of the light rays by water droplets suspended in the atmosphere Nussenzweig (1969a,b).

In acoustics, the presence of a vector field in the description of wave propagation in elastic objects complicates the analysis to some extent (Knopoff, 1959). The reflection-refraction phenomena are accompanied by transitions from longitudinal to transverse modes and vice versa. This possibility of mode conversion, which does not exist in electromagnetics, renders an expansion into Debye series more difficult for elastic media. In spite of this, a partial wave expansion has been obtained in the exterior region for a fluid sphere enclosed in an elastic space by Scholte (1956), and independently by Nagase (1956), by a division of the determinants contained in the exact solution of the problem. The opposite case (elastic object imbedded in a fluid) has been solved for the cylindrical geometry by Brill and Überall (1970, 1971). Their methods consist in separating the exact solution in the fluid into two terms the first of which represents the specular reflection on the cylinder. The other term is expanded, by using the binomial theorem, in a series in which the various elements are subsequently identified as the reflection and refraction coefficients of the cylinder. The application of these various methods is complicated, representing a considerable task that could not be envisaged for multilayered objects. However, for many experimental problems it would be desirable to have a physical interpretation

[1] No partial waves because each term are an elementary solution of the boundary conditions (continuity) at each interface of the scatterer.

of the exact solution available. Another method has been proposed (Gérard, 1979, 1980, 1983) which, starting out from a matrix notation, uses directly the notion of operators and linear algebra. It furnishes the exact solution in all the layers of an elastic multilayered sphere in terms of a generalized Debye series (Gérard, 1987). This method uses the concept of outgoing and ingoing waves together with a particular organization of the system of continuity conditions for the fields at the various interfaces of the model in terms of an appropriate matrix representation. This method has also been used for the case of plates (Conoir et al, 1991; Gérard, 1987), or of the solid elastic or fluid (non multilayered) cylinder immersed in a fluid (Conoir, 1988; Derem, 1982). In all these applications, a judicious choice of notation leads to a form of solution that is independent of the geometry of the model under consideration (cylindrical or spherical). Only the analytic expressions for the reflection and transmission coefficients are specific to the geometry considered. One may thus consider it possible to formulate and solve in a unified fashion the problem of the diffraction of a plane wave by an elastic sphere or cylinder (of circular cross section) immersed in a fluid.

In the present review, we shall first show that our approach can be formulated for the case of submerged objects of linear behavior (whether elastic or not), as long as their surfaces are cylinders of circular cross section, or spheres. We then develop our method for obtaining the generalized Debye series. This method, which we call the GDST (Generalized Debye Series Theory), is subsequently explored in various ways for different examples of targets insonified by external or internal sources. The GDST then appears as a useful complement of the RST (Resonance Scattering Theory) developed by Flax et al (1978); Fiorito et al (1979); Murphy et al (1979a).

18.2 Generalized Debye Series

18.2.1 Formulation of the Problem

The scattering object is assumed to be bounded by a surface that is part of the coordinate surface of a curvilinear system. If the object is multilayered, the interfaces of the layering are also assumed surfaces of the same curvilinear system.

The curvilinear coordinates are designated ξ, η, ζ where ξ is along the exterior normal $\vec{n} = (\xi, 0, 0)$. We assume that source and receiver are located in a plane normal to the direction ζ, which is an axis of revolution of the problem. According to the geometry considered, we have $\xi = r$, $\eta = \theta$, and $\zeta = z$ for the cylinder (circular cross section of radius r and of axis $\zeta = z$), and $\xi = r, \eta = \theta$, $\zeta = \varphi$ for the sphere centered at the origin of coordinates. For the case most frequently treated, and in order to simplify, we assume the source to be exterior to the obstacle. However, the present theory is equally valid if the source is interior to the object, as we shall see later on. For that latter case, an application to the elastic cylinder imbedded in air is given in Sect. 18.6.

We assume that the object, for which we defined the geometry, has linear and isotropic elastic behavior. For the developments to follow, we note that linear elastic behavior plays a fundamental role, while the isotropic character is not essential, but is chosen here solely in order to simplify the calculations. We shall utilize the potential theory of Helmholtz, which for the elastic medium 2 permits to decompose the displacement field (or velocity field) in two potentials (Pao and Mow, 1973). The scalar potential ϕ (or vectorial potential $\overrightarrow{\chi}$) is associated with longitudinal (transverse) waves which propagate in the (ξ,η) plane at speed Cp (Cs). From the basic laws of dynamics and of the behavior of medium 2, these potentials satisfy, respectively, the decoupled wave equations:

$$C_p{}^2\Delta\phi - \frac{\partial^2\phi}{\partial t^2} = 0, \quad C_s{}^2\Delta\chi_r - \frac{\partial^2\chi_r}{\partial t^2} = 0, \tag{18.1}$$

where χ_r designates the radial component of the potential vector.

The modal solution of these equations is then searched for by the classical method of separation of variables. The elementary solutions of order n (n being integer) are obtained as the product $f_n(\xi)g_n(\eta)$ of two functions of a real variable. For any of our chosen geometries (cylinder on sphere), $f_n(\xi)$ is the Hankel function of first on second kind of order n for the cylinder, and of order $n+1/2$ for the sphere. In contrast, for the cylindrical geometry $g_n(\eta)$ is a trigonometric function ($sinn\theta$ or $cosn\theta$) while it is a Legendre polynomial $P_n(cos\theta)$ for the sphere. With a time dependence $exp(-i\omega t)$ of the fields where ω is the circular frequency, the concept of ingoing (or convergent) waves is associated with the Hankel function of second kind, $H^{(2)}$, and that of outgoing (or divergent) waves with the Hankel function of first kind, $H^{(1)}$ (van-der-Pol and Bremmer, 1937; Nussenzweig, 1969a,b). To simplify, we shall use a subscript + (or -) with all quantities associated with outgoing waves proportional to $H^{(1)}$ (or ingoing waves proportional to $H^{(2)}$, respectively), such that the two independent elementary solutions can be written as $f_n^{\pm}(\xi)g_n(\eta)$.

Thus, the modal decomposition of the potential associated with an incident plane wave, or with a longitudinal wave interior to the obstacle, is of the form:

$$\phi = \sum_{n=0}^{\infty}(a_n\phi_n^+ + b_n\phi_n^-)e^{-\omega t}, \quad \phi_n^{\pm} = f_n^{\pm}(\xi)g_n(\eta), \tag{18.2}$$

where the coefficients b_n (a_n) are known (unknown) for the incident wave (the wave reemitted by the obstacle), while these same coefficients a_n and b_n are unknown for the waves propagating in the interior of the obstacle. These coefficients are determined by solving the system arising from the boundary conditions, which express the continuity of the components of displacement and of the stress vector. These quantities are obtained starting from the expression of type Eq. (18.2) by linear derivative operations, which conserve the structure of Eq. (18.2) for the components of displacement and stress. More precisely, we call medium 1 the ambient fluid and medium 2 the submerged elastic object, and then the scalar potential at the boundary of medium 1 is:

$$\phi_1 = \sum_{n=0}^{\infty} (\phi_1^-(1) + a_n \phi_n^+(1)), \tag{18.3}$$

where $1 = k_1 \xi_0$ k_1 being the wave number in medium 1 and ξ_0 the racius of the surface of the object. On the boundary of medium 2, denoting for simplicity χ_2 as the radial component of the vector potential of medium 2, the two potentials necessary for the description of the displacements and stresses are:

$$\phi_2 = \sum_{n=0}^{\infty} (\phi_n^-(p) + \phi_n^+(p)) b_n, \quad \chi_2 = \sum_{n=0}^{\infty} (\chi_n^-(s) + \chi_n^+(s)) c_n, \tag{18.4}$$

where, for the p and s waves coming from medium 2 we have set $q = kq\xi_0$, kq being the wave number of the wave with polarization q ($q = p, s$). Incidentally, there appears in Eq. (18.4) only one constant to be determined for ϕ_2, and also for χ_2, because we have taken into account the fact that the fields must remain finite at $\xi = 0$. This condition is often interpreted as the passage of the waves through a focal point of the object, or also as the reflection of the waves at $\xi = 0$. We shall designate by $\vec{u}^{(j)}$ the displacement field in medium j ($j = 1, 2$). Using the theorem of Helmholtz (Morse and Feshbach, 1953), the structure of ingoing on outgoing waves is conserved under the operators $\overrightarrow{grad}$ and $\overrightarrow{rot}$ in a way that the displacements in each of the media are given by:

$$u_\xi^{(1)} = u_1^-(1) + u_1^+(1) a_n, \tag{18.5}$$

in the fluid,

$$u_\xi^{(2)} = [u_2^-(p) + u_2^+(p)] b_n + [u_2^-(s) + u_2^+(s)] c_n \tag{18.6}$$

in the solid. Likewise, the law of the behavior being linear elastic, the components of the stress tensor $\sigma_{lm}^{(j)}$ for the medium j ($j = 1, 2$) are obtained by linear combinations of derivatives of the expressions Eqs. (18.3) and (18.4). We thus obtain in a similar fashion the nonvanishing components of the stress at the interface in the form:

$$\sigma_{\xi\xi}^{(1)} = \sigma_{\xi\xi}^-(1) + \sigma_{\xi\xi}^+(1) a_n, \quad \sigma_{\xi n}^{(1)} = 0 \tag{18.7}$$

linear in a_n in the fluid,

$$\sigma_{\xi x}^{(2)} = [\sigma_{\xi x}^-(p) + \sigma_{\xi x}^+(p)] b_n + [\sigma_{\xi x}^-(s) + \sigma_{\xi x}^+(s)] c_n, \quad x = \xi, \eta \tag{18.8}$$

linear in b_n and c_n in the solid. Thus, the system of boundary conditions at the fluid-solid interface results as :

$$u_\xi^{(1)} - u_\xi^{(2)} = 0, \quad \sigma_{\xi\xi}^{(1)} - \sigma_{\xi\xi}^{(2)} = 0, \quad \sigma_{\xi\eta}^{(2)} = 0. \tag{18.9}$$

From this, the operator (which, below, we shall designate by $\underline{D}$), which represents the application of the boundary conditions that contain the elements of Eq. (18.9), linear in a_n b_n and c_n, is easily obtained via Eqs. (18.5) and (18.6) and (18.7 and 18.8). Since these boundary conditions must be satisfied for each mode n separately, we may suppress the index n at all the constants and write $a_n \to a$ etc. This leads to the following concise formulation of the boundary conditions:

$$\vec{f}^{\,-}(1) + \underline{D}\vec{d} = 0 \Leftrightarrow \vec{f}^{\,-}(1) + \underline{D}_L\vec{d} + \vec{f}^{\,+}(p)b + \vec{f}^{\,+}(s)c = 0 \quad (18.10)$$

where the following notations were used:

$$\vec{d} = (a,b,c)^t, \vec{f}^{\,-}(1) = (u_\xi^-(1), \sigma_{\xi\xi}^-(1), 0)^t, \vec{f}^{\,\pm}(q) = (-u_\xi^+(q), -\sigma_{\xi\xi}^+(q), \sigma_{\xi\eta}^+(q))^t$$

and

$$\underline{D} = \begin{pmatrix} u_\xi^+(1) & -u_\xi^-(p) - u_\xi^+(p) & -u_\xi^-(s) - u_\xi^+(s) \\ \sigma_{\xi\xi}^+(1) & -\sigma_{\xi\xi}^-(p) - \sigma_{\xi\xi}^+(p) & -\sigma_{\xi\xi}^-(s) - \sigma_{\xi\xi}^+(s) \\ 0 & \sigma_{\xi\eta}^-(p) + \sigma_{\xi\eta}^+(p), & \sigma_{\xi\eta}^-(s) + \sigma_{\xi\eta}^+(s) \end{pmatrix} \quad (18.11)$$

In these expressions, the subscript "t" designates the transpose, $\vec{f}^{\,-}(1)$ is the incident field and $\underline{D}_L$ the matrix of local interactions (hence the index L) which deduce from Eq. (18.11) substituting column 2 and 3 by $\vec{f}^{\,-}_{(p)}$ and $\vec{f}^{\,-}_{(s)}$ respectively. We introduce the column matrix $\underline{C}$; so the two last terms in Eq. (18.10) can be combined in the simple expression

$$\underline{C} = (\vec{0}, \vec{f}^{\,(+)}(p), \vec{f}^{\,(+)}(s)) \quad \Rightarrow \quad \underline{C}\vec{d} = (\vec{0}, \vec{f}^{\,(+)}(p)b, \vec{f}^{\,(+)}(s)c) \quad (18.12)$$

The global system of boundary conditions Eq. (18.7) is thus written, employing these operators:

$$\vec{f}^{\,-}(1) + (\underline{D}_L + \underline{C})\vec{d} = 0 \quad \text{with} \quad \underline{D} = \underline{D}_L + \underline{C}. \quad (18.13)$$

In the general case of multilayered media one can show that the matrix operator of local interactions $\underline{D}_L$ (of block-diagonal form) dissociates the interfaces while the matrix operator $\underline{C}$ (simultaneously block and banded) represents the coupling; hence the notations $\underline{D}$ and $\underline{C}$. Before solving the "global" system of boundary conditions Eq. (18.10) and Eq. (18.13), we shall first introduce the "local" modal reflection and refraction coefficients.

18.2.2 "Local" Modal Reflection and Refraction Coefficients

18.2.2.1 Reflection and Refraction of a Wave Incident from Medium 1 (Fluid) on Medium 2 (Solid)

In a general fashion, we denote by R_{ij}^{qv} (T_{ij}^{qv}) the reflection (transmission) coefficient for a wave of polarization q (q = p, s) incident from medium i on medium j, which undergoes mode conversion v (v = p, s). The incident field coming from medium 1, the potentials in each domain are expressed on the surface of the object $\xi = \xi_0$ by

$$\phi_1 = \phi_1^- + R_{12}^{pp}\phi_1^+(1), \quad \phi_2 = T_{12}^{pp}\phi_2^-(p), \quad \chi_2 = T_{12}^{ps}\chi_2^-(s), \tag{18.14}$$

these expressions containing the local reflection R_{12}^{pp} and refraction coefficient T_{12}^{pp}. The subscripts designate, successively, the nature of the incident wave, followed by that of the reflected (or transmitted) wave. Writing the boundary conditions for the fields of Eq. (18.14) leads to:

$$\begin{aligned} u_\xi^-(1) + u_\xi^+(1)R_{12}^{pp} - u_\xi^-(p)T_{12}^{pp} - u_\xi^-(s)T_{12}^{ps} &= 0, \\ \sigma_{\xi\xi}^-(1) + \sigma_{\xi\xi}^+(1)R_{12}^{pp} - \sigma_{\xi\xi}^-(p)T_{12}^{pp} - \sigma_{\xi\xi}^-(s)T_{12}^{ps} &= 0, \\ \sigma_{\xi\eta}^-(1)T_{12}^{pp} + \sigma_{\xi\eta}^-(s)T_{12}^{ps} &= 0. \end{aligned} \tag{18.15}$$

This system of boundary conditions becomes, from Eqs. (18.10) and (18.11):

$$\vec{f}^-(1) + \underline{D}_L \vec{a}_L = 0, \vec{a}_L = (R_{12}^{pp}, T_{12}^{pp}, T_{12}^{ps})^t. \tag{18.16}$$

The reflection and refraction coefficients are thus given by a simple inversion of relation Eq. (18.16). The analytic expression of these coefficients is given in Conoir and Gérard (1989) for the elastic cylinder, and in Gérard (1987) for the elastic sphere.

18.2.2.2 Reflection and Refraction of Wave Incident from Medium 2 on Medium 1

In the case of a p wave incident on medium 1, the potentials associated with waves propagating in each of the media are expressed at the interface by:

$$\phi_1 = \phi_1^+(1)T_{21}^{pp}, \quad \phi_2 = \phi_2^+(p) + \phi_2^-(p)R_{21}^{pp}, \quad \chi_2 = \chi_2^-(s)R_{21}^{ps}, \tag{18.17}$$

We introduce the notation $\vec{b}^p = (T_{21}^{pp}, R_{21}^{pp}, R_{21}^{ps})^t$ so that the boundary conditions across the interface ξ_0 can be written as:

$$\vec{f}^+(p) + \underline{D}_L \vec{b}_L = 0 \tag{18.18}$$

leading to the analytic expression of the reflection-refraction coefficients. A quite similar problem arises for the case of an incident s wave. With the above notations, it is sufficient to replace p by s in the foregoing relations. In then follows that for a field of polarization q incident from medium 2, the system of boundary conditions of the fields across the interface can be written as:

$$\vec{f}^+(q) + \underline{D}_L \vec{b}^q = 0, \quad \vec{b}^q = (T_{21}^{qp}, R_{21}^{qp}, R_{21}^{qs})^t, \quad q = p, s \tag{18.19}$$

The various reflection and refraction conditions are thus given by a simple inversion as before. Inserting this inversion of Eq. (18.19) in Eq. (18.12) we thus have

$$C\vec{d} = -\underline{D}_L(\vec{b}^p b + \vec{b}^s c)$$

Introducing the matrix $\underline{R}$ of column vectors

$$\underline{R} = (\overrightarrow{0}, \overrightarrow{b}^p, \overrightarrow{b}^s)$$

we obtain

$$\underline{R}\overrightarrow{d} = \overrightarrow{b}^p + \overrightarrow{b}^s c, \quad \underline{C}\overrightarrow{d} = -\underline{D}_L \underline{R}\overrightarrow{d}$$

Equation 18.13 then becomes

$$\overrightarrow{f}^-(1) + \underline{D}_L(\underline{1} - \underline{R})\overrightarrow{d} = 0 \tag{18.20}$$

where $\underline{1}$ designates the unit matrix. From Eqs. (18.20) and (18.10) follows:

$$\underline{D} = \underline{D}_L(\underline{1} - \underline{R}) \tag{18.21}$$

By a simple inversion, the vector of unknowns can be expressed by

$$\overrightarrow{d} = (\underline{1} - \underline{R})^{-1}\overrightarrow{d}_L$$

If we use the Cayley-Hamilton theorem (Morse and Feshbach, 1953) for expanding $(\underline{1} - \underline{R})^{-1}$, we find:

$$\overrightarrow{d} = \sum_{n-0}^{\infty} \underline{R}^n \overrightarrow{d}_L \tag{18.22}$$

which is the generalized Debye series (generalized with respect to the original formulation (Debye, 1908), valid for a scalar field diffracted by a homogeneous object). This formulation of the Generalized Debye Series Theory (GDST), as reviewed here, will be applied below to various different cases.

18.3 Transmitted Waves

In this section, we work out some natural applications of the present GDST. This first application concerns the physical meaning of the exact solution for waves in a fluid normally incident on a fluid cylinder (Brill and Überall, 1970; Derem, 1982). For this scalar case we have no mode conversion and we can suppress the superscripts p and s in the local reflection-refraction coefficients. Using Eq. (18.22), it is easy to show that the exact solution an of the scattering problem is:

$$a_n = R_{12} + T_{12}(1 - R_{12})^{-1}T_{21} \quad \Rightarrow a_n = R_{12} + T_{12}(\sum_{n=0}^{\infty} R_{12}^n)T_{21}. \tag{18.23}$$

For each term of Eq. (18.23) the application of the Sommerfeld-Watson transformation and the saddle point method (Flax et al, 1978) furnish the expressions for the individual transmitted wave amplitudes in terms of reflection-refraction coefficients. After this the first term in Eq. (18.23) represents the specularly reflected wave while the n-th term of a_n is representative of a wave that, after being refracted into the medium 2 (coef. T_{12}) has undergone $(n-1)$ internal reflections into this medium (coef. R_{12}^n) before being refracted in the medium 1 (coef. T_{21}). Equation (18.22) furnishes again the GDST for an elastic cylinder treated above with

$$a_n = R_{12} + \frac{T_{12}^{pp}}{\Delta}[(1-R_{21}^{ss})T_{21}^{pp} + R_{21}^{ps}T_{21}^{sp}] + \frac{T_{12}^{ps}}{\Delta}[(1-R_{21}^{pp})T_{21}^{sp} + R_{21}^{sp}T_{21}^{pp}] \tag{18.24}$$

$$\Delta = (1-R_{21}^{pp})(1-R_{21}^{ss}) - R_{21}^{ps}R_{21}^{sp}.$$

Considering the case where no critical angle is crossed, one finds that:

$$| R_{21}^{pp} | < 1, \quad | R_{21}^{ss} | < 1 \quad \text{and} \quad | R_{21}^{ps}R_{21}^{sp} | < | 1-R_{21}^{pp} || 1-R_{21}^{ss} | . \tag{18.25}$$

As previously, the asymptotic evaluation of all coefficients in Eq. (18.24), and expansion of Δ^{-1} in the manner of Eq. (18.23), describe the "avalanche" as illustrated by figure 4 in Brill and Überall (1970). Equation (18.22) furnishes a similar interpretation of the solution inside to the obstacle, which is not usually given. This will be developed shortly below regarding the example of the internal source treated in Sect. 18.6.

18.4 Contribution to the Resonance Scattering Theory

18.4.1 Case of Solid Submerged Elastic Objects

Another exploitation of Eq. (18.22) concerns the exact calculation of the frequencies and the widths of the scatterer's resonances. It is well known from the Resonance Scattering Theory (RST) (Flax et al, 1978) that the resonances of an object are given by the roots of the secular equation $det(\underline{D}) = 0$, and thus, using Eq. (18.21), by the roots of the equation $det(\underline{1} - \underline{R}) = 0$ (from the definition of the reflection and refraction coefficients, $det(\underline{D_L}) \neq 0$). Computational results are given below for the case of scattering by an aluminum and brass sphere immersed in water (Gérard and Rousselot, 1983). The numerical evaluation of the roots is accomplished in two steps. After having performed the complex extension ν of the index n, one solves the characteristic equation $det(\underline{1} - \underline{R}) = 0$. For a given frequency, one locates the roots in a quadrant of the complex plane. One thus obtains an approximate value of the roots $\hat{\nu}_0$. Starting from this estimation $\hat{\nu}_0$ one refines the result using an algorithm derived from the Newton-Raphson method. The roots group themselves into families labeled by the integer $l = 1$ whose numbers increase with frequency.

All results are given as a function of the dimensionless frequency parameter $x = k_1 a$ (a radius, k_1 acoustic wavenumber in water). In agreement with the RST, the resonances are labeled by two numbers (n,l). The first index defines the ordinal number (order) of the resonance, and the second determines the type of peripheral

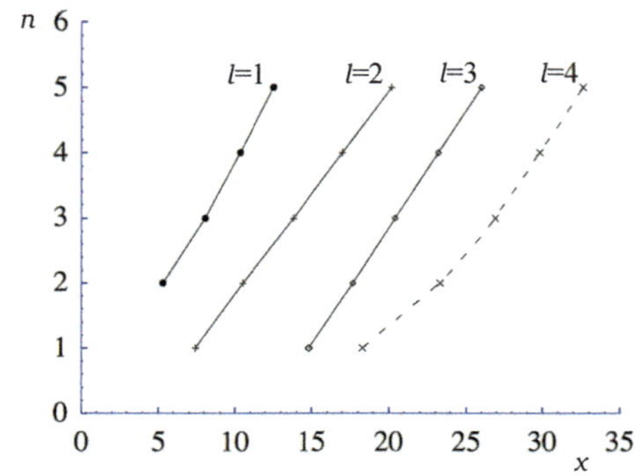

Fig. 18.1 Regge trajectories for a solid aluminum sphere

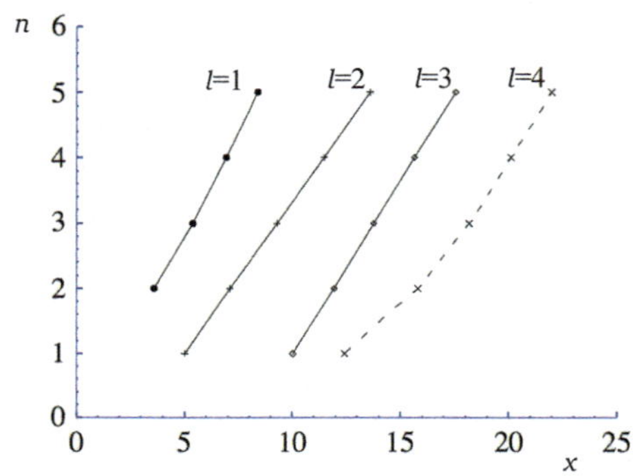

Fig. 18.2 Regge trajectories for a solid brass sphere

wave revolving around the elastic sphere that generates the resonance. The index $l = 1$ corresponds to the "Rayleigh" wave and $l = 2, 3, \ldots$ corresponds to "Whispering Gallery" waves. The curves called "Regge trajectories" (Figs. 18.1 and 18.2) are analogous to what one finds in Maze and Ripoche (1983); Sessarego et al (1987); Maze et al (1981) after a direct search of the roots of the characteristic equation $det\underline{D} = 0$.

18.4.2 Case of Solid Submerged Lossy Elastic Objects

The acoustic scattering by a lossy elastic cylinder at low frequencies has been obtained in the same way (Grosse-Tête et al, 1985). The attenuation is taken into account by introducing the complex wavenumber $k_p = k_1 C_1 C_p^{-1}(1 - i\alpha_p)$, $k_s = k_1 C_1 C_s^{-1}(1 - i\alpha_s)$. To facilitate comparison we have used the following parameters for lucite material (Schuetz and Neubauer, 1977):
$C_1 = 1,468.72\text{m/s}$, $C_p = 2,688.50\text{m/s}$, $C_s = 1,340.0\text{m/s}$,
$\rho_2 = 1.18\text{g/cm}^3$, $\alpha_p = 0.00348\text{m}$ $\alpha_s = 0.00531$
The resonance frequencies are determined by the roots of $det(\underline{1} - \underline{R}) = 0$ and listed

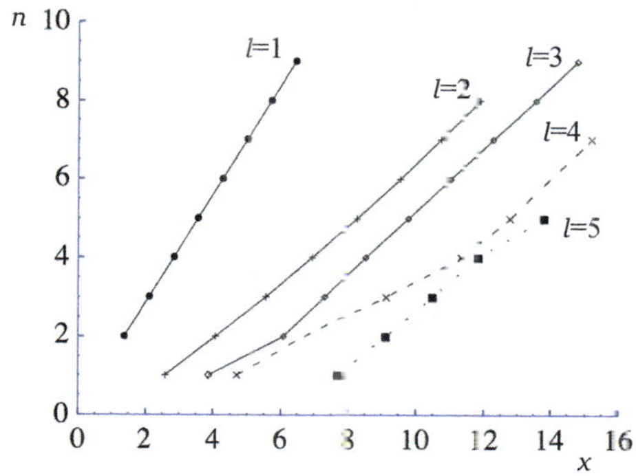

Fig. 18.3 Regge trajectories for a solid brass sphere

in Grosse-Tête et al (1985) and the results of Schuetz and Neubauer (1977) are improved. Here again, the results lead to Regge trajectories, Fig. 18.3, which as a whole resemblance those found in the literature. The slope of these curves is always between 0 and 1. One also notes that the resonances are very large for $l = 1$, but narrower for $l = 2, 3, \ldots$ in the low frequency range $x \leq 16$. Mode conversions appear between the Regge curves $l = 3, 4$ and 5 (Fig. 18.3).

In Grosse-Tête et al (1985) the curves of the pressure field compared with those for lossless materials show that for an isolated resonance the pressure amplitude is weakly attenuated whatever the direction of observation (homogeneous attenuation), whereas when two resonance frequencies are very near the frequency under consideration, the amplitude of scattering pressure is very strongly reduced in specific directions (preferential attenuation).

18.4.3 Case of Submerged Elastic Shells

For a multilayered spherical structure subject to an external point source emitting harmonic elastic waves, the characteristic equation giving the resonance frequencies of the obstacle has been factored by using the present GDST (Gérard, 1983; Gérard and Rousselot, 1984) and still presents itself in the form Eq. (18.22). For example, for the case of shells where the internal fluid medium is numbered by 3, the system of boundary conditions has the form:

1	2	$r = r_1$,	
	2	3	$r = r_2$.

In each block one note the number of the region for which the displacements and stresses must be matched on the surface, and the corresponding radius is written to the right of the block. The outermost (innermost) interface has radius $r = r_1 (r = r_2)$. The solution of the system of boundary conditions proceeds as in Sect. 18.2 and leads to a matrix $\underline{R}$ of the form:

$$\underline{R} = \begin{pmatrix} \underline{0} & \underline{R}_1^u \\ \underline{R}_1^d & \underline{0} \end{pmatrix}, \quad \underline{R}_1^u = \begin{pmatrix} 0 & T_{21}^{pp} & T_{21}^{sp} \\ 0 & R_{21}^{pp} & R_{21}^{sp} \\ 0 & R_{21}^{ps} & R_{21}^{ss} \end{pmatrix}, \quad \underline{R}_1^d = \begin{pmatrix} 0 & T_{23}^{pp} & T_{23}^{sp} \\ 0 & R_{23}^{pp} & R_{23}^{sp} \\ 0 & R_{23}^{ps} & R_{23}^{ss} \end{pmatrix} \tag{18.26}$$

where $\underline{0}$ designates the 3x3 matrix with all elements equal to zero, while R_1^u (or R_1^d) is the 3x3 matrix in which the elements of the first column are zero and those of the two other columns are the reflection and refraction coefficients of the interface at $r = r_1$ or at $r = r_2$, respectively. The characteristic equation thus appears in the simple form

$$det(\underline{1} - \underline{R}_1^u R_1^d) = 0$$

We note that one pass from a 6x6 determinant to a 3x3 determinant.

For an air-filled aluminum and brass spherical shell immersed in water, the theoretical resonance frequencies, and their theoretical resonance widths, have been obtained by us earlier (Gérard and Rousselot, 1984). For these numerical applications we have used the following numerical values of the material properties: $r_2/r_1 = 0.67$ (r_1 outside radius, r_2 inner radius);
aluminum shell: $C_p = 6,420\text{m/s}$, $C_s = 3,040\text{m/s}$;
brass shell: $C_p = 4,350\text{m/s}$, $C_s = 2,050\text{m/s}$
The Regge trajectories, shown in Figs 18.4 and 18.5, also agree with what is usually found in the literature, and demonstrate that the number of resonances increases with frequency more rapidly for shells than for solid objects (Murphy et al, 1979b; Maze and Ripoche, 1983; Maze et al, 1984; Likhodaeva and Shenderov, 1971; Dickey et al, 1983). The slope of these curves always lies between 0 and 1.

The study of a thin spherical shell of Fortal (Rousselot et al, 1988) ($r_2/r_1 = 0.80$) has also brought into evidence a new series of resonances denoted $l = 0$ (Fig. 18.6) which has not been visible for a thicker shell with $r_2/r_1 = 0.67$ (Fig. 18.7). The Regge trajectories displaying these results (Figs. 18.7 and 18.6) are in agreement with experiments (Rousselot et al, 1988). One notes, among other things, that these new resonances appear in a limited frequency band. This series of resonances $l = 0$, unknown a priori but existing also in cylindrical geometry, has recently been related to Scholte-Stoneley type interface waves which encircle the shells at their outer

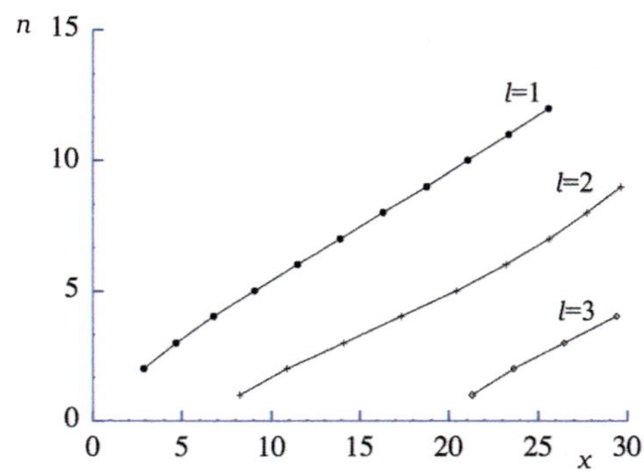

Fig. 18.4 Regge trajectories for a spherical air-filled aluminum shell.

circumference and which exhibit certain frequency limitations (Gérard et al, 1988; Izbicki et al, 1991).

Evidently the simplest separable geometrical shape is the plate. This geometry has been the subject of many studies. Since the first work of Brill and Überall (1970) who have achieved the extension of Debye's series by the binomial theorem, many authors have used the present GDST. Results which have been obtained in recent years concerning the GDST for acoustical interactions, are systematically presented in a thorough manner for the example of an elastic plate immersed in a fluid (Conoir et al, 1991; Gérard, 1983). This special case allows to state clearly the essential principles underlying the way in which GDST is constructed. The characteristic equation of an orthotropic elastic plate is then written in a new form, which uses the local reflection-refraction coefficients linked to the two interfaces of the plate, and the acoustical trajectories of the longitudinal and transverse waves between the interfaces (Deschamps and Hosten, 1992). In Conoir (1991) it is shown that there is a countable infinity of angles (phase velocities), for which the dispersion curves of the symmetric and antisymmetric Lamb waves are crossing. At these angles, there exists a periodicity in the distribution of resonances, and when the dispersion curves are crossing, the resonances result from interferences between all longitudinal and transverse waves refracted in the plate. At high frequencies, situations have also been studied where the resonances result only from transverse wave interferences.

18.5 Non Resonant Background

When making use of Eq. (18.22) that the resonances are intimately linked to the local interactions of the waves in the scatterer, and their combinations. Another interesting consequence of the GDST is the introduction of a new type of intermediate background into the S matrix theory. This theory was transferred into acoustics from nuclear reaction theory by (Flax et al, 1978; Fiorito et al, 1979). In the resonance frequency calculation the interest in the S matrix results from this property of factorization, showing one term of "potential" (or non-resonant) background $S_n^{(p)}$

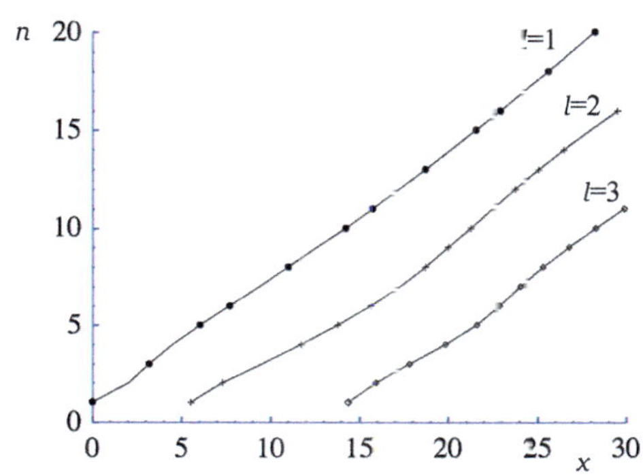

Fig. 18.5 Regge trajectories for a spherical air-filled brass shell.

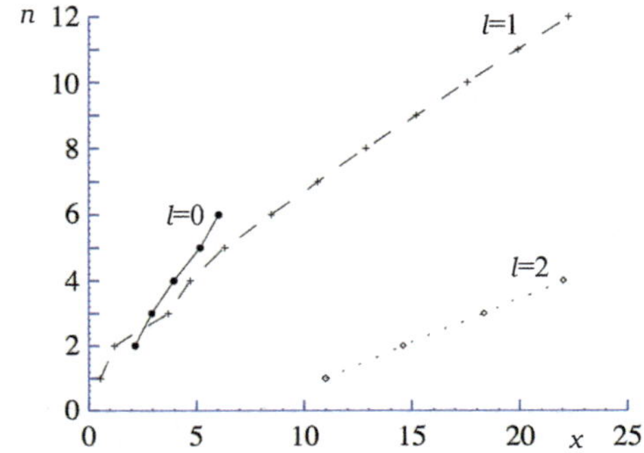

Fig. 18.6 Regge trajectories for a spherical Fortal shell, $r_2/r_1 = 0.80$.

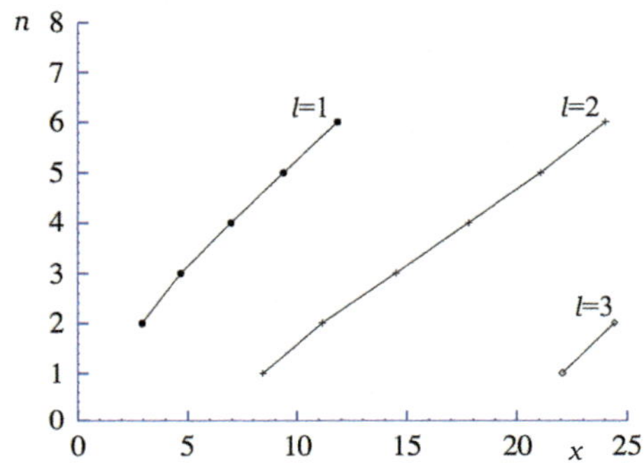

Fig. 18.7 Regge trajectories for a spherical Fortal shell, $r_2/r_1 = 0.67$.

and one term $S_n^*(Sn = S_n^{(p)} S_n^*$, here $S_n = a_n)$ which explicitly displays its resonance character. The idea consists in introducing the transition matrices T_n and $T_n^{(p)}$ defined by $S_n = 1 + 2iT_n$, $S_n^{(p)} = 1 + 2i(T_n^{(p)}$ and observing the behavior of $U_n^p = T_n^- T_n^{(p)}$ as the frequency varies. If the choice made for S_n^p is appropriate, then U_n^p will reach significant values at the resonances, and these are then detected precisely when δ_n^*, the half-phase of S_n^*, passes through $\pi/2$.

Thus the purpose of identifying the background component is to isolate easily the resonance frequencies. Generally the background is to be chosen according to the relative density of the scatterer or the relative thickness of the shell; for an elastic scatterer or a thick shell the rigid background, for some average thickness the intermediate background and for very thin air-filled shells (or air bubbles for instance) the soft background is used (Fiorito et al, 1979). The principal difficulty is to choose the intermediate background. Below, we shall develop a possible factorization of the matrix S which is based on the GDST. In order to simplify the calculations, we take the example of a fluid cylinder immersed in a fluid of different mechanical properties.

Remembering the decomposition of the amplitude of the incident field in a Debye series, Eqs. (18.22) and (18.23), it appears the Sn is the sum of a resonant term – second term of the second expression of Eq. (18.23)– and of a non-resonant term R_{12}. Let us remember that starting from Eqs. (18.21) and (18.23) the characteristic equation of the problem is $1 - R_{12} = 0$. If subsequently we put $\underline{S}_n = R_{12} = 1 +$

$2i\underline{T}_n$ we can evaluate the difference $\underline{U}_n$ between the transition matrix T_n and the transition matrix $\underline{T}_n$ associated with the non-resonant background R_{21}. In this case, we have $\underline{U}_n = (2i)^{-1}T_{12}T_{21}(1-R_{21})^{-1}$ from which follows that $\underline{U}_n$ is likely to assume significant values at the resonant frequencies since $1-R_{12}$ passes though a minimum at the resonance, or is close to a minimum.

For different resonance modes, the development of the modulus of R_{12} as a function of frequency has been observed numerically (Schuetz and Neubauer, 1977; Conoir and Gérard, 1989). It was found that this modulus is smaller than 1 and tends towards the modulus of the reflection coefficient at a plane fluid/fluid interface at normal incidence when $k_1\xi_0$ tends to infinity (i.e. ξ_0, the radius of the cylindrical on spherical object, tends to infinity). It thus appears natural to define the resonant contribution S_n^* of the process by $S_n^* = S_n/R_{12}$ and if we introduce δ_n^* we have

$$S_n^* = \mid R_{12} \mid^{-1} \exp(2i\delta_n^*)$$

since $\mid S_n \mid = 1$. We conclude from this that

$$\mid \underline{U}_n \mid = 2^{-1} \mid \exp(2i\delta_n^*) - \mid R_{12} \mid\mid$$

which does not assure as before that

$$U_n^{(p)} (= \mid \underline{U}_n \mid)$$

is a maximum when δ_n^* passes exactly through $\pi/2$ since R_{12} depends on the frequency. In any case, if we postulate that the variation of the phase is much more rapid than the variation of R_{12} around the resonances, we will not commit an important error. We have thus conserved as a criterion for the detection of resonances the passage of δ_n^* through $\pi/2$ (Gérard and Conoir, 1989).

When the characteristic acoustic impedance

$$Z = \rho_2 C_2 (\rho_1 C_1)^{-1}$$

is high ($Z > 1$) (resp. low ($Z < 1$)) and through the potential background (R_{12}) we find again the results obtained with the rigid background ($S_n^* = S_n^{rigid}$) (resp. soft ($S_n^* = S_n^{soft}$)). For example, the resonance frequency evaluation with R_{12} and S_n^{soft} as potential background (impedance < 1) is given in Conoir and Gérard (1989). We observe that the results globally are in a quite good agreement with an accuracy better than 10^{-1}. The table of resonance frequencies established with R_{12} and S_n^{rigid} as potential background (impedance > 1) gives the same kind of results. This study tends to show that when setting $\underline{S}_n = R_{12}$, one can effect the calculation of resonances equally well for large as well as small impedance ratios. In these conditions, contrary to the preceding cases where two different types of potential background had to be considered (rigid or soft) depending on the impedance ratio, we now only have a single one.

This is not surprising since, if one observes the behavior of the reflection coefficient R_{12} when the density of the scatterer becomes very large $\rho_2 \to -\infty$ or very small $\rho_2 \to 0$, one has according to the results shown in Gérard and Conoir (1989) :

$$R_{12} = -\frac{x_1 H_n^{(2)'}(x_1) H_n^{(2)}(x_2) - \rho_1 \rho_2^{-1} x_2 H_n^{(2)'}(x_2) H_n^{(2)}(x_1)}{x_1 H_n^{(1)'}(x_1) H_n^{(2)}(x_2) - \rho_1 \rho_2^{-1} x_2 H_n^{(2)'}(x_2) H_n^{(1)}(x_1)} \qquad (18.27)$$

from which one deduces the following asymptotic behavior (Conoir and Gérard, 1989):

$$\lim_{\rho_2 \to +\infty} R_{12} = S_n^{rigid} = -\frac{H_n^{(2)'}(x_1)}{H_n^{(1)'}(x_1)}, \quad \lim_{\rho_2 \to 0} R_{12} = S_n^{soft} = -\frac{H_n^{(2)}(x_1)}{H_n^{(1)}(x_1)} \qquad (18.28)$$

This proves that, as stated, the potential background R_{12} adopts the rigid and soft limits for the scatterer when the density of the letter is large or small.

To be rigorous, it remains to be shown that $\underline{S}_n = R_{12}$ is a potential background that remains correct when the impedance ratio is near unity. This latter situation presents a difficulty for a comparison of the results since one does not have any reference values for the resonances. The validation of our hypotheses can thus only be achieved by a comparison with experiments. However, we note that the potential background is supposed to represent the scattering outside the resonances. From this, it appears natural to adopt a rigid or soft background which correspond to the case where no energy penetrates into the scatterer (or, is conserved by the latter). In this sense, the background $\underline{S}_n = R_{12}$ differs from S_n^{rigid} and S_n^{soft}, since it takes into account the penetration (or non-penetration) of the acoustic waves into the scatterer depending on the nature of the later : elastic, rigid or soft. Thus, the coefficient R_{12}, which naturally takes into account the penetration of acoustic energy inside the target, allows us to continuously move from the case in which the insonified object has a rather rigid behavior, to that in which it exhibits a rather soft behavior. As we have verified, these results confirm that the potential R_{12} is well adjusted to obtain the resonance frequencies to a good approximation.

All the examples of the use of the GDST which we have treated here concern the case of the response of an object insonified by an external acoustic source. But as we have already indicated, the methodology developed in Sect. 18.2 equally applies to the case where the acoustic source is inside the object. To illustrate this latter situation, we shall now proceed to evaluate the space-time response to a bounded beam inside an elastic cylindrical waveguide.

18.6 Space-Time Dependence of a Bounded Beam Inside an Elastic Cylindrical Guide

The space-time development of a signal with original time dependence e(t), emitted by a source and received at a space point M, may be described using Fourier analysis. We shall consider here such a development of the signal in a solid elastic medium of cylindrical geometry imbedded in air, and excited by a source centered on the axis. This represents an application of the GDST to the case of an internal source with axial propagation, and no longer in a plane of normal cross section this obliges us to recommence the problem from its beginnings (Danthez et al, 1989a,b). In the appropriate coordinates ($\xi = r$, $\eta = \theta$, $\zeta = z$) where the z direction coincides with the axis of the cylinder, the velocity field is still given, after the theorem of Helmholtz, by the two potentials ϕ (scalar) and $\overrightarrow{\chi}$ (vectorial).

18.6.1 Propagation Equations

The scalar potential and the axial component χ_z of the vector potential satisfy the uncoupled equations while the components χ_r and χ_θ of the vector potential are governed by the coupled wave equations. The vector potential also has to satisfy the condition $div(\overrightarrow{\chi}) = 0$. The axial symmetry entrains the θ independence of the phenomena. The only nonvanishing component of the vector potential is thus along this axis ($\zeta = z$). The expression for the potentials in the solid cylinder is therefore:

$$\phi = \phi_0 J_0(2\pi k_p r)\exp[-2i\pi(\omega t - k_w z)], \tag{18.29}$$

$$\overrightarrow{\chi} = \chi_0 J_1(2\pi k_s r)\exp[-2i\pi(\omega t - k_w z)]\overrightarrow{e_\theta} \tag{18.30}$$

where ϕ_0 and χ_0 are normalization constants, while $k_p, k_s, k_w(\omega)$ the spatial (temporal) frequencies are linked by the dispersion equations:

$$k_p^2 + k_w^2 = \omega^2 C_p^{-2}, \quad k_s^2 + k_w^2 = \omega^2 C_s^{-2}. \tag{18.31}$$

In the following, we shall employ for the Fourier-Bessel transform the following notation:

$$F_k^n(\omega k) = 2\pi \int_0^\infty \omega(k) J_n(2\pi k r) exp(2\pi k_w z) k dk \tag{18.32}$$

in which the kernel can be interpreted in terms of monochromatic cylindrical waves in the r, θ plane, and plane progressive waves in the axial direction z.

18.6.2 Initial Conditions and Limiting Conditions

We assume the field of normal velocities to be known at $z = 0$ in the form of separated variables $v_0(r,t) = e(t)a(r)$. This form permits to separately treat the space part and the time part of the field. In an infinite medium, the spatial development of the field can be written with the help of the scalar potential expressed as a Fourier-Bessel transform

$$\phi(r,z) = F_p^0[\phi(k_p)]$$

a relation in which the regular spectrum of the potential appears, defined by

$$\phi(k_p) = (2i\pi k_w)^{-1} \int_0^\infty a(r) J_0(2\pi k_p r) 2\pi r dr$$

In a bounded medium, it is expedient to introduce in these expression a transfer function that includes the effects of diffraction, and notably the multiple reflections in the (r,θ) plane as well as the effects of diffraction linked to the finite opening of the source, independently of the temporal form of the signal. In order to obtain coherent phases in the reflection coefficients, it is necessary to call upon the Hankel functions of first and second kind, representing divergent and convergent cylindrical waves. Let us recall that these two functions are conjugates of each other and that their half-sum is the Bessel function of first kind. Thus, by a simple permutation on the Hankel functions, the study of propagation with a potential ϕ^+ defined by:

$$\phi^+ = \frac{1}{2}\phi(k_p) H_0^{(1)}(2\pi k_p r) \exp(-2i\pi k_w z) \tag{18.33}$$

permits to derive from it that corresponding to a potential ϕ^- defined by:

$$\phi^- = \frac{1}{2}\phi(k_p) H_0^{(2)}(2\pi k_p r) \exp(-2i\pi k_w z) \tag{18.34}$$

The total response is thus written as a superposition of the propagations Eqs. (18.33) and (18.34). The necessary transfer functions are obtained as before, in the form of an expansion in a GDST as a solution of the global boundary conditions on the fields relevant to the present problem. These limiting conditions are the vanishing of normal and tangential stresses on the outer surface of the cylinder (i. e., a free surface at $r = \xi_0$). The notations of the preceding sections are conserved and a superscript + or - is attached to the quantities associated with divergent ($H^{(1)}$) or convergent ($H^{(2)}$) waves.

18.6.3 Solution of the Problem: Generalized Debye Series

One may thus treat the problem in its global form as before, Sect. 18.2. The incident potential ϕ^+ generates on the internal surface $r = \xi_0$ a reflected scalar potential and a reflected vector potential by mode conversion, and these potentials will in turn get reflected by the cylinder axis (interpretation of the fields being finite at $r = 0$). One denotes thus

$$\overrightarrow{X}^- = (L^-(p,\omega), T^-(p,\omega))$$

as the unknown field amplitudes. Writing the condition of a stress-free surface $r = \xi_0$ furnishes the system

$$\underline{D}\overrightarrow{X}^{-} = \overrightarrow{f_{ext}} \quad (18.35)$$

where $\overrightarrow{f_{ext}}$ denotes the stresses of excitation imposed by the internal source, and where $\underline{D}$ follows from Eq. (18.11) (after having suppressed the first row and column). To solve this system, we first evaluate the reflection coefficients on the internal surface of the cylinder as in Sect. 18.2. For example, the incident potential Eq. (18.33) is reflected as a longitudinal wave and a transverse wave. The associated reflected scalar potential is

$$\phi^{-} = \frac{1}{2}\phi(k_p)R_{pp}^{-}H_0^{(2)}(2\pi k_p r)\exp(-2i\pi k_w z) \quad (18.36)$$

in which there appears R_{pp}^{-}, the longitudinal reflection coefficient without mode conversion associated with a convergent wave, while the vector potential generated by the field Eq. (18.33) is written as a function of the reflection coefficient with mode conversion associated with a convergent wave R_{ps}^{-}:

$$\chi_{\theta}^{-} = \frac{1}{2}\phi(k_p)R_{ps}^{-}H_1^{(2)}(2\pi k_s r)\exp(-2i\pi k_w z) \quad (18.37)$$

The boundary conditions are expressed as a system equivalent to Eq. (18.18), of which the solutions is

$$\underline{\overrightarrow{b}^{p}} = -\underline{D_L}^{-1}\overrightarrow{f_{ext}}, \quad (18.38)$$

with

$$\overrightarrow{b}^{p} = \begin{pmatrix} R_{pp}^{-} \\ R_{ps}^{-} \end{pmatrix}, \quad \underline{D_L} = \begin{pmatrix} \sigma_{rr}^{-}(p) & \sigma_{rr}^{-}(s) \\ \sigma_{rz}^{-}(p) & \sigma_{rz}^{-}(s) \end{pmatrix}$$

Likewise, proceeding as in Eq. (18.19) for an s wave incident on the wall of the cylinder one has:

$$\underline{\overrightarrow{b}^{p}} = \underline{D_L}^{-1}\overrightarrow{f}^{+}(s), \quad \underline{\overrightarrow{b}^{s}} = \begin{pmatrix} R_{21}^{sp} \\ R_{21}^{ss} \end{pmatrix} \quad (18.39)$$

With these notations, the following reflection coefficients for the incident ϕ^{-} or χ_{θ}^{-} fields are obtained from the above relations by permutation of the superscripts + and -. The global system may now easily be solved. It follows from Eqs. (18.20) and (18.21), denoting by$\underline{R}$ the 2x2 matrix formed from the column vectors $\overrightarrow{b}^{p}$ and $\overrightarrow{b}^{s}$, for which analytical expressions are known (Danthez, 1988):

$$\underline{D}\overrightarrow{X}^{-} = \underline{D_L}(\underline{1} - \underline{R})\overrightarrow{X}^{-} = \overrightarrow{f_{ext}}. \quad (18.40)$$

By a simple inversion and after use of the Cayley - Hamilton theorem (Morse and Feshbach, 1953) to the GDST :

$$X^{-} = \sum_{n=0}^{+\infty} \underline{R}^{n}[\overrightarrow{b}^{p}] \quad (18.41)$$

The expression for the unknown coefficients is thus:

$$L^{-}(p,\omega)=[R^{-}_{pp}(1-R^{-}_{ss})+R^{-}_{ps}R^{-}_{sp}][(1-R^{-}_{pp})(1-R^{-}_{ss})-R^{-}_{ps}R^{-}_{sp}]^{-1} \tag{18.42}$$

$$T^{-}(p,\omega)=R^{-}_{ps}[(1-R^{-}_{pp})(1-R^{-}_{ss})-R^{-}_{ps}R^{-}_{sp}]^{-1} \tag{18.43}$$

For the applications, we consider the case where no critical angle has been passed. The inversion of the matrix $(\underline{1}-\underline{R})$ is thus possible, and the convergence of the Debye series is assured. In effect, we can show that the moduli of the coefficients verify the inequalities equivalent to Eq. (18.25). Thus, the following series expansions lead to transfer functions of the following form (Danthez et al, 1989a) :

$$\begin{aligned} L^{-}(p,\omega) &= H^{-}_{p}(p,\omega) \\ &= \sum_{c=1}^{+\infty}(R^{-}_{pp})^{c}+\sum_{n=1}^{+\infty}\sum_{c=n}^{+\infty}\sum_{d=n}^{+\infty}C^{n}_{(c-1)}C_{d}{}^{(n-1)}(R^{-}_{pp})^{c-n+1}(R^{-}_{ss})^{d-n+1}(R^{-}_{ps}R^{-}_{sp})^{n} \end{aligned} \tag{18.44}$$

where there clearly appears a first term representing the ensemble of interactions that have propagated uniquely as longitudinal modes, and a second term containing at least one transverse interaction and one double mode conversion. Likewise, the transverse transfer function is:

$$\begin{aligned} T^{-}(p,\omega) &= H^{-}_{s}(p,\omega) \\ &= (R^{-}_{pp})\sum_{n=0}^{+\infty}\sum_{c=n}^{+\infty}\sum_{d=n}^{+\infty}C^{n}_{c}C^{n}_{d}(R^{-}_{pp})^{c-n}(R^{-}_{ss})^{d-n}(R^{-}_{ps}R^{-}_{sp})^{n} \end{aligned} \tag{18.45}$$

where at least on longitudinal-transverse mode conversion appears. In these conditions, n designates the number of double mode conversion, c the number of longitudinal interactions and d the number of transverse interactions. In this way, the transfer functions Eq. (18.43) can be interpreted in terms of multiple reflections-refractions, and the figure of the avalanche given by Überall (Fig. 4 of Ref. Brill and Überall, 1971), transferred to the present situation, illustrates perfectly the propagation phenomenon considered here.

Since the same decompositions are valid for the inverse situation (divergent waves), on finally has for the response associated with Eqs. (18.44) and (18.45):

$$H_{p}(p,\omega)=H^{+}_{p}(p,\omega)+H^{-}_{p}(p,\omega), \quad H_{s}(s,\omega)=H^{+}_{s}(s,\omega)+H^{-}_{s}(s,\omega) \tag{18.46}$$

Noting the complex form of the reflection coefficients, one has

$$\begin{aligned} &R^{-}_{pp}=(R^{+}_{pp})^{*}, \quad R^{-}_{ss}=(R^{+}_{ss})^{*}, \quad R^{-}_{ps}=-(R^{+}_{ps})^{*}, \\ &R^{-}_{sp}=-(R^{+}_{sp})^{*}, \quad R^{-}_{ps}R^{-}_{sp}=(R^{+}_{ps}R^{+}_{sp})^{*} \end{aligned} \tag{18.47}$$

where $*$ designates the complex conjugate. One verifies that the transfer function $H^{-}_{p}(p,\omega)$ associated with the propagation ϕ^{-} is the conjugate of $H^{+}_{p}(p,\omega)$. Thus, the longitudinal transfer functions is purely real

$$H_p(p,\omega) = 2Re[H_p^-(p,\omega)]$$

In the functions $H_s^+(s,\omega)$ and $H_s^-(s,\omega)$ it appears a term R_{ps}^- as a factor before the combinations, and since

$$R_{ps}^- = -(R_{ps}^+)^*,$$

therefore follows immediately (Danthez et al, 1989a) that the transverse transfer function is thus purely imaginary, $H_s(s,\omega) = 2Im[H_s^-(s,\omega)]$.

18.6.4 Velocity Fields and Simulation

The velocity field is expressed by the Helmholtz potential. Using the preceding results, we have for the radial component:

$$V_r(r,z,\omega) = F_p^1\{-2\pi k_p\phi(k_p)[1+H_p(k_p,\omega)]\} + F_s^1\{2i\pi k_w\phi(k_p)H_s(k_p,\omega)\} \quad (18.48)$$

and for the axial component:

$$V_z(r,z,\omega) = F_p^0\{-2i\pi k_w\phi(p)[1+H_p(p,\omega)]\} + F_q^0\{2\pi k_w\phi(p)H_s(s,\omega)\} \quad (18.49)$$

For a simulation the axial propagation alone, Eq. (18.49), is of interest to us, and using the properties of transfer function this expression becomes

$$V_z(r,z,\omega) = F_p^0\{-2i\pi k_w\phi(p)[1+ReH_p(p,\omega)]\} + F_q^0\{2\pi k_w\phi(p)2ImH_s(s,\omega)\} \quad (18.50)$$

The application of an inverse temporal Fourier transformation to Eq. (18.50) furnishes the space-time response of a cylindrical guide to a bounded beam:

$$v_z(r,z,t) = \int_{-\infty}^{\infty} V_z(r,z,\omega)E(\omega)exp(2i\pi\omega t)d\omega \quad (18.51)$$

where $E(\omega)$ is the temporal spectrum of the original signal $e(t)$. We shall synthesize by numerical calculation (Danthez et al, 1989b) the signal received by a longitudinal transducer placed on the axis at a cross section of a circular cylinder at a distance z from the emitter plane (Fig. 18.8). In most real situations, we can describe the signal delivered by the longitudinal transducer/receiver by the integral of the axial velocity on the surface of the latter, with radius $r = a$. In the axisymmetric case, the answer is expressed by

$$s(z,t) = \int_0^a v_z(r,z,t)2\pi r dr$$

The radial distribution of the emitter is modeled by a Gaussian distribution of rotational symmetry: $a(r) = \exp[(-\pi r^2)/\alpha^2]$ where α is th equivalent radius that can be adjusted appropriate to the experimental situation. The angular spectrum of the potentials, also of rotational symmetry and furnishing the angular velocity spectrum, is

$$\phi(p) = (-2i\pi k_w)^{-1}\alpha^2 \exp(-\pi\alpha^2 k_p^2)$$

Finally, the original signal $e(t)$, depending on the transducer used, is obtained by a spatial deconvolution of the signal received in an infinite medium (Danthez et al, 1989b). The temporal spectrum of the response in an infinite medium is

$$S_\infty(z,\omega) = F_p^1(-2i\pi k_w r/k_p \phi(p)) \,|_{r=a}\, E(\omega)$$

from which one obtains immediately:

$$e(t) = \int_{-\infty}^{+\infty} S_\infty(z,\omega)(F_p^1[-2i\pi k_w r k_p \phi(p)] \,|_{r=a})^{-1} \exp(2i\pi\omega t) d\omega \tag{18.52}$$

In a finite medium, evanescent modes appear only for radial wavenumbers of longitudinal waves $2\pi k_p$ larger than the longitudinal wavenumber k_p. At higher frequencies, no resonance mode is excited and use of the reflection coefficients is sufficient. Furthermore, the contribution of the transverse transfer function is negligible in the expression for the axial velocity. The response of the guide is only given by the longitudinal transfer function. The responses consist of the sum of the signal in an infinite medium and those furnished by the various reflection. In this way, on obtains a succession of separated temporally echoes, whose order is linked to the number of transverse interactions d, cf. Eqs. (18.44) and (18.45).

Thus, the first echo consists of the sum of the signal in an infinite medium and of the modes that have only undergone longitudinal reflections ($d = 0$ and $c \geq 1$). The second echo represents the sum of the modes that have undergone a single transverse interaction ($d = 1$ and $c \geq 1$), and similarly from there on. It has been verified (Danthez, 1988) that for the parameters chosen here, a truncation of order 3 ($1 \leq c \leq 3$) was sufficient for the convergence of the Debye series describing the longitudinal transfer function Eq. (18.44). As an example, Fig. 18.9 shows (Danthez et al, 1989b) the good agreement between experiment and the numerical simulation for a distance (L/ξ_0) of 20. The experimental signal amplitudes, strongly linked to

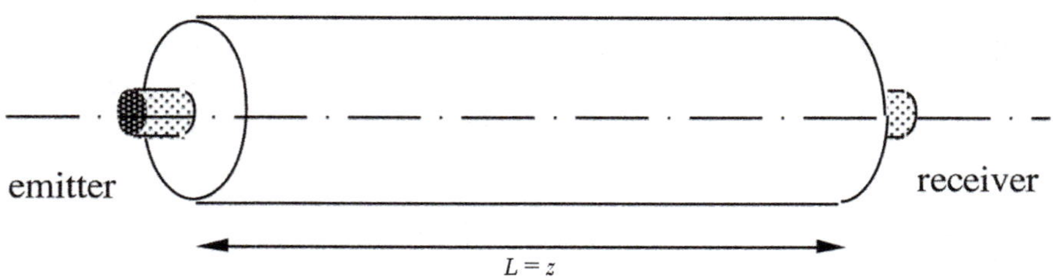

Fig. 18.8: The simulated problem.

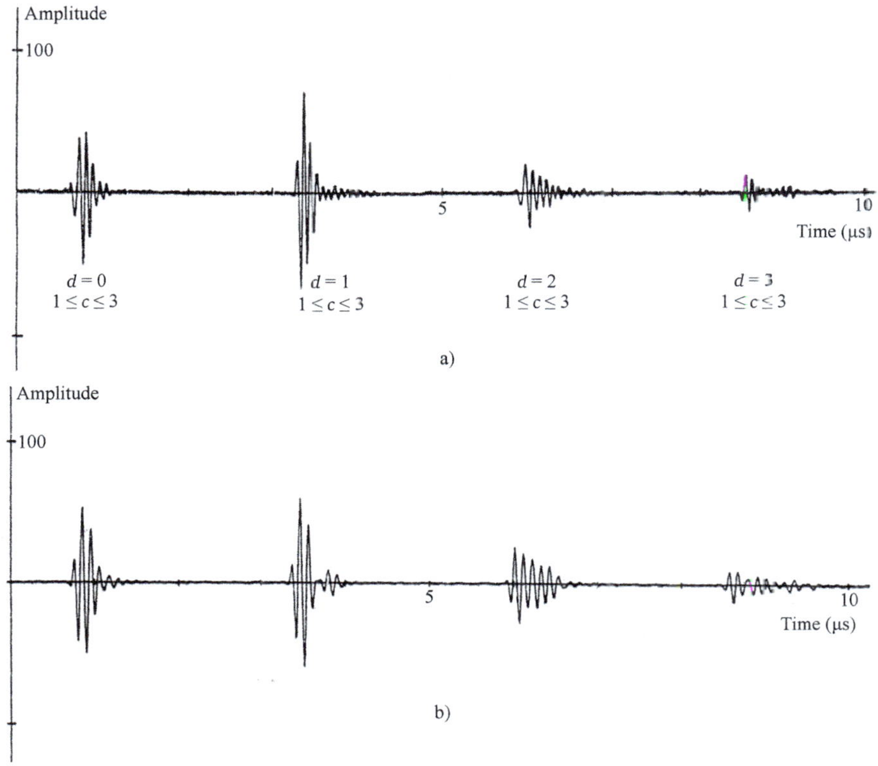

Fig. 18.9: Comparison of signals. a) Theoretical signal for $L = 100$ mm, $\xi_0 = 5$ mm. b) Experimental signal for $L = 100$ mm, $\xi_0 = 5$ mm (after Danthez et al (1989a,b))

the coupling layer, were normalized relative to the first echo of the corresponding numerical simulation. The experimental parameters used for the numerical simulation are: $\omega_0 = 10$ MHz, longitudinal (transverse) wave velocity in the cylinder $C_p = 5{,}900$ m/s ($C_s = 3{,}240$ m/s), dimensionless frequency $F_a = \omega_0 a / C_p = 5.39$, $a = 3.18$ mm (transducer radius). The experimentally observed signals thus confirm the validity of the theoretical model based on the GDST which, among other things, furnishes very well the order of arrival of the mode conversions.

18.7 Conclusions

In acoustics, the doubling of the waves transmitted into an elastic medium into longitudinal and transverse waves renders the scattering problem vectorial, and of a complexity clearly much higher than that of the earlier studies performed in electromagnetic scattering. The original technique used by Debye is here not applicable. The system of continuity conditions of the fields across the interfaces, which is the essence of the problem, has thus been examined in a different fashion. Considering the matrix form of this system, we have established that in its solution, it is perfectly possible to present the ensemble of reflection-refraction coefficients that represent the local elementary interactions which, in fact, are constituents of the exact global solution. Via the Cayley-Hamilton theorem, this solution is then developed in a generalized Debye series. Numerous advantages arise from the fine structure of the thus-formulated solution, first as to the way of treating the problem, but also relative to the intended objectives. Apart from its interest for the study of the reflected and transmitted rays, the GDST is equally well adapted for a search of the resonances, be that for the properties of the S matrix or for the search of the roots of the characteristic equation. In the first case, the GDST furnishes an intermediate potential background, appropriate for a search of resonances for shells that may be both thin -or thick-walled. In the second case, the presence of reflection and refraction coefficients in the characteristic equation shows that the resonances are the consequence of interferences, constructive or destructive, in multiple reflections-refractions taking place inside the insonified object. The several applications presented in this paper show that the GDST is a contribution of interest to the RST as established successfully by Herbert Überall and co-workers. Among other things, the GDST is found to be efficient for the analysis of space-time phenomena.

References

Brill D, Überall H (1970) Transmitted waves in the diffraction of sound from liquid cylinders. J Acoust Soc Am 47:1467–1469

Brill D, Überall H (1971) Acoustic waves transmitted through solid elastic cylinders. J AcoustSoc Am 50:921–939

Conoir JM (1988) Diffusion acoustique par un cylindre. Théorie modale et approximation de l'acoustique géométrique. PhD thesis, University Pierre et Marie Curie, Paris 6

Conoir JM (1991) Interférences périodicités sur les résonances dans une plaque élastique. J Acoust 4:377–412

Conoir JM, Gérard A (1989) Un nouveau fond potentiel pour la matrice S. J Acoust 2:217–227

Conoir JM, Gérard A, Derem A (1991) Ondes acoustiques transmises et séries de Debye généralisées. I Traitement des l'interfaces planes. J Acoust 4:159–200

Danthez JM (1988) Réponse spatio-temporelle d'un guide cylindrique solide à un faisceau ultrasonore. Application à la mesure des tensions résiduelles de serrage dans les assemblages boulonnés. PhD thesis, University Bordeaux 1

Danthez JM, Deschamps M, Gérard A (1989a) Réponse spatio-temporelle d'un guide cylindrique solide à un faisceau borné. I. Partie théorique. J Acoust 2:119–125

Danthez JM, Deschamps M, Roux J (1989b) Réponse spatio-temporelle d'un guide cylindrique solide à un faisceau borné. II. Partie expérimentale. J Acoust 2:127–136

Debye P (1908) Das elektromagnetische Feld um einen Zylinder und die Theorie des Regenbogens. Physik Z 9:775–778

Derem A (1982) Série des ondes transmises pour un cylindre fluide et creux: une solution exacte. Revue du CETHEDEC 73:1–27

Deschamps M, Hosten B (1992) The effects of viscoelasticity on the reflection and transmission of ultrasonic waves by an orthotropic plate. J AcoustSoc Am 91:2007–2015

Dickey JW, Nixon DA, d'Archangelo JM (1983) Acoustic high frequency by elastic cylindrical shells. J Acoust Soc Amer 74:297–304

Fiorito R, Madigosky W, Überall H (1979) Resonance theory of acoustic waves interacting with an elastic plate. J Acoust Soc Am 66:1857–1866

Flax L, Dragonette LR, Überall H (1978) Theory of elastic resonance excitation by sound scattering. J Acoust Soc Am 63:723–761

Gérard A (1979) Champ résultant de l'incidence d'ondes P ou SV sur une sphère élastique. C R Acad Sci Paris 289:237–240

Gérard A (1980) Champ résultant de l'incidence d'ondes P et SV sur un milieu stratifié à symétrie sphérique. C R Acad Sci Paris 290:43–46

Gérard A (1983) Factorisation de l'équation caractéristique d'une sphère élastique multi-couches: interprétation des résonances. C R Acad Sci Paris 297:17–19

Gérard A (1987) Formalisme modal: interprétation. In: Gespa N, Poirée B (eds) La Diffusion Acoustique, Revue du CEDOCAR, Paris, pp 165–287

Gérard A, Conoir JM (1989) S matrix theory: a new intermediate background. In: 13th Int. Cong. of Acoust., Jugoslavia, pp 231–233

Gérard A, Rousselot JL (1983) Solution exacte de diffusion d'une sphère pleine: interprétation et résonances. In: 11th Int. Cong. on Acoustics, Paris

Gérard A, Rousselot JL (1984) Diffusion à basse fréquence d'une coque sphérique immergée. Revue du CETHEDEC 78:121–133

Gérard A, Rousselot JL, Izbicki JL, Maze G, Ripoche J (1988) Résonance des ondes d'interface coques cylindriques minces immergées : détermination et interprétation. Rev Phys Appl 23:289–299

Grosse-Tête A, Gérard A, Rousselot JL (1985) Influence d'un milieu élastique dissipatif sur la diffusion acoustique d'objets cylindriques. Traitement du signal 2:397–400

Izbicki JL, Rousselot JL, Gérard A, Maze G, Ripoche J (1991) Analysis of resonances related to scholte-stoneley waves around circular cylindrical shells. J Acoust Soc Am 90(5):2602–2608

Knopoff L (1959) Scattering of compression waves by spherical obstacles. Geophys 24:30–39

Likhodaeva EA, Shenderov EL (1971) Circumferential waves produced by plane wave diffraction at a cylindrical shell. Sov Phys Acoust 17:63–81

Maze G, Ripoche J (1983) Méthode d'isolement et d'identification de résonances de cylindres et de tubes soumis à une onde acoustique plane dans l'eau. Rev Phys Appl 28:319–326

Maze G, Traconet B, Ripoche J (1981) Influence des ondes de Galerie à écho sur la diffusion d'une onde ultrasonore plane par un cylindre. Phys Lett 84A:309–312

Maze G, Ripoche J, Derem A, Rousselot JL (1984) Diffusion d'une onde ultrasonore par des tubes remplis d'air, immergés dans l'eau. Acustica 55:69–85

Morse PM, Feshbach H (1953) Methods of Theoretical Physics, vol 1. McGraw-Hill, New-York

Murphy JD, George J, Nagl A, Überall H (1979a) Isolation of resonant component in acoustic scattering from fluid loaded elastic spherical shells. J Acoust Soc Am 65:368–373

Murphy JD, George J, Überall H (1979b) Isolation of resonant component in acoustic scattering from fluid loaded cylindrical shells. Wave Motion 1:141–147

Nagase M (1956) Diffraction of elastic waves by a spherical surface. J Phys Soc Japan 11:279–301

Nussenzweig HM (1969a) High-frequency scattering by a transparent sphere. I. Direct reflection and transmission. J Math Phys 10:82–124

Nussenzweig HM (1969b) High-frequency scattering by a transparent sphere. II. Theory of the rainbow and glory. J Math Phys 10:125–176

Pao YH, Mow CC (1973) Diffraction of elastic waves and dynamic stress concentrations. Crane, Russak and Co, New-York

Rousselot JL, Gérard A, Sessarego JP, Sageloli J (1988) Réponse basse fréquence d'une coque sphérique mince immergée. Acustica 66:203–213

Scholte JGJ (1956) On seismic waves in a spherical earth. In: Koninklijk Nederlands Meteorologisch Instituut, Medelingen en Verhandelingen, vol 65

Schuetz LS, Neubauer WG (1977) Acoustic reflexion from cylinders non absorbing and absorbing. J Acoust Soc Am 62:513–517

Sessarego JP, Sageloli J, Gazanhes C (1987) Diffusion acoustique par des sphères élastique. Acustica 64:206–210

Sezawa K (1927) Scattering of elastic waves and some allied problems. Bull Earthquake Res Inst 3:19–42

van-der-Pol B, Bremmer H (1937) The diffraction of electromagnetic waves from an electrical point source round a finitely conducting sphere, with applications to radiotelegraphy and the theory of the rainbow. Phil Mag 24:141–176

Chapter 19
Simplest Linear Homogeneous Reduced Gyrocontinuum as an Acoustic Metamaterial

Elena F. Grekova

Abstract We consider gyrocontinuum, whose each point-body is an infinitesimal rigid body containing inside an axisymmetric rotor, attached to the body but freely rotating about its axis. Point-bodies of the medium may perform independent translations and rotations of general kind. The proper rotation of their rotors does not cause stresses in the medium. We consider the case of infinitesimal density of inertia tensor both of rotor and carrying body and large proper rotation velocity of the rotor, resulting together in a finite dynamic spin. Rotor inside each point body does not interact with anything but its carrying point body, i.e. its existence only contributes into the kinetic energy but not to the strain energy. We suppose that this medium does not react to the gradient of turn of the carrying bodies, therefore we call it "reduced". This yields in zero couple stresses. For simplicity we consider the elastic energy of the medium to be isotropic. This is a medium similar to the reduced Cosserat medium but with the kinetic moment consisting of a gyroscopic term. An example of such an artificially made medium could be a medium consisting of interacting light spheres with light but fast rotating rotors inside them. We consider linear motion of the carrying spheres and investigate harmonic waves in this continuum. We see that, similar to isotropic reduced Cosserat medium, longitudinal wave is non-dispersional, and shear-rotational wave has dispersion and one its branch has a band gap. The band gap depend on the dynamic spin of point bodies and can be controlled via it. Note that all the shear harmonic waves in this medium are not plane waves but have polarization, if the direction of propagation is not orthogonal to the rotor axes.

Elena F. Grekova
Institute for Problems in Mechanical Engineering of the Russian Academy of Sciences, Bolshoy pr., 61, V.O., St. Petersburg, 199178, 195251, Russia
e-mail: elgreco@pdmi.ras.ru

© Springer International Publishing AG, part of Springer Nature 2018
H. Altenbach et al. (eds.), *Generalized Models and Non-classical Approaches in Complex Materials 1*, Advanced Structured Materials 89,
https://doi.org/10.1007/978-3-319-72440-9_19

19.1 Introduction

There exist various media whose particles possess rotational degrees of freedom. The most famous example are Cosserat media. Three-dimensional continuum whose particles are infinitesimal rigid bodies possessing independent rotational and translational degrees of freeedom, was suggested for the first time by brothers Cosserat (Cosserat and Cosserat, 1909). They suggested for description of strains in this medium so-called Cosserat deformation tensor

$$\overset{\circ}{\nabla}\boldsymbol{R}\cdot\boldsymbol{P},$$

where $\overset{\circ}{\nabla}$ is the gradient operator with respect to the reference configuration, $\boldsymbol{R}$ is the position vector in the actual configuration, and $\boldsymbol{P}$ is the turn tensor in the actual configuration. The theory of Cosserat continua was developed in its full variant by Kafadar and Eringen (1971). One of the first works on 3D linear micropolar elasticity is by (Palmov, 1964). For the elastic case one may obtain constitutive equations of the Cosserat continua basing upon the balance of energy (see Altenbach and Zhilin, 1988) where it is done for 2D case and Grekova and Zhilin (2001) where this method is generalized for 3D case. Among various books on elastic Cosserat continua one can mention Eringen and Maugin (2012); Erofeyev (2003); Eremeyev et al (2012).

Strain energy in the elastic full Cosserat medium depends also on the second measure of deformation, equal to

$$-(\overset{\circ}{\nabla}\boldsymbol{P}\cdot\boldsymbol{P}^{\top})\cdot\cdot(\boldsymbol{E}\times\boldsymbol{P})/2$$

(transposed wryness tensor). Reduced Cosserat continuum is a Cosserat continuum which does not work on the gradient of the angular velocity, and therefore couple stress tensor is zero, though stress tensor is asymmetric. For elastic case it means that the strain energy depends only on the Cosserat deformation tensor (Grekova, 2012).

Linear isotropic elastic reduced Cosserat medium was suggested in Schwartz et al (1984) for description of granular media. Waves in this medium and in its various anisotropic versions were investigated in Grekova et al (2009); Grekova (2016, 2018) and some other works. In isotropic case the compression wave is not dispersive, but the shear-rotational wave has a band gap and is highly dispersive close to this domain of frequencies (it is a single negative acoustic metamaterial). If an anisotropic coupling between rotational and translational strains exists, mixed and (if they exist) compression waves are also dispersive and in most cases have a band gap.

Kelvin's medium, a special type of Cosserat continuum consisting of fastly and freely rotating about their axes axisymmetric point bodies, was investigated in Grekova and Zhilin (2001); Grekova and Maugin (2005) and some other works. Its equations are analogous to the equations of ferromagnetic insulators in the approximation of quasimagnetostatics in the state of magnetic saturation by Maugin (1988). Note that both reduced Cosserat medium and Kelvin's medium are specific Cosserat continua with certain restrictions on the strain energy. Constraints on the strain energy yield in specific type of constitutive equations. Symmetry groups and use of different type of strain measures for Cosserat media with constraints are discussed in Eremeyev and Pietraszkiewicz (2012, 2016); Pietraszkiewicz and Eremeyev (2009).

In (Zhilin, 2012) the basic laws for a gyrocontinuum consisting of multispin point bodies (infinitesimal rigid bodes with embedded gyrostats) are suggested. Such a point body is equivalent to a one-rotor gyrostat, a rigid case containing a freely rotating about its axis rotor attached to the case. In Ivanova (2010, 2014, 2015, 2017) Zhilin's ideas were enhanced.

We consider a variant of such a one-rotor elastic gyrocontinuum, consisting of interacting light spheres with light but fast rotating rotors inside them (Fig. 19.1). We consider linear motion of the carrying spheres. We consider a special kind of this gyrocontinuum imposing a restriction that the strain energy does not depend on the gradient of turn of carrying spheres. Therefore we call it *reduced* gyrocontinuum, analogously to the reduced Cosserat medium.

Note that reduced gyrocontinuum considered in this paper is similar but not the same as reduced Kelvin's medium, since Kelvin's medium does not react on the proper rotation of a point body about it axis. Kelvin's medium cannot possess isotropic elastic energy for this reason. On the other hand, it reacts on all other turns of the axis relatively to the background of centres of mass. Reduced gyrocontinuum considered here is rather similar to the reduced isotropic Kelvin's medium but with a specific kinetic moment that yields to the gyroscopic dynamic term in the balance of torque.

19.2 Basic Equations for the Linear Reduced Gyrocontinuum

Define $\boldsymbol{R}$ the position vector of the centre of mass of a point body, $\boldsymbol{R}$ equals $\boldsymbol{r}$ in the reference configuration, $\boldsymbol{P}$ the turn tensor of a carrying point body, $\boldsymbol{P} = \boldsymbol{E}$ in the reference configuration, $\boldsymbol{m}$ the unit vector of its rotor axis in the reference configuration. The rotor performs proper rotation about its axis described by the turn tensor $\boldsymbol{P} \cdot \boldsymbol{Q}$, where $\boldsymbol{Q} = (1 - \cos\varphi)\boldsymbol{mm} + \cos\varphi\boldsymbol{E} + \sin\varphi\boldsymbol{m} \times \boldsymbol{E}$. Note that this rotation does not influence the tensor of inertia of the rotor, neither of the whole point body, since due to the axial symmetry of the rotor it does not change the mass distribution. For this reason Zhilin calls such a point body "quasi-rigid particle" (Zhilin, 2012). The tensor of inertia of the point body will have axial symmetry since the carrying body is a sphere, and the rotor is axisymmetric. Denote ρ the mass density, J the moment of inertia of the carrying sphere, $\boldsymbol{I}_0 = I(\boldsymbol{E} - \boldsymbol{mm}) + I_1\boldsymbol{mm}$ density of tensor of inertia of the rotor in the reference configuration, $\boldsymbol{E}$ the identity tensor. We consider

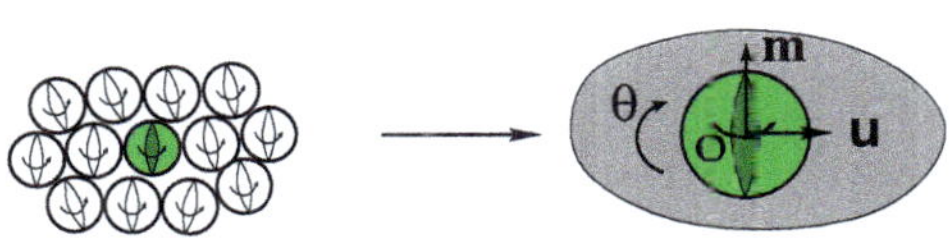

Fig. 19.1 Reduced gyrocontinuum

homogeneous medium with equally oriented axes of symmetry of point bodies in the reference configuration ($\boldsymbol{m}$ does not depend on the space co-ordinates).

Theorem 19.1. *Consider infinitesimal J, I, I_1 and large $\dot{\varphi}$, such that*

$$M \stackrel{def}{=} I_1 \dot{\varphi} = O(1)$$

is a finite quantity. Angular velocity of the carrying point body $\boldsymbol{\omega}$ satisfies the Poisson equation $\dot{\boldsymbol{P}} = \boldsymbol{\omega} \times \boldsymbol{P}$. Angular velocity of the rotor $\boldsymbol{\omega}_1$ equals $\boldsymbol{\omega}_1 = \boldsymbol{\omega} + \dot{\varphi} \boldsymbol{P} \cdot \boldsymbol{m}$. Then the proper kinetic moment $\boldsymbol{K}_2$ of a point body equals $M\boldsymbol{P} \cdot \boldsymbol{m} + o(M)$.

Proof.

$$\begin{aligned}
\boldsymbol{K}_2 &= J\boldsymbol{E}\boldsymbol{\omega} + \boldsymbol{I} \cdot \boldsymbol{\omega}_1 \\
&= J\boldsymbol{\omega} + (I(\boldsymbol{E} - (\boldsymbol{P} \cdot \boldsymbol{m})(\boldsymbol{P} \cdot \boldsymbol{m})) + I_1 (\boldsymbol{P} \cdot \boldsymbol{m})(\boldsymbol{P} \cdot \boldsymbol{m})) \cdot (\boldsymbol{\omega} + \dot{\varphi} \boldsymbol{P} \cdot \boldsymbol{m}) \\
&= (J\boldsymbol{E} + \boldsymbol{I}) \cdot \boldsymbol{\omega} + I_1 \dot{\varphi} \boldsymbol{P} \cdot \boldsymbol{m} = M\boldsymbol{P} \cdot \boldsymbol{m} + o(M)
\end{aligned}$$

The last equality is true since $\boldsymbol{I}, J$ are infinitesimal, and $\boldsymbol{\omega}$ is finite.□

We will consider the linear theory, i.e. $\boldsymbol{u} = \boldsymbol{R} - \boldsymbol{r}$ is an infinitesimal translational displacement, $\boldsymbol{P} = \boldsymbol{E} + \boldsymbol{\theta} \times \boldsymbol{E} + o^2(1)$, $\boldsymbol{\theta}$ is an infinitesimal vector of turn, $\boldsymbol{\omega} = \dot{\boldsymbol{\theta}} + o(\boldsymbol{\theta}) = o(1)$, and we do not need to distinguish between gradients in the reference $\overset{\circ}{\nabla}$ and actual (∇) configurations. In the linear theory, up to the higher order terms, the time derivative of kinetic moment is equal to $M\dot{\boldsymbol{\theta}} \times \boldsymbol{m}$. Consider the case when there is no external torque acting upon the rotor along its axis. In this case $\dot{\varphi} = \text{const}$.

Indeed,

$$\dot{\boldsymbol{K}}_2 = ((J\boldsymbol{E} + \boldsymbol{I}) \cdot \boldsymbol{\omega} + I_1 \dot{\varphi} \boldsymbol{P} \cdot \boldsymbol{m})^{\cdot} = I_1 \dot{\varphi} \dot{\boldsymbol{\theta}} \times (\boldsymbol{E} + \boldsymbol{\theta} \times \boldsymbol{E}) \cdot \boldsymbol{m} + o(\boldsymbol{\theta}) \approx I_1 \dot{\varphi} \dot{\boldsymbol{\theta}} \times \boldsymbol{m} \quad (19.1)$$

We suppose that rotors do not interact with anything outside their carrying point bodies. Therefore the strain energy of this medium is the same as for the reduced Cosserat medium (Grekova et al, 2009), and there are no any other type of stresses caused by presence of rotors. In fact, the equations for this gyrocontinuum are the same as for the reduced Cosserat medium, but with another dynamic term in the balance of kinetic moment. The requirement that the gradient of turn does not cause any stresses in the medium yields in

$$\frac{\partial U}{\partial \nabla \boldsymbol{\theta}} = \boldsymbol{0} \qquad \Longleftrightarrow \qquad \boldsymbol{\mu} = \boldsymbol{0}. \quad (19.2)$$

The stress tensor, however, is asymmetric.

We consider isotropic linear case, then we have

$$\rho U = \frac{1}{2} (\nabla \boldsymbol{u})^S \cdot\cdot \boldsymbol{C} \cdot\cdot (\nabla \boldsymbol{u})^S + \frac{1}{2} \alpha (\boldsymbol{\theta} - \nabla \times \boldsymbol{u}/2)^2, \quad (19.3)$$

where

$$\boldsymbol{C} = \lambda \boldsymbol{E}\boldsymbol{E} + 2\mu (\boldsymbol{i}_m \boldsymbol{i}_n)^S (\boldsymbol{i}_m \boldsymbol{i}_n)^S. \quad (19.4)$$

Constitutive equations look as follows

$$\boldsymbol{\tau} = \frac{\partial \rho U}{\partial(\nabla \boldsymbol{u} + \boldsymbol{\theta} \times \boldsymbol{E})} = \boldsymbol{C} \cdot\cdot \nabla \boldsymbol{u}^S + \alpha(\boldsymbol{\theta} - \nabla \times \boldsymbol{u}/2) \times \boldsymbol{E}$$
$$= \lambda \nabla \cdot \boldsymbol{u} \boldsymbol{E} + 2\mu (\nabla \boldsymbol{u})^S + \alpha(\boldsymbol{\theta} - \nabla \times \boldsymbol{u}/2) \times \boldsymbol{E}. \quad (19.5)$$

We will consider the case when external loads are zero. In fact, the equations for this gyrocontinuum are the same as for the reduced Cosserat medium, but with another dynamic term in the balance of kinetic moment. Then equations of motion can be written as

$$\nabla \cdot \boldsymbol{\tau} = \rho \ddot{\boldsymbol{u}}, \quad (19.6)$$
$$\boldsymbol{\tau}_\times = M \dot{\boldsymbol{\theta}} \times \boldsymbol{m}. \quad (19.7)$$

Substituting here (19.5), we obtain equations of motion in displacements

$$(\lambda + 2\mu)\nabla\nabla \cdot \boldsymbol{u} - \mu \nabla \times (\nabla \times \boldsymbol{u}) + 2\alpha \nabla \times (\boldsymbol{\theta} - \nabla \times \boldsymbol{u}/2) = \rho \ddot{\boldsymbol{u}}, \quad (19.8)$$
$$-4\alpha (\boldsymbol{\theta} - \nabla \times \boldsymbol{u}/2) = M \dot{\boldsymbol{\theta}} \times \boldsymbol{m}. \quad (19.9)$$

19.3 Special Solution in Case $\omega = \omega_0$

Denote

$$\omega_0 = \frac{4\alpha}{M} = \frac{4\alpha}{I_1 \dot{\varphi}}. \quad (19.10)$$

Note that

$$\omega_0 \dot{\varphi} = 4\alpha / I_1$$

(cf. $\omega_0{}^2 = 4\alpha/I$ for the reduced Cosserat medium). We see that if we consider only real $\boldsymbol{\theta}_0$, then $\boldsymbol{\theta} = \boldsymbol{\theta}_0 e^{i\omega_0 t}, \boldsymbol{u} = \boldsymbol{0}$ is not a solution of the problem (19.8)-(19.9), contrary to the case of the reduced Cosserat continuum. However, there is a solution $\boldsymbol{u} = \boldsymbol{0}$, $\boldsymbol{\theta} = \boldsymbol{Q}(-\omega_0 t\boldsymbol{m}) \cdot \boldsymbol{\theta}_0$, corresponding to the regular precession of the axes of point bodies about their initial direction with the angular velocity $-\omega_0 \boldsymbol{m}$ (directed contrary to the proper rotation velocity). Here $\boldsymbol{Q}(-\omega_0 t\boldsymbol{m}) = (1 - \cos\omega_0 t)\boldsymbol{mm} + \cos\omega_0 t\boldsymbol{E} - \sin\omega_0 t\boldsymbol{m} \times \boldsymbol{E}$ is a turn tensor about axis $\boldsymbol{m}$ at angle $\omega_0 t$. We also may write down it as $\boldsymbol{\theta} = \boldsymbol{\theta}_0 e^{i\omega_0 t}$ if $\boldsymbol{\theta}_0 = \theta_0(\boldsymbol{i}_1 + i\boldsymbol{i}_2)$, where $\boldsymbol{i}_{1,2} \cdot \boldsymbol{m} = 0$. This solution exists if $\boldsymbol{\theta}_0$ does not depend on space co-ordinates, and therefore $\nabla \times \boldsymbol{\theta} = \boldsymbol{0}$.

19.4 Longitudinal Waves and Spectral Problem for the Shear-Rotational Wave

Let us look for the harmonic wave solution of (19.8) - (19.9):

$$\boldsymbol{u} = \boldsymbol{u}_0 e^{i(\boldsymbol{k}\cdot\boldsymbol{r}+\omega t)}, \qquad \boldsymbol{\theta} = \boldsymbol{\theta}_0 e^{i(\boldsymbol{k}\cdot\boldsymbol{r}+\omega t)}.$$

Denote

$$\hat{\boldsymbol{k}} = \boldsymbol{k}/k, \qquad k = |\boldsymbol{k}|.$$

The spectral problem takes form

$$\begin{aligned}&-(\lambda+2\mu)k^2\hat{\boldsymbol{k}}\hat{\boldsymbol{k}}\cdot\boldsymbol{u}\\&-(\mu+\alpha)k^2(\boldsymbol{E}-\hat{\boldsymbol{k}}\hat{\boldsymbol{k}})\cdot\boldsymbol{u}+\alpha k^2(\hat{\boldsymbol{k}}\times\boldsymbol{m}\hat{\boldsymbol{k}}\times\boldsymbol{m})\cdot\boldsymbol{u}+2i\alpha\boldsymbol{k}\times\boldsymbol{\theta}+\rho\omega^2\boldsymbol{u}=\boldsymbol{0},\end{aligned} \tag{19.11}$$

$$2i\alpha(\boldsymbol{k}\times\boldsymbol{u}) = 4\alpha\boldsymbol{\theta} + iM\omega\boldsymbol{\theta}\times\boldsymbol{m}. \tag{19.12}$$

Since $\boldsymbol{k}\cdot\boldsymbol{u}$ does not enter into (19.12), the equation for the longitudinal plane wave $\boldsymbol{u}_0 = u_0\hat{\boldsymbol{k}}$ separates from the system and its dispersion relation is classical:

$$\omega = \sqrt{\frac{\lambda+2\mu}{\rho}}\,k. \tag{19.13}$$

For $\boldsymbol{\theta}$ we have in this case trivial solution if we look for the harmonic wave. Also, the combination of the longitudinal plane wave for $\boldsymbol{u}$ and special solution for $\boldsymbol{\theta}$ (see section 19.3) is possible.

For the shear-rotational wave we obtain

$$(\rho\omega^2-(\mu+\alpha)k^2)(\boldsymbol{E}-\hat{\boldsymbol{k}}\hat{\boldsymbol{k}})\cdot\boldsymbol{u}_0+\alpha k^2(\hat{\boldsymbol{k}}\times\boldsymbol{m}\hat{\boldsymbol{k}}\times\boldsymbol{m})\cdot\boldsymbol{u}_0+2i\alpha\boldsymbol{k}\times\boldsymbol{\theta}_0=0, \tag{19.14}$$

$$\frac{i}{2}\boldsymbol{k}\times\boldsymbol{u}_0 = (\boldsymbol{E}-i\frac{\omega}{\omega_0}\boldsymbol{m}\times\boldsymbol{E})\cdot\boldsymbol{\theta}_0. \tag{19.15}$$

In future consider $\omega \neq \omega_0$.

19.5 Shear-Rotational Wave. Reduced Spectral Problem

Let us multiply (19.15) from the left side on

$$(\boldsymbol{E}-i\frac{\omega}{\omega_0}\boldsymbol{m}\times\boldsymbol{E})^\top = \boldsymbol{E}+i\frac{\omega}{\omega_0}\boldsymbol{m}\times\boldsymbol{E}.$$

This will not add parasite roots to the solution, since $\det\boldsymbol{A} = \det\boldsymbol{A}^\top$ for any tensor $\boldsymbol{A}$. We obtain

$$((1-\frac{\omega^2}{\omega_0^2})(\boldsymbol{E}-\boldsymbol{mm})+\boldsymbol{mm})\cdot\boldsymbol{\theta}_0 = \frac{i}{2}\boldsymbol{k}\times\boldsymbol{u}_0 - \frac{\omega}{2\omega_0}\boldsymbol{m}\times(\boldsymbol{k}\times\boldsymbol{u}_0). \tag{19.16}$$

We consider in this section $\omega \neq \omega_0$, therefore we may multiply this equation on $((1-\omega^2/\omega_0^2)^{-1}(\boldsymbol{E}-\boldsymbol{mm})+\boldsymbol{mm})$ and express $\boldsymbol{\theta}_0$ via $\boldsymbol{u}_0$:

$$\boldsymbol{\theta}_0 = \frac{i}{2}((1-\omega^2/\omega_0^2)^{-1}(\boldsymbol{E}-\boldsymbol{mm})+\boldsymbol{mm})\cdot(\boldsymbol{k}\times\boldsymbol{u}_0)-(1-\frac{\omega^2}{\omega_0^2})^{-1}\frac{\omega}{2\omega_0}\boldsymbol{m}\times(\boldsymbol{k}\times\boldsymbol{u}_0)$$

$$= \frac{i}{2}(\frac{\omega_0^2}{\omega_0^2-\omega^2}\boldsymbol{k}\times\boldsymbol{u}_0+\frac{\omega^2}{\omega_0^2-\omega^2}\boldsymbol{m}(\boldsymbol{m}\times\boldsymbol{k})\cdot\boldsymbol{u}_0)-\frac{\omega\omega_0}{2(\omega_0^2-\omega^2)}\boldsymbol{m}\times(\boldsymbol{k}\times\boldsymbol{u}_0)$$

$$= \frac{\omega_0^2}{2(\omega^2-\omega_0^2)}(-i\boldsymbol{k}\times\boldsymbol{u}_0-i\frac{\omega^2}{\omega_0^2}\boldsymbol{m}(\boldsymbol{k}\times\boldsymbol{m})\cdot\boldsymbol{u}_0+\frac{\omega}{\omega_0}\boldsymbol{m}\times(\boldsymbol{k}\times\boldsymbol{u}_0)) \quad (19.17)$$

Thus we obtain

$$2i\alpha\boldsymbol{k}\times\boldsymbol{\theta}_0 = \frac{\alpha\omega_0^2k^2}{\omega_0^2-\omega^2}((\boldsymbol{E}-\hat{\boldsymbol{k}}\hat{\boldsymbol{k}})\cdot\boldsymbol{u}_0-\frac{\omega^2}{\omega_0^2}(\hat{\boldsymbol{k}}\times\boldsymbol{m}\hat{\boldsymbol{k}}\times\boldsymbol{m})\cdot\boldsymbol{u}_0+i\frac{\omega}{\omega_0}\hat{\boldsymbol{k}}\cdot\boldsymbol{m}\hat{\boldsymbol{k}}\times\boldsymbol{u}_0). \quad (19.18)$$

Now, substituting this equation into (19.14), we obtain reduced spectral problem for the shear-rotational wave

$$(\rho\omega^2-(\mu+\alpha\frac{\omega^2}{\omega^2-\omega_0^2})k^2)(\boldsymbol{E}-\hat{\boldsymbol{k}}\hat{\boldsymbol{k}})\cdot\boldsymbol{u}_0+\alpha\frac{\omega^2k^2}{\omega^2-\omega_0^2}(\hat{\boldsymbol{k}}\times\boldsymbol{m}\hat{\boldsymbol{k}}\times\boldsymbol{m})\cdot\boldsymbol{u}_0$$
$$-i\frac{\alpha\omega_0\omega k^2}{\omega^2-\omega_0^2}\hat{\boldsymbol{k}}\cdot\boldsymbol{m}\hat{\boldsymbol{k}}\times\boldsymbol{u}_0=\boldsymbol{0}, \quad (19.19)$$

which can be rewritten as

$$(\rho\omega^2-(\mu+\alpha(\hat{\boldsymbol{k}}\cdot\boldsymbol{m})^2\frac{\omega^2}{\omega^2-\omega_0^2})k^2)\boldsymbol{e}_1\boldsymbol{e}_1$$
$$+(\rho\omega^2-(\mu+\alpha\frac{\omega^2}{\omega^2-\omega_0^2})k^2)\boldsymbol{e}_2\boldsymbol{e}_2$$
$$-i\frac{\alpha\omega_0\omega k^2}{\omega^2-\omega_0^2}\hat{\boldsymbol{k}}\cdot\boldsymbol{m}\hat{\boldsymbol{k}}\times\boldsymbol{E})\cdot\boldsymbol{u}_0=\boldsymbol{0}, \quad (19.20)$$

where $\boldsymbol{e}_1=\hat{\boldsymbol{k}}\times\boldsymbol{m}/|\hat{\boldsymbol{k}}\times\boldsymbol{m}|$, $\boldsymbol{e}_2=\boldsymbol{m}\cdot(\boldsymbol{E}-\hat{\boldsymbol{k}}\hat{\boldsymbol{k}})/|\boldsymbol{m}\cdot(\boldsymbol{E}-\hat{\boldsymbol{k}}\hat{\boldsymbol{k}})|$, $-\hat{\boldsymbol{k}}$ form an orthonormal right-handed basis.

19.6 Shear-Rotational Wave Propagating Perpendicular to the Rotors' Axes ($\hat{\boldsymbol{k}}\cdot\boldsymbol{m}=0$).

We see that if the wave propagates orthogonal to $\boldsymbol{m}$, the last term in the equation (19.19) disappears, and there are two shear waves, one is non-dispersive classical shear wave $\omega=\sqrt{\mu/\rho}k$ with $\boldsymbol{u}=u_0\boldsymbol{e}_1$, and another one ($\boldsymbol{u}_0=u_0\boldsymbol{m}$, in this case $\boldsymbol{m}=\boldsymbol{e}_2$) has the same dispersion relation as in the isotropic reduced Cosserat continuum (Grekova et al, 2009)

Fig. 19.2 Dispersion curves for the shear-rotational waves for $\boldsymbol{m}\cdot\hat{\boldsymbol{k}}=0$, $\boldsymbol{u}_0=u_0\boldsymbol{m}$. Velocities at low and high frequencies $c_s=\sqrt{\mu/\rho}$, $c_{s\alpha}=\sqrt{(\mu+\alpha)/\rho}$, respectively, $\omega_0=4\alpha/M$, $\omega_1^2=\omega_0^2\mu/(\mu+\alpha)$.

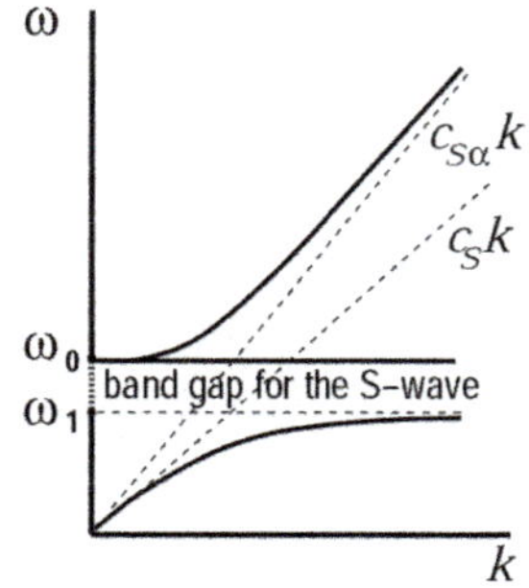

$$k^2=\frac{\rho\omega^2}{\mu+\alpha\frac{\omega^2}{\omega^2-\omega_0^2}}=\frac{\omega^2}{c_s^2}\frac{\omega^2-\omega_0^2}{\omega^2-\omega_1^2},\tag{19.21}$$

where $c_s=\sqrt{\mu/\rho}$, $c_{s\alpha}=\sqrt{(\mu+\alpha)/\rho}$, $\omega_1^2/\omega_0^2=\mu/(\mu+\alpha)$ (see Fig. 19.2). Curiously, in this particular case the gyrostatic nature of the continuum results (only) in the effective anisotropy for shear-rotational waves, though rotors do not interact with anything outside.

19.7 Shear-Rotational Wave Propagating Parallel to the Rotors' Axes ($\hat{\boldsymbol{k}}\times\boldsymbol{m}=\boldsymbol{0}$).

If $\hat{\boldsymbol{k}}\times\boldsymbol{m}=\boldsymbol{0}$, the middle term in (19.19) vanishes, and we obtain

$$(\rho\omega^2-(\mu+\alpha\frac{\omega^2}{\omega^2-\omega_0^2})k^2)(\boldsymbol{E}-\hat{\boldsymbol{k}}\hat{\boldsymbol{k}})\mp i\frac{\alpha\omega_0\omega k^2}{\omega^2-\omega_0^2}\hat{\boldsymbol{k}}\times\boldsymbol{E})\cdot\boldsymbol{u}_0=\boldsymbol{0},\tag{19.22}$$

where $\mp$ takes the value opposite to the sign of $\hat{\boldsymbol{k}}\cdot\boldsymbol{m}$. In this case rotation does not yield the effective anisotropy, but we clearly see the influence of the gyroscopic term. We find in this case circular polarized harmonic waves.

Since

$$i\hat{\boldsymbol{k}}\times(\boldsymbol{e}_1+i\boldsymbol{e}_2)=i(i\boldsymbol{e}_1-\boldsymbol{e}_2)=-(\boldsymbol{e}_1+i\boldsymbol{e}_2)\quad\text{and}\quad i\hat{\boldsymbol{k}}\times(\boldsymbol{e}_1-i\boldsymbol{e}_2)=\boldsymbol{e}_1-i\boldsymbol{e}_2,$$

it is easy to check that $\boldsymbol{u}_0=u_0(\boldsymbol{e}_1\pm i\boldsymbol{e}_2)$ are the eigen vectors of (19.22). Similar to the section 19.3, this corresponds to the circular waves

$$\boldsymbol{u}=u_0(\cos(\omega t+\boldsymbol{k}\cdot\boldsymbol{r})\boldsymbol{e}_1\pm\sin(\omega t+\boldsymbol{k}\cdot\boldsymbol{r})\boldsymbol{e}_2).$$

The dispersion relation for these waves can be obtained from (19.22)

$$(\rho\omega^2-(\mu+\alpha\frac{\omega^2}{\omega^2-\omega_0^2})k^2)\mp\frac{\alpha\omega_0\omega k^2}{\omega^2-\omega_0^2}=0.\tag{19.23}$$

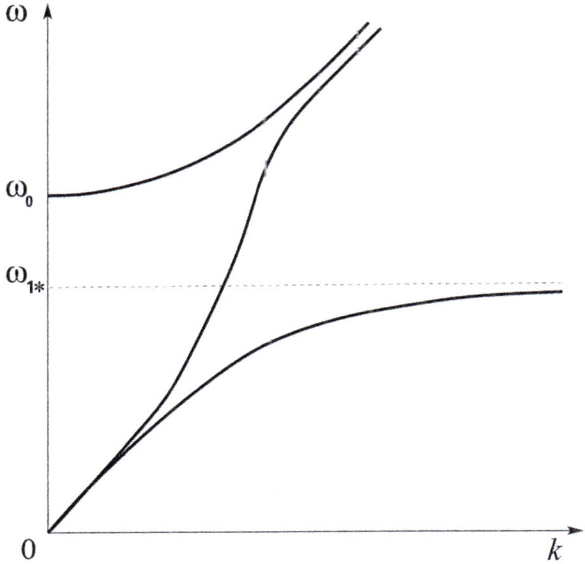

Fig. 19.3: Dispersion curves for shear–rotational waves, $\hat{\boldsymbol{k}} \times \boldsymbol{m} = \boldsymbol{0}$.

This dispersion relation can be written also as

$$k^2 = \frac{\omega^2}{c_{s\alpha}^2} \frac{\omega \pm \omega_0}{(\omega \pm \omega_{1*})}, \tag{19.24}$$

where $\omega_{1*} = \omega_0 \mu/(\mu + \alpha)$. The branch with the minus sign has a band gap $[\omega_{1*}; \omega_0)$, a horizontal asymptote at $\omega = \omega_{1*}$, the lower part starts at zero with the group velocity c_s and the upper one starts with zero group velocity at its cut-off frequency ω_0, changing the velocity to $c_{s\alpha}$ at large ω. The branch with the plus sign has no band gap, its group velocity tends to c_s at small ω and to $c_{s\alpha}$ at large ω. Its typical graph is presented in Fig. 19.3. Thus we see that our reduced gyrocontinuum is a single negative acoustic metamaterial (with respect to some waves), since there are band gaps where certain waves cannot propagate.

19.8 Shear-Rotational Wave Directed in General Way with Respect to the Rotors' Axes

Now let us look for the solution of (19.20) in form of a harmonic polarized wave. We will see that polarization will depend not only on the direction of the wave propagation with respect to the axes of rotors of point bodies, but also on frequency. Let $\boldsymbol{u}_0 = u_0(\boldsymbol{e}_1 + i\beta \boldsymbol{e}_2)$. We will look for such β that $\boldsymbol{u}_0$ is an eigen vector of (19.20).

Calculate

$$-i\hat{\boldsymbol{k}} \times \boldsymbol{u}_0 = -i(-\boldsymbol{e}_2 + i\beta \boldsymbol{e}_1)u_0 = -i(-\boldsymbol{e}_2 + i\beta \boldsymbol{e}_1)u_0 = (\beta \boldsymbol{e}_1 + i\boldsymbol{e}_2)u_0. \qquad (19.25)$$

Equation (19.20) gives us

$$\boldsymbol{e}_1((\rho\omega^2 - (\mu + \alpha(\hat{\boldsymbol{k}}\cdot\boldsymbol{m})^2 \frac{\omega^2}{\omega^2-\omega_0^2})k^2) + \frac{\alpha\omega_0\omega k^2}{\omega^2-\omega_0^2}\hat{\boldsymbol{k}}\cdot\boldsymbol{m}\,\beta)$$
$$+ i\beta\boldsymbol{e}_2(\rho\omega^2 - (\mu + \alpha\frac{\omega^2}{\omega^2-\omega_0^2})k^2 + \frac{\alpha\omega_0\omega k^2}{\omega^2-\omega_0^2}\frac{\hat{\boldsymbol{k}}\cdot\boldsymbol{m}}{\beta}) = \boldsymbol{0}. \qquad (19.26)$$

We have to find such β that coefficients at $\boldsymbol{e}_1$ and $i\beta\boldsymbol{e}_2$ in (19.26) would coincide, then $\boldsymbol{e}_1 + i\beta\boldsymbol{e}_2$ will be an eigen vector of the spectral problem, and if we require this coefficient to be zero, we will obtain the dispersion relation. To determine β we write down

$$\rho\omega^2 - (\mu + \alpha(\hat{\boldsymbol{k}}\cdot\boldsymbol{m})^2 \frac{\omega^2}{\omega^2-\omega_0^2})k^2 + \frac{\alpha\omega_0\omega k^2}{\omega^2-\omega_0^2}\hat{\boldsymbol{k}}\cdot\boldsymbol{m}\,\beta$$
$$= \rho\omega^2 - (\mu + \alpha\frac{\omega^2}{\omega^2-\omega_0^2})k^2 + \frac{\alpha\omega_0\omega k^2}{\omega^2-\omega_0^2}\frac{\hat{\boldsymbol{k}}\cdot\boldsymbol{m}}{\beta} \qquad (19.27)$$

Simplifying this relation, we obtain

$$\frac{1-(\hat{\boldsymbol{k}}\cdot\boldsymbol{m})^2}{\hat{\boldsymbol{k}}\cdot\boldsymbol{m}}\frac{\omega}{\omega_0} + \beta - \frac{1}{\beta} = 0 \qquad (19.28)$$

Thus if we choose $\boldsymbol{u}_0 = u_0(\boldsymbol{e}_1 + i\beta\boldsymbol{e}_2)$, where

$$\beta = -\xi\frac{\omega}{\omega_0} \pm \sqrt{(\xi\frac{\omega}{\omega_0})^2 + 1}, \quad \xi = \frac{(\hat{\boldsymbol{k}}\times\boldsymbol{m})^2}{2\hat{\boldsymbol{k}}\cdot\boldsymbol{m}}, \qquad (19.29)$$

it will be an eigenvector corresponding to the frequency ω. For small ω we have $\beta \approx \pm(1 + (\xi\omega/\omega_0)^2/2) - \xi\omega/\omega_0$, for large ω we obtain

$$\beta \approx \frac{\xi\omega}{\omega_0}(-1 \pm (1 + (\xi\omega/\omega_0)^{-2}/2)),$$

for $\omega \to \omega_0$ we have an exact equality $\beta = -\xi \pm \sqrt{\xi^2+1}$.

The dispersion relations corresponding to these polarized waves are

$$\rho\omega^2 = (\mu + \alpha\frac{\omega^2}{\omega^2-\omega_0^2} - \frac{\alpha\omega_0\omega}{\omega^2-\omega_0^2}\frac{\hat{\boldsymbol{k}}\cdot\boldsymbol{m}}{\beta_\pm(\omega)})k^2. \qquad (19.30)$$

They can be rewritten as

$$k^2 = \frac{\omega^2}{c_{s\alpha}^2} \frac{\omega^2 - \omega_0^2}{\omega^2 - 2\omega_2 \hat{\boldsymbol{k}} \cdot \boldsymbol{m}\, \omega/\beta(\omega) - \omega_1^2}. \quad (19.31)$$

This relation is enough sophisticated. We see that at small ω its group velocity tends to c_s, at large ω to $c_{s\alpha}$, there is a band gap below ω_0 for the wave corresponding to β_-, and the branch corresponding to β_+ does not have a band gap. These curves look qualitatively analogous to Fig. 19.3, though their equations are different from that shown in the figure.

19.9 Conclusions

We discussed harmonic waves in the simplest linear reduced elastic homogeneous gyrocontinuum. Though its strain energy was chosen to be isotropic, gyroscopic effects in some cases yield in effective anisotropy of the medium. Compression wave is not affected and it coincides with the classical one. We have plane shear wave propagation in the direction perpendicular to the axes of rotors. In this case we observe two waves, one is classical and another one is dispersive and has a band gap. Shear-rotational harmonic waves propagating in other directions are polarized. One branch has a band gap below characteristic frequency of the gyrocontinuum, and another branch is dispersive but has no band gap. Thus this gyrocontinuum is a kind of single negative acoustic metamaterial, though only with respect to certain type of waves.

If the propagation of the shear-rotational wave is parallel to the axes of rotors, the wave is circular, and its eigen vectors do not depend on the frequency. For other directions of shear-rotational wave propagation (not parallel and not orthogonal to the axes of rotors) the polarization depends on the frequency. We have found the dispersion relations and the eigen vectors for this spectral problem.

We may control characteristic frequency and the band gap changing the velocity of proper rotation of rotors. Also, in band gaps one may expect localisation phenomena if a heterogeneity is present, and we may create it changing the velocity of the proper rotation of a rotor or changing noticeably its initial position. The other, easier way to control the properties of this acoustic metamaterial would be via magnetic field, but this problem, as well as localisation phenomena, are a subject of future investigation.

Acknowledgements The author had a pleasure to meet Prof. Gérard Maugin in 1999. He was interested in her work on Kelvin's medium where also the analogy between equations of magnetic continua by Prof. Maugin and equations of Kelvin's medium was established. Even now she remembers vividly this interesting discussion. Later she collaborated on gyrocontinua with Prof. Maugin during her postdoc stay in Laboratoire en Modélisation en Mécanique, Jussieu, Paris. It is an honour for her to dedicate her work on reduced gyrocontinuum to the memory of Prof. Maugin.

This work was supported by the Russian Foundation for Basic Research (grant 17-01-00230), by Spanish Government Agency Ministerio de Economía y Competitividad (project No. FIS2014-54539-P) and by Andalusian Government (Junta de Andalucía), support for research group FQM-253.

References

Altenbach H, Zhilin PA (1988) The theory of elastic simple shells (in Russ.). Advances in Mechanics 11(4):107–147

Cosserat E, Cosserat F (1909) Théorie des corps déformables. Hermann, Paris

Eremeyev VA, Pietraszkiewicz W (2012) Material symmetry group of the non-linear polar-elastic continuum. International Journal of Solids and Structures 49(14):1993–2005

Eremeyev VA, Pietraszkiewicz W (2016) Material symmetry group and constitutive equations of micropolar anisotropic elastic solids. Mathematics and Mechanics of Solids 21(2):210–221

Eremeyev VA, Lebedev LP, Altenbach H (2012) Foundations of micropolar mechanics. Springer Science & Business Media, Heidelberg, New York, Dordrecht, London

Eringen AC, Maugin GA (2012) Electrodynamics of Continua, vol I: Foundations and Solid Media. Springer Science & Business Media

Erofeyev VI (2003) Wave Processes in Solids with Microstructure, Series on Stability, Vibration and Control of Systems, Series A, vol 8. World Scientific

Grekova E, Zhilin P (2001) Basic equations of Kelvin's medium and analogy with ferromagnets. Journal of Elasticity and the Physical Science of Solids 64(1):29–70

Grekova EF (2012) Nonlinear isotropic elastic reduced Cosserat continuum as a possible model for geomedium and geomaterials. Spherical prestressed state in the semilinear material. Journal of Seismology 16(4):695–707

Grekova EF (2016) Plane waves in the linear elastic reduced Cosserat medium with a finite axially symmetric coupling between volumetric and rotational strains. Mathematics and Mechanics of Solids 21(1):73–93

Grekova EF (2018) Waves in elastic reduced Cosserat medium with anisotropy in the term coupling rotational and translational strains or in the dynamic term. In: Dell'Isola F, Eremeyev V, Porubov A (eds) Advances in Mechanics of Microstructured Media and Structures, Springer, Singapore, Advanced Structured Materials, vol 87

Grekova EF, Maugin GA (2005) Modelling of complex elastic crystals by means of multi-spin micromorphic media. International Journal of Engineering Science 43(5):494–519

Grekova EF, Kulesh MA, Herman GC (2009) Waves in linear elastic media with microrotations, Part 2: Isotropic reduced Cosserat model. Bulletin of the Seismological Society of America 99(2B):1423–1428

Ivanova EA (2010) Derivation of theory of thermoviscoelasticity by means of two-component medium. Acta Mechanica 215(1):261–286

Ivanova EA (2014) Description of mechanism of thermal conduction and internal damping by means of two-component Cosserat continuum. Acta Mechanica 225(3):757–795

Ivanova EA (2015) A new model of a micropolar continuum and some electromagnetic analogies. Acta Mechanica 226(3):697–721

Ivanova EA (2017) Description of nonlinear thermal effects by means of a two-component Cosserat continuum. Acta Mechanica 228(6):2299–2346

Kafadar CB, Eringen AC (1971) Micropolar media - I the classical theory. International Journal of Engineering Science 9(3):271–305

Maugin GA (1988) Continuum Mechanics of Electromagnetic Solids, North-Holland Series in Applied Mathematics and Mechanics, vol 33. Elsevier

Palmov V (1964) Fundamental equations of the theory of asymmetric elasticity. Journal of Applied Mathematics and Mechanics 28(3):496–505

Pietraszkiewicz W, Eremeyev VA (2009) On natural strain measures of the non-linear micropolar continuum. International Journal of Solids and Structures 46(3):774–787

Schwartz LM, Johnson DL, Feng S (1984) Vibrational modes in granular materials. Phys Rev Lett 52:831–834

Zhilin PA (2012) Rational Continuum Mechanics (in Russ.). St. Petersburg Polytechnic University

Chapter 20
A Mathematical Model of Nucleic Acid Thermodynamics

Sonia Guarguagli and Franco Pastrone

Abstract The DNA denaturation, the double-stranded DNA unwinding process, is a vital process for cells. The percentage of the denaturation is used as an index of organism complexity and it is the base of the DNA hybridisation technique, which provides a great deal of information. It can be detected by observing the increase in the ability of a DNA solution to absorb ultraviolet light at a wavelength of 260 nm. Based on experimental data, we found a mathematical model capable of predicting the behaviour of a general DNA, given the melting temperature T_m.

20.1 Introduction

Gérard Maugin has shown great interest for mathematical models applied to medicine and biology, especially in the latest years of his work (see, for instance Ciarletta et al, 2013, 2012). Thus, we have decided to present this paper, where a mathematical model capable of predicting the behaviour of DNA during its denaturation process is discussed. The two main characteristics of the DNA denaturation curve are its inflection point and its slope, which in turn represent melting temperature and heterogeneity (the percentage of guanine and cytosine) (Nelson and Cox, 2005). At the end of this paper, a relation between the two parameters is found and a single variable α is used to characterise strand behaviour.

Sonia Guarguagli
Università degli Studi di Torino, Via Giuseppe Verdi 8, Torino - Italy
e-mail: sonia.guarguagli@gmail.com

Franco Pastrone
Accademia delle Scienze di Torino, Via Accademia delle Scienze 6, Torino - Italy
e-mail: franco.pastrone@unito.it

© Springer International Publishing AG, part of Springer Nature 2018

H. Altenbach et al. (eds.), *Generalized Models and Non-classical Approaches in Complex Materials 1*, Advanced Structured Materials 89,
https://doi.org/10.1007/978-3-319-72440-9_20

20.2 Denaturation of DNA

Nucleic acid thermodynamics studies how the nucleic acid structure of double-stranded DNA (dsDNA) is affected by temperature (see, for instance Nelson and Cox, 2005). The helical structure of nucleic acids is determined by thermodynamically favoured stacking between adjacent bases in the same strand and the double-stranded structure of the helix is maintained by hydrogen-bonding between the base pairs.

When heated, the double-stranded DNA unwinds because the hydrogen bonds holding the two strands together break. A macromolecule in which the molecules are in a nearly random conformation, is said to be denatured and the process of double helix separation is called DNA denaturation or DNA melting. The measurement of some properties of the molecule changing as denaturation proceeds provides a great deal of information about the structure of DNA. For example, denaturation can be detected by observing the absorbance of ultraviolet light at a wavelength of 260 nm. In double stranded DNA - when bases are highly ordered - A_{260} is lower than that for the less ordered state in single stranded DNA.

If A_{260} is measured at various temperatures while the DNA solution is slowly heated, a melting curve, such as that shown in Fig. (20.1), is obtained (Rastogi, 2003). The temperature at which the rise in A_{260} is halfway through (strands are half double-stranded, half single-stranded) is called the melting temperature (T_m).

T_m is a convenient parameter to characterise the melting transition: the percentage of guanine (G) and cytosine (C) in the DNA molecule has a significant effect on its T_m: the higher the percentage of guanine and cytosine, the higher T_m. The ratio of G to C and A (adenine) to T (thymine) is fixed in an organism's DNA, but the GC content can vary noticeably from one DNA to another. This is because G-C pairs form three hydrogen bonds, while A-T pairs form only two and thus a double-stranded DNA rich in G and C needs more energy to be broken than one rich in A and T; this results in a higher melting temperature.

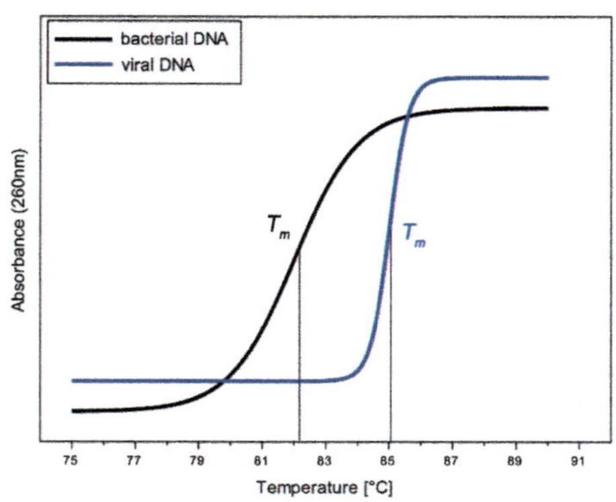

Fig. 20.1 A comparison between bacterial and viral DNA

The study of the ability of a DNA solution to absorb ultraviolet light is a comparative method to quantify differences among organisms. Bacterial double-stranded DNA starts to unwind at a lower temperature than viral DNA, but the process takes longer. This shows that bacterial DNA is more extensive and complex than viral DNA. Moreover, it can be deduced that bacterial DNA contains both AT and GC pairs, whereas viral DNA is shorter, simpler and mainly made of CG pairs that split at higher temperatures.

DNA denaturation is also the base of DNA hybridisation, a technique for selectively binding specific segments of single-stranded DNA or RNA by base pairing to complementary sequences on single-stranded DNA molecules. The field of application of this technique is wide: for example, DNA-DNA hybridisation is commonly used to determine the degree of sequence identity between DNAs of different species, while DNA-RNA hybridisation is used to select those molecules that are complementary to a specific DNA from a heterogeneous population of RNAs (SantaLucia, 1998).

20.3 Mathematical Model

Based on experimental data, our aim is to find a model capable of predicting the behaviour of a general DNA with known T_m. From general analysis, it seems that logistic equations are suitable to represent such a behaviour.

As well known, a logistic function is a sigmoid curve, similar to an exponential function on the left side of the inflection point and with a slow growth on the right of the inflection point. This function can be represented as:

$$x(T) = \frac{q}{m + cqe^{-qT}}, \tag{20.1}$$

where q and m are parameters and c is always positive (the function we are trying to represent is monotonically increasing). Here and further T denotes temperature [°C]. The function is thus continuous at every time t.

We recall here the first and the second derivatives of the function:

$$x' = \frac{cq^3 e^{-qT}}{m + cqe^{-qT}}, \tag{20.2}$$

$$x'' = \frac{cq^4 e^{-qT}(cqe^{-qT} - m)}{(m + cqe^{-qT})^2}. \tag{20.3}$$

Expression (20.3) vanishes only at T_m (the inflection point). Thus, it can be deduced that

$$T_m = \frac{1}{q} \ln \frac{cq}{m}. \tag{20.4}$$

It is obvious from Expression (20.4) that the horizontal asymptote ($x = q/m$) – for t approaching infinity – corresponds to 100% denaturation.

The inflection point of the function is for $T = T_m$ and, as mentioned previously, it represents a denaturation of 50%. It can be calculated by assuming that $x'' = 0$. Hence we obtain:

$$T = \frac{1}{q} \ln \frac{cq}{m} = T_m. \tag{20.5}$$

In addition, we have

$$x(T_m) = \frac{q}{2m} = 0.5. \tag{20.6}$$

Consequently, $q = m$. This result can also be obtained by calculating the limit for $T \to +\infty$ in Expression (20.1).

$$\lim_{T \to +\infty} x(T) = \lim_{T \to +\infty} \frac{q}{m + cqe^{-qT}} = \frac{q}{m}. \tag{20.7}$$

Expression (20.4) is simplified:

$$T_m = \frac{\ln(c)}{q}. \tag{20.8}$$

This means that the inflection point of denaturation curves has a fixed T value and the peculiarity of the DNA is due to a different T_m.

Expression (20.1) is simplified as well:

$$x(T) = \frac{q}{m + cqe^{-qT}} = \frac{1}{1 + ce^{-qT}}. \tag{20.9}$$

Considering the parameter q and trying to define the curve with the only characteristic known, T_m, we can say that:

$$T_m = \frac{\ln(c)}{q}, \tag{20.10}$$

and consequently,

$$q = \frac{1}{T_m} \ln(c). \tag{20.11}$$

Thus, Expression (20.9) becomes:

$$x(T) = \frac{1}{1 + ce^{-Tq}}. \tag{20.12}$$

To simplify Expression (20.12), we introduce $\tau = T/T_m$ and $b = \ln(c)$. We obtain:

$$x(\tau) = \frac{1}{1 + e^{b(1-\tau)}}. \tag{20.13}$$

20.4 The Role of Parameter b

First of all, to study parameter b, we calculate the first derivative of Expression (20.13).

$$x'(\tau) = \frac{be^{b(1-\tau)}}{\left(1+e^{b(1-\tau)}\right)^2}, \tag{20.14}$$

and consequently,

$$x'(T) = \frac{b}{T_m} \frac{e^{\frac{b}{T_m}(T_m-T)}}{\left(1+e^{\frac{b(T_m-T)}{T_m}}\right)^2}. \tag{20.15}$$

Approximating the curve in the vicinity of T_m to a straight line, its slope m is $x'(T_m)$

$$x'(T_m) = \frac{b}{4T_m}. \tag{20.16}$$

Here T_m is fixed and established by a DNA type; as a result, the slope depends exclusively on b. The equation of this straight line is:

$$y = \frac{b}{4T_m}x + q. \tag{20.17}$$

Let us consider as an example, the bacterial DNA shown in Fig. 20.1 (see Rastogi, 2003), where A(80,0) and B(85,100) can be used to identify the slope of the curve. It is thus possible to calculate m and to estimate the value of b to be 660. The equation that represent the denaturation curve of bacterial DNA shown in Fig. 20.1 is:

$$y(T) = \frac{1}{1+e^{\left(660\frac{82.5-T}{82.5}\right)}}. \tag{20.18}$$

The slope, and therefore parameter b, is an indirect index of the complexity of a genome: the more dishomogeneous the DNA, the slower the denaturation process and vice versa.

20.5 Discussion

As shown in Fig. 20.2, the model obtained perfectly conforms to laboratory results, where T_m (representing denaturation temperature) is an index of DNA guanine and cytosine percentage and the slope b is an index of DNA heterogeneity (and thus of genoma complexity): the broader the denaturation range, the more heterogeneous the system. As we know from literature (see, for instance, Mandel and Marmur, 1968) the DNA content of cytosine and guanine can be extimated with T_m. It thus appears

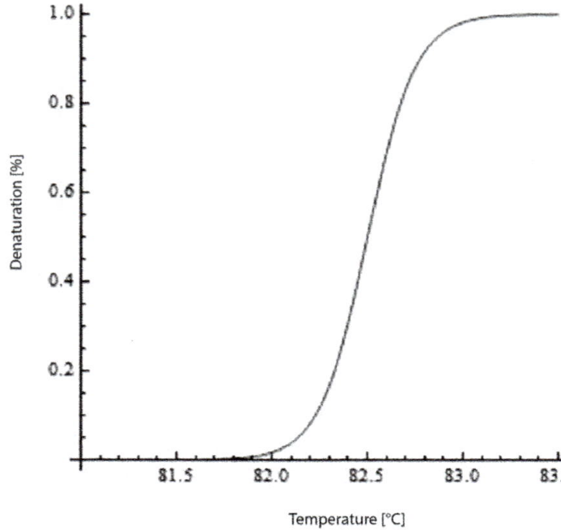

Fig. 20.2 Curve expressed by Expression (20.18)

possible to further simplify the model found, by uniting the two parameters b and T_m into one.

The model obtained perfectly conforms to laboratory results, where T_m (representing denaturation temperature) is an index of DNA guanine and cytosine percentage and the slope b is an index of DNA heterogeneity (and thus of genoma complexity): the broader the denaturation range, the more heterogeneous the system. As we know from literature the DNA content of cytosine and guanine can be estimated with T_m. It thus appears possible to further simplify the model found, by uniting the two parameters b and T_m into one.

To do so, we introduce the variable α ($0 < \alpha < 1$), representing the percentage of guanine and cytosine. As said, T_m is directly proportional to α, while b depends on a function of α as shown in Eq. (20.20):

$$T_m \propto \alpha, \tag{20.19}$$

$$b \propto \frac{1}{\alpha(1-\alpha)}. \tag{20.20}$$

The function $\alpha(1-\alpha)$ vanishes for $\alpha = 0$ and for $\alpha = 1$ and it reaches its maximum for $\alpha = 0.5$, so when heterogeneity is greatest (see Fig. 20.3 for a visual representation).

Introducing α in Expression (20.13) we obtain Expression (20.21):

$$x(T) = \frac{1}{1 + e^{g(\alpha)}(1 - \frac{T}{f(\alpha)})}. \tag{20.21}$$

Fig. 20.3 Curve expressed by Expression (20.20)

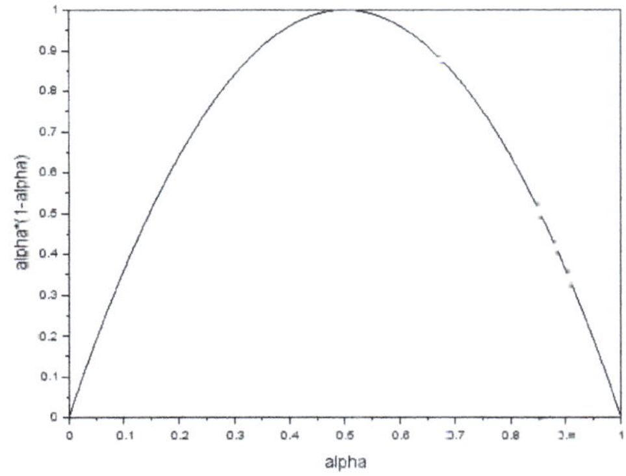

where

$$g(\alpha) = \frac{\lambda}{\mu + \alpha(1-\alpha)}, \tag{20.22}$$

$$f(\alpha) = (T_m(1) - T_m(0)) \cdot \alpha + T_m(0). \tag{20.23}$$

From experimental data, we know that

- $T_m(0) = 80$,
- $T_m(1) = 85$,
- $g(0) = \frac{\lambda}{\mu} = 20$,
- $g(\frac{1}{2}) = \frac{\lambda}{\mu + \frac{1}{4}} = 1.8$

Hence we obtain the graphs in Fig (20.4) for a set of alpha.

The curve on the left side in Fig. 20.4 represents a DNA with adenine and thymine only, thus having a lower T_m and a very steep slope. The same slope is found for the curve on the right hand side, representing a DNA with guanine and cytosine only and a higher T_m (as said, the three hydrogen bonds of guanine-cytosine pairs require a larger amount of energy to be broken than the double hydrogen bonds of adenine-thymine pairs). All curves in between these two show the behaviour of the function with steps of α every 10%: the slope is lowest where heterogeneity is maximum, so when $\alpha = 0.5$.

20.6 Conclusion

The aim of this study was to find a mathematical model to represent the empirical curve of DNA denaturation. Expression (20.21) satisfyingly reaches this purpose,

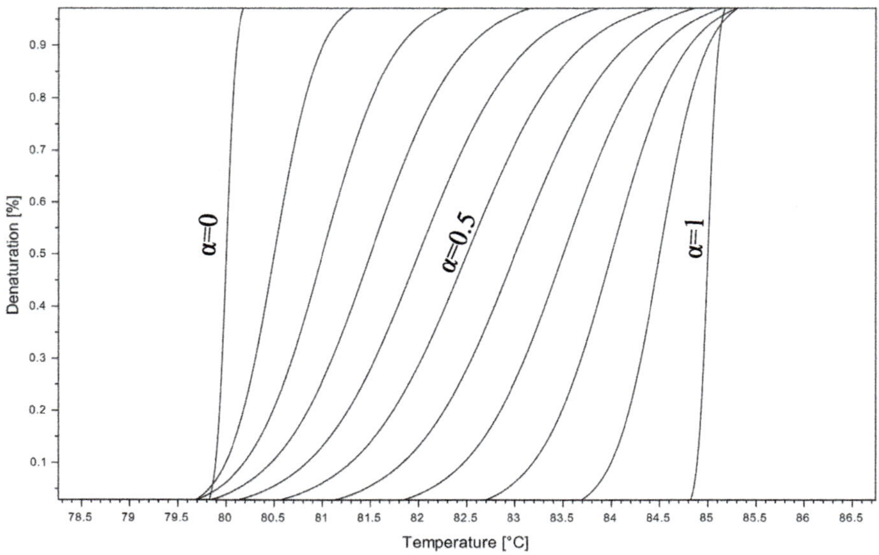

Fig. 20.4: Curve expressed by Expression (20.21) in the interval $0 < \alpha < 1$ (with intervals of 0.1).

where the guanine-cytosine content in the DNA (α) is the main variable to characterise the strands behaviour.

References

Ciarletta P, Preziosi L, Maugin GA (2012) Thermo-mechanics of growth and mass transfer: morphogenesis of seashells. Computer Methods in Biomechanics and Biomedical Engineering 15(sup1):110–112

Ciarletta P, Ambrosi D, Maugin GA, Preziosi L (2013) Mechano-transduction in tumour growth modelling. The European Physical Journal E 36(3):23

Mandel M, Marmur J (1968) [109] use of ultraviolet absorbance-temperature profile for determining the guanine plus cytosine content of dna. Methods in Enzymology 12(Part B):195–206

Nelson DL, Cox MM (2005) Lehninger Principles of Biochemistry. WH Freeman, New York

Rastogi SC (2003) Cell and Molecular Biology, 2nd edn. New Age International (P) Ltd.

SantaLucia JJ (1998) A unified view of polymer, dumbbell, and oligonucleotide DNA nearest-neighbor thermodynamics. Proceedings of the National Academy of Sciences USA 95(4):1460–1465

Chapter 21
Bulk Nonlinear Elastic Strain Waves in a Bar with Nanosize Inclusions

Igor A. Gula and Alexander M. Samsonov (†)

Abstract We propose a mathematical model for propagation of the long nonlinearly elastic longitudinal strain waves in a bar, which contains nanoscale structural inclusions. The model is governed by a nonlinear doubly dispersive equation (DDE) with respect to the one unknown longitudinal strain function. We obtained the travelling wave solutions to DDE, and, in particular, the strain solitary wave solution, which was shown to be significantly affected by parameters of the inclusions. Moreover we found some critical inaccuracies, committed in papers by others in the derivation of a constitutive equation for the long strain waves in a microstructured medium, revised them, and showed an importance of improvements for correct estimation of wave parameters.

21.1 Introduction

Studies of physical properties of composite materials (composites) are of a great importance due to variety of applications in physics and technology. A special class of composites is formed by the so called micro- and nanocomposites, which are multicomponent materials, consisting of a matrix and fillers of extraordinary small size. A comprehensive understanding of dynamical behaviour of composites is substantial for studies of the elastic properties, robustness and strength of materials for various applications, e.g., the non-destructive testing, since the latter allows one to detect defects in structural elements.

Igor A. Gula
The Ioffe Institute of the Russian Academy of Sciences, 194021, St. Petersburg, Russia (current Department of Physics, Chemistry and Pharmacy, University of Southern Denmark (SDU), Campusvej 55, Odense M 5230, Denmark)
e-mail: wswggg@gmail.com

Alexander M. Samsonov (†)
The Ioffe Institute of the Russian Academy of Sciences, 194021, St. Petersburg, Russia

© Springer International Publishing AG, part of Springer Nature 2018
H. Altenbach et al. (eds.), *Generalized Models and Non-classical Approaches in Complex Materials 1*, Advanced Structured Materials 89,
https://doi.org/10.1007/978-3-319-72440-9_21

Physical properties of nanocomposites differ dramatically from those of the matrix homogeneous isotropic materials. The local deformation in nanocomposites affects significantly their elastic properties and mechanical behaviour, however, the classical elasticity theory fails in attempts to describe these phenomena. As a result, nonlocal theories appeared in the beginning of last century and thoroughly elaborated later (Cosserat and Cosserat, 1909; Mindlin, 1964; Eringen and Suhubi, 1964a,b; Capriz, 1989), in which a material point was considered as a finite volume with additional degrees of freedom.

One of the linear nonlocal theories of elastic solids with microstructures (also called "microelastic" or "microstructured" solids) is the Mindlin theory of microstructured medium (Mindlin, 1964), which is based on two hypotheses. In accordance with the first one each material point of solids is considered as solids also and, simultaneously, as an elementary cell, while both macro- and microdisplacements are evaluated separately. The second hypothesis states that u'_i components of microdisplacement vector are linear in local coordinates, i.e. can be represented in the form $u'_i = x'_k\,\psi_{ki}(x_j,t)$, where x'_k – microcoordinates, and ψ_{ki} are microdeformation tensor components, depending only on macrocoordinates x_j. The function of potential energy density Π was chosen in the form of homogeneous quadratic function, depending on tensor components ε_{ij}, γ_{ij} and χ_{ijk}:

$$\begin{aligned}\Pi = &\frac{1}{2} c_{ijkl}\,\varepsilon_{ij}\,\varepsilon_{kl} + \frac{1}{2} b_{ijkl}\,\gamma_{ij}\,\gamma_{kl} + \frac{1}{2} a_{ijklmn}\,\chi_{ijk}\,\chi_{lmn} \\ &+ d_{ijklm}\,\gamma_{ij}\,\chi_{klm} + f_{ijklm}\,\chi_{ijk}\,\varepsilon_{lm} + g_{ijkl}\,\gamma_{ij}\,\varepsilon_{kl}\ ,\end{aligned} \qquad (21.1)$$

where $\varepsilon_{ij} = (\partial_i u_j + \partial_j u_i)/2$ – linear macrodeformation tensor, $\gamma_{ij} = \partial_i u_j - \psi_{ij}$ – a relative deformation tensor, $\chi_{ijk} = \partial_i \psi_{jk}$ – microdeformation gradient (in Mindlin's notation).

Considering an isotropic medium with central symmetry, Mindlin neglected terms, containing the coefficient tensors of 5th rank: d_{ijklm} and f_{ijklm}. Author's explanation was "there are no isotropic tensors of an odd rank". We follow this proposition since it makes the model much simpler, however, it has to be noted, the proposition is to be justified, since, e.g., the Levi-Civita symbols is known to constitute the 3rd order isotropic tensor. Moreover, according to Weisstein (2017), there are 6 isotropic tensors of 5th order.

The Mindlin theory has been applied to many problems of long strain waves in micro- and nanostructured solids, which have been studied during the last two decades. This may be explained mainly by relative simplicity of the mathematical models formulation in partial differential equations (PDE). In the linear problem of unidirectional long elastic longitudinal strain waves in a medium with microstructures (Porubov and Pastrone, 2004) the linear Mindlin theory was applied via formal addition both physical and geometrical nonlinearities, and the obtained results were used in their following work devoted to $1+2$D localized strain waves (Porubov et al, 2004). It was shown in Casasso et al (2010) that consideration of two scales of microstructures in a composite leads to a PDE of 6th order for long nonlinear strain waves. The problem of plane strain waves in microstructured solids was studied in Porubov et al (2012). The general model of solids with vectorial microstructures was proposed in Pastrone (2005) as a generalization of the Mindlin theory. An overview of various models of nonlocal media, exploited in the solid state mechanics, was presented in Engelbrecht and Braun (1998) from the viewpoint of wave propagation theory. Main theoretical results that have been obtained in the framework of the generalized continuum mechanics in the past 40 years are collected in a famous paper by Maugin (2011). A detailed review of mathematical models for strain waves in elastic microstructured solids can be found in Engelbrecht and Berezovski (2015).

As it has been noted, the Mindlin theory (Mindlin, 1964), being even the linear one, is often used as a background for models of nonlinear wave processes in composite solids. Many authors proposed various models for potential energy of elastic deformation, however, the linear parts of the models were inconsistent with the Mindlin model in most cases, and nonlinear terms were not well justified. Moreover, up to date there are no models for description of the nonlinear strain waves in any of three basic types of structural elements (rod, plate and shell), containing microstructures.

Nonlinear longitudinal strain waves in isotropic homogeneous rods have been also studied experimentally in the Ioffe Institute since 80s, see Samsonov (2001) and the studies are in progress up to now, e.g., Semenova et al (2011); Samsonov et al (2017). It should be noted that to interprete experimental data on bulk strain solitons propagation in a cylindrical rod it is of interest to obtain the solution in the form of strain soliton to the problem for a rectangular rod (a bar). It is known (Samsonov, 2001; Semenova et al, 2011; Samsonov et al, 2017), that for the holographic method of registration of such waves a specimen should have surfaces perpendicular to an incident laser light. Consequently, small parts are to be removed from an initial cylindrical rod. Results of estimations of bulk solitons in a bar would allow one to estimate an error, initiated by the necessary shape modification of a cylindrical rod.

The aim of the present work is to derive and to analyze the constitutive equation of the model for the long nonlinearly elastic longitudinal strain waves in a microstructured bar. To derive the equation we proposed the generalization of the Mindlin model and established the one-to-one correspondence between the generalization and the Murnaghan model for potential energy of an elastic deformation Murnaghan (1951) of solids. Besides we found and corrected some critical inaccuracies in recent paper (Porubov and Pastrone, 2004), where the Mindlin theory was applied. Moreover, the model for the long nonlinearly elastic longitudinal strain waves for an isotropic bar was refined to satisfy the condition of a free boundary with a given precision.

21.2 Refinement of the Model of a Continuous Microstructured Medium

The problem of the long nonlinearly elastic longitudinal strain waves in a microstructured *medium* was considered in Porubov and Pastrone (2004). Authors focused their attention on the $1+1$D case to simplify analysis and studied general properties of the nonlinear strain waves in a microstructured medium. The main result of the work is that presence of microscopic inclusions in a medium formally provides formation of the nonlinear localized strain waves (solitons, kinks). As it is known, energy canalization and, as a result, the nonlinear unidirectional strain solitary waves are impossible in an isotropic homogeneous medium without inclusions. However, it is to be noted that authors of Porubov and Pastrone (2004) assumed implicitly that the medium under consideration contains microstructures oriented rather specifically than in a chaotic way. Only the assumption about unidirectionality of microinclusions allowed the authors to derive the constitutive equation of the model for strain wave propagation in a microstructured medium in the form of the long nonlinearly elastic longitudinal localized unidirectional waves. Nevertheless, authors committed few inaccuracies, which led, in particular, to the conclusion that even the linear wave velocity was independent of inclusion presence and properties.

To improve the statement of the problem and correct some mistakes we reconsider the derivation of the governing equation for waves in microstructured medium. We introduce the Cartesian coordinate system $Oxyz$ in an elastic medium and consider propagation of a nonlinear strain wave along Ox axis. The geometrical nonlinearity is described by a single component of the Cauchy-Green macrostrain tensor $C_{11} = U_x + U_x^2/2$, where $U(x,t)$ is the macrodisplacements vector component, and the tensor $\boldsymbol{C}$ invariants may be written as follows: $I_1(\boldsymbol{C}) = tr\boldsymbol{C} = C_{11} = U_x + U_x^2/2$, $I_2(\boldsymbol{C}) = (tr\boldsymbol{C})^2/2 - tr\left(\boldsymbol{C}^2\right) = 0$, $I_3(\boldsymbol{C}) = \det \boldsymbol{C} = 0$. The microdisplacements vector also has a single non-zero component.

The potential energy density Π in Porubov and Pastrone (2004) was chosen to keep its macroscopic part consistent with the Murnaghan model and its microscopic part - with the Mindlin model for an isotropic material with central symmetry. Instead of an expression for Π only its partial derivatives were written with respect to the arguments C_{11}, γ_{11}, χ_{111} with reference to Mindlin (1964) and Murnaghan (1951):

$$\frac{\partial \Pi}{\partial C_{11}} = (\lambda + 2\mu) U_x + \beta U_x^2, \quad \frac{\partial \Pi}{\partial \gamma_{11}} = D(U_x - \psi), \quad \frac{\partial \Pi}{\partial \chi_{111}} = G \psi_x, \tag{21.2}$$

where λ, μ are the Lamé coefficients, l, m, n are the Murnaghan moduli, and $\beta = 3/2\,(\lambda + 2\mu) + l + m + 3n/2$. However, if one writes the precise expression for potential energy density Π in accordance with Mindlin (1964) and Murnaghan (1951):

$$\Pi = \frac{\lambda + 2\mu}{2} I_1^2(\boldsymbol{C}) + \frac{l + 2m}{3} I_1^3(\boldsymbol{C}) + \frac{D}{2} \gamma_{11}^2 + \frac{G}{2} \chi_{111}^2 + H C_{11} \gamma_{11}, \tag{21.3}$$

where D, G, H are elastic constants of microstructures, then the following relations will be derived instead of (21.2):

$$\frac{\partial \Pi}{\partial C_{11}} = (\lambda+2\mu)U_x + \left(\frac{\lambda+2\mu}{2}+l+2m\right)U_x^2 + O(U_x^3) + H\,(U_x-\psi)\,,$$
$$\frac{\partial \Pi}{\partial \gamma_{11}} = D(U_x-\psi) + H\left(U_x+\frac{U_x^2}{2}\right),\ \frac{\partial \Pi}{\partial \chi_{111}} = G\psi_x\,, \tag{21.4}$$

and, comparing (21.4) with (21.2), we conclude, that in the problem of waves in a nonlinearly elastic *medium* the coefficient n in $\partial\Pi/\partial C_{11}$ is absent, since it is multiplied by $I_3(\boldsymbol{C}) = 0$ in the Murnaghan model for the potential energy density. Besides, the term with $C_{11}\,\gamma_{11}$ was discarded for unknown reasons in Porubov and Pastrone (2004).

The wave motion equations in Porubov and Pastrone (2004) were written with the reference to Mindlin (1964) in the form:

$$\rho u_{tt} = \tau_x + \sigma_x\,, \qquad I\psi_{tt} = \sigma + \eta_x, \tag{21.5}$$

where ρ is material's density, I is microinclusions' inertia, $\tau = \partial\Pi/\partial C_{11}$, $\sigma = \partial\Pi/\partial\gamma_{11}$, $\eta = \partial\Pi/\partial\chi_{111}$. Mindlin derived (21.5) for the linear strain tensor ε_{ij}, and, consequently, defined τ as $\partial\Pi/\partial\varepsilon$, however, as mentioned in Porubov and Pastrone (2004), the geometrical nonlinearity was taken into account Therefore the wave motion equations are to be written in the following form:

$$\rho U_{tt} = (\tau+\sigma+\tau U_x)_x\,, \qquad I\psi_{tt} = \sigma + \eta_x. \tag{21.6}$$

This can be argued easily. Neglecting the microinclusions influence, the motion equations in a 3D medium, with the physical and geometrical nonlinearities being taken into account, can be written as:

$$\rho\boldsymbol{U}_{tt} = \nabla\boldsymbol{P}\,, \qquad \boldsymbol{P} = \frac{\partial\Pi}{\partial\boldsymbol{C}}\,(\boldsymbol{E}+\nabla\boldsymbol{U}) \tag{21.7}$$

where $\boldsymbol{U}(x,y,z,t) = U(x,y,z,t)\boldsymbol{i} + V(x,y,z,t)\boldsymbol{j} + W(x,y,z,t)\boldsymbol{k}$ is the displacement vector, $\boldsymbol{P}$ is the Piola-Kirchhoff strain tensor, $\boldsymbol{E}$ is the unit tensor. Then the first equation in (21.7) for $1+1$D problem is reduced to:

$$\rho U_{tt} = (\tau + \tau U_x)_x,$$

that differs from the first equation in (21.5) by the term $(\tau U_x)_x$. This term was absent in Porubov and Pastrone (2004), obviously, because the term $(\partial\Pi/\partial\boldsymbol{C})\,\nabla\boldsymbol{U}$ in the second equation (21.7) was neglected in Porubov and Pastrone (2004), and, as a result, the physical nonlinearity was *not taken* into account. The rigorous derivation of the motion equations based on the Hamilton principle leads to the same refinements.

For convenience a refined derivation of the general equation for the long nonlinearly elastic longitudinal strain damped waves in a microstructured medium is presented in the Supplement 21.5.1 in a dimensionless form:

$$\begin{aligned}&u_{tt}-u_{xx}\,\alpha_0-\left(u^2\right)_{xx}\varepsilon\,\alpha_1-u_{xxt}\,\gamma\,\alpha_2+u_{xxxx}\,\delta\,\alpha_3-u_{xxtt}\left[\delta\,\alpha_4+\gamma^2\,\alpha_7\right]\\&+\gamma\delta\left[u_{xxxxt}\,\alpha_5+u_{xxttt}\,\alpha_6\right]+\delta\,\varepsilon\left[\left(2u\,u_{xx}+u_x^2\right)_{xx}\overline{\alpha}_5-\left(2u\,u_{tt}+u_t^2\right)_{xx}\overline{\alpha}_6\right]\\&+\left(u\,u_t\right)_{xx}\gamma\varepsilon\,\overline{\alpha}_4+\delta^2\left[u_{xxxxxx}\,\alpha_8-u_{xxxxtt}\,\alpha_9+u_{xxtttt}\,\alpha_{10}\right]+\left(u^3\right)_{xx}\varepsilon^2\,\alpha_{11}=0\end{aligned} \tag{21.8}$$

along with definition of the parameters ε, γ, δ. For simplicity we restrict here our consideration by the model of solitary waves without damping, i.e. dissipative terms are neglected, $\gamma=0$, and assume that nonlinearity and dispersion terms are balanced, $\delta=O(\varepsilon)$. Leaving the terms of order $O(\varepsilon)$, we obtain the doubly dispersive equation (DDE) from (21.8) in the form:

$$u_{tt}-u_{xx}\,\alpha_0-\varepsilon\left[\left(u^2\right)_{xx}\alpha_1-u_{xxxx}\,\alpha_3+u_{xxtt}\,\alpha_4\right]=0, \tag{21.9}$$

where $u=U_x$, $\alpha_0=1-H^2/(D(\lambda+2\mu))$, $\alpha_1=(l+2m)/(\lambda+2\mu)+3\,\alpha_0/2$, $\alpha_4=I^*\,(1+H/D)^2$, $\alpha_3=G^*\,(1+H/D)^2/(\lambda+2\mu)$, $G^*=G/p^2$, $I^*=I/\left(\rho\,p^2\right)$, and p is a characteristic size of a microstructural element (Porubov and Pastrone, 2004). The equation (21.9) after introduction of the phase variable $\theta=x\pm\mathrm{v}t$ can be reduced to the Weierstrass equation, which general solution can be found in terms of the elliptic Weierstrass $\wp$-function (Samsonov, 2001). In particular, the solution can be written in the form of solitary strain waves with velocity v:

$$u(\theta)=\frac{3\left(\mathrm{v}^2-\alpha_0\right)}{2\varepsilon\,\alpha_1}\cosh^{-2}\left(\theta\sqrt{\frac{\alpha_0-\mathrm{v}^2}{4\varepsilon\left(\alpha_3-\mathrm{v}^2\,\alpha_4\right)}}\right),\quad\theta=x\pm\mathrm{v}t, \tag{21.10}$$

which exists, obviously, if the following conditions for a wave velocity are satisfied:

$$\left[\begin{array}{l}\mathrm{v}^2>\max\left\{\alpha_0;\,\dfrac{\alpha_3}{\alpha_4}\right\}\\ \mathrm{v}^2<\min\left\{\alpha_0;\,\dfrac{\alpha_3}{\alpha_4}\right\}\end{array}\right.\Leftrightarrow\left[\begin{array}{l}\mathrm{v}^2>\max\left\{(\lambda+2\mu)-\dfrac{H^2}{D};\,\dfrac{\rho\,G}{I}\right\}\\ \mathrm{v}^2<\min\left\{(\lambda+2\mu)-\dfrac{H^2}{D};\,\dfrac{\rho\,G}{I}\right\}\end{array}\right.$$

It can be seen from (21.9) and (21.10), that the linear wave velocity in a microstructured medium is defined by the expression $c_0\sqrt{\alpha_0}$, where $c_0=(\lambda+2\mu)/\rho$, and turns out to be dependent on the microstructure parameters also. In fact potential influence of microstructures on behaviour of linear strain waves in a medium was shown earlier (see Engelbrecht and Pastrone, 2003; Engelbrecht et al, 2005). Nevertheless it is useful to notice that if authors of the two mentioned papers strictly follow the Mindlin theory they would not have revealed this phenomena. Anyway this effect could not be taken into account in Porubov and Pastrone (2004) because the term $H\,C_{11}\,\gamma_{11}$ from (21.3) was neglected. Besides, the additional term $H\,C_{11}\,\gamma_{11}$ in Π also contributes to the condition of the existence of the solitary strain wave.

21.3 Nonlinear Strain Waves in a Bar

We consider firstly the long nonlinear wave propagation in an isotropic bar and generalize the results to take into consideration the microinclusions in the material. It allows us to underline details arrived due to rectangle cross section, and, separately, due to microsctructure of the material.

21.3.1 The Model for Wave Propagation in an Isotropic Bar

Let us consider the problem of the long nonlinearly elastic longitudinal strain waves in a rod with rectangular cross section (a bar). The dimensional DDE was derived in Khusnutdinova and Samsonov (2008) in the form:

$$u_{tt} - \frac{E}{\rho} u_{xx} = \frac{\beta_0}{2\rho} (u^2)_{xx} - \frac{\nu^2 R^2}{3} \left[u_{tt} - \frac{E}{2\rho(\nu+1)} u_{xx} \right]_{xx}, \qquad (21.11)$$

where $u = U_x$, $U(x,t)$ is the longitudinal displacement function, E is the Young modulus, ν is the Poisson ratio, l, m, n are the 3rd order Murnaghan moduli, $\beta_0 = 3E + 2l(1-2\nu)^3 + 4m(\nu+1)^2(1-2\nu) + 6n\nu^2$, $R^2 = (H_y^2 + H_z^2)/4$, H_y and H_z are the bar cross section sizes along the Oy and Oz axes resp. The Eq. (21.11) is based on the plane section hypothesis $U(x,y,z,t) = U_0(x,t)$ and the Love hypothesis for transversal displacements $V(x,y,z,t) = -\nu y U_{0,x}$, $W(x,y,z,t) = -\nu z U_{0,x}$. However, the hypotheses does not lead to the free boundary condition with the necessary accuracy. Our aim is to generalize the hypotheses in such a way for the condition of the bar free boundary to be satisfied more accurately.

The bar geometry is presented in Fig. 21.1. The bar material is assumed to be the homogeneous isotropic one without inclusions, and the bar length is much greater than the cross section sizes. Let us introduce the scales for x, t and u as Λ, T and A correspondingly and dimensionless variables $x' = x/\Lambda$, $t' = t/\mathrm{T}$, $u' = u/A$. We select the wave solutions with velocities close to velocity of the linear longitudinal strain wave in the bar $c_0 = \sqrt{E/\rho}$, and suppose that $\Lambda = c_0\mathrm{T}$. Now (21.11) can be rewritten in the dimensionless form (primes are omitted):

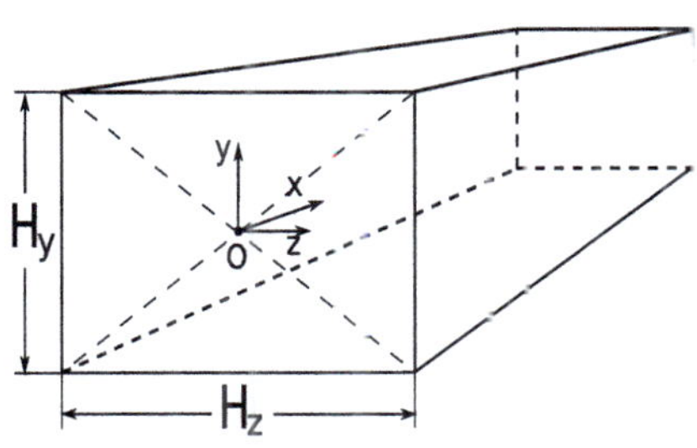

Fig. 21.1 A rod with rectangular cross-section (a bar). The bar length is supposed to be much greater compared to its cross section sizes H_y and H_z. A coordinate system is introduced with the origin placed in the centre of one of the bar edges. The bar is oriented by its longest edge parallel to the abscissa axis

$$u_{tt}-u_{xx}=\frac{\beta\,\mathrm{A}}{2E}\,(u^2)_{xx}+\frac{\nu^2 R^2}{3\Lambda^2}\left[u_{tt}-\frac{1}{2(\nu+1)}\,u_{xx}\right]_{xx}. \tag{21.12}$$

Deformations are supposed to be elastic, i.e., small enough to assume that $\mathrm{A} \ll 1$, therefore the small parameter $\varepsilon = \mathrm{A}$ can be introduced. Let us consider the long longitudinal waves, $R^2/\Lambda^2 \ll 1$, and suppose that the nonlinear and the dispersive terms are in balance, $\beta\,\mathrm{A}/(2E) = \nu^2 R^2/(6\Lambda^2) \Rightarrow \mathrm{A} \sim R^2/\Lambda^2$. Then (21.12) can be rewritten with the first order accuracy with respect to ε in the form:

$$u_{tt}-u_{xx}=\varepsilon\left[\frac{\beta}{2E}\,(u^2)_{xx}+\frac{\nu^2}{3}\left(u_{tt}-\frac{1}{2(\nu+1)}\,u_{xx}\right)_{xx}\right]. \tag{21.13}$$

We may write the dimensionless components P_{yx}, P_{yy}, P_{yz}, P_{zx}, P_{zy}, P_{zz} of the Piola-Kirchhoff tensor $\boldsymbol{P} = \partial\Pi_{3\mathrm{D}}/\partial\boldsymbol{C}(E+\nabla\boldsymbol{U})$ upon an introduction the additional scales for transversal coordinates $[y] = H_y$, $[z] = H_z$ and for displacements $[V] = \mathrm{A}H_y$, $[W] = \mathrm{A}H_z$, and a definition of dimensionless variables $y' = y/H_y$, $z' = z/H_z$, $V' = V/(\mathrm{A}H_y)$, $W' = W/(\mathrm{A}H_z)$. The ratio of the bar cross section sizes is denoted as $H_z/H_y = k$.

Let us expand the displacements U, V, W in a series with respect to ε. Collecting terms with the same powers of ε in the dimensionless components of $\boldsymbol{P}$ and successfully setting them to zero on the bar boundaries $y = \pm 1/2$, $z = \pm 1/2$, we find the generalized expressions for the displacements with an accuracy of $O(\varepsilon^2)$ in the form:

$$U(x,y,z,t)=U_0(x,t)+\varepsilon\,\frac{2\nu}{1+k^2}\,(y^2+k^2z^2)\,U_{0,xx}, \tag{21.14}$$

$$V(x,y,z,t)=-\nu\,y\,U_{0,x}-\varepsilon\,\frac{2\nu}{1+k^2}\left[\frac{y^3}{3}+\frac{y}{4}\,((1+k^2)\,\nu-1)\right]U_{0,xxx}-$$
$$-\varepsilon\left[\frac{\nu(1+\nu)}{2}+\frac{(1-2\nu)(1+\nu)}{E}\,\left(l(1-2\nu)^2+2m\nu(1+\nu)-n\nu\right)\right]yU_{0,x}^2,$$

$$W(x,y,z,t)=-\nu\,z\,U_{0,x}-\varepsilon\,\frac{2\nu}{1+k^2}\left[\frac{k^2z^3}{3}+\frac{z}{4}\,((1+k^2)\,\nu-k^2)\right]U_{0,xxx}-$$
$$-\varepsilon\left[\frac{\nu(1+\nu)}{2}+\frac{(1-2\nu)(1+\nu)}{E}\,\left(l(1-2\nu)^2+2m\nu(1+\nu)-n\nu\right)\right]zU_{0,x}^2$$

or in dimensional form (useful for interpretation of experimental data):

$$U(x,y,z,t)=U_0(x,t)+\frac{\nu}{2}\,(y^2+z^2)\,U_{0,xx}, \tag{21.15}$$

$$V(x,y,z,t)=-\nu\,y\,U_{0,x}-\frac{\nu}{2}\left[\frac{y^3}{3}+\frac{yH_y^2}{4}\,((1+k^2)\,\nu-1)\right]U_{0,xxx}-$$
$$-\left[\frac{\nu(1+\nu)}{2}+\frac{(1-2\nu)(1+\nu)}{E}\,\left(l(1-2\nu)^2+2m\nu(1+\nu)-n\nu\right)\right]yU_{0,x}^2,$$

$$W(x,y,z,t) = -\nu z U_{0,x} - \frac{\nu}{2}\left[\frac{z^3}{3} + \frac{zH_z^2}{4}\left(\frac{1+k^2}{k^2}\nu - 1\right)\right] U_{0,xxx} -$$

$$-\left[\frac{\nu(1+\nu)}{2} + \frac{(1-2\nu)(1+\nu)}{E}\left(l(1-2\nu)^2 + 2m\nu(1+\nu) - n\nu\right)\right] zU_{0,x}^2.$$

Application of the generalized expressions (21.14, 21.15) leads to the refined DDE in a dimensionless form with the small parameter ε:

$$u_{tt} - u_{xx} = \varepsilon\left[\frac{\beta}{2E}(u^2)_{xx} - \frac{\nu(1-\nu)}{3} u_{xxtt} + \frac{\nu}{3} u_{xxxx}\right], \qquad (21.16)$$

or in dimensional form:

$$u_{tt} - c_0^2 u_{xx} = \frac{\beta}{2\rho}(u^2)_{xx} - \frac{\nu(1-\nu)}{3} R^2 u_{xxtt} + \frac{\nu}{3} R^2 c_0^2 u_{xxxx}, \qquad (21.17)$$

where $u(x,t) = U_{0,x}(x,t)$. The comparison of (21.16), (21.17) with (21.13), (21.11), shows that the refined hypotheses (21.14), (21.15) led to the signs changes in the dispersion terms. This important result is a consequence of taking into account the highest degrees corrections of the transversal displacements.

The particular solution to the DDE with constant coefficients in the form of the solitary bell-shaped wave (soliton) (Samsonov, 2001) is well known. To interpret physical experiments it is useful to write the solutions to the dimensional equations (21.11) for conventional (subindex usl) and refined (subindex rfn) statements:

$$u_{\mathrm{usl}}(\theta) = \mathrm{A}\cosh^{-2}\left(\frac{\theta}{B_{\mathrm{usl}}(\mathrm{A})}\right) = \frac{3(\rho\,\mathrm{v}^2 - E)}{\beta}\cosh^{-2}\left(\frac{\theta}{B_{\mathrm{usl}}(\mathrm{v})}\right), \qquad (21.18)$$

$$B_{\mathrm{usl}}(\mathrm{A}) = 2\nu R\sqrt{\frac{1}{3} + \frac{E}{\mathrm{A}\beta}\frac{2\nu+1}{2(\nu+1)}}, \qquad B_{\mathrm{usl}}(\mathrm{v}) = 2\nu R\sqrt{\frac{\mathrm{v}^2 - \frac{c_0^2}{2(1+\nu)}}{3(\mathrm{v}^2 - c_0^2)}},$$

and (21.17):

$$u_{\mathrm{rfn}}(\theta) = \mathrm{A}\cosh^{-2}\left(\frac{\theta}{B_{\mathrm{rfn}}(\mathrm{A})}\right) = \frac{3(\rho\,\mathrm{v}^2 - E)}{\beta}\cosh^{-2}\left(\frac{\theta}{B_{\mathrm{rfn}}(\mathrm{v})}\right), \qquad (21.19)$$

$$B_{\mathrm{rfn}}(\mathrm{A}) = 2\nu R\sqrt{\frac{E}{\mathrm{A}\beta} - \frac{1-\nu}{3\nu}}, \qquad B_{\mathrm{rfn}}(\mathrm{v}) = 2\nu R\sqrt{\frac{c_0^2 - \mathrm{v}^2(1-\nu)}{3\nu(\mathrm{v}^2 - c_0^2)}}.$$

In (21.18), (21.19) the phase variable $\theta = x \pm \mathrm{v}t$ was introduced, together with $\mathrm{A} = 3(\rho\,\mathrm{v}^2 - E)/\beta$ as the amplitude of the solitary wave, and v as the solitary wave velocity.

The refined DDE (21.17) differs from the refined model for the nonlinearly elastic longitudinal strain waves in an isotropic cylindrical rod, proposed in Samsonov

(2001), only by the dispersion terms coefficients. Denominators of the coefficients of our model contain 3 instead of 2. Considering propagation of solitons with the same amplitudes in a bar with cross section sizes H_y and H_z and in a cylindrical rod of radius R, one can conclude that the soliton velocities will be the same. However, if the rod and the bar cross sections are related as $R^2 = \left(H_y^2 + H_z^2\right)/4$, then the soliton in the bar will have the width of $\sqrt{3/2} \approx 1.23$ times less than that of the soliton in the cylindrical rod.

The condition for the existence of the solution in the form (21.18) has the following form:

$$\left[\begin{array}{l} \mathrm{v} > c_0, \\ \mathrm{v} < \dfrac{c_0}{\sqrt{2(\nu+1)}} \end{array}\right.,$$

whereas the condition of existence of the solution (21.19) to the refined DDE is:

$$c_0 < \mathrm{v} < \frac{c_0}{\sqrt{1-\nu}}$$

and provides a limit to the maximal velocity of the solitary waves in a bar. Since $0 < \nu < 0.5$ for most isotropic materials, then both the existence conditions are relevant.

To draw the graphs for comparison and estimation of quantitative characteristics of the solitary *compression* waves the polystyrene (PS) elastic moduli were taken from Hughes and Kelly (1953), whereas the elastic properties for plexiglas (polymethylmethacrylate, PMMA) were found in Semenova et al (2011), see Table 21.1. The bar cross section sizes were equal to 10×10mm along axes y, and z, and the strain soliton amplitudes for PS and PMMA were equal, correspondingly, to -3.2×10^{-4} and -2.3×10^{-4}, and taken from Samsonov et al (2017). Quantitative estimations for velocities c_0 of linear waves and velocities v of the nonlinear waves, the values of the nonlinear term coefficients β and the solitons full width at the half maximum of an amplitude (FWHM) for (21.18) and (21.19) are presented in Table 21.2. The results of numerical estimations allow one to conclude that the solution to the refined equation defines the solitary wave with a greater width. Besides, the

Table 21.1: Elastic moduli of PS and PMMA (Hughes and Kelly, 1953; Semenova et al, 2011)

	density	Lamé coefficients		Young's modulus	Poisson's ratio	3rd order Murnaghan moduli		
	ρ kg/m^3	λ N/m^2 10^{10}	μ N/m^2 10^{10}	E N/m^2 10^{10}	ν	l N/m^2 10^{10}	m N/m^2 10^{10}	n N/m^2 10^{10}
PS	1060	0.289 ±0.001	0.138 ±0.001	0.369 ±0.003	0.338 ±0.001	−1.89 ±0.32	−1.33 ±0.29	−1.0 ±0.14
PMMA	1160	0.39	0.186	0.498	0.339	−1.09	−0.77	−0.14

Table 21.2: Calculation results

	linear wave velocity	nonlinear solitary wave velocity	nonlinear term coefficient	width at amplitude half max, usual	width at amplitude half max., refin.
	c_0, m/s	v, m/s	β, N/m^2 10^{10}	$FWHM_{usl}$, cm	$FWHM_{rfn}$, cm
PS	1866.8 ±7.6	1867.4 ±7.7	−2.786 ±0.799	13.6 ±2.7	17.1 ±3.4
PMMA	2071.8	2071.9	−0.458	46.2	58.4

values of c_0 and v differ insignificantly, and at the same time, since the solitons have the amplitudes laying in admissible limits then the deformations do not transfer a bar to the plastic zone. The graphs of (21.18) and (21.19) are presented in Fig. 21.2.

21.3.2 The Refined Model Application for a Bar with Nanosize Inclusions

To derive the governing equations we use the Mindlin theory Mindlin (1964) and the refinements of the Love hypothesis and the plane section hypothesis, obtained in previous subsection.

Let us assume that a bar is made of a nanocomposite containing a matrix with small scale structural inclusions embedded (nanostructures). According to the Mindlin theory each inclusion is a parallelepiped, and its edges are oriented parallel to the bar edges. We shall call these inclusions as nanobars. For each nanobar the local

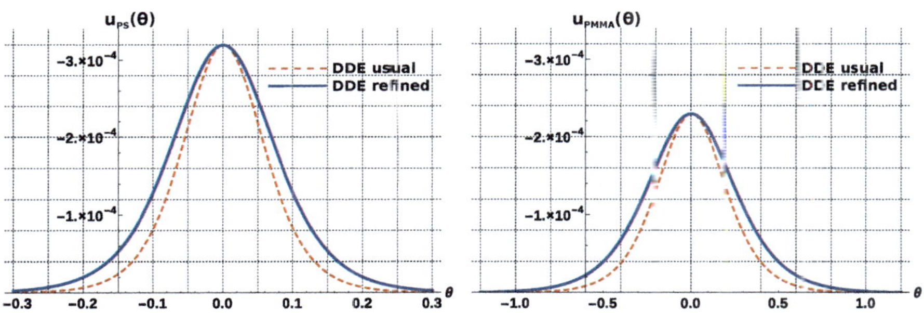

Fig. 21.2: Comparison graphs of the solutions to the DDE (21.18) and to the refined DDE (21.19) for the PS bar (left graphs) and the PMMA bar (right graphs)

Cartesian coordinate system is introduced with its origin placed in the centre of one of the nanobar end surfaces. The Ox' axis of the system is perpendicular to the end surface, while axes Oy' and Oz' are parallel to it. The axes of all of the local nanocoordinate systems are codirectional to the axes of the macrocoordinate system. Let us introduce the 3D microdisplacement vector in each nanobar

$$\boldsymbol{U'}(x',y',z',t) = U'(x',y',z',t)\boldsymbol{i} + V'(x',y',z',t)\boldsymbol{j} + W'(x',y',z',t)\boldsymbol{k}.$$

We assume additionally that all nanobars have lengths much greater than their cross section linear sizes, which, in turn, are equal to H'_y and H'_z along the corresponding axes. Consequently, the Love hypothesis and the plane section hypothesis can be admitted for nanodisplacements:

$$\begin{aligned} U'(x',y',z',t) &= U'_0(x',t), \\ V'(x',y',z',t) &= -\kappa y' U'_{0,x'}, \\ W'(x',y',z',t) &= -\kappa z' U'_{0,x'}, \end{aligned} \tag{21.20}$$

where κ is a nanoscopic analog of the Poisson ratio of the bar matrix. The refined hypotheses obtained previously are not suitable for nanodisplacements since it would be impossible to eliminate the local coordinates x', y', z' in the motion equations. The Mindlin hypothesis is used in a simple form $U'_0(x',t) = x'\,\varphi(x,t)$, choice of $\varphi(x,t)$ instead of $\varphi(x,y,z,t)$ will be analyzed later. The gradient tensor of the nanodisplacements $\nabla'\boldsymbol{U'}$ is denoted as Ψ.

The potential energy density $\Pi_{3\text{D}}$, according to Mindlin (1964), is a quadratic function of components of the Cauchy-Green strain tensor $\boldsymbol{C} = C_{ij}$, of the relative deformation tensor $\Gamma = \nabla\boldsymbol{U} - \Psi = \gamma_{ij}$ and of the nanodeformation gradient $\mathrm{X} = \nabla\Psi = \chi_{ijk}$. Assuming the materials of the bar (both matrix and filler) are isotropic and taking into account that $\boldsymbol{C}$ is the symmetrical tensor, one can write $\Pi_{3\text{D}}$ in the form:

$$\begin{aligned} \Pi_{3\text{D}} = {} & \frac{A_1}{2} C_{ii}C_{jj} + A_2 C_{ij}C_{ij} + \frac{B_1}{2}\gamma_{ii}\gamma_{jj} + \frac{B_2}{2}\gamma_{ij}\gamma_{ij} + \frac{B_3}{2}\gamma_{ij}\gamma_{ji} \\ & + D_1 C_{ii}\gamma_{jj} + D_2 C_{ij}\,(\gamma_{ij} + \gamma_{ji}) + G_1\chi_{iik}\chi_{kjj} + G_2\chi_{iik}\chi_{jkj} \\ & + \frac{G_3}{2}\chi_{iik}\chi_{jjk} + \frac{G_4}{2}\chi_{ijj}\chi_{ikk} + G_5\chi_{ijj}\chi_{kik} + \frac{G_6}{2}\chi_{iji}\chi_{kjk} + \frac{G_7}{2}\chi_{ijk}\chi_{ijk} \\ & + G_8\chi_{ijk}\chi_{jki} + \frac{G_9}{2}\chi_{ijk}\chi_{ikj} + \frac{G_{10}}{2}\chi_{ijk}\chi_{jik} + \frac{G_{11}}{2}\chi_{ijk}\chi_{kji}. \end{aligned} \tag{21.21}$$

Note that the macroscopic physical nonlinearity is not taken into acount in (21.21). To fix this, let us add to $\Pi_{3\text{D}}$ an expression, cubic with respect to the $\boldsymbol{C}$ components in the form $A_{ijklmn}C_{ij}C_{kl}C_{mn}$. It contains 729 terms in general, whereas only 56 are independent due to symmetry of $\boldsymbol{C}$. Since the bar is isotropic, than the tensor A_{ijklmn} has to be isotropic as well, and, consequently, it has to be a linear function of the Kronecker delta productions. It is known, that there are 15 independent productions of three Kronecker deltas, providing isotropic tensors of 6th rank (Mindlin, 1964):

$$\begin{aligned} A_{ijklmn} &= A_3\,\delta_{ij}\,\delta_{kl}\,\delta_{mn} + A_4\,\delta_{ij}\,\delta_{km}\,\delta_{ln} + A_5\,\delta_{ij}\,\delta_{kn}\,\delta_{lm} + A_6\,\delta_{ik}\,\delta_{jl}\,\delta_{mn} \\ &+ A_7\,\delta_{ik}\,\delta_{jm}\,\delta_{ln} + A_8\,\delta_{ik}\,\delta_{jn}\,\delta_{lm} + A_9\,\delta_{il}\,\delta_{jk}\,\delta_{mn} + A_{10}\,\delta_{il}\,\delta_{jm}\,\delta_{kn} \\ &+ A_{11}\,\delta_{il}\,\delta_{jn}\,\delta_{km} + A_{12}\,\delta_{im}\,\delta_{kl}\,\delta_{jn} + A_{13}\,\delta_{im}\,\delta_{jk}\,\delta_{ln} + A_{14}\,\delta_{im}\,\delta_{kn}\,\delta_{jl} \\ &+ A_{15}\,\delta_{in}\,\delta_{kl}\,\delta_{jm} + A_{16}\,\delta_{in}\,\delta_{jk}\,\delta_{lm} + A_{17}\,\delta_{in}\,\delta_{km}\,\delta_{jl}. \end{aligned}$$

Convolution of $C_{ij}C_{kl}C_{mn}$ with each term of the written above shows that $A_4 = A_5 = A_6 = A_9 = A_{12} = A_{15}, A_7 = A_8 = A_{10} = A_{11} = A_{13} = A_{14} = A_{16} = A_{17}$, therefore it can be concluded that $A_{ijklmn}C_{ij}C_{kl}C_{mn} = A_3\,C_{ii}C_{jj}C_{kk} + A_4\,C_{ii}C_{jk}C_{jk} + A_7\,C_{ij}C_{ik}C_{jk}$. The obtained expression is equal to the third order correction in the Landau model for elastic energy of isotropic solid up to the coefficients, see Landau and Lifshitz (1986). The physical nonlinearity resulted from (21.21) has the form:

$$\frac{A_3}{3}\,C_{ii}C_{jj}C_{kk} + A_4\,C_{ii}C_{jk}C_{jk} + A_5\,C_{ij}C_{ik}C_{jk}.$$

Let us write the relations between the Mindlin and Murnaghan model coefficients, establishing one-to-one correspondence between the models:

$$A_1 = \lambda,\; A_2 = \mu,\; A_3 = l - m + \frac{n}{2},\; A_4 = m - \frac{n}{2},\; A_5 = \frac{n}{3}, \tag{21.22}$$

this, in turn, allows one to use the Lamé coefficients and the 3rd order Murnaghan moduli instead of the A_i coefficients of the Mindlin model.

Let us admit the hypotheses (21.15) for the displacements U, V, W and integrate the Lagrangian volume density $\mathscr{L}_{3\mathrm{D}}$ to eliminate the transversal coordinates y and z. The integration is possible to be done because the function φ depends only on the variable x. Applying the Hamilton's principle, we write the system of two Euler-Lagrange differential equations:

$$\begin{cases} \left(\dfrac{\partial \mathscr{L}_{1\mathrm{D}}}{\partial U_{0,t}}\right)_t = -\left(\dfrac{\partial \mathscr{L}_{1\mathrm{D}}}{\partial U_{0,x}}\right)_x + \left(\dfrac{\partial \mathscr{L}_{1\mathrm{D}}}{\partial U_{0,xx}}\right)_{xx} + \left(\dfrac{\partial \mathscr{L}_{1\mathrm{D}}}{\partial U_{0,xt}}\right)_{xt} \\ \qquad - \left(\dfrac{\partial \mathscr{L}_{1\mathrm{D}}}{\partial U_{0,xxx}}\right)_{xxx} - \left(\dfrac{\partial \mathscr{L}_{1\mathrm{D}}}{\partial U_{0,xxt}}\right)_{xxt} + \left(\dfrac{\partial \mathscr{L}_{1\mathrm{D}}}{\partial U_{0,xxxx}}\right)_{xxxx} + \left(\dfrac{\partial \mathscr{L}_{1\mathrm{D}}}{\partial U_{0,xxxt}}\right)_{xxxt}, \\ \left(\dfrac{\partial \mathscr{L}_{1\mathrm{D}}}{\partial \varphi_t}\right)_t = \left(\dfrac{\partial \mathscr{L}_{1\mathrm{D}}}{\partial \varphi}\right) - \left(\dfrac{\partial \mathscr{L}_{1\mathrm{D}}}{\partial \varphi_x}\right)_x \end{cases} \tag{21.23}$$

where

$$\mathscr{L}_{1\mathrm{D}} = \int\limits_{-H_y/2}^{H_y/2}\int\limits_{-H_z/2}^{H_z/2} \mathscr{L}_{3\mathrm{D}}\,dy\,dz$$

is the Lagrangian linear density.

To write (21.23) in dimensionless form, we introduce the scale A for φ and the new scale for time Λ/c_0, where c_0 is a characteristic velocity of a strain wave propagation that in general differs from that one defined in Sect. 21.3. Introducing the dimensionless function $\varphi' = \varphi/\mathrm{A}$ and the same simplifications about deformations and scale relations as those from Sect. 21.3, we multiply the first equation in (21.23) by $\Lambda^2/\left(\rho\, c_0^2 \mathrm{A} H_y H_z\right)$ and the second one by $1/\left(\mathrm{A} H_y H_z\right)$. The result is written up to the first order terms with respect to ε:

$$\begin{cases} u_{tt} = \dfrac{E+\alpha_1}{\rho\, c_0^2}\, u_{xx} + \dfrac{\alpha_2}{\rho\, c_0^2}\, \varphi_{xx} + \varepsilon \left\{ \dfrac{\beta_0+\alpha_3}{2\rho\, c_0^2} \left(u^2\right)_{xx} - \dfrac{\nu(1-\nu)}{3}\, u_{xxtt} \right. \\ \qquad \left. + \dfrac{\alpha_4}{\rho\, c_0^2} (u\,\varphi)_{xx} + \dfrac{1}{\rho\, c_0^2} \left(\dfrac{\nu E}{3} + \alpha_5 \right) u_{xxxx} + \dfrac{\alpha_6}{\rho\, c_0^2}\, \varphi_{xxxx} \right\}, \\ \beta_1\, \varphi = -\alpha_2\, u + \varepsilon \left\{ -\dfrac{\alpha_4}{2} u^2 - \dfrac{I c_0^2}{R^2}\, \varphi_{tt} - \alpha_6\, u_{xx} + \dfrac{\beta_2}{R^2}\, \varphi_{xx} \right\}, \end{cases} \tag{21.24}$$

where $u(x,t) = U_{0,x}(x,t)$, $R^2 = \left(H_y^2 + H_z^2\right)/4$, $\varepsilon = \mathrm{A} = R^2/\Lambda^2$, and the coefficients α_i, β_i are presented in the Supplement 21.5.2. Let us define $c_0^2 = E/\rho$ to present the coefficient at the u_{xx} in (21.24) in the form $1 + \alpha_1/\left(\rho\, c_0^2\right)$.

We apply the slaving principle, assuming $\varphi(x,t)$ is expanded with respect to ε as $\varphi = \varphi_0 + \varepsilon\, \varphi_1 + \ldots$. Substitution of the expansion for $\varphi(x,t)$ to the second equation of (21.23) results in:

$$\varphi_0 = -\frac{\alpha_2}{\beta_1} u, \ \varphi_1 = -\left(\frac{\alpha_6}{\beta_1} + \frac{\alpha_2\, \beta_2}{\beta_1^2 R^2} \right) u_{xx} + \frac{I c_0^2\, \alpha_2}{\beta_1^2 R^2}\, u_{tt} - \frac{\alpha_4}{2\beta_1} u^2 .$$

Let us substitute φ_0, φ_1 in the first equation of (21.23). Leaving the first order terms with respect to ε inclusively, we derive the equation for $u(x,t)$:

$$\begin{aligned} & u_{tt} - \left(1 + \frac{\alpha_1}{E} - \frac{\alpha_2^2}{E\, \beta_1} \right) u_{xx} \\ & = \varepsilon \left[\left(\frac{\beta_0 + \alpha_3}{2E} - \frac{3\, \alpha_2\, \alpha_4}{2E\, \beta_1} \right) \left(u^2\right)_{xx} - \left(\frac{\nu(1-\nu)}{3} - \frac{\alpha_2^2 I}{\beta_1^2 \rho R^2} \right) u_{xxtt} \right. \\ & \left. + \left(\frac{\nu}{3} + \frac{\alpha_5}{E} - \frac{2\, \alpha_2\, \alpha_6}{E\, \beta_1} - \frac{\alpha_2^2\, \beta_2}{E\, \beta_1^2 R^2} \right) u_{xxxx} \right] \end{aligned} \tag{21.25}$$

or in the dimensional form

$$u_{tt} - a_0\, u_{xx} = a_1 \left(u^2\right)_{xx} + a_2\, u_{xxtt} + a_3\, u_{xxxx}\,, \tag{21.26}$$

where:

$$a_0 = c_0^2 + \frac{\alpha_1}{\rho} - \frac{\alpha_2^2}{\rho\,\beta_1}, \qquad a_1 = \frac{\beta_0 + \alpha_3}{2\rho} - \frac{3\,\alpha_2\,\alpha_4}{2\rho\,\beta_1},$$
$$a_2 = -\frac{\nu(1-\nu)R^2}{3} + \frac{\alpha_2^2 I}{\rho\,\beta_1^2},\; a_3 = R^2\left(\frac{\nu\,c_0^2}{3} + \frac{\alpha_5}{\rho} - \frac{2\,\alpha_2\,\alpha_6}{\beta_1\,\rho}\right) - \frac{\alpha_2^2\,\beta_2}{\beta_1^2\,\rho}.$$

The equations above are similar to the DDE derived and investigated in Samsonov (2001) for the isotropic cylindrical rod. Moreover it can be found in Porubov and Pastrone (2004), where the problem of the long nonlinearly elastic longitudinal strain waves in *a medium* with nanoinclusions was considered, as well as in Sect. 21.3 of this paper (Eq. (21.17)) in the context of problem of the strain waves in an isotropic bar. The exact solutions to the DDE were derived in Samsonov (2001) in the form of a travelling wave, expressed via the elliptical Weierstrass $\wp$-function and, in particular, in the form of the solitary wave:

$$u_{nano}(\theta) = \mathrm{A}\cosh^{-2}\left[\theta\sqrt{\frac{a_1\,\mathrm{A}}{6\left(a_3 + \frac{2}{3}a_1\,a_2\,\mathrm{A} + a_0\,a_2\right)}}\right], \quad \theta = x \pm \mathrm{v}t, \qquad (21.27)$$

which amplitude A and velocity v are related as:

$$\mathrm{A} = \frac{3}{2}\,\frac{\mathrm{v}^2 - a_0}{a_1}.$$

The condition of existence of the solitary wave solution can be obtained straightforward:

$$\left[\begin{array}{l} \mathrm{A} \geq \max\{0; -\dfrac{a_3}{a_1\,a_2} - \dfrac{a_0}{a_1}\}, \\ \mathrm{A} \leq \min\{0; -\dfrac{a_3}{a_1\,a_2} - \dfrac{a_0}{a_1}\}. \end{array}\right.$$

The coefficient a_0 at the u_{xx} in (21.26) allows one to calculate the velocity of the linear strain waves propagation $c_1 = \sqrt{a_0}$, moreover a_0 defines the "effective" Young modulus $E_1 = \rho\,c_1^2$. It is clear that various values of the coefficients B_{1-3}, $D_{1,2}$, G_{1-11} may lead to a value of E_1 either less or greater than the Young modulus E of a matrix material of the bar.

To characterize the solution (21.27) quantitatively and draw its graph we took the same values for the amplitude A, density ρ and elastic properties of PS and PMMA as those in the Sect. 21.3. Additionally for simplicity we assumed that $\kappa = \nu$, $I = 10^{-3}$. We are hardly be able to derive analytically the conditions of the solution existense in form of the solitary wave due to a lot of coefficients. Instead, the inverse problem of mathematical modelling of the coefficients B_{1-3}, $D_{1,2}$, G_{1-11}, characterizing nanoinclusion properties, was solved with the help of the function *NMinimize*, embedded in the computer algebra system *Wolfram Mathematica* (Wolfram Research, Inc, 2017). As an example, we found the coefficients in such a way that the value L_{FWHM} of a solitary wave full width at the half maximum of an amplitude calculated as:

$$L_{\mathrm{FWHM}} = 2\,\mathrm{arcosh}(\sqrt{2})\sqrt{\frac{6\left(\rho\, a_3 + \frac{2}{3} a_1\, a_2\, \mathrm{A} + a_0\, a_2\right)}{a_1\, \mathrm{A}}}\,, \tag{21.28}$$

which was close to 4 cm, 3 cm and 2 cm, resp., while the linear wave propagation velocities c_1 were, correspondingly, within the ranges $[2000; 4000]$ m/sec., $[1200; 2000]$ m/sec. and $[800; 1200]$ m/sec., see Table 21.3. The values for L_{FWHM} and c_1 were chosen to demonstrate values of the nanostructure coefficients, necessary for the existence of a bulk soliton with parameters, suitable for observation in future physical experiments with nanocomposite bars.

Figure 21.3 presents the graph of the solution to the DDE (21.26) for a nanostructured bar with L_{FWHM} value equal to 3 cm compared with the solution to the refined DDE (21.17). It can be seen, that inclusions may formally provide a solitary strain pulse in a nanostructured bar with a width, essentially (in times) less than a width of similar soliton in an isotropic homogeneous bar.

Quantitative estimations of soliton velocities v_1 and linear strain wave velocities c_1 in nanostructured bars, given in the Table 21.4, show that inclusions affect significantly both the velocity of the linear waves as well as of the nonlinear ones. Formally they can provide both waves acceleration and deceleration compared to waves velocities in isotropic homogeneous materials, correspondingly, c_0 and v.

Table 21.3: Coefficients of the potential energy density function (21.21) that provides the value of a solitary wave L_{FWHM} equal to 4 cm, 3 cm or 2 cm, resp.

	Polystyrene			Polymethylmetacrylate		
L_{FWHM}	0.04 m	0.03 m	0.02 m	0.04 m	0.03 m	0.02 m
B_1, N/m^2 $\times 10^9$	−0.114	0.287	0.241	6.219	−0.559	−0.168
B_2, N/m^2 $\times 10^9$	−0.294	−0.489	0.248	−3.565	0.338	0.269
B_3, N/m^2 $\times 10^9$	−0.288	−0.489	0.237	−3.262	0.509	0.212
D_1, N/m^2 $\times 10^9$	−0.109	−0.460	0.603	4.177	−0.158	−0.382
D_2, N/m^2 $\times 10^9$	1.183	−0.204	−0.510	−4.105	0.619	0.624
G_1, N $\times 10^7$	−1.279	2.216	−2.12	−1.014	3.803	14.330
G_2, N $\times 10^7$	−1.243	2.67	−1.083	1.753	−5.798	−5.728
G_3, N $\times 10^7$	1.626	2.334	1.758	1.469	−6.613	−6.663
G_4, N $\times 10^7$	1.87	−1.951	2.48	1.251	−7.348	−3.88
G_5, N $\times 10^7$	−1.619	2.214	−1.41	1.584	3.848	−4.786
G_6, N $\times 10^7$	−1.785	2.334	−1.302	1.974	4.187	−5.168
G_7, N $\times 10^7$	1.554	2.411	1.834	−1.253	4.298	−7.998
G_8, N $\times 10^7$	1.506	2.671	1.787	−1.116	5.061	−4.694
G_9, N $\times 10^7$	1.374	−1.561	1.733	−1.323	−6.381	17.670
G_{10}, N $\times 10^7$	−1.139	−9.543	−2.129	−1.024	4.163	−3.979
G_{11}, N $\times 10^7$	−1.146	−9.51	−2.095	−1.024	−1.876	19.010

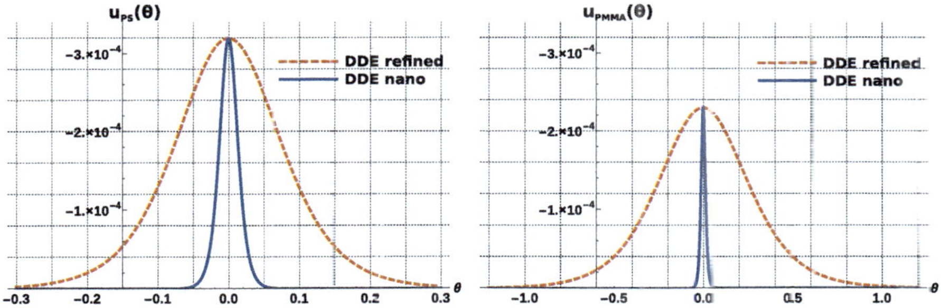

Fig. 21.3: Comparison graphs of the solutions to the DDE (21.26) for a nanostructured bar, and to the refined DDE (21.17) for an isotropic bar without nanostructures in the form of solitary waves for PS (left graphs) and PMMA (right graphs)

Table 21.4: Estimations of the propagation velocities of the linear waves c_1 and nonlinear longitudinal strain waves v_1 in nanostructured bars made of PS and PMMA with the corresponding values of the nanostructures coefficients from the Table 21.3

	Polystyrene			Polymethylmethacrylate		
$FWHM_{rfn}$, m	0.17			0.58		
c_0, m/s	1866.8			2071.8		
v, m/s	1867.4			2071.9		
$FWHM_{nano}$, m	0.04 m	0.03 m	0.02 m	0.04 m	0.03 m	0.02 m
c_1, m/s	3792.6	1930.5	1155.1	3830.7	1517.3	969.9
v_1, m/s	3793.2	1931.3	1155.7	3830.8	1517.4	970.1

Besides, it is clear from (21.25), (21.26) that among the coefficients characterizing nanoinclusion properties only B_i, D_i affect c_1 and v_1, whereas G_i contribute only to a nonlinear pulse width (21.28).

21.4 Conclusions

We propose the mathematical model for the long nonlinearly elastic longitudinal wave propagation in a rod with rectangular cross section (a bar), containing nanoscale inclusions. The model is based on the Mindlin theory of microstructured solids (Mindlin, 1964). We found and corrected some critical inaccuracies, committed by other authors in derivation of similar equation, see Porubov and Pastrone (2004). For the problem of the nonlinear strain wave propagation in an isotropic bar the formulae were obtained, which generalized the Love hypothesis and the plane section hypothesis and allowed one to satisfy the free boundary conditions with desired accuracy.

We note that inaccuracies committed in Porubov and Pastrone (2004) led to the model equations, which were incorrect, in particular, in expressions for the velocity of linear waves propagation in a material with small scale structural inclusions, that was equal to the one in a matrix material. We showed, that the presence of inclusions affected even the velocity of linear strain waves propagation.

The doubly dispersive equation (DDE) for the long nonlinearly elastic longitudinal strain waves propagation in a bar was refined. The derived model satisfies the free boundary condition with desired accuracy. The comparison of the model with the one for the long nonlinearly elastic longitudinal strain waves propagation in a cylindrical isotropic rod Samsonov (2001) showed, that when the compression strain solitons in a bar and in a rod have the same amplitudes, then their velocities are equal. Besides, if the bar and the rod cross sections are related as $R^2 = (H_y^2 + H_z^2)/4$, then the soliton in the bar has the width which is $\sqrt{3/2} \approx 1.23$ times less than that of the soliton in the rod.

The results obtained for the bar with nanoinclusions demonstrate that inclusions *can* define the admissible limits of amplitudes and propagation velocities of the solitary waves in elastic wave guides. Moreover it is shown that inclusions *can* also affect significantly the qualitative wave characteristics, e.g., an ability to focus strain solitons or spread them.

If no microstructure inclusions are presented, i.e. when $B_{1-3} = 0$, $D_{1,2} = 0$, $G_{1-11} = 0$, the coefficients α_{1-6} and β_{1-2} in (21.25) are equal to 0 and, as a consequence, (21.26) reduces to the refined DDE (21.17). It can be seen from (21.26) that reduction of R in the dispersion terms makes the nanostructure influence (terms $\alpha_2^2 I/(\rho\,\beta_1^2)$ and $\alpha_2^2 \beta_2/(\rho\,\beta_1^2)$ in the coefficients at the u_{xxtt} and u_{xxxx}, resp.) more significant in comparison with the macroscopic one. This conclusion seems to be predictable as the reduction of the rod size should lead to more significant effects of the nanostructure. However, from the formal viewpoint, an exact estimation depends on the relation between parameters of the nanostructure.

Acknowledgements The support of AMS by the Russian Science Foundation via grant no. 17-72-20201 is gratefully acknowledged. Additionally we sincerely thank I.V. Semenova for the useful comments and kind help during the paper preparation.

21.5 Supplement

21.5.1 The Model for Longitudinal Nonlinearly Elastic Damping Waves Propagation in a Microstructured Medium

The expressions containing the dissipative terms:

$$\tau = \tau_{eq} + A\,u_{xt} + a\,\psi_t\,,\ \sigma = \sigma_{eq} + B\,u_{xt} + b\,\psi_t\,,\ \eta = \eta_{eq} + F\,u_{xxt} + f\,\psi_{xt}\,, \quad (21.29)$$

are substituted to the motion equations:

$$\rho U_{tt} = (\tau + \sigma + \tau U_x)_x\,, \quad I\psi_{tt} = \sigma + \eta_x \tag{21.30}$$

as it was done in Porubov and Pastrone (2004). Here A, a, B, b, F, f are the dissipation coefficients,

$$\tau_{eq} = (\lambda + 2\mu)U_x + \left(\frac{\lambda + 2\mu}{2} + l + 2m\right) U_x^2 + O(U_x^3) + H\,(U_x - \psi)\,,$$

$$\sigma_{eq} = D\,(U_x - \psi) + H\left(U_x + \frac{U_x^2}{2}\right),\; \eta_{eq} = G\,\psi_x\,,$$

and D, G, H are the microstructure elastic constants. Let us introduce the strain function $u = U_x$ and differentiate the first equation of (21.30) with respect to x. Collecting the terms up to the second order, one can derive:

$$\begin{aligned}
\rho u_{tt} &= (D + 2H + \lambda + 2\mu)\,u_{xx} - (D + H)\,\psi_{xx} - H\,(u\,\psi)_{xx} \\
&+ \left(\frac{3H}{2} + \frac{3}{2}\,(\lambda + 2\mu) + l + 2m\right)\left(u^2\right)_{xx} \\
&+ (A + B)\,u_{xxt} + (a + b)\,\psi_{xxt} + \frac{A}{2}\left(u^2\right)_{xxt} + a\,(u\,\psi_t)_{xx}\,, \\
I\psi_{tt} &= (D + H)\,u - D\,\psi + G\,\psi_{xx} + \frac{H}{2}u^2 + B\,u_t + b\,\psi_t + F\,u_{xxt} + f\,\psi_{xxt}
\end{aligned} \tag{21.31}$$

The correction of the potential energy density function Π affected only the coefficients in (21.31) (e.g., all terms with the coefficient H as well as the coefficient of $\left(u^2\right)_{xx}$). However, the correct description of the geometrical and physical nonlinearities led to appearance of two additional nonlinear terms $A/2\,\left(u^2\right)_{xxt}$ and $a\,(u\,\psi_t)_{xx}$ in the first equation in (21.31).

Let us introduce the scales Λ for the x coordinate, Λ/c_0 for time, where $c_0^2 = (\lambda + 2\mu)/\rho$ is the propagation velocity of the linear waves in an isotropic homogeneous medium, and A for both macro- and microdeformations. Let us define the dimensionless variables $x' = x/\Lambda$, $t' = c_0 t/\Lambda$ and dimensionless functions $u' = u/\mathrm{A}$, $\psi' = \psi/\mathrm{A}$. We assume the following simplifications: the long waves with characteristic lengths $\Lambda \gg 1$ are under consideration, strains are small $\mathrm{A} \ll 1$, microinertia depend on some squared characteristic size p of microstructural element as $I = \rho\,p^2 I^*$, $G = p^2 G^*$, the coefficients at the dissipative terms can be written as $A = A^*\,d$, $a = a^*\,d$, $B = B^*\,d$, $b = b^*\,d$, $F = F^*\,d\,p^2$, $f = f^*\,d\,p^2$, where the parameter d has the length dimension. We introduce the following positive dimensionless parameters $\varepsilon = \mathrm{A} \ll 1$ to characterize the deformation scale, $\delta = p^2/\Lambda^2 \ll 1$ to characterize the microstructure size in comparison with the wave length and $\gamma = d/\Lambda \ll 1$ to characterize the dissipation influence. Let us rewrite (21.31) in the dimensionless form (primes are omitted)

$$u_{tt}-\left(1+\frac{D+2H}{\lambda+2\mu}\right)u_{xx}+\frac{D+H}{\lambda+2\mu}\psi_{xx}=\gamma\left[\frac{c_0\,(A^*+B^*)}{\lambda+2\mu}u+\frac{c_0\,(a^*+b^*)}{\lambda+2\mu}\psi\right]_{xxt}$$
$$+\varepsilon\left[\frac{3\lambda+6\mu+l+2m+3H}{2\,(\lambda+2\mu)}u^2-\frac{H}{\lambda+2\mu}u\,\psi\right]_{xx}$$
$$+\varepsilon\gamma\left[\frac{c_0A^*}{2(\lambda+2\mu)}\left(u^2\right)_t+\frac{c_0\,a^*}{\lambda+2\mu}u\,\psi_t\right]_{xx}, \tag{21.32}$$

$$D\,\psi=(D+H)\,u+\gamma\,(c_0\,B^*\,u+c_0\,B^*\,\psi)_t+\delta\,(G^*\,\psi_{xx}$$
$$-I^*\,(\lambda+2\mu)\,\psi_{tt})+\varepsilon\frac{H}{2}u^2+\delta\,\gamma\,(c_0\,F^*\,u+c_0\,f^*\,\psi)_{xxt}\,. \tag{21.33}$$

Expanding ψ in power series with respect to the introduced small parameters, one can derive:

$$\psi=\psi_0+\gamma\psi_1+\delta\,\psi_2+\varepsilon\,\widetilde{\psi}_2+\delta\,\gamma\,\psi_3+\gamma^2\,\psi_4+\delta\,\varepsilon\,\widetilde{\psi}_3+\gamma\varepsilon\,\widetilde{\psi}_4+\delta^2\,\overline{\psi}_3+\varepsilon^2\,\overline{\psi}_4. \tag{21.34}$$

Substitution of (21.34) to the second equation of (21.33) allows one to find ψ_i. After substitution of ψ_i to the first equation of (21.33), leaving the terms up to the second order with respect to the small parameters, we obtain the following nonlinear wave propagation equation:

$$u_{tt}-u_{xx}\,\alpha_0-\left(u^2\right)_{xx}\varepsilon\,\alpha_1-u_{xxt}\,\gamma\alpha_2+u_{xxxx}\,\delta\,\alpha_3-u_{xxtt}\,\left[\delta\,\alpha_4+\gamma^2\,\alpha_7\right]$$
$$+\gamma\delta\,\left[u_{xxxxt}\,\alpha_5+u_{xxttt}\,\alpha_6\right]+\delta\,\varepsilon\,\left[\left(2\,u\,u_{xx}+u_x^2\right)_{xx}\overline{\alpha}_5-\left(2\,u\,u_{tt}+u_t^2\right)_{xx}\overline{\alpha}_6\right]$$
$$+\left(u\,u_t\right)_{xx}\gamma\varepsilon\,\overline{\alpha}_4+\delta^2\,\left[u_{xxxxxx}\,\alpha_8-u_{xxxxtt}\,\alpha_9+u_{xxtttt}\,\alpha_{10}\right]+\left(u^3\right)_{xx}\varepsilon^2\,\alpha_{11}=0.$$

21.5.2 Coefficients of the Coupled Equations in (21.24)

$$\begin{aligned}
\beta_0 &= 3E+2l(1-2\nu)^3+4m(\nu+1)^2(1-2\nu)+6n\,\nu^2,\\
\alpha_1 &= (1-2\nu)^2(B_1+2D_1)+(1+2\nu^2)\,(B_2+B_3+4D_2)\,,\\
\alpha_2 &= -((1-2\nu)\,(1-2\kappa)\,(B_1+D_1)+(1+2\kappa\nu)\,(B_2+B_3+2D_2))\,,\\
\alpha_3 &= 12\,\omega(-1+2\nu)B_1+12\,\omega\,\nu\,(B_2+B_3)\\
&\quad -3(-1+2\nu)(1-8\,\omega+2\nu^2)D_1+6(1+8\,\omega\,\nu-2\nu^3)D_2,\\
\alpha_4 &= -(4\,\omega\,(2\kappa-1)B_1+4\,\omega\,\kappa\,(B_2+B_3)+\\
&\quad +(1-2\kappa)(1-4\,\omega+2\nu^2)D_1+2(1+4\,\omega\,\kappa-2\kappa\nu^2)D_2)\,,\\
\alpha_5 &= \nu(1-2\nu)^2\,(B_1+2D_1)+\frac{\nu}{3}\left[(1+6\nu^2)\,(B_2+B_3+4D_2)-4\nu\,(B_2+2D_2)\right],\\
\alpha_6 &= -\frac{\nu}{2}\,(1-2\nu)\,(1-2\kappa)\,(B_1+D_1)-\frac{\nu}{6}\,(1-2\kappa+6\nu\kappa)\,(B_2+B_3+2D_2)\,,
\end{aligned}$$

$$\begin{aligned}\beta_1 &= (1-2\kappa)^2 B_1 + \left(1+2\kappa^2\right)(B_2+B_3),\\ \beta_2 &= 2(1-2\kappa)(G_1+G_5)+2(G_2+G_8)+G_3+(1-2\kappa)^2 G_4\\ &\quad + G_6 + \left(1+2\kappa^2\right)(G_7+G_9)+G_{10}+G_{11},\\ \omega &= \frac{\nu(1+\nu)}{2} + \frac{(1-2\nu)(1+\nu)}{E}\left(l(1-2\nu)^2+2m\nu(1+\nu)-n\nu\right).\end{aligned}$$

References

Capriz G (1989) Continua with Microstructure. Springer

Casasso A, Pastrone F, Samsonov AM (2010) Travelling waves in microstructure as the exact solutions to the 6th order nonlinear equation. Acoustical Physics 56(6):871–876

Cosserat E, Cosserat F (1909) Théorie des Corps Déformables. Hermann, Paris

Engelbrecht J, Berezovski A (2015) Reflections on mathematical models of deformation waves in elastic microstructured solids. Mathematics and Mechanics of Complex Systems 3(1):43 – 82

Engelbrecht J, Braun M (1998) Nonlinear waves in nonlocal media. Appl Mech Reviews 51(8):475 – 487

Engelbrecht J, Pastrone F (2003) Waves in microstructured solids with nonlinearities in microscale Proceedings of the Estonian Academy of Sciences Physics Mathematics 52(1):12–20

Engelbrecht J, Berezovski A, Pastrone F, Braun M (2005) Waves in microstructured materials and dispersion. Philosophical Magazine 85:4127–4141

Eringen A, Suhubi E (1964a) Nonlinear theory of simple micro-elastic solids - I International Journal of Engineering Science 2(2):189 – 203

Eringen A, Suhubi E (1964b) Nonlinear theory of simple micro-elastic solids - II International Journal of Engineering Science 2(4):389 – 404

Hughes DS, Kelly JL (1953) Second-order elastic deformation of solids. Physical Review 92(5):1145–1149

Khusnutdinova KR, Samsonov AM (2008) Fission of a longitudinal strain solitary wave in a delaminated bar. Physical Review E 77:066,603

Landau LD, Lifshitz EM (1986) Theoretical Physics, vol VII. Theory of Elasticity, 3rd edn. Butterworth-Heinemann, Oxford

Maugin GA (2011) A historical perspective of generalized continuum mechanics. In: Altenbach H Maugin GA, Erofeev VI (eds) Mechanics of Generalized Continua, Springer, Berlin, Heidelberg, Advanced Structured Materials, vol 7, pp 3–19

Mindlin RD (1964) Micro-structure in linear elasticity. Archive for Rational Mechanics and Analysis 16:51–78

Murnaghan FD (1951) Finite Deformation of an Elastic Solid. Chapman & Hall

Pastrone F (2005) Wave propagation in microstructured solids. Mathematics and Mechanics of Solids 10:349 – 357

Porubov A, Andrianov I, Danishevs'kyy V (2012) Nonlinear strain wave localization in periodic composites. International Journal of Solids and Structures 49(23):3381 – 3387

Porubov AV, Pastrone F (2004) Non-linear bell-shaped and kink-shaped strain waves in microstructured solids. International Journal of Non-Linear Mechanics 39(8):1289 – 1299

Porubov AV, Pastrone F, Maugin GA (2004) Selection of two-dimensional nonlinear strain waves in micro-structured media. Comptes Rendus Mécanique 332(7):513 – 518

Samsonov AM (2001) Strain Solitons in Solids and How to Construct Them. Chapman & Hall/CRC

Samsonov AM, Semenova IV, Belashov AV (2017) Direct determination of bulk strain soliton parameters in solid polymeric waveguides. Wave Motion 71:120 – 126

Semenova IV, Dreiden GV, Samsonov AM (2011) Strain solitary waves in polymeric nanocomposites. In: Proulx T (ed) Dynamic Behavior of Materials, Springer, New York, vol 1, pp 261–267

Weisstein EW (2017) "Isotropic Tensor". From *MathWorld.* A Wolfram Web Resource
Wolfram Research, Inc (2017) Mathematica, Version 11.1. URL `http://www.wolfram.com`, Champaign, IL

Chapter 22
On the Deformation of Chiral Piezoelectric Plates

Dorin Ieşan and Ramon Quintanilla

Abstract The paper is concerned with the linear theory of piezoelectricity for isotropic chiral Cosserat elastic solids. The behavior of chiral bodies is of interest for the investigation of auxetic materials, carbon nanotubes, bones, honeycomb structures, as well as composites with inclusions. First, we establish the basic equations which govern the behavior of thin plates. It is shown that, in contrast with the theory of achiral plates, the stretching and flexure cannot be treated independently of each other. Then, we present a uniqueness result with no definiteness assumption on elastic constitutive coefficients. A reciprocity theorem is also established. Then, we present the conditions on the constitutive coefficients which guarantee that the energy of the system is positive definite and we give a continuous dependence result. In the case of stationary theory we derive a uniqueness result for the Neumann problem. Finally, the effects of a concentrated charge density in an unbounded plate are investigated.

22.1 Introduction

The interaction of electromagnetic fields with elastic continua has been studied in many works (see, e.g., Toupin, 1963; Mindlin, 1974; Maugin, 1988; Eringen and Maugin, 1990; Yang, 2006; Eringen, 1999, 2004, and references therein). In recent years there has been much interest in the study of piezoelectric plates (see,

Dorin Ieşan
Department of Mathematics, Al. I. Cuza University and Octav Mayer Institute of Mathematics (Romanian Academy), Bd. Carol I, nr. 8, 700508 Iaşi, Romania
e-mail: iesan@uaic.ro

Ramon Quintanilla
Department of Mathematics, ESEIAAT, Polytechnic University of Catalonia, Colón, 11, 08222 Terrassa, Barcelona, Spain
e-mail: ramon.quintanilla@upc.edu

© Springer International Publishing AG, part of Springer Nature 2018
H. Altenbach et al. (eds.), *Generalized Models and Non-classical Approaches in Complex Materials 1*, Advanced Structured Materials 89,
https://doi.org/10.1007/978-3-319-72440-9_22

e.g., Mindlin, 1984; Maugin and Attou, 1990; Tiersten, 1993; Bisegna and Maceri, 1996; Krommer and Irschik, 2000; Irschik, 2002; Batra and Vidoli, 2002, and the literature cited therein). In this paper we use the results of Mindlin (1951); Eringen (1999, 1967); Naghdi (1972) in order to derive a theory of chiral piezoelectric thin plates. The behavior of chiral materials is of interest for the investigation of auxetic materials, carbon nanotubes, bones, honeycomb structures, as well as composites with inclusions. It is known that the deformation of chiral elastic materials cannot be described within classical elasticity. Various authors have studied the behavior of chiral materials by using the theory of Cosserat elasticity (see, e.g., Lakes, 2001; Park and Lakes, 1986; Lakes, 1987; Healey, 2002; Natroshvili et al, 2006; Chandraseker et al, 2009; De Cicco and Ieşan, 2013, and references therein). The Cosserat theory studies continua with oriented particles, in which each material point has the six degrees of freedom of a rigid body. Lakes (2001) has shown that a chiral plate bent to hyperbolic shape is predicted to exhibit size effects from the Cosserat characteristic length, and a shear force from the chirality. The general theory of piezoelectric Cosserat elastic solids has been established by Eringen (1999). We use this theory to derive the equations of isotropic piezoelectric plates. This work is motivated by the recent interest in the study of piezoelectric carbon nanotubes, bones and composites (see, e.g., Telega and Wojnar, 2002; Ray and Batra, 2007; Ha et al, 2016).

The deformation of achiral Cosserat elastic plates has been investigated in many papers. A detailed analysis of the results established in this theory have been presented in Eringen (1999); Altenbach and Eremeyev (2013). In this paper we derive a theory of piezoelectric chiral Cosserat elastic plates. In contrast with the theory of achiral plates, the stretching and flexure cannot be treated independently of each other. We assume that on the upper and lower faces of the plate there are prescribed the surface traction, the surface moment and the normal component of the electrical displacement. In Sect. 22.2 we present the basic equations of homogeneous and isotropic chiral Cosserat piezoelectric solids. Section 22.3 is devoted to the deriving of the theory of thin plates. In Sect. 22.4 we present a uniqueness result with no definiteness assumption on elastic constitutive coefficients in the dynamic theory. A reciprocity theorem is also established. In Sect. 22.5 we establish the conditions on constitutive coefficients which guarantee that the energy of the system is positive and we give a continuous dependence result under these conditions. Section 22.6 is concerned with a uniqueness result for the Neumann problem in the case of stationary theory. In Sect. 22.7 we investigate the effects of a concentrated charge density in an unbounded plate.

22.2 Basic Equations

We consider a body that at time t_0 occupies the region B of Euclidean three-dimensional space and is bounded by the piecewise smooth surface ∂B. The motion of the body is referred to a fixed system of rectangular cartesian axes $Ox_k (k = 1,2,3)$. We denote by $\boldsymbol{n}$ the outward unit normal of ∂B. We shall employ the usual summation and differentiation conventions: Greek subscripts are understood to range over integers $(1,2)$ whereas Latin subscripts to the range $(1,2,3)$; summation over repeated subscripts is implied and subscripts preceded by a comma denote partial differentiation with respect to the corresponding cartesian coordinates. The partial derivative with respect to time t is denoted by a superposed dot. We assume that B is occupied by a homogeneous and isotropic chiral piezoelectric material. The fundamental system of field equations consists of the equations of motion

$$\begin{gathered} t_{ji,j}+f_i=\rho\ddot{u}_i, \\ m_{ji,j}+\varepsilon_{ijk}t_{jk}+g_i=J\ddot{\varphi}_i, \end{gathered} \tag{22.1}$$

the equations of the electric field

$$D_{j,j}=q, \quad E_k=-\psi_{,k}, \tag{22.2}$$

the constitutive equations (cf. Eringen, 1999; Lakes, 2001)

$$\begin{gathered} t_{ij}=\lambda e_{rr}\delta_{ij}+(\mu+\kappa)e_{ij}+\mu e_{ji}+C_1\kappa_{ss}\delta_{ij}+C_2\kappa_{ji}+C_3\kappa_{ij}+\lambda_1\varepsilon_{ijk}E_k, \\ m_{ij}=\alpha\kappa_{ss}\delta_{ij}+\beta\kappa_{ji}+\gamma\kappa_{ij}+C_1e_{rr}\delta_{ij}+C_2e_{ji}+C_3e_{ij}-\lambda_2\varepsilon_{ijk}E_k, \\ D_k=-\lambda_1\varepsilon_{ijk}e_{ij}-\lambda_2\varepsilon_{ijk}\kappa_{ij}+\chi E_k, \end{gathered} \tag{22.3}$$

and the geometrical equations

$$e_{ij}=u_{j,i}+\varepsilon_{jik}\varphi_k, \quad \kappa_{ij}=\varphi_{j,i}. \tag{22.4}$$

Here we have used the following notations: t_{ij} is the stress tensor, f_i is the body forces, ρ is the reference mass density, u_i is the displacement vector, m_{ij} is the couple stress tensor, ε_{ijk} is the alternating symbol, g_i is the body couple, J is the coefficient of microinertia, φ_i is the microrotation vector, D_j is the dielectric displacement vector, q is the volume charge density, E_j is the electric field vector, ψ is the electrostatic potential, e_{ij} and κ_{ij} are kinematic strain measures, and $\lambda,\mu,\kappa,\alpha,\beta,\gamma,C_1,C_2,C_3,\lambda_1,\lambda_2$ and χ are constitutive constants. In the case of an achiral material the coefficients C_1,C_2,C_3 and λ_1 are equal to zero

The components of surface traction, the components of the surface moment, and the normal component of the electrical displacement at a regular point of ∂B are given by

$$t_i=t_{ji}n_j, \quad m_i=m_{ji}n_j, \quad \sigma=D_kn_k, \tag{22.5}$$

respectively.

Let M and N be non-negative integers and $\mathscr{T}$ a given interval of time. We say that f is of class $C^{M,N}$ on $B\times\mathscr{T}$ if f is continuous on $B\times\mathscr{T}$ and the functions

$$\frac{\partial^m}{\partial x_i\partial x_j\ldots\partial x_p}\left(\frac{\partial^n f}{\partial t^n}\right), m\in\{0,1,2,\ldots,M\}, n\in\{0,1,2,\ldots,N\}, m+n\leq\max\{M,N\},$$

exist and are continuous on $B\times\mathscr{T}$. We write C^M for $C^{M,M}$.

We say that $p = (u_i, \varphi_i, \psi)$ is an admissible state provided: (i) u_k and φ_k are of class C^2 on $B \times \mathscr{T}$; (ii) u_i, φ_i and ψ are of class C^1 on $\overline{B} \times \mathscr{T}$; (iii) ψ is of class $C^{2,0}$ on $B \times \mathscr{T}$. We assume that $\boldsymbol{f}, \boldsymbol{g}$ and q are continuous on $B \times \mathscr{T}$ and that ρ and J are positive constants.

22.3 Chiral Piezoelectric Plates

We assume now that the region B refers to the interior of a right cylinder of length $2h$ with open cross-section Σ and the lateral boundary II. The cartesian coordinate frame consists of the orthonormal basis $\{\boldsymbol{e}_1, \boldsymbol{e}_2, \boldsymbol{e}_3\}$ and the origin O. The coordinate frame is supposed to be chosen in such a way that the plane x_1Ox_2 is middle plane. We denote by Γ the boundary of Σ. Clearly,

$$B = \{\boldsymbol{x} : (x_1, x_2) \in \Sigma, -h < x_3 < h\}, \Pi = \{\boldsymbol{x}; (x_1, x_2) \in \Gamma, -h < x_3 < h\}.$$

In what follows we present a theory of thin plates assuming that

$$\begin{aligned} &u_\alpha = w_\alpha(x_1, x_2, t) + x_3 v_\alpha(x_1, x_2, t),\ u_3 = w_3(x_1, x_2, t), \\ &\varphi_j = \Phi_j(x_1, x_2, t), \psi = U(x_1, x_2, t) + x_3 V(x_1, x_2, t), \\ &(x_1, x_2, x_3) \in B, t \in \mathscr{T}. \end{aligned} \tag{22.6}$$

In Eringen (1999, 1967) a theory of achiral Cosserat elastic plates was established. It was shown that in this case the stretching and flexure of plates can be treated independently of each other. Following Mindlin (1951); Eringen (1967); Naghdi (1972), to establish a plate theory, we perform the following integrations: (i) we integrate equation of balance of momenta with respect to x_3 over the thickness of the plate; (ii) we take the cross product of the equations of the balance of linear momentum with $x_3\boldsymbol{e}_3$ and integrate over the thickness of the plate; (iii) we integrate the Gauss equation $(22.2)_1$ over x_3 between the limits $-h$ and h; (iv) we multiply equation $(22.2)_1$ by x_3 and integrate over the thickness of the plate. The results of (i) are

$$\begin{aligned} &\tau_{\beta k,\beta} + F_k = \rho \ddot{w}_k, \quad \mu_{\beta\alpha,\beta} + \varepsilon_{3\rho\alpha}(\tau_{3\rho} - \tau_{\rho 3}) + G_\alpha = J\ddot{\Phi}_\alpha, \\ &\mu_{\alpha 3,\alpha} + \varepsilon_{3\rho\alpha}\tau_{\rho\alpha} + G_3 = J\ddot{\Phi}_3, \end{aligned} \tag{22.7}$$

where

$$\tau_{ij} = \frac{1}{2h}\int_{-h}^{h} t_{ij}dx_3,\ \mu_{ij} = \frac{1}{2h}\int_{-h}^{h} m_{ij}dx_3, \tag{22.8}$$

$$F_i = \frac{1}{2h}\int_{-h}^{h} f_i dx_3 + \frac{1}{2h}[t_{3i}]_{-h}^{h},\ G_i = \frac{1}{2h}\int_{-h}^{h} g_i dx_3 + \frac{1}{2h}[m_{3i}]_{-h}^{h}.$$

We assume that the functions t_i, m_i and σ are prescribed on the surfaces $x_3 = \pm h$. To the Eqs. (22.7) we add the result of (ii), i.e.,

$$M_{\beta\alpha,\beta} - 2h\tau_{3\alpha} + H_\alpha = \rho I \ddot{v}_\alpha, \tag{22.9}$$

where we have used the notations

$$M_{\alpha\beta} = \int_{-h}^{h} x_3 t_{\alpha\beta} dx_3, \; I = \frac{2}{3}h^3, \; H_\alpha = \int_{-h}^{h} x_3 f_\alpha dx_3 + [x_3 t_{3\alpha}]_{-h}^{h}. \tag{22.10}$$

If we integrate the Eq. $(22.2)_1$ with respect to x_3 between the limits $-h$ and h, then we obtain the following equation

$$\sigma_{\alpha,\alpha} = P, \tag{22.11}$$

where

$$\sigma_k = \frac{1}{2h}\int_{-h}^{h} D_k dx_3, \; P = \frac{1}{2h}\int_{-h}^{h} q dx_3 - \frac{1}{2h}[D_3]_{-h}^{h}. \tag{22.12}$$

The equation which results from the multiplication of Eq. $(22.2)_1$ by x_3 and integration over x_3 from $x_3 = -h$ to $x_3 = h$ can be written in the form

$$d_{\alpha,\alpha} - 2h\sigma_3 = Q, \tag{22.13}$$

where we have used the notations

$$d_\alpha = \int_{-h}^{h} x_3 D_\alpha dx_3, \; Q = \int_{-h}^{h} x_3 q dx_3 - [x_3 D_3]_{-h}^{h}. \tag{22.14}$$

The functions F_j, G_j, H_α, P and Q are prescribed. From (22.4), (22.2) and (22.6) we obtain

$$\begin{aligned} &e_{\alpha\beta} = \gamma_{\alpha\beta} + x_3\xi_{\alpha\beta}, \; e_{\alpha 3} = \gamma_{\alpha 3}, \; e_{33} = 0, \; e_{3\alpha} = \gamma_{3\alpha}, \\ &\kappa_{\alpha j} = \eta_{\alpha j}, \; \kappa_{3j} = 0, \; E_\alpha = e_\alpha + x_3\zeta_\alpha, \; E_3 = -V, \end{aligned} \tag{22.15}$$

where

$$\begin{aligned} &\gamma_{\alpha j} = w_{j,\alpha} + \varepsilon_{j\alpha k}\Phi_k, \; \gamma_{3\alpha} = v_\alpha + \varepsilon_{3\beta\alpha}\Phi_\beta, \; \eta_{\alpha k} = \Phi_{k,\alpha}, \\ &\xi_{\alpha\beta} = v_{\beta,\alpha}, e_\alpha = -U_{,\alpha}, \; \zeta_\alpha = -V_{,\alpha}. \end{aligned} \tag{22.16}$$

It follows from (22.3), (22.8), (22.10), (22.12), (22.14) and (22.15) that

$$\begin{aligned}
\tau_{\alpha\beta} &= \lambda\gamma_{\rho\rho}\delta_{\alpha\beta}+(\mu+\kappa)\gamma_{\alpha\beta}+\mu\gamma_{\beta\alpha}+C_1\eta_{\rho\rho}\delta_{\alpha\beta}+C_2\eta_{\beta\alpha}+C_3\eta_{\alpha\beta}-\lambda_1\varepsilon_{3\alpha\beta}V,\\
\tau_{\alpha 3} &= (\mu+\kappa)\gamma_{\alpha 3}+\mu\gamma_{3\alpha}+C_3\eta_{\alpha 3}+\lambda_1\varepsilon_{3\beta\alpha}e_\beta,\\
\tau_{3\alpha} &= (\mu+\kappa)\gamma_{3\alpha}+\mu\gamma_{\alpha 3}+C_2\eta_{\alpha 3}+\lambda_1\varepsilon_{3\alpha\beta}e_\beta,\\
\mu_{\nu\kappa} &= \alpha\eta_{\rho\rho}\delta_{\nu\kappa}+\beta\eta_{\kappa\nu}+\gamma\eta_{\nu\kappa}+C_1\gamma_{\rho\rho}\mathrm{d}_{\nu\kappa}+C_2\gamma_{\kappa\nu}+C_3\gamma_{\nu\kappa}-\lambda_2\varepsilon_{3\nu\kappa}V,\\
\mu_{\alpha 3} &= \beta\eta_{3\alpha}+\gamma\eta_{\alpha 3}+C_2\gamma_{3\alpha}+C_3\gamma_{\alpha 3}+\lambda_2\varepsilon_{3\beta\alpha}e_\beta,\\
M_{\alpha\beta} &= I[\lambda\xi_{\rho\rho}\mathrm{d}_{\alpha\beta}+(\mu+\kappa)\xi_{\alpha\beta}+\mu\xi_{\beta\alpha}],\\
\sigma_\alpha &= \chi e_\alpha-\lambda_1\varepsilon_{ij\alpha}\gamma_{ij}-\lambda_2\varepsilon_{\rho 3\alpha}\eta_{\rho 3},\ \sigma_3=-\lambda_1\varepsilon_{3\alpha\beta}\gamma_{\alpha\beta}-\lambda_2\varepsilon_{3\alpha\beta}\eta_{\alpha\beta}-\chi V,\\
d_\alpha &= I\chi\zeta_\alpha.
\end{aligned}\tag{22.17}$$

Thus, the basic equations of the theory of chiral plates consist of the equations of motion (22.7) and (22.9), the equations of the electric fields (22.11) and (22.13), the constitutive equations (22.17) and the geometrical equations (22.16). The field equations can be expressed in terms of the functions w_k, Φ_k, v_α, U and V. We obtain the following equations

$$\begin{aligned}
&(\mu+\kappa)\Delta w_\alpha+(\lambda+\mu)w_{\rho,\rho\alpha}+C_3\Delta\Phi_\alpha+(C_1+C_2)\Phi_{\rho,\rho\alpha}+\kappa\varepsilon_{3\alpha\beta}\Phi_{3,\beta}\\
&\quad-\lambda_1\varepsilon_{3\beta\alpha}V_{,\beta}+F_\alpha=\rho\ddot{w}_\alpha,\\
&(\mu+\kappa)\Delta w_3+C_3\Delta\Phi_3+\kappa\varepsilon_{3\alpha\beta}\Phi_{\beta,\alpha}+\mu v_{\rho,\rho}+F_3=\rho\ddot{w}_3,\\
&C_3\Delta w_\alpha+(C_1+C_2)w_{\rho,\rho\alpha}+\gamma\Delta\Phi_\alpha+(\alpha+\beta)\Phi_{\rho,\rho\alpha}+\kappa\varepsilon_{3\alpha\rho}(w_{3,\rho}-v_\rho)\\
&\quad+2(C_3-C_2)\varepsilon_{3\alpha\beta}\Phi_{3,\beta}-2\kappa\Phi_\alpha-\lambda_2\varepsilon_{3\beta\alpha}V_{,\beta}-2\lambda_1U_{,\alpha}+G_\alpha=J\ddot{\Phi}_\alpha,\\
&C_3\Delta w_3+\gamma\Delta\Phi_3+\kappa\varepsilon_{3\rho\alpha}w_{\alpha,\rho}+2\varepsilon_{3\alpha\beta}(C_3-C_2)\Phi_{\beta,\alpha}\\
&\quad+C_2v_{\alpha,\alpha}-2\kappa\Phi_3-2\lambda_1V+G_3=J\ddot{\Phi}_3,\\
&I[(\mu+\kappa)\Delta v_\alpha+(\lambda+\mu)v_{\rho,\rho\alpha}]-2h[\mu w_{3,\alpha}+C_2\Phi_{3,\alpha}+\kappa\varepsilon_{3\beta\alpha}\Phi_\beta\\
&\quad+(\mu+\kappa)v_\alpha]+2h\lambda_1\varepsilon_{3\alpha\beta}U_{,\beta}+H_\alpha=\rho I\ddot{v}_\alpha,\\
&\chi\Delta U-2\lambda_1\Phi_{\rho,\rho}-\lambda_1\varepsilon_{3\alpha\rho}v_{\rho,\alpha}=-P,\\
&I\chi\Delta V-2h(\lambda_1\varepsilon_{3\alpha\beta}w_{\beta,\alpha}-2\lambda_1\Phi_3+\lambda_2\varepsilon_{3\alpha\beta}\Phi_{\beta,\alpha}+\chi V)=-Q,
\end{aligned}\tag{22.18}$$

where Δ is the two-dimensional Laplacian.

To the field equations we must adjoin initial conditions and boundary conditions. The initial conditions are

$$\begin{aligned}
&w_j(x_1,x_2,0)=w_j^0(x_1,x_2),\ \ \Phi_j(x_1,x_2,0)=\Phi_j^0(x_1,x_2),\\
&v_\alpha(x_1,x_2,0)=v_\alpha^0(x_1,x_2),\dot{w}_j(x_1,x_2,0)=\vartheta_j^0(x_1,x_2),\\
&\dot{\Phi}_j(x_1,x_2,0)=\omega_j^0(x_1,x_2),\dot{v}_\alpha(x_1,x_2,0)=\eta_\alpha^0(x_1,x_2),\ \ (x_1,x_2)\in\Sigma,
\end{aligned}\tag{22.19}$$

where the functions w_j^0, $\Phi_j^0, v_\alpha^0, \vartheta_j^0, \omega_j^0$ and η_α^0 are given. The Neumann problem is characterized by the following boundary conditions

$$\tau_{\beta j}n_\beta=\widetilde{\tau}_j,\ \mu_{\beta j}n_\beta=\widetilde{\mu}_j,\ M_{\beta\alpha}n_\beta=\widetilde{M}_\alpha,\ \sigma_\alpha n_\alpha=\widetilde{\sigma},\ d_\alpha n_\alpha=\widetilde{d}\ \text{ on }\Gamma\times\mathscr{T},\tag{22.20}$$

where the functions $\widetilde{\tau}_j, \widetilde{\mu}_j, \widetilde{M}_\alpha, \widetilde{\sigma}$ and $\widetilde{d}$ are prescribed. In the case of Dirichlet problem the boundary conditions are

$$w_j = \widetilde{w}_j,\ \Phi_j = \widetilde{\Phi}_j,\ v_\alpha = \widetilde{v}_\alpha,\ \ U = \widetilde{U},\ \ V = \widetilde{V}\ \text{ on } \Gamma \times \mathscr{T}, \tag{22.21}$$

where the functions $\widetilde{w}_j, \widetilde{\Phi}_j, \widetilde{v}_\alpha, \widetilde{U}$ and $\widetilde{V}$ are given.

22.4 General Theorems

In this section we establish some basic theorems in the dynamic theory. First, following Brun (1965), we derive a uniqueness theorem without using positive definiteness assumptions on the elastic constitutive coefficients. We assume that the boundary conditions have the form (22.20). Let us consider two data systems of loading

$$\mathscr{L}^{(\alpha)} = \{F_k^{(\alpha)}, G_k^{(\alpha)}, H_\beta^{(\alpha)}, P^{(\alpha)}, Q^{(\alpha)}, w_k^{0(\alpha)}, \Phi_k^{0(\alpha)}, v_\rho^{0(\alpha)},$$
$$\vartheta_j^{0(\alpha)}, \omega_j^{0(\alpha)}, \eta_\beta^{0(\alpha)}, \widetilde{\tau}_j^{(\alpha)}, \widetilde{\mu}_j^{(\alpha)}, \widetilde{M}_\rho^{(\alpha)}, \widetilde{\sigma}^{(\alpha)}, \widetilde{d}^{(\alpha)}\},\ (\alpha = 1,2),$$

and denote by $A^{(\alpha)} = \{w_k^{(\alpha)}, \Phi_k^{(\alpha)}, v_\beta^{(\alpha)}, U^{(\alpha)}, V^{(\alpha)}, \gamma_{ij}^{(\alpha)}, \eta_{\beta k}^{(\alpha)}, \xi_{\beta\alpha}^{(\alpha)}, e_\rho^{(\alpha)}, \zeta_\rho^{(\alpha)}, \tau_{ij}^{(\alpha)}, \mu_{ij}^{(\alpha)}, M_{\rho\nu}^{(\alpha)}, \sigma_k^{(\alpha)}, d_\alpha^{(k)}\}$, a solution corresponding to $\mathscr{L}^{(\alpha)}$. We define the following functions associated to the solution $A^{(\alpha)}$

$$\tau_k^{(\alpha)} = \tau_{\beta k}^{(\alpha)} n_\beta,\ \mu_k^{(\alpha)} = \mu_{\beta k}^{(\alpha)} n_\beta,\ M_\rho^{(\alpha)} = M_{\beta\rho}^{(\alpha)} n_\beta,\ \sigma^{(\alpha)} = \sigma_\rho^{(\alpha)} n_\rho,\ S^{(\alpha)} = d_\rho^{(\alpha)} n_\rho. \tag{22.22}$$

Let us introduce the notations

$$\begin{aligned} J_{\kappa\nu}(r,s) = &\int_\Gamma [2h\tau_j^{(\kappa)}(r)w_j^{(\nu)}(s) + 2h\mu_j^{(\kappa)}(r)\Phi_j^{(\nu)}(s) + 2h\sigma^{(\kappa)}(r)U^{(\nu)}(s) \\ &+ M_\beta^{(\kappa)}(r)v_\beta^{(\nu)}(s) + S^{(\kappa)}(r)V^{(\nu)}(s)]dl \\ &+ \int_\Sigma [2hF_j^{(\kappa)}(r)w_j^{(\nu)}(s) + 2hG_j^{(\kappa)}(r)\Phi_j^{(\nu)}(s) - 2hP^{(\kappa)}(r)U^{(\nu)}(s) \\ &+ H_\alpha^{(\kappa)}(r)v_\alpha^{(\nu)}(s) - Q^{(\kappa)}(r)V^{(\nu)}(s)]da, \end{aligned} \tag{22.23}$$

$$\begin{aligned} K_{\kappa\nu}(r,s) = &\int_\Sigma [2h\rho\ddot{w}_j^{(\kappa)}(r)w_j^{(\nu)}(s) + 2hJ\ddot{\Phi}_j^{(\kappa)}(r)\Phi_j^{(\nu)}(s) \\ &+ \rho I\ddot{v}_\alpha^{(\kappa)}(r)v_\alpha^{(\nu)}(s)]da, \end{aligned}$$

for all $r, s \in \mathscr{T}$. Here, for convenience, we have suppressed the argument $\boldsymbol{x}$.

In the next theorem we present a reciprocity relation which involves two processes at different instants. This relation forms the basis of a uniqueness result.

Theorem 22.1. *Let*

$$D_{\alpha\beta}(r,s)=J_{\alpha\beta}(r,s)-K_{\alpha\beta}(r,s), \tag{22.24}$$

for all $r,s\in\mathscr{T}$. *Then*

$$D_{\alpha\beta}(r,s)=D_{\beta\alpha}(s,r),\ \ (\alpha,\beta=1,2). \tag{22.25}$$

Proof. Let us denote

$$\begin{aligned}W_{\kappa\nu}(r,s)=2h[\tau^{(\kappa)}_{\beta j}(r)\gamma^{(\nu)}_{\beta j}(s)+\tau^{(\kappa)}_{3\alpha}(r)\gamma^{(\nu)}_{3\alpha}(s)+\mu^{(\kappa)}_{\beta j}(r)\eta^{(\nu)}_{\beta j}(s)-\\ -\sigma^{(\kappa)}_{\alpha}(r)e^{(\nu)}_{\alpha}(s)+\sigma^{(\kappa)}_{3}(r)V^{(\nu)}(s)]+M^{(\kappa)}_{\alpha j}(r)\xi^{(\nu)}_{\alpha j}(s)-d^{(\kappa)}_{\alpha}(r)\zeta^{(\nu)}_{\alpha}(s).\end{aligned} \tag{22.26}$$

In view of the constitutive equations (22.17) we find that

$$W_{\kappa\nu}(r,s)=2hW^{(1)}_{\kappa\nu}(r,s)+IW^{(2)}_{\kappa\nu}(r,s), \tag{22.27}$$

where

$$\begin{aligned}W^{(1)}_{\kappa\nu}(r,s)=\lambda\gamma^{(\kappa)}_{\rho\rho}(r)\gamma^{(\nu)}_{\eta\eta}(s)+(\mu+\kappa)[\gamma^{(\kappa)}_{\alpha j}(r)\gamma^{(\nu)}_{\alpha j}(s)+\gamma^{(\kappa)}_{3\alpha}(r)\gamma^{(\nu)}_{3\alpha}(s)]\\ +\mu[\gamma^{(\kappa)}_{\beta j}(r)\gamma^{(\nu)}_{j\beta}(s)+\gamma^{(\kappa)}_{3\alpha}(r)\gamma^{(\nu)}_{\alpha 3}(s)]+C_1[\eta^{(\kappa)}_{\rho\rho}(r)\gamma^{(\nu)}_{\beta\beta}(s)+\gamma^{(\kappa)}_{\rho\rho}(r)\eta^{(\nu)}_{\beta\beta}(s)]\\ +C_2[\eta^{(\kappa)}_{\alpha j}(r)\gamma^{(\nu)}_{j\alpha}(s)+\gamma^{(\kappa)}_{j\beta}(r)\eta^{(\nu)}_{\beta j}(s)]+C_3[\eta^{(\kappa)}_{\alpha j}(r)\gamma^{(\kappa)}_{\alpha j}(s)+\gamma^{(\kappa)}_{\alpha j}(r)\eta^{(\nu)}_{\alpha j}(s)]\\ +\alpha\eta^{(\kappa)}_{\rho\rho}(r)\eta^{(\nu)}_{\beta\beta}(s)+\beta\eta^{(\kappa)}_{\alpha\rho}(r)\eta^{(\nu)}_{\rho\alpha}(s)+\gamma\eta^{(\kappa)}_{\alpha j}(r)\eta^{(\nu)}_{\alpha j}(s)-\chi e^{(\kappa)}_{\alpha}(r)e^{(\nu)}_{\alpha}(s)\\ -\lambda_1\varepsilon_{3\alpha\beta}[V^{(\kappa)}(r)\gamma^{(\nu)}_{\alpha\beta}(s)+V^{(\nu)}(s)\gamma^{(\kappa)}_{\alpha\beta}(r)]\\ -\lambda_2\varepsilon_{3\alpha\beta}[V^{(\kappa)}(r)\eta^{(\nu)}_{\alpha\beta}(s)+V^{(\nu)}(s)\eta^{(\kappa)}_{\alpha\beta}(r)]\\ +\lambda_1\varepsilon_{3\alpha\beta}[e^{(\kappa)}_{\alpha}(r)\gamma^{(\nu)}_{\beta 3}(s)+\gamma^{(\kappa)}_{\beta 3}(r)e^{(\nu)}_{\alpha}(s)+e^{(\kappa)}_{\beta}(r)\gamma^{(\nu)}_{3\alpha}(s)+\gamma^{(\kappa)}_{3\alpha}(r)e^{(\nu)}_{\beta}(s)]\\ +\lambda_2\varepsilon_{3\alpha\beta}[e^{(\kappa)}_{\alpha}(r)\eta^{(\nu)}_{\beta 3}(s)+\eta^{(\kappa)}_{\beta 3}(r)e^{(\nu)}_{\alpha}(s)]-\chi V^{(\kappa)}(r)V^{(\nu)}(s),\end{aligned} \tag{22.28}$$

and

$$\begin{aligned}W^{(2)}_{\kappa\nu}(r,s)\ =\ &\lambda\xi^{(\kappa)}_{\rho\rho}(r)\xi^{(\nu)}_{\alpha\alpha}(s)+(\mu+\kappa)\xi^{(\kappa)}_{\alpha\rho}(r)\xi^{(\nu)}_{\alpha\rho}(s)\\ &+\ \mu\xi^{(\kappa)}_{\beta\alpha}(r)\xi^{(\nu)}_{\alpha\beta}(s)-\chi\zeta^{(\kappa)}_{\alpha}(r)\zeta^{(\nu)}_{\alpha}(s).\end{aligned} \tag{22.29}$$

It follows from (22.27)-(22.29) that

$$W_{\kappa\nu}(r,s)=W_{\nu\kappa}(s,r). \tag{22.30}$$

On the other hand, if we use the equations (22.7), (22.9), (22.11), (22.13) and (22.16) we find that

$$
\begin{aligned}
W_{\kappa\nu}(r,s) &= 2h[F_k^{(\kappa)}(r)w_k^{(\nu)}(s) + G_k^{(\kappa)}(r)\Phi_k^{(\nu)}(s) - P^{(\kappa)}(r)U^{(\nu)}(s)] \\
&+ H_\alpha^{(\kappa)}(r)v_\alpha^{(\nu)}(s) - Q^{(\kappa)}(r)V^{(\nu)}(s) - 2h[\rho\ddot{w}_k^{(\kappa)}(r)w_k^{(\nu)}(s) \\
&+ I\ddot{\Phi}_k^{(\kappa)}(r)\Phi^{(\nu)}(s)] - \rho I\ddot{v}_\alpha^{(\kappa)}(r)v_\alpha^{(\nu)}(s) \\
&+ \{2h[\tau_{\beta k}^{(\kappa)}(r)w_k^{(\nu)}(s) + \mu_{\beta k}^{(\kappa)}(r)\Phi_k^{(\nu)}(s) + \sigma_\beta^{(\kappa)}(r)U^{(\nu)}(s)] \\
&+ M_{\beta\alpha}^{(\kappa)}(r)v_\alpha^{(\nu)}(s) + d_\beta^{(\kappa)}(r)V^{(\nu)}(s)\}_{,\beta}.
\end{aligned} \tag{22.31}
$$

If we integrate (22.31) over Σ and use the divergence theorem and (22.22)-(22.24), then we obtain

$$
\int_\Sigma W_{\kappa\nu}(r,s)da = D_{\kappa\nu}(r,s). \tag{22.32}
$$

In view of (22.30) we obtain the desired result.□

We introduce the notations

$$
\tau_k = \tau_{\alpha k}n_\alpha,\ \mu_k = \mu_{\alpha k}n_\alpha,\ M_\alpha = M_{\beta\alpha}n_\beta,\ \ \sigma = \sigma_\alpha n_\alpha,\ S = d_\alpha n_\alpha. \tag{22.33}
$$

The following theorem is a consequence of Theorem 22.1.

Theorem 22.2. *Let*

$$
A = \{w_k, \Phi_k, v_\beta, U, V, \gamma_{ij}, \eta_{\beta k}, \xi_{\beta\rho}, e_\rho, \zeta_\rho, \tau_{ij}, \mu_{ij}, M_{\alpha\beta}, \sigma_k, d_\alpha\}
$$

be a solution corresponding to the system of loading

$$
\{F_i, G_i, H_\alpha, P, Q, w_i^0, \Phi_i^0, v_\alpha^0, \vartheta_j^0, \omega_j^0, \eta_\alpha^0, \widetilde{\tau}_j, \widetilde{\mu}_j, \widetilde{M}_\alpha, \widetilde{\sigma}, \widetilde{d}\}
$$

and let

$$
\begin{aligned}
\Pi(r,s) &= \int_\sigma [2hF_j(r)w_j(s) + 2hG_j(r)\Phi_j(s) \\
&- 2hP(r)U(s) + H_\alpha(r)v_\alpha(s) - Q(r)V(s)]da \\
&+ \int_L [2h\tau_j(r)w_j(s) + 2h\mu_j(r)\Phi_j(s) + 2h\sigma(r)U(s) + M_\beta(r)v_\beta(s) + S(r)V(s)]dl,
\end{aligned} \tag{22.34}
$$

for all $r,s \in \mathscr{T}$. *Then*

$$
\begin{aligned}
\frac{d}{dt}\{\int_\Sigma (2h\rho w_j w_j + 2hJ\Phi_j\Phi_j + \rho I v_\alpha v_\alpha)da\} &= \int_0^t [\Pi(t-s,t+s) - \Pi(t+s,t-s)]ds \\
+ \int_\Sigma \{2h\rho[\dot{w}_j(0)w_j(2t) + \dot{w}_j(2t)w_j(0)] &+ 2hJ[\dot{\Phi}_j(0)\Phi_j(2t) + \dot{\Phi}_j(2t)\Phi_j(0)] \\
&+ \rho I[\dot{v}_\alpha(0)v_\alpha(2t) + \dot{v}_\alpha(2t)v_\alpha(0)]\}da.
\end{aligned} \tag{22.35}
$$

Proof. From (22.25) we get

$$\int_0^t D_{11}(t+s,t-s)ds = \int_0^t D_{11}(t-s,t+s)ds. \tag{22.36}$$

Let us apply this relation to the solution $A^{(1)} = A$. From (22.23), (22.24) and (22.34) we obtain

$$\int_0^t D_{11}(t+s,t-s)ds = \int_0^t \Pi(t+s,t-s)ds - \int_0^t \int_\Sigma [2h\rho \ddot{w}_j(t+s)w_j(t-s)$$
$$+2hJ\ddot{\Phi}_j(t+s)\Phi_j(t-s) + \rho I\ddot{v}_\alpha(t+s)v_\alpha(t-s)]dsda, \tag{22.37}$$

and

$$\int_0^t D_{11}(t-s,t+s)ds = \int_0^t \Pi(t-s,t+s)ds - \int_0^t \int_\Sigma [2h\ddot{w}_j(t-s)w_j(t+s)$$
$$+2hJ\ddot{\Phi}_j(t-s)\Phi_j(t+s) + \rho I\ddot{v}_\alpha(t-s)v_\alpha(t+s)]dsda. \tag{22.38}$$

With the help of the relations

$$\int_0^t \ddot{f}(t+s)g(t-s)ds = \dot{f}(2t)g(0) - \dot{f}(t)g(t) + \int_0^t \dot{f}(t+s)\dot{g}(t-s)ds,$$

$$\int_0^t \ddot{g}(t-s)f(t+s)ds = \dot{g}(t)f(t) - \dot{g}(0)f(2t) + \int_0^t \dot{g}(t-s)\dot{f}(t+s)ds,$$

we can present (22.37) and (22.38) in an alternative form. Thus, we have

$$\int_0^t D_{11}(t+s,t-s)ds = \int_0^t \Pi(t+s,t-s)ds - \int_\Sigma \{2h\rho[\dot{w}_j(2t)w_j(0)$$
$$-\dot{w}_j(t)w_j(t) + \int_0^t \dot{w}_j(t+s)\dot{w}_j(t-s)ds] + 2hJ[\dot{\Phi}_j(2t)\Phi_j(0)$$
$$-\dot{\Phi}_j(t)\Phi_j(t) + \int_0^t \dot{\Phi}_j(t+s)\dot{\Phi}_j(t-s)ds] \tag{22.39}$$
$$+\rho I[\dot{v}_\alpha(2t)v_\alpha(0) - \dot{v}_\alpha(t)v_\alpha(t) + \int_0^t \dot{v}_\alpha(t+s)\dot{v}_\alpha(t-s)ds]\}da.$$

In a similar way we can transform (22.38). With the help of (22.36) and (22.38) we obtain the desired result.□

In what follows we shall use the following relations

$$\int_{\Gamma} \sigma_\alpha n_\alpha U dl = \int_{\Sigma} (PU + \sigma_\alpha U_{,\alpha}) da, \int_{\Gamma} d_\alpha n_\alpha V dl = \int_{\Sigma} (QV + 2h\sigma_3 V + d_\alpha V_{,\alpha}) da, \tag{22.40}$$

which are consequences of (22.11) and (22.13) .

Theorem 22.2 leads to the following uniqueness theorem.

Theorem 22.3. *Assume that ρ, J and χ are strictly positive. Let*

$$A^* = \{w_k^*, \Phi_k^*, v_\beta^*, U^*, V^*, \gamma_{ij}^*, \eta_{\beta k}^*, \xi_{\beta\rho}^*, e_\rho^*, \zeta_\rho^*, \tau_{ij}^*, \mu_{ij}^*, M_{\alpha\beta}^*, \sigma_k^*, d_\alpha^*\}$$

be the difference of any two solutions of the problem corresponding to the loading system $\mathscr{L}^{(\alpha)}$. Then

$$w_j^* = 0,\ \Phi_j^* = 0,\ v_\alpha^* = 0,\ U^* = const.,\ V^* = 0 \ \text{on}\ \Sigma \times \mathscr{T}.$$

Proof. The solution A^* corresponds to null data so that (22.35) implies that

$$\frac{d}{dt}\{\int_{\Sigma} (2h\rho w_j^* w_j^* + 2hJ\Phi_j^*\Phi_j^* + \rho I v_\alpha^* v_\alpha^*) da\} = 0.$$

If we take into account that w_j^*, Φ_j^* and v_α^* vanish initially, we obtain

$$\int_{\Sigma} (2h\rho w_j^* w_j^* + 2hJ\Phi_j^*\Phi_j^* + \rho I v_\alpha^* v_\alpha^*) da = 0.$$

In view of hypotheses we find

$$w_j^* = 0,\ \Phi_j^* = 0,\ v_\alpha^* = 0 \ \text{on}\ \Sigma \times \mathscr{T}.$$

It follows from (22.17) that

$$\sigma_\alpha^* = -\chi U_{,\alpha}^*,\ d_\alpha^* = -I\chi V_{,\alpha}^*,\ \sigma_3^* = -\chi V^*.$$

With the help of the equations (22.40) we arrive at the desired result.□

In what follows we denote by $f * g$ the convolution of f and g,

$$f * g(\mathbf{x}, t) = \int_0^t f(x, t-\tau) g(x, \tau) d\tau,\ x \in \Sigma, t \in \mathscr{T},$$

where f and g are continuous functions on $\Sigma \times \mathscr{T}$.

Theorem 22.1 forms the basis of the following reciprocity theorem

Theorem 22.4. *Let $A^{(\alpha)}$ be a solution corresponding to the external data system $\mathscr{L}^{(\alpha)}$, $(\alpha = 1, 2)$. Then*

$$\int_{\Sigma}[2h\mathscr{F}_i^{(1)} * w_i^{(2)} + 2h\mathscr{G}_i^{(1)} * \Phi_i^{(2)} + \mathscr{H}_\alpha^{(1)} * v_\alpha^{(2)} - k * (2hP^{(1)} * U^{(2)} + Q^{(1)} * V^{(2)})]da$$
$$+ \int_{\Gamma} k * [2h\tau_j^{(1)} * w_j^{(2)} + 2h\mu_j^{(1)} * \Phi_j^{(2)} + 2h\sigma^{(1)} * U^{(2)} + M_\beta^{(1)} * v_\beta^{(2)} \tag{22.41}$$
$$+ S^{(1)} * V^{(2)}]dl = \int_{\Sigma}[2h\mathscr{F}_i^{(2)} * w_i^{(1)} + 2h\mathscr{G}_i^{(2)} * \Phi_i^{(1)}$$
$$+ \mathscr{H}_\alpha^{(2)} * v_\alpha^{(1)} - k * (2hP^{(2)} * U^{(1)} + Q^{(2)} * V^{(1)})]da$$
$$+ \int_{\Gamma} k * [2h\tau_j^{(2)} * w_j^{(1)} + 2h\mu_j^{(2)} * \Phi_j^{(1)} + 2h\sigma^{(2)} * U^{(1)} + M_\beta^{(2)} * v_\beta^{(1)} + S^{(2)} * V^{(1)}]dl,$$

where

$$\mathscr{F}_i^{(\alpha)} = k * F_i^{(\alpha)} + \rho(t\vartheta_i^{0(\alpha)} + w_i^{0(\alpha)}),\ k(t) = t,\ t \in [0,\infty), \tag{22.42}$$
$$\mathscr{G}_i^{(\alpha)} = k * G_i^{(\alpha)} + J(t\omega_i^{0(\alpha)} + \Phi_i^{0(\alpha)}),\ \mathscr{H}_\rho^{(\alpha)} = k * H_\rho^{(\alpha)} + \rho I(t\eta_\rho^{0(\alpha)} + v_\rho^{0(\alpha)}).$$

Proof. We take in (22.24), $r = \tau$ and $s = t - \tau$ and integrate from 0 to t. With the aid of (22.25) we obtain

$$\int_{\Sigma}[2hF_j^{(1)} * w_j^{(2)} + 2hG_j^{(1)} * \Phi_j^{(2)} - 2hP^{(1)} * U^{(2)} + H_\alpha^{(1)} * v_\alpha^{(2)} - Q^{(1)} * V^{(2)}]da$$
$$+ \int_{\Gamma}[2h\tau_j^{(1)} * w_j^{(2)} + 2h\mu_j^{(1)} * \Phi_j^{(2)} + 2h\sigma^{(1)} * U^{(2)} + M_\beta^{(1)} * v_\beta^{(2)} + S^{(1)} * V^{(2)}]dl$$
$$- \int_{\Sigma}[2h\rho\ddot{w}_j^{(1)} * w_j^{(2)} + 2hI\ddot{\Phi}_j^{(1)} * \Phi_j^{(2)} + \rho I\ddot{v}_\alpha^{(1)} * v_\alpha^{(2)}]da \tag{22.43}$$
$$= \int_{\Sigma}[2hF_j^{(2)} * w_j^{(1)} + 2hG_j^{(2)} * \Phi_j^{(1)} - 2hP^{(2)} * U^{(1)} + H_\alpha^{(2)} * v_\alpha^{(1)} - Q^{(2)} * V^{(1)}]da$$
$$+ \int_{\Gamma}[2h\tau_j^{(2)} * w_j^{(1)} + 2h\mu_j^{(2)} * \Phi_j^{(1)} + 2h\sigma^{(2)} * U^{(1)} + M_\beta^{(2)} * v_\beta^{(1)} + S^{(2)} * V^{(1)}]dl$$
$$- \int_{\Sigma}[2h\rho\ddot{w}_j^{(2)} * w_j^{(1)} + 2hJ\ddot{\Phi}_j^{(2)} * \Phi_j^{(1)} + \rho I\ddot{v}_\alpha^{(2)} * v_\alpha^{(1)}]da.$$

Let us note that

$$k * \ddot{w}_j^{(\alpha)} = w_j^{(\alpha)} - t\vartheta_j^{0(\alpha)} - w_j^{0(\alpha)}. \tag{22.44}$$

If we take the convolution of the relation (22.43) with k, then with the help of (22.42) and (22.44) we conclude that the relation (22.41) holds.□

The method to obtain this reciprocal theorem has been established in Ieşan and Quintanilla (2007).

Let us study now the continuous dependence of solutions upon initial data and body loads. We restrict our attention to Dirichlet boundary conditions. Therefore in the remains of this section when we will consider a system of loading the corresponding boundary conditions will be of Dirichlet type. Let $A^{(1)}$ and $A^{(2)}$ be solutions corresponding to the data systems of loading

$$\mathscr{L}^{(1)} = \{F_k^{(1)}, G_k^{(1)}, H_\beta^{(1)}, P^{(1)}, Q^{(1)}, w_k^{0(1)}, \Phi_k^{0(1)}, v_\rho^{0(1)}, \vartheta_j^{0(1)}, \omega_j^{0(1)}, \eta_\beta^{0(1)}, \widetilde{w}_j, \widetilde{\Phi}_j, \widetilde{v}_\rho, \widetilde{U}, \widetilde{V}\}$$

and

$$\mathscr{L}^{(2)} = \{F_k^{(2)}, G_k^{(2)}, H_\beta^{(2)}, P^{(2)}, Q^{(2)}, w_k^{0(2)}, \Phi_k^{0(2)}, v_\rho^{0(2)}, \vartheta_j^{(2)}, \omega_j^{0(2)}, \eta_\beta^{0(1)}, \widetilde{w}_j, \widetilde{\Phi}_j, \widetilde{v}_\rho, \widetilde{U}, \widetilde{V}\},$$

respectively. We denote

$$w_i = w_i^{(1)} - w_i^{(2)}, \Phi_i = \Phi_i^{(1)} - \Phi_i^{(2)}, v_i = v_i^{(1)} - v_i^{(2)}, U = U^{(1)} - U^{(2)}, V = V^{(1)} - V^{(2)},$$

$$\gamma_{ij} = \gamma_{ij}^{(1)} - \gamma_{ij}^{(2)}, \eta_{\beta k} = \eta_{\beta k}^{(1)} - \eta_{\beta k}^{(2)}, \xi_{\beta\alpha} = \xi_{\beta\alpha}^{(1)} - \xi_{\beta\alpha}^{(2)}, e_\rho = e_\rho^{(1)} - e_\rho^{(2)}, \zeta_\rho = \zeta_\rho^{(1)} - \zeta_\rho^{(2)},$$

$$\tau_{ij} = \tau_{ij}^{(1)} - \tau_{ij}^{(2)}, \mu_{ij} = \mu_{ij}^{(1)} - \mu_{ij}^{(2)}, M_{\rho\nu} = M_{\rho\nu}^{(1)} - M_{\rho\nu}^{(2)}, \sigma_k = \sigma_k^{(1)} - \sigma_k^{(2)}, d_\alpha = d_\alpha^{(1)} - d_\alpha^{(2)}.$$

Clearly,

$$A = \{w_k, \Phi_k, v_\beta, U, V, \gamma_{ij}, \eta_{\beta k}, \xi_{\beta\rho}, e_\rho, \zeta_\rho, \tau_{ij}, \mu_{ij}, M_{\alpha\beta}, \sigma_k, d_\alpha\}$$

is a solution of the problem corresponding to the system of loading

$$\mathscr{L} = \{F_i, G_i, H_\alpha, P, Q, w_i^0, \Phi_i^0, v_\alpha^0, \vartheta_j^0, \omega_j^0, \eta_\alpha^0, 0, 0, 0, 0, 0\},$$

where

$$F_i = F_i^{(1)} - F_i^{(2)}, G_i = G_i^{(1)} - G_i^{(2)}, H_\alpha = H_\alpha^{(1)} - H_\alpha^{(2)}, P = P^{(1)} - P^{(2)}, Q = Q^{(1)} - Q^{(2)},$$

$$w_i^0 = w_i^{0(1)} - w_i^{0(2)}, \Phi_i^0 = \Phi_i^{0(1)} - \Phi_i^{0(2)}, v_\alpha^0 = v_\alpha^{0(1)} - v_\alpha^{0(2)}, \vartheta_j^0 = \vartheta_j^{(1)} - \vartheta_j^{(2)},$$

$$\omega_j^0 = \omega_j^{0(1)} - \omega_j^{0(2)}, \eta_\alpha^0 = \eta_\alpha^{0(1)} - \eta_\alpha^{0(2)}.$$

Let us denote this problem by $(\mathscr{P})$. We introduce the function E on $[0, t_1]$ by

$$E = \frac{1}{2} \int_\Sigma (2h\rho \dot{w}_j \dot{w}_j + 2hJ\dot{\Phi}_j \dot{\Phi}_j + \rho I \dot{v}_\alpha \dot{v}_\alpha + 4h\mathscr{N}_1 + 2I\mathscr{N}_2) da, \tag{22.45}$$

where

$$\begin{aligned} 2\mathscr{N}_1 = {} & \lambda \gamma_{\alpha\alpha} \gamma_{\rho\rho} + (\mu + \kappa)(\gamma_{\alpha j} \gamma_{\alpha j} + \gamma_{3\alpha} \gamma_{3\alpha}) + \mu(\gamma_{\beta j} \gamma_{j\beta} + \gamma_{3\alpha} \gamma_{\alpha 3}) \\ & + \alpha \eta_{\rho\rho} \eta_{\nu\nu} + \beta \eta_{\alpha\rho} \eta_{\rho\alpha} + \gamma \eta_{\alpha j} \eta_{\alpha j} + 2C_1 \gamma_{\rho\rho} \eta_{\beta\beta} + 2C_2 \eta_{\beta j} \gamma_{j\beta} \\ & + 2C_3 \gamma_{\alpha j} \eta_{\alpha j} + \chi(e_\alpha e_\alpha + V^2), \\ 2\mathscr{N}_2 = {} & \lambda \xi_{\alpha\alpha} \xi_{\beta\beta} + (\mu + \kappa) \xi_{\alpha\beta} \xi_{\alpha\beta} + \mu \xi_{\beta\alpha} \xi_{\alpha\beta} + \chi \zeta_\alpha \zeta_\alpha. \end{aligned} \tag{22.46}$$

Lemma 22.1. *Let A be a solution of the problem* $(\mathscr{P})$. *Then*

$$\dot{E} = \int_{\Sigma} [2h(F_j \dot{w}_j + G_j \dot{\Phi}_j + \dot{P}U) + H_\alpha \dot{v}_\alpha + \dot{Q}V]da. \tag{22.47}$$

Proof. In view of (22.17) and (22.46), we find that

$$2h(\tau_{\alpha j}\dot{\gamma}_{\alpha j} + \tau_{3\alpha}\dot{\gamma}_{3\alpha} + \mu_{\alpha j}\dot{\eta}_{\alpha j} + \dot{\sigma}_\alpha e_\alpha - \dot{\sigma}_3 V) + M_{\alpha\beta}\dot{\xi}_{\alpha\beta} + \dot{d}_\alpha \zeta_\alpha = 2h\dot{\mathscr{N}_1} + I\dot{\mathscr{N}_2}. \tag{22.48}$$

On the other hand, by using (22.7), (22.11) and (22.13)) we obtain

$$\begin{aligned} &2h(\tau_{\alpha j}\dot{\gamma}_{\alpha j} + \tau_{3\alpha}\dot{\gamma}_{3\alpha} + \mu_{\alpha j}\dot{\eta}_{\alpha j} + \dot{\sigma}_\alpha e_\alpha - \dot{\sigma}_3 V) + M_{\alpha\beta}\dot{\xi}_{\alpha\beta} \\ &+\dot{d}_\alpha \zeta_\alpha = \{2h(\tau_{\beta j}\dot{w}_j + \mu_{\beta j}\dot{\Phi}_j - \dot{\sigma}_\beta U) + M_{\beta\alpha}\dot{v}_\alpha - \dot{d}_\beta V\}_{,\beta} \\ &+2h(F_j \dot{w}_j + G_j \dot{\Phi}_j + \dot{P}U) + H_\alpha \dot{v}_\alpha + \dot{Q}V - 2h(\rho \ddot{w}_j \dot{w}_j + J\ddot{\Phi}_j \dot{\Phi}_j) - \rho I \ddot{v}_\alpha \dot{v}_\alpha. \end{aligned} \tag{22.49}$$

If we integrate (22.48) over Σ and use the divergence theorem, the boundary conditions and (22.47), then we obtain the desired result.□

We assume that $\mathscr{N}_1$ and $\mathscr{N}_2$ are positive definite quadratic forms. Thus, there exist the positive constants $\kappa_j, (j = 1,2,3,4)$, such that

$$\begin{aligned} \kappa_1(\gamma_{\alpha j}\gamma_{\alpha j} + \gamma_{3\alpha}\gamma_{3\alpha} + e_\alpha e_\alpha + V^2) \le \mathscr{N}_1 \le \kappa_2(\gamma_{\alpha j}\gamma_{\alpha j} + \gamma_{3\alpha}\gamma_{3\alpha} + e_\alpha e_\alpha + V^2), \\ \kappa_3(\xi_{\alpha\beta}\xi_{\alpha\beta} + \zeta_\alpha\zeta_\alpha) \le \mathscr{N}_2 \le \kappa_4(\xi_{\alpha\beta}\xi_{\alpha\beta} + \zeta_\alpha\zeta_\alpha), \end{aligned} \tag{22.50}$$

for all the variables and any $t \in [0,t_1]$.

We define the functions Ψ and M on $[0,t_1]$ by

$$\begin{aligned} \Psi = [\int_{\Sigma} (\dot{w}_j \dot{w}_j + \dot{\Phi}_j \dot{\Phi}_j + \dot{v}_\alpha \dot{v}_\alpha + \gamma_{\alpha j}\gamma_{\alpha j} + \gamma_{3\alpha}\gamma_{3\alpha} + \xi_{\alpha\beta}\xi_{\alpha\beta}, \\ +e_\alpha e_\alpha + \zeta_\alpha\zeta_\alpha + U^2 + V^2)da]^{1/2}, \\ M = \{\int_{\Sigma} [F_i F_i + G_i G_i + H_\alpha H_\alpha + (\dot{P})^2 + (\dot{Q})^2]da\}^{1/2}. \end{aligned} \tag{22.51}$$

Lemma 22.1 leads to the following continuous dependence result.

Theorem 22.5. *Assume that* ρ *and* J *are strictly positive and that* $\mathscr{N}_1$ *and* $\mathscr{N}_2$ *are positive definite quadratic forms. Let A be a solution of the problem* $(\mathscr{P})$. *Then, there exist the positive constants* ρ_1 *and* ρ_2 *such that*

$$\Psi(t) \le \rho_1 \Psi(0) + \rho_2 \int_0^t M(s)ds, \quad t \in [0,t_1]. \tag{22.52}$$

Proof. With the help of the Schwarz inequality, from (22.47) we find that

$$\dot{E} \leq Mk_0\Big[\int_{\Sigma}(\dot{w}_j\dot{w}_j+\dot{\Phi}_j\dot{\Phi}_j+\dot{v}_\alpha\dot{v}_\alpha+U^2+V^2)da\Big]^{1/2}, \tag{22.53}$$

where $k_0=\max(1,2h)$. It follows from (22.51) and (22.53) that

$$\dot{E} \leq Mk_0\Psi, \tag{22.54}$$

so that we get

$$E(t) \leq E(0)-k_0\int_0^t M(s)\Psi(s)ds,\ \ t\in[0,t_1] \tag{22.55}$$

In view of (22.45) and (22.50) we have

$$E(t)\geq \omega_1\Psi^2(t),\ \ E(0)\leq \omega_2\Psi^2(0), \tag{22.56}$$

where

$$\begin{aligned}
\omega_1 &= \frac{1}{2}\min(2h\rho,2hJ,\rho I,4h\kappa_1(1-\epsilon),4h,\kappa_1\epsilon C^{-1},2I\kappa_3),\\
\omega_2 &= \frac{1}{2}\max(2h,2hJ,\rho I,4\kappa_2 h,2I\kappa_4).
\end{aligned}$$

Here ϵ is an arbitrary constant between zero and one and C is the Poincaré constant for the domain Σ.

The relations (22.55) and (22.56) imply that

$$\Psi^2(t)\leq \rho_1^2\Psi^2(0)+2\rho_2\int_0^t M(s)\psi(s)ds,\ t\in[0,t_1], \tag{22.57}$$

where

$$\rho_1=(\omega_2/\omega_1)^{1/2},\ \ 2\rho_2=k_0/\omega_1.$$

In view of the Gronwall inequality, from (22.57) we obtain the desired result.□

We can prove that the quadratic forms $\mathscr{N}_1$ and $\mathscr{N}_2$ are simultaneously positive definite, if and only if

$$\begin{gathered}
2\lambda+2\mu+\kappa>0,2\mu+\kappa>0,\kappa>0,\gamma+\kappa>\beta,\chi>0,\\
\beta+\gamma+\kappa+2\mu>0,2\alpha+\beta+\gamma+2\lambda+2\mu+\kappa>0,\\
\kappa(\gamma-\beta)>(C_2-C_3)^2,\ (\beta+\gamma)(\kappa+2\mu)>(C_2+C_3)^2,\\
(2\alpha+\beta+\gamma)(2\lambda+2\mu+\kappa)>(2C_1+C_2+C_3)^2,\\
\kappa\gamma(2\mu+\kappa)>\kappa(C_2^2+C_3^2)+\mu(C_2-C_3)^2.
\end{gathered} \tag{22.58}$$

To prove this we introduce the matrices $\mathscr{A}$ and $\mathscr{B}$ defined by

$$\begin{gathered}
\mathscr{A}=||a_{ij}||_{8\times 8},\ \mathscr{B}=||b_{mn}||_{6\times 6},\ a_{ij}=a_{ji},\ b_{mn}=b_{nm},\\
a_{11}=a_{22}=\lambda+2\mu+\kappa,\ a_{33}=a_{44}=\mu+\kappa,\ a_{55}=a_{66}=\alpha+\beta+\gamma,\\
a_{77}=a_{88}=\gamma,\ a_{12}=\lambda,\ a_{13}=a_{14}=a_{17}=a_{18}=0,\\
a_{15}=a_{26}=C_1+C_2+C_3,\ a_{16}=a_{25}=C_1, a_{23}=a_{24}=a_{27}=a_{28}=0,\\
a_{34}=\mu, a_{35}=a_{36}=0,\ a_{37}=a_{48}=C_3, a_{38}=a_{47}=C_2, \qquad (22.59)\\
a_{45}=a_{46}=0, a_{56}=\alpha, a_{57}=a_{58}=0, a_{67}=a_{68}=0, a_{78}=\beta,\\
b_{11}=b_{22}=b_{33}=b_{44}=\mu+\kappa, b_{55}=b_{66}=\gamma, b_{12}=b_{14}=b_{16}=0,\\
b_{13}=b_{24}=\mu,\ b_{15}=b_{26}=C_3,\ b_{23}=b_{25}=0,\ b_{34}=b_{36}=0,\\
b_{35}=b_{46}=C_2,\ b_{45}=b_{56}=0.
\end{gathered}$$

Clearly, the quadratic form $\mathscr{N}_1$ is positive definite if and only if the matrices $\mathscr{A}$ and $\mathscr{B}$ are positive definite and $\chi>0$. The matrix $\mathscr{A}$ has six distinct eigenvalues ε_k defined by

$$\begin{gathered}
2\varepsilon_{1,2}=\gamma+\kappa-\beta\pm[(\beta+\kappa-\gamma)^2+4(C_2-C_3)^2]^{1/2},\\
2\varepsilon_{3,4}=2\mu+\kappa+\beta+\gamma\pm[(\beta+\gamma-\kappa-2\mu)^2+4(C_2+C_3)^2]^{1/2}, \qquad (22.60)\\
2\varepsilon_{5,6}=2\lambda+2\mu+\kappa+2\alpha+\beta+\gamma\\
\pm[(2\alpha+\beta+\gamma-2\lambda-2\mu-\kappa)^2+4(2C_1+C_2+C_3)^2]^{1/2}.
\end{gathered}$$

If b is positive, then the numbers $a\pm b^{1/2}$ are both positive if and only if a and a^2-b are positive. Therefore, the eigenvalues of the matrix $\mathscr{A}$ are positive if and only if the following conditions

$$\begin{gathered}
2\mu+\kappa+\beta+\gamma>0,\ \gamma+\kappa>\beta,\ \kappa(\gamma-\beta)>(C_2-C_3)^2,\\
2\lambda+2\mu+\kappa+2\alpha+2\beta+\kappa>0, (\beta+\gamma)(2\mu+\kappa)>(C_2+C_3)^2, \qquad (22.61)\\
(2\alpha+\beta+\gamma)(2\lambda+2\mu+\kappa)>(2C_1+C_2+C_3)^2,
\end{gathered}$$

are satisfied. By using Sylvester theorem we find that the matrix $\mathscr{B}$ is positive definite if and only if the following hold true

$$2\mu+\kappa>0,\ \kappa>0,\ \kappa\gamma(2\mu+\kappa)>\mu(C_2-C_3)^2+\kappa(C_2^2+C_3^2). \qquad (22.62)$$

Similarly, we see that the form $\mathscr{N}_2$ is positive definite if and only if following inequalities

$$2\lambda+2\mu+\kappa>0,\ 2\mu+\kappa>0,\ \kappa>0, \qquad (22.63)$$

hold. With the help of (22.60), (22.61) and (22.62) we obtain (22.73).

22.5 Equilibrium Theory

In this section we derive a uniqueness result for the Neumann problem in the stationary theory. In the case of the equilibrium theory the basic system of field equations consists of the geometrical equations (22.16), the equations of the electric field (22.11) and (22.13), the constitutive equations (22.17) and the following equations of equilibrium

$$\tau_{\beta k,\beta}+F_k=0,\ \mu_{\beta\alpha,\beta}+\varepsilon_{3\rho\alpha}(\tau_{3\rho}-\tau_{\rho 3})+G_\alpha=0,$$
$$\mu_{\alpha 3,\alpha}+\varepsilon_{3\rho\alpha}\tau_{\rho\alpha}+G_3=0,\ M_{\beta\alpha,\beta}-2h\tau_{3\alpha}+H_\alpha=0 \qquad (22.64)$$

on Σ. To the field equations we must adjoin boundary conditions. We consider the Neumann problem which is characterized by the following conditions

$$\tau_{\beta j}n_\beta=\widetilde{\tau}_j,\ \mu_{\beta j}n_\beta=\widetilde{\mu}_j,\ M_{\beta\alpha}n_\beta=\widetilde{M}_\alpha,\ \ \sigma_\alpha n_\alpha=\widetilde{\sigma},\ d_\alpha n_\alpha=\widetilde{d}\ \text{ on } \Gamma, \qquad (22.65)$$

where $\widetilde{\tau}_j,\widetilde{\mu}_j,\widetilde{M}_\alpha,\widetilde{\sigma}$ and $\widetilde{d}$ are prescribed functions on Γ.

We introduce the function W by

$$W=2h[\lambda\gamma_{\alpha\alpha}\gamma_{\beta\beta}+(\mu+\kappa)(\gamma_{\alpha j}\gamma_{\alpha j}+\gamma_{3\alpha}\gamma_{3\alpha})+\mu(\gamma_{\beta j}\gamma_{j\beta}+\gamma_{3\alpha}\gamma_{\alpha 3})+\alpha\eta_{\rho\rho}\eta_{\nu\nu}$$
$$+\beta\eta_{\alpha\rho}\eta_{\rho\alpha}+\gamma\eta_{\alpha j}\eta_{\alpha j}+2C_1\gamma_{\rho\rho}\eta_{\beta\beta}+2C_2\eta_{\beta j}\gamma_{j\beta}+2C_3\gamma_{\alpha k}\eta_{\alpha k}+\chi(e_\alpha e_\alpha+V^2)]$$
$$+I[\lambda\xi_{\alpha\alpha}\xi_{\beta\beta}+(\mu-\kappa)\xi_{\alpha\beta}\xi_{\alpha\beta}+\mu\xi_{\beta\alpha}\xi_{\alpha\beta}+\chi\zeta_\alpha\zeta_\alpha]. \qquad (22.66)$$

We note that the strain measures $\gamma_{\alpha j},\gamma_{3\alpha},\eta_{\alpha j}$ and $\xi_{\alpha\beta}$ defined by (22.16) are equal to zero if and only if $w_j=w'_j$, $\Phi_j=\Phi'_j$, $v_\alpha=v'_\alpha$, where w'_j, Φ'_j, v'_α are defined by

$$w'_j=a_j+\varepsilon_{jk\beta}x_\beta b_k,\ \ \Phi'_j=b_j,\ v'_\alpha=\varepsilon_{3\alpha\beta}b_\beta \qquad (22.67)$$

and a_j and b_j are arbitrary constants.

Theorem 22.6. *Assume that the function W is a positive definite quadratic form. Let $A^*=\{w^*_j,\Phi^*_j,v^*_\beta,U^*,V^*,\gamma^*_{ij},\eta^*_{\beta i},\xi^*_{\beta\rho},e^*_\alpha,\zeta^*_\alpha,\tau^*_{ij},\mu^*_{ij},M^*_{\alpha\beta},\sigma^*_i,d^*_\alpha\}$ be the difference of any two solutions of the Neumann problem. Then*

$$w^*_j=w'_j,\ \Phi^*_j=\Phi'_j,\ v^*_\alpha=v'_\alpha,\ U^*=const.,V^*=0\ \ on\ \Sigma,$$

where w'_j,Φ'_j and v'_α are defined by (22.67).

Proof. With the help of constitutive equations (22.17) we find that

$$W=2h(\tau_{\alpha k}\gamma_{\alpha k}+\tau_{3\alpha}\gamma_{3\alpha}+\mu_{\alpha k}\eta_{\alpha k}+\sigma_\alpha e_\alpha-\sigma_3 V)+M_{\alpha\beta}\xi_{\alpha 3}+d_\alpha\zeta_\alpha. \qquad (22.68)$$

On the other hand, by using the relations (22.16) and Eqs. (22.11), (22.13) and (22.63), from (22.68) we find that

$$W=2h(F_kw_k+G_k\Phi_k+PU)+H_\alpha v_\alpha+QV$$
$$+[2h(\tau_{\beta k}w_k+\mu_{\beta k}\Phi_k-\sigma_\beta U)+M_{\beta\alpha}v_\alpha-d_\beta V]_{,\beta}. \qquad (22.69)$$

If we use (22.68), (22.69) and the divergence theorem we obtain

$$\int_{\Sigma} W da = \int_{\Sigma} [2h(F_k w_k + G_k \Phi_k + PU) + H_\alpha v_\alpha + QV] da$$
$$+ \int_{\Gamma} [2h(\tau_{\beta k} w_k + \mu_{\beta k} \Phi_k - \sigma_\beta U) + M_{\beta\alpha} v_\alpha - d_\beta V] n_\beta dl. \quad (22.70)$$

Let us assume that the Neumann problem has two solutions. If A^* is the difference of any two solutions of the problem, then A^* satisfies the Neumann problem corresponding to null external data. From (22.70) we find that

$$\int_{\Sigma} W^* da = 0, \quad (22.71)$$

where W^* denotes the function W associated to the solution A^*. Since W^* is positive definite, from (22.71) we obtain the desired result.□

We note that in the three dimensional theory the restrictions imposed on the constitutive coefficients by the positive definiteness of the elastic potential have been established in Ieşan and Quintanilla (2007).

22.6 Effects of a Concentrated Charge Density

In this section we present an application of the stationary theory of chiral piezoelectric plates. We study the deformation of an infinite plate subjected to a concentrated charge density. We assume that

$$F_j = 0,\ G_j = 0,\ H_\alpha = 0,\ Q = 0. \quad (22.72)$$

We consider that the concentrated charge acts in the point (y_1, y_2), and introduce the notation $r = [(x_1 - y_1)^2 + (x_2 - y_2)^2]^{1/2}$. We try to solve the problem assuming that

$$w_\alpha = F_{,\alpha},\ w_3 = 0,\ \Phi_\alpha = G_{,\alpha},\ \Phi_3 = 0,\ v_\alpha = \varepsilon_{3\alpha\beta} H_{,\beta},\ U = \Lambda,\ V = 0, \quad (22.73)$$

where F, G, H and Λ are unknown functions of r. The equations (22.18) are satisfied if the functions F, G, H and Λ satisfy the equations

$$\begin{aligned} &a\Delta F + b\Delta G = 0, \\ &b\Delta F + (c\Delta - 2\kappa)G + \kappa H - 2\lambda_1 \Lambda = 0, \\ &(\mu + \kappa)(\Delta - v^2)H + \kappa v^2 G + \lambda_1 v^2 \Lambda = 0, \\ &\chi\Delta\Lambda - 2\lambda_1 \Delta G + \lambda_1 \Delta H = -P, \quad \text{on } \Sigma \end{aligned} \quad (22.74)$$

where

$$a=\lambda+2\mu+\kappa,\ b=C_1+C_2+C_3,\ c=\alpha+\beta+\gamma,\ \nu^2=2h/I. \tag{22.75}$$

We introduce the operator $\mathscr{D}$ defined by

$$\mathscr{D}=\eta_1\Delta\Delta-\eta_2\Delta+\eta_3, \tag{22.76}$$

where we have used the notations

$$\begin{gathered}\eta_1=\chi(ac-b^2)(\mu+\kappa),\\ \eta_2=(ac-b^2)\nu^2[\chi(\lambda+\mu)+\lambda_1^2]+2a(\mu+\kappa)(\kappa\chi+2\lambda_1^2),\\ \eta_3=a\nu^2(\kappa\chi+2\lambda_1^2)(2\mu+\kappa).\end{gathered} \tag{22.77}$$

Let Ω be a function of class C^6 on Σ, and let us consider the representation

$$\begin{gathered}F=-b\lambda_1[2(\mu+\kappa)\Delta-(2\mu+\kappa)\nu^2]\Omega,\\ G=a\lambda_1[2(\mu+\kappa)\Delta-(2\mu+\kappa)\nu^2]\Omega,\\ H=-\nu^2\lambda_1(ac-b^2)\Delta\Omega,\\ \Lambda=[(ac-b^2)(\mu+\kappa)\Delta(\Delta-\nu^2)-2\kappa a(\mu+\kappa)(\Delta-\nu^2)-a\kappa^2\nu^2]\Omega.\end{gathered} \tag{22.78}$$

Then, the functions F,G,H and Λ satisfy the system (22.74) if the function Ω satisfies the equation

$$\Delta\mathscr{D}\Omega=-P. \tag{22.79}$$

We can prove this assertion by substituting F,G,H and Λ from (22.78) into Eqs. (22.74). The operator $\mathscr{D}$ can be expressed in the form

$$\mathscr{D}=\eta_1(\Delta-k_1^2)(\Delta-k_2^2),$$

where k_α^2 are the roots of the equation

$$\eta_1 y^2-\eta_2 y+\eta_3=0.$$

The function Ω satisfies the equation

$$(\Delta-k_1^2)(\Delta-k_2^2)\Delta\Omega=-\xi P, \tag{22.80}$$

where $\xi=1/\eta_1$. In what follows we assume that k_α are distinct positive constants. Let S_j be functions that satisfy the equations

$$(\Delta-k_1^2)S_1=-\xi P,\ (\Delta-k_2^2)S_2=-\xi P,\ \Delta S_3=-\xi P. \tag{22.81}$$

The solution of the equation (22.80) can be written in the form

$$\Omega=\sum_{j=1}^{3}d_jS_j, \tag{22.82}$$

where

$$d_1^{-1}=k_1^2(k_1^2-k_2^2),\ d_2^{-1}=k_2^2(k_2^2-k_1^2),\ d_3^{-1}=k_1^2k_2^2. \tag{22.83}$$

We now assume that $P=P^*\delta(x-y)$, where $\delta(\cdot)$ is the Dirac measure and P^* is a given constant. In this case we have

$$S_\alpha=-\frac{1}{2\pi}\xi P^* K_0(k_\alpha r),\ S_3=\frac{1}{2\pi}\xi P^*\ln r, \tag{22.84}$$

where K_p denotes the modified Bessel function of the third kind and order p. Thus, we get

$$\Omega=-\frac{1}{2\pi}\xi P^*[d_1K_0(k_1r)+d_2K_0(k_2r)-d_3\ln r]. \tag{22.85}$$

The functions, F,G,H and Λ can be determined from (22.78) and (22.85). The function F is given by

$$F=-\frac{1}{2\pi}\xi P^* b\lambda_1[d_1A_1K_0(k_1r)+d_2A_2K_0(k_2r)+d_3A_3\ln r],$$

where

$$A_1=(2\mu+\kappa)\nu^2-2(\mu+\kappa)k_1^2,\ A_2=(2\mu+\kappa)\nu^2-2(\mu+\kappa)k_2^2,\ A_3=-(2\mu+\kappa)\nu^2.$$

The displacements w_α produced by the concentrated charge density are given by

$$w_\alpha=\frac{1}{2\pi r}\xi P^* b\lambda_1 x_\alpha[d_1A_1k_1K_1(k_1r)+d_2A_2k_2K_2(k_2r)-d_3A_3r^{-1}].$$

Similarly, we find that the function v_α have the form

$$v_\alpha=\nu^2(ac-b^2)\lambda_1\xi P^*\varepsilon_{3\alpha\beta}x_\beta(2\pi r)^{-1}[d_1k_1^3K_0(k_1r)+d_2k_2^3K_1(k_2r)].$$

We see that, in contrast with the theory of achiral materials, the stretching and flexure of the plate cannot be treated independently of each other.

22.7 Conclusions

The original results presented in this paper can be summarized as follows:

a) We present a theory of chiral piezoelecric elastic plates in the framework of the Cosserat theory. In contrast with the case of achiral plates, the stretching and flexure cannot be treated independently of each other.
b) We establish a uniqueness result with no definiteness assumption on the constitutive elastic coefficients. A reciprocity theorem is also presented.
c) We give the suitable conditions on the parameters to guarantee that the energy of the system is positive and we present a continuous dependence result under these assumptions.
d) We establish a uniqueness result in the stationary theory.

e) We study the effects of a concentrated charge density acting in an unbounded plate.

Acknowledgements R.Q. is supported by the Project "Análisis Matemático de Problemas de la Termomecánica"(MTM2016-74934-P) (AEI/FEDER, UE) of the Spanish Ministry of Economy and Competitiveness.

References

Altenbach H, Eremeyev VA (2013) Cosserat-type shells. In: Altenbach H, Eremeyev VA (eds) Generalized Continua from the Theory to Engineering Applications, Springer Vienna, Vienna, pp 131–178

Batra RC, Vidoli S (2002) Higher-order piezoelectric plate theory derived from a three-dimensional variational principle. AIAA J 40:91–104

Bisegna P, Maceri F (1996) A consistent theory of thin piezoelectric plates. Journal of Intelligent Material Systems and Structures 7(4):372–389

Brun L (1965) Sur l'unicité en thermoélasticité dynamique et diverses expressions analogues à la formule de Clapeyron. C R Acad Sci 261 A:2584–2587

Chandraseker K, Mukherjee S, Paci JT, Schatz GC (2009) An atomistic-continuum Cosserat rod model of carbon nanotubes. Journal of the Mechanics and Physics of Solids 57(5):932–958

De Cicco S, Ieşan D (2013) A theory of chiral Cosserat elastic plates. Journal of Elasticity 111(2):245–263

Eringen AC (1967) Theory of micropolar plates. Zeitschrift für angewandte Mathematik und Physik ZAMP 18(1):12–30

Eringen AC (1999) Microcontinuum Field Theories, vol I. Foundations and Solids. Springer, New York

Eringen AC (2004) Electromagnetic theory of microstretch elasticity and bone modeling. International Journal of Engineering Science 42(3):231–242

Eringen AC, Maugin GA (1990) Electrodynamics of Continua. Springer, New York

Ha CS, Plesha ME, Lakes RS (2016) Chiral three-dimensional lattices with tunable Poisson's ratio. Smart Materials and Structures 25(5):054,005

Healey TJ (2002) Material symmetry and chirality in nonlinearly elastic rods. Mathematics and Mechanics of Solids 7(4):405–420

Ieşan D, Quintanilla R (2007) Some theorems in the theory of microstretch thermopiezoelectricity. International Journal of Engineering Science 45(1):1–16

Irschik H (2002) A review on static and dynamic shape control of structures by piezoelectric actuation. Engineering Structures 24(1):5–11

Krommer M, Irschik H (2000) A Reissner-Mindlin-type plate theory including the direct piezoelectric and the pyroelectric effect. Acta Mechanica 141(1):51–69

Lakes R (1987) Foam structures with a negative poisson's ratio. Science 235(4792):1038–1040

Lakes R (2001) Elastic and viscoelastic behavior of chiral materials. International Journal of Mechanical Sciences 43(7):1579–1589

Maugin GA (1988) Continuum Mechanics of Electromagnetic Solids, North-Holland Series in Applied Mathematics and Mechanics, vol 33. North-Holland, Amsterdam, New York, Tokio

Maugin GA, Attou D (1990) An asymptotyc theory of thin piezoelectric plates. The Quarterly Journal of Mechanics and Applied Mathematics 43(3):347–362

Mindlin RD (1951) Influence of rotatory inertia and shear on the flexural motions of isotropic elastic plates. Trans ASME J Appl Mech 18:31–38

Mindlin RD (1974) Equations of high frequency vibrations of thermopiezoelectric crystal plates. International Journal of Solids and Structures 10(6):625–637

Mindlin RD (1984) Frequencies of piezoelectrically forced vibrations of electroded, doubly rotated, quartz plates. International Journal of Solids and Structures 20(2):141–157
Naghdi PM (1972) Theory of shells and plates. In: Truesdell C (ed) Handbuch der Physik, Springer, Berlin, vol VIa/2, pp 425–640
Natroshvili D, Giorgashvili L, Stratis I (2006) Representation formulae of general solutions in the theory of hemitropic elasticity. The Quarterly Journal of Mechanics and Applied Mathematics 59:451–474
Park HC, Lakes RS (1986) Cosserat micromechanics of human bone: Strain redistribution by a hydration sensitive constituent. Journal of Biomechanics 19(5):385–397
Ray MC, Batra RC (2007) A single-walled carbon nanotube reinforced 1-3 piezoelectric composite for active control of smart structures. Smart Materials and Structures 16(5):1936–1947
Telega JJ, Wojnar R (2002) Piezoelectric effects in biological tissues. J Theoretical Appl Mech 40:723–759
Tiersten HF (1993) Equations for the extension and flexure of relatively thin electrostatic plates undergoing electric fields. In: Lee JS, Maugin GA, Shindo Y (eds) Mechanics of Electromagnetic Materials and Structures, ASMA, New York, pp 21–34
Toupin RA (1963) A dynamical theory of elastic dielectrics. International Journal of Engineering Science 1(1):101–126
Yang J (2006) The Mechanics of Piezoelectric Structures. World Scientific, Singapore

Chapter 23
Non-Equilibrium Temperature and Reference Equilibrium Values of Hidden and Internal Variables

David Jou and Liliana Restuccia

Abstract In previous papers the concepts of caloric and entropic temperatures were outlined and illustrated in a few examples (ideal gas, ideal systems with two or three energy levels, solids with defects or dislocations). In equilibrium states, all degrees of freedom are at the same temperature, but out equilibrium they have different non-equilibrium temperatures. In this work, using a systematic methodology of classical irreversible thermodynamics, we take into consideration an undeformable medium in which the contributions of microscopic phenomena to the macroscopic specific internal energy U can be described by introducing two internal variables and one hidden variable. Internal variables are measurable (from the thermal point of view, they exchange directly heat with the system acting as thermometer) but not controllable, whereas (in our proposal) hidden variables are also not controllable but, in addition, they do not exchange directly heat with the thermometer, but only with other variables. The aim of this paper is to explore the difference between internal and hidden variables and to establish connections and relations among their corresponding non-equilibrium temperatures and the equilibrium temperature of the medium under consideration.

23.1 Introduction

In equilibrium states, all degrees of freedom are at the same temperature. Out of equilibrium, there are different definitions of temperature and, in any of such defini-

David Jou
Universitat Autònoma de Barcelona, Grup de Física Estadística, 08193 Bellaterra, Catalonia, Spain
e-mail: David.Jou@uab.cat

Liliana Restuccia
University of Messina, Department of Mathematical and Computer Sciences, Physical Sciences and Earth Sciences, Contrada Papardo, Viale Ferdinando Stagno d'Alcontres, 98166 Messina, Italy
e-mail: lrestuccia@unime.it

© Springer International Publishing AG, part of Springer Nature 2018
H. Altenbach et al. (eds.), *Generalized Models and Non-classical Approaches in Complex Materials 1*, Advanced Structured Materials 89
https://doi.org/10.1007/978-3-319-72440-9_23

tions, the different degrees of freedom may have different temperatures. In previous papers (Jou and Restuccia, 2013, 2016; Criado-Sancho et al, 2006) we outlined the concepts of caloric and entropic temperatures and emphasized their differences in a few examples (ideal gas, solids with defects of dislocations, ideal systems with two or three energy levels). The aim of this work is to consider how these temperatures differ for hidden and internal variables, depending on the caloric or entropic restrictions considered in the definition, or, in other words, on the equilibrium state taken as reference state. We will specify in more detail the differences between internal variables and hidden variables, which have special consequences on their respective temperatures. In Sect. 23.2 we describe the physical situation, in Sect. 23.3 we deal with non-equilibrium steady states, and in Sect. 23.4 we comment on unsteady states.

One of the motivations of this work is to go deeper into the thermal interaction of systems and thermometers. In general, thermometers are sensitive only to a given number of degrees of freedom of the system. This is not a problem in equilibrium thermodynamics, because all degrees of freedom have the same temperature, but this become a problem in non-equilibrium steady states, where each degree of freedom may have a different "temperature". Thus, the situation of variables which do not have a direct exchange of heat with the thermometer is not a purely academic issue but also a practical topic of reflexion.

23.2 Internal Variables and Hidden Variables

In this paper, using a systematic methodology of classical irreversible thermodynamics (CIT) (de Groot, 1951; de Groot and Mazur, 1962; Jou et al, 2010; Meixner and Reik, 1959), we show that different nonequilibrium temperatures arise when in the thermodynamic vector state of thermodynamic variables hidden and internal thermodynamic variables, that influence thermal behaviour of the material under consideration, must be introduced.

The introduction of internal variables permits to describe the internal structure of the body or some physical aspects of it, which influence the behaviour of the observed variables (Maugin and Muschik, 1994a,b; Maugin, 1999, 2013, 2015). The use of internal variables in nonequilibrium thermodynamics was done, for the first time, by Eckart (1940a,b) that introduced scalar internal variables; later Kluitenberg used vectorial and tensorial internal variables for describing physical relaxation mechanical, magnetic and dielectric phenomena (Kluitenberg, 1962, 1977, 1981, 1984). Internal variables were used also in Ciancio (1989); Ciancio and Restuccia (1990); Ciancio et al (1990); Ciancio and Restuccia (2016); Dolfin et al (2010); Restuccia and Kluitenberg (1987, 1988, 1989, 1990); Jou and Restuccia (2011); Restuccia (2010, 2014).

Internal variables and hidden variables share the fact that they are not directly measured, but they influence the behaviour of the system, as expressed in terms of a reduced set of directly measured variables. The difference between hidden and internal variables is that the former ones are much more difficult to be measured and controlled than the internal variables, or, in some occasions, hidden variables are in fact not known. Internal variables are measurable (from the thermal point of view, they exchange directly heat with the system acting as thermometer) but not controllable, whereas (in our proposal) hidden variables are also not controllable but, in addition, they do not exchange directly heat with the thermometer, but only with other variables. Hidden variables cannot still be related to any particular internal aspect of the system, because the latter is insufficiently known, at least at the moment in which the system is being considered. This does not imply, of course, that in the future these aspects may, eventually, be identified, measured and controlled. In spite of this lack of direct knowledge and control hidden variables influence the behaviour of the observable variables. This influence must be sufficiently mild, otherwise, the results of each experiment would be too different from those of the other experiments, but it must be perceptible. A possibility is, for instance, that these variables have a very slow or a very fast dynamics as compared to that of the variables being considered. We do not mean that the hidden variables must necessarily be much slower or faster than the other ones, but we only want to stress that hidden variables (in our proposal) not only do not exchange directly heat with the thermometer, but also they could have a very different dynamics than the directly observable variables. This last feature is not necessary for our proposal of hidden variables, and it is only examined in Sect. 23.4, but it is not used in Sect 23.3.

In Jou and Restuccia (2013, 2016); Criado-Sancho et al (2006) we discussed three aspects of temperature: on *empirical* temperature, based on the zeroth law, *caloric* temperature, based on the first law and related to the *energy content* of the system, and *entropic* temperature, based on the second law and related to the capacity of *energy exchange* with other systems. The meaning of the caloric and the entropic temperatures is thus deeply different, but, despite the conceptual differences, their respective values coincide in equilibrium states.

There are also other definitions of temperature based on statistical physics, on fluctuation theory, on fluctuation-dissipation theory, and other theories (see Casas-Vázquez and Jou, 2003; Criado-Sancho et al, 2006; Hoover and Hoover, 2008; Jou and Restuccia, 2011; Luzzi et al, 1997; Powles et al, 2005). In equilibrium states, all these definitions lead to the same value for the temperature, but in non-equilibrium steady states (where the material system is crossed by fluxes of energy, matter, electric current, ...) they lead to different values and give information on different aspects of the system, as for instance on different degrees of freedom, or on kinds of thermometers (for example, the intensity of several spectral lines in a radiation spectrum gives information on vibrational, rotational and electronic temperatures of excited atoms or molecules) (Puglisi et al, 2017). Thus, to try to establish connections between them and relations with the equilibrium temperature is a task for non-equilibrium thermodynamics.

In this paper we consider an undeformable medium in which the contributions of microscopic phenomena to the macroscopic specific internal energy U (energy per unit mass) can be described by introducing two internal variables (x_1, x_3), and one hidden variable (x_2), with their corresponding contributions to the internal energy denoted as U_1, U_2, U_3. Now, we consider that degrees of freedom 1 and 3, associated to U_1 and U_3, are measurable and controllable from the outside, i.e. they may directly exchange energy with the environment or with other systems; in contrast, degree 2, associated to U_2, is a truly "hidden" degree of freedom and it cannot be directly controlled. *It does not exchange directly energy with the environment, but only indirectly, through the other two degrees of freedom 1 and 3.*

Then, let us suppose that the total specific internal energy U is additively composed of three mentioned contributions

$$U = U_1 + U_2 + U_3$$

and we introduce the state space C

$$C = C(U_1, U_2, U_3). \tag{23.1}$$

Now, let us assume that the specific entropy (the entropy per unit of mass) S is a constitutive function of the state space C, and in particular

$$S = S(U_1, U_2, U_3). \tag{23.2}$$

We shall define the non-equilibrium entropic temperatures T_1, T_2, T_3, that represent the thermodynamic affinities conjugate to the internal variables U_1, U_2, U_3 as follows

$$T_1^{-1} = \frac{\partial S}{\partial U_1}, \qquad T_2^{-1} = \frac{\partial S}{\partial U_2}, \qquad T_3^{-1} = \frac{\partial S}{\partial U_3}. \tag{23.3}$$

Considering very small deviations with respect to an equilibrium state state, we expand the entropy (23.2) into Taylor's series with respect to this state and, confining our consideration to the linear terms, we have

$$S = S_{eq} + \frac{\partial S}{\partial U_1} dU_1 + \frac{\partial S}{\partial U_2} dU_2 + \frac{\partial S}{\partial U_3} dU_3. \tag{23.4}$$

From (23.3) we obtain the following Gibbs' relation

$$dS = T_1^{-1} dU_1 + T_2^{-1} dU_2 + T_3^{-1} dU_3. \tag{23.5}$$

Note that one could also consider

$$S = S\left(U_1, U_2, U_3, X_1', X_2', X_3'\right),$$

with X_i' related to work variables (as mechanical, electrical, magnetic or chemical work). In more general terms, we could also have

$$dS = \sum_i T_i^{-1} dU_i + \sum_i T_i^{-1} Y_i dX_i', \tag{23.6}$$

with

$$T_i^{-1} Y_i = \frac{\partial S}{\partial X_i'}. \tag{23.7}$$

For instance, if X_1' is an electrical polarization and X_2' a magnetic polarization, the Y_i would be related to the electric field and the magnetic field, respectively. Here we will deal with (23.5).

23.3 Temperatures in Steady States

The temperatures T_1 and T_3 of the internal variables x_1 and x_3, related to U_1 and U_3, may be measured from outside. The open problem is to find the temperature T_2 of the hidden variable related to U_2. In equilibrium this is not a problem because $T_1 = T_2 = T_3$, but in non-equilibrium this equality is no longer true.

23.3.1 Asymptotic Equilibrium Expressions for Caloric and Entropic Temperatures

There are at least two different possibilities to relate T_2 to T_1 and T_3, depending on whether we are using a caloric or entropic reference state. The caloric reference state is that which has the same total internal energy than the non-equilibrium state, and its temperature is denoted as T_U. Then

$$c_1 T_1' + c_2 T_2' + c_3 T_3' = c_{tot} T_U, \tag{23.8}$$

where T_U is the equilibrium temperature in the caloric reference state, the c_i $(i = 1, 2, 3)$ are the specific heats corresponding to the respective variables, and

$$c_{tot} = (c_1 + c_2 + c_3)$$

is the total specific heat. In other words, T_U will be the temperature of the equilibrium state, which is eventually reached a sufficiently long time after isolating the system. Accordingly, T_2' will be given by

$$T_2' = \frac{c_{tot} T_U - c_1 T_1' - c_3 T_3'}{c_2}. \tag{23.9}$$

The integration path in this case is to isolate the system at a given time and let it reach the final total equilibrium. Being the system isolated, its total internal energy will remain constant.

In the entropic point of view, we select as equilibrium reference state that having the same total entropy of our non-equilibrium state, i.e.

$$S' - S_{eq} = 0.$$

But, to do so, one should bring the degrees of freedom 1 and 3 in contact with very slowly varying thermal reservoirs slowly approaching T_1' and T_3' to their common eventual equilibrium value. Hopefully, this very slow process will not increase the total entropy of the system (unless the internal degree of freedom is analogous to dry friction, in which care slowness will not avoid some entropy production) so that

$$S' - S_{eq} = c_1 \ln \frac{T_1'}{T_S} + c_2 \ln \frac{T_2'}{T_S} + c_3 \ln \frac{T_3'}{T_S}, \tag{23.10}$$

where T_S is the equilibrium temperature of the entropic reference state. This may be obtained (see (23.5)) from

$$dS = \frac{dU_1}{T_1} + \frac{dU_2}{T_2} + \frac{dU_3}{T_3}, \tag{23.11}$$

with $dU_i = c_i dT_i$ and assuming, for the the sake of simplicity, c_i constant. Equation (23.10) may be rewritten in a parallel way to (23.8) as

$$c_1 \ln T_1' + c_2 \ln T_2' + c_3 \ln T_3' = c_{tot} \ln T_S. \tag{23.12}$$

In this case, the temperature T_2' in the initial non-equilibrium state will be related to T_1', T_3' and T_S as

$$\ln T_2' = \frac{c_{tot} \ln T_S - c_1 \ln T_1' - c_3 \ln T_3'}{c_2}. \tag{23.13}$$

The integration path in this case is different from the previous one leading to (23.9). In (23.9) the total internal energy was kept constant; indeed, here the total entropy is kept constant. This could be achieved, for instance, by using reversible heat engines acting between the several degrees of freedom. Therefore, the temperature T_2' related to the hidden degree of freedom will depend on the selected reference state. For small separations from equilibrium, i.e. if

$$| T_1' - T_S | \ll T_S \text{ and } | T_3' - T_S | \ll T_S \text{ and } | T_1' - T_U | \ll T_U \text{ and } | T_3' - T_U | \ll T_U,$$

both T_S and T_U will be very close to each other and also the caloric and the entropic values of T_2' will be very close to each other. Note that these operational definitions of T_2' do not yield the instantaneous temperature of x_2 at given time, because reaching the total equilibrium may require a considerably long time.

23.3.2 Dynamical Steady-State Expressions for Caloric and Entropic Temperatures

The expressions (23.9) and (23.13) are very general but they do not allow us to relate T_i' to the heat fluxes with the external world. A way to achieve this more detailed knowledge and to be able to estimate T_2' without need to let the system reach the

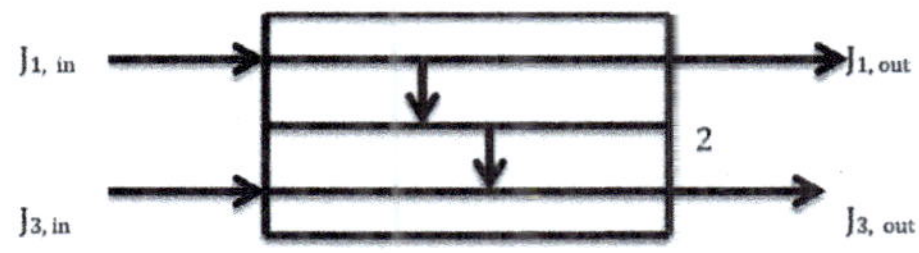

Fig. 23.1 Three degrees of freedom and their fluxes $J_{i,in}$ and $J_{i,out}$ $(i = 1,2)$.

caloric or the entropic equilibrium reference states would be to assume, for instance, that the evolution equations for T_1, T_2 and T_3 are (see Fig. 23.1)

$$\begin{aligned} \dot{U}_1 &= c_1\dot{T}_1 = J_{1,in} - J_{1,out} - \sigma_{12}, \\ \dot{U}_2 &= c_2\dot{T}_2 = \sigma_{12} - \sigma_{23}, \\ \dot{U}_3 &= c_3\dot{T}_3 = J_{3,in} - J_{3,out} + \sigma_{23}, \end{aligned} \tag{23.14}$$

with U_i internal energy, c_i heat capacity, T_i temperature of the degree i $(1 = 1,2,3)$. $J_{i,in}$ and $J_{i,out}$ (i=1,3) the rate of energy exchange of degrees 1 and 3 with the external world, and σ_{ij} rate of energy transfer from degree i to degree j inside the system.

We assume that degree 2 is not directly accessible, and we want to know how to measure its temperature in non-equilibrium steady state. In equilibrium,

$$J_{1,in} = J_{1,out} = J_{3,in} = J_{3,out} = 0.$$

Furthermore, we assume

$$\sigma_{12} = \alpha_{12}(T_1 - T_2); \quad \sigma_{23} = \alpha_{23}(T_2 - T_3), \tag{23.15}$$

with α_{12} and α_{23} being energy exchange coefficients.

In the steady state $\dot{T}_1 = \dot{T}_2 = \dot{T}_3 = 0$, under a given net supply

$$J_1' \equiv J_{1,in} - J_{1,out},$$

with

$$J_{1,in} + J_{3,in} = J_{1,out} + J_{3,out}$$

(corresponding to the steady state), the temperatures T_1' and T_3' will be, by solving Eqs. $(23.14)_1$, $(23.14)_3$ and (23.15),

$$T_1' = T_2' + \frac{J_1'}{\alpha_{12}}, \quad T_3' = T_2' - \frac{|J_3'|}{\alpha_{23}}. \tag{23.16}$$

Now, the results for T_1' and T_3' will depend on the values of T_2'. As we did above, we may relate T_2' to a caloric or an entropic reference state. In the first case, using (23.8) and taking into account (23.16), one obtains

$$
\begin{aligned}
T_3' &= T_U - \frac{c_1}{c_{tot}}\left[\frac{J_1'}{\alpha_{12}} + \frac{|J_3'|}{\alpha_{23}}\right] - \frac{c_2}{c_{tot}}\frac{|J_3'|}{\alpha_{23}}, \\
T_2' &= T_U - \frac{c_1}{c_{tot}}\frac{J_1'}{\alpha_{12}} + \frac{c_3}{c_{tot}}\frac{|J_3'|}{\alpha_{23}}, \\
T_1' &= T_U + \frac{c_2+c_3}{c_{tot}}\frac{J_1'}{\alpha_{12}} + \frac{c_3}{c_{tot}}\frac{|J_3'|}{\alpha_{23}}.
\end{aligned}
\tag{23.17}
$$

The interest of this expression is that it shows in a direct way how the temperatures of three different degrees of freedom differ from the temperature T_U. It is seen that the temperatures depend on the specified net fluxes J_i', as external quantities, and on the internal quantities c_i and on α_{ij}.

In the second case, if we express $T_1' = T_3' + \Delta T_1'$ and $T_2' = T_3' + \Delta T_2'$, relation (23.13) may be rewritten as

$$
\left[T_3' + \Delta T_1'\right]^{c_1}\left[T_3' + \Delta T_2'\right]^{c_2} T_3'^{c_3} = T_S^{c_{tot}}. \tag{23.18}
$$

From here, T_3' as a function of T_S may be obtained and when this result is combined with (23.16), one would get the expressions for T_2' and T_1' as well, and we would have the expressions analogous to (23.17). To have an explicit illustration, assume that $c_1 = c_2 = c_3$. In this case

$$
\begin{aligned}
T_3' &= T_S - \frac{1}{3}\frac{J_1'}{\alpha_{12}} - \frac{2}{3}\frac{|J_3'|}{\alpha_{23}}, \\
T_2' &= T_S - \frac{1}{3}\frac{J_1'}{\alpha_{12}} + \frac{1}{3}\frac{|J_3'|}{\alpha_{23}}, \\
T_1' &= T_S + \frac{2}{3}\frac{J_1'}{\alpha_{12}} + \frac{1}{3}\frac{|J_3'|}{\alpha_{23}}.
\end{aligned}
\tag{23.19}
$$

In general, the temperatures T_1', T_2' and T_3' will depend on whether the caloric or the entropic equilibrium reference state are used.

In (23.17) and (23.19), T_2' is not an univocally defined, "objective", quantity, but it yields the values of T_2' without need to wait for the system to reach the corresponding equilibrium state. Thus, the status of T_2' depends on the observer - not only through the choice of the kind of thermometer - but also by choosing the strategy (caloric or entropic) to reach the eventual equilibrium state.

In Fig. 23.2 we sketch as a continuous curve the manifold of equilibrium states of the system, the non-equilibrium state as the dot *neq*, and with discontinuous horizontal and vertical lines the respective entropic and caloric projections of the non-equilibrium state on the equilibrium manifold, with respective temperatures T_S and T_U.

The role of different "projections" of a non-equilibrium steady state on the manifold of equilibrium states of the system has been emphasized in Muschik (1990, 1977). The corresponding equilibrium state is sometimes called the "accompanying equilibrium state" (Muschik, 1977). When the fluxes are small the non-equilibrium state is "close" to the equilibrium manifold and the caloric and entropic reference states are close to each other, and T_S and T_U are almost the same, but if the fluxes are high, the differences between T_S and T_U and, correspondingly, between (23.17) and (23.19), are significant.

23.4 A Model for System's Aging

We have assumed that variable 2 cannot directly exchange energy with the environment, in contrast to the variables 1 and 3. This is a conceptual problem for the measurement of T_2, since there will not be a "thermometer" able to exchange energy with it. Thus, the measurements of T_2 should be of "dynamical" nature, i.e., through measurements of T_1 as of T_3 since, according to (23.16), in the steady state

$$T_2 - T_3 = \frac{|J_3'|}{\alpha_{23}}, \qquad \text{as} \qquad T_1 - T_2 = \frac{|J_1'|}{\alpha_{12}}.$$

Another problem illustrated by variable 2 is found in the case that c_2 is much higher than c_1 and c_3. Indeed, the characteristic time of temperatures T_1 and T_3 may be found writing $(23.14)_1$ and $(23.14)_3$ as

$$\begin{aligned} c_1\delta\dot{T}_1 &= J_{1,in} - J_{1,out} - \alpha_{12}(\delta T_1 - T_2), \\ c_3\delta\dot{T}_3 &= J_{1,in} - J_{3,out} + \alpha_{23}(T_2 - \delta T_3), \end{aligned} \tag{23.20}$$

which imply for the respective characteristic time scale

$$\tau_1 = \frac{c_1}{\alpha_{12}}, \qquad \tau_3 = \frac{c_3}{\alpha_{23}}. \tag{23.21}$$

The corresponding characteristic time for T_2 found from

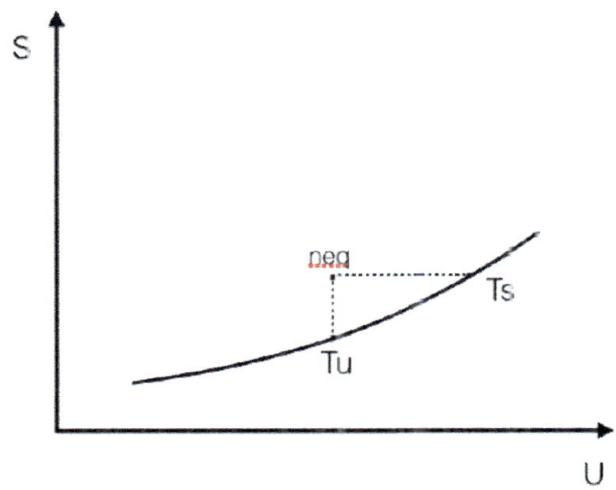

Fig. 23.2 A sketch of the manifold of equilibrium states of the system as a continuous curve and of a non-equilibrium state as the dot *neq*.

$$c_2\delta\dot{T}_2 = \alpha_{12}(T_1 - \delta T_2) - \alpha_{23}(\delta T_2 - T_3) \tag{23.22}$$

is

$$\tau_2 = \frac{c_2}{\alpha_{12}+\alpha_{23}}. \tag{23.23}$$

Thus, time τ_2 will be very long. This implies that variable 2 could act as an energy storage system. In the case that $\sigma_{12} - \sigma_{23}$ is not zero, but small, i.e. $\sigma_{12} - \sigma_{23} = \sigma'$, with σ' small, this would mean that T_2 changes in time as

$$T_2(t) = T_2(0) + \frac{\sigma'}{c_2 t}. \tag{23.24}$$

In this case, the system would exhibit "aging" in temperature. This means that research measuring T_1 and T_3 would find a different value at a different moment. Indeed, the measurement of T_1 as T_3 should take a time of the order of, let us say, $3\tau_1$ as $3\tau_3$, in order that T_1 and T_3 take a steady value. If this time is much shorter than c_2/σ', the changes in T_2 will not be perceptible according to (23.24) and the system will seem to be in a truly steady state. However, if an observation after an interval of the order of c_2/σ' is made, the internal T_2 will be changed considerably. Since, the measured temperature is

$$T_1(t) = T_2(t) + \frac{J_1'}{\alpha_{12}}, \tag{23.25}$$

with J_1'/α_{12} constant (in the simplest case), the measured temperature for T_1 will be different than that measured in an earlier time.

23.5 Concluding Remarks

We have presented a very simplified model for a system with three degrees of freedom. One of them, degree 2, acts as a hidden variable U which cannot be directly manipulated from the outside, in contrast with the other two internal degrees of freedom 1 and 3. Degree 2 does not exchange directly energy with the environment, but only indirectly, through the other two degrees of freedom 1 and 3. We have considered two conceptual problems related to the temperature of such internal variable:

1. First, we have obtained the values of T_1, T_2 and T_3 in a nonequilibrium steady state. Expression (23.16) shows how they depend on the internal energy exchange given by
$$J_i' = J_{i,in} - J_{i,out},$$
and of internal constitutive quantities c_i, α_{ij}.
2. Second, we have related these three temperatures to the single value of temperature of an equilibrium state having the same energy U or the same entropy S than the non-equilibrium system. These temperatures are called T_U and T_S, which are the caloric and entropic temperatures, respectively. It is seen that the temper-

atures of the degrees of freedom, given by (23.17) and (23.19), respectively, will be different in both cases.

3. Third, we have seen that if the internal variable 2 has associated a high heat capacity, it may act as an internal energy reservoir, changing its temperature very slowly. Thus, the values for the measured T_1 and T_3 variables will change with time in a very slow way, showing a thermal aging of the true system. Equilibrium states will be obtained only after a very long time that the perturbation occurs.

Acknowledgements D. Jou knowledges the financial support of the Spanish Ministry of Economy and Competitiveness under grant TEC 2015-67462-C2-2-R, and of the Direccio' General de Recerca of the Generalitat of Catalunya under grant 2009 SGR 164. L. R. acknowledges the hospitality of the Physics Department of the Universitat Autonoma de Barcelona from December 9th to December 16th 2016.

References

Casas-Vázquez J, Jou D (2003) Temperature in non-equilibrium states: a review of open problems and current proposals. Reports on Progress in Physics 66(11):1937–2023

Ciancio V (1989) On the generalized Debye equation for media with dielectric relaxation phenomena described by vectorial internal variables. Journal of Non-equilibrium Thermodynamics 14:239–250

Ciancio V, Restuccia L (1990) On the invariance of Onsager's reciprocal relations in the thermodynamic theory of dielectric relaxation phenomena. Physica A: Statistical Mechanics and its Applications 162(3):489–498

Ciancio V, Restuccia L (2016) On heat equation in the framework of classic irreversible thermodynamics with internal variables. International Journal of Geometric Methods in Modern Physics 13:1640,003

Ciancio V, Restuccia L, Kluitenberg GA (1990) A thermodynamic derivation of equations for dielectric relaxation phenomena in anisotropic polarizable media. Journal of Non-equilibrium Thermodynamics 15:151–171

Criado-Sancho M, Jou D, Casas-Vázquez J (2006) Nonequilibrium kinetic temperatures in flowing gases. Physics Letters A 350(5):339–341

de Groot SR (1951) Thermodynamics of Irreversible Processes. North-Holland Publishing Company, Amsterdam

de Groot SR, Mazur P (1962) Non-Equilibrium Thermodynamics. North-Holland Publishing Company, Amsterdam

Dolfin M, Francaviglia M, Preston S, Restuccia L (2010) Material element model and the geometry of the entropy form. International Journal of Geometric Methods in Modern Physics 7(6):1021–1042

Eckart C (1940a) The thermodynamics of irreversible processes. I. The simple fluid. Phys Rev 58:267–269

Eckart C (1940b) The thermodynamics of irreversible processes. II. Fluid mixtures. Phys Rev 58:269–275

Hoover WG, Hoover CG (2008) Nonequilibrium temperature and thermometry in heat-conducting ϕ^4 models. Phys Rev E 77:041,104

Jou D, Restuccia L (2011) Mesoscopic transport equations and contemporary thermodynamics: an introduction. Contemporary Physics 52(5):465–474

Jou D, Restuccia L (2013) Non-equilibrium temperatures in systems with internal variables. In: Pilotelli M, Beretta GP (eds) Proceedings of the 12th Joint European Thermodynamics Conference, University of Brescia, Brescia

Jou D, Restuccia L (2016) Caloric and entropic temperatures in non-equilibrium steady states. Physica A: Statistical Mechanics and its Applications 460:246–253

Jou D, Casas-Vazquez J, Lebon G (2010) Extended Irreversible Thermodynamics, 4th edn. Springer, Berlin

Kluitenberg GA (1962) On rheology and thermodynamics of irreversible processes. Physica 28(11):1173–1183

Kluitenberg GA (1977) On dielectric and magnetic relaxation phenomena and vectorial internal degrees of freedom in thermodynamics. Physica A: Statistical Mechanics and its Applications 87(2):302–330

Kluitenberg GA (1981) On vectorial internal variables and dielectric and magnetic relaxation phenomena. Physica A: Statistical Mechanics and its Applications 109(1):91–122

Kluitenberg GA (1984) Plasticity and non-equilibrium thermodynamics. In: Lehmann T (ed) The Constitutive Law in Thermoplasticity, Springer, Vienna, pp 157–250

Luzzi R, Vasconcellos AR, Casas-Vazquez J, Jou D (1997) Characterization and measurement of a nonequilibrium temperature-like variable in irreversible thermodynamics. Physica A: Statistical Mechanics and its Applications 234(3):699–714

Maugin GA (1999) The thermomechanics of nonlinear irreversible behaviours. An introduction. World Scientific, Singapore

Maugin GA (2013) Continuum mechanics through the twentieth century: a concise historical perspective. Springer, Berlin

Maugin GA (2015) The saga of internal variables of state in continuum thermo-mechanics (1893-2013). Mechanics Research Communications 69:79–86

Maugin GA, Muschik W (1994a) Thermodynamics with internal variables. Part I. General concepts. J Non-Equilib Thermodyn 19:217–249

Maugin GA, Muschik W (1994b) Thermodynamics with internal variables. Part II. Applications. J Non-Equilib Thermodyn 19:250–289

Meixner J, Reik HG (1959) Thermodynamik der Irreversiblen Prozesse. In: Flügge S (ed) Handbuch der Physik, Springer, Berlin, vol III/2

Muschik W (1977) Empirical foundation and axiomatic treatment of non-equilibrium temperature. Archive for Rational Mechanics and Analysis 66(4):379–401

Muschik W (1990) Aspects of Non-Equilibrium Thermodynamics. Word Scientific, Singapore

Powles JG, Rickayzen G, Heyes DM (2005) Temperatures: old, new and middle aged. Molecular Physics 103(10):1361–1373

Puglisi A, Sarracino A, Vulpiani A (2017) Temperature in and out of equilibrium: A review of concepts, tools and attempts. Physics Reports 709-710:1–60

Restuccia L (2010) On a thermodynamic theory for magnetic relaxation phenomena due to n microscopic phenomena described by n internal variables. J Non-Equilib Thermodyn 35:379–413

Restuccia L (2014) Generalizations of Snoek equation for anisotropic media with magnetic relaxation. Archive of Applied Mechanics 84(9):1539–1563

Restuccia L, Kluitenberg GA (1987) On possible interactions among dielectric relaxation, magnetic relaxation, heat conduction, electric conduction, diffusion phenomena, viscous flow and chemical reactions in fluid mixtures. Atti della Accademia Peloritana dei Pericolanti LXV:309–336

Restuccia L, Kluitenberg GA (1988) On generalizations of the Debye equation for dielectric relaxation. Physica A: Statistical Mechanics and its Applications 154(1):157–182

Restuccia L, Kluitenberg GA (1989) On generalizations of the Snoek equation for magnetic relaxation phenomena. Atti della Accademia Peloritana dei Pericolanti LXVII:141–194

Restuccia L, Kluitenberg GA (1990) Hidden vectorial variables as splitting operators for the polarization vector in the thermodynamic theory of dielectric relaxation. J Non-Equilib Thermodyn 15:335–346

Chapter 24
On the Foundation of a Generalized Nonlocal Extensible Shear Beam Model from Discrete Interactions

Attila Kocsis and Noël Challamel

Abstract In this paper a generalized discrete elastica model including bending, normal and shear interactions is developed. Nonlinear static analysis of the discrete model is accomplished, its buckling and post-buckling behavior are thoroughly studied. It is revealed that based on what finite strain theory is used, the discrete model yields a generalized (extensible) Engesser elastica, or a generalized (extensible) Haringx elastica. The local continuum counterparts of these models are also obtained. Then nonlocal models are developed from the introduced flexural, extensible, shearable discrete systems using a continualization technique. Analytical and numerical solutions are given for the discrete and nonlocal models, and it is shown that the scale effects of the discrete models are well captured by the continualized nonlocal models.

24.1 Introduction

Bridging the gap between molecular mechanics (formulated as a discrete lattice problem) and continuum mechanics is an old topic, which dates, at least from the XIXth century (see the initial works of Navier, 1823; Cauchy, 1823, 1828). This question was extensively debated during the XIXth century, for elaborating a consistent molecular-based three-dimensional elasticity theory (see Timoshenko, 1953; Foce, 1995) for a historical overview of this question). In case of local linear elasticity,

Attila Kocsis
Department of Structural Mechanics, Budapest University of Technology and Economics and Engineering Center Budapest, Robert Bosch Kft., 1111, Budapest, Hungary
e-mail: kocsis@ep-mech.me.bme.hu

Noël Challamel
Université de Bretagne Sud, EA 4250, Institut de Recherche Dupuy de Lôme (IRDL), Centre de Recherche, Rue de Saint Maudé-BP 92116, F-56100 Lorient, France
e-mail: noel.challamel@univ-ubs.fr

© Springer International Publishing AG, part of Springer Nature 2018
H. Altenbach et al. (eds.), *Generalized Models and Non-classical Approaches in Complex Materials 1*, Advanced Structured Materials 89,
https://doi.org/10.1007/978-3-319-72440-9_24

Cauchy derived expressions of the elastic moduli from the atomistic potentials (see also the complete paper of Stakgold (1950) on this point). Modern treatment of the problem of relating lattice and local continuum elasticity can be found in the treatise of Born and Huang (1954), who included both monatomic and multiatomic cases. The development of nonlocal elasticity theories from a lattice formulation emerged more recently, in the 60's by researchers such as Krumhansl (1965); Kröner and Datta (1966); Kunin (1966) (see the extensive analysis of Eringen, 2002, which includes a consistent thermodynamically-based framework of nonlocal theories including the lattice-based calibration). As pointed out by Dell'Isola et al (2014, 2015), the idea of nonlocality introduced from particles interaction was already probably anticipated by *Piola* during the XIXth century. Strain-based integral nonlocal models generally lead to the resolution of a complex integro-differential problem, whose mathematical treatment in the nonlinear range may be mathematically quite difficult. Among the nonlocal theories, the stress gradient model introduced by Eringen (1983) avoids the computation of such an integro-differential problem, by preserving the differential nature of the problem to be solved. This efficient engineering model, which may be classified as a phenomenological nonlocal model, can be calibrated from the wave dispersive properties of the axial lattice (Eringen, 1983). Alternatively, low frequency calibration of such a nonlocal model may be also proposed by Challamel et al (2015b). These results mainly concern three-dimensional or, as a particular case, uniaxial lattice and nonlocal media. The nonlocal models investigated in this paper may be also classified as differential type nonlocal models, in the sense that the mathematical problem to be solved also avoids integro-differential operators and mainly used additional gradient operators for introducing the additional length scale effects. The development of nonlocal beam theories and their fundamental discrete (or lattice) background is much more recent. An extensible local beam theory, which includes both bending and shear contributions was developed by Bresse (1859) for curved elements. Bresse developed a fully consistent bending-shear local beam theory, based on independent kinematic variables, namely the deflection, the axial displacement and the rotation for the in- plane behavior of curved beam elements. The shear correction factor, later introduced by Engesser (1891); Föppl (1897) was implicitly equal to unity in Bresse's theory. In his paper dated from 1922 Timoshenko calibrated the shear correction factor for an elastic beam composed of rectangular cross section, from the exact two- dimensional wave length solution, also labelled as Rayleigh-Lamb equation, giving rise to the so-called Bresse-Timoshenko beam theory (Timoshenko, 1922). The lattice foundation of these engineering beam models, especially formulated in a nonlocal framework, is still under debate. When restricted to the bending contribution, Hencky's model (Hencky, 1920) can be considered as the paradigmatic lattice model, which may asymptotically converge towards the bending continuum beam model, namely the Euler-Bernoulli beam model. It has been shown that this lattice model behaves as a nonlocal Euler-Bernoulli beam model in its linear range (Challamel et al, 2014)) but also in its nonlinear range, for the post-bifurcation regime (Challamel et al, 2015a)). A bending/shear lattice model was also shown to behave as a nonlocal Bresse-Timoshenko beam element (Zhang et al, 2013; Duan et al, 2013). These results valid for bending/shear lattices have been recently extended in the nonlinear range by Kocsis et al (2017). The aim of the present paper is to generalize these results and to develop a fully consistent axial/shear/bending lattice, with its nonlocal counterpart, based on a rational introduction of the extensibility effect. Shear effect will be introduced through both Engesser's approach (Engesser, 1891) and Haringx's approach (Haringx, 1942). Analytical solutions will be presented for the nonlocal extensible shear/beam model and its lattice initial formulation. A closely related study is the paper of Turco et al (2016) who considered a generalized two-dimensional Hencky-type system with bending, shear and extensional springs. Alibert et al (2017) recently presented an extensible Hencky model in the nonlinear range and investigated in detail the convergence property of the associated difference scheme, even in the nonlinear range. As an interesting point concerning the mathematical structure of the difference equations of lattice mechanics, Maugin (1999) pointed out the link between lattice mechanics and finite difference methods (as a numerical method) for the associated continuous medium, a property which is confirmed for beam elements as already mentioned by Silverman (1951) for Hencky beams (restricted to the bending contribution). We will show that this property is valid as well for the fully consistent extensible/shear/bending lattice beam. The fully consistent extensible/shear/bending lattice, which is developed in this paper, can be viewed as the lattice foundation of a general nonlinear and nonlocal (Bresse-Timoshenko) beam kinematics model.

24.2 The Mechanical Model

The generalized *Hencky* bar-chain model is shown in Fig. 24.1 (a). It consists of N links connected by frictionless hinges. In the unloaded undeformed state, the links form a rectangle of longer side a_0, their longer axes form a straight line which is the undeformed axis of the linkage of length $L = N \times a_0$. The leftmost hinge 0 of the linkage is fixed, the rightmost hinge N is equipped with a roller. The linkage is loaded axially by a compressive force P. The links are extensible and shearable, i.e. they can be elongated and distorted. The tensile stiffness of the links is r, the shear stiffness is k. The elongated length of the ith link is $a_i = (1 + \varepsilon_i)a_0$, with ε_i the normal strain in the link, which is constant along the link. The link distortion is measured by the shear angle γ_i. The rotation of the axis of link i is

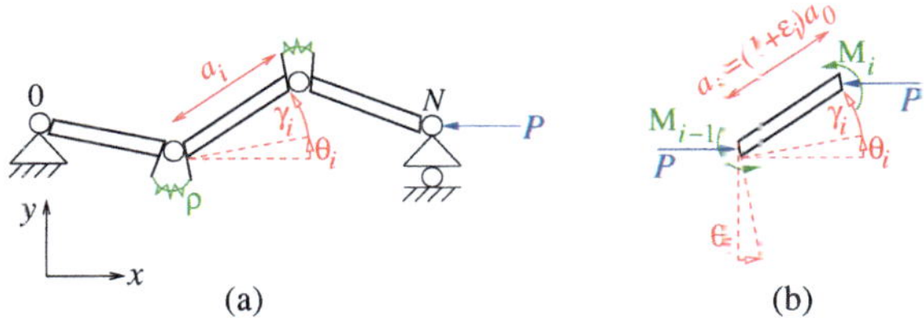

Fig. 24.1 (a) Generalized *Hencky* bar-chain under axial compression with pinned-pinned ends. (b) A deformed link with moments in the bending spring and the resultant of the normal and shear forces.

$$\Psi_i = \theta_i + \gamma_i \tag{24.1}$$

with θ_i the rotation of the end sections of the link. The ends of the neighboring links are interconnected by a rotational spring. If the linkage is a discrete model of a homogeneous, prismatic beam of bending stiffness EI, shear stiffness κGA, and normal stiffness EA, then the stiffness of the bending springs is EI/a_0, the shear stiffness of the links is κGAa_0, and their tensile stiffness is EAa_0.

The material law for the rotational (bending) springs is $M_i = EI/a_0(\theta_{i+1} - \theta_i)$, with M_i the moment in the ith spring. The material law for elongation is $N_i = EA\varepsilon_i$, with N_i the normal force, and the material law for shearing (considering small displacements at this point) is $V_i = \kappa GA\gamma_i$, with V_i the shear force in link i. This latter law will be generalized for large displacements depending on the applied finite shear strain theory later. A deformed link with moments in the bending spring and the resultant of the normal and shear forces are shown in Fig. 24.1 (b).

In Sect. 24.3 we follow *Timoshenko*'s finite strain theory and analyse the equilibrium states, buckling and post-buckling behavior of the discrete system for large displacements. The continuum counterpart and the continualized nonlocal analogue of the discrete model are derived and it is revealed that these are the discrete, local and nonlocal extensible *Engesser* elastica.

In Sect. 24.4 we do the same analyses using *Love*'s finite strain theory, leading us to the discrete, local and nonlocal extensible *Haringx* elastica.

24.3 Extensible Engesser Elastica

In this section, we implement *Timoshenko*'s finite strain theory in the discrete model. First the static behavior of the discrete model is analysed, then the continuum counterpart of the system is determined and a quasi-continuum is also developed from the lattice system, following a continualization technique.

24.3.1 Discrete Extensible Engesser Elastica

Figure 24.2 shows a deformed link of the model. The link axis, which is of initial length a_0, becomes $a_i = (1+\varepsilon_i)a_0$ upon stretching, and rotates by $\Psi_i = \theta_i + \gamma_i$. Here θ_i can be thought of as a rigid-body-like rotation of the lengthened link, and γ_i, the shear strain, measures the distortion of the link. The normal and shear forces of the ith link, N_i and V_i, are aligned and perpendicular, respectively, to the link axis.

The total potential energy of the structure is the sum of internal and external potentials:

Fig. 24.2 Deformation of the shearable, extensible link according to *Timoshenko*'s finite strain theory

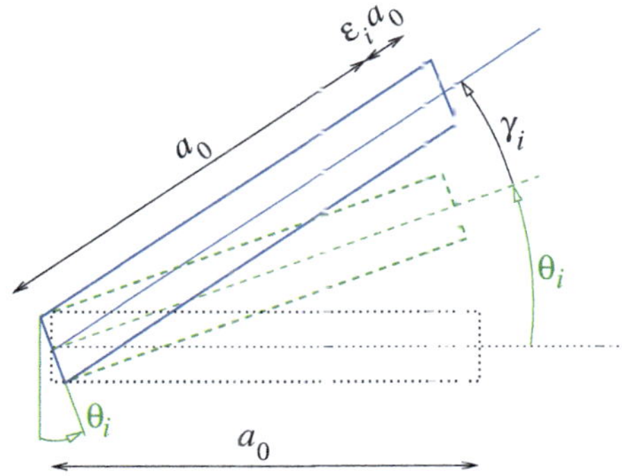

$$\Pi_{\text{tot}} = \frac{1}{2}EAa_0\sum_{i=1}^{N}\varepsilon_i^2 + \frac{1}{2}\kappa GAa_0\sum_{i=1}^{N}\gamma_i^2 + \frac{1}{2}\cdot\frac{EI}{a_0}\sum_{i=1}^{N-1}(\theta_{i+1}-\theta_i)^2 + Pa_0\sum_{i=1}^{N}(1+\varepsilon_i)\cos(\theta_i+\gamma_i)\,. \tag{24.2}$$

The dimensionless total potential energy is obtained by dividing Eq. (24.2) by EI/a_0:

$$\Pi = \frac{\chi}{2N^2}\sum_{i=1}^{N}\varepsilon_i^2 + \frac{\alpha}{2N^2}\sum_{i=1}^{N}\gamma_i^2 + \frac{1}{2}\sum_{i=1}^{N-1}(\theta_{i+1}-\theta_i)^2 + \frac{\beta}{N^2}\sum_{i=1}^{N}(1+\varepsilon_i)\cos(\theta_i+\gamma_i) \tag{24.3}$$

Here the following *stiffness and load parameters* are introduced:

$$\alpha = \frac{\kappa GAL^2}{EI}, \quad \chi = \frac{AL^2}{I}, \quad \beta = \frac{PL^2}{EI}. \tag{24.4}$$

The first derivatives of this potential with respect to $\varepsilon = \{\varepsilon_1, \varepsilon_2, \ldots, \varepsilon_N\}$, $\gamma = \{\gamma_1, \gamma_2, \ldots, \gamma_N\}$, and $\boldsymbol{\theta} = \{\theta_1, \theta_2, \ldots, \theta_N\}$ are:

$$\frac{\partial \Pi}{\partial \varepsilon_i} = \frac{\chi}{N^2}\varepsilon_i + \frac{\beta}{N^2}\cos(\theta_i+\gamma_i) = 0, \quad i = 1,2,\ldots,N, \tag{24.5}$$

$$\frac{\partial \Pi}{\partial \gamma_i} = \frac{\alpha}{N^2}\gamma_i - \frac{\beta}{N^2}(1+\varepsilon_i)\sin(\theta_i+\gamma_i) = 0, \quad i = 1,2,\ldots,N, \tag{24.6}$$

$$\frac{\partial \Pi}{\partial \theta_i} = -(\theta_{i+1} - 2\theta_i + \theta_{i-1}) - \frac{\beta}{N^2}(1+\varepsilon_i)\sin(\theta_i+\gamma_i) = 0, \quad i = 1,2,\ldots,N. \tag{24.7}$$

The trivial equilibrium state is a straight linkage under simple compression: $\theta_i = \gamma_i = 0$, $\varepsilon_i = -\beta/\chi$, $i = 1,\ldots,N$, and β is arbitrary.

Note that Eq (24.5) gives the normal strain of link i,

$$\varepsilon_i = -\frac{\beta}{\chi}\cos\Psi_i. \tag{24.8}$$

Here $\beta\cos(\Psi_i)$ is the compressive (nondimensional) normal force. Similarly, Eq. (24.6) gives the shear strain,

$$\gamma_i = \frac{\beta}{\alpha}\left(1-\frac{\beta}{\chi}\cos\Psi_i\right)\sin\Psi_i \tag{24.9}$$

with $\beta\sin(\Psi_i)$ the (nondimensional) shear force.

Hence, the equation system Eq. (24.5)–(24.7) can be written also by using the rotation of the link axes Ψ_i (see Eq. (24.1)), by using Eq. (24.7) and the normal and shear strain equations:

$$\begin{aligned}&\Psi_{i+1}-2\Psi_i+\Psi_{i-1}-\frac{\beta}{\alpha}\left\{\left(1-\frac{\beta}{\chi}\cos\Psi_{i+1}\right)\sin\Psi_{i+1}\right.\\&\left.-2\left(1-\frac{\beta}{\chi}\cos\Psi_i\right)\sin\Psi_i+\left(1-\frac{\beta}{\chi}\cos\Psi_{i-1}\right)\sin\Psi_{i-1}\right\}\\&+\frac{\beta}{N^2}\left(1-\frac{\beta}{\chi}\cos\Psi_i\right)\sin\Psi_i=0,\quad i=1,2,\ldots,N.\end{aligned} \tag{24.10}$$

The pinned-pinned boundary conditions are: $\Psi_0=\Psi_1$ and $\Psi_N=\Psi_{N+1}$.

Summing Eq. (24.10) for all i yields:

$$\frac{\beta}{N^2}\sum_{i=1}^{N}\left(1-\frac{\beta}{\chi}\cos\Psi_i\right)\sin\Psi_i=0. \tag{24.11}$$

24.3.1.1 Buckling Loads

The N–by–N *Jacobian* of Eq. (24.10) around the trivial equilibrium state is:

$$\boldsymbol{J}=\left\{1-\frac{\beta}{\alpha}\left(1-\frac{\beta}{\chi}\right)\right\}\boldsymbol{C}-\frac{\beta}{N^2}\left(1-\frac{\beta}{\chi}\right)\boldsymbol{I}. \tag{24.12}$$

Here $\boldsymbol{I}$ is the N–by–N identity matrix, while $\boldsymbol{C}$ is the N–by–N modified continuant matrix defined as $C_{i,i}=2$ (except for $C_{1,1}=C_{N,N}=1$), $C_{i,i+1}=C_{i,i-1}=-1$, and zero otherwise. Based on Eq. (24.12) one can reformulate the linearized difference equation as:

$$\left[1-\frac{\beta}{\alpha}\left(1-\frac{\beta}{\chi}\right)\right](\Psi_{i+1}-2\Psi_i+\Psi_{i-1})+\frac{\beta}{N^2}\left(1-\frac{\beta}{\chi}\right)\Psi_i=0. \tag{24.13}$$

In a critical equilibrium state an eigenvalue of Eq. (24.12) is zero. Hence its determinant is zero, which holds if

$$\frac{\beta}{N^2}\left(1-\frac{\beta}{\chi}\right)/\left\{1-\frac{\beta}{\alpha}\left(1-\frac{\beta}{\chi}\right)\right\}$$

is an eigenvalue of $\boldsymbol{C}$. According to Rózsa (1991), the eigenvalues of $\boldsymbol{C}$ are

$$\lambda_i = 4\sin^2\frac{i\pi}{2N}, \quad i=0,1,\ldots,N-1. \tag{24.14}$$

It leads to the nondimensional buckling loads of the discrete extensible *Engesser* elastica:

$$\beta^{\mathrm{cr}}_{2i,2i+1} = \frac{\chi}{2}\left(1\pm\sqrt{1-4\frac{\frac{4N^2\sin^2\frac{i\pi}{2N}}{\chi}}{1+\frac{4N^2\sin^2\frac{i\pi}{2N}}{\alpha}}}\right), \quad i=0,1,\ldots,N-1 \tag{24.15}$$

For $i=0$, there are two critical loads. One is $\beta=\chi$, which means the linkage gets fully compressed under the axial load, while the other, $\beta=0$, corresponds to free rigid-body-like rotation of the structure. However, the latter solution is prevented by the supports. For $i=1,2,\ldots,N-1$ there are $2(N-1)$ compressive buckling loads if the discriminant in Eq. (24.15) is positive.

The fundamental buckling load is:

$$\beta_{\mathrm{deE}} = \frac{\chi}{2}\left(1-\sqrt{1-4\frac{\frac{4N^2\sin^2\frac{\pi}{2N}}{\chi}}{1+\frac{4N^2\sin^2\frac{\pi}{2N}}{\alpha}}}\right), \tag{24.16}$$

which is a new result, to the best of our knowledge. (The subscript "deE" refers to "discrete extensible Engesser".) Based on Eq. (24.13) one can also obtain the fundamental buckling mode:

$$\Psi_j = \Phi\cos\left[\frac{\pi}{N}\left(j-\frac{1}{2}\right)\right], \quad j=1,2,\ldots,N. \tag{24.17}$$

Note that this fundamental buckling load exists if the discriminant is non-negative, i.e. if

$$\chi\left(1+\frac{4N^2\sin^2\frac{\pi}{2N}}{\alpha}\right) \geq 16N^2\sin^2\frac{\pi}{2N}. \tag{24.18}$$

For large N it yields $\chi(1+\pi^2/\alpha)\geq 4\pi^2$.

For an unshearable column, i.e. for a discrete extensible elastica, the buckling load is

$$\beta_{\mathrm{dee}} = \frac{\chi}{2}\left(1-\sqrt{1-\frac{16N^2\sin^2\frac{\pi}{2N}}{\chi}}\right), \tag{24.19}$$

and it exists if $\chi \geq 4\pi^2$. (Here the subscript "dee" refers to "discrete extensible elastica".) For the inextensible case, i.e. for $\chi \to \infty$, we asymptotically find:

$$\beta_{\text{dE}} = \frac{4N^2 \sin^2 \frac{\pi}{2N}}{1 + \frac{4N^2 \sin^2 \frac{\pi}{2N}}{\alpha}}, \tag{24.20}$$

which was already obtained by Zhang et al (2013) for a discrete *Timoshenko* column (see also Kocsis et al, 2017). For the inextensible and unshearable case ($\chi \to \infty$ and $\alpha \to \infty$), the solution of Wang (1951, 1953) is found for the *Hencky* bar-chain model:

$$\beta_{\text{Hencky}} = 4N^2 \sin^2 \frac{\pi}{2N}. \tag{24.21}$$

24.3.1.2 Analytical Solution for Short Linkages

For $N = 2$ analytical solution can be given, at least for reflection symmetric post-buckled configurations. The equilibrium equations are Eq (24.10) with $N = 2$:

$$\begin{aligned} &\Psi_2 - \Psi_1 - \frac{\beta}{\alpha}\left\{\left(1 - \frac{\beta}{\chi}\cos\Psi_2\right)\sin\Psi_2 - \left(1 - \frac{\beta}{\chi}\cos\Psi_1\right)\sin\Psi_1\right\} \\ &\quad + \frac{\beta}{N^2}\left(1 - \frac{\beta}{\chi}\cos\Psi_1\right)\sin\Psi_1 = 0, \quad \text{and} \end{aligned} \tag{24.22}$$

$$\begin{aligned} &-\Psi_2 + \Psi_1 - \frac{\beta}{\alpha}\left\{-\left(1 - \frac{\beta}{\chi}\cos\Psi_2\right)\sin\Psi_2 + \left(1 - \frac{\beta}{\chi}\cos\Psi_1\right)\sin\Psi_1\right\} \\ &\quad + \frac{\beta}{N^2}\left(1 - \frac{\beta}{\chi}\cos\Psi_2\right)\sin\Psi_2 = 0. \end{aligned} \tag{24.23}$$

The sum of these equations, Eq. (24.11) at $N = 2$, is solved by any reflection symmetric configuration:

$$\Psi_2 = -\Psi_1 - 2k\pi. \tag{24.24}$$

Equation (24.23) with the above solution reads:

$$\Psi_1 + k\pi - \frac{\beta}{\alpha}\left(1 - \frac{\beta}{\chi}\cos\Psi_1\right)\sin\Psi_1 - \frac{\beta}{2N^2}\left(1 - \frac{\beta}{\chi}\cos\Psi_1\right)\sin\Psi_1 = 0. \tag{24.25}$$

Finally, the equilibrium paths are given by:

$$\beta = \frac{\chi}{2\cos\psi_1}\left(1 \pm \sqrt{1 - \frac{\frac{32}{\chi}}{1 + \frac{8}{\alpha}}(\Psi_1 + k\pi)\cot\Psi_1}\right). \tag{24.26}$$

Note that the above equation is singular at $\Psi_1 = 0$. An asymptotic expansion yields

$$\beta = \frac{\chi}{2\cos\psi_1}\left(1 \pm \sqrt{1 - \frac{\frac{32}{\chi}}{1+\frac{8}{\alpha}} \cdot \frac{\cos\Psi_1}{1+\sum\limits_{m=1}^{\infty}(-1)^m \frac{\Psi_1^{2m}}{(2m+1)!}}}\right) \tag{24.27}$$

for $k = 0$, which at $\Psi_1 = 0$ gives the buckling loads:

$$\beta_{\text{deE}}^{N=2} = \frac{\chi}{2}\left(1 \pm \sqrt{1 - \frac{\frac{32}{\chi}}{1+\frac{8}{\alpha}}}\right). \tag{24.28}$$

This coincides with Eq. (24.15) at $N = 2$, $i = 1$.

The buckling load exists if the discriminant is positive, i.e. if $8\chi/\alpha/(32-\chi) > 1$. By fixing $\alpha = 10$, the above condition fulfills if $\chi > 160/9 = 17.7778$. This case is shown on top left of Fig. 24.3, by plotting Eq. (24.26) for $k = -5, \cdots, 4$. Here the two critical loads coincide at $\beta = \chi/2$. If the normal stiffness parameter is slightly larger, for example $\chi = 160/8 = 20$, then these two critical loads separate, which can be well observed on top right of Fig. 24.3. If the normal stiffness parameter is slightly smaller, for example $\chi = 160/10 = 16$, then both of these two critical loads

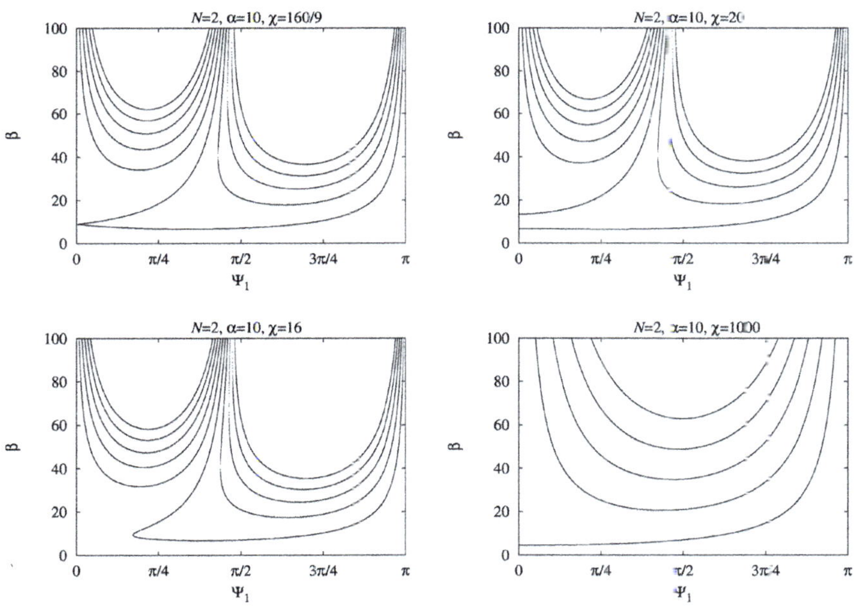

Fig. 24.3: Equilibrium paths of the discrete, extensible Engesser elastica of $N = 2$ links, for $k = -5, \cdots, 4$. Values of α and χ are written on the top of the figures.

vanish, as shown on bottom left of Fig. 24.3. The solution for a linkage with large tensile stiffness ($\chi = 1000$) is shown on bottom right of Fig. 24.3.

On the two top figures of Fig. 24.3, it can be seen that the first post-buckling path is not stable. We can determine the condition for the stability of the first post-buckling path as follows. We assume that the fundamental buckling load exists (i.e. the discriminant in Eq. (24.28) is positive). The first derivative of Eq. (24.27) with respect to Ψ_1, evaluated at $\Psi_1 = 0$, yields zero, hence the bifurcations are symmetric. The second derivative of Eq. (24.27), with subtraction, with respect to Ψ_1, evaluated at $\Psi_1 = 0$ yields:

$$\frac{\partial^2 \beta}{\partial \Psi_1^2}(\Psi_1 = 0) = \frac{\chi}{2}\left(1 - \sqrt{1 - \frac{32/\chi}{1+8/\alpha}}\right) - \frac{16}{1+8/\alpha} \cdot \frac{1}{3\sqrt{1 - \frac{32/\chi}{1+8/\alpha}}}. \tag{24.29}$$

If the above curvature of the first post-buckling path is positive, then the post-buckling path is *stable*. If it is negative, the first post-buckling path is *unstable*. The separator between the two states can be obtained by finding the roots of the above equation (i.e., by solving $\partial^2\beta/\partial\Psi_1^2(\Psi_1 = 0) = 0$). This yields

$$\chi\left(1 + \frac{8}{\alpha}\right) = \frac{128}{3}. \tag{24.30}$$

If $\chi\,(1+8/\alpha) > 128/3$, then the first post-buckling path is stable, if $\chi\,(1+8/\alpha) < 128/3$, then it is unstable. For the top left figure of Fig. 24.3, $\chi(1+8/\alpha) = 32$, and for the top right figure it is 36. Both are smaller than $128/3 = 42.667$, and the first post buckling path is unstable in both cases. It can be clearly seen that snapping can occur in these two cases. For the bottom right figure of Fig. 24.3 $\chi(1+8/\alpha) = 1800$, which is larger than $128/3$, and the first post-buckling path is stable.

For $N = 3$ a similar approach can be used. The geometrically exact equilibrium equation of the discrete *Engesser* elastica of $N = 3$ links is obtained from Eq. (24.10) with $N = 3$. Equation (24.11) with $N = 3$ is satisfied by any reflection symmetric configuration:

$$\Psi_3 = -\Psi_1 - 2k\pi, \quad \Psi_2 = -m\pi. \tag{24.31}$$

The geometrically exact solution for the equilibrium states can be developed from the equilibrium equations:

$$\beta = \frac{\chi}{2\cos\psi_1}\left(1 \pm \sqrt{1 - \frac{\frac{36}{\chi}}{1+\frac{9}{\alpha}}(\Psi_1 + m\pi)\cot\Psi_1}\right). \tag{24.32}$$

With an asymptotic expansion the buckling loads are obtained:

$$\beta_{\text{deE}}^{N=3} = \frac{\chi}{2}\left(1 \pm \sqrt{1 - \frac{\frac{36}{\chi}}{1+\frac{9}{\alpha}}}\right). \tag{24.33}$$

This coincides with Eq. (24.15) at $N = 3$, $i = 1$. The buckling load exists if the discriminant is positive, i.e. if $9\chi/\alpha/(36-\chi) > 1$. By fixing $\alpha = 10$, the above condition fulfills if $\chi > 360/19 = 18.9474$. This case is shown on top left of Fig. 24.4, by plotting Eq. (24.32) for $k = -5, \cdots, 4$.

Here, the two critical loads coincide at $\beta = \chi/2$. If the normal stiffness parameter is slightly larger, for example $\chi = 360/18 = 20$, then these two critical loads separate, which can be well observed on top right of Fig. 24.4. If the normal stiffness parameter is slightly smaller, for example $\chi = 360/20 = 18$, then both of these two critical loads vanish, as shown on bottom left of Fig. 24.4. Equilibrium paths for a linkage with large tensile stiffness ($\chi = 1000$) are shown on bottom right of Fig. 24.4.

On the two top figures of Fig. 24.4, it can be seen that the first post-buckling path is not stable. The condition for stability of the first post-buckling path can be developed. The first derivative of Eq. (24.32) with respect to Ψ_1, expanded up to the second order terms, and evaluated at $\Psi_1 = 0$ yields zero, hence the bifurcations are symmetric. The second derivative of Eq. (24.32) with respect to Ψ_1, expanded up to the second order terms, with subtraction, evaluated at $\Psi_1 = 0$ yields:

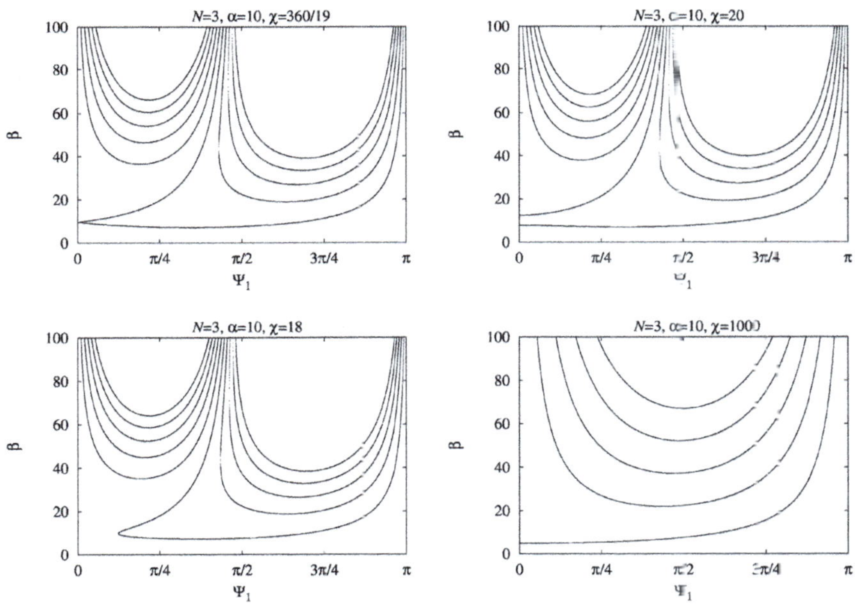

Fig. 24.4: Equilibrium paths of the discrete, extensible Engesser elastica of $N = 3$ links, for $k = -5, \cdots, 4$. Values of α and χ are noted on the top of the figures.

$$\frac{\partial^2 \beta}{\partial \Psi_1^2}(\Psi_1 = 0) = \frac{\chi}{2}\left(1 - \sqrt{1 - \frac{36/\chi}{1+9/\alpha}}\right) - \frac{6}{1+9/\alpha} \cdot \frac{1}{\sqrt{1 - \frac{36/\chi}{1+9/\alpha}}}. \tag{24.34}$$

If the above curvature of the first post-buckling path is positive, then the post-buckling path is *stable*. If it is negative, the first post-buckling path is *unstable*. The separator between the two states can be obtained by finding the roots of the above equation. It yields

$$\chi\left(1 + \frac{9}{\alpha}\right) = 48. \tag{24.35}$$

If $\chi\,(1+9/\alpha) > 48$, then the first post-buckling path is stable, if $\chi\,(1+9/\alpha) < 48$, then it is unstable. For the top left figure of Fig. 24.4, $\chi(1+9/\alpha) = 36$, and for the top right figure, it is 38.Both are smaller than 48, and the first post buckling path is unstable in both cases. It can be clearly seen that snapping can occur in these two cases. For the bottom right figure of Fig. 24.4, $\chi(1+9/\alpha) = 1900$, which is larger than 48, and the first post-buckling path is stable. (The second post-buckling path is well above $\beta = 100$.)

We can conjecture that the stability limit of the first post-buckling path of the discrete *Engesser* elastica is

$$\left[1 + \frac{4N^2}{\alpha}\sin^2\left(\frac{\pi}{2N}\right)\right]\frac{3\chi}{16} = 4N^2\sin^2\left(\frac{\pi}{2N}\right). \tag{24.36}$$

24.3.1.3 Numerical Solution

Numerically computed equilibrium paths of the discrete extensible *Engesser* elastica are shown in Fig. 24.5 for $N = 2$, and in Fig. 24.6 for $N = 3$ in gray color. For completeness the analytical solutions for symmetric configurations are also shown by black curves. It can be seen that there are a vast amount of equilibrium states which are not symmetric, hence not found with the analytical approach. The numerical results are obtained with the simplex scanning algorithm (Gáspár et al, 1997).

24.3.2 Asymptotic Limit: the Local Extensible Engesser Elastica

Equation (24.10) in the asymptotic limit, $N \to \infty$ and $a_0 \to 0$, yields the governing equation of the local (continuum) extensible *Engesser* elastica:

$$\frac{\mathrm{d}^2}{\mathrm{d}\xi^2}\left(\Psi - \frac{\beta}{\alpha}\sin\Psi + \frac{\beta^2}{2\alpha\chi}\sin 2\Psi\right) + \beta\sin\Psi - \frac{\beta^2}{2\chi}\sin 2\Psi = 0. \tag{24.37}$$

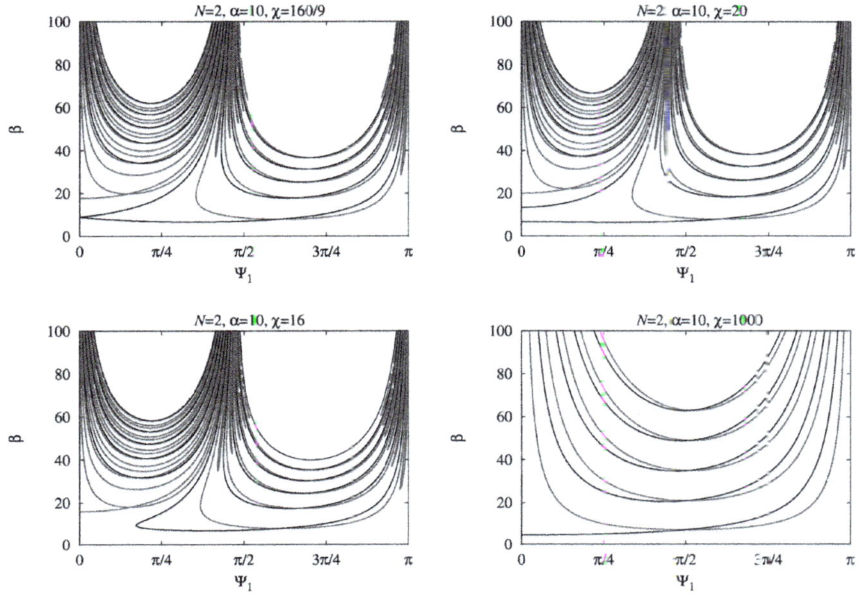

Fig. 24.5: Equilibrium paths of the discrete, extensible Engesser elastica of $N = 2$ links. Analytical results are shown by black curves (for $k = -5, \cdots, 4$), numerical results are shown by gray color. Scanned domain of the angle Ψ_2 is $[-10\pi, 10\pi]$ for the numerical algorithm. Values of α and χ are written on the top of the figures.

Here $\xi = s/L$ is the nondimensional arc-length parameter and $\Psi(\xi) = \theta(\xi) + \gamma(\xi)$. Equation (24.37) can also be written as:

$$\begin{aligned}&\left[1 - \frac{\beta}{\alpha}\left(\cos\Psi - \frac{\beta}{\chi}\cos(2\Psi)\right)\right]\frac{\mathrm{d}^2\Psi}{\mathrm{d}\xi^2}\\ &- \frac{\beta}{\alpha}\left(\frac{2\beta}{\chi}\sin(2\Psi) - \sin\Psi\right)\left(\frac{\mathrm{d}\Psi}{\mathrm{d}\xi}\right)^2 + \beta\left(1 - \frac{\beta}{\chi}\cos\Psi\right)\sin\Psi = 0.\end{aligned} \tag{24.38}$$

This geometrically exact differential equation of the continuous extensible *Engesser* elastica has not been presented yet in the literature, to the best of our knowledge.

In order to obtain the buckling loads of the local continuous extensible *Engesser* elastica, Eq. (24.37) is linearized:

$$\left(1 - \frac{\lambda}{\alpha}\right)\frac{\mathrm{d}^2\Psi}{\mathrm{d}\xi^2} + \lambda\Psi = 0, \quad \text{with} \quad \lambda = \beta\left(1 - \frac{\beta}{\chi}\right), \tag{24.39}$$

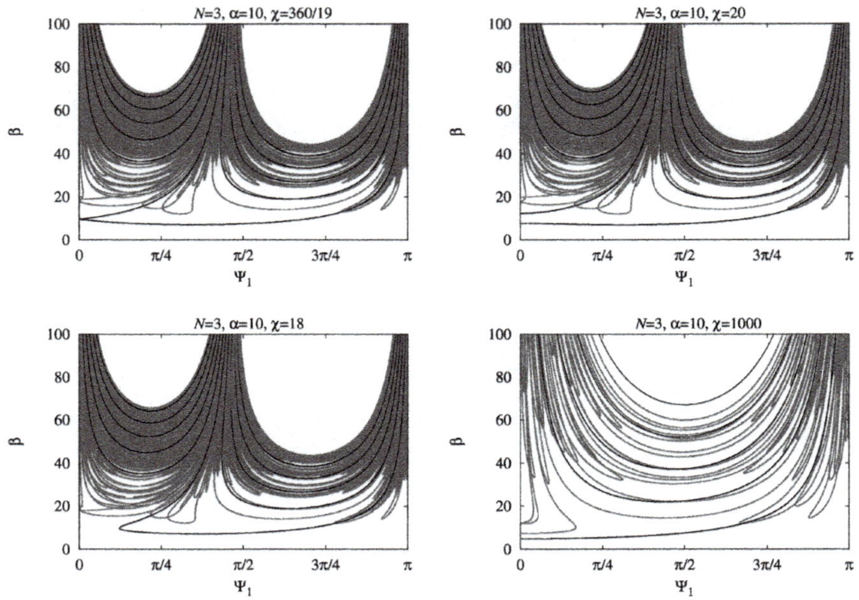

Fig. 24.6: Equilibrium paths of the discrete, extensible Engesser elastica of $N = 3$ links. Analytical results are shown by black curves (for $k = -5, \cdots, 4$), numerical results are shown by gray color. Scanned domain of the angles Ψ_2, Ψ_3 is $[-7\pi, 7\pi]$ for the numerical solution. Values of α and χ are noted on the top of the figures.

with pinned-pinned boundary conditions $\mathrm{d}\Psi/\mathrm{d}\xi(0) = 0$ and $\mathrm{d}\Psi/\mathrm{d}\xi(1) = 0$. It leads to the buckling modes

$$\Psi_r(\xi) = A_r \cos(r\pi\xi), \tag{24.40}$$

and to the nondimensional buckling load parameters:

$$\beta_{\mathrm{leE},r} = \frac{\chi}{2}\left(1 \pm \sqrt{1 - 4\frac{\frac{r^2\pi^2}{\chi}}{1 + \frac{r^2\pi^2}{\alpha}}}\right). \tag{24.41}$$

(With subscript "leE" referring to "local extensible Engesser".) Here $r^2\pi^2$ are the buckling load parameters of the local elastica (i.e. the nondimensional *Euler* forces). The fundamental buckling load is obtained with $r = 1$,

$$\beta_{\mathrm{leE}} = \frac{\chi}{2}\left(1 \pm \sqrt{1 - 4\frac{\frac{\pi^2}{\chi}}{1 + \frac{\pi^2}{\alpha}}}\right). \tag{24.42}$$

This equation was first obtained by *Ziegler* (see Eq. (5.14) of Ziegler, 1982).

This fundamental buckling load exists if the discriminant is non-negative, i.e. if

$$\chi\left(1+\frac{\pi^2}{\alpha}\right) \geq 4\pi^2, \tag{24.43}$$

which can also be obtained from Eq. (24.18) with $N \to \infty$. Considering Eq. (24.4) the (dimensional) fundamental buckling load of the extensible *Engesser* elastica is:

$$P_{\text{leE}} = \frac{EA}{2}\left(1-\sqrt{1-4\frac{\frac{P_{\text{E}}}{EA}}{1+\frac{P_{\text{E}}}{\kappa GA}}}\right) \tag{24.44}$$

Here P_{E} is the *Euler* force.

For infinite normal stiffness ($\chi \to \infty$) Eq. (24.38) yields

$$\left(1-\frac{\beta}{\alpha}\cos\Psi\right)\frac{\mathrm{d}^2\Psi}{\mathrm{d}\xi^2}+\frac{\beta}{\alpha}\sin\Psi\left(\frac{\mathrm{d}\Psi}{\mathrm{d}\xi}\right)^2+\beta\sin\Psi = 0. \tag{24.45}$$

It is the equation of the (inextensible) *Engesser* elastica, already obtained by *Atanackovic* (see Eq. (3.3.57) of Atanackovic, 1997) and also given by Kocsis et al (2017). The *Engesser* solution (Engesser, 1891) can be obtained from Eq. (24.42) in the inextensible limit $\chi \to \infty$:

$$\beta_{\text{lE}} = \frac{\pi^2}{1+\frac{\pi^2}{\alpha}}. \tag{24.46}$$

This load always exists. (The subscript “lE” refers to “local Engesser”.)

For infinite shear stiffness ($\alpha \to \infty$) Eq. (24.38) yields

$$\frac{\mathrm{d}^2\Psi}{\mathrm{d}\xi^2}+\beta\left(1-\frac{\beta}{\chi}\cos\Psi\right)\sin\Psi = 0. \tag{24.47}$$

This extensible elastica equation has been obtained by Pflüger (1964); Atanackovic (1997); Magnusson et al (2001). See also special cases of Koiter (2009); Goto et al (1990); Attard (2003); Humer (2013). The fundamental buckling load parameter of the extensible elastica is obtained in the limit $\alpha \to \infty$:

$$\beta_{\text{lee}} = \frac{\chi}{2}\left(1\pm\sqrt{1-\frac{4\pi^2}{\chi}}\right). \tag{24.48}$$

(With subscript “lee” referring to “local extensible elastica”.) The buckling load Eq. (24.48) exists if $\chi \geq 4\pi^2$. *Euler*'s critical force can be obtained in the inextensible, unshearable limit ($\chi \to \infty$ and $\alpha \to \infty$).

Asymptotic solutions in the post-buckling range can be derived based on a *Taylor* expansion of the load parameter and the rotation field, as extensively used by Koiter

(1963, 2009). The derivation is given in Appendix A. Based on that, the limit between the stable and the unstable post-bifurcation branch is

$$\left(1+\frac{\pi^2}{\alpha}\right)\frac{3\chi}{16}=\pi^2. \tag{24.49}$$

Note that it is the continuum limit of Eq. (24.36). For the unshearable case (extensible elastica), the critical parameter for normal stiffness, $\chi = 16\pi^2/3$, has been obtained by Atanackovic (1997) and Magnusson et al (2001), which validates our results in the limit of $\alpha \to \infty$.

24.3.3 Continualized Nonlocal Extensible Engesser Elastica

Introducing the *Taylor* expansion of a rotational field $\omega(\xi)$ of nondimensional coordinate $\xi = x/L$,

$$\omega\left(\xi+\frac{1}{N}\right)=\sum_{k=0}^{\infty}\frac{\frac{\mathrm{d}^k}{\mathrm{d}\xi^k}\omega(\xi)}{k!}\left(\frac{1}{N}\right)^k=e^{\frac{1}{N}\frac{\mathrm{d}}{\mathrm{d}\xi}}\omega(\xi), \tag{24.50}$$

and using a Padé approximant of order [2,2], the following identity can be written:

$$\begin{aligned}\omega_{i+1}-2\omega_i+\omega_{i-1}&=\omega\left(\xi+\frac{1}{N}\right)-2\omega(\xi)+\omega\left(\xi-\frac{1}{N}\right)\\&=4\sinh^2\left(\frac{1}{2N}\frac{\mathrm{d}}{\mathrm{d}\xi}\right)\omega(\xi)=\frac{1}{N^2}\cdot\frac{\frac{\mathrm{d}^2}{\mathrm{d}\xi^2}}{1-\frac{1}{12N^2}\frac{\mathrm{d}^2}{\mathrm{d}\xi^2}}\omega(\xi)+\dots.\end{aligned} \tag{24.51}$$

Similarly,

$$\begin{aligned}\sin\omega_{i+1}-2\sin\omega_i+\sin\omega_{i-1}&=4\sinh^2\left(\frac{1}{2N}\frac{\mathrm{d}}{\mathrm{d}\xi}\right)\sin\omega(\xi)\\&=\frac{1}{N^2}\cdot\frac{\frac{\mathrm{d}^2}{\mathrm{d}\xi^2}}{1-\frac{1}{12N^2}\frac{\mathrm{d}^2}{\mathrm{d}\xi^2}}\sin\omega(\xi)+\dots.\end{aligned} \tag{24.52}$$

Based on this formulae, the difference equation system Eq. (24.10) can be continualized. It yields the geometrically exact differential equation of the continualized nonlocal extensible *Engesser* elastica:

$$\frac{\mathrm{d}^2}{\mathrm{d}\xi^2}\left\{\Psi-\beta\left(\frac{1}{\alpha}+\frac{1}{12N^2}\right)\sin\Psi+\frac{\beta^2}{2\chi}\left(\frac{1}{\alpha}+\frac{1}{12N^2}\right)\sin 2\Psi\right\}$$
$$+\beta\sin\Psi-\frac{\beta^2}{2\chi}\sin 2\Psi=0. \qquad (24.53)$$

The linearization of the above nonlinear differential equation leads to:

$$\left\{1-\lambda\left(\frac{1}{\alpha}+\frac{1}{12N^2}\right)\right\}\frac{\mathrm{d}^2\Psi}{\mathrm{d}\xi^2}+\lambda\Psi=0, \quad \text{with} \quad \lambda=\beta\left(1-\frac{\beta}{\chi}\right). \qquad (24.54)$$

In order to determine the buckling loads, we can use either local or nonlocal boundary conditions. See for example Challamel et al (2015a) for a discussion on local and nonlocal boundary conditions of the nonlocal *elastica*.

By using local pinned-pinned boundary conditions, i.e. $\Psi'(0)=\Psi'(1)=0$, the buckling modes are the same as Eq. (24.40) and the nondimensional buckling load parameters of the nonlocal extensible *Engesser* elastica are:

$$\beta_{\mathrm{nleE},r}=\frac{\chi}{2}\left(1\pm\sqrt{1-4\frac{\frac{r^2\pi^2}{\chi}}{1+\left(\frac{1}{\alpha}+\frac{1}{12N^2}\right)r^2\pi^2}}\right). \qquad (24.55)$$

Here "nleE" refers to "nonlocal extensible Engesser". The nonlocal pinned-pinned boundary conditions are: $\Psi(0)=\Psi(1/N)$ and $\Psi(1)=\Psi(1+1/N)$.

24.3.3.1 Numerical Solution: Discrete Versus Nonlocal Extensible Engesser Elastica

The first post-buckling paths of the discrete and the nonlocal *Engesser* elastica are shown in Fig. 24.7 for various numbers of links N. The results are obtained numerically with the simplex path-following algorithm (Domokos and Gáspár, 1995). It can be seen that the path computed for the nonlocal model follows well the path of the corresponding discrete model. Hence the scale effect of the lattice system is well captured by the nonlocal model also for large displacements, at least as far as the first post-buckling path is considered, which has the greatest importance in practice.

24.4 Extensible Haringx Elastica

In this section we implement *Love*'s finite strain theory in the discrete model. First the static behavior of the discrete model is analysed, then the continuum counterpart of the system is determined and a quasi-continuum is also developed from the lattice system following a continualization technique.

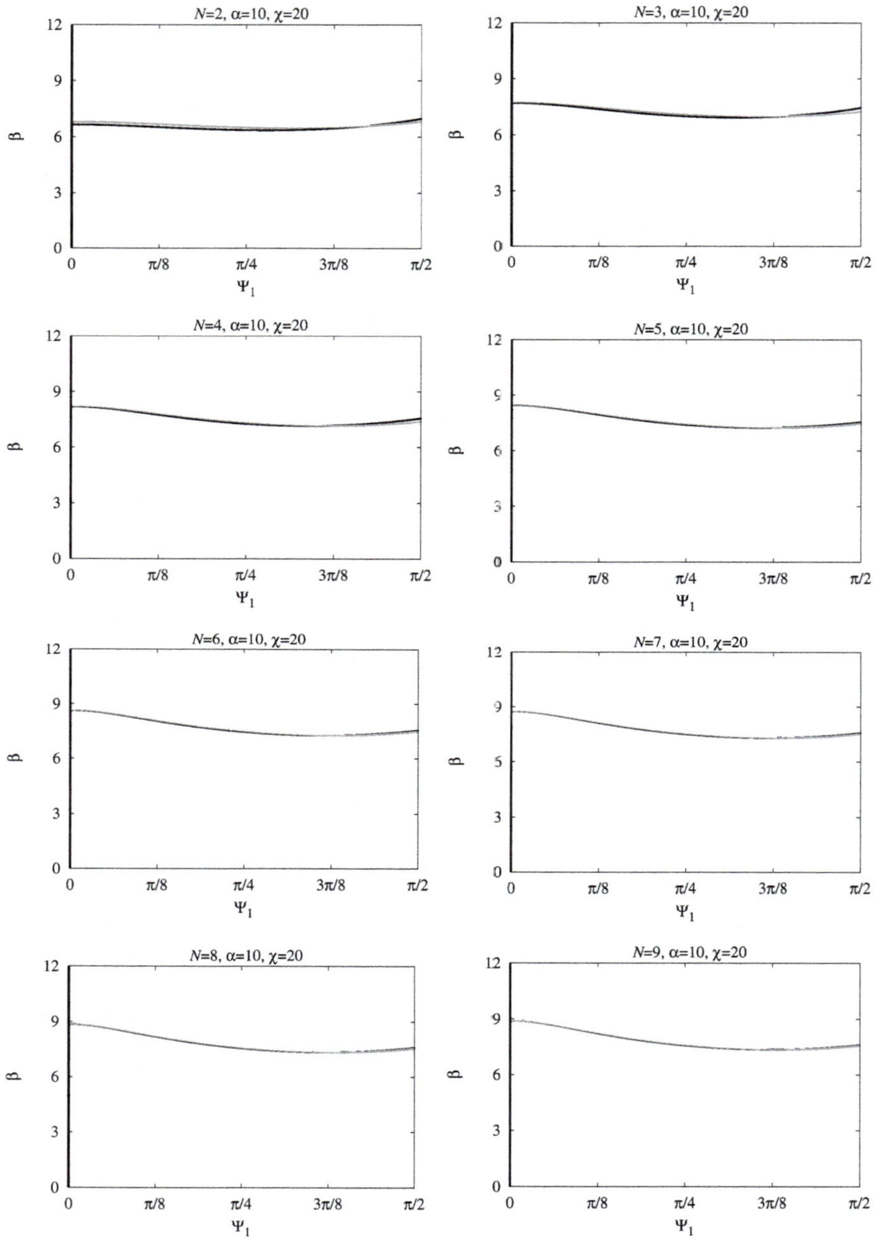

Fig. 24.7: First post-buckling equilibrium path of the nonlocal versus discrete extensible Engesser elastica of $N = 2,3,4,5,7,8,9$ links ($\alpha = 10$, $\chi = 20$).

24.4.1 Discrete Extensible Haringx Elastica

Figure 24.8 sketches a deformed link of the generalized *Hencky* bar-chain model. The link is rotated as a rigid body by the angle θ_i, then distorted by the shear angle γ_i. According to *Love*'s finite shear strain theory (Love, 1944), the area of the link is preserved under shearing. The elongation of the link occurs perpendicular to its end sections, and not along the link axis as in the case of *Timoshenko*'s finite strains. Hence, a link gets longer by $a_0(1+\varepsilon_i)/\cos\widetilde{\gamma}_i$, where $\widetilde{\gamma}_i$ is obtained from the following equation: $\tan\gamma_i = (1+\varepsilon_i)\tan\widetilde{\gamma}_i$. Here ε_i and γ_i are the normal and shear strains of the ith link, respectively.

Hence, the nondimensional potential energy of the discrete model is:

$$
\begin{aligned}
\Pi = {} & \frac{\chi}{2N^2}\sum_{i=1}^{N}\varepsilon_i^2 + \frac{\alpha}{2N^2}\sum_{i=1}^{N}\tan^2\gamma_i + \frac{1}{2}\sum_{i=1}^{N-1}(\theta_{i+1}-\theta_i)^2 \\
& + \frac{\beta}{N^2}\sum_{i=1}^{N}\{(1+\varepsilon_i)\cos\theta_i - \tan\gamma_i\sin\theta_i\}\,.
\end{aligned}
\tag{24.56}
$$

Here the same load and stiffness parameters are used as in Eq. (24.4). The first derivatives of this potential with respect to $\boldsymbol{\theta}$, γ, and ε are:

$$
\frac{\partial\Pi}{\partial\varepsilon_i} = \frac{\chi}{N^2}\varepsilon_i + \frac{\beta}{N^2}\cos\theta_i = 0, \quad i = 1,2,\ldots,N, \tag{24.57}
$$

$$
\frac{\partial\Pi}{\partial\gamma_i} = \frac{\frac{\alpha}{N^2}\tan\gamma_i - \frac{\beta}{N^2}\sin\theta_i}{\cos^2\gamma_i} = 0, \quad i = 1,2,\ldots,N, \tag{24.58}
$$

$$
\frac{\partial\Pi}{\partial\theta_i} = -(\theta_{i+1}-2\theta_i+\theta_{i-1}) - \frac{\beta}{N^2}\{(1+\varepsilon_i)\sin\theta_i + \tan\gamma_i\cos\theta_i\} = 0, \; i = 1,2,\ldots,N. \tag{24.59}
$$

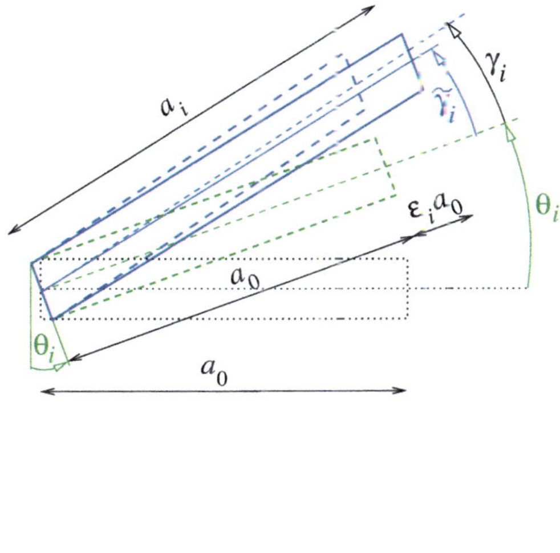

Fig. 24.8 Deformation of a shearable, extensible link according to *Love*'s finite strain theory

The trivial equilibrium state is a straight linkage under simple compression: $\theta_i = \gamma_i = 0$, $\varepsilon_i = -\beta/\chi$ ($i = 1,2,\ldots,N$), and β is arbitrary.

Note that (24.57) gives the normal strain, ε_i, due to the compressive normal force $\beta \cos\theta_i$:

$$\varepsilon_i = -\frac{\beta}{\chi}\cos\vartheta_i, \tag{24.60}$$

while the shear strain γ_i can be computed from Eq. (24.58):

$$\gamma_i = \arctan\left(\frac{\beta}{\alpha}\sin\theta_i\right). \tag{24.61}$$

By using these strains the equilibrium equations can be rewritten from Eq. (24.59) for the rotations θ_i:

$$\theta_{i+1} - 2\theta_i + \theta_{i-1} + \frac{\beta}{N^2}\left\{1+\beta\left(\frac{1}{\alpha}-\frac{1}{\chi}\right)\cos\theta_i\right\}\sin\theta_i = 0, \quad i = 1,2,\ldots,N. \tag{24.62}$$

Summing the above equation for all i, and taking into account the boundary condition $\theta_0 = \theta_1$ and $\theta_n = \theta_{n+1}$ yield:

$$\frac{\beta}{N^2}\sum_{i=1}^{N}\left\{1+\beta\left(\frac{1}{\alpha}-\frac{1}{\chi}\right)\cos\theta_i\right\}\sin\theta_i = 0. \tag{24.63}$$

24.4.1.1 Buckling Loads

The N–by–N *Jacobian* of Eq. (24.62) around the trivial equilibrium state is:

$$\boldsymbol{J} = -\boldsymbol{C} + \frac{\beta}{N^2}\left(1-\frac{\beta}{\chi}+\frac{\beta}{\alpha}\right)\boldsymbol{I}. \tag{24.64}$$

Based on Eq. (24.64), one can reformulate the linearized difference equation as:

$$\theta_{i+1} - 2\theta_i + \theta_{i-1} + \frac{\beta}{N^2}\left[1+\beta\left(\frac{1}{\alpha}-\frac{1}{\chi}\right)\right]\theta_i = 0. \tag{24.65}$$

In a critical equilibrium state the determinant of Eq. (24.64) is zero, i.e $\frac{\beta}{N^2}\left(1-\frac{\beta}{\chi}+\frac{\beta}{\alpha}\right)$ is an eigenvalue of $\boldsymbol{C}$ (given already by Eq. (24.14)). The nondimensional buckling loads of the discrete, extensible *Haringx* elastica is:

$$\beta^{\mathrm{cr}}_{2i,2i+1} = \frac{\chi\,\alpha}{2(\chi-\alpha)}\left(\pm\sqrt{1+16\frac{\chi-\alpha}{\chi\,\alpha}N^2\sin^2\frac{i\pi}{2N}}-1\right), \quad i = 0,1,\ldots,N-1. \tag{24.66}$$

The above formula at $i = 0$, with subtraction, leads to a *pure shear buckling* mode:

$$\beta_{\mathrm{deH}}^{\mathrm{shear}} = \frac{\chi\,\alpha}{\alpha - \chi}. \tag{24.67}$$

(With "deH" referring to "discrete extensible Haringx".) Note that this load may be negative, hence the buckling occurs *under tension*, if $\chi > \alpha$, i.e. if the links are easier to shear than to shorten[1]. Meanwhile, the shear buckling occurs under compression, if the links are easier to shorten than to shear. However, this compressive pure shear buckling cannot be achieved, since it is physically inadmissible: it requires a larger force than needed to the total compression of the links.

Besides this shear buckling mode, there are at most $2(N-1)$ buckling loads, given by Eq. (24.66) with $i = 1, 2, \ldots, N-1$. At most, because the discriminant can be negative. If $\chi > \alpha$, then half of these buckling loads are tensile, the other half is compressive. If $\chi < \alpha$, then all the buckling loads are compressive (but not necessarily physically admissible). Possessing tensile buckling loads is a unique feature of *Haringx*-type columns, see for example Kocsis (2016); Kocsis et al (2017). (Note that Genovese (2017) also found a tension shear buckling phenomenon.)

The fundamental compressive buckling load is Eq. (24.66) at $i = 1$ with addition:

$$\beta_{\mathrm{deH}}^{\mathrm{mixed}} = \frac{\chi\,\alpha}{2(\chi - \alpha)}\left(\sqrt{1 + 16\frac{\chi - \alpha}{\chi\,\alpha}N^2 \sin^2\frac{\pi}{2N}} - 1\right), \tag{24.68}$$

which exists if

$$1 + 16N^2\frac{\chi - \alpha}{\chi\,\alpha}\sin^2\frac{\pi}{2N} \geq 0. \tag{24.69}$$

Based on Eq. (24.65), one can also derive the fundamental buckling mode:

$$\theta_{\mathrm{deH},j} = \Phi\cos\left[\frac{\pi}{N}\left(j - \frac{1}{2}\right)\right], \quad j = 1, 2, \ldots, N. \tag{24.70}$$

For the inextensible case ($\chi \to \infty$), we asymptotically find

$$\beta_{\mathrm{dH}} = -\frac{\alpha}{2}\left(1 \pm \sqrt{1 + 16N^2\frac{1}{\alpha}\sin^2\frac{\pi}{2N}}\right), \tag{24.71}$$

which has been obtained by Kocsis et al (2017). For the unshearable case ($\alpha \to \infty$), we find the same buckling load of a discrete extensible elastica as Eq. (24.19). For the inextensible, unshearable case ($\chi \to \infty$ and $\alpha \to \infty$) we obtain Eq. (24.21), the solution of Wang (1951, 1953) for *Hencky*-chains (Hencky, 1920).

[1] Note that Domokos (1992) also found a pure shear buckling mode under tension for a discrete model of a cord.

24.4.1.2 Analytical Solution for Short Linkages

For $N = 2$ analytical solution can be given, at least for equilibrium configurations possessing symmetric post-buckling states. The equation system of the discrete extensible *Engesser* elastica is:

$$\theta_2 - \theta_1 + \frac{\beta}{4}\left\{\left(1 - \frac{\beta}{\chi}\cos\theta_1\right) + \frac{\beta}{\alpha}\cos\theta_1\right\}\sin\theta_1 = 0, \tag{24.72}$$

$$-\theta_2 + \theta_1 + \frac{\beta}{4}\left\{\left(1 - \frac{\beta}{\chi}\cos\theta_2\right) + \frac{\beta}{\alpha}\cos\theta_2\right\}\sin\theta_2 = 0. \tag{24.73}$$

The sum of these equations, Eq. (24.63) at $N = 2$, is solved by

$$\left\{\left(1 - \frac{\beta}{\chi}\cos\theta_1\right) + \frac{\beta}{\alpha}\cos\theta_1\right\}\sin\theta_1 + \left\{\left(1 - \frac{\beta}{\chi}\cos\theta_2\right) + \frac{\beta}{\alpha}\cos\theta_2\right\}\sin\theta_2 = 0. \tag{24.74}$$

A solution set of this equation is

$$\theta_2 = -\theta_1 - 2k\pi, \tag{24.75}$$

Eq. (24.72) with the above solution reads:

$$-\theta_1 - k\pi + \frac{\beta}{16}\left\{2\sin\theta_1 + \beta\left(\frac{1}{\alpha} - \frac{1}{\chi}\right)\sin 2\theta_1\right\} = 0. \tag{24.76}$$

Finally, the equilibrium paths are given by:

$$\beta = \frac{\chi\,\alpha}{2(\chi - \alpha)\cos\theta_1}\left(-1 \pm \sqrt{1 + 32\frac{\chi - \alpha}{\chi\,\alpha}\cdot(\theta_1 + k\pi)\cot\theta_1}\right). \tag{24.77}$$

Note that the above equation is singular at $\Psi_1 = 0$. An asymptotic expansion yields

$$\beta = \frac{\chi\,\alpha}{2(\chi - \alpha)\cos\theta_1}\left(-1 \pm \sqrt{1 + 32\frac{\chi - \alpha}{\chi\,\alpha}\cdot\frac{\cos\theta_1}{1 + \sum\limits_{m=1}^{\infty}(-1)^m\frac{(\theta_1)^{2m}}{(2m+1)!}}}\right). \tag{24.78}$$

Equation (24.78) at $k = 0$ and $\theta_1 = 0$ gives the fundamental buckling loads,

$$\beta_{\text{deH}}^{N=2} = \frac{\chi\,\alpha}{2(\chi - \alpha)}\left(-1 \pm \sqrt{1 + 32\frac{\chi - \alpha}{\chi\,\alpha}}\right), \tag{24.79}$$

which coincide with Eq. (24.66) at $N = 2$, $i = 1$. These buckling loads exist if the discriminant is positive, i.e. if $\alpha\chi > 32(\alpha - \chi)$. By fixing $\alpha = 10$, this condition fulfills if $\chi > 320/42 = 7.6190$. This case is shown on top left of Fig. 24.9, which is obtained by plotting Eq. (24.77) for $k = -5, \cdots, 4$.

Here the two critical loads coincide at $\beta = \chi\alpha/2/(\alpha-\chi)$. If the normal stiffness parameter is slightly larger, for example $\chi = 8$, then these two critical loads separate, which can be well observed on top right of Figure 24.9. If the normal stiffness parameter is slightly smaller, for example $\chi = 7$, then both of these two critical loads vanish, as shown on bottom left of Figure 24.9. Finally, if the tensile stiffness parameter χ is larger than the shear stiffness parameter α, then one critical force is tensile and the other one is compressive. The compressive post-buckling branch and other higher-order equilibrium paths can be seen on bottom right of Figure 24.9 for $\alpha = 10$, $\chi = 20$. The second derivative of Eq. (24.78) with respect to θ_1, expanded up to the second order, and evaluated at $\theta_1 = 0$ yields:

$$\frac{\partial^2\beta}{\partial\theta_1^2}(\theta_1=0) = \frac{p}{6(p+32)}\left\{(3p+64)\sqrt{\frac{p+32}{p}} - 3(p+32)\right\}, \quad p = \frac{\chi\alpha}{\chi-\alpha}. \tag{24.80}$$

If the above curvature of the first post-buckling path is positive, the post-buckling path is *stable*, if it is negative, the post-buckling path is *unstable*. The separator between the two states can be obtained by finding the roots of the above equation, which yields

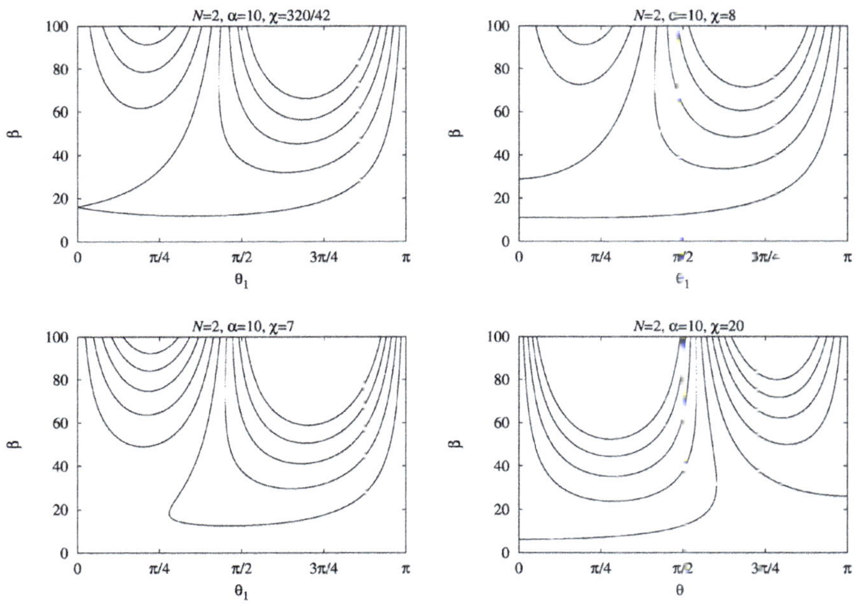

Fig. 24.9: Equilibrium paths of the discrete, extensible Haringx elastica of $N = 2$ links, for $k = -5, \cdots, 4$. Values of α and χ are written on the top of the figures.

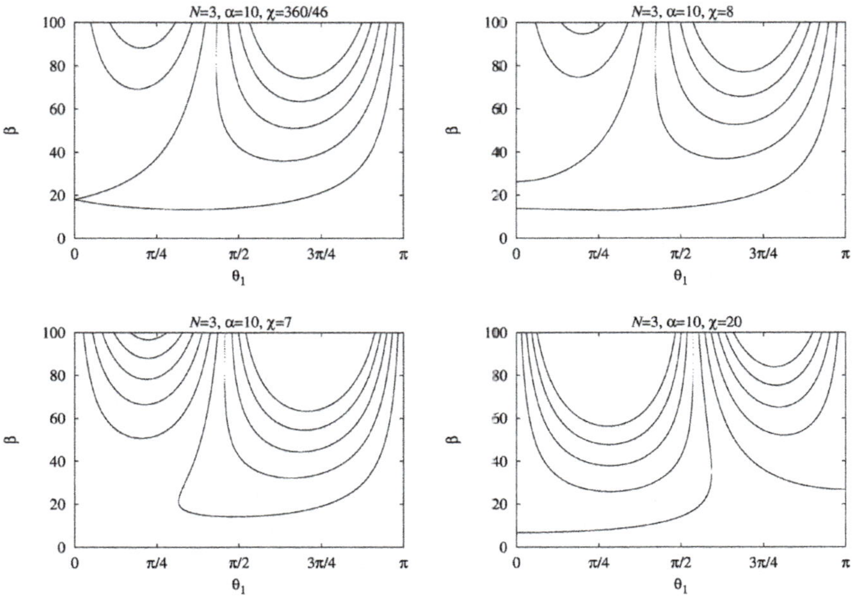

Fig. 24.10: Equilibrium paths of the discrete, extensible Haringx elastica of $N=3$ links, for $k=-5,\cdots,4$. Values of α and χ are given on the top of the figures.

$$\frac{\chi\alpha}{\alpha-\chi}=\frac{128}{3}. \tag{24.81}$$

For $N=3$ a similar procedure can be followed. Eq. (24.63) at $N=3$ is solved by

$$\theta_3=-\theta_1-2k\pi, \quad \theta_2=-m\pi. \tag{24.82}$$

Then the equilibrium paths can be derived from the equilibrium equations (24.62):

$$\beta=\frac{\chi\,\alpha}{2(\chi-\alpha)\cos\theta_1}\left(-1\pm\sqrt{1+36\frac{\chi-\alpha}{\chi\,\alpha}\cdot(\theta_1+k\pi)\cot\theta_1}\right). \tag{24.83}$$

For $k=0$ and $\theta_1=0$, after an asymptotic expansion, the fundamental buckling loads can be obtained,

$$\beta_{\mathrm{deH}}^{N=3}=\frac{\chi\,\alpha}{2(\chi-\alpha)}\left(-1\pm\sqrt{1+36\frac{\chi-\alpha}{\chi\,\alpha}}\right), \tag{24.84}$$

which coincide with Eq. (24.66) at $N=3$, $i=1$. The buckling load exists if the discriminant is positive, i.e. if $\alpha\chi>36(\alpha-\chi)$. By fixing $\alpha=10$, this condition fulfills if $\chi>360/46=7.8261$. Figure 24.10 shows the equilibrium paths for various values of α and χ.

The second derivative of Eq. (24.83) at $\theta_1 = 0$ leads to:

$$\frac{\partial^2 \beta}{\partial \theta_1^2}(\theta_1 = 0) = \frac{p}{2(p+36)} \left\{ (p+24)\sqrt{\frac{p+36}{p}} - (p+36) \right\}, \quad p = \frac{\chi\alpha}{\chi - \alpha}. \tag{24.85}$$

The separator between stable and unstable first post-buckling paths can be found by finding the roots of the above equation:

$$\frac{\chi\alpha}{\alpha - \chi} = 48. \tag{24.86}$$

We conjecture that the stability limit of the first post-buckling path of the discrete *Haringx* elastica is

$$\frac{\chi\alpha}{\alpha - \chi} = \frac{64}{3} N^2 \sin^2\left(\frac{\pi}{2N}\right). \tag{24.87}$$

24.4.1.3 Numerical Solution

Numerically computed equilibrium paths of the discrete extensible *Haringx* elastica are shown in Fig. 24.11 for $N = 2$, and in Fig. 24.12 for $N = 3$ in gray color.

For completeness, the analytical solutions obtained for symmetric configurations are also shown by black curves. It can be seen that there are a vast amount of equilibrium states which are not symmetric, hence not found by the analytical approach. The numerical results are obtained with the simplex scanning algorithm (Gáspár et al, 1997).

24.4.2 Asymptotic Limit: the Local Extensible Haringx Elastica

The local *Haringx* elastica can be asymptotically obtained from Eq. (24.62):

$$\frac{\mathrm{d}^2\theta}{\mathrm{d}\xi^2} + \beta \left\{ 1 + \beta \left(\frac{1}{\alpha} - \frac{1}{\chi} \right) \cos\theta \right\} \sin\theta = 0 \tag{24.88}$$

with $\xi = s/L$ being the nondimensional arc-length parameter. The continuous extensible *Haringx* elastica has already been studied by Goto et al (1990); Atanackovic (1997); Attard (2003); Koiter (2009); Humer (2013).

Now we discuss special cases of Eq. (24.88). For infinite normal stiffness, i.e. for $\chi \to \infty$, Eq. (24.88) yields

$$\frac{\mathrm{d}^2\theta}{\mathrm{d}\xi^2} + \beta \left(1 + \frac{\beta}{\alpha} \cos\theta \right) \sin\theta = 0, \tag{24.89}$$

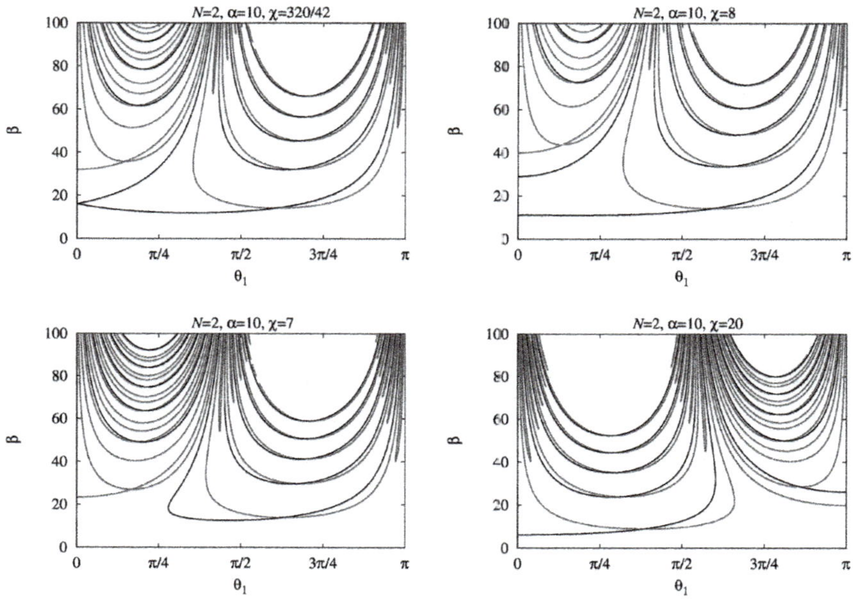

Fig. 24.11: Equilibrium paths of the discrete, extensible Engesser elastica of $N = 2$ links. Analytical results are shown by black curves (for $k = -5, \cdots, 4$), numerical results are shown by gray color. Scanned domain of the angle Ψ_2 is $[-10\pi, 10\pi]$ for the numerics. Values of α and χ are noted on the top of the figures.

the *Haringx* elastica equation, which has been already investigated by Goto et al (1990); Atanackovic (1997); Huang and Kardomateas (2002); Attard (2003); Koiter (2009); Humer (2013) (see also the discussion after Eq. (58) of Kocsis et al (2017)). For infinite shear stiffness, i.e. for $\alpha \to \infty$, Eq. (24.88) leads to

$$\frac{\mathrm{d}^2\theta}{\mathrm{d}\xi^2} + \beta\left(1 - \frac{\beta}{\chi}\cos\theta\right)\sin\theta = 0, \tag{24.90}$$

the extensible elastica equation, which has been obtained by Pflüger (1964); Goto et al (1990); Atanackovic (1997); Magnusson et al (2001). The *Euler* elastica (Euler, 1744) equation can be asymptotically obtained from Eq. (24.88) as $\chi \to \infty$ and $\alpha \to \infty$:

$$\frac{\mathrm{d}^2\theta}{\mathrm{d}\xi^2} + \beta\sin\theta = 0. \tag{24.91}$$

The linearized differential equation of the extensible *Haringx elastica*, Eq. (24.88), is:

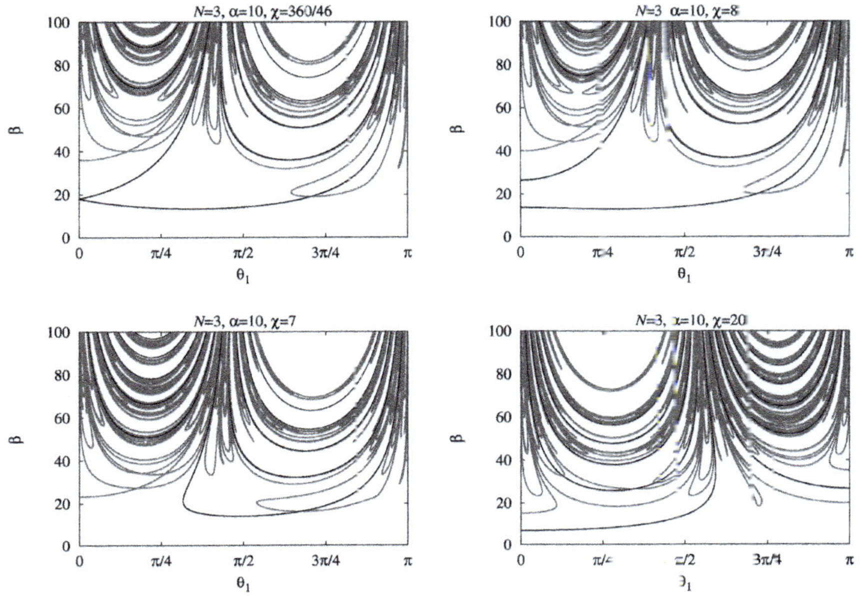

Fig. 24.12: Equilibrium paths of the discrete, extensible Haringx elastica of $N = 3$ links. Analytical results are shown by black curves (for $k = -5 \cdots, 4$), numerical results are shown by gray color. Scanned domain of the angles Ψ_2, Ψ_3 is $[-7\pi, 7\pi]$ for the numerical solution. Values of α and χ are given on the top of the figures.

$$\frac{d^2\theta}{d\xi^2} + \beta\left\{1 + \beta\left(\frac{1}{\alpha} - \frac{1}{\chi}\right)\right\}\theta = 0, \tag{24.92}$$

with pinned-pinned boundary conditions $d\theta/d\xi(0) = 0$ and $d\theta/d\xi(1) = 0$. The buckling modes are

$$\theta_r(\xi) = A_r \cos(r\pi\xi), \tag{24.93}$$

and the nondimensional buckling load parameters of the continuous extensible *Haringx* elastica are:

$$\beta_{\mathrm{leH},r} = \frac{\chi\alpha}{2(\chi-\alpha)}\left(-1 \pm \sqrt{1 + 4\frac{\chi-\alpha}{\chi\alpha} r^2\pi^2}\right). \tag{24.94}$$

(With subscript "leH" referring to "local extensible Haringx".) Here $r^2\pi^2$ is the buckling load parameter of the *Euler* elastica. For $r = 0$ a pure shear buckling load of the local continuum is obtained:

$$\beta_{\mathrm{leH,shear}} = \frac{\chi\alpha}{\alpha-\chi}. \tag{24.95}$$

This is the same as for the discrete version, with the same conclusions: it physically exists if the column is softer for shearing than for tension/compression, and it occurs then under tension. For $r = 1$ the fundamental buckling load parameter of mixed (flexural, normal and shear) buckling modes can be obtained:

$$\beta_{\text{leH}} = \frac{\chi\alpha}{2(\chi-\alpha)}\left(-1 \pm \sqrt{1+4\frac{\chi-\alpha}{\chi\alpha}\pi^2}\right). \tag{24.96}$$

These loads exist if the discriminant is positive, i.e. if $1+4\pi^2(\chi-\alpha)/(\chi\alpha) \geq 0$, and they can be both compressive, or one can be tensile and one compressive (if the column is softer for shearing than for tension/compression). Eq. (24.96) was first obtained by Reissner (1982). Considering Eq. (24.4), the dimensional buckling load of pure shear buckling mode of the continuous extensible *Haringx* elastica is

$$P_{\text{leH}}^{\text{shear}} = \frac{GA \cdot EA}{GA - EA} \tag{24.97}$$

while its fundamental compressive buckling load of mixed deformations is:

$$P_{\text{leH}} = \frac{\kappa GA}{2} \cdot \frac{EA}{EA-\kappa GA}\left(\sqrt{1+4\frac{EA-\kappa GA}{EA}\cdot\frac{P_{\text{E}}}{\kappa GA}} - 1\right). \tag{24.98}$$

Here P_{E} is *Euler*'s critical force. Eq. (24.98) coincides with the results of Kocsis (2016) for discrete planar *Cosserat* rods.

For the unshearable case, $\alpha \to \infty$, the same buckling load is found as Eq. (24.48). The original solution of Haringx (1942) is obtained in the inextensible limit, $\chi \to \infty$:

$$\beta_{\text{lH}} = -\frac{\alpha}{2}\left(1 \pm \sqrt{1+\frac{4\pi^2}{\alpha}}\right). \tag{24.99}$$

Finally, *Euler*'s critical force is obtained in the unshearable, inextensible limit.

Asymptotic solutions in the post-buckling range are derived in Appendix B.

Based on that derivation we can conclude that for $\chi\alpha/(\alpha-\chi) \in \left[4\pi^2, 16\pi^2/3\right]$ the post-bifurcation path is unstable, and for $\chi\alpha/(\alpha-\chi) \geq 16\pi^2/3$ it is stable. Hence the limit between stable and unstable post-bifurcation branch of the continuous *Haringx* elastica is

$$\frac{\chi\alpha}{\alpha-\chi} = \frac{16\pi^2}{3}, \tag{24.100}$$

which is the asymptotic limit of Eq. (24.87). For unshearable columns, this critical parameter is $\chi = 16\pi^2/3$, as already obtained by Magnusson et al (2001). In summary, the critical parameters at the boundary of stable and unstable responses are given for the extensible *Engesser* elastica by

$$\left(1+\frac{\pi^2}{\alpha}\right)\chi=\frac{16\pi^2}{3}. \tag{24.101}$$

and for the extensible *Haringx* elastica by

$$\frac{\alpha\chi}{\alpha-\chi}=\frac{16\pi^2}{3}. \tag{24.102}$$

In both cases, for the unshearable limit $\alpha\to\infty$, we find the value of Atanackovic (1997); Magnusson et al (2001)

$$\chi=\frac{16\pi^2}{3}. \tag{24.103}$$

24.4.3 Continualized Nonlocal Extensible Haringx Elastica

The governing difference equation, Eq. (24.62), can be continualized based on Eq. (24.51):

$$\frac{\mathrm{d}^2\theta}{\mathrm{d}\xi^2}+\beta\left(1-\frac{1}{12N^2}\frac{\mathrm{d}^2}{\mathrm{d}\xi^2}\right)\cdot\left\{1+\beta\left(\frac{1}{\alpha}-\frac{1}{\chi}\right)\cos\theta\right\}\sin\theta=0. \tag{24.104}$$

The linearization of the above nonlinear differential equation is

$$\left(1-\frac{\hat{\lambda}}{12N^2}\right)\frac{\mathrm{d}^2\theta}{\mathrm{d}\xi^2}+\hat{\lambda}\theta=0, \quad \text{with} \quad \hat{\lambda}=\beta\left(1+\beta\left(\frac{1}{\alpha}-\frac{1}{\chi}\right)\right). \tag{24.105}$$

The buckling loads can be computed based on local or nonlocal boundary conditions (Challamel et al, 2015a). With local pinned-pinned boundary conditions, i.e. $\theta'(0)=\theta'(1)=0$, the buckling modes are the same as Eq. (24.93) and the nondimensional buckling load parameters of the nonlocal extensible *Haringx* elastica are:

$$\beta_{\mathrm{nleH},r}=\frac{\chi\alpha}{2(\chi-\alpha)}\left(-1\pm\sqrt{1+4\frac{\chi-\alpha}{\chi\alpha}\frac{r^2\pi^2}{1+\frac{r^2\pi^2}{12N^2}}}\right). \tag{24.106}$$

(With subscript "nleH" referring to "nonlocal extensible Haringx".) Here $r^2\pi^2/(1+r^2\pi^2/(12N^2))$ are the buckling load parameters of the nonlocal elastica (Challamel et al, 2015a). The nonlocal boundary conditions would be $\theta(0)=\theta(1/N)$ and $\theta(1)=\theta(1+1/N)$.

Note that the local and the nonlocal extensible *Haringx* elasticavobey pure shear buckling mode, same as Eq. (24.67). Besides, they possess a countable infinity of tensile and compressive buckling loads. Note also that the discriminant in Eq. (24.94) and (24.106) can be negative, meaning that some critical loads do not exist.

24.4.3.1 Numerical Solution: Discrete Versus Nonlocal Extensible Haringx Elastica

The first post-buckling paths of the discrete and the nonlocal, extensible *Haringx* elastica are shown in Fig. 24.13 for various numbers of links N. The results are obtained numerically with the simplex path-following algorithm (Domokos and Gáspár, 1995). It can be seen that the path computed for the nonlocal model follows well the path of the corresponding discrete mode. Hence the scale effect of the lattice system is well captured by the nonlocal model also for large displacements.

24.5 Conclusions

In this paper lattice systems with shear, normal and bending interactions were developed. The buckling and post-buckling analysis of these systems were studied using exact geometric nonlinearities. *Timoshenko*'s finite strain theory was included in one lattice system, which led to the discrete counterpart of the extensible *Engesser* elastica, and *Love*'s finite strain theory was embedded in the other lattice system, resulting in a discrete extensible *Haringx* elastica model. Analytical and numerical solutions were given for large displacements and high forces and these solutions were compared on bifurcation diagrams.

It was shown that in the continuum limit these systems lead to the local continous *Engesser* and *Haringx* elastica solutions. In the inextensible and unshearable limits the obtained continuum solutions were verified to results available in the literature. Analytical solutions for the post-buckling paths of these continuum models were derived.

Through a continualization technique, nonlocal counterparts of the extensible *Engesser* and *Haringx* elastica models were developed starting from our original lattice systems. Analytical formulas for the buckling loads of the nonlocal models were given. The first post-buckling paths of the nonlocal models were computed numerically and they were compared to those of the source lattice systems. It was shown that these nonlocal models can capture the scale effects of the lattice systems.

Acknowledgements The work of A. Kocsis was supported by the János Bolyai Research Scholarship of the Hungarian Academy of Sciences.

Appendix A

The following *Taylor* expansions are used:

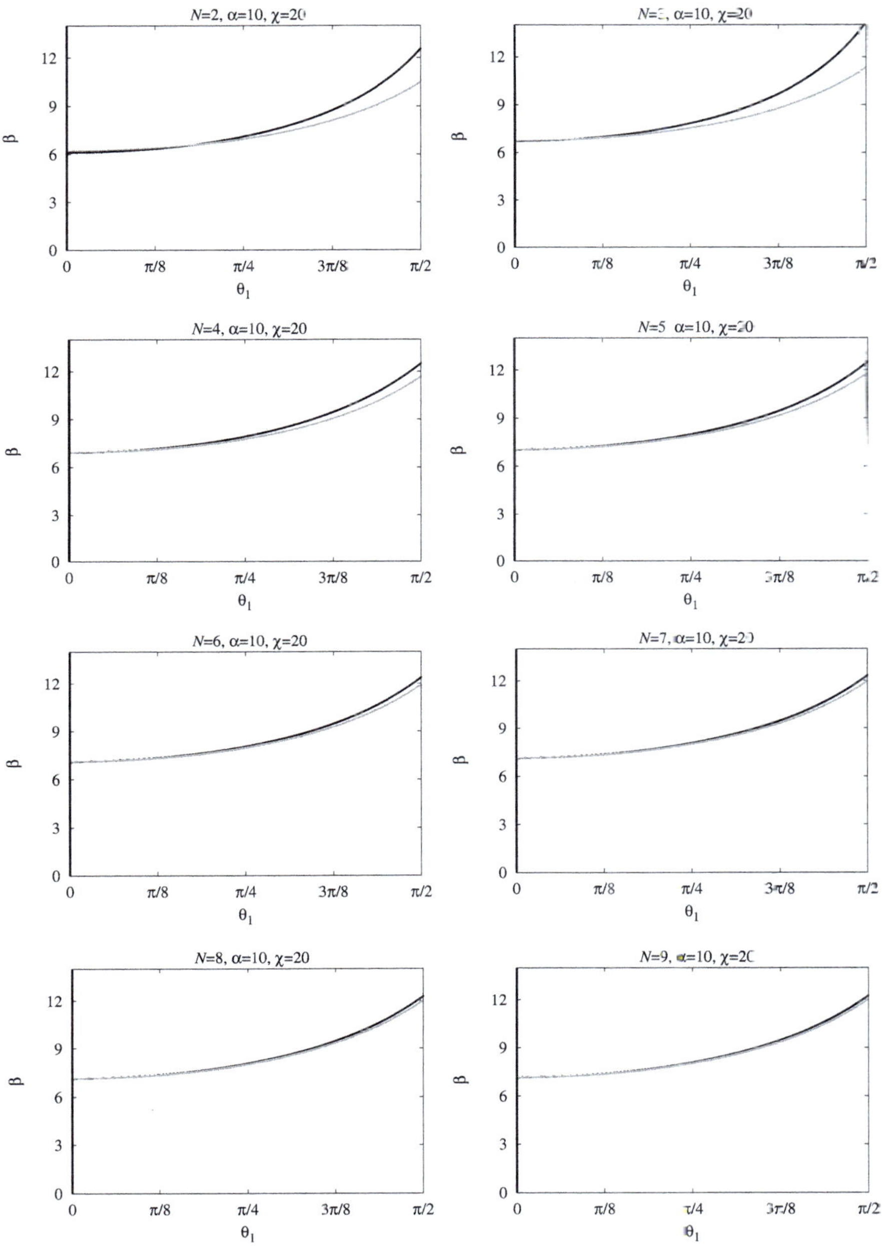

Fig. 24.13: First post-buckling equilibrium path of the nonlocal versus discrete extensible Haringx elastica of $N = 2, 3, 4, 5, 7, 8, 9$ links ($\alpha = 10$, $\chi = 20$).

$$\beta = \beta_0 + \epsilon\beta_1 + \epsilon^2\beta_2 + \epsilon^3\beta_3 + \ldots \tag{24.107}$$

$$\Psi(\xi) = v_0(\xi) + \epsilon v_1(\xi) + \epsilon^2 v_2(\xi) + \epsilon^3 v_3(\xi) + \ldots \quad \text{with} \quad \epsilon = \Psi(0) = \Psi_0. \tag{24.108}$$

Here ϵ is a small parameter related to the amplitude of the post-buckling mode. The fundamental (or trivial) path is characterized by no rotation, $v_0 = 0$. Furthermore, for reasons related to symmetry, it can be shown that some terms vanish in the asymptotic expansion:

$$\text{for} \quad k \geq 1, \quad v_{2k}(\xi) = 0 \quad \text{and} \quad \beta_{2k-1} = 0, \tag{24.109}$$

leading to the third-order asymptotic expansion:

$$\beta = \beta_0 + \epsilon^2\beta_2 + \ldots \tag{24.110}$$

$$\Psi(\xi) = \epsilon v_1(\xi) + \epsilon^3 v_3(\xi) + \ldots \quad \text{with} \quad \epsilon = \Psi(0) = \Psi_0, \tag{24.111}$$

Using this expansion in Eq. (24.38), and considering up to the third power of small parameter ϵ (note that the coefficient of $\epsilon^2 = 0$), one obtains the following two differential equations:

$$\begin{aligned}
&\left[1 - \frac{\beta_0}{\alpha}\left(1 - \frac{\beta_0}{\chi}\right)\right]\frac{\mathrm{d}^2 v_1}{\mathrm{d}\xi^2} + \beta_0\left(1 - \frac{\beta_0}{\chi}\right)v_1 = 0,\\
&\left[1 - \frac{\beta_0}{\alpha}\left(1 - \frac{\beta_0}{\chi}\right)\right]\frac{\mathrm{d}^2 v_3}{\mathrm{d}\xi^2} + \beta_0\left(1 - \frac{\beta_0}{\chi}\right)v_3 = \frac{\beta_0}{\alpha}\left(-\frac{v_1^2}{2} + \frac{2\beta_0}{\chi}v_1^2 - \frac{\beta_2}{\chi}\right)\frac{\mathrm{d}^2 v_1}{\mathrm{d}\xi^2}\\
&+\frac{\beta_2}{\alpha}\left(1 - \frac{\beta_0}{\chi}\right)\frac{\mathrm{d}^2 v_1}{\mathrm{d}\xi^2} + \left(\frac{\mathrm{d}v_1}{\mathrm{d}\xi}\right)^2\frac{\beta_0}{\alpha}\left(\frac{4\beta_0}{\chi} - 1\right)v_1\\
&+\left(1 - \frac{\beta_0}{\chi}\right)\left(\beta_0\frac{v_1^3}{6} - \beta_2 v_1\right) - \frac{\beta_0 v_1}{\chi}\left(-\beta_2 + \frac{\beta_0}{2}v_1^2\right).
\end{aligned} \tag{24.112}$$

with boundary conditions

$$v_i'(0) = 0, v_i'(1) = 0. \tag{24.113}$$

The first differential equation of Eq. (24.112) gives the linearized fundamental buckling mode

$$v_1(\xi) = \cos(\pi\xi) \tag{24.114}$$

with β_0 equals the buckling load given by Eq. (24.42). Here the normalization criterion $v_1(0) = 1$ was used. Using this buckling mode in the second differential equation of Eq. (24.112) yields

$$\left[1-\frac{\beta_0}{\alpha}\left(1-\frac{\beta_0}{\chi}\right)\right]\frac{d^2 v_3}{d\xi^2}+\beta_C\left(1-\frac{\beta_0}{\chi}\right)v_3 = A\cos(\pi\xi)+B\cos(3\pi\xi) \quad (24.115)$$

with

$$\begin{aligned}
A &= \frac{\pi^2}{\alpha}\beta_0\left(\frac{4\beta_0}{\chi}-1\right)+\frac{2\beta_0\beta_2}{\chi}\left(1+\frac{\pi^2}{\alpha}\right) \\
&\quad -\beta_2\left(1+\frac{\pi^2}{\alpha}\right)+\frac{3\beta_0}{4}\left(\frac{1}{6}+\frac{3\pi^2}{2\alpha}\right)-\frac{3\beta_0^2}{4}\left(\frac{6\pi^2}{\alpha\chi}+\frac{2}{3\chi}\right), \\
B &= \frac{\beta_0}{4}\left(\frac{1}{6}+\frac{3\pi^2}{2\alpha}\right)-\frac{\beta_C^2}{4}\left(\frac{6\pi^2}{\alpha\chi}+\frac{2}{3\chi}\right).
\end{aligned} \quad (24.116)$$

Due to the boundary conditions Eq. (24.113), the solution of this differential equation for the second-order buckling load factor β_2 is

$$\beta_2 = \frac{\beta_0}{1-\frac{2\beta_0}{\chi}}\left(\frac{1}{8}-\frac{\beta_0}{2\chi}\right). \quad (24.117)$$

β_2 measures the curvature of the post-bifurcation branch. The limit between the stable and the unstable post-bifurcation branch is associated with $\beta_2 = 0$ and $\beta_0 = \chi/4$, so that

$$\beta_0 = \frac{\chi}{4} \quad \Rightarrow \quad \left(1+\frac{\pi^2}{\alpha}\right)\frac{3\chi}{16} = \pi^2. \quad (24.118)$$

Appendix B

The following expansion of the load parameter and the rotation field can be used:

$$\begin{aligned}
\beta &= \beta_0+\epsilon^2\beta_2+\ldots \\
\theta(\xi) &= \epsilon v_1(\xi)+\epsilon^3 v_3(\xi)+\ldots.
\end{aligned} \quad (24.119)$$

Substituting these expansions in the nonlinear differential equation, Eq. (24.88), and considering the first and third powers of the small parameter ϵ (since the coefficient of ϵ^2 is zero), one obtains the following system of two differential equations:

$$\begin{aligned}
&\frac{d^2 v_1}{d\xi^2}+\beta_0(1-\mu\beta_0)v_1 = 0, \\
&\frac{d^2 v_3}{d\xi^2}+\beta_0(1-\mu\beta_0)v_3 = \beta_0(1-\mu\beta_0)\frac{v_1^3}{6}-\beta_2(1-\mu\beta_0)v_1-\mu\beta_0\left(-\beta_2-\beta_0\frac{v_1^2}{2}\right)v_1
\end{aligned} \quad (24.120)$$

with

$$\mu = \frac{1}{\chi} - \frac{1}{\alpha}.$$

The first differential equation gives the linearized fundamental buckling mode

$$v_1(\xi) = \cos(\pi\xi) \quad \text{with} \quad \beta_0 = \frac{1 \pm \sqrt{1 - 4\mu\pi^2}}{2\mu} \tag{24.121}$$

where the normalization criteria $v_1(0) = 1$ was used. Inserting this first order buckling mode in the second differential equation of Eq. (24.120) leads to:

$$\begin{aligned} &\frac{d^2 v_3}{d\xi^2} + \pi^2 v_3 = A\cos(\pi\xi) + B\cos(3\pi\xi) \\ &\text{with} \quad A = -\beta_2(1 - 2\mu\beta_0) + \frac{\beta_0}{8}(1 - 4\mu\beta_0),\ B = \frac{\beta_0}{24}(1 - 4\mu\beta_0). \end{aligned} \tag{24.122}$$

In view of the boundary conditions, the solution of this differential equation (which requires A to vanish) yields the second-order buckling load factor:

$$\beta_2 = \frac{\beta_0}{8}\frac{1 - 4\mu\beta_0}{1 - 2\mu\beta_0}. \tag{24.123}$$

Since β_2 measures the curvature of the bifurcated equilibrium path at the bifurcation point, the limit between stable and unstable post-bifurcation branches can be obtained by solving for $\beta_2 = 0$:

$$\beta_0 = \frac{1}{4\mu} \quad \Rightarrow \quad \mu = \frac{3}{16\pi^2} \quad \Rightarrow \quad \frac{\alpha\chi}{\alpha - \chi} = \frac{16\pi^2}{3}. \tag{24.124}$$

References

Alibert J, Della Corte A, Giorgio I, Battista A (2017) Extensional elastica in large deformation as Γ-limit of a discrete 1D mechanical system. Z Angew Math Phys 68(42):1–19

Atanackovic TM (1997) Stability theory of elastic rods, Series on Stability, Vibration and Control of Systems, Series A: Volume 1. World Scientific, Singapore

Attard M (2003) Finite strain-beam theory. Int J Solids Structures 40:4563–4584

Born M, Huang K (1954) Dynamical theory of crystal lattices. Oxford University Press, Oxford

Bresse JAC (1859) Cours de mécanique appliquée – Résistance des matériaux et stabilité des constructions. Gauthier-Villars, Paris

Cauchy A (1823) Recherches sur l'équilibre et le mouvement intérieur des corps solides ou fluides, élastiques ou non élastiques. Bulletin des sciences par la Société Philomatique de Paris pp 9–13

Cauchy A (1828) Sur l'équilibre et le mouvement d'un système de points matériels sollicités par des forces d'attraction ou de répulsion mutuelle. Exercices de Mathématiques 3:188–212

Challamel N, Wang CM, Elishakoff I (2014) Discrete systems behave as nonlocal structural elements: Bending, buckling and vibration analysis. European Journal of Mechanics - A 44:125–135

Challamel N, Kocsis A, Wang CM (2015a) Discrete and non-local elastica. International Journal of Non-Linear Mechanics 77:128 – 140

Challamel N, Picandet V, Collet B, Michelitsch T, Elishakoff I, M WC (2015b) Revisiting finite difference and finite element methods applied to structural mechanics within enriched continua. Eur J Mech A/Solids 53:107–120

Dell'Isola F, Maier G, Perego U, Andreaus U, Esposito R, Forest S (2014) The complete works of Gabrio Piola: Volume i commented english translation. In: Advanced Structured Materials, Springer, vol 38

Dell'Isola F, Andreaus U, Placidi L (2015) At the origins and in the vanguard of peridynamics, non-local and higher-gradient continuum mechanics: An underestimated and still topical contribution of Gabrio Piola. Mathematics and Mechanics of Solids 20(8):887–928

Domokos G (1992) Buckling of a cord under tension. Acta Technica Academiae Scientiarum Hungaricae Civil Engineering 104:63–73

Domokos G, Gáspár Z (1995) A global, direct algorithm for path-following and active static control of elastic bar structures. Mechanics of Structures and Machines 23:549–571

Duan W, Challamel N, Wang CM, Ding Z (2013) Development of analytical vibration solutions for microstructured beam model to calibrate length scale coefficient in nonlocal Timoshenko beams. J Applied Physics 114(104312):1–11

Engesser F (1891) Die Knickfestigkeit gerader Stäbe. Zentralblatt der Bauverwaltung 11:483

Eringen A (1983) On differential equations of nonlocal elasticity and solutions of screw dislocation and surface waves. J Appl Phys 54:4703–4710

Eringen AC (2002) Nonlocal Continuum Field Theories. Springer-Verlag

Euler L (1744) Methodus inveniendi lineas curvas maximi minimive proprietate gaudentes, sive solutio problematis isoperimetrici lattissimo sensu accepti. Reprint in Opera Omnia I 24:231–297

Foce F (1995) The theory of elasticity between molecular and continuum approach in the xix century. In: Radelet de Grave P, Benvenuto E (eds) Between Mechanics and Architecture, Birkhaüser-Verlag

Föppl A (1897) Vorlesungen über Technische Mechanik – Dritter Band. Festigkeitslehre, Lepizig

Gáspár Z, Domokos G, Szeberényi I (1997) A parallel algorithm for the global computation of elastic bar structures. Computer Assisted Mechanics and Engineering Sciences 4:55–68

Genovese D (2017) Tensile buckling in shear deformable rods. Int J Struct Stab Dyn 17(6):1750,063

Goto Y, Yoshimitsu T, Obata M (1990) Elliptic integral solutions of plane elastica with axial and shear deformations. Int J Solids Structures 26(4):375–390

Haringx JA (1942) On the Buckling and Lateral Rigidity of Helical Springs. Proc Konink Ned Akad Wet 45:533

Hencky H (1920) Über die angenäherte Lösung von Stabilitätsproblemen im Raummittels der elastischen Gelenkkette. Der Eisenbau 11:437–452, in German

Huang H, Kardomateas G (2002) Buckling and initial post-buckling behavior of sandwich beams including transverse shear. AIAA Journal 40(11):2331–2335

Humer A (2013) Exact solutions for the buckling and post-buckling of shear-deformable beams. Acta Mechanica 224:1493–1525

Kocsis A (2016) Buckling analysis of the discrete planar cosserat rod. International Journal of Structural Stability and Dynamics 16(3):1450,111/1–29

Kocsis A, Challamel N, Károlyi G (2017) Discrete and nonlocal models of Engesser and Haringx elastica. International Journal of Mechanical Sciences 130:571–585

Koiter WT (1963) Elastic stability and post-buckling. In: Langer RE (ed) Proc. Symp. Nonlinear Problems, University of Wisconsin Press

Koiter WT (2009) Elastic stability of solids and structures (Ed. van der Heijden AMA). Cambridge University Press, Cambridge

Kröner E, Datta B (1966) Nichtlokale elastostatik: Ableitung aus der gittertheorie. Z Phys 196:203–211

Krumhansl J (1965) Generalized continuum field representations for lattice vibrations. In: Wallis R (ed) Lattice Dynamics, Proceedings of the International Conference held at Copenhagen, Denmark 5-9 August 1963, Pergamon Press, Oxford

Kunin I (1966) Model of elastic medium with simple structure and space dispersion. Prikl Mat Mekh 30:542–550, In Russian

Love AEH (1944) A treatise on the mathematical theory of elasticity, 4th edn. Dover Publications, New York

Magnusson A, Ristinmaa M, Ljung C (2001) Behaviour of the extensible elastica solution. International Journal of Solids and Structures 38:8441–8457

Maugin GA (1999) Nonlinear waves in elastic crystals. Oxford University Press

Navier L (1823) Sur les lois de l'équilibre et du mouvement des corps solides élastiques. Bulletin des sciences par la Société Philomatique de Paris pp 177–181

Pflüger A (1964) Stabililätsprobleme der Elastostatik. Springer–Verlag, Berlin Heidelberg

Reissner E (1982) Some remarks on the problem of Euler buckling. Ingenieur–Archiv 52:115–119

Rózsa P (1991) Linear Algebra and its Applications (in Hungarian). Tankönyvkiadó, Budapest

Silverman IK (1951) Discussion on the paper of "Salvadori M.G., Numerical computation of buckling loads by finite differences". Trans ASCE 116:625–626

Stakgold I (1950) The cauchy relations in a molecular theory of elasticity. Quart Appl Math 8:169–186

Timoshenko SP (1922) On the transverse vibration of bars with uniform cross-section. Philosophical Magazine 43:125–131

Timoshenko SP (1953) History of strength of materials with a brief account of the history of theory of elasticity and theory of structures. McGraw-Hill

Turco E, Dell'Isola F, Cazzani A, Rizzi L (2016) Hencky-type discrete model for pantographic structures: numerical comparison with second gradient continuum models. Z Angew Math Phys 67(85):1–28

Wang CT (1951) Discussion on the paper of "Salvadori M.G., Numerical computation of buckling loads by finite differences". Trans ASCE 116:629–631

Wang CT (1953) Applied Elasticity. McGraw-Hill, New York

Zhang Z, Challamel N, Wang CM (2013) Eringen's small length scale coefficient for buckling of nonlocal Timoshenko beam based on microstructured beam model. Journal of Applied Physics 114:114,902/1–6

Ziegler H (1982) Arguments for and against Engesser's buckling formulas. Ingenieur–Archiv 52:105–113

Chapter 25
A Consistent Dynamic Finite-Strain Plate Theory for Incompressible Hyperelastic Materials

Yuanyou Li and Hui-Hui Dai

Abstract In this chapter, a dynamic finite-strain plate theory for incompressible hyperelastic materials is deduced. Starting from nonlinear elasticity, we present the three-dimensional (3D) governing system through a variational approach. By series expansion of the independent variables about the bottom surface, we deduce a 2D vector dynamic plate system, which preserves the local momentum-balance structure. Then we propose appropriate position and traction boundary conditions. The 2D plate equation guarantees that each term in the variation of the generalized potential energy functional attains the required asymptotic order. We also consider the associated weak formulations of the plate model, which can be applied to different types of practical edge conditions.

25.1 Introduction

Plate structures are defined as plane elements with one small thickness dimension compared with the other two planar dimensions. The theory of plates has been widely studied by scientists in both mathematical and engineering communities since the nineteenth century. The literature in this field is extremely plentiful, including theories based on engineering intuitions and assumptions, derived theories from three-dimensional elasticity, as well as direct theories (Timoshenko and Woinowsky-Krieger, 1959; Naghdi, 1972; Reddy, 2007; Altenbach et al, 2010). We also refer the readers to Dai and Song (2014) for a review of some selected works. For derived plate theories, we usually focus on how to reduce the original 3D elasticity theory to a two-dimensional (2D) approximate model while the fundamental mechanical properties of plate structures can be appropriately captured.

Yuanyou Li · Hui-Hui Dai
Department of Mathematics, City University of Hong Kong, 83 Tat Chee Avenue, Kowloon Tong, Hong Kong
e-mail: yuanyouli2-c@cityu.edu.hk, mahhdai@cityu.edu.hk

© Springer International Publishing AG, part of Springer Nature 2018
H. Altenbach et al. (eds.), *Generalized Models and Non-classical Approaches in Complex Materials 1*, Advanced Structured Materials 89,
https://doi.org/10.1007/978-3-319-72440-9_25

Early attempts on plate theories relied on a priori hypotheses, either of geometrical and mechanical nature or on the specific form of the solution (displacements and stresses). Classical plate theories derived in this way include the Kirchhoff-Love theory (Kirchhoff, 1850; Love, 1888), the von Kármán theory (von Kármán, 1910), the Mindlin-Reissner theory (Mindlin, 1951). Despite the success of these theories to various specific situations, their applicability for relatively thick plates and general loadings (say, under shear tractions) may be limited.

One approach involves no explicit kinematic assumptions on displacements (or deformed positions) except general power (or other function) series expansions

$$\boldsymbol{x}(\boldsymbol{r},Z) = \sum_{j=0}^{N} Z^j \boldsymbol{x}^j(\boldsymbol{r}). \tag{25.1}$$

Normally, all the coefficients $\boldsymbol{x}^j$ ($j \geqslant 1$) are treated as independent unknowns. By first integrating out the Z variable and conducting a truncation, the 2D potential energy of the plate is formulated. Then the governing equations are derived from the two-dimensional variational or virtual work principle. Such an approach has been adopted by Kienzler (2002) based on linear elasticity. Based on nonlinear elasticity, Meroueh (1986); Steigmann (2007) adopted Legendre polynomials of Z in (25.1) and formulated a system in terms of generalized (high order) stress resultants for finite-strain problems. It is worth mentioning that by imposing restrictions on the high-order coefficients in (25.1) instead of treating them as independent unknowns, Steigmann (2013) derived more proper plate and shell models which incorporate both stretching and bending.

Based on a priori scalings between the plate thickness and the deformations (or applied loads), some consistent mathematical approaches are utilized for deriving asymptotically correct plate theories. The method of Gamma convergence (Friesecke et al, 2002) is concerned with the two-dimensional variational problem in the limit of small thickness, but it cannot be used to study dynamic problems and derive plate theories incorporating both bending and stretching. The method of asymptotic analysis, which aims at developing the leading-order weak formulation by formal expansions with the thickness as the small parameter, was used to derive the von Kármán plate equations from the 3D weak formulation in Ciarlet (1980). In Millet et al (2001), based on the 3D differential formulation, a hierarchy of leading-order plate equations were derived.

Most of the works in the literature consider compressible materials, the existing plate theories for incompressible materials are much fewer. With the Gamma convergence method, Trabelsi (2005) formulated a nonlinear elastic thin membrane model for incompressible materials, while Conti and Dolzmann (2008) extended the plate theory derived in Friesecke et al (2002) to the case of incompressible materials. By using the principle of virtual work, Batra (2007) proposed a compatible shear and normal deformable theory for a plate made of an incompressible linear elastic material, in which the orthonormal Legendre polynomials were adopted to derive the high-order plate theory.

Nowadays, soft materials and biological materials have attracted attentions of researchers of different fields. It happens that most of soft materials are incompressible. In this chapter, we intend to provide a dynamic plate theory for incompressible materials. Taking the incompressibility constraint into account, we extend the consistent plate theory proposed in Song and Dai (2016) to the case of incompressible hyperelastic materials.

We organize this chapter as follows. In Sect. 25.2, the 3D governing system of incompressible materials is derived through conventional variational approach. In Sect. 25.3, according to the criterion of consistency we derive the 2D vector plate equation and propose some proper edge boundary conditions as well. In Sect. 25.4, we consider the associated weak formulations of the 2D plate equation and adopt them to distinct types of boundary conditions. Finally, we make some concluding remarks.

25.2 The 3D Governing Equations

In this section, we consider a homogeneous thin plate of constant thickness, which is composed of an incompressible hyperelastic material. A material point of the plate in the reference configuration $\kappa = \Omega \times [0, 2h]$ is denoted by $\boldsymbol{X} = (\boldsymbol{r}, Z)$, where the thickness $2h$ of the plate is small compared with the planar dimensions of the top (or bottom) surface Ω. The coordinates of a material point in the current configuration κ_t is denoted as $\boldsymbol{x}$. Throughout the paper, symbols with typefaces $\mathbb{a}$, $\boldsymbol{a}$, $\mathbb{A}$, $\mathscr{A}$ represent scalar, vector, second-order tensor (matrix) and higher-order tensor, respectively. In component forms, we adopt the convention that Latin indices run from 1 to 3 whereas Greek indices run from 1 to 2, repeated summation convention is used and the index after the comma indicates differentiation.

The deformation gradient tensor of a material point in the plate can then be represented by

$$\mathbb{F} = \frac{\partial \boldsymbol{x}}{\partial \boldsymbol{X}} = \frac{\partial \boldsymbol{x}}{\partial \boldsymbol{r}} + \frac{\partial \boldsymbol{x}}{\partial Z} \otimes \boldsymbol{k} = \nabla \boldsymbol{x} + \frac{\partial \boldsymbol{x}}{\partial Z} \otimes \boldsymbol{k}, \tag{25.2}$$

where ∇ is the in-plane two-dimensional gradient and $\boldsymbol{k}$ is the unit outward normal vector of the reference top surface Ω. More precisely, with rectangular Cartesian coordinates $\boldsymbol{r} = X_1 \boldsymbol{E}_1 + X_2 \boldsymbol{E}_2$, we have

$$\nabla \boldsymbol{x} = \frac{\partial \boldsymbol{x}}{\partial X_1} \otimes \boldsymbol{E}_1 + \frac{\partial \boldsymbol{x}}{\partial X_2} \otimes \boldsymbol{E}_2.$$

Besides, we consider the following incompressibility constraint equation

$$R(\mathbb{F}) = \mathrm{Det}(\mathbb{F}) - 1 = 0, \quad \text{in } \Omega \times [0, 2h]. \tag{25.3}$$

Suppose the material has the strain-energy density function $\Phi(\mathbb{F})$, the associated first and second order elastic moduli are defined by

$$\mathscr{A}^1(\mathbb{F}) = \frac{\partial^2 \Phi}{\partial \mathbb{F} \partial \mathbb{F}} \left(\mathscr{A}^1_{ijkl} = \frac{\partial^2 \Phi}{\partial F_{ji} \partial F_{lk}} \right), \quad \mathscr{A}^2(\mathbb{F}) = \frac{\partial^3 \Phi}{\partial \mathbb{F} \partial \mathbb{F} \partial \mathbb{F}}. \tag{25.4}$$

It is assumed that the strain energy function for the deformations concerned satisfies the strong-ellipticity condition

$$\boldsymbol{a} \otimes \boldsymbol{b} : \mathscr{A}^1(\mathbb{F})[\boldsymbol{a} \otimes \boldsymbol{b}] > 0, \quad \text{for all } \boldsymbol{a} \otimes \boldsymbol{b} \neq 0, \tag{25.5}$$

where the colon between second-order tensors means a scalar tensor product defined by $\mathbb{A} : \mathbb{B} = A_{kl} B_{lk}$ and the square bracket after a higher-order modulus tensor represents the operations

$$\{\mathscr{A}^1[\mathbb{A}]\}_{ij} = \mathscr{A}^1_{ijkl} A_{lk}, \quad \{\mathscr{A}^2[\mathbb{A}, \mathbb{B}]\}_{ij} = \mathscr{A}^2_{ijklmn} A_{lk} B_{nm}. \tag{25.6}$$

For the dynamic case with dead-loading on the boundary, suppose $\boldsymbol{q}_b$ is the body force, $\boldsymbol{q}^{\pm}$ are the applied tractions on the top and bottom surfaces, and $\boldsymbol{q}$ is the applied traction on $\partial \Omega_q$. The kinetic energy, the strain energy and the load potential are given by

$$\begin{aligned}
\underline{K} &= \int_{\kappa} \frac{1}{2} \rho \dot{\boldsymbol{x}} \cdot \dot{\boldsymbol{x}} d\boldsymbol{X}, \\
\underline{\Phi} &= \int_{\kappa} \Phi(\mathbb{F}) d\boldsymbol{X}, \\
\underline{V} &= -\int_{\kappa} \boldsymbol{q}_b(\boldsymbol{X}) \cdot \boldsymbol{x}(\boldsymbol{X}) d\boldsymbol{X} - \int_{\Omega} \boldsymbol{q}^-(\boldsymbol{r}) \cdot \boldsymbol{x}(\boldsymbol{r}, 0) + \boldsymbol{q}^+(\boldsymbol{r}) \cdot \boldsymbol{x}(\boldsymbol{r}, 2h) d\boldsymbol{r} \\
&\quad - \int_{\partial \Omega_q} \int_0^{2h} \boldsymbol{q}(s, Z) \cdot \boldsymbol{x}(s, Z) dZ ds,
\end{aligned}$$

where ρ is the mass density of the plate material and the overhead dot means time derivative. The lateral surface of the plate, which is denoted as $\partial \Omega$, is composed of the position boundary $\partial \Omega_0$ and the traction boundary $\partial \Omega_q$. All the quantities are defined in the reference configuration.

In order to calculate the minimum of the potential energy functional under the constraint condition (25.3), we consider the following generalized potential energy functional

$$\Psi(\boldsymbol{x}, p; \boldsymbol{X}) = \int_{t_1}^{t_2} \left\{ \underline{\Phi} + \underline{V} - \underline{K} - \int_{\kappa} p(\boldsymbol{X}) R(\mathbb{F}) d\boldsymbol{X} \right\} dt,$$

where $p(\boldsymbol{X})$ plays the role of the Lagrangian multiplier. Next, the governing system of the current plate model will be derived by calculating the variations of Ψ with respect to the independent variables $\boldsymbol{x}$ and p.

First, from the Hamilton's principle and upon using the divergence theorem, we obtain the variation of Ψ with respect to $\boldsymbol{x}$

$$\frac{\delta\Psi}{\delta\boldsymbol{x}} = \int_{t_1}^{t_2}\Big\{\int_{\kappa}(-\mathrm{Div}\mathbb{S} - \boldsymbol{q}_b + \rho\ddot{\boldsymbol{x}})\cdot\delta\boldsymbol{x}d\boldsymbol{X} - \int_{\Omega}(\mathbb{S}^T\boldsymbol{k}|_{Z=0} + \boldsymbol{q}^-)\cdot\delta\boldsymbol{x}(\boldsymbol{r},0)d\boldsymbol{r}$$
$$+ \int_{\Omega}(\mathbb{S}^T\boldsymbol{k}|_{Z=2h} - \boldsymbol{q}^+)\cdot\delta\boldsymbol{x}(\boldsymbol{r},2h)d\boldsymbol{r} + \int_{\partial\Omega_0}\int_0^{2h}\mathbb{S}^T\boldsymbol{N}\cdot\delta\boldsymbol{x}(s,Z)dZds$$
$$+ \int_{\partial\Omega_q}\int_0^{2h}(\mathbb{S}^T\boldsymbol{N} - \boldsymbol{q})\cdot\delta\boldsymbol{x}(s,Z)dZds\Big\}dt, \tag{25.7}$$

where

$$\mathbb{S} = \frac{\partial\Phi}{\partial\mathbb{F}} - p\frac{\partial R}{\partial\mathbb{F}}, \tag{25.8}$$

is the nominal stress tensor of the incompressible material (Ogden, 1984)), $\boldsymbol{N}$ is the unit outward normal to the lateral surface, and $\dot{\boldsymbol{x}}\delta\boldsymbol{x}$ is assumed to vanish at both t_1 and t_2. Due to the arbitrariness of $\delta\boldsymbol{x}$ in (25.7), the equations of motion for any $t \in (t_1, t_2)$ together with boundary conditions are

$$\begin{array}{ll} \mathrm{Div}\mathbb{S} + \boldsymbol{q}_b = \rho\ddot{\boldsymbol{x}}, & \text{in } \Omega\times[0,2h], \\ \mathbb{S}^T\boldsymbol{k}|_{Z=0} = -\boldsymbol{q}^-, & \text{in } \Omega, \\ \mathbb{S}^T\boldsymbol{k}|_{Z=2h} = \boldsymbol{q}^+, & \text{in } \Omega, \\ \boldsymbol{x} = \boldsymbol{b}(s,Z), & \text{on } \partial\Omega_0\times[0,2h], \\ \mathbb{S}^T\boldsymbol{N} = \boldsymbol{q}(s,Z), & \text{on } \partial\Omega_q\times[0,2h], \end{array} \tag{25.9}$$

where $\boldsymbol{b}$ is the prescribed position on the boundary $\partial\Omega_0$, and we omit the argument t in all the above quantities. Next, we obtain the variation of Ψ with respect to p

$$\frac{\delta\Psi}{\delta p} = -\int_{t_1}^{t_2}\int_{\kappa} R(\mathbb{F})\,\delta p d\boldsymbol{X}dt. \tag{25.10}$$

We can obtain the constraint Eq.(25.3) from the above equation. Now we have formulated the 3D governing system Eq.(25.3) and Eq.(25.9), which contains two independent variables $\boldsymbol{x}$ and p.

25.3 The 2D Dynamic Plate Theory

In this section, we derive the 2D plate theory for incompressible materials from the previous 3D governing partial differential equations system, including the consistent dynamic plate equations, the boundary conditions as well as the associated weak formulation. It is a general approach to make approximations to eliminate the Z variable. Here we use the same consistency criterion proposed in Dai and Song (2014):

For all loadings that satisfy some smooth requirements, each term in the first variation of the energy functional should be of a required asymptotic order (say, $O(h^4)$) separately for the plate approximation.

Without loss of generality, it is assumed that all the spatial variables are scaled by the typical dimension of the in-plane surface, then $2h$ in fact represents the thickness ratio of the plate. The derivation follows the similar lines proposed in Song and Dai (2016); Wang et al (2016), but here we take into account the incompressibility constraint. We start from the series expansion of the independent variables with respect to Z.

25.3.1 Dynamic 2D Vector Plate Equation

Suppose that both $\boldsymbol{x}(\boldsymbol{X})$ and $p(\boldsymbol{X})$ are C^5 functions in their arguments, then we obtain the following series expansions:

$$\begin{aligned}\boldsymbol{x}(\boldsymbol{X}) &= \boldsymbol{x}^{(0)}(\boldsymbol{r}) + Z\boldsymbol{x}^{(1)}(\boldsymbol{r}) + \frac{1}{2}Z^2\boldsymbol{x}^{(2)}(\boldsymbol{r}) \\ &\quad + \frac{1}{6}Z^3\boldsymbol{x}^{(3)}(\boldsymbol{r}) + \frac{1}{24}Z^4\boldsymbol{x}^{(4)}(\boldsymbol{r}) + O(Z^5), \qquad (25.11)\\ p(\boldsymbol{X}) &= p^{(0)}(\boldsymbol{r}) + Zp^{(1)}(\boldsymbol{r}) + \frac{1}{2}Z^2p^{(2)}(\boldsymbol{r}) \\ &\quad + \frac{1}{6}Z^3p^{(3)}(\boldsymbol{r}) + \frac{1}{24}Z^4p^{(4)}(\boldsymbol{r}) + O(Z^5), \qquad (25.12)\end{aligned}$$

where $(\cdot)^{(n)} = \partial^n(\cdot)/\partial Z^n|_{Z=0}$ $(n = 1,\ldots,4)$. According to the expansion of $\boldsymbol{x}$, the deformation gradient tensor can also be expanded as

$$\mathbb{F}(\boldsymbol{X}) = \mathbb{F}^{(0)}(\boldsymbol{r}) + Z\mathbb{F}^{(1)}(\boldsymbol{r}) + \frac{1}{2}Z^2\mathbb{F}^{(2)}(\boldsymbol{r}) + \frac{1}{6}Z^3\mathbb{F}^{(3)}(\boldsymbol{r}) + O(Z^4). \qquad (25.13)$$

By substituting (25.11) into (25.2) and comparing with (25.13), we obtain the following relations

$$\mathbb{F}^{(n)} = \nabla\boldsymbol{x}^{(n)} + \boldsymbol{x}^{(n+1)} \otimes \boldsymbol{k}, \quad n = 0,1,2,3. \qquad (25.14)$$

An observation from (25.14) is that the dependence of $\mathbb{F}^{(n)}$ on $\boldsymbol{x}^{(n+1)}$ is linearly algebraic. We also suppose the strain energy $\Phi(\mathbb{F})$ is C^5 functions in their arguments, so the nominal stress tensor can be expanded as

$$\mathbb{S}(\boldsymbol{X}) = \mathbb{S}^{(0)}(\boldsymbol{r}) + Z\mathbb{S}^{(1)}(\boldsymbol{r}) + \frac{1}{2}Z^2\mathbb{S}^{(2)}(\boldsymbol{r}) + \frac{1}{6}Z^3\mathbb{S}^{(3)}(\boldsymbol{r}) + O(Z^4). \qquad (25.15)$$

Besides, from (25.8) and by using the chain rule, we obtain

$$
\begin{aligned}
\mathbb{S} = {} & \mathbb{A}^{(0)}(\mathbb{F}^{(0)}) + \mathscr{A}^{(1)}(\mathbb{F}^{(0)})[\mathbb{F}-\mathbb{F}^{(0)}] + \frac{1}{2}\mathscr{A}^{(2)}(\mathbb{F}^{(0)})[\mathbb{F}-\mathbb{F}^{(0)},\mathbb{F}-\mathbb{F}^{(0)}] \\
& + \frac{1}{6}\mathscr{A}^{(3)}(\mathbb{F}^{(0)})[\mathbb{F}-\mathbb{F}^{(0)},\mathbb{F}-\mathbb{F}^{(0)},\mathbb{F}-\mathbb{F}^{(0)}] + \cdots \\
& - \left\{ p^{(0)}(\boldsymbol{r}) + Z p^{(1)}(\boldsymbol{r}) + \frac{1}{2}Z^2 p^{(2)}(\boldsymbol{r}) + \frac{1}{6}Z^3 p^{(3)}(\boldsymbol{r}) + \cdots \right\} \\
& \times \left\{ \mathbb{R}^{(0)}(\mathbb{F}^{(0)}) + \mathscr{R}^{(1)}(\mathbb{F}^{(0)})[\mathbb{F}-\mathbb{F}^{(0)}] + \frac{1}{2}\mathscr{R}^{(2)}(\mathbb{F}^{(0)})[\mathbb{F}-\mathbb{F}^{(0)},\mathbb{F}-\mathbb{F}^{(0)}] \right. \\
& \left. + \frac{1}{6}\mathscr{R}^{(3)}(\mathbb{F}^{(0)})[\mathbb{F}-\mathbb{F}^{(0)},\mathbb{F}-\mathbb{F}^{(0)},\mathbb{F}-\mathbb{F}^{(0)}] + \cdots \right\},
\end{aligned}
\tag{25.16}
$$

where $\mathscr{A}^{(i)}$ $(i=1,2,3)$ are elastic moduli associated with Φ, the similar moduli $\mathscr{R}^{(i)}$ $(i=1,2,3)$ are defined by replacing Φ with the constraint R, and

$$
\mathbb{A}^{(0)}(\mathbb{F}^{(0)}) = \left.\frac{\partial \Phi}{\partial \mathbb{F}}\right|_{\mathbb{F}=\mathbb{F}^{(0)}},
$$
$$
\mathbb{R}^{(0)}(\mathbb{F}^{(0)}) = \left.\frac{\partial R}{\partial \mathbb{F}}\right|_{\mathbb{F}=\mathbb{F}^{(0)}} = \mathrm{Det}(\mathbb{F}^{(0)})(\mathbb{F}^{(0)})^{-1}.
$$

By substituting (25.13) into (25.16) and comparing with (25.15), we obtain the following expressions for $\mathbb{S}^{(n)}$

$$
\begin{aligned}
\mathbb{S}^{(0)}(\boldsymbol{r}) = {} & \mathbb{A}^{(0)} - p^{(0)}\mathbb{R}^{(0)}, \\
\mathbb{S}^{(1)}(\boldsymbol{r}) = {} & \underline{\mathscr{A}}[\mathbb{F}^{(1)}] - p^{(1)}\mathbb{R}^{(0)}, \\
\mathbb{S}^{(2)}(\boldsymbol{r}) = {} & \underline{\mathscr{A}}[\mathbb{F}^{(2)}] - p^{(2)}\mathbb{R}^{(0)} + \mathscr{A}^{(2)}[\mathbb{F}^{(1)},\mathbb{F}^{(1)}] - p^{(0)}\mathscr{R}^{(2)}[\mathbb{F}^{(1)},\mathbb{F}^{(1)}] - 2p^{(1)}\mathscr{R}^{(1)}[\mathbb{F}^{(1)}], \\
\mathbb{S}^{(3)}(\boldsymbol{r}) = {} & \underline{\mathscr{A}}[\mathbb{F}^{(3)}] - p^{(3)}\mathbb{R}^{(0)} + 3\mathscr{A}^{(2)}[\mathbb{F}^{(1)},\mathbb{F}^{(2)}] + \mathscr{A}^{(3)}[\mathbb{F}^{(1)},\mathbb{F}^{(1)},\mathbb{F}^{(1)}] \\
& - 3p^{(0)}\mathscr{R}^{(2)}[\mathbb{F}^{(1)},\mathbb{F}^{(2)}] - p^{(0)}\mathscr{R}^{(3)}[\mathbb{F}^{(1)},\mathbb{F}^{(1)},\mathbb{F}^{(1)}] - 3p^{(1)}\mathscr{R}^{(1)}[\mathbb{F}^{(2)}] \\
& - 3p^{(1)}\mathscr{R}^{(2)}[\mathbb{F}^{(1)},\mathbb{F}^{(1)}] - 3p^{(2)}\mathscr{R}^{(1)}[\mathbb{F}^{(1)}],
\end{aligned}
\tag{25.17}
$$

where the function of the new combined modulus

$$
\underline{\mathscr{A}} = \mathscr{A}^{(1)} - p^{(0)}\mathscr{R}^{(1)}
$$

in this incompressible case resembles $\mathscr{A}^{(1)}$ in the compressible case, and the argument $\mathbb{F}^{(0)}$ in $\mathbb{A}^{(0)}$, $\mathbb{R}^{(0)}$, $\mathscr{A}^{(i)}$ and $\mathscr{R}^{(i)}$ $(i=1,\ldots,3)$ is omitted for brevity. Due to the series expansions (25.11) and (25.12), we obtain totally 9 unknowns in the governing system (including five vectors $\boldsymbol{x}^{(n)}$ $(n=0,\ldots,4)$ and four scalars $p^{(n)}$ $(n=0,\ldots,3)$), which are necessary in formulating a closed system by some consistent truncations of the 3D system. In addition, some equations in (25.9) serves to eliminate most of the unknowns, leading to a single vector plate equation. From

(25.14) and (25.17), we can observe that $\mathbb{S}^{(i)}$ depend linearly on $\boldsymbol{x}^{(i+1)}$ $(i=1,2,3)$ as well, which plays a fundamental role in deriving the following recursion relations.

First, by substituting $(25.17)_1$ into the bottom traction condition $(25.9)_2$, we obtain

$$\begin{aligned}
\{\mathbb{S}^{(0)}\}^T\boldsymbol{k} &= \left\{\mathbb{A}^{(0)}(\mathbb{F}^{(0)}) - p^{(0)}\mathbb{R}^{(0)}(\mathbb{F}^{(0)})\right\}^T\boldsymbol{k} \\
&= \left\{\mathbb{A}^{(0)}(\nabla\boldsymbol{x}^{(0)}+\boldsymbol{x}^{(1)}\otimes\boldsymbol{k}) - p^{(0)}\mathbb{R}^{(0)}(\nabla\boldsymbol{x}^{(0)}+\boldsymbol{x}^{(1)}\otimes\boldsymbol{k})\right\}^T\boldsymbol{k} \\
&= -\boldsymbol{q}^-.
\end{aligned} \tag{25.18}$$

Equation (25.18) provides three algebraic equations for the unknowns $\boldsymbol{x}^{(1)}$ and $p^{(0)}$. In order to ease the sequel derivations, we define

$$\begin{aligned}
\boldsymbol{g}(\boldsymbol{x}^{(0)}) \triangleq \mathbb{R}^{(0)T}\boldsymbol{k} &= \mathrm{Det}(\mathbb{F}^{(0)})(\mathbb{F}^{(0)})^{-T}\boldsymbol{k} = \mathbb{F}^{(0)^*}\boldsymbol{k} = \mathbb{F}^{(0)^*}(\boldsymbol{E}_1\wedge\boldsymbol{E}_2) \\
&= (\mathbb{F}^{(0)}\boldsymbol{E}_1)\wedge(\mathbb{F}^{(0)}\boldsymbol{E}_2) = \boldsymbol{x}^{(0)}_{,1}\wedge\boldsymbol{x}^{(0)}_{,2},
\end{aligned}$$

where '$*$' represents the adjugate and '$\wedge$' means the cross product (Chadwick, 1999). So (25.18) reduces to

$$\mathbb{A}^{(0)T}(\mathbb{F}^{(0)})\boldsymbol{k} - p^{(0)}\boldsymbol{g}(\boldsymbol{x}^{(0)}) = -\boldsymbol{q}^-. \tag{25.19}$$

Next, vanishing of the coefficients of Z^n from $(25.9)_1$ yields that

$$\nabla\cdot\mathbb{S}^{(n)} + \mathbb{S}^{(n+1)T}\boldsymbol{k} + \boldsymbol{q}_b^{(n)} = \rho\ddot{\boldsymbol{x}}^{(n)}, \quad n=0,1,2. \tag{25.20}$$

Equation (25.20) provides three linear algebraic equations for the unknowns $\boldsymbol{x}^{(n+1)}$ and $p^{(n)}$ $(n=1,2,3)$.

Furthermore, by substituting the series expansion (25.13) into the constraint Eq. (25.3), we obtain

$$\begin{aligned}
&R(\mathbb{F}^{(0)}) + \mathbb{R}^{(0)} : \left\{Z\mathbb{F}^{(1)} + \frac{1}{2}Z^2\mathbb{F}^{(2)} + \frac{1}{6}Z^3\mathbb{F}^{(3)}\right\} \\
&\quad + \frac{1}{2}\left\{Z\mathbb{F}^{(1)} + \frac{1}{2}Z^2\mathbb{F}^{(2)}\right\} : \mathscr{R}^{(1)}\left[Z\mathbb{F}^{(1)} + \frac{1}{2}Z^2\mathbb{F}^{(2)}\right] \\
&\quad + \frac{1}{6}Z\mathbb{F}^{(1)} : \mathscr{R}^{(2)}[Z\mathbb{F}^{(1)}, Z\mathbb{F}^{(1)}] + O(Z^4) = 0.
\end{aligned}$$

The vanishing of the coefficients of Z^n $(n=0,1,2,3)$ in the above equation leads to

$$\begin{aligned}
&R(\mathbb{F}^{(0)}) = (\boldsymbol{x}^{(0)}_{,1} \wedge \boldsymbol{x}^{(0)}_{,2}) \cdot \boldsymbol{x}^{(1)} - 1 = \boldsymbol{g} \cdot \boldsymbol{x}^{(1)} - 1 = 0,\\
&\mathbb{R}^{(0)} : \mathbb{F}^{(1)} = \boldsymbol{g} \cdot \boldsymbol{x}^{(2)} + \mathbb{R}^{(0)} : \nabla \boldsymbol{x}^{(1)} = 0,\\
&\mathbb{R}^{(0)} : \mathbb{F}^{(2)} + \mathbb{F}^{(1)} : \mathscr{R}^{(1)}[\mathbb{F}^{(1)}] = 0,\\
&\mathbb{R}^{(0)} : \mathbb{F}^{(3)} + 3\mathbb{F}^{(2)} : \mathscr{R}^{(1)}[\mathbb{F}^{(1)}] + \mathbb{F}^{(1)} : \mathscr{R}^{(2)}[\mathbb{F}^{(1)}, \mathbb{F}^{(1)}] = 0,
\end{aligned} \tag{25.21}$$

which provide the additional (linear) equations for the unknowns $\boldsymbol{x}^{(n)}$ $(n = 1,2,3,4)$ and $p^{(n)}$ $(n = 1,2,3)$.

These equations can be used to derive the recursion relations for $\boldsymbol{x}^{(n+1)}$ and $p^{(n)}$ $(n = 1,2,3)$. For instance, substituting $(25.17)_2$ into (25.20) $(n = 0)$ furnishes

$$\mathbb{B}\boldsymbol{x}^{(2)} + \boldsymbol{f}^{(2)} - p^{(1)}\boldsymbol{g} = \rho \ddot{\boldsymbol{x}}^{(0)},$$

where the second-order (acoustic) tensor $\mathbb{B}$ and the vector $\boldsymbol{f}^{(2)}$ are defined as

$$\begin{aligned}
&\mathbb{B}\boldsymbol{x} = \{\underline{\mathscr{A}}[\boldsymbol{x} \otimes \boldsymbol{k}]\}^T \boldsymbol{k}, \quad \Rightarrow (\mathbb{B})|_{ij} = \underline{\mathscr{A}}_{3i3j},\\
&\boldsymbol{f}^{(2)} = \left\{\underline{\mathscr{A}}[\nabla \boldsymbol{x}^{(1)}]\right\}^T \boldsymbol{k} + \nabla \cdot \mathbb{S}^{(0)} + \boldsymbol{q}_b^{(0)}.
\end{aligned}$$

By the strong-ellipticity condition in (25.5), $\mathbb{B}$ is invertible and positive-definite and we obtain

$$\boldsymbol{x}^{(2)} = -\mathbb{B}^{-1}\boldsymbol{f}^{(2)} + p^{(1)}\mathbb{B}^{-1}\boldsymbol{g} + \mathbb{B}^{-1}\rho\ddot{\boldsymbol{x}}^{(0)}. \tag{25.22}$$

By substituting (25.22) into $(25.21)_2$, we easily derive the expression of $p^{(1)}$:

$$p^{(1)} = \frac{1}{g}\left(\boldsymbol{g} \cdot \mathbb{B}^{-1}\boldsymbol{f}^{(2)} - \boldsymbol{g} \cdot \mathbb{B}^{-1}\rho\ddot{\boldsymbol{x}}^{(0)} - \mathbb{R}^{(0)} : \nabla \boldsymbol{x}^{(1)}\right), \text{ with } g = \boldsymbol{g} \cdot \mathbb{B}^{-1}\boldsymbol{g}, \tag{25.23}$$

where $g > 0$ due to the positive-definiteness of $\mathbb{B}$. Similarly, we obtain the following expressions of $\boldsymbol{x}^{(3)}$ and $p^{(2)}$

$$\boldsymbol{x}^{(3)} = -\mathbb{B}^{-1}\boldsymbol{f}^{(3)} + p^{(2)}\mathbb{B}^{-1}\boldsymbol{g} + \mathbb{B}^{-1}\rho\ddot{\boldsymbol{x}}^{(1)}, \tag{25.24}$$

$$p^{(2)} = \frac{1}{g}\left(\boldsymbol{g} \cdot \mathbb{B}^{-1}\boldsymbol{f}^{(3)} - \boldsymbol{g} \cdot \mathbb{B}^{-1}\rho\ddot{\boldsymbol{x}}^{(1)} - \mathbb{R}^{(0)} : \nabla \boldsymbol{x}^{(2)} - \mathbb{F}^{(1)} : \mathscr{R}^{(1)}[\mathbb{F}^{(1)}]\right), \tag{25.25}$$

where

$$\begin{aligned}
\boldsymbol{f}^{(3)} = &\left\{\underline{\mathscr{A}}[\nabla \boldsymbol{x}^{(2)}]\right\}^T \boldsymbol{k} + \nabla \cdot \mathbb{S}^{(1)} + \boldsymbol{q}_b^{(1)}\\
&+ \left\{(\mathscr{A}^{(2)} - p^{(0)}\mathscr{R}^{(2)})[\mathbb{F}^{(1)}, \mathbb{F}^{(1)}] - 2p^{(1)}\mathscr{R}^{(1)}[\mathbb{F}^{(1)}]\right\}^T \boldsymbol{k}.
\end{aligned}$$

The recursion relations for $p^{(3)}$ and $\boldsymbol{x}^{(4)}$ are not needed in the following derivations, however the relation (25.20) with $n = 2$ as a whole will be used to eliminate them.

Finally, the top traction condition in $(25.9)_3$ states

$$\mathbb{S}^{(0)T}\boldsymbol{k} + 2h\mathbb{S}^{(1)T}\boldsymbol{k} + 2h^2\mathbb{S}^{(2)T}\boldsymbol{k} + \frac{4}{3}h^3\mathbb{S}^{(3)T}\boldsymbol{k} + O(h^4) = \boldsymbol{q}^+, \tag{25.26}$$

which contains all the unknowns $\boldsymbol{x}^{(n)}$ $(n=0,\ldots,4)$ and $p^{(n)}$ $(n=0,\ldots,3)$.

As for $\boldsymbol{x}^{(1)}$ and $p^{(0)}$, it can be seen from (25.19) and $(25.21)_1$ that they only depend on $\nabla\boldsymbol{x}^{(0)}$. However, (25.19) and $(25.21)_1$ are nonlinear algebraic equations of $\boldsymbol{x}^{(1)}$ and $p^{(0)}$, which can only be solved when the concrete form of the strain-energy density $\Phi(\mathbb{F})$ is given. The strong-ellipticity condition and the implicit function theorem imply that they can uniquely determined in terms of $\nabla\boldsymbol{x}^{(0)}$, as shown in Wang et al (2016).

Finally, by subtracting (25.18) from (25.26) and further using (25.20), we can obtain the dynamic 2D vector plate equation

$$\nabla\cdot\underline{\mathbb{S}}+\underline{\boldsymbol{q}}=\rho\ddot{\underline{\boldsymbol{x}}}, \tag{25.27}$$

where

$$\begin{aligned}\underline{\boldsymbol{q}}&=\frac{\boldsymbol{q}^{+}+\boldsymbol{q}^{-}}{2h}+\underline{\boldsymbol{q}_b},\\ \underline{\boldsymbol{x}}&=\frac{1}{2h}\int_0^{2h}\boldsymbol{x}dZ=\boldsymbol{x}^{(0)}+h\boldsymbol{x}^{(1)}+\frac{2}{3}h^2\boldsymbol{x}^{(2)}+O(h^3),\end{aligned}$$

and $\underline{\mathbb{S}}$ and $\underline{\boldsymbol{q}_b}$ are defined in the same way as $\underline{\boldsymbol{x}}$. The dynamic plate equation, after substituting the recursion relations, becomes a fourth-order differential equation for $\boldsymbol{x}^{(0)}$ with an error of $O(h^3)$.

25.3.2 *Edge Boundary Conditions*

Besides the vector plate equation, we shall also reduce the original 3D lateral surface conditions to appropriate boundary conditions for 2D equation. Since the plate equation is of fourth order in spatial derivatives, on either the position boundary $\partial\Omega_0$ or the traction boundary $\partial\Omega_q$ two conditions regarding $\boldsymbol{x}^{(0)}$ or its derivatives are required. Some conditions might involve time-derivative of $\boldsymbol{x}^{(0)}$, which is different from the boundary conditions proposed in Dai and Song (2014).

25.3.2.1 Case 1. Prescribed Position in the 3D Formulation

Suppose that on $\partial\Omega_0\times[0,2h]$ the position $\boldsymbol{b}$ is prescribed, then we adopt the following two boundary conditions

$$\begin{aligned}&\boldsymbol{x}^{(0)}=\boldsymbol{b}^{(0)}(s),\\ &\underline{\boldsymbol{x}}=\underline{\boldsymbol{b}}\quad\text{on }\partial\Omega_0\\ \Leftrightarrow\;&\boldsymbol{x}^{(1)}+\frac{2}{3}h\boldsymbol{x}^{(2)}+\frac{1}{3}h^2\boldsymbol{x}^{(3)}+O(h^3)=\frac{1}{h}(\underline{\boldsymbol{b}}-\boldsymbol{b}^{(0)}),\end{aligned} \tag{25.28}$$

where

$$b^{(0)} = b|_{Z=0}$$

and $\underline{b}$ represents the prescribed averaged position. The second condition contains up to the third-order spatial-derivatives and third-order time-derivatives of $x^{(0)}$ upon using the recursion relations.

25.3.2.2 Case 2. Prescribed traction in the 3D formulation

Suppose that on $\partial\Omega_q \times [0, 2h]$ the traction q is specified and is C^4 in Z, then we adopt the following two boundary conditions

$$\begin{aligned} &\mathbb{S}^{(0)T} N = q^{(0)}, \\ &\underline{\mathbb{S}}^T N = \frac{1}{2h}\int_0^{2h} \mathbb{S}^T N dZ = \frac{1}{2h}\int_0^{2h} q dZ = q_0 \quad \text{on } \partial\Omega_q \\ \Leftrightarrow &\left(\mathbb{S}^{(0)} + h\mathbb{S}^{(1)} + \frac{2}{3}h^2\mathbb{S}^{(2)} + \frac{1}{3}h^3\mathbb{S}^{(3)} + O(h^4)\right)^T N = q_0, \end{aligned} \tag{25.29}$$

where

$$q^{(0)} = q|_{Z=0}$$

and q_0 is the averaged traction along the thickness of the plate. As for the second condition, we may utilize the traction condition at an arbitrary Z, or alternatively, use the specified moment about the middle line

$$\begin{aligned} &\frac{1}{2h}\int_0^{2h} (Z-h)\mathbb{S}^T N dZ = \frac{1}{2h}\int_0^{2h} (Z-h) q dZ = m(s) \\ \Leftrightarrow &\frac{1}{3}\mathbb{S}^{(1)T} N + \frac{1}{3}h\mathbb{S}^{(2)T} N + \frac{1}{5}h^2\mathbb{S}^{(3)T} N + O(h^3) = \frac{m(s)}{h^2}, \end{aligned} \tag{25.30}$$

where $m(s)$ can be expressed in terms of $q^{(i)}$ in the same way as the left-hand side.

25.3.3 Examination of the Consistency

According to the criterion introduced in Sect. 25.3, in order to examine the consistency of the derived 2D dynamic vector plate equation system, we analyze the asymptotic orders of the terms in the variations (25.7) and (25.10).

For the first term on the right hand side (r.h.s.) of (25.7), we consider the series expansions of Div$\mathbb{S}$ in terms of Z. As the first three terms in the series expansion (25.20) have been used together with $(25.21)_{(2-4)}$ to obtain the recursion relations of $\boldsymbol{x}^{(i)}$ $(i=2,3,4)$ and $p^{(i)}$ $(i=1,2,3)$, we have $\mathrm{Div}\mathbb{S}=O(Z^3)$. Thus the first term in (25.7) is of $O(h^4)$. The second term on the r.h.s of (25.7) is exactly equal to zero because Eq. (25.19) together with $(25.21)_1$ have been used to derive the expressions of $\boldsymbol{x}^{(1)}$ and $p^{(0)}$. In Eq. (25.26), $\mathbb{S}^T\boldsymbol{k}$ on the top surface has been expanded to $O(h^4)$, which implies that the third term in (25.7) is $O(h^4)$.

Next, we examine the asymptotic order of the fourth term in (25.7) which involves the prescribed position boundary condition on $\partial\Omega_0\times[0,2h]$. We rewrite the integrand in the following form

$$\begin{aligned}\int_0^{2h}\mathbb{S}^T\boldsymbol{N}\cdot\delta\boldsymbol{x}dZ=&\int_0^{2h}\mathbb{S}^{(0)T}\boldsymbol{N}\cdot\left(\delta\boldsymbol{x}^{(0)}+Z\delta\boldsymbol{x}^{(1)}+\frac{1}{2}Z^2\delta\boldsymbol{x}^{(2)}\right)dZ\\&+\int_0^{2h}Z\mathbb{S}^{(1)T}\boldsymbol{N}\cdot\left(\delta\boldsymbol{x}^{(0)}+Z\delta\boldsymbol{x}^{(1)}\right)dZ\\&+\int_0^{2h}\frac{1}{2}Z^2\mathbb{S}^{(2)T}\boldsymbol{N}\cdot\delta\boldsymbol{x}^{(0)}dZ+O(h^4).\end{aligned}\tag{25.31}$$

From the position boundary conditions (25.28), it can be obtained that $\delta\boldsymbol{x}^{(0)}=0$, $\delta\boldsymbol{x}^{(1)}=O(h)$ and $\delta\boldsymbol{x}^{(1)}+2/3h\delta\boldsymbol{x}^{(2)}=O(h^2)$. By substituting these results into (25.31), it is easy to check that all the three terms on the r.h.s. of (25.31) are of $O(h^4)$. Thus, the fourth term in (25.7) also satisfies the consistency condition.

To examine the asymptotic order of the fifth term in (25.7), we denote

$$\tilde{\boldsymbol{q}}=\mathbb{S}^T\boldsymbol{N}-\boldsymbol{q}.$$

The coefficients of the series expansion of $\tilde{\boldsymbol{q}}$ are represented as $\tilde{\boldsymbol{q}}^{(i)}$ $(i=0,\dots,3)$. Then the integration of the fifth term in (25.7) can be rewritten as

$$\begin{aligned}\int_0^{2h}\tilde{\boldsymbol{q}}\cdot\delta\boldsymbol{x}dZ=&\int_0^{2h}\left(\tilde{\boldsymbol{q}}^{(0)}+Z\tilde{\boldsymbol{q}}^{(1)}+\frac{1}{2}Z^2\tilde{\boldsymbol{q}}^{(2)}\right)\cdot\delta\boldsymbol{x}^{(0)}dZ\\&+\int_0^{2h}Z\left(\tilde{\boldsymbol{q}}^{(0)}+Z\tilde{\boldsymbol{q}}^{(1)}\right)\cdot\delta\boldsymbol{x}^{(1)}dZ\\&+\int_0^{2h}\frac{1}{2}Z^2\tilde{\boldsymbol{q}}^{(0)}\cdot\delta\boldsymbol{x}^{(2)}dZ+O(h^4).\end{aligned}\tag{25.32}$$

From the traction boundary conditions (25.29), it can be obtained that $\tilde{\boldsymbol{q}}^{(0)}=0$, $\tilde{\boldsymbol{q}}^{(1)}=O(h)$ and $\tilde{\boldsymbol{q}}^{(1)}+2/3h\tilde{\boldsymbol{q}}^{(2)}=O(h^2)$. By substituting these results into (25.32), it is easy to check that all the three terms on the r.h.s. of (25.32) are of $O(h^4)$. Thus, the fifth term in (25.7) also satisfies the consistency condition.

For the variation (25.10), we have considered the series expansion of $R(\mathbb{F})$ in terms of Z, where the coefficients of Z^i $(i=0,1,2,3)$ are set to be zero to derive the recursion relations of $\boldsymbol{x}^{(i+1)}$ $(i=0,1,2,3)$. Thus the variation (25.10) also attains $O(h^4)$, which satisfies the consistency condition. To sum up, the 2D vector plate equation (25.27) and the edge boundary conditions (25.28) and (25.29) ensure each term in the variations to be of an asymptotic order of $O(h^4)$, which satisfies the consistency criterion.

25.4 The Associated Weak Formulations

In this section, we shall derive the associated weak formulations of the previous 2D vector plate system to be prepared for future numerical calculations. Furthermore, when the 3D edge conditions are unknown, suitable boundary conditions can be proposed for practical loading cases according to the weak form.

First, by multiplying both sides of the plate equation (25.27) with $\boldsymbol{\xi}=\delta\boldsymbol{x}^{(0)}$ and calculating the integrations over the region Ω, we obtain

$$\begin{aligned}&\int_{\Omega}(\nabla\cdot\underline{\mathbb{S}})\cdot\boldsymbol{\xi}\,d\boldsymbol{r}=-\int_{\Omega}\underline{\boldsymbol{q}}\cdot\boldsymbol{\xi}\,d\boldsymbol{r}+\int_{\Omega}\rho\underline{\ddot{\boldsymbol{x}}}\cdot\boldsymbol{\xi}\,d\boldsymbol{r}\\ \Rightarrow&\int_{\partial\Omega}(\underline{\mathbb{S}}^T\boldsymbol{N})\cdot\boldsymbol{\xi}\,ds-\int_{\Omega}\underline{\mathbb{S}}:\nabla\boldsymbol{\xi}\,d\boldsymbol{r}=-\int_{\Omega}\underline{\boldsymbol{q}}\cdot\boldsymbol{\xi}\,d\boldsymbol{r}+\int_{\Omega}\rho\underline{\ddot{\boldsymbol{x}}}\cdot\boldsymbol{\xi}\,d\boldsymbol{r}.\end{aligned}\tag{25.33}$$

Generally, the weak formulation associated with the fourth-order plate equation (25.27) should only contain up to the second-order derivative of $\boldsymbol{x}^{(0)}$. However, the weak formulation (25.33) involves the third-order derivatives, which originates from the terms $\mathbb{F}^{(2)}$ and $p^{(2)}$ in $\mathbb{S}^{(2)}$ and should be eliminated.

By substituting (25.23) and (25.25) into (25.14) and through some manipulations, we decompose $p^{(2)}$ and $\mathbb{F}^{(2)}$ into two parts

$$p^{(2)}=p_1^{(2)}+p_2^{(2)},\quad \mathbb{F}^{(2)}=\mathbb{F}_1^{(2)}+\mathbb{F}_2^{(2)},$$

where

$$\begin{aligned}p_1^{(2)}=&\frac{1}{g}\Big\{\boldsymbol{g}\cdot\mathbb{B}^{-1}\left\{(\mathscr{A}^{(2)}-p^{(0)}\mathscr{R}^{(2)})[\mathbb{F}^{(1)},\mathbb{F}^{(1)}]\right\}^T\boldsymbol{k}-2p^{(1)}\boldsymbol{g}\cdot\mathbb{E}^{-1}\left(\mathscr{R}^{(1)}[\mathbb{F}^{(1)}]\right)^T\boldsymbol{k}\\&+\boldsymbol{g}\cdot\mathbb{B}^{-1}\boldsymbol{q}_b^{(1)}-\boldsymbol{g}\cdot\mathbb{B}^{-1}\rho\ddot{\boldsymbol{x}}^{(1)}-\mathbb{F}^{(1)}:\mathscr{R}^{(1)}[\mathbb{F}^{(1)}]\Big\},\\ p_2^{(2)}=&\frac{1}{g}\Big\{\boldsymbol{g}\cdot\mathbb{B}^{-1}\left(\nabla\cdot\mathbb{S}^{(1)}\right)+\boldsymbol{g}\cdot\mathbb{E}^{-1}\left\{\underline{\mathscr{A}}[\nabla\boldsymbol{x}^{(2)}]\right\}^T\boldsymbol{k}-\mathbb{R}^{(0)}:\nabla\boldsymbol{x}^{(2)}\Big\},\end{aligned}$$

$$\begin{aligned}\mathbb{F}_1^{(2)} =& p_1^{(2)}\mathbb{B}^{-1}\boldsymbol{g}\otimes\boldsymbol{k}-\mathbb{B}^{-1}\left\{(\mathscr{A}^{(2)}-p^{(0)}\mathscr{R}^{(2)})[\mathbb{F}^{(1)},\mathbb{F}^{(1)}]\right\}^T\boldsymbol{k}\otimes\boldsymbol{k}\\ &+2p^{(1)}\mathbb{B}^{-1}\left(\mathscr{R}^{(1)}[\mathbb{F}^{(1)}]\right)^T\boldsymbol{k}\otimes\boldsymbol{k}-\mathbb{B}^{-1}\boldsymbol{q}_b^{(1)}\otimes\boldsymbol{k}+\mathbb{B}^{-1}\rho\ddot{\boldsymbol{x}}^{(1)}\otimes\boldsymbol{k},\\ \mathbb{F}_2^{(2)} =& \nabla\boldsymbol{x}^{(2)}-\mathbb{B}^{-1}\left\{\underline{\mathscr{A}}[\nabla\boldsymbol{x}^{(2)}]\right\}^T\boldsymbol{k}\otimes\boldsymbol{k}-\mathbb{B}^{-1}(\nabla\cdot\mathbb{S}^{(1)})\otimes\boldsymbol{k}+p_2^{(2)}\mathbb{B}^{-1}\boldsymbol{g}\otimes\boldsymbol{k}.\end{aligned}$$

It can be found that only $p_2^{(2)}$ and $\mathbb{F}_2^{(2)}$ involve the third-order derivative of $\boldsymbol{x}^{(0)}$. Correspondingly, we consider the following decomposition

$$\underline{\mathbb{S}}:\nabla\boldsymbol{\xi}=\mathscr{W}_1(\nabla\nabla\boldsymbol{x}^{(0)},\nabla\boldsymbol{\xi})+\mathscr{W}_2(\nabla\nabla\nabla\boldsymbol{x}^{(0)},\nabla\boldsymbol{\xi}),\tag{25.34}$$

where

$$\begin{aligned}\mathscr{W}_1 =& (\mathbb{S}^{(0)}+h\mathbb{S}^{(1)}):\nabla\boldsymbol{\xi}+\frac{2}{3}h^2\left\{\underline{\mathscr{A}}[\mathbb{F}_1^{(2)}]-p_1^{(2)}\mathbb{R}^{(0)}\right\}:\nabla\boldsymbol{\xi}\\ &+\frac{2}{3}h^2\left\{(\mathscr{A}^{(2)}-p^{(0)}\mathscr{R}^{(2)})[\mathbb{F}^{(1)},\mathbb{F}^{(1)}]-2p^{(1)}\mathscr{R}^{(1)}[\mathbb{F}^{(1)}]\right\}:\nabla\boldsymbol{\xi},\\ \mathscr{W}_2 =& \frac{2}{3}h^2\left\{\underline{\mathscr{A}}[\mathbb{F}_2^{(2)}]-p_2^{(2)}\mathbb{R}^{(0)}\right\}:\nabla\boldsymbol{\xi}.\end{aligned}\tag{25.35}$$

In order to eliminate the third-order derivative terms of $\boldsymbol{x}^{(0)}$, we then substitute the expressions of $\mathbb{F}_2^{(2)}$ and $p_2^{(2)}$ into (25.35). Further manipulations yield the following result

$$\mathscr{W}_2=\frac{2}{3}h^2\left\{\mathscr{S}_0:\nabla\boldsymbol{x}^{(2)}+\boldsymbol{\eta}\cdot(\nabla\cdot\mathbb{S}^{(1)})\right\},$$

and

$$\begin{aligned}\mathscr{S}_0 &= \underline{\mathscr{A}}[\nabla\boldsymbol{\xi}+\boldsymbol{\eta}\otimes\boldsymbol{k}]-\zeta\mathbb{R}^{(0)},\\ \boldsymbol{\eta} &= -\mathbb{B}^{-1}\{\underline{\mathscr{A}}[\nabla\boldsymbol{\xi}]\}^T\boldsymbol{k}+\zeta\mathbb{B}^{-1}\boldsymbol{g},\\ \zeta &= \frac{1}{g}\left(\underline{\mathscr{A}}[\mathbb{B}^{-1}\boldsymbol{g}\otimes\boldsymbol{k}]-\mathbb{R}^{(0)}\right):\nabla\boldsymbol{\xi}\triangleq\mathbb{P}:\nabla\boldsymbol{\xi}.\end{aligned}\tag{25.36}$$

In fact, it can be proved that

$$\delta p^{(0)}=\zeta,\quad \delta\boldsymbol{x}^{(1)}=\boldsymbol{\eta},\quad \delta\mathbb{S}^{(0)}=\mathscr{S}_0.$$

Then, integration by parts leads to

$$\begin{aligned}\int_\Omega\mathscr{W}_2 d\boldsymbol{r} &= \frac{2}{3}h^2\int_{\partial\Omega}(\mathscr{S}_0^T\boldsymbol{N})\cdot\boldsymbol{x}^{(2)}+(\mathbb{S}^{(1)T}\boldsymbol{N})\cdot\boldsymbol{\eta}ds+\int_\Omega\mathscr{W}_3 d\boldsymbol{r},\\ \mathscr{W}_3 &= -\frac{2}{3}h^2\left\{(\nabla\cdot\mathscr{S}_0)\cdot\boldsymbol{x}^{(2)}+\mathbb{S}^{(1)}:\nabla\boldsymbol{\eta}\right\}.\end{aligned}\tag{25.37}$$

It can be seen from (25.37) that the third-order derivatives of $\boldsymbol{x}^{(0)}$ have been eliminated. Combining the results (25.33), (25.34) and (25.37), the following 2D weak formulation can be derived

$$\int_{\Omega} (\mathscr{W}_1 + \mathscr{W}_3 + (\rho\ddot{\underline{x}} - \underline{q})\cdot\boldsymbol{\xi})\,d\boldsymbol{r} = \int_{\partial\Omega} \left(\underline{\mathbb{S}}^T\boldsymbol{N}\cdot\boldsymbol{\xi} - \frac{2}{3}h^2\mathbb{S}^{(1)T}\boldsymbol{N}\cdot\boldsymbol{\eta} - \frac{2}{3}h^2\mathscr{S}_0^T\boldsymbol{N}\cdot\boldsymbol{x}^{(2)}\right) ds. \tag{25.38}$$

In the following, by considering the boundary conditions, the 2D weak formulation (25.38) can be further simplified. We adapt it to two distinct types of boundary conditions, i.e., the previous cases in Sect. 25.3.2 and other practical cases when 3D edge conditions are unclear.

25.4.0.1 Case 1. Edge position and traction in the 3D formulation are known

From (25.28), it is easy to deduce that $\boldsymbol{\xi} = \delta\boldsymbol{x}^{(0)} = \boldsymbol{0}$ and $\boldsymbol{\eta} = \delta\boldsymbol{x}^{(1)} = O(h)$ on $\partial\Omega_0$, which together with (25.36) further implies that $\nabla\boldsymbol{\xi} = O(h)$ and $\mathscr{S}_0 = O(h)$. Consequently, the boundary integral on $\partial\Omega_0$ in (25.38) is of $O(h^3)$ and can be neglected.

While on $\partial\Omega_q$, it follows from conditions (25.29), (25.30) that

$$\mathscr{S}_0^T\boldsymbol{N} = \delta[\mathbb{S}^{(0)T}\boldsymbol{N}] = O(h^2).$$

Thus, the third term in the boundary integral can be neglected. Besides, replacing $\mathbb{S}^{(1)T}\boldsymbol{N}$ by the condition in (25.30) only causes a higher-order correction. Then, the 2D weak formulation (25.38) reduces to

$$\int_{\Omega} (\mathscr{W}_1 + \mathscr{W}_3 + (\rho\ddot{\underline{x}} - \underline{q})\cdot\boldsymbol{\xi})\,d\boldsymbol{r} = \int_{\partial\Omega_q} \left(\underline{\mathbb{S}}^T\boldsymbol{N}\cdot\boldsymbol{\xi} - \frac{2}{3}h^2\mathbb{S}^{(1)T}\boldsymbol{N}\cdot\boldsymbol{\eta}\right) ds$$
$$= \int_{\partial\Omega_q} \boldsymbol{q}_0\cdot\boldsymbol{\xi} - 2\boldsymbol{m}\cdot\boldsymbol{\eta}\,ds.$$

25.4.0.2 Case 2. Edge position and traction in the 3D formulation are unknown

In many practical situations, where the edge traction distribution (e.g. a pinned edge) or displacement distribution (e.g. a clamped edge) is unknown, we should propose the so-called natural boundary conditions according to the weak formulation. For this purpose, we shall recast the boundary integral in (25.38) in terms of $\boldsymbol{\xi}$ and its normal derivative $\boldsymbol{\xi}_{,N}$.

For convenience, considering the last two terms in the boundary integral of (25.38), we introduce a third-order tensor $\mathscr{M}$ through

$$
\begin{aligned}
& -\frac{2}{3}h^2\left(\mathbb{S}^{(1)T}\boldsymbol{N}\cdot\boldsymbol{\eta}+\mathscr{S}_0^T\boldsymbol{N}\cdot\boldsymbol{x}^{(2)}\right)\\
& =\frac{2}{3}h^2\left\{\underline{\mathscr{A}}\left[\boldsymbol{t}\otimes\boldsymbol{k}-\boldsymbol{x}^{(2)}\otimes\boldsymbol{N}\right]+\left(\boldsymbol{x}^{(2)}\cdot\mathbb{R}^{(0)T}\boldsymbol{N}-\boldsymbol{t}\cdot\boldsymbol{g}\right)\mathbb{P}\right\}:\nabla\boldsymbol{\xi}\triangleq\left(\mathscr{M}[\boldsymbol{N}]\right)^T:\nabla\boldsymbol{\xi},
\end{aligned}
\tag{25.39}
$$

where

$$
\boldsymbol{t}=\mathbb{B}^{-1}\left(\mathbb{S}^{(1)T}\boldsymbol{N}+\mathbb{B}_1\boldsymbol{x}^{(2)}\right),\quad (\mathbb{B}_1)_{ij}=\underline{\mathscr{A}}_{3i\alpha j}N_\alpha.
$$

Furthermore, we introduce the decomposition

$$
\nabla\boldsymbol{\xi}=\boldsymbol{\xi}_{,s}\otimes\boldsymbol{T}+\boldsymbol{\xi}_{,N}\otimes\boldsymbol{N},
\tag{25.40}
$$

where $\boldsymbol{T}$ is the unit tangential vector, and $\boldsymbol{\xi}_{,s}$ is tangential derivative on $\partial\Omega$. Substituting (25.39) and (25.40) into the boundary integral in (25.38) and a simple integration by parts leads to

$$
\int_{\Omega}\left(\mathscr{W}_1+\mathscr{W}_3+(\rho\ddot{\underline{x}}-\underline{q})\cdot\boldsymbol{\xi}\right)d\boldsymbol{r}=\int_{\partial\Omega}\left\{\underline{\mathbb{S}}^T\boldsymbol{N}-\left(\mathscr{M}[\boldsymbol{N}]\boldsymbol{T}\right)_{,s}\right\}\cdot\boldsymbol{\xi}+\left\{\mathscr{M}[\boldsymbol{N}]\boldsymbol{N}\right\}\cdot\boldsymbol{\xi}_{,N}ds.
\tag{25.41}
$$

If we regard

$$
\mathscr{W}:=\mathscr{W}_1+\mathscr{W}_3
$$

as the variation of the plate stress work due to the virtual displacement $\boldsymbol{\xi}$, the weak formulation (25.41) can be rewritten as

$$
\int_{\Omega}\left(\mathscr{W}+(\rho\ddot{\underline{x}}-\underline{q})\cdot\boldsymbol{\xi}\right)d\boldsymbol{r}=\int_{\partial\Omega}\hat{\boldsymbol{q}}(s)\cdot\boldsymbol{\xi}+\hat{\boldsymbol{m}}(s)\cdot\boldsymbol{\xi}_{,N}ds,
\tag{25.42}
$$

where $\hat{\boldsymbol{q}}$ and $\hat{\boldsymbol{m}}$ are respectively the applied generalized traction and bending moment at the edge. Based on (25.41) or (25.42), boundary conditions can be suitably proposed for various practical cases (e.g., clamped, pinned, simply-supported).

25.5 Conclusions

In this chapter, we propose a consistent dynamic finite-strain plate theory for incompressible hyperelastic materials with no special restrictions on loadings or the order of deformations. The developed plate theory follows similar lines as the previous work, except that the dynamic terms and an additional constraint equation are involved into the recursion relations and the final dynamic system. It is consistent with the 3D weak formulation since each term in the variations of the generalized potential energy functional attains the asymptotic order of $O(h^4)$. The current plate theory can recover the 3D displacement and stress fields. For the convenience of numerical calculations, we also derive the weak formulation of the 2D vector plate equation together with the position or traction edge boundary conditions. In comparison with other plate theories, the present one takes into account dynamic, finite-strain, bending and stretching effects together with incompressibility constraint, so it may provide a general framework for studying mechanical behaviour of soft-material plates under various loading conditions.

References

Altenbach J, Altenbach H, Eremeyev VA (2010) On generalized Cosserat-type theories of plates and shells: a short review and bibliography. Archive of Applied Mechanics 80(1):73–92

Batra RC (2007) Higher-order shear and normal deformable theory for functionally graded incompressible linear elastic plates. Thin-Walled Structures 45(12):974–982

Chadwick P (1999) Continuum Mechanics: Concise Theory and Problems, 2nd edn. Dover Publications, Inc., Mineola, New York

Ciarlet PG (1980) A justification of the von Kármán equations. Archive for Rational Mechanics and Analysis 73(4):349–389

Conti S, Dolzmann G (2008) Γ-convergence for incompressible elastic plates. Calculus of Variations and Partial Differential Equations 34(4):531–551

Dai HH, Song Z (2014) On a consistent finite-strain plate theory based on three-dimensional energy principle. Proceedings of the Royal Society of London A: Mathematical, Physical and Engineering Sciences 470(2171):20140,494

Friesecke G, James RD, Müller S (2002) A theorem on geometric rigidity and the derivation of nonlinear plate theory from three-dimensional elasticity. Communications on Pure and Applied Mathematics 55(11):1461–1506

Kienzler R (2002) On consistent plate theories. Archive of Applied Mechanics 72(4):229–247

Kirchhoff G (1850) Über das Gleichgewicht und die Bewegung einer elastischen Scheibe. Journal für die reine und angewandte Mathematik 40:51–88

Love AEH (1888) The small free vibrations and deformation of a thin elastic shell. Philosophical Transactions of the Royal Society of London A 179:491–546

Meroueh KA (1986) On a formulation of a nonlinear theory of plates and shells with applications. Computers & Structures 24(5):691–705

Millet O, Hamdouni A, Cimetière A (2001) A classification of thin plate models by asymptotic expansion of non-linear three-dimensional equilibrium equations. International Journal of Non-Linear Mechanics 36(1):165–186

Mindlin RD (1951) Influence of rotary inertia and shear on flexural motions of isotropic, elastic plates. Trans ASME J of Appl Mech 18(1):31–38

Naghdi PM (1972) Theory of shells and plates. In: Truesdell C (ed) Handbuch der Physik, Springer, Berlin, vol VIa/2, pp 425–640

Ogden R (1984) Non-Linear Elastic Deformations. Ellis Horwood, New York

Reddy JN (2007) Theory and Analysis of Elastic Plates and Shells. CRC Press, Taylor and Francis

Song Z, Dai HH (2016) On a consistent dynamic finite-strain plate theory and its linearization. Journal of Elasticity 125(2):149–183

Steigmann DJ (2007) Thin-plate theory for large elastic deformations. International Journal of Non-Linear Mechanics 42(2):233–240

Steigmann DJ (2013) Koiter's shell theory from the perspective of three-dimensional nonlinear elasticity. Journal of Elasticity 111(1):91–107

Timoshenko S, Woinowsky-Krieger S (1959) Theory of Plates and Shells. McGraw-Hill, New York

Trabelsi K (2005) Incompressible nonlinearly elastic thin membranes. Comptes Rendus Mathematique 340(1):75–80 von Kármán T (1910) Festigkeitsprobleme im Maschinenbau. In: Encyklopädie der mathematischen Wissenschaften, Teubner, vol 4-4/ Heft 3, pp 311–385

Wang J, Song Z, Dai HH (2016) On a consistent finite-strain plate theory for incompressible hyperelastic materials. International Journal of Solids and Structures 78-79:101–109

Chapter 26
A One-Dimensional Problem of Nonlinear Thermo-Electroelasticity with Thermal Relaxation

Wael Mahmoud, Moustafa S. Abou-Dina, Amr R. El Dhaba, Ahmed F. Ghaleb, and Enaam K. Rawy

Abstract We investigate a nonlinear, one-dimensional problem of thermo-electroelasticity with thermal relaxation and in quasi-electrostatics. The system of basic equations is a restriction to one spatial dimension of that proposed earlier in Abou-Dina et al (2017). This model is based on the introduction of the heat flow vector as an additional state variable, thus leading to a Cattaneo-type evolution equation. It includes several non-linear couplings and may be useful in studying problems of elastic dielectrics at low temperatures, as well as in problems of high-temperature heat conduction in dielectric solids subjected to strong high-frequency laser beams. For the present investigation, however, only a few nonlinearities have been retained in the equations for conciseness. A numerical solution is presented for the system of nonlinear equations using an iterative, quasilinearization scheme by finite differences. The numerical results are discussed.

26.1 Introduction

An extensive and rigorous presentation of the general theory of continuum mechanics of electromagnetic media with numerous applications may be found in Maugin and Eringen (1977); Maugin (1988). A state-of-the-art overview of smart materials, including theoretical and computational aspects, is given in Ogden and Steigmann (2011). The theories of electroelasticity and electro-thermoelasticity were treated by numerous authors, among whom we cite Tiersten (1971); Maugin (1985); Dorfmann

Wael Mahmoud · Moustafa S. Abou-Dina · Ahmed F. Ghaleb · Enaam K. Rawy
Department of Mathematics, Faculty of Science, Cairo University, Giza 12613, Egypt
e-mail: mwael@sci.cu.edu.eg, moustafa_aboudina@hotmail.com, afghaleb@sci.cu.edu.eg, enaamkhalifaq@yahoo.com

Amr R. El Dhaba
Department of Mathematics, Faculty of Science, Damanhour University, Egypt
e-mail: amrramadaneg@gmail.com

© Springer International Publishing AG, part of Springer Nature 2018

H. Altenbach et al. (eds.), *Generalized Models and Non-classical Approaches in Complex Materials 1*, Advanced Structured Materials 89,
https://doi.org/10.1007/978-3-319-72440-9_26

and Ogden (2006); Maugin et al (1992). A numerical treatment of nonlinear thermo-electroelasticity for smart materials by finite elements is given by Mehnert et al (2017). The general models of electro-thermoelasticity involve rather complicated interactions between the mechanical, thermal and electric fields. Nonlinearity is relevant in most cases for technological applications. As Maugin (1985) notes in his preface: "On certain occasions, one wants to benefit from the nonlinearities. On other occasions, one wants to avoid them". In any case, for a successful treatment of the subject, the laws of thermodynamics must be respected.

The theory of extended thermoelasticity mainly addresses the problem of finiteness of the speed of propagation of the heat disturbances, or second sound. This leads to hyperbolic systems of equations, instead of the classical hyperbolic-parabolic systems of thermoelasticity. Many approaches to extended thermodynamics exist in the literature. They all agree on a central requirement: The need for an enrichment of the basic set of thermodynamical variables describing the system. Most of them rely on rigorous thermodynamics, meaning that the Clausius-Duhem inequality is satisfied. One of the existing approaches to extended thermodynamics relies on the assignment of a new definition to entropy outside thermodynamical equilibrium by the addition of a dissipative component. As noted by Maugin (2015, footnote on p. 81), this may lead to a hyperbolic equation for the propagation of heat. Along this trend one cites the pioneering work by Coleman et al (1982); Coleman and Newman (1988) who proposed a theory for rigid conductors which is compatible with thermodynamics only if the specific internal energy has a dissipative part exhibiting a quadratic dependence on the heat flux. The same was adopted by Ghaleb (1986) and others. Kuang (2014, cf. p. 23) adds a dissipative part to entropy, calling it inertial entropy, and relates it to the time rate of change of temperature. Adding such a time rate to the basic set of thermodynamical variables raised some concern, but was shown to be consistent with the basic principles of thermodynamics in earlier work by Coleman et al (1982). The laws of thermodynamics have been re-visited in certain cases (cf. Green and Naghdi, 1991; Barletta and Zanchini, 1997) in order to find a solution for the paradox of infinite speed of propagation of the thermal disturbances. The limits of applicability of Fourier law for heat conduction is questioned nowadays in many relevant practical problems, for example those related to laser treatment of surfaces, under high power ultra-short energy pulses, when extreme rapidly changing temperature fields are formed (cf. Zudin, 2017).

The theory of electro-thermoelasticity in generalized thermodynamics has witnessed a growing interest due to experimental work by Rybalko and others (Rybalko et al, 2007; Pashitskii and Ryabchenko, 2007; Pashitskii et al, 2009, cf.), in which experimental evidence was given that the propagation of heat waves at low temperature is accompanied by an electrical activity. A multitude of investigations have appeared since. Montanaro (2011) develops a model of nonlinear thermoelasticity in extended thermodynamics for electrically polarizable, heat conducting elastic continua. Ghaleb (2014) presents a fully nonlinear model for electrically polarizable, heat conducting elastic continuous media in the quasi-electrostatic approximation. Kuang (2014) considers wave propagation in pyroelectrics and other materials with complex structure in extended thermodynamics within the linearized theory. Montanaro (2015); Giorgi and Montanaro (2016) treat the case of electrical continuous media within Green and Naghdi thermoelasticity theory. Abou-Dina et al (2017) present a model of nonlinear thermo-electroelasticity within the frame of extended thermodynamics and in the quasi-electrostatic regime, including several couplings between the mechanical, thermal and electric fields. In particular, the free energy of the system has a quadratic dependence on the heat flow vector, with a multiplicative coefficient which does not depend on temperature. In this case, the entropy retains its classical form, and the objection raised by Barletta and Zanchini (1997) that Coleman's entropy does not satisfy the highest-entropy principle does not apply here. The model proposed in Abou-Dina et al (2017) involves several nonlinearities, and is thus adequate for the description of dielectric materials, as the nonlinear effects are often relevant in technological applications involving these materials, and the control of such effects is of primordial importance.

Here we investigate a one-dimensional nonlinear problem of thermo-electro-elasticity in extended thermodynamics and in the quasi-electrostatic regime. Thus, the magnetic field effects are discarded from the outset. The model is a restriction to one spatial dimension of the one recently presented by some of the authors Abou-Dina et al (2017) Solutions to nonlinear problems of thermoelasticity within the frame of extended thermodynamics are few in the literature. One notices here the numerical treatments by Pulvirenti et al (1998) for materials with temperature dependent heat conduction properties, and by Yen and Wu (2003) in which the authors consider a hyperbolic heat problem with nonlinear boundary condition. To the authors knowledge, solutions to problems of nonlinear thermo-electroelasticity are almost inexistant. A numerical treatment of the nonlinear equations by finite differences is proposed for the half-space. This is an iterative, quasi-linearization scheme. For this reason, a brief discussion of the associated linear system is given. It puts in evidence the existence of two speeds of wave propagation as expected, one for the usual coupled thermoelastic wave, and the other for second sound. Only a few nonlinearities are kept for conciseness, but the model may be extended further without difficulty. The obtained results are represented in two-, and three-dimensional plots. They illustrate the main features of the solution, more precisely the existence of two velocities of propagation of disturbances, and the effect of an important constant expressing the quadratic dependence of the free energy on heat flux. Future work is under progress to study nonlinear wave propagation in two dimensions.

26.2 The Nonlinear Equations

The following equations are a restriction to one spatial dimension of a more general system of equations of electro-thermoelasticity based on the introduction of the heat flow vector as an additional state variable in the free energy density of the medium and introduced in Ghaleb (2014). The equations are in dimensionless form, the velocity of coupled thermoelastic waves being taken as unity:

$$-\epsilon\phi_{xx}+\gamma' u_{xx}+\beta^{\star}\theta_x=0, \tag{26.1}$$

$$u_{tt}-u_{xx}=-\beta_1\theta_x+f(x,t), \tag{26.2}$$

$$(\theta+au_x)_t=-q_x+\frac{1}{\theta_0}q\,\theta_x+Nq^2+r(x,t), \tag{26.3}$$

$$(\theta_0+\theta)\,q_t=-q-\frac{1}{N}\theta_x. \tag{26.4}$$

Here, u denotes the mechanical displacement component, θ is the absolute temperature as measured from a reference temperature θ_0, q is the heat flow vector and ϕ is the electric potential. These four relations represent respectively:

i the equation of electrostatics expressing the vanishing of the divergence of the electric induction, in which only two linear coupling terms have been retained;
ii the equation of motion with only a linear thermoelastic coupling term;
iii the equation of heat conduction transformed using the constitutive equation for entropy, and
iv Vernotte-Cattaneo law which replaces the usual Fourier law for heat conduction.

The coefficients appearing in the equations denote material constants having obvious meaning. In particular, ϵ represents the dielectric constant, $\beta^{\star}$ is a thermoelectric coupling constant, β_1 is the thermoelastic coefficient, N is an important parameter illustrating the quadratic dependence of the free energy on the heat flow vector, related to the thermal relaxation time. Functions f and r are, respectively, the volume force and the heat supply. In the subsequent calculations, both of them will be taken equal to zero. It may be noticed that the classical Fourier law for heat conduction can be obtained from the Vernotte-Cattaneo law (26.4) in the limit of vanishing thermal relaxation time as explained in Abou-Dina et al (2017). In the above equations, only few nonlinear terms have been retained for conciseness, as compared to the more ample equations proposed in Abou-Dina et al (2017). Among others, terms expressing the dependence of material parameters on strain, and the quadratic electric terms have been neglected.

In what follows we consider a one-dimensional problem of thermo-electro-elasticity for the half-space $0 \le x < \infty$ based on the above governing equations. The present formulation necessitates initial and boundary conditions for the new variable of state q, to be considered side by side with the boundary condition for temperature. This is in distinction from other models of generalized thermoelasticity in which time derivatives are imbedded in the basic equations.

26.3 The Associated System of Linear Equations

Investigation of the associated linear system is necessary, as the adopted method of solution to deal with the nonlinear equations involves an iterative scheme with quasi-linearization. The system of linear equations may be written as a system of first-order partial differential equations in six unknown functions $\{u,\theta,q,E,P,R\}$, where E is the electric field component given by

$$E = -\frac{\partial \phi}{\partial x},$$

while P, R are defined below. One has

$$\zeta \frac{\partial E}{\partial t} + \gamma' \frac{\partial u}{\partial x} = -\epsilon E - \beta^{\star}\theta, \tag{26.5}$$

$$\frac{\partial P}{\partial t} - \frac{\partial P}{\partial x} + \beta_1 \frac{\partial \theta}{\partial x} = f(x,t), \tag{26.6}$$

$$\frac{\partial \theta}{\partial t} + \frac{\partial q}{\partial x} + a \frac{\partial R}{\partial t} = r(x,t), \tag{26.7}$$

$$N\theta_0 \frac{\partial q}{\partial t} + \frac{\partial \theta}{\partial x} = -Nq, \tag{26.8}$$

$$\frac{\partial u}{\partial t} + \frac{\partial u}{\partial x} = P, \tag{26.9}$$

$$\eta \frac{\partial R}{\partial t} + \frac{\partial u}{\partial x} = R. \tag{26.10}$$

The two unknown functions P and R are defined from the last two equations of the above system. The parameters ζ, η are two positive small parameters introduced artificially for convenience. Subsequently, they will be made to tend to zero during the computations. The other parameters appearing in the equations are dimensionless quantities involving the physical parameters of the medium. In matrix form, this system reads:

$$T \frac{\partial Z}{\partial t} + W \frac{\partial Z}{\partial x} = F, \tag{26.11}$$

with

$$T = \begin{pmatrix} 1 & 0 & 0 & 0 & 0 & 0 \\ 0 & 1 & 0 & 0 & 0 & a \\ 0 & 0 & N\theta_0 & 0 & 0 & 0 \\ 0 & 0 & 0 & \zeta & 0 & 0 \\ 0 & 0 & 0 & 0 & 1 & 0 \\ 0 & 0 & 0 & 0 & 0 & \eta \end{pmatrix}, \quad W = \begin{pmatrix} 1 & 0 & 0 & 0 & 0 & 0 \\ 0 & 0 & 1 & 0 & 0 & 0 \\ 0 & 1 & 0 & 0 & 0 & 0 \\ \gamma' & 0 & 0 & 0 & 0 & 0 \\ 0 & \beta_1 & 0 & 0 & -1 & 0 \\ 1 & 0 & 0 & 0 & 0 & 0 \end{pmatrix},$$

$$F = \begin{pmatrix} -\epsilon E - \beta^{\star}\theta \\ f \\ r \\ -Nq \\ P \\ R \end{pmatrix}, \quad Z = \begin{pmatrix} u \\ \theta \\ q \\ E \\ P \\ R \end{pmatrix}.$$

Matrix T is upper triangular. It is invertible with inverse

$$T^{-1} = \begin{pmatrix} 1 & 0 & 0 & 0 & 0 & 0 \\ 0 & 1 & 0 & 0 & 0 & -\frac{a}{\eta} \\ 0 & 0 & \frac{1}{N\theta_0} & 0 & 0 & 0 \\ 0 & 0 & 0 & \frac{1}{\zeta} & 0 & 0 \\ 0 & 0 & 0 & 0 & 1 & 0 \\ 0 & 0 & 0 & 0 & 0 & \frac{1}{\eta} \end{pmatrix}.$$

The set of eigenvalues of T is

$$\{1, 1, 1, N\theta_0, \zeta, \eta\}.$$

Thus the linear system under consideration is of *t-hyperbolic* type. Such systems are known to have many common features with the hyperbolic systems. Multiplying (26.11) by T^{-1}, the linear system takes the standard form:

$$\frac{\partial Z}{\partial t} + W_1 \frac{\partial Z}{\partial x} = F_1, \tag{26.12}$$

where

$$W_1 = T^{-1} W$$

and

$$F_1 = T^{-1} F.$$

One may now determine the speeds v of wave propagation in the present model by looking for autonomous solutions of the homogeneous system in (26.12) in the form

$$Z(x,t) = Z(t - \frac{x}{v}) = Z(\xi).$$

The condition for the existence of non-trivial solutions is

$$\det(I - \frac{1}{v} W_1) = 0,$$

yielding two values

$$v_1 = 1, \qquad v_2 = \frac{1}{\sqrt{N\theta_0}}.$$

The first speed is for the coupled thermoelastic wave as expected, while the second one is for second sound. Clearly, both speeds in the linear approximation are not affected by the electric field.

26.4 Numerical Scheme

In what follows, we use a three-step, iterative quasi-linearization finite-difference scheme to solve the set of nonlinear equations (26.1) - (26.4). For the computational work, one considers finite intervals on the x- and t-axes:

$$0 \leq x \leq X, \qquad 0 \leq t \leq T,$$

The domain in the (x,t) plane is discretized by a grid with step length

$$\triangle x = h$$

and time step

$$\triangle t = k$$

in such a way so as to secure stability of the computational scheme. Details about the scheme may be found in Mahmoud et al (2014). The following figures were calculated with $h = 0.5$, $k = 0.005$ and $X = T = 150$. For this same value of k, we noticed instability of the numerical scheme for $h \leq 0.1$.

The system of equations is solved for the values of the constants

$$\theta_0 = 1, \quad \epsilon = 0.9, \quad \beta_1 = 0.5, \quad a = 5, \quad \gamma' = 0.001, \quad \beta^* = 0.01$$

and under the following initial-boundary conditions:

$$\begin{aligned}
u(x,0) &= u_t(x,0) = \theta(x,0) = \theta_t(x,0) = q(x,0) = \phi(x,0) = 0,\\
u(0,t) &= u_0(1-\cos t),\\
\theta_x(0,t) &= Bi\,\theta(0,t),\\
q(0,t) &= q_0(1-\cos t),\\
\phi(0,t) &= 0,\\
u(X,t) &= \theta(X,t) = q(X,t) = \phi(X,t) = 0,\\
u_0 &= q_0 = 0.1,\\
Bi &= 2.
\end{aligned}$$

Zeroing all the variables at $x = X$ introduces an artificial boundary, necessary for the application of the shooting method. The choice of the values for X and T provides the possibility to detect the reflected waves at this far boundary travelling with velocity $v > 1$. On the other hand, zeroing the electric potential at $x = X$ induces some error, as the used quasi-electrostatic approximation allows for non-zero values of the electric potential at any location of space at any positive time value. This error, however, can be made smaller by considering larger values for X.

The solutions for the mechanical displacement, temperature, heat flow vector and electric potential for five values of time $\{t = 30, 60, 90, 120, 150\}$ and two values of the constant N $\{N = 0.01, 0.1\}$ are represented on the 3D-figures (26.1) - (26.4), in which one clearly notices the effect of nonlinearity, growing with time, on the distributions of the main unknown functions.

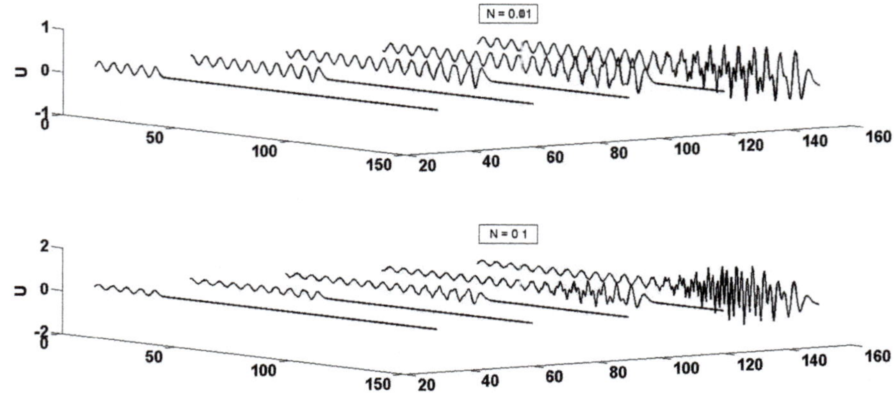

Fig. 26.1: Displacement for five values of time and two values of the constant N.

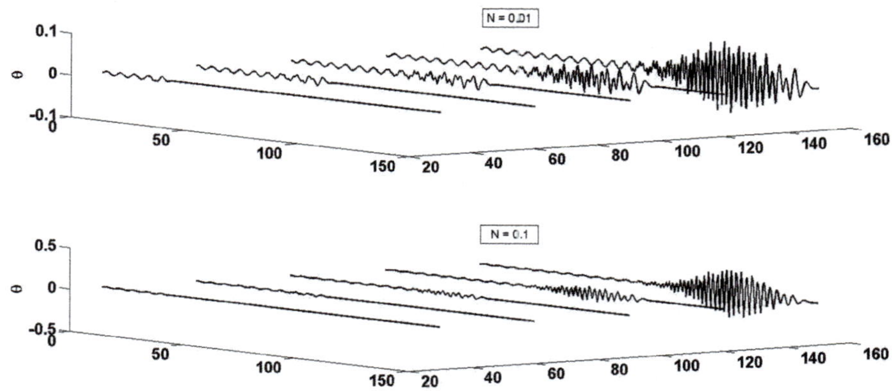

Fig. 26.2: Temperature for five values of time and two values of the constant N.

We have drastically decreased the value of constant N. Comparison is carried out for two values $\{N = 0.0001, 0.1\}$ on Fig. (26.5), in which we have taken $u_0 = 0.01$ and $q_0 = 0.01$. Understandably, the heat wave is invisible for $N = 0.0001$, i.e. for relatively large values of the speed of second sound, since the effects of extended irreversible in this case may be disregarded.

Figure (26.6) shows the electric potential as function of time at a certain location of the half-space $x = 100$. The fast wave is clearly visible for times $t < 100$, while the classical coupled thermoelastic wave starts to appear at $t = 100$. We have omitted the curves for the displacement, temperature and heat flux because the rapid wave is not visible on them, due to the smallness of its amplitude, compared to that of the slower, classical wave.

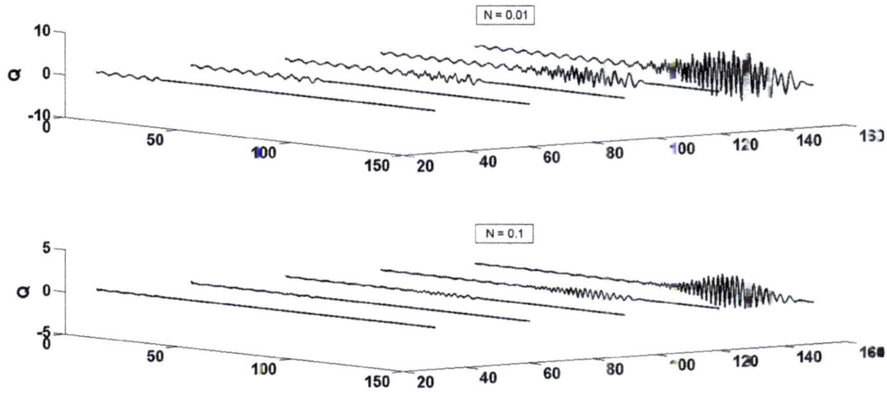

Fig. 26.3: Heat flux for five values of time and two values of the constant N.

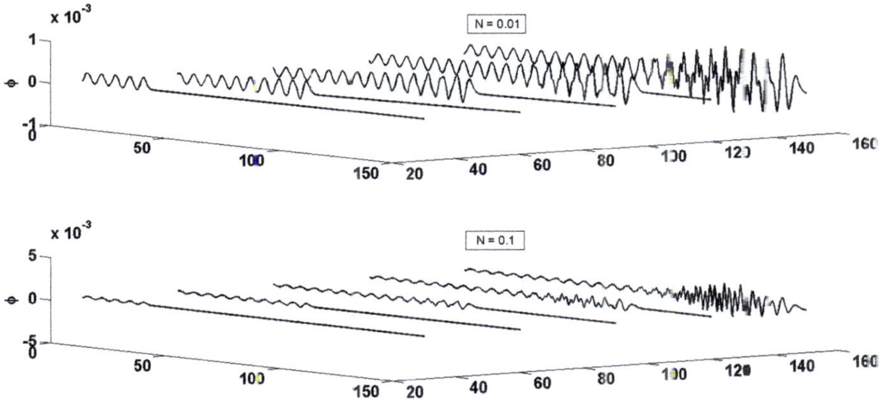

Fig. 26.4: Electric potential for five values of time and two values of the constant N.

26.5 Conclusions

The solution to a problem of nonlinear thermo-electroelasticity in extended thermodynamics and in the quasi-electrostatic approximation has been presented. The main ingredient in the used model is a Cattaneo-type evolution equation, with the inclusion of the heat flow vector as an additional state variable, independent of temperature. This requires considering an additional boundary condition for the heat flux, independent of the usual thermal boundary condition. Dimension analysis and a study of the characteristic curves has revealed the mixed "parabolic-hyperbolic" character of the proposed system of linear equations of electro-thermoelasticity. The parabolic component here is due to the basic hypothesis of quasi-electrostatic approximation

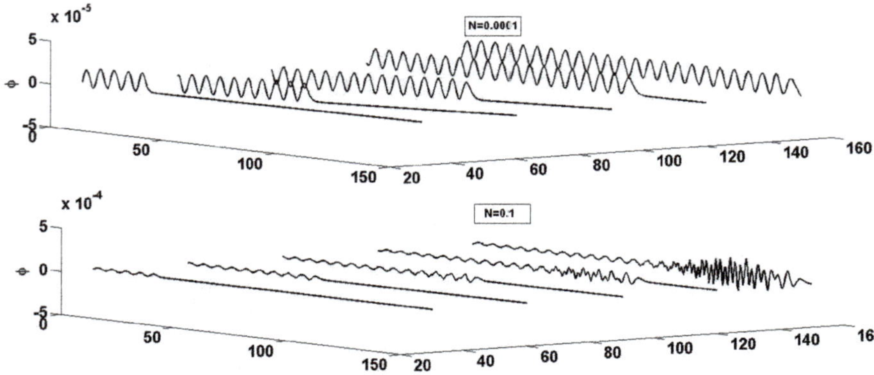

Fig. 26.5: Comparison between two cases $N = 0.0001$ and $N = 0.1$.

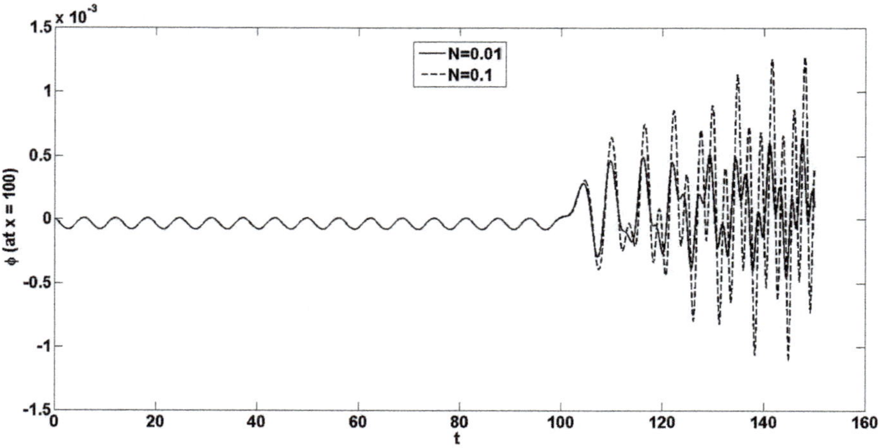

Fig. 26.6: Electric potential at $x = 100$ for two values of N.

of the electric field. The numerical treatment involves periodic boundary regimes for the mechanical displacement and the heat flux. The obtained results clearly show two types of waves propagating in the medium, the usual coupled thermoelastic wave, and the second sound wave. Thus the used model successfully illustrates the studied phenomenon, showing at the same time the electrical activity that accompanies the propagation of thermal and mechanical disturbances. In the almost total absence of solutions to nonlinear problems of thermo-electroelasticity, the obtained results may be of interest in investigating the nonlinear thermo-electromechanical interactions in dielectrics at low temperature, or in problems involving surface treatment in dielectrics by means of strong high-frequency laser beams.

References

Abou-Dina MS, Dhaba ARE, Ghaleb AF, Rawy EK (2017) A model of nonlinear thermo-electroelasticity in extended thermodynamics. International Journal of Engineering Science 119:29 – 39

Barletta A, Zanchini E (1997) Hyperbolic heat conduction and local equilibrium: a second law analysis. International Journal of Heat and Mass Transfer 40(5):1007 – 1016

Coleman BD, Newman DC (1988) Implications of a nonlinearity in the theory of second sound in solids. Phys Rev B 37:1492–1498

Coleman BD, Fabrizio M, Owen DR (1982) On the thermodynamics of second sound in dielectric crystals. Archive for Rational Mechanics and Analysis 80(2):135–158

Dorfmann A, Ogden RW (2006) Nonlinear electroelastic deformations. Journal of Elasticity 82(2):99–127

Ghaleb AF (1986) A model of continuous, thermoelastic media within the frame of extended thermodynamics. International Journal of Engineering Science 24(5):765–771

Ghaleb AF (2014) Coupled thermoelectroelasticity in extended thermodynamics. In: Hetnarski RB (ed) Encyclopedia of Thermal Stresses, Springer, vol c, pp 767–774

Giorgi C, Montanaro A (2016) Constitutive equations and wave propagation in Green-Naghdi type II and III thermoelectroelasticity. Journal of Thermal Stresses 39(9):1051–1073

Green AE, Naghdi PM (1991) A re-examination of the basic postulates of thermomechanics. Proceedings of the Royal Society of London A: Mathematical, Physical and Engineering Sciences 432(1885):171–194

Kuang ZB (2014) Theory of Electroelasticity. Springer

Mahmoud W, Ghaleb AF, Rawy EK, Hassan HAZ, Mosharafa AA (2014) Numerical solution to a nonlinear, one-dimensional problem of thermoelasticity with volume force and heat supply in a half-space. Archive of Applied Mechanics 84(9):1501–1515

Maugin GA (1985) Nonlinear Electromechanical Effects and Applications, Series in Theoretical and Applied Mechanics (ed. RKT Hsieh), vol 1. World Scientific, Singapore

Maugin GA (1988) Continuum Mechanics of Electromagnetic Solids. North-Holland Series in Applied Mathematics and Mechanics (eds. J. D. Achenbach, B. Budiansky, W. T. Koiter, H. A. Lauwerier, and L. van Wijngaarden), Elsevier, Amsterdam

Maugin GA (2015) The saga of internal variables of state in continuum thermo-mechanics (1893-2013). Mechanics Research Communications 69:79 – 86

Maugin GA, Eringen AC (1977) On the equations of the electrodynamics of deformable bodies of finite extent. Journal de Mécanique 16:101–147

Maugin GA, Cillet B, Drouot R, Pouget J (1992) Nonlinear Electromechanical Couplings. Nonlinear Science, John Wiley and Sons Inc.

Mehnert M, Pelteret JP, Steinmann P (2017) Numerical modelling of nonlinear thermo-electro-elasticity. Mathematics and Mechanics of Solids 22(11):2196–2213

Montanaro A (2011) On the Constitutive Relations for Second Sound in Thermo-Electroelasticity. Arch Mech 63(3):225–254

Montanaro A (2015) A Green-Naghdi approach for thermo-electroelasticity. Journal of Physics: Conference Series 633(1):012,129

Ogden RW, Steigmann DJ (eds) (2011) Mechanics and Electrodynamics of Magneto- and Electro-Elastic Materials, CISM Courses and Lectures, vol 527. Springer, Wien

Pashitskii EA, Ryabchenko SM (2007) On the cause of the electrical activity of superfluid helium upon excitation of a second sound wave and normal-component velocity oscillations in it. Low Temperature Physics 33(1):8–14

Pashitskii EA, Tkachenko OM, Grygoryshyn KV, Lev BI (2009) On the nature of electrical activity in superfluid helium at second sound excitations (in Ukrain.). Ukrayins'kij Fyizichnij Zhurnal 54(1-2):89–93

Pulvirenti B, Barletta A, Zanchini E (1998) Finite-difference solution of hyperbolic heat conduction with temperature-dependent properties. Numerical Heat Transfer, Part A: Applications 34(2):169–183

Rybalko A, Rudavskii E, Rubets S, Tikhiy V, Derkach V, Tarapov S (2007) Electric induction in He II. Journal of Low Temperature Physics 148(5):527–534

Tiersten HF (1971) On the nonlinear equations of thermo-electroelasticity. International Journal of Engineering Science 9(7):587 – 604

Yen CC, Wu CY (2003) Modelling hyperbolic heat conduction in a finite medium with periodic thermal disturbance and surface radiation. Applied Mathematical Modelling 27(5):397 – 408

Zudin YB (2017) Hyperbolic Heat Conduction Equation, Springer, Berlin, Heidelberg, pp 183–200

Chapter 27
Analysis of Mechanical Response of Random Skeletal Structure

María-Belén Martínez-Pavetti and Shoji Imatani

Abstract In order to discuss the transmission of forces and moments in cellular materials, a computational scheme by use of rod-beam element is adopted to represent open-celled skeletal materials with random configuration. The target domain is obtained by so-called the Voronoi tessellation technique in which we consider the line segments of the polyhedra as the substantial beam-like members, and these members are interconnected with each other at the corners. Finite element analyses by rod-beam elements are then carried out to examine the characteristics of the complex structures. We discuss the transition from microscopic deformation in member beams to macroscopic response of such structures.

27.1 Introduction

Cellular structures are very common in nature; plant tissues, bones, honeycombs, and so on; some are made of so-called closed-cell structures and others are of open-celled structures. A particular configuration by beam-like members with hollow portions reveals outstanding characteristics of extremely light weight and high energy absorption, which are of our great interest for engineering application. Such materials are, needless to say, extremely inhomogeneous; most of the portion is just hollow, i.e. neither force nor moment is transmitted there, while the substantial members carry forces, moments, and any other kinds of signals from one position to others. When we touch a sponge in dairy life, for example, we feel a softness for a sponge

María-Belén Martínez-Pavetti
Kyoto University, Sakyo-ku, Kyoto Japan (currently Universidad Nacional de Asunción, San Lorenzo, Paraguay)
e-mail: bmartinez.py@gmail.com

Shoji Imatani
Kyoto University, Sakyo-ku, Kyoto Japan
e-mail: imatani@energy.kyoto-u.ac.jp

© Springer International Publishing AG, part of Springer Nature 2018

H. Altenbach et al. (eds.), *Generalized Models and Non-classical Approaches in Complex Materials 1*, Advanced Structured Materials 89,
https://doi.org/10.1007/978-3-319-72440-9_27

with fine foams while, in contrast, we feel something hard for another one with rough foams, even if they have the same volume fraction and the same ingredients. A question may arise: why is the sense of touch different from each other even with the same density at the macroscopic level? or how do mechanical signals transmit in the porous/skeletal structure? Such a primitive question is also of importance when we formulate a macroscopic material model for porous materials.

Mechanical behavior of porous materials have been discussed for a long time (Gibson, 1989). Among them we can readily refer the celebrated monographs, for example, by Ashby et al (2000); Gibson and Ashby (1997); Weaire and Hutzler (2009). Stimulated by these pioneering works, experimental investigations as well as numerical simulations can be found in literatures, e.g. Gong et al (2005); Luxner et al (2007). One of the authors also reported the mechanical behavior of closed-cell foams "ALPORAS" both from experimental and computational approaches (Imatani, 2012b). The complexity of dealing with the porous materials lies in the randomness in geometrical configuration. So far as we employ stresses and strains in the macroscopic level, this kind of statistical variation should be discussed somewhere in the process of constitutive modeling.

Recent development of production technology enables us to fabricate complex structures quite easily. Figure 27.1 demonstrates a porous/skeletal structure by use of a 3D printer. The advantage of using the 3-D printer is that we can generate mostly any kind of geometrical data as we expect, and the data are readily converted to the input file for numerical analysis schemes. Then we can exactly compare the response with computations. The volume fraction of the above example takes about 10%, i.e. the rest of 90% is hollow in the rectangular body. The cross-section of the bridge members is controlled to achieve the volume. The resolution of accuracy of the machine nowadays reaches less than 0.1mm, and so we expect fabrications of structures in smaller size, through which the real size effect can be discussed. In order to formulate a macroscopic constitutive model taking account of what really takes place in the structures, we are now going on the experimental work on the skeletal structures with this technique. Here we present a computational scheme to evaluate the mechanical behavior of random skeletal structures and discuss some of the characteristics to be taken into account in the macroscopic models.

The objective domain, i.e. skeletal structures, should have specific features, and they must be realized in the computational scheme. The most characteristic point for our purpose is the regularity and randomness not only in shape but also in dimension. The real scale effect of the skeletal structure is out of our consideration for

Fig. 27.1 A sample structure fabricated by use of 3-D printer. The volume fraction of the substantial portion is about 10%, and it is easily controlled by setting the thickness of line segments of Voronoi polyhedra.

the moment, and it is reserved for our future work. Here we limit our discussion to the effect of configuration of skeletal members, which carry forces and moments. Problems become simple when adopting a regular shape for pores to be generated in the way that the structures are composed of regular polyhedron such as tetrahedron, cube, octahedron and so on. We can apply so-called the homogenization method (Cioranescu and Donato, 1999; Sanchez-Palencia and Zaoui, 1987) to the representative unit cell. Then the homogenized stiffness is immediately obtained, and the local responses in any state can be evaluated, at least, within the linear elasticity theory. However, a specific anisotropy may be induced in the structural response of such regular structures, and this would make our discussion complicated in the macroscopic level. And such an treatment may be far from the reality of our targetted materials. We take an opposite way; starting from the shape and the volume of pores sufficiently to be various and random, and focusing to the specific phenomena. The Voronoi tessellation algorithm (Aurenhammer and Klein, 2000; Brostow et al, 1978; Finney, 1979) is the one of the promising techniques to generate such a skeletal structure in the three dimensional space. Here we apply this procedure with an extension in hyperspace (Imatani, 2012a) while regarding the line segments of edges as the substantial members of skeletal structures.

27.2 Material Characterization

27.2.1 Extended Voronoi Tessellation

In the usual three dimensional Voronoi tessellation technique, we first set many points (seeds) at random. And we seek the corner which is equi-distant from four seeds and is the closest from all the other seeds, i.e. no other seed exists within the distance from that specific corner. When we repeat this procedure for all the seeds, the domain surrounded by closer planes from the specific seed comprises a convex polyhedron call "Voronoi" polyhedron. The objective domain is then occupied by many polyhedra generated in such a way. The algorithm is just simple and this is widely used to analyze the crystallisation processes; overall discussions are presented in Aurenhammer and Klein (2000). Since we set many seeds at random in the three dimensional space, the polyhedra in fact reveal various shape with various number of planes, e.g. from 18 to 22, planes. It should be noted, however, that the volume of polyhedra is quite homogeneous. When we set, for example, 100 seeds in a unit cube, the average of the volume of polyhedra is of course 0.01, but the standard deviation of the volumes takes less than 0.002! This indicates that the largest polyhedron takes at most 0.014 or so, while the smallest 0.006 in the cells assembly. As we observe various volume distribution of skeletal structures in nature, this could be far from reality. This is the reason why we extend the tessellation technique to the hyperspace.

Let (x_i, y_i, z_i, w_i) be the coordinate of a seed in the four dimensional space, where i takes from $1, 2, \ldots, N$. We set the value for the fourth parameter w_i to be relatively smaller than the other components, and the component w_i indicates the additional parameter through which the randomness of volume is controlled. The seed with the negative of w_i, i.e. a seed at the mirror position $(x_i, y_i, z_i, -w_i)$, is also added in the list of seeds. The key point of setting the mirror seed is to cut the four dimensional polytope at $w = 0$, which is in fact a mid-point of w_i and $-w_i$, and to get the projection in three dimensional space. The four dimensional Voronoi polytope can be specified from the following steps:

1. We solve the linear equations by five seeds $i,\ j,\ \cdots, m$ to the interconnected point (x_p, y_p, z_p, w_p) as

$$\begin{cases} (x_i - x_p)^2 + (y_i - y_p)^2 + (z_i - z_p)^2 + (w_i - w_p)^2 = R^2 \\ (x_j - x_p)^2 + (y_j - y_p)^2 + (z_j - z_p)^2 + (w_j - w_p)^2 = R^2 \\ \vdots \\ (x_m - x_p)^2 + (y_m - y_p)^2 + (z_m - z_p)^2 + (w_m - w_p)^2 = R^2 \end{cases} \tag{27.1}$$

   while eliminating the radius R. This is a simple extension of the three dimensional search.
2. In the above equations, the mirror seed $(x_i, y_i, z_i, -w_i)$ should be included either in $j, \ldots, m$ for the seed i, and then $w_p = 0$ must be the solution. This indicates the projection from the four dimensional space to the three dimension at $w = 0$. The candidate $(x_p, y_p, z_p, 0)$ is reserved if it is the closest point (corner) from any other kind of combinations. If it does not fit this condition, e.g. $w \neq 0$, the candidate is disregarded.
3. The mid-plane is composed of the corners $(x_p, y_p, z_p, 0)$ around the position (x_i, y_i, z_i, w_i). The rest of the procedure just follows the usual three dimensional technique. The position $(x_i, y_i, z_i, 0)$ is regarded as the center of the polyhedron. A sample of Voronoi polyhedra is shown in Fig. 27.2 where each polyhedron is displayed in separate. All these polyhedra are strictly closed and convex because the original one in the four dimension is closed and convex.
4. In order to construct the open-celled structure from the above polyhedra, we use the edges, i.e. the line segments between two corner points, as the substantial material portion. For the experimental sample in Fig. 27.1, we reserve the portion with triangular cross-section. In the following simulation, however, each line

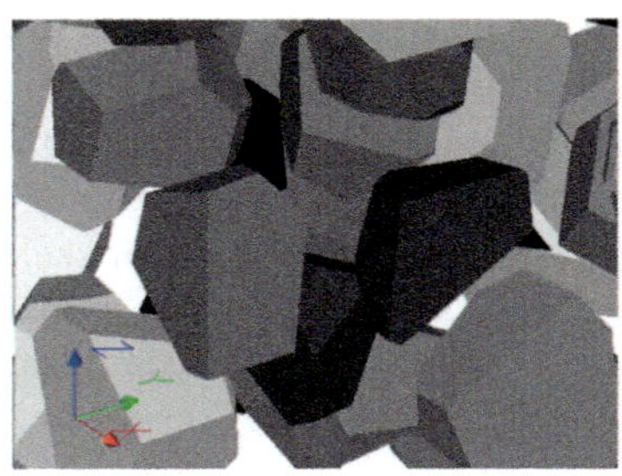

Fig. 27.2 Various shape of Voronoi polyhedra in an assembly example. The polyhedra are displayed, just for convention, in separate with each other. In order to create the skeletal structure, the edges of the polyhedra are regarded as the members of the substantial portion.

segment is assumed to be a rod/beam-like member with a constant circular cross-section. The diameter d is settled being proportional to the length of each line segment. This simplification remarkably reduces the computational costs while violates the equivalence between the experiment and simulation because the warping and other related issues may occur in the experiment. Another option with fully three dimensional mesh generation is also possible. The material parameters are introduced into this member material. Each line segment is sub-divided into several finite elements.

27.2.2 Basic Equations

We assume that the substantial member portion is composed of a linear elastic material. All the governing equations are well-known. The equation of motion in a body, i.e. the substantial portion $\mathscr{R}$, is given by

$$\rho \dot{\vec{v}} = \operatorname{div} \vec{\sigma} + \rho \vec{b}. \tag{27.2}$$

Here the standard nomenclatures are used; ρ is the mass density, $\vec{v} = \dot{\vec{u}}$ is the velocity and therefore $\dot{\vec{v}}$ means the acceleration, $\vec{\sigma}$ stands for the stress tensor, and $\vec{b}$ is the body force per unit mass. The conventional linear elasticity material model reads

$$\vec{\sigma} = \vec{E}\, \vec{\varepsilon}. \tag{27.3}$$

The elasticity coefficient $\vec{E}$ is a fourth-rank tensor, and we set Young's modulus E and the shear modulus G. The kinematic variable $\vec{\varepsilon}$ is the infinitesimal strain which is related to the displacement $\vec{u}$ as

$$\vec{\varepsilon} = \frac{1}{2} \left(\operatorname{grad} \vec{u} + \operatorname{grad}^{\mathrm{T}} \vec{u} \right). \tag{27.4}$$

In a part of the surrounded region, i.e. a surface $\partial \mathscr{R}_u$, the displacement $\vec{u}$ is prescribed as $\underline{\vec{u}}$ at the fundamental boundary, while the surface force $\underline{\vec{t}}$ is given at the complementary boundary $\partial \mathscr{R}_t$.

The Lagrangian $\mathscr{L}$ of the continuum body is given by the kinematic energy $\mathscr{K}$ subtracted by the potential energy $\mathscr{U}$ as $\mathscr{L} = \mathscr{K} - \mathscr{U}$:

$$\begin{cases} \mathscr{K} = \displaystyle\int_{\mathscr{R}} \frac{1}{2} \rho \vec{v} \cdot \vec{v}\, dv, \\ \mathscr{U} = \displaystyle\int_{\mathscr{R}} \frac{1}{2} \vec{\varepsilon} \cdot (\vec{E}\, \vec{\varepsilon})\, dv - \int_{\mathscr{R}} \rho \vec{b} \cdot \vec{u}\, dv - \int_{\partial \mathscr{R}} \vec{t} \cdot \vec{u}\, ds. \end{cases} \tag{27.5}$$

The Lagrangian is integrated over the time. Taking the first variation on the functional, we obtain the virtual work principle. Making use of some integration techniques, we have the following virtual work principle at any instant:

$$\int_{\mathscr{R}} \delta \vec{u} \cdot \rho \ddot{\vec{u}}\, dv + \int_{\mathscr{R}} \delta \vec{\varepsilon} \cdot \vec{E}\, \vec{\varepsilon}\, dv = \int_{\mathscr{R}} \delta \vec{u} \cdot \rho \vec{b}\, dv + \int_{\partial \mathscr{R}_t} \delta \vec{u} \cdot \underline{\vec{t}}\, ds. \quad (27.6)$$

Here the virtual displacement $\delta \vec{u}$ does not violate the fundamental boundary, i.e. $\delta \vec{u} = 0,\ \vec{x} \in \partial \mathscr{R}_u$. The static version of this principle is easily reduced or recovered when the dynamic term is neglected. This gives the basis for the finite element equation (Hughes, 1987).

27.2.3 Finite Element Discretization

Let an element e be composed of node-i and node-j, and the z-axis lies along the longitudinal direction of the element. The other axes x and y are set perpendicular to the z-axis. The components in this coordinate can be transformed from the global coordinate X-Y-Z, as schematically illustrated in Fig. 27.3. The discretization procedure is just straightforward as is well recognized (Bathe, 1995). Almost no particular technique is employed. The displacement $\vec{u}$ in the element e is interpolated by the shape function $\phi(z)$. We employ so-called the Timoschenko type element together with longitudinal tension and twisting. Each component of the displacement in the local system is expressible as

$$\begin{cases} u_x^{\mathrm{e}} = [\phi]\{u_x^n\} - y[\phi]\{\theta_z^n\}, \\ u_y^{\mathrm{e}} = [\phi]\{u_y^n\} + x[\phi]\{\theta_z^n\}, \\ u_z^{\mathrm{e}} = [\phi]\{u_z^n\} + y[\phi]\{\theta_x^n\} - x[\phi]\{\theta_y^n\}. \end{cases} \quad (27.7)$$

The first terms in the right hand side of Eq. (27.7) represent the translation, and the second terms for u_x and u_y contribute to the twisting around z-axis. The second and third term in the component u_z are to express the shear in the Timoshenko beam theory.

The infinitesimal rotation vector $\vec{\theta}$ is arranged around the axes, and so there are totally six degrees of freedom at the node-n as

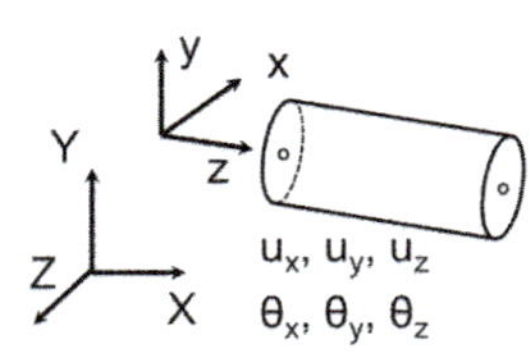

Fig. 27.3 A schematic model of two-node rod-beam element. The line segments of Voronoi polyhedra are composed of this type of element. There are totally six degrees of freedom per node.

$$\{u^n\} = [u_x^n \ u_y^n \ u_z^n \ \theta_x^n \ \theta_y^n \ \theta_z^n]^{\mathrm{T}}. \tag{27.8}$$

Equation (27.7) is simply denoted by $\{u^{\mathrm{e}}\} = [\Phi]\{u^n\}$. Then we obtain the non-zero strain components of ε_z for tension/bending and γ_{yz} and γ_{zx} for shear, which are related to the nodal displacement by $[B]$-matrix. Non-zero stress components are also σ_z for tension/bending and σ_{yz} and σ_{zx} for shear. Setting $\{\varepsilon\} = [\varepsilon_z \ \gamma_{yz} \ \gamma_{zx}]^{\mathrm{T}}$ and $\{\sigma\} = [\sigma_z \ \sigma_{yz} \ \sigma_{zx}]^{\mathrm{T}}$ in vector table, the strain-displacement relation and the stress-strain relation are respectively expressed by

$$\begin{cases} \{\varepsilon\} = [B]\{u^n\}, \\ \{\sigma\} = [D]\{\varepsilon\}. \end{cases} \tag{27.9}$$

The 3×3 stiffness matrix $[D]$ is a diagonal matrix whose components are E and G. The shear correction factor is chosen to be 0.9 for this type of Timoshenko beam. The surface force is also discretized by the standard technique, and the body force is neglected for simplicity. In this case we can prescribe the rotation angle as the fundamental boundary while surface moment can also be given as the natural boundary condition.

Introducing these equations from (27.7) to (27.9) into Eq.(27.6), we find the equation of motion in discrete system. For simplicity we prepare the elemental mass matrix $[M^{\mathrm{e}}]$ and the elemental stiffness matrix $[K^{\mathrm{e}}]$ as

$$[M^{\mathrm{e}}] = \int_{\mathrm{e}} \rho[\Phi]^{\mathrm{T}}[\Phi]\,dv, \qquad [K^{\mathrm{e}}] = \int_{\mathrm{e}} [B]^{\mathrm{T}}[D][B]\,dv. \tag{27.10}$$

This is written in the local coordinate system. Here we employ the consistent mass matrix technique rather than the lumped matrix which is well used in structural analysis. Converting from the local system to the global system, we have the following equation of motion:

$$[\overrightarrow{M}]\{\ddot{\overrightarrow{U}}^n\} + [\overrightarrow{K}]\{\overrightarrow{U}^n\} = \{\overrightarrow{F}^n\}. \tag{27.11}$$

$[\overrightarrow{M}]$ and $[\overrightarrow{K}]$ are the total mass matrix and the stiffness matrix respectively. For step-by-step analysis, the nodal acceleration $\{\ddot{\overrightarrow{U}}^n\}$ is approximated between the time τ and $\tau+\Delta\tau$ by use of the implicit Newmark-β algorithm with a proper parameter

$$1/6 \le \beta \le 1/4.$$

The FE equation to be solved for the nodal displacement $\{\overrightarrow{U}^n_{\tau+\Delta\tau}\}$ is given by

$$[\hat{\overrightarrow{K}}]\{\overrightarrow{U}^n_{\tau+\Delta\tau}\} = \{\hat{\overrightarrow{F}}^n_{\tau+\Delta\tau}\} \tag{27.12}$$

where the matrices are given below:

$$
\begin{cases}
[\hat{\vec{K}}] &= [\vec{K}] + \frac{1}{\beta(\Delta\tau)^2}[\vec{M}], \\
\{\hat{\vec{F}}^{n}_{\tau+\Delta\tau}\} &= \{\vec{F}^{n}_{\tau+\Delta\tau}\} \\
&\quad + [\vec{M}]\left\{\frac{1}{\beta(\Delta\tau)^2}\{\vec{U}^{n}_{\tau}\} + \frac{1}{\beta\Delta\tau}\{\dot{\vec{U}}^{n}_{\tau}\} + \left(\frac{1}{2\beta} - 1\right)\{\ddot{\vec{U}}^{n}_{\tau}\}\right\}.
\end{cases}
\tag{27.13}
$$

We notice that there is no damping in this system, and that we have no steady state to be approached. By virtue to the Helmholtz theorem, we have two kinds of characteristic wave speed for one kind of vector field; usually the longitudinal wave and shear wave in terms of the displacement $\vec{u}$. However, it should be pointed out that there is one more wave pattern through rot $\vec{\theta}$ (N.B. div $\vec{\theta} = 0$), since we have two kinds of vector fields $\vec{u}$ and $\vec{\theta}$. For the static analysis, the matrix $[\hat{\vec{K}}]$ is replaced by $[\vec{K}]$ and the vector $\{\hat{\vec{F}}^{n}_{\tau+\Delta\tau}\}$ by $\{\vec{F}^{n}\}$. Then the equation (27.12) is recovered to the static stiffness equation;

$$
[\vec{K}]\{\vec{U}^{n}\} = \{\vec{F}^{n}\}\,. \tag{27.14}
$$

The results of $\{\vec{U}^{n}_{\tau+\Delta\tau}\}$ for dynamic analysis and $\{\vec{U}^{n}\}$ for static analysis are utilized for the evaluation of elemental stress/strain response.

27.3 Analyses and Discussions

27.3.1 Preparation of Skeletal Model

Figure 27.4 shows two types of skeletal structures of unit cube $(X,Y,Z) \in [0,1]$. The bold frames in grey color, which are not valid in the practical computation, are added to distinguish the domain. The internal structure is completely different due to the seeds distributed in the domain as is given in Sect. 27.2.1. The left side of the figure indicates a regular structure composed of the line segments of face centered cubic (f.c.c.) cells; i.e. rhombic dodecahedron. The cross-section of each member is

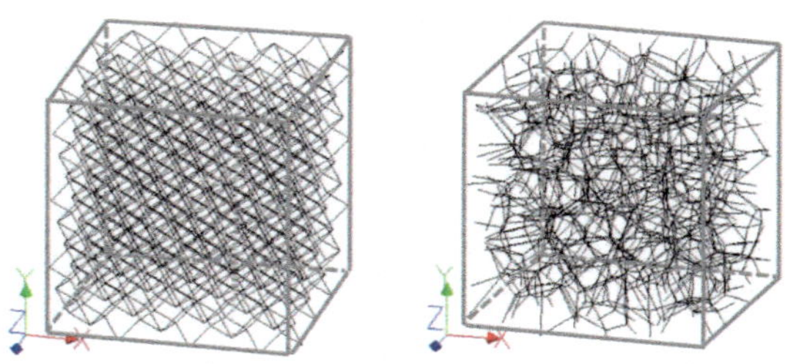

Fig. 27.4 Two types of skeletal structures. The left side model is a regular structure composed of the line segments of f.c.c. dodecahedral cells, and the right side model is a random skeletal structure. Two-nodes rod-beam elements are utilized to compose the structures, and thus the substantial portion is displayed in line.

constant and uniform. In contrast the right side of the figure indicates a unit-cube structure composed of random skeletal members. The cross-section of this structure depends upon the length of the line segments, and the total volume of the substantial portion is set to be the same; about 10% in the above example. In the following simulation, we set the nodes at the left end of $X = 0$ to be fixed and the displacements of the nodes at the right end of $X = 1$ are prescribed. The number of interconnected points is about 600 to 800, and the total number of elements and nodes is about 20,000 to 24,000, both of which depend on the number of seeds. We prepare several patterns for the random structure with different sets of seeds. The material properties and other parameters related to the simulation are, mainly for simplicity, chosen to be simple enough; the mass density ρ to be 1.0, the shear modulus G to be 1.0, and Young's modulus is set 2.6.

When a signal of axial force, shear forces, axial (twisting) moment, and bending moments leaves from an interconnected point (corner), it propagates along the line segments, which are subdivided by several elements, and reaches the other corner. Since all the line segments are identical in the regular structure, the reflection and the transmission at the corner occur at the same time. From this meaning the signal may be kept coherent in the structure, while the bending strain has some distribution along the member. In contrast, the length of the line segments in the random structure is so various that the reflection and the transmission take place almost at random. And this may affect the macroscopic behavior of mechanical response of the structure. It should be noticed that the forces and the moments change the role at the interconnected point, and so a simple push-pull signal does not keep the same signal pattern.

27.3.2 Static Tension

As the first examination, a simple tension by static analysis is carried out for the structures. The displacement at the right end $\underline{U}_X$ in Fig. 27.4 is stretched to 0.1% at the X-direction. For homogeneous solids, this simply induces a uniform strain over the whole domain. However the results are far different from the uniform stretch. Figure 27.5 represents the displacement of some of the representative nodes. The members exposed to the outside of the domain are free, and therefore they do not contribute to any deformation. In order to exclude such a free end effect, the sample nodes located on the inside of the domain are chosen. The open circles are the plots by the regular structure and the triangular marks are the result by the random structure, while some of the marks are overlapped. Both plots of displacement are almost linear on the distance from the origin, which implies that the structures as a whole undergo the prescribed stretch, i.e. $\varepsilon_X \approx 0.1\%$. And this also indicates the uniformity in the macroscopic response of strain, since the macroscopic strain is given by the gradient of displacement at the macroscopic level. However, this does not mean that the response is uniform in the structures at the microscopic level. Even in the regular structure a periodic distribution may occur in each line segment.

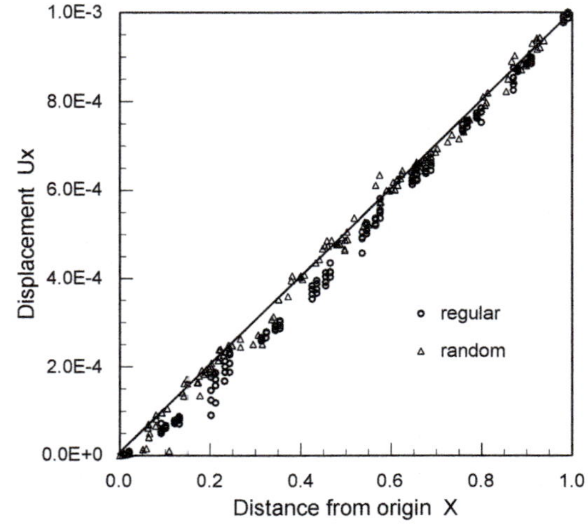

Fig. 27.5 Displacement along the structures. Some of the representative nodes near the center are plotted in order that peculiar responses due to the free ends around the side surface are to be excluded. Since the nodes at the origin $X = 0$ are strictly fixed not only at the X-direction but also at Y- and Z-direction as well as the angles, the edge effect is observed around the origin O.

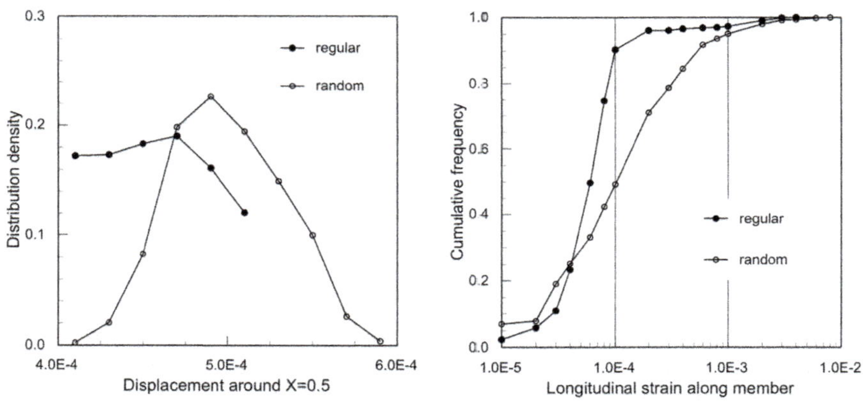

Fig. 27.6: Statistical properties of displacement and strain within the middle part of the domain under tension test. The regular structure shows a flat distribution in displacement while the random structure reveals a normal distribution due to the randomness of the skeletal members. The local strain shows a wide variety and sometimes 10 times larger than the average strain.

Lest us discuss the statistical aspect of the microscopic deformation. Figure 27.6 demonstrates the distribution profiles around the middle part, i.e. $X \in [0.4, 0.6]$, of the structures. The samples are also chosen on the inside of the domain $(Y, Z) \in [0.4, 0.6]$ to exclude the nodes and elements around the side surface. Notice that the number of plotted data in this figure is much more than the previous one in $X \in [0.4, 0.6]$ of Fig. 27.5. The solid circles are the plot for the regular structure and the open ones for the random structure. The left side of the figure indicates the distribution profile of displacement U_X in the global coordinate system, and the right side stands for the cumulative frequency, i.e. the sum of plotted data, of the longitudinal strain ε_z in each local coordinate system in the elements. Both figures are normalized by the sum of the event data. The magnitude of the local longitudinal strain ε_z is in general much smaller than the macroscopic strain ε_X, because the deformation is achieved by the structural response in which the local members are deflected mainly by bending and twisting. It follows that a uniform stretch in homogeneous materials is replaced by the structural rotation in skeletal materials.

Characteristic differences are observed between these structures in the distribution of displacement in the left side of the figure. A flat form like a step-function is obtained in the regular structure, whereas a normal distribution is predicted in the random structure. If a homogeneous solid is applied, we have an exact step-function within $[4\times10^{-4}, 6\times10^{-4}]$ for the finite range of $X \in [0.4, 0.6]$, corresponding to the 0.1%-stretch. This step function may approach steeper and steeper, i.e. finally like the Dirac-delta, when the sampling range becomes narrower. From this point of view, the regular structure keeps this kind of regularity in displacement. This assures, in a sense, the validity of the homogenization technique when we deal with the regular structure. In contrast, the normal distribution is induced from the randomness in geometry. When we take narrower sampling range for the random structure, the normal distribution may also approach steeper distribution, but not the Dirac-delta function in final. Furthermore the gradient of this kind of distribution does not reduce any regularity. Imatani (2007) reported that the stress/strain response, but not the displacement, in a polycrystalline model also reveals a transition to the normal distribution, although each crystal grain possesses the cubic anisotropy at random orientation. In such a way the randomness in geometry gives more and more dominant influence in the gradient of displacement, i.e. the strain.

The local strain ε_z reveals much complicated distributions in the right figure. The slope of the cumulative frequency is equivalent to the event density, i.e. the local strain level, while the horizontal axis is displayed in logarithmic scale. In the regular structure, the local longitudinal strain distribution is mostly located in the range from 4×10^{-5} to 1×10^{-4}, and the density is similar to the normal distribution. In contrast, the strain response in random structure shows larger variation from extremely small stain of less than 1×10^{-5} to relatively large strain up to 1×10^{-3}. This is because we set many kinds of beams with various cross-section as well as length. We have found a rough tendency such that larger strain is generated in thinner members, but this tendency does not always hold since the geometrical configuration gives much more dominant influence on the location strain response.

27.3.3 Dynamic Loading

The dynamic response is examined next, in order to discuss the transmission of force and moment. Being similar to the static test, we use the cubic structures in Fig. 27.4, and the left end at $X = 0$ is fixed while the right end at $X = 1$ is stretched. The prescribed displacement given to the structures is

$$\underline{U}_X|_{X=1} = \begin{cases} \underline{U}_0 \left(1 - \cos \dfrac{\pi t_n}{20}\right) & \cdots \; t_n < 20 \, , \\ 2\underline{U}_0 & \cdots \; t_n \geq 20 \text{ (hold)} \, . \end{cases} \tag{27.15}$$

The time t_n indicates the normalized scale. Starting from $X = 1$, the deformation makes progress or propagates from the right end to the left. However, the propagation itself is very complex because of the reflection at interconnecting points. Recall that the equation of motion (27.11) does not involve any damping mechanism, and therefore the structures do not show any stationary state even when a long time passes while holding the displacement at the right end. Here we pick up deformation profiles at the time $t_n = 100$ and $t_n = 400$ and compare the progress of deformation. Since we set the elasticity parameters $E = 2.6$ and $G = 1.0$ while the mass density is $\rho = 1.0$, the wave speed, whether it is a longitudinal wave or a shear wave, is in the order of 1. The sampling times at 100 and 400 should sufficiently be large.

Figure 27.7 demonstrates the variation of microscopic deformation, in which the left figure shows the local strain distribution ε_z and the right figure corresponds to the rotation $|\omega|$. The infinitesimal rotation $\vec{\omega}$ is evaluated as the skew-symmetric part of the displacement gradient, and it is also related to the axial vector. In spite that the prescribed displacement is held constant with $\underline{U}_0$ at $t_n = 100$ and 400, the deformation is concentrated near the right end and it is not yet uniform from the macroscopic viewpoint. In the regular structure, all the members have the same span and the same connecting configuration, the signal traveling is coherent. This coherence is advantageous in transmission of signals across the interconnecting point,

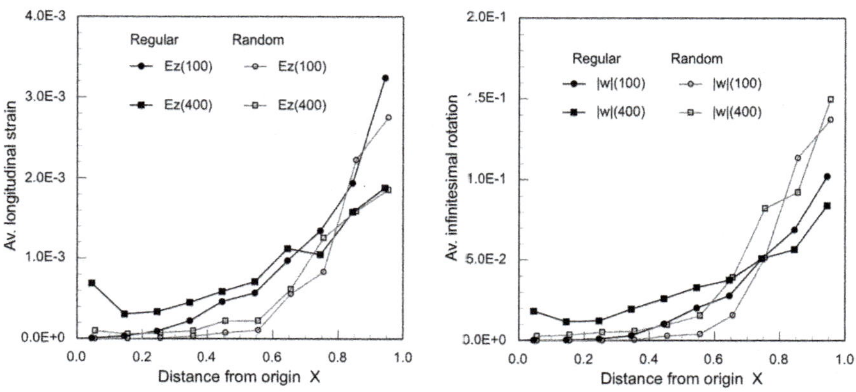

Fig. 27.7: Distributions of longitudinal strain and rotation under dynamic tension. The signals generally move from the right end to the left, while the reflections at the interconnecting point play an important role as the structural damping.

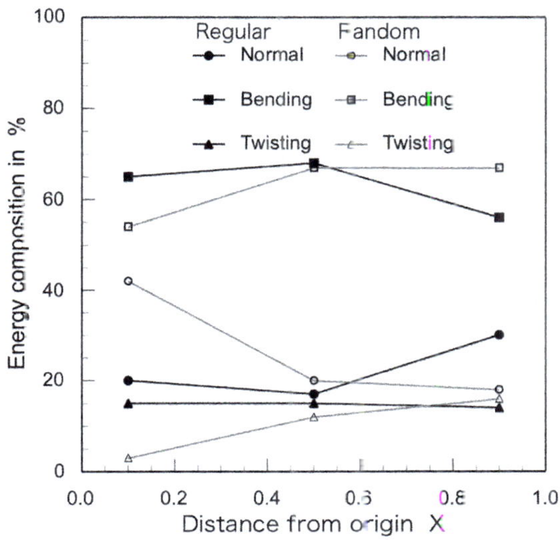

Fig. 27.8 Composition of strain energy along the axis over the simulation time. In spite of the remarkable difference of mechanical response between the regular and random structures, the composition of strain energy stored in the body is not so different. When the member is thinner, The bending energy becomes more dominant in the storage of strain energy.

and the local strain arrives at the left end. In the random structure, in contrast, the deformation almost ceases the propagation at the middle of the cube. This may be due to the simultaneous recurrence of reflection of signals at the interconnecting points, and this behavior seems a structural damping while no damping mechanism exists. Such a tendency is common for both the strain and the rotation.

With reference to the interpolation of displacement in Eq. (27.7), the strain components can be decomposed into normal strain, bending/shear strains, and twisting strain. Then the total strain energy can also be split into these terms. Figure 27.8 clarifies the contribution of the strain energy stored in the member elements. The energy composition means the relative ratio of each term with respect to the total strain energy at that location, and so the values at $X = 0.1$ are overestimated in a sense because the deformation is so small there. It should also be pointed out that the nodal displacements at both ends are fixed, and then further constraints give influences on the evaluation of energy. However, the bending energy is always the highest among all, and it is dominant on the deformation of skeletal structures.

27.4 Concluding Remarks

Skeletal structures with random configuration are discussed from computational approaches. Here we obtained the following remarks:

1. Based upon a computational geometry technique, skeletal structures with random configuration is generated. The advantage of this procedure is that we can

reproduce the model in practice with a 3-D printer while the model is also applied to the numerical analysis scheme.

2. It is found that both the displacement and the strain should be treated as a sort of distribution density function when we formulate a macroscopic constitutive model for skeletal materials. The homogenization technique would be a promising method for a material with some regularity in geometry, while further technique must be applied to a material with random configuration.
3. The bending mechanism plays a dominant role in the deformation of each member. And so the couple stress concept or higher order stress should taken into account in the macroscopic constitutive model.
4. The reflection of plural waves at the interconnecting point decelerates the travel of signals, and this effect appears as a sort of mechanical damping at the macroscopic level. The high energy absorption in skeletal materials may be caused by this effect.

Acknowledgements A part of this work was carried out when the first author stayed as a student in the International Course of Energy Science, Kyoto University. The financial support by the Ministry of Education, Sports, and Culture, Japan is highly appreciated. The authors wish to thank Dr. H. Miyauchi, a technical staff at Kyoto Prefectural Technology Center for Small and Medium Enterprises, for his help in preparation of materials by use of 3-D printer,

References

Ashby MF, Evans A, Fleck N, Gibson L, Hutchinson JW, Wadley H (2000) Metal foams: A design guide. Butterworth-Heinemann, Boston

Aurenhammer F, Klein R (2000) Voronoi diagrams. In: Sack JR, Urrutia J (eds) Handbook of Computational Geometry, North-Holland, Amsterdam, pp 201–290

Bathe KJ (1995) Finite Element Procedures. Prentice Hall, New Jersey

Brostow W, Dussault JP, Fox BL (1978) Construction of Voronoi polyhedra. Journal of Computational Physics 29(1):81–92

Cioranescu D, Donato P (1999) An Introduction to Homogenization. Oxford Univ. Press, Oxford

Finney JL (1979) A procedure for the construction of Voronoi polyhedra. Journal of Computational Physics 32(1):137–143

Gibson LJ (1989) Modelling the mechanical behavior of cellular materials. Materials Science and Engineering: A 110:1–36

Gibson LJ, Ashby FA (1997) Cellular Solids: Structure and Properties, 2nd edn. Cambridge University Press, Cambridge

Gong L, Kyriakides S, Jang WY (2005) Compressive response of open-cell foams. Part I: Morphology and elastic properties. International Journal of Solids and Structures 42(5):1355–1379

Hughes TJR (1987) The Finite Element Method: Linear Static and Dynamic Finite Element Analysis. Dover Publications, New York

Imatani S (2007) Development of microscopic inhomogeneities in polycrystalline aggregate. In: Bruhns OT, Meyers A (eds) Plasticity and Impact Mechanics - IMPLAST 2007, The Univ. Press Bochum, Bochum, pp 531–538

Imatani S (2012a) Material characterization of polycrystals through hyperspace. In: Proceedings of 23rd International Congress on Theoretical and Applied Mechanics, pp SM04–049

Imatani S (2012b) Numerical evaluation of compressible plasticity behavior of metal foams. Technische Mechanik 32:265–272

Luxner MH, Stampfl J, Pettermann HE (2007) Numerical simulations of 3d open cell structures: influence of structural irregularities on elasto-plasticity and deformation localization. International Journal of Solids and Structures 44(9):2990–3003
Sanchez-Palencia E, Zaoui A (1987) Homogenization Techniques for Composite Media. Springer, Vienna
Weaire D, Hutzler S (2009) The Physics of Foams. Oxford University Press, New York

Chapter 28
On the Influence of the Coupled Invariant in Thermo-Electro-Elasticity

Markus Mehnert, Tiphaine Mathieu-Pennober, and Paul Steinmann

Abstract Electro-active polymers (EAPs) are a comparatively new class of smart materials that can change their properties and undergo large deformations as a result of an external electric excitation. These characteristics make them promising candidates in a wide range of applications, for example in sensor and actuator technology. As the experimental testing is both expensive and time consuming, simulation methods are developed in order to predict the material behavior. These simulations are based on well established energy formulation that are amended by additional coupling terms, often times in form of an invariant description. While the form of the purely mechanical and purely electric invariant quantities does not vary among the contributions of the electro-mechanical community, two different formulations for the coupling invariant can be found. In this contribution we demonstrate the influence of the selected coupling invariant on the material response. Therefore a thermo-electro-mechanically coupled constitutive model is derived based on the frequently used total energy approach. We devise the relevant constitutive equations starting from the basic laws of thermodynamics. Two distinctively different non-homogeneous boundary value problems are solved analytically in order to demonstrate the influence of the selected coupling invariant.

Markus Mehnert · Paul Steinmann
Chair of Applied Mechanics, University of Erlangen-Nuremberg, Paul-Gordan-Str. 3, 91054 Erlangen, Germany
e-mail: markus.mehnert@fau.de, paul.steinmann@fau.de

Tiphaine Mathieu-Pennober
Ecole Polytechnique, Paris, France
e-mail: tiphaine.mathieu-pennober@polytechnique.org

© Springer International Publishing AG, part of Springer Nature 2018
H. Altenbach et al. (eds.), *Generalized Models and Non-classical Approaches in Complex Materials 1*, Advanced Structured Materials 89,
https://doi.org/10.1007/978-3-319-72440-9_28

28.1 Introduction

Increased standards on design and efficiency, the demand for multi-functionality and small scale applications while simultaneously ensuring low production costs and robustness of the final product pose major challenges for engineers in the development of state-of-the-art technology. This has lead to an increased interest in smart materials that possess the ability to change their shape and their mechanical features in response to the application of a non-mechanical field. One particularly interesting example of this material class are electro-active polymers (EAPs) which can react to the application of an electric field with large deformations and a change in their material properties, like stiffness or viscosity. In the design of actuators for example, EAPs have the ability to outperform traditional technologies like shape memory alloys, piezoelectrics or electro-magnetic motors, especially when low weight, small production costs and a fast reaction time are are the dominating factors (Pelrine et al, 2000a,b). But also in a wide range of other applications such as artificial muscles, sensors, generators, microfluidic pumping systems and adaptable optics (O'Halloran et al, 2008; Bar-Cohen, 2002, 2004; Vertechy et al, 2014; Böse and Fuß, 2014; Koh et al, 2011) electro-active polymers have been employed successfully.

We can differentiate EAPs based on their deformation mechanism (Bar-Cohen, 2002; Brochu and Pei, 2010). In ionic electro-active polymers (IEAPs), intramolecular forces lead to electrostriction whereas the mechanical response of electronic electro-active polymers (EEAPs) is due to the Maxwell stress originating from electrostatic forces between electric charges. If used as an actuator, a thin film of an EEAP is sandwiched between two compliant electrodes. Upon the application of an electric potential difference the thin-film contracts in the direction of the resulting field while simultaneously expanding in the transverse planar direction.

A group of smart materials that shows a very similar behavior as EAPs are magneto-rheological elastomers (MREs) which consist of a soft elastomer matrix filled with magnetizable micro particles (Kankanala and Triantafyllidis, 2004). When a magnetic field is applied, MREs can undergo large deformations and change their elastic properties (Danas et al, 2012). This makes magneto-rheological elastomers interesting candidates in the design of e.g. controllable stiffness devices or tactile interfaces for the visually impaired (Bodelot et al, 2016). For the latter, the magneto-rheological material is applied in form of a thin layer on a non-magnetic substrate. When the combination is subjected to a transverse magnetic field buckling of the material occurs (Danas and Triantafyllidis, 2014).

The basic principles of the mathematical modelling of electro- and magneto-elasticity were established for example in the works of Voltairas et al (2003); Maugin and Eringen (1977); Maugin (1976); Maugin and Eringen (1972); Maugin (2013); Eringen and Maugin (2012); Dorfmann and Ogden (2005a). The work of Maugin was furthermore extended by contributing studies on defects in electro-active polymers (Maugin, 1993) using the concept of material forces (Epstein and Maugin, 1990b,a). In the modelling of magneto-active materials the interface between the micro particles and the elastomer matrix poses additional challenges that were investigated for example in Pössinger et al (2014). Just as conventional polymeric materials, EAPs and MREs are sensitive to changes in temperature which often times may be impossible to avoid during their operation due to the high electric potential differences needed for the activation but also due to external thermal loading. Therefore it is crucial not only to analyze the electro- and magneto-mechanical coupling but also to consider the influence of temperature changes. A thermo-electro-elastic framework was introduced in a preceding contribution (Mehnert et al, 2016) which presented a phenomenological approach to the simulation of EAPs. In contrast to micro-mechanical models, which are derived from the statistical analysis of networks of idealized chain molecules, phenomenological approaches are formulated in terms of strain invariants or principal stretches (Steinmann et al, 2012; Hossain and Steinmann, 2013). One common approach to capture the interaction between the non-mechanical field and the deformation in EAPs or MREs is to amend a classical phenomenological model such as a realization of the class of Ogden models (Ogden, 1972) or models of Mooney-Rivlin type (Rivlin, 1948, 1949a,b) with additional coupling invariants. Throughout the literature two different formulations for this coupling invariant have been predominant whereas both fulfill the requirements for the formulation of an isotropic material. In this contribution we want to present how the selection of the coupling invariant may change the material response based on two analytical examples. This is done in the context of electro-active polymers but can in principle also be translated to the magnetic case for MREs.

This paper is organized as follows. In Subsect. 28.1.1, the finite strain theory of nonlinear electro-elasticity is presented by deriving the relevant nonlinear kinematics and balance laws for the spatial and material configuration. Furthermore this chapter contains constitutive relations based on a total free energy function and the derivation of a modified heat equation. The chapter is concluded by the derivation of a thermo-electro-mechanical coupled framework. In Sect. 28.2 two non-homogeneous boundary problems are solved analytically in order to present the different material behavior due to the selection of the coupling invariant. The final Sect. 28.3 concludes the paper with a summary and an outlook for future works.

28.1.1 Kinematics

The position vector $\boldsymbol{X}$ defines the position of a point in the reference configuration $\mathscr{B}_0$ of a nonlinearly deforming body $\mathscr{B}$. The corresponding position vector in the deformed configuration $\mathscr{B}_t$ is denoted $\mathbf{x}$, which is connected to the reference configuration by the nonlinear deformation map $\boldsymbol{\varphi}$ in the form $\boldsymbol{x} = \boldsymbol{\varphi}(\mathbf{X})$. By calculating the gradient of the deformation map with respect to the material coordinates we introduce the deformation gradient $\boldsymbol{F} = \text{Grad}\,\boldsymbol{\varphi}$ with the Jacobian determinant $J = \det(\boldsymbol{F})$. The deformation gradient $\mathbf{F}$ can be used to define the right Cauchy-Green tensor $\boldsymbol{C} = \boldsymbol{F}^T\boldsymbol{F}$. As the considered deformation is a combination of thermal and electro-mechanical loading, we introduce a multiplicative decomposition of the deformation gradient following Lu and Pister (1975) into an electro-mechanical deformation $\boldsymbol{F}_{EM}$ and a thermal part $\boldsymbol{F}_{\Theta} = \exp(\alpha\Delta\Theta)\boldsymbol{I}$, c.f. Erbts et al (2015), that captures the thermal expansion

$$\boldsymbol{F} = \boldsymbol{F}_{EM}\boldsymbol{F}_{\Theta} \tag{28.1}$$

In this context α is the thermal expansion coefficient and $\Delta\Theta$ is the temperature difference. The Jacobian determinant can be decomposed accordingly as

$$\begin{aligned} J &= \det\boldsymbol{F} = \det\boldsymbol{F}_{EM}\ \det\boldsymbol{F}_{\Theta} = J_{EM}J_{\Theta}, \\ \text{with}\quad J_{\Theta} &= \exp(3\alpha\Delta\Theta) \quad \text{and} \quad J_{EM} = J\exp(-3\alpha\Delta\Theta). \end{aligned} \tag{28.2}$$

28.1.2 Balance Laws in Electrostatics

We will now present the well established electro-mechanical balance equations. For further details on these relations the reader is referred to the works of e.g. Voltairas et al (2003); Maugin (2013); Griffiths (1989) or Dorfmann and Ogden (2005a).

28.1.2.1 Spatial Configuration

In vacuum an electric field $\mathbb{e}$ induces the electric displacement $\mathbb{d}^{\varepsilon} = \varepsilon_0\mathbb{e}$, where $\varepsilon_0 = 8.85\times10^{-12}$ F/m is the vacuum electric permittivity, whereas in the presence of matter, this relationship is amended by the electric polarization $\mathbb{p}$ which results in the constitutive relation for the electric displacement $\mathbb{d}$ (Eringen, 1963; Trimarco and Maugin, 2001)

$$\mathbb{d} =: \mathbb{d}^{\varepsilon} + \mathbb{p}, \quad \text{in } \mathscr{B}_t. \tag{28.3}$$

The governing equations of the electric problem in the spacial configuration are summarized by the Maxwell equations that, in the absence of free currents and free electric charges, take the form

$$\operatorname{div}\mathbb{d} = 0, \quad \operatorname{curl}\mathbb{e} = \boldsymbol{0} \quad \text{in } \mathscr{B}_t. \tag{28.4}$$

Here div and curl denote the corresponding differential operators with respect to the position vectors $\boldsymbol{x}$ in $\mathscr{B}_t$. We can ensure that Equation $(28.4)_2$ is satisfied by defining the electric field $\mathbb{e}$ as the gradient of the scalar electric potential ϕ with respect to the spatial coordinates Vu et al (2007); Vogel (2015)

$$\mathbb{e} = -\operatorname{grad}\phi. \tag{28.5}$$

The balance of linear momentum governs the mechanical response of the body. In electro-mechanics it states that the divergence of the non-symmetric mechanical Cauchy stress $\boldsymbol{\sigma}$ has to be in equilibrium with the mechanical body forces $\boldsymbol{b}_t$ and the ponderomotive body forces $\boldsymbol{b}_t^{\text{pon}}$

$$\operatorname{div}\boldsymbol{\sigma} + \boldsymbol{b}_t^{\text{pon}} + \boldsymbol{b}_t := \operatorname{div}\boldsymbol{\sigma}^{\text{tot}} + \boldsymbol{b}_t = \boldsymbol{0} \quad \text{in } \mathscr{B}_t. \tag{28.6}$$

These ponderomotive body forces capture the interaction between matter and the electric field in terms of the polarization and the gradient of the electric field

$$\boldsymbol{b}_t^{\text{pon}} := \text{grad}\, \mathbb{e} \cdot \mathbb{p}, \tag{28.7}$$

and can be expressed as the divergence of a corresponding ponderomotive stress $\boldsymbol{\sigma}^{\text{pon}}$

$$\boldsymbol{b}_t^{\text{pon}} = \text{div}\, \boldsymbol{\sigma}^{\text{pon}}. \tag{28.8}$$

The ponderomotive stress can be decomposed into the symmetric Maxwell stress $\boldsymbol{\sigma}^{\text{max}}$ and a non-symmetric polarization stress $\boldsymbol{\sigma}^{\text{pol}}$ (Vogel et al, 2014; Steinmann, 2011; Eringen, 1980; Eringen and Maugin, 2012)

$$\boldsymbol{\sigma}^{\text{max}} = -\frac{1}{2}\varepsilon_0[\mathbb{e}\cdot\mathbb{e}]\boldsymbol{i} + \varepsilon_0\mathbb{e}\otimes\mathbb{e}, \quad \boldsymbol{\sigma}^{\text{pol}} = \mathbb{e}\otimes\mathbb{p}, \tag{28.9}$$

where $\boldsymbol{i}$ represents the second order spatial identity tensor. If we combine the mechanical stress $\boldsymbol{\sigma}$ with the ponderomotive stress $\boldsymbol{\sigma}^{\text{pon}}$ we can define a total Cauchy-type symmetric stress tensor (Dorfmann and Ogden, 2005a; Bustamante, 2009b; Dorfmann and Ogden, 2005b, 2003, 2004)

$$\boldsymbol{\sigma}^{\text{tot}} = \boldsymbol{\sigma} + \boldsymbol{\sigma}^{\text{pon}}. \tag{28.10}$$

In order to complete the description of the electro-mechanical balance equations, we will now define the behavior at the boundary of the body $\partial\mathscr{B}_t$. For this we define $[\![\bullet]\!] = [\bullet]^{\text{out}} - [\bullet]^{\text{in}}$ as the jump of a quantity at the interface between the bulk material and the free space surrounding the body. The total stress tensor $\boldsymbol{\sigma}^{\text{tot}}$ must satisfy

$$-[\![\boldsymbol{\sigma}^{\text{tot}}]\!]\cdot\boldsymbol{n} = \boldsymbol{t}_t^{\text{p}}, \quad \text{on } \partial\mathscr{B}_t, \tag{28.11}$$

where $\boldsymbol{t}_t^{\text{p}}$ are the imposed mechanical tractions and $\boldsymbol{n}$ is the surface normal on $\partial\mathscr{B}_t$ pointing outwards. This formulation simplifies to the standard boundary condition if we neglect the influence of the free space. In this case the boundary $\partial\mathscr{B}_t$ is split into a Dirichlet boundary $\partial\mathscr{B}_t^{\varphi}$ and a Neumann boundary $\partial\mathscr{B}_t^{\text{t}}$ where mechanical tractions $\boldsymbol{t}_t^{\text{p}}$ are imposed

$$\boldsymbol{\varphi} = \boldsymbol{\varphi}^{\text{p}}, \quad \text{on } \partial\mathscr{B}_t^{\varphi}, \quad \boldsymbol{\sigma}^{\text{tot}}\cdot\boldsymbol{n} = \boldsymbol{t}_t^{\text{p}}, \quad \text{on } \partial\mathscr{B}_t^{\text{t}}. \tag{28.12}$$

For the spatial electric field vector $\mathbb{e}$ and the electric displacement $\mathbb{d}$ we can formulate similar jump conditions at the boundary of the body in the form (Eringen, 1980; Eringen and Maugin, 2012; Kovetz, 2000; Kost, 1994)

$$[\![\mathbb{e}]\!]\times\boldsymbol{n} = \boldsymbol{0} \quad \text{and} \quad [\![\mathbb{d}]\!]\cdot\boldsymbol{n} = \hat{\rho}_t^f, \quad \text{on } \partial\mathscr{B}_t, \tag{28.13}$$

with the density of free surface charges per deformed area $\hat{\rho}_t^f$ (Vogel et al, 2014). If we consider the influence of the free space as negligible these jump conditions can be transformed into classical boundary conditions. Recalling the relationship in Eq. (28.5), i.e. the electric field $\mathbb{e}$ can be defined as the gradient of a scalar potential ϕ

we split the boundary $\partial\mathscr{B}_t$ into a part $\partial\mathscr{B}_t^\phi$ on which Dirichlet boundary conditions for the electric potential are prescribed and $\partial\mathscr{B}_t^\rho$ where Neumann conditions have to be satisfied

$$\phi = \phi^{\mathrm{p}}, \quad \text{on } \partial\mathscr{B}_t^\phi \quad \text{and} \quad \mathbb{d}\cdot\boldsymbol{n} = \hat{\rho}_t^f, \quad \text{on } \partial\mathscr{B}_t^\rho. \tag{28.14}$$

28.1.2.2 Material Configuration

The relations presented in the previous section are now transformed into the material configuration, starting with the electric field, the electric displacement and the polarization

$$\mathbb{E} = \boldsymbol{F}^T\cdot\mathbb{e}, \quad \mathbb{D} = J\boldsymbol{F}^{-1}\cdot\mathbb{d}, \quad \mathbb{P} = J\boldsymbol{F}^{-1}\cdot\mathbb{p}. \tag{28.15}$$

With these quantities in the undeformed configuration the Maxwell equations (28.4) take the form

$$\mathrm{Div}\,\mathbb{D} = 0, \quad \mathrm{Curl}\,\mathbb{E} = \boldsymbol{0} \quad \text{in } \mathscr{B}_0. \tag{28.16}$$

Here Div and Curl are the corresponding differential operators defined with respect to the material position vector $\boldsymbol{X}$. If we derive the electric field $\mathbb{E}$ from a scalar potential such that

$$\mathbb{E} = -\mathrm{Grad}\,\phi, \quad \text{in } \mathscr{B}_0, \tag{28.17}$$

we ensure that Eq. $(28.16)_2$ is satisfied. When we define the vacuum electric displacement as $\mathbb{D}^\varepsilon := \varepsilon_0 J\boldsymbol{C}^{-1}\cdot\mathbb{E}$ the electric displacement $\mathbb{D}$ in matter can be formulated as

$$\mathbb{D} = \mathbb{D}^\varepsilon + \mathbb{P} \quad \text{in } \mathscr{B}_0, \tag{28.18}$$

which follows directly from a pull-back of the spatial format given in Eq. (28.3). We can derive the total Piola stress tensor $\boldsymbol{P}^{\mathrm{tot}}$ as the material counterpart of the total Cauchy stress as $\boldsymbol{\sigma}^{\mathrm{tot}}$ as

$$\boldsymbol{P}^{\mathrm{tot}} = J\boldsymbol{\sigma}^{\mathrm{tot}}\cdot\boldsymbol{F}^{-T}. \tag{28.19}$$

This stress measure can be decomposed into a mechanical part $\boldsymbol{P}$ and a ponderomotive part $\boldsymbol{P}^{\mathrm{pon}}$ containing the polarization stress $\boldsymbol{P}^{\mathrm{pol}}$ and the Maxwell stress $\boldsymbol{P}^{\max}$

$$\boldsymbol{P}^{\mathrm{tot}} = \boldsymbol{P} + \boldsymbol{P}^{\mathrm{pon}} = \boldsymbol{P} + \boldsymbol{P}^{\max} + \boldsymbol{P}^{\mathrm{pol}} \tag{28.20}$$

with

$$\boldsymbol{P}^{\mathrm{pol}} = \mathbb{e}\otimes\mathbb{P}, \; \boldsymbol{P}^{\max} = -\frac{1}{2}\varepsilon_0 J\boldsymbol{C}^{-1} : [\mathbb{E}\otimes\mathbb{E}]\boldsymbol{F}^{-T} + \mathbb{e}\otimes\mathbb{D}^\varepsilon. \tag{28.21}$$

The balance of linear momentum (28.6) in the undeformed configuration can therefore be translated to

$$\mathrm{Div}\,\boldsymbol{P}^{\mathrm{tot}} + \boldsymbol{b}_0 = \boldsymbol{0}. \tag{28.22}$$

To conclude the material description we transform the jump conditions into

$$[\![\mathbb{D}]\!]\cdot\boldsymbol{N} = \hat{\rho}_0^f, \quad \text{and} \quad [\![\boldsymbol{P}^{\mathrm{tot}}]\!]\cdot\boldsymbol{N} = \boldsymbol{t}_0^p, \quad \text{on } \partial\mathscr{B}_0, \tag{28.23}$$

which renders the standard boundary conditions in the case that the free space is not considered

$$\mathbb{D} \cdot \boldsymbol{N} = \hat{\rho}_0^f, \quad \text{on } \partial \mathscr{B}_0^\rho, \\ \boldsymbol{P}^{\text{tot}} \cdot \boldsymbol{N} = \boldsymbol{t}_0^p, \quad \text{on } \partial \mathscr{B}_0^t. \tag{28.24}$$

28.1.3 Heat Equation

We will now present various basic equations in thermo-electro-elasticity. The reader is referred to the works of Holzapfel (2000); Erbts et al (2015); Miehe (1995) and Vogel (2015) for a description in further detail.

The starting point is the first law of thermodynamics. It states that the rate of the energy $\mathscr{E}$, that can be decomposed into the internal part $\mathscr{U}$ and a kinetic energy contribution $\mathscr{K}$, has to be equal to the rate of work done on a continuum body which comprises of the sum of the external mechanical power $\mathscr{P}_{\text{ext}}$ and the non-mechanical power $\mathscr{Q}$ combining thermal power $\mathscr{Q}^{\text{thm}}$ and ponderomotive power density $\mathscr{Q}^{\text{pon}}$ (Holzapfel, 2000)

$$\dot{\mathscr{E}} = \dot{\mathscr{U}} + \dot{\mathscr{K}} = \mathscr{P}_{\text{ext}} + \mathscr{Q} = \mathscr{P}_{\text{ext}} + \mathscr{Q}^{\text{thm}} + \mathscr{Q}^{\text{pon}}. \tag{28.25}$$

Following Vogel (2015) for the quasi static case, we can derive the local form of the balance of energy in the material configuration as

$$\dot{\mathscr{U}} = \boldsymbol{P} : \dot{\boldsymbol{F}} - \text{Div}\, \boldsymbol{Q} + \mathscr{R} + \mathbb{E} \cdot \dot{\mathbb{P}} + \boldsymbol{P}^{\text{pol}} : \dot{\boldsymbol{F}} \tag{28.26}$$

Here we have introduced the change in the internal energy density per unit undeformed volume $\dot{\mathscr{U}}$, the heat sources $\mathscr{R}$ and the heat flux vector $\boldsymbol{Q}$, expressed in the undeformed configuration. The heat flux can be calculated from the gradient of the absolute temperature Θ using the relation

$$\boldsymbol{Q} := -\kappa J \boldsymbol{C}^{-1} \cdot \text{Grad}\, \Theta \tag{28.27}$$

with the isotropic heat conductivity κ. This can be transformed to the Clausius-Duhem inequality by introducing the entropy H, which gives a formulation of the dissipation power density $\mathscr{D} = \mathscr{D}(\boldsymbol{X}, t) \geq 0$ in the form

$$\mathscr{D} = \Theta \dot{H} - \frac{\boldsymbol{Q}}{\Theta} \cdot \text{Grad}\, \Theta - \dot{\mathscr{U}} + \boldsymbol{P} : \dot{\boldsymbol{F}} + \mathbb{E} \cdot \dot{\mathbb{P}} + \boldsymbol{P}^{\text{pol}} : \dot{\boldsymbol{F}} \geq 0. \tag{28.28}$$

The dissipative term consists of two contributions, the dissipation power density due to heat conduction

$$\mathscr{D}^{\text{con}} = -\frac{\boldsymbol{Q}}{\Theta} \cdot \text{grad}\, \Theta \geq 0$$

and the local dissipation power density $\mathscr{D}^{\text{loc}}$ Vu (2014), that can be expressed in form of the Clausius-Planck inequality as

$$\mathscr{D}^{\text{loc}} = \Theta \dot{H} - \dot{\mathscr{U}} + \boldsymbol{P} : \dot{\boldsymbol{F}} + \mathbb{E} \cdot \dot{\mathbb{P}} + \boldsymbol{P}^{\text{pol}} : \dot{\boldsymbol{F}} \geq 0. \tag{28.29}$$

This inequality holds for any irreversible process whereas the local dissipation term vanishes for a reversible process. We introduce the free energy density Ψ that depends on the current state of deformation, the electric field and the temperature, resulting from a double Legendre transformation (Coleman and Noll, 1963) in the form

$$\Psi(\boldsymbol{F}, \Theta, \mathbb{E}) = \mathscr{U} - \Theta H - \mathbb{E} \cdot \mathbb{P}, \tag{28.30}$$

rendering the Clausius-Planck inequality

$$\mathscr{D}^{\text{loc}} = -\dot{\Psi} - \dot{\Theta} H + [\boldsymbol{P} + \boldsymbol{P}^{\text{pol}}] : \dot{\boldsymbol{F}} - \dot{\mathbb{E}} \cdot \mathbb{P} \geq 0. \tag{28.31}$$

When we finally choose to take into account the energy stored in the electric field itself, the free energy density $\Psi(\boldsymbol{F}, \Theta, \mathbb{E})$ has to be amended by the term

$$E(\boldsymbol{F}, \mathbb{E}) = -\frac{1}{2} \varepsilon_0 J [\mathbb{E} \otimes \mathbb{E}] : \boldsymbol{C}^{-1}$$

(Dorfmann and Ogden, 2003, 2004) leading to the amended total free energy density per unit volume in $\mathscr{B}_0$

$$\Omega(\boldsymbol{F}, \Theta, \mathbb{E}) = \Psi(\boldsymbol{F}, \Theta, \mathbb{E}) + E(\boldsymbol{F}, \mathbb{E}). \tag{28.32}$$

The Clausius-Duhem inequality is transformed accordingly to a formulation containing the total Piola stress

$$\mathscr{D}^{\text{loc}} = -\dot{\Omega} - \dot{\Theta} H + \boldsymbol{P}^{\text{tot}} : \dot{\boldsymbol{F}} - \dot{\mathbb{E}} \cdot \mathbb{D} \geq 0. \tag{28.33}$$

From this formulation we can derive the constitutive relations for the total Piola stress, the electric displacement and the entropy (Mehnert et al, 2016)

$$\boldsymbol{P}^{\text{tot}} = \frac{\partial \Omega}{\partial \boldsymbol{F}}, \quad \text{with} \quad \boldsymbol{P}^{\text{max}} = \frac{\partial E}{\partial \boldsymbol{F}}, \quad \mathbb{D} = -\frac{\partial \Omega}{\partial \mathbb{E}}, \quad H = -\frac{\partial \Omega}{\partial \Theta}. \tag{28.34}$$

If we combine Eq. (28.26) with the Clausius-Planck inequality (28.31) we find the first law of thermodynamics in entropy form

$$\Theta \dot{H} = \mathscr{R} - \text{Div} \boldsymbol{Q} + \mathscr{D}^{\text{loc}}, \tag{28.35}$$

The term $\Theta \dot{H}$ can be expanded with the total free energy density (28.32) using the chain rule, c.f. Mehnert et al (2016),

$$\Theta \dot{H} = -\Theta \frac{\partial^2 \Omega}{\partial \Theta \partial \Theta} \dot{\Theta} - \Theta \frac{\partial^2 \Omega}{\partial \boldsymbol{F} \partial \Theta} : \dot{\boldsymbol{F}} - \Theta \frac{\partial^2 \Omega}{\partial \mathbb{E} \partial \Theta} \cdot \dot{\mathbb{E}}. \tag{28.36}$$

If we recall the definition of the heat capacity $c_0 = -\Theta \dfrac{\partial^2 \Omega}{\partial \Theta^2}$, c.f. Holzapfel and Simo (1996); Elahinia et al (2013); Vertechy et al (2010); Santapuri et al (2013);

Santapuri (2012), the first law of thermodynamics can be transformed into the heat conduction equation

$$c_0\dot{\Theta} = \mathscr{R} - \mathrm{Div}\boldsymbol{Q} + \underbrace{\Theta\partial_\Theta\left[\boldsymbol{P}^{\mathrm{tot}} : \dot{\boldsymbol{F}} - \mathbb{D}\cdot\dot{\mathbb{E}}\right]}_{\mathscr{H}} + \mathscr{D}^{\mathrm{loc}}. \tag{28.37}$$

The structural thermo-mechanical and thermo-electric heating/cooling effect are here combined to the term $\mathscr{H}$. To finalize the formulation we split the boundary of the body into a part $\partial\mathscr{B}_0^\Theta$ with Dirichlet boundary conditions for the temperature, i.e. $\Theta = \Theta^p$ on $\partial\mathscr{B}_0^\Theta$, and a part $\partial\mathscr{B}_0^q$ with Neumann boundary conditions for the heat flux, i.e. $\boldsymbol{Q}\cdot\boldsymbol{N} = \overline{Q}$ on $\partial\mathscr{B}_0^q$, in such a way that $\partial\mathscr{B}_C = \partial\mathscr{B}_C^\Theta \bigcup \partial\mathscr{B}_0^q$ and $\partial\mathscr{B}_0^\Theta \bigcap \partial\mathscr{B}_0^q = \emptyset$.

28.1.4 Energy Function

We will now derive a thermo-electro-mechanically coupled free energy function. As a starting point we use the heat capacity $c_{\boldsymbol{F},\mathbb{E}}$ at constant deformation and constant electric field that is assumed to be constant as a simple first attempt to model the thermo-electro-mechanical behavior of elastomers

$$c_{\boldsymbol{F},\mathbb{E}}(\Theta) = c_{\boldsymbol{F},\mathbb{E}}(\Theta_0) = c_0. \tag{28.38}$$

Here Θ_0 is a reference temperature. With the definition of the heat capacity c_0 used in the previous section we can derive a thermodynamically consistent formulation for the free energy density Ψ by integrating the expression twice as presented in Mehnert et al (2016). This leads to the formulation

$$\Psi(\boldsymbol{F},\Theta,\mathbb{E}) = c_0\left[\Theta - \Theta_0 - \Theta\ln\left(\frac{\Theta}{\Theta_0}\right)\right] - \left[\Theta - \Theta_0\right]M_1(\boldsymbol{F},\mathbb{E}) + \mathscr{W}(\boldsymbol{F},\mathbb{E}). \tag{28.39}$$

The first term on the right side of Eq. (28.39) can be identified as the capacitive contribution to the free energy density, whereas the second term is the coupled contribution and the last is an expression for the isothermal energy contribution. Following Mehnert et al (2016) we can transform $\Psi(\boldsymbol{F},\Theta,\mathbb{E})$ into

$$\Psi(\boldsymbol{F},\Theta,\mathbb{E}) = c_0\left[\Theta - \Theta_0 - \Theta\ln\left(\frac{\Theta}{\Theta_0}\right)\right] - \left[\Theta - \Theta_0\right]M(J) + \frac{\Theta}{\Theta_0}W(\boldsymbol{F},\mathbb{E}). \tag{28.40}$$

In this context, the term $M(J)$ describes the volumetric contribution to the free energy. This expression can be introduced into the definition of the total free energy density (28.32). We consider the electro-elastic material to be isotropic which is closely connected to the theory of isotropic tensor functions Vu (2014) and can therefore be expressed using the representation theorem (Spencer, 2013; Wang, 1970) that states that a scalar-valued function $\mathscr{F} = \mathscr{F}(\boldsymbol{A},\boldsymbol{b})$ is invariant with respect to the proper orthogonal group $SO(3)$ if and only if it can be expressed as a function of the six principal invariants of $\boldsymbol{A}$ and $\boldsymbol{b}$

$$\mathscr{F}(\boldsymbol{A},\boldsymbol{b}) = \mathscr{F}(I_{1\boldsymbol{A}}, I_{2\boldsymbol{A}}, I_{3\boldsymbol{A}}, I_{4\boldsymbol{b}}, I_{5\boldsymbol{Ab}}, I_{6\boldsymbol{Ab}}). \tag{28.41}$$

Thus we define the isothermal energy function $W(\boldsymbol{F},\mathbb{E})$ as an isotropic function depending on the right Cauchy-Green tensor $\boldsymbol{C}$ and the electric field vector $\mathbb{E}$. As presented in e.g. Dorfmann and Ogden (2005a); Vu (2014); Spencer (2013) the first three purely mechanical invariants take the form

$$I_1 = \text{tr}\,(\boldsymbol{C}), \qquad I_2 = \frac{1}{2}\left[[\text{tr}\,(\boldsymbol{C})]^2 - \text{tr}\,(\boldsymbol{C}^2)\right], \qquad I_3 = \det\,(\boldsymbol{C}), \tag{28.42}$$

and the fourth invariant, which is a purely electric quantity, can be expressed as

$$I_4 = [\mathbb{E}\otimes\mathbb{E}] : \boldsymbol{I}. \tag{28.43}$$

This invariant can be physically interpreted as the square of the electric field strength in the undeformed configuration. While these four invariants are basically consistent throughout the literature, different formulations can be found for the remaining two invariants describing the coupling between the electric field and the mechanical deformation. On the one hand in a number of contributions (Vu, 2014; Ask et al, 2012; Bustamante, 2010, 2009a; Vogel, 2015) the coupled invariants are formulated using the right Cauchy-Green tensor

$$I_5 = [\mathbb{E}\otimes\mathbb{E}] : \boldsymbol{C}, \qquad I_6 = [\mathbb{E}\otimes\mathbb{E}] : \boldsymbol{C}^2. \tag{28.44}$$

The physical interpretation of this formulation is not quite straight forward. Following the notion in Dorfmann and Ogden (2005a) that the electric field plays a similar role in electro-elasticity as the preferred direction for transversely isotropic elastic materials, one attempt of a physical interpretation can be made analogously to Holzapfel and Ogden (2010). We assume the electric field vector as a normal vector to an area-like quantity comparable to an area element $\mathrm{d}a$ in the spatial configuration and $\mathrm{d}A$ in the material configuration. Using Nanson's formula it can be shown that I_5 is a measure for the ratio $\mathrm{d}a^2/\mathrm{d}A^2$, hence a measure of the changes of this area-like quantity normal to the electric field.

In contrast to this in Dorfmann and Ogden (2005a); Holzapfel and Ogden (2010); Merodio and Ogden (2002); Dorfmann and Ogden (2014) a formulation based on the inverse of the right Cauchy-Green tensor can be found which will be labeled $\widetilde{I}_5$ in the context of this work

$$\widetilde{I}_5 = [\mathbb{E}\otimes\mathbb{E}] : \boldsymbol{C}^{-1}, \qquad \widetilde{I}_6 = [\mathbb{E}\otimes\mathbb{E}] : \boldsymbol{C}^{-2}. \tag{28.45}$$

A nice feature of this formulation is that the fifth invariant can be physically interpreted as the square of the electric field strength in the deformed configuration as

$$\widetilde{I}_5 = [\mathbb{E}\otimes\mathbb{E}] : \boldsymbol{C}^{-1} = \mathbb{E}\cdot\boldsymbol{C}^{-1}\cdot\mathbb{E}^T = |\mathbb{e}|^2. \tag{28.46}$$

Both of these formulation are invariant to the proper orthogonal group $SO(3)$ and are therefore valid options for the free energy function of an isotropic material. Nevertheless the choice of the coupled invariant can have a significant influence on the material response as we will demonstrate in Sect. 28.2.

28.2 Non-Homogeneous Boundary Value Problems

In the following section we will present two different boundary value problems as examples of how the choice of the coupled invariant can influence the material response. The isothermal energy function used will take one of the following forms

$$\begin{aligned} W(\boldsymbol{C},\mathbb{E}) &= \mu(I_4)[I_1-3]+c_1 I_4+c_2 I_5, \\ \widetilde{W}(\boldsymbol{C},\mathbb{E}) &= \mu(I_4)[I_1-3]+c_1 I_4+\widetilde{c}_2\widetilde{I}_5. \end{aligned} \tag{28.47}$$

In both cases the material response is based on a Neo-Hookean type material with a field dependent shear modulus $\mu(I_4)$. For the sake of simplicity in the analytical calculations this field dependency is assumed to be linear (Mehnert et al, 2016; Bustamante, 2010), i.e. $\mu(I_4)=\mu_0+g_1 I_4$, with the zero field shear modulus μ_0. For $g_1<0$ this leads to a softening whereas $g_1>0$ results in a hardening of the material when an electric field is applied. The coupling between the electric field and the mechanical deformation is implemented by the fifth invariant with the coupling parameter c_2 and $\widetilde{c}_2$ respectively. The parameter c_1 vanishes in the following analytical calculations, it should be noted though that we would have to select $c_1<0$ to ensure that the free energy is concave in $\mathbb{E}$.

As the focus of this contribution lies on the effect of the electro-mechanical coupling invariant on the material behavior, we will neglect the deformation due to thermal expansion in the following calculations. This is done because i) the expected deformation from thermal expansion will be significantly smaller when compared to the deformations caused by the electric field or the prescribed displacement and ii) the effect of the thermal expansion on the results does not influence the conclusion on the role of the selected coupling invariant. With this in mind we assume the material to be incompressible at constant temperature. Thus it holds that the contribution $M(J)$ in (28.40) vanishes and the definition of the total Cauchy stress can be expressed for

- $I_5=[\mathbb{E}\otimes\mathbb{E}]:\boldsymbol{C}$ as

$$\begin{aligned} \boldsymbol{\sigma}^{\text{tot}} &= \boldsymbol{b}\Omega_1+[[\boldsymbol{C}:\boldsymbol{I}]\boldsymbol{b}-\boldsymbol{b}^2]\Omega_2+[\boldsymbol{b}\mathbb{e}\otimes\boldsymbol{b}\mathbb{e}]\Omega_5 \\ &\quad -\frac{1}{2}\varepsilon_0[\mathbb{E}\otimes\mathbb{E}]:[\boldsymbol{F}^{-T}\overline{\otimes}\boldsymbol{F}^{-1}]-p\boldsymbol{I}. \end{aligned} \tag{28.48}$$

- $\widetilde{I}_5=[\mathbb{E}\otimes\mathbb{E}]:\boldsymbol{C}^{-1}$ as

$$\begin{aligned}\widetilde{\boldsymbol{\sigma}}^{\text{tot}} =& \boldsymbol{b}\Omega_1 + [[\boldsymbol{C}:\boldsymbol{I}]\boldsymbol{b} - \boldsymbol{b}^2]\Omega_2 - [\mathbb{e}\otimes\mathbb{e}]\widetilde{\Omega}_5 \\ &- \frac{1}{2}\varepsilon_0[\mathbb{E}\otimes\mathbb{E}] : [\boldsymbol{F}^{-T}\overline{\otimes}\boldsymbol{F}^{-1}] - p\boldsymbol{I}.\end{aligned} \tag{28.49}$$

Here the parameter p is the Lagrange multiplier connected to the incompressibility constraint and Ω_i is the derivative of the augmented energy function Ω with respect to the invariant I_i, whereas $\widetilde{\Omega}_5$ is the derivative of the free energy $\widetilde{\Omega}$ with respect to $\widetilde{I}_5$

$$\begin{aligned}&\Omega_1 = \frac{\Theta}{2\Theta_0}[\mu_0 + g_1 I_4], \quad \Omega_2 = \Omega_6 = 0, \quad \Omega_4 = \frac{\Theta}{\Theta_0}c_1 + \frac{\Theta}{\Theta_0}\frac{g_1}{2}[I_1 - 3], \\ &\Omega_5 = c_2\frac{\Theta}{\Theta_0}, \quad \widetilde{\Omega}_5 = \widetilde{c}_2\frac{\Theta}{\Theta_0}.\end{aligned} \tag{28.50}$$

Furthermore we have introduced the special dyadic product (Wriggers, 2008; Holzapfel, 2000)

$$\boldsymbol{A}\overline{\otimes}\boldsymbol{B} = [A_{ij}\boldsymbol{e}_i\otimes\boldsymbol{e}_j]\overline{\otimes}[B_{kl}\boldsymbol{e}_k\otimes\boldsymbol{e}_l] = A_{ik}B_{kl}\boldsymbol{e}_i\otimes\boldsymbol{e}_j\otimes\boldsymbol{e}_k\otimes\boldsymbol{e}_l. \tag{28.51}$$

28.2.1 Deformation of a Cube with a Uniaxially Applied Electric Field

As a first example we will focus on a cube consisting of electro-active material that deforms due to a uniaxially applied electric field resulting from a potential difference $\Delta\phi$. The sides with normal vector in direction $\boldsymbol{e}_1$ are labeled as top/bottom, with normal vector in direction $\boldsymbol{e}_2$ as left/right and with normal vector in direction $\boldsymbol{e}_3$ as front/back. We assume that the bottom surface of the cube is fixed in $\boldsymbol{e}_1$ direction whereas all other sides are free to deform. In this example for the sake of simplicity we will restrict ourself to the isothermal case and also neglect the field sensitivity of the shear modulus, therefore we choose $g_1 = 0$. The zero field shear modulus is selected as $\mu_0 = 5$ MPa, similar to Vu et al (2007) . Using the incompressibility condition $\lambda_1\lambda_2\lambda_3 = 1$, where λ_i is the stretch in the respective coordinate direction, and assuming a symmetric deformation we can formulate the deformation gradient depending exclusively on the stretch $\lambda = \lambda_1$ in the direction of the electric field $\mathbb{E}$ as

$$\boldsymbol{F} = \begin{bmatrix}\lambda & 0 & 0\\ 0 & \lambda^{-0.5} & 0\\ 0 & 0 & \lambda^{-0.5}\end{bmatrix}, \quad \boldsymbol{C} = \boldsymbol{b} = \begin{bmatrix}\lambda^2 & 0 & 0\\ 0 & \lambda^{-1} & 0\\ 0 & 0 & \lambda^{-1}\end{bmatrix}, \quad \mathbb{E} = \begin{bmatrix}E_0\\ 0\\ 0\end{bmatrix}. \tag{28.52}$$

This deformation gradient in combination with the derived equations for the total Cauchy stress (28.48) and (28.49) results in the value σ_{11}^{tot} and respectively $\widetilde{\sigma}_{11}^{\text{tot}}$ in the direction of the electric field as

$$\begin{aligned} \sigma_{11}^{\text{tot}} &= \mu\lambda^2 + 2c_2 E_0^2 \lambda^2 - \frac{1}{2}\varepsilon_0 E_0^2 \lambda^2 - p, \\ \widetilde{\sigma}_{11}^{\text{tot}} &= \mu\lambda^2 - 2\widetilde{c}_2 E_0^2 \lambda^{-2} - \frac{1}{2}\varepsilon_0 E_0^2 \lambda^2 - p. \end{aligned} \tag{28.53}$$

The Lagrange multiplier can be identified as $p = \mu\lambda^{-1}$ using the definition of the stress in direction perpendicular to the applied electric field. When we assume that the upper surface of the cube is stress free in normal direction we can calculate the resulting stretch from the applied electric field $\mathbb{E}$ which is depicted in Fig. 28.1 for c_2 and various values of $\widetilde{c}_2$. The validity of these results is checked with numerical results obtained from an in-house FEM-Code and are therefore deemed to be correct.

We expect the electro-active material to contract in the direction of the applied electric field which is the case for the invariant $I_5 = [\mathbb{E} \otimes \mathbb{E}] : \boldsymbol{C}$ with a positive value for c_2. In contrast to this the material does not show the expected response to the electric loading for $\widetilde{I}_5 = [\mathbb{E} \otimes \mathbb{E}] : \boldsymbol{C}^{-1}$ with a positive value for $\widetilde{c}_2$ as the resulting stretch λ is larger than one. By simply switching the sign of $\widetilde{c}_2$ we can achieve a physically sensible material response and by selecting a value of $\widetilde{c}_2 = -4.9799 N/V^2$ we can fit the material response with $\widetilde{I}_5$ to the one calculated with I_5. It should be

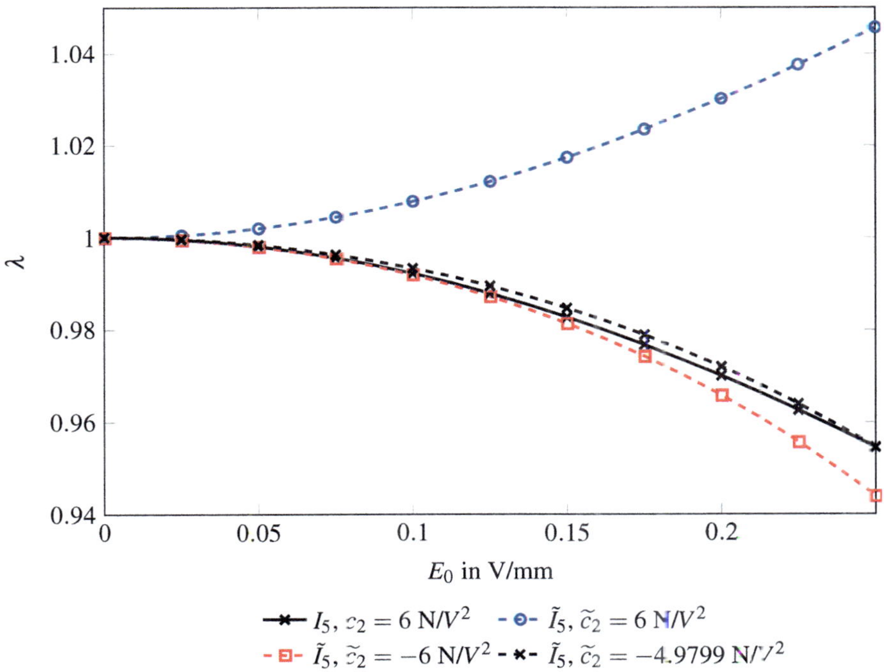

Fig. 28.1: Plot of the stretch λ resulting from the applied electric field with the field strength E_0

noted though that the selection of a negative value for $\widetilde{c}_2$ comes with the downside that the expression for $\widetilde{\sigma}_{11}^{\text{tot}}$ found in Eq. (28.53) only gives real valued results until a threshold value of E_0 is reached. After this we will find only complex solutions. For this simple example the selection of $\widetilde{I}_5$ as the coupled invariant therefore is only sensible for a very restricted amplitude of the electric field, whereas there is no restriction if we choose I_5.

28.2.2 Extension and Torsion of a Cylindrical Tube

As a second example we are going to investigate a cylindrical tube under thermo-electro-mechanical loading. Due to the small value of the vacuum permittivity the energy stored in the electric field $E(\boldsymbol{F},\mathbb{E})$ will be neglected, c.f. Vu et al (2007). In the considered case it is reasonable to work in the cylindrical coordinates (R,Φ,Z) with the material unit basis vectors $(\boldsymbol{E}_R,\boldsymbol{E}_\Phi,\boldsymbol{E}_Z)$. The corresponding quantities in the spatial configuration are defined as (r,ϕ,z) and $(\boldsymbol{e}_r,\boldsymbol{e}_\phi,\boldsymbol{e}_z)$. The geometry of the cylinder in the deformed configuration can be described by

$$a_i \leq r \leq a_e, \quad 0 \leq \phi \leq 2\pi, \quad -\infty < z < \infty, \tag{28.54}$$

where the internal radius is denoted by a_i and the external radius by a_e. The length of the cylinder is assumed to be infinite in order to avoid problems with the end conditions of a finitely long tube. In the undeformed configuration the geometry of the tube can be expressed by

$$A_i \leq R \leq A_e, \quad 0 \leq \Phi \leq 2\pi, \quad -\infty < Z < \infty, \tag{28.55}$$

We introduce the ratio of the undeformed external radius to the undeformed internal radius $\zeta = \frac{A_e}{A_i}$ as a measure for the wall thickness of the cylinder. The prescribed deformation that is considered in this example is a combination of an axial stretch characterized by the parameter λ_z and a torsion of the angle τ around the cylinder axis. Thus the transformation of the undeformed to the deformed coordinates reads

$$r = \lambda_z^{-1/2} R, \quad \phi = \Phi + \lambda_z \tau Z, \quad z = \lambda_z Z. \tag{28.56}$$

This mechanical loading results in a deformation gradient in the form

$$\boldsymbol{F} = \begin{pmatrix} \lambda_z^{-1/2} & 0 & 0 \\ 0 & \lambda_z^{-1/2} & \tau r \lambda_z \\ 0 & 0 & \lambda_z \end{pmatrix} = \begin{pmatrix} \lambda_z^{-1/2} & 0 & 0 \\ 0 & \lambda_z^{-1/2} & \gamma \lambda_z \\ 0 & 0 & \lambda_z \end{pmatrix}, \tag{28.57}$$

where we have introduced the definition of $\gamma = r\tau$ and used the incompressibility constraint $\det(\boldsymbol{F}) = 1$. Furthermore we will apply a purely axial electric field characterized by the field strength E_0 and a radial temperature gradient. With the given deformation gradient and an axial electric field we can use the relations in Eqs.

(28.48) and (28.49) to calculate the respective non-zero entries of the total Cauchy stress. For the coupling invariant I_5 this results in

$$\begin{aligned}
\sigma_{rr}^{\text{tot}} &= 2\Omega_1 \lambda_z^{-1} - p, \\
\sigma_{\phi\phi}^{\text{tot}} &= 2\Omega_1 [\lambda_z^{-1} + \gamma^2 \lambda_z^2] + 2\Omega_5 E_0^2 \gamma^2 \lambda_z^2 - p, \\
\sigma_{zz}^{\text{tot}} &= 2\Omega_1 \lambda_z^2 + 2\Omega_5 E_0^2 \lambda_z^2 - p, \\
\sigma_{z\phi}^{\text{tot}} &= \sigma_{\phi z}^{\text{tot}} = 2\Omega_1 \gamma \lambda_z^2 + 2\Omega_5 E_0^2 \gamma \lambda_z^2,
\end{aligned} \tag{28.58}$$

whereas the selection of $\widetilde{I}_5$ leads to the stress entries

$$\begin{aligned}
\widetilde{\sigma}_{rr}^{\text{tot}} &= 2\Omega_1 \lambda_z^{-1} - p, \\
\widetilde{\sigma}_{\phi\phi}^{\text{tot}} &= 2\Omega_1 [\lambda_z^{-1} + \gamma^2 \lambda_z^2] - p, \\
\widetilde{\sigma}_{zz}^{\text{tot}} &= 2\Omega_1 \lambda_z^2 - 2\Omega_5 E_0^2 \lambda_z^{-2} - p, \\
\widetilde{\sigma}_{z\phi}^{\text{tot}} &= \sigma_{\phi z}^{\text{tot}} = 2\Omega_1 \gamma \lambda_z^2.
\end{aligned} \tag{28.59}$$

As we apply a radial temperature gradient we need to find a solution to the heat conduction equation (28.37). In order to find an analytical solution first we have to simplify this equation. We will assume a steady state temperature field without heat sources. Furthermore we neglect the thermo-mechanical and thermo-electric heating effect $\mathscr{H}$ and local dissipation effects. All these assumptions simplify the heat conduction equation to the Laplace equation

$$\Delta\Theta = 0 \tag{28.60}$$

with Δ as the Laplace operator. For an axial symmetric problem as the torsion of a hollow cylinder, this equation can be transformed to the cylindrical coordinates (r, ϕ, z) as

$$\frac{\partial^2 \Theta(r)}{\partial r^2} + \frac{1}{r}\frac{\partial \Theta(r)}{\partial r} = 0. \tag{28.61}$$

One possible analytical solution to this equation found for example in Bland (1956); Rajagopal and Huang (1994) is

$$\Theta(r) = k_1 + k_2 \ln(r) \tag{28.62}$$

with the constants k_1 and k_2 depending on the deformed internal and external radius of the cylinder and on the respective temperature

$$k_1 = \frac{\Theta(a_i)\ln(a_e) - \Theta(a_e)\ln(a_i)}{\ln(a_e) - \ln(a_i)}, \quad k_2 = \frac{\Theta(a_e) - \Theta(a_i)}{\ln(a_e) - \ln(a_i)}. \tag{28.63}$$

We will investigate the influence of the coupling invariant by analyzing the torque $\mathscr{M}$ that has to be applied in order to achieve the prescribed deformation. The torque can defined as the integral over the cross section of the cylinder of the mechanical stress in azimuthal direction

$$\mathscr{M} = 2\pi \int_{a_i}^{a_e} \sigma_{z\phi}\ r^2 \mathrm{d}r = 2\pi \int_{a_i}^{a_e} (\sigma_{z\phi}^{\mathrm{tot}} - \sigma_{z\phi}^{\mathrm{max}})\ r^2 \mathrm{d}r. \tag{28.64}$$

As the Maxwell stress $\sigma_{z\phi}^{\mathrm{max}}$ for the applied electric field vanishes, this definition reduces to

$$\mathscr{M} = 2\pi \int_{a_i}^{a_e} \sigma_{z\phi}^{\mathrm{tot}}\ r^2 \mathrm{d}r. \tag{28.65}$$

Using the definition of the Cauchy stress (28.58) and (28.59) combined with the derivatives of the energy function (28.50) and the solution of the heat equation (28.62) this results in the following expression for the torque

- for the invariant I_5

$$\begin{aligned}\mathscr{M} =& \frac{4\pi\tau\lambda_z^2}{\Theta_0}[0.5[g_0 + g_1 E_0^2] + c_2 E_0^2] \times \\ & \left[\frac{k_1}{4}[a_e^4 - a_i^4] + \frac{k_2}{16}[a_e^4[4\ln(a_e) - 1] - a_i^4[4\ln(a_i) - 1]]\right],\end{aligned} \tag{28.66}$$

- for the invariant $\widetilde{I}_5$

$$\begin{aligned}\widetilde{\mathscr{M}} =& \frac{4\pi\tau\lambda_z^2}{\Theta_0}[0.5[g_0 + g_1 E_0^2]] \times \\ & \left[\frac{k_1}{4}[a_e^4 - a_i^4] + \frac{k_2}{16}[a_e^4[4\ln(a_e) - 1] - a_i^4[4\ln(a_i) - 1]]\right].\end{aligned} \tag{28.67}$$

It is obvious that the formulation for the torque in the case of $\widetilde{I}_5$ does not contain the coupling parameter $\widetilde{c}_2$. Therefore, in contrast to the previous example, the expected material response in the case of $\widetilde{I}_5$ will inevitably be different from the one with I_5 independent from the selected material parameter.

A similar example can be found in Bustamante (2010); Mehnert et al (2016), therefore we choose the zero-field shear modulus as $\mu_0 = 0.1$ MPa accordingly. The parameters $g_1 = -0.01$ N/V^2 and $c_2 = 0.05$ N/V^2 are selected in such a way that the influence of the coupled invariant becomes intuitive.

We will start by investigating the electro-mechanical loading case. Figure 28.2 illustrates the behavior of the torque $\mathscr{M}$ (solid lines) and $\widetilde{\mathscr{M}}$ (dashed lines) depending on the angle of torsion (a) and the axial stretch (b) in the case of an axially applied electric field for selected values of the initial wall thickness ζ.

Figure 28.2 shows a linear dependency of the resulting torque on both the angle τ and the axial stretch λ_z. For $\tau = 0$ the torque vanishes independently from the applied electric field and the axial stretch λ_z. In Fig. 28.2(b) the torque never vanishes as we assume a constant torsion of the cylinder of $\tau = \pi/4$. It can be seen in both plots that the incline of $\mathscr{M}$ is larger than $\widetilde{\mathscr{M}}$ and if we look at the final formulation of the torque in Eqs. (28.66) and (28.67) it is obvious that this is due to the influence of the

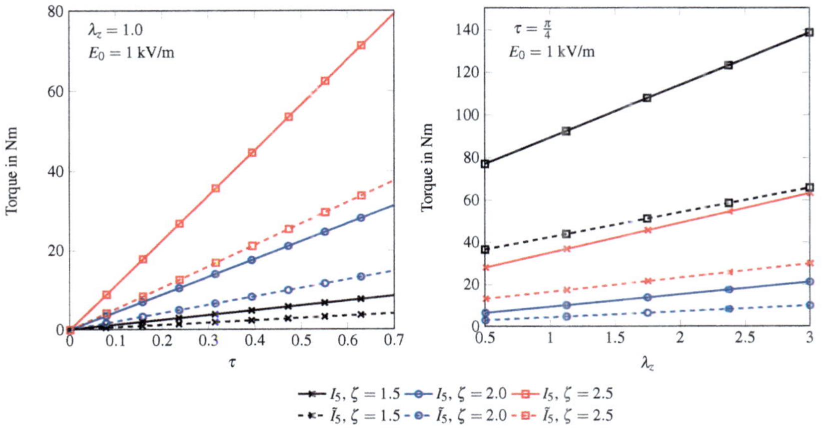

Fig. 28.2: Plot of the torque with respect to (a) τ and (b) λ_z under a electro-mechanically coupled load for selected values of ζ

coupling parameter c_2 that results in an increased stiffness of the material when an electric field is applied.

Figure 28.3 shows the dependency of the torque on the thickness of the tube (a) and on the applied electric field (b). In the first plot it is visible that the torque increases when the wall thickness of the cylinder is increased. When comparing the results with respect to the two different coupled invariants a similar effect as in Fig. 28.2 can be identified. The material response for an electro-mechanical loading in the case of I_5 is stiffer when compared to the response for $\widetilde{I}_5$.

The difference between the two formulations becomes even more distinct when we analyze the material behavior depending on the electric field depicted in Fig. 28.3(b). It becomes visible that for the invariant I_5 the torque increases with the electric field whereas the effect is the opposite for the invariant $\widetilde{I}_5$. This can once again be explained by the final formulation of the torque in Eqs. (28.66) and (28.67). Regardless of the selected coupling invariant the influence of the electric field is only present in the first term in brackets and is coupled in both cases by the selected material parameter $g_1 < 0$, which results in a softening of the material. As the expression for $\mathscr{M}$ additionally contains the positive parameter $c_2 > g_1$ the result is an increase of the torque with the electric field despite the decrease of the material parameter $\mu(I_4)$. It should be restated at this point, that in the final expression for $\widetilde{\mathscr{M}}$ the coupling parameter $\widetilde{c}_2$ is not present. Therefore independently from the selected value of $\widetilde{c}_2$ the torque will always show a softening of the material when an electric field is applied.

Next we apply an additional radial temperature gradient by keeping the temperature at the internal surface of the tube fixed at the reference temperature

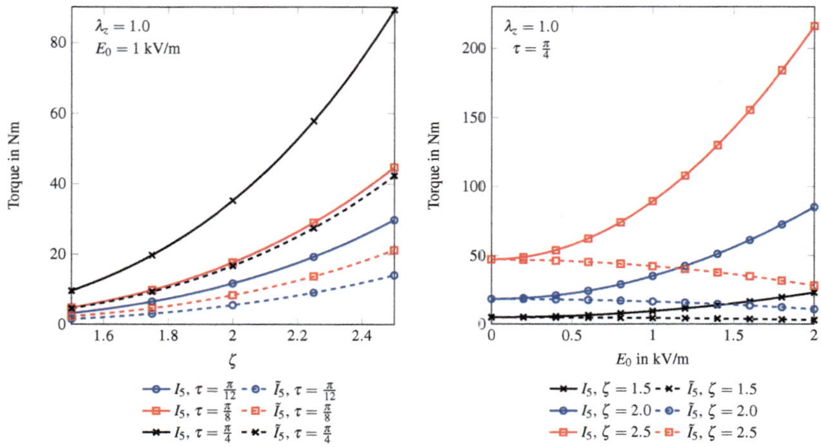

Fig. 28.3: Plot of the torque with respect to ζ for selected values of τ (a) and with respect to E_0 for selected values of ζ (b) under a electro-mechanically coupled load

$\Theta(a_i) = \Theta_i = 293$ K and varying the temperature $\Theta(a_e) = \Theta_e$ on the external surface of the tube. Figure 28.4 depicts the material response with respect to the applied electric field (a) and the temperature Θ_e (b).

The plot on the left side of Fig. 28.4 shows a similar behavior of the torque with respect to the electric field when compared to the isothermal case. For I_5 the torque increases with the field strength whereas a decrease is visible when $\widetilde{I}_5$ is selected. The effect is scaled by the temperature gradient as the torque becomes larger/smaller for an increased/decreased value of Θ_e, independent of the selected coupling invariant. Figure 28.4(b) confirms this observation further. If no electric field is applied the resulting torque is equal in both cases whereas the application of an electric field leads to an increase of the torque when I_5 is selected and a decrease for $\widetilde{I}_5$. In both cases the torque increases linearly with the temperature on the external cylinder surface although this increase is more distinct for $\mathscr{M}$.

As mentioned before the results of $\widetilde{\mathscr{M}}$ are independent of the selected parameter $\widetilde{c}_2$. Thus the only way to achieve a similar material behavior in this example would mean to select the coupling parameter $c_2 = 0$ and therefore reducing the electro-mechanical coupling to the softening effect of the field sensitive shear modulus.

28.3 Conclusions

While the isothermal description of electro-elasticity is well documented in the literature, the influence of the temperature on the response of electro-active materials has yet to reach a comparable level of understanding. With our previous work (Mehnert et al, 2016) we introduced a coupled thermo-electro-mechanical framework that might be used to fill this gap as it expands the formulations commonly used for the description of isothermal electro-elasticity to include thermal contributions. Both the isothermal formulations and the presented framework have in common that they employ an invariant approach, i.e. the free energy function used is formulated using a number of purely mechanical, purely electric and coupled invariant quantities, the latter of which vary in form throughout the literature. Thus in this contribution, we have focused on the influence of this coupling invariant in a phenomenological simulation approach of thermo-electro-elasticity. By the investigation of the analytical solution of two non-homogeneous boundary value problems we have presented that the selection of the coupling invariant can have a significant influence on the results. While in the first example, the displacement of a cube of electro-active material due to an uniaxially applied electric field, the difference between the two formulations could be minimized by the proper selection of material parameters, the second example proved that the material response might differ independently from the material parameters, due to the selected coupling invariant. In forthcoming contributions the behavior of the coupled thermo-electro-mechanical framework will be investigated further by using a finite element implementation with which we may model for

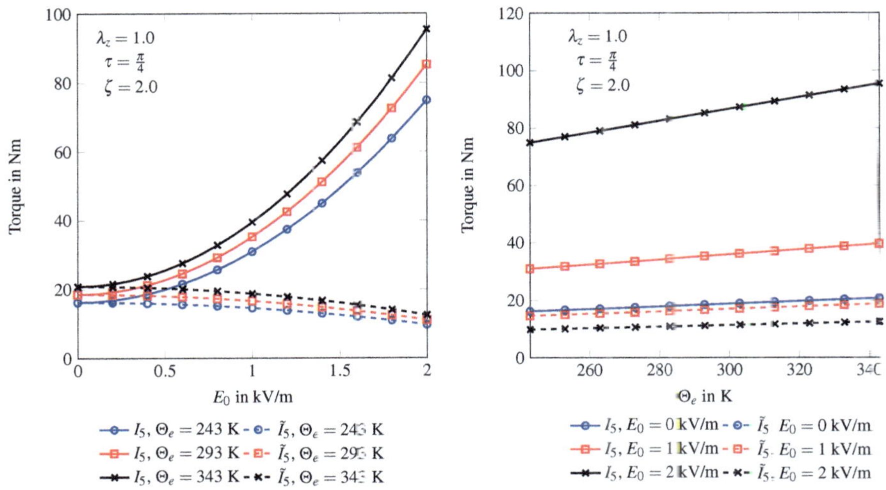

Fig. 28.4: Plot of the torque with respect to ζ for selected values of τ (a) and with respect to E_0 for selected values of ζ (b) under a electro-mechanically coupled load

example the influence of the surrounding free space on the material response. Future work will include experimental studies in order to further investigate the best possible selection of the coupling invariants and the addition of essential time dependent effects, such as viscoelasticity of the underlying polymeric material.

Acknowledgements The authors acknowledge the funding within the DFG project No. STE 544/52-1 and by the ERC advanced grant MOCOPOLY.

References

Ask A, Menzel A, Ristinmaa M (2012) Electrostriction in electro-viscoelastic polymers. Mechanics of Materials 50:9–21

Bar-Cohen Y (2002) Electroactive polymers: current capabilities and challenges. In: SPIE's 9th Annual International Symposium on Smart Structures and Materials, International Society for Optics and Photonics, pp 1–7

Bar-Cohen Y (2004) Electroactive Polymer (EAP) Actuators as Artificial Muscles: Reality, Potential, and Challenges, vol 136. SPIE press

Bland D (1956) Elastoplastic thick-walled tubes of work-hardening material subject to internal and external pressures and to temperature gradients. Journal of the Mechanics and Physics of Solids 4(4):209–229

Bodelot L, Pössinger T, Danas K, Triantafyllidis N, Bolzmacher C (2016) Magnetorheological elastomers: Experimental and modeling aspects. In: Mechanics of Composite and Multi-functional Materials, Volume 7, Springer, pp 251–256

Böse H, Fuß E (2014) Novel dielectric elastomer sensors for compression load detection. In: SPIE Smart Structures and Materials+ Nondestructive Evaluation and Health Monitoring, International Society for Optics and Photonics, pp 905,614–905,614

Brochu P, Pei Q (2010) Advances in dielectric elastomers for actuators and artificial muscles. Macromolecular Rapid Communications 31(1):10–36

Bustamante R (2009a) Transversely isotropic non-linear electro-active elastomers. Acta Mechanica 206(3):237–259

Bustamante R (2009b) A variational formulation for a boundary value problem considering an electro-sensitive elastomer interacting with two bodies. Mechanics Research Communications 36(7):791–795

Bustamante R (2010) Transversely isotropic nonlinear magneto-active elastomers. Acta Mechanica 210(3-4):183–214

Coleman BD, Noll W (1963) The thermodynamics of elastic materials with heat conduction and viscosity. Archive for Rational Mechanics and Analysis 13(1):167–178

Danas K, Triantafyllidis N (2014) Instability of a magnetoelastic layer resting on a non-magnetic substrate. Journal of the Mechanics and Physics of Solids 69:67–83

Danas K, Kankanala SV, Triantafyllidis N (2012) Experiments and modeling of iron-particle-filled magnetorheological elastomers. Journal of the Mechanics and Physics of Solids 60(1):120–138

Dorfmann A, Ogden R (2003) Magnetoelastic modelling of elastomers. European Journal of Mechanics-A/Solids 22(4):497–507

Dorfmann A, Ogden R (2004) Nonlinear magnetoelastic deformations of elastomers. Acta Mechanica 167(1):13–28

Dorfmann A, Ogden R (2005a) Nonlinear electroelasticity. Acta Mechanica 174(3-4):167–183

Dorfmann A, Ogden RW (2005b) Some problems in nonlinear magnetoelasticity. Zeitschrift für angewandte Mathematik und Physik (ZAMP) 56(4):718–745

Dorfmann L, Ogden RW (2014) Nonlinear mechanics of soft fibrous materials. Springer

Elahinia M, Vertechy R, Berselli G, Parenti Castelli V, Bergamasco M (2013) Continuum thermo-electro-mechanical model for electrostrictive elastomers. Journal of Intelligent Material Systems and Structures 24(6):761–778
Epstein M, Maugin GA (1990a) Energy-momentum tensor and J-integral in electrodeformable bodies. International Journal of Applied Electromagnetics in Materials 2(2):141–145
Epstein M, Maugin GA (1990b) Inhomogeneities, Eshelby's tensor and J-integral in electroelasticity. In: Hsieh RKT (ed) Mechanical Modelling of New Electromagnetic Materials, pp 253–258
Erbts P, Hartmann S, Düster A (2015) A partitioned solution approach for electro-thermo-mechanical problems. Archive of Applied Mechanics 85(8):1075–1101
Eringen AC (1963) On the foundations of electroelastostatics. International Journal of Engineering Science 1(1):127–153
Eringen AC (1980) Mechanics of Continua. Robert E. Krieger Publishing Co., Huntington, NY
Eringen AC, Maugin GA (2012) Electrodynamics of Continua I: Foundations and Solid Media. Springer Science & Business Media
Griffiths DJ (1989) Introduction to Electrodynamics, 2nd edn. Prentice Hall, Englewood Cliffs
Holzapfel GA (2000) Nonlinear Solid Mechanics, vol 24. Wiley, Chichester
Holzapfel GA, Ogden RW (2010) Constitutive modelling of arteries. Proceedings of the Royal Society of London A: Mathematical, Physical and Engineering Sciences 466(2118):1551–1597
Holzapfel GA, Simo JC (1996) Entropy elasticity of isotropic rubber-like solids at finite strains. Computer Methods in Applied Mechanics and Engineering 132(1-2):17–44
Hossain M, Steinmann P (2013) More hyperelastic models for rubber-like materials: consistent tangent operators and comparative study. Journal of the Mechanical Behavior of Materials 22(1-2):27–50
Kankanala SV, Triantafyllidis N (2004) On finitely strained magnetorheological elastomers. Journal of the Mechanics and Physics of Solids 52(12):2869–2908
Koh SJA, Keplinger C, Li T, Bauer S, Suo Z (2011) Dielectric elastomer generators: How much energy can be converted? IEEE/ASME Transactions on mechatronics 16(1):33–41
Kost A (1994) Numerische Methoden in der Berechnung elektromagnetischer Felder. Springer
Kovetz A (2000) Electromagnetic Theory. Oxford University Press, Oxford
Lu S, Pister K (1975) Decomposition of deformation and representation of the free energy function for isotropic thermoelastic solids. International Journal of Solids and Structures 11(7-8):927–934
Maugin GA (1976) A continuum theory of deformable ferrimagnetic bodies. I. Field equations. Journal of Mathematical Physics 17(9):1727–1738
Maugin GA (1993) Material Inhomogeneities in Elasticity, vol 3. CRC Press
Maugin GA (2013) Continuum Mechanics of Electromagnetic Solids, vol 33. Elsevier
Maugin GA, Eringen AC (1972) Deformable magnetically saturated media. I. Field equations. Journal of Mathematical Physics 13(2):143–155
Maugin GA, Eringen AC (1977) On the equations of the electrodynamics of deformable bodies of finite extent. Journal de Mécanique 16:101–147
Mehnert M, Hossain M, Steinmann P (2016) On nonlinear thermo-electro-elasticity. Proc R Soc A 472(2190)
Merodio J, Ogden RW (2002) Material instabilities in fiber-reinforced nonlinearly elastic solids under plane deformation. Archives of Mechanics 54(5-6):525–552
Miehe C (1995) Entropic thermoelasticity at finite strains. aspects of the formulation and numerical implementation. Computer Methods in Applied Mechanics and Engineering 120(3-4):243–269
Ogden R (1972) Large deformation isotropic elasticity-on the correlation of theory and experiment for incompressible rubberlike solids. Proceedings of the Royal Society of London A: Mathematical, Physical and Engineering Sciences 326(1567):565–584
O'Halloran A, O'Malley F, McHugh P (2008) A review on dielectric elastomer actuators, technology, applications, and challenges. Journal of Applied Physics 104(7):9
Pelrine R, Kornbluh R, Kofod G (2000a) High-strain actuator materials based on dielectric elastomers. Advanced Materials 12(16):1223–1225
Pelrine R, Kornbluh R, Pei Q, Joseph J (2000b) High-speed electrically actuated elastomers with strain greater than 100%. Science 287(5454):836–839

Pössinger T, Bolzmacher C, Bodelot L, Triantafyllidis N (2014) Influence of interfacial adhesion on the mechanical response of magneto-rheological elastomers at high strain. Microsystem Technologies 20(4-5):803–814

Rajagopal KR, Huang YN (1994) Finite circumferential shearing of nonlinear solids in the context of thennoelasticity. IMA Journal of Applied Mathematics 53(2):111–125

Rivlin RS (1948) Large elastic deformations of isotropic materials. IV. Further developments of the general theory. Philosophical Transactions of the Royal Society of London A: Mathematical, Physical and Engineering Sciences 241(835):379–397

Rivlin RS (1949a) Large elastic deformations of isotropic materials. V. The problem of flexure. Proceedings of the Royal Society of London A: Mathematical, Physical and Engineering Sciences 195(1043):463–473

Rivlin RS (1949b) Large elastic deformations of isotropic materials. VI. Further results in the theory of torsion, shear and flexure. Philosophical Transactions of the Royal Society of London A: Mathematical, Physical and Engineering Sciences 242(845):173–195

Santapuri S (2012) Unified continuum modeling of fully coupled thermo-electro-magneto-mechanical behavior, with applications to multifunctional materials and structures. PhD thesis, The Ohio State University

Santapuri S, Lowe RL, Bechtel SE, Dapino MJ (2013) Thermodynamic modeling of fully coupled finite-deformation thermo-electro-magneto-mechanical behavior for multifunctional applications. International Journal of Engineering Science 72:117–139

Spencer AJM (2013) Part III. Theory of invariants. In: Eringen AC (ed) Continuum Physics, vol 1, pp 239–353

Steinmann P (2011) Computational nonlinear electro-elasticity—getting started—. In: Mechanics and Electrodynamics of Magneto-and Electro-elastic Materials, Springer, pp 181–230

Steinmann P, Hossain M, Possart G (2012) Hyperelastic models for rubber-like materials: consistent tangent operators and suitability for Treloar's data. Archive of Applied Mechanics 82(9):1183–1217

Trimarco C, Maugin GA (2001) Material mechanics of electromagnetic solids. In: Configurational Mechanics of Materials, Springer, pp 129–171

Vertechy R, Berselli G, Parenti Castelli V, Vassura G (2010) Optimal design of lozenge-shaped dielectric elastomer linear actuators: mathematical procedure and experimental validation. Journal of Intelligent Material Systems and Structures 21(5):503–515

Vertechy R, Fontana M, Papini GPR, Forehand D (2014) In-tank tests of a dielectric elastomer generator for wave energy harvesting. In: SPIE Smart Structures and Materials+ Nondestructive Evaluation and Health Monitoring, International Society for Optics and Photonics, pp 90,561G–90,561G

Vogel F (2015) On the Modeling and Computation of Electro- and Magneto-active Polymers. Lehrstuhl für Technische Mechanik, Universität Erlangen-Nürnberg

Vogel F, Göktepe S, Steinmann P, Kuhl E (2014) Modeling and simulation of viscous electro-active polymers. European Journal of Mechanics-A/Solids 48:112–128

Voltairas P, Fotiadis D, Massalas C (2003) A theoretical study of the hyperelasticity of electro-gels. Proceedings of the Royal Society of London A: Mathematical, Physical and Engineering Sciences 459(2037):2121–2130

Vu D, Steinmann P, Possart G (2007) Numerical modelling of non-linear electroelasticity. International Journal for Numerical Methods in Engineering 70(6):685–704

Vu DK (2014) A study on nonlinear electro-elastostatics: Theory and numerical simulation. Habilitation, Friedrich-Alexander University of Erlangen-Nürnberg: Erlangen, Bayern, Germany

Wang CC (1970) A new representation theorem for isotropic functions: An answer to Professor G.F. Smith's criticism of my papers on representations for isotropic functions. Archive for Rational Mechanics and Analysis 36(3):166–197

Wriggers P (2008) Nonlinear Finite Element Methods. Springer Science & Business Media

Chapter 29
On Recurrence and Transience of Fractional Random Walks in Lattices

Thomas Michelitsch, Bernard Collet, Alejandro Perez Riascos, Andrzej Nowakowski, and Franck Nicolleau

Abstract The study of random walks on networks has become a rapidly growing research field, last but not least driven by the increasing interest in the dynamics of online networks. In the development of fast(er) random motion based search strategies a key issue are first passage quantities: How long does it take a walker starting from a site p_0 to reach 'by chance' a site p for the first time? Further important are recurrence and transience features of a random walk: A random walker starting at p_0 will he *ever* reach site p (ever return to p_0)? How often a site is visited? Here we investigate Markovian random walks generated by fractional (Laplacian) generator matrices $L^{\frac{\alpha}{2}}$ $(0 < \alpha \leq 2)$ where L stands for 'simple' Laplacian matrices. This walk we refer to as 'Fractional Random Walk' (FRW). In contrast to classical Pólya type walks where only local steps to next neighbor sites are possible, the FRW allows nonlocal long-range moves where a remarkably rich dynamics and new features arise. We analyze recurrence and transience features of the FRW on infinite d-dimensional simple cubic lattices. We deduce by means of lattice Green's function (probability generating functions) the mean residence times (MRT) of the walker at preselected sites. For the infinite 1D lattice (infinite ring) we obtain for the transient regime $(0 < \alpha < 1)$ closed form expressions for these characteristics. The lattice Green's function on infinite lattices existing in the transient regime fulfills Riesz

Thomas Michelitsch, Bernard Collet
Sorbonne Université, Centre National de la Recherche Scientifique, UMR 7190, Institut Jean Le Rond d'Alembert, F-75005 Paris, France
e-mail: michel@lmm.jussieu.fr, bernard.collet@upmc.fr

Alejandro Perez Riascos
Department of Civil Engineering, Universidad Mariana San Juan de Pasto, Colombia
e-mail: aaappprrr@gmail.com

Andrzej Nowakowski, Franck Nicolleau
Sheffield Fluid Mechanics Group, Department of Mechanical Engineering, University of Sheffield, Mappin Street, Sheffield S1 3JD, United Kingdom
e-mail: a.f.nowakowski@sheffield.ac.uk, f.nicolleau@sheffield.ac.uk

© Springer International Publishing AG, part of Springer Nature 2018
H. Altenbach et al. (eds.), *Generalized Models and Non-classical Approaches in Complex Materials 1*, Advanced Structured Materials 89,
https://doi.org/10.1007/978-3-319-72440-9_29

potential asymptotics being a landmark of anomalous diffusion, i.e. random motion (Lévy flights) where the step lengths are drawn from a Lévy α-stable distribution.

29.1 Introduction

The scientific activity of Gérard Maugin was not limited and focused to only one discipline. He by his nature indeed was a generalist with fine sense for the 'big picture' and interest in interdisciplinary problems and approaches where the broad spectrum of his contributions is well documented in his last book (Maugin, 2017). Inspired over many years by his sprit, the present paper is concerned to the interdisciplinary subject of random walks.

Many stochastic processes and transport phenomena in very different contexts such as the diffusion of particles, the foraging of animals, the spread of plant seeds, pandemic diseases, rumors, migration, the transitions in chemical reactions, the time evolution of stock market prices, and many further natural processes can be attributed to random walks. Random walks may take place on continuous spaces or on discrete finite or infinite sets of points (graphs) (Newman, 2010; Albert and Barabási, 2002; Noh and Rieger, 2004; Gonçalves et al, 2011; Mieghem, 2011). The notion of 'random walk' first was introduced by Karl Pearson in 1905 (Pearson, 1905). A fundamental question arising in many contexts such as information spread in a network: How long does it take for news headlines emitted somewhere in the web to hit a preselected set of target sites for the *first time*? Indeed the study of expected 'first passage times' of a random walk strategy is of crucial interest in many interdisciplinary problems for instance in contexts of survival models or to describe the efficiency of random walk based search strategies, the speed of chemical reactions, and one easily can enumerate further examples (Metzler et al, 2009).

Closely related to the problem how many steps are needed until a target is hit, is the question whether a target is hit at all, and whether there exists a finite probability of never hitting a target. In the random walk picture this question is related to the problem of recurrence (transience) of a walk: Will two random walkers starting their walk from the same site at the same time performing independent walks of the same type ever meet again somewhere on the network? Indeed these features are well understood for 'classical' random walks where the walker in one time increment can only make steps with equal probability to next neighbor sites. This 'simple' type of random walk performed on the infinite d-dimensional lattice was first analyzed by Pólya (1921) formulating the celebrated 'recurrence theorem' ('Pólya theorem') for such random walks: The random walker returns for sure to the site of departure for lattice dimensions $d = 1,2$ (recurrence of the walk), whereas there is a finite escape probability of never return in dimensions $d = 3,4$ (transience of the random walk) (Pólya, 1921; Montroll, 1956; Montroll and Weiss, 1965; Hudges, 1995; Spitzer, 1976). The question on whether or not the walker returns to its departure site, i.e. recurrence or transience of a walk indeed is crucial for the development of random walk based search strategies (Viswanathan et al, 2008). In the meantime a multitude of random walks with different features were introduced. Hughes and Shlesinger (1982) have analyzed random walks on simple cubic lattices with asymptotic power-law behavior of the vibrational dispersion relation of the lattice and derived conditions for recurrence of these walks. They demonstrated that the recurrence features of this type of walk are different from the Pólya walk.

The very rich dynamics of this new type of random walk model has turned out to be suitable to describe phenomena of 'anomalous diffusion', i.e. stochastic motions with self-similar (heavy tailed inverse power law) distribution of step lengths where long-range steps appear which are drawn from Lévy α-stable ('heavy tailed') distributions. This type of anomalous diffusion which also is referred to as Lévy motions or Lévy flights during the last two decades and at present is extensively being studied (Metzler and Klafter, 2000; Chechkin et al, 2008; Palyulin et al, 2016; Klages, 2016; Dybiec et al, 2017). The classical Pólya walk which allows only local steps to 'neighbor sites' is a model for Brownian motion and describes well phenomena of classical diffusion where the step length are drawn from a normal distribution.

In the present paper our goal is to generalize Pólya walks on regular networks in such a way to become a suitable model exhibiting long-range steps with asymptotic emergence of Lévy motions. We will demonstrate subsequently that an appropriate discrete model to achieve this is the Fractional Random Walk (FRW), a random walk which is generated by a fractional power $L^{\frac{\alpha}{2}}$ $(0 < \alpha \leq 2)$ where the Laplacian matrix L itself generates a Pólya type walk (Michelitsch et al, 2017b,a). It has been found in various models that search efficiency may be increased for sparse targets by using search strategies based on random walks with long-range Lévy flight characteristics rather than for 'classical' Brownian Pólya type random walks (Palyulin et al, 2016; Klages, 2016; Metzler et al, 2009; Metzler and Klafter, 2004; Riascos and Mateos, 2012; Michelitsch et al, 2017b,a; Bénichou et al, 2011, and references therein).

We define the FRW as a Markovian random walk generated by fractional powers $L^{\frac{\alpha}{2}}$ of 'simple' Laplacian matrices L. The index range is restricted $0 < \alpha \leq 2$ to maintain the good properties of the Laplacian (Michelitsch et al, 2017b) where $\alpha = 2$ recovers the classical Pólya type walk allowing only next neighbor steps. We will demonstrate some of the universal features of the FRW such as Lévy flight dynamics on sufficiently large lattices which is a consequence of the asymptotic scale free non-locality of fractional Laplacian matrices. The latter take asymptotic representations of Riesz fractional derivatives (continuous fractional Laplacian operator kernels) on sufficiently large lattices depending only on the dimension d of the lattice and index α of the fractional Laplacian. Asymptotic scale free non-locality indeed is an 'universal' feature, in the sense that it is independent of the spectral details (apart of some constraints mentioned below) of the Laplacian matrix L.

The present paper is organized as follows: First we evoke basic features of time-discrete Markavian random walks and define the FRW on a regular undirected networks and lattices. We introduce for the FRW transition matrices and deduce by employing the method of probability generating functions the Green's functions of the network containing the entire statistical information on occupation probabilities, first passage probabilities and recurrence (transience) of the random walk on the network. We then analyze the FRW on d-dimensional cubic lattices: We deduce the lattice Green's function of the FRW which determines the average number of visits of the lattice points (nodes) and contains the probabilities that a node (n)ever is visited. In this way we uncover the universal recurrence (transience) features for a FRWs performed on infinite d-dimensional lattices. Finally we analyze the asymptotic behavior of long-range steps. We demonstrate that the probability of long-range moves of distance $l \gg 1$ obeys an universal inverse power law scaling $\sim l^{-(\alpha+d)}$ of Riesz fractional derivative kernel form independently of the choice of Laplacian L. This asymptotic universal scaling law on sufficiently large lattices is the landmark of the emergence of Lévy flights, that is anomalous diffusion drawn from Lévy α-stable (heavy tailed) distribution. For a more detailed analysis, we refer to our recent paper (Michelitsch et al, 2017a).

29.2 Time Discrete Markovian Random Walks on Undirected Networks

We consider here Markovian random walks on regular undirected connected networks (graphs) consisting of N nodes denoted by $p = 0, .., N-1$ (see Fig. 29.1). In a network nodes p, q are either connected by an edge or disconnected. Whether or not the nodes p, q are connected is described by the $N \times N$ adjacency matrix A with elements $A_{pq} = 1$ if the nodes p, q are connected, and $A_{pq} = 0$ if they are disconnected. We do not permit self-connections thus $A_{pp} = 0$. Nodes p, q which are connected $A_{pq} = 1$ are also referred to as 'adjacent nodes' or 'next neighbor nodes'. In undirected networks which uniquely are considered in the present paper, the connections (edges) between nodes have no direction thus the adjacency matrix is symmetric $A_{pq} = A_{qp}$. In a network of N nodes an important quantity is the (symmetric) distance matrix $D_{pq} = D_{qp} \geq 0$ ($D_{pq} = D_{qp} > 0$ when $p \neq q$ and $D_{pp} = 0$). The elements D_{pq} of the distance matrix indicate the shortest possible numbers of edges needed to generate a path linking the nodes p, q. It follows that adjacent nodes have distance $D_{pq} = 1$. In Fig. 29.1 is represented a connected network where different colors indicate different distances of nodes from node 1. Closely related is the degree K_p of a node p, representing the number of adjacent nodes of node p. The degree is given by

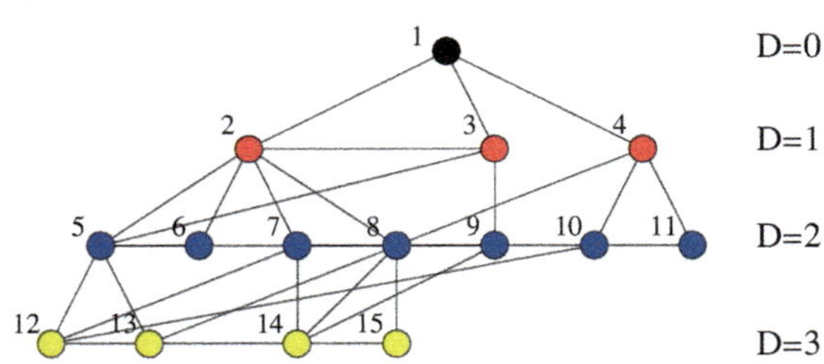

Fig. 29.1 Representation of a connected network. The colors indicate different distances from node 1 (black) on the network: Red nodes can be reached from node 1 in one step (distance $D = 1$), blue nodes in two steps, they have distance $D = 2$ from node 1, and yellow nodes are $D = 3$ steps away from node 1.

$$K_p = \sum_{q=0}^{N-1} A_{pq}. \tag{29.1}$$

We restrict us here uniquely to 'regular' networks where all nodes p have constant degree $K_p = K \, \forall p = 0,..,N-1$. This especially is true in the lattice structures considered in this paper, such as simple cubic lattices ('integer lattices'). The properties of the network are taken into account by the Laplacian matrix which has the canonic representation (Newman, 2010; Mieghem, 2011)

$$L_{pq} = \delta_{pq} K_p - A_{pq} = \sum_{j=1}^{N} \mu_j \langle p|\Psi_j\rangle\langle\Psi_j|q\rangle \tag{29.2}$$

where $L_{pq} = L_{qp}$ is symmetric in undirected networks reflecting the symmetry $A_{pq} = A_{qp}$ of the adjacency matrix. We adopt here for our convenience Dirac's bracket notation. For constant degree $K_p = K$ the Laplacian matrix has the form $L = K\hat{1} - A$ where $\hat{1}$ indicates the unity matrix. Due to the symmetry of the Laplacian matrix the set of eigenvectors constitutes a complete N-dimensional ortho-normal canonic basis[1]. It follows from (29.1) and (29.2) that the constant vector

$$|\Psi_1\rangle = \frac{1}{\sqrt{N}}(1,\ldots,1)$$

is eigenvector of the Laplacian matrix L to the vanishing eigenvalue $\mu_1 = 0$.

Generally the Laplacian matrix is positive-semidefinite and in connected networks the zero eigenvalue μ_1 appears uniquely together with $N-1$ positive eigenvalues $\mu_1 = 0 < \mu_2 \leq ..,\leq \mu_N$ (Mieghem, 2011). This positive-semidefinite structure of the Laplacian matrix follows from the quadratic form

$$\mathscr{V}(\xi_0,..,\xi_{N-1}) = \frac{1}{2}\sum_{p=0}^{N-1}\sum_{q=0}^{N-1} L_{pq}\xi_p\xi_q = \frac{1}{4}\sum_{p=0}^{N-1}\sum_{q=0}^{N-1} A_{pq}(\xi_p - \xi_q)^2 \geq 0 \tag{29.3}$$

where $\mathscr{V}(\xi_0,..,\xi_{N-1}) = 0$ only when $\xi_p = \xi_q = c \forall p,q$ is constant (which corresponds to the constant eigenvector to $\mu_1 = 0$) and $\mathscr{V} > 0$ otherwise. We further observe in (29.2) that the off-diagonal elements of the Laplacian matrix are non-positive $L_{pq} = -A_{pq} \leq 0 \; (p \neq q)$. Both, (i) the positive semi-definiteness together with (ii) the property of non-positiveness of the off diagonal elements define the 'good' properties of a Laplacian matrix which are necessary to generate random walks.

In order to characterize the time-evolution of a random walk it is convenient to introduce the occupation probability vector $\boldsymbol{P}_t = (P_t(0),..,P_t(p),..,P_t(N-1))$ where $P_t(p)$ indicates the occupation probability that the walker occupies node p at time t. We consider time-discrete random walks at integer times $t = 0,1,2,..$ where the walk starts at $t = 0$. Time continuous Fractional Random Walks we analyzed elsewhere (Michelitsch et al, 2017b). The normalization condition

[1] where $\langle\Psi_i|\Psi_j\rangle = \delta_{ij}$ and $\sum_{m=1}^{N} \langle i|\Psi_m\rangle\langle\Psi_m|j\rangle = \delta_{ij}$

$$\sum_{p=0}^{N-1} P_t(p) = 1$$

indicates that the walker at any time is on the network during an infinite time of observation $0 \leq t < \infty$. For a Markovian walk which uniquely is considered here the time evolution of the occupation probabilities is described by a discrete master equation (Hudges, 1995)

$$P_{t+1}(p) = \sum_{q=0}^{N-1} \mathscr{W}_{pq} P_t(q), \qquad \boldsymbol{P}_{t+1} = \mathscr{W} \cdot \boldsymbol{P}_t \tag{29.4}$$

where we utilize alternatively matrix and index notations. In (29.4) is introduced the constant $N \times N$ matrix $\mathscr{W} =: W(\delta t = 1)$ which is referred to as *transition matrix* constituting a linear relationship between the occupation probabilities $\boldsymbol{P}_{t+1}$ and $\boldsymbol{P}_t$ (Spitzer, 1976). The time-evolution (29.4) of the transition matrix iterating t time-steps writes

$$P_t(p) = \sum_{q=0}^{N-1} \langle p|\mathscr{W}^t q\rangle P_0(q) \tag{29.5}$$

where $\mathscr{W}^0 = \hat{1}$ denotes the $N \times N$ unity matrix. We employ the equivalent notations $\langle p|\mathscr{W}^t q\rangle = W_{pq}(t)$ for the elements of the transition matrix $\mathscr{W}^t = W(t)$ at time t. The transition matrix fulfills the normalization condition

$$\sum_{p=0}^{N-1} W_{pq}(t) = 1, \qquad 0 \leq W_{pq}(t) \leq 1 \tag{29.6}$$

as a consequence of the normalization condition for the occupation probabilities. We also have the restriction $0 \leq W_{pq}(t) \leq 1$ allowing the probability interpretation to be maintained for the entire observation time $0 \leq t = n < \infty$. For the analysis to follow it is worthy to consider the spectral properties of the transition matrix. As the transition matrix is symmetric (self-adjoint), it can be expressed by its (purely real) eigenvalues λ_m and eigenvectors $|\Psi_m\rangle$ as

$$\begin{aligned} \mathscr{W} &= W(t=1) = \sum_{m=1}^{N} \lambda_m |\Psi_m\rangle\langle\Psi_m|, \\ W(t)_{pq} &= (\mathscr{W}^t)_{pq} = \sum_{m=1}^{N} (\lambda_m)^t \langle p|\Psi_m\rangle\langle\Psi_m|q\rangle. \end{aligned} \tag{29.7}$$

In connected ergodic networks with constant degree the stationary distribution is constituted by the equal-distribution (Mieghem, 2011; Riascos and Mateos, 2012)

$$W_{pq}(t \to \infty) = \langle p|\Psi_1\rangle\langle\Psi_1|q\rangle = \frac{1}{N} \qquad \forall p,q = 0,..,N-1 \tag{29.8}$$

From $\lim_{t\to\infty}\mathscr{W}^t = |\Psi_1\rangle\langle\Psi_1|$ follows that eigenvalue $\lambda_1 = 1$ has in connected networks multiplicity one and the remaining $N-1$ eigenvalues fulfill $|\lambda_m| < 1$ $m = 2,..,N$. All eigenvalues λ_m of the transition matrix are real due the symmetry $\mathscr{W}_{pq} = \mathscr{W}_{qp}$ in regular networks[2].

Consider now a time-discrete random walk where the walker in one time increment is allowed to step only to adjacent nodes $q \to p$ with $A_{pq} = 1$, i.e. to perform steps of distance $D_{pq} = 1$. This is realized when the transition matrix of one time increment of (29.4) has the following form, e.g. Noh and Rieger (2004)

$$\mathscr{W}_{pq} = \delta_{pq} - \frac{1}{K_q}L_{pq} = \frac{1}{K}A_{pq}, \qquad K = \frac{1}{N}tr(L) \tag{29.9}$$

where $tr(..)$ denotes the trace

$$\sum_{p=0}^{N-1}(..)_{pp}$$

of a matrix with

$$\frac{1}{K_q}L_{pq} = \frac{1}{K}L_{pq}.$$

As a consequence in regular networks the transition matrix $\mathscr{W}_{pq} = \mathscr{W}_{qp}$ is symmetric with the same canonic basis of eigenvectors $\{|\Psi_j\rangle\}$ $(j = 1,2,..,N)$ as the Laplacian matrix. It follows also that the eigenvalues of transition matrix and Laplacian matrix then are related by

$$\lambda_m = 1 - \frac{\mu_m}{K}, \qquad m = 1,2,..,N. \tag{29.10}$$

The constitutive relation (29.9) connects the random walk dynamics with the network properties. We notice that $\mathscr{W}_{pp} = 0$, that is the walker at each time-step has to move to an adjacent node. The transition matrix (29.9) in a regular network thus defines a Pólya type random walk where only the K adjacent nodes can be reached in a single time-step with equal probability $1/K$. This 'locality' of moves is a typical feature of Pólya walks and is in contrast to the 'nonlocality' of the subsequently analyzed FRWs.

29.3 Probability Generating Functions - Green's Functions

For the subsequent analysis of the FRW it is convenient to employ the powerful method of probability generating functions and network Green's functions. For general properties and discussions we refer to Mieghem (2011); Montroll (1956); Montroll and Weiss (1965); Hudges (1995); Spitzer (1976); Doyle and Laurie Snell (1984); Kemeny and Laurie Snell (1976); Feller (1950). To this end the following definitions are required. $F_t(p,q)$ denotes the probability that the random walker starting at $t = 0$ at node q reaches node p at time-step t *for the first time*. The first passage

[2] We define here 'regular' networks as networks with constant degree $K_p = K$.

probabilities in an undirected network define symmetric matrices $F_t(p,q) = F_t(q,p)$ with respect to departure and arrival nodes. The same is true for occupation probabilities which we denote here as $P_t(p,q) = P_t(q,p)$ indicating the probability that a walker starting at $t = 0$ from departure node q reaches node p at time t (not necessarily for the first time). It follows that $P_t(p,q) = (\mathscr{W}^t)_{pq} = W_{pq}(t)$ can be identified with the entries of transition matrix (29.5).

For a time sequence of probabilities $\{Q_t(p,q)\}$ $(t = 0,1,2,...)$ the $N \times N$ matrix $Q(\xi) = (Q(p,q,\xi))$ of probability generating function is defined by the power series (Hudges, 1995)

$$Q(p,q,\xi) = \sum_{n=0}^{\infty} Q_n(p,q)\xi^n \qquad |\xi| < 1 \tag{29.11}$$

where the (non-negative) probabilities $0 \leq Q_n(p,q) \leq 1$ appear as coefficients. (29.11) has according to Abel's theorem (at least) the radius of convergence $\xi = 1$. By putting $\xi = e^{-s}$ (29.11) can be read as discrete Laplace transform $Q(\xi = e^{-s}) = \hat{Q}(s)$ of the probabilities Q_t (converging at least for $\Re(s) > 0$). The $Q(p,q,\xi) = (F(p,q,\xi), P(p,q,\xi))$ stand in the following analysis for the generating functions of the first passage- and occupation probabilities, respectively. For Markovian walks we can establish the following relation between first passage probabilities and occupation probabilities

$$P_t(p,q) = \delta_{t0}\delta_{pq} + \sum_{k=0}^{t} F_{t-k}(p,q)P_k(0,0) \tag{29.12}$$

with $P_0(p,q) = \delta_{pq}$ where $F_0(p,q) = 0$ and $F_1(p,q) = P_1(p,q)$ as at $t = 1$ only next neighbor nodes can be visited for the first time.

From this relation (29.12) which constitutes the nth-order in ξ of the functional identity $P(p,q,\xi) - \delta_{pq} = F(p,q,\xi)P(0,0,\xi)$, one can establish a classical relationship between the first passage probability generating functions and the occupation probability generating functions holding for Markovian walks, namely Montroll and Weiss (1965); Hudges (1995); Spitzer (1976); Michelitsch et al (2017a)

$$F(p,q,\xi) = \frac{P(p,q,\xi) - \delta_{pq}}{P(0,0,\xi)}, \qquad F(\xi) = \frac{1}{P(0,0,\xi)}\left(P(\xi) - \hat{1}\right) \tag{29.13}$$

where the second relation represents matrical representation. For the analysis to follow the following physical interpretations of the probability generating functions are important (Hudges, 1995; Spitzer, 1976):
Firstly the cumulated first passage probability

$$F(p,q,\xi = 1) = \sum_{t=0}^{\infty} F_t(p,q) \quad (0 \leq F(p,q,\xi = 1) \leq 1)$$

indicates the probability that a node p *ever* is visited by the walker for a walk starting at node q during an infinite time of observation $t \to \infty$. For $p = q$ this quantity

$F(p,p,\xi=1)$ indicates the probability that the walker ever returns to the departure node. If $F(p,p,\xi=1)=1$ then the walker returns for sure to the departure node, thus the walk is *recurrent*. In contrast if $F(p,p,\xi=1)<1$ there is a nonzero probability of non-return $1-F(p,p,\xi=1)>0$ thus the walk then is *transient*.
Secondly the cumulated occupation probabilities

$$P(p,q,\xi=1)=\sum_{t=0}^{\infty}P_t(p,q)$$

indicate the average number of time-steps (the mean residence time - MRT) the walker occupies node p when starting the walk at node q during an infinite time of observation. Since the walker per construction changes at any time-step the node (as $\mathscr{W}_{pp}=0$, see (29.9)), the MRT $P(p,q,\xi=1)$ measures the average number of visits of a node p for a walk starting at q during an infinite time of observation (Spitzer, 1976; Dybiec et al, 2017; Michelitsch et al, 2017a). For networks with constant degree $K_p=K$ the MRTs $P(p,q,\xi=1)=P(q,p,\xi=1)$ constitute a symmetric $N\times N$ matrix. This matrix $P(\xi=1)=(P(q,p,\xi=1))$ also is referred to as network Green's function or lattice Green's function if the network constitutes a lattice (Montroll and Weiss, 1965; Hudges, 1995; Spitzer, 1976)[3].

These definitions are sufficient for our goal to analyze recurrence and transience features of the FRW. The Green's function contains the full statistical information whether or not the walker (n)ever returns to the departure node and (n)ever reaches any preselected destination node (Mieghem, 2011; Pólya, 1921; Montroll, 1956; Montroll and Weiss, 1965; Hudges, 1995; Spitzer, 1976; Doyle and Laurie Snell, 1984; Kemeny and Laurie Snell, 1976; Feller, 1950). The following statistical interpretations of the Green's function are important.

a) A random walk is recurrent if the MRT (i.e. the matrix elements of the Green's function) $\lim_{\xi\to 1}P(p,q,\xi)\to\infty$ diverge. A sufficient criteria for recurrence of a walk is when the diagonal element (being constant for regular networks) $P(p,p,1)$ diverges, which is equivalent that the walker is expected to return infinitely often to the departure node $P(q,q,\xi)-1\to\infty$ and hence to any ever visited node.
In view of the spectral representation of the transition matrix (29.7) the probability generating function $P(p,q,\xi)$ for a finite network of N nodes of the occupation probabilities is obtained by the geometrical series

$$P(\xi)=\sum_{n=0}^{\infty}\mathscr{W}^n\xi^n=[\hat{1}-\xi\mathscr{W}]^{-1}=\frac{|\Psi_1\rangle\langle\Psi_1|}{(1-\xi)}+\sum_{m=2}^{N}|\Psi_m\rangle\langle\Psi_m|\frac{1}{(1-\lambda_m\xi)},\quad |\xi|<1. \tag{29.14}$$

We see that $P(p,q,\xi\to 1-0)\to\infty$ due to the contribution which corresponds to $\lambda_1=1$ namely

$$\frac{|\Psi_1\rangle\langle\Psi_1|}{(1-\xi)}$$

[3] We subsequently see that the Green's function exists only for transient walks on infinite networks

is diverging for N finite. One can conclude due to the presence of $\lambda_1 = 1$ in finite networks, that random walks on *finite connected networks always are recurrent*, i.e. any node of the network is sure to be (infinitely often) visited (during an infinite time of observation) (Hudges, 1995). This feature may change for walks on *infinite* networks ($N \to \infty$) where the stationary (equal-) distribution $N \to \infty$ due to

$$\langle p|\Psi_1\rangle\langle\Psi_1|q\rangle = \frac{1}{N} \to 0$$

is suppressed (Michelitsch et al, 2017a).

b) In contrast, if a node p for an infinite observation time in the average is visited only for a finite number of times where the MRT

$$\lim_{\xi\to 1} P(p,q,\xi) < \infty$$

is finite, the random walk is *transient*. Transient walks may occur *only* on *infinite networks* ($N \to \infty$) since then

$$\lim_{N\to\infty} \frac{|\Psi_1\rangle\langle\Psi_1|}{(1-\xi)} \to 0$$

is suppressed, thus only the contributions of the relaxing modes $|\lambda_m| < 1$ contribute to (29.14).

Whether or not a walk is transient or recurrent in an infinite network depends on the network properties which are crucial for the convergence of the infinite network limit $N \to \infty$ of (29.14) which writes with (29.9)

$$\begin{aligned} r_{pq}(\xi = 1) &= \lim_{\xi\to 1-0}(\lim_{N\to\infty} P(\xi)) = K\sum_{m=2}^{\infty} \langle p|\Psi_m\rangle\langle\Psi_m|q\rangle(\mu_m)^{-1}, \\ &= \int_0^{\mu_{max}} D(\mu)\mu^{-1}\langle p|\tilde{\Psi}(\mu)\rangle\langle\tilde{\Psi}(\mu)|q\rangle \mathrm{d}\mu, \end{aligned} \tag{29.15}$$

containing only relaxing modes

$$|\lambda_m| = |1 - \frac{\mu_m}{K}| < 1 \quad (\mu_m > 0).$$

In the second line we have introduced the density of (eigen)states $D(\mu)$ where $\mu_2 \leq .. \leq \mu_{N-1} =: \mu_{max}$ becomes a continuous eigenvalue spectrum for $N \to \infty$ and the $\tilde{\Psi}(\mu)$ denotes the appropriately renormalized continuous set of eigenmodes. The general representation (29.15) for the Green's function will be useful to explore subsequently recurrence and transience features for the FRW.

29.4 The Fractional Random Walk

Here our goal is to generalize the 'simple' Pólya type walk to introduce the 'Fractional Random Walk' (FRW) (Michelitsch et al, 2017b; Riascos and Mateos, 2015, 2014). In order to define a random walk transition matrix (29.9), it is important to notice that only Laplacian matrices L fulfilling the following properties are admissible (Michelitsch et al, 2017b; Riascos et al, 2017):

i) The eigenvalue structure of the Laplacian matrix in connected networks is positive-semidefinite with (in connected networks) one vanishing $\mu_1 = 0$ and $N-1$ positive eigenvalues $\mu_1 = 0 < \mu_2 \leq .. \leq \mu_N$. This positive-semidefinite structure is a consequence of (29.3).

ii) The off-diagonal elements of the Laplacian matrix are non-positive

$$L_{pq} = -A_{pq} \leq 0 \quad (p \neq q)$$

allowing probability interpretation

$$0 \leq \mathscr{W}_{pq} = \frac{A_{pq}}{K} \leq 1$$

of the transition matrix elements. The vanishing eigenvalue of the Laplacian matrix $\mu_1 = 0$ guarantees

$$K = \sum_{q=0}^{N-1} A_{pq} > 0$$

the normalization condition with eigenvalue $\lambda_1 = 1$ of the transition matrix together with the stationary (equal-) distribution

$$\langle p|\Psi_1\rangle\langle\Psi_1|q\rangle = \frac{1}{N}$$

representing the state of maximum entropy (ergodicity of the network, see relation (29.8)).

The properties **i)**, **ii)** constitute the necessary 'good properties' of a Laplacian matrix to generate a random walk as defined in (29.9). We notice that in a context of elasticity (29.3) can be interpreted as harmonic elastic potential where the adjacency matrix elements play the role of harmonic spring constants where the conditions **i)**, **ii)** guarantee elastic stability. We emphasize that when we replace in (29.9) Laplacian matrix L by a matrix function $g(L)$ only functions $g(L)$ are admissible that conserve the structure **i)**, **ii)**. Indeed this is the case for the fractional Laplacian matrix $g(L) = L^{\frac{\alpha}{2}}$ in the index range $0 < \alpha \leq 2$. For a demonstration see Michelitsch et al (2017b).

In above relation (29.9) we defined Pólya type random walks where the walker can step with equal probabilities $1/K$ only to any next neighbor (adjacent) node. The Pólya walk serves as a model for Brownian normal diffusive motion (Pólya, 1921;

Metzler and Klafter, 2000; Michelitsch et al, 2017a) appearing as a special case in FRWs.

To define below the FRW we introduce the fractional Laplacian matrix as a power matrix function of (29.2) in its canonic representation

$$L^{\frac{\alpha}{2}} = \hat{1}K^{(\alpha)} - A^{(\alpha)} = \sum_{m=2}^{N} (\mu_m)^{\frac{\alpha}{2}} |\Psi_m\rangle\langle\Psi_m|, \qquad 0 < \alpha \leq 2. \tag{29.16}$$

We define the transition matrix of one time-step for the FRW corresponding to (29.9) (Michelitsch et al, 2017a; Riascos and Mateos, 2015, 2014)

$$\mathscr{W}_{pq}^{(\alpha)} = \delta_{pq} - \frac{1}{K^{(\alpha)}}(L^{\frac{\alpha}{2}})_{pq} =: \frac{1}{K^{(\alpha)}}(A^{(\alpha)})_{pq}, \qquad 0 < \alpha \leq 2. \tag{29.17}$$

We introduced in (29.16) the fractional degree $K^{(\alpha)}$ which is the diagonal element of the fractional Laplacian matrix. This quantity which is constant in regular networks is given by

$$K^{(\alpha)} = [L^{\frac{\alpha}{2}}]_{pp} = \frac{1}{N} tr(L^{\frac{\alpha}{2}}) = \frac{1}{N}\sum_{m=1}^{N} (\mu_m)^{\frac{\alpha}{2}}, \qquad 0 < \alpha \leq 2. \tag{29.18}$$

The diagonal element of the transition matrix is vanishing $\mathscr{W}_{pp}^{(\alpha)} = 0$ as for the Pólya walk (see (29.9)). So as for the Pólya walk, the fractional random walker changes the node at any time-step. Further we introduced in relation (29.17) the fractional adjacency matrix

$$A_{pq}^{(\alpha)} = \delta_{pq}K^{(\alpha)} - (L^{\frac{\alpha}{2}})_{pq} \geq 0, \qquad 0 < \alpha \leq 2. \tag{29.19}$$

We observe analogous properties as in the non-fractional case

$$K^{(\alpha)} = \sum_{q=0}^{N-1} A_{pq}^{(\alpha)}$$

for the constant fractional degree reflecting conservation of eigenvalue zero and corresponding eigenvector $|\Psi_1\rangle$ of the fractional Laplacian matrix. We notice further per construction (29.19) the diagonal elements $A_{pp}^{(\alpha)} = 0$ are vanishing and (29.17) are normalized probabilities

$$\sum_{p=0}^{N-1} \mathscr{W}_{pq}^{(\alpha)} = \sum_{p=0}^{N-1} \frac{1}{K^{(\alpha)}}(A^{(\alpha)})_{pq} = 1.$$

The fractional adjacency matrix $A_{pq}^{(\alpha)}$ has uniquely non-negative off diagonal elements

$$A_{pq}^{(\alpha)} = -(L^{\frac{\alpha}{2}})_{pq} \geq 0 \quad (p \neq q)$$

which are especially nonzero and positive in the fractional range $0 < \alpha < 2$ reflecting the nonlocality of the FRW: In contrast to the Pólya type walk, for $0 < \alpha < 2$ the FRW walker can reach in one single time-step *any node* including far distant nodes ($D_{pq} >> 1$) in the network with jump probabilities $\mathscr{W}_{pq}^{(\alpha)} > 0$ where any jump distances may occur[4]. The nonlocal characteristics of the FRW makes its dynamics remarkably more rich as compared to Pólya type walks. The FRW has the capacity to describe phenomena of anomalous transport and diffusion where the large world property of a network seen by a Pólya walker is transformed by the FRW dynamics into a small world. We can demonstrate here only a few of these newly emerging phenomena generated by the FRW (Michelitsch et al, 2017b,a Riascos and Mateos, 2014, 2015).

29.5 Universality of Fractional Random Walks

The goal of the present section which is the principal part of this paper, is to elaborate some of the universal properties of FRWs generated by a fractional Laplacian $L^{\frac{\alpha}{2}}$ ($0 < \alpha \leq 2$). We refer these properties to as 'universal' as they appear independently from the spectral details of the Laplacian matrix.

29.5.1 Universal Behavior in the Limit $\alpha \to 0$

The first of these properties to be discussed here emerges when we consider the limit $\alpha \to 0+$ where α remains infinitesimally positive: In this limiting case the $N-1$ positive eigenvalues $\mu_m > 0$ take asymptotically independent of their values

$$(\mu_m)^{\frac{\alpha}{2}} \to 1$$

and where only the zero eigenvalue

$$\mu_1^{\frac{\alpha}{2}} = 0$$

is maintained. We obtain hence

$$\lim_{\alpha \to 0+} L^{\frac{\alpha}{2}} = \sum_{m=2}^{N} |\Psi_m\rangle\langle\Psi_m| = \hat{1} - |\Psi_1\rangle\langle\Psi_1|. \tag{29.20}$$

The universality of this relation manifests itself in the sense as it only requires one vanishing and $N-1$ positive eigenvalues, i.e. good structure i) ii) of the fractional Laplacian. Consider now a rescaled version of (29.20) generating the same walk (see

[4] With characteristic inverse power law asymptotics for long range jumps as demonstrated subsequently.

eq. (29.9))

$$\lim_{\alpha\to 0+} N\langle p|L^{\frac{\alpha}{2}}|q\rangle = \delta_{pq}(N-1) - (1-\delta_{pq}). \tag{29.21}$$

which coincides with the Laplacian of a *fully connected network* where all nodes are connected to each other $A_{pq} = (1-\delta_{pq}) = 1 \forall p \neq q$ and where the degree of any node is

$$\lim_{\alpha\to 0+} NK^{(\alpha)} = K_N = N-1,$$

i.e. each node has the maximum number of $N-1$ adjacent nodes. From (29.17) we obtain for $\alpha \to 0+$ the transition matrix

$$\lim_{\alpha\to 0+} \mathscr{W}^{(\alpha)} = \frac{1}{K^{(\alpha\to 0)}} A_{pq}^{(\alpha\to 0)} = \frac{1}{N-1}(1-\delta_{pq}) \tag{29.22}$$

coinciding with the transition matrix of a Pólya walk on a fully connected network where the walker can reach in one time-step any destination node different from the departure node in only one time-step with equal probability $1/(N-1)$. In this limit $\alpha \to 0+$ the FRW reaches its maximum speed. This holds for finite networks when N is finite and not necessarily large. We come back to the limit $\alpha \to 0+$ in subsequent paragraph in the context of our analysis of recurrence and transience features of the FRW on infinite lattices.

29.5.2 Recurrence Theorem for the Fractional Random Walk on Infinite Simple Cubic Lattices

Here we analyze universal behavior of the FRW which emerges in d-dimensional infinite simple cubic lattices (Z^d-integer lattices) where $d = 1,2,3,4,...$ denotes the dimension of the lattice. The recurrence behavior of the FRW to be analyzed in this paragraph can also be considered as an 'universal' property in the sense as it does not depend on the spectral details of Laplacian matrix L (apart from above discussed requirements for L of positive semi-definiteness **i**), and non-positive off diagonal elements **ii**)).

For the sake of simplicity of the present demonstration we consider now the special cases of regular undirected networks constituted by d-dimensional simple cubic lattices ($d = 1,2,3,4,..$) where we assume infinite boundary conditions in any dimension $j = 1,..,d$. The lattice points can be identified with the nodes of the network and are denoted by the lattice vectors $\boldsymbol{p} = (p_1,..,p_d)$ having integer valued components where $p_j = 0, \pm 1, \pm 2, .. \pm \infty \in Z$ may take any integer value. The Laplacian matrix $\mathscr{L}$ in this lattice with only next neighbor connections have the matrix elements (Michelitsch et al, 2017a,b)

$$
\begin{aligned}
\mathscr{L}(\boldsymbol{p}-\boldsymbol{q}) = \mathscr{L}_{p_1,..p_n|q_1,..,q_n} &= 2d \prod_{j=1}^{d} \delta_{p_j q_j} - \sum_{j=1}^{d} \left(\delta_{p_{j+1} q_j} + \delta_{p_{j-1} q_j}\right) \prod_{s \neq j}^{n} \delta_{p_s q_s} \\
&= \frac{1}{(2\pi)^d} \int_{\boldsymbol{\kappa}} e^{i\boldsymbol{\kappa}\cdot(\boldsymbol{p}-\boldsymbol{q})} \mu(\boldsymbol{\kappa}) \mathrm{d}^q \boldsymbol{\kappa}, \qquad \mu(\boldsymbol{\kappa}) = 2d - 2\sum_{j=1}^{d} \cos(\kappa_j) \\
&\quad -\pi \leq \kappa_j \leq \pi, \qquad j = 1,..,d
\end{aligned}
\tag{29.23}
$$

and constant degree $K = 2d$. The second line indicates the spectral representation in terms of 2π-periodic Bloch eigenfunctions

$$
\frac{e^{i\kappa_j p_j}}{\sqrt{2\pi}}
$$

(see (29.2)) where $\boldsymbol{\kappa} = (\kappa_1, .., \kappa_j, .., \kappa_d)$. In (29.23) we have introduced the abbreviation

$$
\int_{\boldsymbol{\kappa}} h(\boldsymbol{\kappa}\cdot(\boldsymbol{p}-\boldsymbol{q}))\mathrm{d}\boldsymbol{\kappa} = \int_{-\pi}^{\pi} \mathrm{d}\kappa_1 .. \int_{-\pi}^{\pi} \mathrm{d}\kappa_d h(\kappa_1(p_1-q_1) + .. + \kappa_d(p_d - q_d)) \tag{29.24}
$$

which indicates integration over the d-dimensional first Brillouin zone $(2\pi)^d$. Further we introduced the unity matrix

$$
\delta_{pq} \rightarrow \delta_{\boldsymbol{p}-\boldsymbol{q}} = \prod_{j=1}^{d} \delta_{p_j q_j}
$$

of the d-dimensional lattice. Laplacian (29.23) with transition matrix (29.9) defines the classical Pólya walk (Pólya, 1921). Now to define the FRW on the cubic lattice we introduce the fractional Laplacian (fractional power of (29.23)) which has the canonic representation

$$
[\mathscr{L}^{\frac{\alpha}{2}}](\boldsymbol{p}-\boldsymbol{q}) = \frac{1}{(2\pi)^d} \int_{\boldsymbol{\kappa}} e^{i\boldsymbol{\kappa}\cdot(\boldsymbol{p}-\boldsymbol{q})} (\mu(\boldsymbol{\kappa}))^{\frac{\alpha}{2}} \mathrm{d}^q \boldsymbol{\kappa}, \qquad 0 < \alpha \leq 2 \tag{29.25}
$$

where throughout this analysis we confine us on the admissible range $0 < \alpha \leq 2$ and for $\alpha = 2$ (29.25) recovers (29.23). The transition matrix of the FRW on the lattice is defined by (29.17) and can then be written as

$$
\mathscr{W}^{(\alpha)}(\boldsymbol{p}-\boldsymbol{q}) = \frac{1}{(2\pi)^d} \int_{\boldsymbol{\kappa}} e^{i\boldsymbol{\kappa}\cdot(\boldsymbol{p}-\boldsymbol{q})} \lambda^{(\alpha)}(\boldsymbol{\kappa}) \mathrm{d}^q \boldsymbol{\kappa}, \qquad \lambda^{(\alpha)}(\boldsymbol{\kappa}) = 1 - \frac{(\mu(\boldsymbol{\kappa}))^{\frac{\alpha}{2}}}{K^{(\alpha)}} \tag{29.26}
$$

where the fractional degree $K^{(\alpha)}$ being the constant diagonal element of (29.25) is given by

$$
K^{(\alpha)} = [\mathscr{L}^{\frac{\alpha}{2}}](\mathbf{0}) = \frac{1}{(2\pi)^d} \int \mu^{\frac{\alpha}{2}}(\boldsymbol{\kappa}) \mathrm{d}^d \kappa, \qquad 0 < \alpha \leq 2. \tag{29.27}
$$

We introduce now the Green's function of the FRW on the infinite simple cubic d-dimensional lattice. Using general relation (29.15) yields

$$r^{(\alpha)}_{\boldsymbol{p}-\boldsymbol{q}}(\xi=1) = \frac{K^{(\alpha)}}{(2\pi)^d}\int e^{i(\boldsymbol{p}-\boldsymbol{q})\cdot\boldsymbol{\kappa}}\mu^{-\frac{\alpha}{2}}(\boldsymbol{\kappa})\mathrm{d}^d\kappa \tag{29.28}$$

where the integrations always are performed over the first Brillouin zone (see (29.24)). As already mentioned the matrix elements of the Green's function

$$r^{(\alpha)}_{\boldsymbol{p}-\boldsymbol{q}}(\xi=1)$$

have the statistical interpretation of the mean residence time (MRT), that is the average number of time-steps the walker is visiting a node $\boldsymbol{p}$ during an infinite time of observation where the walk always is starting at node $\boldsymbol{q}$. The Green's function is a symmetric matrix with respect of exchange of departure- and arrival node. Its diagonal element of (29.28) indicates then the average number of time-steps the walker spends in the departure node

$$r^{(\alpha)}_0(\xi=1) = \frac{K^{\alpha}}{(2\pi)^d}\int \mu^{-\frac{\alpha}{2}}(\boldsymbol{\kappa}')\mathrm{d}^d\kappa'. \tag{29.29}$$

Since the walker for $t=0$ per definition always is present in the departure node for one time-step, the inequality $r^{(\alpha)}_0(\xi=1)\geq 1$ is fulfilled. Thus $r^{(\alpha)}_0(\xi=1)-1$ indicates the expected number of returns to the departure node measured during an infinite time of observation. It is a sufficient criteria for recurrence of the walk when this quantity (and hence the Green's function (29.28)) diverges, indicating that the walker returns to the departure node infinitely often. On the other hand if $r^{(\alpha)}_0(\xi=1)-1$ is convergent, i.e. when the walker returns during an infinite time of observation only a finite number of times to the departure node, then the walk necessarily is transient.

Let us now analyse the convergence behavior of Green's function (29.28) for the FRW more closely. We notice in (29.27) that the fractional degree $K^{(\alpha)}$ for $0<\alpha\leq 2$ always converges. The question of whether or not the Green's function converges thus depends on the behavior of the eigenvalues $\mu^{\frac{\alpha}{2}}(\boldsymbol{\kappa})$ of the fractional Laplacian (29.25) close to the origin $|\boldsymbol{\kappa}|\to 0$. Since in this limit $\mu^{\frac{\alpha}{2}}(\boldsymbol{\kappa})\sim|\boldsymbol{\kappa}|^{\alpha}$ (see relation (29.23)), integral (29.29) can be written as

$$\begin{aligned}\frac{r^{(\alpha)}_0(\xi=1)}{K^{(\alpha)}} &= \frac{1}{(2\pi)^d}\int \mu^{-\frac{\alpha}{2}}(\boldsymbol{\kappa}')\mathrm{d}^d\kappa' \\ &= \frac{1}{(2\pi)^d}\left\{\frac{2\pi^{\frac{d}{2}}}{\Gamma(\frac{d}{2})}\lim_{\epsilon\to 0}\int_{\epsilon}^{\kappa_0}\kappa^{d-1-\alpha}\mathrm{d}\kappa + \int_{V_c}\mu^{-\frac{\alpha}{2}}(\boldsymbol{\kappa}')\mathrm{d}^d\kappa'\right\} \\ &\sim \lim_{\epsilon\to 0} a(\epsilon) + C(\kappa_0).\end{aligned} \tag{29.30}$$

In (29.30) $0 < \kappa_0 \ll 1$ is sufficiently small that $(\mu(\boldsymbol{\kappa_0}))^{-\frac{\alpha}{2}} \approx \kappa_0^{-\alpha}$ and $C(\kappa_0)$ is the contribution of the integral of $\mu^{-\frac{\alpha}{2}}(\boldsymbol{\kappa}')$ over V_c which is the cube $-\pi < \kappa_j < \pi$ without the d-sphere of radius $\kappa = \kappa_0$.

The first integral in $(29.30)_2$ is crucial for the divergence or convergence of $r_0^{(\alpha)}(1)$: It behaves as

$$a(\epsilon) \sim -\frac{\epsilon^{d-\alpha}}{d-\alpha}$$

for $d \neq \alpha$ and

$$a(\epsilon) \sim -\log(\epsilon)$$

when $d = \alpha$ where $\epsilon \to 0+$. Hence (29.30) diverges for $d \leq \alpha$ and as a consequence the FRW then is recurrent. On the other hand integral (29.30) is finite for $d > \alpha$, i.e. Green's function (29.28) then is convergent where the walker in the average visits any node only a finite number of times. That is for $d > \alpha$ ($0 < \alpha \leq 2$) the FRW is transient with a finite probability

$$\frac{1}{r_0^{(\alpha)}(\xi = 1)} > 0$$

of never return to the departure node.

The generalized recurrence theorem for Fractional Random Walks can then be formulated as follows (For a more thorough analysis and further comments, we refer to Michelitsch et al (2017a)). The FRW is recurrent for lattice dimensions $d \leq \alpha$ and transient for $d > \alpha$ where always $0 < \alpha \leq 2$. We emphasize that the recurrence behavior of the FRW is an universal feature, that is, it does not depend on the spectral details of the generating Laplacian L. This behavior is represented in Fig. 29.2: Lattice dimensions of transient FRWs are indicated by bullet points. The FRW-recurrence theorem remains true for the entire class of random walks on infinite networks with the same power law asymptotics for long-range steps (analyzed in subsequent section) as Lévy flights. The recurrence behavior for Lévy flights in the *continuous* d-dimensional infinite space was considered earlier (Ferraro and Zaninetti, 2006; Sato, 1999) and generally recurrence of symmetric stable processes was analyzed in Blumenthal et al (1961); Getoor (1961) and see also the Hughes and Shlesinger (1982) where some recurrence features of random walks with power-law asymptotics have been obtained.

29.5.3 Universal Asymptotic Scaling: Emergence of Lévy Flights

The asymptotic power law behavior of the eigenvalues of the fractional Laplacian as $|\boldsymbol{\kappa}|^{\alpha}$ for $|\boldsymbol{\kappa}| \to 0$ have as further consequence an universal scaling behavior of long-range steps. To demonstrate this briefly, let us consider the probability determined by (29.26) that the walker makes a long-range move $|\boldsymbol{p}-\boldsymbol{q}| \gg 1$. Using $(\mu(\boldsymbol{\kappa}))^{\frac{\alpha}{2}} \sim |\boldsymbol{\kappa}|^{\alpha}$ for $|\boldsymbol{\kappa}| \to 0$ we obtain as leading contribution ($A^{(\alpha)} = K^{(\alpha)} W^{(\alpha)}$)

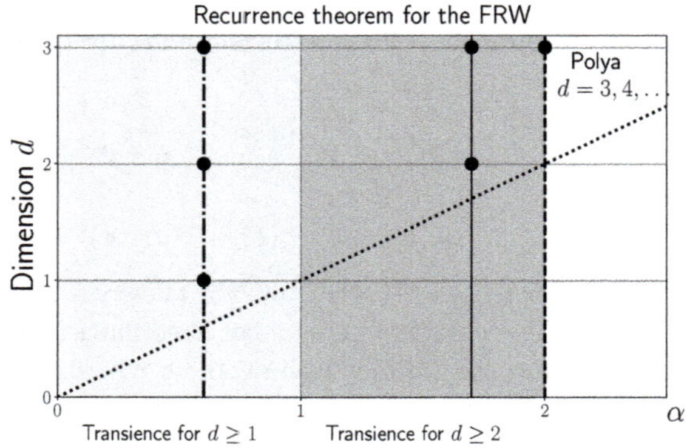

Fig. 29.2 Representation of the recurrence theorem for the Fractional Random Walk for the admissible range $0 < \alpha \leq 2$ (Michelitsch et al, 2017a): The plot shows for two values of α (within $0 < \alpha < 2$) the lattice dimensions of transience (indicated as bullet points). Recurrent FRWs exist for $d = 1$ within $1 \leq \alpha < 2$. For $\alpha = 2$ representing the Pólya walk, Pólya's classical recurrence theorem is recovered where the walk is recurrent for lattice dimensions $d = 1,2$ and transient for $d \geq 3$.

$$A^{(\alpha)}(\boldsymbol{p}-\boldsymbol{q})) \approx -\frac{1}{(2\pi)^d}\int |\boldsymbol{\kappa}|^{\alpha} e^{i(\boldsymbol{p}-\boldsymbol{q})\cdot\boldsymbol{\kappa}} = -\left(-\Delta_{(\boldsymbol{p}-\boldsymbol{q})}\right)^{\frac{\alpha}{2}} \delta^d(\boldsymbol{p}-\boldsymbol{q}) = \frac{C_{\alpha,d}}{|\boldsymbol{p}-\boldsymbol{q}|^{d+\alpha}} \tag{29.31}$$

with the positive constant (Michelitsch et al, 2014)

$$C_{\alpha,d} = \frac{2^{\alpha-1}\alpha\Gamma(\frac{\alpha+d}{2})}{\pi^{\frac{d}{2}}\Gamma(1-\frac{\alpha}{2})}$$

where the inverse power law kernel holds for $0 < \alpha < 2$ ($\alpha \neq 2$) having the form of the kernel of the fractional Laplacian (Riesz fractional derivative) in the d-dimensional infinite space (Michelitsch et al, 2014, 2017b). Returning to master equation (29.4), the time evolution of the occupation probabilities with (29.31) is for $|\boldsymbol{p}-\boldsymbol{q}| \gg 1$ asymptotically described by [5]

$$\frac{dP_t(\boldsymbol{p}-\boldsymbol{q})}{dt}|_{t=0} \approx -\frac{(L^{\frac{\alpha}{2}})_{\boldsymbol{p}-\boldsymbol{q}}}{K^{(\alpha)}} \approx -\frac{1}{K^{(\alpha)}}\left(-\Delta_{(\boldsymbol{p}-\boldsymbol{q})}\right)^{\frac{\alpha}{2}} \delta^d(\boldsymbol{p}-\boldsymbol{q}) \tag{29.32}$$

This relation is the evolution equation of a (time-continuous) Lévy flight in the d-dimensional infinite space with Lévy index $0 < \alpha < 2$ where $\delta^d(\boldsymbol{p}-\boldsymbol{q})$ denotes the d-dimensional Dirac's δ-function (Michelitsch et al, 2017b). Asymptotic relation (29.32) recovers for $\alpha = 2$ the conventional diffusion equation with the Brownian nature of the Pólya walk.

By a similar consideration the asymptotic representation of the fractional lattice Green's function (29.28) of the d-dimensional infinite lattice for $|\boldsymbol{p}-\boldsymbol{q}| >> 1$ for the transient regime $d-\alpha > 0$ is obtained in Riesz potential form (see also Sato, 1999, pp. 261)

[5] $P_{t+1}(\boldsymbol{p}-\boldsymbol{q}) - P_t(\boldsymbol{p}-\boldsymbol{q}) \approx \frac{dP_t(\boldsymbol{p}-\boldsymbol{q})}{dt}$

$$\frac{1}{K^{(\alpha)}} r^{(\alpha)}(\boldsymbol{p}-\boldsymbol{q}) \approx \left(-\Delta_{(\boldsymbol{p}-\boldsymbol{q})}\right)^{-\frac{\alpha}{2}} \delta^d(\boldsymbol{p}-\boldsymbol{q}) = \frac{(-C_{-\alpha,d})}{|\boldsymbol{p}-\boldsymbol{q}|^{d-\alpha}} > 0 \tag{29.33}$$

evanescent at infinity since existing only in the transient regime $d > \alpha$ ($0 < \alpha \leq 2$). This expression formally is obtained when replacing $\alpha \to -\alpha$ (and adding multiplyer (-1)) in (29.31). It is worth noticing that

$$-C_{-\alpha,d} = 2^{-\alpha} \pi^{-\frac{d}{2}} \frac{\Gamma(\frac{d-\alpha}{2})}{\Gamma(\frac{\alpha}{2})} > 0$$

occurring in (29.33) indeed is a positive constant as required for probabilities. We mention also that asymptotic relation (29.33) holds also for the transient Pólya walks $d > \alpha = 2$, i.e. $d \geq 3$. For a Pólya walk on a $d = 3$-dimensional lattice, the Green's function (29.33) takes the representation of a Newtonian potential

$$\frac{1}{4\pi|\boldsymbol{p}-\boldsymbol{q}|} = \left(-\Delta_{(\boldsymbol{p}-\boldsymbol{q})}\right)^{-1} \delta^3(\boldsymbol{p}-\boldsymbol{q})$$

as a landmark of the Brownian nature of the Pólya walk. It is a very interesting observation that the lattice Green's functions existing in the transient regime and static potentials of lattice dynamics have their correspondence in both contexts, namely for random walks and lattice dynamics. In this context we refer especially to the seminal works (Montroll, 1956; Montroll and Weiss, 1965; Spitzer, 1976) where this general correspondence is analyzed.

29.6 Transient Regime $0 < \alpha < 1$ for the Inifinite One-Dimensional Chain

The FRW is transient for all lattice dimensions $d = 1, 2, 3, 4, ..$ in the index range $0 < \alpha < 1$. Here we discuss the transient regime for the one-dimensional infinite lattice $d = 1$. For this case explicit expressions for the fractional Laplacian matrix (Michelitsch et al, 2015, 2016; Zoia et al, 2007), and for the lattice Green's function (Michelitsch et al, 2017a) have been obtained. We assume an undirected network of an infinite chain where the nodes are located at the integer lattice points $p = 0, \pm 1, \pm 2 \in \mathbf{Z}$. The Laplacian matrix (29.23) of the one-dimensional lattice with only next neighbor connections and constant degree $K = 2$ has the simple representation

$$L_{pq} = 2\delta_{pq} - \delta_{p,q+1} - \delta_{p,q-1} \tag{29.34}$$

of a symmetric second difference operator where the eigenvalues of the Laplacian are

$$\mu(\kappa) = 2(1 - \cos(\kappa)) = 4\sin^2\left(\frac{\kappa}{2}\right), \qquad -\pi \leq \kappa \leq \pi. \tag{29.35}$$

The spectral representation of the fractional Laplacian matrix $L^{\frac{\alpha}{2}}$ then is defined by

$$(L^{\frac{\alpha}{2}})_{|p|} = \frac{1}{2\pi}\int_{-\pi}^{\pi} e^{i\kappa p}\left(4\sin^2\frac{\kappa}{2})\right)^{\frac{\alpha}{2}} \mathrm{d}\kappa \tag{29.36}$$

depending only on $|p-q| \to |p| = p$ and is obtained in explicit form as (Michelitsch et al, 2015, 2016; Zoia et al, 2007)

$$(L^{\frac{\alpha}{2}})_{|p|} = (-1)^p \frac{\alpha!}{(\frac{\alpha}{2}+p)!(\frac{\alpha}{2}-p)!} = -\frac{\alpha!}{\pi}\sin\left(\frac{\alpha\pi}{2}\right)\frac{(|p|-1-\frac{\alpha}{2})!}{(\frac{\alpha}{2}+|p|)!}. \tag{29.37}$$

We utilize here the same notation as in Michelitsch et al (2015, 2016) in terms of generalized binomial coefficients with the notation

$$\zeta! = \Gamma(\zeta+1) \tag{29.38}$$

where $\Gamma(..)$ denotes the Γ-function. Note that as discussed above the fractional Laplacian maintains its good properties for $0 < \alpha \leq 2$, where $\alpha = 2$ in (29.36), (29.37) recovers the Laplacian (29.34). The fractional degree (diagonal element of the fractional Laplacian matrix (29.37)) yields

$$K^{(\alpha)} = (L^{\frac{\alpha}{2}})_0 = \frac{2^\alpha}{\pi}\frac{(\frac{\alpha-1}{2})!(-\frac{1}{2})!}{\frac{\alpha}{2}!} = \frac{\alpha!}{\frac{\alpha}{2}!\frac{\alpha}{2}!}. \tag{29.39}$$

Expression (29.28) for the Green's function with $d = 1$ takes the representation (Michelitsch et al, 2017a)

$$r_{|p|}^{(\alpha)}(\xi = 1) = \frac{K^{(\alpha)}}{2\pi}\int_{-\pi}^{\pi} e^{i\kappa p}\left(4\sin^2\frac{\kappa}{2})\right)^{-\frac{\alpha}{2}} \mathrm{d}\kappa, \qquad 0<\alpha<1. \tag{29.40}$$

We emphasize that for $d = 1$ the lattice Green's function converges only in the transient regime $0 < \alpha < 1$, whereas diverges in the recurrent regime $1 \leq \alpha \leq 2$. The Green's function (29.40) then yields an explicit expression which can be obtained when replacing in (29.37) $\alpha \to -\alpha$, namely

$$r_{|p|}^{(\alpha)}(\xi = 1) = K^{(\alpha)}(L^{-\frac{\alpha}{2}})_{|p|} = \frac{\alpha!}{\frac{\alpha}{2}!\frac{\alpha}{2}!}\frac{(-1)^p(-\alpha)!}{(\frac{-\alpha}{2}+p)!(\frac{-\alpha}{2}-p)!}, \qquad 0<\alpha<1. \tag{29.41}$$

For $p = 0$ this relation yields

$$r_0^{(\alpha)}(1) = \frac{\Gamma(1+\alpha)\Gamma(1-\alpha)}{\Gamma^2(1+\frac{\alpha}{2})\Gamma^2(1-\frac{\alpha}{2})} \geq 1, \qquad 0<\alpha<1 \tag{29.42}$$

which in view of our above interpretation as MRT and (29.40) can only take values greater than one (Fig. 29.3) where the minimum value

$$r_0^{(\alpha\to 0+)}(1) = 1$$

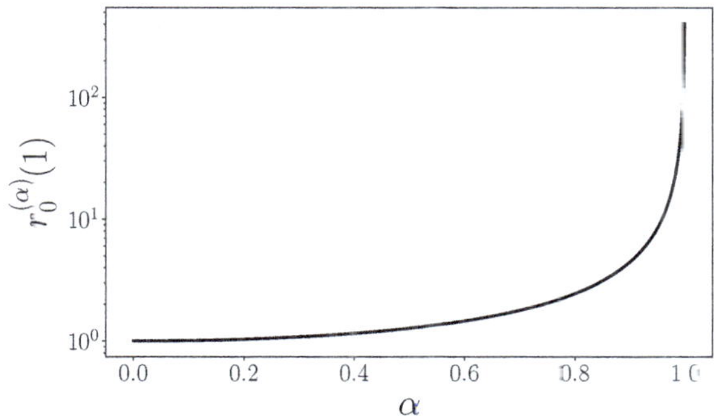

Fig. 29.3 Diagonal element $r_0^{(\alpha)}(1)$ of the Green's function (29.42) versus α in the transient regime $0 < \alpha < 1$, indicating the expected number of time units the walker is present in the departure node. The smaller α the more rare the walker returns to the departure node where for $\alpha \to 0+$ (extreme transience) the walker never returns to the departure node. With increasing α the MRT $r_0^{(\alpha)}(1)$ increases monotonously diverging at the recurrent limit $\alpha \to 1-0$ where the walker returns infinitely often to the departure node.

is taken for $\alpha \to 0+$. For $|p| >> 1$ (29.41) yields the Riesz potential asymptotic formula

$$\frac{r^{(\alpha)}_{|p|>>1}(\xi = 1)}{K^{(\alpha)}} \approx \frac{(-\alpha)!}{\pi} \sin\left(\frac{\pi\alpha}{2}\right) \frac{1}{|p|^{1-\alpha}}, \qquad 0 < \alpha < 1 \tag{29.43}$$

which is in accordance with (29.33) for $d = 1$ ($0 < \alpha < 1$).

Let us now discuss more closely some characteristics of the FRW where we invoke the statistical meaning of the components of the Green's function as the mean residence time (MRT). Due to the construction of the walk (29.9) where the walker at any time increment has to move to another node, (29.41) thus counts the expected number of visits of nodes $\pm p$ during an infinitely long FRW. For visits of the departure node holds

$$r_0^{(\alpha)}(\xi = 1) \geq 1$$

since the walker per definition visits the departure node (at least) once, namely at $t = 0$.

In Fig. 29.3 is drawn the diagonal element $r_0^{(\alpha)}(1)$. Taking into account that $r_0^{(\alpha)}(1) - 1$ indicates the average number of *returns* to the departure node, Fig. 29.3 shows the following behavior: The smaller α the more rare become returns to the departure node. A physical interpretation is that the smaller α, the mode frequent are long-range steps (see (29.31) with $d = 1$) and the higher the average speed of the FRW. This tendency is most pronounced in the limit of extreme transience $\alpha \to 0+$ where the walker for sure never returns to the departure node where

$$r_0^{(\alpha \to 0+)}(1) - 1 = 0.$$

For increasing α the number of returns to the departure node increases monotonously, diverging in the limit recurrence $\alpha = 1-0$ (see Fig. 29.3 and Eq. (29.42) where

$$\lim_{\alpha \to 1-0} \Gamma(1-\alpha) = \infty),$$

where the walker infinitely often returns to the departure node. In view of relation (29.13) we mention that $1/r_0^{(\alpha)}(1)$ indicates the probability of escape (never return) to the departure node: for $\alpha \to 0+$ the probability that walker never returns to the departure node tends to 1 whereas in the recurrent limit $\alpha = 1$ the probability of escape

$$\frac{1}{r_0^{(\alpha \to 1-0)}(1)} = 0$$

is vanishing.

In Fig. 29.4 are drawn the matrix elements of Green's function (closed form expression (29.41)) for destination nodes different form the departure node ($p \neq 0$). We observe that the smaller α, the smaller $r_{|p|}^{(\alpha)}(\xi = 1)$. That is the smaller α, the more rarely a preselected target node is visited. In terms of search efficiency this means that a FRW searcher is less successful for smaller α. This tendency is the most pronounced at $\alpha = 0+$ where the expected number of visits of any preselected destination node tends to zero. A FRW searcher on the infinite lattice in the limit of extreme transience $\alpha \to 0$ becomes inefficient [6]. This effect can be understood in terms of more frequent long-range jumps occuring for smaller α where any preselected destination node tends to be overleaped. This overshooting effect was already found in the context of Lévy flight search (Klages, 2016; Palyulin et al, 2014). We also observe in Fig. 29.4 when α is kept constant that the average number of visits is increased for nodes closer to the departure node.

On the other hand we also see in Fig. 29.4 the expected number of visits of a node p increases with increasing α and diverges when approaching the recurrent limit $\alpha = 1$. That is in terms of search efficiency the FRW searcher becomes the more efficient the closer the recurrent limit $\alpha = 1$ is approached. Similar results were reported for the case of Lévy flight based search strategies (Klages, 2016; Palyulin et al, 2014) (and references therein), however, this subject is beyond of the scope of the present paper. For a further discussion and analysis of the present results in terms of ever passage probabilities we refer to Michelitsch et al (2017a).

In Fig. 29.5 is plotted the Green's function (29.41) versus $|p|$ in log-log representation. Equal values are indicated by lines of the same color. The linear characteristic of the same colored lines indicate the inverse power law (Riesz potential) characteristics which is obtained for $|p| \gg 1$ in (29.43). We observe that, the more the recurrent limit $\alpha = 1$ is approached, the more horizontal the equal colored lines become which is confirmed by the Riesz potential (29.43) behaving as $\sim |p|^{\alpha-1}$.

[6] This statement is not any more true on *finite* lattices where any walk is recurrent.

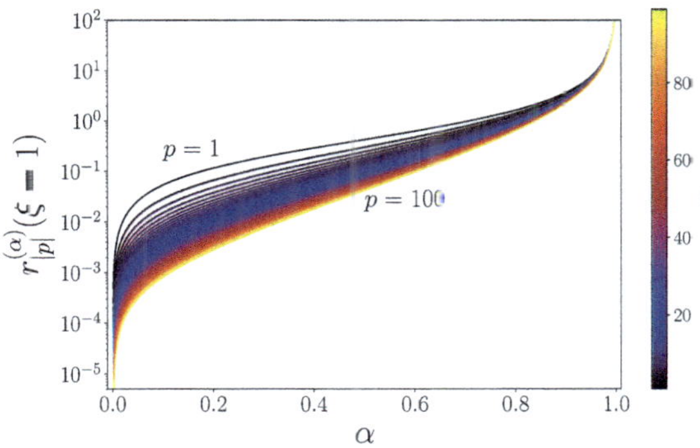

Fig. 29.4 Expected number of visits given by the elements of Green's function (29.41) for nodes $p \neq 0$ different from the departure node. The larger α the more frequent a destination node is visited (diverging at the recurrent limit $\alpha = 1$). For the same α a node is more often visited the closer located to the departure node. In the limit of extreme transience $\alpha \to 0+$ caused by the extreme occurence of long range jumps any preselected node p is overleaped and therefore in the average is never visited.

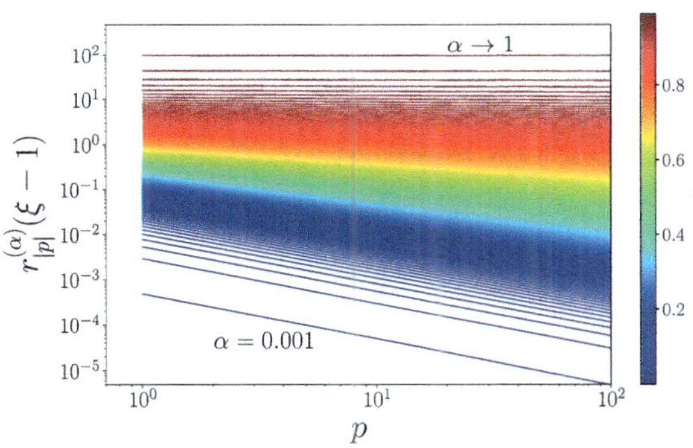

Fig. 29.5 Plot $r^{(\alpha)}_{|p|}(\xi = 1)$ vs. $\log p$ from (29.41) taking Riesz potential inverse power law characteristics $\log(r^{(\alpha)}_{|p|}(\xi = 1)) \sim (\alpha - 1)\log|p| + \text{const}$ (see Eq. (29.43)). In rainbow colors from small $\alpha = 0.001$ (blue: extreme transience) to $\alpha = 1 - 0$ (red: recurrent limit).

29.7 Conclusions

In the present paper we have presented a brief analysis of some universal features of Fractional Random Walks on regular networks, especially simple cubic lattices. Among these universal features are the recurrence and transience behavior of the FRW: The FRW on *infinite lattices* is transient for lattice dimensions $d > \alpha$ (where always $0 < \alpha \leq 2$) otherwise recurrent (Michelitsch et al, 2017a). On finite lattices the FRW (as any Markovian walk) *always* is recurrent (Hughes, 1995).

Another universal feature of the FRW emerging on infinite networks is the inverse power law asymptotics for the probability of long-range moves where the master equation governing the time evolution of the occupation probabilities asymptotically takes the form of a fractional Lévy flight diffusion equation involving Riesz fractional derivatives (relation (29.32) together with (29.31)). As a consequence the infinite lattice Green's function of the FRW indicating the average number of visits of nodes (converging only in the transient regime $d > \alpha$) takes the asymptotic form of the inverse of the Riesz fractional derivative, namely the Riesz potential for nodes far from the departure node (expression (29.33)).

All these mentioned features are universal in the sense that they do not depend on the spectral details of the Laplacian matrix L (where only 'good' properties of the Laplacian are required). They emerge on infinite lattices and depend only on lattice dimension d and index α. In the limit $\alpha \to 0$ the FRW appears as a Pólya walk on a fully connected network when the network is finite. In the infinite network the limit $\alpha \to 0+$ exhibits the phenomena of extreme transience and appears for any lattice dimension d where the walker for sure never returns to the departure node (see (29.28) together with (29.27) and (29.41) for $d = 1$). For a more thorough analysis and discussions of FRW we again refer to Michelitsch et al (2017a).

The FRW offers interesting perspectives in the development for faster random walk based search strategies as demonstrated for Lévy flights (Palyulin et al, 2016, 2014; Bénichou et al, 2011). The FRW due to his universal characteristics with relation to Lévy α stable distributions and capacity to describe anomalous transport phenomena in networks deserves further investigations: For instance it would be desirable to analyse MRTs for finite observation times t_{ob} and to characterize how the $MRT(t_{ob})$ diverge in the recurrent limit with respect to t_{ob}.

Further important questions calling for deeper analysis are: What is the relationship of the FRW to *fractal* networks? Is the FRW performed on a regular lattice equivalent to a walk performed on a fractal network and what is then the relation of its fractal dimension with the index α of the FRW? In Hughes and Shlesinger (1982) random walks with asymptotic power law behavior were considered and a relationship between scaling index and (fractal) lattice dimensions were analysed. With a similar reasoning Pólya type walks on fractal networks could be related to FRWs: Consider a path in the d-dimensional lattice realized during a FRW. This transcience/recurrence behavior depends on $d - \alpha$. Take now a network of a 'dimension' d^* containing the same path where successively visited points are adjacent. The FRW on this unknown network thus is of Pólya type where only steps to adjacent nodes are performed. We consider the identical walk in both networks, appearing as a FRW in the d-dimensional lattice as well as a Pólya type walk in the unknown network of dimension d^*. Since identical walks have the same recurrence behavior, we obtain $d^* - 2 = d - \alpha$ which is a condition that determines the dimension $d^* = d + 2 - \alpha$ of the unknown network. For $0 < \alpha < 2$ dimension d^* generally is (except for $\alpha = 1$) a non-integer and hence a fractal dimension $d^* > d$. For the Pólya walk case $\alpha = 2$ both dimensions $d^* = d$ and hence both lattices coincide. In other words the FRW on regular lattice seem to have Pólya type walk counterparts taking place on fractal network. This rough consideration may give an idea of the potential of the FRW concept to open new directions towards approaches that describe the dynamics of random motions on fractal sets and complex networks.

References

Albert R, Barabási AL (2002) Statistical mechanics of complex networks Rev Mod Phys 74:47–97
Bénichou O, Loverdo C, Moreau M, Voituriez R (2011) Intermittent search strategies. Rev Mod Phys 83:81–129
Blumenthal RM, Getoor RK, Ray DB (1961) On the distribution of first hits for the symmetric stable processes. Transactions of the American Mathematical Society 99(3):540–554
Chechkin AV, Metzler R, Klafter J, Gonchar VY (2008) Introduction to the theory of lévy flights. In: Klages R, Radons G, Sokolov IM (eds) Anomalous Transport: Foundations and Applications, Wiley, pp 129–162
Doyle PG, Laurie Snell J (1984) Random Walks and Electric Networks, Carus Mathematical Monographs, vol 22. Mathematical Association of America
Dybiec B, Gudowska-Nowak E, Barkai E, Dubkov AA (2017) Lévy flights versus Lévy walks in bounded domains. Phys Rev E 95(5):052,102
Feller W (1950) An Introduction to Probability Theory and its Applications. John Wiley & Sons Inc., New York, London
Ferraro M, Zaninetti L (2006) Mean number of visits to sites in Levy flights. Phys Rev E 73(5):057,102
Getoor RK (1961) First passage times for symmetric stable processes in space. Transactions of the American Mathematical Society 101(1):75–90
Gonçalves B, Perra N, Vespignani A (2011) Modeling users' activity on twitter networks: Validation of dunbar's number. PLOS ONE 6(8):e22,656
Hudges BD (1995) Random Walks and Random Environments. Cambridge University Press, New York
Hughes BD, Shlesinger MF (1982) Lattice dynamics, random walks, and nonintegral effective dimensionality. Journal of Mathematical Physics 23(9):1688–1692
Kemeny JG, Laurie Snell J (1976) Finite Markov Chains. Springer, New York, Berlin, Tokyo
Klages R (2016) Search for food of birds, fish and insects. In: Bunde A, Caro J, Kaerger J, Vogl G (eds) Diffusive Spreading in Nature, Technology and Society, Springer, Berlin, pp 129–162
Maugin GA (2017) Nonclassical Continuum Mechanics. Springer
Metzler R, Klafter J (2000) The random walk's guide to anomalous diffusion: a fractional dynamics approach. Physics Reports 339(1):1–77
Metzler R, Klafter J (2004) The restaurant at the end of the random walk: recent developments in the description of anomalous transport by fractional dynamics. Journal of Physics A: Mathematical and General 37(31):R161–R208
Metzler R, Koren T, van den Broek B, Wuite GJL, Lomholt MA (2009) And did he search for you, and could not find you? Journal of Physics A: Mathematical and Theoretical 42(43):434,005
Michelitsch T, Collet BA, Riascos AP, Nowakowski AF, Nicolleau F (2017a) On recurrence of random walks with long-range steps generated by fractional Laplacian matrices on regular networks and simple cubic lattices. Journal of Physics A: Mathematical and Theoretical 50:505,004
Michelitsch TM, Maugin GA, Derogar S, Rahman M (2014) A regularized representation of the fractional Laplacian in n dimensions and its relation to Weierstrass-Mandelbrot-type fractal functions. IMA Journal of Applied Mathematics 79(5):753–777
Michelitsch TM, Collet B, Nowakowski AF, Nicolleau FCGA (2015) Fractional Laplacian matrix on the finite periodic linear chain and its periodic Riesz fractional derivative continuum limit. Journal of Physics A: Mathematical and Theoretical 48(29):295,202
Michelitsch TM, Collet B, Nowakowski AF, Nicolleau FCGA (2016) Lattice fractional laplacian and its continuum limit kernel on the finite cyclic chain. Chaos, Solitons & Fractals 82:38–47
Michelitsch TM, Collet BA, Riascos AP, Nowakowski AF, Nicolleau FCGA (2017b) Fractional random walk lattice dynamics. Journal of Physics A: Mathematical and Theoretical 50(5):055,003
Mieghem PV (2011) Graph Spectra for Complex Networks. Cambridge University Press, New York, NY, USA

Montroll EW (1956) Random walks in multidimensional spaces, especially on periodic lattices. Journal of the Society for Industrial and Applied Mathematics 4(4):241–260

Montroll EW, Weiss GH (1965) Random walks on lattices. II. Journal of Mathematical Physics 6(2):167–181

Newman MEJ (2010) Networks: An Introduction. Oxford University Press, Oxford

Noh JD, Rieger H (2004) Random Walks on Complex Networks. Physical Review Letters 92(11):118,701

Palyulin VV, Chechkin AV, Metzler R (2014) Lévy flights do not always optimize random blind search for sparse targets. Proc Nat Acad Sci USA 111(8):2931–2936

Palyulin VV, Chechkin AV, Klages R, Metzler R (2016) Search reliability and search efficiency of combined Lévy-Brownian motion: long relocations mingled with thorough local exploration. Journal of Physics A: Mathematical and Theoretical 49(39):394,002

Pearson K (1905) The problem of the random walk. Nature 72(294):318–342

Pólya G (1921) Über eine Aufgabe der Wahrscheinlichkeitsrechnung betreffend die Irrfahrt im Straßennetz. Mathematische Annalen 83:149–160

Riascos AP, Mateos JL (2012) Long-range navigation on complex networks using Lévy random walks. Phys Rev E 86:056,110

Riascos AP, Mateos JL (2014) Fractional dynamics on networks: Emergence of anomalous diffusion and Lévy flights. Phys Rev E 90:032,809

Riascos AP, Mateos JL (2015) Fractional diffusion on circulant networks: emergence of a dynamical small world. Journal of Statistical Mechanics: Theory and Experiment 2015(7):P07,015

Riascos AP, Michelitsch TM, Collet B, Nowakowski AF, Nicolleau FCGA (2017) Random walks defined in terms bof functions of the laplacian matrix: Emergence of long-range transport on networks. in preparation

Sato K (1999) Lévy processes and Infinitely Divisible Distributions, Cambridge Studies in Advanced Mathematics, vol 68. Cambridge University Press

Spitzer F (1976) Principles of Random Walks. Springer, Heidelberg

Viswanathan GM, Raposo EP, da Luz MGE (2008) Lévy flights and superdiffusion in the context of biological encounters and random searches. Physics of Life Reviews 5(3):133–150

Zoia A, Rosso A, Kardar M (2007) Fractional Laplacian in bounded domains. Phys Rev E 76(2):021,116

Chapter 30
Micropolar Theory with Production of Rotational Inertia: A Rational Mechanics Approach

Wolfgang H. Müller and Elena N. Vilchevskaya

Abstract The aim of this paper is a review on recently found new aspects in the theory of micropolar media. For this purpose the necessary theoretical framework for a micropolar continuum is initially presented. Here the standard macroscopic equations for mass, linear and angular momentum, and energy are extended in two ways. First, the aspect of coupling linear and angular rotational kinetic energies is emphasized. Second, the equations are complemented by a recently proposed kinetic equation for the moment of inertia tensors containing a production term. We then continue to explore the possibilities of this new term for the case of micropolar media encountering a change of moment of inertia during a thermomechanical process. Particular emphasis is put on the general form of the production of moment of inertia for a transversally isotropic medium and its potential to describe, for example, structural changes from a transversally isotropic state to an isotropic one. In order to be able to comprehend and to study the influence of the various material parameters the production term is interpreted mesoscopically and various other examples are solved in closed form. Moreover, in context with the presented example problems it will also become clear that the traditional Lagrangian way of describing the motion of solids might sometimes no longer be adequate and must then be replaced by a Eulerian approach.

Wolfgang H. Müller
Institute of Mechanics, Chair of Continuum Mechanics and Constitutive Theory, Technische Universität Berlin, Einsteinufer 5, 10587 Berlin, Germany
e-mail: `wolfgang.h.mueller@tu-berlin.de`

Elena N. Vilchevskaya
Institute for Problems in Mechanical Engineering of the Russian Academy of Sciences Bolshoy pr., 61, V.O., St. Petersburg, 199178 & Peter the Great St.Petersburg Polytechnic University, Polytechnicheskaya st., 29, St.Petersburg, 195251, Russia
e-mail: `vilchevska@gmail.com`

© Springer International Publishing AG, part of Springer Nature 2018
H. Altenbach et al. (eds.), *Generalized Models and Non-classical Approaches in Complex Materials 1*, Advanced Structured Materials 89,
https://doi.org/10.1007/978-3-319-72440-9_30

30.1 Review of the Current State-of-the-Art

Generalized Continuum Theories (GCTs) have proven to be most useful for the materials science community, because they are needed for computational modeling of new technical designs with high performance materials carrying an inner structure. Such materials appear in large and small scale applications ranging from light-weight aerospace and automotive applications down to micromechanics and microelectronic gadgets. A particular form of GCT is the so-called micropolar theory. It emphasizes the aspect of inner rotational degrees of freedom of a material consisting of rigid particles on a mesoscopic scale, see Eringen and Kafadar (1976), Eringen (2002, Section 13), or Eremeyev et al (2012). Hence this theory seems particularly promising in context with the description of soils, polycrystalline and composite matter, granular and powder-like materials, and also for porous media and foams.

Recall that the tensor of the moment of inertia of a continuum particle, $\boldsymbol{J}$, plays an important role in context with its rotational degree of freedom, specifically in combination with the angular velocity vector, $\boldsymbol{\omega}$, assigned to the continuum element. Traditionally $\boldsymbol{J}$ is conserved and known *a priori*. However, more recently, it has been emphasized by Ivanova and Vilchevskaya (2016) that the moment of inertia tensor deserves to be treated as an independent structural field variable. The situation is completely analogous to the inertia associated with linear momentum, namely the field of mass density, ρ, of compressible media, which, as it is commonly known, is governed by a kinetic or (better) balance equation of its own. However, in contrast to the balance of mass, moment of inertia is not conserved. Rather its governing equation contains a production term, $\boldsymbol{\chi}_J$, which in terms of continuum theory must be considered as a new constitutive quantity. As we shall see, it allows to model additional features of materials, *e.g.*, processes with considerable structural changes, such as transitions from an anisotropic to an isotropic state, and *vice versa*.

We therefore need to extend the original goals of micropolar theory. In fact, in what follows we will initially formulate the corresponding equations in a very general manner, so that they are ready for future investigations of various kinds of boundary value problems, may they be static or dynamic, and with or without coupling of linear and angular momentum. After that we shall successively specialize to the needs of this paper. In this spirit we proceed as follows:

In its most general form the objective of micropolar theory is, mathematically speaking, to determine the following primary fields: (a) the scalar field of mass density, $\rho(\boldsymbol{x},t)$; (b) the symmetric, second rank, and positive definite specific moment of inertia tensor field, $\boldsymbol{J}(\boldsymbol{x},t)$; (c) the specific moment of inertia coupling tensor field, $\boldsymbol{B}(\boldsymbol{x},t)$; (d) the vector field of linear velocity, $\boldsymbol{v}(\boldsymbol{x},t)$; (e) the spin (a.k.a. angular velocity) field, $\boldsymbol{\omega}(\boldsymbol{x},t)$; and the temperature field, $T(\boldsymbol{x},t)$, in all points, $\boldsymbol{x}$, and at all times, t, within a region of space, $\mathscr{B}$, which can be either a material volume, *i.e.*, it consists of the same matter at all times, or be a region through which matter is flowing.

The determination of these fields relies on field equations for the primary fields. The field equations are based on balance laws and need to be complemented by suitable constitutive relations. In spatial description, putting the substantial time derivatives of the balancable quantities exclusively on the left hand side, these continuum scale balances read as follows:

- balance of mass:

$$\frac{\delta\rho}{\delta t} = -\rho\nabla\cdot\boldsymbol{v}, \tag{30.1}$$

with the substantial (a.k.a. material) derivative,

$$\frac{\delta(\cdot)}{\delta t} = \frac{\mathrm{d}(\cdot)}{\mathrm{d}t} + (\boldsymbol{v}-\boldsymbol{w})\cdot\nabla(\cdot), \tag{30.2}$$

$\mathrm{d}(\cdot)/\mathrm{d}t$ being the total time derivative including the mapping velocity $\boldsymbol{w}$ of the observational point;

- balances for the moment of inertia and for the coupling moment of inertia tensors,

$$\frac{\delta \boldsymbol{J}}{\delta t} = \boldsymbol{\omega}\times\boldsymbol{J} - \boldsymbol{J}\times\boldsymbol{\omega} + \boldsymbol{\chi}_J\,, \quad \frac{\delta \boldsymbol{B}}{\delta t} = \boldsymbol{\omega}\times\boldsymbol{B} - \boldsymbol{B}\times\boldsymbol{\omega} + \boldsymbol{\chi}_B\,; \tag{30.3}$$

- balance of linear momentum,

$$\rho\frac{\delta}{\delta t}(\boldsymbol{v}+\boldsymbol{B}\cdot\boldsymbol{\omega}) = \nabla\cdot\boldsymbol{\sigma} + \rho\boldsymbol{f} + \rho\,\boldsymbol{\chi}_B\cdot\boldsymbol{\omega}; \tag{30.4}$$

- balance of spin,

$$\rho\frac{\delta}{\delta t}(\boldsymbol{J}\cdot\boldsymbol{\omega}+\boldsymbol{v}\cdot\boldsymbol{B}) = \nabla\cdot\boldsymbol{\mu} + \boldsymbol{\sigma}_\times + \rho\boldsymbol{m} - \rho\boldsymbol{v}\times\boldsymbol{B}\cdot\boldsymbol{\omega} + \rho\,(\boldsymbol{v}\cdot\boldsymbol{\chi}_B + \boldsymbol{\chi}_J\cdot\boldsymbol{\omega}); \tag{30.5}$$

- and the balance of internal energy,

$$\rho\frac{\delta u}{\delta t} = -\nabla\cdot\boldsymbol{q} + \rho r + \boldsymbol{\sigma} : (\nabla\boldsymbol{v} + \mathbf{I}\times\boldsymbol{\omega}) + \boldsymbol{\mu} : \nabla\boldsymbol{\omega}. \tag{30.6}$$

The symbol ":" denotes the outer double scalar product, see Eq. (30.69) of the Appendices. We will elaborate more on the important issue of spatial description below. For the time being it suffices to say that $\boldsymbol{\sigma}$ denotes the (non-symmetric) Cauchy stress tensor, $\boldsymbol{f}$ is the specific body force, $\boldsymbol{\chi}_J$ (a symmetric tensor of second rank) and $\boldsymbol{\chi}_B$ (a general tensor of second rank) are the productions related to the moment of inertia tensor, $\boldsymbol{J}$, and to the coupling moment of inertia tensor, $\boldsymbol{B}$, respectively; $\boldsymbol{\mu}$ is the couple stress tensor, $\boldsymbol{\sigma}_\times$ is the Gibbsian cross (see Eremeyev et al, 2012, pg. 34) or (Naumenko and Altenbach, 2007, pg. 176) applied to the (non-symmetric) Cauchy stress tensor, $\boldsymbol{m}$ are specific volume couples, u is the specific internal energy, $\boldsymbol{q}$ is the heat flux, and r is the specific heat supply.

In context with Eq. (30.3) several comments are in order. It has already been mentioned that in this form such relations were first introduced in a recent paper by one of the authors (Ivanova and Vilchevskaya, 2016). There is a precedent to the equation for the inertia tensor, $\boldsymbol{J}$, namely what is called "conservation of microinertia" in Eringen and Kafadar (1976), pg. 15. However, being characterized as a conserved quantity it cannot contain a production term, χ_J, *a priori*. On the continuum level this term must be interpreted as a new constitutive quantity. One of the purposes of this paper is to specify possible forms of this constitutive relation and to illustrate its potential for modeling materials with higher inner degrees of freedom by various example problems, which in hindsight confirm the necessity for extending the theory.

However, if there exists a production of moment of inertia, we face another problem, as follows: Continuum mechanics of solids is typically formulated in the Lagrangian way, a.k.a. material description, which is based on the concept of an indestructible "material particle." This particle is identifiable by its reference position vector, $\boldsymbol{X}$, which can then be used in a bijective mapping for describing uniquely the motion, $\boldsymbol{x} = \hat{\boldsymbol{x}}(\boldsymbol{X},t)$, of the particle through three-dimensional space in time. Note that this requires the neighboring material particles to remain "close" to each other during the motion. Furthermore note that a material particle in the continuum sense is composed of myriads of atoms or molecules, so that statistical fluctuations play no role in a macroscopic continuum. Moreover, there is no exchange of atoms and molecules between material particles. A material particle is indestructible and its mass is simply conserved.

Traditionally, micropolar theory was not any different in this respect. One may say that the corresponding material particle consists of a statistically significant number of subunits on a mesoscopic scale, which are often also called "particles," an unfortunate terminology, which may give rise to confusion. Now, if the Lagrangian idea of a material particle is followed, material particles must stay together during motion. There should be no exchange of subunits between them. Particles are neither destroyed nor generated. In other words, this concept cannot handle production. Consequently, under such assumptions the field equations of mass and of the inertia tensor, Eqs. (30.1) and $(30.3)_1$, can be integrated in time. In Eringen and Kafadar (1976, pp. 15/16) we find:

$$\rho = \frac{\rho_0}{\det\boldsymbol{F}}, \ \boldsymbol{J} = \boldsymbol{Q}\cdot\boldsymbol{J}_0\cdot\boldsymbol{Q}^\top, \tag{30.7}$$

where all functions of the current configuration depend on $(\boldsymbol{X},t)$ and the ones in the reference placement (identifiable by the subscript 0) on $(\boldsymbol{X},t_0)$. The first result for the mass density is well known in continuum mechanics, $\boldsymbol{F} = \nabla_X\hat{\boldsymbol{x}}(\boldsymbol{X},t)$ being the deformation gradient. The second one is less frequently mentioned, $\boldsymbol{Q}$ being the so-called tensor of micromotion or microrotation, as it is called in modern literature (*e.g.* Eremeyev and Pietraszkiewicz, 2012) an orthogonal rotation tensor connecting the directors of the reference configuration with those of the current placement.

However, this means that within this framework certain processes and effects in materials can simply not be modeled. Consider, for example, a granular medium, which is milled. The milling will affect the material particle, because its subunits will be crushed. They will change their mass as well as their moment of inertia and, what is more, during the milling process there might even be an exchange of crushed subunits between neighboring material particles, which are then no longer material in the original sense of the word. Consequently, on a continuum scale, the moment of inertia will change in space and time as dictated by the production term, χ_J. As another example consider a medium, which on a mesoscale consists of electrically charged particles. If an electrostatic field is applied externally, is will affect their orientation as well as their shape. On a continuum scale this means that the moment of inertia tensor field is affected.

Also recall that the Eulerian description is based and widely used in fluid mechanics. It does not impose strict constraints on the motion of mass-conserved material points. Rather it embraces the idea of an open system, allowing a priori for exchange of mass, momentum, energy, moment of inertia, *etc.*, between and within the cells of the Eulerian grid.

For such reasons (Ivanova and Vilchevskaya, 2016) have decided to consider an independent balance for $\boldsymbol{J}$ (and for $\boldsymbol{B}$), and this is why they depart from the Lagrangian description and turn to the Eulerian perspective (a.k.a. spatial description) instead.

A second remark concerns Eqs. (30.4) and (30.5). Note that the specific moment of inertia coupling tensor, $\boldsymbol{B}$, originally stems from rigid body dynamics (see, *e.g.* Hibbeler, 2010, Section 21.3). It is skewsymmetric and different from zero, if we choose some arbitrary "geometric" center as a reference point, which does not coincide with the center of mass. In Ivanova and Vilchevskaya (2016, Section 3.2) a mesoscopic interpretation of the continuum field $\boldsymbol{B}$ was given. From their explanations it follows that if the microparticles in an RVE (Representative Volume Element in an Eulerian grid) are homogeneously distributed within, then, due to symmetry considerations, the skewsymmetric tensor $\boldsymbol{B}$ must be zero. Consequently, if we intend to use $\boldsymbol{B}$ on the continuum level, we have to abandon the above approach and accept that the $\boldsymbol{B}$ tensor of the microparticle should no longer be skewsymmetric but an arbitrary tensor of second rank characterizing an additional physical quality of the microparticle. It was exactly for this reason that Zhilin (2015) introduced it in the continuum world. We proceed to explain this issue.

By coupling linear and angular velocities, *i.e.*, $\boldsymbol{v} \cdot \boldsymbol{B} \cdot \boldsymbol{\omega}$, it contributes toward the specific kinetic energy contained in a spatial element:

$$\mathscr{K} = \tfrac{1}{2}\boldsymbol{v} \cdot \boldsymbol{v} + \boldsymbol{v} \cdot \boldsymbol{B} \cdot \boldsymbol{\omega} + \tfrac{1}{2}\boldsymbol{\omega} \cdot \boldsymbol{J} \cdot \boldsymbol{\omega}. \tag{30.8}$$

The kinetic energy serves as a potential for the linear and for the angular momentum:

$$\boldsymbol{\mathscr{K}}_{\text{lin}} = \frac{\partial \mathscr{K}}{\partial \boldsymbol{v}} = \boldsymbol{v} + \boldsymbol{B} \cdot \boldsymbol{\omega}, \; \boldsymbol{\mathscr{K}}_{\text{ang}} = \frac{\partial \mathscr{K}}{\partial \boldsymbol{\omega}} = \boldsymbol{v} \cdot \boldsymbol{B} + \boldsymbol{J} \cdot \boldsymbol{\omega}. \tag{30.9}$$

This way $\boldsymbol{B}$ also couples the linear balance of momentum and the spin balance, Eqs. (30.4) and (30.5), which is also a new feature when compared to the usual formulations found in (for example) Eringen and Kafadar (1976, pg. 15) or Eremeyev et al (2012, pg. 30).

In Zhilin (2015, Section 3.2) a first fascinating application of $\boldsymbol{B}$ was presented. Zhilin specialized the equations from above, which are valid for the continuum, to the case of a point particle carrying mass m and isotropic rotational inertia. The coupling tensor was also assumed to be isotropic:

$$\boldsymbol{B} = B\mathbf{I}\,, \; \boldsymbol{J} = J\mathbf{I}, \tag{30.10}$$

where B and J are constants. The point particle was initially assigned a constant translational speed, $\boldsymbol{v}_0$, and a constant angular velocity, $\boldsymbol{\omega}_0$. Then, in the absence of external forces and moments, linear and angular momentum are conserved and the angular velocity develops according to:

$$\boldsymbol{\omega}(t) = [1 - \cos(\alpha t)]\boldsymbol{n} \cdot \boldsymbol{\omega}_0 \boldsymbol{n} + \cos(\alpha t)\boldsymbol{\omega}_0 + \sin(\alpha t)\boldsymbol{n} \times \boldsymbol{\omega}_0, \tag{30.11}$$

where $\alpha = \frac{Ba}{B^2-J}$ and $\boldsymbol{v}_0 + B\boldsymbol{\omega}_0 = \boldsymbol{a} = a\boldsymbol{n}$, $\boldsymbol{n}$ being a unit vector indicating a direction.

The translational speed and the path of the particle are given by

$$\boldsymbol{v}(t) = \boldsymbol{v}_0 + B\big[(1 - \cos(\alpha t))(\mathbf{I} - \boldsymbol{n}\otimes\boldsymbol{n}) - \sin(\alpha t)\boldsymbol{n} \times \mathbf{I}\big] \cdot \boldsymbol{\omega}_0 \tag{30.12}$$

and

$$\begin{aligned} \boldsymbol{x}(t) = {} & \boldsymbol{x}_0 + \boldsymbol{v}_0 t + \\ & B\Big[\big(t - \tfrac{1}{\alpha}\sin(\alpha t)\big)(\mathbf{I} - \boldsymbol{n}\otimes\boldsymbol{n}) - \tfrac{1}{\alpha}\big(1 - \cos(\alpha t)\big)\boldsymbol{n} \times \mathbf{I}\Big] \cdot \boldsymbol{\omega}_0, \end{aligned} \tag{30.13}$$

respectively.

The first two terms on the right hand side are the classical result according to Newton's First Law: Not driven by forces a mass point continues to move uniformly along a straight line. Clearly, the terms on the second line show the difference between Newtonian and Eulerian mechanics, where the balance of angular momentum is truly independent and a coupling between translational and angular parts of the kinetic energy is assumed: The mass point moves along a curved path. The initial angular velocity influences its path considerably. However, note that all of these differences to the classical point of view arise only if we allow the coupling coefficient B to be different from zero.

In this paper, we are going to consider extensively structural changes from transversal anisotropy with no translational motion. In these examples there is a need for a non-spherical inertia tensor, $\boldsymbol{J}(\boldsymbol{x},t)$, but not for a coupling inertia tensor field, $\boldsymbol{B}(\boldsymbol{x},t)$. A first examination of the modeling possibilities contained in the latter was just presented: Zhilin's idea of induced helical motion of a particle without external forces and moments originally moving translationally at a constant speed. However, a more detailed examination of the capabilities of the coupling tensor is left to future research.

We shall now proceed to discuss possible rather general forms of the production χ_J and then illustrate the general theory by various examples.

30.2 Productions of Microinertia and the Coupling Tensor for Transversally Isotropic Media

In this and in the next section we will concentrate on the balance for the inertia tensor $(30.3)_1$ in combination with the balance of spin (30.5). We will study purely static problems, where the linear and the angular velocity vanish, $\boldsymbol{\omega} = \mathbf{0}$, $\boldsymbol{v} = \mathbf{0}$, or problems of purely angular motion, $\boldsymbol{\omega} \neq \mathbf{0}$, $\boldsymbol{v} = \mathbf{0}$. In particular, we will focus on a transversally isotropic medium, such that the plane of isotropy is characterized by a unit normal vector $\boldsymbol{n}$ perpendicular to it. In a first cut to the problem[1] we will assume that the production $\boldsymbol{\chi}_J$ is a function of the microinertia and of the unit vector alone:

$$\boldsymbol{\chi}_J = \boldsymbol{\chi}_J(\boldsymbol{J}, \boldsymbol{n}). \tag{30.14}$$

Recall that the production $\boldsymbol{\chi}_J$ is a symmetric second rank tensor. The most general representation for the production under these circumstances was derived in the Appendices:

$$\begin{aligned}\boldsymbol{\chi}_J(\boldsymbol{J}, \boldsymbol{n}) = \chi_1 \mathbf{I} + \chi_2 \boldsymbol{n} \otimes \boldsymbol{n} + \chi_3 \boldsymbol{J} + \chi_4 \boldsymbol{J}^2 + \\ \chi_5 (\boldsymbol{n} \otimes \boldsymbol{J} \cdot \boldsymbol{n} + \boldsymbol{J} \cdot \boldsymbol{n} \otimes \boldsymbol{n}) + \chi_6 (\boldsymbol{n} \otimes \boldsymbol{J}^2 \cdot \boldsymbol{n} + \boldsymbol{J}^2 \cdot \boldsymbol{n} \otimes \boldsymbol{n}),\end{aligned} \tag{30.15}$$

where the coefficients χ_i, $i = 1, \cdots, 6$ are arbitrary scalar functions of five invariants, $\mathrm{Tr}(\boldsymbol{J}), \mathrm{Tr}(\boldsymbol{J}^2), \mathrm{Tr}(\boldsymbol{J}^3), \boldsymbol{n} \cdot \boldsymbol{J} \cdot \boldsymbol{n}, \boldsymbol{n} \cdot \boldsymbol{J}^2 \cdot \boldsymbol{n}$.[2]

We consider a few special cases. Let us assume that, on a mesoscopic level, the material consists of (not necessarily rigid) spheroids with (initial) minor axes a_0 and major axis $c_0 > a_0$. Let us further assume that as part of a production process the spheroids are all oriented in the direction of the major axis, $\boldsymbol{n}$. Then, on a continuum level the homogenized tensor of inertia field of a transversally isotropic medium is initially given by:

$$\boldsymbol{J}_0 = \tfrac{1}{5}(a_0^2 + c_0^2)(\mathbf{I} - \boldsymbol{n} \otimes \boldsymbol{n}) + \tfrac{2}{5} a_0^2 \boldsymbol{n} \otimes \boldsymbol{n}. \tag{30.16}$$

This result calls for a short explanation. The inertia tensor of a single spheroidal particle within the RVE is (initially) of the form shown in Eq. (30.16). Following the averaging procedure outlined in Subsection 3.2 of Ivanova and Vilchevskaya (2016) this is then also the homogenized initial field of the moment of inertia tensor on the continuum scale.

If the material is purely isotropic ($a_0 = c_0$, *i.e.*, spherical particles) the moment of inertia turns into a purely spherical tensor:

$$\boldsymbol{J}_0 = \tfrac{2}{5} a_0^2 \mathbf{I}. \tag{30.17}$$

If the particles are hollow spheres with (initial) inner and outer radii $r_{\mathrm{i},0}$ and $r_{\mathrm{o},0}$, respectively, we may write:

$$\boldsymbol{J}_0 = \frac{2}{5} \frac{r_{\mathrm{o},0}^5 - r_{\mathrm{i},0}^5}{r_{\mathrm{o},0}^3 - r_{\mathrm{i},0}^3} \mathbf{I}. \tag{30.18}$$

---

[1] This is supposed to mean that in a complete phenomenological theory further variables should appear, such as deformation or stress tensors, temperature and, in particular, electromagnetic ones, such as polarization or magnetization, or their duals, the electric and the magnetic field. However, the detailed description is left to future research.

[2] Also observe the previous footnote. The invariant $\boldsymbol{n} \cdot \boldsymbol{n}$ does not appear because $\boldsymbol{n}$ is a unit vector.

In the isotropic case the most general form for the moment of inertia and its production are then given by:

$$\boldsymbol{J} = J\mathbf{I}\,, \ \boldsymbol{\chi}_J = \chi_J\mathbf{I}, \tag{30.19}$$

where the scalar χ_J is an arbitrary function of the three invariants of $\boldsymbol{J}$:

$$\chi_J = \chi_J(I_1, I_2, I_3)\ ,\ I_1 = 3J\ ,\ I_2 = 3J^2\ ,\ I_3 = 3J^3. \tag{30.20}$$

Moreover, for later purpose, we now also decompose the moment of a general inertia tensor $\boldsymbol{J}$ into volumetric (spherical) and deviatoric parts:

$$\boldsymbol{J} = \boldsymbol{J}^{\text{vol}} + \boldsymbol{J}^{\text{dev}}\ ,\ \boldsymbol{J}^{\text{vol}} = \tfrac{1}{3}\,\mathsf{Tr}\boldsymbol{J}\,\mathbf{I}\,,\ \boldsymbol{J}^{\text{dev}} = \boldsymbol{J} - \tfrac{1}{3}\,\mathsf{Tr}\boldsymbol{J}\,\mathbf{I}, \tag{30.21}$$

i.e., similarly as in the case of Eq. (30.16):

$$\boldsymbol{J}^{\text{vol}} = \tfrac{2}{15}\left[2a(t)^2 + c(t)^2\right]\mathbf{I},\ \boldsymbol{J}^{\text{dev}} = \tfrac{1}{15}\left[c(t)^2 - a(t)^2\right](\mathbf{I} - 3\boldsymbol{n}\otimes\boldsymbol{n}). \tag{30.22}$$

The first two terms on the right hand side of Eq. $(30.3)_1$, which characterize the change of the moment of inertia tensor due to rigid body rotation, always vanish in the case of a sphere for physical reasons, *i.e.*, in the latter case, for the isotropic part, $\boldsymbol{J}^{\text{vol}}$. Of course, this can also be demonstrated mathematically:

$$\boldsymbol{\omega}\times\mathbf{I} - \mathbf{I}\times\boldsymbol{\omega} = \omega_k\epsilon_{mkj}(\boldsymbol{e}_m\otimes\boldsymbol{e}_j + \boldsymbol{e}_j\otimes\boldsymbol{e}_m) = \mathbf{0}, \tag{30.23}$$

where ϵ_{mik} denotes the Levi-Civita symbol and $\boldsymbol{e}_i$ is a Cartesian base. Moreover, if we now assume that the angular velocity points exclusively in the direction of transversal anisotropy, $\boldsymbol{\omega} = \omega\boldsymbol{n}$, it is easily seen that the $\boldsymbol{n}$-dependent parts in $\boldsymbol{J}^{\text{dev}}$ will also lead to vanishing rigid body rotation terms on the right hand side of Eq. $(30.3)_1$.

Hence, under the requirements discussed in context with Eqs. (30.19) and (30.21) the balance of the moment of inertia tensor simplifies and reads:

$$\frac{\delta \boldsymbol{J}}{\delta t} = \boldsymbol{\chi}_J. \tag{30.24}$$

30.3 Discussion of Special Cases for the Production Term for the Moment of Inertia, $\boldsymbol{\chi}_J$

In previous papers some solutions of the temporal development of the field for the moment of inertia have already been studied. One of the purposes of this paper is to revisit these examples and to discuss them from a more general point of view. Moreover, some new cases will be presented, which have not been investigated so far.

30.3.1 Examples for the Isotropic Case

As a first example for the case of an isotropic structural change consider the situation depicted in Fig. 30.1. On the inset of the upper left we see a Representative Volume Element (RVE) or Eulerian volume element containing hollow spherical particles filled with an ideal gas. We assume that the shell of these particles is incompressible with a constant volume V_0. The inner pressure is initially at the level p_0. The outer pressure can change as a function time, $p(t)$.

Depending on how we model the shell the inner pressure will then adjust differently. If the shell is modeled as a linear elastic solid it will be subjected to a stress level p_0 inside and carry the difference to the outer pressure by developing internal stresses according to the Lamè solution. On the other hand, if we model the shell as a rubber balloon, the inner pressure will adjust and equal the outer one. In both cases the volume of the spheres will change over time (right inset on the top of the figure). The moment of inertia tensors of all spheres in an RVE are equal and given by an isotropic tensor expression analogous to Eq. (30.18). If homogenized (see the two insets on the bottom of the figure) the inertia tensor field on the continuum level is also isotropic and it stays that way.

Hence Eq.(30.3)$_1$ implies that the production $\boldsymbol{\chi}_J$ is isotropic as well. As it was shown in Ivanova and Vilchevskaya (2016) and in Müller et al (2017) it consists in both cases only of the first term on the right hand side of Eq. (30.15), where the coefficient χ_1 is independent of invariants of $\boldsymbol{J}$. With some modeling effort it can be expressed in terms of mesoscopic parameters. For example, in the case of the Lamè model it is given by:

$$\chi_1 = -\frac{r_{o,0}^2}{2\mu}\frac{\beta^3\left(1-\beta^2\right)}{\left(1-\beta^3\right)\left(1-\beta^5\right)}\frac{\mathrm{d}p(t)}{\mathrm{d}t}\,,\quad \beta = \frac{r_{c,0}}{r_{i,0}}, \tag{30.25}$$

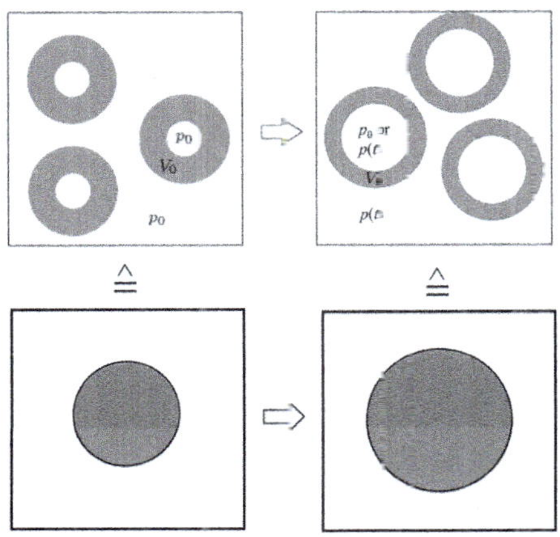

Fig. 30.1 Structural shape change and corresponding homogenization of pressurized spheres.

μ being the shear modulus of the mesoscopic particle. For further details the reader should turn to the two references.

Another structural change governed by an isotropic moment of inertia and an inevitably isotropic production is the case of particles isotropically distributed on the mesoscale being milled in a so-called crusher. The assumption is that the distribution stays isotropic during the milling process. This situation is also treated extensively in the two references. There it is also motivated that the production is given by the following expression:

$$\boldsymbol{\chi}_J = -\alpha\,(J - J_*)\mathbf{I}, \tag{30.26}$$

where α and J_* are positive constants, which can be interpreted as being related to the minimum grain size the particles are made of, to the particle toughness, and to the crusher action. Obviously this kind of an expression is also contained in the first term on the right hand side of the general equation (30.15) if we assume χ_1 to depend only on $\mathrm{Tr}\boldsymbol{J}$ and then perform a linear Taylor expansion.

30.3.2 Structural Change I: Purely Deviatoric Production

Consider now the case of a material, which, on a mesoscale, consists of an assembly of rigid spheroids with equatorial axes a_0 and polar axis c_0, respectively. Initially, at times $t < 0$, as consequence of an external force acting as part of a "production process," all particles are uniformly aligned in $\boldsymbol{n}$-direction (Fig. 30.2, top left). The initial, homogenized field of micropolar inertia is then given by Eq. (30.16) and illustrated in Fig. 30.2 on the bottom left. In order to realize such an ordered structure one could think of an electret powder. Electrets are materials that possess a quasi-permanent electric charge or electric dipole polarization. One way of obtaining a (coarse) electret powder consists of irradiating a fluorine-containing resin sheet with an electron beam in order to convert the fluorine-containing resin sheet into an electret sheet, and then pulverize it. The resulting powder particles are more or less rigid. An external electric field orients them as indicated in the figure, where the positive and negative electric charges are indicated by white and black bullets, respectively (also

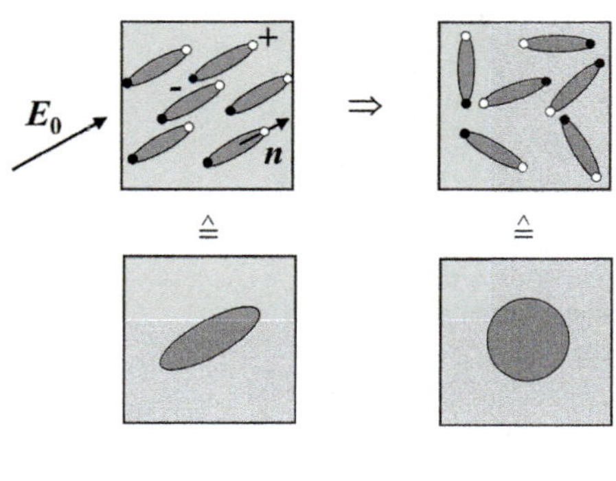

Fig. 30.2 Structural shape change and corresponding homogenization.

see Kestelman et al, 2013, Fig. 1.1). The net charge in the representative volume element is equal to zero and it is for that reason that the depicted equivalent spheroids on the bottom of the figure do not contain symbols for charges.

At $t = 0$ the electric field is then suddenly switched off. Stochastic (thermal) motion sets in and the particles start to realign arbitrarily (Fig. 30.2, top right). For physical reasons we expect the resulting field of microinertia eventually to degenerate into a spherical tensor. This final state is illustrated in Fig. 30.2 on the bottom right. It is important to note that the electric field is not running the transition. It was used only for motivation how the initial state of order could be achieved and is then simply switched off. This will be different in the next example where it drives the structural change actively.

The objective is now to study the transition process. For this purpose we argue that the transition is nothing else but a "shape" change. It is known that shape changes are characterized by deviatoric expressions. Hence we choose for the production of microinertia shown in Eq. (30.15) only the linear term in $\boldsymbol{J}$ and specialize it to its deviatoric part according to Eq. (30.21). Moreover, we assume that the parameter χ_3 in that equation is a negative constant, $-\beta$, and write:

$$\boldsymbol{\chi}_J = -\beta \boldsymbol{J}^{\text{dev}}(t)\,,\ \ \beta > 0. \tag{30.27}$$

As we shall see this assumption is physically reasonable. The tensor of microinertia is notably time dependent. In order to state its explicit mathematical form we generalize Eq. (30.16) by assuming the semi-axes to be time dependent and write

$$\boldsymbol{J}(t) = \tfrac{1}{5}\left[a^2(t) + c^2(t)\right](\mathbf{I} - \boldsymbol{n}\otimes\boldsymbol{n}) + \tfrac{2}{5}a^2(t)\boldsymbol{n}\otimes\boldsymbol{n}. \tag{30.28}$$

Analogously to Eq. (30.22) we have:

$$\begin{aligned} \boldsymbol{J}^{\text{vol}}(t) &= \tfrac{2}{15}\left[2a^2(t) + c^2(t)\right]\mathbf{I}\,, \\ \boldsymbol{J}^{\text{dev}}(t) &= \tfrac{1}{15}\left[c^2(t) - a^2(t)\right](\mathbf{I} - 3\boldsymbol{n}\otimes\boldsymbol{n}). \end{aligned} \tag{30.29}$$

We do not expect that the shape transition is accompanied by macroscopic translational movement, *i.e.*, $\boldsymbol{v} = \mathbf{0}$ at all times. Eq. (30.24) then becomes

$$\frac{\partial \boldsymbol{J}(t)}{\partial t} = -\beta \boldsymbol{J}^{\text{dev}}(t) \quad \Rightarrow \quad \frac{\mathrm{d}\boldsymbol{J}(t)}{\mathrm{d}t} = -\beta \boldsymbol{J}^{\text{dev}}(t), \tag{30.30}$$

the latter being an ordinary time differentiation,[3] because of the homogeneity of the problem.

Recall the volumetric (spherical) and the deviatoric projectors defined by:

$$\boldsymbol{P}^{\text{vol}} = \frac{1}{3}\mathbf{I}\otimes\mathbf{I}\,,\ \boldsymbol{P}^{\text{dev}} = \boldsymbol{e}_i\otimes\boldsymbol{e}_j\otimes\boldsymbol{e}_i\otimes\boldsymbol{e}_j - \boldsymbol{P}^{\text{vol}}. \tag{30.31}$$

We apply them successively to Eq. (30.30) and obtain:

[3] not to be confused with the total time derivative of Eq. (30.2)

$$\frac{\mathrm{d}\boldsymbol{J}^{\mathrm{vol}}}{\mathrm{d}t} = \mathbf{0}\,, \quad \frac{\mathrm{d}\boldsymbol{J}^{\mathrm{dev}}}{\mathrm{d}t} = -\beta \boldsymbol{J}^{\mathrm{dev}}. \tag{30.32}$$

We conclude that the volumetric part is conserved and the deviatoric one decreases exponentially in time:

$$\boldsymbol{J}^{\mathrm{vol}}(t) = \boldsymbol{J}_0^{\mathrm{vol}}\,, \quad \boldsymbol{J}^{\mathrm{dev}}(t) = \boldsymbol{J}_0^{\mathrm{dev}} \exp(-\beta t)\,. \tag{30.33}$$

In order to find out how the axes $a(t)$ and $c(t)$ "adjust" during the shape change we insert Eqs. (30.29) and (30.22) into this result. Two algebraic equations for $a(t)$ and $c(t)$ are obtained, which can be solved easily:

$$\begin{aligned} a(t) &= \tfrac{1}{\sqrt{3}}\sqrt{3a_0^2 + \left(c_0^2 - a_0^2\right)\left[1 - \exp(-\beta t)\right]}, \\ c(t) &= \tfrac{1}{\sqrt{3}}\sqrt{c_0^2 + 2a_0^2 + 2\left(c_0^2 - a_0^2\right)\exp(-\beta t)}. \end{aligned} \tag{30.34}$$

We conclude that for very long times t the semi axes $a(t)$ and $c(t)$ become equal and that the average axis in all directions is given by:

$$a(\infty) = c(\infty) = \tfrac{1}{\sqrt{3}}\sqrt{c_0^2 + 2a_0^2}. \tag{30.35}$$

Moreover, if we insert these results into Eq. (30.28) we can explicitly see how the moment of inertia field turns exponentially into a spherical tensor:

$$\boldsymbol{J}(t) = \frac{2}{15}\left(2a_0^2 + c_0^2\right)\mathbf{I} + \frac{1}{15}\left(c_0^2 - a_0^2\right)\left(\mathbf{I} - 3\boldsymbol{n}\otimes\boldsymbol{n}\right)\exp(-\beta t). \tag{30.36}$$

Obviously the factor β has the physical meaning of a rate parameter: The greater it is, the faster the transition from order to disorder will be achieved. In view of our mesoscopic interpretation we may say that β characterizes how fast the particles within the RVE "rotate." It is thus tempting to relate it to a temperature, but we leave it with this remark.

At this point our mathematical analysis of the shape change to an isotropic configuration is complete. It is exactly what we expect to happen for physical reasons and it supports in hindsight our choice of the minus sign in Eq. (30.27). However, we may also argue that shape change is a transition from order to disorder due to fluctuations on the mesoscopic scale. Obviously the state of order is initially much higher than after the structural change. This calls for an examination of the irreversibility associated with the process. For this purpose it was shown in the Appendices that the minus sign in front of β guarantees consistency with the second law. We can even calculate the entropy production and the change in entropy associated with the process. To this end we start from the local entropy balance shown in Eq. (30.74), observe that the substantial time derivative degenerates for matter not moving translationally in the homogeneous case to a partial time derivative, which can then be replaced by a standard differentiation w.r.t. time, ignore the heat terms and rewrite the entropy production by observing Eqs. (30.79) and (30.80) in context with the production of

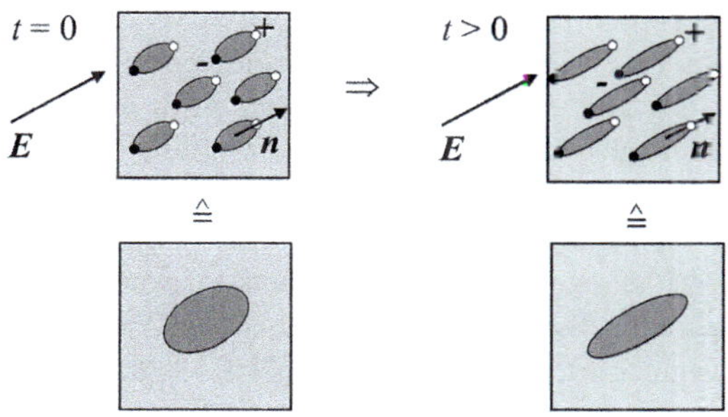

Fig. 30.3 Structural shape change and corresponding homogenization.

interest shown in (30.27):[4]

$$\rho \frac{\mathrm{d}s}{\mathrm{d}t} = \sigma = -\rho \frac{1}{T} \frac{\partial f}{\partial \boldsymbol{J}} : \boldsymbol{\chi}_J = \rho \frac{\beta_3 \beta}{T} \boldsymbol{J}^{\mathrm{dev}} : \boldsymbol{J}^{\mathrm{dev}}. \tag{30.37}$$

We now observe the solution shown in Eqs. (30.28) and (30.34) and find:

$$\frac{\mathrm{d}s}{\mathrm{d}t} = \frac{6\beta_3\beta}{15^2 T} \left(c_0^2 - a_0^2\right)^2 \exp(-2\beta t). \tag{30.38}$$

This can be integrated between $t = 0$ and $t = \infty$ to yield the change in specific entropy during the structural transition:

$$s(t = \infty) - s(t = 0) = \frac{\beta_3}{75T} \left(c_0^2 - a_0^2\right)^2. \tag{30.39}$$

We expect entropy to increase and, indeed, this is the case if β_3 is positive as already discussed in the Appendices.

30.3.3 Structural Change II: Purely Axial Production

We consider a transversally isotropic material characterized by the axis $\boldsymbol{n}$. On a mesoscopic scale it consists of little spheroids of initial axes a_0 in the plane of symmetry and with an axis c_0 in the direction of $\boldsymbol{n}$. The situation is depicted in Fig. 30.3 on the top left. As in the example of Subsection 30.3.2 we suppose that the initial alignment is due to some external force stemming from a production process, for example, by an electric field $\boldsymbol{E}$, as will be detailed below. However, this time we will not assume that the particles are rigid. Rather they shall be able to deform most easily up to a certain limit due to an externally imposed "load." In other words they can be stretched in $\boldsymbol{n}$-direction and will finally turn into little rods of finite length (Fig. 30.3, top right). As a motivation[5] for such a material and process, one can again turn to

[4] Because of the assumed homogeneity there is no difference in partial and regular derivatives, $\partial s/\partial t$ vs. $\mathrm{d}s/\mathrm{d}t$.

electrets for a possible realization. However, an electret powder is not suitable for this purpose. It is simply too stiff. We could envision the powder particles to behave linear elastically, but the deformation resulting from an external electric field would be very small indeed. The situation is similar to the case of permanent magnets, where even enormous external fields result in only very small deformations (*cf.*, Reich et al, 2017). What we need are "membrane-like electrets," similar to the balloons depicted in Fig. 30.1. Most recently materials scientists invented a substance with properties coming close to this, the so-called Ferroelectrets (Wikipedia, 2017). These are polymers, *e.g.*, Fluorinated Ethylene Propylene (FEP) containing lens-shaped voids filled with gas or a highly porous matter consisting of expanded PolyTetraFluorEthylen (ePTFE).

Gross and Xu (2015) calculated the electric field distributions in such structures. In particular, they investigated the configuration of a single "unit" consisting of an lens-shape body made of ePTFE surrounded by an FEP layer, which with a grain of salt we may idealize as an electrically polarized spheroidal balloon. Note that their simulations concentrate on calculating the electric field in this "unit" and that they do not detail the associated deformation, although mechanical stresses and strains are included in their theoretical framework.[6] They claim that such units have also been investigated experimentally. However, after looking at the corresponding references it is fair to say that such experiments have not been performed with single units of microscopic size. With some good will it can be said that porous bodies have been manufactured consisting of such units (of different size) forming a permanently electrically polarizable network.

All of this in mind we can now reinterpret Fig. 30.3 in two ways. First, a flexible ferroelectrete material, pre-polarized in $\boldsymbol{n}$-direction, is subjected to a constant electric field $\boldsymbol{E} = E\boldsymbol{n}$, $E > 0$ at time $t = 0$. This will lead at times $t > 0$ to an elongation of the spheroids within the FEP/ePTFE network. Of course, in reality, the spheroids will not be able to elongate freely. Rather there will be some elastic response, constraints, and interactions between them, which are not so easy to model in the production term for $\boldsymbol{J}$. Here we will simply ignore such effects.

Therefore, second, imagine polarized, independent, ellipsoidal "balloon units" to exist and that a constant electric field $\boldsymbol{E}$ is suddenly acting on them in $\boldsymbol{n}$-direction. We assume that all of these particles will instantaneously realign accordingly and then start elongating under the influence of that electric field. In the figure it was assumed that $c_0 > a_0$ but the inverse is also possible. We conclude that on the continuum scale the (homogenized) initial moment of inertia field is given by Eq. (30.16). This situation is illustrated on the bottom left inset of the figure. If we envision the electric polarization charges to be contained within an inextensible membrane, we expect the spheroidal electrets to be elongated in normal direction, $c(t) > c_0$, and to shrink within the plane of symmetry, $0 \leq a(t) < a_0$. The limit case $a = 0$ corresponds to a rod, so to speak. On a continuum scale the material stays transversally isotropic and

[5] And nothing more but a motivation! Note that the intention of this paper is not to present a proper mesomodel. Rather we wish to demonstrate the possibilities of the various terms for the moment of inertia production according to the representation theorem.

[6] Recall that what Gross and Xu (2015) call "displacement" in their paper is not the mechanical one but the charge potential field in matter, $\boldsymbol{D}$, a.k.a. dielectric displacement.

its (homogenized) moment of inertia field is given by Eq. (30.28). This is illustrated by the inset on the bottom right of the figure.

So far the process of axial elongation. The opposite process, axial shrinkage, is also imaginable, at least in an idealized theoretical manner. To this end consider Fig. 30.4. We start with a flexible ferroelectrete material which, similarly as in Subsection 30.3.2, has been pre-polarized in $\boldsymbol{n}$-direction as a consequence of a suitable production process: See the left inset on the top of Fig. 30.4. At times $t > 0$ it is then subjected to a constant electric field $\boldsymbol{E} = -E\boldsymbol{n}$, $E > 0$. This will lead to a shrinkage of the spheroids within the FEP/ePTFE network (see the top right inset). But again, the spheroids will not be able to shrink freely, and there will be some elastic influence, constraints, and interactions, *etc.*, which we will ignore when modeling the production term for $\boldsymbol{J}$.

In this spirit we take refuge to our idea of polarizable, independent, ellipsoidal "balloon units" which we assume to be pre-polarized in $\boldsymbol{n}$-direction as indicated in the figure. However, if an electric field is instantaneously switched on pointing in $-\boldsymbol{n}$-direction, $\boldsymbol{E} = -|\boldsymbol{E}|\boldsymbol{n}$, the slightest fluctuation would cause the particles to flip, because they are free to rotate. The situation is that of an unstable equilibrium, which we do not want to "disturb," because the intention is to study a shrinkage process instead, due to the electrostatic forces associated with $\boldsymbol{E}$. Hence, in reality some embedding of the particles, as in the case of the FEP/ePTFE network, is required. Moreover, initially rod-like particles will steadily become stouter and finally turn into spheres. After that structural instability might set in, which would let them depart from the spheroidal shape into some more complex form. In other words, the particles assume first an oblong spheroidal shape, then turn into a spheres, and most likely will never be able to assume the shape of a penny. We will take this into account when modeling the temporal transition for the case of axial shrinkage on the continuum scale.

In order to make a long story short: In this paper we will not conceive a more complex mesomechanical model of the FEP/ePTFE network and what follows is more of a mathematical exercise than a true physical model. In this idealistic spirit we will assume that the spheroidal pores behave like a limb inextensible membrane, and that the electric charges, which are so to speak situated at the top points of the axis of symmetry, may glide freely until the shrinkage process toward a sphere, $c = a$,

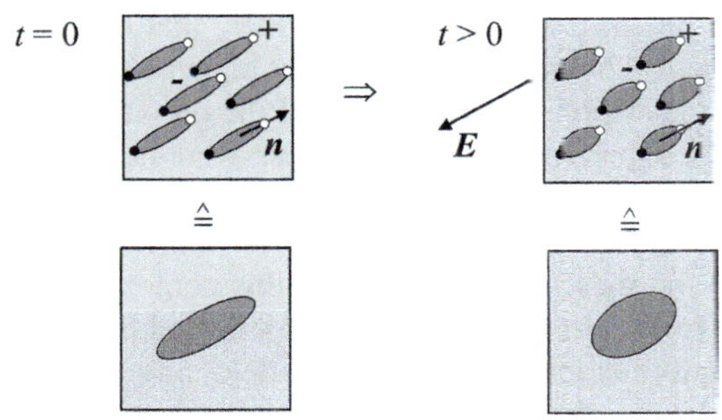

Fig. 30.4 Structural shape change during shrinkage.

is complete, without an interaction between the charges, and without flipping in the direction of the electric field. We also assume that during the shrinkage process the simple spheroidal shape remains and we will therefore not extrapolate our results beyond the formation of a sphere during shrinkage.

We now wish to quantify both processes, elongation as well as shrinkage. For this purpose we put $\boldsymbol{b} = \pm\boldsymbol{E}$ and choose only the term in $\boldsymbol{E}\otimes\boldsymbol{E}$ in the production of microinertia shown in Eq. (30.73). Moreover, the function a_2 will initially simply be replaced by a constant, α, such that we find:

$$\boldsymbol{\chi}_J = \mp\alpha\,\boldsymbol{E}\otimes\boldsymbol{E}. \tag{30.40}$$

The minus and plus signs refer to the case of elongation and to the case of shrinkage, respectively. Correspondingly, the sign of α is, in general, positive, but α can also become zero, as follows. In the case of elongation we write:

$$\alpha > 0 \text{ if } a > 0 \text{ and } \alpha = 0 \text{ if } a = 0. \tag{30.41}$$

This way we make mathematically sure that as long as $a(t)$ is positive there will be a production of moment of inertia and that the production stops abruptly as soon as the ellipsoid has turned into a rod.

In the case of shrinkage we choose as initial condition $c_0 > a_0$ (in order to avoid instability as discussed above) and:

$$\alpha > 0 \text{ if } c > a \text{ and } \alpha = 0 \text{ if } c = a, \tag{30.42}$$

because the moment of inertia shall change only as long as $c(t) > a(t)$. It shall come to a standstill the moment the spheroid turns into a sphere in order to avoid the problem of instability that was mentioned above. Again this is a purely mathematical trick and further below we will attempt to ameliorate the situation somewhat, at least for the case of elongation.

Note that the production is not explicitly time-dependent. This makes sense, because it is the result of the immediate change to a constant electric field $-\boldsymbol{E}$ that stays put. It is also not implicitly time-dependent in terms of the current moment of inertia, $\boldsymbol{J}$, as in Eqs. (30.26) or (30.27). Also, if we gave some more thought to the modeling of the elongation on the mesoscale we would be able to relate the parameter α to mesoscopic or external parameters, similarly as in the case of Eq. (30.25). For example, it is likely that the magnitude of the electric field will influence the production such that higher electric fields will speed up the structural change. Moreover, in the case of the porous network made of FEP/ePTFE elastic stiffness parameters will to appear, whereas in the case of the electric polarization charges contained within an inextensible, limb membrane, α is a true constant, because the charges "glide" freely within the membrane and continuously stretch it to the maximum possible. In this context it should also be mentioned that we ignore the effect of internal work required for separating the electric charges: They do not interact and the external field simply moves each of them separately and independently. This simple case will be considered in what follows, because our intention is to illustrate the various possibilities of the production, χ_J, and not a thorough micromechanical modeling of the spheroidal balloons or of the FEP/ePTFE network. The latter only serves as a vehicle to put our theory in context with reality.

A final word of caution is in order: The elongation or deformation on the mesoscale must not be confused with a deformation of the RVE on the continuum scale. In fact the RVE does not deform at all, even though its substructure changes as described. But on the continuum scale this change has a repercussion only on the $\boldsymbol{J}$-field, at least within the framework of this paper, where continuum field variables like stress and strain are of no concern.

Now if translational motion is ignored we obtain for elongation similarly to the case of Eq. (30.30):

$$\frac{\mathrm{d}\boldsymbol{J}(t)}{\mathrm{d}t} = -\alpha\, \boldsymbol{E} \otimes \boldsymbol{E}. \tag{30.43}$$

In view of Eq. (30.28) we obtain by taking the trace of this equation:

$$\frac{\mathrm{d}}{\mathrm{d}t}\left[2a^2(t) + c^2(t)\right] = -\frac{5}{2}\alpha E^2, \tag{30.44}$$

and by double scalar multiplication with $\boldsymbol{n}$ from the left and from the right:

$$\frac{\mathrm{d}}{\mathrm{d}t}a^2(t) = -\frac{5}{2}\alpha E^2, \tag{30.45}$$

Hence the semi-axis of the plane of isotropy decreases with time:

$$a(\bar{t}) = a_0\sqrt{1-\bar{t}}\ ,\ \bar{t} = \frac{5\alpha E^2}{2a_0^2}t, \tag{30.46}$$

and vanishes completely from $\bar{t} = 1$ onward. Moreover, the semi-axis in the direction of $\boldsymbol{n}$ increases toward a maximum value at time $\bar{t} = 1$ according to:

$$c(\bar{t}) = c_0\sqrt{1 + \frac{a_0^2}{c_0^2}\bar{t}}. \tag{30.47}$$

If we insert this in Eq. (30.28) we learn how the moment of inertia field develops in time:

$$\boldsymbol{J}(\bar{t}) = \frac{1}{5}\left(a_0^2 + c_0^2\right)\mathbf{I} - \frac{1}{5}\left[c_0^2 - (1-2\bar{t})a_0^2\right]\boldsymbol{n}\otimes\boldsymbol{n}. \tag{30.48}$$

Especially at the end of the elongation process, $\bar{t} = 1$, we obtain:

$$\boldsymbol{J}(\bar{t}=1) = \frac{1}{5}\left(a_0^2 + c_0^2\right)\left(\mathbf{I} - \boldsymbol{n}\otimes\boldsymbol{n}\right). \tag{30.49}$$

In comparison with Eq. (30.28) we see that the semi-axis characterizing the plane of isotropy, a, completely disappeared and that the semi-axis in the direction of the axis of anisotropy, c, became $\sqrt{a_0^2 + c_0^2}$: Equivalently, the ellipsoid has turned into a rod of vanishing cross-section with a fixed moment of inertia tensor, the projection of which in $\boldsymbol{n}$-direction, $\boldsymbol{J}\cdot\boldsymbol{n}$, is zero, because a rod has no resistance to rotation about its axis. To be precise, the length of this equivalent rod is given by $l=\sqrt{\frac{12}{5}(a_0^2+c_0^2)}$. Clearly, the production of moment of inertia vanishes when this state is reached. In other words α must then be put equal to zero.

A final comment on the case of elongation is in order. Some readers may feel uncomfortable with the case distinction for the constitutive relation for the production shown in Eq. (30.40). An alternative is the following *ansatz*, which is consistent with our representation principles:

$$\boldsymbol{\chi}_J = -\hat{\alpha}\,\det\boldsymbol{J}\,\boldsymbol{E}\otimes\boldsymbol{E}\,, \tag{30.50}$$

$\hat{\alpha}>0$ being constant. Obviously the determinant takes the volume change explicitly into account and makes sure that there is no production any more as soon as the case of the rod has been reached. Note that in contrast to Eq. (30.40) the production now depends explicitly on time. Following the lines of reasoning explained before we find with $\det\boldsymbol{J}=\frac{2}{125}\left[a^2(t)+c^2(t)\right]^2a^2(t)$:

$$\begin{aligned} a(t) &= a_0\exp\left[-\frac{\hat{\alpha}E^2}{50}\left(a_0^2+c_0^2\right)^2 t\right], \\ c(t) &= \sqrt{a_0^2+c_0^2-a_0^2\exp\left[-\frac{\hat{\alpha}E^2}{25}\left(a_0^2+c_0^2\right)^2 t\right]}\,. \end{aligned} \tag{30.51}$$

It should be noted that in this case the production vanishes at $t\to\infty$, unlike the case represented by Eqs. (30.40) and (30.41), where it vanishes at a finite time, more or less by mathematical requirement. We may want to interpret these results as a relaxation process, which at times $t\to\infty$ yields exactly the results shown previously in Eqs. (30.46) and (30.47) at the finite (dimensionless) time $\bar{t}=1$. It is therefore a matter of taste which *ansatz* for the production one prefers.

In the case of shrinkage we find the following solution for the temporal development of semi-axes simply by switching the signs in Eqs. (30.46) and (30.47):

$$a(\bar{t}) = a_0\sqrt{1+\bar{t}}\,,\ \ c(\bar{t}) = c_0\sqrt{1-\frac{a_0^2}{c_0^2}\bar{t}}\,,\ \ \bar{t}=\frac{5\alpha E^2}{2a_0^2}t. \tag{30.52}$$

Both semi-axes become equal at time $\bar{t}_{\mathrm{iso}}=\frac{1}{2}\left(c_0^2/a_0^2-1\right)$. Then the final state of a sphere is reached with an equivalent radius of $\frac{1}{\sqrt{2}}\sqrt{a_0^2+c_0^2}$, which will stay fixed afterwards: The spheroid has turned into a sphere of constant dimensions and isotropic conditions prevail. This can be confirmed by studying the temporal development of the moment of inertia field until that time. We find by using the solution (30.52) in combination with Eq. (30.28):

$$\boldsymbol{J}(\bar{t}) = \frac{1}{5}\left(a_0^2+c_0^2\right)\mathbf{I}-\frac{1}{5}\left[c_0^2-(1+2\bar{t})a_0^2\right]\boldsymbol{n}\otimes\boldsymbol{n}, \tag{30.53}$$

which, as it should, differs from the solution for elongation shown in Eq. (30.48) only by one change of sign. Especially at time $\bar{t}_{\text{iso}}$, we obtain:

$$\boldsymbol{J}(\bar{t}_{\text{iso}}) = \frac{1}{5}\left(a_0^2 + c_0^2\right)\mathbf{I}, \tag{30.54}$$

which corresponds to the equation for the specific moment of inertia tensor of a sphere with radius $\frac{1}{\sqrt{2}}\sqrt{a_0^2 + c_0^2}$

30.4 Dynamics of Micropolar Media with Time-Varying Micro-Inertia

30.4.1 General Remarks

Recall the general spin balance for micropolar media shown in Eq. (30.5). If there is no coupling tensor $\boldsymbol{B}$, it can be rewritten while observing the balance of microinertia of Eq. $(30.3)_1$:

$$\rho \boldsymbol{J} \cdot \frac{\delta \boldsymbol{\omega}}{\delta t} + \boldsymbol{\omega} \times \boldsymbol{J} \cdot \boldsymbol{\omega} = \nabla \cdot \boldsymbol{\mu} + \boldsymbol{\sigma}_{\times} + \rho \boldsymbol{m}. \tag{30.55}$$

We want to study the evolution of angular velocity for some of the situations of temporally changing microinertia tensor discussed in Section 30.3. In the context of axial elongation and shrinkage based upon the action of an electric field it seems reasonable to assume that the angular velocity vector points in the direction of the normal, $\boldsymbol{\omega} = \omega \boldsymbol{n}$. Then for microinertia tensors of the form shown in Eq. (30.28) the second term on the left hand side of Eq. (30.55) drops out completely. Moreover, we will assume that the spheroidal particles behave like dust, *i.e.*, the spin and the stress tensors vanish, $\boldsymbol{\mu} = \mathbf{0}$ and $\boldsymbol{\sigma} = \mathbf{0}$. This assumption is somewhat critical because the particles within the RVE are electrically charged and, as we shall see now, must also carry a magnetic moment in order to enable rotation. It is also critical because of the embedding medium we had to assume in case of particles oriented in opposite direction to $\boldsymbol{n}$.

For the moment couple density we assume that it is constant and points in normal direction, $\boldsymbol{m} = m_0 \boldsymbol{n}$. In order to provide a physical interpretation one may imagine that it is created by a constant external magnetic field acting on elementary magnetic dipoles: If the particles of the RVE on the mesolevel carry in addition to the electric charges a magnetic moment oriented in $\boldsymbol{n}$-direction this will lead to a rotational motion about their $\boldsymbol{n}$-axis, similar to a DC electric engine, provided that the original electric field $\boldsymbol{E}$ is accompanied by an additional external (constant) magnetic field, $\boldsymbol{B}$.[7] Intuitively speaking, we may imagine the magnetic moment of a spheroidal particle to result from an elementary Ampère current loop (see, *e.g*, Bansal, 2006, Section 3.2.8) circulating about its equator. Note that both, the elementary magnetic

[7] which is not to be confused with the coupling tensor

moments as well as the external magnetic field, are not shown in Figs. 30.3 and 30.4 in order not to overload the cartoons. This is all we say at this point and, in particular, we will not attempt to find a real material with such microscopic properties.

Finally we assume that there is no translational motion and homogeneous conditions prevail. Then Eq. (30.55) simplifies to a first order ODE:[8]

$$\boldsymbol{J}(t)\cdot\boldsymbol{n}\frac{\mathrm{d}\omega(t)}{\mathrm{d}t} = m_0\boldsymbol{n}. \tag{30.56}$$

30.4.2 Axial Elongation and Shrinkage

We will first examine Eq. (30.56) for the case of axially elongation, where the tensor of inertia is given by Eq. (30.48). In this case we have to solve for $0 \leq \bar{t} < 1, \bar{t} = \frac{5\alpha E^2}{2a_0^2}t$:

$$\frac{\mathrm{d}\overline{\omega}(\bar{t})}{\mathrm{d}\bar{t}} = \frac{1}{1-\bar{t}}\,,\quad \overline{\omega} = \frac{\omega}{\omega_0}\,,\quad \omega_0 = \frac{m_0}{\alpha E^2} \tag{30.57}$$

with $\overline{\omega}(0) = 0$, which yields:

$$\overline{\omega}(\bar{t}) = \ln\frac{1}{1-\bar{t}}. \tag{30.58}$$

At time $\bar{t} = 1$ we end up with a singularity, which makes sense, because then the spheroid turns into a rod. Recall that a rod has no resistance against rotation about its axis. Hence the slightest m_0 will result in infinite angular velocity. Note that the curvature of the temporal development is positive: It becomes easier and easier to accelerate about the $\boldsymbol{n}$-axis because the resisting moment of inertia becomes smaller and small. Also note that this behavior is independent as to whether c_0/a_0 is less or greater one.

If we turn to the case of the alternative constitutive equation for elongation detailed in Eqs. (30.50) and (30.51) we have to write:

$$\frac{\mathrm{d}\overline{\omega}(\bar{t})}{\mathrm{d}\bar{t}} = \exp\bar{t}\,,\quad \overline{\omega} = \frac{\omega}{\omega_0}, \tag{30.59}$$
$$\omega_0 = \frac{125 m_0}{2\hat{\alpha}E^2 a_0^2(a_0^2+c_0^2)^2}\,,\quad \bar{t} = \frac{\hat{\alpha}E^2}{25}(a_0^2+c_0^2)^2 t.$$

Using the initial condition $\overline{\omega}(0) = 0$ yields:

$$\overline{\omega}(\bar{t}) = \exp\bar{t} - 1. \tag{30.60}$$

[8] The material derivative of Eq. (30.55) degenerates into a partial derivative because the observational point does not move and there is no translational velocity on the continuum scale. Moreover, because of the assumed homogeneity there is no difference in partial and regular derivatives, $\partial\omega/\partial t$ *vs.* $\mathrm{d}\omega/\mathrm{d}t$.

This means the occurrence of an infinite angular velocity is postponed untill infinity.

We will finally examine Eq. (30.56) for the case of axially shrinkage where the tensor of inertia is given by Eq. (30.53). In this case we have to solve the ODE in two steps. First we consider times $0 \leq \bar{t} \leq \bar{t}_{\mathrm{iso}}$, $\bar{t} = \frac{5\alpha E^2}{2a_0^2} t$. Here we seek the solution to:

$$\frac{\mathrm{d}\overline{\omega}(\bar{t})}{\mathrm{d}\bar{t}} = \frac{1}{1+\bar{t}} \, , \quad \overline{\omega} = \frac{\omega}{\omega_0} \, , \quad \omega_0 = \frac{m_0}{\alpha E^2} \tag{30.61}$$

with $\overline{\omega}(0) = 0$, which yields:

$$\overline{\omega}(\bar{t}) = \ln(1+\bar{t}) \, , \quad 0 \leq \bar{t} \leq \bar{t}_{\mathrm{iso}}. \tag{30.62}$$

At time $\bar{t} = \bar{t}_{\mathrm{iso}}$ the spheroid has turned into a sphere, and by brute mathematical force we have made sure that the moment of inertia will then not change any more. We solve:

$$\frac{\mathrm{d}\overline{\omega}(\bar{t})}{\mathrm{d}\bar{t}} = \frac{1}{1+\bar{t}_{\mathrm{iso}}} \tag{30.63}$$

with $\overline{\omega}(\bar{t}_{\mathrm{iso}}) = \ln(1+\bar{t}_{\mathrm{iso}})$, which yields:

$$\overline{\omega}(\bar{t}) = \frac{\bar{t}-\bar{t}_{\mathrm{iso}}}{1+\bar{t}_{\mathrm{iso}}} + \ln(1+\bar{t}_{\mathrm{iso}}) \, , \quad \bar{t}_{\mathrm{iso}} \leq \bar{t} < \infty. \tag{30.64}$$

In contrast to the previous case the angular velocity now develops initially with a negative curvature. This makes sense because the spheroid turns into a sphere with a greater radius than the initial axis in the plane of isotropy, $\frac{1}{\sqrt{2}}\sqrt{a_0^2+c_0^2}$ *vs.* $\sqrt{\frac{2}{5}}a_C$. Thus the resistance against revolution about the $\boldsymbol{n}$-axis grows. After the sphere has formed the angular velocity continues to grow linearly as to be expected by a constant driving moment with constant resistance of revolution.

30.5 Conclusions and Outlook

In this paper several tasks were accomplished:

- The features of an extended micropolar theory were presented, which allows for coupling of translational and angular kinetic energies as well as changes of microinertia.
- An attempt based on representation theory was made to present a general form for the recently introduced production of inertia of a transversally isotropic material.
- Based on the representation for the production of microinertia various solutions of its kinetic or balance equation were discussed: isotropic productions, deviatoric production, and a production purely related to the axis of symmetry.
- Closed form solutions for the moment of inertia tensor and for the spin as functions of time were obtained.

For an outlook into future work recall first the so-called Einstein-de Haas effect: Consider a suspended cylinder made of magnetizable material. The cylinder is initially motionless. Now a magnetic field is applied in axial direction whereupon the cylinder starts turning. This effect is interpreted as follows: Initially the angular momentum was zero. Aligning the electron spins means that they create an effective angular momentum, which must be counterbalanced macroscopically leading to an observable rotation of the material. One may also say that the spins of the electrons are initially isotropically oriented. After the magnetic field has been applied the electrons form a transversally isotropic medium. In other words, there is a structural change from isotropic to transversally isotropic.

In fact, our example of axially elongated spheroids shows similar features. Let us assume that the axes a_0 and c_0 are initially equal. This means that in the left upper inset of Fig. 30.3 we would have to draw polarized spheres, which are then extended to rods. According to Eq. (30.48) we have to write:

$$\boldsymbol{J}(\bar{t}) = \frac{2}{5} a_0^2 (\mathbf{I} - \boldsymbol{n} \otimes \boldsymbol{n}) + \frac{2}{5} a_0^2 (1 - \bar{t}) \boldsymbol{n} \otimes \boldsymbol{n}. \tag{30.65}$$

Note that this tensor is clearly isotropic at $\bar{t} = 0$ and its transversal isotropy evolves with increasing values of time, $\bar{t}$. It reaches the highest level of transversal isotropy at $\bar{t} = 1$, namely:

$$\boldsymbol{J}(\bar{t} = 1) = \frac{2}{5} a_0^2 (\mathbf{I} - \boldsymbol{n} \otimes \boldsymbol{n}), \tag{30.66}$$

where the resistance against rotation about the $\boldsymbol{n}$-axis becomes zero. In future work we shall explore if and how such thoughts could lead to a phenomenological modeling of the Einstein-de Haas effect. Here it will become necessary to include the coupling tensor, $\boldsymbol{B}$, because changes of the angular velocity field must generate a translational (circumferential) velocity.

A second potential field of application of the presented theory are liquid crystals and, in particular, the transition from an orderly anisotropic structure to an isotropic one of the Navier-Stokes type, if temperature is increased. The structural tensor parameters introduced in the work of de Gennes (*e.g.*, de Gennes, 1971; de Gennes and Prost, 1995) bear a certain similarity to those used in context with the examples presented in Subsections 30.3.2 and 30.3.3. It can be expected that the continuum point of view is able to shed a different light on that problem, too.

Acknowledgements E.N.V. wishes to gratefully acknowledge support of her work by a grant from the Russian Foundation for Basic Research (17-51-12055).

Appendices

Below general representations for the production term and for the free energy density will be derived and discussed.

Representation of the Production of Moment of Inertia

In the case of a transversally isotropic material we assume that the production of the moment of inertia, $\boldsymbol{\chi}_J$ is a symmetric isotropic tensor function of second rank depending on a symmetric second order tensor, $\boldsymbol{B}$[9], and a vector vector $\boldsymbol{b}$. Note that in the examples of structural change presented in this paper the vector $\boldsymbol{b}$ will either be the direction of the spheroidal axis, $\boldsymbol{n}$, or the applied (constant) electric field, $\pm\boldsymbol{E}$, and the tensor $\boldsymbol{B}$ is replaced by the tensor of the moment of inertia, $\boldsymbol{J}$. We write:

$$\boldsymbol{\chi}_J = \boldsymbol{\chi}_J(\boldsymbol{B},\boldsymbol{b}). \tag{30.67}$$

Without loss of generality we turn this isotropic tensor function into an isotropic scalar function $\phi = \phi(\boldsymbol{A},\boldsymbol{B},\boldsymbol{b})$ by outer scalar multiplication with a symmetric tensor second rank, $\boldsymbol{A}$, as follows:

$$\phi(\boldsymbol{A},\boldsymbol{B},\boldsymbol{b}) = \boldsymbol{A} : \boldsymbol{\chi}_J(\boldsymbol{B},\boldsymbol{b}). \tag{30.68}$$

The symbol ":" refers to the outer double scalar product (see Holzapfel, 2000, pg. 14):

$$\boldsymbol{C} : \boldsymbol{D} = C_{ij}D_{ij}, \tag{30.69}$$

$\boldsymbol{C}$ and $\boldsymbol{D}$ being arbitrary tensors of second rank.

We now turn to the tables of Smith (1965), where we examine the case of a scalar depending on two symmetric tensors of second rank, $\boldsymbol{A}$ and $\boldsymbol{B}$, and one vector, $\boldsymbol{b}$. We restrict ourselves to linear dependencies in the "dummy tensor," $\boldsymbol{A}$. Then the most general polynomial form for the isotropic scalar $\phi(\boldsymbol{A},\boldsymbol{B},\boldsymbol{b})$ is given by the following expression based on six terms:

$$\begin{aligned}\phi(\boldsymbol{A},\boldsymbol{B},\boldsymbol{b}) = \boldsymbol{A} : \Big(&a_1\mathbf{I} + a_2\boldsymbol{b}\otimes\boldsymbol{b} + a_3\boldsymbol{B} + a_4\boldsymbol{B}^2 + \\ &a_5(\boldsymbol{b}\otimes\boldsymbol{B}\cdot\boldsymbol{b} + \boldsymbol{B}\cdot\boldsymbol{b}\otimes\boldsymbol{b}) + a_6(\boldsymbol{b}\otimes\boldsymbol{B}^2\cdot\boldsymbol{b} + \boldsymbol{B}^2\cdot\boldsymbol{b}\otimes\boldsymbol{b})\Big),\end{aligned} \tag{30.70}$$

where the necessity for symmetry of the tensorial expression in parentheses, namely for $\boldsymbol{\chi}_J$, has explicitly been taken into account. The coefficients a_i, $i = 1,\cdots,6$ are arbitrary scalar functions of the following six invariants:

$$a_i = a_i\Big(\mathrm{Tr}(\boldsymbol{B}), \mathrm{Tr}(\boldsymbol{B}^2), \mathrm{Tr}(\boldsymbol{B}^3), \boldsymbol{b}\cdot\boldsymbol{b}, \boldsymbol{b}\cdot\boldsymbol{B}\cdot\boldsymbol{b}, \boldsymbol{b}\cdot\boldsymbol{B}^2\cdot\boldsymbol{b}\Big). \tag{30.71}$$

Note that the dependence on the invariant, $\boldsymbol{b}\cdot\boldsymbol{b}$, can be omitted, if $\boldsymbol{b}$ is a unit vector. Also note that if physical interpretation is sought, then it is advisable to replace $\mathrm{Tr}(\boldsymbol{B}^3)$ by $\det\boldsymbol{B}$, which is characteristic of volume changes, since $\det\boldsymbol{B} = \frac{1}{6}\left(I_1^3 - 3I_1I_2 + 2I_3\right)$ with $I_1 = \mathrm{Tr}(\boldsymbol{B})$, $I_2 = \mathrm{Tr}(\boldsymbol{B}^2)$, $I_3 = \mathrm{Tr}(\boldsymbol{B}^3)$.

This result is in agreement with investigations on tensorial representations performed in Müller (1985), pg. 88, or Naumenko and Altenbach (2007), Section C2,

---

[9] not to be confused with the coupling tensor

where the former reference uses more or less intuitive arguments. The latter one is obviously inspired by the work of Zhilin, specifically Zhilin (2003, 2005). In there we find a very helpful formula for predicting the to-be-expected number of independent invariants, N_*, namely:

$$N_* = N - 3\ ,\ \ N > 3, \tag{30.72}$$

where N is the "coordinate number" or (better) number of independent components of the participating objects, in our case $N = 6+3 = 9$, hence $N_* = 6$ as it was claimed above.

Finally we conclude that the most general structure of the production of moment of inertia for transversally isotropic media is given by:

$$\begin{aligned}\boldsymbol{\chi}_J(\boldsymbol{B},\boldsymbol{b}) = {} & a_1\mathbf{I} + a_2\boldsymbol{b}\otimes\boldsymbol{b} + a_3\boldsymbol{B} + a_4\boldsymbol{B}^2 + \\ & a_5(\boldsymbol{b}\otimes\boldsymbol{B}\cdot\boldsymbol{b} + \boldsymbol{B}\cdot\boldsymbol{b}\otimes\boldsymbol{b}) + a_6(\boldsymbol{b}\otimes\boldsymbol{B}^2\cdot\boldsymbol{b} + \boldsymbol{B}^2\cdot\boldsymbol{b}\otimes\boldsymbol{b}),\end{aligned} \tag{30.73}$$

where $\boldsymbol{b}$ will either be the unit normal $\boldsymbol{n}$ or the electric field $\boldsymbol{E}$, and $\boldsymbol{B}$ will be replaced by the moment of inertia tensor, $\boldsymbol{J}$.

Restrictions on the Production of Moment of Inertia by the Second Law

We start from the standard entropy balance and from the second law of thermodynamics:

$$\rho\frac{\delta s}{\delta t} = -\nabla\cdot\frac{\boldsymbol{q}}{T} + \frac{\rho r}{T} + \sigma\ ,\ \ \sigma \geq 0. \tag{30.74}$$

In other words, we assume that the entropy flux and the entropy supply are given by $\boldsymbol{q}/T$ and $\rho r/T$, respectively, where $\boldsymbol{q}$ is the heat flux and r the volumetric heat supply known from the balance of internal energy of micropolar media shown in Eq. (30.6). Finally σ refers to the entropy production density. If both heat terms in these equations are eliminated we arrive at the Clausius-Duhem inequality for micropolar media:

$$\rho\frac{\delta f}{\delta t} + \rho s\frac{\delta T}{\delta t} - \boldsymbol{\sigma} : (\nabla\boldsymbol{v} + \mathbf{I}\times\boldsymbol{\omega}) - \boldsymbol{\mu} : \nabla\boldsymbol{\omega} + \frac{1}{T}\boldsymbol{q}\cdot\nabla T = -T\sigma \leq 0. \tag{30.75}$$

For a more general examination according to Coleman-Noll we would now assume that the specific free energy, $f = u - Ts$, is a scalar function of:

$$f = f(\rho, T, \nabla\boldsymbol{v} + \mathbf{I}\times\boldsymbol{\omega}, \nabla\boldsymbol{\omega}, \nabla T, \boldsymbol{J}, \boldsymbol{n}). \tag{30.76}$$

However, in this paper we are interested in restrictions of the form of the production of the microinertia, $\boldsymbol{\chi}_J$, in the absence of mechanical and thermal influences. Hence we concentrate on the dependencies on $\boldsymbol{J}$ and $\boldsymbol{n}$ alone and focus on the following reduced Clausius-Duhem inequality:

$$\rho \frac{\delta f(\boldsymbol{J},\boldsymbol{n})}{\delta t} = -T\sigma \leq 0 \quad \Rightarrow \tag{30.77}$$
$$\rho \left(\frac{\partial f}{\partial \boldsymbol{J}} : \frac{\delta \boldsymbol{J}}{\delta t} + \frac{\partial f}{\partial \boldsymbol{n}} \cdot \frac{\delta \boldsymbol{n}}{\delta t} \right) = \rho \frac{\partial f}{\partial \boldsymbol{J}} : \boldsymbol{\chi}_J = -T\sigma \leq 0,$$

if we restrict ourselves to situations where Eq. (30.24) holds, *i.e.*, $\delta \boldsymbol{J}/\delta t = \boldsymbol{\chi}_J$, this is to say, where the rigid body rotation, $\boldsymbol{\omega} \times \boldsymbol{J} - \boldsymbol{J} \times \boldsymbol{\omega}$, vanishes and if $\delta \boldsymbol{n}/\delta t = \mathbf{0}$. Both is the case for the microinertia field of interest in this paper, *i.e.*, for a transversally isotropic one of the form shown in Eq. (30.28). Now we consider a representation for f analogous to (30.71):

$$f(\boldsymbol{J},\boldsymbol{n}) = f\Big(\mathsf{Tr}(\boldsymbol{J}), \mathsf{Tr}(\boldsymbol{J}^2), \mathsf{Tr}(\boldsymbol{J}^3), \boldsymbol{n}\cdot\boldsymbol{J}\cdot\boldsymbol{n}, \boldsymbol{n}\cdot\boldsymbol{J}^2\cdot\boldsymbol{n}\Big). \tag{30.78}$$

The partial differentiation in (30.77) can then be executed by using the chain rule. If we restrict ourselves to a linear irreversible theory in $\boldsymbol{J}$ then the following terms remain:

$$\rho\left(\beta_1 \mathbf{I} + (\beta_2 + \tfrac{1}{3}\beta_3)\,\mathsf{Tr}\boldsymbol{J}\mathbf{I} + \beta_3 \boldsymbol{J}^{\mathrm{dev}} + \beta_4 \boldsymbol{n}\otimes\boldsymbol{n}\right) : \boldsymbol{\chi}_J = -T\sigma \leq 0, \tag{30.79}$$

where β_i, $i = 1, \cdots, 4$ are true constants, namely:

$$\beta_1 = \left.\frac{\partial f}{\partial\,\mathsf{Tr}\boldsymbol{J}}\right|_{\boldsymbol{J}=0}, \quad \beta_2 = \left.\frac{\partial^2 f}{\partial\,\mathsf{Tr}\boldsymbol{J}\,\partial\,\mathsf{Tr}\boldsymbol{J}}\right|_{\boldsymbol{J}=0}, \tag{30.80}$$
$$\beta_3 = 2\left.\frac{\partial f}{\partial\,\mathsf{Tr}\boldsymbol{J}^2}\right|_{\boldsymbol{J}=0}, \quad \beta_4 = \frac{\partial f}{\partial \boldsymbol{n}\cdot\boldsymbol{J}\cdot\boldsymbol{n}}{}_{\boldsymbol{J}=0}.$$

Let us investigate as to whether the choice of a purely deviatoric production according to Eq. (30.27) might lead to a conflict with the second law. By inserting the expression into Eq. (30.79) when specialized to deviators we obtain:

$$\rho\beta_3 \boldsymbol{J}^{\mathrm{dev}} : \boldsymbol{\chi}_J = -\rho\beta_3\beta\boldsymbol{J}^{\mathrm{dev}} : \boldsymbol{J}^{\mathrm{dev}} = -T\sigma \leq 0. \tag{30.81}$$

Since β is positive this leads to a constraint for the free energy:

$$\beta_3 = 2\left.\frac{\partial f}{\partial\,\mathsf{Tr}\boldsymbol{J}^2}\right|_{\boldsymbol{J}=0} \geq 0. \tag{30.82}$$

References

Bansal R (2006) Fundamentals of engineering electromagnetics. CRC press

Eremeyev VA, Pietraszkiewicz W (2012) Material symmetry group of the non-linear polar-elastic continuum. International Journal of Solids and Structures 49(14):1993–2005

Eremeyev VA, Lebedev LP, Altenbach H (2012) Foundations of micropolar mechanics. Springer Science & Business Media, Heidelberg, New York, Dordrecht, London

Eringen AC, Kafadar CB (1976) Polar field theories. In: Continuum physics IV. Academic Press, London

Eringen C (2002) Nonlocal continuum field theories. Springer Science & Business Media, New York

de Gennes PG (1971) Short range order effects in the isotropic phase of nematics and cholesterics. Molecular Crystals and Liquid Crystals 12(3):193–214

de Gennes PG, Prost J (1995) The physics of liquid crystals, second edition. Oxford University Press, Oxford

Gross D, Xu B (2015) Micromechanical modeling of cellular ferroelectrets by using internal variables. Procedia IUTAM 12:62–72

Hibbeler RC (2010) Dynamics. 12th ed. Prentice Hall, Upper Saddle River (NJ)

Holzapfel GA (2000) Nonlinear Solid Mechanics: A Continuum Approach for Engineering. John Wiley & Sons, Ltd., West Sussex, England

Ivanova EA, Vilchevskaya EN (2016) Micropolar continuum in spatial description. Continuum Mechanics and Thermodynamics 28(6):1759–1780

Kestelman VN, Pinchuk LS, Goldade VA (2013) Electrets in Engineering: Fundamentals and Applications. Springer Science & Business Media

Müller I (1985) Thermodynamics. Pitman

Müller WH, Vilchevskaya EN, Weiss W (2017) A meso-mechanics approach to micropolar theory: A farewell to material description. Physical Mesomechanics 20(3):13–24

Naumenko K, Altenbach H (2007) Modeling of creep for structural analysis. Springer Science & Business Media, Singapore

Reich FA, Rickert W, Stahn O, Müller WH (2017) Magnetostriction of a sphere: Stress development during magnetization and residual stresses due to the remanent field. Continuum Mechanics and Thermodynamics 29(2):535–557

Smith GF (1965) On isotropic integrity bases. Archive for rational mechanics and analysis 18(4):282–292

Wikipedia (2017) Ferroelectret. https://en.wikipedia.org/wiki/Ferroelectret

Zhilin PA (2003) Модифицированная теория симметрии тензоров и тензорных инвариантов (Modified theory of the theory of tensors and tensor invariants, in Russian). Изв высш учеб зав Северо - Кавказский регион Естественные науки (S):176–195

Zhilin PA (2005) Symmetries and orthogonal invariants in oriented space. In: Proceedings of the XXXII Summer School - Conference "Advanced Problems in Mechanics", St. Petersburg, pp 470–483

Zhilin PA (2015) Динамика твердого тела (Dynamics of the Solid Body, in Russian). Санкт-Петербург Издательство Политехнического университета, St. Petersburg

Chapter 31
Contact Temperature as an Internal Variable of Discrete Systems in Non-Equilibrium

Wolfgang Muschik

Abstract State space and entropy rate of a discrete non-equilibrium system are shortly considered including internal variables and the contact temperature. The concept of internal variables in the context of non-equilibrium thermodynamics of a closed discrete system is discussed. The difference between internal variables and degrees of freedom are repeated, and different types of their evolution equations are mentioned in connection with Gérard A. Maugin's numerous papers on applications of internal variables. The non-equilibrium contact temperature is recognized as an internal variable and its evolution equation is presented.

31.1 Introduction

Temperature is a quantity which can be measured easily, but whose theoretical foundation is complicated. There is a huge variety of different thermometers (Kestin, 1966; Lieneweg, 1976) all measuring "temperature", but the concept of temperature is first of all only properly defined in equilibrium. For elucidating this fact, we consider the simple example of a thermometer whose surface[1] has different heat conductivities. Contacted with a non-equilibrium system, the measured "temperature" depends at the same position on the orientation of such a "thermometer". It is clear, that this orientation sensitivity of the thermometer vanishes in equilibrium (if one knows what equilibrium is). A second example is a thermometer which measures the intensity of radiation which is composed of different parts of the spectrum. The measured "temperature' depends on the sensitivity distribution of the thermometer

Wolfgang Muschik
Institut für Theoretische Physik, Technische Universität Berlin, Hardenbergstr. 36, D-10623 Berlin, Germany
e-mail: muschik@physik.tu-berlin.de

[1] The thermometer is here a discrete system which has a volume and a surface, as small as ever.

© Springer International Publishing AG, part of Springer Nature 2018
H. Altenbach et al. (eds.), *Generalized Models and Non-classical Approaches in Complex Materials 1*, Advanced Structured Materials 89,
https://doi.org/10.1007/978-3-319-72440-9_31

over the spectrum with the result, that different thermometers measure different "temperatures" at the same object.

For escaping these "thermometer induced" difficulties, a theoretical definition of temperature is considered as a remedy . We define (Kestin, 1966)

$$\begin{aligned} \text{discrete systems: } \frac{1}{T} &:= \frac{\partial S}{\partial U}, \\ \text{field formulation: } \frac{1}{T} &:= \frac{\partial s}{\partial u}. \end{aligned} \tag{31.1}$$

But also these definitions have their drawbacks: First of all, a state space is needed, because the partial derivatives have no sense without it. Then entropy S or entropy density s and internal energy U or internal energy density u are needed in equilibrium or out of it. And finally, the open question is, if there exists a thermometer which measures the temperature T.

To avoid all these uncertainties, a simple idea is the following: why not define a general concept of temperature which is valid independently of equilibrium or non-equilibrium and which is introduced into the theoretical framework by defining the RHSs of (31.1)

$$\begin{aligned} \frac{1}{\Theta} &=: \frac{\partial S}{\partial U}, \\ \frac{1}{\Theta} &=: \frac{\partial s}{\partial u} ? \end{aligned} \tag{31.2}$$

If additionally Θ is connected with a measuring instruction which "defines" the temperature Θ experimentally, temperature comes from the outside into the theoretical framework and not vice versa. How to realize this idea in connection with internal variables is the objective of this paper.

31.2 Contact Temperature

31.2.1 Definition

We consider a closed discrete non-equilibrium system[2] which is contacted with an equilibrium environment of thermostatic temperature T^*. The heat exchange per time between the considered system and its environment is $\overset{\bullet}{Q}$. We now introduce a temperature Θ which satisfies the inequality

$$\overset{\bullet}{Q}\left(\frac{1}{\Theta} - \frac{1}{T^*}\right) \geq 0. \tag{31.3}$$

[2] For the sake of a minimum of formalism, we consider here closed systems. This choice does not influence the definition of the contact temperature below, that means, closing an open system does not change its contact temperature. More details in Muschik (2009).

According to this inequality, we obtain[3]

$$\overset{\bullet}{Q} > 0 \Longrightarrow T^* > \Theta,$$
$$\overset{\bullet}{Q} < 0 \Longrightarrow T^* < \Theta, \qquad (31.4)$$
$$T^* = \Theta \Longleftrightarrow \overset{\bullet}{Q} = 0.$$

Consequently, we have the following definition.

Definition 31.1. The system's contact temperature is that thermostatic temperature of the system's equilibrium environment for which the net heat exchange between the system and this environment through an inert[4] partition vanishes by change of sign[5].

The contact temperature is defined for discrete systems in non-equilibrium embracing the case of equilibrium (Muschik, 1977; Muschik and Brunk, 1977; Muschik and Berezovski, 2004). In both cases, the net heat exchange vanishes, if the thermostatic temperature of the controlling environment T^* is equal to the contact temperature Θ in non-equilibrium according to $(31.4)_3$, or if T^* is equal to the thermostatic temperature T of the system in equilibrium. If the system is in non-equilibrium at the contact temperature $\Theta = T^*$, the sum of the non-vanishing partial heat exchanges between system and heat reservoir of T^* vanishes. If the system is in equilibrium, all these partial heat exchanges vanish.

The contact temperature is not defined by (31.2), but it is a basic quantity similar as the energy. Entropy and internal energy in connection with a suitable state space have to be defined so that (31.2) is satisfied. This item is treated in Sect. 31.3.

31.2.2 Contact Temperature and Internal Energy

The contact temperature Θ is independent of the internal energy U of the system[6]. The proof of this statement runs as follows: we consider the energy balance equation of a closed discrete system with a rigid power impervious partition

$$\overset{\bullet}{U} = \overset{\bullet}{Q} + \overset{\bullet}{W}, \qquad \overset{\bullet}{W} = 0, \text{ the power.} \qquad (31.5)$$

The process taking place in the non-equilibrium system generates a time dependent contact temperature $\Theta(t)$ and a time dependent heat exchange $\overset{\bullet}{Q}(t)$ which also depends on the temperature $T^*(t)$ of the equilibrium environment which controls the system.

[3] A more detailed proof is represented in first subsection of Appendices.

[4] inert means: the partition does not absorb or emit energy and/or material

[5] Do not take the vanishing net heat exchange for an adiabatic condition there are positive and negative heat exchanges through partial surfaces between system and environment.

[6] see Muschik (1990a, 4.1.2)

We now choose the environment's temperature for all times equal to the contact temperature of the system, and we obtain

$$T^*(t) \doteq \Theta(t) \quad \Longrightarrow \quad \overset{\bullet}{Q}(t) = 0 \quad \Longrightarrow \quad \overset{\bullet}{U}(t) = 0. \tag{31.6}$$

The last implication is due to (31.5). Because Θ is time dependent and U is constant, both quantities are independent of each other.

31.3 State Space and Entropy Rate

We consider the state space of a closed discrete non-equilibrium system which contains the contact temperature as an independent variable according to Sect. 31.2.2

$$Z := (U, \boldsymbol{a}, \Theta, \boldsymbol{\xi}). \tag{31.7}$$

Here, $\boldsymbol{a}$ are the work variables

$$\overset{\bullet}{W} = \boldsymbol{A} \cdot \overset{\bullet}{\boldsymbol{a}} \tag{31.8}$$

and $\boldsymbol{\xi}$ the internal variables –"measurable but not controllable"[7]– which are discussed in Sect. 31.5.

A process $Z(t)$ is represented by a trajectory $\mathscr{T}$ on the non-equilibrium space (31.7). According to this state space, the time rate of the non-equilibrium entropy becomes along $\mathscr{T}$ by inserting the first law (31.5)

$$\overset{\bullet}{S} := \frac{1}{\Theta}\Big(\overset{\bullet}{U} - \boldsymbol{A} \cdot \overset{\bullet}{\boldsymbol{a}}\Big) + \alpha \overset{\bullet}{\Theta} + \boldsymbol{\beta} \cdot \overset{\bullet}{\boldsymbol{\xi}} = \frac{1}{\Theta} \overset{\bullet}{Q} + \alpha \overset{\bullet}{\Theta} + \boldsymbol{\beta} \cdot \overset{\bullet}{\boldsymbol{\xi}}. \tag{31.9}$$

By definition, the entropy rate of an isolated system is the entropy production Σ which is non-negative according to the second law

$$\overset{\bullet}{U} \equiv 0, \quad \overset{\bullet}{\boldsymbol{a}} \equiv 0 \longrightarrow \overset{\bullet}{S}{}^{isol} =: \Sigma = \alpha \overset{\bullet}{\Theta} + \boldsymbol{\beta} \cdot \overset{\bullet}{\boldsymbol{\xi}} \geq 0. \tag{31.10}$$

Because the contact temperature is independent of the other internal variables, we can decompose $(31.10)_4$

$$\alpha \overset{\bullet}{\Theta} \geq 0, \qquad \boldsymbol{\beta} \cdot \overset{\bullet}{\boldsymbol{\xi}} \geq 0, \tag{31.11}$$

and using (31.9) and (31.3), we obtain the inequalities

$$\overset{\bullet}{S} \geq \frac{\overset{\bullet}{Q}}{\Theta} \geq \frac{\overset{\bullet}{Q}}{T^*}. \tag{31.12}$$

[7] see Maugin (1999a, 4.1); Maugin (2013a, 5.6)

If we presuppose that a state function $S(U, \boldsymbol{a}, \Theta, \boldsymbol{\xi})$ exists[8] the integration along a cyclic trajectory on (31.7) results in the extended Clausius inequality of closed systems

$$0 \geq \oint \frac{\overset{\bullet}{Q}}{\Theta} dt \geq \oint \frac{\overset{\bullet}{Q}}{T^*} dt, \tag{31.13}$$

and $(31.2)_1$ becomes an integrability condition

$$\frac{1}{\Theta} = \left(\frac{\partial S}{\partial U}\right)_{\boldsymbol{a},\Theta,\boldsymbol{\xi}} \quad \longrightarrow \quad S(U, \boldsymbol{a}, \Theta, \boldsymbol{\xi}) = \frac{1}{\Theta} U + K(\boldsymbol{a}, \Theta, \boldsymbol{\xi}), \tag{31.14}$$

and $-\Theta K(\boldsymbol{a}, \Theta, \boldsymbol{\xi})$ is the free energy[9]. Because of the two last terms in (31.9), the contact temperature takes the placing of an internal variable which are discussed in Sect. 31.5.

31.4 Equilibrium and Reversible "Processes"

Equilibrium in thermally homogeneous systems satisfies the following equilibrium conditions:

$$\text{no time dependence:} \quad \overset{\bullet}{\boxtimes}{}^{eq} \doteq 0, \tag{31.15}$$

$$\text{thermostatic temperature:} \quad \Theta^{eq} \doteq T(U, \boldsymbol{a}) = T^* \tag{31.16}$$

Consequently, we obtain

$$\overset{\bullet}{U}{}^{eq} = 0, \quad \overset{\bullet}{\boldsymbol{a}}{}^{eq} = \boldsymbol{0}, \quad \overset{\bullet}{Q}{}^{eq} = 0, \quad \overset{\bullet}{\Theta}{}^{eq} = 0, \quad \overset{\bullet}{\boldsymbol{\xi}}{}^{eq} = \boldsymbol{0} \longrightarrow \Sigma^{eq} = 0 \tag{31.17}$$

according to (31.10). We now have to distinguish two kinds of equilibria concerning the affinities $\boldsymbol{\beta}$

$$\boldsymbol{\beta}\Big(U, \boldsymbol{a}, T(U, \boldsymbol{a}), \boldsymbol{\xi}^{eq}\Big) = \boldsymbol{0} \quad \text{and} \quad \boldsymbol{\beta}\Big(U, \boldsymbol{a}, T(U, \boldsymbol{a}), \boldsymbol{\xi}_f\Big) \neq \boldsymbol{0}. \tag{31.18}$$

Consequently, we obtain

$$\text{unconstrained equilibrium:} \quad \overset{\bullet}{\boldsymbol{\xi}}{}^{eq} = \boldsymbol{0} \wedge \boldsymbol{\beta} = \boldsymbol{0}, \tag{31.19}$$

$$\text{constrained equilibrium:} \quad \overset{\bullet}{\boldsymbol{\xi}}_f = \boldsymbol{0} \wedge \boldsymbol{\beta} \neq \boldsymbol{0}. \tag{31.20}$$

Unconstrained equilibrium means that the internal variables are fixed at their equilibrium values

$$\boldsymbol{\xi}^{eq} = \boldsymbol{\xi}(U, \boldsymbol{a}) \neq \boldsymbol{\xi}_f \tag{31.21}$$

---

8 That is the case in large state spaces, if the system is adiabatically unique (Muschik, 2009).

9 more details in Muschik (2009, 2014)

according to the solution of $(31.18)_1$. Constraint equilibrium means that the internal variables $\boldsymbol{\xi}_f$ are according to $(31.20)_1$ "frozen in" at a value which is different from its unconstrained equilibrium value and which is not determined by $(U,\boldsymbol{a})$. "Mixed equilibria" are possible in which one part $\boldsymbol{\zeta}$ of the internal variables is in unconstrained equilibrium, whereas the other part $\boldsymbol{\chi}$ of them is in constraint equilibrium:

$$\boldsymbol{\xi} = (\boldsymbol{\zeta},\boldsymbol{\chi}) \longrightarrow \boldsymbol{\xi}^{eq} = \Big(\boldsymbol{\zeta}^{eq}(U,\boldsymbol{a}),\boldsymbol{\chi}^{eq} = \boldsymbol{\chi}_f\Big). \tag{31.22}$$

The individual parts $\boldsymbol{\zeta}$ and $\boldsymbol{\chi}$ of a mixed equilibrium may depend on the time for which the system is isolated making the relaxation to equilibrium possible.

The dimension of the state space (31.7) shrinks at equilibrium according to (31.16) and $(31.22)_1$

$$Z^{eq} := \Big(U,\boldsymbol{a},\Theta^{eq}(U,\boldsymbol{a}),\boldsymbol{\zeta}^{eq}(U,\boldsymbol{a}),\boldsymbol{\chi_f}\Big) \longrightarrow \mathscr{Z}^{eq} = (U,\boldsymbol{a},\boldsymbol{\chi}_f). \tag{31.23}$$

Such as the entropy rate of a non-equilibrium process $\mathscr{T}$ is defined on (31.7), we define the reversible entropy "rate" along a *reversible "process"* $\mathscr{R}$ on the equilibrium sub-space $(31.23)_2$ $\mathscr{Z}^{eq}(t)$

$$\begin{aligned} \overset{\bullet}{S}{}^{rev} &:= \frac{1}{T}\Big(\overset{\bullet}{U} - \boldsymbol{A}^{rev}\cdot\overset{\bullet}{\boldsymbol{a}}\Big) + \frac{\partial S^{rev}}{\partial \boldsymbol{\chi}_f}\cdot\overset{\bullet}{\boldsymbol{\chi}}_f, \\ \overset{\bullet}{U} - \boldsymbol{A}^{rev}&\cdot\overset{\bullet}{\boldsymbol{a}} =: \overset{\bullet}{Q}{}^{rev}. \end{aligned} \tag{31.24}$$

The process parameter "t" along $\mathscr{R}$ is not the real time because real processes are not possible in the equilibrium sub-space. The time parameter is formally generated by a projection of $\mathscr{T}$ onto $\mathscr{R}$

$$\mathscr{P}\mathscr{T}(t) = \mathscr{P}(U,\boldsymbol{a},\Theta,\boldsymbol{\xi})(t) = (U,\boldsymbol{a},\boldsymbol{\chi}_f)(t) = \mathscr{R}(t), \tag{31.25}$$

where the reversible *accompanying "process"* $(U,\boldsymbol{a},\boldsymbol{\chi}_f)(t)$ takes place[10] which belongs to the original one $(U,\boldsymbol{a},\Theta,\boldsymbol{\xi})(t)$ by projection (Muschik, 1993).

To connect (31.9) with $(31.24)_1$, we apply the *embedding axiom* (Muschik, 1990a)

$$S_{B(eq)} - S_{A(eq)} = \mathscr{T}\int_A^B \overset{\bullet}{S}(t)dt \doteq \mathscr{R}\int_A^B \overset{\bullet}{S}{}^{rev}(t)dt \tag{31.26}$$

which by use of $(31.9)_2$ and (31.24) results in

$$(\mathscr{T}/\mathscr{R})\int_A^B \Big(\frac{\overset{\bullet}{Q}}{\Theta} - \frac{\overset{\bullet}{Q}{}^{rev}}{T} + \alpha\,\overset{\bullet}{\Theta} + \boldsymbol{\beta}\cdot\overset{\bullet}{\boldsymbol{\xi}} - \frac{\partial S^{rev}}{\partial \boldsymbol{\chi}_f}\cdot\overset{\bullet}{\boldsymbol{\chi}}_f\Big)dt = 0. \tag{31.27}$$

---

[10] t is the "slaved time" according to (31.25)

We obtain according to (31.3) and (31.24)$_2$, and that $T(U,\boldsymbol{a}) = T^*$ is valid along $\mathscr{R}$

$$\frac{\overset{\bullet}{Q}}{\Theta} - \frac{\overset{\bullet}{Q}{}^{rev}}{T} \geq \frac{\overset{\bullet}{Q}}{T^*} - \frac{\overset{\bullet}{Q}{}^{rev}}{T} = \frac{1}{T^*}\Big(\overset{\bullet}{Q} - \overset{\bullet}{Q}{}^{rev}\Big) = \frac{1}{T^*}\Big(\boldsymbol{A}^{rev} - \boldsymbol{A}\Big)\cdot\overset{\bullet}{\boldsymbol{a}}. \quad (31.28)$$

Taking (31.10)$_3$ and (31.28) into account, we obtain from (31.27)

$$\mathscr{T}/\mathscr{R}\int_A^B \Big[\frac{1}{T^*}\Big(\boldsymbol{A}^{rev} - \boldsymbol{A}\Big)\cdot\overset{\bullet}{\boldsymbol{a}} - \frac{\partial S^{rev}}{\partial \boldsymbol{\chi}_f}\cdot\overset{\bullet}{\boldsymbol{\chi}}_f\Big]dt \leq 0. \quad (31.29)$$

If the system under consideration has only unconstrained equilibria, (31.29) yields

$$\mathscr{R}\int_A^B \boldsymbol{A}^{rev}\cdot\overset{\bullet}{\boldsymbol{a}}\,dt \leq \mathscr{T}\int_A^B \boldsymbol{A}\cdot\overset{\bullet}{\boldsymbol{a}}\,dt, \quad (31.30)$$

and we obtain for the volume work the well known inequality

$$\boldsymbol{A}\cdot\overset{\bullet}{\boldsymbol{a}} \equiv -p\overset{\bullet}{V} \longrightarrow \mathscr{R}\int_A^B p^{rev}dV \geq \mathscr{T}\int_A^B p\overset{\bullet}{V}\,dt. \quad (31.31)$$

The non-equilibrium entropy (31.9) has to be complemented by evolution laws for the internal variables Θ and $\boldsymbol{\xi}$ in the next section.

31.5 Brief Overview of Internal Variables

Historically, the concept of internal variables can be traced back to Bridgeman (1943); Meixner (1943) and many others. The introduction of internal variables makes possible to use large state spaces, that means, material properties can be described by mappings defined on the state space variables (including the internal ones), thus avoiding the use of their histories which appear in small state spaces (Muschik, 1990b). Those are generated, if the internal variables are eliminated. Consequently, internal variables allow to use the methods of Irreversible and/or Extended Thermodynamics (Maugin, 1999a).

Internal variables cannot be chosen arbitrarily: there are concepts which restrict their introduction (Muschik, 1990b). The most essential ones are:

1. Internal variables need a model or an interpretation,
2. Beyond the constitutive and balance equations, internal variables require rate equations which can be adapted to different situations, making the use of internal variables flexible and versatile,
3. The time rates of the internal variables do not occur in the work differential of the First Law,

4. An isolation of the discrete system does not influence the internal variables,
5. In equilibrium, the internal variables become dependent on the variables of the equilibrium sub-space, if the equilibrium is unconstrained.

Satisfying these concepts, the internal variables involve an ambiguous relationship with the microstructure and internal degrees of freedom (Maugin and Muschik, 1994a,b). But internal variables and internal degrees of freedom represent different concepts for extending the state space: both are included in the state space, both need evolution laws, but whereas internal variables do not occur in the work differential of the First Law according to 3., degrees of freedom appear in the time rate of the internal energy. Consequently, the question "internal variables or degrees of freedom?"[11] can be answered clearly.

As the last term of (31.9) shows, internal variables must be complemented by an evolution law[12] which may have the shape

$$\overset{\bullet}{\boldsymbol{\xi}} = \boldsymbol{f}(U,\boldsymbol{a},\boldsymbol{\xi}) + \boldsymbol{g}(U,\boldsymbol{a},\boldsymbol{\xi})\,\overset{\bullet}{U} + \boldsymbol{h}(U,\boldsymbol{a},\boldsymbol{\xi})\cdot\overset{\bullet}{\boldsymbol{a}}\,. \tag{31.32}$$

Special one-dimensional cases are (Muschik, 1990a)

$$\text{relaxation type:}\quad \overset{\bullet}{\xi}(t) = -\frac{1}{\tau(U,\boldsymbol{a},\Theta)}\Big(\xi(t)-\xi^{eq}\Big), \tag{31.33}$$

$$\text{reaction type:}\quad \overset{\bullet}{\xi}(t) = \gamma(U,\boldsymbol{a},\Theta)\Big[1-\exp\Big(-\mu(t)\beta(U,\boldsymbol{a},\Theta)\Big)\Big]. \tag{31.34}$$

If for a special degree of freedom an evolution criterion exists (Muschik and Papenfuß, 1993)

$$\frac{d}{dt}\int\limits_{G(t)} \mathscr{L}(...)dV \geq 0, \tag{31.35}$$

we obtain a variational problem at equilibrium

$$\Big(\int\limits_{G(t)} \mathscr{L}(...)dV\Big)^{eq} \longrightarrow \max, \tag{31.36}$$

and the Euler-Lagrange equations of $\mathscr{L}$ result in the Landau-Ginzburg equations for the considered degree of freedom at equilibrium[13].

---

[11] This question was discussed during Gérard's stay at the Wissenschaftskolleg zu Berlin, 1991/92, (Maugin, 1999a, sect.4.7 B), (Maugin and Muschik, 1994a,b).

[12] Maugin (1999a), 3.5, 4.7.B, Maugin and Muschik (1994a)

[13] An example for a degree of freedom is the second order alignment tensor of liquid crystal theory (Muschik and Papenfuß, 1993).

31.6 Contact Temperature as an Internal Variable

When at the University of Calgary (Canada) on the first week of August, 1979, Gérard delivered a lecture on "Electromagnetic internal variables in ferroelectric and ferromagnetic continua", he started a series of papers on internal variables (Maugin, 1993, 1981a,b, 1990; Maugin and Drouot, 1983; Maugin and Morro, 1989; Maugin, 1999b) which comes to its end in 2013 (Maugin, 2013b). The concept of contact temperature is mentioned in Sect. 4.3 of Maugin (1999a), but without any connection to internal variables. In the sequel, Gérard replaced the contact temperature in the Clausius-Duhem inequality by the thermostatic temperature, thus blurring the differences between these two concepts of temperature. Beyond that, nobody was aware at that time, that despite its appearance in the entropy rate (31.9) the contact temperature may be an internal variable . This knowledge came into consideration in the course of 2012/14 (Muschik, 2014).

Starting out with the defining inequality of the contact temperature (31.3), we obtain the following constitutive equation

$$\overset{\bullet}{Q} = \kappa(T^* - \Theta), \quad \kappa = \mathcal{K}[T^* - \Theta], \quad \kappa > 0. \tag{31.37}$$

The heat exchange number κ is positive and depends on the temperature difference between the non-equilibrium system of contact temperature Θ and the equilibrium environment of thermostatic temperature T^*.

We now explain in four steps why the contact temperature is an internal variable (Muschik, 2014):

1: We consider the pure thermal contact between the non-equilibrium system and its equilibrium environment –marked by *– through an inert partition. The First Law of the equilibrium environment –the heat reservoir– is

$$\overset{\bullet}{U}{}^* = \overset{\bullet}{Q}{}^* = -\overset{\bullet}{Q}, \qquad \overset{\bullet}{\boldsymbol{a}}{}^* \equiv \mathbf{0}. \tag{31.38}$$

The caloric equation of the heat reservoir is

$$T(U^*, \boldsymbol{a}^*) = T^* \longrightarrow \frac{\partial T}{\partial U^*}(-\overset{\bullet}{Q}) = \overset{\bullet}{T}{}^* \longrightarrow \overset{\bullet}{Q} = -\frac{\overset{\bullet}{T}{}^*}{\partial T/\partial U^*}. \tag{31.39}$$

Consequently, the heat exchange can be measured by calorimetry using the heat reservoir.

2: If according to $(31.4)_3$, the net heat exchange between the non-equilibrium system and the heat reservoir vanishes, the non-equilibrium systems has by definition the contact temperature Θ. Consequently, the contact temperature is measurable, but not controllable by the heat reservoir.

3: Because T^*, $\overset{\bullet}{Q}$ and Θ are measurable quantities, so also is the function $\mathcal{K}$ is according to $(31.37)_1$ known by measurement.

4: We obtain from (31.37) the time rate of the heat exchange

$$\partial_t \overset{\bullet}{Q} = \Big(\mathscr{K}'[T^* - \Theta](T^* - \Theta) + \kappa\Big)(\overset{\bullet}{T}{}^* - \overset{\bullet}{\Theta}) \tag{31.40}$$

resulting in an evolution equation for the contact temperature

$$\overset{\bullet}{\Theta} = \overset{\bullet}{T}{}^* - \frac{\kappa \partial_t \overset{\bullet}{Q}}{\mathscr{K}' \overset{\bullet}{Q} + \kappa^2}. \tag{31.41}$$

Thus, because of **#2** and (31.41), the contact temperature Θ is an internal variable.

For the sake of simplicity, the special case of a closed discrete system is here considered. The general case of open discrete systems and of field formulation is treated in Muschik (2014), a paper which was dedicated to Gérard on the occasion of his 70th birthday in 2014[14].

Appendices

Heat Exchange and Contact Temperature

The heat exchange $\overset{\bullet}{Q}$ between the considered non-equilibrium system and the equilibrium environment of the thermostatic temperature T^* represent a two-place one-to-one mathematical correlation $\mathscr{R}$ satisfying the two statements

$$(T^*, \overset{\bullet}{Q}) \in \mathscr{R} \wedge (T^*, \overset{\bullet}{Q}_0) \in \mathscr{R} \Longrightarrow \overset{\bullet}{Q} = \overset{\bullet}{Q}_0, \tag{31.42}$$

$$(T^*, \overset{\bullet}{Q}) \in \mathscr{R} \wedge (T_0^*, \overset{\bullet}{Q}) \in \mathscr{R} \Longrightarrow T^* = T_0^*. \tag{31.43}$$

In a more physical language this means: To each temperature of the environment belongs a unique heat exchange between system and environment and vice versa.

We now introduce a temperature Θ with the property

$$(T^*, \overset{\bullet}{Q}) \in \mathscr{R} \wedge \overset{\bullet}{Q} \geq 0 \Longrightarrow \Theta \leq T^*, \tag{31.44}$$

$$(T^*, \overset{\bullet}{Q}) \in \mathscr{R} \wedge \Theta \geq T^* \Longrightarrow \overset{\bullet}{Q} \leq 0. \tag{31.45}$$

Especially for $T^* = \Theta$ follows

$$(\Theta, \overset{\bullet}{Q}) \in \mathscr{R} \wedge \overset{\bullet}{Q} \geq 0 \Longrightarrow \Theta = \Theta, \tag{31.46}$$

$$(\Theta, \overset{\bullet}{Q}) \in \mathscr{R} \wedge \Theta = \Theta \Longrightarrow \overset{\bullet}{Q} \leq 0, \tag{31.47}$$

resulting in

[14] Other papers dedicated to him are Muschik and v. Borzeszkowski (2014) and Muschik (2004).

$$(\Theta, 0) \in \mathscr{R} \tag{31.48}$$

which means: the heat exchange $\overset{\bullet}{Q}$ vanishes, if the environment has the *contact temperature* Θ and vice versa because of the one-to-one correlation $\mathscr{R}$. Although the contact temperature is by definition a thermostatic one of the equilibrium environment, we attach it to the non-equilibrium system as a non-equilibrium temperature which satisfy the defining inequality (31.3). Of course, the value of the contact temperature depends on the properties of the partition generating the contact between non-equilibrium system and heat reservoir. The terminology "contact temperature" stems from this contact depending which disappears in equilibrium.

We now consider two non-equilibrium systems –1 and 2– into contact with each other. The corresponding contact temperatures are Θ_1 and Θ_2, and the correlations are

$$(T_1^*, \overset{\bullet}{Q}_1) \in \mathscr{R}_1, \qquad (T_2^*, \overset{\bullet}{Q}_2) \in \mathscr{R}_2. \tag{31.49}$$

The defining inequalities (31.3) of the contact temperatures are

$$\overset{\bullet}{Q}_1 \left(\frac{1}{\Theta_1} - \frac{1}{T_1^*}\right) \geq 0, \qquad \overset{\bullet}{Q}_2 \left(\frac{1}{\Theta_2} - \frac{1}{T_2^*}\right) \geq 0. \tag{31.50}$$

The heat exchange $\overset{\bullet}{Q}_{12}$ between these two non-equilibrium systems takes place through an inert partition[15] resulting in

$$\overset{\bullet}{Q}_1 \equiv \overset{\bullet}{Q}_{12} \doteq -\overset{\bullet}{Q}_{21} \equiv -\overset{\bullet}{Q}_2\,. \tag{31.51}$$

Because the correlations are one-to-one, we obtain from (31.49) and (31.50)

$$(T_{12}^*, \overset{\bullet}{Q}_{12}) \in \mathscr{R}_1, \qquad (T_{21}^*, \overset{\bullet}{Q}_{21}) \in \mathscr{R}_2, \tag{31.52}$$

$$\overset{\bullet}{Q}_{12} \left(\frac{1}{\Theta_1} - \frac{1}{T_{12}^*}\right) \geq 0, \qquad -\overset{\bullet}{Q}_{12} \left(\frac{1}{\Theta_2} - \frac{1}{T_{21}^*}\right) \geq 0. \tag{31.53}$$

Addition of $(31.53)_1$ and $(31.53)_2$ results in

$$\overset{\bullet}{Q}_{12} \left(\frac{1}{\Theta_1} - \frac{1}{\Theta_2}\right) \geq \overset{\bullet}{Q}_{12} \left(\frac{1}{T_{12}^*} - \frac{1}{T_{21}^*}\right). \tag{31.54}$$

Consequently, the reservoir temperature T_1^* in the defining inequality $(31.50)_1$ cannot be replaced by the contact temperature of system #2, because the RHS of (31.54) does not vanish. The endoreversible case is defined by replacing the contact temperatures by thermostatic ones. That means, reservoirs are contacting to each other in the endoreversible case:

$$\Theta_1 \doteq T_1, \qquad \Theta_2 \doteq T_2, \tag{31.55}$$

$$T_{12}^* \doteq T_2, \qquad T_{21}^* \doteq T_1. \tag{31.56}$$

---

[15] which is impervious to work and material

Consequently, we obtain from (31.54) to (31.56)

$$\overset{\bullet}{Q}_{12}\left(\frac{1}{T_1}-\frac{1}{T_2}\right) \geq \overset{\bullet}{Q}_{12}\left(\frac{1}{T_2}-\frac{1}{T_1}\right) \quad \longrightarrow \quad \overset{\bullet}{Q}_{12}\left(\frac{1}{T_1}-\frac{1}{T_2}\right) \geq 0, \tag{31.57}$$

an inequality which is well known in thermostatics. These considerations show that the contact temperature is a useful tool in non-equilibrium thermodynamics.

Contact Temperature and Efficiency

We consider a cyclic, power-producing process of a closed discrete non-equilibrium system which works between two heat reservoirs of constant thermostatic temperatures $T_H^* > T_L^*$ (Muschik, 2009). The contact temperatures of the two contacts between the system and the corresponding reservoirs are $\Theta_H(t)$ and $\Theta_L(t)$, the heat exchanges through the inertial contacts are

$$\overset{\bullet}{Q}{}_H^*(t) < 0$$

and

$$\overset{\bullet}{Q}{}_L^*(t) > 0,$$

relative to the heat reservoirs. According to the defining inequality (31.3), we obtain for the two heat reservoirs

$$\begin{aligned} \left(\frac{1}{T_H^*}-\frac{1}{\Theta_H}\right)\overset{\bullet}{Q}{}_H^* &\geq 0, \\ \left(\frac{1}{T_L^*}-\frac{1}{\Theta_L}\right)\overset{\bullet}{Q}{}_L^* &\geq 0, \end{aligned} \tag{31.58}$$

resulting in

$$T_H^* \geq \Theta_H, \qquad \Theta_L \geq T_L^*. \tag{31.59}$$

Integration over the cycle time yields

$$\oint \frac{\overset{\bullet}{Q}{}_H^*}{\Theta_H}dt \leq \frac{1}{T_H^*}\oint \overset{\bullet}{Q}{}_H^* dt \;=:\; \frac{1}{T_H^*}Q_H^*, \tag{31.60}$$

$$\oint \frac{\overset{\bullet}{Q}{}_L^*}{\Theta_L}dt \leq \frac{1}{T_L^*}\oint \overset{\bullet}{Q}{}_L^* dt \;=:\; \frac{1}{T_L^*}Q_L^*. \tag{31.61}$$

The mean value theorem applied to $(31.60)_1$ and $(31.61)_1$ results in

$$\oint \frac{\overset{\bullet}{Q}{}^*_H}{\Theta_H} dt = \frac{Q^*_H}{[\Theta_H]} \leq \frac{1}{T^*_H} Q^*_H \longrightarrow T^*_H \geq [\Theta_H], \tag{31.62}$$

$$\oint \frac{\overset{\bullet}{Q}{}^*_L}{\Theta_L} dt = \frac{Q^*_L}{[\Theta_L]} \leq \frac{1}{T^*_L} Q^*_L \longrightarrow T^*_L \leq [\Theta_L], \tag{31.63}$$

Here, the square brackets denote mean values over the cyclic process which are defined by $(31.62)_1$ and $(31.63)_1$. We obtain the following estimation of the Carnot efficiency according to $(31.62)_3$ and $(31.63)_3$

$$\eta_{Car} = 1 - \frac{T^*_L}{T^*_H} \geq 1 - \frac{[\Theta_L]}{[\Theta_H]} =: \eta_{neq}. \tag{31.64}$$

The non-equilibrium efficiency η_{neq} is smaller or equal to the Carnot efficiency which belongs to reversible processes in contrast to η_{neq} which is a more realistic efficiency.

An essential presupposition for the considerations above is according to (31.59)

$$T^*_H \geq \Theta_H(t), \qquad \Theta_L(t) \geq T^*_L, \tag{31.65}$$

that the contact temperatures during the non-equilibrium process satisfy (31.65) for all times.

We now consider the First and the Second Law with respect to the contact temperature. The First Law per cycle of the considered power-producing non-equilibrium system runs as follows

$$Q_H + Q_L + W = 0 \quad \longrightarrow \quad Q^*_H + Q^*_L = W < 0 \quad \longrightarrow \quad Q^*_H = W - Q^*_L < -Q^*_L, \tag{31.66}$$

and the Clausius inequality (Second Law) $(31.13)_1$ becomes by use of the mean value theorem

$$0 \leq \oint \frac{\overset{\bullet}{Q}{}^*}{\Theta} dt = \oint \frac{\overset{\bullet}{Q}{}^*_H}{\Theta_H} dt + \oint \frac{\overset{\bullet}{Q}{}^*_L}{\Theta_L} dt = \frac{Q^*_H}{[\Theta_H]} + \frac{Q^*_L}{[\Theta_L]}. \tag{31.67}$$

Taking $(31.66)_3$ into account, we obtain an inequality of the mean values of the contact temperatures belonging to the cycle

$$0 \leq \Big(-\frac{1}{[\Theta_H]} + \frac{1}{[\Theta_L]}\Big) Q^*_L \quad \longrightarrow \quad [\Theta_H] > [\Theta_L] \quad \longrightarrow \quad \eta_{Car} \geq \eta_{neq} > 0, \tag{31.68}$$

and the positive definiteness of η_{neq} which is according to $(31.68)_3$ a more realistic efficiency in comparison with that of Carnot.

References

Bridgeman PW (1943) The Nature of Thermodynamics. Harvard University Press, Cambridge (Mass)

Kestin J (1966) A Course in Thermodynamics. Blaisdell, Waltham Mass.

Lieneweg F (1976) Handbuch der Technischen Temperaturmessung. Vieweg, Braunschweig

Maugin G, Drouot R (1983) Internal variables and the thermodynamics of macromolecule solutions. International Journal of Engineering Science 21(7):705–724

Maugin G, Morro A (1989) Viscoelastic materials with internal variables and dissipation functions. Acta Phys Hung 66:69–78

Maugin GA (1981a) Electromagnetic internal variables in electromagnetic continua. Arch Mech 33:927–935

Maugin GA (1981b) The notion of internal variable in fluid mechanics. In: Complex Fluids and Structurally Complex Flows, Theoretical and Applied Mechanics, Bulgarian Acad.Sci., Sofia, pp 671–676

Maugin GA (1990) Internal variables and dissipative structures. Journal of Non-Equilibrium Thermodynamics 15(2):173–192

Maugin GA (1993) Vectorial internal variables in magnetoelasticity. J Méchanique 18:541–563

Maugin GA (1999a) The Thermomechanics of Nonlinear Irreversible Behaviors, An Introduction. World Scientific, Singapore

Maugin GA (1999b) The Thermomechanics of Nonlinear Irreversible Behaviors. Chap. 4. World Scientific, Singapore

Maugin GA (2013a) Continuum Mechanics Through the Twentieth Century. Springer, Dordrecht

Maugin GA (2013b) Continuum Mechanics Through the Twentieth Century. Chap. 5.6. Springer, Dordrecht

Maugin GA, Muschik W (1994a) Thermodynamics with internal variables. I. General Concepts. Journal of Non-Equilibrium Thermodynamics 19(3):217–249

Maugin GA, Muschik W (1994b) Thermodynamics with internal variables. II. Applications. Journal of Non-Equilibrium Thermodynamics 19(3):250–289

Meixner J (1943) Zur Thermodynamik der irreversiblen Prozesse. Z Phys Chem 538:235–263

Muschik W (1977) Empirical foundation and axiomatic treatment of non-equilibrium temperature. Archive for Rational Mechanics and Analysis 66(4):379–401

Muschik W (1990a) Aspects of Non-Equilibrium Thermodynamics. World Scientific, Singapore

Muschik W (1990b) Internal Variables in Non-Equilibrium Thermodynamics. Journal of Non-Equilibrium Thermodynamics 15(2):127–137

Muschik W (1993) Fundamentals of nonequilibrium thermodynamics. In: Muschik W (ed) Non-Equilibrium Thermodynamics with Applications to Solids, Springer, Wien, CISM Courses and Lectures, vol 336

Muschik W (2004) Gérard A. Maugin, 60 years young. Journal of Non-Equilibrium Thermodynamics 29(2):205–208

Muschik W (2009) Contact Quantities and Non-Equilibrium Entropy of Discrete Systems. Journal of Non-Equilibrium Thermodynamics 34:75–92

Muschik W (2014) Contact temperature and internal variables: A glance back, 20 years later. Journal of Non-Equilibrium Thermodynamics 39(3):113–121

Muschik W, Berezovski A (2004) Thermodynamic interaction between two discrete systems in non-equilibrium. Journal of Non-Equilibrium Thermodynamics 29(3):237–255

Muschik W, Brunk G (1977) A concept of non-equilibrium temperature. International Journal of Engineering Science 15(6):377–389

Muschik W, Papenfuß C (1993) An evolution criterion of nonequilibrium thermodynamics and its application to liquid crystals. Physica A: Statistical Mechanics and its Applications 201(4):515–526

Muschik W, v Borzeszkowski HH (2014) Exploitation of the dissipation inequality in general relativistic continuum thermodynamics. Archive of Applied Mechanics 84(9):1517–1531

Chapter 32
Angular Velocities, Twirls, Spins and Rotation Tensors in the Continuum Mechanics Revisited

Konstantin Naumenko and Holm Altenbach

Abstract In the classical continuum mechanics several quantities related to angular velocity of rotation are introduced. Examples include vorticity vector, twirl tensors and logarithmic spin. Furthermore the corresponding rotation tensors can be defined to capture the orientation of triads. All of these quantities are measures of accompanying rotational motion and can be related to the deformation and velocity gradient. Such relationships are crucial for constitutive modeling of material behavior. The aim of this contribution is to recall classical definitions of rotation-like quantities and to present several new relationships between them.

32.1 Introduction

In the classical continuum mechanics several quantities related to angular velocity of rotation are introduced. Examples include vorticity vector, twirl tensors and logarithmic spin. Furthermore the correspondent rotation tensors can be defined to capture the orientation of triads. Since all of these quantities are derived from motion and deformation, relationships between them exist. Such relationships are crucial for constitutive modeling of complex material behavior. For example, in finite plasticity, the multiplicative decomposition of the deformation gradient into the elastic and inelastic parts is frequently applied. Here several rotation tensors and spins corresponding to elastic and plastic parts can be introduced, e.g. Altenbach and Eremeyev (2014); Bertram (2012); Maugin (1992, 2013); Naumenko and Altenbach (2016). Physically sound definitions satisfying the objectivity requirements are only available for specific constitutive theories, e.g. crystal plasticity.

Konstantin Naumenko · Holm Altenbach
Institut für Mechanik, Otto-von-Guericke-Universität Magdeburg, Universitätsplatz 2, D-39106 Magdeburg, Germany
e-mail: konstantin.naumenko@ovgu.de, holm.altenbach@ovgu.de

© Springer International Publishing AG, part of Springer Nature 2018
H. Altenbach et al. (eds.), *Generalized Models and Non-classical Approaches in Complex Materials 1*, Advanced Structured Materials 89,
https://doi.org/10.1007/978-3-319-72440-9_32

Another example is related to anisotropic materials with evolving microstructure, e.g. short-fiber suspensions and composites (Altenbach et al, 2003a,b, 2007). Here additional tensorial state variables for microstructure orientation states are defined by means of evolution equations. The rotation tensor and the angular velocity vector of rigid particles in a suspension must be defined in order to derive the evolution equations for state variables. The aim of this contribution is to recall all classical definitions of rotation-like quantities in the continuum mechanics and to present several relationships between them.

32.2 Rotation Tensor and Angular Velocity Vector

In order to fix our notation let us recall several definitions from rigid body kinematics. Let $\boldsymbol{N}_k, k = 1,2,3$, $(\boldsymbol{N}_1 \times \boldsymbol{N}_2)\cdot\boldsymbol{N}_3 = 1$ be a triple of orthogonal unit vectors attached to a point of a body in a reference configuration. Let $\boldsymbol{n}_k(t), k = 1,2,3$ be the actual triple of orthogonal unit vectors obtained from $\boldsymbol{N}_k$ by a rigid body rotation, i.e. the length and the angles between the vectors does not change during the motion and $(\boldsymbol{n}_1 \times \boldsymbol{n}_2)\cdot\boldsymbol{n}_3 = 1$. To describe such a rotation of the triad the proper orthogonal tensor $\boldsymbol{R}$ can be introduced as follows

$$\boldsymbol{R}(t) = \boldsymbol{n}_1(t)\otimes\boldsymbol{N}_1 + \boldsymbol{n}_2(t)\otimes\boldsymbol{N}_2 + \boldsymbol{n}_3(t)\otimes\boldsymbol{N}_3, \quad \boldsymbol{n}_k(t) = \boldsymbol{R}(t)\cdot\boldsymbol{N}_k$$

$$\boldsymbol{R}\cdot\boldsymbol{R}^{\mathrm{T}} = \boldsymbol{R}^{\mathrm{T}}\cdot\boldsymbol{R} = \boldsymbol{I}, \quad \det\boldsymbol{R} = 1,$$

where $\boldsymbol{I}$ is the second rank unit tensor. For the rotation tensor $\boldsymbol{R}$ the angular velocity vector $\boldsymbol{\Omega}_R$ and the spin tensor $\boldsymbol{\Omega}_R \times \boldsymbol{I}$ are introduced as follows. According to the definition of the orthogonal tensor we obtain

$$\boldsymbol{R}\cdot\boldsymbol{R}^{\mathrm{T}} = \boldsymbol{I} \quad \Rightarrow \quad \dot{\boldsymbol{R}}\cdot\boldsymbol{R}^{\mathrm{T}} + \boldsymbol{R}\cdot\dot{\boldsymbol{R}}^{\mathrm{T}} = \boldsymbol{0} \quad \Rightarrow \quad \dot{\boldsymbol{R}}\cdot\boldsymbol{R}^{\mathrm{T}} = -(\dot{\boldsymbol{R}}\cdot\boldsymbol{R}^{\mathrm{T}})^{\mathrm{T}}$$

The skew-symmetric tensor $\boldsymbol{W}_R = \dot{\boldsymbol{R}}\cdot\boldsymbol{R}^{\mathrm{T}}$ is called the left spin tensor or simply spin tensor. In the rigid body kinematics the associated vector $\boldsymbol{\Omega}_R$ called the (left) angular velocity vector is introduced as follows

$$\boldsymbol{W}_R = \boldsymbol{\Omega}_R \times \boldsymbol{I} = \boldsymbol{I} \times \boldsymbol{\Omega}_R, \quad \boldsymbol{\Omega}_R = -\frac{1}{2}(\boldsymbol{W})_\times, \tag{32.1}$$

where $(\ldots)_\times$ denotes the vector invariant or the Gibbs cross operation applied to a second rank tensor, see Subsect. 32.4. The axial vector $\boldsymbol{\Omega}_R$ is widely used in the rigid body dynamics, e.g. Altenbach et al (2007); Zhilin (1996). If the rotation tensor $\boldsymbol{R}$ is given, then the corresponding angular velocity of rotation is determined as follows

$$\boldsymbol{\Omega}_R = -\frac{1}{2}(\dot{\boldsymbol{R}}\cdot\boldsymbol{R}^{\mathrm{T}})_\times \tag{32.2}$$

For the given angular velocity of rotation $\boldsymbol{\Omega}_{\boldsymbol{R}}(t)$ the rotation tensor can be computed by solving the Darboux problem

$$\dot{\boldsymbol{R}} = \boldsymbol{\Omega}_{\boldsymbol{R}} \times \boldsymbol{R}, \quad \boldsymbol{R}(0) = \boldsymbol{R}_0, \tag{32.3}$$

where $\boldsymbol{R}_0$ is the initial condition.

32.3 Rotation Tensors and Spins in the Classical Continuum Mechanics

In the Cosserat-type theories of continuum (Maugin, 2014, 2016; Eremeyev et al, 2013) and in theories of shells (Altenbach et al, 2005; Antman, 1995; Libai and Simmonds, 1998) the rotation tensor and the angular velocity of rotation are introduced as independent quantities to capture the motion of triads. In the classical continuum mechanics rotation tensors and angular velocities are measures of accompanying rotational motion and can be related to the deformation and velocity gradient. The spatial velocity gradient $\boldsymbol{L}$ is defined as follows

$$\boldsymbol{L} = (\nabla \boldsymbol{v})^{\mathrm{T}} = \dot{\boldsymbol{F}} \cdot \boldsymbol{F}^{-1}, \tag{32.4}$$

where $\boldsymbol{v}$ is the velocity vector and $\boldsymbol{F}$ is the deformation gradient. The tensor $\boldsymbol{L}$ can be additively decomposed into the symmetric and skew symmetric parts

$$\boldsymbol{L} = \boldsymbol{D} + \boldsymbol{\omega} \times \boldsymbol{I}, \tag{32.5}$$

where the symmetric part

$$\boldsymbol{D} = \frac{1}{2}\left[\nabla \boldsymbol{v} + (\nabla \boldsymbol{v})^{\mathrm{T}}\right]$$

is called the deformation rate tensor while

$$\boldsymbol{\omega} = -\frac{1}{2}\nabla \times \boldsymbol{v}$$

is called vorticity vector.

Applying the polar decomposition theorem relations

$$\boldsymbol{F} = \boldsymbol{R} \cdot \boldsymbol{U} = \boldsymbol{V} \cdot \boldsymbol{R} \quad \Rightarrow \quad \dot{\boldsymbol{F}} = \dot{\boldsymbol{R}} \cdot \boldsymbol{U} + \dot{\boldsymbol{U}} \cdot \boldsymbol{R} \quad \text{and} \quad \boldsymbol{F}^{-1} = \boldsymbol{U}^{-1} \cdot \boldsymbol{R}^{\mathrm{T}},$$

where $\boldsymbol{U}$ is the right stretch tensor, $\boldsymbol{V}$ is the left stretch tensor and $\boldsymbol{R}$ is the rotation tensor the velocity gradient can be given as follows

$$\boldsymbol{L} = \dot{\boldsymbol{F}} \cdot \boldsymbol{F}^{-1} = \dot{\boldsymbol{R}} \cdot \boldsymbol{R}^{\mathrm{T}} + \boldsymbol{R} \cdot \dot{\boldsymbol{U}} \cdot \boldsymbol{U}^{-1} \cdot \boldsymbol{R}^{\mathrm{T}} \tag{32.6}$$

With Eq. (32.5) Eq. (32.6) takes the form

$$\boldsymbol{D}+\boldsymbol{\omega}\times\boldsymbol{I}=\boldsymbol{\Omega}_{\boldsymbol{R}}\times\boldsymbol{I}+\boldsymbol{R}\cdot\dot{\boldsymbol{U}}\cdot\boldsymbol{U}^{-1}\cdot\boldsymbol{R}^{\mathrm{T}}, \tag{32.7}$$

where $\boldsymbol{\Omega}_{\boldsymbol{R}}$ is the angular velocity vector corresponding to $\boldsymbol{R}$. The tensor $\boldsymbol{\Omega}_{\boldsymbol{R}}\times\boldsymbol{I}$ is sometimes called body spin, e.g. Reinhardt and Dubey (1996). By analogy let us call $\boldsymbol{\Omega}_{\boldsymbol{R}}$ the body angular velocity. The vector invariant of Eq. (32.7) provides the vorticity vector as follows

$$\boldsymbol{\omega}=\boldsymbol{\Omega}_{\boldsymbol{R}}-\frac{1}{2}(\boldsymbol{R}\cdot\dot{\boldsymbol{U}}\cdot\boldsymbol{U}^{-1}\cdot\boldsymbol{R}^{\mathrm{T}})_{\times} \tag{32.8}$$

The symmetric part of Eq (32.7) is

$$\boldsymbol{D}=\frac{1}{2}\boldsymbol{R}\cdot\boldsymbol{U}^{-1}\cdot\dot{\boldsymbol{U}}\cdot\boldsymbol{R}^{\mathrm{T}}=\frac{1}{2}\boldsymbol{R}\cdot(\dot{\boldsymbol{U}}\cdot\boldsymbol{U}^{-1}+\boldsymbol{U}^{-1}\cdot\dot{\boldsymbol{U}})\cdot\boldsymbol{R}^{\mathrm{T}} \tag{32.9}$$

Equation (32.9) can be put in the following form

$$\boldsymbol{R}^{\mathrm{T}}\cdot\boldsymbol{D}\cdot\boldsymbol{R}=\frac{1}{2}(\dot{\boldsymbol{U}}\cdot\boldsymbol{U}^{-1}+\boldsymbol{U}^{-1}\cdot\dot{\boldsymbol{U}})$$

or

$$\boldsymbol{F}^{\mathrm{T}}\cdot\boldsymbol{D}\cdot\boldsymbol{F}=\frac{1}{2}\frac{\mathrm{d}}{\mathrm{d}t}(\boldsymbol{U}^2)=\frac{1}{2}\dot{\boldsymbol{C}}, \tag{32.10}$$

where $\boldsymbol{C}$ is the right Cauchy-Green tensor. Let us take the time derivatives of stretch tensors applying the spectral representations[1]

$$\begin{aligned}\dot{\boldsymbol{U}}&=\sum_{i=1}^{3}\left(\dot{\lambda}_i\overset{U}{\boldsymbol{N}}_i\otimes\overset{U}{\boldsymbol{N}}_i+\lambda_i\frac{\mathrm{d}}{\mathrm{d}t}\overset{U}{\boldsymbol{N}}_i\otimes\overset{U}{\boldsymbol{N}}_i+\lambda_i\overset{U}{\boldsymbol{N}}_i\otimes\frac{\mathrm{d}}{\mathrm{d}t}\overset{U}{\boldsymbol{N}}_i\right),\\ \dot{\boldsymbol{V}}&=\sum_{i=1}^{3}\left(\dot{\lambda}_i\overset{V}{\boldsymbol{n}}_i\otimes\overset{V}{\boldsymbol{n}}_i+\lambda_i\frac{\mathrm{d}}{\mathrm{d}t}\overset{V}{\boldsymbol{n}}_i\otimes\overset{V}{\boldsymbol{n}}_i+\lambda_i\overset{V}{\boldsymbol{n}}_i\otimes\frac{\mathrm{d}}{\mathrm{d}t}\overset{V}{\boldsymbol{n}}_i\right),\end{aligned} \tag{32.11}$$

where λ_i are principal stretches, $\overset{U}{\boldsymbol{N}}_i$ are eigenvectors of $\boldsymbol{U}$ and $\overset{V}{\boldsymbol{n}}_i$ are eigenvectors of the tensor $\boldsymbol{V}$.

32.3.1 Rotations of Principal Directions and Twirls

Consider a triple of fixed orthogonal unit vectors $\boldsymbol{e}_i$ and the rotation tensor $\boldsymbol{P}_U$ such that

$$\overset{U}{\boldsymbol{N}}_i=\boldsymbol{P}_U\cdot\boldsymbol{e}_i$$

Hence

$$\overset{V}{\boldsymbol{n}}_i=\boldsymbol{R}\cdot\boldsymbol{P}_U\cdot\boldsymbol{e}_i$$

[1] In the sequel we assume that the tensors $\boldsymbol{U}$ and $\boldsymbol{V}$ have distinct eigenvalues. The cases of coincident eigenvalues should be analyzed separately.

or

$$\overset{V}{\boldsymbol{n}}_i = \boldsymbol{P}_V \cdot \boldsymbol{e}_i, \quad \boldsymbol{P}_V = \boldsymbol{R} \cdot \boldsymbol{P}_U \tag{32.12}$$

For the rotation tensors $\boldsymbol{P}_U$ and $\boldsymbol{P}_V$ the spin tensors and the angular velocity vectors can be introduced as follows

$$\begin{aligned} \dot{\boldsymbol{P}}_U \cdot \boldsymbol{P}_U^{\mathrm{T}} &= \boldsymbol{\Omega}_U \times \boldsymbol{I}, \quad \dot{\boldsymbol{P}}_U = \boldsymbol{\Omega}_U \times \boldsymbol{P}_U, \\ \dot{\boldsymbol{P}}_V \cdot \boldsymbol{P}_V^{\mathrm{T}} &= \boldsymbol{\Omega}_V \times \boldsymbol{I}, \quad \dot{\boldsymbol{P}}_V = \boldsymbol{\Omega}_V \times \boldsymbol{P}_V \end{aligned} \tag{32.13}$$

The tensors $\boldsymbol{\Omega}_U \times \boldsymbol{I}$ and $\boldsymbol{\Omega}_V \times \boldsymbol{I}$ are called Lagrangian and Eulerian twirls, respectively (Reinhardt and Dubey, 1996; Zhong-Heng et al, 1992). By analogy let us call the vectors $\boldsymbol{\Omega}_U$ and $\boldsymbol{\Omega}_V$ twirl angular velocities.

The time derivative of Eq $(32.12)_2$ yields

$$\begin{aligned} \dot{\boldsymbol{P}}_V &= \dot{\boldsymbol{R}} \cdot \boldsymbol{P}_U - \boldsymbol{R} \cdot \dot{\boldsymbol{P}}_U \\ &= \boldsymbol{\Omega}_R \times \boldsymbol{R} \cdot \boldsymbol{P}_U + \boldsymbol{R} \cdot (\boldsymbol{\Omega}_U \times \boldsymbol{I}) \cdot \boldsymbol{R}^{\mathrm{T}} \cdot \boldsymbol{R} \cdot \boldsymbol{P}_U \\ &= (\boldsymbol{\Omega}_R + \boldsymbol{R} \cdot \boldsymbol{\Omega}_U) \times \boldsymbol{P}_V \end{aligned}$$

Hence the following relationship between the angular velocity vectors can be established

$$\boldsymbol{\Omega}_V = \boldsymbol{\Omega}_R + \boldsymbol{R} \cdot \boldsymbol{\Omega}_U \tag{32.14}$$

With Eqs. (32.13) the rates of change of principal directions can be computed as follows

$$\frac{\mathrm{d}}{\mathrm{d}t} \overset{U}{\boldsymbol{N}}_i = \boldsymbol{\Omega}_U \times \overset{U}{\boldsymbol{N}}_i, \quad \frac{\mathrm{d}}{\mathrm{d}t} \overset{V}{\boldsymbol{n}}_i = \boldsymbol{\Omega}_V \times \overset{V}{\boldsymbol{n}}_i$$

Consequently the rates of change of stretch tensors (32.11) take the following form

$$\begin{aligned} \dot{\boldsymbol{U}} &= \sum_{i=1}^{3} \dot{\lambda}_i \overset{U}{\boldsymbol{N}}_i \otimes \overset{U}{\boldsymbol{N}}_i + \boldsymbol{\Omega}_U \times \boldsymbol{U} - \boldsymbol{U} \times \boldsymbol{\Omega}_U, \\ \dot{\boldsymbol{V}} &= \sum_{i=1}^{3} \dot{\lambda}_i \overset{V}{\boldsymbol{n}}_i \otimes \overset{V}{\boldsymbol{n}}_i + \boldsymbol{\Omega}_V \times \boldsymbol{V} - \boldsymbol{V} \times \boldsymbol{\Omega}_V \end{aligned} \tag{32.15}$$

With Eq. (32.15) and

$$\boldsymbol{U}^{-1} = \sum_{i=1}^{3} \frac{1}{\lambda_i} \overset{U}{\boldsymbol{N}}_i \otimes \overset{U}{\boldsymbol{N}}_i,$$

one may compute

$$\boldsymbol{R} \cdot \dot{\boldsymbol{U}} \cdot \boldsymbol{U}^{-1} \cdot \boldsymbol{R}^{\mathrm{T}} = \boldsymbol{R} \cdot \left[\sum_{i=1}^{3} \dot{\lambda}_i \lambda_i^{-1} \overset{U}{\boldsymbol{N}}_i \otimes \overset{U}{\boldsymbol{N}}_i + \boldsymbol{\Omega}_U \times \boldsymbol{I} - (\boldsymbol{U} \times \boldsymbol{\Omega}_U) \cdot \boldsymbol{U} \right] \cdot \boldsymbol{R}^{\mathrm{T}}$$

Applying Eqs. (32.12) and (32.14) we obtain

$$\begin{aligned}\boldsymbol{R}\cdot\dot{\boldsymbol{U}}\cdot\boldsymbol{U}^{-1}\cdot\boldsymbol{R}^{\mathrm{T}} &= \sum_{i=1}^{3}\dot{\lambda}_i\lambda_i^{-1}\overset{V}{\boldsymbol{n}}_i\otimes\overset{V}{\boldsymbol{n}}_i \\ &+ (\boldsymbol{\Omega}_V-\boldsymbol{\Omega}_R)\times\boldsymbol{I}-\boldsymbol{V}\cdot[(\boldsymbol{\Omega}_V-\boldsymbol{\Omega}_R)\times\boldsymbol{I}]\cdot\boldsymbol{V}^{-1}\end{aligned} \tag{32.16}$$

With identities (32.38) and (32.39) the vector invariant of Eq. (32.16) yields

$$(\boldsymbol{R}\cdot\dot{\boldsymbol{U}}\cdot\boldsymbol{U}^{-1}\cdot\boldsymbol{R}^{\mathrm{T}})_\times = -2(\boldsymbol{\Omega}_V-\boldsymbol{\Omega}_R)-\boldsymbol{A}_V\cdot(\boldsymbol{\Omega}_V-\boldsymbol{\Omega}_R), \tag{32.17}$$

where

$$\boldsymbol{A}_V=\sum_{i_1}^{3}\lambda_i\overset{V}{\boldsymbol{n}}_i\times\boldsymbol{V}^{-1}\times\overset{V}{\boldsymbol{n}}_i=\sum_{i=1}^{3}\sum_{j=1}^{3}\frac{\lambda_i}{\lambda_j}\overset{V}{\boldsymbol{n}}_i\times\overset{V}{\boldsymbol{n}}_j\otimes\overset{V}{\boldsymbol{n}}_j\times\overset{V}{\boldsymbol{n}}_i$$

According to (32.39) and the Cayley-Hamilton theorem the tensor $\boldsymbol{A}_V$ has the following representations

$$\boldsymbol{A}_V=J^{-1}\boldsymbol{V}\cdot[\boldsymbol{V}^2-(\operatorname{tr}\boldsymbol{V}^2)\boldsymbol{I}]=\boldsymbol{I}+\frac{J_{2_V}-J_{1_V}^2}{J}\boldsymbol{V}+\frac{J_{1_V}}{J}\boldsymbol{V}^2$$

where J_{1_V}, J_{2_V} and $J=J_{3_V}$ are principal invariants of the tensor $\boldsymbol{V}$. The spectral form of the tensor $\boldsymbol{A}_V$ is

$$-\boldsymbol{A}_V=\frac{\lambda_2^2+\lambda_3^2}{\lambda_2\lambda_3}\overset{V}{\boldsymbol{n}}_1\otimes\overset{V}{\boldsymbol{n}}_1+\frac{\lambda_3^2+\lambda_1^2}{\lambda_3\lambda_1}\overset{V}{\boldsymbol{n}}_2\otimes\overset{V}{\boldsymbol{n}}_2+\frac{\lambda_1^2+\lambda_2^2}{\lambda_1\lambda_2}\overset{V}{\boldsymbol{n}}_3\otimes\overset{V}{\boldsymbol{n}}_3 \tag{32.18}$$

With Eqs (32.8) and (32.17) the following relationship between the angular velocities can be obtained

$$\boldsymbol{\omega}=\boldsymbol{\Omega}_V+\frac{1}{2}\boldsymbol{A}_V\cdot(\boldsymbol{\Omega}_V-\boldsymbol{\Omega}_R) \tag{32.19}$$

The relationship (32.19) can also be derived from the polar decomposition

$$\boldsymbol{F}=\boldsymbol{V}\cdot\boldsymbol{R} \quad\Rightarrow\quad \boldsymbol{F}^{-1}=\boldsymbol{R}^{\mathrm{T}}\cdot\boldsymbol{V}^{-1}$$

such that the velocity gradient takes the form

$$\boldsymbol{L}=\dot{\boldsymbol{F}}\cdot\boldsymbol{F}^{-1}=\dot{\boldsymbol{V}}\cdot\boldsymbol{V}^{-1}+\boldsymbol{V}\cdot\dot{\boldsymbol{R}}\cdot\boldsymbol{R}^{\mathrm{T}}\cdot\boldsymbol{V}^{-1} \tag{32.20}$$

With Eqs (32.1) and (32.15), Eq. (32.20) takes the form

$$\boldsymbol{L}=\sum_{i=1}^{3}\dot{\lambda}_i\lambda_i^{-1}\overset{V}{\boldsymbol{n}}_i\otimes\overset{V}{\boldsymbol{n}}_i+\boldsymbol{\Omega}_V\times\boldsymbol{I}+\boldsymbol{L}_\Omega, \tag{32.21}$$

where

$$\boldsymbol{L}_\Omega=\boldsymbol{V}\cdot(\tilde{\boldsymbol{\Omega}}\times\boldsymbol{I})\cdot\boldsymbol{V}^{-1}, \quad \tilde{\boldsymbol{\Omega}}=\boldsymbol{\Omega}_R-\boldsymbol{\Omega}_V$$

The vector invariant of Eq. (32.21) provides the relationship (32.19).

With the identity $(32.36)_2$ the tensor $\boldsymbol{L}_\Omega$ can be represented as follows

$$\begin{aligned} \boldsymbol{L}_{\boldsymbol{\Omega}} &= \boldsymbol{a} \times \boldsymbol{V}^{-2} = \boldsymbol{V}^2 \times \boldsymbol{b}, \\ \boldsymbol{a} &= J\boldsymbol{V}^{-1} \cdot \tilde{\boldsymbol{\Omega}}, \quad \boldsymbol{b} = J^{-1}\boldsymbol{V} \cdot \tilde{\boldsymbol{\Omega}} \end{aligned} \tag{32.22}$$

The right dot product of Eq. (32.21) with $\boldsymbol{V}^2$ yields

$$\boldsymbol{L} \cdot \boldsymbol{V}^2 = \sum_{i=1}^{3} \dot{\lambda}_i \lambda_i \overset{V}{\boldsymbol{n}}_i \otimes \overset{V}{\boldsymbol{n}}_i + \boldsymbol{\Omega}_V \times \boldsymbol{V}^2 + \boldsymbol{a} \times \boldsymbol{I}, \tag{32.23}$$

With the decomposition of the velocity gradient (32.5), Eq. (32.23) takes the following form

$$\boldsymbol{D} \cdot \boldsymbol{V}^2 = \sum_{i=1}^{3} \dot{\lambda}_i \lambda_i \overset{V}{\boldsymbol{n}}_i \otimes \overset{V}{\boldsymbol{n}}_i + (\boldsymbol{\Omega}_V - \boldsymbol{\omega}) \times \boldsymbol{V}^2 + \boldsymbol{a} \times \boldsymbol{I} \tag{32.24}$$

Taking the vector invariant of Eq. (32.24) yields

$$\frac{1}{2J} \boldsymbol{V} \cdot (\boldsymbol{D} \cdot \boldsymbol{V}^2)_\times = \left(\boldsymbol{I} + \frac{1}{2} \boldsymbol{A}_V \right) \cdot \boldsymbol{\Omega}_V - \frac{1}{2} \boldsymbol{A}_V \cdot \boldsymbol{\omega} - \boldsymbol{\Omega}_R \tag{32.25}$$

From Eqs. (32.19) and (32.25) we obtain

$$\frac{1}{2J} \boldsymbol{V} \cdot (\boldsymbol{D} \cdot \boldsymbol{V}^2)_\times = \left(\boldsymbol{I} - \frac{1}{2} \boldsymbol{A}_V \right) \cdot (\boldsymbol{\omega} - \boldsymbol{\Omega}_R)$$

With Eq. (32.18) one may verify the tensor $\boldsymbol{I} - \frac{1}{2}\boldsymbol{A}_V$ is non-singular. Hence

$$\boldsymbol{\omega} = \boldsymbol{\Omega}_R + \boldsymbol{K}_V \cdot (\boldsymbol{D} \cdot \boldsymbol{V}^2)_\times, \quad \boldsymbol{K}_V = \frac{1}{2J} \left(\boldsymbol{I} - \frac{1}{2} \boldsymbol{A}_V \right)^{-1} \cdot \boldsymbol{V} \tag{32.26}$$

Applying Eq. (32.18) the following spectral representation of the tensor $\boldsymbol{K}_V$ can be established

$$\boldsymbol{K}_V = \frac{1}{(\lambda_2 + \lambda_3)^2} \overset{V}{\boldsymbol{n}}_1 \otimes \overset{V}{\boldsymbol{n}}_1 + \frac{1}{(\lambda_3 + \lambda_1)^2} \overset{V}{\boldsymbol{n}}_2 \otimes \overset{V}{\boldsymbol{n}}_2 + \frac{1}{(\lambda_1 + \lambda_2)^2} \overset{V}{\boldsymbol{n}}_3 \otimes \overset{V}{\boldsymbol{n}}_3 \tag{32.27}$$

Equation (32.26) is the relationship between the vorticity vector and the body angular velocity vector. It shows that if the vector $(\boldsymbol{D} \cdot \boldsymbol{V}^2)_\times = \boldsymbol{0}$, i.e. the tensors $\boldsymbol{D}$ and $\boldsymbol{V}$ are coaxial, the vorticity vector coincides with the body angular velocity. To the best of our knowledge Eq. (32.26) is new and not introduced in the classical continuum mechanics literature.

32.3.2 Logarithmic Spin

Let us relate the tensor $\boldsymbol{D}$ to the time derivative of the spatial Hencky tensor $\boldsymbol{h}$. To this end consider the symmetric part of Eq. (32.21)

$$\boldsymbol{D} = \sum_{i=1}^{3} \dot{\lambda}_i \lambda_i^{-1} \overset{V}{\boldsymbol{n}}_i \otimes \overset{V}{\boldsymbol{n}}_i + \frac{1}{2}(\boldsymbol{V}^2 \times \boldsymbol{b} - \boldsymbol{b} \times \boldsymbol{V}^2) \tag{32.28}$$

The time derivative of the spatial Hencky strain tensor $\boldsymbol{h} = \ln \boldsymbol{V}$ can be computed as follows

$$\dot{\boldsymbol{h}} = \sum_{i=1}^{3} \dot{\lambda}_i \lambda_i^{-1} \overset{V}{\boldsymbol{n}}_i \otimes \overset{V}{\boldsymbol{n}}_i + \boldsymbol{\Omega}_{\boldsymbol{V}} \times \boldsymbol{h} - \boldsymbol{h} \times \boldsymbol{\Omega}_{\boldsymbol{V}} \tag{32.29}$$

With Eqs. (32.28) and (32.29) we obtain

$$\boldsymbol{D} = \dot{\boldsymbol{h}} - \boldsymbol{\Omega}_{\boldsymbol{V}} \times \boldsymbol{h} + \boldsymbol{h} \times \boldsymbol{\Omega}_{\boldsymbol{V}} + \frac{1}{2}(\boldsymbol{V}^2 \times \boldsymbol{b} - \boldsymbol{b} \times \boldsymbol{V}^2) \tag{32.30}$$

The tensor

$$\boldsymbol{D}_{\boldsymbol{\Omega}} = \frac{1}{2}(\boldsymbol{V}^2 \times \boldsymbol{b} - \boldsymbol{b} \times \boldsymbol{V}^2)$$

has the following representation

$$2\boldsymbol{D}_{\boldsymbol{\Omega}} = \sum_{i=1}^{3} \sum_{j=1}^{3} (\lambda_i^2 - \lambda_j^2) \boldsymbol{b} \cdot (\overset{V}{\boldsymbol{n}}_i \times \overset{V}{\boldsymbol{n}}_j)(\overset{V}{\boldsymbol{n}}_i \otimes \overset{V}{\boldsymbol{n}}_j + \overset{V}{\boldsymbol{n}}_j \otimes \overset{V}{\boldsymbol{n}}_i) \tag{32.31}$$

Assuming that the tensor $\boldsymbol{V}$ has distinct principal values λ_i let us consider the following identity

$$\begin{aligned} 2\boldsymbol{D}_{\boldsymbol{\Omega}} &= \sum_{i=1}^{3} \sum_{j=1}^{3} (\ln \lambda_i - \ln \lambda_j) \frac{\lambda_i^2 - \lambda_j^2}{(\ln \lambda_i - \ln \lambda_j)} \boldsymbol{b} \cdot (\overset{V}{\boldsymbol{n}}_i \times \overset{V}{\boldsymbol{n}}_j)(\overset{V}{\boldsymbol{n}}_i \otimes \overset{V}{\boldsymbol{n}}_j + \overset{V}{\boldsymbol{n}}_j \otimes \overset{V}{\boldsymbol{n}}_i) \\ &= \sum_{i=1}^{3} \sum_{j=1}^{3} (\ln \lambda_i - \ln \lambda_j) \boldsymbol{c} \cdot (\overset{V}{\boldsymbol{n}}_i \times \overset{V}{\boldsymbol{n}}_j)(\overset{V}{\boldsymbol{n}}_i \otimes \overset{V}{\boldsymbol{n}}_j + \overset{V}{\boldsymbol{n}}_j \otimes \overset{V}{\boldsymbol{n}}_i) \\ &= \boldsymbol{h} \times \boldsymbol{c} - \boldsymbol{c} \times \boldsymbol{h}, \quad i \neq j, \end{aligned} \tag{32.32}$$

where the components of vector $\boldsymbol{c}$ are related to the components of vector $\boldsymbol{b}$ as follows

$$\boldsymbol{c} \cdot (\overset{V}{\boldsymbol{n}}_i \times \overset{V}{\boldsymbol{n}}_j) = \frac{\lambda_i^2 - \lambda_j^2}{(\ln \lambda_i - \ln \lambda_j)} \boldsymbol{b} \cdot (\overset{V}{\boldsymbol{n}}_i \times \overset{V}{\boldsymbol{n}}_j), \quad i \neq j$$

Hence

$$\boldsymbol{c} \cdot \sum_{i=1}^{3} \sum_{j=1}^{3} \overset{V}{\boldsymbol{n}}_i \times \overset{V}{\boldsymbol{n}}_j \otimes \overset{V}{\boldsymbol{n}}_j \times \overset{V}{\boldsymbol{n}}_i = \boldsymbol{b} \cdot \sum_{i=1}^{3} \sum_{j=1}^{3} \frac{\lambda_i^2 - \lambda_j^2}{(\ln \lambda_i - \ln \lambda_j)} \overset{V}{\boldsymbol{n}}_i \times \overset{V}{\boldsymbol{n}}_j \otimes \overset{V}{\boldsymbol{n}}_j \times \overset{V}{\boldsymbol{n}}_i, \; i \neq j$$

Applying the identity (32.37) we obtain

$$\sum_{i=1}^{3} \sum_{j=1}^{3} \overset{V}{\boldsymbol{n}}_i \times \overset{V}{\boldsymbol{n}}_j \otimes \overset{V}{\boldsymbol{n}}_j \times \overset{V}{\boldsymbol{n}}_i = \sum_{i=1}^{3} \overset{V}{\boldsymbol{n}}_i \times \boldsymbol{I} \times \overset{V}{\boldsymbol{n}}_i = \sum_{i=1}^{3} (\overset{V}{\boldsymbol{n}}_i \otimes \overset{V}{\boldsymbol{n}}_i - \overset{V}{\boldsymbol{n}}_i \cdot \overset{V}{\boldsymbol{n}}_i \boldsymbol{I}) = -2\boldsymbol{I}$$

Consequently

$$2\boldsymbol{c} = -\boldsymbol{b} \cdot \sum_{i=1}^{3} \sum_{j=1}^{3} \frac{\lambda_i^2 - \lambda_j^2}{(\ln \lambda_i - \ln \lambda_j)} \overset{V}{\boldsymbol{n}}_i \times \overset{V}{\boldsymbol{n}}_j \otimes \overset{V}{\boldsymbol{n}}_j \times \overset{V}{\boldsymbol{n}}_i, \quad i \neq j \tag{32.33}$$

With Eqs. (32.22), (32.30), (32.32) and (32.33) the tensor $\boldsymbol{D}$ is related to the rate of the Hencky tensor $\boldsymbol{h}$ as follows

$$\boldsymbol{D} = \dot{\boldsymbol{h}} - \boldsymbol{\Omega}_h \times \boldsymbol{h} + \boldsymbol{h} \times \boldsymbol{\Omega}_h, \quad \boldsymbol{\Omega}_h = \boldsymbol{\Omega}_V + \boldsymbol{A}_h \cdot (\boldsymbol{\Omega}_R - \boldsymbol{\Omega}_V), \tag{32.34}$$

where

$$\boldsymbol{A}_h = -\frac{1}{4J} \boldsymbol{V} \cdot \sum_{i=1}^{3} \sum_{j=1}^{3} \frac{\lambda_i^2 - \lambda_j^2}{(\ln \lambda_i - \ln \lambda_j)} \overset{V}{\boldsymbol{n}}_i \times \overset{V}{\boldsymbol{n}}_j \otimes \overset{V}{\boldsymbol{n}}_j \times \overset{V}{\boldsymbol{n}}_i, \quad i \neq j$$

The tensor $\boldsymbol{A}_h$ has the following spectral representation

$$2\boldsymbol{A}_h = \frac{\lambda_2^2 - \lambda_3^2}{\lambda_2 \lambda_3 \ln \frac{\lambda_2}{\lambda_3}} \overset{V}{\boldsymbol{n}}_1 \otimes \overset{V}{\boldsymbol{n}}_1 + \frac{\lambda_3^2 - \lambda_1^2}{\lambda_3 \lambda_1 \ln \frac{\lambda_3}{\lambda_1}} \overset{V}{\boldsymbol{n}}_2 \otimes \overset{V}{\boldsymbol{n}}_2 + \frac{\lambda_1^2 - \lambda_2^2}{\lambda_1 \lambda_2 \ln \frac{\lambda_1}{\lambda_2}} \overset{V}{\boldsymbol{n}}_3 \otimes \overset{V}{\boldsymbol{n}}_3$$

In Xiao et al (1997) the tensor $\boldsymbol{\Omega}_h \times \boldsymbol{I}$ is called logarithmic spin. By analogy let us call the vector $\boldsymbol{\Omega}_h$ the logarithmic angular velocity. With Eqs. (32.14), (32.26) and (32.34) the vector $\boldsymbol{\Omega}_h$ can be computed as follows

$$\boldsymbol{\Omega}_h = \boldsymbol{\omega} + \boldsymbol{K}_h \cdot (\boldsymbol{D} \cdot \boldsymbol{V}^2)_\times, \tag{32.35}$$

where

$$2\boldsymbol{K}_h = \sum_{i=1}^{3} \sum_{j=1}^{3} \frac{1}{\lambda_i^2 - \lambda_j^2} \left(\frac{\lambda_i^2 + \lambda_j^2}{\lambda_i^2 - \lambda_j^2} - \frac{1}{\ln \frac{\lambda_i}{\lambda_j}} \right) \overset{V}{\boldsymbol{n}}_i \times \overset{V}{\boldsymbol{n}}_j \otimes \overset{V}{\boldsymbol{n}}_j \times \overset{V}{\boldsymbol{n}}_i, \quad i \neq j$$

The tensor $\boldsymbol{K}_h$ has the following spectral representation

$$\begin{aligned} \boldsymbol{K}_h = &\frac{1}{\lambda_2^2 - \lambda_3^2} \left(\frac{1}{\ln \frac{\lambda_2}{\lambda_3}} - \frac{\lambda_2^2 + \lambda_3^2}{\lambda_2^2 - \lambda_3^2} \right) \overset{V}{\boldsymbol{n}}_1 \otimes \overset{V}{\boldsymbol{n}}_1 \\ &+ \frac{1}{\lambda_3^2 - \lambda_1^2} \left(\frac{1}{\ln \frac{\lambda_3}{\lambda_1}} - \frac{\lambda_3^2 + \lambda_1^2}{\lambda_3^2 - \lambda_1^2} \right) \overset{V}{\boldsymbol{n}}_2 \otimes \overset{V}{\boldsymbol{n}}_2 \\ &+ \frac{1}{\lambda_3^2 - \lambda_1^2} \left(\frac{1}{\ln \frac{\lambda_1}{\lambda_2}} - \frac{\lambda_1^2 + \lambda_2^2}{\lambda_1^2 - \lambda_2^2} \right) \overset{V}{\boldsymbol{n}}_3 \otimes \overset{V}{\boldsymbol{n}}_3 \end{aligned}$$

With Eqs. (32.26) and (32.35) one may relate the logarithmic angular velocity to the vectors $\boldsymbol{\omega}$ and $\boldsymbol{\Omega}_R$. To the best of our knowledge Eq. (32.34) is new. Equation (32.35) in a different notation is firstly derived by Xiao et al (1997). It indicates that if tensors $\boldsymbol{D}$ and $\boldsymbol{V}$ are co-axial, the logarithmic angular velocity coincides with the vorticity vector. Therefore the deformation rate tensor $\boldsymbol{D}$ is the Jaumann co-rotational rate of the spatial Hencky tensor.

32.4 Conclusions

The derived equations relate twirl vectors and the logarithmic angular velocity vector to combinations of the vorticity vector and the body angular velocity vector. If the deformation rate tensor is co-axial with the left stretch tensor, then the vorticity vector coincides with the body angular velocity vector. In our view the notions of angular velocities is more convenient if compared to the skew-symmetric tensors for the following reasons. First the derived relations are more transparent as similar relations discussed in the literature, for example for twirls, e.g. Reinhardt and Dubey (1996). Second, one may use foundations of rigid body dynamics to give interpretations to angular velocities. Furthermore, rotation tensors corresponding to angular velocities, for example the logarithmic one can be introduced.

Appendix: Some Operations with Second Rank Tensors

Dot Products of a Second Rank Tensor and a Vector

The right dot product of a second rank tensor $\boldsymbol{A}$ and a vector $\boldsymbol{c}$ is defined by

$$\boldsymbol{A}\cdot\boldsymbol{c} = \left(\sum_{i=1}^{3}\boldsymbol{a}_{(i)}\otimes\boldsymbol{b}_{(i)}\right)\cdot\boldsymbol{c} = \sum_{i=1}^{3}(\boldsymbol{b}_{(i)}\cdot\boldsymbol{c})\boldsymbol{a}_{(i)} = \sum_{i=1}^{3}\alpha_{(i)}\boldsymbol{a}_{(i)}$$

with $\alpha_{(i)} \equiv \boldsymbol{b}_{(i)}\cdot\boldsymbol{c}$. The left dot product is defined by

$$\boldsymbol{c}\cdot\boldsymbol{A} = \boldsymbol{c}\cdot\left(\sum_{i=1}^{3}\boldsymbol{a}_{(i)}\otimes\boldsymbol{b}_{(i)}\right) = \sum_{i=1}^{3}(\boldsymbol{c}\cdot\boldsymbol{a}_{(i)})\boldsymbol{b}_{(i)} = \sum_{i=1}^{3}\beta_{(i)}\boldsymbol{b}_{(i)}$$

with $\beta_{(i)} \equiv \boldsymbol{c}\cdot\boldsymbol{a}_{(i)}$. The results of these operations are vectors. One can verify that

$$\boldsymbol{A}\cdot\boldsymbol{c} \neq \boldsymbol{c}\cdot\boldsymbol{A}, \quad \boldsymbol{A}\cdot\boldsymbol{c} = \boldsymbol{c}\cdot\boldsymbol{A}^{\mathrm{T}}$$

Cross Products of a Second Rank Tensor and a Vector

The right cross product of a second rank tensor $\boldsymbol{A}$ and a vector $\boldsymbol{c}$ is defined by

$$\boldsymbol{A} \times \boldsymbol{c} = \left(\sum_{i=1}^{3} \boldsymbol{a}_{(i)} \otimes \boldsymbol{b}_{(i)} \right) \times \boldsymbol{c} = \sum_{i=1}^{3} \boldsymbol{a}_{(i)} \otimes (\boldsymbol{b}_{(i)} \times \boldsymbol{c}) = \sum_{i=1}^{3} \boldsymbol{a}_{(i)} \otimes \boldsymbol{d}_{(i)}$$

with $\boldsymbol{d}_{(i)} \equiv \boldsymbol{b}_{(i)} \times \boldsymbol{c}$. The left cross product is defined by

$$\boldsymbol{c} \times \boldsymbol{A} = \boldsymbol{c} \times \left(\sum_{i=1}^{3} \boldsymbol{a}_{(i)} \otimes \boldsymbol{b}_{(i)} \right) = \sum_{i=1}^{3} (\boldsymbol{c} \times \boldsymbol{a}_{(i)}) \otimes \boldsymbol{b}_{(i)} = \sum_{i=1}^{3} \boldsymbol{e}_{(i)} \otimes \boldsymbol{b}_{(i)}$$

with $\boldsymbol{e}_{(i)} \equiv \boldsymbol{b}_{(i)} \times \boldsymbol{c}$. The results of these operations are second rank tensors. It can be shown that

$$\boldsymbol{A} \times \boldsymbol{c} = -[\boldsymbol{c} \times \boldsymbol{A}^{\mathrm{T}}]^{\mathrm{T}}$$

The following identities valid for a non-singular tensor $\boldsymbol{A}$ and vectors $\boldsymbol{a}$ and $\boldsymbol{b}$ (Zhilin, 2001)

$$\begin{aligned} \det \boldsymbol{A} \boldsymbol{A}^{-T} \cdot (\boldsymbol{a} \times \boldsymbol{b}) &= (\boldsymbol{A} \cdot \boldsymbol{a}) \times (\boldsymbol{A} \cdot \boldsymbol{b}), \\ \det \boldsymbol{A} \boldsymbol{A}^{-T} \cdot (\boldsymbol{a} \times \boldsymbol{I}) \cdot \boldsymbol{A}^{-1} &= (\boldsymbol{A} \cdot \boldsymbol{a}) \times \boldsymbol{I} \end{aligned} \tag{32.36}$$

Vector Invariant

The vector invariant or the Gibbs cross of a second rank tensor $\boldsymbol{A}$ is defined by

$$\boldsymbol{A}_{\times} = \left(\sum_{i=1}^{3} \boldsymbol{a}_{(i)} \otimes \boldsymbol{b}_{(i)} \right)_{\times} = \sum_{i=1}^{3} \boldsymbol{a}_{(i)} \times \boldsymbol{b}_{(i)}$$

The result of this operation is a vector. The vector invariant of a symmetric tensor is the zero vector. The following identities can be verified

$$\begin{aligned} (\boldsymbol{a} \times \boldsymbol{I})_{\times} &= -2\boldsymbol{a}, \\ \boldsymbol{a} \times \boldsymbol{I} \times \boldsymbol{b} &= \boldsymbol{b} \otimes \boldsymbol{a} - (\boldsymbol{a} \cdot \boldsymbol{b})\boldsymbol{I} \end{aligned} \tag{32.37}$$

For any vector $\boldsymbol{a}$ and any second rank tensor $\boldsymbol{B}$ the following identity is valid

$$(\boldsymbol{a} \times \boldsymbol{B})_{\times} = \boldsymbol{B} \cdot \boldsymbol{a} - (\mathrm{tr}\, \boldsymbol{B})\boldsymbol{a}, \quad (\boldsymbol{B} \times \boldsymbol{a})_{\times} = \boldsymbol{B}^{T} \cdot \boldsymbol{a} - (\mathrm{tr}\, \boldsymbol{B})\boldsymbol{a} \tag{32.38}$$

For any vector $\boldsymbol{a}$ and symmetric tensors $\boldsymbol{A}, \boldsymbol{B}$ the following identity can be established

$$\begin{aligned} &[(\boldsymbol{A} \times \boldsymbol{a}) \cdot \boldsymbol{B}]_{\times} = [\boldsymbol{A} \cdot (\boldsymbol{a} \times \boldsymbol{B})]_{\times} = \boldsymbol{C} \cdot \boldsymbol{a}, \\ &\boldsymbol{C} = \sum_{i=1}^{3} \lambda_i^A \overset{A}{\boldsymbol{n}}_i \times \boldsymbol{B} \times \overset{A}{\boldsymbol{n}}_i = \sum_{i=1}^{3} \lambda_i^B \overset{B}{\boldsymbol{n}}_i \times \boldsymbol{A} \times \overset{B}{\boldsymbol{n}}_i \\ &\quad = [\mathrm{tr}\, (\boldsymbol{A} \cdot \boldsymbol{B}) - \mathrm{tr}\, \boldsymbol{A} \mathrm{tr}\, \boldsymbol{B}]\boldsymbol{I} + (\mathrm{tr}\, \boldsymbol{B})\boldsymbol{A} + (\mathrm{tr}\, \boldsymbol{A})\boldsymbol{B} - \boldsymbol{A} \cdot \boldsymbol{B} - \boldsymbol{B} \cdot \boldsymbol{A}, \end{aligned} \tag{32.39}$$

where λ_i^A and λ_i^B are principal values of the tensors $\boldsymbol{A}$ and $\boldsymbol{B}$, respectively. The othonormal vectors $\overset{A}{\boldsymbol{n}}_i$ and $\overset{B}{\boldsymbol{n}}_i$ are corresponding principal directions.

References

Altenbach H, Eremeyev V (2014) Strain rate tensors and constitutive equations of inelastic micropolar materials. International Journal of Plasticity 63:3 – 17

Altenbach H, Naumenko K, L'vov G, Pilipenko S (2003a) Numerical estimation of the elastic properties of thin-walled structures manufactured from short-fiber-reinforced thermoplastics. Mechanics of composite materials 39(3):221–234

Altenbach H, Naumenko K, Zhilin P (2003b) A micro-polar theory for binary media with application to phase-transitional flow of fiber suspensions. Continuum Mechanics and Thermodynamics 15:539 – 570

Altenbach H, Naumenko K, Zhilin PA (2005) A direct approach to the formulation of constitutive equations for rods and shells. In: Pietraszkiewicz W, Szymczak C (eds) Shell Structures: Theory and Applications, Taylor & Francis, Leiden, pp 87 – 90

Altenbach H, Naumenko K, Pylypenko S, Renner B (2007) Influence of rotary inertia on the fiber dynamics in homogeneous creeping flows. ZAMM-Journal of Applied Mathematics and Mechanics/Zeitschrift für Angewandte Mathematik und Mechanik 87(2):81 – 93

Antman S (1995) Nonlinear Problems of Elasticity. Springer, Berlin

Bertram A (2012) Elasticity and Plasticity of Large Deformations, 3rd edn. Springer, Berlin

Eremeyev VA, Lebedev LP, Altenbach H (2013) Foundations of micropolar mechanics. Springer Science & Business Media

Libai A, Simmonds JG (1998) The Nonlinear Theory of Elastic Shells. Cambridge University Press, Cambridge

Maugin G (2013) Continuum Mechanics Through the Twentieth Century: A Concise Historical Perspective. Solid Mechanics and Its Applications, Springer

Maugin GA (1992) The Thermomechanics of Plasticity and Fracture. Cambridge University Press, Cambridge

Maugin GA (2014) Continuum Mechanics Through the Eighteenth and Nineteenth Centuries. Springer

Maugin GA (2016) Non-Classical Continuum Mechanics: A Dictionary. Advanced Structured Saterials, vol 51. Springer

Naumenko K, Altenbach H (2016) Modeling High Temperature Materials Behavior for Structural Analysis: Part I: Continuum Mechanics Foundations and Constitutive Models. Advanced structured materials, vol 28. Springer

Reinhardt W, Dubey R (1996) Coordinate-independent representation of spins in continuum mechanics. Journal of Elasticity 42(2):133–144

Xiao H, Bruhns O, Meyers A (1997) Logarithmic strain, logarithmic spin and logarithmic rate. Acta Mechanica 124(1-4):89–105

Zhilin PA (1996) A new approach to the analysis of free rotations of rigid bodies. ZAMM-Journal of Applied Mathematics and Mechanics/Zeitschrift für Angewandte Mathematik und Mechanik 76(4):187 – 204

Zhilin PA (2001) Vektory i tensory vtorogo ranga v trekhmernom prostranstve (Vectors and second rank tensors in three-dimensional space, in Russ.). Nestor, St. Petersburg

Zhong-Heng G, Lehmann T, Haoyun L, Man CS (1992) Twirl tensors and the tensor equation AX- XA= C. Journal of Elasticity 27(3):227–245

Chapter 33
Towards Continuum Mechanics with Spontaneous Violations of the Second Law of Thermodynamics

Martin Ostoja-Starzewski and Bharath V. Raghavan

Abstract As dictated by modern statistical physics, the second law is to be replaced by the fluctuation theorem (FT) on very small length and/or time scales. This means that the deterministic continuum thermomechanics must be generalized to a stochastic theory allowing randomly spontaneous violations of the Clausius-Duhem inequality to take place anywhere in the material domain. This paper outlines a formulation of stochastic continuum thermomechanics, where the entropy evolves as a submartingale while the dissipation function is consistent with the FT. A summary is then given of the behavior of an atomic fluid in Couette flow, studied using a combination of kinetic theory, hydrodynamic theory, and molecular dynamics. Overall, the developed framework may be applied in many fields involving fluid flow and heat conduction on very small spatial scales.

33.1 Dissipation Function in Thermomechanics within Second Law

In the version of continuum mechanics called *thermomechanics with internal variables* (TIV), the Second Law is written in terms of the reversible ($\dot{s}^{(r)}$) and irreversible ($\dot{s}^{(i)}$) parts of entropy production rate ($\dot{s}$) (Maugin, 1999; Ziegler, 2012; Ziegler and Wehrli, 1987)

$$\dot{s} = \dot{s}^{(r)} + \dot{s}^{(i)} \quad \text{with} \quad \dot{s}^{(r)} = -\frac{\partial (q_i/\theta)}{\partial x_i} \quad \text{and} \quad \dot{s}^{(i)} \geq 0. \tag{33.1}$$

Here θ is the absolute temperature while q_i ($\boldsymbol{q}$) is the heat flux [Hereinafter we interchangeably, whichever is more convenient, use the symbolic ($\boldsymbol{f}$) and subscript

Martin Ostoja-Starzewski · Bharath V. Raghavan
University of Illinois at Urbana-Champaign, Urbana, IL 61801, USA
e-mail: martinos@illinois.edu, braghav2@illinois.edu

© Springer International Publishing AG, part of Springer Nature 2018

H. Altenbach et al. (eds.), *Generalized Models and Non-classical Approaches in Complex Materials 1*, Advanced Structured Materials 89,
https://doi.org/10.1007/978-3-319-72440-9_33

($f_{i...}$) notations for tensors; an overdot means the material time derivative.]. This form of writing the Second Law naturally involves the dissipation function $\phi = \dot{s}^{(i)}$, which provides a constitutive law for the irreversible part of response. Typically, $\phi = \boldsymbol{Y} \cdot \boldsymbol{V}$ is taken as a functional of the velocity $\boldsymbol{V}$ so as to obtain the dissipative force $\boldsymbol{Y}$:

$$Y_i = \lambda \frac{\partial \phi (\boldsymbol{V})}{\partial V_i} \quad \text{where} \quad \lambda = \left(V_i \frac{\partial \phi (\boldsymbol{V})}{\partial V_i} \right)^{-1} \phi . \tag{33.2}$$

Effectively, this means that, provided the dissipative force $\boldsymbol{Y}$ is prescribed, the actual velocity maximizes the dissipation rate $\dot{s}^{(i)} = \boldsymbol{Y} \cdot \boldsymbol{V}$ subject to the side condition (33.1)

$$\phi(\boldsymbol{V}) = \boldsymbol{Y} \cdot \boldsymbol{V} = \dot{s}^{(i)} \geq 0 \tag{33.3}$$

This is not the only extremum-type interpretation of (33.2), others being e.g. the principle of least velocity, the principle of least dissipative force.

While (33.2) applies to a very wide range of linear and nonlinear material behaviors, it does not cover all of them. The most general solution of the inequality (33.3) is based on a decomposition theorem of Edelen (1973, 1974): assuming $\boldsymbol{Y} = \boldsymbol{Y}(\boldsymbol{V})$, there always exist functions $\varphi (\boldsymbol{V})$ and $\boldsymbol{U} (\boldsymbol{V})$ such that

$$Y_i = \frac{\partial \varphi (\boldsymbol{V})}{\partial V_i} + U_i \tag{33.4}$$

with

$$\varphi (\boldsymbol{V}) \equiv \int_0^1 V_i Y_i (\tau \boldsymbol{V}) \, d\tau \tag{33.5}$$

and

$$U_i (\boldsymbol{V}) \equiv \int_0^1 \tau V_j \left[\frac{\partial Y_j (\tau \boldsymbol{V})}{\partial (\tau V_i)} - \frac{\partial Y_i (\tau \boldsymbol{V})}{\partial (\tau V_j)} \right] d\tau \tag{33.6}$$

Since $\boldsymbol{Y} \cdot \boldsymbol{U} = 0$, $\boldsymbol{U} (\boldsymbol{V})$ is called the *non-dissipative* (or *powerless*) vector; also $\boldsymbol{U}(\mathbf{0}) = \mathbf{0}$. The Maxwell-Cattaneo heat conduction is an example of a process derivable by this approach.

By analogy to the role played by the free energy ψ for quasi-conservative processes (such as hyperelasticity), ϕ or φ plays the role of a potential for all the dissipative processes, which are then called *hyperdissipative*. Effectively, the functional $\phi (\boldsymbol{V})$ or $\phi (\boldsymbol{Y})$ is employed to derive the constitutive laws of continua. In general, there is no perfectly-established (and uniformly agreed-upon) rule to decide whether $\boldsymbol{V}$ or $\boldsymbol{Y}$ should be the argument of the functional φ or ϕ.

Note that ϕ, but not φ, is directly equal to the dissipation $s^{*(i)}$. Clearly, in the linear response regime, $\varphi = \phi/2$ providing $\boldsymbol{U} = \mathbf{0}$.

33.2 Dissipation Function in Statistical Physics beyond Second Law

While the inequality (33.3) is assumed to hold instantaneously (i.e. for $\forall t$), contemporary statistical physics (e.g. Evans and Searles, 2002; Searles and Evans, 2001) the Second Law is replaced by the *fluctuation theorem* which gives the relative probability of observing processes that have positive (A) and negative ($-A$) total dissipation in non-equilibrium systems:

$$\frac{\mathsf{P}(\phi_t = A)}{\mathsf{P}(\phi_t = -A)} = e^{At}. \tag{33.7}$$

Here ϕ_t is the total dissipation for a trajectory $\boldsymbol{\Gamma} \equiv \{q_1, p_1, ..., q_N, p_N\}$ of N particles originating at $\boldsymbol{\Gamma}(0)$ and evolving for a time t:

$$\phi_t(\boldsymbol{\Gamma}(0)) = \int_0^t \phi(\boldsymbol{\Gamma}(s))\, ds. \tag{33.8}$$

The integral in (33.8) involves an instantaneous dissipation function:

$$\phi(\boldsymbol{\Gamma}(t)) = \frac{d\phi_t(\boldsymbol{\Gamma}(0))}{dt}. \tag{33.9}$$

The Second Law of thermodynamics is recovered upon ensemble averaging, time averaging, or upscaling.

In view of (33.7) above, the dissipation function is a stochastic (not deterministic) quantity which possibly and spontaneously takes negative values, so that the positive-definiteness does not absolutely hold. Therefore, we write (33.3) as

$$\phi(\boldsymbol{V}, \omega) = \boldsymbol{Y}(\omega) \cdot \boldsymbol{V} = s^{*(i)}, \quad \omega \in \Omega, \tag{33.10}$$

where $\boldsymbol{Y}(\omega)$ are the dissipative forces conjugate to $\boldsymbol{V}$, while Ω is the set of all possible outcomes. Thus, the argument ω indicates that $\phi(\boldsymbol{V}, \omega)$ is a stochastic functional, while $\boldsymbol{Y}(\omega)$ is a random quantity for a non-random (prescribed) velocity $\boldsymbol{V}$. An analogous picture holds for $\boldsymbol{Y}$ being prescribed and $\boldsymbol{V}$ being the random outcome. It is tacitly assumed that Ω is equipped with a σ-algebra of observable events $\mathscr{A}$ and a probability measure P defined on $(\Omega, \mathscr{A})$.

The fluctuation theorem as expressed by (33.7) states that (i) positive dissipation is exponentially more likely to be observed than negative dissipation, and (ii) ensemble averaging of ϕ_t leads to

$$\langle \Delta\phi_t \mid \mathscr{F}_t \rangle \geq 0. \tag{33.11}$$

Here $\mid \mathscr{F}_t$ indicates the conditioning on the past history and is discussed below, while $\langle f \rangle := \int f\, d\mathsf{P}$. Thus, the entropy production rate is non-negative on average. In view of the random fluctuations, ϕ_t is a stochastic process with a specific type of memory effect: a *submartingale* (Doob, 1953; Ostoja-Starzewski and Malyarenko, 2014).

Treating time as a continuous parameter and noting that ϕ_t is deterministic for a forward evolution, we have

$$\langle \phi_{t+dt} \mid \mathscr{F}_t \rangle \geq \phi_t. \tag{33.12}$$

Clearly, this is a weaker statement than a purely deterministic one

$$\phi_{t+dt} \mid \mathscr{F}_t \geq \phi_t \quad \text{or, equivalently,} \quad \frac{\phi_{t+dt} - \phi_t}{\Delta t} \geq 0. \tag{33.13}$$

The latter inequality is the same as writing (33.1)$_3$ in a finite difference form.

Next, we recall the Doob–Meyer decomposition to write ϕ_t as a sum of a martingale (M) and a "drift" process (G):

$$\phi_t(\boldsymbol{V}, \omega) = M + G \quad \text{and} \quad \phi(\boldsymbol{V}, \omega) = \dot{M} + \dot{G}. \tag{33.14}$$

Thus, $M \neq 0$ reflects the fluctuations of entropy production about the zero level $\left\langle s^{(i)} \right\rangle = 0$. The four different cases depending on whether $M = 0$ or $M \neq 0$ and $G = 0$ or $G > 0$ have been discussed in Ostoja-Starzewski and Malyarenko (2014). Overall, the deterministic continuum mechanics is smoothly recovered as the time and/or spatial scale increases (so that $M \to 0$) or via ensemble averaging, a result that is consistent with intuition. A corresponding re-examination of axioms of continuum mechanics has been given in Ostoja-Starzewski (2016).

Observe that one might also work with a discrete time formulation, making the mathematical analysis of martingales simpler. It is fitting here to note that mathematical physics – also in classical, i.e. non-quantum, regime – may be formulated from the standpoint of discrete, rather than continuous, time (Jaroszkiewicz, 2014).

There are three types of phenomena in classical physics where the fluctuation theorem is applicable: viscous, thermal, and electrical (Evans and Searles, 2002; Searles and Evans, 2001). If we concern ourselves with the first two, a contact with continuum thermomechanics is made by writing the scalar product $\boldsymbol{Y} \cdot \boldsymbol{V}$ as one involving the intrinsic mechanical dissipation (viscous effects) and thermal dissipation in spatial (Eulerian) description:

$$\phi(\boldsymbol{V}, \omega) = \phi_{th}(\boldsymbol{V}_1, \omega) + \phi_{mech}(\boldsymbol{V}_2, \omega), \quad \boldsymbol{V} \equiv (\boldsymbol{V}_1, \boldsymbol{V}_2) = \left(\frac{-\boldsymbol{\nabla}\theta}{\theta}, \boldsymbol{d} \right). \tag{33.15}$$

Thus, the generalized velocity vector $\boldsymbol{V}$ is made up of two parts: the negative temperature gradient divided by the temperature (i.e., $-\boldsymbol{\nabla}\theta/\theta$) and the deformation rate $\boldsymbol{d}$. The reason we take the former as the argument of ϕ_{th} is that the fluctuation theorem for heat flow was derived for controllable temperature differences (Searles and Evans, 2001), with the heat flux being the stochastic outcome. Analogously, the fluctuation theorem for Couette and Poiseuille flows was derived for controllable velocities (Evans and Searles, 2002), with the Cauchy stress being the stochastic outcome. Thus, the dissipative force corresponding to $\boldsymbol{V}$ is made up of the heat flux and the dissipative stress

$$\boldsymbol{Y} \equiv (\boldsymbol{Y}_1, \boldsymbol{Y}_2) = \left(\boldsymbol{q}, \boldsymbol{\sigma}^{(d)}\right). \tag{33.16}$$

It is largely a matter of convenience whether $-\boldsymbol{\nabla}\theta/\theta$ or $\boldsymbol{q}$ should be taken as a velocity or a dissipative force. In the section on thermoviscous fluids we work with the setup outlined above, while in the section on inviscid thermoelastic solids we invert the roles of $-\boldsymbol{\nabla}\theta/\theta$ and $\boldsymbol{q}$.

There are two basic possibilities here:

- both processes in (33.15) may independently exhibit spontaneous random violations of the Second Law;
- both processes in (33.15) are coupled implying that the Clausius-Duhem inequality holds for thermal and viscous violations jointly; the relevant statistical physics has not yet been studied.

In Ostoja-Starzewski (2017b,a), we considered the first possibility focusing on: (i) thermoviscous fluids with parabolic or hyperbolic type heat conduction, (ii) thermoelasticity with parabolic or hyperbolic type heat conduction, and (iii) poromechanics with dissipation within the skeleton, the fluid, and the temperature field. The reason we considered parabolic or hyperbolic cases is that the statistical physics has established the spontaneous violations of the Fourier type law, but a hyperbolic heat conduction in fluids and solids can still be modeled in continuum mechanics providing two relaxation times - one in the mechanical and another in the entropy constitutive law - are introduced (recall the "thermoelasticity with two relaxation times"). The theoretical developments hinge on the fact that the balance laws apply irrespective of the conventional Second Law being obeyed or not. At the same time, we are interested in formulating models which are hyperelastic and hyperdissipative in ensemble average sense (or, for long time averages), thereby extending such class beyond the deterministic media fully obeying the Second Law (Edelen, 1974; Evans and Searles, 2002).

33.3 Stochastic Dissipation Function

33.3.1 Basics

In view of the preceding section, the constitutive relation linking $\boldsymbol{Y}$ with $\boldsymbol{V}$ should be stochastic. Therefore, we replace the deterministic picture by a stochastic one so the internal energy density u and the entropy s are real-valued random fields over the material ($\mathscr{D}$) and time (T) domains. For example, in the case of heat conduction in a rigid (undeformable) conductor,

$$u : \mathscr{D} \times T \times \Omega \to \mathbb{R}, \quad s : \mathscr{D} \times T \times \Omega \to \mathbb{R} \tag{33.17}$$

The randomness disappears as the time and/or spatial scales become large: the field quantities simplify to deterministic functions of a homogeneous continuum. Focusing on the thermal dissipation in (33.15), we have

$$\phi_{th}\left(\frac{-\nabla\theta}{\theta},\omega\right) = -q_k\frac{\theta_{,k}}{\theta} \equiv -\boldsymbol{q}\cdot\frac{\nabla\theta}{\theta}. \tag{33.18}$$

Given the stochastic violations of the Second Law, $\phi_{th}(\boldsymbol{q},\omega)$ takes the form:

$$\phi_{th}(\boldsymbol{q},\omega) = \dot{G}(\boldsymbol{q}) + \dot{M}(\boldsymbol{q},\omega). \tag{33.19}$$

For the linear Fourier-type conductivity, it becomes more explicit with

$$\dot{G}(\boldsymbol{q}) = \frac{1}{\theta}q_i\kappa_{ij}q_j,\ \dot{M}(\boldsymbol{q},\omega) = \frac{1}{\theta}q_i\mathscr{M}_{ij}(\omega)\,q_j. \tag{33.20}$$

Here $\dot{G}(\boldsymbol{q})$ involves the thermal resistivity λ_{ij} which is positive definite, and $\dot{M}(\boldsymbol{q},\omega) = dM(\boldsymbol{q},\omega)/dt$, with M being the martingale modeling the random fluctuations according to (33.2). Clearly, the randomness residing in $M(\boldsymbol{d},\omega)$ allows the total thermal conductivity $\kappa_{ij}+\mathscr{M}_{ij}$ to become negative since $\mathscr{M}_{ij}$ is not required to be positive definite, thus signifying the violations of the Second Law. More specifically, $\mathscr{M}_{ij}:\mathscr{V}\to\mathscr{V}$ (where $\mathscr{V}$ is a real vector space) is a second-order rank 2 tensor random field (Malyarenko and Ostoja-Starzewski, 2014, 2016)

$$\mathscr{M}_{ij}:\mathscr{D}\times\Omega\to\mathscr{V}^2. \tag{33.21}$$

In view of the Gaussian character of nanoscale fluctuations, $\mathscr{M}_{ij}$ is a Gaussian tensor random field. A departure from Gaussianity has been studied through the atomic fluid model below.

33.3.2 Atomic Fluid in Couette Flow

In Raghavan and Ostoja-Starzewski (2017), based on a kinetic theory and Non-Equilibrium Molecular Dynamics (NEMD) we have derived a shear-thinning model equation of state for an atomic fluid with interactions of Lennard-Jones type. This, in turn, leads to a dissipation function $\varphi(\boldsymbol{Y})$, in which the dissipative force $\boldsymbol{Y}$ is the symmetric Cauchy stress tensor $\sigma_{ij}^{(d)}$ and $\boldsymbol{V}$ is the deformation rate d_{ij}. For a uniform shear flow, the dissipative force is just the shear stress $\sigma_{12}^{(d)}$ corresponding to the applied deformation rate $\dot{\gamma}$, so that

$$\sigma_{ij}^{(d)}(d_{ij})\,d_{ij} = \phi(\boldsymbol{d}) = \frac{\eta_0\dot{\gamma}^2}{1+\frac{2}{3}(\dot{\gamma}/\nu)^2}, \tag{33.22}$$

where $\nu = p/\eta_0$ with η_0 being the Newtonian viscosity, and p being the pressure. Adopting thermodynamic orthogonality (33.2), on account of the side conditions ($d_{(1)}=0$, $d_{(2)}=\dot{\gamma}^2$, and $d_{(3)}=0$) and several intermediate steps, we obtain a quasi-linear fluid model

$$\sigma_{ij}^{(d)} = \lambda \frac{\partial \phi}{\partial d_{(2)}} d_{ij} = 2\eta \left(d_{(2)}\right) d_{ij}, \tag{33.23}$$

with the constant of proportionality in (33.23) being

$$\lambda = \frac{1 + \frac{2}{3}\left(\dot{\gamma}/\nu\right)^2}{2}. \tag{33.24}$$

The fluid viscosity η needs to at least depend on $d_{(2)} = \dot{\gamma}^2$.

Continuing with NEMD in Raghavan et al (2018), we obtained a connection between a local density contrast and temporal fluctuations in the shear stress, which arise naturally through the equivalence between the dissipation function and entropy production according to the fluctuation theorem. The bispectral density of the shear stress was used to measure the degree of departure from the Gaussian model and the degree of nonlinearity induced in the system owing to the applied strain rate. The information theory was used to account for the departure from Gaussian statistics and to develop a more general probability distribution function. By accounting for negative shear stress increments, this distribution (i) preserves the violations of the Second Law of thermodynamics observed in planar Couette flow of atomic fluids and (ii) captures the non-Gaussian nature of the system by allowing for non-zero higher moments. It was also demonstrated how the temperature affects the band-width of the shear-stress and how the density affects its power spectral density, thus determining the conditions under which the shear-stress acts is a narrow-band or wide-band random process. The changes in the statistical characteristics of the parameters of interest occur at a critical deformation rate at which an ordering transition occurs in the fluid causing shear thinning and affecting its stability.

33.4 Closure

There are myriad directions in which this research can be continued. For instance, we have already suggested a formulation of inelastic materials, thermoviscous fluids, and micropolar fluid mechanics accounting for the lack of symmetry of stress tensor on molecular scales. Overall, the situations where violations of the Second Law are relevant occur on very small length scales (extremely thin wavefronts, nano-channels, nano-rods, ...) and short time scales (although times up to 3 seconds have been observed for cholesteric liquids). Thus, a natural setting in which to expect these phenomena is that of permeability and poromechanics involving nanoscale channels in random (and possibly fractal) media.

Acknowledgements This material is based upon work partially supported by the NSF under grants CMMI-1462749 and IIP-1362146 (I/UCRC on Novel High Voltage / Temperature Materials and Structures).

References

Doob JL (1953) Stochastic Processes. Wiley New York

Edelen D (1974) Primitive thermodynamics: a new look at the Clausius-Duhem inequality. Int J Eng Sc 12(2):121–141

Edelen DGB (1973) On the existence of symmetry relations and dissipation potentials. Arch Rational Mech Anal 51(3):218–227

Evans DJ, Searles DJ (2002) The fluctuation theorem. Adv Phys 51(7):1529–1585

Jaroszkiewicz G (2014) Principles of Discrete Time Mechanics. Cambridge University Press

Malyarenko A, Ostoja-Starzewski M (2014) Statistically isotropic tensor random fields: correlation structures. Math Mech Complex Sys (MEMOCS) 2(2):209–231

Malyarenko A, Ostoja-Starzewski M (2016) Spectral expansions of homogeneous and isotropic tensor-valued random fields. ZAMP 67(3):59

Maugin GA (1999) The Thermomechanics of Nonlinear Irreversible Behaviors: An Introduction. World Scientific

Ostoja-Starzewski M (2016) Second law violations, continuum mechanics, and permeability. Cont Mech Thermodyn 28(1-2):489

Ostoja-Starzewski M (2017a) Admitting spontaneous violations of the second law in continuum thermomechanics. Entropy 19(2):78

Ostoja-Starzewski M (2017b) Continuum physics with violations of the second law of thermodynamics. In: Math. Model. Sol. Mech., Springer, pp 181–192

Ostoja-Starzewski M, Malyarenko A (2014) Continuum mechanics beyond the second law of thermodynamics. In: Proc. R. Soc. A, p 20140531

Raghavan BV, Ostoja-Starzewski M (2017) Shear-thinning of molecular fluids in Couette flow. Phys Fluids 29(2):023,103

Raghavan BV, Karimi P, Ostoja-Starzewski M (2018) Stochastic characteristics and Second Law violations of atomic fluids in Couette flow. Physica A: Statistical Mechanics and its Applications 496:90 – 107

Searles DJ, Evans DJ (2001) Fluctuation theorem for heat flow. Int J Thermophys 22(1):123–134

Ziegler H (2012) An Introduction to Thermomechanics. Elsevier

Ziegler H, Wehrli C (1987) The derivation of constitutive relations from the free energy and the dissipation function. Adv Appl Mech 25:183–238

Chapter 34
Nonlocal Approach to Square Lattice Dynamics

Alexey V. Porubov, Alena E. Osokina, and Thomas M. Michelitsch

Abstract The algorithm is developed to model two-dimensional dynamic processes in a nonlocal square lattice on the basis of the shift operators. The governing discrete equations are obtained for local and nonlocal models. Their dispersion analysis reveals important differences in the dispersion curve and in the sign of the group velocity caused by nonlocality. The continuum limit allows to examine possible auxetic behavior of the material described by the nonlocal discrete model.

34.1 Introduction

Studies of the waves propagation in various crystal lattices are associated, first of all, with the name of M. Born Born and Huang (1954), whose works are dated from the beginning of the 20th century and don't lose relevance up to this day in connection with the development of nanotechnologies and nanoelectronics. Recent achievements may be found, e.g., in Askar (1985); Kunin (1975); Maugin (1999); Maugin et al (1992); Manevich and Manevitch (2005); Ostoja-Starzewski (2002); Andrianov et al (2010); Kuzkin et al (2016); Eringen (1972); Metrikine and Askes (2006); Askes and Metrikine (2005); Kosevich and Savotchenko (1999); Zabusky and Deem (1967). The problem of linear waves propagation in one-dimensional monoatomic chain, where particles are represented by material points with a nonzero mass, and the inter-

Alexey V. Porubov · Alena E. Osokina
Institute for Problems in Mechanical Engineering of the Russian Academy of Sciences, Bolshoy pr., 61, V.O., St. Petersburg, 199178 & Peter the Great St.Petersburg Polytechnic University, Polytechnicheskaya st., 29, St.Petersburg, 195251, Russia
e-mail: alexey.porubov@gmail.com, aeosokina@gmail.com

Thomas M. Michelitsch
Sorbonne Université, Centre National de la Recherche Scientifique, UMR 7190, Institut Jean Le Rond d'Alembert, F-75005 Paris, France
e-mail: michel@lmm.jussieu.fr

© Springer International Publishing AG, part of Springer Nature 2018

H. Altenbach et al. (eds.), *Generalized Models and Non-classical Approaches in Complex Materials 1*, Advanced Structured Materials 89,
https://doi.org/10.1007/978-3-319-72440-9_34

atomic bonds are described by elastic springs, has become classical (Askar, 1985; Ostoja-Starzewski, 2002; Andrianov et al, 2010; Metrikine and Askes, 2006; Zabusky and Deem, 1967). The particles with different masses or the springs with various rigidities were also considered, as well as nonlinear and nonlocal generalizations of interatomic interactions. First of all, linear approximation problems were solved. Structures with monatomic crystal lattices (triangular, square, cubic, FCC and BCC) (Askar, 1985; Metrikine and Askes, 2006; Askes and Metrikine, 2005) and diatomic crystal lattices (Born and Huang, 1954; Askar, 1985; Manevich and Manevitch, 2005; Ostoja-Starzewski, 2002) were examined. Discrete and continuum models both possess analytical solutions in the linear case, which allows complex analysis of the mechanical phenomena from micro- and macroscopic points of view (Kuzkin et al, 2016). In the monograph (Askar, 1985), perfect and imperfect lattices dynamics was considered and their connection with the continuum theories of four main types: elasticity, piezoelectricity, viscoelasticity and plasticity in the framework of continuum mechanics. Furthermore, two-dimensional problems were considered for the triangular, square, and hexagonal lattices in Askar (1985); Manevich and Manevitch (2005); Metrikine and Askes (2006); Askes and Metrikine (2005); Porubov and Berinskii (2014).

The importance of nonlinear interactions has risen after the report of Fermi, Pasta and Ulam in 1955, which presented the model of one-dimensional nonlinear lattice that did not show energy equipartition, i.e. practically the entire energy was concentrated in the first mode, and thus, nonlinearity doesn't guarantee an equal energy distribution between modes, see Zabusky and Deem (1967); Ablowitz and Segur (1981). An outstanding contribution to the nonlinear wave theory of crystals has been done by Maugin (1999); Maugin et al (1992); Sayadi and Pouget (1991). In Maugin (1999), aspects of deformable solids (also known as inelastic crystals) nonlinear dynamics were considered, where nonlinear effects combine or compete with each other. Various models were investigated - both discrete and continuum, in particular, the effects of thermal, electric or magnetic nature in the crystal structure, and were analyzed utilizing the equations of rational mechanics. More complicated highly nonlinear discrete systems were studied in Maugin et al (1992); Sayadi and Pouget (1991) giving rise to new coupled nonlinear wave equations in the continuum limit.

A nonlocal approach to the study of the crystalline structures allows to take into consideration the influence of more distant particles onto discrete system dynamics (Kunin, 1975; Eringen, 1972; Xu et al, 2005). The interaction between non-neighboring particles in the lattice has attracted considerable interest because of their influence on the dispersion features of the waves (Maugin, 1999; Manevich and Manevitch, 2005; Ostoja-Starzewski, 2002; Andrianov et al, 2010; Metrikine and Askes, 2006; Askes and Metrikine, 2005; Kosevich and Savotchenko, 1999; Michelitsch et al, 2014). This is also important for a modeling of a microstructure of the material. Dynamic processes in one-dimensional lattices are studied more widely both in linear and nonlinear cases (Askar, 1985; Maugin, 1999; Ostoja-Starzewski, 2002) while two-dimensional lattices are mainly considered in the linear case (Ostoja-Starzewski, 2002; Metrikine and Askes, 2006; Askes and Metrikine, 2005). Some of the two-dimensional processes can be studied in the one-dimensional approximation, for example, propagation of plane waves (Porubov and Berinskii, 2014), while their transverse instability requires a two-dimensional analysis.

Dynamic processes in lattices are investigated using both discrete and continuum modelling (Askar, 1985; Maugin, 1999; Ostoja-Starzewski, 2002; Andrianov et al, 2010). In the linear case, one can solve analytically both the discrete and the continuum equations. However, only certain discrete nonlinear equations, such as, for example, the Toda lattice equation, have exact solutions, see, e.g. Ablowitz and Segur (1981). Therefore, in order to obtain continuous equations in the nonlinear case, an approach based on the continuum long-wave limit of the original discrete equation is needed (Askar, 1985; Maugin, 1999; Ostoja-Starzewski, 2002; Andrianov et al, 2010) that corresponds to a discrete model only for small wave numbers.

Lattice structural features are usually taken into account when describing the negative Poisson ratio (Lakes, 1991; Prawoto, 2012; Vasiliev, 2013; Dirrenberger et al, 2013; Erofeev and Pavlov, 2015). It was obtained in Baughman et al (1998), that metals with cubic structure possess negative value of Poisson's ratio or auxetic behavior exactly because of their crystal lattice features. Relations for the Poisson ratio in materials with cubic lattice can be found in Turley and Sines (1971). Possible applications of auxetics in various branches of industry and technology may be found, e.g., in Stavroulakis (2005). In particular, they may be useful for creating semiconductors with low thermal conductivity and ultra-high hole mobility Zhang et al (2016), or in aerospace engineering (Alderson and Alderson, 2007). Also, in medicine, such materials are used to create blood vessels, which tend to increase their walls' thickness (rather than decrease) in response to the blood pulsation to prevent the vessel's rupture (Evans and Alderson, 2000). The range of applications varies from sport industry (Sanami et al, 2014) to the military-industrial complex (Underhill, 2014). Among the models of auxetics, one considers the particles which are no longer represented as material points, but as bodies possessing shape and mass, with additional degrees of freedom (Erofeev and Pavlov 2015) The system of anisotropic particles is considered, and the couple stress interactions are taken into account: a square lattice is modeled as a set of rigid circular particles possessing two translational and one rotational degrees of freedom.

In this paper, a new formalism is introduced to simplify the derivation of nonlocal equations. To do this, we extend the method based on the usage of the shift operators developed in Michelitsch et al (2014), up to a two-dimensional problem. The dispersion relation analysis is used to reveal additional effects that non-local interactions bring into the model changing the shape of the dispersion curve and the sign of the group velocity. In addition, the continuum limit of the discrete equations allows us to see how non-locality affects elastic constants and to analyze the effect of additional long-range interactions on ta possible auxetic behavior of the model.

34.2 Linear Local Model

Let us consider the square lattice with particles whose interactions are modeled by linearly elastic springs, see Fig. 34.1. The particles have equal masses M, the lattice constant in horizontal and vertical directions is denoted by h, the rigidities of the horizontal and vertical springs are equal to C_1 , while rigidity of the diagonal springs is C_2. We introduce the displacement vector for the particle with the number $m,\ n$ in the following form: $\overrightarrow{U}_{m,n} = u_{m,n}\overrightarrow{i} + v_{m,n}\overrightarrow{j}$, where $u_{m,n}, v_{m,n}$ are the horizontal and vertical components of the displacement vector, and n_i are the unit vectors corresponding to the directions from the particle m,n to its neighbour as shown in Fig. 34.1,

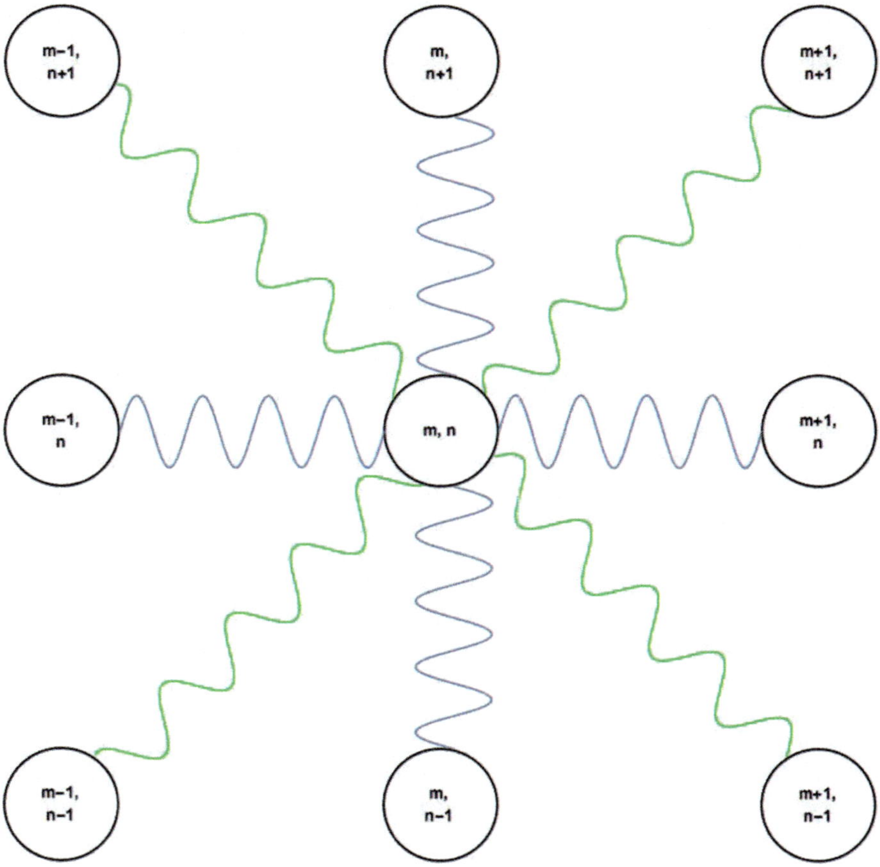

Fig. 34.1: Local model of square lattice.

$$\begin{aligned}
&\vec{n}_1=\vec{i}, \quad \vec{n}_2=-\vec{i}, \quad \vec{n}_3=\vec{j}, \quad \vec{n}_4=-\vec{j},\\
&\vec{n}_5=\frac{1}{\sqrt{2})}(\vec{i}+\vec{j}), \quad \vec{n}_6=\frac{1}{\sqrt{2}}(-\vec{i}+\vec{j}),\\
&\vec{n}_7=\frac{1}{\sqrt{2}}(\vec{i}-\vec{j}), \quad \vec{n}_8=\frac{1}{\sqrt{2}}(-\vec{i}-\vec{j}).
\end{aligned} \tag{34.1}$$

The shift operators,

$$\begin{aligned}
D_1 &= \sum_{m=0}^{\infty}\frac{h^m}{m!}\frac{d^m}{dx^m}=e^{h\frac{d}{dx}}, \quad D_2=\sum_{n=0}^{\infty}\frac{h^n}{n!}\frac{d^n}{dy^n}=e^{h\frac{d}{dy}},\\
D_1^\dagger &= \sum_{m=0}^{\infty}\frac{(-h)^m}{m!}\frac{d^m}{dx^m}=e^{-h\frac{d}{dx}}, \quad D_2^\dagger=\sum_{n=0}^{\infty}\frac{(-h)^n}{n!}\frac{d^n}{dy^n}=e^{-h\frac{d}{dy}},\\
D_1D_2 &= \sum_{m=0}^{\infty}\sum_{n=0}^{\infty}\frac{h^{m+n}}{m!n!}\frac{d^m}{dx^m}\frac{d^n}{dy^n}=e^{h(\frac{d}{dx}+\frac{d}{dy})},\\
D_1D_2^\dagger &= \sum_{m=0}^{\infty}\sum_{n=0}^{\infty}\frac{(-1)^n h^{m+n}}{m!n!}\frac{d^m}{dx^m}\frac{d^n}{dy^n}=e^{h(\frac{d}{dx}-\frac{d}{dy})},\\
D_1^\dagger D_2 &= \sum_{m=0}^{\infty}\sum_{n=0}^{\infty}\frac{(-1)^m h^{m+n}}{m!n!}\frac{d^m}{dx^m}\frac{d^n}{dy^n}=e^{h(-\frac{d}{dx}+\frac{d}{dy})},\\
D_1^\dagger D_2^\dagger &= \sum_{m=0}^{\infty}\sum_{n=0}^{\infty}\frac{(-1)^{m+n} h^{m+n}}{m!n!}\frac{d^m}{dx^m}\frac{d^n}{dy^n}=e^{h(-\frac{d}{dx}-\frac{d}{dy})},
\end{aligned}$$

act on the function $u(x_m, y_n)$ as

$$\begin{aligned}
D_1u_{m,n} &= u(x_m+h,y_n)=u(x_{m+1},y_n)=u_{m+1,n},\\
D_1^\dagger u_{m,n} &= u(x_m-h,y_n)=u(x_{m-1},y_n)=u_{m-1,n},\\
D_2u_{m,n} &= u(x_m,y_n+h)=u(x_m,y_{n+1})=u_{m,n+1},\\
D_2^\dagger u_{m,n} &= u(x_m,y_n-h)=u(x_m,y_{n-1})=u_{m,n-1},\\
D_1*D_2u_{m,n} &= u(x_m+h,y_n+h)=u(x_{m+1},y_{n+1})=u_{m+1,n+1},\\
D_1*D_2^\dagger u_{m,n} &= u(x_m+h,y_n-h)=u(x_{m+1},y_{n-1})=u_{m+1,n-1},\\
D_1^\dagger*D_2u_{m,n} &= u(x_m-h,y_n+h)=u(x_{m-1},y_{n+1})=u_{m-1,n+1},\\
D_1^\dagger*D_2^\dagger u_{m,n} &= u(x_m-h,y_n-h)=u(x_{m-1},y_{n-1})=u_{m-1,n-1}.
\end{aligned} \tag{34.2}$$

Similar expressions hold for the action on the function $v(x_m, y_n)$.

Let the potential energy $V_{m,n}$ contains only the local part, $V_1(u_m, v_n)$, responsible for the interaction of the central particle (m,n) with the neighboring particles shown in Fig. Fig. 34.1. The interactions should take into account the geometry of the lattice characterized by the unit vectors n_i introduced above, while the shift operators allow us to write the expression for the energy in a compact form,

$$V_l=\frac{C_1}{2}\Big(((D_1-1)\vec{U}_{m,n}\cdot\vec{n_1})^2-((D_1^\dagger-1)\vec{U}_{m,n}\cdot\vec{n_2})^2+((D_2-1)\vec{U}_{m,n}\cdot\vec{n_3})^2+$$

$$((D_2^\dagger - 1)\overrightarrow{U}_{m,n} \cdot \overrightarrow{n_4})^2\Big) + \frac{C_2}{2}\Big(9(D_1 D_2 - 1)\overrightarrow{U}_{m,n} \cdot \overrightarrow{n_5})^2 + ((D_1^\dagger D_2^\dagger - 1)\overrightarrow{U}_{m,n} \cdot \overrightarrow{n_6})^2 +$$

$$((D_1^\dagger D_2 - 1)\overrightarrow{U}_{m,n} \cdot \overrightarrow{n_7})^2 + ((D_1 D_2^\dagger - 1)\overrightarrow{U}_{m,n} \cdot \overrightarrow{n_8})^2\Big), \tag{34.3}$$

To obtain discrete equations of motion, it is necessary to write down the Lagrangian,

$$L_{m,n} = T_{m,n} - V_{m,n} = \frac{M}{2}(\dot{u}_{m,n}^2 + \dot{v}_{m,n}^2) - V_1, \tag{34.4}$$

where $T_{m,n}$ is the kinetic energy of the particle with the number m,n. Then the discrete equations of motion are obtained by utilizing the Hamilton-Ostrogradsky variational principle. They are

$$\begin{aligned} M\,\ddot{u}_{m,n} &= C_1(D_1 + D_1^\dagger - 2)u_{m,n} \\ &+ \frac{C_2}{2}\Big((D_1 * D_2 + D_1 * D_2^\dagger + D_1^\dagger * D_2 + D_1^\dagger * D_2^\dagger - 4)u_{m,n} \\ &\quad + (D_1 * D_2 - D_1 * D_2^\dagger + D_1^\dagger * D_2 - D_1^\dagger * D_2^\dagger)v_{m,n}\Big) \end{aligned} \tag{34.5}$$

$$\begin{aligned} M\,\ddot{v}_{m,n} &= C_1(D_2 + D_2^\dagger - 2)v_{m,n} \\ &+ \frac{C_2}{2}\Big((D_1 * D_2 + D_1 * D_2^\dagger + D_1^\dagger * D_2 + D_1^\dagger * D_2^\dagger - 4)v_{m,n} \\ &\quad + (D_1 * D_2 - D_1 * D_2^\dagger + D_1^\dagger * D_2 - D_1^\dagger * D_2^\dagger)u_{m,n}\Big) \end{aligned} \tag{34.6}$$

Nonlocal interactions result in variations in the equations of motion.

34.3 Nonlocal Linear Model

Consider the lattice model shown in Fig. 34.2 where the central particle m,n interacts with additional non-neighboring particles. The rigidities of the springs modeling interactions with four vertical and horizontal non-neighbors is assumed to be C_3, while for four diagonal particles the rigidity of the springs is equal to C_4.

For further investigation it's necessary to choose a "generator" function, which, on the basis of the shift operators formalism and the expression for energy in the local case, would describe the nonlocal behaviour of the lattice, and would generate energy expressions for any indicated order of non-locality. The power function is the most suitable in this case Michelitsch et al (2014), the exponent of which determines the order of non-locality. Consequently, different orders of non-locality of the model correspond to different exponents. Thus, for the second order of non-locality shown in Fig. 34.2 the non-local interactions are described by the energy V_n in the form

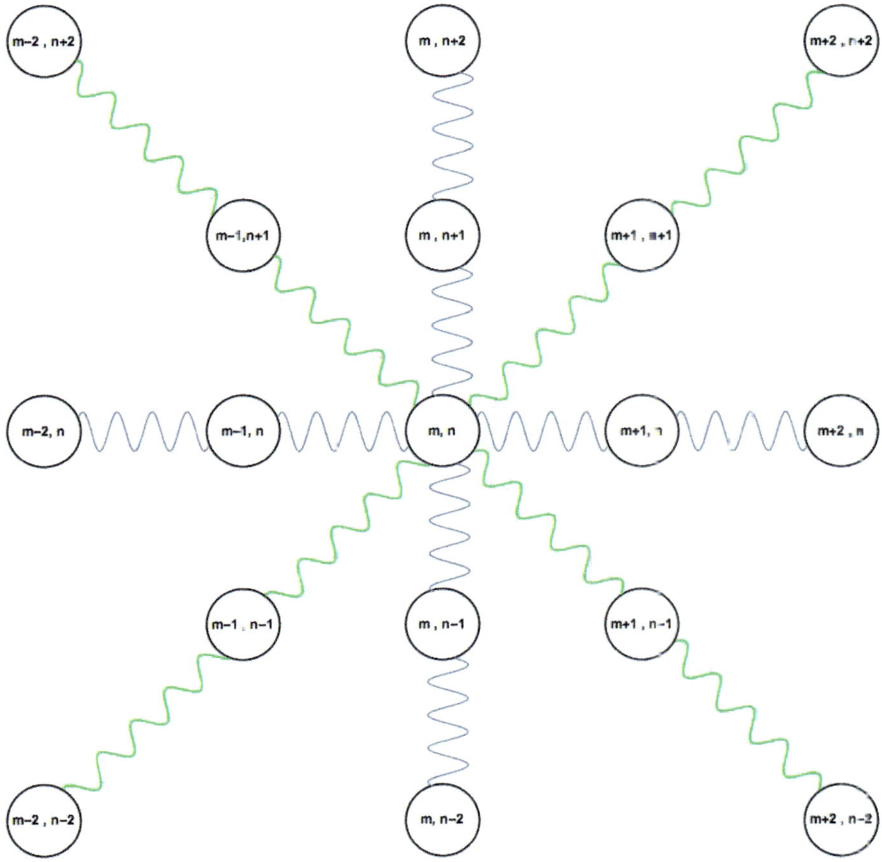

Fig. 34.2: Lattice with the second order non-locality.

$$V_n = \frac{C_3}{2}\Big[((D_1^2-1)\vec{U}_{m,n}\cdot\vec{n_1})^2 + ((D_1^{\dagger 2}-1)\vec{U}_{m,n}\cdot\vec{n_2})^2 + ((D_2^2-1)\vec{U}_{m,n}\cdot\vec{n_3})^2 + ((D_2^{\dagger 2}-1)\vec{U}_{m,n}\cdot\vec{n_4})^2\Big] + \frac{C_4}{2}\Big[((D_1^2D_2^2-1)\vec{U}_{m,n}\cdot\vec{n_5})^2 + ((D_1^{\dagger 2}D_2^{\dagger 2}-1)\vec{U}_{m,n}\cdot\vec{n_6})^2 + ((D_1^{\dagger 2}D_2^2-1)\vec{U}_{m,n}\cdot\vec{n_7})^2 + ((D_1^2D_2^{\dagger 2}-1)\vec{U}_{m,n}\cdot\vec{n_8})^2\Big]. \quad (34.7)$$

Then the whole potential energy V is

$$V = V_l + V_n \quad (34.8)$$

where the local energy V_l is defined by Eq. (34.3). The utilization of the variational principle for the Lagrangian (34.4) with the potential energy defined by Eq. (34.8) results in the discrete equations of motion,

$$M\,\ddot{u}_{m,n} = C_1(D_1 + D_1^\dagger - 2)u_{m,n} + C_3(D_1^2 + D_1^{\dagger 2} - 2)u_{m,n} +$$
$$\frac{C_2}{2}\Big[(D_1D_2 + D_1^\dagger D_2 + D_1D_2^\dagger + D_1^\dagger D_2^\dagger - 4)u_{m,n} +$$
$$(D_1D_2 - D_1^\dagger D_2 - D_1D_2^\dagger + D_1^\dagger D_2^\dagger)v_{m,n}\Big] +$$
$$\frac{C_4}{2}\Big[(D_1^2D_2^2 + D_1^{\dagger 2}D_2^2 + D_1^2D_2^{\dagger 2} + D_1^{\dagger 2}D_2^{\dagger 2} - 4)u_{m,n} +$$
$$\frac{C_4}{2}(D_1^2D_2^2 - D_1^{\dagger 2}D_2^2 - D_1^2D_2^{\dagger 2} + D_1^{\dagger 2}D_2^{\dagger 2})v_{m,n}\Big], \tag{34.9}$$
$$M\,\ddot{v}_{m,n} = C_1(D_2 + D_2^\dagger - 2)v_{m,n} + C_3(D_2^2 + D_2^{\dagger 2} - 2)v_{m,n} +$$
$$\frac{C_2}{2}\Big[(D_1D_2 + D_1^\dagger D_2 + D_1D_2^\dagger + D_1^\dagger D_2^\dagger - 4)v_{m,n} +$$
$$(D_1D_2 - D_1^\dagger D_2 - D_1D_2^\dagger + D_1^\dagger D_2^\dagger)u_{m,n}\Big] +$$
$$\frac{C_4}{2}\Big[(D_1^2D_2^2 + D_1^{\dagger 2}D_2^2 + D_1^2D_2^{\dagger 2} + D_1^{\dagger 2}D_2^{\dagger 2} - 4)v_{m,n} +$$
$$\frac{C_4}{2}(D_1^2D_2^2 - D_1^{\dagger 2}D_2^2 - D_1^2D_2^{\dagger 2} + D_1^{\dagger 2}D_2^{\dagger 2})u_{m,n}\Big]. \tag{34.10}$$

The influence of the additional nonlocal terms in Eqs. (34.9), (34.10) will be studied using the dispersion relation analysis.

34.4 Dispersion Relations Analysis

Let us consider the propagation of longitudinal plane waves in the horizontal direction. In this case, the transverse displacement is absent, $v_{m,n} = 0$, and no variations by n occur, i.e., the corresponding shift operators are equal to zero. Then the local model (34.5), (34.6) is reduced to only one equation of motion,

$$M\,\ddot{u}_m = (C_1 + C_2)(2(1 - D_1 + D_1^\dagger)u_m). \tag{34.11}$$

In the case of the non-local model (34.9), (34.10) one obtains

$$M\,\ddot{u}_m = (C_1 + C_2)(2(1 - D_1 + D_1^\dagger)u_m + (C_3 + C_4)(2(1 - D_1^2 + D_1^{\dagger 2})u_m). \tag{34.12}$$

The solution of both equations is sought in the form

$$u_m = A\,\exp\,\imath(k\,m - \omega\,t), \tag{34.13}$$

where k is normalized wave number, ω is frequency.

Substitution of Eq. (34.13) into the local model equation (34.11) gives rise to the dispersion relation,

$$\omega^2 = \frac{4}{M}((C_1 + C_2)\sin^2(k/2), \tag{34.14}$$

while substitution to Eq. (34.12) results in nonlocal dispersion relation,

$$\omega^2 = \frac{4}{M}\left((C_1 + C_2)\sin^2(k/2) + (C_3 + C_4)\sin^2(k) \right). \tag{34.15}$$

The first derivative of ω^2 in Eq.(34.15) with respect to k reveals additional values of k where the first derivative is zero. Besides $k = \pi N$, N is integer, from the local model, there are additional extrema defined by $C_1 + C_2 + 4(C_3 + C_4)\cos(k) = 0$. Comparison of the dispersion curves is shown Fig. 34.3, where an appearance additional extrema for nonlocal model is seen. The difference in the shapes for nonlocal model results in variations in the shape of the curve of the group velocity a shown in Fig. 34.4. In particular, it provides different signs of the group velocity for one and the same k.

34.5 Continuum Equations

The transition from discrete equations to continuum ones is provided by the following algorithm: the discrete displacements $u_{m,n}, v_{m,n}$ for small values of the wave number are associated with smooth differentiable functions $u(x,y,t)$ and $v(x,y,t)$, describing the displacements in the continuum medium to be obtained. The continuum displacements for the particles with neighboring numbers the Taylor series expansion around the position m,n is used.

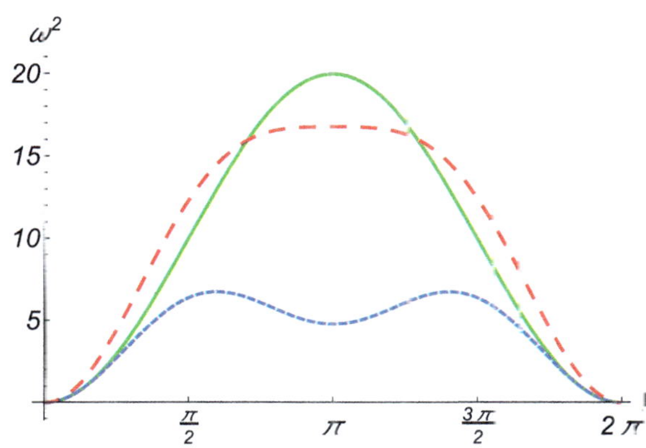

Fig. 34.3 Dispersion curves for local and nonlocal model. Solid line corresponds to local relation (34.14), dashed line corresponds to nonlocal relation (34.15) at $C_1 + C_2 > 4(C_3 + C_4)$, dotted line corresponds to the curve (34.15) at $C_1 + C_2 < 4(C_3 + C_4)$.

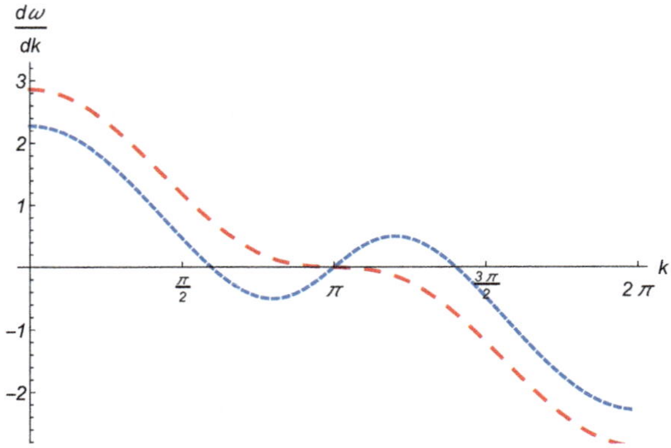

Fig. 34.4 Group velocities for nonlocal model: dashed line corresponds to nonlocal relation (34.15) at $C_1 + C_2 > 4(C_3 + C_4)$, dotted line corresponds to the curve (34.15) at $C_1 + C_2 < 4(C_3 + C_4)$.

34.5.1 Local Model

Discrete local equations (34.5), (34.6) in the long wave continuum limit are

$$\begin{aligned} M\ u_{tt} &= h^2[C_2(u_{yy} + 2v_{xy}) + (C_1 + C_2)u_{xx}], \\ M\ v_{tt} &= h^2[C_2(v_{xx} + 2u_{xy}) + (C_1 + C_2)v_{yy}]. \end{aligned} \tag{34.16}$$

Equations (34.16) are related to the equations of motion of the cubic crystals provided that the elastic cubic constants C_{11}, C_{12} and C_{44} are connected with our constants C_1, C_2 as

$$C_{11} = \frac{C_1 + C_2}{a}, \quad C_{12} = \frac{C_2}{a}, \quad C_{44} = \frac{C_2}{a} \tag{34.17}$$

It can be seen from Eqs. (34.17) that such relations occur only if the Cauchy condition ($C_{12} = C_{44}$) is fulfilled, i.e., when central forces act between uniformly deformed regions of the lattice. The Poisson ratio is not the same in different directions for cubic crystals. Using the expressions for the main crystallographic directions of the cubic lattice (Turley and Sines, 1971), we obtain:

$$\nu_{<110,001>} = \frac{4C_2^2}{C1^2 + 5C_1C_2 + 2C_2^2}, \quad \nu_{<110,1\underline{1}0>} = \frac{C_1^2 + C_1C_2 - 2C_2^2}{C_1^2 + 5C_1C_2 + 2C_2^2} \tag{34.18}$$

Only Poisson's ratio in the $< 110, 1\underline{1}0 >$ direction may possess negative values if $C_1 < C_2$. The designation corresponds to the measurement of Poisson's ratio in the direction $< 1\underline{1}0 >$ when stretched along $< 110 >$, see Figs. 34.5 and 34.6. The figures show the crystallographic directions in the cubic lattice.

34.5.2 Nonlocal Model

Discrete nonlocal equations (34.9), (34.10) in the long wave continuum limit are

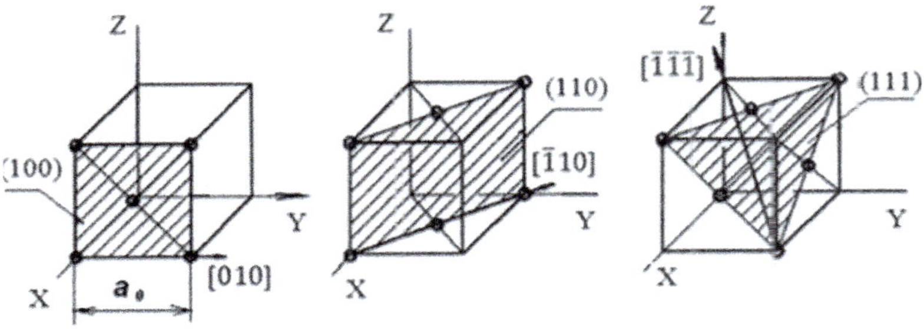

Fig. 34.5: Crystallographic directions

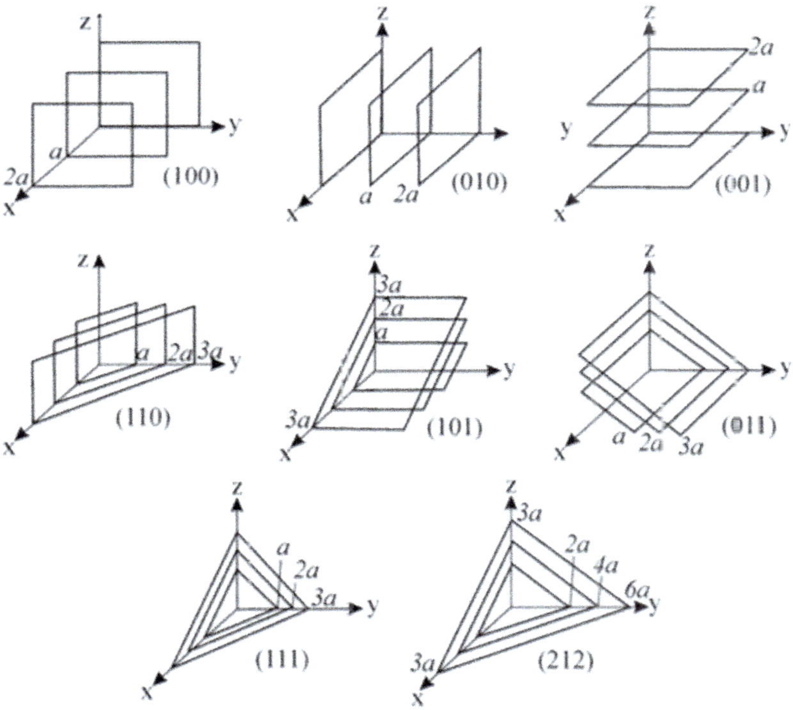

Fig. 34.6: Crystallographic directions

$$
\begin{aligned}
M\, u_{tt} &= a^2[(C_2+4C_4)u_{yy}+2(C_2+4C_4)v_{xy}+(C_1+C_2+4(C_3+C_4))u_{xx}],\\
M\, v_{tt} &= a^2[(C_2+4C_4)v_{xx}+2(C_2+4C_4)u_{xy}+((C_1+C_2+4(C_3+C_4))v_{yy}]
\end{aligned}
$$

Cubic elastic constants are connected with the rigidities of the model as

$$
C_{11} = \frac{C_1+C_2+4(C_3+C_4)}{a}, \quad C_{12} = \frac{C_2+4C_4}{a}, \quad C_{44} = \frac{C_2+4C_4}{a} \tag{34.19}
$$

Then anisotropic Poisson ratios are

$$\nu_{<100,001>} = \frac{C_2 + 4C_4}{C_1 + 2C_2 + 4C_3 + 8C_4} \tag{34.20}$$

$$\nu_{<110,1\underline{1}0>} = \frac{(C_1 - C_2 + 4C_3 - 4C_4)(C_1 + 2C_2 + 4C_3 + 8C_4)}{(2C_2 + 4C_4)(C_1 + C_2 + 4(C_3 + C_4)) + (C_1 + 4C_3)(C_1 + 3C_2 + 4(C_3 + 3C_4))} \tag{34.21}$$

One of the coefficients may be negative,

$$\nu_{<110,1\underline{1}0>} < 0 \tag{34.22}$$

if $C_1 + 4C_3 < C_2 + 4C_4$. Hence, the probability of manifestation of auxetic properties may be provided by the greater contribution of the diagonal interactions. This may be done by nonlocal diagonal interaction described by the rigidity C_4. The longitudinal nonlocal interaction supports positive values of the Poisson ratio.

34.5.3 Nonlinear Interaction

The proposed algorithm of obtaining discrete equations using the shift operators may be generalized for a weak nonlinearity. In particular, it is done by the following expression for energy in the local model

$$\begin{aligned}
V_{nloc} = &\frac{1}{2}[C_1((D_1-1)\vec{U}_{m,n}\cdot\vec{n_1})^2 + C_1((D_1^\dagger - 1)\vec{U}_{m,n}\cdot\vec{n_2})^2 + C_1((D_2 - 1)\vec{U}_{m,n}\cdot\vec{n_3})^2 \\
&+ C_1((D_2^\dagger - 1)\vec{U}_{m,n}\cdot\vec{n_4})^2 + C_2(((D_1D_2 - 1)\vec{U}_{m,n}\cdot\vec{n_5})^2 + ((D_1^\dagger D_2^\dagger - 1)\vec{U}_{m,n}\cdot\vec{n_6})^2 + \\
&((D_1^\dagger D_2 - 1)\vec{U}_{m,n}\cdot\vec{n_7})^2 + ((D_1 D_2^\dagger - 1)\vec{U}_{m,n}\cdot\vec{n_8})^2) + \frac{Q_1}{3}(((D_1 - 1)\vec{U}_{m,n}\cdot\vec{n_1})^3 + \\
&((D_1^\dagger - 1)\vec{U}_{m,n}\cdot\vec{n_2})^3 + ((D_2 - 1)\vec{U}_{m,n}\cdot\vec{n_3})^3 + ((D_2^\dagger - 1)\vec{U}_{m,n}\cdot\vec{n_4})^3) \\
&+ \frac{Q_2}{3}(((D_1D_2 - 1)\vec{U}_{m,n}\cdot\vec{n_5})^3 + ((D_1^\dagger D_2^\dagger - 1)\vec{U}_{m,n}\cdot\vec{n_6})^3 + \\
&((D_1^\dagger D_2 - 1)\vec{U}_{m,n}\cdot\vec{n_7})^3 + ((D_1 D_2^\dagger - 1)\vec{U}_{m,n}\cdot\vec{n_8})^3]
\end{aligned} \tag{34.23}$$

where $Q - i$ account for nonlinear rigidity. More detailed consideration of nonlinear problem will be done elsewhere.

34.6 Conclusion

The algorithm is developed to obtain two-dimensional discrete equations of motion for different orders of non-locality in the square lattice. Owing to the introduced mathematical formalism, the equations can be obtained in a simpler way because the nonlocal model is obtained from the local by raising the operators included in the expression for the internal energy. Dispersion relation analysis of the plane wave in the discrete system shows the role of nonlocality in the appearance of additional extrema in the dispersion curve and in the sign of the group velocity. In addition, it is shown that the Poisson coefficients for a square lattice in the nonlocal model can take negative values and, accordingly, the lattice demonstrates auxetic behavior along one of the crystalline directions. Also, the lattice rigidities relation at which the auxetic behavior will definitely manifest itself has been derived.

The approach to describe non-local interactions in crystals that was presented in this work, can also be generalized to any type of crystal lattices. Further development is connected, first of all, with the development of a nonlinear model based on the shift operators. Taking into account nonlinear terms greatly complicates the two-dimensional problem, which makes it necessary to make reasonable simplifications with respect to the nature of the wave propagation, the characteristic scale for the weak nonlinearity and its correlation to the scale for changes in the direction perpendicular to the direction of nonlinear strain wave propagation. Another possible direction in the development of the algorithm based on the usage of shift operators is to consider the construction of continuous models in the short-wave limit, just as was done for the triangular lattice in Porubov and Berinskii (2014).

Acknowledgements The work of AVP and AEO has been supported by the Russian Foundation for Basic Researches, grant No 17-01-00230-a.

References

Ablowitz M, Segur H (1981) Solitons and the Inverse Scattering Transform. Society for Industrial and Applied Mathematics

Alderson A, Alderson KL (2007) Auxetic materials. Proceedings of the Institution of Mechanical Engineers, Part G: Journal of Aerospace Engineering 221(4):565–575

Andrianov IV, Awrejcewicz J, Weichert D (2010) Improved continuous models for discrete media. Mathematical Problems in Engineering 2010:35

Askar A (1985) Lattice Dynamical Foundations of Continuum Theories, Series in Theoretical and Applied Mechanics, vol 2. World Scientific, Singapore

Askes H, Metrikine AV (2005) Higher-order continua derived from discrete media: continualisation aspects and boundary conditions. International Journal of Solids and Structures 42(1):187–202

Baughman RH, Shacklette JM, Zakhidov AA, Stafstroem S (1998) Negative poisson's ratios as a common feature of cubic metals. Nature 392(26):362–365

Born M, Huang K (1954) Dynamical Theory of Crystal Lattices. Clarendon Press, Oxford

Dirrenberger J, Forest S, Jeulin D (2013) Effective elastic properties of auxetic microstructures: anisotropy and structural applications. International Journal of Mechanics and Materials in Design 9(1):21–33

Eringen A (1972) Linear theory of nonlocal elasticity and dispersion of plane waves. International Journal of Engineering Science 10(5):425–435

Erofeev VI, Pavlov IS (2015) Parametric identification of crystals having a cubic lattice with negative Poisson's ratios. Journal of Applied Mechanics and Technical Physics 56(6):1015–1022

Evans KE, Alderson A (2000) Auxetic materials: Functional materials and structures from lateral thinking! Advanced Materials 12(9):617–628

Kosevich AM, Savotchenko SE (1999) Peculiarities of dynamics of one-dimensional discrete systems with interaction extending beyond nearest neighbors, and the role of higher dispersion in soliton dynamics. Low Temperature Physics 25(7):550–557

Kunin IA (1975) Elastic Media with Microstructure. Nonlocal Theory of Material Media (in Russ.). Nauka, Moscow

Kuzkin VA, Krivtsov AM, Podolskaya EA, Kachanov ML (2016) Lattice with vacancies: elastic fields and effective properties in frameworks of discrete and continuum models. Philosophical Magazine 96(15):1538–1555

Lakes R (1991) Deformation mechanisms in negative Poisson's ratio materials: structural aspects. Journal of Materials Science 26(9):2287–2292

Manevich AI, Manevitch LI (2005) The Mechanics of Nonlinear Systems with Internal Resonances. Imperial College Press, London

Maugin GA (1999) Nonlinear Waves in Elastic Crystals. Oxford University Press, Oxford

Maugin GA, Pouget J, Drouot R, Collet B (1992) Nonlinear Electromechanical Couplings. John Wiley & Sons, New York

Metrikine AV, Askes H (2006) An isotropic dynamically consistent gradient elasticity model derived from a 2D lattice. Philosophical Magazine 86(21-22):3259–3286

Michelitsch TM, Collet B, Wang X (2014) Nonlocal constitutive laws generated by matrix functions: Lattice dynamics models and their continuum limits. International Journal of Engineering Science 80(Supplement C):106–123

Ostoja-Starzewski M (2002) Lattice models in micromechanics. Appl Mech Rev 55:35–60

Porubov AV, Berinskii IE (2014) Non-linear plane waves in materials having hexagonal internal structure. International Journal of Non-Linear Mechanics 67(Supplement C):27–33

Prawoto Y (2012) Seeing auxetic materials from the mechanics point of view: A structural review on the negative Poisson's ratio. Computational Materials Science 58(Supplement C):140–153

Sanami M, Ravirala N, Alderson K, Alderson A (2014) Auxetic materials for sports applications. Procedia Engineering 72(Supplement C):453–458

Sayadi MK, Pouget J (1991) Soliton dynamics in a microstructured lattice model. Journal of Physics A: Mathematical and General 24(9):2151–2172

Stavroulakis GE (2005) Auxetic behaviour: appearance and engineering applications. physica status solidi (b) 242(3):710–720

Turley J, Sines G (1971) The anisotropy of Young's modulus, shear modulus and Poisson's ratio in cubic materials. Journal of Physics D: Applied Physics 4(2):264–271

Underhill RS (2014) Defense applications of auxetic materials. Defense Applications of Auxetic Materials 1:7–13

Vasiliev A (2013) Analysis of auxetic properties of the cell having finite size particles. Letters on Materials 3(4):288–291

Xu Z, Kartashov YV, Torner L (2005) Soliton mobility in nonlocal optical lattices. Phys Rev Lett 95:113,901

Zabusky NJ, Deem GS (1967) Dynamics of nonlinear lattices I. Localized optical excitations, acoustic radiation, and strong nonlinear behavior. Journal of Computational Physics 2(2):126–153

Zhang LC, Qin G, Fang WZ, Cui HJ, Zheng QR, Yan QB, Su G (2016) Tinselenidene: a two-dimensional auxetic material with ultralow lattice thermal conductivity and ultrahigh hole mobility. Scientific Reports 6:1–9

Chapter 35
A New Class of Models to Describe the Response of Electrorheological and Other Field Dependent Fluids

Vít Průša and Kumbakonam R. Rajagopal

Abstract We propose a new class of models for electrorheological fluids. While the standard constitutive relations for electrorheological fluids are based on the assumption that the stress is a function of the symmetric part of the velocity gradient and the intensity of the electric field, we formulate constitutive relations in an implicit way. The stress, the symmetric part of the velocity gradient and the intensity of the electric field are linked via a tensorial implicit equation. The potential benefit of the new class of models is investigated by the analysis of a simple shear flow in a transverse electric field.

35.1 Introduction

Ever since the pioneering work of Winslow (1949) concerning suspensions that can exhibit significant changes in their response characteristics due to the application of electrical or magnetic fields, field dependent fluids such as magnetorheological fluids, see Rosenweig (1985), and electrorheological fluids, see Conrad et al (1991) or Halsey et al (1992), are finding ever increasing applications in a variety of technological areas such as vibration dampers and absorbers, see Gavin et al (1996), brakes and clutches in automobiles, lubricants, exercise machines, in the fine polishing of surfaces and so forth, see for example Choi et al (2005) and the list of references therein.

Vít Průša
Faculty of Mathematics and Physics, Charles University, Sokolovská 83, Praha 8 – Karlín, CZ 186 75, Czech Republic
e-mail: prusv@karlin.mff.cuni.cz

Kumbakonam R. Rajagopal
Texas A&M University, Department of Mechanical Engineering, 3123 TAMU, College Station TX 77843-3123, United States of America
e-mail: krajagopal@tamu.edu

© Springer International Publishing AG, part of Springer Nature 2018
H. Altenbach et al. (eds.), *Generalized Models and Non-classical Approaches in Complex Materials 1*, Advanced Structured Materials 89,
https://doi.org/10.1007/978-3-319-72440-9_35

While a great deal of effort has been expended in the manufacture of such field dependent fluids, see Liu and Choi (2012) and references therein, the same cannot be said with respect to the development of models to describe the response of such materials and the solution of initial-boundary value problems relevant to the problems that arise in technological applications wherein such materials are employed. While there is interest in understanding the response characteristics of both magnetorheological and electrorheological fluids[1] this study is devoted to the modeling of electrorheological fluids. In particular, we speculate about a new methodology for the development of mathematical models for these fluids.

Electrorheological fluids are essentially suspensions. While some suspensions might be describable by the Navier–Stokes constitutive relation, most cannot be described by the Navier–Stokes model. In fact, electrorheological fluids exhibit a variety of non-Newtonian characteristics, see Abu-Jdayil and Brunn (1997); Kollias and Dimarogonas (1993); Zukoski (1993); Krztoń-Maziopa et al (2005); Martin et al (1994); Gast and Zukoski (1989) or Belza et al (2008) to name a few. Concerning the rheological behaviour of suspensions and colloids, an important characteristic exhibited by many of them is the non-monotone relationship between the shear stress and the shear rate in simple shear flow, see for example Boltenhagen et al (1997); David and Filip (2004) or Galindo-Rosales et al (2011). Such behavior can be described by *implicit constitutive relations* that have been introduced recently within the context of field-independent fluids, see Rajagopal (2003, 2006) and Perlácová and Průša (2015). The question is, whether the same methodology can be fruitfully applied in the context of field-dependent fluids in general, and electrorheological fluids in particular.

The models that are currently in use to describe the response of electrorheological fluids provide explicit expression for the Cauchy stress tensor $\mathbb{T}$ in terms of the density, deformation that the fluid undergoes, temperature, and the electrical field, see Gamota and Filisko (1991); Conrad et al (1991); Rajagopal and Wineman (1992); Rajagopal and Růžička (1996, 2001) and Růžička (2000). While this allows one to incorporate effects associated for instance the invariants of the symmetric part of the velocity gradient in the expression for the stress, whereby providing the ability to model shear thinning, it does not allow for the material properties to depend for instance on the mean value of the stress or other invariants of the stress or on mixed invariants that depend on stress as well as an appropriate measure of the deformation and so forth. Further, if the electrorheological fluid exhibited S-shaped response in the shear stress/shear rate plot, as many suspensions do, see Perlácová and Průša (2015) and Janečka and Průša (2015) and references therein, the standard approach based on expressing the Cauchy stress as a function of the symmetric part of the velocity gradient would be inapplicable.

[1] Field dependent fluids have to be studied within the full scope of an electromagnetic theory and the stress response of such fluids would depend on both the electric as well as the magnetic fields. However, most such fluids seems to primarily respond to either the electric or magnetic fields and hence one can simplify their study by restricting one's interest to one or the other field. By electrorheological fluids we refer to fluids whose stress response depends primarily on the electric field. For such fluids, the effects of the magnetic field are secondary and can be ignored.

In order to address these issues, it is necessary to develop *constitutive relations* wherein one has an implicit relationship between the stress, and the other quantities. This study concerns the development of such implicit constitutive relationship. Note that while we restrict ourselves to electrorheological fluids in order to illustrate the issues involved, a similar approach applies for other field dependent fluids as well.

In order to pose the initial-boundary value problem for the motion of electrorheological fluid fully, one needs to formulate the balance equations for mass, linear and angular momentum, energy, as well as the full Maxwell's equations. However, this is too daunting and challenging problem and hence we shall, as it is commonly practiced, simplify it drastically. The approximations that we shall employ parallel the standard development of explicit models for the stress in terms of the symmetric part of the velocity gradient and the electric field that has worked reasonably well. We will completely ignore Maxwell's equations and consider the electric field as a known and controllable quantity that need not be determined, see Rajagopal and Wineman (1992); Ceccio and Wineman (1994) or Růžička (2004) and references therein for such an approach. The only departure from the previous studies in electrorheology is that we consider the class of models wherein the stress, symmetric part of the velocity gradient and the electric field bear an implicit relation.

Initially we shall define a very general class of implicit constitutive relations that can be used to model electrorheological field dependent fluids. We then particularize to a simpler class of models wherein the symmetric part of the velocity gradient depends on the Cauchy stress and the applied fields. We shall restrict ourselves to what we shall refer to as *algebraic models*. (Constitutive relation is formulated as a relation between pointwise values of the quantities such as Cauchy stress tensor and the symmetric part of the velocity gradient. Neither time not space derivatives of these quantities enter the constitutive relation.) It is quite straightforward to generalize the modeling to include *rate type* models such as the Maxwell, Oldroyd or Burgers fluid models as well as integral type fluid models.

35.2 Preliminaries

A study of the mechanics of field dependent materials, even in the absence of thermal effects, requires the study of the usual balance equations for mass, linear and angular momentum, as well as the equations governing the field in question. If electromagnetic fields are involved, one has to also consider Maxwell's equations. However, in many instances one does not consider the full set of governing equations but simplifies the system, as in the case of magnetohydrodynamics, see for example Chapman and Cowling (1990).

An extreme and useful approximation that is often used in the study of the behavior of electrorheological fluids in particular and field dependent materials in general is to merely consider the balance equations for the mass and balance of linear and angular momentum and consider the field as a variable that is given, that is one does not solve the coupled equations for the balance of mass, linear and angular momentum and Maxwell's equations. This is the approach that we shall adopt[2]. Also, we shall restrict ourselves to the case of *incompressible electrorheological fluids*, but it is straightforward to extend the current approach to the case of compressible fluids, and to other field-dependent fluids as well.

For the electromagnetic field we can use a quasistatic approximation of Maxwell equations

$$\operatorname{div}\left(\varepsilon_0\varepsilon_r\overrightarrow{E}\right) = 0, \tag{35.1a}$$

$$\operatorname{rot}\overrightarrow{E} = 0, \tag{35.1b}$$

where $\overrightarrow{E}$ denotes the intensity of the electric field, ε_r is the relative permittivity (a constant) and ϵ_0 denotes the vacuum permittivity (a constant). For the fluid (no external forces) we get the equations

$$\rho\frac{\mathrm{d}\overrightarrow{v}}{\mathrm{d}t} = \operatorname{div}\mathbb{T} + \epsilon_0\varepsilon_r\left[\nabla\overrightarrow{E}\right]\overrightarrow{E}, \tag{35.2a}$$

$$\mathbb{T} = \mathbb{T}^\top, \tag{35.2b}$$

$$\operatorname{div}\overrightarrow{v} = 0, \tag{35.2c}$$

where ρ is the density of the fluid, $\overrightarrow{v}$ is the velocity field, $\mathbb{T}$ is the Cauchy stress tensor, $\mathbb{D}$ is the symmetric part of the velocity gradient $\mathbb{D} =_{\text{def}} \frac{1}{2}\left(\nabla\overrightarrow{v} + \nabla\overrightarrow{v}^\top\right)$, and $\frac{\mathrm{d}\overrightarrow{v}}{\mathrm{d}t} =_{\text{def}} \frac{\partial\overrightarrow{v}}{\partial t} + \left(\overrightarrow{v}\bullet\nabla\right)\overrightarrow{v}$ denotes the material time derivative. In (35.1) and (35.2) we have already made a constitutive assumption that the polarization $\overrightarrow{P}$ is proportional to the intensity of the electric field, thus $\overrightarrow{P} = \epsilon_0\chi_e\overrightarrow{E}$, where χ_e denotes the electric susceptibility, $\varepsilon_r =_{\text{def}} \chi_e + 1$. Another constitutive assumption that is in fact tacitly introduced in (35.2) is the absence of internal body couples that leads to the symmetry (35.2b) of the Cauchy stress tensor.

The electric field $\overrightarrow{E}$ enters the balance of linear momentum (35.2a) in two ways, "implicitly" in the constitutive equation for the Cauchy stress $\mathbb{T}$, and "explicitly" as a volume force, $\left[\nabla\overrightarrow{E}\right]\overrightarrow{E}$. Obviously, the latter term vanishes if one considers a homogeneous electric field, for example an electric field between two infinite plates, but this is not the case in more complicated geometries.

System (35.1) can be reduced to a single equation for the scalar potential φ,

$$\Delta\varphi = 0, \tag{35.3}$$

where the intensity of the electric field $\overrightarrow{E}$ is related to the potential ϕ through the formula $\overrightarrow{E} = -\nabla\varphi$. Obviously, the equation for the potential can be solved without knowledge of the pressure and velocity field. Consequently, the intensity of the

[2] See Rajagopal and Wineman (1992); Rajagopal and Růžička (1996); Růžička (2000); Rajagopal and Růžička (2001) and Růžička (2004) for a discussion of this approximation. More involved approaches that are necessary to follow in other flow regimes are described for example in Pao (1978); Penfield and Haus (1967) or Eringen and Maugin (1990a,b).

electric field enters system (35.2) as a known "parameter", which justifies the notion of "field dependent fluid".

In order to get a closed system of equations, one needs to complement (35.2) with a constitutive relation that provides a link between the Cauchy stress tensor and the other quantities, namely the electric field $\vec{E}$ and the symmetric part of the velocity gradient $\mathbb{D}$.

35.3 Constitutive Relation

In this section we shall develop *implicit constitutive relations* to describe the response of electrorheological fluids. Within the purview of the classical approach, the Cauchy stress is expressed in terms of the symmetric part of the velocity gradient $\mathbb{D}$ and the intensity of the electric field $\vec{E}$ as

$$\mathbb{T} = \mathfrak{f}\left(\mathbb{D}, \vec{E}\right), \tag{35.4}$$

where $\mathfrak{f}$ denotes a tensorial function.

If the fluid of interest is *isotropic*, then one can appeal to standard representation theorems in continuum mechanics, see Spencer (1971), to obtain the most general expression for the stress in terms of the symmetric part of the velocity gradient and the electric field. If one requires invariance with respect to the full proper orthogonal group, it follows that

$$\begin{aligned} \mathbb{T} = {} & \beta_0 \mathbb{I} + \beta_1 \mathbb{D} + \beta_2 \left(\vec{E} \otimes \vec{E}\right) + \beta_3 \mathbb{D}^2 \\ & + \beta_4 \left(\vec{E} \otimes \mathbb{D}\vec{E} + \mathbb{D}\vec{E} \otimes \vec{E}\right) + \beta_5 \left(\vec{E} \otimes \mathbb{D}^2\vec{E} - \mathbb{D}^2\vec{E} \otimes \vec{E}\right), \end{aligned} \tag{35.5}$$

see also Rajagopal and Wineman (1992). The scalar coefficients $\{\beta_i\}_{i=0}^{5}$ depend on the invariants $\mathrm{Tr}\,\mathbb{D}$, $\mathrm{Tr}\,\mathbb{D}^2$, $\mathrm{Tr}\,\mathbb{D}^3$, $\mathrm{Tr}\left(\vec{E} \otimes \vec{E}\right)$, $\mathrm{Tr}\left(\mathbb{D}\vec{E} \otimes \vec{E}\right)$ and $\mathrm{Tr}\left(\mathbb{D}^2\vec{E} \otimes \vec{E}\right)$.

We shall consider constitutive relations of the form

$$\mathfrak{g}\left(\mathbb{T}, \mathbb{D}, \vec{E}\right) = \mathbb{0}, \tag{35.6}$$

where $\mathfrak{g}$ denotes a tensorial function. The generalisation from (35.4) to (35.5) is motivated by the ideas introduced by Rajagopal (2003, 2005), that have been later fruitfully applied in the mechanics of solids, electroelastic solids, magnetoleastic solids and non-Newtonian fluids, see Rajagopal and Srinivasa (2009, 2007); Málek et al (2010); Průša and Rajagopal (2012b); Le Roux and Rajagopal (2013); Ferlácová and Průša (2015); Kulvait et al (2013); Bustamante and Rajagopal (2013b,a, 2015); Rajagopal and Srinivasa (2015, 2016); Bustamante and Rajagopal (2017) and Arvanitakis (2017) to name a few.

Appealing to standard representation theorems for isotropic tensorial functions, see Smith (1971); Spencer (1971) and Zheng (1994), we find that (35.6) leads to

$$\begin{aligned}
\alpha_0 \mathbb{I} + \alpha_1 \mathbb{T} + \alpha_2 \mathbb{T}^2 + \alpha_3 \mathbb{D} + \alpha_4 \mathbb{D}^2 + \alpha_5 (\mathbb{T}\mathbb{D} + \mathbb{D}\mathbb{T}) + \alpha_6 (\mathbb{T}^2\mathbb{D} + \mathbb{D}\mathbb{T}^2) \\
+ \alpha_7 (\mathbb{D}^2\mathbb{T} + \mathbb{T}\mathbb{D}^2) + \alpha_8 (\mathbb{D}^2\mathbb{T}^2 + \mathbb{T}^2\mathbb{D}^2) + \alpha_9 \left(\vec{E} \otimes \vec{E}\right) \\
+ \alpha_{10} \left(\vec{E} \otimes \mathbb{T}\vec{E} + \mathbb{T}\vec{E} \otimes \vec{E}\right) + \alpha_{11} \left(\vec{E} \otimes \mathbb{D}\vec{E} + \mathbb{D}\vec{E} \otimes \vec{E}\right) \\
+ \alpha_{12} \left(\vec{E} \otimes \mathbb{T}^2\vec{E} + \mathbb{T}^2\vec{E} \otimes \vec{E}\right) + \alpha_{13} \left(\vec{E} \otimes \mathbb{D}^2\vec{E} + \mathbb{D}^2\vec{E} \otimes \vec{E}\right) \\
+ \alpha_{14} \left(\vec{E} \otimes \mathbb{T}\mathbb{D}\vec{E} + \mathbb{T}\mathbb{D}\vec{E} \otimes \vec{E}\right) + \alpha_{15} \left(\mathbb{T}\vec{E} \otimes \mathbb{D}\vec{E} + \mathbb{D}\vec{E} \otimes \mathbb{T}\vec{E}\right) = \mathbb{O}, \quad (35.7)
\end{aligned}$$

where the material moduli $\{\alpha_i\}_{i=0}^{15}$ scalar functions of the invariants

$$I_1 =_{\mathrm{def}} \mathrm{Tr}\,\mathbb{T}, \qquad I_2 =_{\mathrm{def}} \mathrm{Tr}\,\mathbb{T}^2, \qquad I_3 =_{\mathrm{def}} \mathrm{Tr}\,\mathbb{T}^3, \quad (35.8a)$$

$$I_4 =_{\mathrm{def}} \mathrm{Tr}\,\mathbb{D}, \qquad I_5 =_{\mathrm{def}} \mathrm{Tr}\,\mathbb{D}^2, \qquad I_6 =_{\mathrm{def}} \mathrm{Tr}\,\mathbb{D}^3, \quad (35.8b)$$

$$I_7 =_{\mathrm{def}} \mathrm{Tr}\,(\mathbb{T}\mathbb{D}), \qquad I_8 =_{\mathrm{def}} \mathrm{Tr}\left(\mathbb{T}^2\mathbb{D}\right), \qquad I_9 =_{\mathrm{def}} \mathrm{Tr}\left(\mathbb{T}\mathbb{D}^2\right), \quad (35.8c)$$

$$I_{10} =_{\mathrm{def}} \mathrm{Tr}\left(\mathbb{T}^2\mathbb{D}^2\right) \quad (35.8d)$$

$$I_{11} =_{\mathrm{def}} \vec{E} \bullet \vec{E}, \quad (35.8e)$$

$$I_{12} =_{\mathrm{def}} \vec{E} \bullet \mathbb{T}\vec{E}, \qquad I_{13} =_{\mathrm{def}} \vec{E} \bullet \mathbb{T}^2\vec{E}, \quad (35.8f)$$

$$I_{14} =_{\mathrm{def}} \vec{E} \bullet \mathbb{D}\vec{E}, \qquad I_{15} =_{\mathrm{def}} \vec{E} \bullet \mathbb{D}^2\vec{E}, \quad (35.8g)$$

$$I_{16} =_{\mathrm{def}} \mathbb{T}\vec{E} \bullet \mathbb{D}\vec{E} \quad (35.8h)$$

see also Bustamante and Rajagopal (2013b) and comments therein.

It is well-nigh impossible to develop an experimental protocol that can identify the numerous material moduli that are functions of the numerous invariants that characterize the general model. We thus look for a sensible simple sub-class, and we point out some features that make the models of the type (35.7) of interest in the modelling of the response of electrorheological fluids.

First, we decompose the Cauchy stress tensor to the spherical and traceless part,

$$\mathbb{T} = -m\mathbb{I} + \mathbb{S}, \quad (35.9a)$$

where $\mathbb{S}$ is a traceless extra stress tensor, $\mathrm{Tr}\,\mathbb{S} = 0$, and m is the mean normal stress. Now instead of specification of the constitutive relation for the full Cauchy stress tensor $\mathbb{T}$, we will search for a constitutive relation for the extra stress tensor $\mathbb{S}$ only. That is we replace (35.6) by

$$\mathfrak{h}\left(\mathbb{S}, \mathbb{D}, \vec{E}\right) = \mathbb{O}. \quad (35.9b)$$

Doing so, we have lost the opportunity to model phenomena such as viscosity dependent on the mean normal stress. The benefit is that we will be dealing with a simpler but yet very rich class of constitutive relations. (In particular the incompressibility of the material is much easier to handle if we use restricted form (35.9) instead of (35.6).) If we use the representation theorem for function $\mathfrak{h}$ in (35.9b), we get

$$
\begin{aligned}
&\widetilde{\alpha}_1 \mathbb{S} + \widetilde{\alpha}_2 \left(\mathbb{S}^2\right)_\delta + \widetilde{\alpha}_3 \mathbb{D} + \widetilde{\alpha}_4 \left(\mathbb{D}^2\right)_\delta + \widetilde{\alpha}_5 (\mathbb{S}\mathbb{D} + \mathbb{D}\mathbb{S})_\delta + \widetilde{\alpha}_6 \left(\mathbb{S}^2\mathbb{D} + \mathbb{D}\mathbb{S}^2\right)_\delta \\
&\quad + \widetilde{\alpha}_7 \left(\mathbb{D}^2\mathbb{S} + \mathbb{S}\mathbb{D}^2\right)_\delta + \widetilde{\alpha}_8 \left(\mathbb{D}^2\mathbb{S}^2 + \mathbb{S}^2\mathbb{D}^2\right)_\delta + \widetilde{\alpha}_9 \left(\vec{E} \otimes \vec{E}\right)_\delta \\
&\quad + \widetilde{\alpha}_{10} \left(\vec{E} \otimes \mathbb{S}\vec{E} + \mathbb{S}\vec{E} \otimes \vec{E}\right)_\delta + \widetilde{\alpha}_{11} \left(\vec{E} \otimes \mathbb{D}\vec{E} + \mathbb{D}\vec{E} \otimes \vec{E}\right)_\delta \\
&\quad + \widetilde{\alpha}_{12} \left(\vec{E} \otimes \mathbb{S}^2\vec{E} + \mathbb{S}^2\vec{E} \otimes \vec{E}\right)_\delta + \widetilde{\alpha}_{13} \left(\vec{E} \otimes \mathbb{D}^2\vec{E} + \mathbb{D}^2\vec{E} \otimes \vec{E}\right)_\delta \\
&\quad + \widetilde{\alpha}_{14} \left(\vec{E} \otimes \mathbb{S}\mathbb{D}\vec{E} + \mathbb{S}\mathbb{D}\vec{E} \otimes \vec{E}\right)_\delta + \widetilde{\alpha}_{15} \left(\mathbb{S}\vec{E} \otimes \mathbb{D}\vec{E} + \mathbb{D}\vec{E} \otimes \mathbb{S}\vec{E}\right)_\delta = \mathbb{C},
\end{aligned}
\tag{35.10}
$$

where $\{\widetilde{\alpha}_i\}_{i=1}^{16}$ are functions of combined invariants of $\mathbb{S}$, $\mathbb{D}$ and $\vec{E}$. (These are essentially the same as those listed in (35.8a), it suffices to replace $\mathbb{T}$ by $\mathbb{S}$.) The symbol $\mathbb{A}_\delta$ denotes the traceless part of the corresponding tensor, that is $\mathbb{A}_\delta = \mathbb{A} - \frac{1}{3}\operatorname{Tr}\mathbb{A}$.

Out of the plethora of particular models in the class (35.9), we shall first focus on *linear models*. This includes the standard models wherein the extra stress tensor is a linear function of the symmetric part of the velocity gradient, and the complementary models wherein the symmetric part of the velocity gradient is a linear function of the extra stress tensor. In the former case the representation theorem implies that

$$
\mathbb{S} = a_3 \mathbb{D} + a_9 \operatorname{Tr}\left(\mathbb{D}\vec{E} \otimes \vec{E}\right) \left(\vec{E} \otimes \vec{E}\right)_\delta + a_{11} \left(\mathbb{D}\vec{E} \otimes \vec{E} + \vec{E} \otimes \mathbb{D}\vec{E}\right)_\delta, \tag{35.11}
$$

where the scalar coefficients a_3, a_9 and a_{11} can be functions of $\vec{E} \bullet \vec{E}$. (Recall that the fluid is assumed to be incompressible, hence the linear invariant $\operatorname{Tr}\mathbb{D}$ vanishes.) On the other hand, if we assume the opposite, that is if we assume that the symmetric part of the velocity gradient $\mathbb{D}$ is a linear function of the extra stress tensor $\mathbb{S}$, then the representation theorem implies that

$$
\mathbb{D} = b_1 \mathbb{S} + b_9 \operatorname{Tr}\left(\mathbb{S}\vec{E} \otimes \vec{E}\right) \left(\vec{E} \otimes \vec{E}\right)_\delta + b_{10} \left(\vec{E} \otimes \mathbb{S}\vec{E} + \mathbb{S}\vec{E} \otimes \vec{E}\right)_\delta, \tag{35.12}
$$

where the scalar coefficients b_1, b_9 and b_{10} can be functions of $\vec{E} \bullet \vec{E}$.

After discussing linear models, we move forward to *nonlinear* models. If one insists on the standard approach, that is if one requires the extra stress tensor $\mathbb{S}$ to be a function of the symmetric part of the velocity gradient $\mathbb{D}$ and the intensity of the electric field $\vec{E}$, then, according to the representation theorem, the most general isotropic tensorial function of $\mathbb{D}$ and $\vec{E}$ has the form

$$\mathbb{S} = a_3\mathbb{D} + a_4\left(\mathbb{D}^2\right)_\delta + a_9\left(\vec{E}\otimes\vec{E}\right)_\delta + a_{11}\left(\mathbb{D}\vec{E}\otimes\vec{E} + \vec{E}\otimes\mathbb{D}\vec{E}\right)_\delta + a_{13}\left(\mathbb{D}^2\vec{E}\otimes\vec{E} + \vec{E}\otimes\mathbb{D}^2\vec{E}\right)_\delta \tag{35.13}$$

where a_3, a_4, a_9 and a_{13} can be functions of the corresponding combined invariants of $\mathbb{D}$ and $\vec{E}$. In particular we will focus on the case where the material coefficients a_3, a_4, a_9 and a_{13} are constants.

Finally, we introduce a fully implicit model that includes all the *bilinear tensorial terms* involving $\mathbb{S}$ and $\mathbb{D}$, that is we assume that the constitutive relation is given by the formula

$$c_1\mathbb{S} + c_3\mathbb{D} + c_5(\mathbb{S}\mathbb{D} + \mathbb{D}\mathbb{S})_\delta + c_9\left(\vec{E}\otimes\vec{E}\right)_\delta + c_{10}\left(\vec{E}\otimes\mathbb{S}\vec{E} + \mathbb{S}\vec{E}\otimes\vec{E}\right)_\delta + c_{11}\left(\vec{E}\otimes\mathbb{D}\vec{E} + \mathbb{D}\vec{E}\otimes\vec{E}\right)_\delta + c_{14}\left(\vec{E}\otimes\mathbb{S}\mathbb{D}\vec{E} + \mathbb{S}\mathbb{D}\vec{E}\otimes\vec{E}\right)_\delta + c_{15}\left(\mathbb{S}\vec{E}\otimes\mathbb{D}\vec{E} + \mathbb{D}\vec{E}\otimes\mathbb{S}\vec{E}\right)_\delta = \mathbb{O}, \tag{35.14}$$

where c_1, c_3, c_5, c_9, c_{10}, c_{11}, c_{14} and c_{15} can be functions of the complete set of invariants.

In what follows, the models will *not assessed with respect to their consistency with the laws of thermodynamics*. Such a task would require one to deal with intricated nonlinear models, see for example Perlácová and Průša (2015) for a preliminary study in the case of non-Newtonian fluids, and also Rajagopal and Srinivasa (2008) for a general approach to the same. Here the aim is different. We want to discuss the potential of the additional tensorial terms especially form the perspective of their qualitative features such as the presence of normal stress differences[3], which is an important qualitative characteristics of complex fluids, see for example Coleman et al (1966); Barnes et al (1989) or Tanner and Walters (1998).

35.4 Simple Shear Flow

In order to assess qualitative features of the models, we investigate the simple shear flow under a constant transverse electric field, see Fig. 35.1. The fluid flows in between two infinite parallel plates kept at constant electric potential φ_{top} and φ_{bottom}. The flow is driven by the motion of the top plate that moves with the velocity $\vec{V}$, and the distance between the plates is $2h$. No-slip boundary condition for the velocity field $\vec{v}$ is assumed on both plates.

Given the simple geometry we are interested in, it is straightforward to solve the field equation (35.3). Intensity of the electric field $\vec{E}$ is then given by the formula

[3] See Sect. 35.4 for a precise definition of the concept of normal stress differences.

$$\vec{E} = E\vec{e}_y = \begin{bmatrix} 0 \\ E \\ 0 \end{bmatrix}, \tag{35.15}$$

where $E = \frac{\varphi_B - \varphi_T}{2h}$.

Cauchy stress tensor is assumed to take the form

$$\mathbb{T} = \begin{bmatrix} T_{\hat{x}\hat{x}} & 0 & 0 \\ 0 & T_{\hat{y}\hat{y}} & T_{\hat{y}\hat{z}} \\ 0 & T_{\hat{y}\hat{z}} & T_{\hat{z}\hat{z}} \end{bmatrix}, \tag{35.16}$$

where the components are constants. Further, the velocity field is assumed to take the form of a steady unidirectional flow

$$\vec{v} = v^{\hat{z}}(y)\vec{e}_{\hat{z}}, \tag{35.17}$$

and the magnitude of the velocity is a linear function of y, $v^{\hat{z}} = V_{\text{top}}\left(\frac{y}{h}+1\right)$. If the velocity field takes the form (35.17) and the Cauchy stress field is given by the formula (35.16), then the equations (35.2) with no-slip boundary condition for the velocity are fulfilled. It remains to check whether the corresponding constitutive relation is also satisfied. In other words, using the corresponding constitutive relation, we need to express the components of the stress tensor in terms of the problem data, that is in terms of the gradient $\frac{dv^{\hat{z}}}{dy}$.

A more convenient way of writing the formula for the Cauchy stress tensor is

$$\mathbb{T} = \begin{bmatrix} C & 0 & 0 \\ 0 & C+A & T \\ 0 & T & C+B \end{bmatrix}. \tag{35.18}$$

If the Cauchy stress tensor has the form (35.18), then it is easy to keep track on the normal stress differences. If the coordinate systems is chosen as above, then the *first and second normal stress differences* are defined as follows

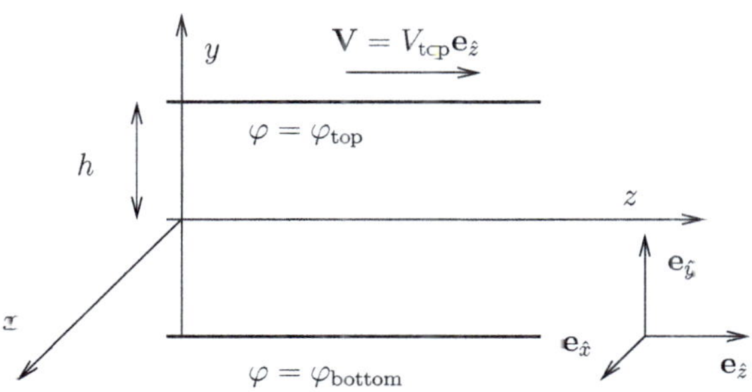

Fig. 35.1 Simple shear flow in a transverse electric field.

$$N_1 = \mathrm{T}_{\hat{z}\hat{z}} - \mathrm{T}_{\hat{y}\hat{y}}, \tag{35.19a}$$

$$N_2 = \mathrm{T}_{\hat{y}\hat{y}} - \mathrm{T}_{\hat{x}\hat{x}}, \tag{35.19b}$$

see for example Coleman et al (1966), hence we see that $N_1 = \mathrm{B} - \mathrm{A}$, and $N_2 = \mathrm{A}$.

The traceless part of $\mathbb{T}$ then reads

$$\mathbb{S} = \begin{bmatrix} -\frac{\mathrm{A}+\mathrm{B}}{3} & 0 & 0 \\ 0 & \frac{2\mathrm{A}-\mathrm{B}}{3} & \mathrm{T} \\ 0 & \mathrm{T} & \frac{2\mathrm{B}-\mathrm{A}}{3} \end{bmatrix}. \tag{35.20}$$

Further, the symmetric part of the velocity gradient is given by the formula

$$\mathbb{D} = \frac{1}{2}\frac{\mathrm{d}v^{\hat{z}}}{\mathrm{d}y}\begin{bmatrix} 0 & 0 & 0 \\ 0 & 0 & 1 \\ 0 & 1 & 0 \end{bmatrix}, \tag{35.21}$$

and the formulae for some of the tensorial terms in the representation theorem read

$$\vec{E}\otimes\vec{E}=E^2\begin{bmatrix}0&0&0\\0&1&0\\0&0&0\end{bmatrix}, \tag{35.22a}$$

$$\left(\vec{E}\otimes\vec{E}\right)_\delta=E^2\begin{bmatrix}-\frac{1}{3}&0&0\\0&\frac{2}{3}&0\\0&0&-\frac{1}{3}\end{bmatrix}, \tag{35.22b}$$

$$\mathbb{D}^2=\frac{1}{4}\left(\frac{\mathrm{d}v^{\hat{z}}}{\mathrm{d}y}\right)^2\begin{bmatrix}0&0&0\\0&1&0\\0&0&1\end{bmatrix}, \tag{35.22c}$$

$$(\mathbb{D}^2)_\delta=\frac{1}{4}\left(\frac{\mathrm{d}v^{\hat{z}}}{\mathrm{d}y}\right)^2\begin{bmatrix}-\frac{2}{3}&0&0\\0&\frac{1}{3}&0\\0&0&\frac{1}{3}\end{bmatrix}, \tag{35.22d}$$

$$\mathbb{D}\mathbb{S}+\mathbb{S}\mathbb{D}=\frac{1}{2}\frac{\mathrm{d}v^{\hat{z}}}{\mathrm{d}y}\begin{bmatrix}0&0&0\\0&2\mathrm{T}&\frac{\mathrm{A+B}}{3}\\0&\frac{\mathrm{A+B}}{3}&2\mathrm{T}\end{bmatrix} \tag{35.22e}$$

$$(\mathbb{D}\mathbb{S}+\mathbb{S}\mathbb{D})_\delta=\frac{1}{3}\frac{\mathrm{d}v^{\hat{z}}}{\mathrm{d}y}\begin{bmatrix}-2\mathrm{T}&0&0\\0&\mathrm{T}&\frac{\mathrm{A+B}}{2}\\0&\frac{\mathrm{A+B}}{2}&\mathrm{T}\end{bmatrix} \tag{35.22f}$$

$$\mathbb{D}\vec{E}\otimes\vec{E}+\vec{E}\otimes\mathbb{D}\vec{E}=\frac{1}{2}E^2\frac{\mathrm{d}v^{\hat{z}}}{\mathrm{d}y}\begin{bmatrix}0&0&0\\0&0&1\\0&1&0\end{bmatrix}, \tag{35.22g}$$

$$\left(\mathbb{D}\vec{E}\otimes\vec{E}+\vec{E}\otimes\mathbb{D}\vec{E}\right)_\delta=\frac{1}{2}E^2\frac{\mathrm{d}v^{\hat{z}}}{\mathrm{d}y}\begin{bmatrix}0&0&0\\0&0&1\\0&1&0\end{bmatrix}, \tag{35.22h}$$

$$\mathbb{D}^2\vec{E}\otimes\vec{E}+\vec{E}\otimes\mathbb{D}^2\vec{E}=E^2\left(\frac{\mathrm{d}v^{\hat{z}}}{\mathrm{d}y}\right)^2\begin{bmatrix}0&0&0\\0&1&0\\0&0&0\end{bmatrix}, \tag{35.22i}$$

$$\left(\mathbb{D}^2\vec{E}\otimes\vec{E}+\vec{E}\otimes\mathbb{D}^2\vec{E}\right)_\delta=E^2\left(\frac{\mathrm{d}v^{\hat{z}}}{\mathrm{d}y}\right)^2\begin{bmatrix}-\frac{1}{3}&0&0\\0&\frac{2}{3}&0\\0&0&-\frac{1}{3}\end{bmatrix}, \tag{35.22j}$$

$$\mathbb{S}\vec{E}\otimes\vec{E}+\vec{E}\otimes\mathbb{S}\vec{E}=E^2\begin{bmatrix}0&0&0\\0&\frac{2}{3}(2\mathrm{A}-\mathrm{B})&\mathrm{T}\\0&\mathrm{T}&0\end{bmatrix}, \tag{35.22k}$$

$$\left(\mathbb{S}\vec{E}\otimes\vec{E}+\vec{E}\otimes\mathbb{S}\vec{E}\right)_\delta=E^2\begin{bmatrix}-\frac{2}{9}(2\mathrm{A}-\mathrm{B})&0&0\\0&\frac{4}{9}(2\mathrm{A}-\mathrm{B})&\mathrm{T}\\0&\mathrm{T}&-\frac{2}{9}(2\mathrm{A}-\mathrm{B})\end{bmatrix}, \tag{35.22l}$$

$$\mathbb{S}\mathbb{D}\vec{E}\otimes\vec{E}+\vec{E}\otimes\mathbb{S}\mathbb{D}\vec{E}=\frac{1}{2}\frac{\mathrm{d}\mathrm{v}^{\hat{z}}}{\mathrm{d}y}E^2\begin{bmatrix}0&0&0\\0&2\mathrm{T}&\frac{2\mathrm{B}-\mathrm{A}}{3}\\0&\frac{2\mathrm{B}-\mathrm{A}}{3}&0\end{bmatrix}, \tag{35.22m}$$

$$\left(\mathbb{S}\mathbb{D}\vec{E}\otimes\vec{E}+\vec{E}\otimes\mathbb{S}\mathbb{D}\vec{E}\right)_{\delta}=\frac{1}{2}\frac{\mathrm{d}\mathrm{v}^{\hat{z}}}{\mathrm{d}y}E^2\begin{bmatrix}-\frac{2}{3}\mathrm{T}&0&0\\0&\frac{4}{3}\mathrm{T}&\frac{2\mathrm{B}-\mathrm{A}}{3}\\0&\frac{2\mathrm{B}-\mathrm{A}}{3}&-\frac{2}{3}\mathrm{T}\end{bmatrix}, \tag{35.22n}$$

$$\mathbb{S}\vec{E}\otimes\mathbb{D}\vec{E}+\mathbb{D}\vec{E}\otimes\mathbb{S}\vec{E}=\frac{1}{2}\frac{\mathrm{d}\mathrm{v}^{\hat{z}}}{\mathrm{d}y}E^2\begin{bmatrix}0&0&0\\0&0&\frac{2\mathrm{A}-\mathrm{B}}{3}\\0&\frac{2\mathrm{A}-\mathrm{B}}{3}&2\mathrm{T}\end{bmatrix}, \tag{35.22o}$$

$$\left(\mathbb{S}\vec{E}\otimes\mathbb{D}\vec{E}+\mathbb{D}\vec{E}\otimes\mathbb{S}\vec{E}\right)_{\delta}=\frac{1}{2}\frac{\mathrm{d}\mathrm{v}^{\hat{z}}}{\mathrm{d}y}E^2\begin{bmatrix}-\frac{2}{3}\mathrm{T}&0&0\\0&-\frac{2}{3}\mathrm{T}&\frac{2\mathrm{A}-\mathrm{B}}{3}\\0&\frac{2\mathrm{A}-\mathrm{B}}{3}&\frac{4}{3}\mathrm{T}\end{bmatrix}. \tag{35.22p}$$

Let us now use the explicit formulae in the specific constitutive relations (35.11), (35.12), (35.13) and (35.14), and let us see the implications of the given constitutive relation with respect to qualitative properties such as the normal stress differences.

35.4.1 Extra Stress Tensor $\mathbb{S}$ is a Linear Function of the Symmetric Part of the Velocity Gradient $\mathbb{D}$

If we substitute (35.22) into (35.11), we get

$$\begin{bmatrix}-\frac{\mathrm{A}+\mathrm{B}}{3}&0&0\\0&\frac{2\mathrm{A}-\mathrm{B}}{3}&\mathrm{T}\\0&\mathrm{T}&\frac{2\mathrm{B}-\mathrm{A}}{3}\end{bmatrix}=a_3\frac{1}{2}\frac{\mathrm{d}\mathrm{v}^{\hat{z}}}{\mathrm{d}y}\begin{bmatrix}0&0&0\\0&0&1\\0&1&0\end{bmatrix}+a_{11}\frac{1}{2}E^2\frac{\mathrm{d}\mathrm{v}^{\hat{z}}}{\mathrm{d}y}\begin{bmatrix}0&0&0\\0&0&1\\0&1&0\end{bmatrix}. \tag{35.23}$$

(The invariant $\mathrm{Tr}\left(\mathbb{D}\vec{E}\otimes\vec{E}\right)$ vanishes in the simple shear flow setting we are interested in, see (35.22g).) Consequently, we see that

$$\mathrm{A}=0, \tag{35.24a}$$

$$\mathrm{B}=0, \tag{35.24b}$$

$$\mathrm{T}=\left(a_3+a_{11}E^2\right)\frac{1}{2}\frac{\mathrm{d}\mathrm{v}^{\hat{z}}}{\mathrm{d}y}. \tag{35.24c}$$

This means that no normal stress differences are generated in the simple shear flow. In fact, the fluid described by the constitutive relation (35.11) effectively behaves, in the simple shear flow, as a fluid with constitutive relation

$$\mathbb{S}=\left(a_3+a_{11}E^2\right)\mathbb{D}, \tag{35.25}$$

which means that the extra tensorial term $\left(\mathbb{D}\vec{E}\otimes\vec{E}+\vec{E}\otimes\mathbb{D}\vec{E}\right)_{\delta}$ in (35.11) has no impact on the structure of the extra stress tensor matrix.

The fact that A and B vanish is a natural consequence of the fact that the normal stress differences are second order effects, hence they can not be captured by a linear constitutive relation.

35.4.2 Symmetric Part of the Velocity Gradient $\mathbb{D}$ is a Linear Function of the Extra Stress Tensor $\mathbb{S}$

Let us now consider the other linear constitutive relation. If we substitute into (35.12), we get

$$\frac{1}{2}\frac{\mathrm{d}v^{\hat{x}}}{\mathrm{d}y}\begin{bmatrix}0&0&0\\0&0&1\\0&1&0\end{bmatrix}=b_1\begin{bmatrix}-\frac{\mathrm{A}+\mathrm{B}}{3}&0&0\\0&\frac{2\mathrm{A}-\mathrm{B}}{3}&\mathrm{T}\\0&\mathrm{T}&\frac{2\mathrm{B}-\mathrm{A}}{3}\end{bmatrix}+b_9\frac{1}{3}E^2\left(2\mathrm{A}-\mathrm{B}\right)\begin{bmatrix}0&0&0\\0&1&0\\0&0&0\end{bmatrix}$$
$$+b_{10}E^2\begin{bmatrix}-\frac{2}{9}\left(2\mathrm{A}-\mathrm{B}\right)&0&0\\0&\frac{4}{9}\left(2\mathrm{A}-\mathrm{B}\right)&\mathrm{T}\\0&\mathrm{T}&-\frac{2}{9}\left(2\mathrm{A}-\mathrm{B}\right)\end{bmatrix}, \tag{35.26}$$

which implies that

$$\mathrm{A}=0, \tag{35.27a}$$

$$\mathrm{B}=0, \tag{35.27b}$$

$$\left(b_1+b_{10}E^2\right)\mathrm{T}=\frac{1}{2}\frac{\mathrm{d}v^{\hat{x}}}{\mathrm{d}y}. \tag{35.27c}$$

Again, as expected, no normal stress differences are generated in the simple shear flow, and the fluid effectively behaves as a fluid with constitutive relation

$$\left(b_1+b_{10}E^2\right)\mathbb{S}=\mathbb{D}, \tag{35.28}$$

hence the extra tensorial term $\left(\vec{E}\otimes\mathbb{S}\vec{E}+\mathbb{S}\vec{E}\otimes\vec{E}\right)_{\delta}$ plays no important role regarding the structure of the extra stress tensor matrix.

35.4.3 Extra stress tensor $\mathbb{S}$ is a function of the symmetric part of the velocity gradient

Substituting for $\mathbb{S}$, $\mathbb{D}$ and $\vec{E}$ into (35.13) yields

$$
\begin{bmatrix} -\frac{\mathrm{A}+\mathrm{B}}{3} & 0 & 0 \\ 0 & \frac{2\mathrm{A}-\mathrm{B}}{3} & \mathrm{T} \\ 0 & \mathrm{T} & \frac{2\mathrm{B}-\mathrm{A}}{3} \end{bmatrix} = a_3 \frac{1}{2} \frac{\mathrm{d}v^{\hat{z}}}{\mathrm{d}y} \begin{bmatrix} 0 & 0 & 0 \\ 0 & 0 & 1 \\ 0 & 1 & 0 \end{bmatrix} + a_4 \frac{1}{4} \left(\frac{\mathrm{d}v^{\hat{z}}}{\mathrm{d}y} \right)^2 \begin{bmatrix} -\frac{2}{3} & 0 & 0 \\ 0 & \frac{1}{3} & 0 \\ 0 & 0 & \frac{1}{3} \end{bmatrix}
$$
$$
+ a_9 E^2 \begin{bmatrix} -\frac{1}{3} & 0 & 0 \\ 0 & \frac{2}{3} & 0 \\ 0 & 0 & -\frac{1}{3} \end{bmatrix} + a_{11} \frac{1}{2} E^2 \frac{\mathrm{d}v^{\hat{z}}}{\mathrm{d}y} \begin{bmatrix} 0 & 0 & 0 \\ 0 & 0 & 1 \\ 0 & 1 & 0 \end{bmatrix} + a_{13} E^2 \left(\frac{\mathrm{d}v^{\hat{z}}}{\mathrm{d}y} \right)^2 \begin{bmatrix} -\frac{1}{3} & 0 & 0 \\ 0 & \frac{2}{3} & 0 \\ 0 & 0 & -\frac{1}{3} \end{bmatrix}. \tag{35.29}
$$

This implies that the following three independent equations must be satisfied,

$$
\mathrm{A} + \mathrm{B} = \left(\frac{a_4}{2} + a_{13} E^2 \right) \left(\frac{\mathrm{d}v^{\hat{z}}}{\mathrm{d}y} \right)^2 + a_9 E^2, \tag{35.30a}
$$

$$
\mathrm{A} - \mathrm{B} = a_9 E^2 + a_{13} E^2 \left(\frac{\mathrm{d}v^{\hat{z}}}{\mathrm{d}y} \right)^2, \tag{35.30b}
$$

$$
\mathrm{T} = \frac{1}{2} \left(a_3 + a_{11} E^2 \right) \frac{\mathrm{d}v^{\hat{z}}}{\mathrm{d}y}. \tag{35.30c}
$$

(The first equation follows from comparison of $\hat{x}\hat{x}$ elements of the matrices, the second equation follows by subtracting equations that follow comparison of $\hat{z}\hat{z}$ and $\hat{y}\hat{y}$ elements in the matrices. The last equation follows from comparison of $\hat{y}\hat{z}$ elements of the matrices.) This means that

$$
\mathrm{A} = \left(\frac{a_4}{4} + a_{13} E^2 \right) \left(\frac{\mathrm{d}v^{\hat{z}}}{\mathrm{d}y} \right)^2 + a_9 E^2, \tag{35.31a}
$$

$$
\mathrm{B} = \frac{a_4}{4} \left(\frac{\mathrm{d}v^{\hat{z}}}{\mathrm{d}y} \right)^2, \tag{35.31b}
$$

$$
\mathrm{T} = \frac{1}{2} \left(a_3 + a_{11} E^2 \right) \frac{\mathrm{d}v^{\hat{z}}}{\mathrm{d}y}, \tag{35.31c}
$$

hence we get *nonzero* normal stress differences.

In particular, if we want the material parameters in (35.13) to be constants, and if we require the extra stress tensor to vanish if $\mathbb{D} = 0$ and $\overrightarrow{E} = \overrightarrow{0}$, then we have to fix $a_9 = 0$. In such a case, we see that the constitutive relation does not allow us to describe a fluid where $\mathrm{A} + \mathrm{B} = 0$ for all shear rates and intensities of the electric field, and simultaneously have $\mathrm{A} \neq 0$ and $\mathrm{B} \neq 0$.

35.4.4 Fully Implicit Constitutive Relation – Constitutive Relation with Bilinear Tensorial Terms

Substituting for $\mathbb{S}$, $\mathbb{D}$ and $\overrightarrow{E}$ into (35.14) yields

$$c_1\begin{bmatrix}-\frac{A+B}{3} & 0 & 0\\ 0 & \frac{2A-B}{3} & T\\ 0 & T & \frac{2B-A}{3}\end{bmatrix}+c_3\frac{1}{2}\frac{dv^{\hat{z}}}{dy}\begin{bmatrix}0&0&0\\0&0&1\\0&1&0\end{bmatrix}+c_5\frac{1}{3}\frac{dv^{\hat{z}}}{dy}\begin{bmatrix}-2T & 0 & 0\\ 0 & T & \frac{A+B}{2}\\ 0 & \frac{A+B}{2} & T\end{bmatrix}$$
$$+c_9E^2\begin{bmatrix}-\frac{1}{3}&0&0\\0&\frac{2}{3}&0\\0&0&-\frac{1}{3}\end{bmatrix}+c_{10}E^2\begin{bmatrix}-\frac{2}{9}(2A-B) & 0 & 0\\ 0 & \frac{4}{9}(2A-B) & T\\ 0 & T & -\frac{2}{9}(2A-B)\end{bmatrix}$$
$$+c_{11}\frac{1}{2}E^2\frac{dv^{\hat{z}}}{dy}\begin{bmatrix}0&0&0\\0&0&1\\0&1&0\end{bmatrix}+c_{14}\frac{1}{2}\frac{dv^{\hat{z}}}{dy}E^2\begin{bmatrix}-\frac{2}{3}T & 0 & 0\\ 0 & \frac{4}{3}T & \frac{2B-A}{3}\\ 0 & \frac{2B-A}{3} & -\frac{2}{3}T\end{bmatrix}$$
$$+c_{15}\frac{1}{2}\frac{dv^{\hat{z}}}{dy}E^2\begin{bmatrix}-\frac{2}{3}T & 0 & 0\\ 0 & -\frac{2}{3}T & \frac{2A-B}{3}\\ 0 & \frac{2A-B}{3} & \frac{4}{3}T\end{bmatrix}=\begin{bmatrix}0&0&0\\0&0&0\\0&0&0\end{bmatrix}. \tag{35.32}$$

This implies that the following three independent equations must be satisfied

$$c_1(A+B)+c_5 2T\frac{dv^{\hat{z}}}{dy}+c_9E^2+c_{10}\frac{2}{3}E^2(2A-B)$$
$$+\frac{dv^{\hat{z}}}{dy}E^2T(c_{14}+c_{15})=0, \tag{35.33a}$$

$$c_1(A-B)+c_9E^2+c_{10}\frac{2}{3}E^2(2A-B)+\frac{dv^{\hat{z}}}{dy}E^2T(c_{14}-c_{15})=0, \tag{35.33b}$$

$$c_1T+c_3\frac{1}{2}\frac{dv^{\hat{z}}}{dy}+c_5\frac{dv^{\hat{z}}}{dy}\frac{A+B}{6}+c_{10}E^2T$$
$$+\frac{E^2}{2}\frac{dv^{\hat{z}}}{dy}\left[c_{11}+c_{14}\frac{2B-A}{3}+c_{15}\frac{2A-B}{3}\right]=0. \tag{35.33c}$$

(The first equation follows from comparison of $\hat{x}\hat{x}$ elements of the matrices, the second equation follows by subtracting equations that follow from comparison of $\hat{z}\hat{z}$ and $\hat{y}\hat{y}$ elements in the matrices. The last equation follows from comparison of $\hat{y}\hat{z}$ elements of the matrices.)

Constitutive relation (35.14) leads to interesting results even if we put most of the material parameters equal to zero. Indeed, let us set $c_{10}=0$, $c_5=0$, $c_{11}=0$ and $c_9=0$. This means that we do not consider tensorial terms that do not involve $\overrightarrow{E}$, the terms that are linear in $\mathbb{S}$ and $\mathbb{D}$, and the tensorial term that depends exclusively on $\overrightarrow{E}$. If we do so, and if we further fix

$$c_{14} =_{\text{def}} -c_{15}, \tag{35.34}$$

then (35.33) reduces to

$$\mathrm{A}+\mathrm{B}=0, \tag{35.35a}$$

$$\mathrm{A}-\mathrm{B}=2c_{15}\frac{\mathrm{d}\mathrm{v}^{\hat{z}}}{\mathrm{d}y}E^2\mathrm{T}, \tag{35.35b}$$

$$c_1\mathrm{T}+c_3\frac{1}{2}\frac{\mathrm{d}\mathrm{v}^{\hat{z}}}{\mathrm{d}y}+\frac{E^2}{2}\frac{\mathrm{d}\mathrm{v}^{\hat{z}}}{\mathrm{d}y}c_{15}\frac{\mathrm{A}-\mathrm{B}}{3}=0, \tag{35.35c}$$

which yields

$$\mathrm{A}=c_{15}\frac{\mathrm{d}\mathrm{v}^{\hat{z}}}{\mathrm{d}y}E^2\mathrm{T}, \tag{35.36a}$$

$$\mathrm{B}=-\mathrm{A}, \tag{35.36b}$$

$$\mathrm{T}=-\frac{\frac{c_3}{2}}{c_1+\frac{c_{15}^2}{3}E^4\left(\frac{\mathrm{d}\mathrm{v}^{\hat{z}}}{\mathrm{d}y}\right)^2}\frac{\mathrm{d}\mathrm{v}^{\hat{z}}}{\mathrm{d}y}, \tag{35.36c}$$

or, in the explicit form,

$$\mathrm{A}=-\frac{1}{2}\frac{c_3c_{15}}{c_1+\frac{c_{15}^2}{3}E^4\left(\frac{\mathrm{d}\mathrm{v}^{\hat{z}}}{\mathrm{d}y}\right)^2}\left(\frac{\mathrm{d}\mathrm{v}^{\hat{z}}}{\mathrm{d}y}\right)^2E^2, \tag{35.37a}$$

$$\mathrm{B}=-\mathrm{A}, \tag{35.37b}$$

$$\mathrm{T}=-\frac{\frac{c_3}{2}}{c_1+\frac{c_{15}^2}{3}E^4\left(\frac{\mathrm{d}\mathrm{v}^{\hat{z}}}{\mathrm{d}y}\right)^2}\frac{\mathrm{d}\mathrm{v}^{\hat{z}}}{\mathrm{d}y}. \tag{35.37c}$$

Consequently, wee see that in the case of constant material coefficients the simple constitutive relation (35.14) allows us to describe the behaviour that is beyond the reach of constitutive relation (35.13) and *vice versa*.

35.5 Conclusion

We have seen that the implicit type constitutive relations substantially expand the class of models the can be used to characterise the response of electrorheological fluids, especially regarding genuine three dimensional effects such as normal stress differences. This is apparent even in the simplest setting of simple shear flow and transverse electric field. It would be worthwhile to test the influence of the additional tensorial terms in more complex geometries and inhomogeneous electric fields, in particular in geometries where one can still have an analytic formula for the electric field, see for example Love (1924); Jeffery (1921) and Bewley (1948). Some partial results in this direction can be found for example in Průša and Rajagopal (2012a).

An important qualitative feature of electrorheological fluids, namely the yield stress behaviour, is not captured by the specific models discussed in Sect. 35.3. However, in the framework of implicit constitutive relations, it is easy to design models that are capable of capturing the yield stress phenomenon, and amend the constitutive relations for electrorheological fluids correspondingly. In fact, the yield stress phenomenon has been studied from the perspective of implicit constitutive relations in the context of non-Newtonian fluids, see especially Rajagopal and Srinivasa (2008) and follow-up works such as Bulíček et al (2009, 2012); Diening et al (2013) and Maringová and Žabenský (2018) focused on the corresponding mathematical issues.

The outlined approach to the constitutive relations for electrorheological fluids fits into a larger framework of the so-called implicit constitutive relations, that has been found useful both in modelling the response of solids and fluids. The reader interested in the benefits of using implicit constitutive relations is kindly referred to recent review (Rajagopal and Saccomandi, 2016).

References

Abu-Jdayil B, Brunn PO (1997) Study of the flow behavior of electrorheological fluids at shear- and flow-mode. Chem Eng Process 36(4):281–289

Arvanitakis A (2017) On implicit constitutive relations in elastic ferroelectrics. Z angew Math Phys 68(5)

Barnes HA, Hutton JF, Walters K (1989) An Introduction to Rheology. Elsevier Amsterdam

Belza T, Pavlínek V, Sáha P, Quadrat O (2008) Effect of field strength and temperature on viscoelastic properties of electrorheological suspensions of urea-modified silica particles. Colloid Surf A-Physicochem Eng Asp 316(1-3):89–94

Bewley LV (1948) Two-Dimensional Fields in Electrical Engineering. Macmillan, New York

Boltenhagen P, Hu Y, Matthys EF, Pine DJ (1997) Observation of bulk phase separation and coexistence in a sheared micellar solution. Phys Rev Lett 79:2359–2362

Bulíček M, Gwiazda P, Málek J, Świerczewska-Gwiazda A (2009) On steady flows of incompressible fluids with implicit power-law-like rheology. Adv Calc Var 2(2):109–136

Bulíček M, Gwiazda P, Málek J, Świerczewska-Gwiazda A (2012) On unsteady flows of implicitly constituted incompressible fluids. SIAM J Math Anal 44(4):2756–2801

Bustamante R, Rajagopal KR (2013a) On a new class of electro-elastic bodies. II. Boundary value problems. Proc R Soc A: Math Phys Eng Sci 469(2155)

Bustamante R, Rajagopal KR (2013b) On a new class of electroelastic bodies. I. Proc R Soc A: Math Phys Eng Sci 469(2149)

Bustamante R, Rajagopal KR (2015) Implicit constitutive relations for nonlinear magnetoelastic bodies. Proc R Soc A: Math Phys Eng Sci 471(2175)

Bustamante R, Rajagopal KR (2017) Implicit equations for thermoelastic bodies. Int J Non-Linear Mech 92:144–152

Ceccio SL, Wineman AS (1994) Influence of orientation of electric field on shear flow of electrorheological fluids. J Rheol 38(3):453–463

Chapman S, Cowling TG (1990) The Mathematical Theory of Nonuniform Gases, 3rd edn. Cambridge Mathematical Library, Cambridge University Press, Cambridge, an account of the kinetic theory of viscosity, thermal conduction and diffusion in gases, In co-operation with D. Burnett, With a foreword by Carlo Cercignani

Choi YT, Cho JU, Choi SB, Wereley NM (2005) Constitutive models of electrorheological and magnetorheological fluids using viscometers. Smart Mater Struct 14(5):1025–1036

Coleman BD, Markovitz H, Noll W (1966) Viscometric Flows of Non-Newtonian Fluids. Theory and Experiment. Springer-Verlag, Berlin

Conrad H, Sprecher AF, Choi Y, Chen Y (1991) The temperature dependence of the electrical properties and strength of electrorheological fluids. J Rheol 35(7):1393–1410
David J, Filip P (2004) Phenomenological modelling of non-monotonous shear viscosity functions. Appl Rheol 14(2):82–88
Diening L, Kreuzer C, Süli E (2013) Finite element approximation of steady flows of incompressible fluids with implicit power-law-like rheology. SIAM J Numer Anal 51(2):984–1015
Eringen AC, Maugin GA (1990a) Electrodynamics of Continua, vol I - Foundations and Solid Media. Springer-Verlag, New York
Eringen AC, Maugin GA (1990b) Electrodynamics of Continua, vol II- Fluids and Complex Media. Springer-Verlag, New York
Galindo-Rosales FJ, Rubio-Hernández FJ, Sevilla A (2011) An apparent viscosity function for shear thickening fluids. J Non-Newton Fluid Mech 166(5–6):321–325
Gamota DR, Filisko FE (1991) Dynamic mechanical studies of electrorheological materials – moderate frequencies. J Rheol 35(3):399–425
Gast AP, Zukoski CF (1989) Electrorheological fluids as colloidal suspensions. Adv Colloid Interface Sci 30:153–202
Gavin HP, Hanson RD, Filisko FE (1996) Electrorheological dampers 1. Analysis and design. J Appl Mech-Trans ASME 63(3):669–675
Halsey TC, Martin JE, Adolf D (1992) Rheology of electrorheological fluids. Phys Rev Lett 68(10):1519–1522
Janečka A, Průša V (2015) Perspectives on using implicit type constitutive relations in the modelling of the behaviour of non-Newtonian fluids. AIP Conference Proceedings 1662:020003
Jeffery GB (1921) Plane stress and plane strain in bipolar co-ordinates. Philos Trans R Soc A-Math Phys Eng Sci 221:265–293
Kollias A, Dimarogonas A (1993) Properties of zeolite- and cornstarch-based electrorheological fluids at high shear strain rates. J Intell Mater Syst Struct 4(4):519–526
Krztoń-Maziopa A, Wyciślik H, Płocharski J (2005) Study of electrorheological properties of poly(p-phenylene) dispersions. J Rheol 49(6):1177–1192
Kulvait V, Málek J, Rajagopal KR (2013) Anti-plane stress state of a plate with a V-notch for a new class of elastic solids. Int J Frac 179(1–2):59–73
Le Roux C, Rajagopal KR (2013) Shear flows of a new class of power-law fluids. Appl Math 58(2):153–177
Liu YD, Choi HJ (2012) Electrorheological fluids: smart soft matter and characteristics. Soft Matter 8(48):11,961–11,978
Love AEH (1924) Some electrostatic distributions in two dimensions. Proceedings of the London Mathematical Society s2-22(1):337–369
Málek J, Průša V, Rajagopal KR (2010) Generalizations of the Navier–Stokes fluid from a new perspective. Int J Eng Sci 48(12):1907–1924
Maringová E, Žabenský J (2018) On a Navier–Stokes–Fourier-like system capturing transitions between viscous and inviscid fluid regimes and between no-slip and perfect-slip boundary conditions. Nonlinear Anal - Real World Appl 41:152–178
Martin JE, Adolf D, Halsey TC (1994) Electrorheology of a model colloidal fluid. J Colloid Interface Sci 167(2):437–452
Pao YH (1978) Mechanics Today, vol 4, Pergamon Press, New York, chap Electromagnetic forces in deformable continua, pp 209–305
Penfield P, Haus HA (1967) Electrodynamics of Moving Media. MIT Press, Cambridge, Massachusetts
Perlácová T, Průša V (2015) Tensorial implicit constitutive relations in mechanics of incompressible non-Newtonian fluids. J Non-Newton Fluid Mech 216:13–21
Průša V, Rajagopal KR (2012a) Flow of an electrorheological fluid between eccentric rotating cylinders. Theor Comput Fluid Dyn 26(1-4):1–21
Průša V, Rajagopal KR (2012b) On implicit constitutive relations for materials with fading memory. J Non-Newton Fluid Mech 181–182:22–29
Rajagopal KR (2003) On implicit constitutive theories. Appl Math 48(4):279–319

Rajagopal KR (2006) On implicit constitutive theories for fluids. J Fluid Mech 550:243–249
Rajagopal KR, Růžička M (1996) On the modeling of electrorheological materials. Mech Res Commun 23(4):401–407
Rajagopal KR, Růžička M (2001) Mathematical modeling of electrorheological materials. Contin Mech Thermodyn 13(1):59–78
Rajagopal KR, Saccomandi G (2016) A novel approach to the description of constitutive relations. Frontiers in Materials 3:36
Rajagopal KR, Srinivasa AR (2007) On the response of non-dissipative solids. Proc R Soc Lond, Ser A, Math Phys Eng Sci 463(2078):357–367
Rajagopal KR, Srinivasa AR (2008) On the thermodynamics of fluids defined by implicit constitutive relations. Z angew Math Phys 59(4):715–729
Rajagopal KR, Srinivasa AR (2009) On a class of non-dissipative materials that are not hyperelastic. Proc R Soc Lond Ser A Math Phys Eng Sci 465(2102):493–500
Rajagopal KR, Srinivasa AR (2015) Inelastic response of solids described by implicit constitutive relations with nonlinear small strain elastic response. Int J Plast 71:1 – 9
Rajagopal KR, Srinivasa AR (2016) An implicit three-dimensional model for describing the inelastic response of solids undergoing finite deformation. Z angew Math Phys 67(4)
Rajagopal KR, Wineman AS (1992) Flow of electrorheological materials. Acta Mech 91(1-2):57–75
Rosenweig RE (1985) Ferrohydrodynamics. Cambridge Monographs on Mechanics, Cambridge University Press
Růžička M (2000) Electrorheological Fluids: Modeling and Mathematical Theory, Lecture Notes in Mathematics, vol 1748. Springer-Verlag, Berlin
Růžička M (2004) Modeling, mathematical and numerical analysis of electrorheological fluids. Appl Math, Praha 49(6):565–609
Smith GF (1971) On isotropic functions of symmetric tensors, skew-symmetric tensors and vectors. Int J Eng Sci 9(10):899–916
Spencer AJM (1971) Theory of invariants. In: Eringen AC (ed) Continuum Physics, vol I, Academic Press, New York, pp 239–353
Tanner RI, Walters K (1998) Rheology: an Historical Perspective, Rheology Series, vol 7. Elsevier, Amsterdam
Winslow WM (1949) Induced fibration of suspensions. J Appl Phys 20(12):1137–1140
Zheng QS (1994) Theory of representations for tensor functions – A unified invariant approach to constitutive equations. Applied Mechanics Reviews 47(11):545–587
Zukoski CF (1993) Material properties and the electrorheological response. Annu Rev of Mater Sci 23(1):45–78

Chapter 36
Second Gradient Continuum: Role of Electromagnetism Interacting with the Gravitation on the Presence of Torsion and Curvature

Lalaonirina R. Rakotomanana

Abstract The goal of this paper is to link the geometric variables of strain gradient continuum with electromagnetic fields. For that purpose, we derive the electromagnetic wave equation within a Riemann-Cartan continuum where curvature and torsion are present. We also show that the gravitational and electromagnetic fields are respectively identified as geometric objects of such a continuum, namely the curvature $\mathfrak{R}^{\gamma}_{\alpha\beta\lambda}$ for gravitation which is a classical result, and the torsion $\aleph^{\gamma}_{\alpha\beta}$ as source of electromagnetism.

36.1 Introduction

First, interaction of Einstein gravitation and electromagnetism was often considered in a curved Riemannian continuum e.g. Fernandez-Nunez and Bulashenko (2016). It is usually assumed the case where the electromagnetic field is of the order of small perturbation of the spacetime metric. Only the influence of the metric on electromagnetic field is mostly accounted for, not the converse. Second, the influence of the Riemann-Cartan geometry on the electromagnetic field is not so easy. A free electromagnetic field is suggested to not produce torsion e.g. Hehl et al (1976), and there is in principle no contribution from torsion in Maxwell equations. When a strong magnetic field coexists with matter distribution, there is however a possibility to induce spin polarization of individual particles composing the continuum matter e.g. Prasanna (1975). Some authors have even suggested that torsion play a keyrole in electromagnetism when considering electromagnetic field within twisted continuum e.g. Hammond (1989). They propose that the electromagnetic potential is represented by the torsion vector $A_\alpha := \aleph_\alpha = \aleph^{\beta}_{\alpha\beta}$. The influence of torsion tensor as cosmic

Lalaonirina R. Rakotomanana
Institut de Recherche Mathématique de Rennes, Campus Beaulieu, 35042 Rennes, France
e-mail: lalaonirina.rakotomanana-ravelonarivo@univ-rennes1.fr

© Springer International Publishing AG, part of Springer Nature 2018

H. Altenbach et al. (eds.), *Generalized Models and Non-classical Approaches in Complex Materials 1*, Advanced Structured Materials 89,
https://doi.org/10.1007/978-3-319-72440-9_36

dislocation (singularity of the curvature tensor) was investigated in e.g. Dias and Moraes (2005), or some material defects as screw dislocations (Fumeron et al, 2015), or fluids with spin density e.g. Schutzhold et al (2002). They have included the 2-form torsion with one non-vanishing component as $2\pi\beta\,\delta^2(r)\;dr\wedge d\theta$ in the metric of the spacetime, then derive Maxwell's equations in a cylindrical coordinates to solve two interesting cases: the electric field of a line charge, and the magnetic field of the line current. However, they seemed to assume a connection based only a metric but do not consider the contortion tensor for covariant derivation of Maxwell's equations in the framework of Riemann-Cartan geometry.

In the context of nonhomogeneous continuum matter, a question arised some years ago in Maugin (1993) about the couplings of cracks and more generally inhomogeneities with the electromagnetic fields within the matter. A main difficulty arises with the derivation of governing wave equations, namely the formulation of constitutive laws and conservation laws for strain gradient continuum which requires the identification of physical quantities with geometrical variables e.g. Futhazar et al (2014). A convenient tool for deriving constitutive laws and conservation laws from a Lagrangian density lies on the concept of variation of an action, namely the Lagrangian variation and the Eulerian variation (Poincaré invariance). Covariance requires the use of metric, torsion, and curvature as arguments of Lagrangian function for a second gradient continuum e.g. Antonio Tamarasselvame and Rakotomanana (2011). For the present paper, we are interested in the wave propagation and relative gravitation inspired by the pioneering work of Maugin (1978). We attempt in this paper to define some link between the electromagnetic fields and the second gradient continuum geometric variables. Electromagnetic waves, including light wave propagation, are described by Maxwell's equations within Minkowskian, Riemannian or Riemann-Cartan continuum. We consider in this work some elements of the theory of interaction between gravitation and electromagnetism based on the classical Hilbert-Einstein action.

36.2 Electromagnetism in Minkowski Spacetime

36.2.1 Maxwell's 3D Equations in Vacuum

Electromagnetism theory is built upon electric field $\boldsymbol{E}$, and magnetic field $\boldsymbol{B}$. Both of them depend on the space coordinate and the time in the general case. The general form of the Maxwell's equations is intimately linked to the geometry of the spacetime vacuum characterized by the Minkowski metric, they constitute the fundamental basis of classical electrodynamics. Vacuum Maxwell's equations are derived in the local coordinates of flat Minkowski spacetime $\mathscr{M}$ endowed with the metric

$$\hat{g}_{\mu\nu} := \{+1,-1,-1,-1\}$$

Coordinates of the spacetime are denoted

$$x^\mu := (x^0 = ct, x^1, x^2, x^3)$$

Conservation laws are rigorously derived with the derivative with respect to the proper time which is a particular case of the objective derivative defined from the concept of integral invariance of Poincaré e.g. Rakotomanana (2003). Before going into the derivation of Maxwell's equation, it is worth to remind the notion of proper time in the framework of special relativity. Consider a body / or a referential frame moving with a uniform velocity v with respect to $\mathscr{M}$. The proper time is given by: $d\tau := \sqrt{1-(v^2/c^2)}dt$, where the proper time τ along a timelike world line in the spacetime $\mathscr{M}$ is the laps of time measured by a clock following that line. The derivative with respect to the proper time τ is then be calculated by applying the transitivity rule:

$$\frac{\partial}{\partial \tau} = \frac{\partial x^0}{\partial \tau}\,\partial_0 = c\frac{dt}{d\tau}\partial_0 \simeq c\,\partial_0$$

when the velocity is small compared to the light speed $v \ll c$. Therefore, classical three-dimensional Maxwell's equations take the form of, the connection of 3D vacuum space is denoted $\hat{\nabla}$, e.g. Kleinert (2008); Kovetz (2000):

$$\begin{cases} \hat{\nabla}\cdot \boldsymbol{D} = \rho \\ \hat{\nabla}\times \boldsymbol{H} - c\,\partial_0 \boldsymbol{D} = \boldsymbol{J} \end{cases}, \qquad \begin{cases} \hat{\nabla}\cdot \boldsymbol{B} = 0 \\ \hat{\nabla}\times \boldsymbol{E} + c\,\partial_0 \boldsymbol{B} = 0 \end{cases} \tag{36.1}$$

in which $\boldsymbol{E}$ and $\boldsymbol{H}$ are the electric and magnetic field intensities, whereas $\boldsymbol{D}$ (displacement) and $\boldsymbol{B}$ (magnetic induction) are the electric and magnetic flux densities. ρ and $\boldsymbol{J}$ are the volume charge density and the electric current density respectively. The fluxes $\boldsymbol{D}$ and $\boldsymbol{B}$ are related to the field intensities via the electromagnetic constitutive laws for the vacuum e.g. Kovetz (2000):

$$\boldsymbol{D} = \epsilon_0 \boldsymbol{E}, \qquad \boldsymbol{B} = \mu_0 \boldsymbol{H} \tag{36.2}$$

Constants ϵ_0 and μ_0 are called electric permittivity and magnetic permeability of the vacuum space, respectively, and $c^2 := (\epsilon_0\mu_0)^{-1}$. As such, the Maxwell's equations are the conservation laws, and then by accounting for constitutive laws, we obtain the partial differential equations where the electric and magnetic intensities are the unknowns. From the equations (36.1), an alternative expression of the electric and magnetic fields holds:

$$\boldsymbol{E} := -c\,\partial_0 \boldsymbol{A} - \hat{\nabla}\phi, \qquad \text{and} \qquad \boldsymbol{B} := \hat{\nabla}\times \boldsymbol{A} \tag{36.3}$$

where $\boldsymbol{A}$ and ϕ are the vector and scalar potentials of electromagnetism. Fluxes may be re-written as magnetic $\boldsymbol{B} := B_{jk}\,dx^j \wedge dx^k$ and $B^i := (1/2)\epsilon^{ijk}\,B_{jk}$ and electric variables $\boldsymbol{D} := D_{jk}\,dx^j \wedge dx^k$ and $D^i := (1/2)\epsilon^{ijk}\,D_{jk}$. By using the form notation, we remind the electromagnetic strength or also Faraday tensor as a 2-form:

$$\begin{aligned} \mathscr{F} &= E_1 dx^1 \wedge dx^0 + E_2 dx^2 \wedge dx^0 + E_3 dx^3 \wedge dx^0 \\ &+ B^1 dx^2 \wedge dx^3 + B^2 dx^3 \wedge dx^1 + B^3 dx^1 \wedge dx^2 \end{aligned} \tag{36.4}$$

from which we can easily check the correspondence of the components $\mathscr{F}_{\mu\nu}$ with the components of $\boldsymbol{E}$ and $\boldsymbol{B}$. The Maxwell's equations in terms of in terms ϕ and $\boldsymbol{A}$ do not uniquely determine the electric potential and magnetic potential. If ϕ_0 and $\boldsymbol{A}_0$ are solutions, then so are the pair:

$$\phi = \phi_0 + c\partial_0\Lambda, \qquad \boldsymbol{A} = \boldsymbol{A}_0 - \hat{\nabla}\Lambda \tag{36.5}$$

by checking $\boldsymbol{E} = -c\partial_0(\boldsymbol{A}_0 - \hat{\nabla}\Lambda) - \hat{\nabla}(\phi_0 + c\partial_0\Lambda)$ and $\boldsymbol{B} = \hat{\nabla} \times (\boldsymbol{A}_0 + \hat{\nabla}\Lambda)$ where $\Lambda(x^\mu)$ is an arbitrary scalar function assumed to be of class $\mathscr{C}^2$, which is called gauge-transformation function. However, the gauge invariance is satisfied since we assumed: $-c\partial_0(\hat{\nabla}\Lambda) - \hat{\nabla}t(c\partial_0\Lambda) = 0$, and $\hat{\nabla} \times (\hat{\nabla}\Lambda) = 0$.

36.2.2 Covariant Formulation of Maxwell's Equations

Maxwell's equations in vacuum include both conservation laws and the constitutive equations. The invariant formulation of electromagnetism theory in the Minkowski spacetime $\mathscr{M}$ is obtained by considering the Faraday tensor with components (36.4):

$$\mathscr{F}_{\mu\nu} := \partial_\mu A_\nu - \partial_\nu A_\mu, \qquad \mathscr{F}_{\mu\nu} = \begin{bmatrix} 0 & -E_1 & -E_2 & -E_3 \\ E_1 & 0 & B^3 & -B^2 \\ E_2 & -B^3 & 0 & B^1 \\ E_3 & B^2 & -B^1 & 0 \end{bmatrix} \tag{36.6}$$

where the combined electromagnetic field $(\boldsymbol{E}, \boldsymbol{B})$ do not transform as 3-vectors but as the six components of the skew-symmetric tensor $\mathscr{F}_{\mu\nu}$. We conform here to the convention in e.g. Hehl (2008); Obukhov (2008). In this way the skew symmetric tensor $\mathscr{F}_{\mu\nu}$ is chosen as primal variables of the theory. Let us now define the dual variable $\mathscr{H}^{\mu\nu}$ constructed from the displacement and the magnetic field. Independently on the constitutive laws, the classical electromagnetism theory considers the electromagnetic excitation as a two-form $\mathscr{H}_{\mu\nu}$:

$$\begin{aligned} \mathscr{H} := & -H_1 dx^1 \wedge dx^0 - H_2 dx^2 \wedge dx^0 - H_3 dx^3 \wedge dx^0 \\ & + D^3 dx^1 \wedge dx^2 + D^1 dx^2 \wedge dx^3 + D^1 dx^2 \wedge dx^3 \end{aligned} \tag{36.7}$$

in the same way as the definition of the electromagnetic strength (36.4). The variable we are interested in is in fact the dual variable $\mathscr{H}^{\mu\nu}$ in order to be able to link it with the primal variable $\mathscr{F}_{\alpha\beta}$. From (36.7), we easily obtain the two contravariant components as exactly as for the linear case:

$$\mathscr{H}^{\mu\nu} = \begin{bmatrix} 0 & D^1 & D^2 & D^3 \\ -D^1 & 0 & H_3 & -H_2 \\ -D^2 & -H_3 & 0 & H_1 \\ -D^3 & H_2 & -H_1 & 0 \end{bmatrix} \tag{36.8}$$

owing again that $\epsilon_0\mu_0 = 1$ (the choice of coordinate system where $x^0 := ct$ allows us to act as if $c = 1$ and the metric of the flat Minkowski spacetime holds

$$g^{\mu\nu} := \{+1, -1, -1, -1\}$$

It conforms to the linear constitutive law but it should again be stressed that in the general case this is in fact considered as a definition of the dual variable independently on the constitutive law. Previous 3*D* equations (36.1) may be recast in 4-dimensional covariant Maxwell equations by using the Minkowski spacetime connection:

$$\hat{\nabla}_\mu \mathscr{H}^{\mu\nu} = J^\nu, \qquad \hat{\nabla}_\mu \mathscr{F}^{*\mu\nu} = 0 \tag{36.9}$$

where $\mathscr{H}^{\mu\nu}$ denotes the electromagnetic tensor including the electric displacement field and the magnetic field, and $\mathscr{F}^{*\mu\nu} := (1/2)\epsilon^{\mu\nu\kappa\sigma}\mathscr{F}_{\kappa\sigma}$ is the dual of $\mathscr{F}^{\mu\nu}$ ($\epsilon^{\mu\nu\kappa\sigma}$ being the Levi-Civita tensor $\epsilon^{0123} := +1$). Covariant formulation of constitutive laws may be derived by means of the electromagnetic Lagrangian:

$$\mathscr{L} := -\frac{1}{4}\mathscr{H}^{\mu\nu}\mathscr{F}_{\mu\nu} \tag{36.10}$$

By introducing the definitions (36.6) and (36.8) the three-dimensional formulation of the Lagrangian density function reduces to:

$$\mathscr{L} = \frac{1}{2}(\boldsymbol{D}\cdot\boldsymbol{E} - \boldsymbol{B}\cdot\boldsymbol{H})$$

Note 36.1. The covariance of the Maxwell's 3D equations with respect to rotations, space reflection, time reversal, and charge conjugation (modification of positive charge to negative charge) may be checked by means of a four-dimensional covariance analysis with respect to Lorentz group of transformations.

36.3 Electromagnetism in Curved Continuum

36.3.1 Variational Method and Covariant Maxwell's Equations

Various phenomena may have effects on electromagnetic waves such as the presence of gravitational field. The constitutive equations should be worthily changed to account for either the modification of the spacetime environment in presence of gravity for instance (curvature), or the propagation of electromagnetic fields within continuum matter. Here, we thus consider the electromagnetic constitutive laws $\boldsymbol{D} = \epsilon\boldsymbol{E}$ and $\boldsymbol{B} = \mu\boldsymbol{H}$ where ϵ and μ are the electric and magnetic parameters of the continuum replacing ϵ_0 and μ_0. For the sake of the simplicity, let us consider the simplest example of action for free electromagnetic field without sources and occurring within a Riemann curved continuum:

$$\mathscr{S} := \int_{\mathscr{M}} \mathscr{L}\omega_n \quad \text{with} \quad \mathscr{L} := -\frac{1}{4}\mathscr{F}^{\mu\nu}\,\mathscr{F}_{\mu\nu} + \frac{1}{2\chi}\overline{\mathscr{R}} \tag{36.11}$$

where the Faraday tensor and the scalar curvature are defined by the relationships:

$$\mathscr{F}_{\mu\nu} := \overline{\nabla}_\mu A_\nu - \overline{\nabla}_\nu A_\mu, \qquad \overline{\mathscr{R}} := g^{\mu\nu}\overline{\mathfrak{R}}_{\mu\nu} \tag{36.12}$$

where the Faraday tensor $\mathscr{F}^{\mu\nu}$ is calculated with the connection with zero torsion. First, the Lagrangian variation of the action (36.11) allows us to obtain the expression:

$$\begin{aligned}\Delta\mathscr{S} = \int_{\mathscr{M}} & \left\{ -\frac{1}{2}\mathscr{F}^{\mu\nu}\Delta\mathscr{F}_{\mu\nu} + \frac{1}{4}\mathscr{F}_{\mu\nu}\left(g^{u\lambda}\mathscr{F}^{\rho\nu} + \mathscr{F}^{\mu\rho}g^{\lambda\nu}\right)\Delta g_{\lambda\rho} \right. \\ & + \frac{1}{2\chi}\left(\mathfrak{R}^{\lambda\rho} - \frac{\mathscr{R}}{2}g^{\lambda\rho}\right)\Delta g_{\lambda\rho} + \frac{1}{8}\mathscr{F}^{\mu\nu}\,\mathscr{F}_{\mu\nu}\,g^{\lambda\rho}\Delta g_{\lambda\rho} \\ & \left. + \frac{1}{2\chi}g^{\mu\nu}\left[\overline{\nabla}_\lambda\left(\Delta\overline{\Gamma}^\lambda_{\mu\nu}\right) - \overline{\nabla}_\mu\left(\Delta\overline{\Gamma}^\lambda_{\lambda\nu}\right)\right]\right\}\omega_n\end{aligned}$$

For the variation of the Lagrangian (36.11), it is worth to remind the independent variations of the metric and the four-potential vector. The Lagrangian variation of the Faraday tensor takes the form of :

$$\Delta\mathscr{F}_{\mu\nu} = \overline{\nabla}_\mu(\Delta A_\nu) - \overline{\nabla}_\nu(\Delta A_\mu) \tag{36.13}$$

This relation is obtained by directly writing:

$$\begin{aligned}\Delta\mathscr{F}_{\mu\nu} &= \Delta\left(\partial_\mu A_\nu - \overline{\Gamma}^\rho_{\mu\nu}A_\rho\right) - \Delta\left(\partial_\nu A_\mu - \overline{\Gamma}^\rho_{\nu\mu}A_\rho\right) \\ &= \left(\partial_\mu\Delta A_\nu - \overline{\Gamma}^\rho_{\mu\nu}\Delta A_\rho - \Delta\overline{\Gamma}^\rho_{\mu\nu}A_\rho\right) - \left(\partial_\nu\Delta A_\mu - \overline{\Gamma}^\rho_{\nu\mu}\Delta A_\rho - \Delta\overline{\Gamma}^\rho_{\nu\mu}A_\rho\right)\end{aligned}$$

accounting for that the connection variation $\Delta\overline{\Gamma}^\rho_{\mu\nu}$ induces a variation of the field $\Delta\mathscr{F}_{\mu\nu}$. Second, the two systems of conservation laws associated to the unknown primal variables (say the 4-vector potential A_μ, and the Riemannian metric $g_{\mu\nu}$) are derived by varying the Lagrangian along the Lie-derivative variations $\mathscr{L}_\xi A_\mu$, and $\mathscr{L}_\xi g_{\mu\nu}$. Shifting the divergence terms at the boundary of the continuum and assuming a zero divergence at this boundary allow us to obtain the conservation laws. We can rearrange the Lagrangian variation of the action to give:

$$\begin{aligned}\Delta\mathscr{S} = \int_{\mathscr{M}} \overline{\nabla}_\nu\mathscr{F}^{\mu\nu}\,\Delta A_\mu\,\omega_n + \int_{\mathscr{M}} & \left[\frac{1}{2\chi}\left(\overline{\mathfrak{R}}^{\lambda\rho} - \frac{\overline{\mathscr{R}}}{2}g^{\lambda\rho}\right)\right. \\ & \left. + \frac{1}{8}\mathscr{F}^{\mu\nu}\mathscr{F}_{\mu\nu}\,g^{\lambda\rho} + \frac{\mathscr{F}_{\mu\nu}}{4}\left(g^{\mu\lambda}\mathscr{F}^{\rho\nu} + \mathscr{F}^{\mu\rho}g^{\lambda\nu}\right)\right]\Delta g_{\lambda\rho}\,\omega_n\end{aligned} \tag{36.14}$$

owing that the Faraday tensor is *in fine* expressed in terms of the potential A_μ by means of equation (36.12). Due to the arbitrariness of the metric and potential variations, we obtain the classical (and covariant) Einstein-Maxwell's equations:

$$\begin{cases} \overline{\nabla}_{\nu}\mathscr{F}^{\mu\nu} = 0 \\ \frac{1}{2\chi}\left(\overline{\mathfrak{R}}^{\lambda\rho} - \frac{\overline{\mathscr{R}}}{2}g^{\lambda\rho}\right) + \frac{1}{8}\mathscr{F}^{\mu\nu}\mathscr{F}_{\mu\nu}\,g^{\lambda\rho} + \frac{\mathscr{F}_{\mu\nu}}{4}\left(g^{\mu\lambda}\,\mathscr{F}^{\rho\nu} + \mathscr{F}^{\mu\rho}g^{\lambda\nu}\right) = 0 \end{cases} \tag{36.15}$$

where the first equation is the covariant Maxwell's equations in a Riemann continuum, such as second gradient continuum. The second equation governs the interaction of electromagnetism with gravitation. The unknowns in first term of the second equation are the continuum metric. The electromagnetic source (including both the second and the third terms) in the second equation constitutes the energy-momentum tensor. They influence the gravitation field and *vice versa* the metric field has also some influence on the electromagnetic field via the Levi-Civita covariant derivative $\overline{\nabla}$. Maxwell's equations (36.15) (first row) are used to analyze the electromagnetic wave within a Riemann continuum. The first equation may be re-written:

$$\begin{aligned} \overline{\nabla}_{\nu}\mathscr{F}^{\mu\nu} &= \overline{\nabla}_{\nu}\left(g^{\mu\alpha}g^{\alpha\nu}\mathscr{F}_{\alpha\beta}\right) = \overline{\nabla}_{\nu}\left[g^{\mu\alpha}\overline{\nabla}_{\alpha}A^{\nu} - g^{\nu\beta}\overline{\nabla}_{\beta}A^{\mu}\right] \\ &= g^{\mu\alpha}\left[\overline{\nabla}_{\alpha}\overline{\nabla}_{\nu}A^{\nu} + \mathfrak{R}^{\nu}_{\nu\alpha\gamma}A^{\gamma}\right] - g^{\nu\beta}\overline{\nabla}_{\nu}\overline{\nabla}_{\beta}A^{\mu} = 0 \end{aligned}$$

where we have used the Schouten's relations e.g. Rakotomanana (2003) with a zero torsion. Maxwell's equations include a classical wave part, a divergence term, and the contribution of the Ricci curvature of the continuum:

$$-g^{\nu\beta}\overline{\nabla}_{\nu}\overline{\nabla}_{\beta}A^{\mu} + g^{\mu\alpha}\overline{\nabla}_{\alpha}\overline{\nabla}_{\nu}A^{\nu} + g^{\mu\alpha}\overline{\mathfrak{R}}_{\alpha\gamma}A^{\gamma} = 0 \tag{36.16}$$

The first term expresses a D'Alembertian operator. The second term may be dropped if we assume a Lorenz gauge $\overline{\nabla}_{\nu}A^{\nu} = 0$. We then obtain the electromagnetic wave propagation equation within curved continuum:

$$-g^{\nu\beta}\overline{\nabla}_{\nu}\overline{\nabla}_{\beta}A^{\mu} + g^{\mu\alpha}\overline{\mathfrak{R}}_{\alpha\gamma}A^{\gamma} = 0 \tag{36.17}$$

in which we notice the direct influence of the gravitation (represented by Ricci curvature) on the electromagnetic wave propagation. In the following we will consider an extension of the equation (36 17) in the framework of Riemann-Cartan continuum.

36.3.2 Field Equations and Conservation Laws

The second row of system (36.15) is the field equation which extends the Einstein equation for vacuum spacetime, where the term represents the energy momentum analogous of the Maxwell energy-momentum for the space part, and with nonsymmetric property when considering the timelike part:

$$T^{\lambda\rho} := -\frac{1}{4}\mathscr{F}^{\mu\nu}\mathscr{F}_{\mu\nu}\, g^{\lambda\rho} - \frac{\mathscr{F}_{\mu\nu}}{2}\left(g^{\mu\lambda}\mathscr{F}^{\rho\nu} + \mathscr{F}^{\mu\rho} g^{\lambda\nu}\right) \tag{36.18}$$

It is the Minkowski energy-momentum tensor due to electromagnetic field. It modifies the gravitational field as source whereas the spacetime modifies the electromagnetic field according to (36.17). The temporal component of the energy-momentum (36.18) holds:

$$T^{00} = \frac{1}{2}\left(D^i E_i + B^i H_i\right) \tag{36.19}$$

which is exactly the electromagnetic energy density

$$T^{00} = (1/2)\left(\boldsymbol{D}\cdot\boldsymbol{E} + \boldsymbol{B}\cdot\boldsymbol{H}\right) := \mathscr{E}$$

in a three-dimensional formulation. By introducing the electromagnetic tensors (36.6) and (36.8) into the expression of the energy-momentum tensor, we have the following particular cases:

$$\begin{cases} T^{01} = E_2 H_3 - E_3 H_2 \\ T^{02} = E_3 H_1 - E_1 H_3 \\ T^{03} = E_1 H_2 - E_2 H_1 \end{cases}, \qquad \begin{cases} T^{10} = D^2 B^3 - D^3 B^2 \\ T^{20} = D^3 B^1 - D^1 B^3 \\ T^{30} = D^1 B^2 - D^2 B^1 \end{cases}$$

showing again that the $\{T^{01}, T^{02}, T^{03}\}$ are the components of the vector $\boldsymbol{E}\times\boldsymbol{H}$, whereas $\{T^{10}, T^{20}, T^{30}\}$ are the components of the vector $\boldsymbol{D}\times\boldsymbol{B}$, also called Minkowski momentum density e.g. Milonni and Boyd (2010), in the three-dimensional formulation. This highlights that the Minkowski tensor $T^{\mu\nu}$ is not symmetric when considering the time index 0. It is worth to express the energy momentum as:

$$T_M^{\mu\nu} = \begin{bmatrix} \mathscr{E} & \boldsymbol{E}\times\boldsymbol{H} \\ \boldsymbol{D}\times\boldsymbol{B} & \boldsymbol{T}^M \end{bmatrix} \tag{36.20}$$

where $\mathscr{E}$ is the energy, and $\boldsymbol{T}^M$ is the Maxwell tensor with contraviant components T^{ij}. The (nonsymmetric) energy-momentum such defined is called Minkowski energy momentum.

Note 36.2. The Poynting vector $\boldsymbol{S} := \boldsymbol{E}\times\boldsymbol{H}$ (originally discovered by JH Poynting in 1884) represents the rate of energy in the i-direction.

Note 36.3. Due to the nonsymmetry of the Minkowski energy-momentum, other tensors have been considered in the past, the most known is the symmetric Abraham energy momentum tensor which can be written as:

$$T_A^{\mu\nu} = \begin{bmatrix} \mathscr{E} & \boldsymbol{E}\times\boldsymbol{H} \\ \boldsymbol{E}\times\boldsymbol{H} & \boldsymbol{T}^M \end{bmatrix} \tag{36.21}$$

$T_M^{\mu\nu}$ and $T_A^{\mu\nu}$ are by far the most cited and considered. Among the numerous results for supporting the choice of the one or the other, experimental measurements of radiation pressure of light on matter have shown the complementarity of these two energy-momenta, see for instance Milonni and Boyd (2010) for physical interpretations of the two energy-momentum on the radiation pressure of light on a dielectric.

For completeness, the spatial components of the Maxwell energy-momentum take the form of for diagonal contributions:

$$\begin{cases} T^{11} = -\frac{1}{2}\left(+D^1E_1 - D^2E_2 - D^3E_3\right) - \frac{1}{2}\left(+B^1H_1 - B^2H_2 - B^3H_3\right) \\ T^{22} = -\frac{1}{2}\left(-D^1E_1 + D^2E_2 - D^3E_3\right) - \frac{1}{2}\left(-B^1H_1 + B^2H_2 - B^3H_3\right) \\ T^{33} = -\frac{1}{2}\left(-D^1E_1 - D^2E_2 + D^3E_3\right) - \frac{1}{2}\left(-B^1H_1 - B^2H_2 + B^3H_3\right) \end{cases}$$

For off-diagonal spatial terms we obtain (remind that $T^{\mu\nu}$ is symmetric):

$$T^{12} = -\left(D^2E_1 + B^2H_1\right),\ T^{23} = -\left(D^3E_2 + B^3H_2\right),\ T^{31} = -\left(D^1E_3 + B^1H_3\right)$$

By considering the Eulerian variation (Lie derivative variation $\mathscr{L}_\xi g_{\lambda\rho}$) on the Riemann continuum we obtain the conservation equation as follows:

$$\overline{\nabla}_\rho(\overline{\mathfrak{R}}^{\lambda\rho} - (\overline{\mathscr{R}}/2)g^{\lambda\rho}) = 0 \qquad \Longrightarrow \qquad \overline{\nabla}_\rho T^{\lambda\rho} = 0$$

where the covariant derivative of the Einstein tensor vanishes by using the Bianchi relationships e.g. Rakotomanana (2003). It is now essential to introduce the energy-momentum tensor in the conservation laws. The conservation laws take the form of:

$$\begin{cases} \overline{\nabla}_\rho T^{0\rho} = \overline{\nabla}_0 T^{00} + \overline{\nabla}_i T^{0i} \\ \overline{\nabla}_\rho T^{i\rho} = \overline{\nabla}_0 T^{i0} + \overline{\nabla}_j T^{ij} \end{cases} \tag{36.22}$$

We recognize that the first row reduces to the Poynting's theorem when considering an isotropic and homogeneous continuum, where the Joule effect is not present because we did not introduce it in the Lagrangian function $\mathscr{L}$. The second row represents the "force equilibrium" where the time derivative of the Poynting's vector compensates the divergence of the Maxwell's stress tensor. Going back to the system of equations (36.15), it is interesting to re-formulate the second row to give:

$$\frac{1}{\chi}\left(\overline{\mathfrak{R}}^{\lambda\rho} - \frac{\overline{\mathscr{R}}}{2}g^{\lambda\rho}\right) = T_M^{\lambda\rho} = \frac{1}{2}\left(T_M^{\lambda\rho} + T_M^{\rho\lambda}\right) + \frac{1}{2}\left(T_M^{\lambda\rho} - T_M^{\rho\lambda}\right) \tag{36.23}$$

where the left-hand side of the equation is symmetric whereas the right-hand side is not. This induces that the skew-symmetric part of the Minkowski energy-momentum $T_M^{\lambda\rho}$ does not contribute to bend the spacetime and then has no influence on the gravitation field. In a vacuum Minkowski spacetime remind however that $\epsilon_0\mu_0 = c^2 = 1$, for we adopt a coordinate system $(x^0 := ct, x^1, x^2, x^3)$. In such a case the two energy-momenta merge and the Abraham energy-momentum coincides with the Minkowski energy-momentum, problems only arise when electromagnetism interact with continuum matter. Starting from the equation (36.23) which relates the electromagnetic fields $T^{\lambda\rho}$ as source of the bending of the continuum, we can multiply this equation by the covariant components of the metric $g_{\lambda\rho}$ to obtain without difficulty the Ricci curvature and then the curvature of the continuum:

$$-\overline{\mathscr{R}} = \mathscr{T} := g_{\lambda\rho} T_M^{\lambda\rho} = -2\chi \mathscr{F}_{\lambda\rho} \mathscr{F}^{\lambda\rho} = \chi \left(\boldsymbol{D} \cdot \boldsymbol{E} - \boldsymbol{B} \cdot \boldsymbol{H} \right) \tag{36.24}$$

which is exactly χ times twice of the electromagnetic part of the Lagrangian.

Note 36.4. The debate of physicists and mathematicians about the choice of energy-momentum draw back to the beginning of relativistic theory more than 100 years ago. At least two factors may constitute the reasons of its revival nowadays: the increasing role of optics in modern communication technology, and the legitimate seek of consistent physics theory. Previous studies suggest that the two energy-momenta (36.20) and (36.21) are in fact correct but in different circumstances e.g. Milonni and Boyd (2010). They have their own physical interpretations.

36.4 Electromagnetism in Twisted and Curved Continuum

Analysis of electromagnetic fields in presence of extremely massive gravitation remains a relevant topic in relativistic astrophysics. Propagation of electromagnetic waves governed by Maxwell's equations within a curved continuum constitutes a fundamental basis for studying signals received from neutron stars and black holes to name but a few in astrophysics. Other methods consist in measuring the signal due to gravitational waves. It is now admitted that the influence of the non-Minkowskian metric of the curved spacetime is much stronger on the electromagnetic field $\mathscr{F}_{\mu\nu}(x^\alpha)$ than the influence of this field on the bending of the spacetime $\mathscr{M}$. In this section we consider the gravitation electromagnetism interaction within a Riemann-Cartan continuum endowed with metric $g_{\alpha\beta}(x^\mu)$ and connection $\Gamma_{\alpha\beta}^{\gamma}(x^\mu)$.

36.4.1 Faraday Tensor in Twisted Continuum

Formulation of Maxwell's equations by means of differential forms may be not equivalent to formulation by means connection in Riemann-Cartan continuum e.g Vandyck (1996). In a Riemann-Cartan continuum, the Faraday tensor is calculated as follows e.g. Prasanna (1975); Smalley (1986): $\mathscr{F}_{\mu\nu} := \nabla_\mu A_\nu - \nabla_\nu A_\mu = \partial_\mu A_\nu - \partial_\nu A_\mu + \aleph_{\mu\nu}^{\rho} A_\rho$. It is rather different if calculated by means of an exterior derivative of the 1-form $\boldsymbol{A} = (A_\mu)$ e.g. Prasanna (1975) :

$$\mathscr{F} := d\boldsymbol{A} \quad \Longrightarrow \quad \mathscr{F}_{\mu\nu} = \partial_\mu A_\nu - \partial_\nu A_\mu \tag{36.25}$$

where, in such a case, we have exactly the same form of Faraday tensor in either Minkowski spacetime or Riemann continuum. In this framework, two of the Maxwell's equations $d\mathscr{F} = 0$ would be expected since the Faraday tensor 2-form $\mathscr{F}$ is exact, say $\mathscr{F} := d\boldsymbol{A}$, and hence closed, $d\mathscr{F} = d(d\boldsymbol{A}) = 0$. To investigate electromagnetic waves within twisted and curved continuum matter (which may be considered as a Riemann-Cartan manifold), it is assumed that the electromagnetic field is described by an electromagnetic 2-form $\mathscr{F}_{\mu\nu}$. It constitutes an extended model of electromagnetism within curved continuum as earlier as in e.g. Plebanski (1960), and in the framework of differential forms e.g. Prasanna (1975). Prasanna (1975) has derived the Maxwell equations in a Riemann-Cartan continuum. In the following, we would like to derive the Maxwell's equations in a curved and twisted manifold $\mathscr{M}$. By using a formalism based on exterior calculus, Maxwell's equations were established for various continua (Minkowski, Riemann, and almost post-Riemann) (Puntigam et al, 1997) where they considered as basic axioms the conservation of electric charge and the conservation of magnetic flux. This allows them to put aside the connection structure of the continuum. Third, either for metric-based energy, or metric-torsion based energy, it is worth to define Lagrangian $\mathscr{L}(\mathscr{F}_{\mu\nu}, u^{\mu}, \mathscr{M}^{\mu\nu}, \mathscr{J}^{\mu}, \mathscr{A}_{\mu}, \cdots)$ associated to the electromagnetic fields when we face the question of variational formulation. To relate electromagnetism with relativistic gravitation, it is interesting to remind that application of the gauge invariance principle for the group of translation (corresponding to torsion) of the continuum $\mathscr{M}$ with Yang-Mills type Lagrangian, quadratic in the field strengths $\mathscr{F}_{\mu\nu}$ (as for electromagnetism), allows us to deduce the usual Einstein's theory of gravitation, based on the Einstein-Hilbert action e.g. Cho (1976).

36.4.2 Field Equations, Wave Equations

For the sake of the simplicity, let us consider the simplest example of action for free electromagnetic field without sources and occurring within a Riemann-Cartan continuum (curved and twisted):

$$\mathscr{S} := \int_{\mathscr{M}} \mathscr{L}\,\omega_n \quad \text{with} \quad \mathscr{L} := -\frac{1}{4}\mathscr{F}^{\mu\nu}\,\mathscr{F}_{\mu\nu} + \frac{1}{2\chi}\mathscr{R} \tag{36.26}$$

For the variation of (36.26), it is worth to remind that the metric and the torsion are independents primal variables in addition to the electromagnetic four-potential. The Faraday tensor (minimally coupled to the gravitation in a Riemann-Cartan continuum via the torsion) and the scalar curvature are defined by the relationships:

$$\mathscr{F}_{\mu\nu} := \nabla_{\mu}A_{\nu} - \nabla_{\nu}A_{\mu} = \overline{\nabla}_{\mu}A_{\nu} - \overline{\nabla}_{\nu}A_{\mu} + \aleph^{\rho}_{\mu\nu}A_{\rho}, \qquad \mathscr{R} := g^{\mu\nu}\Re_{\mu\nu} \tag{36.27}$$

where the contravariant components of $\mathscr{F}^{\mu\nu}$ are calculated by means the connection with torsion e.g. Smalley and Krisch (1992). Definition (36.25) in Riemann continuum holds for both Euclidean and (pseudo)-Riemannian and also proposed in some post-Riemannian spacetimes e.g. Puntigam et al (1997). As extension the definition (36.27) is valid for both Euclidean, Riemannian and Riemann-Cartan continuum. This again illustrates the fact that the extension of physical variables as $\mathscr{F}_{\mu\nu}$ can be done in many ways (as a 2-form in e.g. Puntigam et al (1997) or as a twice the skew-symmetric part of the gradient in e.g. Smalley and Krisch (1992)). This shows that the Lagrangian variation of this 2-form and curvature include both the variation of the potential A_μ, the variation of the Riemann metric $g_{\alpha\beta}$, and also the variation of the connection $\Gamma^{\gamma}_{\alpha\beta}$. Here, the Lagrangian we consider includes Yang-Mills electromagnetic part and Einstein-Hilbert gravitation part e.g. Charap and Duff (1977). First, the Lagrange variation of the action (36.26) allows us to obtain the expression:

$$\begin{aligned}\Delta\mathscr{S} = \int_{\mathscr{M}} & \left\{ -\frac{1}{2}\mathscr{F}^{\mu\nu}\Delta\mathscr{F}_{\mu\nu} + \frac{1}{4}\mathscr{F}_{\mu\nu}\left(g^{\mu\lambda}\mathscr{F}^{\rho\nu} + \mathscr{F}^{\mu\rho}g^{\lambda\nu}\right)\Delta g_{\lambda\rho}\right. \\ & + \frac{1}{2\chi}\left(\mathfrak{R}^{\lambda\rho} - \frac{\mathscr{R}}{2}g^{\lambda\rho}\right)\Delta g_{\lambda\rho} + \frac{1}{8}\mathscr{F}^{\mu\nu}\,\mathscr{F}_{\mu\nu}\,g^{\lambda\rho}\Delta g_{\lambda\rho} \\ & \left. + \frac{1}{2\chi}g^{\mu\nu}\left[\nabla_\lambda\left(\Delta\Gamma^{\lambda}_{\mu\nu}\right) - \nabla_\mu\left(\Delta\Gamma^{\lambda}_{\lambda\nu}\right) - \aleph^{\rho}_{\lambda\mu}\Delta\Gamma^{\lambda}_{\rho\nu}\right]\right\}\omega_n\end{aligned}$$

where the last line is deduced from Palatini relation e.g. Rakotomanana (2003). The Lagrangian variation of the Faraday tensor takes the form of:

$$\Delta\mathscr{F}_{\mu\nu} = \nabla_\mu(\Delta A_\nu) - \nabla_\nu(\Delta A_\mu) + \Delta\,\aleph^{\rho}_{\mu\nu}A_\rho \tag{36.28}$$

This relation is obtained by directly writing:

$$\begin{aligned}\Delta\mathscr{F}_{\mu\nu} &= \Delta\left(\partial_\mu A_\nu - \Gamma^{\rho}_{\mu\nu}A_\rho\right) - \Delta\left(\partial_\nu A_\mu - \Gamma^{\rho}_{\nu\mu}A_\rho\right) \\ &= \left(\partial_\mu\Delta A_\nu - \Gamma^{\rho}_{\mu\nu}\Delta A_\rho - \Delta\Gamma^{\rho}_{\mu\nu}A_\rho\right) - \left(\partial_\nu\Delta A_\mu - \Gamma^{\rho}_{\nu\mu}\Delta A_\rho - \Delta\Gamma^{\rho}_{\nu\mu}A_\rho\right)\end{aligned}$$

accounting for that the variation of the geometric structure, say $\Delta\,\aleph^{\rho}_{\mu\nu}$, induces a variation of the field $\Delta\mathscr{F}_{\mu\nu}$. At a second step, the three systems of conservation laws associated to the unknown primal variables (say the 4-vector potential A_μ, the Riemannian metric $g_{\mu\nu}$, and the torsion $\aleph^{\rho}_{\mu\nu}$) are derived by varying the Lagrangian along the Lie-derivative variations $\mathscr{L}_\xi A_\mu$, $\mathscr{L}_\xi g_{\mu\nu}$, and $\mathscr{L}_\xi\,\aleph^{\rho}_{\mu\nu}$. Now we factorize the variation with respect to the Lagrangian variations of the electromagnetic potential ΔA_μ, the metric $\Delta g_{\lambda\rho}$, and the connection $\Delta\Gamma^{\lambda}_{\mu\nu}$ respectively. The presence of the term $\Delta\Gamma^{\lambda}_{\mu\nu}$ means that the torsion and curvature may evolve since they are independent primal variables of the theory. By shifting divergence terms at the boundary of the continuum $\mathscr{M}$ we can rearrange the Lagrangian variation of the action to give:

$$\begin{aligned}\Delta\mathscr{S} &= \int_{\mathscr{M}} \nabla_\nu \mathscr{F}^{\mu\nu}\, \Delta A_\mu\, \omega_n + \int_{\mathscr{M}} \left[\frac{1}{2\chi}\left(\Re^{\lambda\rho} - \frac{\mathscr{R}}{2} g^{\lambda\rho}\right)\right. \\ &\quad \left. + \frac{1}{8}\mathscr{F}^{\mu\nu}\mathscr{F}_{\mu\nu}\, g^{\lambda\rho} + \frac{\mathscr{F}_{\mu\nu}}{4}\left(g^{\mu\lambda}\mathscr{F}^{\rho\nu} + \mathscr{F}^{\mu\rho} g^{\lambda\nu}\right)\right] \Delta g_{\lambda\rho}\, \omega_n \\ &\quad - \int_{\mathscr{M}} \left(\left(\mathscr{F}^{\mu\nu} - \mathscr{F}^{\nu\mu}\right) A_\lambda + \frac{1}{\chi} g^{\rho\nu}\, \aleph^{\mu}_{\lambda\rho}\right) \Delta\Gamma^{\lambda}_{\mu\nu}\, \omega_n\end{aligned} \tag{36.29}$$

owing that the Faraday tensor is *in fine* expressed in terms of the potential A_u by means of (36.27). Due to the arbitrariness of the variation of primal variables, we deduce the system of partial differential equations:

$$\begin{cases} \nabla_\nu \mathscr{F}^{\mu\nu} = 0 \\ \frac{1}{2\chi}\left(\Re^{\lambda\rho} - \frac{\mathscr{R}}{2} g^{\lambda\rho}\right) + \frac{1}{8}\mathscr{F}^{\mu\nu}\mathscr{F}_{\mu\nu}\, g^{\lambda\rho} + \frac{\mathscr{F}_{\mu\nu}}{4}\left(g^{\mu\lambda}\mathscr{F}^{\rho\nu} + \mathscr{F}^{\mu\rho} g^{\lambda\nu}\right) = 0 \\ \left(\mathscr{F}^{\mu\nu} - \mathscr{F}^{\nu\mu}\right) A_\lambda + \frac{1}{\chi} g^{\rho\nu}\, \aleph^{\mu}_{\lambda\rho} = 0 \end{cases} \tag{36.30}$$

where we notice a slightly extension of the fields equations in Charap and Duff (1977) for Riemann-Cartan continuum. The first row of the system (36.30) expresses the Maxwell's equations in Riemann-Cartan continuum, and it should be stressed that in this Lagrangian (model), the potential A_μ may be apparently calculated independently on the gravitation (except eventual coupling at the boundary $\partial\mathscr{M}$). The connection approach is equivalent to the differential form approach when the continuum is Riemannian without torsion (Vandyck, 1996), or when the non metricity of the connection is traceless. The Maxwell's equations of the system (36.30) show that the connection approach with torsion is "naturally" deduced from a variation principle, and the same form as the connection approach is obtained. The Maxwell's equations (36.30) (first row) may be used to analyze the electromagnetic wave propagation within a twisted and curved continuum. Let consider a continuum $\mathscr{M}$ endowed with a metric $g_{\alpha\beta}$ and a connected with $\Gamma^{\gamma}_{\alpha\beta}$, this later is compatible with the metric. Maxwell's equations may be re-written as follows:

$$\begin{aligned}\nabla_\nu \mathscr{F}^{\mu\nu} &= \nabla_\nu\left(g^{\mu\alpha} g^{\alpha\nu}\mathscr{F}_{\alpha\beta}\right) = \nabla_\nu\left[g^{\mu\alpha}\nabla_\alpha A^\nu - g^{\nu\beta}\nabla_\beta A^\mu\right] \\ &= g^{\mu\alpha}\left[\nabla_\alpha\nabla_\nu A^\nu - \aleph^{\gamma}_{\nu\alpha}\nabla_\gamma A^\nu + \Re^{\nu}_{\nu\alpha\gamma}A^\gamma\right] - g^{\nu\beta}\nabla_\nu\nabla_\beta A^\mu = 0\end{aligned}$$

where we have used the Schouten's relations e.g. Rakotomanana (2003). By arranging the previous relationships, we notice that the Maxwell's equations include, as for elastic wave propagation, a classical wave part, a divergence term, and the contribution of the twisting and the Ricci curvature of the continuum:

$$-g^{\nu\beta}\nabla_\nu\nabla_\beta A^\mu + g^{\mu\alpha}\nabla_\alpha\nabla_\nu A^\nu - g^{\mu\alpha}\aleph^{\gamma}_{\nu\alpha}\nabla_\gamma A^\nu + g^{\mu\alpha}\Re_{\alpha\gamma}A^\gamma = 0 \tag{36.31}$$

The first term expresses a D'Alembertian operator. The second term may be dropped if we assume a null divergence as a gauge condition. The condition $\nabla_\nu A^\nu \equiv 0$ extends the Lorenz condition in the framework of Riemann-Cartan geometry, more specifically in the way of Gauss units system. For a non twisted and non curved continuum, the electromagnetic wave propagation equation reduces to $\Box A^\mu = 0$. The third term introduces a first covariant derivative which leads to a diffusion of the wave (spacetime attenuation), and the last term points out a breathing mode whenever the boundary conditions allow it e.g. Futhazar et al (2014). What should be observed too is that the torsion and Ricci curvature influence the wave propagation linearly. It should be stressed that the continuum geometry and *in fine* the gravitation is in fact tightly linked to the electromagnetism phenomenon. This may not be perceived at a first sight. In sum, the second row of the system (36.30) gives the coupling equation of the electromagnetic field and the gravitational field one. The electromagnetic terms act as a source-term for the gravitation. They act as a kind of electromagnetic energy generating evolution of the continuum metric. We recognize the Einstein field equation in the absence of the electromagnetic field. Despite its apparent relative simplicity, the system of partial differential equations (36.30) remains complex since the connection, and by the way the Ricci and total curvatures, includes both the (gravitational) metric $g_{\mu\nu}$ and the contortion tensor $\mathscr{T}^\gamma_{\mu\nu}$.

36.4.3 Electromagnetism and Continuum Defects

The third row of (36.30) gives the equation to calculate the torsion field. It is striking its analogy with the result obtained by Fernado et al (2012) by considering a particular Riemann-Cartan continuum and working with contortion tensor. It is a link between electromagnetic fields and the twisting of the continuum. What is interesting is that the electromagnetic field allows us to calculate with an algebraic explicit formula the torsion field by means of the third row. Once the torsion is obtained, we can apply covariant derivative within Riemann-Cartan geometry. By multiplying with $g_{\nu\sigma}$, the explicit formula for calculating the torsion is obtained accordingly by means of an algebraic relation:

$$\aleph^{\mu}_{\lambda\sigma} = -\chi\, g_{\sigma\nu}\,(\mathscr{F}^{\mu\nu} - \mathscr{F}^{\nu\mu})A_\lambda = -2\chi\, g_{\sigma\nu}\left(g^{\mu\alpha}g^{\nu\beta} - g^{\mu\beta}g^{\nu\alpha}\right)A_\lambda \nabla_\alpha A_\beta \tag{36.32}$$

owing the expression of the electromagnetic strength in terms of potential. It may be noticed that the contribution of the electromagnetic potential to the torsion field is of second order "$A_\lambda \nabla_\alpha A_\beta$".

Note 36.5. The investigation of the interaction of electromagnetic masses with Riemann-Cartan continuum was done by numerous authors for charged and spinning "static" dust (static means here no displacement of the center of mass), for perfect fluids with spin density e.g. Smalley and Krisch (1992). It was shown that by analyzing the solutions of Maxwell's equations, the torsion field together with the spin of Einstein-Cartan gravitation theory may be suggested as produced by the electromagnetic field e.g. Tiwari and Ray (1997). Paraphrasing these authors, it was concluded that in the absence of electromagnetic fields, the body has a vanishing spin density which itself is associated to the continuum torsion e.g. Hehl and von der Heyde (1973). The third algebraic equation of (36.30) conforms this conclusion concerning the torsion field.

Note 36.6. What could be relevant is that the torsion field of the Riemann-Cartan continuum may be explicitly calculated. In other words and from the physical point of view, torsion is thought to be generated by the electromagnetic field demonstrated here by assuming the simplest Lagrangian of the type (36.26).

Note 36.7. From the system of equations (36.30), we notice that the electromagnetic energy-momentum in a vacuum has the same shape as for as for electromagnetic within a continuum matter (Obukhov and Hehl, 2003):

$$T^{\lambda\rho} = \frac{1}{4}\mathscr{F}^{\mu\nu}\mathscr{F}_{\mu\nu}\,g^{\lambda\rho} - \frac{\mathscr{F}_{\mu\nu}}{2}\left(g^{\mu\lambda}\mathscr{F}^{\rho\nu} + \mathscr{F}^{\mu\lambda}g^{\rho\nu}\right) \tag{36.33}$$

This is a non symmetric Minkowski (canonical) energy-momentum tensor e.g. Obukhov (2008) for the free electromagnetic field occurring within spacetime. For electromagnetic fields without sources occurring within a Riemann-Cartan continuum (curved and twisted), the Lagrangian (36.26) is slightly extended to:

$$\mathscr{S} := \int_{\mathscr{M}} \mathscr{L}\,\omega_n \quad \text{with} \quad \mathscr{L} := -\frac{1}{4}\mathscr{H}^{\mu\nu}\mathscr{F}_{\mu\nu} + \frac{1}{2\chi}\mathscr{R} \tag{36.34}$$

where the constitutive laws are given by the equation (36.8). The same developments as previously may be conducted to analyze the behavior of the matter electromagnetic interaction. There is a controversy between the version of Minkowski and that of Abraham, not deduced from a Lagrangian. We do not enter into this long last debate, which was done in the past. We have just to remind that the Minkowski version is defined in the framework of Lagrange-Noether conforming to the invariance approach we adopt in this work. Obukhov and Hehl (2003) suggested the adoption of the Minkowski like version (36.33) which is motivated by the Lagrangian axiomatic approach, and by the experimental evidence conducted in the past by Walker & Walker (which is based on experimental measurements of dielectric disk placed in a crossed oscillating radial electric and longitudinal magnetic fields), and James (which is based on a similar experimental jig but with radial electric field and azimuthal magnetic field), see Obukhov and Hehl (2003).

Note 36.8. In the previous equations, the torsion does not propagate. In order to account for the torsion propagation, i.e., a well-known method would be to add a scalar bilinear term of the covariant derivatives of the torsion e.g. Hammond (1987) where the trace of the torsion $\aleph_\nu := \aleph^{\mu}_{\nu\mu}$ can be considered as the electromagnetic four-potential, and the skew-symmetric part of the Ricci curvature tensor as proportional to the electromagnetic Faraday tensor. For that purpose, he has considered the

Lagrangian: $\mathscr{S} := \int_{\mathscr{M}} (\frac{\mathscr{R}}{\chi} + a\, \mathscr{G}_{\mu\nu}\mathscr{G}^{\mu\nu})$ with $\mathscr{G}_{\mu\nu} := \partial_\mu \aleph_\nu - \partial_\nu \aleph_\mu$ where arbitrariness of the metric and the torsion variations hold. In his approach the electromagnetic variables are deduced from continuum geometry.

To go further let us consider the first and third equations of the system of equations relating gravitation and electromagnetism. As for the Maxwell's equations within vacuum spacetime, the above equation may be formulated by means of four-potential vector A^μ by introducing properties of curvature tensor, the metricity of the connection and the Lorenz gauge ($\nabla_\nu A^\nu \equiv 0$). The third equation may be re-arranged to isolate the torsion. The first and third equations thus give, by assuming a null divergence for the potential $\nabla_\nu A^\nu = 0$,:

$$\begin{cases} -g^{\nu\beta}\nabla_\nu\nabla_\beta A^\mu - g^{\mu\alpha}\aleph^{\gamma}_{\nu\alpha}\nabla_\gamma A^\nu + g^{\mu\alpha}\Re_{\alpha\gamma}A^\gamma = 0 \\ 2\epsilon\chi\,(\nabla^\gamma A_\alpha - \nabla_\alpha A^\gamma)A_\beta = \aleph^{\gamma}_{\alpha\beta} \end{cases} \tag{36.35}$$

where, in the Maxwell's equations, the first term represents a wave equation, the second term a diffusion contribution due to the torsion field, and the last term with the Ricci curvature introduces a "breathing" mode due to the non vanishing of curvature tensor. It should be pointed out that the torsion field is of second-order with respect to the potential A^α.

Note 36.9. By proposing a particular Riemann-Cartan spacetime structure, and working with contortion tensor rather than with the torsion tensor, a similar and interesting relation as the second row of the system (36.35) is obtained in Fernado et al (2012) without dealing with a variational formulation.

Note 36.10. We define the Faraday tensor as $\mathscr{F}_{\alpha\beta} := \nabla_\alpha A_\beta - \nabla_\beta A_\alpha$ where the connection $\Gamma^{\gamma}_{\alpha\beta}$ has torsion. The $U(1)$ gauge invariance of Maxwell's equations may be violated without cautions with this choice. Indeed by modifying the potential as $A_\beta \to A_\beta + g_{\beta\gamma}\nabla^\gamma\Lambda$ where $\Lambda(x^\mu)$ is an arbitrary function, we get: $\mathscr{F}_{\alpha\beta} = \nabla_\alpha A_\beta - \nabla_\beta A_\alpha - \aleph^{\gamma}_{\alpha\beta}\nabla_\gamma\Lambda$ where the last term vanishes if and only if the torsion is zero or the function Λ is covariantly uniform. Some previous authors propose to define $\mathscr{F}_{\alpha\beta} := \overline{\nabla}_\alpha A_\beta - \overline{\nabla}_\beta A_\alpha$ as Faraday tensor even in Riemann-Cartan continuum e.g. Smalley (1986); de Andrade and Pereira (1999). Further analyses are required in the future. Results in the framework of Riemann-Cartan Gravitation (e.g. Sotiriou and Liberati, 2007) may highlight some problems of gauge invariance since the electromagnetic tensor does not satisfy the Lorenz gauge invariance (say $U(1)$ gauge invariance) e.g. Puntigam et al (1997). Choosing the Faraday tensor as $\mathscr{F} := d\boldsymbol{A}$ or equivalently $\mathscr{F}_{\mu\nu} := \overline{\nabla}_\mu A_\nu - \overline{\nabla}_\nu A_\mu = \nabla_\mu A_\nu - \nabla_\nu A_\mu - \aleph^{\rho}_{\mu\nu}A_\rho$ allows to obtain a $U(1)$ invariant model but induces the following field equations: $\nabla_\nu \mathscr{F}^{\mu\nu} - 2\aleph^{\mu}_{\rho\nu}\mathscr{F}^{\rho\nu} = 0$ and $\aleph^{\rho}_{\mu\nu} = 0$ replacing the first Maxwell's equations and the third equation coupling the torsion within continuum and the electromagnetic fields, meaning that the torsion is identically zero within the continuum. Further investigations should be done about the definition of the Faraday tensor, there is yet a lot to be done in this domain. This may hurt at first sight, however, more investigations should be conducted since the concept of magnetic monopole enters into the discussion because the Gauss law on magnetic flux should be re-analyzed in such a case e.g. Fernado et al (2012).

36.5 Concluding Remarks

Electromagnetism interacting with gravity is source of change of Riemann continuum to Riemann-Cartan continuum. We have considered in the present study the Hilbert-Einstein action (which is the simplest case among numerous gravitation theories) to relate electromagnetism and gravitation with the extended continuum with torsion. In the present paper, we mainly consider the intimate link between the electrodynamics and the geometry of the continuum where the electromagnetic waves are propagating. Considering a very simple shape of the Lagrangian (the same form for all the models), we extend the geometry structure from the flat spacetime to curved and then twisted and curved continua (second gradient continua). In order to analyze the interaction of electromagnetism and gravitation, the development of the Maxwell's equations within curved continuum shows the electromagnetism-gravitation mutual influence by means of the geometry characterized by metric, Levi-Civita connection, and associated Ricci curvature. When dealing with second gradient continuum, where abrupt gradients of physical properties may occur, the extension of the Maxwell's equations, namely the resulting wave propagation, is necessary to account for the non zero torsion $\aleph^{\gamma}_{\alpha\beta} \neq 0$ and non zero curvature $\Re^{\gamma}_{\alpha\beta\lambda} \neq 0$. Among numerous approaches, the use of Riemann-Cartan manifold as underlying geometrical structure seems worth. The tables below give overview of interaction of electromagnetic waves with various continuua. The table displays most classical approaches for analyzing the interaction of the electromagnetic waves with gravitation and more generally the geometric structure of continua. For the first two continua, the basic geometric variable is the continuum metric tensor: uniform for Minkowski (Special Relativity) and depending on the coordinates x^{μ} for Riemann spacetime (Relativistic Gravitation). We observe that the Minimal Coupling Procedure induces the interaction of the electromagnetic waves with the continuum curvature, by means of the Ricci curvature. This is pointed out by the shape of the field equation e.g de Andrade and Pereira (1999); Smalley (1986). It is striking that the equation (36.30) we obtained, is analogous to the particular contortion tensor $\mathscr{T}^{\gamma}_{\alpha\beta} = -(G/c^4)\mathscr{F}^{\gamma}_{\beta}A_{\alpha}$ found in a paper by Fernado et al (2012). Indeed, they have deduced that a particular connection defined by $\Gamma^{\gamma}_{\alpha\beta} := \overline{\Gamma}^{\gamma}_{\alpha\beta} + \mathscr{T}^{\gamma}_{\alpha\beta}$ in Riemann-Cartan continuum allowed them to derive Maxwell's equations:

$$\nabla_{\beta}\mathscr{F}^{\alpha\beta} = \overline{\nabla}_{\beta}\mathscr{F}^{\alpha\beta} = 0, \qquad \nabla_{[\beta}\mathscr{F}_{\alpha\beta]} = \overline{\nabla}_{[\beta}\mathscr{F}_{\alpha\beta]} = \partial_{[\beta}\mathscr{F}_{\alpha\beta]} = 0$$

For instance, following another path Poplawski suggested to define the four-potential as a part of the trace of the torsion itself e.g. Poplawski (2010). All these aspects will certainly constitute future research topics. We observe that both the torsion and the curvature influence the electromagnetic wave propagation in a Riemann-Cartan continua. Despite the crucial point on the Lorenz gauge invariance, this model seems to extend and thus include all previous models. First, geometrization of gravity developed by Einstein by considers the dependence of the Lagrangian on the curvature tensor as the starting point for deriving the field equations (Einstein equations). Second, introducing the tensor Faraday including electric and magnetic fields - within the Lagrangian, it is recognized that the light, a particular case of electromagnetic waves, bends if viewed from a uniformly accelerating frame and then accordingly that the gravity would therefore bend the light. The interaction of gravitation and electromagnetic waves are described Einstein-Maxwell's equations. The geometrization of the electromagnetic fields constitutes the third step when these fields are present in the continuum. For that purpose, we have considered a strain gradient continuum where curvature and torsion are present, Riemann-Cartan continuum. By observing the fields equation, we propose that the gravitational and electromagnetic fields are respectively identified as geometric objects of such a

Table 36.1: Theories of electromagnetism interacting with gravitation in curved continua : Minkowski (flat spacetime), and Riemann (curved continuum).

Minkowski		Special Relativity
Spacetime metric	$g_{\alpha\beta} := \{+,-,-,-\}$	$g := \sqrt{\lvert \mathrm{Det} g_{\alpha\beta} \rvert}$
Electromagnetic tensor	$\Gamma^{\gamma}_{\alpha\beta} \equiv 0$	$\mathscr{F}_{\alpha\beta} = \partial_\alpha A_\beta - \partial_\beta A_\alpha$
Constitutive laws	$\mathscr{L} := -\frac{1}{4} \mathscr{F}^{\alpha\beta} \mathscr{F}_{\alpha\beta}\, g$	$\mathscr{F}^{\alpha\beta} = \epsilon_0\, g^{\alpha\mu} g^{\beta\nu} \mathscr{F}_{\mu\nu}$
Conservation laws	$\partial_\beta \mathscr{F}^{\alpha\beta} = 0$	$g^{\nu\beta} \partial_\nu \partial_\beta A^\mu = 0$
Riemann		Einstein Gravitation
Continuum metric	$g_{\alpha\beta} := g_{\alpha\beta}(x^\mu)$	$g := \sqrt{\lvert \mathrm{Det} g_{\alpha\beta} \rvert}$
Electromagnetic tensor	$\overline{\Gamma}^{\gamma}_{\alpha\beta}$ (Levi – Civita)	$\mathscr{F}_{\alpha\beta} = \overline{\nabla}_\alpha A_\beta - \overline{\nabla}_\beta A_\alpha$
Constitutive laws	$\mathscr{L} := -\frac{1}{4} \mathscr{F}^{\alpha\beta} \mathscr{F}_{\alpha\beta}\, g$	$\mathscr{F}^{\alpha\beta} = \epsilon_0\, g^{\alpha\mu} g^{\beta\nu} \mathscr{F}_{\mu\nu}$
Conservation laws	$\overline{\nabla}_\beta \mathscr{F}^{\alpha\beta} = 0$	$g^{\nu\beta} \overline{\nabla}_\nu \overline{\nabla}_\beta A^\mu - g^{\mu\alpha} \mathfrak{R}_{\alpha\gamma} A^\gamma = 0$

Table 36.2: Theories of electromagnetism interacting with gravitation in twisted and curved continua : Weitzenböck (twisted, not curved) and Riemann-Cartan (twisted and curved)

Weitzenböck		Tele Parallel Gravitation
Continuum metric	$g_{\alpha\beta} := g_{ij}F^i_\alpha F^j_\beta$	$g := \sqrt{\lvert \mathrm{Det}\, g_{\alpha\beta}\rvert}$
Electromagnetic tensor	$\Gamma^\gamma_{\alpha\beta} := F^\gamma_i \partial_\alpha F^i_\beta = \overline{\Gamma}^\gamma_{\alpha\beta} + \mathscr{T}^\gamma_{\alpha\beta}$	$\mathscr{F}_{\alpha\beta} = \overline{\nabla}_\alpha A_\beta - \overline{\nabla}_\beta A_\alpha$
Constitutive laws	$\mathscr{L} := -\frac{1}{4}\mathscr{F}^{\alpha\beta}\mathscr{F}_{\alpha\beta}\, g$	$\mathscr{F}^{\alpha\beta} = \epsilon_0\, g^{\alpha\mu} g^{\beta\nu}\mathscr{F}_{\mu\nu}$
Conservation laws	$\overline{\nabla}_\beta \mathscr{F}^{\alpha\beta} = 0$	$g^{\nu\beta}\overline{\nabla}_\nu\overline{\nabla}_\beta A^\mu - g^{\mu\alpha}\mathscr{K}_{\alpha\gamma}A^\gamma = 0$
Riemann-Cartan		Einstein-Cartan Gravitation
Continuum metric	$g_{\alpha\beta} := g_{\alpha\beta}(x^\mu)$	$g := \sqrt{\lvert \mathrm{Det}\, g_{\alpha\beta}\rvert}$
Electromagnetic tensor	$\Gamma^\gamma_{\alpha\beta}$	$\mathscr{F}_{\alpha\beta} = \nabla_\alpha A_\beta - \nabla_\beta A_\alpha$
Constitutive laws	$\mathscr{L} := -\frac{1}{4}\mathscr{F}^{\alpha\beta}\mathscr{F}_{\alpha\beta}\, g$	$\mathscr{F}^{\alpha\beta} = \epsilon_0\, g^{\alpha\mu} g^{\beta\nu}\mathscr{F}_{\mu\nu}$
Conservation laws	$\nabla_\beta \mathscr{F}^{\alpha\beta} = 0$	$g^{\nu\beta}\nabla_\nu\nabla_\beta A^\mu + g^{\mu\alpha}\aleph^{\gamma}_{\nu\alpha}\nabla_\gamma A^\nu - g^{\mu\alpha}\Re_{\alpha\gamma}A^\gamma = 0$

continuum, namely the curvature $\Re^\gamma_{\alpha\beta\lambda}$ and the torsion $\aleph^\gamma_{\alpha\beta}$. Further studies are required for the invariance aspects.

References

Antonio Tamarasselvame N, Rakotomanana L (2011) On the form-invariance of Lagrangean function for higher gradient continuum. In: Altenbach H, Maugin GA, Erofeev V (eds) Mechanics of Generalized Continua, Heidelberg, ASM, vol 7, pp 291–322

Charap JM, Duff MJ (1977) Gravitational effects on Yang-Mills topology. Physics Letters 69B(4):445–447

Cho YM (1976) Einstein Lagrangian as the translational Yang-Mills Lagrangian. Physical Review D 14(10):2521–2525

de Andrade VC, Pereira JG (1999) Torsion and the electromagnetic field. International Journal of Modern Physics D 8(2):141–151

Dias L, Moraes F (2005) Effects of torsion on electromagnetic fields. Brazilian Journal of Physics 35(3A):636–640

Fernado J, Giglio T, Rodrigues Jr WA (2012) Gravitation and electromagnetism as geometrical objects of Riemann-Cartan spacetime structure. Advanced Applied Clifford Algebras 22:640–664

Fernandez-Nunez I, Bulashenko O (2016) Anisotropic metamaterial as an anlogue of a black hole. Physics Letters A 380:1–8

Fumeron S, Pereira E, Moraes F (2015) Generation of optical vorticity from topological defects. Physica B 476:19–23

Futhazar G, Le Marrec L, Rakotomanana LR (2014) Covariant gradient continua applied to wave propagation within defected material. Archive for Applied Mechanics 84(9–11):1339–1356

Hammond RT (1987) Gravitation, Torsion, and Electromagnetism. General Relativity and Gravitation 20(8):813–827

Hammond RT (1989) Einstein-Maxwell theory from torsion. Classical Quantum and Gravitation 6:195–198

Hehl FW (2008) Maxwell's equations in Minkowski's world: their premetric generalization and the electromagnetic energy-momentum tensor. Annalen der Physik 17(9-10):691–704

Hehl FW, von der Heyde P (1973) Spin and the structure of spacetime. Annales de l'Institut Henri Poincaré, section A 19(2):179–196

Hehl FW, von der Heyde P, Kerlick JM G D ans Nester (1976) General relativity with spin and torsion: Foundations and prospects. Reviews of Modern Physics 48(3):393–416

Kleinert H (2008) Multivalued Fields: in Condensed matter, Electromagnetism, and Gravitation. World Scientific, Singapore

Kovetz A (2000) Electromagnetic Theory. Oxford University Press, Oxford

Maugin G (1993) Material Inhomogeneities in Elasticity. Chapman and Hall, London

Maugin GA (1978) Exact relativistic theory of wave propagation in prestressed nonlinear elastic solids. Annales de l'Institut Henri Poincaré, section A 28(2):155–185

Milonni PW, Boyd RW (2010) Momentum of light in a dielectric medium. Adv Opt Photon 2(4):519–553

Obukhov YN (2008) Electromagnetic energy and momentum in moving media. Annalen der Physik 17(9-10):830–851

Obukhov YN, Hehl FW (2003) Electromagnetic energy-momentum and forces in matter. Physics Letters A 311:277–284

Plebanski J (1960) Electromagnetic waves in gravitational fields. Physical Review 118(5):1396–1408

Poplawski NJ (2010) Torsion as electromagnetism and spin. International Journal of Theoretical Physics 49(7):1481–1488

Prasanna AR (1975) Maxwell's equations in Riemann-Cartan space U_4. Physics Letters A 54(1):17–18

Puntigam RA, L¨ammerzahl C, Hehl FW (1997) Maxwell's theory on a post-Riemannian spacetime and the equivalence principle. Classical and Quantum Gravitation 14:1347–1356

Rakotomanana RL (2003) A Geometric Approach to Thermomechanics of Dissipating Continua. Progress in Mathematical Physics Series, Birkhaüser, Boston

Schutzhold R, Plunien G, Soff G (2002) Dielectric black hole analogs. Physical Review Letters 88(6):061,101/1–061,101/4

Smalley LL (1986) On the extension of geometric optics from Riemaniann to Riemann-Cartan spacetime. Physics Letters A 117(6):267–269

Smalley LL, Krisch JP (1992) Minimal coupling of electromagnetic fields in Riemann-Cartan spacetimes for perfect fluids with spin density. Journal of Mathematical Physics 33(3):1073–1081

Sotiriou TP, Liberati S (2007) Metric-affine $f(R)$ theories of gravity. Annals of Physics 322:935–966

Tiwari RN, Ray S (1997) Static spherical charged dust electromagnetic mass models in Einstein-Cartan theory. General Relativity and Gravitation 29(6):683–690

Vandyck MA (1996) Maxwell's equations in spaces with non-metricity and torsion. J Physics A : Math Gen 29:2245–2255

Chapter 37
Optimal Calculation of Solid-Body Deformations with Prescribed Degrees of Freedom over Smooth Boundaries

Vitoriano Ruas

Abstract One of the reasons why the finite element method became the most used technique in Computational Solid Mechanics is its versatility to deal with bodies having a curved shape. In this case method's isoparametric version for meshes consisting of curved triangles or tetrahedra has been mostly employed to recover the optimal approximation properties known to hold for standard elements in the case of polygonal or polyhedral domains. However isoparametric finite elements helplessly require the manipulation of rational functions and the use of numerical integration. This can be a brain teaser in many cases, especially if the problem at hand is non linear. We consider a simple alternative to deal with boundary conditions commonly encountered in practical applications, that bypasses these drawbacks, without eroding the quality of the finite-element model. More particularly we mean prescribed displacements or forces in the case of solids. Our technique is based only on polynomial algebra and can do without curved elements. Although it can be applied to countless types of problems in Continuum Mechanics, it is illustrated here in the computation of small deformations of elastic solids.

37.1 Introduction

This work deals with a new technique for the numerical solution of boundary value problems posed in a two- or three-dimensional domain, with a boundary consisting of disconnected smooth curved portions of arbitrary shape. Although it can be used for solving different types of problems in Continuum Mechanics and beyond, here we focus its application to Solid Mechanics. This is because the technique is strongly related to the finite element method, which turns out to be by far the most popular

Vitoriano Ruas
Sorbonne Université, Centre National de la Recherche Scientifique, UMR 7190, Institut Jean Le Rond d'Alembert, F-75005 Paris, France
e-mail: vitoriano.ruas@upmc.fr

© Springer International Publishing AG, part of Springer Nature 2018

H. Altenbach et al. (eds.), *Generalized Models and Non-classical Approaches in Complex Materials 1*, Advanced Structured Materials 89,
https://doi.org/10.1007/978-3-319-72440-9_37

numerical method employed in deformable solid-body modeling. Actually the new technique is aimed at taking the best advantage of equally smooth data, prescribed all over each smooth disconnected boundary portion as essential boundary conditions. In the case of the finite element method these correspond to prescribed boundary degrees of freedom, which can be displacements for a standard minimum-energy formulation, or forces for a complementary energy formulation.

The method presented in this paper attempts to tackle the true boundary of the solid body, akin to the *interpolated boundary condition method* described in Brenner and R.Scott (2008) introduced almost simultaneously in Nitsche (1972); Scott (1973), and the method proposed in Zlámal (1973) extended in Ženíšek (1978), for two-dimensional problems. Although the latter methods are rather intuitive and are known since the seventies, they have been of limited use so far. Among the reasons for this we could quote their difficult implementation, the lack of extensions to three-dimensional problems, or yet restrictions on the choice of nodal points where boundary degrees of freedom are to be prescribed in order to reach optimal convergence rates. In contrast our method is simple to implement in both two- and three-dimensional geometries. Moreover optimality is attained very naturally in both cases for various choices of boundary nodal points.

Without loss of essential aspects, our methodology can be perfectly described taking as a model simple linear partial differential equations with Dirichlet boundary conditions. We can consider for example the Poisson equation solved by different N-simplex based methods, incorporating degrees of freedom other than function values at the mesh vertices. If standard quadratic Lagrange finite elements are employed, it is well-known that approximations of an order not greater than 1.5 in the energy norm are generated (cf. Ciarlet, 1978), in contrast to the second order ones that apply to the case of a polygonal or polyhedral shape, assuming that the solution is sufficiently smooth. If we are to recover the optimal second order approximation property, something different has to be done. Since long the isoparametric version of the finite element method for meshes consisting of curved triangles or tetrahedra (cf. Zienkiewicz, 1971), has been considered as the ideal way to achieve this. It turns out that, besides a more elaborated description of the mesh, the isoparametric technique inevitably leads to the integration of rational functions to compute the system matrix, which raises the delicate question on how to choose the right numerical quadrature formula in the master element. In contrast, in the technique to be considered in this paper exact integration can always be used for this purpose, since we only have to deal with polynomial integrands. Moreover the element geometry remains the same as in the case of polygonal or polyhedral domains. It is noteworthy that both advantages are conjugated with the fact that no erosion of qualitative approximation properties results from the application of our technique, as compared to the equivalent isoparametric one. We should also emphasize that this approach is particularly handy, whenever the finite element method under consideration has normal components or normal derivatives as degrees of freedom. Indeed in this case the definition of isoparametric finite element analogs is not always so straightforward (cf. Bertrand et al, 2014).

An outline of the paper is as follows. In Section 2 we describe our method to solve a model Poisson problem with Dirichlet boundary conditions in a smooth curved two-dimensional domain with conforming Lagrange finite elements based on meshes with straight triangles, in connection with a standard Galerkin formulation. We recall the error estimates established in Ruas (2017b) for this method. In Sect. 37.2 we assess its approximation properties, by solving linear elasticity problems posed in a two-dimensional solid body a smooth curved shape using piecewise quadratic functions. More specifically a circular membrane problem is considered in Subsect. 37.3.1 and in Subsect. 37.3.2we use our methodology to represent the displacement field of a homogeneous isotropic elastic annulus in small plane deformations caused by a prescribed rotation of its outer boundary. We conclude in Sect. 37.4 with a few comments.

37.2 Method Description

As a model we consider the Poisson equation with Dirichlet boundary conditions in an N-dimensional smooth domain Ω with boundary $\partial\Omega$ for $N = 2$ or $N = 3$, that is, $-\Delta u = f$ in Ω, $u = d$ on Γ, where f and d are given functions defined in Ω and on Γ, having suitable regularity properties. We shall be dealing with approximations of u of order k for $k > 1$ in the standard energy norm, assuming that f, d and Ω are sufficiently smooth. Although the method to be described below applies to any spatial dimension, for the sake of simplicity we confine its description to the two-dimensional case. Let us then be given a partition $\mathscr{T}_h$ of Ω into straight triangles satisfying the usual compatibility conditions (see Ciarlet (1978)). $\mathscr{T}_h$ is assumed to belong to a uniformly regular family of partitions. Let Ω_h be the polygonal domain formed by the union of the triangles in $\mathscr{T}_h$ and Γ_h be its boundary. Further h_T being the diameter of a triangle $T \in \mathscr{T}_h$, we denote by h the maximum of the h_Ts as T sweeps $\mathscr{T}_h$. Notice that if Ω is convex Ω_h is a proper subset thereof. We make the more than reasonable assumptions on the mesh that no element in $\mathscr{T}_h$ has more than one edge on Γ_h. We also need some definitions regarding the skin comprised between Γ_h and Γ. First of all, in order to avoid non essential technicalities, we assume that the mesh is constructed in such a way that convex and concave portions of Γ correspond to convex and concave portions of Γ_h. This property is guaranteed if the points separating such portions of Γ are vertices of polygon Γ_h. In doing so, let $\mathscr{S}_h$ be the subset of $\mathscr{T}_h$ consisting of triangles having one edge on Γ_h Now for every T in the subset $\mathscr{S}_h$ we denote by Δ_T the skin portion delimited by Γ and the edge e_T of T whose end-points belong to Γ and by T' the sets indicated in $T \cup \Delta_T$ (see Fig. 37.1).

Next we introduce two sets of functions V_h and W_h^d associated with $\mathscr{T}_h$. V_h is the standard Lagrange finite element space consisting of continuous functions v defined in Ω_h that vanish on Γ_h, whose restriction to every triangle T in $\mathscr{T}_h$ is a polynomial of degree less than or equal to k for $k > 1$. For convenience we extend by zero every

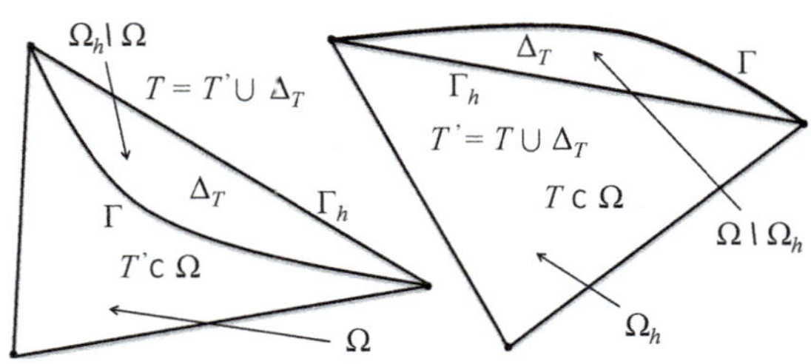

Fig. 37.1 Skin portion Δ_T comprised between e_T and Γ at the level of a mesh triangle T belonging to the subset $\mathscr{S}_h$, next to a convex (right) or a concave (left) portion of Γ; e_T is the edge of T contained in Γ_h.

function $v \in V_h$ to $\Omega \setminus \Omega_h$. W_h^d in turn is the set of functions defined in Ω_h having the properties listed below.

1. The restriction of $w \in W_h^d$ to every $T \in \mathscr{T}_h$ is a polynomial of degree less than or equal to k ;
2. Every w in W_h^d is continuous in Ω_h and equals d at the vertices of Γ_h;
3. A function w in W_h^d is also defined in $\Omega \setminus \Omega_h$, in such a way that its polynomial expression in a triangle T belonging to $\mathscr{S}_h$ also applies to points in the skin portion Δ_T ;
4. For all T belonging to $\mathscr{S}_h$, $w(P) = d(P)$ at every point P of Γ located on the lines passing through the vertex O_T of T not belonging to Γ and the $k-1$ points M of e_T different from a vertex of T lying among those that subdivide this edge into k segments of equal length (cf. Fig. 37.2).

Note 37.1. The construction of the nodes associated with W_h^d located on Γ_h advocated in item 4. is not mandatory. Notice that it differs from the intuitive construction of such nodes lying on normals to edges of Γ_h commonly used in the isoparametric technique. The main advantage of this proposal is an easy determination of boundary node coordinates by linearity, using a supposedly available analytical expression of Γ. Actually the choice of boundary nodes ensuring our method's optimality is absolutely flexible, in contrast to the restrictions inherent to the interpolated boundary condition method (cf. Brenner and R.Scott, 2008).

The fact that W_h^d is a non empty finite-dimensional manifold was established in Ruas (2017b).

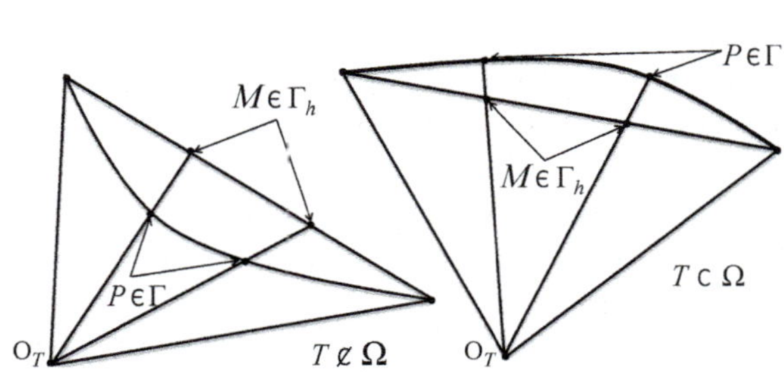

Fig. 37.2 Construction of nodes P located on Γ for the set W_h^d corresponding to nodes M located on Γ_h, exemplified in the particular case of the cubic Lagrange interpolation (i.e. for $k = 3$)

Next we set the problem associated with the sets of functions V_h and W_h^d, whose solution is an approximation u_h of the solution u to the Poisson equation. Taking any regular extension of f to $\Omega_h \setminus \Omega$ and still denoting the resulting function by f, u_h is determined as the solution of the following variational problem:

$$u_h \in W_h^d \text{ fulfills for all } v \in V_h : \int_{\Omega_h} grad\, u_h \cdot grad\, v\, dxdy = \int_{\Omega_h} f\, v\, dxdy \quad (37.1)$$

According to Ruas (2017b), provided h is sufficiently small, problem (37.1) has a unique solution. Moreover the underlying bilinear form is uniformly stable on $W_h^0 \times V_h$ in the sense of Babuska-Brezzi (cf. Ruas, 2017b)) where W_h^0 is W_h^d for $d \equiv 0$. This leads to the conclusion that the approximation method associated with (37.1) is a k-th order method in the natural energy norm, as long as u is sufficiently smooth and all the $\mathscr{T}_h$s under consideration belong to a regular family of partitions of Ω in the sense of Ciarlet (1978). In short

$$[\int_{\Omega_h} |grad(u_h - u)|^2 dx]^{1/2}$$

is an $O(h^k)$ under such assumptions (cf. Ruas (2017b)).

37.3 Method Experimentation

In order to show the effectiveness of the technique studied in this work to take into account Dirichlet boundary conditions prescribed on curved boundaries, we report in this section significant numerical results for some academic test-problems with known exact solution, taking $k = 2$. We selected two classical applications in Solid Mechanics, namely, the small deflexions undergone by a homogeneous elastic membrane under the action of forces orthogonal to its plane, and the small plane deformations of a homogeneous isotropic elastic plate.

37.3.1 Deflections of an Elastic Membrane

In the small strain regime a homogeneous elastic membrane whose edge is kept fixed, subjected to a load distribution g applied perpendicularly to its plane Π is governed by the Poisson equation with homogeneous Dirichlet boundary conditions, i.e. with $d \equiv 0$. More specifically, in appropriate dimensionless form the deflexion u in the direction orthogonal to Π satisfies $-\Delta u = f$ in the region Ω of the plane occupied by the membrane in Π, where $f = cg$ for a suitable constant c. For the sake of simplicity we consider that Ω is the unit disk with center at the origin O of a cartesian coordinate system (O,x,y) attached to Π. Taking g of the conical form αr where $r = (x^2+y^2)^{1/2}$ the exact solution is given by

$$u(x,y) = \frac{\alpha(1-r^3)}{9c}.$$

In our computations we took $\alpha = 9c$, and used meshes with $2M^2$ triangles for $M = 2^m$ and $m = 2,3,4,5,6$, of the quarter disk given by $x > 0$ and $y > 0$ due to symmetry. For better visibility quite abusively we set $h = 1/M$. The resulting errors in the following senses are displayed in Table 37.1. In the third column we show the errors in the energy norm while in the fourth column the mean-square norms of $u - u_h$ in Ω_h are given. For the sake of conciseness we denote the mean-square norm of a function or field $\boldsymbol{a}$ defined in Ω_h by $\| \boldsymbol{a} \|_h$.

In order to make sure that there was no favorable effect owing to the particular shape of the membrane and/or the exact solution, we also computed with the same meshes using the polygonal approach. This consists of prescribing zero boundary values at the mid-points of the edges e_T, besides the vertices of Γ_h. In other words the polygonal approach consists of replacing W_h^0 by V_h in (37.1), thereby generating another numerical solution $\underline{u}_h$ instead of u_h. Corresponding errors are supplied in Table 37.2.

Observation of Tables 37.1 and 37.2 confirms that our method is of the second order in the energy norm, while the polygonal approach yields only $O(h^{1.5})$ approximations in the same sense, as predicted in classical books (cf. Ciarlet (1978)). Even better news come from the errors in the mean-square sense. We observe third order approximations for our method, in contrast to the second order ones for the polygonal approach.

Table 37.1: Errors for the membrane problem solved by the new approach with $k = 2$

M	h	$\| grad(u-u_h) \|_h$	$\| u-u_h \|_h$
4	0.250000	0.14007×10^{-1}	0.43895×10^{-3}
8	0.125000	0.36168×10^{-2}	0.56460×10^{-4}
16	0.062500	0.91850×10^{-3}	0.71709×10^{-5}
32	0.031250	0.23151×10^{-3}	0.90592×10^{-6}
64	0.015625	0.58128×10^{-4}	0.12428×10^{-6}

Table 37.2: Errors for the membrane problem solved by the polygonal approach with $k = 2$

M	h	$\| grad(u - \underline{u}_h) \|_h$	$\| u - \underline{u}_h \|_h$
4	0.250000	0.54344×10^{-1}	0.81891×10^{-2}
8	0.125000	0.19690×10^{-1}	0.19879×10^{-2}
16	0.062500	0.70417×10^{-2}	0.48790×10^{-3}
32	0.031250	0.25026×10^{-2}	0.12070×10^{-3}
64	0.015625	0.88700×10^{-3}	0.29979×10^{-4}

37.3.2 Torsion of an Elastic Annulus

The aim of the experiments reported in this sub-section is to assess the behavior of an obvious extension of our method to a system of linear second order partial differential equations. More specifically we consider the linear elasticity system for a homogeneous, compressible and isotropic solid occupying a bounded region Ω of the plane referred to a cartesian coordinate system (O,x,y), with a smooth curved boundary Γ. Under these circumstances, in case the solid is subjected to body forces $\vec{f}$ and its displacement field $\boldsymbol{u} = (u_x, u_y)$ is prescribed all over its boundary, say $\boldsymbol{u} = \boldsymbol{d}$ on Γ, $\boldsymbol{u}$ is the solution of the system

$$\begin{cases} -2\mu div\, \epsilon(\boldsymbol{u}) - \lambda grad\; div\, \boldsymbol{u} = \boldsymbol{f} \text{ in } \Omega \\ \boldsymbol{u} = \boldsymbol{d} \text{ on } \Gamma, \end{cases} \tag{37.2}$$

where $\epsilon(\boldsymbol{u})$ is the symmetric gradient of $\boldsymbol{u}$ and μ and λ are the Lamé coefficients of the material the solid is made of.

Similarly to the case of the scalar Poisson equation, we will define the set of test-fields to be $\boldsymbol{V}_h := V_h \times V_h$ and the set of trial-fields $\boldsymbol{W}_h^{\boldsymbol{d}} := W_h^{d_x} \times W_h^{d_y}$ with $(d_x, d_y) = \boldsymbol{d}$. Then we determine an approximation $\boldsymbol{u}_h$ of $\boldsymbol{u}$ by solving the discrete variational problem,

$$\begin{cases} \boldsymbol{u}_h \in \boldsymbol{W}_h^{\boldsymbol{d}} \text{ fulfills for all } \boldsymbol{v} \in \boldsymbol{V}_h : \\ \displaystyle\int_{\Omega_h} [2\mu\epsilon(\boldsymbol{u}_h) : \epsilon(\boldsymbol{v}) + \lambda div\, \boldsymbol{u}_h div\, \boldsymbol{v}]\, dxdy = \int_{\Omega_h} \boldsymbol{f} \cdot \boldsymbol{v}\, dxdy. \end{cases} \tag{37.3}$$

We check the performance of the above method in the case where Ω is the annulus delimited by the circles given by $r = r_o$ and $r = r_i$, with $r_i < r_o$. This annulus represents the cross section of the domain comprised between two concentric cylinders with radii r_o and r_i between which lie a homogeneous and isotropic elastic body. This body is assumed to perfectly adhere to both cylinders. The outer cylinder undergoes a rotation about its axis of an angle equal to ω_o while the inner cylinder is kept fixed. We assume deformation conditions under which the resulting strains in the elastic body are small. In this case the displacement field $\boldsymbol{u} = (u_x, u_y)$ referred to a cartesian frame (O,x,y,z) such that the z-axis coincides with the axis of both cylinders is given by

$$u_x(x,y) = -\frac{y u_\theta(r)}{r}$$

and

$$u_y(x,y) = \frac{x u_\theta(r)}{r},$$

where u_θ is the azimuthal displacement of the body at a distance from the z-axis equal to r.

Our experiments consist of approximating by quadratic Lagrange finite elements the components u_x and u_y of the displacement field, which satisfy the equilibrium equation (37.2) in Ω with $\boldsymbol{f} \equiv \boldsymbol{0}$ and $\boldsymbol{d} \equiv \boldsymbol{0}$ for $r = r_i$ and $\boldsymbol{d} = (-x\omega_o, y\omega_o)$ for $x^2 + y^2 = r_o^2$. Then, irrespective of the values of μ and λ, the exact solution is given by

$$\boldsymbol{u} = u_\theta(r)\left(-\frac{y}{r}, \frac{x}{r}\right),$$

where

$$u_\theta(r) = \omega_o r_o^2 \frac{\left(r - \frac{r_i^2}{r}\right)}{(r_o^2 - r_i^2)}.$$

In order to take advantage of symmetry we solve (37.3) in the quarter annulus corresponding to $x > 0$ and $y > 0$, by prescribing $\boldsymbol{u}_h \cdot \boldsymbol{\tau} = 0$ and $\boldsymbol{v} \cdot \boldsymbol{\tau} = 0$ on the axes $\overrightarrow{Ox}$ and $\overrightarrow{Oy}$, where $\boldsymbol{\tau}$ generically represents the unit vector along them. Using dimensionless quantities we take $r_o = 1$ and $r_i = 1/2$, and prescribe $\omega_o = 0.5^{\mathrm{o}}$.

The meshes of the computational domain consist of $4M^2$ triangles constructed by subdividing the range of the radial coordinate r into M equal segments, and its total aperture equal to $\pi/2$ into $2M$ equal angles with $M > 1$. Setting for convenience $h = r_o\pi/(4M)$ we supply in Table 37.3 the errors measured in the gradient mean-square norm, i.e. $\| grad(\boldsymbol{u} - \boldsymbol{u}_h) \|_h$, and the field mean-square norm, i.e. $\| \boldsymbol{u} - \boldsymbol{u}_h \|_h$, for $M = 2, 4, 8, 16, 32$. As one can observe, akin to Sub-section 3.1, the numerical results show that our method is of the second order in the gradient mean-square norm and of the third order in the mean-square sense, instead of being of the order 1.5 and second order, respectively like in the case of the polygonal approach.

Note 37.2. In Table 37.3 $\boldsymbol{u}$ also stands for the extension of $\boldsymbol{u}$ to the skin $\Omega_h \setminus \Omega$ next to the inner boundary given by $r = r_i$, analytically expressed in the same way as in Ω.

37.4 Final Comments

Numerical two-dimensional experiments were further carried out in several author's recent works on the methodology considered in this paper. For instance in Ruas (2017c) this was done again for curved membranes. However in contrast to Subsect. 3.1 in Ruas (2017c) the simulations were also performed with an adaptation of our technique to a Hermite analog of the Raviart-Thomas mixed finite element of the lowest order, allowing for an explicit representation of the stresses (cf. Ruas and Silva Ramos, 2017, and references therein).

Three-dimensional boundary value problems can be dealt with in a very similar manner, even though for obvious reasons our technique becomes slightly more complicated in this case. For further details the author refers to Ruas (2017a). Applications of the method studied in this work to Fluid Mechanics are underway and will be reported in due course.

Acknowledgements This work was partially accomplished while the author was working at PUC-Rio, Brazil, as a CNPq research grant holder.

References

Bertrand F, Münzenmaier S, Starke G (2014) First-order system least squares on curved boundaries: Higher-order raviart–thomas elements. SIAM Journal on Numerical Analysis 52(6):3165–3180

Brenner SC, RScott L (2008) The Mathematical Theory of Finite Element Methods, Texts in Applied Mathematics, vol 15. Springer

Ciarlet PG (1978) The Finite Element Method for Elliptic Problems. North Holland, Amsterdam

Nitsche J (1972) On dirichlet problems using subspaces with nearly zero boundary conditions. In: Aziz A (ed) The Mathematical Foundations of the Finite Element Method with Applications to Partial Differential Equations, Academic Press

Ruas V (2017a) Methods of arbitrary optimal order with tetrahedral finite-element meshes forming polyhedral approximations of curved domains. arXiv Numerical Analysis

Ruas V (2017b) Optimal simplex finite-element approximations of arbitrary order in curved domains circumventing the isoparametric technique. ArXiv e-prints

Table 37.3: Errors for the annulus problem solved by the new approach with $k = 2$

$2M$	h	$\| grad(\boldsymbol{u}-\boldsymbol{u}_h) \|_h$	$\| \boldsymbol{u}-\boldsymbol{u}_h \|_h$
4	$\pi/8$	0.53502×10^{-3}	0.15738×10^{-4}
8	$\pi/16$	0.13977×10^{-3}	0.19913×10^{-5}
16	$\pi/32$	0.35449×10^{-4}	0.25040×10^{-6}
32	$\pi/64$	0.88970×10^{-5}	0.31335×10^{-7}
64	$\pi/128$	0.22264×10^{-5}	0.39163×10^{-8}

Ruas V (2017c) A simple alternative for accurate finite-element modeling in curved domains. In: Comptes-rendus du Congrès Français de Mécanique, Lille

Ruas V, Silva Ramos MA (2017) A hermite fe method for maxwell's equations. AIP Conference Proceedings 1863(1):370,003

Scott LR (1973) Finite element techniques for curved boundaries. Phd, MIT

Ženíšek A (1978) Curved triangular finite c^m-elements. Aplikace Matematiky 23(5):346–377

Zienkiewicz OC (1971) The Finite Element Method in Engineering Science. McGraw-Hill,

Zlámal M (1973) Curved elements in the finite element method. i. SIAM Journal on Numerical Analysis 10(1):229–240

Chapter 38
Toward a Nonlinear Asymptotic Model for Thin Magnetoelastic Plates

Sushma Santapuri and David J. Steigmann

Abstract An asymptotic two-dimensional formulation for the potential energy of a thin magnetoelastic plate is obtained from that for a three-dimensional magnetoelastic body subjected to conservative tractions and an applied magnetic field.

38.1 Introduction

Modern treatments of the theory of elastic plates and shells emphasize asymptotic derivations of the relevant equations from expressions for the potential energy of a three-dimensional nonlinearly elastic body. The availability of such energies in three-dimensional magnetoelasticity theory naturally suggests an avenue for further application of the asymptotic method. We discuss this extension in the present work. In contrast, classical treatments of magnetoelastic plate theory (Maugin, 1988) emphasize through-thickness integration of the three-dimensional equations but do not deliver information about the order of accuracy of the resulting theory *vis a vis* the three-dimensional theory.

We present a summary of three-dimensional magneto-elastostatics in Sect 38.2. A convenient reformulation of the basic theory that facilitates our objectives is outlined in Sect. 38.3. This is augmented in Sect. 38.4 by a derivation of necessary conditions for energy-minimizing states that play a crucial role in securing the eventual two-dimensional model. In Sects. 38.5 and 38.6 we derive the two-dimensional plate energy to an accuracy of order h^3, where h is the small plate thickness; this is the scaling that is classically associated with combined plate bend-

Sushma Santapuri
Department of Applied Mechanics, Indian Institute of Technology Delhi, New Delhi 110016, India
e-mail: ssantapuri@am.iitd.ac.in

David J. Steigmann
Department of Mechanical Engineering, University of California Berkeley, CA 94720, USA
e-mail: dsteigmann@berkeley.edu

© Springer International Publishing AG, part of Springer Nature 2018
H. Altenbach et al. (eds.), *Generalized Models and Non-classical Approaches in Complex Materials 1*, Advanced Structured Materials 89,
https://doi.org/10.1007/978-3-319-72440-9_38

ing and stretching. The resulting expression for the energy inherits certain features from the three-dimensional theory related to the through-thickness derivatives of the three-dimensional deformation and magnetization fields. In Sect. 38.7 we demonstrate that these latter fields may be obtained in terms of the mid-plane deformation and its (two-dimensional) gradients to yield a plate energy functional that is finally determined by the midplane deformation alone, as in the conventional purely elastic theory.

We employ a direct notation throughout so as to convey the underlying ideas as succinctly as possible. Readers preferring more explicit formulae are encouraged to translate these into Cartesian index notation.

38.2 Summary of the Three-Dimensional Theory for Conservative Problems

In the formulation of the three-dimensional static theory of magnetoelasticity adopted here, the independent variables are the deformation field $\mathbf{y}=\boldsymbol{\chi}(\mathbf{x})$, where $\boldsymbol{\chi}$ is the deformation function and $\mathbf{x}$ is the position of a material point in a fixed reference configuration κ; and the magnetization $\boldsymbol{\mu}(\mathbf{x})$ per unit mass.

The basic constitutive function is the strain energy, $W(\mathbf{F},\boldsymbol{\mu})$, per unit volume of κ. This function depends explicitly on $\mathbf{x}$ if the material is not uniform, but we suppress this in the notation for the sake of clarity.

The basic equations of three-dimensional magneto-elastostatics are (Brown, 1966; James, 2002; DeSimone and Podio-Guidugli, 1996; Steigmann, 2004; Dorfmann and Ogden, 2014):

1. The force balance
$$Div(W_{\mathbf{F}})+\mu_0 J(grad\mathbf{h})\mathbf{m}=\mathbf{0}, \tag{38.1}$$
wherein Div is the divergence with respect to $\mathbf{x}$, $grad$ is the gradient with respect to $\mathbf{y}$, $\mathbf{h}$ is the magnetic field, $\mathbf{F}$ is the gradient of $\boldsymbol{\chi}(\mathbf{x})$, $J=\det\mathbf{F}$, $\mathbf{m}$ is the magnetization per unit volume of the current configuration, R say, and μ_0 is the permeability of free space;
2. The magnetostatic equations
$$div\mathbf{b}_s=0; \quad [\mathbf{b}_s]\cdot\mathbf{n}=0, \tag{38.2}$$
where $\mathbf{b}_s$ is the magnetic induction self-field generated by the magnetized body and $[\cdot]\cdot\mathbf{n}$ is the jump in the enclosed quantity across the bounding surface ∂R with exterior unit normal $\mathbf{n}$. In (38.2) the induction self-field is defined by
$$\mu_0^{-1}\mathbf{b}_s=\mathbf{h}_s+\mathbf{m}, \tag{38.3}$$
where $\mathbf{h}_s$ is the magnetic self-field - the part of the field generated by the magnetized body. This is simply the difference between the total and applied magnetic fields; i.e.,

$$h_s = h - h_a. \tag{38.4}$$

The total, self and applied magnetic fields are presumed to be curl free in accordance with Ampére's law. Lastly,

3. The constitutive equation

$$\mu_0 \rho_\kappa h = W_\mu \tag{38.5}$$

connecting the magnetic field to the magnetization and the deformation, where $\rho_\kappa(x)$ is the assigned mass density in κ.

Granted the force balance (38.1), balance of moments is equivalent to the rotational invariance of the energy; i.e., $W(F, \mu) = W(QF, Q\mu)$, where Q is an arbitrary rotation (Steigmann, 2004; Kovetz, 2000). We do not dwell on the consequences of such invariance, which are well known, and, in any case, not germane to the considerations of this work.

In (Brown, 1966; James, 2002; DeSimone and Podio-Guidugli, 1996; Kankanala and Triantafyllidis, 2004) it is shown by a rather involved procedure that, for conservative problems, Eqs. (38.1) and (38.5) are the Euler-Lagrange equations for the potential energy

$$\mathscr{E} = \int_\kappa [W - \frac{1}{2}\mu_0 \rho_\kappa (h + h_a) \cdot \mu] dv - \mathscr{L}, \tag{38.6}$$

where h_a is the applied magnetic field generated by sources remote from the body, and $\mathscr{L}$ is a load potential. This presumes that the magnetostatic Eqs. (38.2)-(38.4) are satisfied *a priori*. The applied field is presumed to be prescribed as a smooth function of y in all space, consisting of the interior and exterior of the body. Further, the variation of the load potential generates the virtual work of the applied Piola tractions p_a; thus,

$$\dot{\mathscr{L}} = \int_{\partial \kappa_p} p_a \cdot \dot{y} da, \tag{38.7}$$

where, here and elsewhere, superposed dots are used to denote variational, or Gateaux, derivatives, and $\partial \kappa_p$ is the complement of the part of $\partial \kappa$ where position y is assigned. For example, in the case of dead loading, $p_a(x)$ is fixed on $\partial \kappa_p$, independently of the deformation, and for $\mathscr{L}$ we may take

$$\mathscr{L} = \int_{\partial \kappa_p} p_a \cdot y da. \tag{38.8}$$

Whether or not the load is dead, the associated natural boundary condition is (Steigmann, 2004)

$$(W_F)N = p_a + \frac{1}{2}\mu_0 (m \cdot n)^2 F^* N, \tag{38.9}$$

where N is the exterior unit normal to $\partial \kappa$ and F^* is the cofactor of F.

38.3 Reformulation

We reformulate the foregoing system to make it more amenable to our present purpose. To this end it proves convenient to define the pull-back

$$\boldsymbol{M} = \rho_\kappa \boldsymbol{F}^{-1}\boldsymbol{\mu} \tag{38.10}$$

of the magnetization field, and to regard $\boldsymbol{F}$ and $\boldsymbol{M}$ as independent variables in the strain-energy function. Accordingly, we define

$$U(\boldsymbol{F},\boldsymbol{M}) = W(\boldsymbol{F},\rho_\kappa \boldsymbol{F}\boldsymbol{M}). \tag{38.11}$$

Then, the potential energy (38.6) may be expressed as

$$\mathscr{E} = \int_\kappa [U - \frac{1}{2}\mu_0(\boldsymbol{H}+\boldsymbol{H}_a)\cdot\boldsymbol{M}]dv - \mathscr{L}, \tag{38.12}$$

where

$$\boldsymbol{H} = \boldsymbol{F}^t\boldsymbol{h} \quad \text{and} \quad \boldsymbol{H}_a = \boldsymbol{F}^t\boldsymbol{h}_a(\boldsymbol{y}). \tag{38.13}$$

It follows easily from (38.5) and the chain rule that

$$U_{\boldsymbol{M}} = \mu_0 \boldsymbol{H}, \tag{38.14}$$

which we impose henceforth. Accordingly, the energy (38.12) becomes

$$\mathscr{E} = \int_\kappa \Xi(\boldsymbol{y},\boldsymbol{F},\boldsymbol{M})dv - \mathscr{L}, \tag{38.15}$$

where

$$\Xi(\boldsymbol{y},\boldsymbol{F},\boldsymbol{M}) = U(\boldsymbol{F},\boldsymbol{M}) - \frac{1}{2}\boldsymbol{M}\cdot U_{\boldsymbol{M}} - \frac{1}{2}\mu_0\boldsymbol{H}_a\cdot\boldsymbol{M}, \tag{38.16}$$

in which the explicit dependence on $\boldsymbol{y}$ is inherited from $(38.13)_2$ via the assigned function $\boldsymbol{h}_a(\boldsymbol{y})$.

Stationarity of the energy with respect to $\boldsymbol{M}$, at fixed $\boldsymbol{\chi}(\boldsymbol{x})$, yields the condition

$$\Xi_{\boldsymbol{M}} = \boldsymbol{0}, \tag{38.17}$$

holding pointwise in κ. Using (38.16), this may be cast as

$$U_{\boldsymbol{M}} = \mu_0\boldsymbol{H}_a + (U_{\boldsymbol{MM}})\boldsymbol{M}, \tag{38.18}$$

and with (38.14) this may be interpreted as

$$\mu_0\boldsymbol{H}_s = (U_{\boldsymbol{MM}})\boldsymbol{M}, \tag{38.19}$$

where

$$H_s = F^t h_s \tag{38.20}$$

is the pull-back of the magnetic self-field (cf. (38.4) and (38.13)) in the region occupied by the material. Accordingly, as expected, the self-field in the body vanishes if the magnetization vanishes; that is, if the body is not magnetized. In the next section we show that U_{MM} is invertible and hence that the self-field vanishes if and only if the magnetization vanishes.

Stationarity of the energy with respect to variation of the deformation yields the differential equation

$$Div(\Xi_F) = \Xi_y, \tag{38.21}$$

holding in κ; and the boundary condition

$$(\Xi_F)N = p_a, \tag{38.22}$$

holding on $\partial\kappa_p$.

38.4 Legendre-Hadamard Conditions

In James (2002); Kankanala and Triantafyllidis (2004) arguments are given in support of regarding stable equilibria as minimizers of the energy $\mathscr{E}$. In the present section we derive pointwise necessary conditions for such states which play an essential role in the considerations that follow. To this end we compute the second variation of the energy (38.15). Suppressing the applied tractions and the applied magnetic field, which in principle are subject to our control, we find the second variation to be

$$\begin{aligned}\ddot{\mathscr{E}} = &\int_\kappa \{\Xi_{FF}[\dot{F}]\cdot\dot{F} + (\Xi_{MM})\dot{M}\cdot\dot{M} + (\Xi_{FM})\dot{M}\cdot\dot{F} + (\Xi_{MF})\dot{F}\cdot\dot{M}\}dv \\ &+ \int_\kappa (\Xi_F\cdot\ddot{F} + \Xi_M\cdot\ddot{M})dv.\end{aligned} \tag{38.23}$$

At equilibrium, the first variation,

$$\dot{\mathscr{E}} = \int_\kappa (\Xi_F\cdot\dot{F} + \Xi_M\cdot\dot{M})dv \tag{38.24}$$

vanishes for all choices of the variations $\dot{F}$ and $\dot{M}$. Therefore the second integral in (38.23) vanishes at equilibrium. If the equilibrium state is an energy minimizer, it is then necessary that

$$\int_\kappa \{\Xi_{FF}[\nabla u]\cdot\nabla u + v\cdot(\Xi_{MM})v + (\Xi_{FM})v\cdot\nabla u + (\Xi_{MF})\nabla u\cdot v\}dv > 0 \tag{38.25}$$

for all vector fields $u(x)$ and $v(x)$ with the proviso that u vanishes on $\partial\kappa\setminus\partial\kappa_p$.

Consider the special variation

$$\boldsymbol{u}(\boldsymbol{x}) = \varepsilon\boldsymbol{\phi}(\boldsymbol{z}) \quad \text{with} \quad \boldsymbol{z} = \varepsilon^{-1}(\boldsymbol{x}-\boldsymbol{x}_0), \quad \varepsilon > 0 \tag{38.26}$$

and $\boldsymbol{x}_0 \in \kappa$. Choose $\boldsymbol{\phi}$ and $\boldsymbol{v}$ to be compactly supported in a strictly interior region D containing the point $\boldsymbol{z} = \boldsymbol{0}$. Then, after division by ε^3 and passage to the limit $\varepsilon \to 0$, we find that (38.25) reduces to

$$\int_D \{\Xi_{\boldsymbol{FF}}[\nabla\boldsymbol{\phi}]\cdot\nabla\boldsymbol{\phi} + (\Xi_{\boldsymbol{MM}})\boldsymbol{v}\cdot\boldsymbol{v} + (\Xi_{\boldsymbol{FM}})\boldsymbol{v}\cdot\nabla\boldsymbol{\phi} + (\Xi_{\boldsymbol{MF}})\nabla\boldsymbol{\phi}\cdot\boldsymbol{v}\}dv > 0, \tag{38.27}$$

in which $\nabla(\cdot)$ is now the gradient with respect to $\boldsymbol{z}$, and the coefficients $\Xi_{\boldsymbol{FF}}$, etc., are evaluated at $\boldsymbol{x}_0$ and hence fixed.

A standard necessary condition for (38.27) is

$$\boldsymbol{c}\cdot(\Xi_{\boldsymbol{MM}})\boldsymbol{c} > 0 \quad \text{for all} \quad \boldsymbol{c} \neq \boldsymbol{0} \tag{38.28}$$

at the arbitrary point $\boldsymbol{x}_0 \in \kappa$; i.e., $\Xi_{\boldsymbol{MM}}$ is pointwise positive definite. To find another necessary condition, we render the left-hand side of inequality (38.27) stationary with respect to $\boldsymbol{v}$. This yields

$$\int_D \dot{\boldsymbol{v}}\cdot\{(\Xi_{\boldsymbol{MM}})\boldsymbol{v} + (\Xi_{\boldsymbol{MF}})\nabla\boldsymbol{\phi}\}dv = 0, \tag{38.29}$$

and the stationary value of $\boldsymbol{v}$ is

$$\boldsymbol{v} = -(\Xi_{\boldsymbol{MM}})^{-1}(\Xi_{\boldsymbol{MF}})\nabla\boldsymbol{\phi} \tag{38.30}$$

in which the existence of the inverse on the right is guaranteed by (38.28). Because (38.27) is satisfied for all $\boldsymbol{v}$, it is satisfied by this choice in particular. Substitution into (38.27) delivers the necessary condition

$$\int_D \mathscr{M}[\nabla\boldsymbol{\phi}]\cdot\nabla\boldsymbol{\phi}\,dv > 0, \tag{38.31}$$

where

$$\mathscr{M} = \Xi_{\boldsymbol{FF}} - \Xi_{\boldsymbol{FM}}(\Xi_{\boldsymbol{MM}})^{-1}\Xi_{\boldsymbol{MF}}. \tag{38.32}$$

The classical pointwise Legendre-Hadamard, or strong-ellipticity, condition

$$\mathscr{M}[\boldsymbol{a}\otimes\boldsymbol{b}]\cdot\boldsymbol{a}\otimes\boldsymbol{b} > 0, \quad \text{for all} \quad \boldsymbol{a}\otimes\boldsymbol{b} \neq 0 \tag{38.33}$$

may be deduced from (38.31) via a well known procedure, detailed, for example in Chapter 4, Section 1.3 of Giaquinta and Hildebrandt (1996). Conditions equivalent to (38.28) and (38.33) are stated without proof in Kankanala and Triantafyllidis (2004). Together they imply a form of the strong-ellipticity condition similar to that arising in the conventional purely mechanical theory; namely,

$$\Xi_{FF}[\boldsymbol{a}\otimes\boldsymbol{b}]\cdot\boldsymbol{a}\otimes\boldsymbol{b}>0, \quad \text{for all} \quad \boldsymbol{a}\otimes\boldsymbol{b}\neq\boldsymbol{0}, \tag{38.34}$$

the necessity of which follows immediately from (38.27) on setting $\boldsymbol{v}=\boldsymbol{0}$ and adapting the procedure used to derive (38.33) from (38.31) (Giaquinta and Hildebrandt, 1996).

38.5 Equations Holding on the Midplane and Small-Thickness Estimates

Let the reference configuration κ of the body be the prismatic cylindrical plate-like region $\Omega\times[-h/2,h/2]$, where Ω is the midplane of the plate and h is its (uniform) thickness. Let $\boldsymbol{k}$ be the (fixed) unit normal to Ω. Using the projection

$$\boldsymbol{1}=\boldsymbol{I}-\boldsymbol{k}\otimes\boldsymbol{k}, \tag{38.35}$$

where $\boldsymbol{I}$ is the identity for three-space, we write the force balance (38.21) in the form

$$div((\Xi_{\boldsymbol{F}})\boldsymbol{1})+(\Xi_{\boldsymbol{F}})'\boldsymbol{k}=\Xi_{\boldsymbol{y}}, \tag{38.36}$$

where $div(\cdot)$ is the two-dimensional divergence on Ω and $(\cdot)'=\partial(\cdot)/\partial\varsigma$, where $\varsigma\in[-h/2,h/2]$ is a convected through-thickness coordinate. This holds at all points of κ and hence on the midplane Ω in particular, whereon it reduces to

$$div((\Xi_{\boldsymbol{F}})_0\boldsymbol{1})+(\Xi_{\boldsymbol{F}})_0'\boldsymbol{k}=(\Xi_{\boldsymbol{y}})_0, \tag{38.37}$$

where, here and henceforth, the subscript $_0$ is used to denote evaluation on the midplane corresponding to $\varsigma=0$.

Further, because (38.17) holds identically in $(-h/2,h/2)$ we may differentiate it with respect to ς. Evaluating the result, together with (38.17), on Ω we find that

$$(\Xi_{\boldsymbol{M}})_0=\boldsymbol{0} \quad \text{and} \quad (\Xi_{\boldsymbol{M}})_0'=\boldsymbol{0}. \tag{38.38}$$

In the sequel we make use of the formulas (Steigmann, 2013)

$$\boldsymbol{F}_0=\nabla\boldsymbol{r}+\boldsymbol{d}\otimes\boldsymbol{k}, \quad \boldsymbol{F}_0'=\nabla\boldsymbol{d}+\boldsymbol{g}\otimes\boldsymbol{k} \quad \text{and} \quad \boldsymbol{F}_0''=\nabla\boldsymbol{g}+\boldsymbol{h}\otimes\boldsymbol{k}, \tag{38.39}$$

where

$$\boldsymbol{r}=\boldsymbol{\chi}_0, \quad \boldsymbol{d}=\boldsymbol{\chi}_0', \quad \boldsymbol{g}=\boldsymbol{\chi}_0'' \quad \text{and} \quad \boldsymbol{h}=\boldsymbol{\chi}_0'''. \tag{38.40}$$

The first of these is the midplane deformation, and the remaining terms are the coefficients in the through-thickness expansion

$$\boldsymbol{\chi}=\boldsymbol{r}+\varsigma\boldsymbol{d}+\frac{1}{2}\varsigma^2\boldsymbol{g}+\frac{1}{6}\varsigma^3\boldsymbol{h}+\ldots \tag{38.41}$$

of the three-dimensional deformation.

Thus far we have merely recast the equations and specialized them to the midplane without making any approximations. We do so now, by approximating the traction boundary conditions at the major surfaces of the plate. Assuming the tractions to vanish on these surfaces, we have cf. (38.22) $(\Xi_{\boldsymbol{F}})^{\pm}\boldsymbol{k}=\boldsymbol{0}$, which may be Taylor-expanded about the midplane to conclude that

$$(\Xi_{\boldsymbol{F}})_0\boldsymbol{k} = O(h^2) \quad \text{and} \quad (\Xi_{\boldsymbol{F}})'_0\boldsymbol{k} = O(h^2). \tag{38.42}$$

38.6 Potential Energy of a Thin Plate

On combining Leibniz' rule with a Taylor expansion, we estimate the first term on the right-hand side of (38.15) as (Steigmann, 2013)

$$\int_{\kappa} \Xi dv = \int_{\Omega} \underline{W} da + o(h^3), \tag{38.43}$$

in which $\underline{W}$ is given by

$$\underline{W} = h\Xi_0 + \frac{1}{24}h^3\Xi''_0, \tag{38.44}$$

where, for uniform materials,

$$\Xi_0 = \Xi(\boldsymbol{r},\boldsymbol{F}_0,\boldsymbol{M}_0) \tag{38.45}$$

and

$$\Xi''_0 = (\Xi_{\boldsymbol{F}})'_0\cdot\boldsymbol{F}'_0 + (\Xi_{\boldsymbol{M}})'_0\cdot\boldsymbol{M}'_0 + (\Xi_{\boldsymbol{F}})_0\cdot\boldsymbol{F}''_0 + (\Xi_{\boldsymbol{M}})_0\cdot\boldsymbol{M}''_0 + (\Xi_{\boldsymbol{y}})'_0\cdot\boldsymbol{d} + (\Xi_{\boldsymbol{y}})_0\cdot\boldsymbol{g}. \tag{38.46}$$

We seek a theory for equilibria and thus invoke partial information about these from the three-dimensional theory. For example, (38.38), (38.39) and (38.42) may be used to simplify (38.46) to

$$\Xi''_0 = (\Xi_{\boldsymbol{F}})'_0\boldsymbol{1}\cdot\nabla\boldsymbol{d} + (\Xi_{\boldsymbol{F}})_0\boldsymbol{1}\cdot\nabla\boldsymbol{g} + (\Xi_{\boldsymbol{y}})'_0\cdot\boldsymbol{d} + (\Xi_{\boldsymbol{y}})_0\cdot\boldsymbol{g}, \tag{38.47}$$

while retaining an order of accuracy consistent with (38.43). The second term on the right is re-arranged and use is made of (38.37) to obtain

$$\begin{aligned}(\Xi_{\boldsymbol{F}})_0\boldsymbol{1}\cdot\nabla\boldsymbol{g} &= div\{[(\Xi_{\boldsymbol{F}})_0\boldsymbol{1}]^t\boldsymbol{g}\} - div[(\Xi_{\boldsymbol{F}})_0\boldsymbol{1}]\cdot\boldsymbol{g}\\ &= div\{[(\Xi_{\boldsymbol{F}})_0\boldsymbol{1}]^t\boldsymbol{g}\} + (\Xi_{\boldsymbol{F}})'_0\boldsymbol{k}\cdot\boldsymbol{g} - (\Xi_{\boldsymbol{y}})_0\cdot\boldsymbol{g}.\end{aligned} \tag{38.48}$$

From (38.42) we then derive the consistent-order estimate

$$\Xi''_0 = (\Xi_{\boldsymbol{F}})'_0\boldsymbol{1}\cdot\nabla\boldsymbol{d} + (\Xi_{\boldsymbol{y}})'_0\cdot\boldsymbol{d} + div\{[(\Xi_{\boldsymbol{F}})_0\boldsymbol{1}]^t\boldsymbol{g}\} \tag{38.49}$$

and thus reduce (38.43), with the aid of the Green-Stokes theorem, to

$$\int_{\kappa} \Xi dv = \int_{\Omega} W da + \frac{1}{24}h^3 \int_{\partial\Omega} \boldsymbol{g} \cdot (\Xi_{\boldsymbol{F}})_0 \mathbf{1}\boldsymbol{\nu} ds + o(h^3), \tag{38.50}$$

in which $\boldsymbol{\nu}$ is the exterior unit normal to the edge $\partial\Omega$ and W (not to be confused with the use of the same symbol in (38.6) is given by

$$W = h\Xi_0 + \frac{1}{24}h^3\{(\Xi_{\boldsymbol{F}})_0'\mathbf{1} \cdot \nabla\boldsymbol{d} + (\Xi_{\boldsymbol{y}})_0' \cdot \boldsymbol{d}\}. \tag{38.51}$$

In the case of dead loading, (38.8) yields (Steigmann, 2013)

$$\mathscr{L} = L + o(h^3), \tag{38.52}$$

with

$$L = \int_{\partial\Omega_p} (\boldsymbol{p}_r \cdot \boldsymbol{r} + \boldsymbol{p}_d \cdot \boldsymbol{d} + \boldsymbol{p}_g \cdot \boldsymbol{g}) da \tag{38.53}$$

where

$$\boldsymbol{p}_r = h(\boldsymbol{p}_a)_0 + \frac{1}{24}h^3(\boldsymbol{p}_a'')_0, \quad \boldsymbol{p}_d = \frac{1}{24}h^3(2\boldsymbol{p}_a')_0 \quad \text{and} \quad \boldsymbol{p}_g = \frac{1}{24}h^3(\boldsymbol{p}_a)_0, \tag{38.54}$$

and where $\partial\Omega_p$ is the intersection of $\partial\kappa_p$ with Ω; i.e, $\partial\kappa_p = \partial\Omega_p \times [-h/2, h/2]$.
Using (38.22) in the form

$$(\boldsymbol{p}_a)_0 = (\Xi_{\boldsymbol{F}})_0 \mathbf{1}\boldsymbol{\nu}, \tag{38.55}$$

we derive the estimate

$$\mathscr{E} = E + O(h^3), \tag{38.56}$$

with

$$E = \int_{\Omega} W da + \frac{1}{24}h^3 \int_{\partial\Omega\setminus\partial\Omega_p} (\Xi_{\boldsymbol{F}})_0 \mathbf{1}\boldsymbol{\nu} \cdot \boldsymbol{g} ds - L \tag{38.57}$$

in which L is now given by

$$L = \int_{\partial\Omega_p} (\boldsymbol{p}_r \cdot \boldsymbol{r} + \boldsymbol{p}_d \cdot \boldsymbol{d}) ds. \tag{38.58}$$

The integral over $\partial\Omega \setminus \partial\Omega_p$ is effectively fixed by the data and thus contributes only a disposable constant to the energy (38.57). This conclusion follows from the fact that three-dimensional position is assigned on $\partial\kappa \setminus \partial\kappa_p$. This fixes the through-thickness derivatives of the deformation thereon and implies that $\boldsymbol{r}$, $\boldsymbol{d}$ and $\boldsymbol{g}$ are assigned on $\partial\Omega \setminus \partial\Omega_p$. Further, $(\Xi_{\boldsymbol{F}})_0$ is controlled by the values of $\boldsymbol{M}_0$, $\boldsymbol{r}$, $\nabla\boldsymbol{r}$ and $\boldsymbol{d}$. We show in the next section that $\boldsymbol{M}_0$ and $\boldsymbol{d}$ are determined by $\boldsymbol{r}$ and $\nabla\boldsymbol{r}$. On a clamped boundary, $\boldsymbol{r}$ and its normal derivative $(\nabla\boldsymbol{r})\boldsymbol{\nu}$ are assigned. Then, since

$$\nabla\boldsymbol{r} = (\nabla\boldsymbol{r})\boldsymbol{\nu} \otimes \boldsymbol{\nu} + (\nabla\boldsymbol{r})\boldsymbol{\tau} \otimes \boldsymbol{\tau},$$

with $\boldsymbol{\tau} = \boldsymbol{k} \times \boldsymbol{v}$, and since the tangential derivative $(\nabla \boldsymbol{r})\boldsymbol{\tau}$ is determined by the assigned function $\boldsymbol{r}$, it follows that $\nabla \boldsymbol{r}$ is fixed on $\partial\Omega \setminus \partial\Omega_p$. The relevant integrand is then fixed by the data and may be suppressed without affecting the mechanics of the system. Accordingly, the approximate plate energy reduces to

$$E = \int_{\Omega} W da - L, \tag{38.59}$$

with W and L as stated.

38.7 Reduction of the Plate Energy

Henceforth we work exclusively with variables defined on Ω and thus suppress the subscript $_0$ for the sake of clarity. Our objective is to show that the midplane values of $\boldsymbol{M}$, $\boldsymbol{M}'$, $\boldsymbol{d}$, $\nabla \boldsymbol{d}$ and $\boldsymbol{g}$ may be found in terms of the midplane deformation $\boldsymbol{r}$ and its first and second midplane gradients and thus that the plate energy may be reduced to a functional of the midplane deformation alone, as in the conventional purely mechanical theory (Steigmann, 2013).

To this end we fix the function $\boldsymbol{r}$ and its gradient $\nabla \boldsymbol{r}$ on Ω. Equation $(38.38)_1$, namely $\Xi_{\boldsymbol{M}} = \boldsymbol{0}$, then imposes a relationship between $\boldsymbol{M}$ and $\boldsymbol{d}$, which yields a unique solution $\boldsymbol{M} = \hat{\boldsymbol{M}}(\boldsymbol{d})$, say, by virtue of the positive definiteness of $\Xi_{\boldsymbol{MM}}$ (cf. (38.28)) and the implicit function theorem. The midplane energy may then be written as a function $\Xi = \Gamma(\boldsymbol{d})$ of $\boldsymbol{d}$ alone, again presuming that $\boldsymbol{r}$ and its gradient are fixed. Consider the one-parameter family $\boldsymbol{d}(u)$. This induces the family $\boldsymbol{M}(u) = \hat{\boldsymbol{M}}(\boldsymbol{d}(u))$. Let $\sigma(u) = \Gamma(\boldsymbol{d}(u))$. The first and second derivatives of this function are

$$\dot{\sigma} = \Xi_{\boldsymbol{M}} \cdot \dot{\boldsymbol{M}} + \Xi_{\boldsymbol{F}} \cdot \dot{\boldsymbol{d}} \otimes \boldsymbol{k} \tag{38.60}$$

and

$$\begin{aligned} \ddot{\sigma} = {} & \Xi_{\boldsymbol{M}} \cdot \ddot{\boldsymbol{M}} + \Xi_{\boldsymbol{F}} \cdot \ddot{\boldsymbol{d}} \otimes \boldsymbol{k} + (\Xi_{\boldsymbol{MM}})\dot{\boldsymbol{M}} \cdot \dot{\boldsymbol{M}} + \dot{\boldsymbol{d}} \otimes \boldsymbol{k} \cdot \Xi_{\boldsymbol{FF}}[\dot{\boldsymbol{d}} \otimes \boldsymbol{k}] \\ & + \dot{\boldsymbol{M}} \cdot (\Xi_{\boldsymbol{MF}})\dot{\boldsymbol{d}} \otimes \boldsymbol{k} + \dot{\boldsymbol{d}} \otimes \boldsymbol{k} \cdot (\Xi_{\boldsymbol{FM}})\dot{\boldsymbol{M}}, \end{aligned} \tag{38.61}$$

where, by construction, $\Xi_{\boldsymbol{M}}(\boldsymbol{d}(u), \boldsymbol{M}(u)) \equiv \boldsymbol{0}$. Accordingly,

$$(\Xi_{\boldsymbol{MM}})\dot{\boldsymbol{M}} + \Xi_{\boldsymbol{MF}}(\dot{\boldsymbol{d}} \otimes \boldsymbol{k}) = \boldsymbol{0}, \tag{38.62}$$

yielding

$$\dot{\boldsymbol{M}} = -(\Xi_{\boldsymbol{MM}})^{-1} \Xi_{\boldsymbol{MF}}(\dot{\boldsymbol{d}} \otimes \boldsymbol{k}) \tag{38.63}$$

and (cf. 38.32)

$$\ddot{\sigma} = \Xi_{\boldsymbol{M}} \cdot \ddot{\boldsymbol{M}} + \Xi_{\boldsymbol{F}} \cdot \ddot{\boldsymbol{d}} \otimes \boldsymbol{k} + \dot{\boldsymbol{d}} \otimes \boldsymbol{k} \cdot \mathscr{M}[\dot{\boldsymbol{d}} \otimes \boldsymbol{k}]. \tag{38.64}$$

Next, recall that (38.42) was used to justify the suppression of $(\Xi_{\boldsymbol{F}})\boldsymbol{k}$ in the order-h^3 term of the energy. Together with the vanishing of $\Xi_{\boldsymbol{M}}$, this furnishes $\dot{\sigma} = 0$

and hence the stationarity of the midplane energy Ξ with respect to $\boldsymbol{d}$. Conversely, granted the vanishing of $\Xi_{\boldsymbol{M}}$, this energy is stationary only if $(\Xi_{\boldsymbol{F}})\boldsymbol{k}$ vanishes, and $\ddot{\sigma}$ then reduces to

$$\ddot{\sigma} = \dot{\boldsymbol{d}} \otimes \boldsymbol{k} \cdot \mathscr{M}[\dot{\boldsymbol{d}} \otimes \boldsymbol{k}], \tag{38.65}$$

which is positive for all non-zero $\dot{\boldsymbol{d}}$ by virtue of (38.33). The energy Ξ is thus a convex function of $\boldsymbol{d}$. Because such functions have unique stationary points, it follows that the conditions $\Xi_{\boldsymbol{M}} = \boldsymbol{0}$ and $(\Xi_{\boldsymbol{F}})\boldsymbol{k} = \boldsymbol{0}$, which were imposed in the course of constructing W, furnish $\boldsymbol{M}$ and $\boldsymbol{d}$ as unique functions of $\boldsymbol{r}$ and $\nabla\boldsymbol{r}$. Importantly, because these functions minimize Ξ absolutely, it is appropriate to incorporate them in the order-h term of the energy also.

It remains to evaluate $\boldsymbol{M}'$ and $\boldsymbol{g}$. To this end we observe, for uniform materials, that $(38.38)_2$ furnishes

$$\boldsymbol{M}' = -(\Xi_{\boldsymbol{MM}})^{-1}\{(\Xi_{\boldsymbol{MF}})\boldsymbol{F}' + (\Xi_{\boldsymbol{My}})\boldsymbol{d}\} \tag{38.66}$$

in which (cf. $(38.39)_2$)

$$\boldsymbol{F}' = \nabla\boldsymbol{d} + \boldsymbol{g} \otimes \boldsymbol{k}. \tag{38.67}$$

Here $\boldsymbol{M}, \boldsymbol{d}$ and $\nabla\boldsymbol{d}$ are fixed in terms of $\boldsymbol{r}$, $\nabla\boldsymbol{r}$ and $\nabla\nabla\boldsymbol{r}$ via the foregoing procedure. For fixed values of the latter, (38.66) thus delivers $\boldsymbol{M}'$ as an explicit function of $\boldsymbol{g}$. To eliminate $\boldsymbol{g}$, and thereby reduce the energy (38.59) to a functional of the midplane position field $\boldsymbol{r}$ alone, recall that $(38.42)_2$ was used to justify the suppression of $(\Xi_{\boldsymbol{F}})'\boldsymbol{k}$ in the course of deriving (38.51). Thus,

$$\boldsymbol{0} = (\Xi_{\boldsymbol{F}})'\boldsymbol{k} = \{\Xi_{\boldsymbol{FF}}[\nabla\boldsymbol{d} + \boldsymbol{g} \otimes \boldsymbol{k}] + (\Xi_{\boldsymbol{FM}})\boldsymbol{M}' + (\Xi_{\boldsymbol{Fy}})\boldsymbol{d}\}\boldsymbol{k}. \tag{38.68}$$

Inserting (38.66), we obtain

$$\boldsymbol{A}\boldsymbol{g} = -(\mathscr{M}[\nabla\boldsymbol{d}])\boldsymbol{k} + \{\Xi_{\boldsymbol{Fy}} - \Xi_{\boldsymbol{FM}}(\Xi_{\boldsymbol{MM}})^{-1}\Xi_{\boldsymbol{My}}\}\boldsymbol{d}, \tag{38.69}$$

where $\boldsymbol{A}$ is the symmetric tensor defined by

$$\boldsymbol{A}\boldsymbol{v} = (\mathscr{M}[\boldsymbol{v} \otimes \boldsymbol{k}])\boldsymbol{k} \tag{38.70}$$

for any three-vector $\boldsymbol{v}$. This is positive definite by virtue of (38.33), and (38.69) yields $\boldsymbol{g}$ uniquely in terms of variables that are uniquely determined by $\boldsymbol{r}$, $\nabla\boldsymbol{r}$ and $\nabla\nabla\boldsymbol{r}$, yielding an energy density W that depends on the latter variables alone.

In principle, this completes the reduction of the plate energy to a functional of the midplane deformation. However, in practice, because $(38.38)_1$ and $(38.42)_1$ do not deliver explicit formulae for $\boldsymbol{M}$ and $\boldsymbol{d}$, it is more feasible to eliminate $\boldsymbol{M}'$ and $\boldsymbol{g}$ using the explicit results (38.66) and (38.69), and then to regard the resulting energy as a functional of $\boldsymbol{r}$, $\boldsymbol{d}$ and $\boldsymbol{M}$, subject to the nonlinear algebraic constraints $(38.38)_1$ and $(38.42)_1$. One may then derive the relevant equilibrium equations as the stationarity conditions for the constrained energy. The details of this procedure are standard and therefore left to the interested reader.

We leave open the challenging question of the well-posedness of the minimization problem for the two-dimensional plate energy. In the purely mechanical context, it is known (Steigmann, 2013) that the energy derived by a procedure like that used here may not be well posed unless certain additional conditions are met. In particular, it may be that truncation of the energy at order higher than h^3 may be needed to secure a well-posed model, in which case the ultimate dimensionally reduced system may well prove to be overly cumbersome *vis à vis* the three-dimensional theory.

Acknowledgements SS thanks the Institute of Functional Nanomaterials at the University of Puerto Rico and the US National Science Foundation, through grant number EPS-1002410, for their support of her Visiting Professorship at UC Berkeley. DJS gratefully acknowledges support provided by the US National Science Foundation through grant number CMMI-1538228.

References

Brown WF (1966) Magnetoelastic Interactions. Springer, Berlin

DeSimone A, Podio-Guidugli P (1996) On the continuum theory of deformable ferromagnetic solids. Arch Rational Mech Anal 136:201–233

Dorfmann L, Ogden RW (2014) Nonlinear Theory of Electroelastic and Magnetoelastic Interactions. Springer, New York

Giaquinta M, Hildebrandt S (1996) Calculus of Variations, vol I. Springer, Berlin

James RD (2002) Configurational forces in magnetism with application to the dynamics of a small-scale ferromagnetic shape memory cantilever. Continuum Mech Thermodyn 14:55–86

Kankanala SV, Triantafyllidis N (2004) On finitely strained magnetorheological elastomers. J Mech Phys Solids 52:2869–2908

Kovetz A (2000) Electromagnetic Theory. Oxford University Press, Oxford

Maugin GA (1988) Continuum Mechanics of Electromagnetic Solids. North-Holland, Amsterdam

Steigmann DJ (2004) Equilibrium theory for magnetic elastomers and magnetoelastic membranes. Int J Non-linear Mech 39:1193–1216

Steigmann DJ (2013) A well-posed finite-strain model for thin elastic sheets with bending stiffness. Math Mech Solids 13:103–112

Chapter 39
Modelling of an Ionic Electroactive Polymer by the Thermodynamics of Linear Irreversible Processes

Mireille Tixier and Joël Pouget

Abstract Ionic polymer-metal composites consist in a thin film of electro-active polymers (Nafion® for example) sandwiched between two metallic electrodes. They can be used as sensors or actuators. The polymer is saturated with water, which causes a complete dissociation and the release of small cations. The strip undergoes large bending motions when it is submitted to an orthogonal electric field and vice versa. We used a continuous medium approach and a coarse grain model; the system is depicted as a deformable porous medium in which flows an ionic solution. We write microscale balance laws and thermodynamic relations for each phase, then for the complete material using an average technique. Entropy production, then constitutive equations are deduced: a Kelvin-Voigt stress-strain relation, generalized Fourier's and Darcy's laws and a Nernst-Planck equation. We applied this model to a cantilever electro-active polymer strip undergoing a continuous potential difference (static case); a shear force may be applied to the free end to prevent its displacement. Applied forces and deflection are calculated using a beam model in large displacements. The results obtained are in good agreement with data published in the literature.

39.1 Introduction

Electro-active polymers (EAP) have attracted much attention from scientists and engineers because of their very promising applications in many areas of science and

---

Mireille Tixier
Département de Physique, Université de Versailles Saint Quentin, 45, Avenue des Etats-Unis, F-78035 Versailles, France
e-mail: mireille.tixier@uvsq.fr

Joël Pouget
Sorbonne Université, Centre National de la Recherche Scientifique, UMR 7190, Institut Jean Le Rond d'Alembert, F-75005 Paris, France
e-mail: joel.pouget@upmc.fr

© Springer International Publishing AG, part of Springer Nature 2018

H. Altenbach et al. (eds.), *Generalized Models and Non-classical Approaches in Complex Materials 1*, Advanced Structured Materials 89,
https://doi.org/10.1007/978-3-319-72440-9_39

the growing market. Their behavior and electro-chemical-mechanical interactions are of great interest and curiosity for research. In particular, the properties of these materials are highly attractive for biomimetic applications (for instance, in robotic mechanisms are based on biologically inspired models), for the rise of artificial muscles (Bar-Cohen, 2001), and for haptic actuators. More recently EAPs are excellent candidates for energy harvesting devices (Brufau-Penella et al, 2008). Promising applications of this smart material consisting of micro-electromechanical systems (MEMS) at the sub-micron scale are also investigated for accurate medical control (Yoon et al, 2007).

The purpose of the present study is to construct step by step a micro-mechanical model which accounts for couplings between the ion transport, electric field and elastic deformation in order to deduce the constitutive equations for this material. Next, an application to the actuation of beam made of thin layer of EAP is presented. Roughly speaking, an electro-active polymer is a polymer that exhibits a mechanical response, such as stretching, contracting, or bending for example, when subject to an electric field (only few volts are needed for actuation). Conversely, the EAP can produce energy in response to a mechanical loading.

The terminology electro-active polymer has very wide meaning and can be applied to a large category of materials. The electro-active polymers are generally divided in two main classes: (i) the electronic EAPs, in which activation is causedby electro-active force between both electrodes which squeezes the polymer and (ii) the ionic EAPs, in which actuation is due to the displacement of ions inside the polymer. Both classes are divided into subfamilies according to the physical or chemical principles of activation. The electronic EAP family encompasses ferroelectric polymers, electrets (PolyVinyliDene Fluoride (PVDF) is an example), dielectric elastomers, electroactive papers, liquide crystal polymers and many others. The ionic EAP category comprises ionic gels, ionic composite (IPMC) (such as Nafion® or Flemion®), ionic conductor polymers (the strong conductivity is due to oxychloreduction process), nanotube of carbone (the electrolyte is modified by additional charges which produce volume change), electrorheologic fluid (fluid with micro particles changing the rheological properties of fluid, viscosity for instance) among others. The reader can refer to Bar-Cohen (2001) for many more details. Each category possesses their own advantages and their drawbacks.

The present work addresses investigation of electro-active polymers belonging to ionic class and more precisely to ionic-exchange polymer-metal composite (IPMC because of the metallic electrodes on the layer faces). The latter consists in an ionic polymer (Nafion®, for instance) sandwiched with two electrodes onto the upper and lower surfaces of the polymer layer. Katchalsky, in 1949 was one of the first investigators to report the ionic chemo-mechanical deformation of polyectrolytes such as polyacrylic acid or polyvinyl chloride systems. More recently a great interest has been devoted to EAPs due to the similarities with biological tissues in terms of achievable stress and they are often called artificial muscles. Moreover, the material has potential applications in the field of robotics, medical technology and so on. Investigation on EAP have been traced to Shahinpoor (1994); Shahinpoor et al (1998) and some many other researchers.

Modeling of EAPs must include complicated electro-chemical-mechanical couplings. Different kinds of approaches have been proposed. Newbury and Leo (2001, 2002) developed empirical and heuristic models to explain sensing and actuating properties of ionic polymer benders. Model based on electrostatic interactions produced by ion motion has been developed by Todokoru et al (2000). A model including the effect of electric field, current, pressure gradient and water flux as state variables has been proposed by de Gennes et al (2000) using the concept of irreversible thermodynamics. A more sophisticated model nonetheless closer to the realistic properties was developed by Nemat-Nasser and Li (2000); Nemat-Nasser (2002). The model is based on the micro-mechanics of ionic polymers including ion transport. Finite element 3D model has been studied by Vokoun et al (2015) to solve the basic governing physical equations for EAPs proposed by Nemat-Nasser with given boundary conditions. A model of electro-viscoelastic polymers as an extension of the nonlinear electro-elasticity theory has been discussed by Ask et al (2012) and finite element numerical simulations were also presented.

The proposed model accounts for electro-mechanical and chemical-electric coupling of ion transport, electric field and elastic deformation to produce the response of the EAP. We first investigate the conservation laws of the different phases at the micro level. An averaging statistical process applied to the different phase quantities and to the equations of the conservation laws at the micro-scale is used to deduce, in a representative elementary volume containing all the phases, the equations of the conservation laws of the polymer at the continuum level. We write down conservation laws (mass, momentum, energy) in the framework of non-equilibrium thermodynamics. The thermodynamics of linear irreversible process allows us to identify the fluxes and generalized forces (Tixier and Pouget, 2014) and the constitutive equations for the continuum model are consequently obtained (Tixier and Pouget, 2016). A generalized Darcy's law and the balance for ion flux (a kind of Nernst-Plank equation) are deduced from the thermodynamics relations and Gibbs' relation. Along with the former equations the stress-strain equation and that of the electric charge conservation are also considered (Tixier and Pouget, 2016).

The paper is organized as follows, the next section is devoted to the description of the EAP giving the main properties and the way of modeling the material. The section reports also the method used for arriving at the continuum model. The Sect. 39.3 concerns the conservation laws, especially the energy balance laws. The fundamental thermodynamic relations as well as the entropy production are reported in Sect. 39.4. Gibbs', Euler's and Gibbs-Duhem's relations are given in order to deduce the rate of entropy production. The constitutive equations are presented in Sect. 39.5, in particular the stress-strain relation, Nernst-Plank equation and generalized Darcy's law are written in terms of concentration, electric field and pressure. The Sect. 39.6 proposes a validation of the model by studying the bending actuation of EAP layer subject to a constant difference of electric potential applied to the upper and lower electrodes. Comparisons to experimental results available in the literature ascertain the present model. The most pertinent results are summarized in the conclusions.

39.2 Description and Modelling of the Material

The system we study is an ionic polymer-metal composite (IPMC); it consists of a membrane of polyelectrolyte coated on both sides with thin metal layers acting as electrodes. The polymer is saturated with water, which results in a quasi-complete dissociation: anions remain bound to the polymer backbone, whereas small cations are released in water (Chabé, 2008). When an electric field perpendicular to the electrodes is applied, cations are attracted by the negative electrode and carry solvent away by osmosis. As a result, the polymer swells near the negative electrode and contracts on the opposite side, leading to the bending of the strip.

To model this system, the polymer chains are assimilated to a deformable porous medium saturated by an ionic solution composed by water and cations. We suppose that the solution is dilute. We depicted the complete material as the superposition of three systems: a deformable solid made up of polymer backbone negatively charged, a solvent (the water) and cations (for schematic representation, see the inset of Fig. 39.1 a and b. The solid and liquid phases are assumed to be incompressible phases separated by an interface whose thickness is supposed to be negligible. We identify the quantities relative to the different components by subscripts: 1 refers to cations, 2 to solvent, 3 to solid, *i* to the interface and 4 to the solution, that is both components 1 and 2; quantities without subscript refer to the whole material. The different components (except 1) as well as the global material are assimilated to continua. We venture the hypothesis that gravity and magnetic field are negligible, so the only external force acting on the system is the electric force (Tixier and Pouget, 2014).

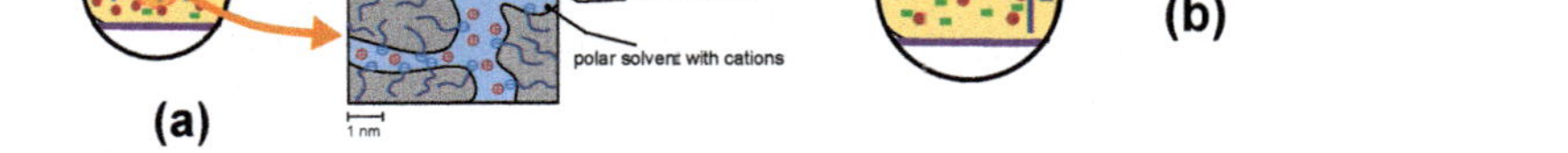

Fig. 39.1: Deformable porous medium: (a) Undeformed strip, (b) Strip bending under an applied electric field

39.2.1 Average Process

We describe this medium using a coarse-grained model developed for two-phase mixtures (Drew, 1983; Drew and Passman, 1998; Ishii and Hibiki, 2006; Lhuillier, 2003; Nigmatulin, 1979, 1990). We use two scales. The microscopic scale must be small enough so that the corresponding volume only contains a single pase (3 or 4), but large enough to use a continuous medium model. For Nafion® completely saturated with water, it is about hundred Angstroms. At the macroscopic scale, we define a representative elementary volume (R.E.V.) which contains the two phases; it must be small enough so that average quantities relative to the whole material can be considered as local, and large enough so that these averages are relevant. Its characteristic length is about micron (Chabé, 2008; Collette, 2008; Gierke et al, 1981).

A microscale Heaviside-like function of presence $\chi_k(\overrightarrow{r},t)$ has been defined for the phases 3 and 4

$$\chi_k = 1 \text{ when phase } k \text{ occupies point } \overrightarrow{r} \text{ at time } t, \quad \chi_k = 0 \text{ otherwise} \tag{39.1}$$

The function of presence of the interface is the Dirac-like function

$$\chi_i = -\overrightarrow{\nabla}\chi_k \cdot \overrightarrow{n_k} \text{ (in m}^{-1}\text{)}$$

where $\overrightarrow{n_k}$ is the outward-pointing unit normal to the interface in the phase k. $\langle\rangle_k$ denotes the average over the phase k of a quantity relative to the phase k only. The macroscale quantities relative to the whole material are obtained by statistically averaging the microscale quantities over the R.E.V., that is by repeating many times the same experiment. We suppose that this average, denoted by $\langle\rangle$, is equivalent to a volume average (ergodic hypothesis) and commutes with the space and time derivatives (Leibniz' and Gauss' rules, Drew, 1983; Lhuillier, 2003). A macroscale quantity g_k verifies

$$g_k = \langle \chi_k g_k^0 \rangle = \phi_k \langle g_k^0 \rangle_k \tag{39.2}$$

where g_k^0 is the corresponding microscale quantity and $\phi_k = \langle \chi_k \rangle$ the volume fraction of the phase k. g_k is relative to the total volume of the whole material. In the following, we use superscript ^{0} to indicate microscale quantities; the macroscale quantities, which are averages defined all over the material, are written without superscript.

39.2.2 Interface Modelling

In practice, contact area between phases 3 and 4 has a certain thickness; extensive physical quantities vary from one bulk phase to the other one. This complicated reality can be modelled by two uniform bulk phases separated by a discontinuity surface Σ whose localization is arbitrary. Let Ω be a cylinder crossing Σ, whose bases are parallel to Σ. We denote by Ω_3 and Ω_4 the parts of Ω respectively included in phases 3 and 4.

The continuous quantities relative to the contact zone are identified by a superscript ^{0} and no subscript. A microscale quantity per surface unit g_i^0 related to the interface is defined by

$$g_i^0 = \lim_{\Sigma \longrightarrow 0} \frac{1}{\Sigma} \left\{ \int_{\Omega} g^0 dv - \int_{\Omega_3} g_3^0 dv - \int_{\Omega_4} g_4^0 dv \right\} \tag{39.3}$$

where Ω_3 and Ω_4 are small enough so that g_3^0 and g_4^0 are constant. Its average over the R.E.V. is the volume quantity g_i defined by

$$g_i = \left\langle g_i^0 \chi_i \right\rangle \tag{39.4}$$

We arbitrarily fix the interface position in such a way that it has no mass density. The different phases do not interpenetrate, thus we can write on the interfaces

$$\overrightarrow{V_1^0} = \overrightarrow{V_2^0} = \overrightarrow{V_3^0} = \overrightarrow{V_4^0} = \overrightarrow{V_i^0} \tag{39.5}$$

where $\overrightarrow{V_k^0}$ denotes the local velocity of the phase k. Moreover, we neglect all the velocities fluctuations on the R.E.V. scale.

39.2.3 Partial Derivatives and Material Derivative

In order to write the balance equations, it is necessary to calculate the variations of the extensive quantities following the material motion. This raises a problem because the different phases do not move with the same velocity: velocities of the solid and the solution are a priori different. Let us consider an extensive quantity of density $g\left(\overrightarrow{r},t\right)$ relative to the whole material. We can define particle derivatives following the motion of the solid ($\frac{d_3}{dt}$), the solution ($\frac{d_4}{dt}$) or the interface ($\frac{d_i}{dt}$)

$$\frac{d_k g}{dt} = \frac{\partial g}{\partial t} + \overrightarrow{\nabla} g \cdot \overrightarrow{V_k} \tag{39.6}$$

According to the theory developed by Biot (1977); Coussy (1989, 1995), we are also able to define a derivative following the motion of the different phases of the medium. We will call it the "material derivative"

$$\rho \frac{D}{Dt}\left(\frac{g}{\rho}\right) = \sum_{k=3,4,i} \rho_k \frac{d_k}{dt}\left(\frac{g_k}{\rho_k}\right) = \sum_{k=3,4\ i} \frac{\partial g_k}{\partial t} + div\left(g_k \overrightarrow{V_k}\right) \tag{39.7}$$

where ρ_k is the mass density of the phase k and g_3, g_4 and g_i the densities relative to the total actual volume attached to the solid, the solution and the interface, respectively

$$g = g_3 + g_4 + g_i \tag{39.8}$$

This derivative must not be confused with the derivative $\frac{d}{dt}$ following the barycentric velocity $\overrightarrow{V}$

$$\rho \frac{D}{Dt}\left(\frac{g}{\rho}\right) = \rho \frac{d}{dt}\left(\frac{g}{\rho}\right) - \sum_k \left[div\left(g_k\left(\overrightarrow{V} - \overrightarrow{V_k}\right)\right)\right] \tag{39.9}$$

39.2.4 Balance Laws

The balance equation of an extensive microscale quantity $g_k^0(\overrightarrow{x},t)$ writes

$$\frac{\partial g_k^0}{\partial t} + div\left(g_k^0\overrightarrow{V_k^0}\right) = -div\overrightarrow{A_k^0} + B_k^0 \tag{39.10}$$

where $\overrightarrow{A_k^0}$ is the flux of g_k^0 linked to phenomena other than convection and B_k^0 the volume production of g_k^0 (source term). At the macroscopic scale

$$\frac{\partial g_k}{\partial t} + div\left(g_k\overrightarrow{V_k}\right) = -div\overrightarrow{A_k} + B_k - \left\langle \overrightarrow{A_k^0}\cdot\overrightarrow{n_k}\chi_i \right\rangle \tag{39.11}$$

in which

$$\overrightarrow{A_k} = \left\langle \chi_k\overrightarrow{A_k^0}\right\rangle \qquad B_k = \left\langle \chi_k B_k^0\right\rangle \tag{39.12}$$

In the case of an interface, the balance law is Ishii and Hibiki (2006)

$$\frac{\partial g_i^0}{\partial t} + div_s\left(g_i^0\overrightarrow{V_i^0}\right) = \sum_{3,4}\left[g_k^0\left(\overrightarrow{V_k} - \overrightarrow{V_i^0}\right)\cdot\overrightarrow{n_k} + \overrightarrow{A_k^0}\cdot\overrightarrow{n_k}\right] - div_s\overrightarrow{A_i^0} - B_i^0 \tag{39.13}$$

where $\overrightarrow{A_i^0}$ is the surface flux of g_i^0 along the interface due to phenomena other than convection, B_i^0 the surface production of g_i^0 and div_s the surface divergence operator. At the macroscopic scale

$$\frac{\partial g_i}{\partial t} + div\left(g_i\overrightarrow{V_i}\right) = \sum_{3,4}\left\langle \chi_i\overrightarrow{A_k^0}\cdot\overrightarrow{n_k}\right\rangle - div\overrightarrow{A_i} + B_i \tag{39.14}$$

Note that g_i, $\overrightarrow{A_i}$ and B_i are volume quantities.

For the complete material, we deduce by summation:

$$\rho\frac{D}{Dt}\left(\frac{g}{\rho}\right) = -div\overrightarrow{A} + B \tag{39.15}$$

where

$$\vec{A} = \sum_{3,4,i} \vec{A_k}, \qquad B = \sum_{3,4,i} B_k \tag{39.16}$$

39.3 Conservation Laws

We now write the balance laws of the following quantities: mass, electric charge, linear momentum, the different energies (potential, kinetic, total and internal), entropy as well as Maxwell's equations.

39.3.1 Conservation of the Mass

In the absence of chemical reaction, there is no volume production of mass and the only flux is due to convection. The macroscale mass continuity equation thus writes

$$\frac{\partial \rho_k}{\partial t} + div\left(\rho_k \vec{V_k}\right) = 0 \qquad (k = 2,3), \qquad \frac{\partial \rho}{\partial t} + div\left(\rho \vec{V}\right) = 0 \tag{39.17}$$

with

$$\rho_1 = \phi_4 C M_1, \qquad \rho_k = \phi_k \rho_k^0 \quad (k = 2,3), \qquad \rho_4 = \rho_2 + \phi_4 C M_1 \tag{39.18}$$

M_k is the molar mass of the component k and C the cations molar concentration relative to the liquid phase. We assume the concentration fluctuations are negligible.

39.3.2 Electric Equations

As in the case of mass, there is, for the electric charge, neither source term nor flux except convection. The electric charge conservation law can be written

$$div\,\vec{I} + \frac{\partial \rho Z}{\partial t} = 0 \tag{39.19}$$

where

$$\rho Z = \sum_{3,4} \rho_k Z_k + Z_i, \qquad \vec{I} = \sum_{3,4,i} \vec{I_k} \tag{39.20}$$

and

$$\vec{I_3} = \rho_3 Z_3 \vec{V_3}, \qquad \vec{I_4} = \rho_1 Z_1 \vec{V_1} \tag{39.21}$$

Z_k denotes the electric charge per unit of mass and $\vec{I_k}$ the current density vector of phase or interface k.

The electric field $\overrightarrow{E_k^0}$ and the electric displacement $\overrightarrow{D_k^0}$ verify Maxwell's equations and their boundary conditions. We assume that the electric field fluctuations are negligible and that macroscale electric fields are identical in all the phases. We thus obtain Maxwell's equations for the complete material

$$rot\,\overrightarrow{E} = \overrightarrow{0} \qquad div\,\overrightarrow{D} = \rho Z \qquad \overrightarrow{D} = \varepsilon\,\overrightarrow{E} \tag{39.22}$$

where the permittivity ε of the whole material writes

$$\varepsilon = \sum_{3,4} \phi_k \left\langle \varepsilon_k^0 \right\rangle_k \tag{39.23}$$

We conclude that the EAP behaves like an isotropic homogeneous linear dielectric. However its permittivity varies over time and space because of variations of the volume fractions ϕ_k.

39.3.3 Linear Momentum Conservation Law

Since the gravity and the magnetic field are negligible, the only force applied is the electric force (source term). The linear momentum flux is related to the stress tensors $\underset{\sim}{\sigma}_k^0$ of the different phases. From (39.5), we deduce that the linear momentum of the interface is zero, which leads to the following linear momentum conservation law for the interfaces

$$\sum_{3,4} \left\langle \underset{\sim}{\sigma}_k^0 \cdot \overrightarrow{n_k} \chi_i \right\rangle = Z_i \overrightarrow{E_i} \tag{39.24}$$

We obtain for the complete material

$$\rho \frac{D\overrightarrow{V}}{Dt} = div\underset{\sim}{\sigma} + \rho Z \overrightarrow{E} = div\left[\underset{\sim}{\sigma} + \varepsilon\left(\overrightarrow{E} \otimes \overrightarrow{E} - \frac{E^2}{2}\underset{\sim}{1}\right)\right] + \frac{E^2}{2} grad\varepsilon \tag{39.25}$$

using Maxwell's equations (39.22), $\underset{\sim}{1}$ denotes the identity tensor,

$$\underset{\sim}{\sigma} = \sum_{k=3,4} \underset{\sim}{\sigma}_k \tag{39.26}$$

is a symmetric tensor.

$$\varepsilon\left(\overrightarrow{E} \otimes \overrightarrow{E} - \frac{E^2}{2}\underset{\sim}{1}\right)$$

is the Maxwell's tensor, which is here symmetric. The additional term

$$\frac{E^2}{2} grad\varepsilon$$

results from the nonhomogeneous material permittivity.

39.3.4 Energy Balance Laws

39.3.4.1 Potential Energy Balance Equation

Potential energy production is equal to the volume power $-\overrightarrow{E_k^0}\cdot\overrightarrow{I_k^0}$ of the force due to the action of the electric field on the electric charges. The two phases are supposed to be non-dissipative isotropic linear media. As a consequence the balance equation for the potential energy (Poynting's theorem) can be written in the integral form (Jackson, 1975; Maugin, 1988)

$$\frac{d}{dt}\int_{\Omega}\frac{1}{2}\left(\overrightarrow{E}\cdot\overrightarrow{D}+\overrightarrow{B}\cdot\overrightarrow{H}\right)dv=-\oint_{\partial\Omega}\left(\overrightarrow{E}\wedge\overrightarrow{H}\right)\cdot\overrightarrow{n}\,ds-\int_{\Omega}\overrightarrow{E}\cdot\overrightarrow{I}\,dv \quad (39.27)$$

The left hand side represents the variation of the potential energy attached to the volume Ω following the charge motion. If the charges are mobile, the associated local equation is written as follows for the phase k, neglecting the magnetic field

$$\frac{\partial E_{pk}^0}{\partial t}+div\left(E_{pk}^0\overrightarrow{V_k^0}\right)=-\overrightarrow{E_k^0}\cdot\overrightarrow{I_k^0} \qquad k=3,4 \quad (39.28)$$

in which

$$E_{pk}^0=\frac{1}{2}\overrightarrow{D_k^0}\cdot\overrightarrow{E_k^0}$$

is the potential energy per unit of volume of the phase k. The potential energy balance equation for the whole material is then

$$\rho\frac{D}{Dt}\left(\frac{E_p}{\rho}\right)=-\overrightarrow{E}\cdot\overrightarrow{I} \quad (39.29)$$

where

$$E_p=\frac{1}{2}\overrightarrow{D}\cdot\overrightarrow{E}$$

39.3.4.2 Kinetic Energy Balance Equation

The relative velocities of the different phases are negligible compared to the velocities measured in the laboratory reference frame. Let us take for example a strip of Nafion® which is 200 μm thick and 1.57 cm long, bending in an electric field; the tip displacement is about 4 mm and is obtained in 1 s (Nemat-Nasser and Li, 2000). Relative velocities of the order of $2\ 10^{-4}$ ms^{-1} and absolute velocities close to $4\ 10^{-3}$ ms^{-1} are thus obtained (Tixier and Pouget, 2016). In a first approximation, we can identify the kinetic energy E_c of the complete material and the sum of the kinetic energies of the constituents. The kinetic energy balance equation derives from the linear momentum balance equation

$$\rho \frac{D}{Dt}\left(\frac{E_c}{\rho}\right) = \sum_{3,4}\left[div\left(\underset{\sim}{\sigma_k}\cdot\vec{V_k}\right) - \underset{\sim}{\sigma_k} : grad\vec{V_k}\right] + \left(\vec{I} - \vec{i}\right)\cdot\vec{E} \tag{39.30}$$

where the diffusion current of the cations in the solution and of the interfaces is

$$\vec{i} = \vec{I} - \sum_{k=3,4}\left(\rho_k Z_k \vec{V_k}\right) - Z_i \vec{V_i} \tag{39.31}$$

39.3.4.3 Total Energy Balance Equation

The total energy E is the sum of internal, potential and kinetic energies. The source term is null, and the flux comes from contact forces work and heat conduction $\vec{Q}$. The total energy conservation law for the whole material is

$$\rho \frac{D}{Dt}\left(\frac{E}{\rho}\right) = div\left(\sum_{k=3,4}\underset{\sim}{\sigma_k}\cdot\vec{V_k}\right) - div\vec{Q} \tag{39.32}$$

39.3.4.4 Internal Energy Balance Equation

Internal energy U is the difference between total energy and potential and kinetic energies

$$U = E - E_c - E_p \tag{39.33}$$

which leads to

$$\rho \frac{D}{Dt}\left(\frac{U}{\rho}\right) = \sum_{3,4}\left(\underset{\sim}{\sigma_k} : grad\vec{V_k}\right) + \vec{i}\cdot\vec{E} - div\vec{Q} \tag{39.34}$$

39.3.4.5 Interpretation of the Equations

The energy balance laws we have written are relative to a thermodynamic closed system because of the use of the material derivative. Source terms correspond to conversion of one kind of energy into another one. Energy exchanges are summarized in Table 39.1. Fluxes can be considered as the rate of variation of the quantity associated with the conduction phenomenon. The kinetic energy flux is equal to the work of the contact forces, the internal energy flux is the heat flux and the total energy flux is the sum of the two. We verify that there is no source term in this last equation. $\left(\vec{I} - \vec{i}\right)\cdot\vec{E}$ results in a motion of the electric charges subject to the electric field and is a conversion of potential energy into kinetic energy. $\vec{i}\cdot\vec{E}$ can be seen as Joule heating, which is a conversion of potential energy into internal energy,

Table 39.1: Energy exchanges

	flux	$E_c \longleftrightarrow E_p$	$U \longleftrightarrow E_p$	$E_c \longleftrightarrow U$
E_p		$-\left(\vec{I}-\vec{i}\right)\cdot\vec{E}$	$-\vec{i}\cdot\vec{E}$	
E_c	$\sum_{3,4}\underset{\sim}{\sigma_k}\cdot\vec{V_k}$	$+\left(\vec{I}-\vec{i}\right)\cdot\vec{E}$		$-\sum_{3,4}\underset{\sim}{\sigma_k}:grad\vec{V_k}$
U	$-\vec{Q}$		$+\vec{i}\cdot\vec{E}$	$+\sum_{3,4}\left(\underset{\sim}{\sigma_k}:grad\vec{V_k}\right)$
E	$\sum_{3,4}\underset{\sim}{\sigma_k}\cdot\vec{V_k}-\vec{Q}$			

$$\sum_{3,4}\left(\underset{\sim}{\sigma_k}:grad\vec{V_k}\right)$$

represents the viscous dissipation, which converts kinetic energy into heat.

39.4 Entropy Production

We shall now write the thermodynamic relations of the electroactive polymer. The thermodynamics of linear irreversible processes will allow us to identify the fluxes and the generalized forces (Tixier and Pouget, 2016).

39.4.1 Entropy Balance Law

The entropy balance law of the whole material is written as

$$\rho\frac{D}{Dt}\left(\frac{S}{\rho}\right)=s-div\vec{\Sigma}, \tag{39.35}$$

where S, $\vec{\Sigma}$ and s denote the entropy, the entropy flux vector and the rate of entropy production, respectively.

39.4.2 Fundamental Thermodynamic Relations

According to de Groot and Mazur (1962), the Gibbs' relations of a single constituent solid and of a two-constituent fluid can be written at the microscopic scale

$$\rho_3^0 \frac{d_3^0}{dt}\left(\frac{U_3^0}{\rho_3^0}\right) = p_3^0 \frac{1}{\rho_3^0}\frac{d_3^0 \rho_3^0}{dt} + \underset{\sim}{\sigma}_3^{0es} : \frac{d_3^0 \underset{\sim}{\varepsilon}_3^{0s}}{dt} + \rho_3^0 T_3^0 \frac{d_3^0}{dt}\left(\frac{S_3^0}{\rho_3^0}\right),$$
$$\frac{d_4^0}{dt}\left(\frac{U_4^0}{\rho_4^0}\right) = T_4^0 \frac{d_4^0}{dt}\left(\frac{S_4^0}{\rho_4^C}\right) - p_4^0 \frac{d_4^0}{dt}\left(\frac{1}{\rho_4^0}\right) + \sum_{k=1,2} \mu_k^0 \frac{d_4^0}{dt}\left(\frac{\rho_k'}{\rho_4^0}\right). \tag{39.36}$$

where T_k^0 denotes the absolute temperature, $\underset{\sim}{\epsilon}_3^0$ and $\underset{\sim}{\sigma}_3^{0e}$ the strain tensor and the equilibrium stress tensor, $\underset{\sim}{\epsilon}_3^{0s}$ and $\underset{\sim}{\sigma}_3^{0es}$ the strain and stress deviator tensors. The solid pressure p_3^0 is related to the stress tensor and verifies the Euler's relation, as well as the liquid pressure p_4^0

$$p_3^0 = -\frac{1}{3} tr\left(\underset{\sim}{\sigma}_3^{0e}\right) = T_3^0 S_3^0 - U_3^0 + \mu_3^0 \rho_3^0,$$
$$U_4^0 - T_4^0 S_4^0 + p_4^0 = \mu_1^0 C M_1 + \mu_2^0 \frac{\rho_2^0 \phi_2}{\phi_4}. \tag{39.37}$$

μ_k^0 is the chemical potential per unit of mass. We venture the hypothesis that the fluctuations over the R.E.V. of the intensive thermodynamic quantities (pressures, temperatures and chemical potentials) and of the strain and equilibrium stress tensors are negligible. We also suppose that the solid deformations are small. Making the hypothesis of local thermodynamic equilibrium, we derive

$$p = p_3 = p_4 = p_3^0 = p_4^0$$
$$T = T_3 = T_4 = T_i = T_3^0 = T_4^0 \tag{39.38}$$

We thus obtain Gibbs's, Euler's and Gibbs-Duhem's relations of the whole material

$$T\frac{D}{Dt}\left(\frac{S}{\rho}\right) = \frac{D}{Dt}\left(\frac{U}{\rho}\right) + p\frac{D}{Dt}\left(\frac{1}{\rho}\right) - \frac{1}{\rho}\underset{\sim}{\sigma}_3^{es} : \frac{d_3 \underset{\sim}{\epsilon}_3^s}{dt} - \sum_{1,2} \mu_k \frac{\rho_4}{\rho}\frac{d_4}{dt}\left(\frac{\rho_k}{\rho_4}\right);$$
$$p = TS - U + \sum_{k=1,2,3} \mu_k \rho_k,$$
$$\frac{dp}{dt} = S\frac{dT}{dt} + \sum_{k=1,2,3} \rho_k \frac{d\mu_k}{dt} - \underset{\sim}{\sigma}^{es} : grad\overrightarrow{V} \tag{39.39}$$

39.4.3 Entropy Production

The stress tensor is composed of the equilibrium stress tensor $\underset{\sim}{\sigma}^e$ and the viscous stress tensor $\underset{\sim}{\sigma}^v$; this second part vanishes at equilibrium

$$\underset{\sim}{\sigma} = \underset{\sim}{\sigma}^e + \underset{\sim}{\sigma}^v = -p\underset{\sim}{1} + \underset{\sim}{\sigma}^{es} + \underset{\sim}{\sigma}^v \tag{39.40}$$

Combining the internal energy and entropy equations with the Gibbs' relation, the rate of entropy production s can be identified

$$s = \frac{1}{T}\underset{\sim}{\sigma}^{v} : grad\overrightarrow{V} + \frac{1}{T}\overrightarrow{i'} \cdot \overrightarrow{E} - \frac{1}{T^2}\overrightarrow{Q'} \cdot gradT + \sum_{k=1,2,3} \rho_k \left(\overrightarrow{V} - \overrightarrow{V_k}\right) \cdot grad\frac{\mu_k}{T} \tag{39.41}$$

with

$$\begin{aligned} \overrightarrow{i'} &= \overrightarrow{I} - \rho Z \overrightarrow{V}, \\ \overrightarrow{Q'} &= \overrightarrow{Q} - \sum_{k=3,4} U_k \left(\overrightarrow{V} - \overrightarrow{V_k}\right) - \sum_{k=3,4} \underset{\sim}{\sigma}_k \cdot \left(\overrightarrow{V_k} - \overrightarrow{V}\right) \end{aligned} \tag{39.42}$$

39.4.4 Generalized Forces and Fluxes

We define the mass diffusion flux of the cations in the solution $\overrightarrow{J_1}$ and the mass diffusion flux of the solution in the solid $\overrightarrow{J_4}$

$$\overrightarrow{J_1} = \rho_1 \left(\overrightarrow{V_1} - \overrightarrow{V_2}\right), \qquad \overrightarrow{J_4} = \rho_4 \left(\overrightarrow{V_4} - \overrightarrow{V_3}\right) \tag{39.43}$$

These two fluxes are linearly independent. The diffusion current $\overrightarrow{i'}$ and the fluxes $\rho_k \left(\overrightarrow{V_k} - \overrightarrow{V}\right)$ can be expressed as functions of these two fluxes. We thus identify the fluxes along with the associated generalized forces (Table 39.2).

Table 39.2: Generalized fluxes and forces

Fluxes	Forces
$\frac{1}{3} tr \underset{\sim}{\sigma}^{v}$	$\frac{1}{T} div \overrightarrow{V}$
$\overrightarrow{Q'}$	$grad \frac{1}{T}$
$\overrightarrow{J_1}$	$\frac{\rho_2}{\rho_4}\left[\frac{1}{T} Z_1 \overrightarrow{E} - grad\frac{\mu_1}{T} + grad\frac{\mu_2}{T}\right]$
$\overrightarrow{J_4}$	$\frac{\rho_3}{\rho}\left[\frac{1}{T}\left(\frac{\rho_1}{\rho_4} Z_1 - Z_3\right)\overrightarrow{E} - \frac{\rho_1}{\rho_4} grad\frac{\mu_1}{T} - \frac{\rho_2}{\rho_4} grad\frac{\mu_2}{T} + grad\frac{\mu_3}{T}\right]$
$\underset{\sim}{\sigma}^{vs}$	$\frac{1}{T}\left[\frac{1}{2}\left(grad\overrightarrow{V} + grad\overrightarrow{V}^{T}\right) - \frac{1}{3}\left(div\overrightarrow{V}\right)\underset{\sim}{1}\right]$

39.5 Constitutive Equations

We venture the hypothesis that the medium is isotropic. According to Curie symmetry principle, there can not be any coupling between fluxes and forces whose tensorial ranks differs from one unit. Moreover, we suppose that coupling between fluxes and different tensorial rank forces are negligible, which is commonly admitted (de Groot and Mazur, 1962). We thus obtain a tensorial law (the rheological equation) and three vectorial constitutive equations.

39.5.1 Rheological Equation

Considering the symmetry of the tensor $\underset{\sim}{\sigma}^v$, the scalar and tensorial fluxes are linear functions of the corresponding forces. Assuming that the complete material satisfies the Hooke's law at equilibrium and the liquid phase is newtonian and stokesian, the pressure verify

$$p = -\frac{1}{3} tr\left(\underset{\sim}{\sigma}^e\right) = \left(\lambda + \frac{2}{3}G\right) tr\underset{\sim}{\epsilon} \qquad (39.44)$$

where λ and G denote the first Lamé constant and the shear modulus of the complete material, respectively, and $\underset{\sim}{\epsilon}$ the material strain

$$\underset{\sim}{\epsilon} = \frac{1}{2}\left(grad\,\vec{u} + grad\,\vec{u}^T\right) \qquad or \qquad \dot{\underset{\sim}{\epsilon}} = \frac{1}{2}\left(grad\,\vec{V} + grad\,\vec{V}^T\right) \qquad (39.45)$$

$\vec{u}$ is the displacement vector. The stress tensor of the complete material thus identifies with a Kelvin-Voigt model

$$\underset{\sim}{\sigma} = \lambda \left(tr\underset{\sim}{\epsilon}\right)\underset{\sim}{1} + 2G\underset{\sim}{\epsilon} + \lambda_v tr\dot{\underset{\sim}{\epsilon}}\underset{\sim}{1} + \mu_v\dot{\underset{\sim}{\epsilon}} \qquad (39.46)$$

λ_v and μ_v are viscoelastic coefficients.

The elastic coefficients have the following values (Bauer et al, 2005; Barclay Satterfield and Benziger, 2009; Silberstein and Boyce, 2010), which are in agreement with the usual ones

$$G \sim 4.5\ 10^7\ \text{Pa}, \qquad \lambda \sim 3\ 10^8\ \text{Pa}, \qquad E \sim 1.3\ 10^8\ \text{Pa} \qquad \nu \sim 0.435 \qquad (39.47)$$

where E is the Young's modulus and ν the Poisson's ratio. Viscoelastic coefficients can be deduced from uniaxial tension tests (Barclay Satterfield and Benziger, 2009; Silberstein and Boyce, 2010; Silberstein et al, 2011) and relaxation times for a traction, which are close to 15 s (Silberstein, 2008; Silberstein and Boyce, 2010; Silberstein et al, 2011)

$$\lambda_v \sim 7\ 10^8\ \text{Pa s}, \qquad \mu_v \sim 10^8\ \text{Pa s} \qquad (39.48)$$

These coefficients depend strongly on the solvent concentration and on the temperature, especially if it is close to the glass transition.

39.5.2 Nafion® Physicochemical Properties

Vectorial constitutive equations require nine phenomenological coefficients, which are a priori second-rank tensors; they can be replaced by scalars because of the isotropy of the medium. The first equation that we obtain is a generalized Fourier's law. We will approximate the two other by restricting ourselves to the isothermal case and focusing on a particular EAP: Nafion® 117 Li^+.

The liquid phase is a dilute solution of strong electrolyte. According to Diu (2007), mass chemical potentials can be written on a first approximation

$$\begin{aligned}
\mu_1(T,p,x) &\simeq \mu_1^0(T,p) + \frac{RT}{M_1}\ln\left(C\frac{M_2}{\rho_2^0}\right),\\
\mu_2(T,p,x) &\simeq \mu_2^0(T,p) - \frac{RT}{\rho_2^0}C,\\
\mu_3(T,p,x) &= \mu_3^0(T),
\end{aligned} \tag{39.49}$$

where $R = 8,314\, J\, \mathrm{K}^{-1}$ is the gas constant. μ_2^0 and μ_3^0 denote the chemical potentials of the single solid and solvent, and μ_1^0 depends on the solvent and the solute. The Gibbs-Duhem's relations for the solid and the liquid phases enable the calculation of $grad\,\mu_k$

$$\begin{aligned}
grad\,\mu_1 &= -\frac{S_1}{\rho_1}gradT + \frac{v_1}{M_1}gradp + \frac{RT\rho_2^0}{M_2M_1C}grad\left(\frac{CM_2}{\rho_2^0}\right)\\
grad\,\mu_2 &= -\frac{S_2}{\rho_2}gradT + \frac{v_2}{M_2}gradp - \frac{RT}{M_2}grad\left(\frac{CM_2}{\rho_2^0}\right)\\
grad\,\mu_3 &= -\frac{S_3}{\rho_3}gradT
\end{aligned} \tag{39.50}$$

where v_k denotes the partial molar volume of the constituent k.

The physicochemical properties of the Nafion® are well documented. Its equivalent weight, that is to say, the weight of polymer per mole of ionic sites is (Gebel, 2000)

$$M_{eq} \sim 1.1\ \mathrm{kg\,eq}^{-1}$$

The solution volume fraction ϕ_4 is close to 38% (Cappadonia et al, 1994; Choi et al, 2005). Other parameters are resumed in Table 39.3 (Heitner-Wirguin, 1996; Nemat-Nasser and Li, 2000). The anions molar concentration, which is equal to the average cations concentration, is

$$C_{moy} = \frac{\rho_3^0 \phi_3}{M_{eq} \phi_4} \sim 3.1\ 10^3 \text{ mol m}^{-3}$$

The dynamic viscosity of water is $\eta_2 = 10^{-3}$ Pa s. We deduce the mass densities of the complete material $\rho \sim 1.7\ 10^3$ kg m^{-3}. We moreover suppose that the temperature is $T = 300$ K. The electric field is typically about 10^4 V m^{-1} (Nemat-Nasser and Li, 2000).

Considering this numerical estimations, we can write in a first approximation

$$Z_1 \gg Z_3 \qquad \rho \sim \rho_2 \sim \rho_3 \gg \rho_1 \qquad \rho_1 Z_1 \sim \rho_4 Z_3 \tag{39.51}$$

39.5.3 Nernst-Planck Equation

It is commonly accepted that the non-diagonal phenomenological coefficients are small compared to the diagonal ones. The expression of the mass diffusion flux of the cations J_1 can be identified with a Nernst-Planck equation (Lakshminarayanaiah, 1969)

$$\vec{V_1} = -\frac{D}{C}\left[grad C - \frac{Z_1 M_1 C}{RT}\vec{E} + \frac{C v_1}{RT}\left(1 - \frac{M_1}{M_2}\frac{v_2}{v_1}\right) grad\, p\right] + \vec{V_2} \tag{39.52}$$

$D \sim 2\ 10^{-9}$m^{2}s^{-1} denotes the mass diffusion coefficient of the cations in the liquid phase Zawodzinski et al (1991). This equation expresses the equilibrium of an ions mole under the action of four forces: the Stokes friction force, which is proportional to $\vec{V_1} - \vec{V_2}$, the pressure force, the electric force and the thermodynamic force $-M_1 grad \mu_1$.

The order of magnitude of the different terms of this equation can be estimated. According to Farinholt and Leo (2004); Nemat-Nasser (2002), the concentration gradient verifies

$$|grad C| \precsim 10^8 \text{ mol m}^{-4} \tag{39.53}$$

Table 39.3: Nafion® 117 Li$^+$ parameters

	Cations	Solvent	Solid
M_k (kg mol^{-1})	$3\ 10^{-3}$	$18\ 10^{-3}$	10^2 - 10^3
ρ_k^0 (kg m^{-3})		10^3	$2.08\ 10^3$
v_k $\left(\text{m}^3 \text{ mol}^{-1}\right)$	$\frac{M_1}{\rho_4^0} \sim 3\ 10^{-6}$	$18\ 10^{-6}$	
ρ_k (kg m^{-3})	3.5	380	$1.3\ 10^3$
Z_k (C kg^{-1})	$3.2\ 10^7$	0	$9\ 10^4$

The pressure gradient can be roughly estimated using the Darcy's law; it is about 10^9 Pa m^{-1}. We thus obtain

$$\begin{aligned} &\frac{M_1 C}{RT} Z_1 \left|\overrightarrow{E}\right| && \sim 1.6\ 10^9 \text{ mol m}^{-4} \\ &\frac{C v_1}{RT}\left(1 - \frac{M_1}{M_2}\frac{v_2}{v_1}\right) |gradp| && \sim 1.1\ 10^3 \text{ mol m}^{-4} \end{aligned} \tag{39.54}$$

The electric field and the mass diffusion have the leading effects; we thereafter neglect the pressure gradient term.

39.5.4 Generalized Darcy's Law

The expression of the mass diffusion flux of the solution in the solid J_4 identifies with a generalized Darcy's law

$$\overrightarrow{V_4} - \overrightarrow{V_3} \simeq -\frac{K}{\eta_2 \phi_4}\left[grad\,p - \rho_4^0 (Z_4 - Z_3)\overrightarrow{E}\right] - \frac{R}{M_1 C \rho_4} L^1 gradC \tag{39.55}$$

where L^1 is a phenomenological coefficient and K the intrinsic permeability of the solid phase, which is on the order of the square of the pore sizes, that is 10^{-16} m^2.

The orders of magnitude of the different terms are

$$\begin{aligned} &\frac{K}{\eta_2 \phi_4} |gradp| && \sim 2.6\ 10^{-4} \text{ m s}^{-1} \\ &\frac{K}{\eta_2 \phi_4} \rho_4^0 (Z_4 - Z_3)\left|\overrightarrow{E}\right| && \sim 0.53 \text{ m s}^{-1} \\ &\frac{R}{M_1 C \rho_4} L^1 |gradC| && \ll 5.9\ 10^{-7} \text{ m s}^{-1} \end{aligned} \tag{39.56}$$

The latter term can therefore be neglected. The first term represents the mass pressure force and the second one is the mass electric force; it expresses the motion of the solution under the action of the electric field and reflects an electro-osmotic phenomenon.

The distribution of cations becomes very heterogeneous (Farinholt and Leo, 2004; Nemat-Nasser, 2002): they gather near the negative electrode, where $Z_4 \gg Z_3$; the expression obtained coincides with that of Biot (1955). Near the positive electrode, where the cation concentration is very low, $Z_4 \ll Z_3$; this corresponds to the result obtained by Grimshaw et al (1990); Nemat-Nasser and Li (2000). In the center of the strip, the mass electric force exerted on the solution is proportional to the net charge $(Z_4 - Z_3)$.

39.6 Validation of the Model: Application to a Cantilevered Strip

In order to validate the model that we have just described, we apply it to the static case of a cantilevered EAP strip bending under the effect of a permanent electric field. In addition, the strip might undergo the action of a shear force exerted on the free end in order to prevent its displacement (blocking force).

39.6.1 Static Equations

A continuous constant voltage φ_0 is applied between the two faces of the strip. As a consequence, the partial derivatives with respect to time and the velocities are zero. We focuse on Nafion® Li^+ strip of length $L = 2$ cm, of thickness $2e = 200\ \mu$m and of width $2l = 5$ mm subject to a potential difference $\varphi_0 = 1$ V. We postulate that the volume fraction ϕ_4 is constant; this hypothesis will be checked a posteriori. As a consequence, the dielectric permittivity of the whole material is a constant too. We assume that it is equal to that measured by Deng and Mauritz (1992) for a material very close to the Nafion®: $\varepsilon \sim 10^{-6}\ \mathrm{F\,m^{-1}}$. Considering the strip dimensions, it is a two-dimensional problem. A coordinate system $Oxyz$ is chosen such that the axis Oz is parallel to the imposed electric field, the axis Ox is along the length of the strip and Oy along its width. We venture the hypothesis that the local electric field axial coordinate E_x is negligible compare to the normal one E_z. On a first approximation, C, E_z, p, the local electric potential φ and the electric charge density ρZ only depend on the z coordinate. Neglecting the pressure term in (39.52), we obtain

$$\begin{aligned} E_z &= -\frac{d\varphi}{dz}, & \varepsilon\frac{dE_z}{dz} &= \rho Z, \\ \frac{dp}{dz} &= \left(CF - \rho_2^0 Z_3\right) E_z, & \frac{dC}{dz} &= \frac{FC}{RT} E_z \\ \rho Z &= \phi_4 F \left(C - C_{moy}\right) \end{aligned} \tag{39.57}$$

$F = 96487\ \mathrm{C\,mol^{-1}}$ is the Faraday's constant. The boundary conditions write

$$\lim_{z\to -e} \varphi = \varphi_0, \qquad \lim_{z\to e} \varphi = 0, \qquad \int_{-e}^{e} \rho Z\, dz = 0 \tag{39.58}$$

The latter condition expresses the electroneutrality condition and is equivalent to $E_z(e) = E_z(-e)$. We introduce the following dimensionless variables:

$$\begin{aligned} \overline{E} &= \frac{E_z e}{\varphi_0}, & \overline{C} &= \frac{C}{C_{moy}}, & \overline{\varphi} &= \frac{\varphi}{\varphi_0}, \\ \overline{\rho Z} &= \frac{\rho Z}{\phi_4 F C_{moy}}, & \overline{p} &= \frac{p}{F \varphi_0 C_{moy}}, & \overline{z} &= \frac{z}{e} \end{aligned} \tag{39.59}$$

which leads to

$$\begin{aligned} \overline{E} &= -\frac{d\overline{\varphi}}{d\overline{z}}, & \frac{d\overline{E}}{d\overline{z}} &= \frac{A_1}{A_2}\overline{\rho Z}, \\ \frac{d\overline{p}}{d\overline{z}} &= \left(\overline{C}+A_3\right)\overline{E}, & \frac{d\overline{C}}{d\overline{z}} &= A_2\overline{CE}, \\ \overline{\rho Z} &= \overline{C}-1, \end{aligned} \tag{39.60}$$

A_1, A_2 and A_3 are dimensionless constants

$$\begin{aligned} A_1 &= \frac{\phi_4 e^2 F^2 C_{moy}}{\varepsilon RT} \sim 4,37\ 10^7, \\ A_2 &= \frac{F\varphi_0}{RT} \sim 38,7, \\ A_3 &= -\frac{\rho_2^0 Z_3 C_{moy}}{FC_{moy}} \sim 0,303 \end{aligned} \tag{39.61}$$

Boundary conditions are

$$\lim_{\overline{z}\to -1} \overline{\varphi} = 1, \qquad \lim_{\overline{z}\to 1} \overline{\varphi} = 0, \qquad \overline{E}(1) = \overline{E}(-1) \tag{39.62}$$

We deduce

$$\frac{d}{d\overline{z}}\left(\frac{d\overline{C}}{\overline{C}d\overline{z}}\right) = A_1\left(\overline{C}-1\right) \tag{39.63}$$

and

$$\overline{C} \simeq A_2 \exp\left(-A_2\overline{\varphi}\right), \qquad \overline{p} = \frac{\overline{C}}{A_2} - A_3\overline{\varphi} + Cte \tag{39.64}$$

The hydrated polymer can be assimilated to a conductive material. As a consequence, the electric field is zero throughout the strip except near the boundaries. We deduce the values of the different quantities on the sides and at the center of the strip (Table 39.4). Equation (39.63) can be solved using Matlab. We deduce $\overline{\rho Z}$, $\overline{\varphi}$, $\overline{p}$ and $\overline{E}$ by (39.60) and (39.64). An evaluation of the pressure term of the equation (39.52)

Table 39.4: Boundary values of the physical quantities

	-1	Center	1
$\overline{C}$	$A_2 e^{-A_2} \simeq 0$	1	A_2
$\overline{\varphi}$	1	$\frac{\ln A_2}{A_2}$	0
$\overline{E}$	$\sqrt{\frac{2A_1}{A_2}\left[1-\frac{1}{A_2}(1+\ln A_2)\right]}$	0	$\sqrt{\frac{2A_1}{A_2}\left[1-\frac{1}{A_2}(1+\ln A_2)\right]}$
$\overline{p}$	$-A_3$	$\frac{1}{A_2}(1-A_3\ln A_2)$	1

shows that it does not exceed 2% of the second term of the equation with the nominal conditions chosen.

39.6.2 Beam Model on Large Displacements

Given the high deflection values, we used a large displacement beam model to determine forces, stress and strain. We consider a straight beam clamped at the end O; the other end A is either free or subject to a shear force bringing the deflection to a zero value (blocking force $\overrightarrow{F^p}$). The polymer is subject to a distributed electric force $\overrightarrow{p^p}$ independent of the x coordinate and orthogonal to the strip. Moreover, the swelling of the strip on the side of the negative electrode generates, through the pressure

$$p = \frac{\sigma_{xx}}{3},$$

a bending moment $\overrightarrow{M_A^p}$ at the free end of the beam, see (39.2), Considering the electroneutrality condition, the distributed force and the bending moment verify

$$p^p = \int_{-l}^{l}\int_{-e}^{e} \rho Z E_z \, dz \, dy = 2l \int_{-e}^{e} \rho Z E_z \, dz = 2l\varepsilon \left[\frac{E_z^2}{2}\right]_{-e}^{e} = 0 \tag{39.65}$$

$$M_A^p = \int_{-l}^{l}\int_{-e}^{e} \sigma_{xx} \, z \, dz \, dy = A_4 \int_{-1}^{1} \overline{p} \, \overline{z} \, d\overline{z} \tag{39.66}$$

with $A_4 = 6le^2 F \varphi_0 C_{r\!-\!oy} \sim 0{,}045\text{N m}$.

We assume that the Bernoulli and Barré Saint Venant hypotheses are verified. Let s and $\overline{s}$ be the curvilinear abscissas along the beam respectively at the rest and deformed configurations, $\overrightarrow{t}$ and $\overrightarrow{n}$ the vectors tangent and normal to the beam, θ the angle of rotation of a cross-section and R^p the radius of curvature (geometric parameters are given in Fig. 39.3). It will be assumed that the elongation

$$\Lambda = \frac{d\overline{s}}{ds}$$

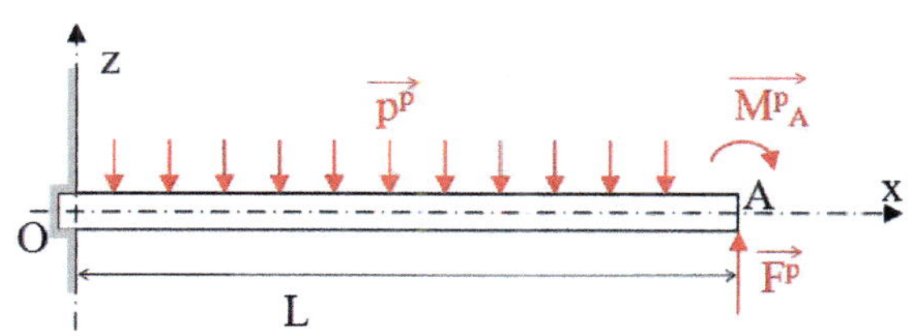

Fig. 39.2 Forces exerted on the beam

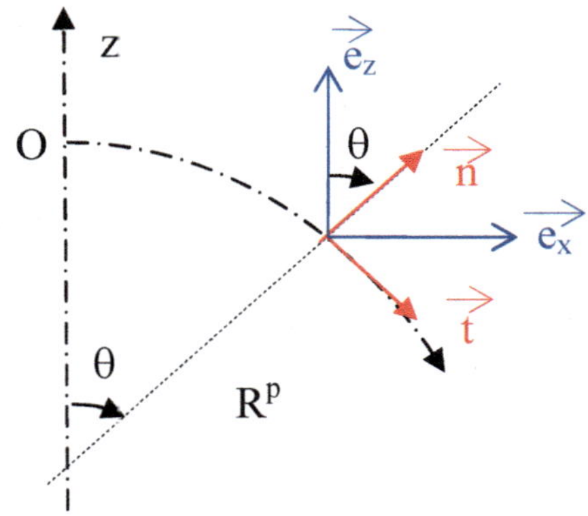

Fig. 39.3 Beam on large displacements

is equal to 1. The bending moment in the current section is

$$M^p = F^p\,(x-L) + M_A^p \tag{39.67}$$

The strain tensor is given by

$$\underset{\sim}{\epsilon} = \frac{1}{2}\left[\left(grad\,\vec{u} + \underset{\sim}{1}\right)^T \left(grad\,\vec{u} + \underset{\sim}{1}\right) - \underset{\sim}{1}\right] \tag{39.68}$$

In the reference frame linked to the undeformed beam, it follows

$$\epsilon_{xx} = -\frac{\overline{n}}{R^p}\left(1 - \frac{\overline{n}}{2R^p}\right) \simeq -\frac{\overline{n}}{R^p} \tag{39.69}$$

where $\overline{n}$ designates the coordinate in the $\vec{n}$ direction. Indeed, the beam being thin, $|\overline{n}| \ll R^p$. In the case of pure bending, the strain is given by

$$\epsilon_{xx} = \frac{M^p}{EI^p}\overline{n}$$

where

$$I^p = \frac{4le^3}{3}$$

is the area moment of inertia, with respect to the Oy axis. The shear force has a negligible effect on the deflection. We deduce the radius of curvature and the angle of rotation

$$\frac{1}{R^p} = \frac{d\theta}{d\overline{s}} = \frac{F^p}{EI^p}\,(L-\overline{s}) - \frac{M_A^p}{EI^p}, \qquad \theta = \frac{F^p}{2EI^p}\overline{s}\,(2L-\overline{s}) - \frac{M_A^p}{EI^p}\overline{s} \tag{39.70}$$

by choosing the point O as the origin of the curvilinear abscissas. The deflection w is obtained by integrating the relation

$$\frac{dz}{d\bar{s}} = \sin\theta$$

In the case of a cantilever beam ($F^p = 0$), the radius of curvature is constant; the beam is circle shaped and at the free end

$$\theta = -\frac{M_A^p}{EI^p}L = -\frac{2A_5}{L}\int_{-1}^{1}\bar{p}\,\bar{z}\,d\bar{z}$$

$$w = \frac{EI^p}{M_A^p}\left[\cos\left(\frac{M_A^p}{EI^p}L\right)-1\right] = \frac{L^2}{2A_5\int_{-1}^{1}\bar{p}\,\bar{z}\,d\bar{z}}\left[\cos\left(\frac{2A_5}{L}\int_{-1}^{1}\bar{p}\,\bar{z}\,d\bar{z}\right)-1\right] \tag{39.71}$$

with

$$A_5 = \frac{9}{4}\frac{L^2 F\varphi_0 C_{moy}}{eE} \sim 20,59\ \text{m}$$

The blocking force is the end loading such that the deflection is zero

$$w(\bar{s}=L) = -\int_0^L \sin\left[\frac{F^p}{2EI^p}\bar{s}(\bar{s}-2L) + \frac{M_A^p}{EI^p}\bar{s}\right]d\bar{s} = 0 \tag{39.72}$$

Using the Fresnel functions to compute this integral, we obtain for the blocking force the same result as in small displacements

$$F^p = \frac{3M_A^p}{2L} = \frac{3A_4}{2L}\int_{-1}^{1}\bar{p}\,\bar{z}\,d\bar{z} \tag{39.73}$$

39.6.3 Simulations Results

The values obtained for the deflection and the blocking force are in good agreement with the experimental values reported in the literature (Nemat-Nasser 2002; Newbury, 2002; Newbury and Leo, 2002, 2003). The variation of cation concentration in the thickness of the strip is shown in Fig. 39.4. This quantity is constant throughout the central part of the strip but varies very strongly in the vicinity of the electrodes, especially near the negative electrode on which the cations accumulate. On the contrary, it is noticed near the positive electrode an aerea of the order of $0,1\ \mu m$ without cations. This result is in good agreement with that of Nemat-Nasser (2002).

$$\int_{-1}^{1}\bar{p}\,\bar{z}\,d\bar{z}$$

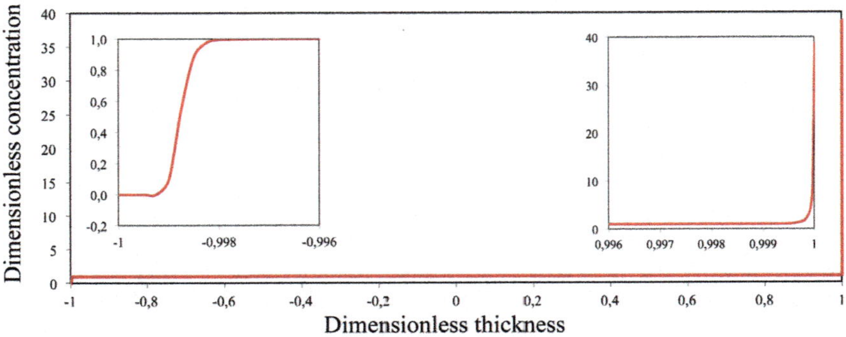

Fig. 39.4: Variation of cation concentration in the thickness of the strip; the distributions at the vicinity of the lower and upper faces (electrodes) are detailed in insets.

depends only on the imposed potential φ_0, the thickness e and the cation chosen. The deflection w is therefore independent of the width and approximately proportional to length square, which corresponds to Shahinpoor measurements (Shahinpoor, 1999). The blocking force is proportional to l and inversely proportional to L, which is in good agreement with Newbury and Leo (2003).

The deflection varies linearly with the imposed potential difference φ_0, a result in agreement with the experiments of Mojarrad and Shahinpoor (1997); Shahinpoor et al (1998). The blocking force follows the same tendency (Fig. 39.5).

39.7 Conclusion

We have modeled the behavior of an ionic electroactive polymer saturated with water and subject to an orthogonal electric field. The presence of water causes a complete dissociation of the polymer and the release of small cations. We have depicted this medium as the superposition of three systems with different velocities fields : the cations, the solvent and the solid assimilated to a deformable porous medium. We

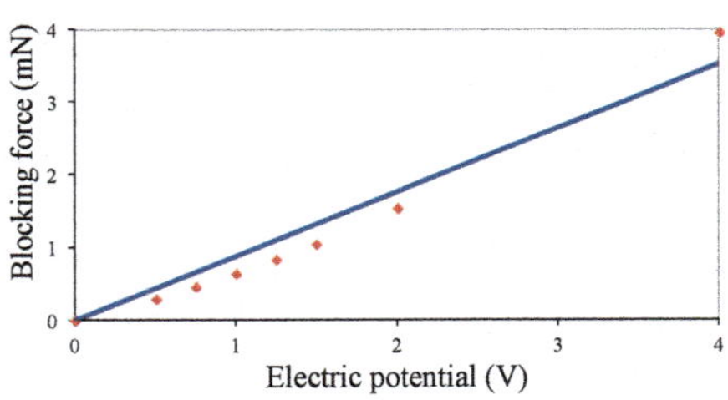

Fig. 39.5 Influence of the potential difference on the blocking force

have used a continuous medium approach and a coarse grain model. We have established the microscale balance equations of mass, linear momentum, energies, entropy and the Maxwell's equations as well as thermodynamic relations for each phase (solid and liquid). We have derived the macroscale equations relative to the whole material using an average technique and the material derivative concept. Thermodynamics of linear irreversible processes have provided the entropy production and the constitutive equations of the complete material. We have obtained a rheological law of the Kelvin-Voigt type, generalised Fourier's and Darcy's laws and a Nernst-Planck equation.

We have applied this model to a cantilevered Nafion® strip subject to a continuous voltage between its two faces, which is a static case. The other end may be either free or subject to a shear force preventing its displacement (blocking force). We have used a beam model in large displacement. We have drawn the profile of cations concentration and evaluated the deflection and the blocking force. We have observed that the concentration varies very strongly in the vicinity of the electrodes, which is characteristic of a conductive material behavior. Our results are in good agreement with the experimental data published in the literature.

To improve this model, we intend to take into account the variations of the permittivity with the cation concentration. Another way of improvement is to replace the rheological law with a Zener model that is better suited to viscoelastic polymers.

39.8 Notations

$k = 1, 2, 3, 4, i$ subscripts respectively represent cations, solvent, solid, solution (water and cations) and interface; quantities without subscript refer to the whole material. Superscript ^{0} denotes a local quantity; the lack of superscript indicates average quantity at the macroscopic scale. Microscale volume quantities are relative to the volume of the phase, average quantities to the volume of the whole material. Superscript ^{s} indicates the deviatoric part of a second-rank tensor, and ^{T} its transpose.

C, C_{moy} : cations molar concentrations (relative to the liquid phase);
D : mass diffusion coefficient of the cations in the liquid phase;
$\vec{D}$: electric displacement field;
e : half-thickness of the strip;
E, G, λ, ν : Young's and shear modulus, first Lamé constant, Poisson's ratio.
$\vec{E}$: electric field;
E, E_p (E^0_{pk}), E_c, U (U_k, U^0_k) : total, potential, kinetic and internal energy densities;
$F = 96487\ \mathrm{C\,mol^{-1}}$: Faraday's constant ;
$\vec{F^b}$: blocking force;
$\vec{i}$ ($\vec{i'}$) : diffusion current;
$\vec{I}$ ($\vec{I_k}$, $\vec{I^0_k}$) : current density vectors;
I^p : area moment of inertia;

$\overrightarrow{J_k}$: mass diffusion flux;
K : intrinsic permeability of the solid phase;
l : half-width of the strip;
L : length of the strip;
M_k : molar mass of component k;
M_{eq} : equivalent weight (weight of polymer per mole of sulfonate groups);
$\overrightarrow{M^p}$ ($\overrightarrow{M_A^p}$) : bending moment;
$\overrightarrow{n_k}$: outward-pointing unit normal of phase k;
p (p_k, p_k^0) : pressure;
$\overrightarrow{Q}$ ($\overrightarrow{Q'}$) : heat flux;
$R = 8,314\,\mathrm{J\,K^{-1}}$: gas constant;
R^p : radius of curvature of the beam;
s : rate of entropy production;
S (S_k^0, S_k) : entropy density;
T (T_k, T_k^0) : absolute temperature;
$\overrightarrow{u}$: displacement vector;
v_k : partial molar volume of component k (relative to the liquid phase);
$\overrightarrow{V}$ ($\overrightarrow{V_k}$, $\overrightarrow{V_k^0}$) : velocity;
w : deflection of the beam;
Z (Z_k) : total electric charge per unit of mass;
ε : permittivity;
$\underset{\sim}{\epsilon}$ ($\underset{\sim}{\epsilon_k^0}$) : strain tensor;
η_2 : dynamic viscosity of water;
θ : angle of rotation of a beam cross section;
λ_v, μ_v : viscoelastic coefficients;
μ_k (μ_k^0) : mass chemical potential;
ρ (ρ_k, ρ_k^0) : mass density;
$\underset{\sim}{\sigma}$ ($\underset{\sim}{\sigma_k}$, $\underset{\sim}{\sigma_k^0}$), $\underset{\sim}{\sigma^v}$, $\underset{\sim}{\sigma^e}$ ($\underset{\sim}{\sigma_k^e}$, $\underset{\sim}{\sigma_k^{0e}}$): stress tensor, dynamic stress tensor, equilibrium stress tensor;
$\boldsymbol{\Sigma}$: entropy flux vector;
ϕ_k : volume fraction of phase k;
φ (φ_0): electric potential;
χ_k : function of presence of phase k ;

References

Ask A, Menzel A, Ristinmaa M (2012) Electrostriction in electro-viscoelastic polymers. Mechanics of Materials 50:9 – 21

Bar-Cohen Y (2001) Electroactive Polymers (EAP) Actuators Artificial Muscles: Reality, Potential and Challenges. SPIE Press

Barclay Satterfield M, Benziger JB (2009) Viscoelastic properties of Nafion at elevated temperature and humidity. J Polym Sci Pol Phys 47(1):11–24

Bauer F, Denneler S, Willert-Porada M (2005) Influence of temperature and humidity on the mechanical properties of Nafion® 117 polymer electrolyte membrane. Journal of Polymer Science Part B: Polymer Physics 43(7):786–795
Biot MA (1955) Theory of elasticity and consolidation for a porous anisotropic solid. Journal of Applied Physics 26(2):182–185
Biot MA (1977) Variational Lagrangian-thermodynamics of nonisothermal finite strain mechanics of porous solids and thermomolecular diffusion. International Journal of Solids and Structures 13(6):579–597
Brufau-Penella J, Puig-Vidal M, Giannone P, Graziani S, Strazzeri S (2008) Characterization of the harvesting capabilities of an ionic polymer metal composite device. Smart Materials and Structures 17(1):015,009.1–15
Cappadonia M, Erning JW, Stimming U (1994) Proton conduction of Nafion® 117 membrane between 140 K and room temperature. Journal of Electroanalytical Chemistry 376(1):189 –193
Chabé J (2008) Etude des interactions moléculaires polymère-eau lors de l'hydratation de la membrane nafion, électrolyte de référence de la pile à combustible. PhD thesis, Université Joseph Fourier Grenoble I
Choi P, Jalani NH, Datta R (2005) Thermodynamics and proton transport in nafion: II. Proton diffusion mechanisms and conductivity. Journal of The Electrochemical Society 152(3):E123–E130
Collette F (2008) Vieillissement hygrothermique du Nafion. PhD thesis, Ecole Nationale Supérieure d'Arts et Métiers
Coussy O (1989) Dielectric relaxation studies of water-containing short side chain perfluorosulfonic acid membranes. Eur J Mech A 8(1):1–14
Coussy O (1995) Mechanics of Porous Continua. Wiley, Chichester
de Groot SR, Mazur P (1962) Non-Equilibrium Thermodynamics. North-Holland, Amsterdam
Deng ZD, Mauritz KA (1992) Dielectric relaxation studies of water-containing short side chain perfluorosulfonic acid membranes. Macromolecules 25(10):2739–2745
Diu B (2007) Thermodynamique. Hermann, Paris
Drew DA (1983) Mathematical modeling of two-phase flow. Annual Review of Fluid Mechanics 15(1):261–291
Drew DA, Passman SL (1998) Theory of Multicomponents Fluids. Springer, New York
Farinholt K, Leo DJ (2004) Modeling of electromechanical charge sensing in ionic polymer transducers. Mechanics of Materials 36(5):421–433
Gebel G (2000) Structural evolution of water swollen perfluorosulfonated ionomers from dry membrane to solution. Polymer 41(15):5829–5838
de Gennes PG, Okumura K, Shahinpoor M, Kim KJ (2000) Mechanoelectric effects in ionic gels. EPL (Europhysics Letters) 50(4):513–518
Gierke TD, Munn GE, Wilson FC (1981) The morphology in nafion perfluorinated membrane products, as determined by wide- and small-angle x-ray studies. Journal of Polymer Science: Polymer Physics Edition 19(11):1687–1704
Grimshaw PE, Nussbaum JH, Grodzinsky AJ, Yarmush ML (1990) Kinetics of electrically and chemically induced swelling in polyelectrolyte gels. The Journal of Chemical Physics 93(6) 4462–4472
Heitner-Wirguin C (1996) Recent advances in perfluorinated ionomer membranes: structure, properties and applications. Journal of Membrane Science 120(1):1–33
Ishii M, Hibiki T (2006) Thermo-Fluid Dynamics of Two-Phase Flow. Springer, New York
Jackson JD (1975) Classical Electrodynamics. Wiley & Sons, New York
Lakshminarayanaiah N (1969) Transport Phenomena in Membranes. Academic Press, New York
Lhuillier D (2003) A mean-field description of two-phase flows with phase changes. International Journal of Multiphase Flow 29(3):511–525
Maugin GA (1988) Continuum Mechanics of Electromagnetic Solids. North-Holland, Amsterdam
Mojarrad M, Shahinpoor M (1997) Ion-exchange-metal composite sensor films. Proc SPIE 3042:52–60

Nemat-Nasser S (2002) Micromechanics of actuation of ionic polymer-metal composites. Journal of Applied Physics 92(5):2899–2915

Nemat-Nasser S, Li J (2000) Electromechanical response of ionic polymer-metal composites. Journal of Applied Physics 87(7):3321–3331

Newbury KM (2002) Characterization, modeling and control of ionic-polymer transducers. PhD thesis, Faculty of the Virginia Polytechnic Institute and State University, Blacksburg, Virginia

Newbury KM, Leo DJ (2001) Mechanical work and electromechanical coupling in ionic polymer bender actuators. In: ASME Adative Structures and Materials Symposium

Newbury KM, Leo DJ (2002) Electromechanical modeling and characterization of ionic polymer benders. Journal of Intelligent Material Systems and Structures 13(1):51–60

Newbury KM, Leo DJ (2003) Linear electromechanical model of ionic polymer transducers - Part II: Experimental validation. Journal of Intelligent Material Systems and Structures 14(6):343–357

Nigmatulin RI (1979) Spatial averaging in the mechanics of heterogeneous and dispersed systems. Int J Multiphase Flow 5:353–385

Nigmatulin RI (1990) Dynamics of Multiphase Media, vol 1 and 2. Hemisphere, New York

Shahinpoor M (1994) Continuum electromechanics of ionic polymeric gels as artificial muscles for robotic applications. Smart Materials and Structures 3:367–372

Shahinpoor M (1999) Electromechanics of ionoelastic beams as electrically controllable artificial muscles. Proc of SPIE 3669

Shahinpoor M, Bar-Cohen Y, Simpson JO, Smith J (1998) Ionic polymer-metal composites (IPMCs) as biomimetic sensors, actuators and artificial muscles - a review. Smart Materials and Structures 7(6):R15–R30

Silberstein MN (2008) Mechanics of proton exchange membranes: Time, temperature and hydration dependence of the stress-strain behavior of persulfonated polytetrafluoroethylene. PhD thesis, Massachusetts Institut of Technology, Cambridge, MA

Silberstein MN, Boyce MC (2010) Constitutive modeling of the rate, temperature, and hydration dependent deformation response of Nafion to monotonic and cyclic loading. J Power Sources 195(17):5692–5706

Silberstein MN, Pillai PV, Boyce MC (2011) Biaxial elastic-viscoplastic behavior of Nafion membranes. Polymer 52(2):529–539

Tixier M, Pouget J (2014) Conservation laws of an electro-active polymer. Continuum Mechanics and Thermodynamics 26(4):465–481

Tixier M, Pouget J (2016) Constitutive equations for an electroactive polymer. Continuum Mechanics and Thermodynamics 28(4):1071–1091

Todokoru S, Yamagami S, Takamori T, Oguro K (2000) Modeling of Nafion-Pt composite actuators (ICPF) by ionic motion. Proc of SPIE 3987

Vokoun D, Qingsong H, Heller L, Min Y, Zhen DD (2015) Modeling of IPMC cantilever's displacements and blocking forces. Journal of Bionic Engineering 12(1):142–151

Yoon WJ, Reinhall PG, Seibel EJ (2007) Analysis of electro-active polymer bending: A component in a low cost ultrathin scanning endoscope. Sensors and Actuators A: Physical 133(2):506–517

Zawodzinski TA, Neeman M, Sillerud LO, Gottesfeld S (1991) Determination of water diffusion coefficients in perfluorosulfonate ionomeric membranes. The Journal of Physical Chemistry 95(15):6040–6044

Chapter 40
Weakly Nonlocal Non-Equilibrium Thermodynamics: the Cahn-Hilliard Equation

Péter Ván

Abstract The Cahn-Hilliard and Ginzburg-Landau (Allen-Cahn) equations are derived from the second law. The intuitive approach of separation of full divergences is supported by a more rigorous method, based on Liu procedure and a constitutive entropy flux. Thermodynamic considerations eliminate the necessity of variational techniques and explain the role of functional derivatives.

40.1 Introduction

In continuum theories the extension of classical, well known evolution equations is one of the most exciting theoretical question, where several, completely different approaches compete. When compared to the kinetic theory or nonequilibrium statistical physics the advantage of pure phenomenological ideas is their universality. This way one can analyse the consequences of general requirements, like the basic balances and the second law of thermodynamics without assuming and introducing particular microscopic structures or mechanisms (Gurtin, 1996; Giorgi, 2009; Ván, 2013). In this respect the so called phase-field theories are particularly interesting, because there the influence of microstructure is introduced mostly indirectly, through fluctuating field quantities. The apparent universality of these descriptions is often attributed either to second order phase transitions or background linear instabilities (see e.g. Hohenberg and Halperin, 1977; Hohenberg and Krekhov, 2015). On the other hand, from a continuum point of view the obtained macroscopic models are weakly nonlocal extensions of the simplest evolution equations. For example the extension

Peter Ván
Department of Theoretical Physics, Wigner Research Centre for Physics, H-1525 Budapest, Konkoly Thege Miklós u. 29-33.; Department of Energy Engineering, Faculty of Mechanical Engineering, Budapest University of Technology and Economics, 1111 Budapest, Műegyetem rkp. 3.; Montavid Thermodynamic Research Group, Budapest, Hungary
e-mail: van.peter@wigner.mta.hu

© Springer International Publishing AG, part of Springer Nature 2018
H. Altenbach et al. (eds.), *Generalized Models and Non-classical Approaches in Complex Materials 1*, Advanced Structured Materials 89,
https://doi.org/10.1007/978-3-319-72440-9_40

of a relaxation dynamics of a single internal variable leads to the Ginzburg-Landau (Allen-Cahn) equation (Landau and Ginzburg, 1950; Landau and Khalatnikov, 1954; Cahn, 1961; Allen and Cahn, 1979), or the extension of a diffusion-Fourier dynamics results in the Cahn-Hilliard equation (Cahn and Hilliard, 1958). In phase-field models the role of the second law is rarely constructive, it is used more restrictively. Thermostatics is based on free energy and entropy appears only when thermal phenomena is considered (Penrose and Fife, 1990; Alt and Pawlow, 1992; Antanovskii, 1995, 1996), or sometimes when dissipation is calculated (e.g. Grmela, 2008). Some more elaborated analyses introduce extra entropy flux (Fabrizio et al, 2006; Giorgi, 2009). The weakly nonlocal extension is either postulated or due to variational techniques applied together with local equilibrium based thermodynamic considerations (Anders and Weinberg, 2011). In phase-field theories the second law compatibility in the weakly nonlocal sense, for the higher order spatial derivatives, is rarely mentioned, and then it is restricted to the Ginzburg-Landau theory. In the case of the Cahn-Hilliard equation the large constitutive state space leads to technical difficulties.

However, for more complicated constraints, like the classical balances of continuum mechanics of solids, a simple variational approach may become problematic as well. The conceptual problem is the doubled theoretical structure. Usually the relaxation dynamics is introduced without referring to nonequilibrium thermodynamics (e.g. Anderson et al, 1998). The methodology can be extended and combined to theories like GENERIC (Öttinger, 2005; Grmela and Öttinger, 1997; Öttinger and Grmela, 1997; Grmela, 2008), where a clear separation of dissipative and nondissipative parts of the dynamics is based on a bracket formalism with functional derivatives for weak nonlocality. That can keep strict thermodynamic compatibility far beyond the usual phase-field approaches.

This doubled thermodynamical-mechanical structure is due to a lack of a resolution of an ancient question regarding the origin of evolution equations of physics[1]. Should we separate the ideal world from the real one or is there a common origin of nondissipative and dissipative dynamics? Here there are two evident strategies. One either try to extend the variational principles to dissipative dynamics or derive the nondissipative evolution from the second law. The historical roots of the first approach are going back to Helmholtz (Yourgrau and Mandelstam, 1999) and to a strategy to overcome some strict mathematical conditions (Ván and Muschik, 1995; Ván and Nyíri, 1999). In order to obtain dissipative evolution and parabolic partial differential equations from variational principles one need to use some inconventional methods (Gyarmati, 1970; Sieniutycz and Farkas, 2005; Verhás, 2014; Glavatskiy, 2015). However, in most cases the doubled theoretical structure is preserved in a modified form (Matolcsi et al, 2005).

The other possibility is pioneered in nonequilibrium thermodynamics, where a deeper mathematical consistency indicated that the second law can be sufficient to derive evolution equations. Anyway, a nondissipative evolution is a special dissipative one, where the dissipation is zero. There exist an intuitive and a rigorous treatment for weakly nonlocal theories, where the constitutive functions depend on the derivatives

[1] That question goes back at least to the greek philosophers Platon and Aristotle, see and overview in Prigogine and Stengers (1986).

of the state variables (Capriz, 1985, 1989; Mariano, 2002). The intuitive technique, the identification of the entropy flux by separation of full divergences, appears independently in many papers and books (Cahn and Hilliard, 1958; de Groot, 1959; de Groot and Mazur, 1962; Maugin, 1999; Giorgi, 2009; Heida et al, 2012). The rigorous treatments are the Coleman-Noll and the Liu procedures (Coleman and Noll, 1963; Liu, 1972), together with a strict interpretation of the second law (Coleman and Mizel, 1967; Muschik and Ehrentraut, 1996). This rigorous methods are equivalent (Triani et al, 2008) and interpret the second law as a constrained inequality, where the identification of independent variables and the constitutive state space are the key aspects for handling differential equations as constraints.

In principle weak nonlocality can be both in time and space, that is the constitutive functions may depend both on time and space derivatives. However, in the following we use the terminology only for the spatial case if not mentioned otherwise. This is the safe approach because we want to avoid spacetime related objectivity questions: spatial derivatives in nonrelativistic space-time are frame independent (Ván, 2017).

The first applications of the strict mathematical requirements of the second law led to blockingly restrictive results. Coleman and Gurtin has proved that for internal variables only local evolution is possible (Coleman and Gurtin, 1967), and Gurtin has shown that there is no gradient extension of elasticity (Gurtin, 1965). These results effectively prevented to understand the universal background of the Ginzburg-Landau (Allen-Cahn) equation and phase-field theories in general. The thermodynamic investigations of higher grade fluids and solids were slowed down. Additional concepts were necessary to circumvent the restrictive conditions.

These concepts are very different. For example *interstitial work* generalizes the energy balance (Dunn and Serrin, 1985), the *configurational forces* generalize the energy and momentum balances (Gurtin, 1996, 2000), and the *virtual power* is based on a kinematic interpretation of internal variables and a related modification of the energy balance (Germain, 1973; Maugin, 1980; Frémond, 2001; Maugin, 2013)[2]. These interpretational issues elevated further communication barriers, and the pitchforking of thermodynamic theories continued[3]. In principle all these extension are capable to construct reasonable conditions and lead to very similar results.

Here, in this paper we argue that there is a minimal set of assumptions – it is essentially the second law – that can generate the Ginzburg-Landau and Cahn-Hilliard type evolution without any further ado. Moreover, the second law is useful at two levels, there is an intuitive and a rigorous approach. The mentioned sophisticated exploitation procedures of Coleman-Noll and Liu can be applied when the blocking barriers are removed: one need to treat the entropy current density as a constitutive quantity and prolong the constraints according to the nonlocality level of the state space. From this point of view the Cahn-Hilliard evolution is particularly instructive, because there the rigorous methods are seemingly hopeless. They are either restricted

[2] The phase-fields themselves have their particular interpretation contexts like order parameters, interfacial free energy, diffuse interface, etc. All these particular ideas are disguising the general conditions and the universal background.

[3] This is somehow deeply related to the situation described in the first chapter of (Maugin, 1999). The specialization is a logical consequence.

to check the thermodynamic compatibility of the already derived equations (Fabrizio et al, 2006) or the obtained conditions are not direct and too complicated for practical applications (Pawłow, 2006; Cimmelli et al, 2016). At the same time we will show that the intuitive method of *separation of full divergences* leads essentially to the same results. This is the classical tool of irreversible thermodynamics (de Groot and Mazur, 1962), the method used originally by Cahn and Hilliard (1958) and it is the approach of Gerard Maugin for weakly nonlocal internal variables (Maugin, 1999, 2006). Thermodynamic concepts can be used to unite dissipative and nondissipative evolution.

40.2 Variational derivation of Ginzburg-Landau and Cahn-Hilliard equations

In both cases we are looking for the evolution equation of a single scalar internal variable without any constraint in a continuum at rest. In the first case without any additional conditions or constraints for the internal variables, and in the second case with a balance form evolution as a constraint. Here we survey the traditional derivation of both equations which is a characteristic mixture of variational and thermodynamical ideas.

Let us denote the scalar field by ξ. The Helmholtz free energy density, f, depends on this variable and its gradient: $f(\xi, \partial_i \xi)$. For the sake of simplicity we assume the following square gradient form, a Ginzburg–Landau free energy function:

$$f(\xi, \partial_i \xi) = f_0(\xi) + \gamma \partial_i \xi \partial^i \xi / 2, \tag{40.1}$$

where γ is a nonnegative material parameter, which is scalar for isotropic continua. f_0 is the classical, local part of the free energy, that may have particular forms, if ξ is an order parameter of a second order phase transition. ∂_ξ denotes the gradient of ξ and we apply Einstein's summation convention and abstract indices, $i, j, k \in \{1,2,3\}$[4].

Then, following the usual arguments, one assumes, that the rate of ξ in a body with volume V is negatively proportional to the change of the free energy, denoted by δ:

$$\frac{d}{dt} \int_V \xi \mathrm{d}V = -l\delta \int_V f(\xi, \partial_i \xi) dV. \tag{40.2}$$

Assuming that this equality is valid for any V we obtain the general *Ginzburg–Landau (Allen-Cahn) equation* in the following form:

$$\partial_t \xi = -l \frac{\delta f}{\delta \xi} = -l \left[\partial_\xi f - \partial_i \left(\partial_{\partial_i \xi} f \right) \right]. \tag{40.3}$$

---

[4] Please note that these indices are not coordinates (Penrose, 2004).

Here ∂_t is the partial time derivative, $\frac{\delta}{\delta\xi}$ in the functional derivative, and l is a material parameter. With the square gradient free energy, (40.1), one arrives at the classical form of the equation:

$$\partial_t \xi = -l\left[\partial_\xi f_0 - \gamma \partial^i_{\ i}\xi\right]. \tag{40.4}$$

The second most important basic example of phase-field theories is the Cahn–Hilliard equation, used for modelling phase transitions in solid media. Usually it is introduced as a dynamic equation of a conserved order parameter (Hohenberg and Halperin, 1977). Therefore now we are looking for a balance form evolution of an internal variable. Then one may assume a derivation by classical irreversible thermodynamics, where in the thermodynamic force the gradient of the internal variable is substituted by a functional derivative, assuming a nonlocal interaction. The variational origin can be less argumented in this case. Therefore, denoting by j^i the current density of ξ, the balance is written as

$$\partial_t \xi + \partial_i j^i = 0. \tag{40.5}$$

Then, according to classical irreversible thermodynamics the thermodynamic flux is the gradient of the corresponding intensive quantity, A_ξ. Without thermal interaction this intensive quantity is the partial derivative of the free energy by ξ, that is

$$A_\xi = \frac{\partial f}{\partial \xi}.$$

Therefore the constitutive equation for the flux is

$$j^i = -\kappa \partial^i A_\xi. \tag{40.6}$$

In case of a weakly nonlocal free energy, $f(\xi, \partial_i \xi)$, the partial derivative, is to be substituted by a functional derivative,

$$\hat{A}_\xi = \frac{\delta f}{\delta \xi}.$$

Then we obtain the Cahn-Hilliard equation for the evolution of the internal variable as

$$\partial_t \xi - \partial_i\left(\kappa \partial^i \hat{A}_\xi\right) = \partial_t \xi - \partial_i\left(\kappa \partial^i\left[\partial_\xi f - \partial_i\left(\partial_{\partial_i \xi} f\right)\right]\right) = 0. \tag{40.7}$$

These derivations are remarkably simple and sound from a physical point of view also when developed in a more detailed form. On the other hand they combine a variational extremum principle with a frame theory of non-equilibrium thermodynamics. In the next section we will see, that variational considerations are not necessary at all, and they role is simply a separation of surface and bulk contributions for the entropy.

First we will investigate the Ginzburg-Landau equation.

40.3 The Thermodynamic Origin of the Ginzburg-Landau (Allen-Cahn) Equation

40.3.1 Separation of Full Divergences

With this intuitive method one constructs the entropy balance by introducing the constraints through the time derivative of the entropy density and then identify an entropy flux by separating full divergences in the expression. This idea was used independently in several works (Cahn and Hilliard, 1958; Giorgi, 2009; Heida et al, 2012), but as a method was introduced in classical irreversible thermodynamics, where the starting point is the Gibbs relation for the classical extensives (de Groot, 1959; de Groot and Mazur, 1962; Gyarmati, 1970). Gerard Maugin applied the approach in several cases, including internal variables (Maugin, 1999; Maugin and Drouot, 1983; Maugin, 2006; Berezovski and Ván, 2017).

Let us assume that the evolution equation of the internal variable ξ is written in a general form as

$$\partial_t \xi = F. \tag{40.8}$$

As we can see one does not fix the domain of the right hand side yet. The starting point is a first order weakly nonlocal entropy density, $s(\xi, \partial_i \xi)$. Now we investigate the entropy balance. With a free energy the following investigation is less convenient, however, there is no conceptual difference. The relation of the two approaches and a more systematic background with complete continuum mechanics on material manifolds is developed e.g. in Berezovski and Ván (2017).

Let us calculate the time derivative of the entropy density and separate the full divergences with Leibnitz rule:

$$\begin{aligned}\partial_t s(\xi, \partial_i \xi) &= \partial_\xi s \partial_t \xi + \partial_{\partial_i \xi} s \partial_{ti} \xi = -\partial_\xi s F - \partial_{\partial_i \xi} s \partial_i F \\ &= -\partial_i \left(F \partial_{\partial_i \xi} s\right) + F\left(-\partial_\xi s + \partial_i (\partial_{\partial_i \xi} s)\right).\end{aligned} \tag{40.9}$$

Therefore the entropy flux is identified as

$$J^i = F \partial_{\partial_i \xi} s, \tag{40.10}$$

and the entropy production is

$$\partial_t s + \partial_i J^i = \Sigma = F\left(-\partial_\xi s + \partial_i (\partial_{\partial_i \xi} s)\right) \geq 0. \tag{40.11}$$

One can see, that the entropy inequality can be solved with the identification of thermodynamic fluxes and forces and assuming a linear relationship between them

$$F = l\left(-\partial_\xi s + \partial_i (\partial_{\partial_i \xi} s)\right), \tag{40.12}$$

where $l > 0$ is a material relaxation coefficient. The complete evolution of ξ will be given by an entropic Ginzburg-Landau equation:

$$\partial_t \xi = l\left(-\partial_\xi s + \partial_i(\partial_{\partial_i \xi} s)\right). \tag{40.13}$$

One may observe that natural boundary conditions emerge assuming a zero entropy flux. These are convenient for numerical solutions and effectively substitute the natural boundary conditions of variational principles (Giorgi, 2009; Berezovski and Ván, 2017).

40.3.2 Ginzburg-Landau Equation: a More Rigorous Derivation

The method is simple and clear, this is the advantage. The disadvantage is that one cannot fix the constitutive quantities and the constitutive state space in advance, it is determined along the calculations, therefore the results are not unique and require a verification by more rigorous methodology.

In this case one should assume a second order weakly nonlocal state space, spanned by the internal variable field, ξ, and its first and second derivatives, $\partial_i \xi$ and $\partial_{ij}\xi$.

We distinguish the

- *space of basic variables*, spanned by ξ,
- the *consititutive state space*, spanned by $(\xi, \partial_i \xi, \partial_{ij}\xi)$,
- and the constitutive functions are s, J^i and F.

Then the so called *process direction space* (Muschik and Ehrentraut, 1996; Ván, 2008) is spanned by the higher derivatives of the constitutive state space, $(\partial_t \xi, \partial_{ti}\xi, \partial_{tij}\xi, \partial_{ijk}\xi)$. We can observe, that now these derivatives are not independent, both the evolution equation, (40.8), and also its derivative

$$\partial_{ti}\xi + \partial_i F = 0_i, \tag{40.14}$$

is a relation in the process direction space. Therefore both the evolution equation of ξ and also its gradient are constraints for the entropy inequality.

In order to apply Liu procedure we introduce the Lagrange-Farkas multipliers, λ and Λ^i, for the equations (40.8) and (40.14) respectively. Then the application of Liu procedure leads to

$$\begin{aligned}
0 \le\ & \partial_t s + \partial_i J^i - \lambda\,(\partial_t \xi + F) - \Lambda^i\,(\partial_{ti}\xi + \partial_i F) = \\
&= \partial_\xi s\, \partial_t \xi + \partial_{\partial_i \xi} s\, \partial_{it}\xi + \partial_{\partial_{ij}\xi} s\, \partial_{ijt}\xi + \partial_\xi J^i\, \partial_i \xi + \partial_{\partial_j \xi} J^i\, \partial_{ij}\xi + \partial_{\partial_{jk}\xi} J^i\, \partial_{ijk}\xi \\
&- \lambda\,(\partial_t \xi + F) - \Lambda^i \left(\partial_{ti}\xi + \partial_\xi F\, \partial_i \xi + \partial_{\partial_j \xi} F\, \partial_{ij}\xi + \partial_{\partial_{jk}\xi} F\, \partial_{ijk}\xi\right)
\end{aligned}$$

$$= (\partial_\xi s - \lambda)\,\underline{\partial_t \xi} + (\partial_{\partial_i \xi} s - \Lambda^i)\;\underline{\partial_{it}\xi} + \partial_{\partial_{ij}\xi} s\;\underline{\partial_{ijt}\xi} +$$
$$+ \left(\partial_{\partial_{jk}\xi} J^i - \Lambda^i \partial_{\partial_{jk}\xi} F\right) \underline{\partial_{ijk} a} + \partial_\xi J^i\, \partial_i \xi + \partial_{\partial_j \xi} J^i\, \partial_{ij}\xi -$$
$$- \Lambda^i \left(\partial_\xi f\, \partial_i \xi + \partial_{\partial_j \xi} f\, \partial_{ij}\xi\right) - \lambda F. \tag{40.15}$$

The multipliers of the underlined partial derivatives, the members of the process direction space, give the Liu equations as

$$\partial_t \xi: \qquad \partial_\xi s = \lambda, \tag{40.16}$$
$$\partial_{it} \xi: \qquad \partial_{\partial_i \xi} s = \Lambda^i, \tag{40.17}$$
$$\partial_{ijt} \xi: \qquad \partial_{\partial_{ij}\xi} s = 0^{ij}, \tag{40.18}$$
$$\partial_{ijk} \xi: \qquad \partial_{\partial_{(jk}\xi} J^{i)} = \Lambda^{(i} \partial_{\partial_{jk)}\xi} f. \tag{40.19}$$

The first two equations determine the Lagrange–Farkas multipliers as the derivatives of the entropy, a solution of the third one gives that the entropy is independent on the second gradient of the state variable, ξ. Therefore the Lagrange–Farkas multiplier, Λ^i, is also independent of this variable. In the last equation the indexed parenthesis indicate the symmetric part of the related tensorial components. The last equation can be integrated as

$$J^i(\xi, \partial_i \xi, \partial_{ij}\xi) = \partial_{\partial_i \xi} s(\xi, \partial_i \xi)\; F(\xi, \partial_i \xi, \partial_{ij}\xi) + \mathfrak{J}^i(\xi, \partial_i \xi). \tag{40.20}$$

Here the extra entropy current density, $\mathfrak{J}^i$, is an arbitrary constitutive function of the indicated variables and we did not restrict the Liu condition to the symmetric part of the expression. This is a complete solution of this solution, (40.16)–(40.19). Considering these results, the dissipation inequality is reduced to the following form:

$$0 \leq \partial_i \mathfrak{J}^i + \left[\partial_i(\partial_{\partial_i \xi} s) - \partial_\xi s\right] F. \tag{40.21}$$

Assuming that the extra entropy flux, $\mathfrak{J}^i$, is zero, we obtain a product of undetermined constitutive quantity and the derivatives of the entropy, the same thermodynamic force-flux system as in (40.11). The classical linear solution of the inequality results in

$$F = l\,\left[\partial_i(\partial_{\partial_i \xi} s) - \partial_\xi s\right], \qquad l \geq 0. \tag{40.22}$$

Therefore the evolution equation of an internal variable in a second order weakly nonlocal constitutive state space will be the Ginzburg-Landau equation:

$$\partial_t \xi = l\,\left[\partial_\xi s - \partial_i(\partial_{\partial_i \xi} s)\right]. \tag{40.23}$$

This result is a consequence of the second law, independently of any microscopic interpretation. If the entropy density has a square gradient form, (40.1), Bedeaux et al (2003); Johannessen and Bedeaux (2003, 2004); Glavatskiy and Bedeaux (2008), we obtain the classical form of the equation if γ is constant. The concavity of the entropy requires $\gamma > 0$.

40.4 The Thermodynamic Origin of the Cahn–Hilliard Equation

40.4.1 Separation of Full Divergences

In this case the internal variable ξ is conservative, its evolution equation has a balance form

$$\partial_t \xi + \partial_i j^i = 0, \tag{40.24}$$

where j^i is the flux of ξ. Now the constitutive functions are the entropy, its current density and the flux of the state variable: s, J^i and j^i. When looking at the equation (40.7) we can observe, that we need at least a fourth order weakly nonlocal state space. In case of the Coleman-Noll or Liu procedures the large number of composite derivatives, and the resulted nonlinearity in the process direction variables, encumbers to find explicit solutions of the inequality. However, the simple separation of the full divergences in the entropy balance leads to the expected result.

Let us assume, that the entropy density depends on the state variable and also on its gradient, $s(\xi, \partial_i \xi)$. Let us calculate its time derivative, and substitute the balance, (40.24), as a constraint:

$$\begin{aligned}\partial_t s(\xi, \partial_i \xi) &= \partial_\xi s \partial_t \xi + \partial_{\partial_i \xi} s \partial_{ti} \xi = -\partial_\xi s \partial_i j^i - \partial_{\partial_i \xi} s \partial_{ik} j^k \\ &= -\partial_i \left(\partial_\xi s j^i\right) + \partial_i \left(\partial_\xi s\right) j^i - \partial_i \left(\partial_{\partial_i \xi} s \partial_k j^k\right) + \partial_i \left(\partial_{\partial_i \xi} s\right) \partial_k j^k = \\ &= -\partial_i \left[\partial_\xi s j^i + \partial_{\partial_i \xi} s \partial_k j^k \partial_k \left(\partial_{\partial_k \xi} s\right) j^i\right] + \partial_i \left(\partial_\xi s - \partial_k \left(\partial_{\partial_k \xi} s\right)\right) j^i\end{aligned} \tag{40.25}$$

Therefore the entropy flux can be identified as

$$J^i = \left(\partial_\xi s - \partial_k \left(\partial_{\partial_k \xi} s\right)\right) j^i + \partial_{\partial_i \xi} s \partial_k j^k, \tag{40.26}$$

and the entropy production is

$$\Sigma = \partial_i \left(\partial_\xi s - \partial_k \left(\partial_{\partial_k \xi} s\right)\right) j^i \geq 0. \tag{40.27}$$

One can see, that the solution of this inequality is easy with the identification of the thermodynamic force by the gradient of $\partial_i \left(\partial_\xi s - \partial_k \left(\partial_{\partial_k \xi} s\right)\right)$, and the thermodynamic flux as the current density of the internal variable, j^i. For isotropic materials the coefficient is a scalar and we obtain the constitutive equation, (40.6):

$$j^i = -\kappa \partial^i \left(\partial_\xi s - \partial_k \left(\partial_{\partial_k \xi} s\right)\right), \tag{40.28}$$

where $\kappa > 0$, because of the second law. Substituting this expression into the balance (40.24) we obtain the Cahn-Hilliard equation:

$$\partial_t \xi - \partial_i \left[\kappa \partial^i \left(\partial_\xi s - \partial_k \left(\partial_{\partial_k \xi} s\right)\right)\right] = 0. \tag{40.29}$$

One may recognize that according to (40.28) the internal variable flux is a third order weakly nonlocal function, depending on the variables $(\xi, \partial_i \xi, \partial_{ij} \xi, \partial_{ijk} \xi)$.

40.4.2 Cahn-Hilliard Equation: a More Rigorous Derivation

Here we should start from a fourth order weakly nonlocal constitutive state space for a systematic analysis. Therefore, the

- space of basic variables is spanned by ξ,
- the consititutive state space is spanned by $(\xi, \partial_i \xi, \partial_{ij} \xi, \partial_{ijk} \xi, \partial_{ijkl} \xi,)$.
- The constitutive functions are s, J^i and j^i.

In order to apply the Liu procedure we introduce the Lagrange-Farkas multipliers λ for the balance of ξ, (40.24), and Λ^j for the gradient of (40.24):

$$\partial_{tj} \xi + \partial_{ij} j^i = 0. \tag{40.30}$$

With a fourth order weakly nonlocal state space one may wonder, whether higher order spatial derivatives of the constraint are to be applied. For example, Cimmelli (2007) argues that the consistent evaluation of the second law may require the third and fourth derivatives as constraints. However, in that case the constructive character of the derivation is lost. Therefore, let us calculate the constrained inequality, with the Lagrange-Farkas multipliers, as in the previous section, but now with different constraints and in a larger state space.

$$\begin{aligned}
0 \le\ & \partial_t s + \partial_i J^i - \lambda \left(\partial_t \xi + \partial_i j^i\right) - \Lambda^j \left(\partial_{tj} \xi + \partial_{ij} j^i\right) = \\
=\ & \partial_\xi s\, \partial_t \xi + \partial_{\partial_i \xi} s\, \partial_{it} \xi + \partial_{\partial_{ij} \xi} s\, \partial_{ijt} \xi + \partial_{\partial_{ijk} \xi} s\, \partial_{ijkt} \xi + \partial_{\partial_{ijkl} \xi} s\, \partial_{ijklt} \xi + \\
&+ \partial_\xi J^i\, \partial_i \xi + \partial_{\partial_j \xi} J^i\, \partial_{ij} \xi + \partial_{\partial_{jk} \xi} J^i\, \partial_{ijk} \xi + \partial_{\partial_{jkl} \xi} J^i\, \partial_{ijkl} \xi + \partial_{\partial_{jklm} \xi} J^i\, \underline{\partial_{ijklm} \xi} + \\
&- \lambda \Big(\partial_t \xi + \left(\partial_\xi j^i\right) \partial_i \xi + \left(\partial_{\partial_j \xi} j^i\right) \partial_{ij} \xi + \left(\partial_{\partial_{jk} \xi} j^i\right) \partial_{ijk} \xi + \\
&+ \left(\partial_{\partial_{jkl} \xi} j^i\right) \partial_{ijkl} \xi + \left(\partial_{\partial_{jklm} \xi} j^i\right) \underline{\partial_{ijklm} \xi}\Big) \\
&- \Lambda^n \Big(\partial_{nt} \xi + \partial_n \left(\partial_\xi j^i\right) \partial_i \xi + \left(\partial_\xi j^i\right) \partial_{in} \xi + \partial_n \left(\partial_{\partial_j \xi} j^i\right) \partial_{ij} \xi + \left(\partial_{\partial_j \xi} j^i\right) \partial_{ijn} \xi + \\
&+ \partial_n \left(\partial_{\partial_{jk} \xi} j^i\right) \partial_{ijk} \xi + \left(\partial_{\partial_{jk} \xi} j^i\right) \partial_{ijkn} \xi + \partial_n \left(\partial_{\partial_{jkl} \xi} j^i\right) \partial_{ijkl} \xi +
\end{aligned}$$

$$+ \quad \left(\partial_{\partial_{jkl}\xi} j^i\right) \underline{\partial_{ijkln}\xi} + \partial_n \left(\partial_{\partial_{jklm}\xi} j^i\right) \underline{\partial_{ijklm}\xi} + \left(\partial_{\partial_{jklm}\xi} j^i\right) \underline{\partial_{ijklmn}\xi}\Big) \geq 0 \tag{40.31}$$

The multipliers of the partial time derivatives give the following Liu equations:

$$\partial_t \xi \ : \partial_\xi s \ = \lambda, \tag{40.32}$$

$$\partial_{it} \xi \ : \partial_{\partial_i \xi} s \ = \Lambda^i, \tag{40.33}$$

$$\partial_{ijt} \xi \ : \partial_{\partial_{ij} \xi} s \ = 0^{ij}, \tag{40.34}$$

$$\partial_{ijkt} \xi \ : \partial_{\partial_{ijk} \xi} s = 0^{ijk}, \tag{40.35}$$

$$\partial_{ijlkt} \xi \ : \partial_{\partial_{ijkl} \xi} s = 0^{ijkl}, \tag{40.36}$$

The first two equations give the Lagrange–Farkas multipliers as the derivatives of the entropy, and a solution of the last three ones result in that the entropy must depend only on the state variable and its gradient. Therefore the λ and Λ^i Lagrange–Farkas multipliers depend only on these variables, too. In summary the entropy density and its derivatives are $s = s(\xi, \partial_i \xi)$, $\partial_\xi s = \lambda$, and $\partial_{\partial_i \xi} s = \Lambda^i$.

Let us investigate now the multipliers of the underlined sixth and fifth spatial derivatives, the remaining members of the process direction space in the inequality:

$$\partial_{ijklmn}\xi : \partial_{\partial_i \xi} s \partial_{\partial_{jklm}\xi} j^n = 0^{ijklmn}, \tag{40.37}$$

$$\partial_{ijklm}\xi : \partial_{\partial_{ijkl}\xi} J^m = \partial_\xi s \partial_{\partial_{ijkl}\xi} j^m + \partial_{\partial_n \xi} s \left(\partial_n \left[\partial_{\partial_{ijkl}\xi} j^m\right] + \delta^i_n \partial_{\partial_{jkl}\xi} j^m\right) \tag{40.38}$$

If the entropy is first order weakly nonlocal, then a solution of the first equation is

$$\partial_{\partial_{ijkl}\xi} \left(j^m\right) = 0. \tag{40.39}$$

The solution of the second equation will be the following expression for the entropy flux

$$J^i = \partial_\xi s j^i + \partial_{\partial_n \xi} s \partial_n j^i + \mathfrak{J}^i(\xi, \partial_i \xi, \partial_{ij}\xi, \partial_{ijk}\xi), \tag{40.40}$$

where the last term, the extra entropy flux, $\mathfrak{J}^i$, is only third order weakly nonlocal.

In this solution we have used the first among the identities below:

$$\begin{aligned}
\partial_{\partial_{ijkl}\xi} \left[\partial_n j^m\right] &= \partial_n \left[\partial_{\partial_{ijkl}\xi} j^m\right] + \delta^i_n \partial_{\partial_{jkl}\xi} j^m, \\
\partial_{\partial_{ikl}\xi} \left[\partial_n j^m\right] &= \partial_n \left[\partial_{\partial_{ikl}\xi} j^m\right] + \delta^i_n \partial_{\partial_{kl}\xi} j^m, \\
\partial_{\partial_{il}\xi} \left[\partial_n j^m\right] &= \partial_n \left[\partial_{\partial_{il}\xi} j^m\right] + \delta^i_n \partial_{\partial_{l}\xi} j^m, \\
\partial_{\partial_{i}\xi} \left[\partial_n j^m\right] &= \partial_n \left[\partial_{\partial_{i}\xi} j^m\right] + \delta^i_n \partial_{\xi} j^m.
\end{aligned} \tag{40.41}$$

Using also the other identities of (40.41) and also the entropy flux, (40.40), the dissipation inequality reduces to the following simple form:

$$
\begin{aligned}
&\partial_{ijkl}\xi\left(\partial_{\partial_{jkl}\xi}J^i - \partial_\xi s\partial_{\partial_{jkl}\xi}j^i - \partial_{\partial_m\xi}s\partial_{\partial_{jkl}\xi}(\partial_m j^i)\right) + \\
&\partial_{ijk}\xi\left(\partial_{\partial_{jk}\xi}J^i - \partial_\xi s\partial_{\partial_{jk}\xi}j^i - \partial_{\partial_m\xi}s\partial_{\partial_{jk}\xi}(\partial_m j^i)\right) + \\
&\partial_{ij}\xi\left(\partial_{\partial_j\xi}J^i - \partial_\xi s\partial_{\partial_j\xi}j^i - \partial_{\partial_m\xi}s\partial_{\partial_j\xi}(\partial_m j^i)\right) + \\
&\partial_i\xi\left(\partial_\xi J^i - \partial_\xi s\partial_\xi j^i - \partial_{\partial_m\xi}s\partial_\xi\partial_m j^i\right) = \\
&= \partial_i\mathfrak{J}^i + \partial_i(\partial_\xi s)j^i + \partial_i(\partial_{\partial_n\xi}s)\partial_n j^i \geq 0. \qquad (40.42)
\end{aligned}
$$

Then the consequence of (40.39) is, that the last line of the above expression depends linearly on the fourth spatial derivative of the internal variable, $\partial_{ijkl}\xi$. The coefficients of this term, $\partial_{\partial_{ijk}\xi}\mathfrak{J}^i$ and $\partial_i(\partial_{\partial_n\xi}s)\partial_{\partial_{ijk}\xi}j^i$, and also the remaining terms in the inequality are all independent on the fourth spatial derivative itself, they are only third order weakly nonlocal. Therefore the coefficient of $\partial_{ijkl}\xi$ must be zero in the inequality. This is a repeated application of Liu procedure. Therefore:

$$\partial_{ijkl}\xi : \partial_{\partial_{jkl}\xi}\mathfrak{J}^i = \partial_n(\partial_{\partial_n\xi})\partial_{\partial_{jkl}\xi}j^n. \qquad (40.43)$$

The solution of this equation for the extra entropy flux is

$$\mathfrak{J}^i = -\partial_n\left(\partial_{\partial_i\xi}s\right)j^n + \hat{\mathfrak{J}}^i(\xi, \partial_i\xi, \partial_{ij}\xi). \qquad (40.44)$$

Here the remaining extra entropy flux, $\hat{\mathfrak{J}}^i$, is only second order weakly nonlocal. Therefore the final form of the entropy flux is

$$J^i = \left(\partial_\xi s - \partial_n(\partial_{\partial_i\xi}s)\right)j^n + (\partial_{\partial_n\xi}s)\partial_n j^i + \hat{\mathfrak{J}}^i(\xi, \partial_i\xi, \partial_{ij}\xi), \qquad (40.45)$$

This concludes a complete solution of the Liu system of equations (40.32)–(40.37). Considering all these conditions, and assuming a zero extra entropy flux, the dissipation inequality reduces to the following simple expression:

$$0 \leq \partial_i\left[\partial_\xi s - \partial_n(\partial_{\partial_n\xi}s)\right]j^i. \qquad (40.46)$$

Equations (40.45) and (40.46) are identical with the expressions (40.26) and (40.27), that we have obtained with the help of divergence separation method.

The product form of the inequality, (40.46), with the fourth order weakly nonlocal constitutive j^i as a multiplier is a flux-force system, where the constitutive state space ensures a linear solution by Lagrange mean value theorem:

$$j^i = -\kappa\partial^i\left(\partial_\xi s - \partial_k\left(\partial_{\partial_k\xi}s\right)\right), \qquad (40.47)$$

Substituting this expression into the balance (40.24) we obtain the Cahn-Hilliard equation again:

$$\partial_t\xi - \partial_i\left[\kappa\partial^i\left(\partial_\xi s - \partial_k\left(\partial_{\partial_k\xi}s\right)\right)\right] = 0. \qquad (40.48)$$

The particular case of first order weakly nonlocal state space leads to the diffusive evolution of Classical Irreversible Thermodynamics (Ván, 2002). This concludes the thermodynamic derivation.

40.5 Discussion

In this paper we have seen that phase-field evolution equations can be derived from the second law without variational considerations and functional derivatives. The intuitive method of separation of divergences enlights that the background is the separation of surface and bulk contributions. A more rigorous analysis with the application of Liu procedure led essentially to the same results when a constitutive entropy flux was applied with derivative constraints. This derivation of the evolution equations is based only on general principles and therefore the obtained Ginzburg-Landau and Cahn-Hilliard dynamics is universal.

Let us emphasize that the differences between the pure thermodynamic approaches are not fundamental. They are negligible when compared to the difficulties of the doubled structures mentioned in the introduction. For example the simple and elegant approach of Heida, Málek and Rajagopal is based on the separation of divergences and the solution of the dissipation inequality. It is applied to obtain coupled diffusive thermal dynamics of fluid mixtures including Korteweg fluids (Heida et al, 2012). A more rigorous second law analysis, probably could lead to a justification of their results. An independent but similar work of Giorgi emphasizes the universality of the uniform thermodynamic derivation when compared to the configurational force balance of Gurtin and the virtual power approach of Fremond (Giorgi, 2009).

On the other hand a less constructive but rigorous standpoint can lead to general but complicated conditions. These conditions are nevertheless useful to improve a heuristic ansatz and to check its validity (Cimmelli et al, 2016). A similarly instructive analysis is the work of Pawłow, where the constitutive state space is first order weakly nonlocal in time, the entropy flux is constitutive but Pawlow does not apply a derivative prolongation of the constraints. Then the Liu equations cannot be solved in a closed function form, and to obtain the particular Cahn-Hilliard dynamics requires a further restriction. However, the evolution equation of the Lagrange-Farkas multiplier leads to the microforce balance of Gurtin (Pawłow, 2006). That is, sacrificing the constructivity obscures the advantages of a uniform approach, but it still can be well interpreted.

Here the thermodynamic methods are demonstrated with scalar internal variables. The generalization for vectors and tensors is straightforward, as well as for classical thermodynamic variables and more complicated constraints (see e.g. Ván and Fülöp, 2006; Ván, 2008, 2009; Ván and Papenfuss, 2010). We need to mention also, that time nonlocality, the treatment of memory and inertial effects is a different matter. For such systems dual internal variables are suitable for a uniform thermodynamic approach (Ván et al, 2008; Berezovski et al, 2011; Ván et al, 2014; Berezovski and Ván, 2017).

Acknowledgements The work was supported by the grants National Research, Development and Innovation Office - NKFIH 116197 (116375), NKFIH 124366 (124366) and 123815.

References

Allen SM, Cahn JW (1979) A microscopic theory for antiphase boundary motion and its application to antiphase domain coarsening. Acta Metallurgica 27(6):1085–1095

Alt HW, Pawlow I (1992) A mathematical model of dynamics of non-isothermal phase separation. Physica D: Nonlinear Phenomena 59(4):389–416

Anders D, Weinberg K (2011) A variational approach to the decomposition of unstable viscous fluids and its consistent numerical approximation. ZAMM-Journal of Applied Mathematics and Mechanics/Zeitschrift für Angewandte Mathematik und Mechanik 91(8):609–629

Anderson DM, McFadden GB, Wheeler AA (1998) Diffuse-interface methods in fluid mechanics. Annual Rev in Fluid Mechanics 30:139–65

Antanovskii LK (1995) A phase field model of capillarity. Physics of Fluids 7(4):747–753

Antanovskii LK (1996) Microscale theory of surface tension. Physical Review E 54(6):6285

Bedeaux D, Johannessen E, Rojorde A (2003) A nonequilibrium Van der Waals square gradient model. (I). The model and its numerical solution. Physica A 330:329–353

Berezovski A, Ván P (2017) Internal Variables in Thermoelasticity. Springer

Berezovski A, Engelbrecht J, Maugin GA (2011) Generalized thermomechanics with dual internal variables. Archive of Applied Mechanics 81(2):229–240

Cahn JW (1961) On spinodal decomposition. Acta Metallica 9:795–801

Cahn JW, Hilliard JE (1958) Free energy of a nonuniform system I. Interfacial free energy. Journal of Chemical Physics 28:258–267

Capriz G (1985) Continua with latent microstructure. Archive for Rational Mechanics and Analysis 90(1):43–56

Capriz G (1989) Continua with Microstructure. Springer, New York

Cimmelli V, Oliveri F, Pace A (2016) Phase-field evolution in Cahn–Hilliard–Korteweg fluids. Acta Mechanica 227(8):2111–2124

Cimmelli VA (2007) An extension of Liu procedure in weakly nonlocal thermodynamics. Journal of Mathematical Physics 48:113,510

Coleman BD, Gurtin ME (1967) Thermodynamics with internal state variables. The Journal of Chemical Physics 47(2):597–613

Coleman BD, Mizel VJ (1967) Existence of entropy as a consequence of asymptotic stability. Archive for Rational Mechanics and Analysis 25:243–270

Coleman BD, Noll W (1963) The thermodynamics of elastic materials with heat conduction and viscosity. Archive for Rational Mechanics and Analysis 13:167–178

Dunn JE, Serrin J (1985) On the thermomechanics of interstitial working. Archive of Rational Mechanics and Analysis 88:95–133

Fabrizio M, Giorgi C, Morro A (2006) A thermodynamic approach to non-isothermal phase-field evolution in continuum physics. Physica D: Nonlinear Phenomena 214(2):144–156

Frémond M (2001) Non-Smooth Thermomechanics. Springer

Germain P (1973) The method of virtual power in continuum mechanics. Part 2: Microstructure. SIAM Journal of Applied Mathematics 25:556–575

Giorgi C (2009) Continuum thermodynamics and phase-field models. Milan Journal of Mathematics 77(1):67–100

Glavatskiy KS (2015) Lagrangian formulation of irreversible thermodynamics and the second law of thermodynamics. The Journal of Chemical Physics 142(20):204,106

Glavatskiy KS, Bedeaux D (2008) Nonequilibrium properties of a two-dimensionally isotropic interface in a two-phase mixture as described by the square gradient model. Physical Review E 77:061,101

Grmela M (2008) Extensions of classical hydrodynamics. Journal of Statistical Physics 132(3) 581–602

Grmela M, Öttinger HC (1997) Dynamics and thermodynamics of complex fluids. I. Development of a general formalism. Physical Review E 56(6):6620–6632

de Groot S (1959) Thermodynamics of Irreversible Processes. North Holland

de Groot SR, Mazur P (1962) Non-Equilibrium Thermodynamics. North-Holland Publishing Company, Amsterdam

Gurtin ME (1965) Thermodynamics and the possibility of spatial interaction in elastic materials. Archive for Rational Mechanics and Analysis 19:339–352

Gurtin ME (2000) Configurational Forces as Basic Concepts of Continuum Physics. Springer, New York

Gurtin MG (1996) Generalized Ginzburg-Landau and Cahn-Hilliard equations based on a microforce balance. Physica D 92:178–192

Gyarmati I (1970) Non-equilibrium Thermodynamics - Field Theory and Variational Principles. Springer, Berlin

Heida M, Málek J, Rajagopal K (2012) On the development and generalizations of Cahn–Hilliard equations within a thermodynamic framework. Zeitschrift für Angewandte Mathematik und Physik 63(1):145–169

Hohenberg PC, Halperin BI (1977) Theory of dynamic critical phenomena. Reviews of Modern Physics 49(3):435–479

Hohenberg PC, Krekhov A (2015) An introduction to the Ginzburg-Landau theory of phase transitions and nonequilibrium patterns. Physics Reports 572:1–42

Johannessen E, Bedeaux D (2003) A nonequilibrium van der Waals square gradient model. (II) Local equilibrium of the Gibbs surface. Physica A 330:354–372

Johannessen E, Bedeaux D (2004) A nonequilibrium van der Waals square gradient model. (III). Heat and mass transfer coefficients. Physica A 336:252–270

Landau LD, Ginzburg VL (1950) K teorii sverkhprovodimosti. Zhurnal Eksperimentalnoi i Teoreticheskoi Fiziki 20:1064, English translation: On the theory of superconductivity, in: Collected papers of L. D. Landau, ed. D ter Haar, (Pergamon, Oxford, 1965), pp. 546-568

Landau LD, Khalatnikov IM (1954) Ob anomal'nom pogloshchenii zvuka vblizi tochek fazovogo perekhoda vtorogo roda. Dokladu Akademii Nauk, SSSR 96:469–472, English translation: On the anomalous absorption of sound near a second order transition point. in: Collected papers of L. D. Landau, ed. D. ter Haar,(Pergamon, Oxford, 1965), pp. 626-633

Liu IS (1972) Method of Lagrange multipliers for exploitation of the entropy principle. Archive of Rational Mechanics and Analysis 46:131–148

Mariano PM (2002) Multifield theories in mechanics of solids. Advances in Applied Mechanics 38:1–94

Matolcsi T, Ván P, Verhás J (2005) Fundamental problems of variational principles: objectivity, symmetries and construction. In: Sieniutycz S, H F (eds) Variational and Extremum Principles in Macroscopic Problems, Elsevier, Amsterdam-etc., pp 57–74

Maugin G (1999) The Thermomechanics of Nonlinear Irreversible Behaviors (An Introduction). World Scientific, Singapore-New Jersey-London-Hong Kong

Maugin G (2013) The principle of virtual power: from eliminating metaphysical forces to providing an efficient modelling tool. Continuum Mechanics and Thermodynamics 25:127–146

Maugin GA (1980) The principle of virtual power in continuum mechanics. Application to coupled fields. Acta Mechanica 35:1–70

Maugin GA (2006) On the thermomechanics of continuous media with diffusion and/or weak nonlocality. Archive of Applied Mechanics 75:723–738

Maugin GA, Drouot R (1983) Internal variables and the thermodynamics of macromolecule solutions. International Journal of Engineering Science 21(7):705–724

Muschik W, Ehrentraut H (1996) An amendment to the Second Law. Journal of Non-Equilibrium Thermodynamics 21:175–192

Öttinger HC (2005) Beyond Equilibrium Thermodynamics. Wiley-Interscience

Öttinger HC, Grmela M (1997) Dynamics and thermodynamics of complex fluids. II. Illustrations of a general formalism. Physical Review E 56(6):6633–6655
Pawłow I (2006) Thermodynamically consistent Cahn-Hilliard and Allen-Cahn models in elastic solids. Discrete and Continuous Dynamical Systems 15(4):1169–1191
Penrose O, Fife PC (1990) Thermodynamically consistent models of phase-field type for the kinetics of phase transitions. Physica D 43:44–62
Penrose R (2004) The Road to Reality. Jonathan Cape
Prigogine I, Stengers I (1986) La nouvelle alliance. Gallimard, Paris
Sieniutycz S, Farkas H (2005) Progress in variational formulations for macroscopic processes. In: Sieniutycz S, Farkas H (eds) Variational and Extremum Principles in Macroscopic Problems, Elsevier, Amsterdam, pp 3–24
Triani V, Papenfuss C, Cimmelli VA, Muschik W (2008) Exploitation of the Second Law: Coleman-Noll and Liu procedure in comparison. Journal of Non-Equilibrium Thermodynamics 33:47–60
Ván P (2002) Weakly nonlocal irreversible thermodynamics - the Ginzburg-Landau equation. Technische Mechanik 22(2):104–110
Ván P (2008) Internal energy in dissipative relativistic fluids. Journal of Mechanics of Materials and Structures 3(6):1161–1169
Ván P (2009) Weakly nonlocal non-equilibrium thermodynamics - variational principles and Second Law. In: Quak E, Soomere T (eds) Applied Wave Mathematics (Selected Topics in Solids, Fluids, and Mathematical Methods), Springer-Verlag, Berlin-Heidelberg, chap III, pp 153–186
Ván P (2013) Thermodynamics of continua: The challenge of universality. In: Pilotelli M, Beretta GP (eds) Proceedings of the 12th Joint European Thermodynamics Conference, Cartolibreria SNOOPY, Brescia, pp 228–233
Ván P (2017) Galilean relativistic fluid mechanics. Continuum Mechanics and Thermodynamics 29(2):585–610
Ván P, Fülöp T (2006) Weakly nonlocal fluid mechanics - the Schrödinger equation. Proceedings of the Royal Society, London A 462(2066):541–557
Ván P, Muschik W (1995) Structure of variational principles in nonequilibrium thermodynamics. Physical Review E 52(4):3584–3590
Ván P, Nyíri B (1999) Hamilton formalism and variational principle construction. Annalen der Physik (Leipzig) 8:331–354
Ván P, Papenfuss C (2010) Thermodynamic consistency of third grade finite strain elasticity. Proceedings of the Estonian Academy of Sciences 59(2):126–132
Ván P, Berezovski A, Engelbrecht J (2008) Internal variables and dynamic degrees of freedom. Journal of Non-Equilibrium Thermodynamics 33(3):235–254
Ván P, Papenfuss C, Berezovski A (2014) Thermodynamic approach to generalized continua. Continuum Mechanics and Thermodynamics 25(3):403–420, erratum: 421-422
Verhás J (2014) Gyarmati's variational principle of dissipative processes. Entropy 16:2362–2383
Yourgrau W, Mandelstam S (1999) Variational Principles in Dynamics and Quantum Theory, 2nd edn. Pitman, New York-Toronto-London

Printed by Printforce, the Netherlands